Energy Transition in the Oil and Gas Industry

The oil and gas industry is in the midst of a paradigm shift, moving from developing solely petroleum-based energy to producing alternative energy forms, including renewables. *Energy Transition in the Oil and Gas Industry* offers a comprehensive overview of renewables and their applications in the oil and gas industry during the current energy transition period. It includes the latest methods and workflows in renewables and oil and gas processes as well as integrated and hybrid approaches currently used as the industry begins its transition to the production of alternative forms of energy.

- Provides a synopsis of fossil fuel resources, along with the latest technologies, applications, and economics, and offers a general outline for the energy transition
- Details various alternative and renewable energy forms and discusses their advantages, disadvantages, maturity levels, and applications, including solar, geothermal, wind, hydropower, fuel cells, hydrogen, biofuels, ocean energy, and nuclear
- Discusses carbon capture and storage, electric vehicles, and energy storage technologies
- Covers the latest advances and technologies related to digital transformation in the oil and gas industry
- Summarizes future trends and directions of technologies related to renewable energy and energy transition in the oil and gas industry

Addressing energy holistically from a technology and engineering perspective, this book offers engineering professionals in the energy sector a wide-ranging view of current and near future changes taking place in this critical industry.

Energy Transition in the Oil and Gas Industry

Edited by Cenk Temizel, Ali Başer, Onder Saracoglu,
Tolga Tural, Luigi Saputelli, and Ole Torsæter

CRC Press
Taylor & Francis Group
Boca Raton London New York

CRC Press is an imprint of the
Taylor & Francis Group, an **informa** business

Designed cover image: Shutterstock

First edition published 2025
by CRC Press
2385 NW Executive Center Drive, Suite 320, Boca Raton FL 33431

and by CRC Press
4 Park Square, Milton Park, Abingdon, Oxon, OX14 4RN

CRC Press is an imprint of Taylor & Francis Group, LLC

ISBN: 978-1-032-25249-0 (hbk)
ISBN: 978-1-032-25250-6 (pbk)
ISBN: 978-1-003-28227-3 (ebk)

DOI: 10.1201/9781003282273

Typeset in Times
by Apex CoVantage, LLC

Dedication

To my mother, Emine Yüksel Temizel.
–Cenk Temizel

To my amazing wife and wonderful son, you are my greatest blessings. Thank you for your endless love and support.
–Tolga Tural

This book is dedicated to the ones who make life a beautiful adventure: my wife, Johimar, my incredible children, Andres, Gabriella and Emilio, and my ever-supportive sisters, Lucia and Adriana.
–Luigi Saputelli

Contents

Foreword

The world stands at a crossroads. Our reliance on traditional energy sources is increasingly challenged by the urgent need for a sustainable future. This transition demands bold action, and the energy sector is at the heart of this transformation. Balancing Energy Security, Equity (Affordability), and Environmental Sustainability has never been more critical, and the diversity of global challenges is clearer than ever.

With more than 25 years of experience in the energy sector across different continents, I have witnessed a remarkable evolution towards cleaner and more environmentally friendly solutions. This shift has become increasingly vital in recent years. A new energy landscape has unlocked significant opportunities, with companies focusing on energy efficiency, transitioning to greener fuels, and adopting renewables to reduce emissions and meet global climate change initiatives driven by the Paris Agreement. Ambitious global targets and pivotal discussions have emerged from the Conferences of the Parties, fostering innovative ideas and collaborative efforts.

My passion for energy transition has driven me to make substantial contributions in my roles as Manager of Strategy and Excellence in National Oil Companies and Director of New Energies in the Middle East in global consulting and engineering companies. I have had the privilege to shape strategic frameworks that emphasize the end-to-end value chain with multisectoral perspectives through energy hubs and green corridors, along with a transparent approach through ESG reporting. Here, energy efficiency, carbon management, sustainability, and new energies are key pillars guiding our path forward. Additionally, my volunteer role as Chair of Decarbonization & Sustainability (SPE Gaia) at the Society of Petroleum Engineers in the UAE and leading a Hydrogen Working Group in the MENA region has fueled my dedication to supporting experts, new professionals, and students eager to delve into energy management and sustainability, sharing knowledge and inspiring the next generation of leaders.

In my previous publications and research, I have emphasized education and awareness among the private sector, communities, and energy end-users as crucial elements in addressing the energy transition. Education campaigns act as catalysts that can accelerate the transition trajectory in a country or region. Consequently, universities, R&D centers, and knowledge societies are essential partners in building and enabling our future energy roadmap. In my conferences, I always encourage students to study with a holistic perspective and critical thinking, from a traditional to a wider approach, combining oil and gas skills with other energy knowledge in a perfect symbiosis for transitioning to the right energy mix needed in future scenarios.

This book, "Energy Transition in the Oil and Gas Industry," arrives at a pivotal moment. It delves into the complexities of the energy transition, exploring the technologies, policies, and societal shifts that will reshape our energy landscape. Within these pages, you will encounter innovative solutions and practical strategies for navigating this transition, presented by international experts. Whether you are an industry leader, policymaker, or concerned citizen, this book offers valuable insights for understanding the challenges and opportunities that lie ahead.

Adopting renewable energy globally is a key point in the book; it necessitates collaboration and partnership between government agencies and the private sector. This involves incorporating more incentives and eliminating obstacles to green energy adoption. It is well acknowledged that this adoption is hampered not only by technological challenges but also by financial obstacles and the high risk and low return ratio of the technologies. Providing appropriate incentives and regulatory frameworks that enhance the growth and adoption of these technologies is imperative.

The journey towards a clean energy future will not be easy and will not happen overnight, but it is clear that it will happen. It requires collaboration across sectors, a commitment to innovation, and a clear vision for a sustainable future with a crucial role for the oil and gas sector. This book serves as a valuable guidance, illuminating the path toward a cleaner, more secure energy future for generations to come.

It is time for an urgent call to action, encouraging countries, companies, and individuals to engage with the ideas presented in this book. Different energy transitions are happening simultaneously, and different countries are experiencing this transition in various ways. Let us take an active role in these essential transformations.

Suhail Diaz
Director of New Energies – Middle East at Wood

Preface

The world's energy situation is going through changes to tackle climate change and meet sustainability objectives. Switching to renewable energy sources and the technological advancements transform how we generate, consume, and handle energy. The book *Energy Transition in the Oil and Gas Industry* was created to look at aspects of this shift, providing insights into the evolving energy industry's technological, economic, and environmental dimensions.

This book is motivated by the need to achieve zero greenhouse gas emissions by 2050 to control warming within 1.5°C, as outlined in international agreements. This ambitious target carries implications for sectors like oil and gas renewables and the larger energy transition. The chapters in this book aim to shed light on these implications while exploring routes toward an energy future.

The book is divided into sections that explore aspects of the energy transition journey. It starts with an overview of the concepts and factors driving this transition before analyzing the energy landscape and the significance of renewable energies.

The book explores how the shift toward energy sources impacts the production and consumption of petroleum products and their derivatives. It also discusses advancements like carbon capture and storage (CCS), the outlook for oil and gas, and the incorporation of energy technologies. Each chapter examines these innovations, processes, and strategies that are driving the energy transition, supported by research findings and real-world examples.

Furthermore, it covers a range of energy sources, including fossil fuels, nuclear power, and emerging renewable options. The book explores the fundamental and practical applications and the recommended approaches for energy storage, fuel cells, and electric vehicles to underline their importance in moving toward a sustainable economy.

A particular focus is placed on catalysts facilitating the energy transition, by exploring their role in chemical processes such as hydrogenation and dehydrogenation of hydrocarbons. The conclusion offers insights into possibilities in the oil and gas industries, renewable energies, and the broader trends shaping the global energy landscape.

Energy Transition in the Oil and Gas Industry aims to appeal to a readership comprising professionals, researchers, policymakers, and students. Our goal is to provide readers with insight into the challenges and possibilities linked to the shift in energy, empowering them to make choices and drive innovation. We trust that this book will guide individuals involved in guiding the world's energy infrastructure toward a sustainable and adaptable tomorrow.

About the Editors

Cenk Temizel is a seasoned reservoir engineer boasting nearly two decades of extensive experience in reservoir simulation, smart fields, optimization, unconventional resources, field development, and enhanced oil recovery. His career has spanned roles at industry giants such as Saudi Aramco, Schlumberger and Halliburton across the Middle East, the United States, and the UK, and a notable tenure as a reservoir engineer at Aera Energy LLC (a Shell–ExxonMobil affiliate) in Bakersfield, California, USA. Prior to his industry roles, he contributed as a teaching and research assistant at the University of Southern California and Stanford University. Throughout his career, Cenk has garnered numerous accolades, including the SPE 2022 Regional Reservoir Description and Dynamics Award. He secured second place at the SPE Global R&D Competition at SPE ATCE 2014 in Amsterdam and the Technical Excellence Awards at the 22nd World Petroleum Congress in 2017 in Istanbul. In 2012, he received the Halliburton Applause Award for Innovation. He is also a holder of several US patents and has a portfolio of publications in areas of reservoir engineering, enhanced oil recovery processes, machine learning, and smart fields. Cenk holds a BS degree (honors) from the Middle East Technical University, Ankara (2003), and an MS degree (2005) from the University of Southern California (USC), both in petroleum engineering. A committed member of SPE since 2000, he has actively contributed to several program committees.

Ali Başer is the managing partner of Flux Energy Solutions, a firm focused on bringing coupled AI solutions to oil, gas, and geothermal firms. Having hands-on experience in both drilling and reservoir engineering aspects of the industry, he worked as an advisor to major geothermal energy companies in Türkiye, specializing in reservoir modeling, production, and injection optimization. Previously, he worked as a programmer for various firms, creating video games, desktop applications, and web applications. His research interests include machine learning-assisted reservoir engineering methods, time series analysis of production dynamics, and the usage of "serious games" as a training medium in the oil and gas industry. He holds a B.Sc. in petroleum and natural gas engineering from Middle East Technical University, Ankara, and an M.Sc. in computer engineering from Başkent University, Ankara.

Onder Saracoglu is the managing partner at Flux Energy Solutions. He has 27 years of experience working on reservoir simulation, reservoir management, field development, and reservoir engineering studies with NOC, service companies, and international E&P companies in the Middle East, Central Asia, and Africa. He has published numerous publications in reservoir and geothermal engineering. He holds a BS (1996) and MS (1999) degrees from the Middle East Technical University in petroleum and natural gas engineering, Ankara, Türkiye.

Tolga Tural received a B.S. degree in electrical engineering from Yildiz Technical University, Istanbul, Turkey, 1991 an M.S. degree in electrical and computer engineering from Oklahoma State University, Stillwater, Oklahoma, 1995 and MBA in finance and investment management from University of North Carolina, Chapel Hill, NC , 2017. Earlier in his career, Mr. Tural specialized in semiconductors specifically integrated circuit design and development design and development engineering roles for both networking and wireless communication applications in both Canada and the US. Later in his career, he focused in renewable industry to develop renewable projects as well as to build commercial and large scale power plants for power utilities and oil & gas industry. Mr. Tural has 28 years of diversified experience in semiconductors, wireless and data communications, power utilities and oil and gas industries. Mr. Tural is currently working at Saudi Aramco New Energies Strategy & Marketing Consultant. His current role is leading decarbonization development with renewables including project development, portfolio transaction and technology validation. Mr.Tural currently holds several technical publications and patents.

Luigi Saputelli is a reservoir engineering expert advisor with 34 years of experience as reservoir, drilling and production engineer in PDVSA, ADNOC, Hess and Halliburton. Saputelli is a researcher, lecturer, and active volunteer in the Society of Petroleum Engineers (SPE) where he served as JPT editor (2012-2020), Production and Operations Advisory Board since 2010, founding member of Real-time Optimization Interest Group and Petroleum Data-driven Analytics technical sections since 2003. He is the recipient of the 2015 SPE International Production and Operations Award. He has published more than 100 industry papers, 6 patents and contributed with over 5 books on digital oilfields, reservoir management and real-time production optimization. He holds a BSc in electronic engineering from Universidad Simon Bolivar (1990), with a MSc in petroleum engineering from Imperial College (1996), and a PhD in chemical engineering from University of Houston (2003). He is also serving as managing partner in Frontender, an upstream energy services firm in Houston.

Ole Torsaeter is professor emeritus in reservoir engineering at the Norwegian University of Science and Technology (NTNU), research associate at PoreLab a Norwegian Centre of Excellence and former adjunct professor at the University of Oslo. He has been researcher or visiting professor with SINTEF, Phillips Petroleum, ResLab, New Mexico Tech, Texas A & M, University of Bordeaux, and A*STAR, Singapore. Torsæter has supervised 220 Master- and 25 PhD-candidates, and he has published 200 research papers. Ole Torsæter has a doctor degree from NTNU with a thesis on water imbibition in the chalk formations in the Ekofisk Field. Torsæter received the Darcy Technical Achievement Award from the Society of Core Analysts (2014). Distinguished Achievement Award (2016) and Management and Information Award (2018) were received from Society of Petroleum Engineers. Torsæter is member of the Norwegian Academy of Technological Sciences.

Contributors

Mohammed Aliyu
Department of Computer Science
Federal College of Education
Gombe, Nigeria

Sepehr Arbabi
The University of Texas Permian Basin
Midland, Texas, USA

Labaran Jamilu Ari
Department of Physics
Federal University of Lafia
Nasarawa, Nigeria

Narendra Boppana
The University of Texas Permian Basin
Midland, Texas, USA

Coruh Celikbilek
Department of Mechanical Engineering
Istanbul Technical University
Istanbul, Türkiye

Jim Henry
Henry Resources
Midland, Texas, USA

Deniz Yağmur Kayhan
Department of Marine Research
General Directorate of Mineral Research and Exploration
Ankara, Türkiye

Aliyu Mohammed Kaura
Department of Geology
Federal University of Lafia
Nasarawa, Nigeria

Mohammed A. Mohammed
Department of Chemistry
Federal University of Lafia
Nasarawa, Nigeria

Adamu Salihu Rabo
Nigerian Upstream Petroleum Regulatory Commission (NUPRC)
Abuja, Nigeria

Jorge Saldaña
The University of Houston
Houston, Texas, USA

Osman Sen
Department of Geological Engineering
The University of Selcuk
Konya, Türkiye

Mesut Yurukcu
The University of Texas Permian Basin
Midland, Texas, USA

Usman Yusuf
Department of Mathematics
Federal University of Lafia
Nasarawa, Nigeria

Sebahattin Ziyanak
The University of Texas Permian Basin
Midland, Texas, USA

1 Energy Transition, Renewables, and the Oil and Gas Industry

Jorge Saldaña, Mesut Yurukcu, Narendra Boppana, Sepehr Arbabi, Jim Henry, and Sebahattin Ziyanak

1.1 INTRODUCTION

According to the United Nations Theme Report on Energy Transition (2021), energy transformation is pivotal to accelerate economic progress and to help combat the potentially harmful consequences of phenomena like climate change associated with a large amount of fossil fuel usage. With information from Our World in Data, summarized in Figure 1.1, we see that over 80% of global energy comes from conventional sources, like oil, gas, and coal, and only less than 16% comes from low-carbon alternatives. An effective transition process will occur when we move from a system primarily based on conventional sources to one that enhances energy efficiency and relies on renewable energy to replace those sources (IRENA, 2018).

The challenge is, then, to optimize conventional energy sources and allow for flexibility in the processes of generation, distribution, and transmission of energy. This system flexibility permits the incorporation of multiple energy sources, which work in tandem to fulfill the needs of a given population. Thus, a more flexible energy generation system in the grid would welcome solar and wind technologies, making it increasingly easier to implement these new energy sources into the structure. Likewise, more flexibility in energy distribution systems can help reach previously overlooked regions that need the energy to develop and thrive.

There are multiple ways of advancing a strategy for energy transition. First, incentivizing the diversification of energy sources is a step that can be furthered by governments, but also by private firms. In this process, "first movers" – or players that take the initiative in a changing environment, standing to win or lose the most when the scenario changes – can help other players follow suit, which would make waves in generating the change that is sought. Similarly, increasing the number of markets where these processes are taking place can be a crucial step toward a mindset shift from shareholders and firms. Figure 1.2 illustrates a projected electricity generation scenario based on the International Renewable Energy Agency's *Renewable Energy Roadmap* (REmap) for a 2°C setting.

Flexibility is essential not only at the time of generating and distributing energy but also when considering how we incorporate electricity generated from renewable energies into our daily lives. Innovations from the ingenuity of engineers and researchers are a vital component of the virtuous circle brought up by the transition to renewable energy. Figure 1.3 illustrates the adaptation processes and technologies that should be implemented to make renewable energies more efficient. From physical improvements like grid enhancement to technological advancements like electric cars, battery storage, and a more widespread use of artificial intelligence, the energy transition process requires a new understanding of the many interconnections between people and businesses.

This chapter focuses on the eventual transition to a new energy paradigm, led by renewable sources and maintained by a structure of optimal energy storage and constant improvement around carbon management, particularly in the oil and gas industry. In the following sections, we will review the role of renewable sources of energy as the main drivers of this energy transition, which

DOI: 10.1201/9781003282273-1

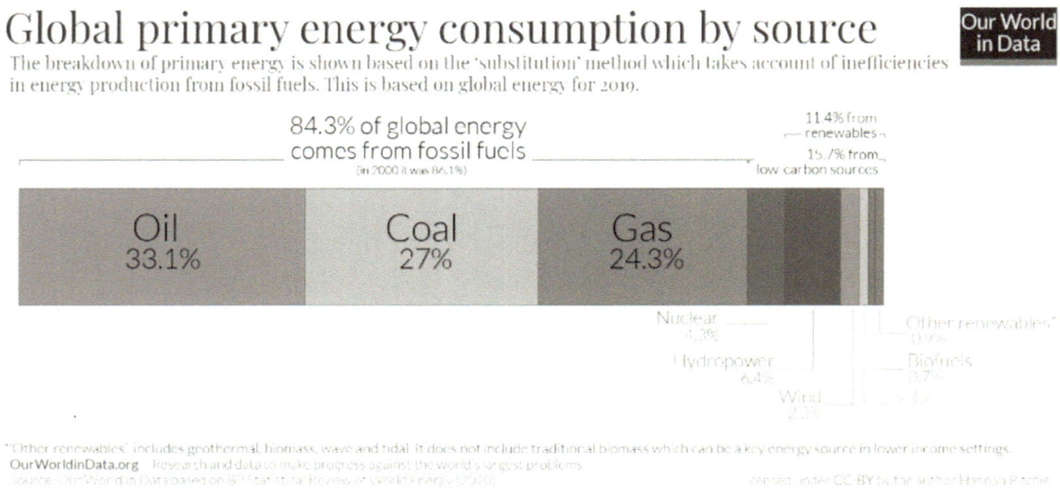

FIGURE 1.1 Global primary energy consumption by source.

Source: Our World in Data (2020).

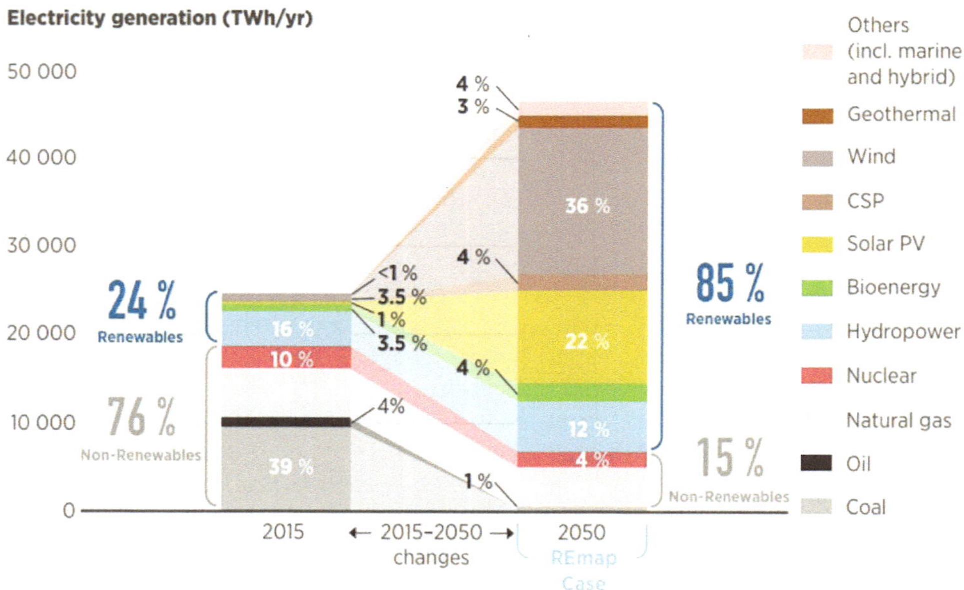

FIGURE 1.2 A 2°C scenario for electricity generation, 2015–2050.

Source: IRENA (2018).

will be supported by a structure of efficient storage facilities, as well as advancements in the fields of battery storage, reutilization of petroleum products, and CO_2 capture, among others. The role of the oil and gas industry will be critical in this transition, as they can remain as prominent players in this process with the appropriate level of research and development, as well as investment for new, greener technologies.

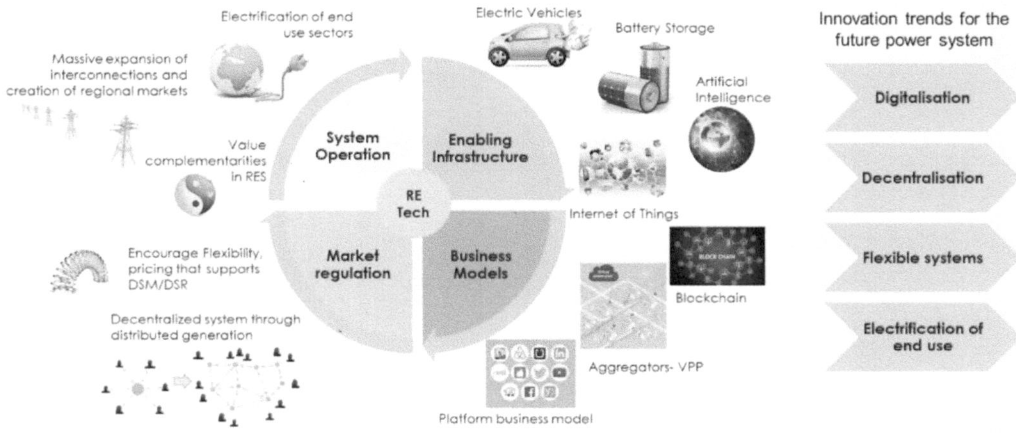

FIGURE 1.3 Power system innovations with renewable energy technologies.

Source: Gielen et al. (2019).

1.2 ENERGY TRANSITION, RENEWABLES, AND THE OIL AND GAS INDUSTRY

The energy transition process can be considered a significant challenge but is also an invaluable opportunity for the oil and gas industry. Thus, it can be regarded as an important test for the sector that produces conventional energy sources to power the world. Oil majors face important challenges to remain competitive in the market while working on solutions and paths toward expanding renewables (Oxford Institute for Energy Studies, 2018; Shojaeddini et al., 2019). However, due to these challenges, the energy transition process can be perceived as an opportunity for the industry to pioneer in the switch to renewables, leading other actors to follow suit and work for a greener future.

This section reviews the energy transition process in greater detail, starting with analyzing the current global energy situation and the steps significant players take for a responsible switch to renewables. Then, the petroleum products and by-products and their contributions to the worldwide energy situation are examined. We balance the vital contribution of these products to the worldwide concern with the challenges that maintaining the status quo can have for the environment. In the end, the process of energy transition involves careful considerations that need to be made by all the relevant actors to diminish the risks of climate change and potential resource scarcity.

1.2.1 Global Energy Outline

The global energy situation is starting to recoup after the worldwide health emergency caused by the COVID-19 pandemic. Figure 1.4 shows the significant drop in the global GDP, energy demand, and CO_2 emissions for 2020 and the expected rebound for 2021 compared to 2019. For a successful energy transition process, we must evaluate the current system's shortcomings in this moment of unprecedented disruptions and analyze the potential benefits of expanding the structure to welcome renewable energy into the mix.

Likewise, Figure 1.5 shows the extensive drop in worldwide carbon emissions by type of fuel in 2020. As one of the only positive points of a health crisis of this magnitude, these almost two years have brought up an opportunity to think about our use of conventional sources of energy and the

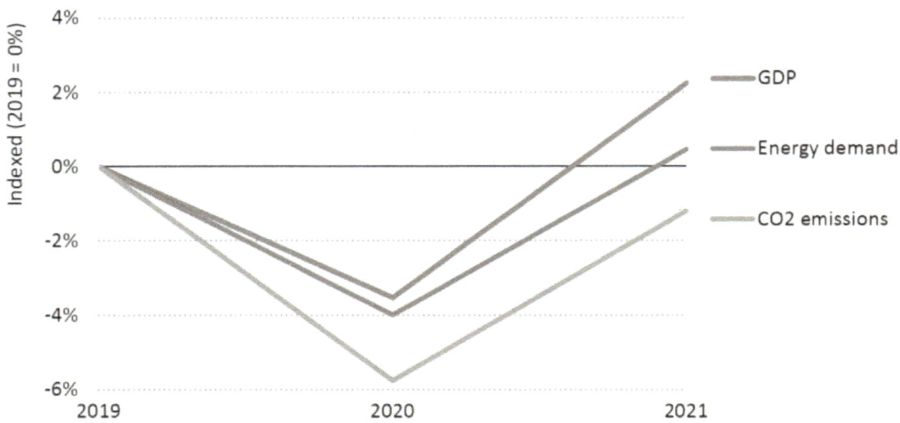

FIGURE 1.4 Evolution of global GDP, total primary energy demand, and energy-related CO_2 emissions relative to 2019.

Source: IEA (2021b).

FIGURE 1.5 Change in CO_2 emissions by fuel, 1990–2020.

Source: IEA (2021a).

potential to scale back on them while we give a larger room to renewables, as well as sources with a lower carbon footprint. It is in points like this in history that we can reflect on the effects of our actions and the potential avenues to remedy these damaging effects. With that said, these remedies also include a series of calculations and externalities (or unintended consequences) that we need to be aware of before implementing swift changes in such a fragile sector as energy.

To avoid having to take two steps back after taking a decisive step forward, researchers and engineers need to study all the branches that these challenges bring, like the different ways to store energy from renewables, the intermittence of these sources of energy, the short-term and long-term replacements that will be needed, the feasibility of hydrogen as the cornerstone of the new green energy sector, as well as the potential of more recent sources of energy, like biofuels, or of older sources which are going through a modernization phase, like nuclear, to power the lives of millions of people worldwide.

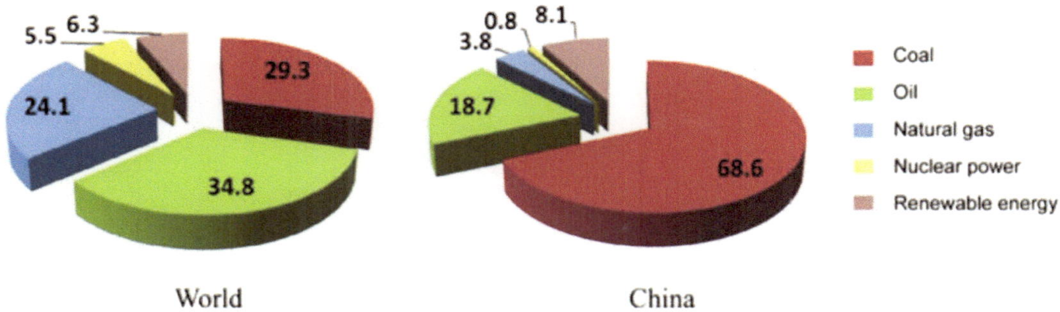

FIGURE 1.6 Structure of primary energy consumption in 2008, world vs. China.

Source: Zhang et al. (2011).

Developed countries have relied mainly on oil and natural gas for their energy needs for years. In contrast, developing countries have relied more on coal and other resources that can release more pollutants, like firewood, and even animal or human waste (Zhang et al., 2011). In the case of China, coal has been the primary energy source for many years, with oil as a distant second (see Figure 1.6). Because of this, the Chinese government has developed a plan focusing on energy conservation, efficiency in primary energy use, and the development of renewable, clean energy. Currently, the Asian giant is one of the leaders in wind energy materials production and the leader in installed capacity of wind farms (Lewis, 2016), which opens the door for them to make this a sustainable endeavor while helping the environment.

Similarly, advanced democracies in Europe and the United States have been developing their own large installations for renewable energy generation for years, and their own phaseout schemes for fossil fuels. Of particular interest is Germany's *Energiewende*, their strategy of energy transition from fossil fuels to renewables, which includes ambitious goals, like an 80% electricity generation by renewable sources by the year 2050 and, perhaps somewhat counterintuitively, a complete nuclear phaseout (Rogge and Johnstone, 2017). Plans like these contribute to the picture in Figure 1.7, with an increasing share of energy capacity being built for renewable energy sources.

1.2.2 Energy Transition

The energy transition process is the switch from conventional energy sources to more sustainable ones, like wind and solar. According to the International Renewable Energy Agency, the goal is to achieve this transformation by the second half of this century. Given the challenges brought up by climate change, moving the energy sector from fossil-based to zero-carbon sources is an urgent matter (IRENA, 2020). This challenge is different for each country due to their different degree of dependence on oil or natural gas, their degree of production, and the stage of development they are currently in. But the thing in common across all countries is that the effort is a global endeavor that reaches and affects all of us.

With that said, this process is bound to take a significant amount of time and will find several vital challenges and trade-offs. For starters, governments must evaluate the need for a transition strategy and weigh it next to all other considerations that can have political ramifications, like welfare or health programs. In some situations, a transition process can be less critical for a government seeking to solve a more urgent problem or one seen more favorably by the people and in the polls. Thus, prescribing a one-size-fits-all path toward energy transition for all countries without considering the context of each actor can be detrimental to the entire process and delay this much-anticipated progress.

In this sense, Chapman and Itaoka (2018) examine the case of Japan, where certain considerations need to be made and understood before an adequate transition process can move forward.

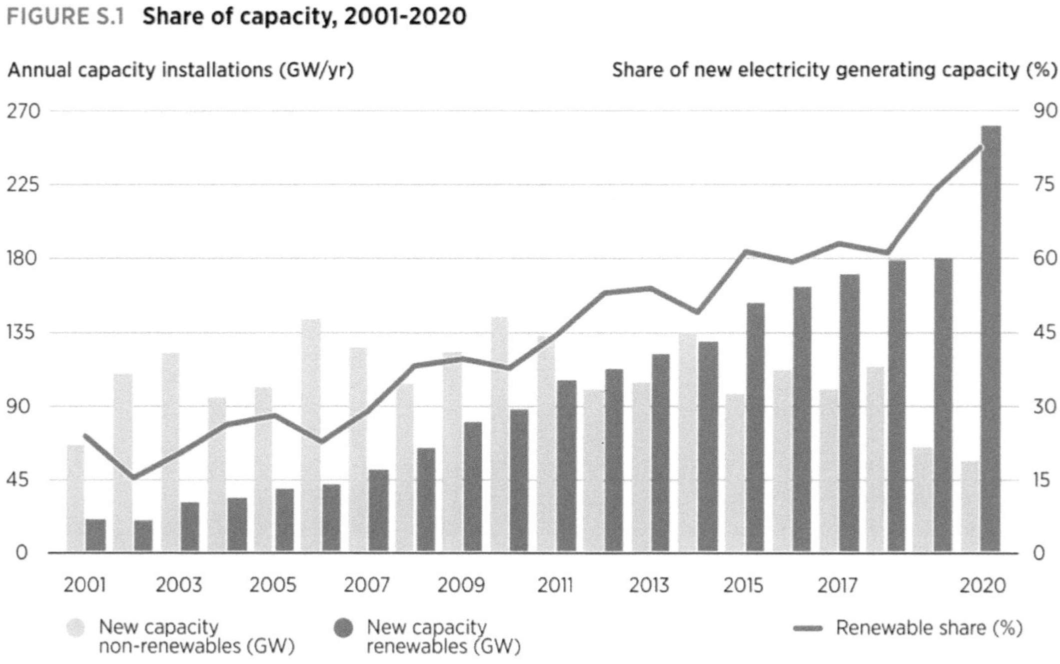

FIGURE 1.7 Share of capacity, 2001–2020.

Source: IRENA (2021).

For example, Japan is heavily dependent on oil imports, and there are some political challenges with the development of nuclear energy, which appears unpalatable for the population, given the Fukushima nuclear plant incident in 2011. These two factors alone have moved authorities in the country to consider a stronger domestic energy supply that also moves away from nuclear energy in their Strategic Energy Plan of 2014.

Therefore, when we start assessing other aspects that can influence transition directly or indirectly, we have enough variability across countries to consider a series of different pathways that can be applied to countries more appropriately. Figure 1.8 shows four different pathways that countries can follow, with some places able to fall into one or more of those pathways, depending on their contexts at a particular moment in time. In the case of Japan, previous experiences suggest a transformation or reconfiguration pathway, where niche innovations like solar PV and wind are reaching a maturity level that allows them to be incorporated into the electricity system (Chapman and Itaoka, 2018).

Given these significant differences at the starting point, it is reasonable to expect different trajectories for countries starting a transition process and varying outcomes once some time has passed. Furthermore, as it is recognized today, these transitions should happen faster than previous shifts, so countries with earlier transition experiences can be at an advantage today. The work of Solomon and Krishna (2011) studies three different cases of energy transition, identifying the most relevant factors that can help or hinder an energy transition process. By looking at successful transition projects between major energy sources in Brazil and France and a failed attempt in the United States, the authors seek to apply learned lessons and recommend paths to follow when initiating this newest worldwide transition endeavor.

The case of Brazil consisted of an expansion of energy options to include that obtained from sugarcane ethanol. Meanwhile, the study of France reviews their choice of evolution to include nuclear power while seeking to diminish oil extraction and production. Lastly, Solomon and Krishna (2011)

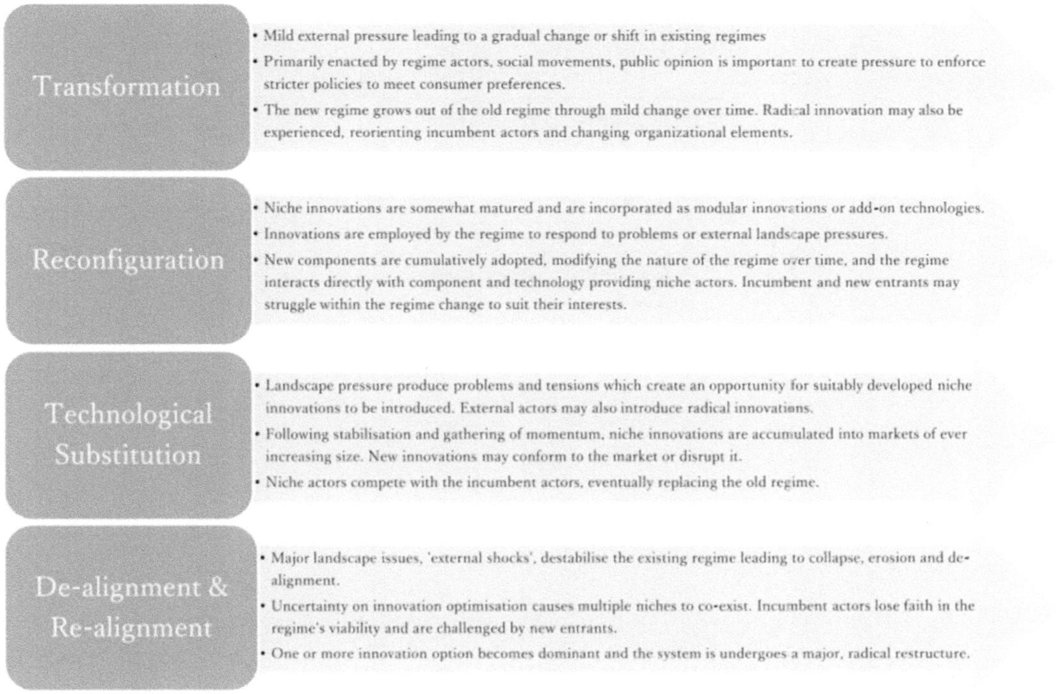

FIGURE 1.8 Potential transition pathways.

Source: Chapman and Itaoka (2018).

highlight the project of the United States to obtain energy independence. In all three cases, ambition and interest were high, but the failure of the American experience can be partly attributed to its much broader focus. Thus, the endeavor generated too many bureaucratic reorganizations, and several government agencies had to participate and meddle with each other to move the project alone, a move that ended up endangering it.

These cases preview what we can expect with a transition process that includes only a few countries but is a global project. The positive aspects center on the fact that this project is achievable for the most part, but timelines should consider a span of at least two to three decades, which can put considerable pressure on many countries with short-term political considerations to make before embarking on such ambitious projects. Suppose governments and relevant actors maintain a focus on the sequence of projects to develop and the citizenry allows for space for these projects to grow and ultimately flourish. In that case, ideally, the world will benefit from moving away from conventional energy sources to power our lives.

These processes are problematic because they involve most people's lives, and some citizens will have to endure higher costs than others. Hence, these processes are highly societal, because transitions of this magnitude involve political institutions, organizations, governments, and industries. The work of Andrews-Speed (2016) establishes some clear parallels between social transitions and the switch from fossil fuels to a low-carbon energy structure. Following the institutional theory, regime transition is another equally lengthy process that requires many actors to influence and modify behaviors when responding to new incentives. Figure 1.9 shows a schematic with a hierarchy of behaviors that actors follow when institutions and norms around them shift.

When governments and market forces meet technological advances and a need to change the game, new incentives are created, leading to behavior change or adaptation and a citizenry that

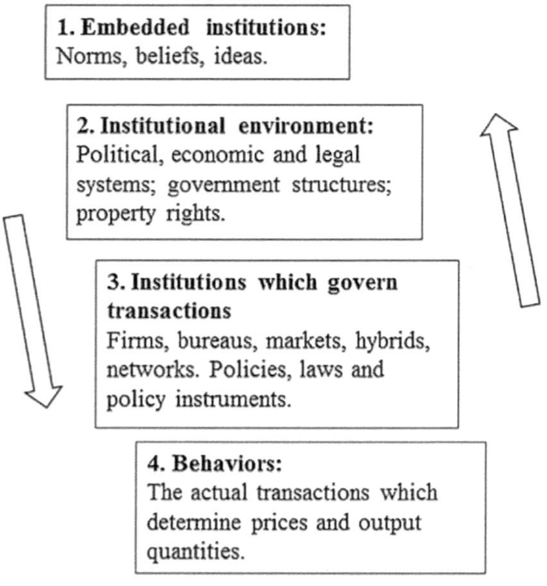

FIGURE 1.9 The three levels of institution which collectively govern actor behaviors.

Source: Andrews-Speed (2016).

can be positively responsive sometimes but not so much at other times. Public policies can help steer those behaviors, but incentives ought to be the appropriate ones. Thus, if an adequate level of investment is secured from a government and the market adapts and generates the necessary paths to development and growth for the most significant number of people, an energy transition process will be less costly. When this optimal point is reached and the transition process is understood and used for the advantage of society, it can be a net positive for many countries.

Regarding public policies governments can implement to incentivize an energy transition process, we can think of limits for greenhouse gas emissions. These are a subset of environmental regulations, such as eco taxes or environmental agreements, that seek to regulate the behavior of private actors (Millot et al., 2020). According to the work of Bailey and Rupp (2004), new forms of environmental regulation have been developed by governments to provide more incentives to reduce impacts on the environment instead of forcing firms to comply through more coercive means. New environmental policy instruments (NEPIs), like eco taxes, tradable permits, and agreements, aim to limit industries' pollution according to international regulations and government expectations.

With more substantial pressure from the international community to respect environmental agreements, NEPIs have become a tool of regulation for polluters. Bailey and Rupp (2004) explore NEPIs in the United Kingdom and Germany in their work. The between-country differences mentioned earlier appear here once again, as the UK has been able to integrate NEPIs more readily than Germany, a country with stronger regulatory traditions. Thus, even within developed nations, the results of regulations that look at tackling the same problems can have varying effects that are region-, country-, or state-dependent.

Additionally, these differences can be noticeable across industries as well. As the analysis performed by Sullivan (2009) suggests, significant differences in response and behavior can be observed between companies that have to operate under environmental regulations that seek to manage greenhouse emissions. From reporting their emissions data to establishing basic management systems and processes to deal with emissions if they become a problem, companies go through several stages of adaptation that start from different points and are affected by resources, time, and both the degree of compliance and as well as enforcement from the regulatory agency. For some companies, the message that "significant reductions in emissions will be required to avert the worst consequences of climate change" is not received equally as for others.

Moreover, we must factor the European Union into the equation in the European case. This supranational organization oversees legislation and implements regulations with different degrees of application locally. In their study of several locations in Italy, Arbolino et al. (2017) find that greenhouse regulations for the transportation sector, tailored for all European Union countries, are applied differently, even within the same country. The cities in the country's north are more incisive and dynamic and focused on reducing emissions. In contrast, the locations in the country's south show more reticence toward the European directives and a slower pace to transition to less-polluting energy.

All these regulations have the ultimate objective of reducing pollution and achieving what is known as "net-zero emissions." This translates into limiting the use of carbon dioxide as much as possible to avoid a potential increase in global temperatures that can be triggered by the indiscriminate use of conventional sources for human consumption. According to the Intergovernmental Panel on Climate Change (IPCC) report for policymakers (2018), reaching net-zero global anthropogenic CO_2 emissions would halt the warming of the Earth and would help keep climate-related risks at bay.

That said, not all sectors can be transformed rapidly and made available for a swift transition. Light-duty transportation, heating, cooling, and lighting can follow relatively quick paths to decarbonization. However, services like aviation, long-distance transport, production of materials like steel or cement, or the implementation of a reliable electricity supply are sectors that need a longer planning time and thus will take a longer time to transition to fewer pollutant forms of energy (Davis et al., 2018). Consequently, even the best-laid plans require time to be realized. Even then, there can be significant challenges to avoid falling back on using conventional energy sources if the transition process suffers a setback.

Its reliance on modeling is a significant challenge to the energy transition process. Statistical techniques that predict scenarios with data that has been input into the models have shortcomings, like omitted variables that translate into a valuable context that could be missing and the possible overdetermination of some variables that are given more significant importance in the modeling than they have in real life. Therefore, to obtain better models that reflect realistic net-zero scenarios, some old ways of modeling must be revised (Pye et al., 2020).

Some of the current tools to include more data in models is using marginal abatement cost (MAC) curves. MAC curves demonstrate the expense of lowering emissions from various sources. In a simple illustration of such a curve, Figure 1.10 shows a series of boxes where each box signifies a different opportunity to reduce emissions (Climate Works Australia, 2019). These boxes are ordered by how costly it becomes to reduce a ton of CO_2 for a given year, so a graph will reflect a series of potential paths to follow, according to their cost, compared to doing nothing at all. Hence, we can see that choosing one measure over another will have marginal costs compared to other strategies, so researchers can have a better view of potential avenues to follow when dealing with cutting emissions.

A significant contribution of MAC curves is that of working with real-world data. By using real-world data, MAC differs a bit from some strains of modeling techniques that rely on simulations or predictions, which are performed from real-world data but can have assumptions that may not always apply to the real world. When considering a reduction of greenhouse gases (GHG), MAC curves can serve the field well, as their reliance on actual data can bridge the gap that appears when the shortcomings of some modeling techniques appear (Ali Almihoub et al., 2013).

Moreover, another vital piece of real-world data is carbon pricing and its variability across world regions. This is an important consideration when factoring in the likelihood and projections of abatement models that seek to reduce GHG worldwide. Carbon pricing serves as an instrument to capture the (sometimes hidden) emissions costs and return it to the actors producing it to signal to

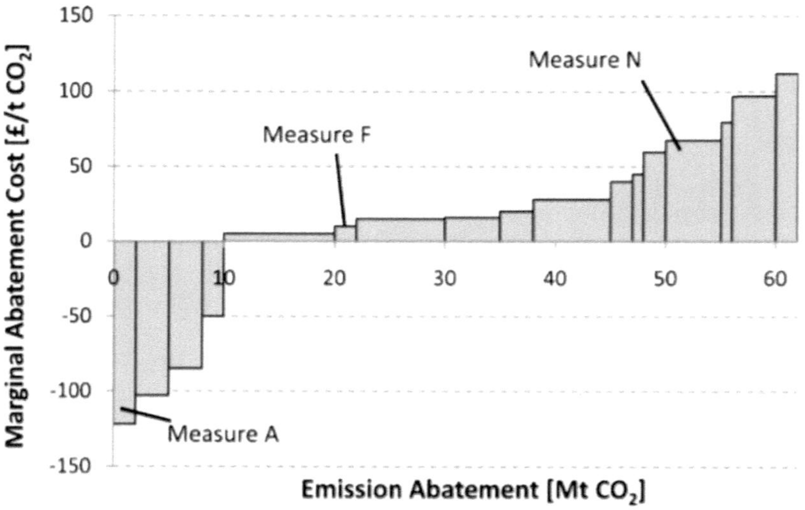

FIGURE 1.10 Sample MAC curve.

Source: Kesicki and Strachan (2011).

them that their production is costly for other actors in society (The World Bank, 2021). In a world that seeks to achieve the commitments made in the Paris Agreement, carbon pricing intends to place a cap on actors that go over their emissions thresholds, planning to keep a fairer distribution of the incentives and the penalties for stepping out of line.

Carbon pricing is seen as the most efficient policy to reduce GHG (Steckel et al., 2021). Still, some equilibrium needs to be found when placing a burden on countries and regions that might benefit from short-term excess emission levels that could lead to more significant economic development. It is more likely that nations will quit relying so heavily on conventional energy sources, like coal, once they have reached a certain level of development. Relatedly, other countries have been able to find alternative energy sources more easily, which has led to lower carbon pricing, but without the risk of losing energy production or risking shortages, like it would happen in less-developed countries. This has been the case in the United States after the shale gas boom of the early 2010s (Kim and Lee, 2015).

Similarly, carbon pricing could affect industrial competitiveness, especially in regions that depend on a particular amount of production. For the United States, Casey et al. (2022) propose a flexible pricing scheme that considers the degree of competitiveness between a plant in a state subject to carbon pricing and plants in other states or countries. Given the differences within jurisdictions, the blind implementation of a carbon pricing strategy is bound to overlook challenges and potential costs to local producers. Thus, a particular price may increase unemployment in the affected region and scare away jobs overseas or in other states.

In summary, many factors influence carbon pricing in different parts of the world, and the potential externalities of making arbitrary changes are another part of the equation. Furthermore, and more generally, the whole energy transition process is limited by geographical, economic, and contextual constraints, which make this path different for every actor seeking to initiate the process. Thus, even though the world has come to the agreement that quick change is needed to revert the potential effects of a warmer climate, there are many steps to this process, and it will be a very different path for all actors that start this endeavor.

The rest of this section explores the creation of products and by-products from petroleum, from elements that power many aspects of our lives to tools and utensils that are present in almost every part of our daily lives. This exploration is performed to highlight the importance of petroleum not only in one pivotal aspect of our lives but also as something that is an integral part of the development of countries and, as such, something that needs a thoughtful process for its decreased use and eventual replacement for more sustainable sources of energy. The ideal strategy for energy transition may not exist. Still, a process that is both positive for the environment and promising for the country that decides to initiate it is possible. We hope that this chapter provides some guidance for it.

1.2.3 Petroleum Products and By-Products

An essential part of the oil extraction and refinement process consists of the many products and by-products obtained from it. Thus, an energy transition process should include manufacturing these products or repurposing some of them to adapt to future needs. Additionally, the demand for these products shows no signs of slowing down (Forman et al., 2014; Speight, 2014), and given the lack of satisfactory alternatives, we are presented with another significant challenge for a dramatic reduction in the extraction and use of petroleum and the manufacturing of products and by-products from crude oil (see Figure 1.11 for an illustration).

One of the main petroleum by-products is the gasoline we put in our cars, but a list of derived products includes many more that encompass multiple areas of our lives. The Colorado Oil & Gas Association (2021) has provided a list of products made from oil and natural gas, including deodorants, lipstick, materials like nylon and polyester, and the smartphones millions of people have in

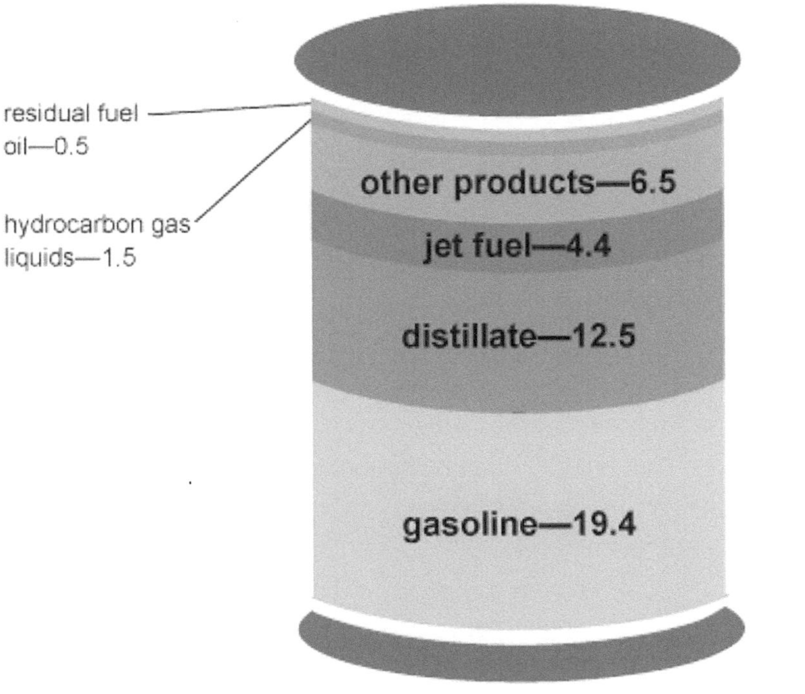

FIGURE 1.11 Petroleum products made from a barrel of crude oil, 2020.

Source: U.S. Energy Information Administration (2021).

their homes. Additionally, many health-related products that are partly made with products derived from petroleum are dentures, artificial limbs, wheelchairs, eyeglasses, and casts (see Figure 1.12).

Meanwhile, in his *Handbook of Petroleum Product Analysis*, Speight (2014) classifies oil's related materials into three categories: materials of natural origin, manufactured materials, and integral fractions derived from the two first categories (see Table 1.1). From the first group, we can think of natural gas, coal, and different kinds of refined oils that power most of our lives today. Then, as noted previously, many manufactured materials cover the everyday needs of millions of people, like wax and asphalt. Lastly, the feedstocks derived from petroleum are used to make daily

FIGURE 1.12 Everyday products made possible by oil and natural gas.

Source: Colorado Oil & Gas Association (2021).

TABLE 1.1
Subdivision of Fossil Fuels into Various Subgroups

Natural Materials	Derived Materials	Manufactured Materials
Natural gas	Saturates	Synthetic crude oil
Petroleum	Aromatics	Distillates
Heavy oil	Resins	Lubricating oils
Bitumen*	Asphaltenes	Wax
Asphaltite	Carbenes†	Residuum
Asphaltoid	Carboids†	Asphalt
Ozocerite (natural wax)		Coke
Kerogen		
Coal		

Source: Speight (2014).

* Bitumen from tar sand deposits.

† Products vof petroleum processing.

products, like plastics, trash bags, paint, pillows, detergents, and even some clothes (Ranken Energy Corporation, 2017).

The current moment invites a rethinking of refineries' work and functioning requirements. Currently, in the United States, refineries are operated strictly to maximize profit (Forman et al., 2014). Suppose one of the global goals of the current couple of decades is to reduce greenhouse emissions and petroleum extraction is seen as a direct contributor to climate warming. In that case, legislation should target these refineries and make them shift their operation objectives to be tailored more to efficiency (Elgowainy et al., 2014). While researchers develop and expand the role of renewable energy, governments can work on creating incentives for refineries for efficiency and innovation that can help move us away from relying strictly on fossil fuels for many of our daily needs.

However, once again, these levers cannot be manipulated without paying close attention to the context around every proposed modification. There are many important variables at play in every interaction, and with the consequences that making radical changes could have on the lives of millions or billions of people, these modifications should follow a closer inspection. Some policies implemented in one country can have positive consequences in that same country but negative consequences in another. For example, the work of Thompson et al. (2011) shows the effects of biofuel policies in the United States on local production and overseas markets.

Conceived to combat the emission of greenhouse gases, these policies can have the best of intentions, but some externalities that appear after the implementation of these regulations can, in fact, lead to increased use of petroleum when some conditions are met. As Figure 1.13 illustrates, the outcomes of regulations and taxes depend on the degree of enforcement of these mandates and other contextual variables (Thompson et al., 2011). As we can see, the change in emissions for one fuel type can be offset by the increase of another. Thus, despite a change, the overall situation may not have changed for the better. This may be especially true when considering the ripple effects these regulations could have on employment or the general state of the economy.

Additionally, refineries have different degrees of energy use according to what they produce from crude oil (Wang et al., 2004). This allocation of resources is another factor in the search for diminishing emissions, as incentivizing less production of something actively being replaced thanks to innovation and research could help significantly reduce emissions without even considering a lower degree of extraction. Table 1.2 summarizes the energy allocations for products derived from 1 kg of crude oil. We can see that the most energy-intensive processes are in producing gasoline and kerosene. Thus, progress in achieving efficiency across refineries is another road to explore on the path to a successful energy transition.

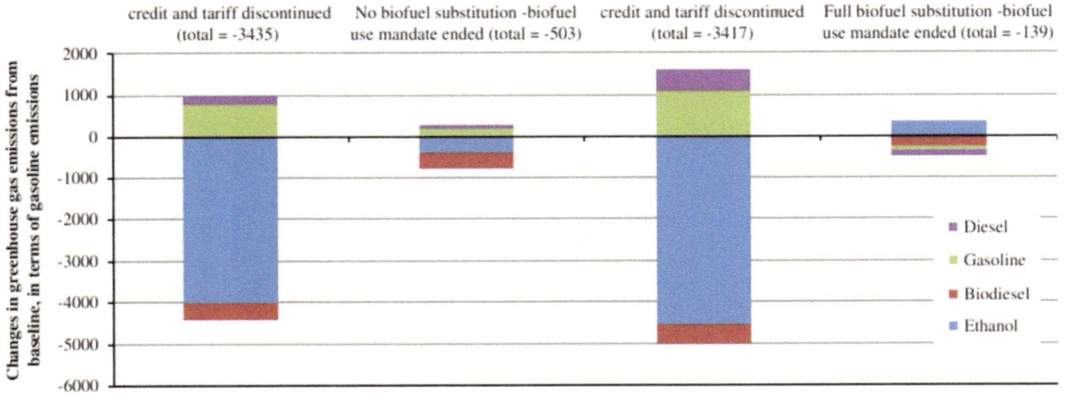

FIGURE 1.13 Policy effects on greenhouse gas emissions, average of 2012/2013–2019/2020.

Source: Thompson et al. (2011).

TABLE 1.2
Energy Content–Based Process Energy Allocation by Final Product: 1 kg of Crude Feed

Final Product	Mass (kg)	Mass Share (%)	Allocated Energy Use				Allocated Energy Use Share (%)				Energy Intensity (%)
			Fuel (kJ)	Electricity (kJ)	Steam (k)	Total (%)	Fuel	Electricity	Steam	Total	
Residual oil	0.004	3.1	0.1	1.6	4.8	0.4	0.1	0.1	0.2	0.2	41.3
Fuel (still) gas	0,044	92.9	6.2	42.1	141.2	4.4	4,2	6.1	6.5	4,8	110.0
Naphtha	0.001	0.5	0.1	0.9	1.5	0.1	0.0	0.1	0.1	0.1	50.9
Diesel	0.094	147.6	9.4	39.9	196.9	9.3	6.7	9.3	6.1	6.7	71.8
Kerosene	0.137	192.5	11.5	45.6	249.6	13.6	8.8	11.4	7.0	8.5	62.4
Gasoline	0.465	1,334.7	53.6	184.4	1,572.6	46.0	60.7	53.2	28.4	53.3	115.9
LPG	0.058	28.4	5.1	44.8	78.3	5.7	1.3	5.0	6.9	2.7	46.2
Gas oil	0.045	126.6	2.9	18.6	148.1	4.5	5.8	2.9	2.9	5.0	112.7
Heavy fuel oil	0.040	58.9	3.1	12.8	74.8	4.0	2.7	3.0	2.0	2.5	64.0
Lube stocks	0.070	111.4	5.3	202.1	318.8	6.9	5.1	5.3	31.1	10.8	156.0
Asphalt	0.020	46.8	1.5	23.5	71.8	2.0	2.1	1.5	3.6	2,4	123.0
Waxes	0.009	14.7	0.7	26.6	42.0	0.9	0.7	0.7	4.1	1.4	160.0
Coke	0.005	12.1	0.3	3.6	16.0	0,5	0,6	0.3	0.6	0.5	109.6
H_2 gas	0.005	21.2	0.8	1.4	23.4	0,5	1.0	0.8	0.2	0.8	160.3
H_2S gas	0.013	6.2	0.3	1.6	8.2	1.3	0.3	0.3	0.2	0.3	21.5
Total	1.010	2,197.8	100.7	649.4	2,947.9	100.00	100.0	100.0	100.0	100.0	100.0

Source: Wang et al. (2004).

We have seen that the energy transition process is located within a context with plenty of moving parts closely related to each other. This means that, despite expecting significant benefits in one aspect after pulling some levers, there is a chance of causing other unexpected effects in other areas, some of which could be negative. Furthermore, these effects can be reflected not only in the oil and gas industry but also in other aspects of the economy. An example scenario could occur when, after emissions-saving modifications are implemented, there is a negative outcome in a town or city that requires energy to continue functioning properly, which makes them rely on more contaminant sources to satisfy that need.

In other words, if the focus is put exclusively on decreasing greenhouse emissions, without much concern about the international context, local production, employment, and the dynamics between markets, we run the risk of policy backfiring. A policy too narrow and strict has the downside of dismissing other contexts and alternative solutions. At the same time, an approach that is too expansive and lax runs the risk of being ignored altogether. If the urgency to combat the potential negative consequences of a warmer climate is manifest, researchers and policymakers need to develop policies that are broad enough to consider alternative dynamics, other markets, and other considerations, like the state of the economy. But at the same time, these regulations need to come with enough credible threats against actors that do not comply; this is not an easy task.

1.2.4 Energy Transition and Its Impact on Petroleum Products and By-Products

As we have reviewed so far, a process of energy transition from conventional energy sources to primarily renewable energy has several steps and moving parts that need to be considered individually and contextually. This section summarizes some strategies that have appeared to replace petroleum-based products so that countries can advance toward a successful energy transition. For example, a few countries have embraced soybean oil to produce rubber materials, and others have instigated the use of biofuel like ethanol to replace diesel, a petroleum-based fuel. Similarly, asphalt binders have been partially replaced by bio-binders made from reconverted swine manure in some places.

Some of these strategies have started a promising endeavor that repurposes a previously discarded element to replace a petroleum-derived product that needs to decrease its circulation. Innovation of this kind is an excellent addition to the list of alternatives to fossil fuel–based products, which is a task that, given its magnitude, needs to be tackled on multiple fronts. Knowing that petroleum has been integrated with the lives of millions of people to such a degree that it touches many of the everyday aspects of their daily lives, going from the plastics they use to the computers and phones they rely on to the cars they drive, a strategy of transition must be deployed across many dimensions of society.

First, biodiesel, as an alternative to petrol diesel, has been a salient topic of discussion for several years. Responding to initial shortages of crude oil but now moved more by the potential dangers that maintaining current oil production levels could have for the environment, oils of plant origin and some animal fats have been considered alternatives (Ajala et al., 2015). Despite having some similarities to petroleum-based fuels and holding a better record in emissions (see Figure 1.14), biofuels are still more expensive to produce. There are some issues with adapting the vehicles and catalysts to work with this new type of oil.

Additionally, were this alternative to gain traction among countries, some of these actors would be faced with a hard choice between adapting soil to produce mostly biofuels or having abundant food to export or keep in their national markets. This is particularly relevant when sustainability issues also appear in this fuel, as water and soil conservation is an essential consideration when countries are to take a step toward the mass production of biofuels (Liska et al., 2014; Payne, 2010). It appears that producing biofuels may not be economically viable for many countries. The trade-offs arising in this new resource's market can leave some actors out of the competition quickly while the root problem is not being solved.

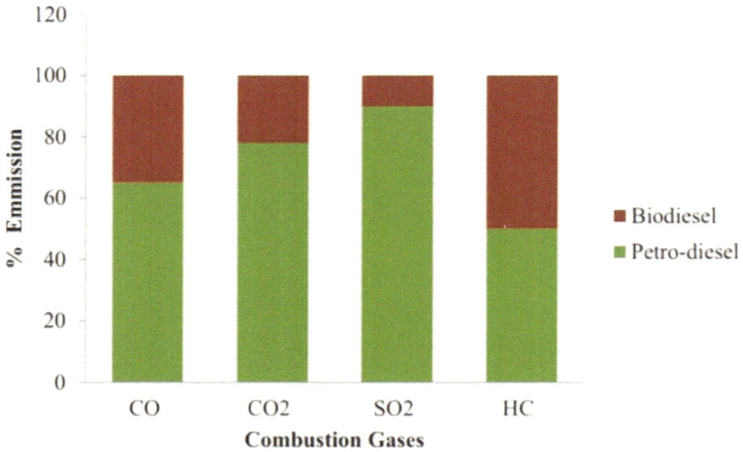

FIGURE 1.14 Comparison of the percentage emission from biodiesel and petrodiesel from compression ignition engines (CIE).

Source: Ajala et al. (2015).

Moreover, some researchers argue that the potential benefits of investing heavily in biofuels may not outweigh the costs. For example, Gomiero (2018) argues that the picture of biofuels functioning as a competitive alternative to fossil fuels has been painted too optimistically. Besides requiring many subsidies, biofuels have poor energy performance while conflicting with food production, altering the soil's quality and health. Alternatively, other researchers have proposed using high-oil-content microalgae to produce biofuel (Gouveia and Oliveira, 2008). All in all, several important questions remain regarding the development of biofuels that need further exploration, especially when externalities could affect millions of vulnerable people.

Then, asphalt is a pivotal element in the development of a country. Advanced roads can serve as an excellent proxy for a population's quality of life, and petroleum can be found in the middle of this development. But because of the preoccupation with the potentially negative consequences of a steady increase in global temperatures worldwide, some alternatives must be found such that high-quality roads continue to be built without relying so much on petroleum products. One of these solutions is located in the development of bio-binders (Fini et al., 2012; Peralta et al., 2012), which are comparable alternatives to asphalt binders placed on a layer of a paved road.

Peralta et al.'s (2012) research examines how a bio-binder constructed of fractionated bio-oil reacted with crumb rubber can be comparable to a conventional asphalt binder used to pave a road. The authors report a high degree of comparability and similar behavior of the biofuel binder with respect to the asphalt binder. Meanwhile, Fini et al. (2012) tested a binder produced from pig manure, finding an acceptable degree of comparability with the asphalt binder. Both works warn of the potential pitfalls of using increasing amounts of biofuel to cover these needs. Still, they are also encouraging this technology, given the current need to reduce CO_2 emissions.

Once again, trade-offs are a significant consideration when governments are confronted with the choice to invest large sums of money while redirecting the use of large amounts of soil that could be used to grow food for millions of people. Country leaders seeking to make these moves are also confronted with at least two other problems. The first consists of the diminishing returns of investing longer in these technologies. This occurs in a context that rewards what is known as "early movers," so every day that authorities in one country decide not to invest in technologies like biofuel, there could be someone else out there that is deciding to do it, and they will start reaping any potential benefits first, ahead of the country that took longer to decide whether to make that investment or not.

And second, there are political and electoral ramifications to making these types of decisions. Some countries have a greater degree of trust in their politicians and will be less likely to reward or punish with their votes when the next election comes. Therefore, keeping all else constant, politicians who see their careers less threatened by a significant social environment change will likely be more inclined to embrace substantial changes that could reap considerable electoral benefits. On the other hand, fragile political majorities are expected to have more reluctance to initiate significant structural changes, particularly on an issue like energy, affecting all electorate members. Thus, a change of this kind constitutes a gamble that could benefit popular politicians but could very well bury unpopular ones.

As we can see, the management and impact of petroleum by-products have ramifications that go beyond an energy sector consideration and touch upon other social areas that affect virtually the entire planet. So for example, if the lack of gas to power cars can be a problem that large, developed countries have to a greater extent than poor and small countries with bad road connections, the long list of petroleum by-products eventually touches all countries almost equally. Therefore, the control and potential phasing out of these products appear to be much more complex than the energy transition process, already a challenging process that requires the collaboration of many sectors of society.

Therefore, in this aspect, more than in others, pulling the levers of government to regulate the management of these by-products can have unexpected consequences on areas not considered relevant initially. With that said, to assess the potential success of the future energy transition, looking at the alternatives to petroleum by-products and the embrace they receive from society is an excellent signal for the advancement of the switch to an energy sector dependent on renewable energy. If the needs of millions of people are met with reliable alternatives to the more polluting current products, the path to the new energy paradigm will start with promising support from those who will need it the most to achieve the long-desired economic development.

1.2.5 Carbon Capture and Storage (CCS) and Applications in the Oil and Gas Industry

To achieve climate change mitigation, one technique that has gained significant momentum in the past few years is carbon capture and storage (CCS), also known as carbon sequestration. This consists of several methods of recovering CO_2 from the atmosphere, involving its capture, transport, utilization, and storage from several perspectives (see Figure 1.15). One subset of CCS techniques is the capture of carbon dioxide at power plants, which can, for example, be injected into reservoirs for long-term storage (Szulczewski et al., 2012). This process of recovery and storage can be achieved through several methods, where the most common are absorption, adsorption, separation by membranes, and cryogenic separation (Pires et al., 2011).

Absorption is a physical or chemical process where atoms, molecules, or ions are dissolved in bulk. This is done to treat industrial gas streams that contain acid gases like hydrogen sulfide (H_2S), nitrogen oxides (NO_x), and carbon dioxide (CO_2). Then, adsorption occurs when molecules in liquid or gaseous mixtures are attached to a solid surface (the adsorbent) through a selective capture process. Meanwhile, gas separation with membranes is a low-cost technology that permeates gases via vacuum conditions. Lastly, cryogenic distillation is an air separation process where different elements are separated through condensation (Boot-Handford et al., 2014; Fernandes Araújo and de Medeiros, 2017; Pires et al., 2011; Tan et al., 2016).

Along with other human-made processes of sequestration and storage of carbon dioxide, another essential way to sequester carbon in the atmosphere centers on enhancing natural processes that increase carbon removal, like forestation processes that increase vegetation (Herzog and Golomb, 2004). Additionally, degraded agricultural soils, pasture management, and soil organic carbon (SOC) are three other methods where the combination of human techniques and nature performing its own work of transforming energy can reap desirable outcomes in the storage of carbon in soils across the world (VijayaVenkataRaman et al., 2012).

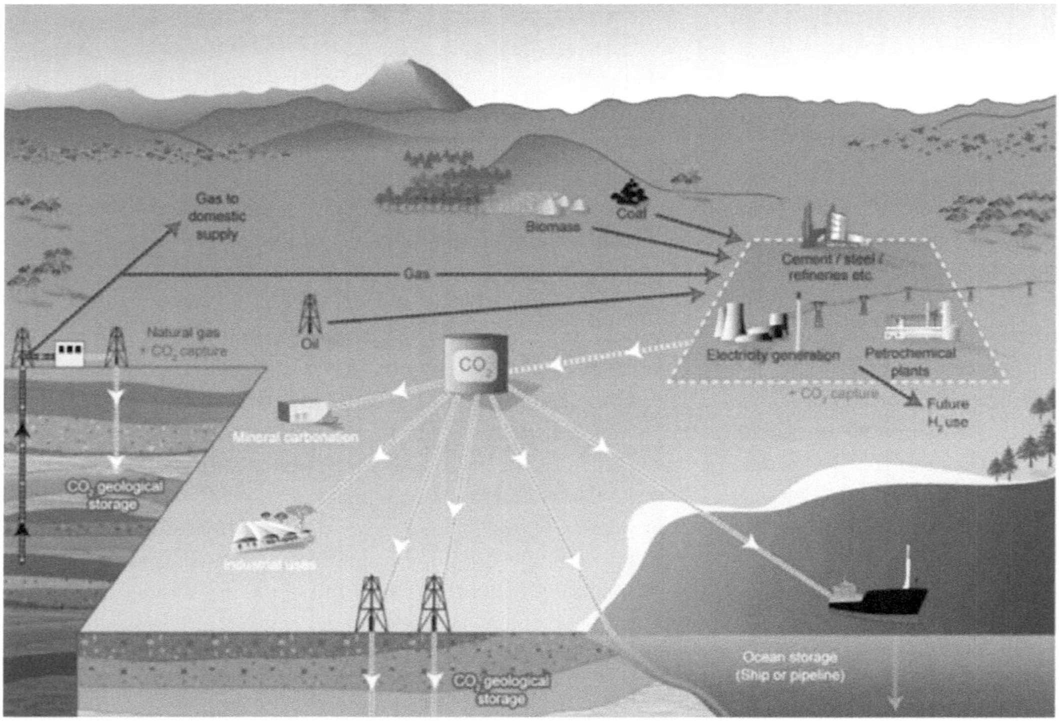

FIGURE 1.15 Schematic view of the carbon capture and storage (CCS) chain.

Source: Pires et al. (2011).

In this endeavor, private actors have also started their capture initiatives in the oil and gas industry. For example, ExxonMobil established a new branch, ExxonMobil Low Carbon Solutions (ELCS), which commercializes CCS technologies and develops CCS initiatives, like the Houston Ship Channel CCS Innovation project, set in a critical industrial cluster with offshore storage in the Gulf of Mexico. Likewise, Siemens and Aker Carbon Capture have joined forces to develop their projects, just like Schlumberger and LaFarge Holcim, and G2, NETPower, and EJM have their own (Global CCS Institute, 2021). Table 1.3 lists carbon capture and storage projects in different stages of development worldwide to gauge how CCS projects have been advancing.

The oil and gas industry might specifically benefit from several CCS storage projects by using the captured CO_2 and injecting it into depleted reservoirs to increase the pressure inside of them and speed up oil recovery (Bui et al., 2018; Pires et al., 2011; Tan et al., 2016). Additionally, the feasibility of offshore power plants with CCS installations that help decarbonize oil and gas platforms and have the potential to decarbonize mainland electricity installations was explored by Roussanaly et al. (2019) with promising results. Figure 1.16 illustrates a cycle CO_2 could follow after we include the need to protect the environment with every energy endeavor we pursue.

Strategies of sequestration and capture of carbon are part of a series of measures compatible with the current state of things in the energy sector, where several actors continue to pollute the environment to worrisome levels. However, these initiatives are also appropriate to pave the way for the energy transition of the 21st century. Thus, CCS initiatives are sorely needed in a world that generates significant greenhouse emissions. However, despite being a welcome improvement for the less-polluting industry, policies to stop the majority of the CO_2 produced in these operations from being released into the atmosphere and causing global warming are still needed.

TABLE 1.3
Largest Contributors to the Growth of Projects in Development, 2021

Plant Early Development	Industry	Country	Mean CO_2 Capture Capacity (Mtpa)
Dave Johnson Plant	Electricity generation	United States	4.00
G2 Net-Zero LNG	Natural gas processing	United States	4.00
Next-Decade Rio Grande LNG	Natural gas processing	United States	5.00
Keadby 3 Power Station	Electricity generation	United Kingdom	2.10
Repsol Sakakemang	Natural gas processing	Indonesia	1.80
Barents Blue Clean Ammonia	Chemical production	Norway	1.60
Shell Refinery Rotterdam CCS	Hydrogen production	Netherlands	1.20
Stockholm Exergi BECCS	Electricity and heat generation	Sweden	0.80
Air Liquide Refinery, Rotterdam CCS	Hydrogen production	Netherlands	0.80
Lawler Biorefinery CCS	Bioethanol production	United States	0.53
Copenhill (Amager Bakke) Waste-to-Energy CCS	Waste processing	Denmark	0.50
Casselton Biorefinery CCS	Bioethanol production	United States	0.47
Marcus Biorefinery CCS	Bioethanol production	United Slates	0.43

Source: Global CCS Institute (2021).

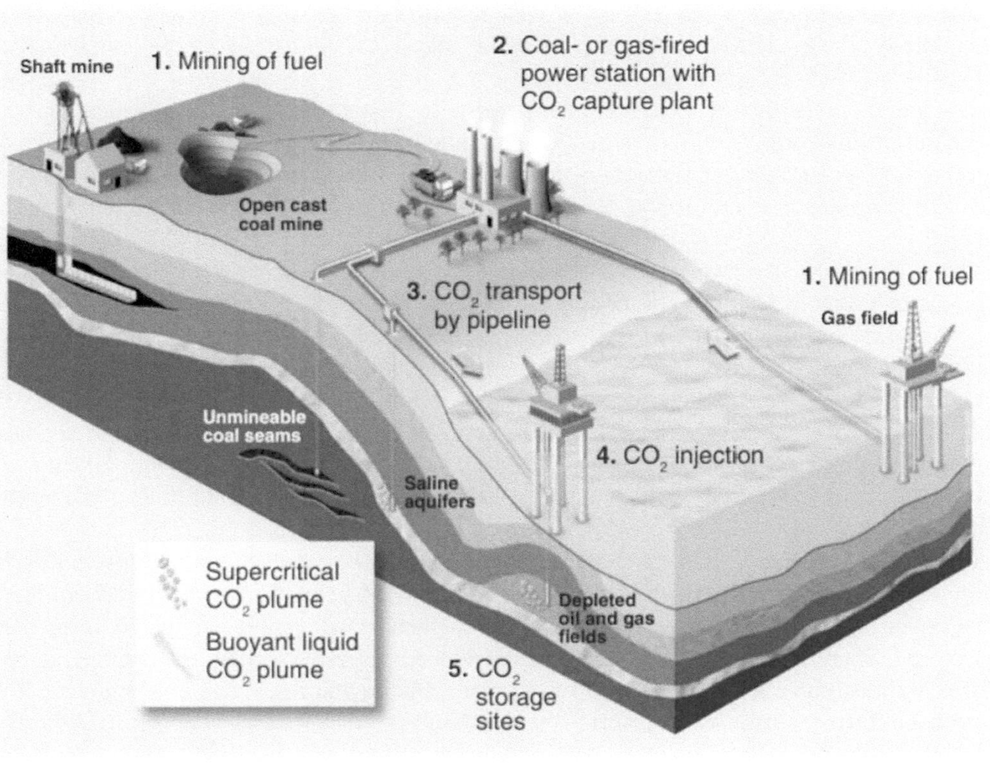

FIGURE 1.16 Diagrammatic representation of the life cycle chain of fossil fuel use. CO_2 separation and capture at power plants enable the storage of CO_2 in porous rocks deep belowground.

Source: Haszeldine (2009).

1.3 FUTURE DIRECTIONS OF OIL AND GAS AND RENEWABLES

In energy terms, what we do today will significantly determine the world's future. We are currently faced with depleting resources and increased global temperatures, closely associated with the amount of carbon dioxide humans produce in powering our lives. This is forcing us to develop solutions that tackle global greenhouse gas emissions to a level that limits that increase. Global warming threatens our way of life, but throughout history, human ingenuity has developed alternative plans and potential solutions that alleviate most of our problems.

Many of these discussions on switching to renewables come from countries that have been able to experience economic development brought by conventional sources of energy, like coal and petroleum. Therefore, it may be considered unjust to subject developing countries to strict environmental agreements which may hinder their economic development. With that said, some developing nations have been able to leapfrog from conventional energy sources to renewables without relying on fossil fuels as much as developed nations did in the past. Ideally, some balance ought to be found between prohibitive orders and environmental considerations, allowing all countries to seek the economic development they desire without compromising the planet.

To achieve a balance between environmental considerations and harsh regulations, we require an immediate communication process between legacy industries, like oil and gas, and the new renewable sectors, as well as with governments and international organizations. Researchers from all these sides have come up with reports and potential solutions to the problems brought by an increased global temperature and recognize the need for close work between all actors. For example, oil companies that have successfully powered the world are now investing in cleaner technologies while inviting governments to finance research investigating biofuel produced from algae in offshore plants.

The oil and gas industry can not only provide research and investment for new technologies but also improve how it handles its business. This way, the sector can help lower emissions, make oil and gas more efficient, and improve recovery techniques to obtain more fuel with less pollution while using more effectively. One example is hydraulic fracturing, or *fracking*, which consists of horizontal drills into gas wellbores aided by a fluid that helps keep a continuous flow of exploited material. Still a controversial method due to its potential externalities, fracking has helped replace coal with natural gas. It shows how the industry has tried to adapt to the current environmental situation.

The industry has also been capable of striking deals with governments and international organizations to promote regional and private investment in renewable energy sources. The United Nations and the European Union are two major players in the search for a cleaner future. The recent Climate Change Conference (COP26) celebrated in Scotland resulted from their work. Itself not lacking controversy, the meeting ended with a compromise from several nations to move away from using or developing projects with coal, one of the main drivers of man-made climate change. At the same time, most countries in attendance pledged to reach zero carbon emissions (albeit without strict deadlines or with many threats if not achieved in a reasonable amount of time) while reversing deforestation by 2030.

To conclude, an energy transition process is akin to a very complex piece of machinery, with multiple moving parts involving many actors across society, going from governments to international organizations, think tanks, researchers, and engineers, and where the citizenry is an integral part as well. Therefore, a successful process requires a considerable number of resources and planning, as well as time to evaluate the results of public policies and research and to study the implementation of further changes or to revert to previous positions due to a bad outcome. Just as previous energy transition processes did not happen from one day to the next, as was the case with the move away from wood into coal, a 21st-century move away from fossil fuels will not occur spontaneously. It will have a few hiccups along the way.

However, the world needs fundamental changes in how energy is collected and consumed. The challenge is driven by increasing temperatures that are associated with higher levels of carbon dioxide in the atmosphere, and unless humans make up for anthropogenic climate change, the problem will only exacerbate. Many technological, political, and social branches need to cooperate and come together to combat this problem, and the oil and gas industry holds an important place in this dynamic. Through innovation, investment, influence, and development, the sector can remain a competitive and proactive player, continuing its work for the development of humanity while striving for a cleaner environment.

REFERENCES

Ajala, O. E., F. Aberuagba, T. E. Odetoye, and A. M. Ajala. 2015. Biodiesel: Sustainable Energy Replacement to Petroleum-Based Diesel Fuel—A Review. *ChemBioEng Reviews* 2(3): 145–156.

Ali Almihoub, A. A., J. M. Mula, and M. Mafizur Rahman. 2013. Marginal Abatement Cost Curves (MACCs): Important Approaches to Obtain (Firm and Sector) Greenhouse Gases (GHGs) Reduction. *International Journal of Economics and Finance* 5(5): 35–54.

Andrews-Speed, P. 2016. Applying Institutional Theory to the Low-Carbon Energy Transition. *Energy Research & Social Science* 13: 216–225.

Arbolino, R., F. Carlucci, A. Cirà, G. Ioppolo, and T. Yigitcanlar. 2017. Efficiency of the EU Regulation on Greenhouse Gas Emissions in Italy: The Hierarchical Cluster Analysis Approach. *Ecological Indicators* 81: 115–123.

Bailey, I., and S. Rupp. 2004. Politics, Industry and the Regulation of Industrial Greenhouse-Gas Emissions in the UK and Germany. *European Environment* 14(4): 235–250.

Boot-Handford, M. E., J. C. Abanades, E. J. Anthony, M. J. Blunt, S. Brandani, N. Mac Dowell, . . ., and P. S. Fennell. 2014. Carbon Capture and Storage Update. *Energy & Environmental Science* 7(1): 130–189.

Bui, M., C. S. Adjiman, A. Bardow, E. J. Anthony, A. Boston, S. Brown, . . . and N. Mac Dowell. 2018. Carbon Capture and Storage (CCS): The Way Forward. *Energy & Environmental Science* 11(5): 1062–1176.

Casey, B., Gray, W. B., Linn, J., & Morgenstern, R. D. 2022. How does state-level carbon pricing in the United States affect industrial competitiveness?. *Environmental and Resource Economics*, 83(3): 831–860.

Chapman, A. J., and K. Itaoka. 2018. Energy Transition to a Future Low-Carbon Energy Society in Japan's Liberalizing Electricity Market: Precedents, Policies and Factors of Successful Transition. *Renewable and Sustainable Energy Reviews* 81: 2019–2027.

Climate Works Australia. 2019. *How to Read a Marginal Abatement Cost Curve*. Available in www.climateworksaustralia.org/resource/how-to-read-a-marginal-abatement-cost-curve/ [accessed October 12th, 2021].

Colorado Oil & Gas Association, 2021. *Everyday Products Made Possible by Oil & Natural Gas*. Available in www.coga.org/factsheets/everyday-products-uses [accessed May 15th, 2022].

Davis, S. J., N. S. Lewis, M. Shaner, S. Aggarwal, D. Arent, I. L. Azevedo, . . ., and K. Caldeira. 2018. Net-Zero Emissions Energy Systems. *Science* 360(6396): eaas9793.

Elgowainy, A., J. Han, H. Cai, M. Wang, G. S. Forman, and V. B. DiVita. 2014. Energy Efficiency and Greenhouse Gas Emission Intensity of Petroleum Products at U.S. Refineries. *Environmental Science & Technology* 48(13): 7612–7624.

Fernandes Araújo, O. D. Q., and J. L. de Medeiros. 2017. Carbon Capture and Storage Technologies: Present Scenario and Drivers of Innovation. *Current Opinion in Chemical Engineering* 17: 22–34.

Fini, E. H., I. L. Al-Qadi, Z. You, B. Zada, and J. Mills-Beale. 2012. Partial Replacement of Asphalt Binder with Bio-Binder: Characterisation and Modification. *International Journal of Pavement Engineering* 13(6): 515–522.

Forman, G. S., V. B. Divita, J. Han, H. Cai, A. Elgowainy, and M. Wang. 2014. U.S. Refinery Efficiency: Impacts Analysis and Implications for Fuel Carbon Policy Implementation. *Environmental Science & Technology* 48(13): 7625–7633.

Gielen, D., F. Boshell, D. Saygin, M. D. Bazilian, N. Wagner, and R. Gorini. 2019. The Role of Renewable Energy in the Global Energy Transformation. *Energy Strategy Reviews* 24: 38–50.

Global CCS Institute. 2021. Global status of CCS report 2021. *Global CCS Institute*. Available in https://www.globalccsinstitute.com/wp-content/uploads/2021/10/2021-Global-Status-of-CCS-Report_Global_CCS_Institute.pdf

Gomiero, T. 2018. Large-Scale Biofuels Production: A Possible Threat to Soil Conservation and Environmental Services. *Applied Soil Ecology* 123(February): 729–736.

Gouveia, L., and A. C. Oliveira. 2008. Microalgae as a Raw Material for Biofuels Production. *Journal of Industrial Microbiology & Biotechnology* 36(2): 269–274.

Haszeldine, R. S. 2009. Carbon Capture and Storage: How Green Can Black Be? *Science* 325(5948): 1647–1652.

Herzog, H., and D. Golomb. 2004. Carbon Capture and Storage from Fossil Fuel Use. *Encyclopedia of Energy*: 277–287.

Intergovernmental Panel on Climate Change. 2018. Summary for Policymakers. In: *Global Warming of 1.5°C. An IPCC Special Report on the Impacts of Global Warming of 1.5°C Above Pre-Industrial Levels and Related Global Greenhouse Gas Emission Pathways, in the Context of Strengthening the Global Response to the Threat of Climate Change, Sustainable Development, and Efforts to Eradicate Poverty* [Masson-Delmotte, V., P. Zhai, H.-O. Pörtner, D. Roberts, J. Skea, P. R. Shukla, . . ., and T. Waterfield (eds.)]. In Press.

International Energy Agency. 2021a. *Change in CO_2 Emissions by Fuel, 1990–2020*. Available in https://www.iea.org/data-and-statistics/charts/change-in-co2-emissions-by-fuel-1990-2020

International Energy Agency. 2021b. *Global Energy Review 2021*. Available in https://www.iea.org/reports/global-energy-review-2021?utm_source=newsletter&utm_medium=email&utm_campaign=newsletter_axiosgenerate&stream=top

IRENA. 2018. *Power System Flexibility for the Energy Transition, Part 1: Overview for Policy Makers*. International Renewable Energy Agency, Abu Dhabi.

IRENA. 2020. *Energy Transition*. Available in www.irena.org/energytransition [accessed October 2nd, 2021].

IRENA. 2021. *World Energy Transitions Outlook: 1.5°C Pathway*. International Renewable Energy Agency, Abu Dhabi.

Kesicki, F., and N. Strachan. 2011. Marginal Abatement Cost (MAC) Curves: Confronting Theory and Practice. *Environmental Science & Policy* 14(8): 1195–1204.

Kim, M.-K., and K. Lee. 2015. Dynamic Interactions between Carbon and Energy Prices in the U.S. Regional Greenhouse Gas Initiative. *International Journal of Energy Economics and Policy* 5(2): 494–501.

Lewis, J. I. 2016. Wind Energy in China: Getting More from Wind Farms. *Nature Energy* 1(6): 16076.

Liska, A. J., H. Yang, M. Milner, S. Goddard, H. Blanco-Canqui, M. P. Pelton, X. X. Fang, H. Zhu, and A. E. Suyker. 2014. Biofuels from Crop Residue Can Reduce Soil Carbon and Increase CO 2 Emissions. *Nature Climate Change* 4(5): 398–401.

Millot, A., A. Krook-Riekkola, and N. Maïzi. 2020. Guiding the Future Energy Transition to Net-Zero Emissions: Lessons from Exploring the Differences Between France and Sweden. *Energy Policy* 139: 111358.

Oxford Institute for Energy Studies. 2018. The Rise of Renewables and Energy Transition: What Adaptation Strategy for Oil Companies and Oil-Exporting Countries? *OIES Paper: MEP* 19 (May).

Payne, W. A. 2010. Are Biofuels Antithetic to Long-Term Sustainability of Soil and Water Resources? *Advances in Agronomy* 105: 1–46.

Peralta, J., R. C. Williams, M. Rover, and H. M. R. D. Silva. 2012. Development of Rubber-Modified Fractionated Bio-Oil for Use as Noncrude Petroleum Binder in Flexible Pavements. *Transportation Research Circular E-C165: Alternative Binders*: 23–36.

Pires, J. C. M., F. G. Martins, M. C. M. Alvim-Ferraz, and M. Simões. 2011. Recent Developments on Carbon Capture and Storage: An Overview. *Chemical Engineering Research and Design* 89(9): 1446–1460.

Pye, S., O. Broad, C. Bataille, P. Brockway, H. E. Daly, R. Freeman, A. Gambhir, O. Geden, F. Rogan, S. Sanghvi, J. Tomei, I. Vorushylo, and J. Watson. 2020. Modelling Net-Zero Emissions Energy Systems Requires a Change in Approach. *Climate Policy*: 1–10.

Ranken Energy Corporation. 2017. *Products Made from Petroleum*. Available in www.ranken-energy.com/index.php/products-made-from-petroleum/ [accessed October 18th, 2021].

Rogge, K., and P. Johnstone. 2017. Exploring the Role of Phase-Out Policies for Low-Carbon Energy Transitions: The Case of the German *Energiewende*. *Energy Research & Social Science* 33: 128–137.

Roussanaly, S., A. Aasen, R. Anantharaman, B. Danielsen, J. Jakobsen, L. Heme-De-Lacotte, . . . and R. Dreux. 2019. Offshore Power Generation with Carbon Capture and Storage to Decarbonise Mainland Electricity and Offshore Oil and Gas Installations: A Techno-Economic Analysis. *Applied Energy* 233: 478–494.

Shojaeddini, E., S. Naimoli, S. Ladislaw, and M. Bazilian. 2019. Oil and Gas Company Strategies Regarding the Energy Transition. *Progress in Energy* 1(1): 012001.

Solomon, B. D., and K. Krishna. 2011. The Coming Sustainable Energy Transition: History, Strategies, and Outlook. *Energy Policy* 39(11): 7422–7431.

Sullivan, R. 2009. The Management of Greenhouse Gas Emissions in Large European Companies. *Corporate Social Responsibility and Environmental Management* 16(6): 301–309.

Speight, J. G. 2014. *Handbook of Petroleum Product Analysis*. John Wiley & Sons, Inc., Hoboken, NJ.

Steckel, J. C., I. I. Dorband, L. Montrone, H. Ward, L. Missbach, F. Hafner, M. Jakob, and S. Renner. 2021. Distributional Impacts of Carbon Pricing in Developing Asia. *Nature Sustainability*. https://doi.org/10.1038/s41893-021-00758-8.

Szulczewski, M. L., C. W. MacMinn, H. J. Herzog, and R. Juanes. 2012. Lifetime of Carbon Capture and Storage as a Climate-Change Mitigation Technology. *Proceedings of the National Academy of Sciences* 109(14): 5185–5189.

Tan, Y., W. Nookuea, H. Li, E. Thorin, and J. Yan. 2016. Property Impacts on Carbon Capture and Storage (CCS) Processes: A Review. *Energy Conversion and Management* 118: 204–222.

Thompson, W., J. Whistance, and S. Meyer. 2011. Effects of US Biofuel Policies on US and World Petroleum Product Markets with Consequences for Greenhouse Gas Emissions. *Energy Policy* 39(9): 5509–5518.

United Nations. 2021. *Theme Report on Energy Transition: Towards the Achievement of SDG 7 and Net-Zero Emissions*. Secretariat of the High-level Dialogue on Energy. Available in https://www.un.org/sites/un2.un.org/files/2021-twg_2-062321.pdf.

U.S. Energy Information Administration. 2021. Petroleum products made from a barrel of crude oil, 2020. *U.S. Department of Energy*. Available in https://www.eia.gov/

VijayaVenkataRaman, S., S. Iniyan, and R. Goic. 2012. A Review of Climate Change, Mitigation and Adaptation. *Renewable and Sustainable Energy Reviews* 16(1): 878–897.

Wang, M., H. Lee, and J. Molburg. 2004. Allocation of Energy Use in Petroleum Refineries to Petroleum Products. *The International Journal of Life Cycle Assessment* 9(1): 34–44.

The World Bank. 2021. *What Is Carbon Pricing?* Available in https://carbonpricingdashboard.worldbank.org/what-carbon-pricing [accessed October 13th, 2021].

Zhang, N., N. Lior, and H. Jin. 2011. The Energy Situation and Its Sustainable Development Strategy in China. *Energy* 36(6): 3639–3649.

2 Energy Resources Overview

Jorge Saldaña, Mesut Yurukcu, Narendra Boppana, Sepehr Arbabi, Jim Henry, and Sebahattin Ziyanak

2.1 ENERGY RESOURCES OVERVIEW

This chapter comprehensively reviews the many energy resources that humans have relied on for the past several decades, and even centuries, to power their lives. From the conventional sources and their evolution and history of use across the world to the renewable sources that are incentivizing these global changes, we will review the advantages and shortcomings of these technologies, the efforts made to achieve more efficiency and sustainability, and the paths forward with each of these resources.

Even though the main goal is an almost-complete transition from conventional sources of energy to those considered "green," like renewables or even nuclear, a strong implication from the first chapter of this book is that this process may not be as simple and it may take many steps that consider several variables. For example, the different realities from many countries in the world, their energy independence, and their ability to rely on coal or natural gas, as opposed to their reliance on nuclear, wind, solar, or more polluting sources like wood or animal manure, are factors which are relevant to the decisions made by policymakers. Also, political and electoral factors may be in the mix, as more popular politicians may favor transition strategies when compared to unpopular politicians with low chances of reaping electoral benefits from this transformation.

2.1.1 Conventional and Fossil Fuels

Humanity has undergone a series of seminal transformations throughout history. In prehistoric times, fire served as one of the first elements to protect humans from the cold and other external threats. In the Middle Ages, more energy sources had been discovered, including charcoal, biomass, and animal power, which included human muscle power. Years later, before the Industrial Revolution, coal was one of the primary energy sources, but this transition from less-reliable sources was slow. London was one of the leading centers of this transition, and with the city's development came a greater demand for coal, mainly imported from the north. Slowly, market demand made coal a cheaper resource than wood, and an energy transition process started (Allen et al., 2013).

A resource first used for industries and blacksmithing, coal was later utilized for heating houses, among other new activities propelled by the Industrial Revolution. From this revolution came the invention of the chimney, a device that proliferated concurrently with coal, becoming a widespread resource for energy generation (Allen et al., 2013). As humanity moved away from using their own bodies to generate small amounts of energy that exert tremendous pressure to rely on weight-bearing animals and, later, to find ways to generate power from resources on the ground, like wood and, later, coal, we see the spirit of innovation and adaptation. This spirit has helped humanity overcome many obstacles, extend the expectancy of our lives, and lift millions of people out of poverty and despair.

However, this spirit has also brought some costs to humanity. Today, the world is warmer, partly attributable to the use of fossil fuels for human consumption. Before this trend becomes dangerous, efforts to facilitate a new energy transition should be considered. This transition must consider the significant advancements in renewable energies like wind and solar and the potential of a clean energy source like nuclear power. Yet all these new resources have their own shortcomings and

DOI: 10.1201/9781003282273-2

issues that researchers are still trying to deal with, which are an apparent roadblock to a quick and painless transition. These new energies will be covered in greater detail in the following sections, but for now, we will talk a little bit more about the traditional energy sources we still use today.

Following the dynamics observed in England, a second big revolution came after the discovery of crude oil in Pennsylvania in 1859. This resource is superior to coal because of its high energy density, flexibility, and ability to be shipped and put in storage. After the First World War, demand for crude oil multiplied (Van Vactor, 2018), and despite ups and downs in global markets, dynamics that have led to innovation, adaptation, and transitions of other types, crude oil has remained one of the essential sources of energy throughout the years.

Figure 2.1 illustrates the trends for the primary sources of energy over time, and of note are the moments when coal replaced wood as a primary source, observed by the end of the 19th century, and when oil replaced coal, around half a century later. After that period, coal was rivaled by gas, which appears to have gained use after the start of the 21st century, particularly in more-developed countries. In general, these countries have also been the first group to invest in renewable and nuclear energies, an advantage they allegedly have because they achieved a level of development using conventional energy sources.

This dynamic opens the debate on the appropriate time to initiate an energy transition process, which can be achievable for some countries but less for others. From the perspective of a developing nation, it can be argued that, to transition from one source of energy to another, the country from which this process is expected should have a minimum level of development, thus being able to provide power for most of its citizens in a reliable manner. Just as countries leading the way for an energy transition to renewables had time to develop up to a certain point before

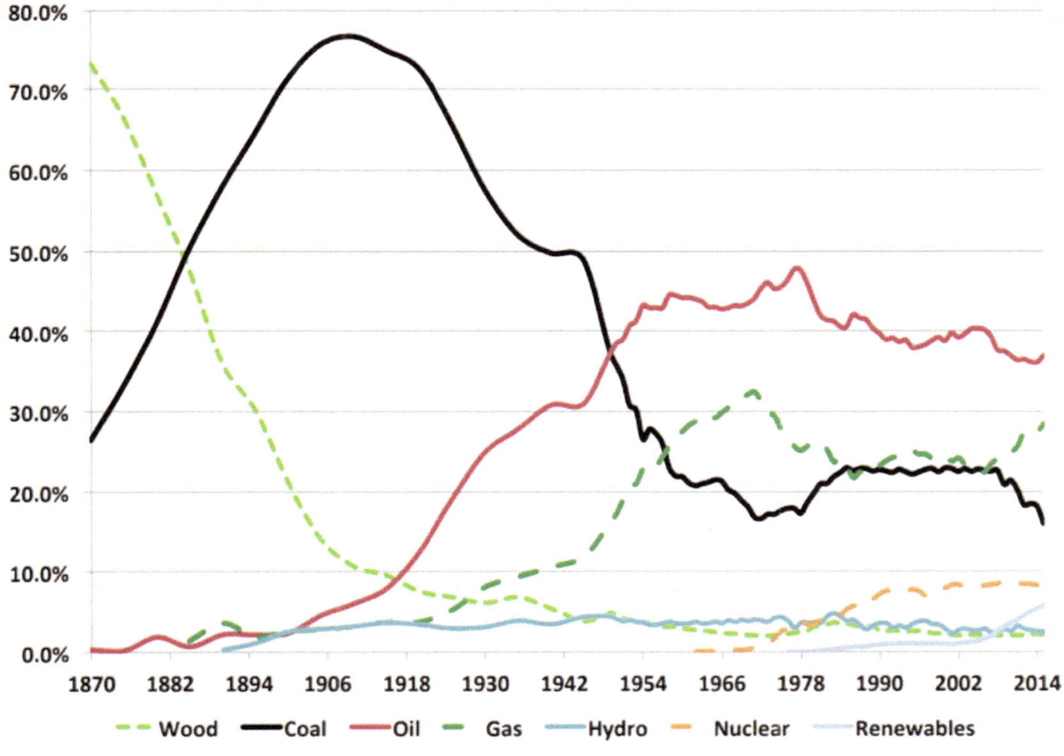

FIGURE 2.1 Market share of US primary energy consumption.

Source: Van Vactor (2018).

the pressure to transition appeared at their door, countries in the middle of the pack should also benefit from it.

As some authors have hypothesized, there is a considerable likelihood that countries with lower levels of economic development, measured in per capita income, have a larger carbon footprint than developed countries. But as Figure 2.2 shows, after these economies get richer, their environmental impacts fall steadily over time, joining the more advanced countries (Dinda, 2004; Stern, 2018). Thus, given that not all countries have reached economic development levels for even considering a move away from the sources that power the lives of their citizens for the most basic needs, like heating and electricity, it is not so simple to mandate an energy transition process with which more developed countries can easily comply.

With that said, this position should be balanced with the global need to slow down carbon emissions, particularly from coal (Li and Hu, 2017). Hence, we are confronted once again with a large-scale problem that affects the planet, but with solutions that need to be gradual and conscious of specific contexts of countries or entire regions. One way to deal with this is to stimulate developing countries, the ones needing more time and space to transition, to invest in alternative ways to power their citizens' lives without incurring massive costs that could backfire the entire endeavor (Marcotullio and Schulz, 2007). For example, after the latest climate meeting, COP26, celebrated in Glasgow, Scotland, many developing countries joined a pledge to stop all investment in new coal projects, which currently produce over one-third of the world's electricity (BBC News, 2021).

Some developing nations have begun investing in various forms of technology to produce more diversified electricity than in the past, aiding the global movement to reduce CO_2 emissions.

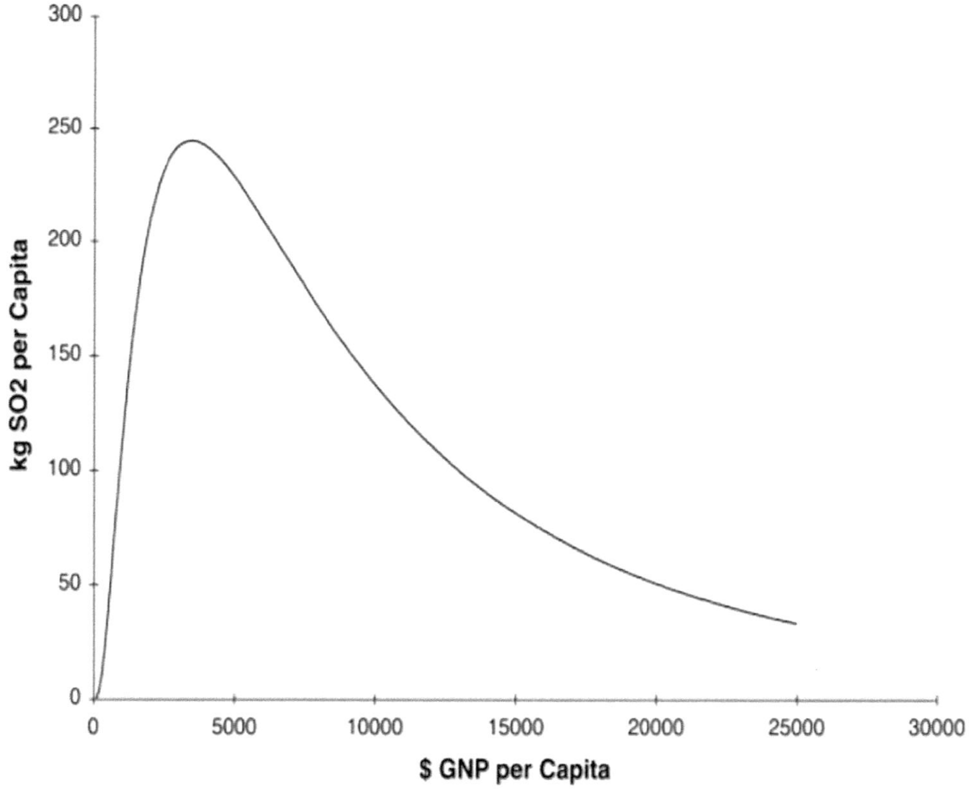

FIGURE 2.2 Environmental Kuznets curve (EKC).

Source: Stern (2018).

At the same time, they seek to comply with agreements and pledges made on the international stage. Figure 2.3 shows the evolution of the capacity for electricity generation from developing countries, and we see that besides solar and wind energy, hydropower and geothermal energies have been two other sources explored to cover the energy needs of the millions of people in developing countries. Therefore, some countries with fewer resources than the big players have found a middle ground between their limitations and the commitments made to international organizations to improve their local energy problems without contributing more emissions to the environment.

Besides, there is still a temporal window where the reliance on fossil fuels will continue to be significant and the preoccupation about their level of emissions will continue. Figure 2.4 illustrates the growth in carbon emissions in the last four decades from gas, oil, and coal in gigatons of carbon dioxide per year. As we can see, emissions from these three sources appear to have peaked close to 2007, with gas having the greatest contribution in emanations. Relatedly, the energy sector is reported to account for close to 70% of all anthropogenic greenhouse gas (GHG) emissions, with a large part generated by fuel combustion (Höök and Tang, 2013).

Along with the potential dangers of continuing the same path regarding emissions, fossil fuels are believed to have limited reserves, making their use at current levels unsustainable over time. In the case of Europe, it has been estimated that fossil fuel energy consumption is still over 60% in most countries, and countries like Germany and the United Kingdom, considered leaders in the push for renewables and clean energy, have fossil fuel consumption values of over 80% (Martins et al., 2019). However, this reliance is limited by resource availability. Table 2.1 shows selected European countries' oil, gas, and coal reserves.

Norway has the largest oil reserves in Europe, with 7,600 million barrels, followed by the United Kingdom, with less than one-third of that. The Nordic country has the largest gas reserves, with almost three times as many billion cubic meters as that of the Netherlands, the second country on the list. Meanwhile, Germany and Poland are far ahead of the rest of the countries shown in the

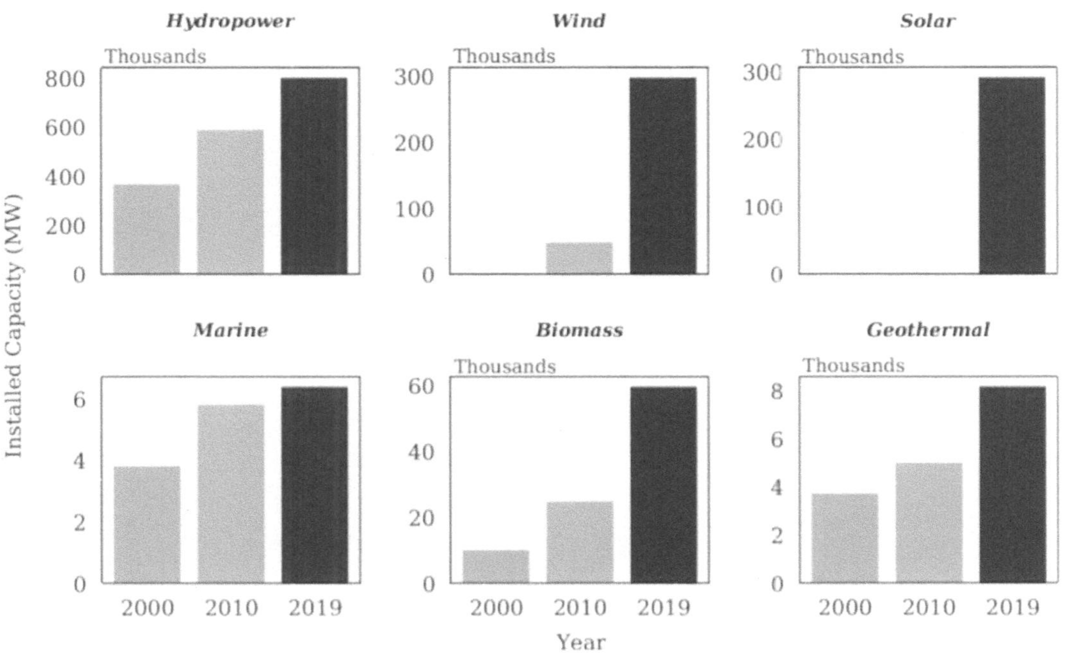

FIGURE 2.3 Evolution of the renewable electricity capacity in developing countries per technology and selected years.

Source: Vanegas Cantarero (2020).

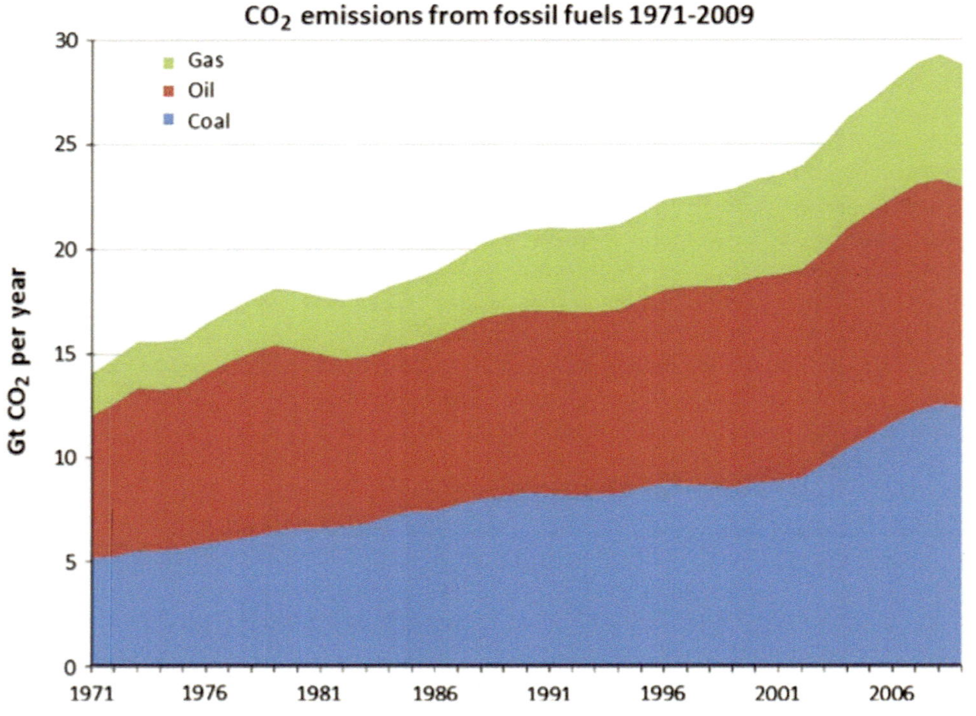

FIGURE 2.4 CO_2 emission trends from 1971 to 2009 by fuel type.

Source: Höök and Tang (2013).

TABLE 2.1
Proven Fossil Fuel Reserves in European Countries

Country	Oil Reserves (Million Barrels)	Coal Reserves (Million Tons)	Gas Reserves (Billion Cubic Meters)
Bulgaria	0	2,366	0
Czech Republic	0	3,676	0
Denmark	400	0	14
Germany	0	36,212	37
Greece	0	2,876	0
Hungary	0	2,909	0
Italy	500	0	37
Netherlands	0	0	700
Norway	7,600	0	1,800
Poland	0	24,161	100
Romania	600	291	100
Spain	0	1,187	0
United Kingdom	2,500	70	200

Source: Martins et al. (2019).

table in coal reserves. The varying levels of resources in European countries show that most of them are confronted with the possibility of energy scarcity or a vicious cycle of importing oil and increasing their carbon emissions after burning coal to power the lives of their citizens.

The fact that some countries are faced with these decisions, whether out of a lack of investment, a robust energy market, or government policies that foster newer energy sources, may contribute to a scenario where we observe higher levels of carbon emissions in the next few decades. Figure 2.5 shows several estimations of CO_2 emissions performed by the Intergovernmental Panel on Climate Change (IPCC) which show where the world could end up by the year 2100. We could observe different emissions levels in only a few decades depending on which policies are undertaken today. Given the close relationship between more carbon emissions and global temperatures, it is essential to coordinate the efforts to decrease emissions if the goal is to remain below the stated limit of 1.5°C established by the Paris Agreement and the United Nations (UNEP, 2021).

Despite the potential shortcomings of models that could be overdetermining the effects of variables to the detriment of others, or that may not be considering the mitigating impact measures like practical adaptation to the current trends can have on the development of anthropogenic climate change (VijayaVenkataRaman et al., 2012), higher world temperatures are still a pivotal issue that researchers and engineers need to tackle with urgency. Therefore, human ingenuity needs to take a predominant role and exploit new technologies that provide new ways to generate energy, relying either on renewable energy sources, alternative sources like nuclear power, or newer sources like shale gas (Kasperson and Ram, 2013). In the following sections, we will review these new sources in more detail and consider their advantages and disadvantages and how they can help humanity in its struggle against a warming climate.

We are in a critical moment in our history as a species. Perhaps more than ever, we have a direct responsibility to the world that future generations will inherit and to the climate in which they will grow. Despite many ways of tackling the problem we face today and the different contexts that each of the countries is subject to, we must develop a common strategy that deals with the warming of the globe and that can advance solutions that are not one-size-fits-all mandates but rather incorporate the needs and realities of the people that are still struggling to improve their quality of life and to which energy sources are still vital for that to happen.

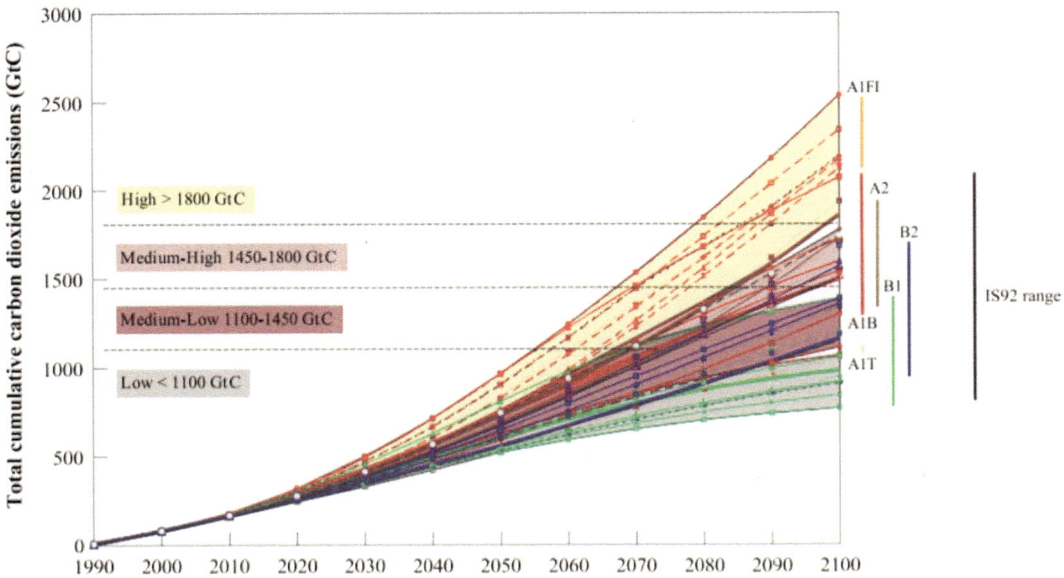

FIGURE 2.5 Estimated scenarios for total global cumulative CO_2 emissions (GtC) from 1990 to 2100.

Source: Intergovernmental Panel on Climate Change (2000).

2.1.2 New Energy and Renewables

One way to tackle the problems outlined in the previous section has been to search and experiment with new energy sources that could complement or replace conventional sources. The most talked about sources of energy are wind and solar, and the technologies derived from them, like photovoltaics (PV). Still, there is also a list of new sources that include biomass and nuclear, in its several applications (Lund, 2006), as well as natural gas and its offshoots, shale gas (Schrag, 2012), and natural gas hydrates (Kvenvolden, 2003). Implementing policies to utilize these sources to their fullest potential will help us improve our current situation and begin a path to lower emissions and more stable global temperatures less susceptible to human action.

This section performs a quick review of these newer sources of energy. We start with natural gas, a hydrocarbon energy source with high energy conversion efficiencies and considerable unexploited resources worldwide (Economides and Wood, 2009). As Figure 2.6 illustrates, natural gas is currently the second energy source in the United States, behind petroleum, surpassing coal and more than tripling the contribution made by renewable sources. As opposed to petroleum sources, natural gas is evenly distributed across the industrial, residential, and commercial sectors and as a source for the electrical grid. Renewable sources have a similar potential, indicating the possibility of these sources replacing petroleum and coal.

Natural gas is a gaseous form of fossil fuels with the lowest carbon intensities and lowest CO_2 emitted per unit of energy generated, and it burns efficiently, almost without any carbon emissions. Given its form, it allows for a greater recovery at a relatively low cost, especially with the new shale technologies we will discuss momentarily. On the other hand, gas has a few shortcomings. One of those is its low energy density, which complicates transmission and storage, caused by its gaseous form. Due to gas being delivered through pipelines, markets have developed regionally rather than globally (Moniz et al., 2011). Thus, an aggressive expansion of this resource is unlikely, and the dependence on oil is still notoriously high.

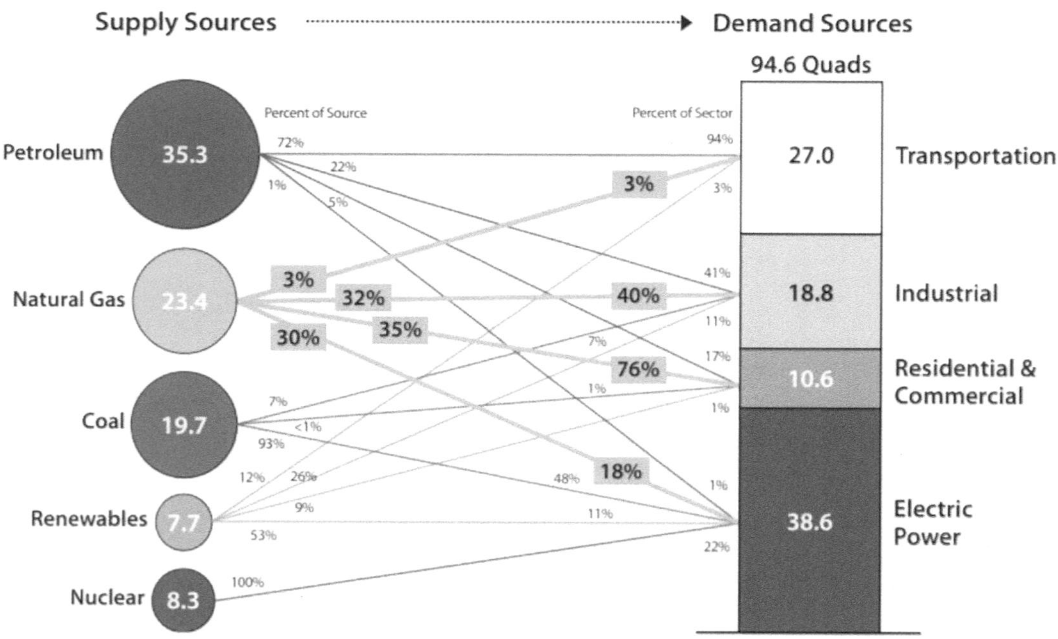

FIGURE 2.6 Sources and use of primary energy sources in the United States, with natural gas highlighted (quadrillion Btu), 2009.

Source: Moniz et al. (2011).

Natural gas has been thought to be a more widespread replacement for oil, especially with its two forms, liquefied natural gas (LNG) and compressed natural gas (CNG). For example, a few countries have embraced transportation systems that feature a significant number of buses that have natural gas engines (Dyr et el., 2019; Pan et al., 2019), both to limit the excessive purchase of cars that use gas and to modernize the systems in countries that are looking to pollute less. Generally, these buses have a longer life than diesel buses while emitting less, thus helping achieve the much-desired clean energy transition.

Relatedly, some regions of the world have developed a market for CNG cars, with Asia and South America being particularly proactive in creating a niche for these vehicles. As Figure 2.7 shows, Europe, Africa, and North America are significantly behind the other regions. Moreover, Pakistan and Argentina appear to have the most significant investment in this type of vehicle. This move was first taken to pollute less in big cities, which is now being kept due to the problem of emissions affecting many countries (Engerer and Horn, 2010). That said, adopting alternative fuel vehicles (AFVs) is highly dependent on the proper incentives given by governments and the policies targeting potential stakeholders (Yeh, 2007).

As noted, natural gas is a relatively new energy source with distinct advantages over older sources like coal and oil. After a few decades of innovation and policies and the profound reforms of the industrial sectors of big countries like China (Dong et al., 2017), natural gas has surpassed coal as an energy source in the developed world but remains behind oil. In the last few years, newer technologies to recover more natural gas from the fields have appeared, and shale gas has garnered some attention. *Shale gas* is natural gas trapped in rock formations below the Earth's surface (Armor, 2013; Kasperson and Ram, 2013). Usually, shale gas is recovered through hydraulic fracturing, or *fracking*. This process utilizes a well stimulant to help flow gas, which, in some cases, is nitroglycerin in solidified form (Montgomery and Smith, 2010). The primarily horizontal drilling performed to obtain gas (seen in Figure 2.8) is boosted by the stimulant, permitting a constant flow to capture and store.

Shale gas is currently considered a "bridge fuel" (Schrag, 2012), thus a source of energy between the conventional sources that humans have relied on for decades and the ones we are establishing as a future goal: renewable sources. Therefore, it is not seen as a permanent replacement for sources like coal but as a complement to the move to solar and wind as primary energy sources, aided in some cases by nuclear power. Relatedly, shale is considered an unconventional form of natural gas

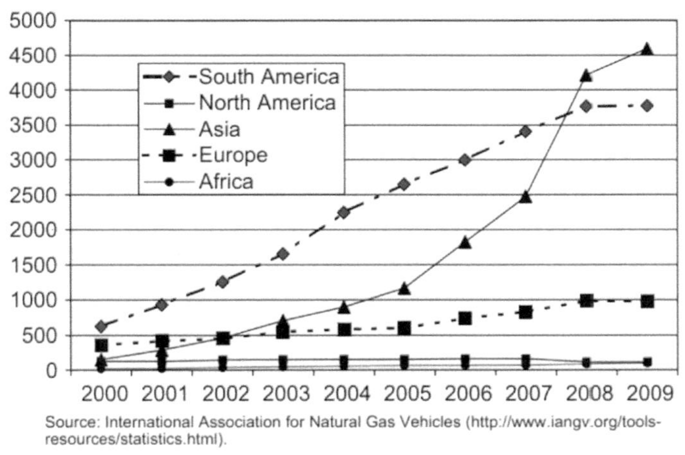

FIGURE 2.7 Number of registered compressed natural gas (CNG) cars per region (in thousands).

Source: Engerer and Horn (2010).

due to being found in shale, a less-permeable rock than sandstone, siltstone, and limestone, where conventional gas is found (London School of Economics, 2018).

Figure 2.9 illustrates the estimated worldwide shale resources, with a significant share in Russia, China, North America, and Brazil. As the largest nations prepare to transition their energy grids, this resource can become of vital importance, both from an environmental and a geopolitical position. Thus, having large amounts of recoverable shale gas, as well as shale oil, can be a beneficial

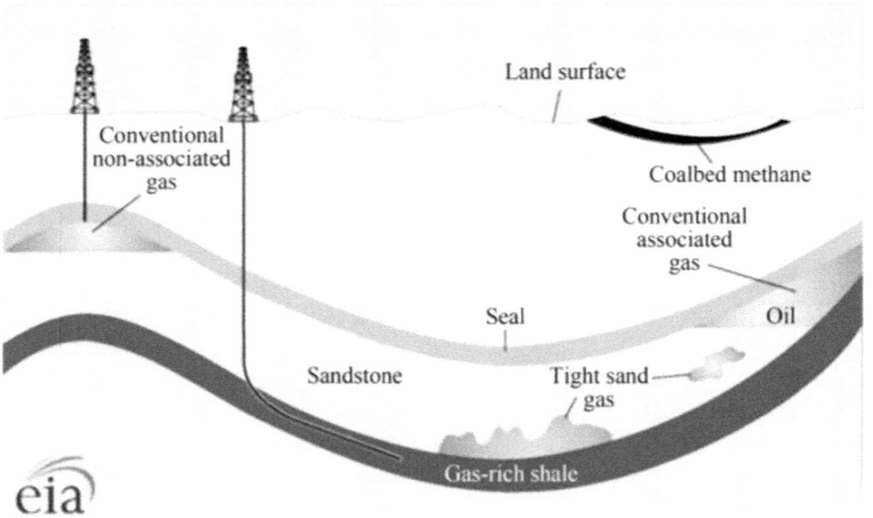

FIGURE 2.8 Types of natural gas deposits.

Source: Armor (2013).

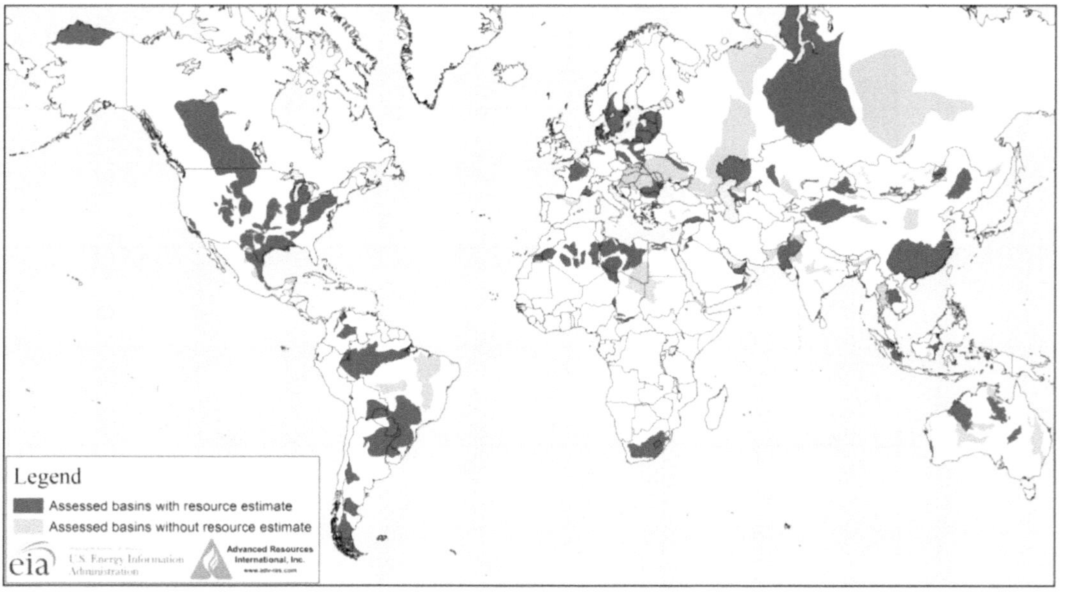

FIGURE 2.9 World map for estimated shale resources.

Source: U.S. Energy Information Administration (2015).

occurrence in those countries' path to cleaner sources of energy while positioning them as global leaders and agents of change, able to establish credible deterrents to countries that fail to comply with international environmental agreements.

Both natural gas and shale gas have been instrumental in reforming the energy sector, allowing for less pollution and adequate preparation for the eventual arrival of renewable energy as the primary source to power the lives of billions of people. On that note, we now turn our attention to renewable energy sources like wind and solar. In what remains of this section, we will discuss the advantages and shortcomings of these sources, as well as their key role in the energy transition of the 21st century. Additionally, we will explore the plans that governments and organizations have developed to integrate the contribution of renewables into electrical grids and the industrial, commercial, and residential sectors.

Newer forms of energy must compete with those resources which have powered our lives for decades. And not only that, but also with markets, stakeholders, investors, and public opinion about the predominant role of oil and gas in modern life. However, it appears that popular support for new forms of energy is growing stronger. The work of Crowe and Li (2020) shows us that even towns and cities with strong ties to conventional sources of energy support renewable sources like wind and solar. Thus, it appears that if the energy transition process is seen as necessary, people not only understand potential switches but also favor changes of this magnitude. We are in the right moment to explore renewables.

The energy that comes from the wind has intrigued people for centuries. Once humanity had the ability to capture this energy, it strived to develop technologies that could do this consistently to power the world. Figure 2.10 shows the newly installed capacity for wind energy, in gigawatts, between 1996 and 2012. As we can see, the new capacity almost doubled between 2008 and 2012, and according to information released by the Global Wind Energy Council, the total capacity has now more than doubled once more, reaching more than 93 GW in 2020. The total capacity now stands at 743 GW globally (Global Wind Energy Council, 2021).

Over the years, we have witnessed an expansion of the reach of wind energy, which has helped with lowering emissions worldwide. But concurrently, this has helped engineers and researchers find new ways to diversify the use of this renewable energy source, expanding the reach of service beyond electricity production. Figure 2.11 shows the innovative ways that wind energy can be

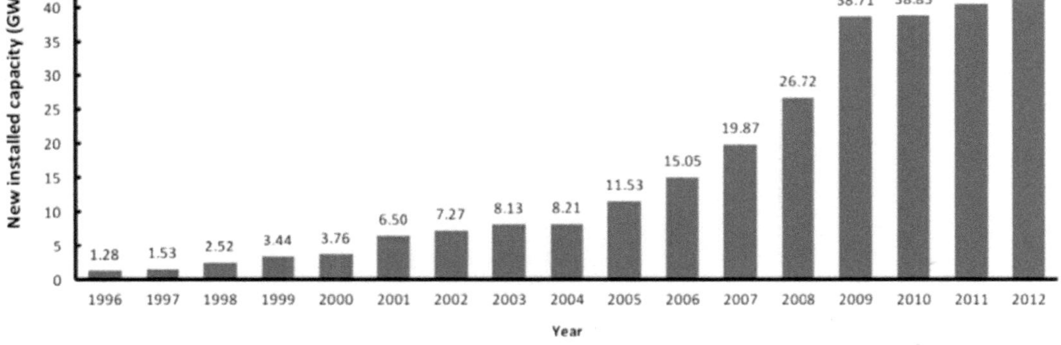

FIGURE 2.10 Global annual installed wind capacity, 1996–2012.

Source: Wang and Wang (2015).

utilized. This development has been brought out by innovation and resource optimization and by perfecting the instruments to capture and store energy. As we can see, wind energy can significantly contribute to the industrial sector, transportation, and buildings.

These innovations in capturing wind energy have been partly boosted by the increasing capabilities to develop offshore wind facilities. Offshore farms can work well for this purpose due to factors like lack of space, varying environments between countries, and/or lower environmental impact from these wind farms compared to onshore installations (Bórawski et al., 2020; Esteban et al., 2011). Additionally, the location of these plants can help engineers study the potential of marine hydrodynamic technologies, like wave, tidal, and current power, to determine their potential role in the future electricity market (Rueda-Bayona et al., 2019).

Regarding offshore exploration, one more potential source of energy that has been systematically studied but has been flying under the radar these past few years, perhaps due to a lack of media coverage, or maybe due to a low level of funding for research, consists of biofuels generated from genetically engineered microalgae. Some bioengineers have stated that some types of microalgae be redesigned to produce oil that is more environmentally friendly and helps us move away from conventional sources of energy (Snow and Smith, 2012). Biofuels derived from this type of source are shown to have little to no impact on the environment, and algae biomass has many applications across distinct sectors (see Figure 2.12), making the potential for this feedstock a very promising one (Adeniyi et al., 2018).

We recognize the vast potential of wind technologies as a source to shape the energy transition of the future significantly. With that said, shortcomings in technology, potential emissions from the production of the required equipment, and some natural limitations still make some people wonder about the ceiling that a source like wind can have, given the current energy environment. Figure 2.13 enumerates a series of potential environmental impacts from wind power, which include the harming of wildlife like birds and flying mammals, the noise from wind turbines, the potential

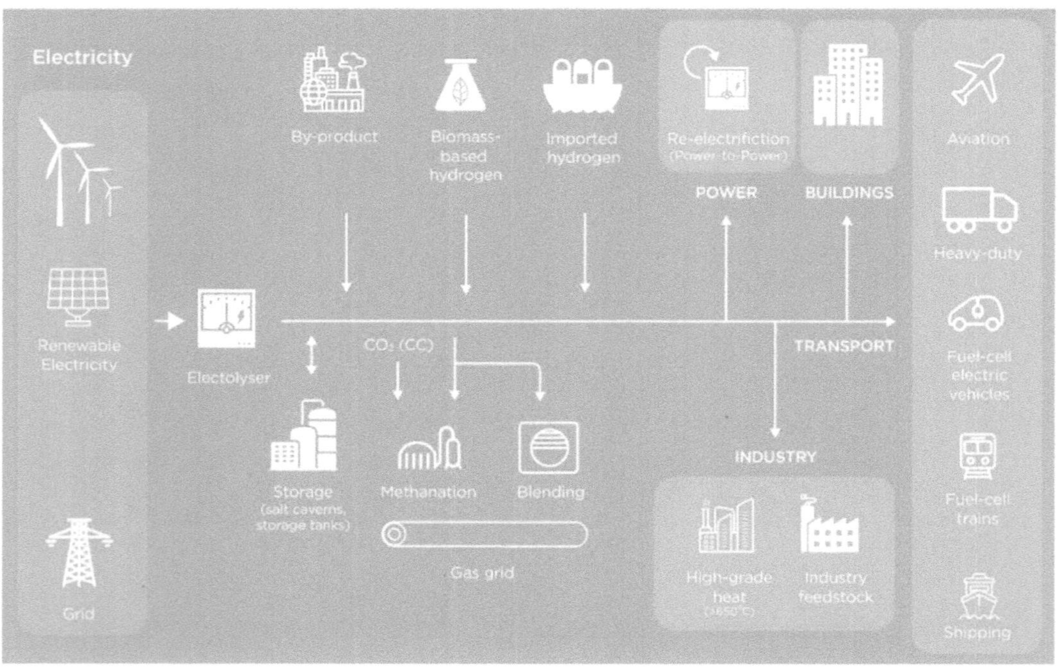

FIGURE 2.11 Innovation in wind energy for multiple end uses.

Source: Global Wind Energy Council (2021).

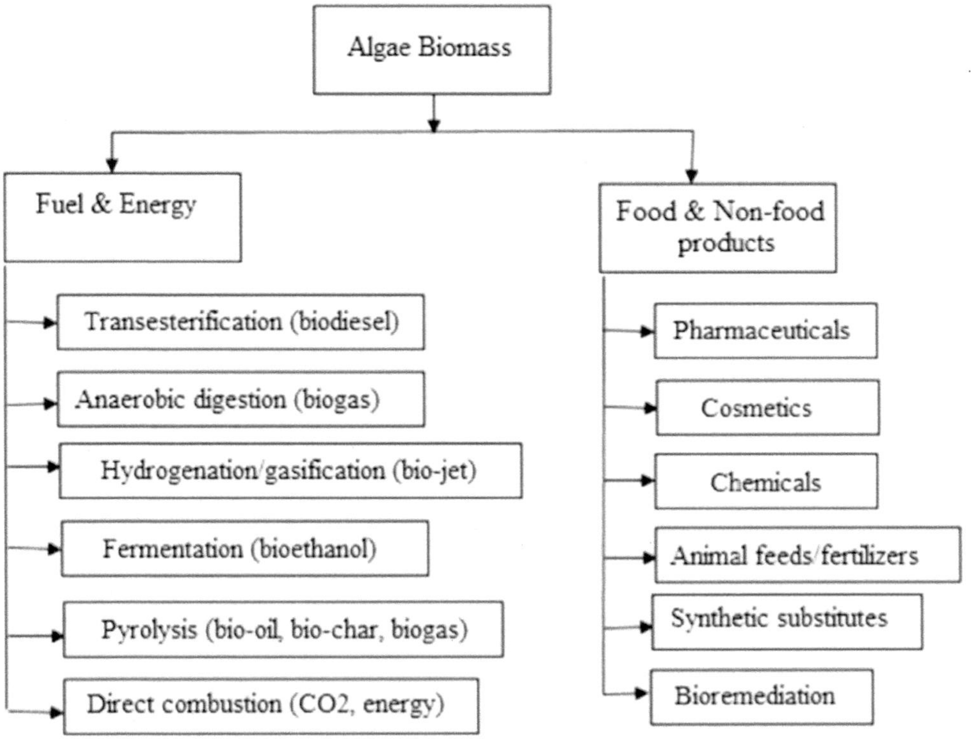

FIGURE 2.12 Diagram of algae biomass products.

Source: Adeniyi et al. (2018).

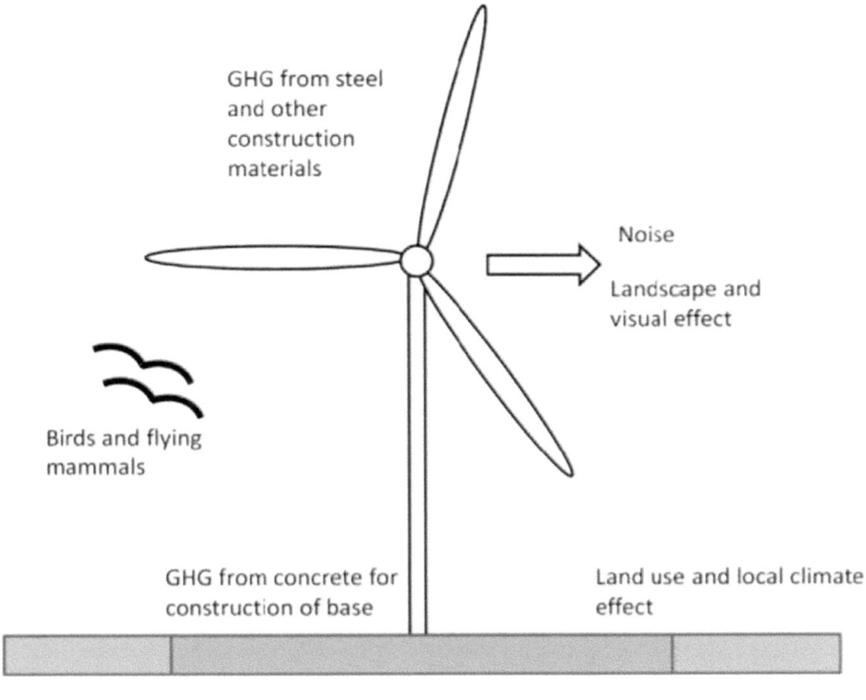

FIGURE 2.13 Environmental impacts of wind power.

Source: Wang and Wang (2015).

greenhouse emissions generated from the creation of the equipment and construction materials, and the visual, land use, and local climate effects.

However, most of these problems have been met with proactive solutions, and some countries have considered the trade-offs acceptable. For example, Lakatos et al. (2011) conclude in their study that Hungary is a perfect spot for implementing wind technologies and solar power installations. Despite the natural limitations of a lack of sunlight or a lack of wind that could help with the energy grid, which sometimes requires the help of conventional sources to complement and satisfy the demand, many countries, particularly in Europe, have developed incentives to develop and fortify a wind energy market in the region (Bórawski et al., 2020).

Similarly, China has embraced the development of its wind market by strongly emphasizing state policies (Ydersbond and Korsnes, 2016). Just like for the European Union (EU), setting up the objectives of reducing greenhouse emissions, securing the energy supply, and creating industries and jobs that are future-oriented have paid off for the Asian giant. Figure 2.14 shows the evolution

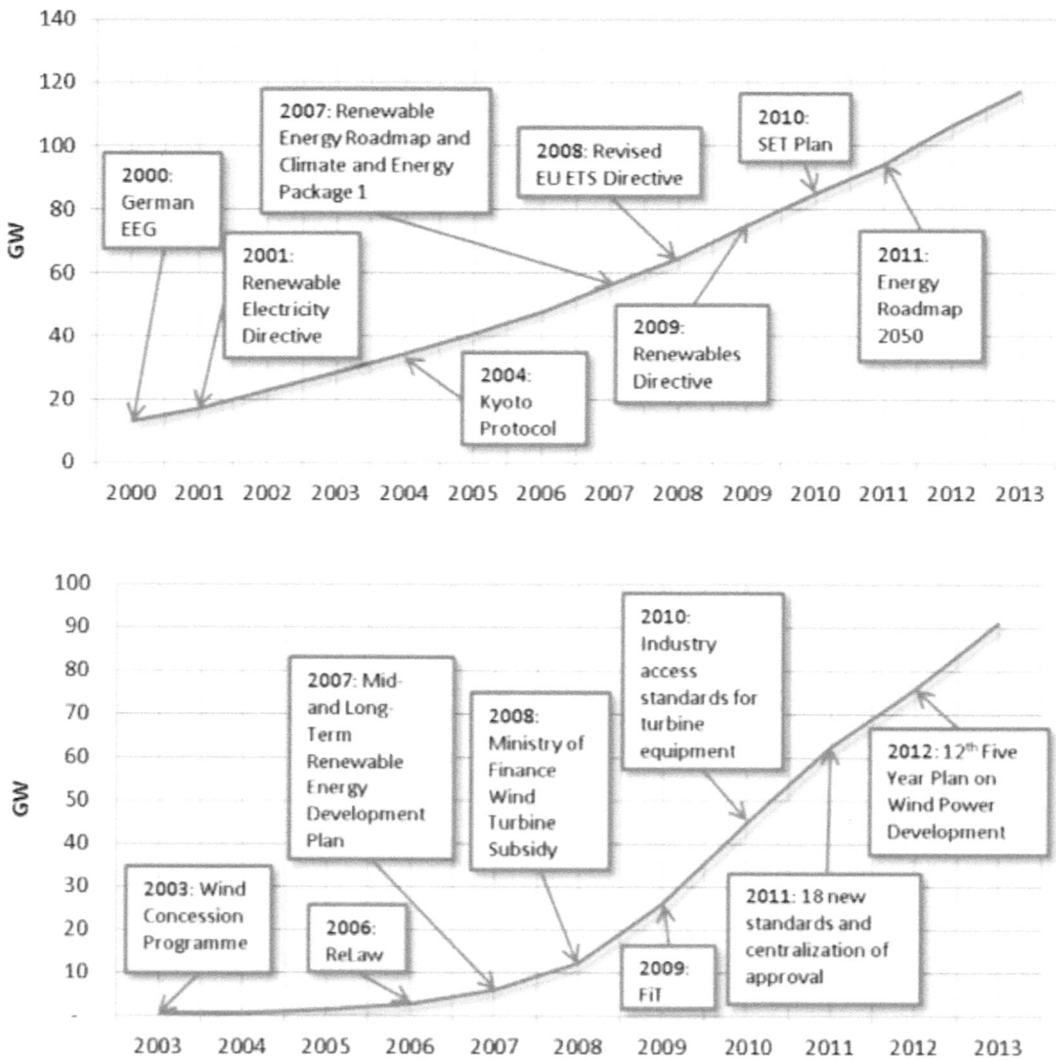

FIGURE 2.14 Installed wind power capacity and policy development in the European Union (2000–2013) and China (2003–2013).

Source: Ydersbond and Korsnes (2016).

of the installed wind capacity in both the EU and China, along with the policy development that has allowed for an expansion of their wind energy markets. Despite not adhering to the Kyoto Protocol, China has embraced renewable energy and is reaching similar capacity levels to the EU, which has committed to follow international agreements and support wind energy generation (Zervos, 2003).

As technology becomes more widespread, trade-offs begin to appear in different areas. Thus, one of the significant challenges the wind energy industry must tackle is sustainability. Because the very point of moving from polluting energy sources to renewable, clean sources is that of lessening the load on the climate through decarbonization, the proper handling and disposal of the materials used in producing wind energy is another aspect that needs to be considered. Figure 2.15 illustrates a cycle for using and reusing wind turbine blades, from sourcing raw materials to repurposing from decommissioned wind farms. This cycle seeks to transform the way the wind industry does business while keeping to international treaties like the United Nations Race to Zero campaign (Global Wind Energy Council, 2021).

At a comparable development pace to wind, solar energy is another renewable energy source touted as a pivotal replacement for technologies like coal and oil. Particularly in developing countries, where their population has increased significantly over the last few decades, energy demands are at an all-time high. But because of the climate challenges already discussed and the depleting

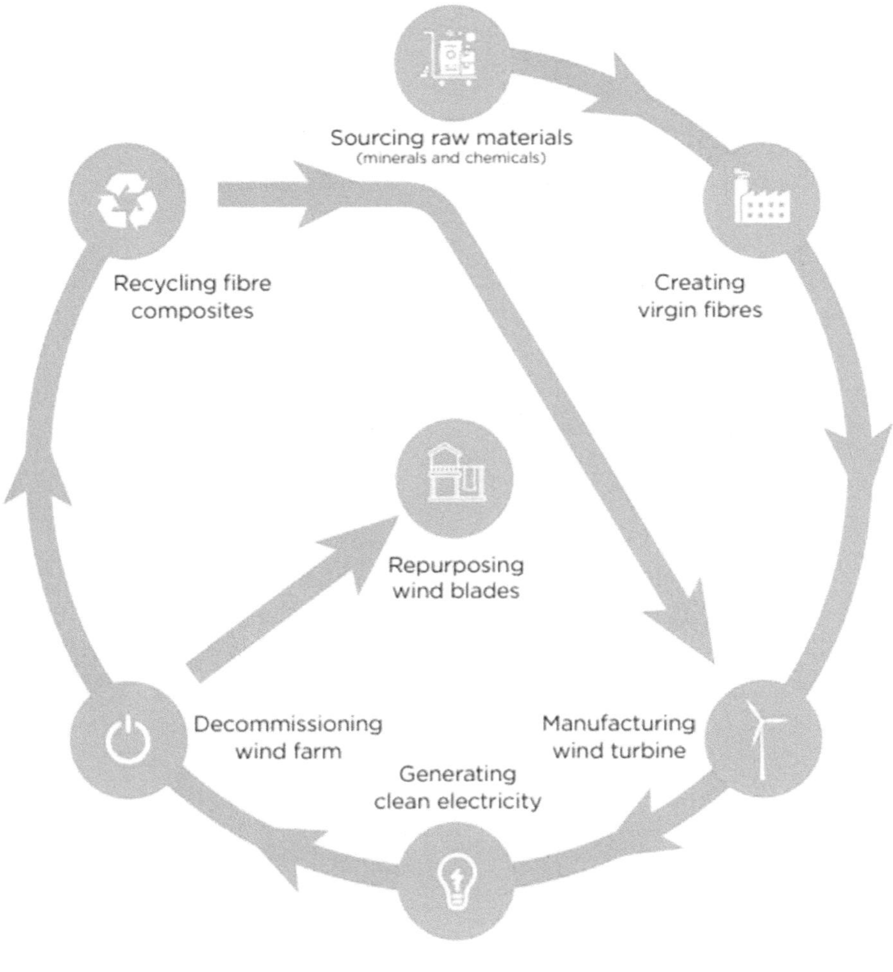

FIGURE 2.15 Circular economy theory of wind turbine blades.

Source: Global Wind Energy Council (2021).

resources, renewables like solar energy are precious (Kannan and Vakeesan, 2016). Figure 2.16 shows a map of the world with the amount of global horizontal irradiation (GHI), where we can see that most of the African continent receives the largest daily sum of GHI in the world.

This advantage could be a starting point for the continent to become energy-independent, and perhaps even an exporter of renewable energy in the future, through what is known as technological *leapfrogging* (Amankwah-Amoah, 2014). This term is understood as skipping older forms of technology by rapidly acquiring and developing the necessary tools to adopt newer forms of technology. An advantage of being a "late bloomer" is that this sort of advancement can help developing countries with large energy needs adopt the latest technologies to exploit cleaner energy sources, reaping comparable benefits as other actors in more developed countries.

Solar technologies can be divided into three main categories: photovoltaic (PV), thermal, and concentrated solar power (CSP). The first technique transforms solar radiation into electricity through materials that conduct the energy, like photosensors and cells (Singh, 2013). Figure 2.17 shows a few of these PV fields in developing countries. Meanwhile, solar thermal is a technology that collects energy from the sun to generate thermal energy, storing it for future use (see Figure 2.18). Lastly, CSP uses mirrors or lenses to concentrate the radiation coming from the sun onto a receiver, collecting it for later use in power-producing heat processes (Barlev et al., 2011).

These three techniques represent different ways of harnessing the sun's power, which is limitless and can help power all our needs – if we have the appropriate tools to collect its energy. Solar PV is better suited for rural or large-scale projects, where land use is not a significant consideration. A potential shortcoming of this technique is its low energy density, which means that one requires several plots of land to produce as much as a conventional energy plant would produce. One proposed solution is using Fresnel lenses (FL), which are mirrorlike concentrators that help obtain larger energy densities for a given plot of land (Sierra and Vázquez, 2005).

Meanwhile, solar thermal systems are better suited for residential and commercial settings to heat water, cook food, or regulate the temperature of a building or apartment. However, some countries have developed plans to build and operate solar thermal plants, like the examples of France (Scolan et al., 2020), Tunisia (Frein et al., 2014), and the island of Cyprus (Poullikkas, 2009). Then, CSP technology can be used for medium-scale projects and as power plants that seek to expand the

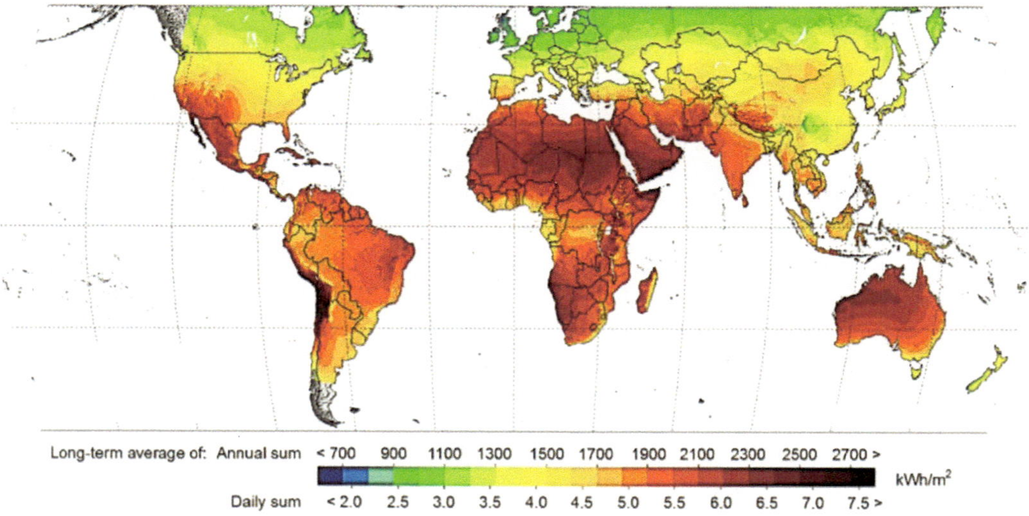

FIGURE 2.16 Maps of global horizontal irradiation (GHI).

Source: Kannan and Vakeesan (2016).

FIGURE 2.17 Solar photovoltaic farms used for rural electrification.

Source: Hoffmann (2006).

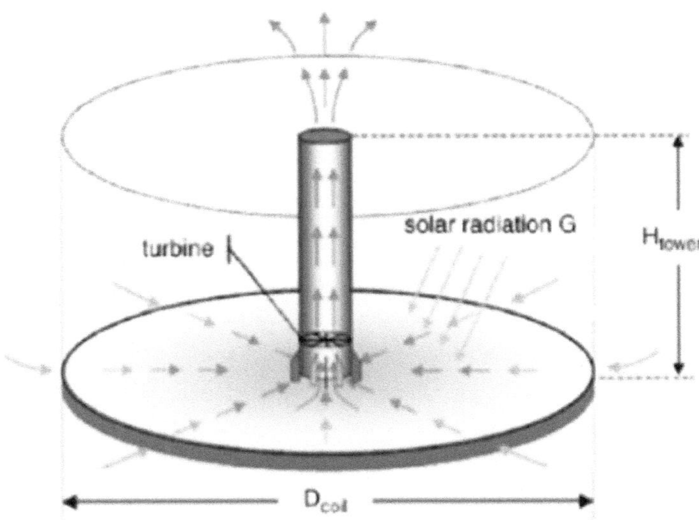

FIGURE 2.18 Solar thermal system with the greenhouse effect and air pressure difference.

Source: Guney (2016).

electricity grid. When placed in a location with favorable solar radiation, and when coupled with sources of energy that can supply in gaps brought by bad weather, like biomass or hydropower, concentrated solar power (with plants like the one illustrated in Figure 2.19) can be the catalyst to a 100% renewable electricity supply (Li et al., 2020).

With each year, the need for a greater share of renewable energy in our grids and industries is growing. Countries set goals and celebrate agreements, and organizations develop projections

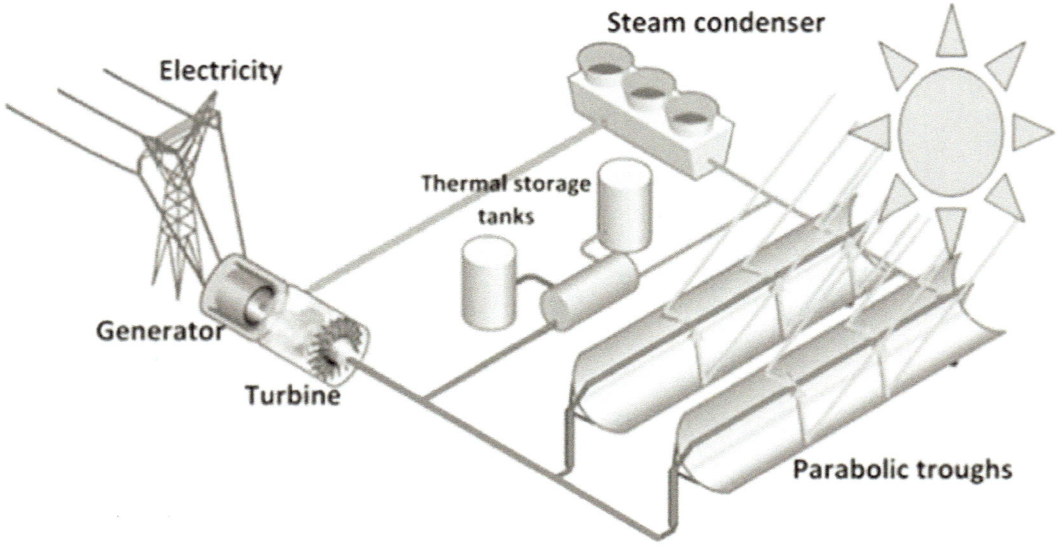

FIGURE 2.19 Concentrating solar thermal power.

Source: Kannan and Vakeesan (2016).

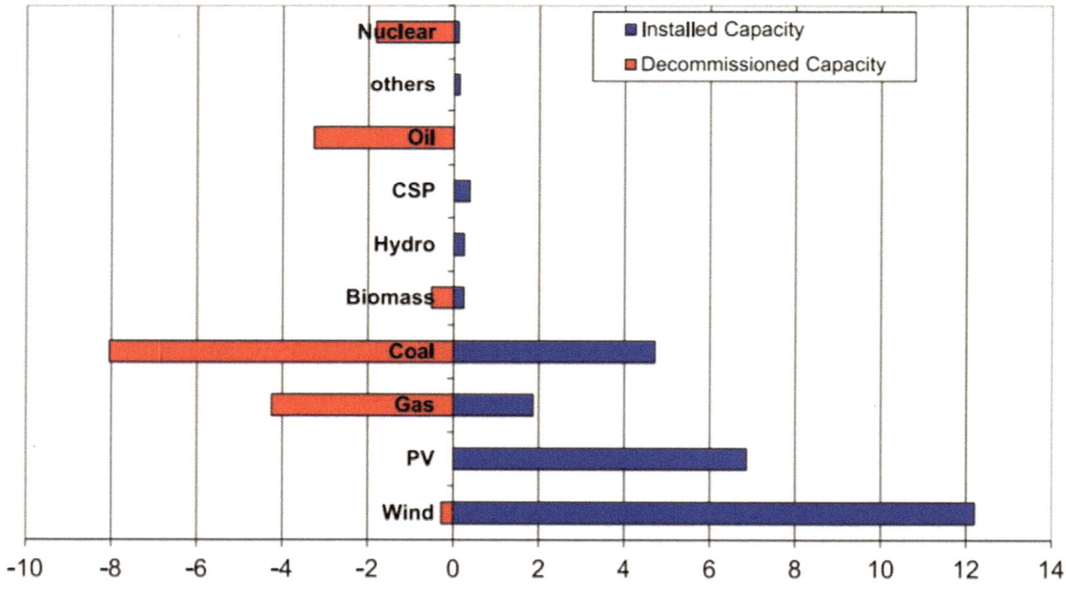

FIGURE 2.20 Newly installed or decommissioned electricity generation capacity in the European Union (in GW) for 2015.

Source: Lacal Arantegui and Jäger-Waldau (2018).

for a sustainable future. One way of fulfilling those accords is to decommission plants that rely on conventional energy sources while approving and building plants that use renewables. As Figure 2.20 shows, the European Union has led the way in connecting capacity that uses renewable energy sources, with wind and solar (primarily through PV farms) contributing almost 20 GW in 2015. And concurrently, the EU has decommissioned many coal, gas, oil, and nuclear capacity installations.

Of note is the decision from the European Union to decommission several nuclear power installations, despite the current academic and political discussion on its potential role in the energy transition of the 21st century. The following section consists of a review of nuclear energy, its challenges and strengths, and the possibilities that this technology, in its latest iteration, can open in the search for a successful energy transition. In the public discussion, nuclear energy brings thoughts of atomic bombs and their destructive power displayed only a few decades ago, so these considerations still hamper the debate about its widespread use.

2.1.3 Nuclear Energy

The energy provided by nuclear reactions can be obtained mainly in two ways: through fission or fusion of atoms. Figure 2.21 shows the process of nuclear fission, currently used to produce electricity around the world, where the nucleus of an atom is split to create energy in a chain reaction that affects other nuclei. The heat released is converted into electricity in a nuclear plant, where the heat warms the cooling agent, which most of the time is water, to produce steam that is later channeled to spin turbines (see Figure 2.22), which in turn produces clean energy (IAEA, 2021). Nuclear fusion is currently in a research and development (R&D) phase; thus, it is not a widespread method to produce energy.

One of the chemical elements best suited for electricity-producing nuclear fission is an enriched form of uranium, called uranium-235, that allows for these chain reactions that provide greater energy (IEIAvideo, 2017). Uranium is mined from open or underground mines or by leaching (reacting the ore with a substance). Then, the uranium is enriched from around 0.7% of ^{235}U to a range between 3% and 4%, an ideal state to produce fuel. After this, nuclear waste that comes from the spent fuel is repurposed to generate the so-called MOX fuel (mixed oxide, $UO_2 + PuO_2$), and the rest is moved into underground repositories or, in some cases, transmuted in generation IV reactors built for this purpose (Ricotti, 2013). An illustration of this cycle is shown in Figure 2.23.

During their operations, nuclear plants produce virtually no CO_2, so at least in theory, nuclear energy appears to fit exceptionally well with the plans from governments and supranational

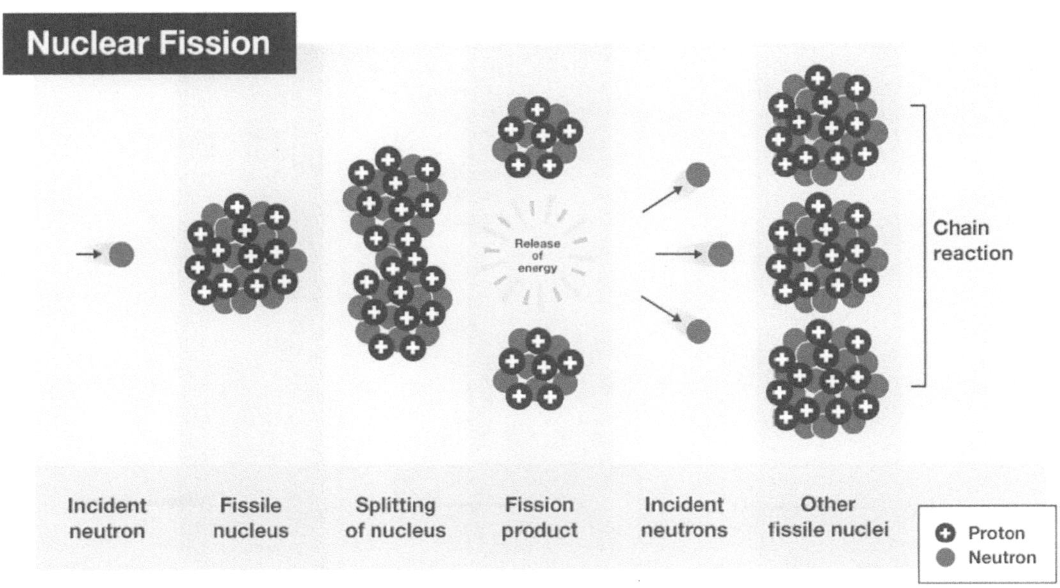

FIGURE 2.21 Process of nuclear fission.

Source: IAEA (2021).

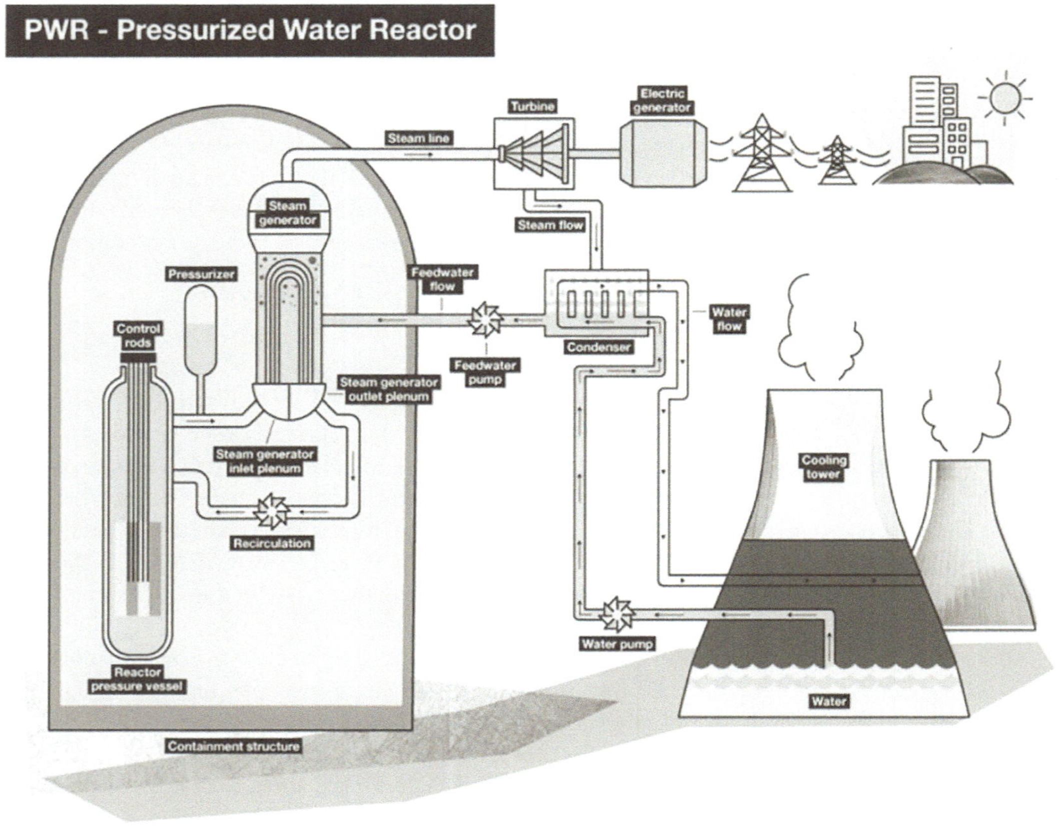

FIGURE 2.22 Pressurized water reactor.

Source: IAEA (2021).

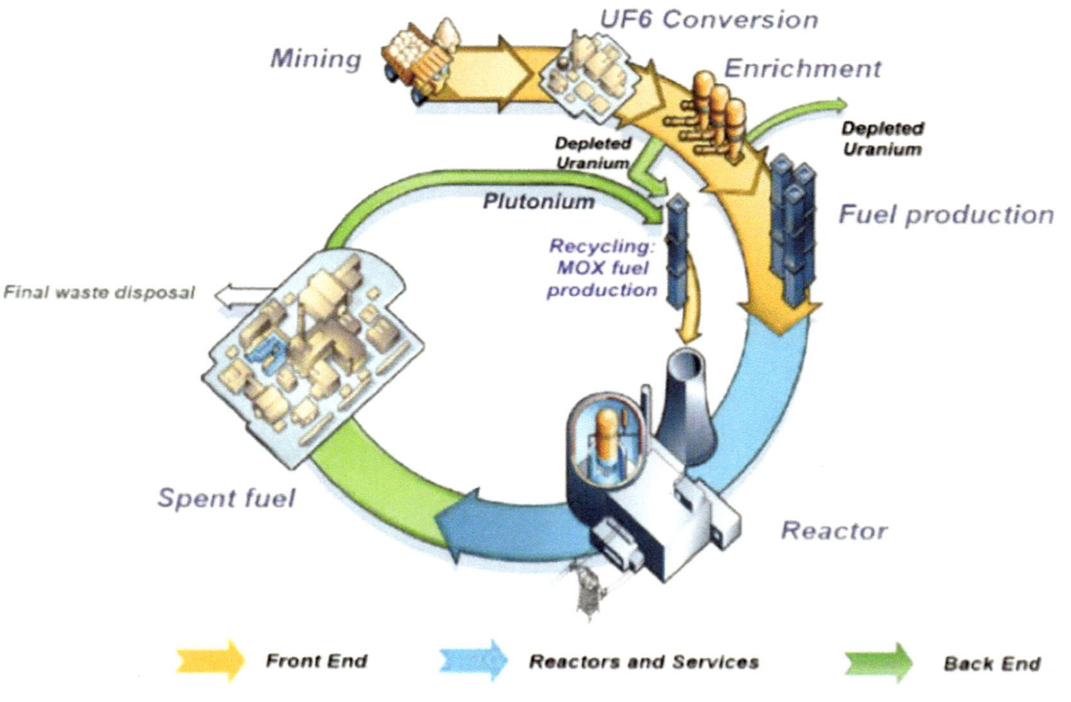

FIGURE 2.23 The main steps of the nuclear fuel cycle.

Source: Ricotti (2013).

organizations to achieve net-zero goals announced at international conferences. Additionally, atomic energy is a very dense form of energy, which means that with small amounts of uranium, plants can produce energy that lasts much longer than the energy produced by multiple acres of wind or solar farms (Cheng and Hammond, 2017; Månsson, 2015). Figure 2.24 shows the vast difference in density between nuclear energy generators and biomass, wind, and PV technologies.

With that said, critics of nuclear energy mention the notorious accidents that have occurred after major nuclear reactor technology failures, like in Fukushima, Three Mile Island, and Chernobyl (Gu, 2018). There is an exciting debate about nuclear accidents and the use of nuclear technology, especially if we consider a broader context. For example, these accidents have remained in the public perception for a long time, despite the low casualty toll of nuclear energy compared to other energy sources. Kharecha and Hansen (2013) estimate that coal and natural gas, besides contributing more greenhouse gases than nuclear, also have a higher death toll per terawatt produced (see Table 2.2).

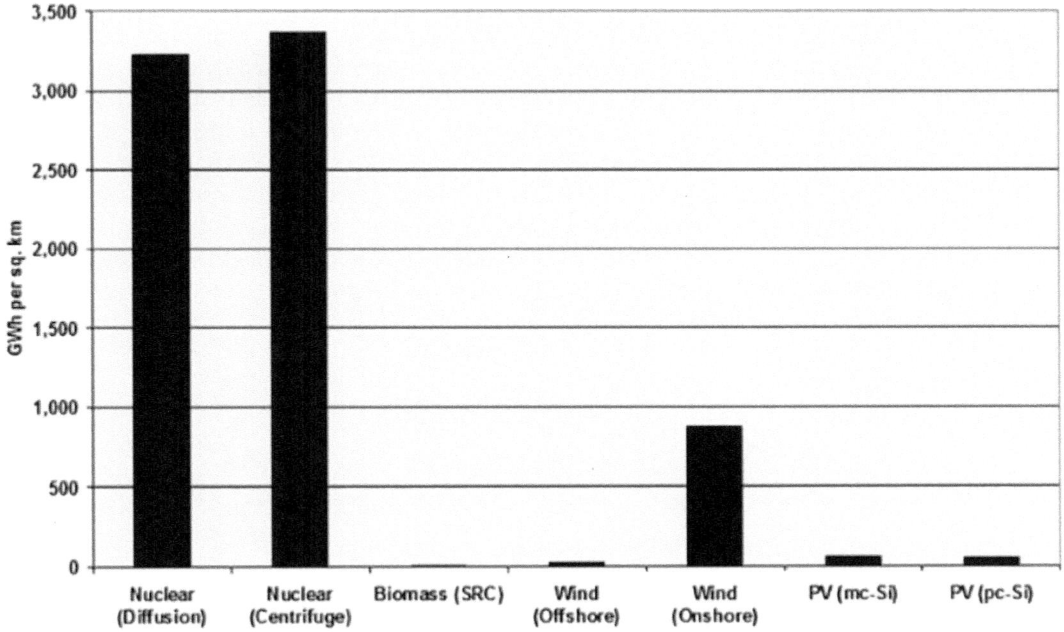

FIGURE 2.24 Energy densities of selected power generators.

Source: Cheng and Hammond (2017).

TABLE 2.2
Mortality and Greenhouse Emissions (GHG) Estimations

Electricity source	Mean Value (Range)	Unit[b]
Coal	28.67 (7.15–114)	Deaths/TWh
	77 (19.25–308)	Deaths/TWh
	1,045 (909–1,182)	tCO_2-eq/GWh
Natural gas	2.821 (0.7–11.2)	Deaths/TWh
	602 (386–818)	tCO_2-eq/GWh
Nuclear	0.074 (range not given)	Deaths/TWh
	65 (10–130)[d]	tCO_2eq/GWh

Source: Kharecha and Hansen (2013).

This discrepancy has some roots in the historical development of advanced countries in the 20th century, some of it appearing after the dropping of the atomic bomb on Japanese territory in 1945. In the United States, caution multiplied after the Three Mile Island accident in 1979, and public perception revealed low levels of trust in the nuclear establishment (Stoutenborough et al., 2013; Whitfield et al., 2009). After the events of March 11, 2011, in Fukushima, where four reactor buildings were severely damaged after an earthquake and subsequent tsunami, Japan revived many of its fears regarding the dangers of nuclear power, taking with it much of the momentum gained by the industry in the past decade (Behling et al., 2019).

A ripple effect from this accident (where one person died from radiation and almost 20 were injured) was that Germany shut down 40% of its nuclear power plants permanently (Liebl, 2019), while Switzerland and Italy undertook some lighter nuclear phaseout plans (Ricotti, 2013), some of which led to instability in energy prices. Moreover, Japan itself suffered some unintended effects from what Neidell et al. (2019) call an "overcaution with the precautionary principle." After they ceased operations of all nuclear power stations and replaced them with fossil fuels, the authors observed an increase in overall mortality during very cold temperatures, which they attribute to lower energy consumption, which was driven by higher energy prices caused by energy scarcity.

Figure 2.25 shows the monthly average price for an energy bill before and after the Fukushima accident, where we see that all Japanese regions experienced an increase in their power bills not long after experiencing a sustained decrease after 2009. The Tokyo region was particularly affected by the consequences of the changes in the electricity grid, but it appears that the supply was affected enough so that prices rose. This lack of alternatives to replace nuclear energy in Japan led to a more significant problem, and this is something that needs close attention before more countries decide to phase out nuclear reactors due to fears of accidents that do occur but are exceedingly rare and do not cause as many fatalities as more conventional sources of energy.

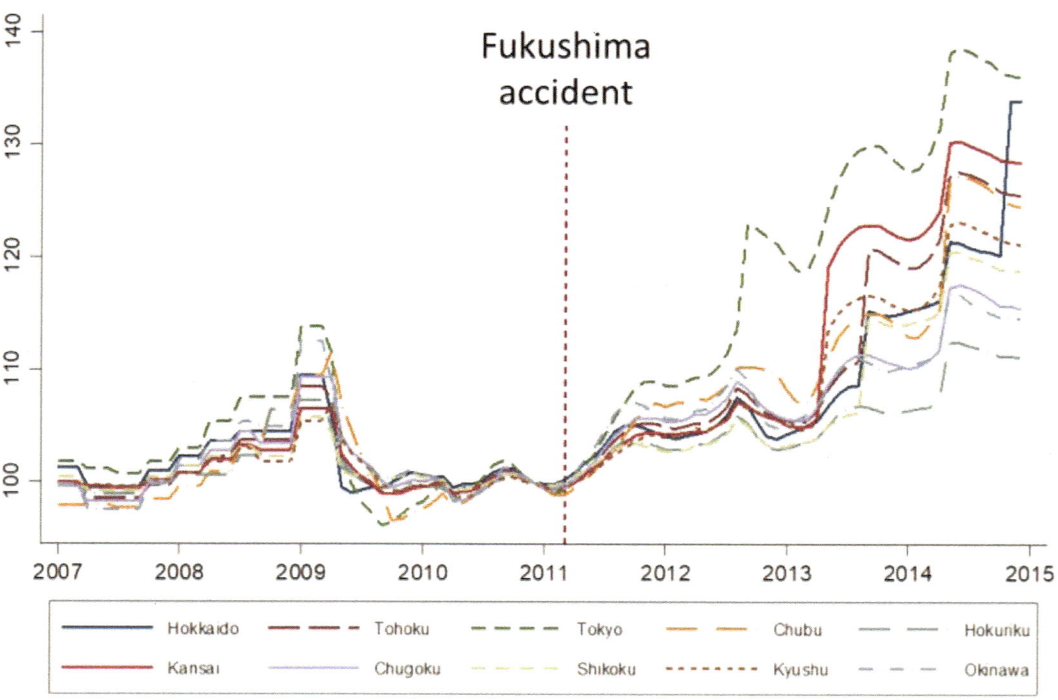

FIGURE 2.25 Monthly average price index of residential electricity by Japanese region, 2007–2014 (2010 = 100).

Source: Neidell et al. (2019).

Some evidence from European Union countries shows relatively low support for nuclear energy, even among those populations most concerned about climate change (Pampel, 2011). Table 2.3 indicates consent for different power sources on a scale from 1 to 7, and we see that nuclear energy is positioned last, with a score of 3.5, while solar, wind, and hydroelectric are in the top three, with scores of at least 6. Before Brexit occurred, two contrasting cases were observed in the cases of Germany and the United Kingdom. On the one hand, Germany has held an anti-nuclear position for a long time, and on the other, the latter country appears to have a "reluctant acceptance" of nuclear energy, yet its population is still divided and uncertain about it (Kiyar and Wittneben, 2012). These differences suggest that varying results across individuals may also be affected by their country, as contexts help individuals shape their perceptions (Wang and Kim, 2018).

Despite these public perceptions, however, some European countries experimenting with nuclear power approved the construction of new reactors and continued the operations of current ones across the continent. Figure 2.26 shows the situation regarding nuclear energy in Europe as of February 2021. France is leading the list by far, with 56 nuclear reactors in operation and one more under construction, followed by Russia, Ukraine, and the United Kingdom, which together add up to 68 operational reactors, with 7 more under construction. The constructed nuclear plants have a net electric capacity of over 150,000 MWe in total (European Nuclear Society, 2021).

An essential part of this move of a few countries toward more nuclear energy has been the development of new generations of nuclear technology. In the year 2000, the US Department of Energy initiated the Generation IV International Forum (GIF), and 13 countries joined to develop new methods of generating nuclear energy that would make the technology safer and more widespread, moving away from second-generation reactors, which constitute most of the reactors in the world (World Nuclear Association, 2020). In their 2020 report, GIF stressed the need to support the necessary research and development strategies to help GEN IV systems that focus on enhanced safety, minimal waste, sustainability, and efficiency (Generation IV International Forum, 2020).

Regarding nuclear reactors' safety, the defense in depth (DiD) concept stemmed from the safety strategy around the first nuclear reactor, inspired by Enrico Fermi. Slowly, this concept developed into a military-style defense mechanism for a nuclear reactor that considers variables touching on quality assurance, automation, monitoring, diagnostics tools, and external hazards (Gu, 2018).

TABLE 2.3
Eurobarometer Items Measuring Support for Nuclear Energy

	Relative Items (With Means and Standard Deviations)	
Are you in favor or opposed to the use of these different sources of energy (in OUR COUNTRY)? Please use a scale from 1 to 7; 1 would mean that you are "strongly opposed to this energy source," and 7 would mean that you are "strongly in favor" of it.		
Solar	Mean = 6.3	*SD* = 1.1
Wind	Mean = 6.0	*SD* = 1.4
Hydroelectric	Mean = 6.0	*SD* = 1.3
Ocean	Mean = 5.9	*SD* = 1.5
Biomass	Mean = 5.6	*SD* = 1.7
Gas	Mean = 5.1	*SD* = 1.6
Oil	Mean = 4.4	*SD* = 1.8
Coal	Mean = 4.2	*SD* = 1.9
Nuclear	Mean = 3.5	*SD* = 2.1

Source: Pampel (2011).

Table 2.4 shows the five layers of this DiD strategy, starting with prevention objectives and moving to controlling abnormal operations, detecting potential failures, proactively assessing plant conditions to avoid accidents, and mitigating the consequences of any possible release of radioactive materials.

In conclusion, many lessons learned the hard way may end up being a blessing for future nuclear energy development. Due to its unparalleled potential, nuclear energy appears to have experienced a new renaissance, seeking a way out of conventional energy sources. With the effect of Gen III and Gen IV nuclear power plants (Locatelli et al., 2013), which have discarded many of the old technologies that were unable to prevent a disaster like the one of Chornobyl, the sector appears poised for a substantial contribution to the energy transition of the 21st century. And if the debate includes positions that at times have remained at the fringes (Ausubel, 2007), safe nuclear technologies could be the backbone of the energy transition of the 21st century.

FIGURE 2.26 Nuclear plants in operation in Europe as of February 2021.

Source: European Nuclear Society (2021).

TABLE 2.4
The Five Continuous Layered Defenses of Defense in Depth (DiD)

DiD Levels	Objectives	Essential Means
Level 1	Prevention of abnormal operation and failure	Conservative design and high quality in construction and operation
Level 2	Control of abnormal operation and detection of failures	Control, limiting, and protection systems and other surveillance features
Level 3	Control of accidents within the design basis	Engineered safety features and accident procedures
Level 4	Control of severe plant conditions, including prevention of accident progression and mitigation of the consequences of severe accidents	Complementary measures and accident management
Level 5	Mitigation of radiological consequences of significant releases of radioactive materials	Off-site emergency response

Source: Gu (2018).

2.1.4 Energy Storage

This section explores the practice of energy storage, which consists of capturing energy produced at one time and storing it so it can be used later. The development of this technique is a crucial aspect of the expansion of renewable energies, as both wind and solar are very unreliable sources of energy, providing it handsomely when conditions are favorable (and sometimes when no more power is needed) but unable to do so when there is no wind, or when the sun illuminates the other half of the world.

Currently, most of the energy that is produced from different sources is consumed almost immediately, so thanks to the technological advancements developed throughout the years, the electricity demand can be covered quickly by the supply provided by power plants, including abnormal moments where demand is higher (but for the extreme case of the Texas cold snap of winter 2021, see Doss-Gollin et al. 2021). However, a successful process of energy transition needs a supply that not only integrates new sources of energy progressively into the grid (Ausfelder et al., 2017) but also necessitates improved and cutting-edge energy storage technologies that can withstand exceptional circumstances (Kittner et al., 2017).

Energy storage systems (ESSs) not only help store unused energy but are also needed for the improvement of distribution and transmission and proper energy management. Table 2.5 shows a list of the most common ESSs and their round-trip efficiency, which is the percentage of electricity stored that is retrieved and used for power (Breeze, 2019). Each of these technologies has its own trade-offs, making them good options at some moments but less so at others. Throughout this section, we will review these methods in more detail and explore current research trends that combine some of these technologies with what is known as "Power-to-X."

Figure 2.27 shows five different sets of applications of ESSs, where some of them help with price arbitration, supply management, physical connection improvements, and voltage control, among other considerations. There are different methods to store energy that vary in their duration, use, and sophistication. Just like how we rely on batteries to power our remote controls at home, researchers have also developed batteries for some future energy uses. But these applications are not only limited to batteries, as we will review next.

First, batteries are one of the most well-known forms of energy storage, helping the proliferation of portable electronic devices a couple of decades ago. Batteries are part of electrochemical energy storage systems, which involve three steps: separation (or ionization), transport, and recombining of

TABLE 2.5
Round-Trip Efficiency of Energy Storage Technologies

Energy Storage Technology	Round-Trip Storage Efficiency (%)
Capacitors	90
Superconducting magnetic energy storage	90
Flow batteries	90
CAES	65
Flywheels	80
Pumped-storage hydropower	75–80
Batteries	75–90

Source: Breeze (2019).

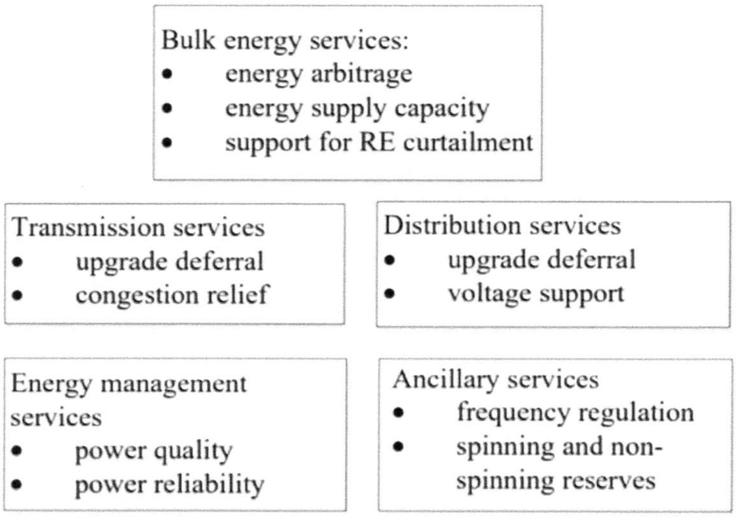

FIGURE 2.27 Applications of energy storage systems (ESSs).

Source: Mohamad et al. (2018).

charge (Gür, 2018). The battery's more widespread version is made from lithium-ion (Li-ion). Still, other iterations are made from lead-acid, redox flow, sodium-sulfur, sodium metal halide, or zinc hybrid cathode (Mongird et al., 2019). Figure 2.28 shows a Ragone diagram with different battery types placed on the spectra of specific energy and specific power.

In general, batteries are a cost-effective solution that offers versatility but has relatively constrained uses. As a response, supercapacitor technology has grown in the past few years. A *supercapacitor* is a high-capacity energy-storing device that can store and deliver energy at very fast rates and tolerate many more charge and discharge cycles than rechargeable batteries, reportedly over 500,000 of these cycles. Supercapacitors have found applications in electric vehicles (more on that later), uninterruptible power supplies (UPS), and memory backups for IT systems (Mensah-Darkwa et al., 2019).

Then, *flywheels* are a mechanical form of energy storage that consists of a rotor (the flywheel) that is accelerated in a vacuum containment, generating electricity for a considerable amount of time. Due to the wheel spinning continuously, without needing added power, the flywheel stores energy in a rotating mass, and the kinetic energy produced is stored as rotational energy (Olabi

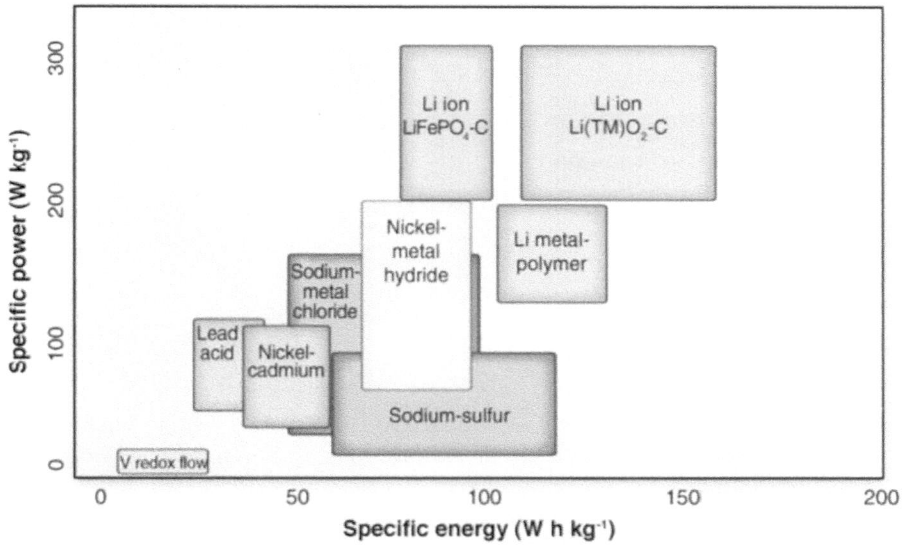

FIGURE 2.28 Ragone diagram of specific energy versus specific power for major rechargeable battery systems.

Source: Gür (2018).

et al., 2021). Figure 2.29 shows a schematic of a basic design for a flywheel, where we see the spinning wheel that works in a vacuum or in a low-pressure environment that is maintained with helium so that friction is diminished. While energy is stored, the flywheel spins at high velocities, and when power is needed, the kinetic energy is converted into electricity as the wheel slows down (Gür, 2018).

Table 2.6 lists the advantages and disadvantages of the flywheel mechanism. Given that it is a mechanical form of energy storage, it has an uncomplicated delivery, is highly efficient, and requires almost no maintenance. However, it has a high self-discharge rate and a low energy density while also having a high capital cost due in part to the materials needed to keep it functioning. Because of these shortcomings, flywheels have not been utilized as much as the next mechanical energy storage technology we will review, hydropower.

Hydropower consists of using the energy of both water and gravity to release stored energy. The general setup of a hydropower plant is shown in Figure 2.30, where we can see two water storage units with different elevations. The top is used for storing purposes, and the bottom unit is used for discharging energy (Gür, 2018). During periods of low energy demand, like what happened throughout the height of lockdowns during the COVID-19 pandemic, pumped storage hydropower (PSH) installations were able to store energy, which was later retrieved in periods of low winds (International Hydropower Association, 2021). For the most part, users overlook the energy source they consume, but PSH systems have been essential to stabilize the electricity grid by supplying daily power to those who need it.

Additionally, pumped hydro technology significantly contributes to the displacement of fossil fuel reliance because it helps eliminate the need to generate backup power through polluting sources. For example, if a town runs out of energy from clean sources and renewables, most grids have backup power plants if needed, but they rely on older technologies like coal. To cover the intermittence of solar and wind energy (Gür, 2018; Javed et al., 2020), at times, it becomes necessary to turn those backup power plants on, a needed move for the well-being of the users, but a decision that could increase greenhouse emissions if it is maintained for extended periods (Fairuz et al., 2013).

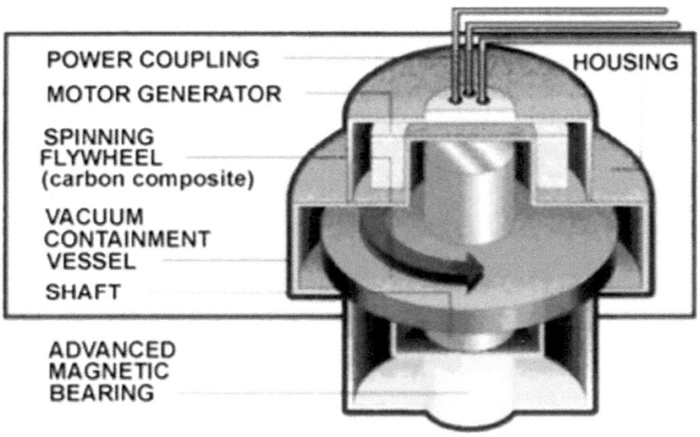

FIGURE 2.29 Schematic cross section illustrating the basic design and operation of a flywheel energy storage system.

Source: Gür (2018).

TABLE 2.6
Advantages and Disadvantages of the Flywheel

Advantages	Disadvantages
• High energy-efficiency	• The need for permanent magnets in the rotor
• Almost immediate delivery	• May require costly cryogenic cooling devices
• Strong power	• Cryogenic cooling also reduces the overall energy storage efficiency
• Requires little maintenance	Deep discharging cannot be achieved
• Long service life	• High capital cost, whether due to the materials' cost for the light rotational mass, i.e., at high rpm, or for the magnetic bearing using heavy rotational masses
• Environmentally friendly	• High self-discharge rate and low energy density
• Simple and safe	
• Flexible in the rate of charging and/or discharging	

Source: Olabi et al. (2021).

These plants are generally complex to build, adding considerable costs to energy supply and grid maintenance. However, many large-scale hydropower storage plants have been developed in several countries over the years, particularly in countries that are installing extensive facilities for renewable energies like wind and solar, like in several European countries (Breeze, 2019). A pumped hydro energy storage plant is an excellent installation to deliver stored energy in peak demand periods, particularly when generating additional power is costly. Thus, coupled with renewable energy in a cleaner grid, these plants can help make the energy transition process as smooth as possible.

Lastly, compressed air energy storage (CAES) methods save energy in the form of air pressurized above atmospheric pressure (Breeze, 2019). CAES is a promising technology due to its high storage capacity, high efficiency, as well as being a safe and environmentally friendly method (Zhang et al., 2019). Additionally, CAES systems are showing encouraging signs of integration with renewable sources so that the intermittent nature of the power coming from the wind and the sun is resolved and we can successfully move away from fossil fuels (Mousavi et al., 2021; Soltani et al., 2020). Figure 2.31 shows an illustration of a simple CAES system.

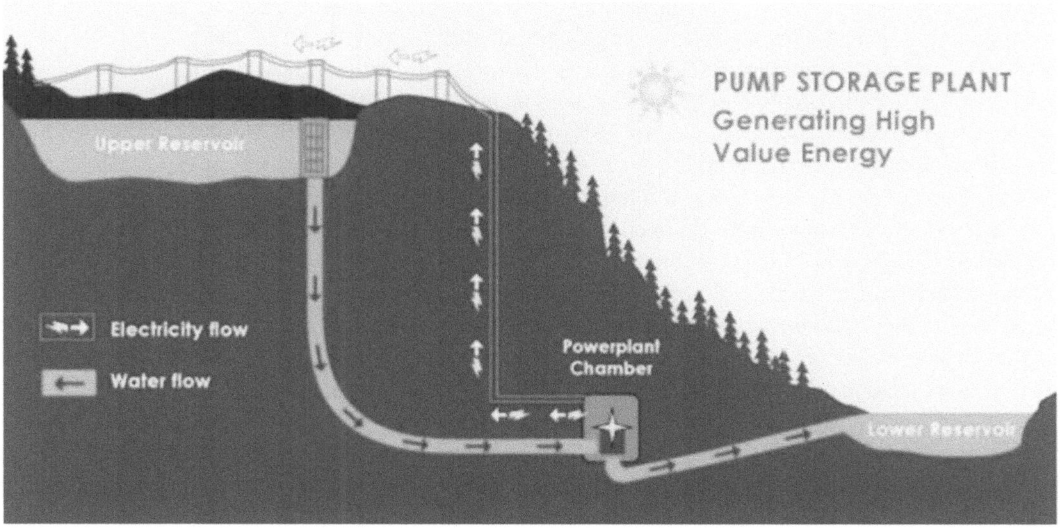

FIGURE 2.30 Configuration of a pumped hydro energy storage (PHES) plant.

Source: Mohamad et al. (2018).

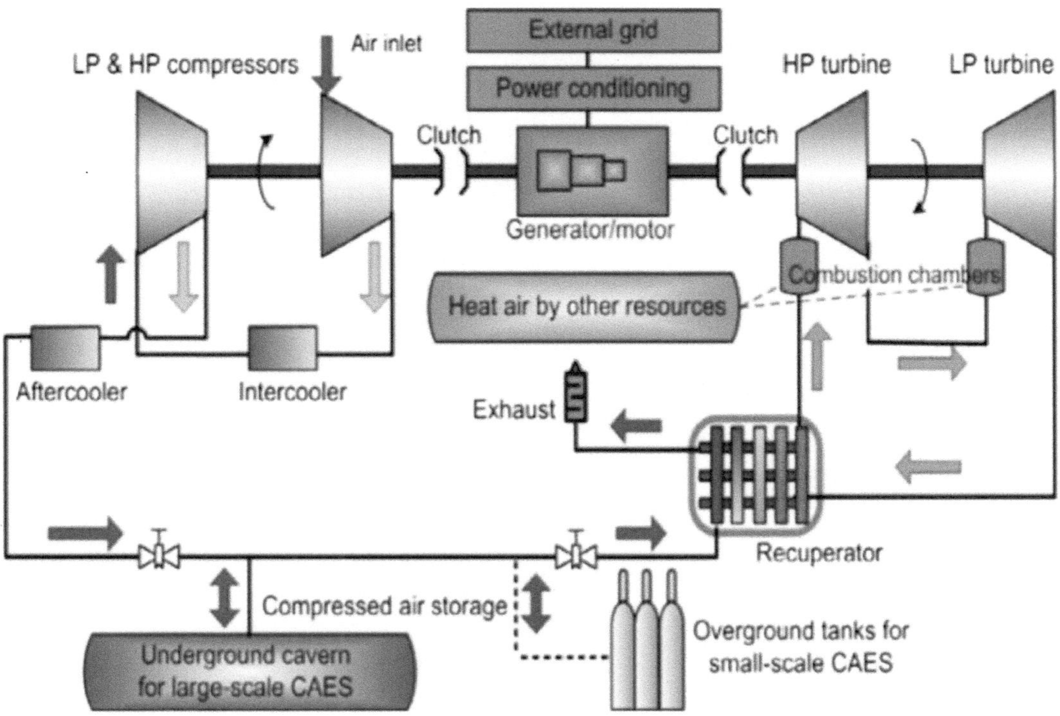

FIGURE 2.31 Schematic of a compressed air energy storage (CAES) system.

Source: Mohamad et al. (2018).

To review the contribution of each method of energy storage, Table 2.7 shows the distribution of different types of energy storage technologies in the world, and we can see that hydropower (noted as *PSH*) takes the most significant share of energy storage use, with almost 170,000 MW deployed, far outpacing all other technologies. The second technology is lithium-ion batteries, with less than one-tenth of the MW contribution from PSH. Then, flywheels are placed third, while compressed air energy storage (CAES) methods are the fourth most used.

In general, these methods serve different purposes at different points in the energy transition process and for various applications. Pumped hydro technologies are ideal when countries are looking for support for a large electricity grid that is intermittent or to discard the support of it by conventional sources of energy. Compressed air methods are like PHS, but discharge times are long, and they require proximity to underground cavities and caverns where gas can be contained for extended periods (Gür, 2018). Meanwhile, batteries appear best suited for smaller tasks or portable endeavors, like cars operated by these storage devices. In turn, flywheels may also help the automotive industry with the storing of energy after braking, to be used as energy when accelerating, in what is known as KERS (kinetic energy recovery system) technology (Farrokhzad Ershad et al., 2020; Milica and Andrei, 2020).

Furthermore, the latest developments in the research of energy storage methods include the "power-to-X" technology, described as the conversion and storage of electrical energy into a carrier (gas, fuel, or raw material) or product (basic material or feedstock). The term is a blanket terminology for different power-to-gas, power-to-liquid, power-to-fuel, power-to-chemicals, and power-to-product techniques (Sterner and Specht, 2021). Figure 2.32 shows a basic concept of the power-to-gas (PtG) technology, which shows the conversion of renewable electricity into renewable gas by replicating the natural photosynthesis process observed in plants. After conversion, the energy can be used for transportation, buildings, or even the industrial sector.

A somewhat overlooked aspect of the energy transition, the storage of copious amounts of energy collected both from renewable sources and the cleaner alternatives developed to fossil fuels is a sector that must advance at a similar pace as the development of power plants and the phasing out of more polluting sources. Given the intermittent nature of renewable energy sources, which means that the sun and the wind will provide large quantities of energy, but only for limited hours of the day, building and maintaining large storage facilities next to power plants is another significant objective of the energy transition.

Equally, processes of clean transformation of different types of renewable energy into, for example, hydrogen to be stored in fuel cells to electric power vehicles, or into gas to provide to the grid for

TABLE 2.7
Worldwide Deployment by Technology Type, 2018

Technology	MW Deployed
Sodium sulfur	189
Lithium-ion	1,629
Lead-acid	75
Sodium metal halide	19
Flow battery	72
PSH	169,557
CAES	407
Flywheels	931
Electrochemical capacitor	49
Total	172,928

Source: Mongird et al. (2019).

heating and cooking needs, are also an essential part of the supporting innovations that are expected for the energy transition process of the new century. As the world continues to develop and finds its footing after the COVID-19 pandemic, power needs are increasing once again. The potential of renewables is virtually endless, but human ingenuity needs to back that potential with structures that can store that energy and convert it through clean processes when other uses become necessary.

2.1.5 Fuel Cells, Electric Vehicles (EV), and Hydrogen Technologies

In a similar category within energy storage, fuel cells, and hydrogen technologies, are a more novel set of methods with the same objective of lowering carbon emissions and contributing to a successful energy transition away from fossil fuels. Fuel cells are energy conversion devices that constantly convert chemical energy into electrical energy if fuel and oxidants are available. One of the elements in this equation is hydrogen. This naturally abundant chemical element is oxidized in a fuel cell system, generating pure water as a by-product without emitting carbon dioxide (Fan et al., 2021). This chapter will review the close relationship between these two components and its relationship to battery-powered vehicles, a category that includes electric vehicles (EVs), a promising avenue for car manufacturers to produce more environmentally friendly vehicles.

Fuel cells work like batteries, but they do not run out if there is fuel being supplied to keep them running. Hydrogen is fed into the negative electrode (or anode), while air is fed into the positive electrode (or cathode). A catalyst at the anode begins separating the hydrogen molecules into protons and electrons, which later go to the cathode through an external circuit that creates an electricity flow. Once in the cathode, protons unite with oxygen and electrons to produce water and heat (Office of Energy Efficiency and Renewable Energy, 2021). Figure 2.33 shows a few applications of fuel cells in our daily lives, where vehicles are an increasingly important commodity. Figure 2.34 shows a basic configuration of a hydrogen-based fuel cell.

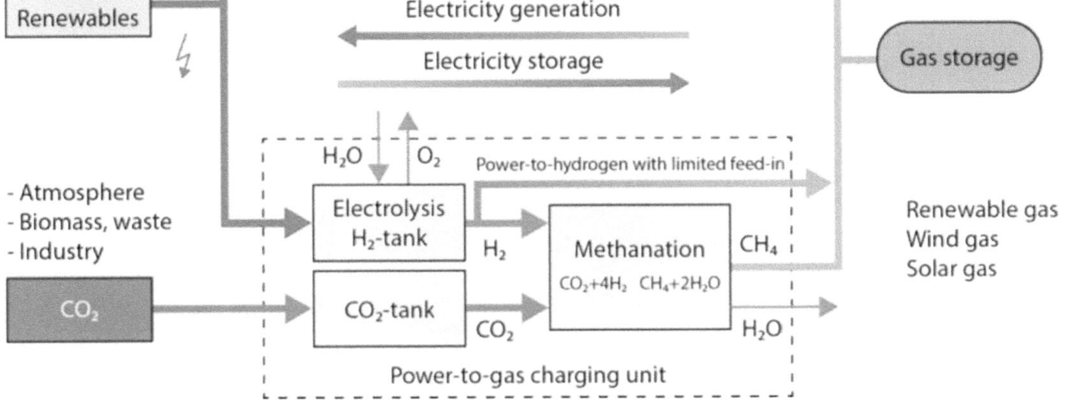

FIGURE 2.32 Basic concept of power-to-gas from 2008.

Source: Sterner and Specht (2021).

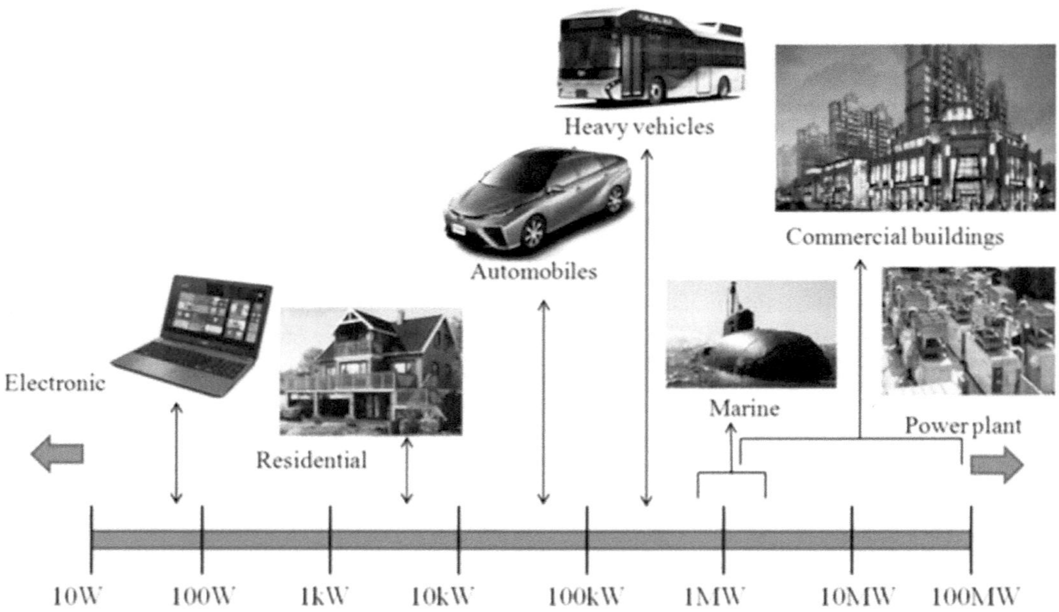

FIGURE 2.33 Typical applications of fuel cells.

Source: Fan et al. (2021).

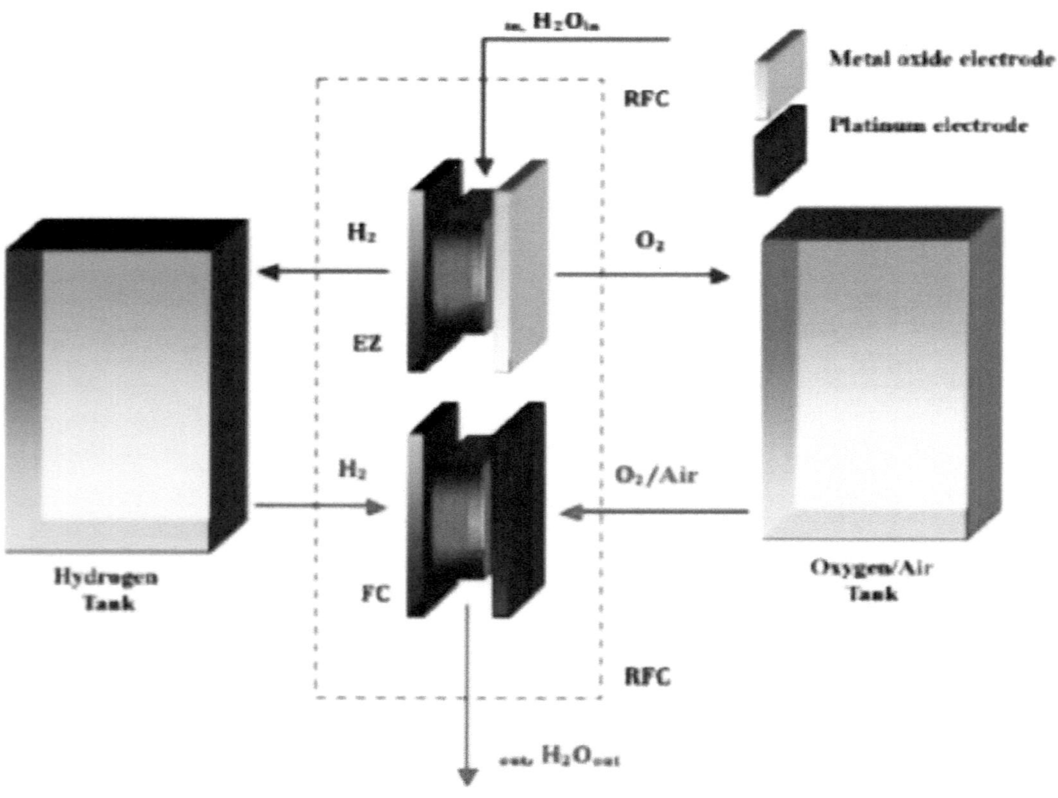

FIGURE 2.34 Configuration of a hydrogen-based fuel cell.

Source: Mohamad et al. (2018).

More generally, hydrogen technologies have been touted as solid replacements for fossil fuels and are pivotal in advancing a low-carbon future (Oliveira et al., 2021). A "hydrogen economy" is an ambitious alternative based on a similar process to the one described earlier, the production of water electrolysis that splits water into hydrogen and oxygen, producing no greenhouse gas emissions, albeit performed on a much larger scale. The strongest proponents of a switch to an energy economy based on hydrogen claim that we are currently surrounded by hydrogen technologies that only need more time to develop, but the move is imminent (Ball and Weeda, 2015; Clark, 2007).

Currently, hydrogen technologies are present in many aspects of the world's energy journey, as hydrogen has been used and produced for industrial uses since around 1920, when it was used for ammonia synthesis, and later as a town gas for street lighting and as a home energy supply for cooking and heating (Ball and Weeda, 2015). Thus, moving toward an energy transition that has this element as one of its main components is relying on a proven part that has stood the test of time and has an almost boundless potential due to its adaptability and efficiency in carrying energy (Li, 2019; Manoharan et al., 2019).

Various hydrogen production technologies are used to obtain hydrogen at a large scale. These techniques include water electrolysis, methanol reforming, ammonia decomposition, water gas, and chlor-alkali industrial tail gas treatment (Fan et al., 2021). The first method, water electrolysis, appears to be the most promising process, with the lowest environmental impact. As advanced at the beginning of this section, water electrolysis consists of the circulation of a current through water to separate its molecules into hydrogen and oxygen, with the former being created at the cathode while the latter is formed at the anode (Ursúa et al., 2012). Table 2.8 shows a detailed summary of some of the typical features of different water electrolysis processes.

Hydrogen is the most efficient energy carrier (Shiva Kumar and Himabindu, 2019), and despite taking more energy to produce than what it provides when converted to useful energy (EIA, 2021), hydrogen has a high energy content per unit of weight, which makes it ideal to be used as, for example, rocket fuel, as well as in fuel cells to power devices of different sizes. This unrealized potential, coupled with the clean process to obtain hydrogen in a world that needs to pollute less, positions this element as a desirable candidate to be exploited for widespread energy use in the future.

Fuel cells are electrochemical devices that convert the energy of a fuel into electrical energy in one step. Comparatively, other processes that require multiple energy conversion steps, such as from chemical to thermal, followed by steps to mechanical and, finally, to electrical energy, have been demonstrated to be more polluting and less efficient (Sharaf and Orhan, 2014). Table 2.9 summarizes fuel cells' key advantages and shortcomings, where the two most significant upsides are their reduced harmful emissions and high efficiency. At the same time, the two most important limitations are their high cost and low durability.

Furthermore, and related to the process of energy transition, fuel cells have a very promising potential for many markets. One of those markets is that of electric vehicles (EVs), where proponents of fuel cell vehicles (FCV) hope that they end up replacing the internal combustion engine (ICE), becoming the principal source of power for vehicles (Hardman et al., 2013). The history of engine-powered vehicles is rich, however. Historically, gas has been the preferred method to run their engines, but batteries have significantly contributed over the last few decades. Batteries are a mass production electrochemical system of energy that can power a vehicle for several hours before being recharged and are another essential branch of a successful move to the energy transition of the 21st century.

Battery-run vehicles have a history dating back as far as the 1830s, when Robert Anderson built the first electric vehicle. A couple of decades later, Gaston Plante invented the first rechargeable lead-acid battery, making the electric car more practical. Interestingly, before the beginning of the 20th century, more automobiles were EVs than gasoline-powered ones. However, with Henry Ford's development of mass production techniques for cars in 1910, which lowered the cost of gas-fueled vehicles, the market for EVs was significantly diminished (Cairns and Albertus, 2010). But by the end of the century, oil shortages and better technologies have positioned EVs to come back just in time for the energy transition of the 21st century.

TABLE 2.8
Typical Characteristics of Main Electrolysis Technologies

	Low-Temperature Electrolysis			High-Temperature Electrolysis		
	Alkaline (OH–) Electrolysis		Proton Exchange (H+) Electrolysis		Oxygen Ion (O_2) Electrolysis	
	Liquid	Polymer Electrolyte Membrane		Solid Oxide Electrolysis (SOE)		
	Conventional	**Solid Alkaline**	**H+–PEM**	**H+–SOE**	**O_2-–SOE**	**Co-Electrolysis**
Operation principles	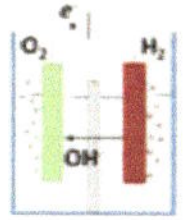	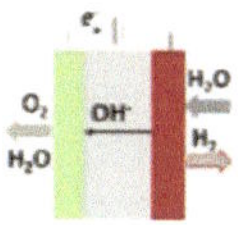	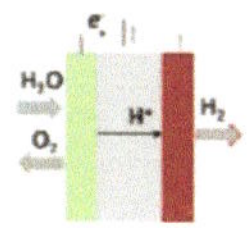	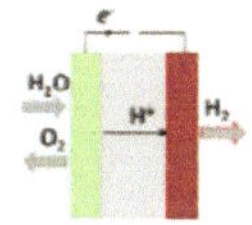	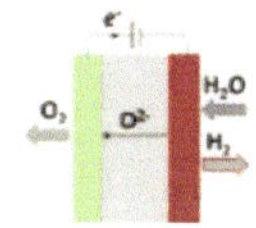	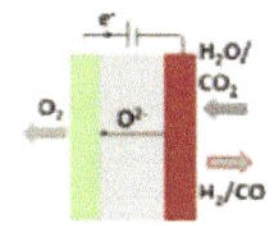
Charge carrier	OH^-	OH^-	H^+	H^+	O_2^-	O_2^-
Temperature	20–80°C	20–200°C	20–200°C	500–1,000°C	500–1,000°C	750–900°C
Electrolyte	Liquid	Solid (polymeric)	Solid (polymeric)	Solid (ceramic)	Solid (ceramic)	Solid (ceramic)
Anodic reaction (OER)	$4OH^- \rightarrow$ $2H_2O + O_2 + 4e^-$	$4OH^- \rightarrow$ $2H_2O + O_2 + 4e^-$	$2H_2O \rightarrow 4H^+ + O_2 + 4e^-$	$2H_2O \rightarrow 4H^+ + 4e^- + O_2$	$O^2 \rightarrow ½O_2 + 2e^-$	$O^2 \rightarrow ½O_2 + 2e^-$
Anodes	Ni > Co > Fe (oxides) Perovskites: $Ba_{0.5}Sr_{0.5}Co_{0.8}Fe_{0.2}O_{3.5}$, $LaCoO_3$	Ni-based	IrO_2 RuO_2, $Ir_xRu_{1-x}O_2$ Supports: TiO_2, ITO, TiC	Perovskites with protonic-electronic conductivity	$La_xSr_{1-x}.MnO_3$) + Y-stabilized ZrO_2 (LSM-YSZ)	La_xSr_{1-x} MnO_3 + Y-stabilized ZrO_2, (LSM-YSZ)
Cathodic reaction (HER)	$2H_2O + 4e \rightarrow$ $4OH + 2H_2$|	$2H_2O + 4e \rightarrow$ $4OH + 2H_2$	$4H^+ + 4e^- \rightarrow 2H_2$,	$4H^+ + 4e^- \rightarrow 2H_2$	$H_2O + 2e \rightarrow$ $H_2 + O^2$	$H_2O + 2e \rightarrow H_2 + O^2$ $CO_2 + 2e \rightarrow CO + O^2$
Cathodes	Ni alloys	Ni, Ni-Fe, $NiFe_2O_4$,	Pt/C M_oS_2	Ni-cermets	Ni-YSZ subst. $LaCrO_3$	Ni-YSZ perovskites
Efficiency	59–70%		65–42%	Up to 100%	Up to 100%	-
Applicability	Commercial	Laboratory-scale	Near-term commercialization	Laboratory-scale	Demonstration	Laboratory-scale

Advantages	Low capital cost, relatively stable, mature technology	Combination of alkaline and H^+-PEM electrolysis	Compact design, fast response/start-up, high-purity H_2	Enhanced kinetics, thermodynamics: lower energy demands, low capital cost	+ Direct production of syngas
Disadvantages	Corrosive electrolyte, gas permeation, slow dynamics	Low OH^- conductivity in polymeric membranes	High-cost polymeric membranes; acidic; noble metals	mechanically unstable electrodes (cracking); safety issues: improper sealing	
Challenges	Improve durability/reliability, and oxygen evolution	Improve electrolyte	Reduce noble metal utilization	Microstructural changes in the electrodes: delamination, blocking of TPBs, passivation	C deposition, microstructural change electrodes

Source: Chi and Yu (2018).

TABLE 2.9
Summary of Fuel Cells' Main Advantages and Disadvantages

Advantages	Disadvantages
Less/no pollution	Immature hydrogen infrastructure
Higher thermodynamic efficiency	Sensitivity to contaminants
Higher part–load efficiency	Expensive platinum catalysts
Modularity and scalability	Delicate thermal and water management
Excellent load response	Dependence on hydrocarbons reformation
Fewer energy transformations	Complex and expensive BoP components
Quiet and static	Long-term durability and stability issues
Water and cogeneration applications	Hydrogen safety concerns
Fuel flexibility	High investment cost per kW
Wide range of applications	Relatively large system size and weight

Source: Sharaf and Orhan (2014).

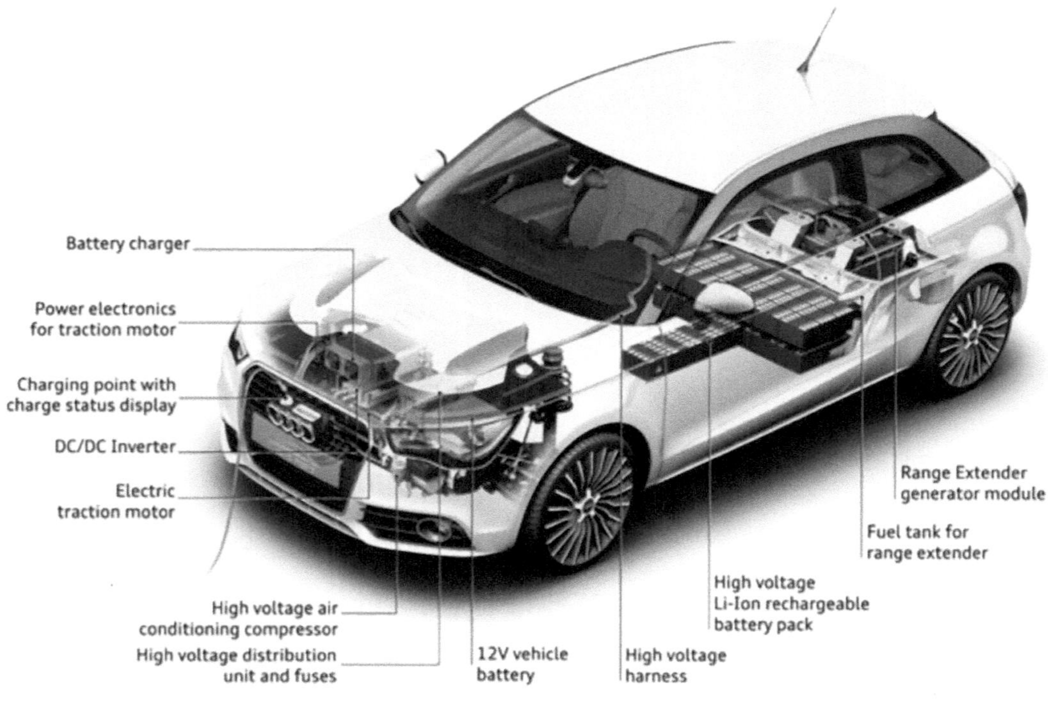

FIGURE 2.35 Battery-charged electric car.

Source: Wilberforce et al. (2017).

Ambitious estimations have FCVs overtaking the vehicle market around the year 2030, where a few expectations are that a hydrogen infrastructure is developed to enable vehicle refueling while engines in the units increase their mileage per tank (Manoharan et al., 2019). Figure 2.35 shows a schematic of a battery-charged electric car. Figure 2.36 displays the estimation of vehicle sales for various types over the next few decades in millions of units per year. As we can see, the peak of

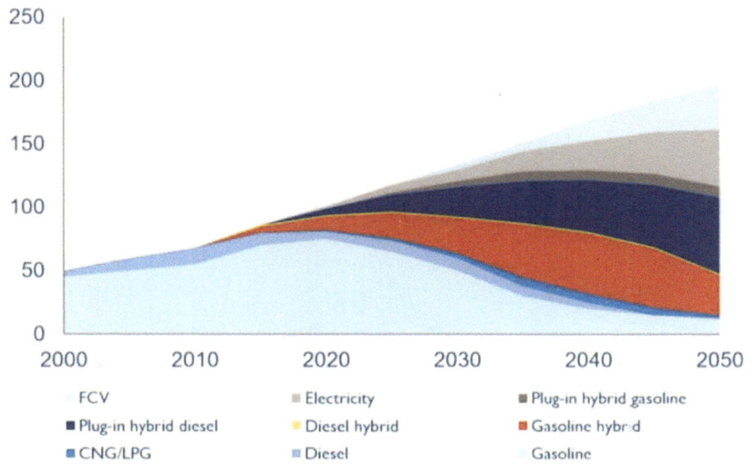

FIGURE 2.36 Fuel cell vehicle (FCV) sales volume, millions of annual units.

Source: Manoharan et al. (2019).

TABLE 2.10
Summary of Commercial FCVs

Vehicle Prototype	Year	Running Range (Miles)	Fuel Economy MPGe at City	Fuel Economy MPGe at Highway	Type
Honda FCX Clarity (Energy, 2016; Anon, 2015c)	2014	231	58	60	PFCV
Honda FCV Concept (Anon, 2015c)	2014	435	/	/	PFCV
Audi A7 Sportback h-tron quattro (Anon, 2016b; Vorsprung durch Technik, 2015; Anon. 2015b)	2014	500	62	62	FCHEV
Roewe 950 Fuel Cell (Wikipedia, 2015)	2014	350	/	/	PRFV
Volkswagen Golf HyMotion (Anon, 2014c, 2014a)	2014	426	/	/	PFCV
Toyota Mirai (Nonobe, 2017; Anon, 2016d)	2020	750	66	66	PFCV
Hyundai Tucson Fuel Cell (America, 2015)	2016	265	49	51	PFCV
Honda Clarity Fuel Cell (Anon, 2014b)	2017	434	/	/	FCHEV

Source: Fan et al. (2021).

gasoline-powered vehicles is placed close to the year 2020, with the following decade marking a significant decline in the sales of this type of vehicle in favor of electrics and FCVs.

Fan et al. (2021) have developed a list of the commercial FCVs currently in the market, with Table 2.10 summarizing their main features. Since 2014, several automakers have released fuel cell models with different characteristics. For example, most of the models are considered "pure" fuel cell vehicles (PFCVs), as opposed to fuel cell hybrid electric vehicles (FCHEV), which combine fuel cells with a different energy storage system. Then, Toyota developed the most efficient range, with their Mirai model able to run approximately 750 mi before running out of energy (Fan et al., 2021).

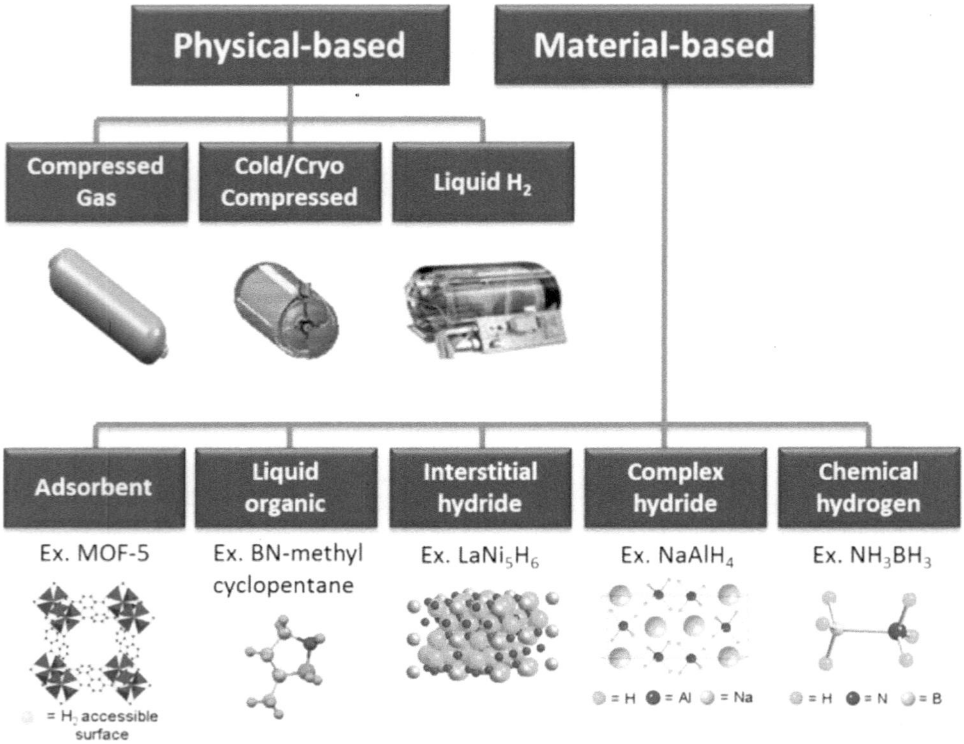

FIGURE 2.37 How is hydrogen stored?

Source: Office of Energy Efficiency and Renewable Energy (2021).

A vital aspect of this efficiency is related to hydrogen storage, a process still in its early stages (Hwang and Varma, 2014).

Whether we are referring to electric vehicles or energy plants, hydrogen storage is a process that is still in its infancy. Yet with environmental pressures and with governments seeking to disincentivize the generation of carbon emissions, renewable energy technologies will become cheaper and more readily available. But as mentioned earlier in this chapter, renewable energies like wind and solar are notoriously unstable in their fluctuation, so a reliable system of transportation and storage of energy will be required for a successful energy transition process. We can again turn our attention to hydrogen, a promising secondary energy vector good for storing, transporting, and distributing vast quantities of energy (Preuster et al., 2018).

For hydrogen energy markets to work, researchers must find efficient storage methods. Generally, hydrogen can be stored as a compressed, liquefied, or cryo-compressed gas or in hydrides (Barthelemy et al., 2017). Figure 2.37 summarizes the few ways that hydrogen can be stored, which are either physical-based or material-based methods. The literature has found that compressed and cryogenic storages are the more developed hydrogen storage methods. The former consists of storing hydrogen in pressure vessels akin to gas cylinders. Meanwhile, cryogenic storage is performed at very low temperatures and kept in tankers or trailers with cryogenic vessels (Moradi and Groth, 2019). Figure 2.38 illustrates some examples of cryogenic carriers for land and sea use.

The potential for hydrogen to replace fossil fuels in a future energy economy is promising, but as we are also witnessing in the review done in this section, a few problems still need attention, like average production efficiency and better ways to store it. However, it is ideal to see hydrogen as an energy storage medium for intermittent renewable sources like wind and solar (Breeze, 2019). For

FIGURE 2.38 Kawasaki heavy industry concept design for liquid hydrogen carriers (left), and cryogenic trailer (right).

Source: Moradi and Groth (2019).

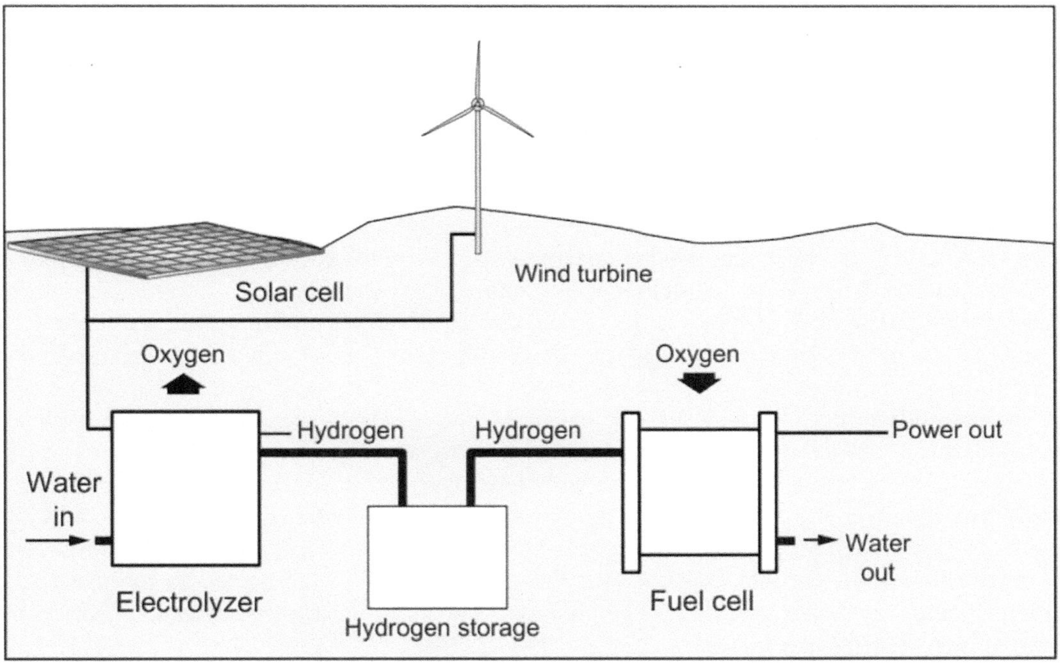

FIGURE 2.39 Schematic of a renewable energy hydrogen energy storage system.

Source: Breeze (2019).

these purposes, Figure 2.39 presents a schematic of an energy storage facility with a built-in electrolyzer that converts the energy from renewable sources (captured in peak energy moments) into hydrogen, which later is stored until the energy is needed (thus, in scarce energy moments).

As the schematic also shows, the process of releasing the power involves a fuel cell, through a process that generates electric energy directly and with a conversion efficiency that is greater than that obtained from the combustion of gasoline, methane, and coal (Nicoletti et al., 2015). Ideally, if this process is replicated all over the world, renewable sources of energy will have a much-needed

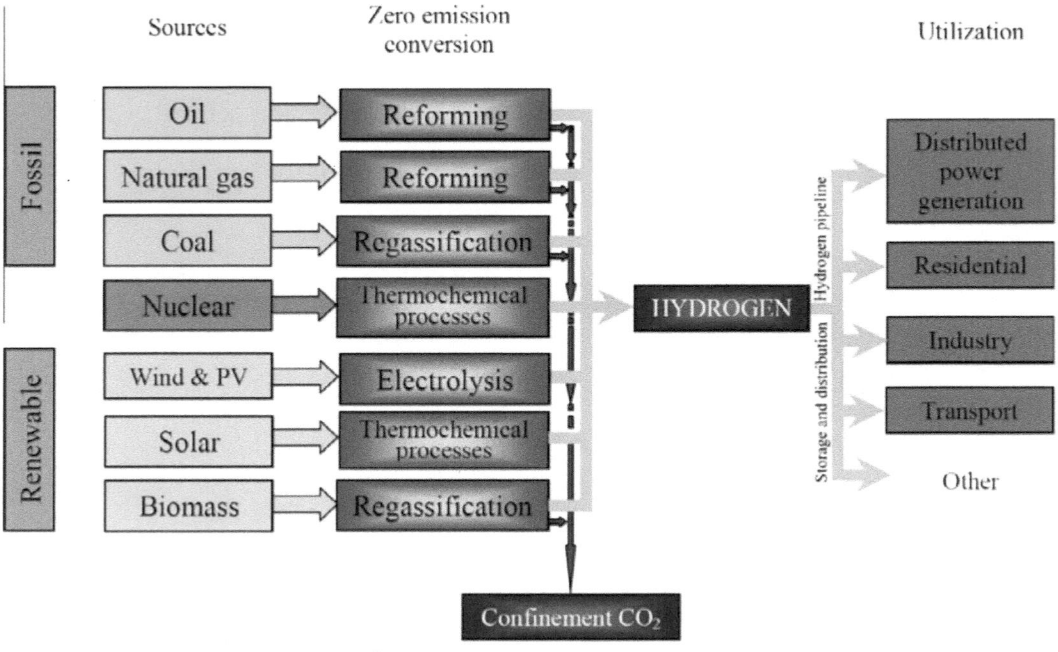

FIGURE 2.40 Possible energy scheme concerning production, transport, accumulation, and use of hydrogen. **Source: Nicoletti et al. (2015).**

aid to store energy collected in abundant times and released in times of scarcity. Only through an explicit embrace of technologies like hydrogen can fossil fuels successfully be phased out over the next couple of decades. Moreover, as Figure 2.40 shows, a zero-emission energy conversion system can be achieved, in which multiple energy sources are utilized efficiently. Therefore, newer energies like biomass, or new iterations of older technologies like fourth-generation nuclear, can also be a part of the energy transition.

Thanks to the work of researchers and engineers, the world has seen unprecedented levels of development. However, as with all massive interventions, some essential costs have been placed. In the case of the energy revolution, these costs have appeared in the form of increasing carbon emissions and a warming climate. Now, it is time for the same community of experts to develop solutions to some of these global problems before they exacerbate them. One of these paths is to rely on natural, renewable energies, coupling them with the results of human ingenuity to generate a new grid that phases out legacy forms of energy, like coal and oil, while keeping coverage and prices to similar levels as the current ones.

The challenge is not easy, but it is necessary to continue to enjoy the world as we have in the past few decades. A "hydrogen economy" may be the right solution at the right time. To implement this, Oliveira et al. (2021) contemplate three main phases, where hydrogen is first used as a chemical feedstock. Thus, in this phase, hydrogen is used as support for large chemical reactions, needed mainly in the industrial sector, which, by itself, produces a large amount of greenhouse gases (GHG). The hydrogen feedstock market has been estimated at around USD 115 billion, with an expected growth of 40 billion dollars more by 2022 (Nicita et al., 2020).

The second phase for developing a hydrogen economy consists of using this element as a transition fuel for transportation and buildings, as well as for heating purposes. As noted previously, the key in this step is the development of more efficient and portable fuel cells, which will help in sectors like transportation (see Figure 2.41). Expanding the reach for hydrogen technologies to daily applications must be accompanied by adequate storage schemes, but this step can be the one that

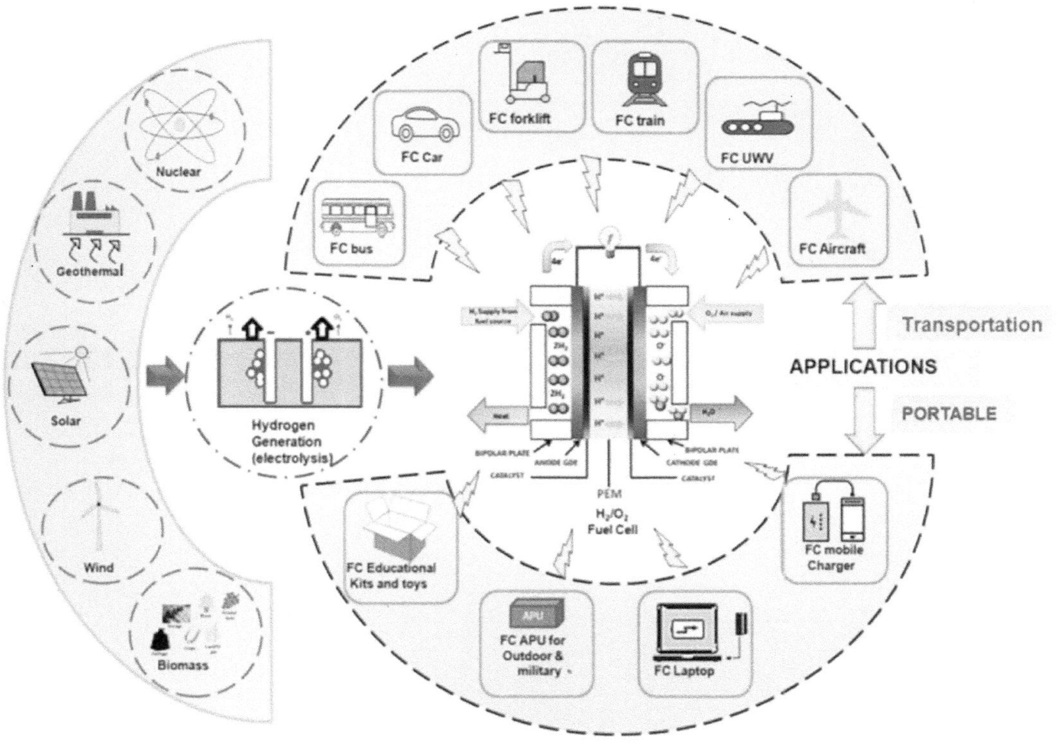

FIGURE 2.41 Specific application of fuel cells in portable and transportation sectors.

Source: Rath et al. (2019).

sentences the beginning of the end for the fossil fuel energy economy. Relatedly, the third phase consists of guaranteeing a degree of energy security through seasonal storage. In this stage, electricity and hydrogen will work as complementary technologies, storing and converting energy between them (Oliveira et al., 2021).

Following these three phases, hydrogen technologies can penetrate the four major sectors of industry – transportation, buildings, heating, and power – to reduce GHG and carbon emissions. This shall open the door to the *green hydrogen economy*, or the clean production of hydrogen from renewable energy sources. This appears as an alternative to how most of the hydrogen is produced today through steam-reforming natural gas, which creates significant amounts of greenhouse gases (Dincer, 2012). As Figure 2.42 advances, the energy sector of the not-too-distant future must contemplate drastic decreases in the amount of carbon emissions generated through all the electrification processes and rely more heavily on cleanly generated hydrogen and on storage facilities that feed every sector.

2.2 CONCLUSION

This chapter focused on the resourcefulness of humankind and the capacity to develop and exploit new energy sources, like those coming from the wind and the sun, as well as nuclear and green hydrogen. These varying energy sources can help diversify the grid and reduce the potential dangers of excessive reliance on fossil fuels. A new energy transition process appears more necessary than ever, and nature itself could provide us with the answer. However, to collect and store the

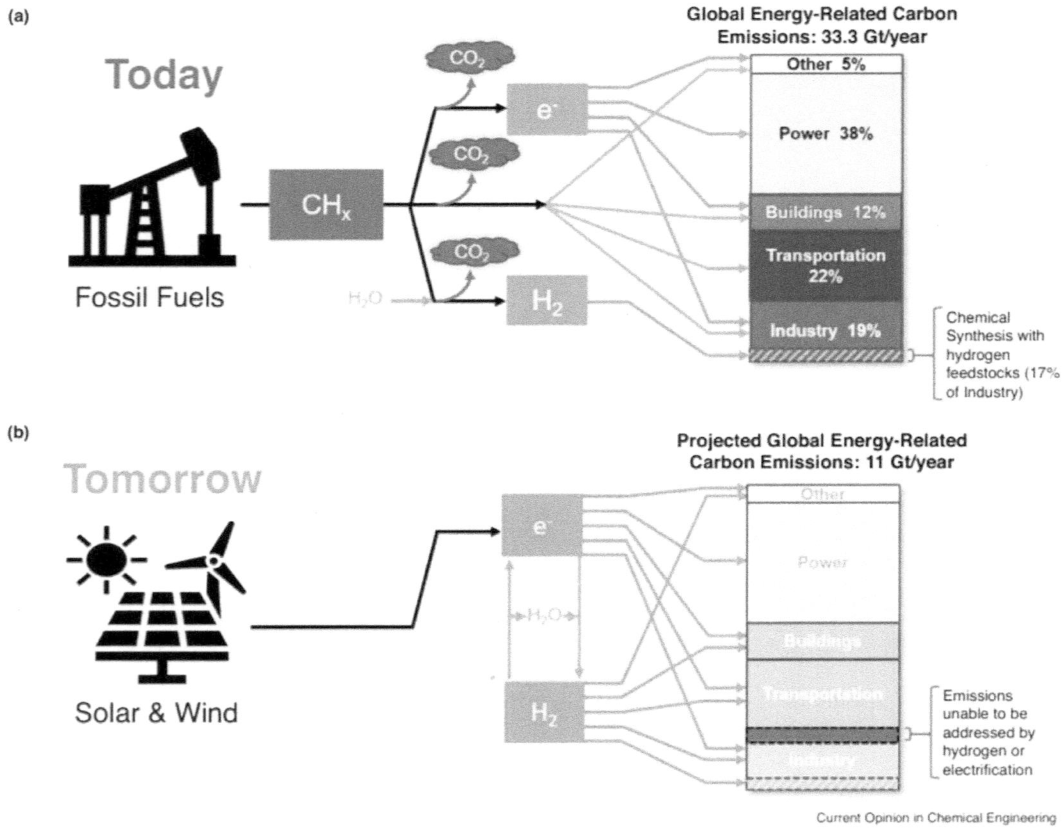

FIGURE 2.42 Energy sectors of today and tomorrow.

Source: Oliveira et al. (2021).

inexhaustible energy from the sun and the wind, we need technological advancements that require more research and the help of engineers and the oil and gas industry.

Similarly, technologies that deal with energy storage, such as batteries, and the research and development of electric vehicles are additional steps in transitioning to cleaner energy sources as the main drivers of the world's power. The goal is to implement these strategies in short to medium term, hoping that the progress made regarding solar, wind, and other clean energy sources replacing conventional sources goes along as expected and delineated by governments and international organizations.

It is important to note that clean technologies also have their own pitfalls and shortcomings. Despite being inexhaustible energy sources in the long term, the contribution of solar and wind to electricity grids is hindered by their absence or scarcity in the immediate moment. Thus, a cloudy day will affect a system that relies heavily on solar energy, something that will not be helped by low winds and batteries that are expensive and cannot save much energy to begin with. Hence, the role of the oil and gas industry is still very relevant and is a crucial connection to any fruitful development that the new sources of energy seek to make in the coming years.

More promising results of the deals and compromises are reflected in the discussion and exploration of a robust hydrogen energy sector, with a zero-emissions production of this element in electrolyzers that separate water into hydrogen and oxygen through electrolysis. Then, hydrogen can power fuel cells that move vehicles and other appliances. If the green production of hydrogen becomes a widespread strategy, moving away from its current production from natural gas, a polluting approach, a significant step toward decarbonization will be achieved.

On this same trend, developing electric vehicles (EVs) that run on fuel cells powered by hydrogen technologies is another significant advancement in the energy paradigm shift. To help with the switch, some countries have already adopted bus transportation systems that run almost entirely on electric energy, and others have funded the development of electric cars to compete with combustion engine cars in their automobile markets. Further, many governments have started offering subsidies for families looking into buying an EV, which is currently more expensive than the more polluting alternatives, while also increasing the availability of charging stations across roads and service stations.

Parallel to these efforts, the oil and gas industry has also researched strategies of carbon sequestration, which consists of capturing and storing CO_2 to impede its release into the atmosphere, where it can contribute to warmer global temperatures. These climate change mitigation strategies can also be used as valuable technologies for the reutilization of depleted reservoirs due to the increased pressure carbon injected into a site, which can lead to enhanced oil recovery (EOR) of fields previously considered inviable. Also, other private actors can incentivize different technologies, like storing carbon in global soils or encouraging forestation, where more trees can absorb the CO_2 and use it for their growth.

At the same time, having the necessary discussions on sources of energy that currently suffer from a tainted public perception, like nuclear power, is another way forward. Due to much-publicized accidents like those of Chernobyl, Three Mile Island, and Fukushima, as well as its being the primary source for a nuclear bomb, nuclear energy has gained a reputation as a destructive source, despite being one of the most energy-dense and cleaner sources known to man, with untapped potential for the improvement of society. Current researchers are working on new generations of nuclear plants and reactors that consider the problems that caused those accidents, so that they are avoided in the future.

REFERENCES

Adeniyi, O. M., U. Azimov, U., and A. Burluka. 2018. Algae biofuel: Current status and future applications. *Renewable and Sustainable Energy Reviews* 90: 316–335.

Allen, D. T., V. M. Torres, J. Thomas, D. W. Sullivan, M. Harrison, A. Hendler, . . . and J. H. Seinfeld. 2013. Measurements of methane emissions at natural gas production sites in the United States. *Proceedings of the National Academy of Sciences* 110(44): 17768–17773.

Amankwah-Amoah, J. 2014. Solar energy in Sub-Saharan Africa: The challenges and opportunities of technological leapfrogging. *Thunderbird International Business Review* 57(1): 15–31.

Arantegui, R. L., and A. Jäger-Waldau. 2018. Photovoltaics and wind status in the European Union after the Paris agreement. *Renewable and Sustainable Energy Reviews* 81: 2460–2471.

Armor, J. N. 2013. Emerging importance of shale gas to both the energy & chemicals landscape. *Journal of Energy Chemistry* 22(1): 21–26.

Ausfelder, F., C. Beilmann, M. Bertau, S. Bräuninger, A. Heinzel, R. Hoer, W. Koch, F. Mahlendorf, A. Metzelthin, M. Peuckert, L. Plass, K. Räuchle, M. Reuter, G. Schaub, S. Schiebahn, E. Schwab, F. Schüth, D. Stolten, G. Teßmer, K. Wagemann, and K.-F. Ziegahn. 2017. Energy storage as part of a secure energy supply. *ChemBioEng Reviews* 4(3): 144–210.

Ausubel, J. H. 2007. Renewable and nuclear heresies. *International Journal of Nuclear Governance Economy and Ecology* 1 (3, January): 229–243.

Ball, M., and M. Weeda. 2015. The hydrogen economy – Vision or reality? *International Journal of Hydrogen Energy* 40: 7903–7919.

Barlev, D., R. Vidu, and P. Stroeve. 2011. Innovation in concentrated solar power. *Solar Energy Materials and Solar Cells* 95(10): 2703–2725.

Barthelemy, H., M. Weber, and F. Barbier. 2017. Hydrogen storage: Recent improvements and industrial perspectives. *International Journal of Hydrogen Energy* 42(11): 7254–7262.

BBC News. 2021. COP26: More than 40 countries pledge to quit coal. *BBC News*. Available in www.bbc.com/news/science-environment-59159018 [accessed November 6th, 2021].

Behling, N., M. C. Williams, and S. Managi. 2019. Regulating Japan's nuclear power industry to achieve zero-accidents. *Energy Policy* 127: 308–319.

Bórawski, P., A. Bełdycka-Bórawska, K. J. Jankowski, B. Dubis, and J. W. Dunn. 2020. Development of wind energy market in the European Union. *Renewable Energy* 161: 691–700.

Breeze, P. 2019. Power system energy storage technologies. *Power Generation Technologies*: 219–249.

Cairns, E. J., and P. Albertus. 2010. Batteries for electric and hybrid-electric vehicles. *Annual Review of Chemical and Biomolecular Engineering* 1(1): 299–320.

Cantarero, M. M. V. 2020. Of renewable energy, energy democracy, and sustainable development: A roadmap to accelerate the energy transition in developing countries. *Energy Research & Social Science* 70: 101716.

Cheng, V. K., and G. P. Hammond. 2017. Life-cycle energy densities and land-take requirements of various power generators: A UK perspective. *Journal of the Energy Institute* 90(2): 201–213.

Chi, J., and H. Yu. 2018. Water electrolysis based on renewable energy for hydrogen production. *Chinese Journal of Catalysis* 39(3): 390–394.

Clark, W. W. 2007. The hydrogen paradigm shift. *CoGeneration and Distributed Generation Journal* 22(2, Spring): 6–38.

Crowe, J. A., and R. Li. 2020. Is the just transition socially accepted? Energy history, place, and support for coal and solar in Illinois, Texas, and Vermont. *Energy Research & Social Science* 59: 101309.

Dincer, I. 2012. Green methods for hydrogen production. *International Journal of Hydrogen Energy* 37(2): 1954–1971.

Dinda, S. 2004. Environmental Kuznets curve hypothesis: A survey. *Ecological Economics* 49(4): 431–455.

Dong, X., G. Pi, Z. Ma, and C. Dong. 2017. The reform of the natural gas industry in the PR of China. *Renewable and Sustainable Energy Reviews* 73(June): 582–593.

Doss-Gollin, J., D. J. Farnham, U. Lall, and V. Modi. 2021. How unprecedented was the February 2021 Texas cold snap? *Environmental Research Letters* 16: 064056.

Dyr, T., P. Misiurski, and K. Ziółkowska. 2019. Costs and benefits of using buses fuelled by natural gas in public transport. *Journal of Cleaner Production* 225: 1134–1146.

Economides, M. J., and D. A. Wood. 2009. The state of natural gas. *Journal of Natural Gas Science and Engineering* 1(1–2, July): 1–13.

Engerer, H., and M. Horn. 2010. Natural gas vehicles: An option for Europe. *Energy Policy* 38(2): 1017–1029.

Esteban, M. D., J. J. Diez, J. S. López, and V. Negro. 2011. Why offshore wind energy? *Renewable Energy* 36(2): 444–450.

European Nuclear Society. 2021. *Nuclear power plants in Europe*. Available in www.euronuclear.org/glossary/nuclear-power-plants-in-europe/ [accessed November 12th, 2021].

Fairuz, S. M. C., M. Y. Sulaiman, C. H. Lim, S. Mat, B. Ali, O. Saadatian, M. H. Ruslan, E. Salleh, and K. Sopian. 2013. Long term strategy for electricity generation in Peninsular Malaysia – Analysis of cost and carbon footprint using MESSAGE. *Energy Policy* 62: 493–502.

Fan, L., Z. Tu, and S. H. Chan. 2021. Recent development of hydrogen and fuel cell technologies: A review. *Energy Reports* 7(November): 8421–8446.

Farrokhzad Ershad, N., R. Tafazzoli Mehrjardi, and M. Ehsani. 2020. High performance 4WD electric powertrain with flywheel kinetic energy recovery. *IEEE Transactions on Power Electronics*: 1–13.

Frein, A., M. Calderoni, and M. Motta. 2014. Solar thermal plant integration into an industrial process. *Energy Procedia* 48: 1152–1163.

Global Wind Energy Council. 2021. *Global Wind Report 2021*. GWEC. Available in https://gwec.net/global-wind-report-2021/.

Gu, Z. 2018. History review of nuclear reactor safety. *Annals of Nuclear Energy* 120: 682–690.

Guney, M. S. 2016. Solar power and application methods. *Renewable and Sustainable Energy Reviews* 57: 776–785.

Gür, T. M. 2018. Review of electrical energy storage technologies, materials and systems: challenges and prospects for large-scale grid storage. *Energy & Environmental Science* 11(10): 2696–2767.

Hardman, S., R. Steinberger-Wilckens, and D. van der Horst. 2013. Disruptive innovations: The case for hydrogen fuel cells and battery electric vehicles. *International Journal of Hydrogen Energy* 38(35): 15438–15451.

Hoffmann, W. 2006. Photovoltaics as a major contributor to the future global energy needs and a 100% renewably powered world. *Photovoltaics for Sustainable Electricity and Buildings* 55–93. https://doi.org/10.1007/978-3-319-39280-6_4

Höök, M., and X. Tang. 2013. Depletion of fossil fuels and anthropogenic climate change—a review. *Energy Policy* 52: 797–809.

Hwang, H. T., and A. Varma. 2014. Hydrogen storage for fuel cell vehicles. *Current Opinion in Chemical Engineering* 5: 42–48.

IEIAvideo on YouTube. 2017. *The nuclear fuel cycle*. Available in www.youtube.com/watch?v=2W4v5_ZVQOA [accessed November 2nd, 2021].

Intergovernmental Panel on Climate Change (IPCC). 2000. *Special report on emissions scenarios*. Cambridge University Press. Available in https://archive.ipcc.ch/pdf/special-reports/emissions_scenarios.pdf

International Atomic Energy Agency. 2021. *What is nuclear energy? The science of nuclear power*. Available in www.iaea.org/newscenter/news/what-is-nuclear-energy-the-science-of-nuclear-power [accessed November 2nd, 2021].

International Hydropower Association. 2021. *Hydropower status report: Sector trends and insights*. Available in https://assets-global.website-files.com/5f749e4b9399c80b5e421384/60c37321987070812596e26a_IHA20212405-status-report-02_LR.pdf [accessed November 2nd, 2021].

Javed, M. S., T. Ma, J. Jurasz, and M. Y. Amin. 2020. Solar and wind power generation systems with pumped hydro storage: Review and future perspectives. *Renewable Energy* 148(April): 176–192.

Kannan, N., and D. Vakeesan. 2016. Solar energy for future world: A review. *Renewable and Sustainable Energy Reviews* 62: 1092–1105.

Kasperson, R. E., and B. J. Ram. 2013. The public acceptance of new energy technologies. *Daedalus* 142(1): 90–96.

Kharecha, P. A., and J. E. Hansen. 2013. Prevented mortality and greenhouse gas emissions from historical and projected nuclear power. *Environmental Science & Technology* 47(9): 4889–4895.

Kittner, N., F. Lill, and D. M. Kammen. 2017. Energy storage deployment and innovation for the clean energy transition. *Nature Energy* 2(9): 17125.

Kiyar, D., and B. F. Wittneben. 2012. Nuclear energy in the European union after Fukushima: Political and economic considerations. *CESifo DICE Report* 10(3): 9–15.

Kvenvolden, K. A. 2003. Natural gas hydrate: Background and history of discovery. *Natural Gas Hydrate in Oceanic and Permafrost Environments*: 9–16.

Lakatos, L., G. Hevessy, and J. Kovács. 2011. Advantages and disadvantages of solar energy and wind-power utilization. *World Futures* 67(6): 395–408.

Li, G. 2019. Spatiotemporal dynamics of ecological total-factor energy efficiency and their drivers in China at the prefecture level. *International Journal of Environmental Research and Public Health* 16(18): 3480.

Li, J., and S. Hu. 2017. History and future of the coal and coal chemical industry in China. *Resources, Conservation and Recycling* 124: 13–24.

Li, M., M. Lenzen, M. Yousefzadeh, and F. A. Ximenes. 2020. The roles of biomass and CSP in a 100 % renewable electricity supply in Australia. *Biomass and Bioenergy* 143: 105802.

Liebl, D. 2019. Nonparametric testing for differences in electricity prices: The case of the Fukushima nuclear accident. *Ann. Appl. Stat.* 13(2): 1128–1146 (June 2019). DOI: 10.1214/18-AOAS1230

Locatelli, G., M. Mancini, and N. Todeschini. 2013. Generation IV nuclear reactors: Current status and future prospects. *Energy Policy* 61: 1503–1520.

London School of Economics. 2018. *What is shale gas, how is it extracted through fracking and what are fracking's impacts?* Available in www.lse.ac.uk/granthaminstitute/explainers/shale-gas-extracted-fracking-frackings-impacts/ [accessed November 7th, 2021].

Lund, J. W. 2006. *Geothermal energy use compared to other renewables*. GHC Quarterly Bulletin, Geo.

Manoharan, Y., S. E. Hosseini, B. Butler, H. Alzhahrani, B. T. F. Senior, T. Ashuri, and J. Krohn. 2019. Hydrogen fuel cell vehicles; current status and future prospect. *Applied Sciences* 9(11): 2296.

Månsson, A. 2015. A resource curse for renewables? Conflict and cooperation in the renewable energy sector. *Energy Research & Social Science* 10: 1–9.

Marcotullio, P. J., and N. B. Schulz. 2007. Comparison of energy transitions in the United States and developing and industrializing economies. *World Development* 35(10): 1650–1683.

Martins, F., C. Felgueiras, M. Smitkova, and N. Caetano. 2019. Analysis of fossil fuel energy consumption and environmental impacts in European countries. *Energies* 12(6): 964.

Mensah-Darkwa, K., C. Zequine, P. K. Kahol, and R. K. Gupta. 2019. Supercapacitor energy storage device using biowastes: A sustainable approach to green energy. *Sustainability* 11(2): 414.

Milica, L., and G. Andrei. 2020. Influence of the flywheel profile on the energy storage capacity of the KERS system. *IOP Conference Series: Materials Science and Engineering* 997: 012123.

Mohamad, F., J. Teh, C.-M. Lai, and L.-R. Chen. 2018. Development of energy storage systems for power network reliability: A review. *Energies* 11(9): 2278.

Mongird, K., V. V. Viswanathan, P. J. Balducci, M. J. E. Alam, V. Fotedar, V. S. Koritarov, and B. Hadjerioua. 2019. *Energy storage technology and cost characterization report* (No. PNNL-28866). Pacific Northwest National Lab (PNNL), Richland, WA.

Moniz, E. J., H. D. Jacoby, A. J. M. Meggs, R. C. Armstrong, D. R. Cohn, S. R. Connors, . . . and G. M. Kaufman. 2011. *The future of natural gas*. Massachusetts Institute of Technology, Cambridge, MA.

Montgomery, C. T., and M. B. Smith. 2010. Hydraulic fracturing: History of an enduring technology. *Journal of Petroleum Technology* 62(12): 26–40.

Moradi, R., and K. M. Groth. 2019. Hydrogen storage and delivery: Review of the state of the art technologies and risk and reliability analysis. *International Journal of Hydrogen Energy* 44(23): 12254–12269.

Mousavi, S. B., P. Ahmadi, A. Pourahmadiyan, and P. Hanafizadeh. 2021. A comprehensive techno-economic assessment of a novel compressed air energy storage (CAES) integrated with geothermal and solar energy. *Sustainable Energy Technologies and Assessments* 47: 101418.

Neidell, M., S. Uchida, and M. Veronesi. 2019. *Be cautious with the precautionary principle: Evidence from Fukushima Daiichi nuclear accident* (No. w26395). National Bureau of Economic Research, Cambridge.

Nicita, A., G. Maggio, A. P. F. Andaloro, and G. Squadrito. 2020. Green hydrogen as feedstock: Financial analysis of a photovoltaic-powered electrolysis plant. *International Journal of Hydrogen Energy*: 1–14.

Nicoletti, G., N. Arcuri, G. Nicoletti, and R. Bruno. 2015. A technical and environmental comparison between hydrogen and some fossil fuels. *Energy Conversion and Management* 89: 205–213.

Olabi, A. G., T. Wilberforce, M. A. Abdelkareem, and M. Ramadan. 2021. Critical review of flywheel energy storage system. *Energies* 14(8): 2159.

Oliveira, A. M., R. R. Beswick, and Y. Yan. 2021. A green hydrogen economy for a renewable energy society. *Current Opinion in Chemical Engineering* 33: 100701.

Pampel, F. C. 2011. Support for nuclear energy in the context of climate change. *Organization & Environment* 24(3): 249–268.

Pan, Y., S. Chen, F. Qiao, S. V. Ukkusuri, and K. Tang. 2019. Estimation of real-driving emissions for buses fueled with liquefied natural gas based on gradient boosted regression trees. *Science of The Total Environment* 660: 741–750.

Poullikkas, A. 2009. Economic analysis of power generation from parabolic trough solar thermal plants for the Mediterranean region – A case study for the island of Cyprus. *Renewable and Sustainable Energy Reviews* 13(9): 2474–2484.

Preuster, P., Q. Fang, R. Peters, R. Deja, L. Blum, D. Stolten, and P. Wasserscheid. 2018. Solid oxide fuel cell operating on liquid organic hydrogen carrier-based hydrogen–making full use of heat integration potentials. *International Journal of Hydrogen Energy* 43(3): 1758–1768.

Rath, R., P. Kumar, S. Mohanty, and S.K. Nayak. 2019. Recent advances, unsolved deficiencies, and future perspectives of hydrogen fuel cells in transportation and portable sectors. *International Journal of Energy Research* 43: 8931–8955.

Ricotti, M. E. 2013. Nuclear energy: Basics, present, future. *EPJ Web of Conferences* 54: 01005.

Rueda-Bayona, J. G., A. Guzmán, J. J. Cabello Eras, R. Silva-Casarín, E. Bastidas-Arteaga, and J. Horrillo-Caraballo. 2019. Renewable energies in Colombia and the opportunity for the offshore wind technology. *Journal of Cleaner Production* 220: 529–543.

Schrag, D. P. 2012. Is shale gas good for climate change? *Daedalus* 141(2): 72–80.

Scolan, S., S. Serra, S. Sochard, P. Delmas, and J. M. Reneaume. 2020. Dynamic optimization of the operation of a solar thermal plant. *Solar Energy* 198: 643–657.

Sharaf, O. Z., and M. F. Orhan. 2014. An overview of fuel cell technology: Fundamentals and applications. *Renewable and Sustainable Energy Reviews* 32: 810–853.

Shiva Kumar, S., and V. Himabindu. 2019. Hydrogen production by PEM water electrolysis – a review. *Materials Science for Energy Technologies* 2: 442–454.

Sierra, C., and A. Vázquez. 2005. High solar energy concentration with a Fresnel lens. *Journal of Materials Science* 40: 1339–1343. http://dx.doi.org/10.1007/s10853-005-0562-6

Singh, G. K. 2013. Solar power generation by PV (photovoltaic) technology: A review. *Energy* 53: 1–13.

Snow, A. A., and V. H. Smith. 2012. Genetically engineered algae for biofuels: A key role for ecologists. *BioScience* 62 (8, August): 765–768.

Soltani, M., M. H. Nabat, A. R. Razmi, M. B. Dusseault, and J. Nathwani. 2020. A comparative study between ORC and Kalina based waste heat recovery cycles applied to a green compressed air energy storage (CAES) system. *Energy Conversion and Management* 222: 113203.

Stern, D. I. 2018. Energy-GDP relationship. *The New Palgrave Dictionary of Economics* 3697–3714. https://www.academia.edu/download/56082075/PalgraveE000340.pdf

Sterner, M., and M. Specht. 2021. Power-to-gas and power-to-X—the history and results of developing a new storage concept. *Energies* 14: 6594.

Stoutenborough, J. W., S. G. Sturgess, and A. Vedlitz. 2013. Knowledge, risk, and policy support: Public perceptions of nuclear power. *Energy Policy* 62: 176–184.

United Nations Environment Programme. 2021. *Emissions gap report 2021: The heat is on – a world of climate promises not yet delivered.* UNEP, Nairobi.

U.S. Department of Energy. 2021. *Office of energy efficiency & renewable energy*. Available in https://www.energy.gov/eere/office-energy-efficiency-renewable-energy

U.S. Energy Information Administration. 2015. World shale gas resources: An initial assessment of 14 regions outside the United States. *U.S. Department of Energy*. Available in https://www.eia.gov/analysis/studies/worldshalegas/

U.S. Energy Information Administration. 2021. Annual energy outlook 2021: Narrative. *U.S. Department of Energy*. EIA, 2021. Available in https://www.eia.gov/outlooks/aeo/pdf/AEO_Narrative_2021.pdf

Ursúa, A., L. M. Gandía, and P. Sanchis. 2012. Hydrogen production from water electrolysis: Current status and future trends. *Proceedings of the IEEE* 100(2): 410–426.

Van Vactor, Samuel A., Historical Perspective on Energy Transitions (May 14, 2018). *USAEE Working Paper*, Available at SSRN: https://ssrn.com/abstract=3178176 or http://dx.doi.org/10.2139/ssrn.3178176

VijayaVenkataRaman, S., S. Iniyan, and R. Goic. 2012. A review of climate change, mitigation and adaptation. *Renewable and Sustainable Energy Reviews* 16(1): 878–897.

Wang, J., and S. Kim. 2018. Comparative analysis of public attitudes toward nuclear power energy across 27 European countries by applying the multilevel model. *Sustainability* 10(5): 1518.

Wang, S., and S. Wang. 2015. Impacts of wind energy on environment: A review. *Renewable and Sustainable Energy Reviews* 49: 437–443.

Whitfield, S. C., E. A. Rosa, A. Dan, and T. Dietz. 2009. The future of nuclear power: Value orientations and risk perception. *Risk Analysis: An International Journal* 29(3): 425–437.

Wilberforce, T., Z. El-Hassan, F. N. Khatib, A. Al Makky, A. Baroutaji, J. G. Carton, and A. G. Olabi. 2017. Developments of electric cars and fuel cell hydrogen electric cars. *International Journal of Hydrogen Energy* 42(40): 25695–25734.

World Nuclear Association. 2020. *Generation IV nuclear reactors*. Available in https://world-nuclear.org/information-library/nuclear-fuel-cycle/nuclear-power-reactors/generation-iv-nuclear-reactors.aspx [accessed November 15th, 2021].

Ydersbond, I. M., and M. S. Korsnes. 2016. What drives investment in wind energy? A comparative study of China and the European Union. *Energy Research & Social Science* 12: 50–61.

Yeh, S. 2007. An empirical analysis on the adoption of alternative fuel vehicles: The case of natural gas vehicles. *Energy Policy* 35(11): 5865–5875.

Zervos, A. 2003. Developing wind energy to meet the Kyoto targets in the European union. *Wind Energy* 6(3): 309–319.

Zhang, X., R. Zeng, Q. Deng, X. Gu, H. Liu, Y. He, K. Mu, X. Liu, H. Tian, and H. Li. 2019. Energy, exergy and economic analysis of biomass and geothermal energy based CCHP system integrated with compressed air energy storage (CAES). *Energy Conversion and Management* 199: 111953.

3 Conventional and Fossil Fuels

Jorge Saldaña, Mesut Yurukcu, Narendra Boppana, Sepehr Arbabi, Jim Henry, and Sebahattin Ziyanak

3.1 OIL AND GAS: INTRODUCTION

In the current energy environment, the tide is moving toward the use of cleaner sources of energy, like renewables, hydrogen, and nuclear. An essential part of this book highlights the energy transition's role and how the oil and gas sector is adapting to it. With that said, conventional resources like coal, oil, and natural gas still play a vital role in the lives of billions of people, powering their homes, businesses, and the transportation systems on which they rely. Additionally, and as we reviewed in Chapter 2, many by-products from oil still offer tremendous benefits to most of humanity, with replacements still in their infancy stages of development or simply offering inferior quality for the user.

Due to factors like population growth, urbanization, and industrialization of formerly underdeveloped nations, the demand for resources like oil, coal, and natural gas has remained relatively high worldwide. In addition, the growing demand for energy in developing countries has increased the demand for conventional resources, particularly as a rebound effect after the COVID-19 pandemic. As a result, oil and gas exploration, production, and distribution have become major industries that drive economic growth and provide employment opportunities across many more nations than in previous decades.

The significant shocks to economies worldwide in the past few years have perhaps moderated expectations regarding a quick turnaround to renewable energy sources. After all, humanity still relies heavily on fossil fuels, as nearly all vehicles require gasoline or a petroleum by-product, if not to run, then to be manufactured. Moreover, the travel demand has returned to levels seen before 2020, which incentivizes the aviation and shipping sectors to expand, thus requiring more fuel to continue functioning. All this does not mean that renewables need not be explored; instead, it reinforces the continuous pivotal role that conventional energy sources have for billions of people.

Furthermore, the conflict between Russia and Ukraine has also sparked a global energy crisis that reminds us of the importance of many economic and societal considerations that must be weighed when considering many energy-related moves. As Figure 3.1 suggests, despite the commitments some countries have made to achieving net-zero emissions in the near future, there are not very noticeable trends in the investment in fossil fuel–driven energy technologies in countries that have made these pledges in the past two years. All the external pressure has likely forced the hands of many countries to continue to rely on conventional sources of energy, at least in the short to medium term.

While shocks of this magnitude may make actors revert to their original positions and try to weather these storms in known and more comfortable standing ground, the environmental impacts of prolonged reliance on fossil fuels can undo years of joint work by governments and international organizations. In this sense, it becomes vital to continue to pursue and push the different mitigation strategies that tackle the various stages in the oil and gas production chain. Therefore, carbon capture and storage strategies should continue to be deployed, and all extraction and transportation improvements should be to lower environmental risks.

Technological advancements have encouraged the development of the concept of Industry 4.0, or Oil and Gas 4.0, where energy generation processes can be streamlined and improved via emerging technologies. As we can see in Figure 3.2, the combination of artificial intelligence, Internet of

 DOI: 10.1201/9781003282273-3

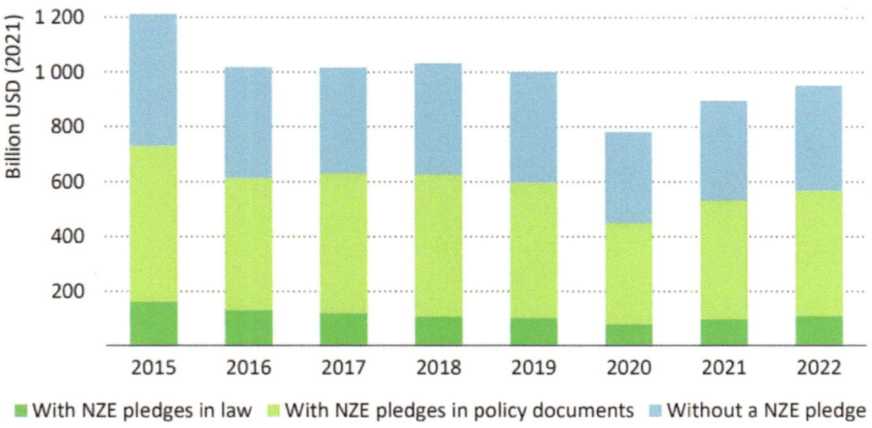

FIGURE 3.1 Fossil fuel investment in countries with and without net-zerc emissions (NZE) pledges, 2015–2022.

Source: IEA (2022).

Things (IoT) enhancements, data analytics, and big data improvements, among other innovations, has helped transform the oil and gas supply chain through more efficiency, safety, and connectivity, all while also improving the industry's environmental consciousness.

These challenges invite essential reflections on the role of conventional energy sources and the trade-offs observed when considering fossil fuels and their alternatives. This chapter reviews the contributions of conventional energy sources, the history of site selection and exploration, and its evolution over time. Amid the energy transition, it also becomes relevant to explore the more novel applications and expected uses for fossil fuels for current technologies. Lastly, the potential of conventional sources in the energy transition context is analyzed to delineate the role that fossil fuels will continue to have in the next few decades.

3.2 SCREENING, SELECTION, AND SITE ECONOMICS FOR OIL AND GAS RESOURCES

Over the years, the selection criteria to pick a site to explore hydrocarbon resources have evolved to include many considerations that go from production expectations to distribution challenges and the most optimal way to consume the extracted resources. With the push toward a cleaner industry, more novel environmental considerations have been added, which, along with technological advancements like machine learning (ML), have helped optimize the potential of the different sites to not only offer plentiful resources but also serve as subsequent CO_2 storage facilities, for example.

Additionally, the prospects for new sites should include at least part of the many considerations and recommendations made by international organizations like the United Nations regarding climate warming mitigation. Figure 3.3 exemplifies one of these dimensions by projecting the different global warming projections, following various degrees of commitment with temperature increase pledges. As we can see, to achieve the more ambitious goal of increasing global temperatures up to 1.5°C, carbon emissions need to decrease considerably from now until 2050. For these goals to be reachable, the current and future site selection for conventional sources must be optimized accordingly.

With more refined analyses, the optimization strategies for site selection become more complex, as they now need to consider not only the economic impacts and benefits for the region they are serving but also the effects on the environment, as well as the social impact that construction,

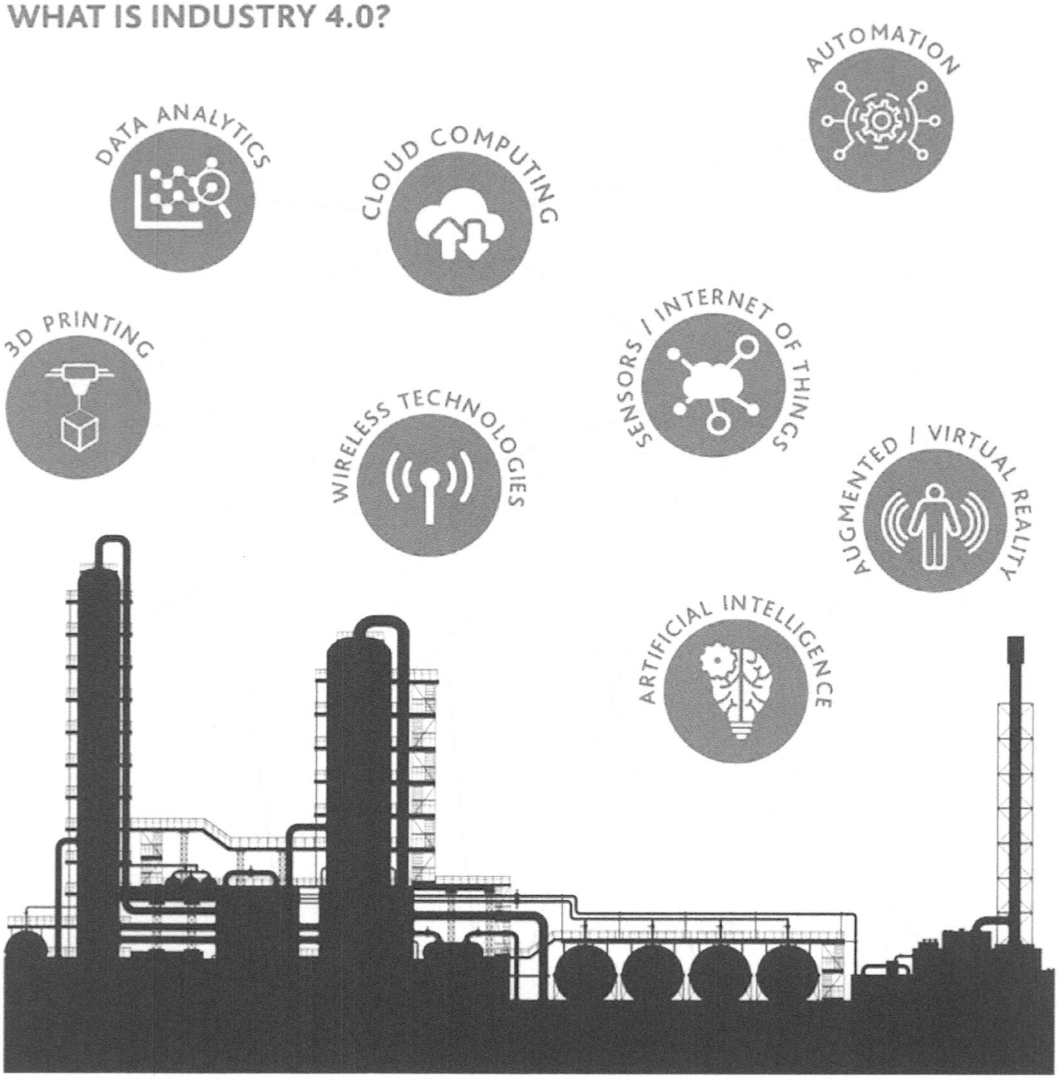

FIGURE 3.2 Illustration of Industry 4.0, or Oil and Gas 4.0.

Source: BDO (2018).

extraction, and distribution will have on communities in the medium to long term (Santibañez-Aguilar et al., 2014). Moreover, the means of transportation for the extracted resources can become a topic of interest when considering infrastructure expenditures and sustainability concerns. For example, Shahriar et al. (2012) explore the complexities of pipeline construction and the risk analyses performed before and during their implementation.

As we can see in Figure 3.4, oil and natural gas exploration and handling can trigger a series of different events occurring as a consequence of their release or spillage, which can go from water pollution in an environmentally related outcome to an evacuation emergency, or even to casualties on the ground when the event collides with life in society. The constant advancement in technology serves to bring efficiency and improve prediction. Still, it is essential to note that in risk calculation analysis, there is always an element of *force majeure* or unforeseeable circumstances that require constant monitoring and optimization strategies from site keepers (Kassem et al., 2019).

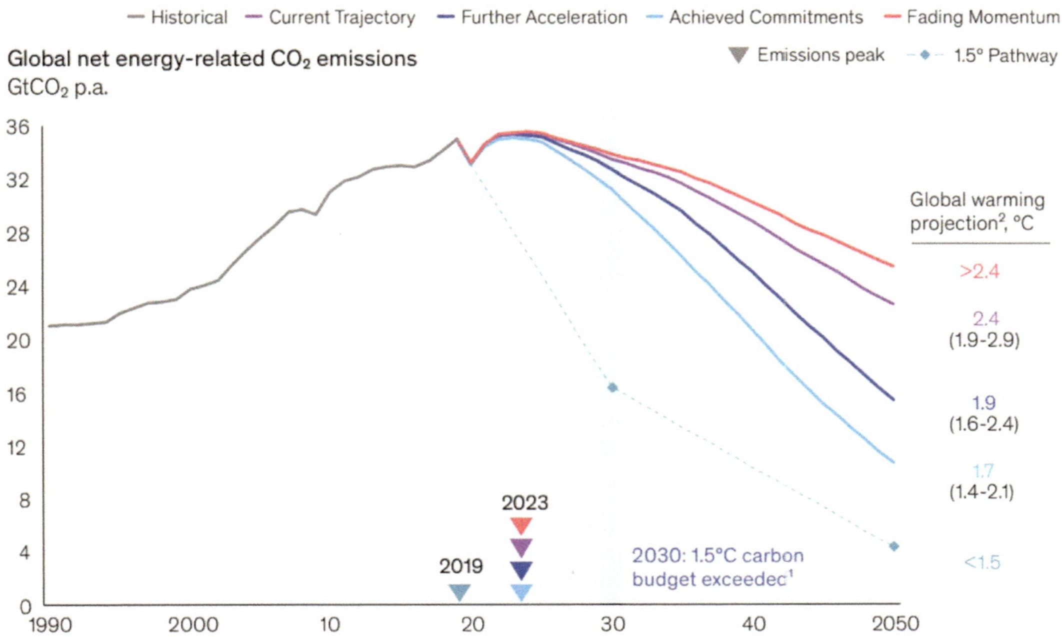

FIGURE 3.3 Global net energy–related CO_2 emissions.

Source: McKinsey & Company (2022).

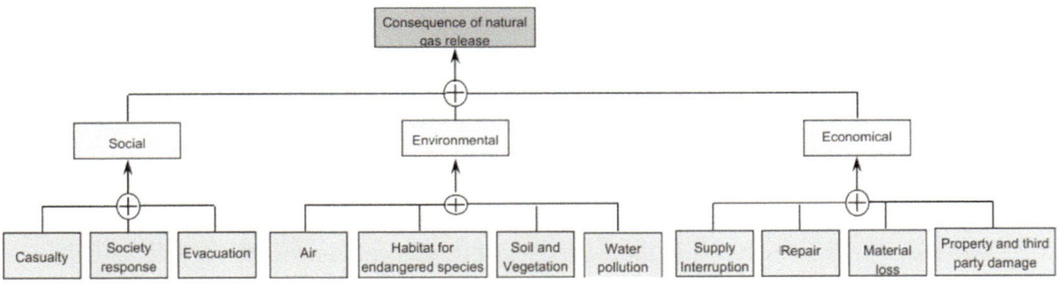

FIGURE 3.4 Hierarchical framework for the consequence of output event (OE).

Source: Shahriar et al. (2012).

One successful approach for site screening has been utilizing geographic information systems (GIS) programs, which provide spatial data to assess variables such as distance from source to market, terrain slope, proximity to population centers, and presence or absence of protected sites. In the work of Abudu and Williams (2015), route selections for oil and gas pipelines in Uganda were determined with the help of a GIS-based technology that evaluated environmental, economic, and security concerns. Thus, any potential interference with wildlife, the proximity to urban centers, and the variation in construction costs were among the considerations being weighed.

Along with geography, information on the climate of the potential location also needs to be addressed, particularly if oil and gas pipelines will be exposed to the elements. Popoola et al. (2013) studied how potential pipeline corrosion concerns affect the behavior of operators, engineers, and designers, who work in conjunction to improve the infrastructure to mitigate problems and delays in hydrocarbon production and delivery. Moreover, in Table 3.1, the authors summarize a handy list of

TABLE 3.1
Recommended Materials for the Oil and Gas Industry

Material Specification	Oil and Gas Applications
Carbon steels	Bulk fluids, crude pipelines, flow lines, water and steam injection lines, production and test separators, KO drums, storage tanks
Low- and medium-alloy steels	Wellhead items, chokes, manifolds, and well components with sour and high-temperature applications
Straight chromium steels (chromium 12% to 18%)	Christmas trees, wellheads, downhole rods, valves, and casing pipes
Chromium-nickel steels (chromium >18%, nickel >8%)	Valve trims, instruments and materials of separators and tanks, low chloride levels
Nickel steels (2.5%, 3.5%, **9%** nickel)	Rarely used in oil and gas sectors, LNG storage tanks, piping, and pumps
Duplex stainless steels (22% chromium duplex, 25% chromium super, duplex)	Piping, vessel, and tank internals, where a very high level of chlorides is present
Nickel-chrome (inconels) Ni-Cr-Fe alloys	Wellhead and flow lines, manifolds with high-sour and high-temperature applications
Nickel-iron (Incoloy) Ni-Fe-Cr alloys	Wellhead and flow lines, manifolds with high-sour and high-temperature applications

Note: KO, "knock out"; *LNG*, "liquefied natural gas."
Source: Popoola et al. (2013).

materials recommended across industry projects while highlighting the role of protective coatings for pipelines that have already been built so that their lifespan can be extended.

The inclusion of more variables and the newfound opportunities to utilize sites once they are depleted invite more sophisticated examinations to arrive at successful site selection strategies. Because of the encouraging prospects regarding site utilization for carbon capture and storage (CCS), engineers now have a new dimension to include in their studies. Shedding some light on this, the work of Davarazar et al. (2020) explores the fuzzy Delphi methodology to select sites for CO_2 sequestration and storage, determining the relative importance of the relevant parameters after collecting the opinions of experts in the field. Their hierarchical analysis found that geology and lithology considerations rank first, while reservoir permeability, depth, and social acceptance fall close behind.

Due to these dynamics, the economics of site selection for fossil fuels has increased in complexity in the past decades. Along with production, distribution, and a convenient geographical location, we must now consider technology's local and global environmental impacts more than ever, as well as the degree of competition or collaboration expected from newer forms of energy, such as renewables or nuclear power. In this sense, a constant feedback loop has been in the works between governments, policymakers, international organizations, and oil and gas players, where all actors have some degree of influence over the other and where decisions can significantly affect supply and demand, as well as prices of the resource.

Additionally, two other vital considerations in the economics of oil and gas resources have been the staggering rate of technological advancements developed in the past few decades, as well as the effects of the globalization efforts to integrate processes across the world, trying to expand the benefits of these energy technologies to the most significant number of people. These two factors are mainly interconnected through the scalability solutions that some technologies have offered, which have facilitated the implementation of some advancements in remote areas of the world at a lower cost than in the past.

It is, in part, thanks to these dynamics that now, for example, blocs of developing countries can aspire to more significant economic growth, assisted by consuming conventional energy sources, such as coal, natural gas, and oil. According to Bildirici and Bakirtas (2014), the set of BRICS countries, or those of Brazil, Russia, India, China, and South Africa, has achieved a significant relationship with conventional energy sources, which has positively impacted their economic growth. Thus, improving energy infrastructure for fossil fuels is found to help create jobs and stimulate the economy, allowing those countries to reach the point where energy diversification can be later explored, just as more developed countries have done in recent decades.

As we can infer, these developments can also facilitate a move toward an energy transition in the future, as economic growth and technological advancements are significant factors for countries to begin the search for cleaner energy sources and more efficient processes altogether. For example, rapid technological developments have helped improve the discovery and exploration of deepwater areas with potential resources (Zhang et al., 2019), increase the role of liquefied natural gas in the energy mix (Litvinenko, 2020), and assist in the shale oil revolution observed in North America (Kilian, 2016).

Therefore, the interconnection of countries achieved by globalization has influenced the economics of oil and gas significantly. Opening new markets in remote areas is possible due to economies of scale, which reduce production costs and the efficiency of the oil and gas sector processes. Meanwhile, technological improvements facilitate the search for alternative decarbonization processes, and institutional mechanisms have been developed to encourage the search for cleaner sources of energy, developed as complements and eventual replacements for fossil fuels.

In the meantime, however, fossil fuel technologies have developed from within to a massive degree, offering cleaner and more efficient processes to power most of the world's population with lower environmental costs as decades pass. Additionally, with the same advancements and the trajectory of current trends, it is possible to envision a future where conventional sources are aided or outright replaced by renewable energy sources. In the next section, we review the leading oil and gas applications, the economics of the sector, the best practices to improve their role in society, and the potential to coexist with sources like nuclear power, as well as solar and wind.

3.3 OIL AND GAS APPLICATIONS AND PRINCIPLES

When we talk about "oil and gas," we mainly refer to technologies developed around exploiting resources commonly characterized as "fossil fuels." These hydrocarbons have been defined as those non-renewable resources formed from living organisms over millions of years, which have gone through different phases of more and less scarcity, depending on new findings of previously unidentified reservoirs. These fossil fuels are identified as coal, oil, and natural gas, and there is ample debate about their extraction, their peak, and the quantity of reserves worldwide (Abas et al., 2015; Department of Energy, 2023; Welsby et al., 2021).

What is less debated, however, is the vital role of these hydrocarbons as catalysts in improving the lives of millions, if not billions, of people worldwide. This is thanks to the high density of oil and gas to produce large quantities of electricity while also serving as components in other consumer products, such as plastics. Additionally, numerous employment opportunities have improved the quality of life of many (Kalair et al., 2020). With that said, governments and international organizations have sounded the alarm concerning the carbon dioxide levels in the atmosphere due to these energy processes, which have been associated with increasing global temperatures. The industry has looked to develop within the oil and gas technologies framework to tackle detriments like these.

Thus, as a way to move forward, researchers and engineers have advanced methods that are applied in the industry but have other objectives in mind. For example, reservoir exploration can now be performed to optimize a site for carbon storage, where CO_2 is placed in depleted sites so that it does not reach the atmosphere and, over time, generates more crude hydrocarbons to be recovered. Similarly, the many procedures to estimate production and recovery have helped develop different types of sites with other resources, like geothermal energy and hydrogen.

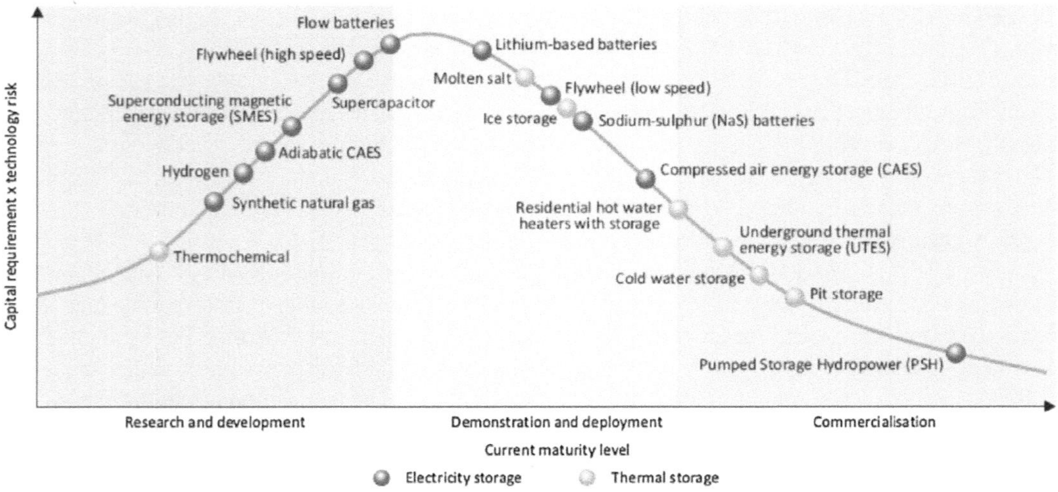

FIGURE 3.5 Maturity phases of future energy systems.

Source: Abas et al. (2015).

There is abundant dynamism in the oil and gas industry, which has seen particularly turbulent and uncertain times in the last half-decade. The COVID-19 pandemic constituted a long reflection moment for all industries, and the oil and gas sector experienced significant fluctuations in the market value of petroleum in the first months of the health emergency. Then, in 2022, the conflict in Ukraine once again tested markets and the ability of many countries to rebound in their use of diverse energy sources. According to the International Energy Agency, the use of coal has seen a revival, particularly in advanced economies, where demand has increased by nearly 10% (IEA, 2022).

Various other market and social dynamics have significantly transformed the sector in the past few decades. Because of this, the term "oil and gas" is increasingly associated with an industry that explores sites and generates optimization strategies, but also a sector that analyzes the potential of an expanding set of energy technologies, funding efforts in both renewable energies and promising alternative resources. As Figure 3.5 illustrates, many energy systems are part of the suite currently covered by oil and gas technologies, with some having varying degrees of exploration and development, while others show different levels of potential, return on investment, and commercialization structures.

One last consideration has been the role of feasibility forecasting as a tool that informs decision-making in the energy sector. In the past couple of decades, there has been a boom in the energy landscape of countries that have explored shale oil and gas resources, with few actors being able to predict such a revolution before it occurs. This oversight is in itself a vital shortcoming that reflects the uncertainty that even experts are sometimes immersed in and the economic paths that are opened up or remain closed down after those analytical tools and models are used to delineate strategies and draw conclusions about the potential of a given energy source (Reed et al., 2019).

In what follows, we review the most common oil and gas technologies, the evolution of these processes over time, and the practices that allow the industry to continue to be a leading sector in improving human prosperity. Additionally, we go through some ways in which the industry has been able to move from a consistent producer of energy using conventional sources and established methods to a proactive player in search of new energy opportunities, as well as a promoter of renewable energies despite their potential to replace the conventional procedures maintained for generations.

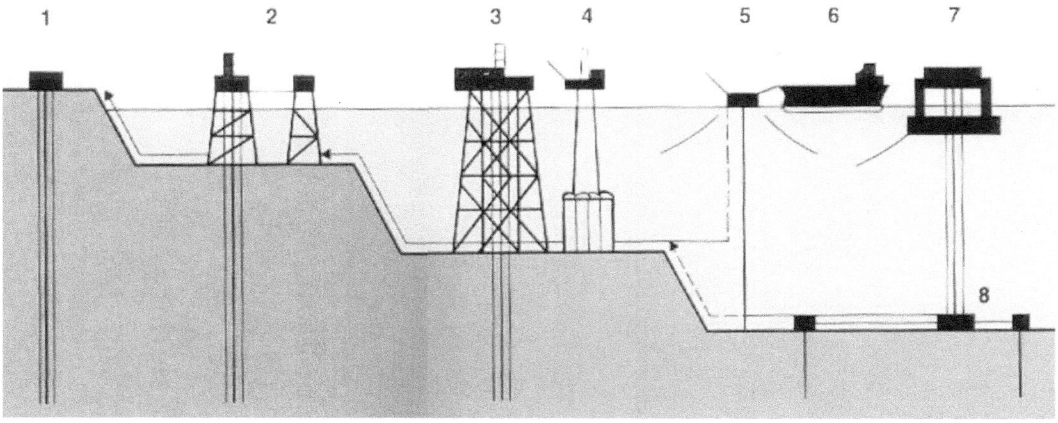

FIGURE 3.6 Oil and gas production facilities: (1) onshore well; (2) offshore, fixed, multi-platforms; (3) offshore, fixed, self-contained platforms; (4) offshore, self-contained, concrete gravity platforms; (5) offshore, floating, single-point mooring; (6) storage/shuttle tanker; (7) offshore, floating, tension leg platforms; (8) subsea manifolds.

Source: Devold (2009).

3.3.1 Oil and Gas Reservoirs

The more common sites for oil and gas are those observed in onshore and offshore reservoirs, where structures are built on land or sea to recover crude that sits below the surface. Figure 3.6 summarizes the types of oil and gas facilities, both on- and offshore, that have been advanced to recover hydrocarbons across varying geographical circumstances. These reservoirs include the first iterations in the infrastructure developed to recover substantial quantities of hydrocarbons and the work of decades of improvement in offshore technologies.

The primary use of these oil and gas facilities is to recover hydrocarbons that are later refined and used as a fuel source to provide heating and transportation and generate electricity for millions of people worldwide. Some petroleum derivatives are chemicals, fertilizers, and cleaning items, like detergents and soap. Other by-products of everyday use include insecticides, plastics, and lubricants for all types of engines. As we can see, oil-derived products are intrinsically linked to many of the habits of consumers, thus constituting a vital part of the global economy.

Even though the oil and gas sector has become a regular staple in the lives of most people on Earth, a couple of worrying trends have it in a constant search for energy alternatives. First, reserves appear to be decreasing in availability, as the total of proven and probable reserves (2P) estimations remains high, but with a somewhat limited time frame to move to other alternatives. As Figure 3.7 suggests, more novel studies have revealed that oil and gas reserves remain healthy and could last a few decades. Still, uncertainty remains about what will happen once humanity reaches those time frames.

In the second place, carbon-heavy processes have been associated with the warming in global temperatures, a trend that has mobilized the industry, along with international organizations and governments, to tackle it before the repercussions become dangerous for humankind. As mentioned previously and touched upon throughout this book, one immediate way to improve the current operations has been to include carbon capture and storage (CCS/CCUS) initiatives to avoid significant quantities of CO_2 from reaching the atmosphere. Additionally, optimized site exploration aided by machine learning (ML) can offer more promising avenues for successful decarbonization efforts.

Moreover, new technologies have allowed engineers to exploit previously unexplored reservoirs, such as those containing heavy oils, bitumen, shale gas, hydrates, and coal bed methane, among

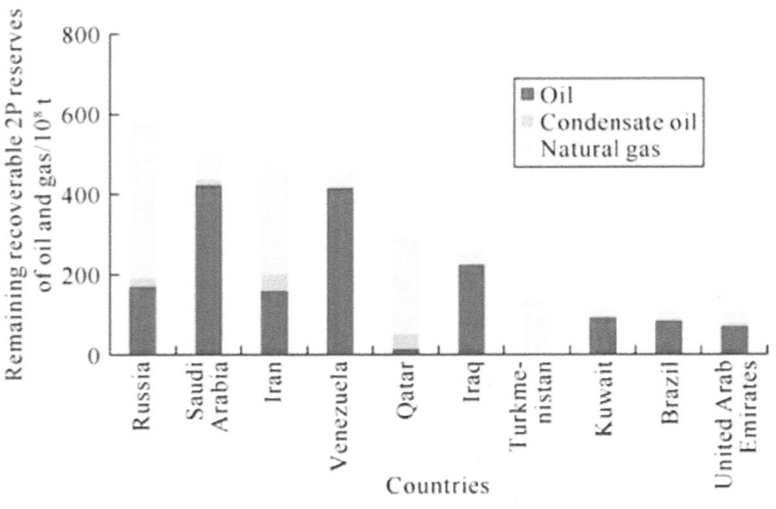

FIGURE 3.7 Histogram of remaining recoverable 2P oil and gas reserves in major countries.

Source: Tong et al. (2018).

other hydrocarbons. Of particular importance, hydraulic fracturing has been a game-changing method to recover unconventionals, which has provided more efficient exploration and recovery, as well as cleaner processes that benefit millions of people worldwide. Thus, while the realization of the transition to renewable energies is not achieved, these unconventional sources can continue to provide reliable energy with a diminished environmental impact.

3.3.2 Heavy Oil and Bitumen Reservoirs

Heavy oil and bitumen are the first unconventional sites to be recovered by new technologies. Heavy oils are those hydrocarbons that are considered more difficult to recover due to their higher degrees of density and viscosity (Chopra et al., 2010), which would require special extraction methods that use newer technologies, like multistage hydraulic fracturing (Al-Fatlawi et al., 2019). Because these heavy oils are rich in metals, sulfur, and carbon while lacking in hydrogen, upgrading technologies like separation, hydrogen addition, and carbon rejection have been developed around their recovery strategies (Ore and Adebiyi, 2021).

Figure 3.8 shows the different technologies that have been advanced to deal with *in situ* upgrading heavy oil that is recovered so that its mobility is improved. In addition, pyrolysis is a low-cost method to remove metals and reduce viscosity, which has varying degrees of outcome quality depending on temperature, heating rate, and other parameters (Ore and Adebiyi, 2021). Relatedly, Zhao et al. (2021) explored other methods to upgrade heavy oils, concentrating on different catalysts that interact with the molecular bonds of rich oils, changing their macromolecules into small molecules through different reactions (see Figure 3.9).

According to the authors, when considering reaction phases, these catalysts are divided into two main types: those where the reactants and the trigger are in the same phase (such as liquid/liquid), and those where the elements are in different phases. Also, depending on their solubility, catalysts can be separated into mineral, water-soluble, oil-soluble, and dispersed (Zhao et al., 2021). These tools decrease the viscosity of heavy oils through an irreversible decomposition process that involves very high temperatures while incurring fewer costs than techniques like steam-assisted gravity drainage (SAGD) or *in situ* combustion (ISC).

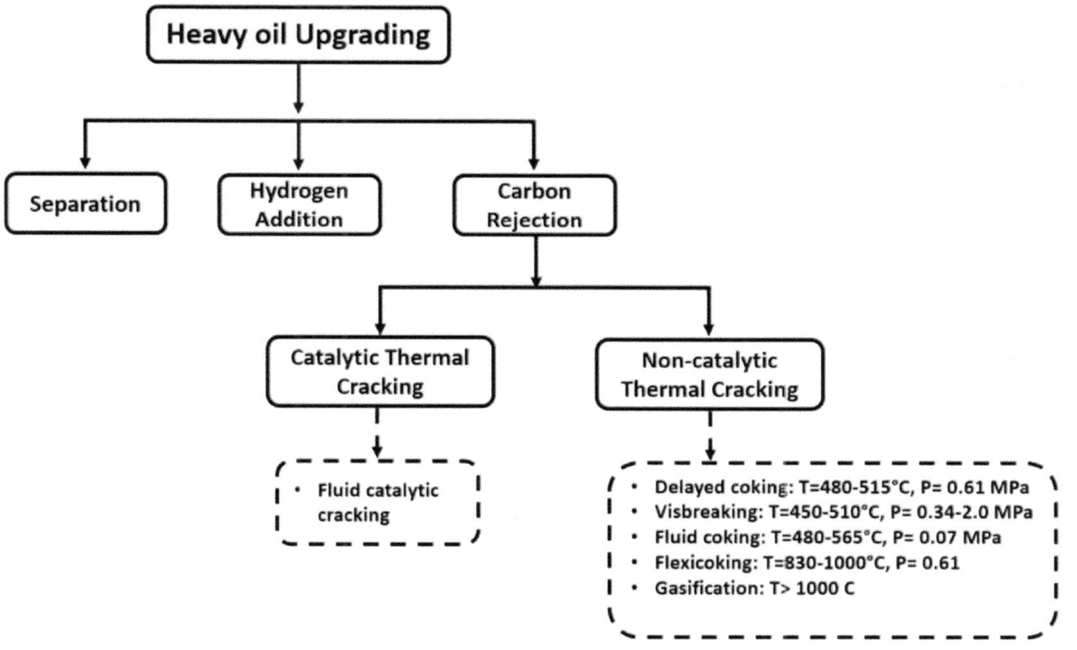

FIGURE 3.8 Classification of the upgrading process of heavy oils.

Source: Ore and Adebiyi (2021).

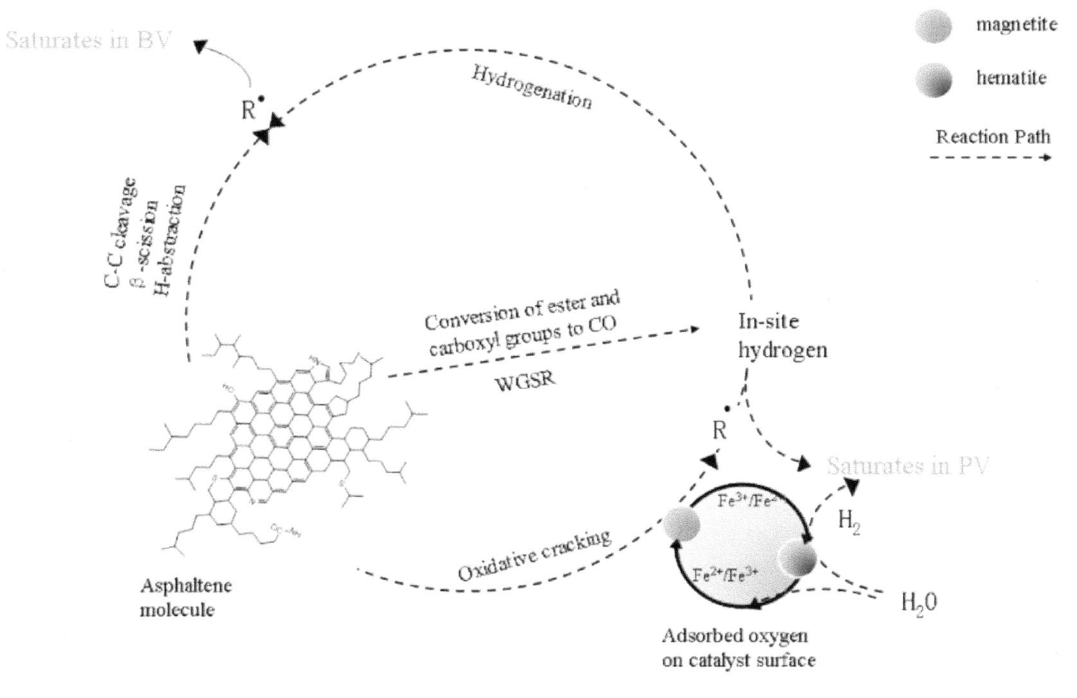

FIGURE 3.9 Reaction pathway of heavy molecules in heavy oil with the supercritical water (SCW) containing the catalyst.

Source: Zhao et al. (2021).

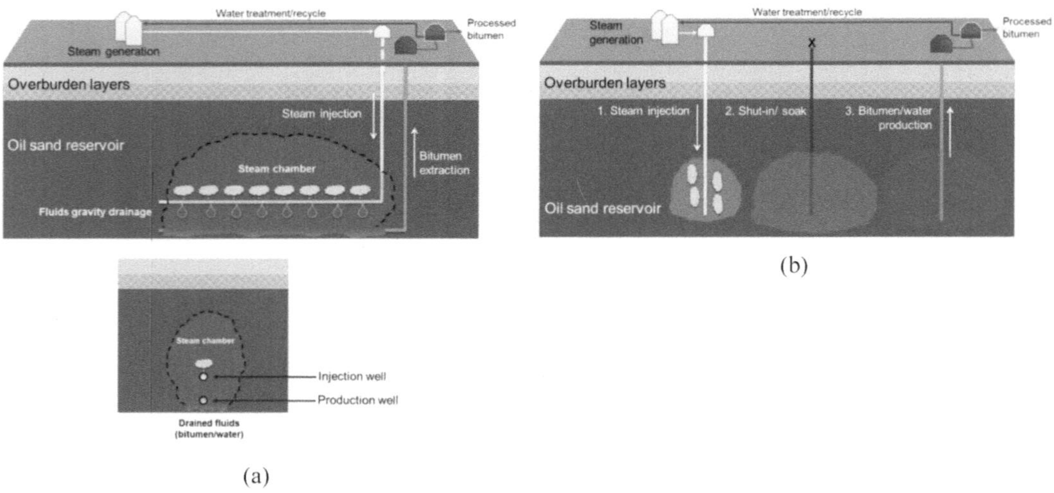

FIGURE 3.10 Schematic representation of (a) a steam-assisted gravity drainage (SAGD) operation, with an illustration of its main elements that show the juxtaposition of injection and production wells in the upper section and a cross-sectional view of the reservoir, with the shape of the steam chamber highlighted in the lower section; and (b) a cyclic steam stimulation (CSS) operation that uses a vertical well for its three phases: steam injection, soak, and production.

Source: Taylor (2018).

Then, bitumen is a similar low-grade heavy oil that is thick and highly viscous and is the main fossil fuel in oil sands. Significant formations have been discovered at oil fields in Iran, Iraq, and Mexico, as well as in Canadian tar fields. Bitumen has strong adhesion properties, which can cause significant accidents, increase the energy consumption to recover the hydrocarbons, and even lead to well abandonment. Because of this, bitumen formations need to be removed by separating oil sands, which can be achieved through improved drilling techniques that increase the density of drilling fluids, add diesel and other emulsifiers, combine chemical plugging, or replace the contaminated pulp (Liu et al., 2019).

SAGD is a well-known thermal method to recover bitumen and heavy oils (see Figure 3.10). It creates a steam chamber below the surface, where the hydrocarbon is located, building two parallel horizontal wells. After a steam injection through the first well, bitumen falls to the second well, known as the production well, which goes in the opposite direction and recovers the separated bitumen. Meanwhile, a second technique, called cyclic steam stimulation (CSS), uses a similar high-pressure steam injection well, followed by a soaking period preceding production. Finally, around one-fifth of the original oil can be recovered, and the CSS process continues until it is not profitable (Taylor, 2018).

Despite ample promise, these methods still show low recovery rates and are highly dependent on the context in where recovery is attempted. In this sense, the reservoir's heterogeneity and permeability can be two critical factors affecting the outcomes of steam-assisted methods. Moreover, the primary separation vessel (PSV) can also make a big difference. This is why some authors have developed improved techniques for these vessels. For example, Shafi et al. (2020) rely on machine learning to create a two-level hierarchical model that simulates the interaction between an agent and the vessel, arriving at an optimal element after many episodic interactions.

Ma et al. (2022) also developed a numerical simulation technique that advances a different recovery method, known as supercritical water flooding (SCWF). Under this framework, a well injects supercritical water (or water above its critical point of 374°C and a pressure of 22.4 MPa) into the heavy oil/bitumen reservoir, which helps heat the hydrocarbon into a recoverable oil that is captured by a production well. Despite showing promise as a simulation method that can lower production costs compared to steam-assisted strategies, SCWF still has no in-field applications. Thus, until practical evidence is not collected, steam injection methods appear to be the most cost-effective procedures for bitumen recovery.

As we can see, more novel techniques that rely on simulations and artificial intelligence have been advanced to recover unconventional and heavy oils efficiently. Coupled with efficiency, however, the environmental impact of heavy oil extraction and processing remains a concern. Because heavy oils contain high levels of sulfur, nitrogen, and heavy metals, potentially affecting air and water quality, *in situ* techniques have gained the momentum they currently hold. According to Guo et al. (2016), catalytic methods that keep hydrocarbons underground while being injected by a catalyst have shown a lower environmental impact and potentially more significant recovery.

3.3.3 Unconventional Sources (Shale Gas and Oil, Gas Hydrates, Coal Bed Methane)

Moving on to other types of recoverable hydrocarbons under newer technologies, shale oil and gas have become two of the most promising fossil fuel sources in the past few decades. Shale gas and oil reservoirs are considered "unconventional" due to the lower degree of permeability where hydrocarbons are trapped, which invites more sophisticated forms of stimulation to help recover them. To help the mudstone rock release the oil and gas trapped inside, hydraulic fracturing has become the preferred method, which consists of an underground horizontal well that releases pressurized fluids to stimulate (or "fracture") shale formations so that the natural gas and oil are eventually released to the surface (Department of Energy, 2013).

As Figure 3.11 shows, the world holds significant reserves of unconventional oil and gas resources. Moreover, an assessment made by Wang et al. (2016) of the seven most common types of unconventional resources finds that there are over 440 billion tons and almost 230 trillion cubic meters of recoverable material. These deposits of heavy oil, oil sand, tight oil, oil shale, shale gas, tight gas, and coal bed methane are distributed predominantly in 476 formations in 363 petroliferous

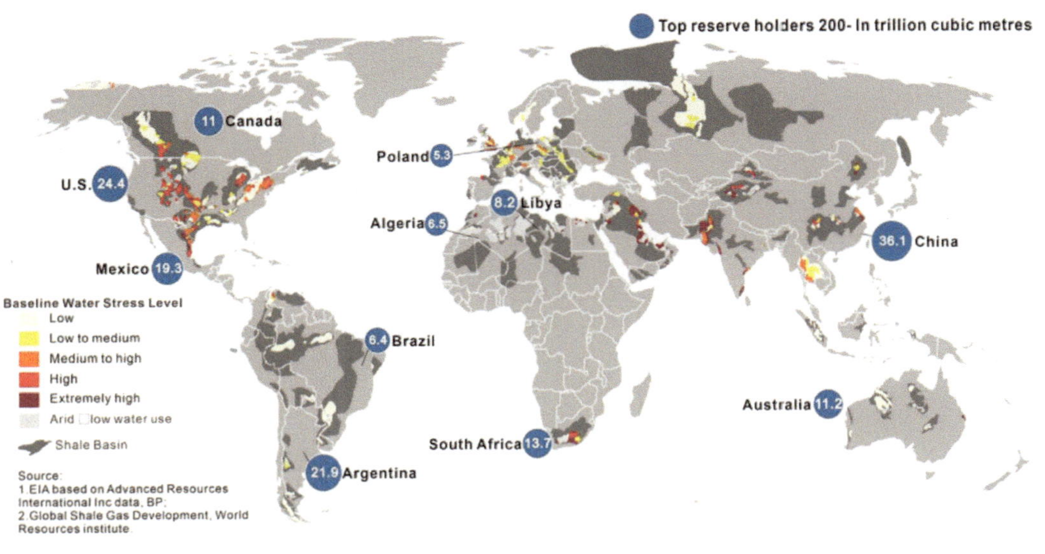

FIGURE 3.11 Estimations of global shale gas resources.

Source: Lyu et al. (2021).

basins, with two of the most studied being the Marcellus Shale Formation in the eastern region of the United States and the Permian Basin in the state of Texas.

With the advent of a more cautious era for oil and gas recovery procedures, more environmentally friendly ways to recover shale oil and gas have been proposed in the past few years. One method that includes carbon capture in shale reservoirs and shale hydrocarbon recovery consists of using supercritical CO_2 (SC-CO_2) as a working fluid. Lyu et al. (2021) explored the potential of this technology, noting its superior efficacy in breaking rocks when compared to water, which leads to improved permeabilities. Additionally, the carbon dioxide injected underground can be sequestered and absorbed on pore surfaces, remaining in place for many decades.

Meanwhile, laboratory techniques have also experienced an improvement and a more widespread use in the field. For example, the work of Chandra and Vishal (2021) follows theoretical frameworks used in digital rock physics (DRP) studies. It develops a framework to analyze images of shale samples, aiming to understand better pore and fracture characteristics that allow a better recovery (see Figure 3.12). Their work reviews different 3D and 2D image processing setups, like X-ray computerized tomography (CT), nano-CT, focused ion beam scanning electron microscopy (FIB-SEM), and confocal laser scanning microscopy (CLSM), and their strengths and weaknesses and various applications.

Additionally, some authors have explored the possibilities that machine learning techniques bring when simulating shale rock characteristics. For example, Anderson et al. (2020) collect a shale sample to create a registered image dataset which is later used as the baseline to predict image cross sections using convolutional neural networks (CNNs) and conditional generative adversarial networks (cGANs). Similarly, Tang and Spikes (2017) rely on neural networks to simulate rock segmentation procedures, which are costly and time-consuming. Lastly, Wu et al. (2019b) built a forest classifier to simulate segmentation procedures after identifying four rock elements in a sample: pore/crack, rock matrix, pyrite, and organic/kerogen components.

Despite these critical advances, some essential costs still remain on the list of factors hindering a widespread replication of methods to recover shale oil gas. First, methane has a considerable risk of being released into the atmosphere. Also considered a greenhouse gas, methane could offset any progress made by CO_2 capture efforts, so researchers should tread carefully when pursuing more shale-related operations. Furthermore, unconventional reservoir development is still in its infancy, and there is an ongoing debate about any potential land and community disruptions that methods like fracking cause. Lastly, current techniques still reach relatively low recovery rates that need to be assisted by subsequent EOR schemes (Lyu et al., 2021).

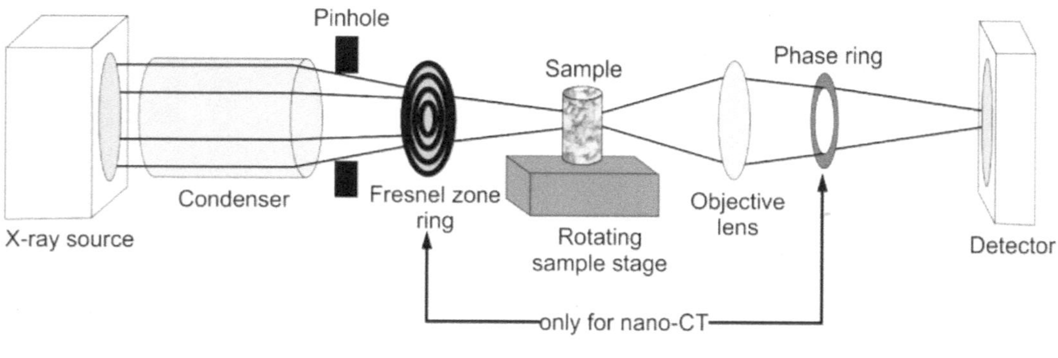

FIGURE 3.12 A schematic of an X-ray micro/nano-CT setup. The Fresnel zone ring and phase ring combination is unique for nano-CT frameworks, which helps in the precise focusing of an X-ray beam that leads to a finer resolution than micro-CT configurations.

Source: Chandra and Vishal (2021).

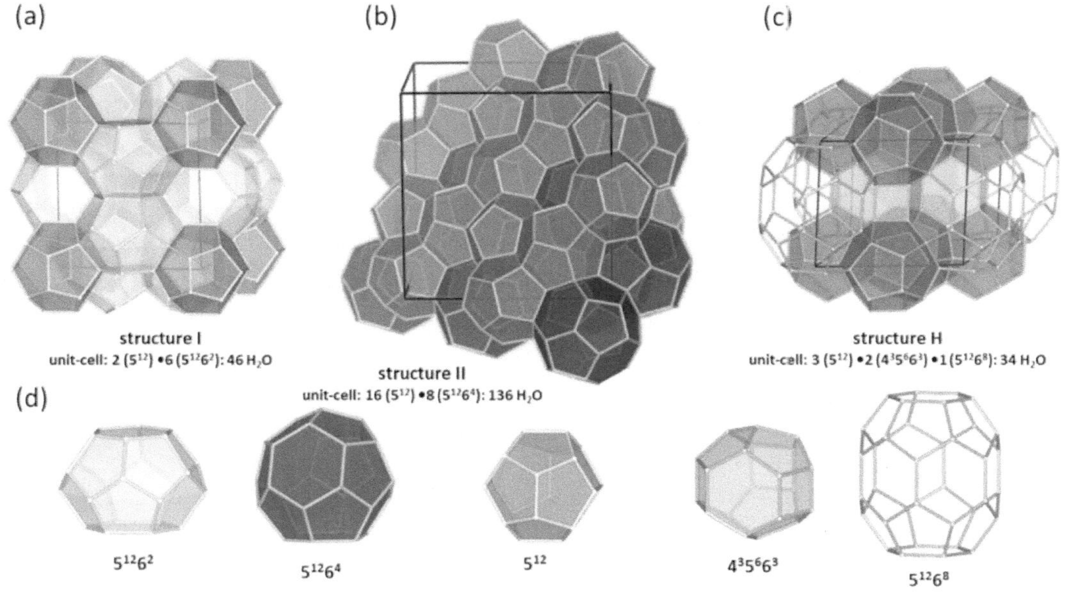

FIGURE 3.13 Three natural gas hydrate structures, (a) structure I, (b) structure II, (c) structure H, and (d) five different host water cages. Solid lines represent the unit cell of each hydrate.

Source: Hassanpouryouzband et al. (2020).

Likewise, the economic cost of many of these strategies continues to be very high for many countries, which has invited the international community to develop various funding plans. This partly explains the enhanced role that supranational organizations like the European Union and the United Nations have had in writing guidelines and potential legislation to be applied across countries that have pledged to develop more environmentally friendly energy processes. These costs will hopefully be brought down by improving technologies that achieve scalability and can be used in a more significant number of projects worldwide.

Another source of hydrocarbons is found in gas hydrates, a potentially vast resource of natural gas that consists of ice-like structures located deep under ocean floors, forming in conditions of elevated pressures and low temperatures (Hassanpouryouzband et al., 2020; Wang et al., 2021). Natural gas hydrates (NGH) are formed when methane gas is trapped within a solid lattice of water molecules, whether through biogenic or thermogenic means (Department of Energy, 2023). Due to their extreme formation conditions in deepwater sediments, gas hydrates are challenging to extract. Still, they possess enormous potential due to their extensive distribution of reserves and high fuel efficiency capacity (Liu et al., 2019).

As Figure 3.13 illustrates, lattice formations are classified into three different hydrate configurations: cubic structure I (sI), cubic structure II (sII), and hexagonal structure H (sH). The first two lattice formations have two distinct cavities, large and small, while the third type of formation has three different cavities. Knowing this, engineers have developed a few approaches to monitor and recover the natural gas in these hydrates. The three most studied methods for gas hydrate production are depressurization, thermal stimulation, and chemical inhibitor injection. These strategies seek to change equilibrium conditions so that the hydrate structure decomposes, producing natural gas (Ruppel, 2018; Wang et al., 2021).

In this sense, a couple of procedures that have the objective of breaking down the lattice frameworks to release natural gas consist of replacing the methane (CH_4) within these structures with a mixture of gases like CO_2 / N_2 and CO_2 / H_2, where the final products are methane and hydrogen, respectively. Additionally, other production methods have been explored in limited field experiments, such as hydraulic fracturing, recovery below freezing point (where the icing effect prevents sand production, which increases recovery), and recovery in solid form (where the seabed hydrate is disintegrated and lifted to a platform in solid state), with encouraging results (Li et al., 2019a).

The abundant reserves of natural gas derived from hydrates estimated at the bottom of the ocean and in permafrost settings have invited researchers to develop novel monitoring and measurement techniques that help enable successful future recovery efforts. Despite remaining mainly in the experimental phase, NGH recovery can still be a commercially viable endeavor, especially in a context where cleaner forms of energy are being sought out, and a phasing out of conventional fossil fuel forms is expected as part of the environmental policy of most advanced countries in the world.

Because of these commitments and policy proposals, countries like Japan, China, the United States, Canada, India, and South Korea have funded gas hydrate exploration and development for several years. Research has found that due to their lower energy consumption and damage to the formation structure, CO_2 replacement strategies appear to have the most significant potential (Wang et al., 2021). Additionally, these procedures allow for the capture of carbon dioxide underground, lowering emissions and combating anthropogenic climate warming. With that said, methane recovery and carbon capture's effectiveness may depend on the system's temperature, with oscillations having interesting potential as an optimization strategy (Zhang et al., 2022).

Then, a third unconventional source of energy that advanced exploration and recovery techniques have boosted is coal bed methane (CBM), a type of natural gas trapped in coal seams. CBM opens a door for countries rich in coal reservoirs but looking for cleaner processes to recover the hydrocarbon. The coal is unique in acting as a source rock and a reservoir rock where light gases are generated inside, and it is possible to explore recovery methods for the gases inside instead of burning the rock and potentially emitting more greenhouse gases in the process. Additionally, coal seams have been found to store up to seven times the CH_4 content than conventional gas reservoirs, which invites commercial exploration and development (Altowilib et al., 2020).

The primary method to recover gases in a CBM process consists of using water and other fluids to stimulate cracks (also known as cleats) in the rock formation, which are connected to a wellbore that brings methane up to the surface after the water inside is pumped out and pressure in the coal seam is reduced (Department of Energy, 2012). Because CBM formations are usually saturated with water, dewatering is required to successfully release the methane adsorbed (or adhered) in the coal seam (Mohamed and Mehana, 2020).

Then, other proposed methods consist of hydraulic flushing, where a high-pressure water jet (HPWJ) load is used to increase permeability by cutting the rock in the coal seam (Zhao and Guo, 2018). Also, supercritical CO_2 flooding in coal beds, where carbon dioxide in high temperatures is injected into the reservoir to generate fractures, is advanced as a technique that surpasses hydraulic fracturing in early experiments. Meanwhile, thermal fracturing by steam injection is another method used to activate the methane molecules, achieving hydrocarbon recovery (Liang et al., 2021). Figure 3.14 summarizes the main injection and transport mechanisms underlying these methods.

While all the previous methods were advanced to recover methane trapped in coal seams, some of those techniques have the added advantage of serving as CO_2 storage strategies, contributing to its capture and sequestration. For example, Pan et al. (2017) explored a CO_2 injection method that helps enhance CBM recovery while also serving as a carbon capture and storage (CCS) endeavor, which allows lower emissions linked to climate change. In this sense, Mukherjee and Misra (2018) identified the primary mechanism for the approach, which consists of three steps: CO_2 adsorption and methane desorption from rock pores, swelling/shrinkage of the core matrix, and flow of gases through the cleats, depending on their degree of compressibility.

Despite few field examples, carbon dioxide injection for enhanced coal bed methane (ECBM) processes has a promising future. One of the first field trials on CO_2-ECBM was done in the San Juan Basin, New Mexico, injecting CO_2 for six years into a deep, unminable coal bed, starting

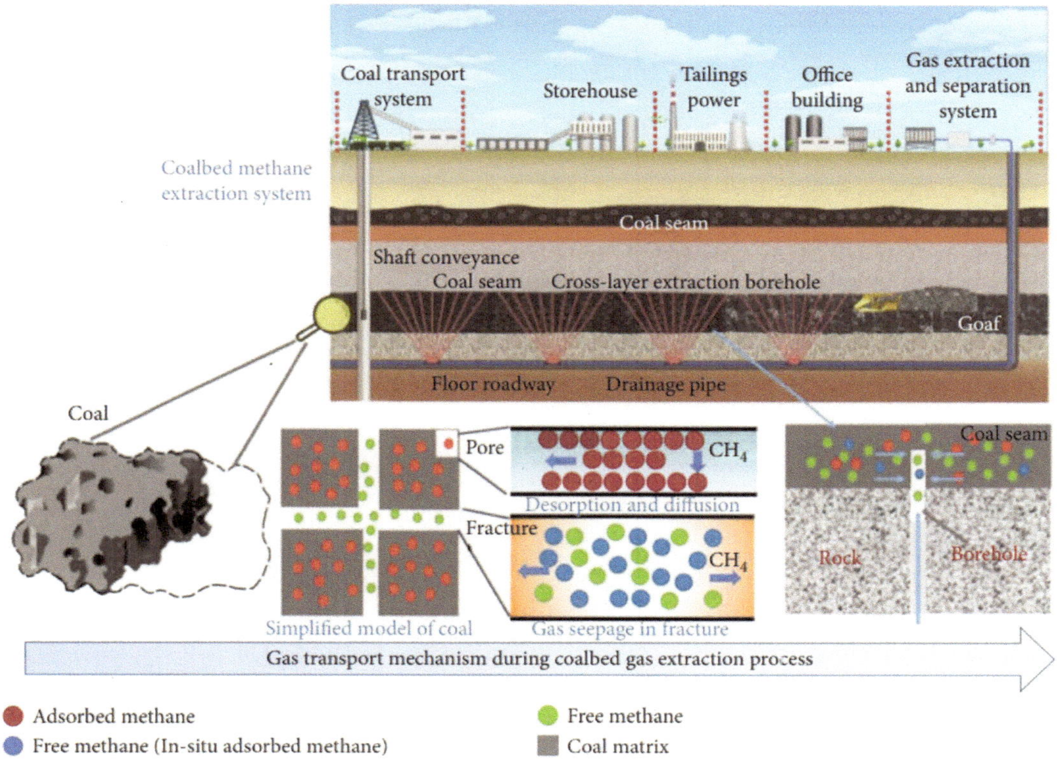

FIGURE 3.14 Gas transport mechanism during the coal bed gas extraction.

Source: Zhang et al. (2020).

in 1995. Shi and Durucan (2005) reported on the results of this field experiment, noting a good agreement between the reservoir simulator and the field data, with discrepancies explained by the lack of a dynamic permeability model. Thus, characterizing the pore fracture accurately, which accounts for the petrophysics features of CBM sites, such as permeability, well heterogeneity, genesis mechanism, and mechanical properties, can significantly optimize these endeavors (Jia et al., 2020).

Additionally, in recent years, methods like cryogenic liquid nitrogen, microwave heating, and high-voltage electrical stimulation techniques have been proposed to improve gas recovery, with positive results (Zhang et al., 2020). Meanwhile, Mu et al. (2019) proposed a coupled hydraulic-mechanical-thermal numerical model that simulates the injection of flue gas, or exhaust gas, from coal-fired power plants into a coal seam. The authors chose the Shaqu coal mine to validate their model, achieving good matching levels between the field data and the simulation results.

In general, CBM methods constitute a cleaner approach to coal exploration. This fossil fuel has significant reserves worldwide and is a safer alternative when dealing with dangerous or unminable reservoirs. In this sense, China is a potential clear beneficiary of the modernization of CBM exploration, given its large number of unexplored coal reserves. However, a clear path toward successful exploration of coal bed methane is still hindered by important challenges, such as the lack of clear government policy support, strict regulations, limited consumer market fostering, and a proactive investment environment (Tao et al., 2019).

As we can see, the advent of new technologies has opened the door to a novel exploration of fossil fuel repositories that allows for the recovery of hydrocarbon located in previously unreachable wells, lowers the dangers for workers, and has advanced cleaner processes that reduce

TABLE 3.2
Enhanced Oil Recovery (EOR) Mechanisms with the Most Potential in Unconventional Reservoirs

Miscible Gas Injection Method (CO_2, N_2, Natural Gas)
- Extraction
- Oil swelling and pressure maintenance
- Reduction of oil viscosity
- Displacement of oil in matrix by high compressibility
- Capillary pressure reduction
- Gas diffusion
- Re-pressurization

Low-Salinity Water Method
- Shale cracking
- Wettability alteration
- Water imbibition
- Clay swelling

Chemical Methods
- Wettability alteration
- Interfacial tension reduction

Source: Elturki and Imqam (2020).

emissions while continuing to provide energy for billions of people worldwide. Applications for heavy oil/bitumen, shale oil and gas, gas hydrates, and coal bed methane are still heavily dependent on factors such as the location of the project, the economic viability and the outside actors willing to invest, the degree of funding support from authorities, and the strictness of regulations, to name a few.

With those constraints, researchers have proposed different techniques to recover hydrocarbons in unconventional wells and improve existing procedures. As mentioned previously, the most common method for enhanced oil recovery (EOR) processes is CO_2 injection. Still, there are a few others, such as low-salinity water flooding and surfactant use, to name a couple (Elturki and Imqam, 2020). Furthermore, Table 3.2 shows a more comprehensive list of mechanisms for EOR in unconventional reservoirs which have shown promising results in experimental and simulation settings, along with those standard techniques with success in field applications.

A CO_2 injection mechanism that has shown promise is known colloquially as "huff-and-puff," where carbon dioxide is injected into a well and a period of soaking follows before resuming production. As Figure 3.15 illustrates, huff-and-puff techniques have three distinct steps: the injection phase, the soaking period, and the production phase. In the first step, the well acts as the injector of carbon dioxide, which goes through at a high rate, significantly increasing the pressure in the reservoir. In the second step, the well is shut in and the soaking period increases the volume of the heavy oil while viscosity decreases. Lastly, the well is opened, and the heavy oil that has been swelled can move through the well and be recovered more easily (Zhou et al., 2018).

Chen et al. (2023) investigated the use of flue gas to perform huff-and-puff operations in tight gas reservoirs, conducting an experiment that builds from a core scanning electron microscopy (SEM) image to create simulations of a well. Findings show that multiple rounds of "huff" help the simulated reservoir gradually improve recovery. Yet a caveat is brought up by Li et al. (2018), who simulated heavy oil retrieval and encountered decreasing returns after the third round, which resulted in poor recovery rates after huff-and-puff. To improve on this, Ding et al. (2021) reviewed the relevant

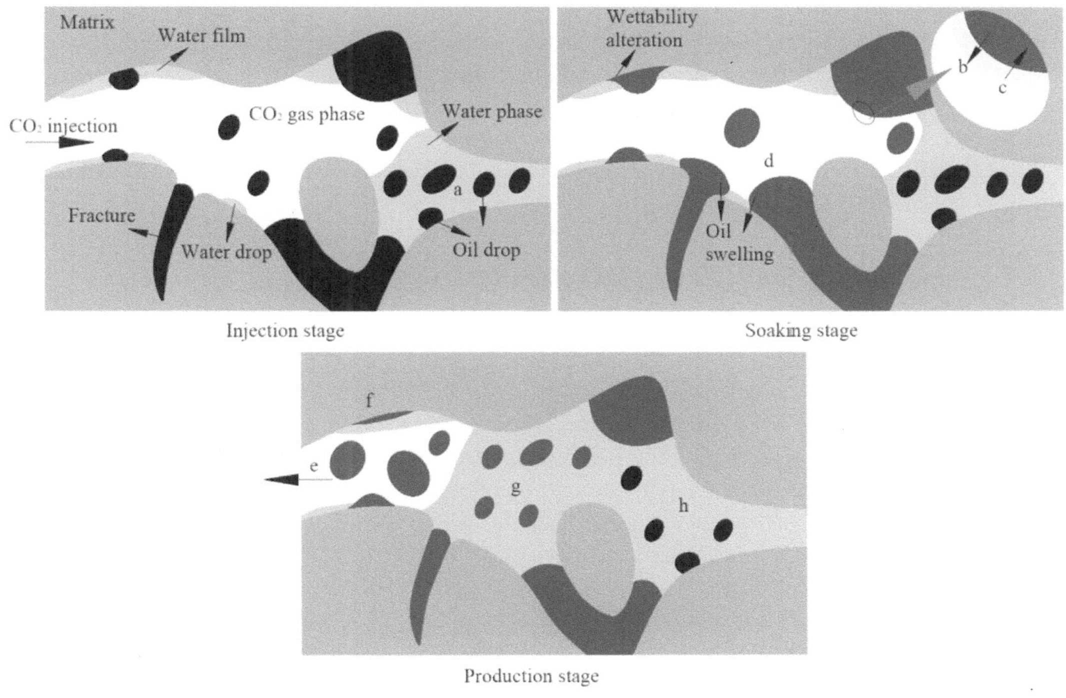

FIGURE 3.15 Schematic of the CO_2 huff-and-puff process in the heavy oil reservoir: (a) oil drops are flushed to a further location in the reservoir by CO_2; (b) light/medium components are extracted by carbon dioxide; (c) CO_2 dissolves into heavy oil; (d) oil drops connect due to oil swelling; (e) fluids flow direction; (f) residual oil is generated due to wettability alteration; (g) oil drops caused from swelled oil in the soaking stage; and (h) oil drops in the water phase are driven from the further location.

Source: Zhou et al. (2018).

parameters that influence EOR in tight oil following CO_2 huff-and-puff methods, concluding that they are more effective at different levels of permeability, with reservoir size also showing significant explaining power.

As mentioned earlier, methods like huff-and-puff have the added feature of sequestering carbon dioxide in underground wells, keeping it from reaching the atmosphere. This potentially positive environmental effect has warranted research into different injection techniques for various shale reservoir types. The work of Du and Nojabaei (2019) reports on field test results performed to achieve EOR while injecting CO_2, showing that huff-and-puff is preferable to other methods, such as gas flooding, mainly due to the quick response that huff-and-puff procedures have to gas injection, as well as the more significant pressure difference observed between measurements down at the reservoir and those at the wellbore.

Meanwhile, an improved mechanism for test experiments is advanced by Mahzari et al. (2021), who observed the lack of live oils that saturate the fractured cores in simulation scenarios. They tested shale cores from the Haynesville formation to tackle this, subject to two huff-and-puff injection experiments. Results show an increased EOR in live oil–saturated cores when compared to dead oil–saturated ones, a finding that invites a close revision of simulation techniques. One avenue for such improvement follows the refinement of operating parameters that optimize huff-and-puff procedures, and here is where the work of Afari et al. (2022) finds that bottom-hole pressure (BHP) and production period are key factors influencing oil recovery.

This review allows us to see that the exploration and recovery of heavy oils and unconventional reservoirs have undergone a process of modernization and improved characterization. The oil and gas industry is not unique in constantly revamping methods and techniques. Still, the sector is certainly keen on exploring new techniques that can help minimize externalities while maximizing economic prospects. This practice is also evident when studying upstream and downstream operations, where a transformation is being led by mathematical models that rely on artificial intelligence. The following section sheds light on the novel AI methods coupled with more conventional techniques, opening promising new avenues for more efficient and cleaner operations for the industry.

3.4 OIL AND GAS APPLICATIONS AND INDUSTRY PROCESSES

The continued exploration of sites and the constantly evolving process of recovery and distribution have led to an identifiable set of industry processes used to systematize oil and gas operations. Generally, these procedures are divided into upstream and downstream sectors and tied to the main oil and gas operations, going from exploration, storage, refining, and distribution to end users. Figure 3.16 shows that the upstream sector involves all exploration and production efforts, including geological studies for site selection and drilling and recovery operations. Meanwhile, downstream operations involve transportation and storage efforts, all the way to refining and distribution. Due to their challenges and advancements, some sections of the literature separate transportation and storage operations into a midstream sector (Library of Congress, 2022; Shafiee et al., 2018).

Upstream processes are the furthest away from final consumers, yet they are vital to guarantee the constant delivery of electricity to the global population. Before production even begins, optimization strategies and site surveillance are pivotal aspects of any potentially successful oil and gas operation. Thus, exploration is the first step to energy distribution. The advent of novel technologies has simplified exploration strategies, with geological surveys performed to assess the necessary parameters to move forward with a given site. Furthermore, deep learning techniques can estimate geological features and predict production without much on-site work, which cuts costs and avoids an essential portion of emissions detrimental to the environment.

The second stage in upstream operations is production, where heavy and expensive machinery and hundreds of skilled workers begin drilling operations, onshore or offshore. Onshore production involves drilling wells into the reservoir and then pumping the oil or gas to the surface using several wells placed together along a field. Meanwhile, offshore reservoirs are more complex, requiring specialized equipment and technologies like radar and sonar to collect information about the seafloors and their potential to be exploited. Once a site is chosen, a large mobile or fixed platform is built, and operations can begin on the coastal shores (Library of Congress, 2022).

Midstream operations involve all modes of transportation and storage of crude oil, going from using pipelines, trucks, and tankers to moving the hydrocarbons to the chosen facilities. In turn, downstream operations include refining and processing, ending with distributing the refined product to businesses, industries, governments, and public consumers (Library of Congress, 2022). Generally speaking, *refining* is converting crude oil into usable products like gasoline, diesel, kerosene, asphalt, and lubricants (Mojarad et al., 2018). Refineries are typically built close to the recovery sites so transportation costs are lowered and centered only on the final distribution stage.

Because of their importance in the global energy supply chain, oil and gas processes are constantly reassessed and perfected. There is currently an intense exploration in the literature to propose general optimization models for upstream and downstream operations that can be applied industry-wide. For example, the work of Shafiee et al. (2018) explores quantitative and qualitative decision-making support (DMS) methods to assist upstream operations. The authors find that tools like cost–benefit analyses, life cycle costing, decision tree analyses, analytic hierarchy processes, and multi-criteria decision analyses have been conventional methods for streamlining. At the same time, novel fuzzy set theories and neural networks have become popular and widespread.

Following a multi-criteria approach, Ghaithan et al. (2017) advanced a downstream oil and gas supply chain (OGSC) model that seeks to minimize total costs and maximize the total revenue

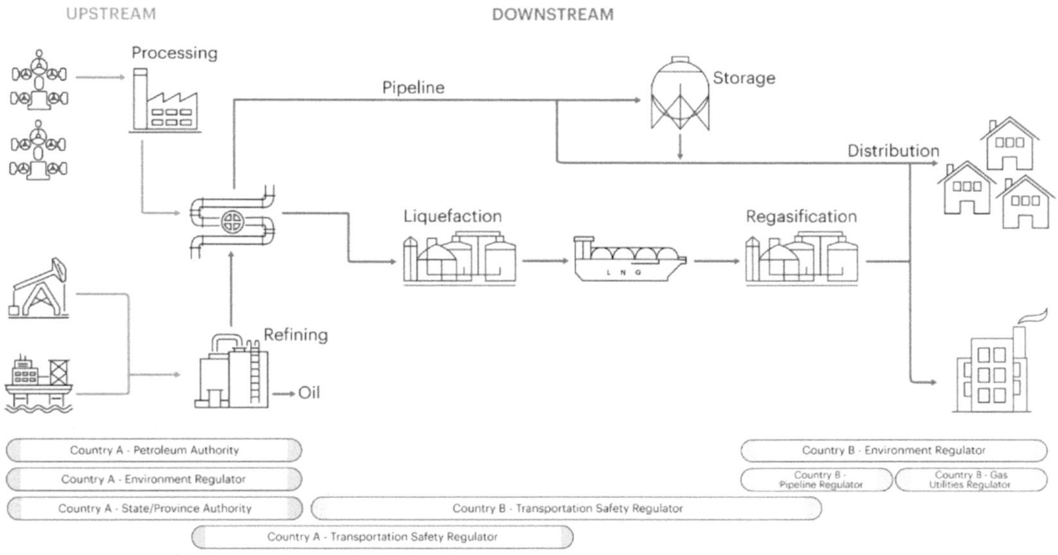

FIGURE 3.16 Oil and gas processes and regulation actors.

Source: IEA (2021).

while maximizing the level of service. Thanks to the mathematical computation of historical data, a curve of predicted scenarios that assesses the multiple objectives established can be drawn, which can then be updated with more variables and data in the future. Then, Patel et al. (2020) explored the tools that big data offers to optimize downstream operations, finding that methods like Apache Hadoop and Cassandra have helped refineries reduce fossil fuel and electricity consumption through the methodization of practices and the pattern recognition granted by big data techniques.

Despite the promising advances regarding conventional oil and gas operations, there continues to be a general concern about the environmental repercussions that exploiting oil and gas sites could have on the climate, as well as the lack of sustainability guidelines that would guarantee a phaseout of some technologies, replacing them with cleaner processes over the next few decades. Because of this, engineers and researchers have investigated a series of improvements that rely on mathematical models and deep learning analysis to tackle the emissions dilemma while delivering energy without significant interruptions to large sections of the world.

A systematic review of data science and machine learning methods used today in the oil and gas industry reveals that the advent of better data collecting procedures has led to an increase of deep learning methods, along with the quality of its results and implications for the environment. As Tariq et al. (2021) find, prediction and classification analyses regarding recovery and economic feasibility that rely on artificial intelligence provide helpful avenues to solve pressing problems, such as climate challenges. For instance, artificial neural networks (ANNs) can help researchers learn patterns in the data and advance novel techniques with outcomes still in the range of accepted solutions from conventional methods.

Likewise, hybrid models are a popular set of techniques that couple the strengths of data-driven models with the positive aspects of physics-based frameworks. Table 3.3 shows that digital twin models have succeeded in drilling applications, while probabilistic approaches with machine learning (ML) have been used for production estimation. Similarly, digital rock analysis and physics coupled with ML workflows are a common staple of reservoir characterization strategies, while surrogate models have been used with similar objectives of site optimization. Despite some important challenges regarding collecting and recording massive databases, big data analytics offers many promising avenues for furthering the oil and gas sector (Mohammadpoor and Torabi, 2020).

TABLE 3.3
A Summary of Selected Hybrid Models in the Oil and Gas Industry

Model	Application	Components
Digital twin	Drilling engineering	Sensor data in near real time (data); synthetic data generated from simulators (physics); humans to interact using avatar (expert); digital siblings for "what if?" scenarios
ML and probabilistic approach	Oil and gas production	Calculated input parameters using existing principles (physics); classification using ML models (data); a probabilistic model to quantify the uncertainty associated with each method; cost model to predict financial impact
ML and digital rock analysis (DRA)	Reservoir characterization	Rock image acquisition; image processing using ML models (data); numerical simulation (physics); result analysis
Surrogate reservoir model (SRM)	Reservoir characterization	Multiple neuro-fuzzy systems; numerical simulation model; spatiotemporal database

Source: Tariq et al. (2021).

The move toward more systematic ways to handle increasing quantities of reservoir data is exciting for the oil and gas sector. The economic viability of hydrocarbon operations and compliance with environmental regulations are also essential parameters that require a closer look. In the next section, we go over some financial considerations for oil and gas operations, along with a series of best practices that have become the standard in the new era of environmentally friendly processes for the industry.

3.5 ECONOMICS AND BEST PRACTICES FOR OIL AND GAS RESOURCES

Regarding the economics of oil and gas processes, the industry is very capital-intensive, with hundreds of thousands of jobs worldwide created to fulfill its needs. As a consequence, a significant boost is created by these dynamics in economies across many countries, which mainly helps developing nations. That said, how those benefits are obtained can put developing countries in conflict with developed states that are looking to expand their decarbonization efforts to combat climate change, because these initiatives may interfere with efforts from developing countries to lift many people from poverty.

Therefore, as a first consideration, the economics of the oil and gas sector need to be contextualized in a world that features actors with varying degrees of development that coordinate in different ways and obtain particular benefits from the way they organize. This unpredictability may help explain why the energy transition process has many moving parts that will likely take several decades to implement fully. Additionally, the boost provided by shale oil and gas reservoirs and recovery techniques like hydraulic fracturing can invite many of the actors considering divesting from fossil fuels back into the fray in the search for economic opportunities with cleaner energy sources.

In this respect, the work of Braungardt et al. (2019) provides a comprehensive list of arguments in favor and against fossil fuel divestment to continue the debate on the implications of their exploration and recovery. On the one hand, ethical considerations regarding the state of the planet and future generations are the most substantial ideas in favor of divestment, while the risk of public opinion backfire and the perceived inconsistency of demanding change of others, yet continuing to consume CO_2, especially in developed countries, are listed as the strongest arguments against divestment.

Investigating the potential effect that fossil fuel exploration and consumption could have on economic growth, Baz et al. (2021) explored the trends that positive and negative shocks to fossil fuel consumption have on short-term and long-term economic prospects for different countries. The authors find asymmetric effects mitigated or accentuated by factors like energy dependency, foreign direct investment (FDI), geographical location, and investment in renewable energies. For the case of Pakistan, highlighted by the authors, renewable energy potential is higher than in other countries. Thus, exploration could be tailored to this sector without expecting a significant hit to their economic growth prospects.

Under this dynamic scenario of incomplete and changing information, actors are required to arrive at energy decisions that affect many people, most times for decades. Because of the high stakes involved in the decision-making process, governments and companies have increasingly relied on international organizations like the United Nations and the European Union to draft agreements on environmental regulations, such as the Paris Agreement or the Kyoto Protocol, a few decades earlier. Using the international stage can make governments and companies less willing to renege on their promises, as their pledges are on the record in front of hundreds of other actors.

This scenario opens the door to different strategies for companies and governments, with transitioning to renewables, allowing for greater energy diversification, or completely resisting being the main paths. In their work, Green et al. (2022) explored the different strategies for oil companies and found that hedging, or mitigating risks through energy diversification, is the preferred strategy by the more ambitious companies. Concurrently, most firms perform political messaging, where they "talk the talk" and announce plans for decarbonization yet do not "walk the walk," failing to initiate processes for energy transition. These lateral moves are partly the result of a market with different players holding different motivations and the continuous advancement of novel technologies to exploit fossil fuels.

The reasons for companies to behave a certain way may be attached to profit margins, changing environments, or even a genuine desire to explore alternative energy sources for the future. All these parameters, along with the professionalization of the sector and the increasing reliance on computing capabilities for simulations and experiments, have helped shape its cost structure over the years. Figure 3.17 shows different breakdowns in the costs incurred by the oil and gas sector, featuring significant allocations for things like health, safety, and environment (HSE) expenses, service contracts, and hydrocarbon transportation. Meanwhile, the intrinsic uncertainty that characterizes the industry is seen in the significant percentage of variable costs.

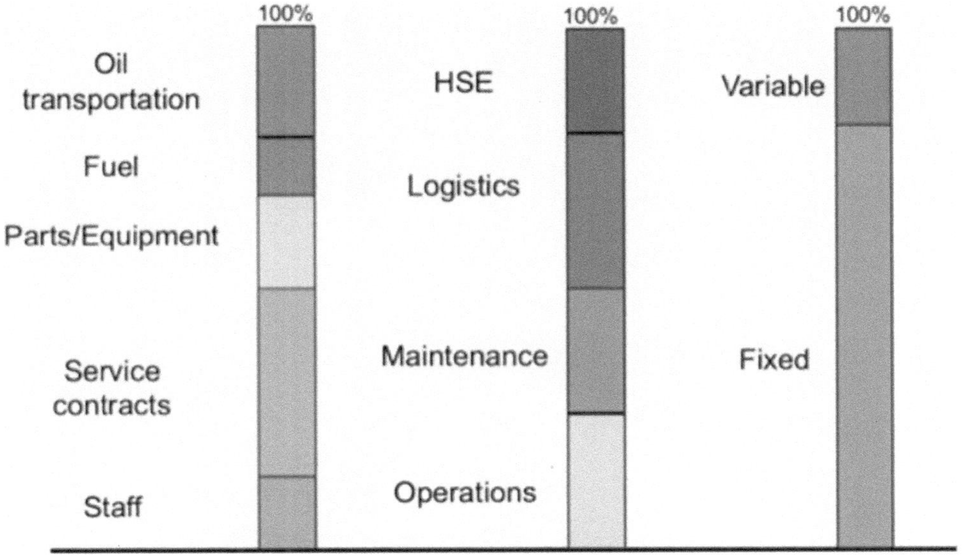

FIGURE 3.17 Cost structures under different types of breakdowns by major cost families.

Source: Bret-Rouzaut (2022).

According to Bret-Rouzaut (2022), the three main actors involved in the economics of oil and gas operations are the states, the contractors, and the oil companies themselves. This does not preclude other players, such as banks, insurance companies, and professional organizations, from involvement. As noted, this framework has allowed for the development of outside regulations enforced by international organizations, but at the same time, it has encouraged oil companies to develop alliances to work in tandem toward improving the climate situation. This has led to the development of the Oil and Gas Climate Initiative. This CEO-led initiative has focused on energy transition matters, such as decarbonization and net-zero pledges (OGCI, 2022).

The development of OGCI has its supporters as well as its detractors. On the one hand, many appreciate the effort from oil and gas companies to deal with matters considered urgent, such as the state of the environment and the effect of CO_2 in the atmosphere, a significant part of it coming from oil and gas processes. On the other hand, some skepticism is raised because oil companies continue to be profit-seeking enterprises, and they may see the energy transition process as an opportunity to develop different strategies to remain as dominant energy players while doing little to commit to treaties and taking decisive action to protect the environment (Bach, 2019).

In turn, governments are also constrained by similar considerations. In choosing their energy suite, countries may stick to what has worked for decades, explore alternatives based on economic incentives granted to new or existing players, or combine strategies to have a diversified grid. In their work, Chudy-Laskowska and Pisula (2022) explored the timeline that has led many European countries to diversify and invest in renewable energies. Joint efforts started around 2009, when member states committed to 20% of the total EU energy consumption to come from renewable sources while each country would consider their starting points and develop strategies to increase their renewable energy shares.

As Figure 3.18 illustrates, EU members vary in their energy shares, with Denmark, Austria, and Luxembourg having the most significant amount of energy coming from renewable sources. On the

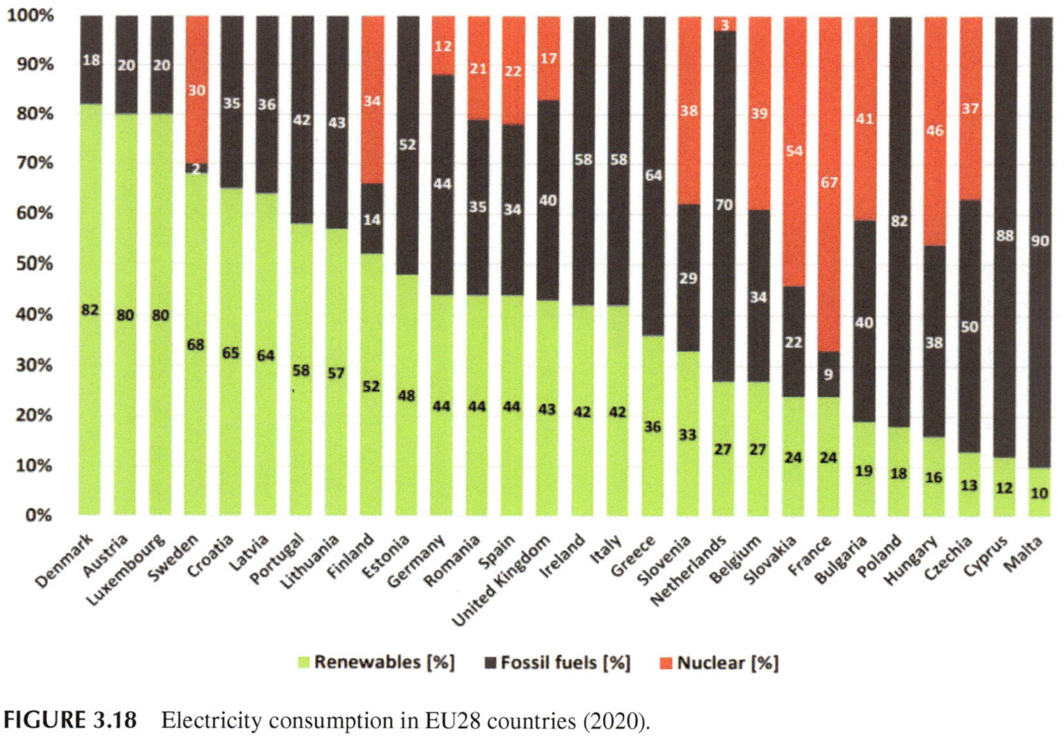

FIGURE 3.18 Electricity consumption in EU28 countries (2020).

Source: Chudy-Laskowska and Pisula (2022).

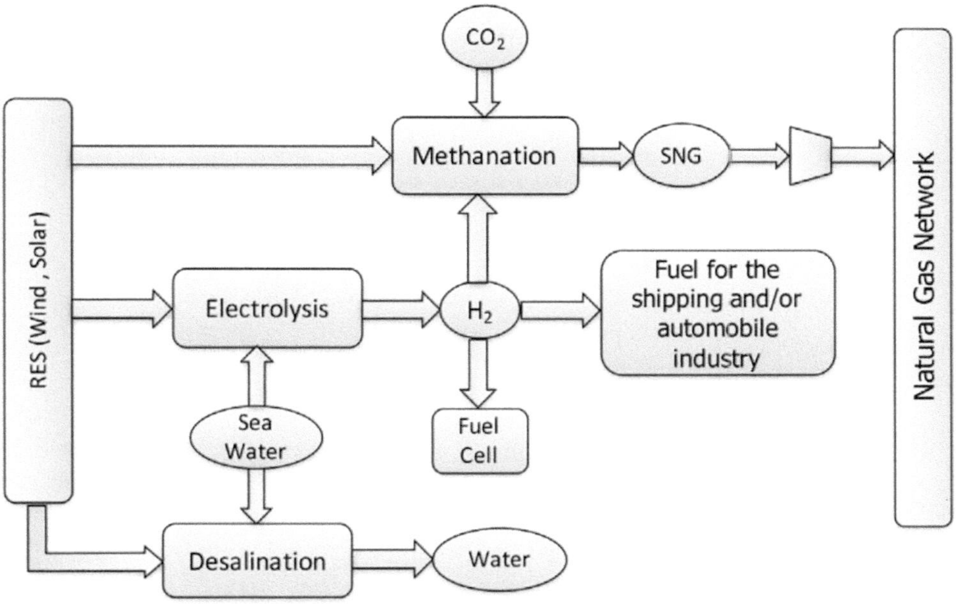

FIGURE 3.19 Global model of the RELife project to reuse offshore oil and gas platforms.

Source: Leporini et al. (2019).

other side of the spectrum, Malta and Cyprus have the largest share of energy produced from fossil fuels. Meanwhile, countries like France and Slovakia hold a more mixed energy suite, with significant shares coming from nuclear power. Currently, the European Commission is behind most of the current environmental pledges as they continue to strive for energy balance from EU members through the 'Fit for 55' plan, which seeks to reduce net emissions by at least 55% by 2030 when compared to 1990 levels and to be the first climate-neutral continent by 2050 (European Commission, 2021).

Along with the supranational policy, governments can also encourage best practice policies regarding the reutilization or decommissioning of oil and gas platforms and infrastructure, improve their efficiency, or foster the creation of markets for the production and transportation of hydrocarbons to incentivize competition. On the former point, Leporini et al. (2019) highlight the role of sub-transitions necessary for the larger energy transition to occur, including the process of oil and gas platform reutilization. With this in mind, the authors developed the Renewable Energy for a New Life of Offshore Platforms (RELife) project, examining all the steps required to repurpose offshore platforms.

As we see in Figure 3.19, the RELife project seeks to convert oil and gas platforms into renewable energy production structures, where a few alternatives can be explored. For example, green hydrogen can be produced through electrolysis and sent through the same pipelines built to transport oil and gas. Alternatively, hydrogen could be used as a catalyst in a methanation process to produce synthetic natural gas (SNG). Then, using energy from the sun or the wind, plants can also be repurposed to generate fresh water from the ocean. Lastly, a more distant alternative requiring further research is generating direct electricity that could be added directly to the national grid.

Meanwhile, Nguyen et al. (2016) proposed a series of measures to improve the efficiency of offshore oil and gas platforms to decrease their CO_2 emissions. With the caveat of great variability across platforms and actors involved in the process, the authors find significant emission reductions when applying their proposed strategy in the four plants they examined. The

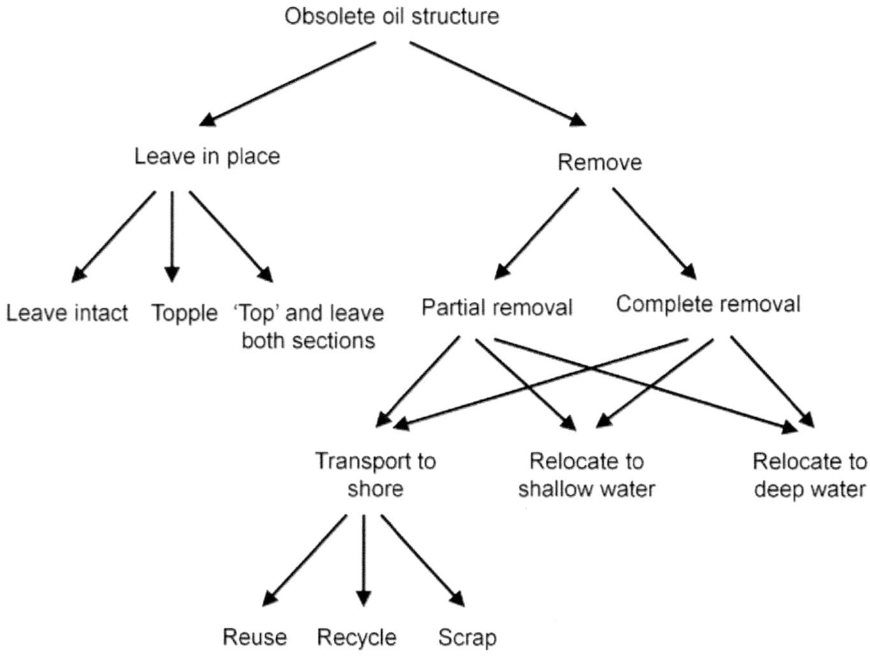

FIGURE 3.20 Decommissioning options for obsolete oil structures.

Source: Fowler et al. (2014).

measures proposed center on optimizing and integrating energy processes and consist of the installation of multiple pressure levels in production manifolds, utilizing multiphase expanders, limiting gas recirculation around compressors, downsizing or replacing existing gas turbines, and using both waste heat from turbine exhausts and low-temperature heat from gas cooling in other processes.

Then, strategies for decommissioning offshore oil and gas infrastructures are explored by Fowler et al. (2014). Noted in Figure 3.20 is a decision tree of several options to decommission obsolete oil and gas structures, a crossroads many countries will face in the short term. The authors developed a multi-criteria framework that helps optimize decisions in a context where a multiplicity of equally essential considerations needs to be assessed. This turn toward more holistic approaches that combine the strengths of different disciplines is required to tackle the unprecedented challenges that the energy transition will bring in the next few decades.

Relatedly, petroleum engineering would gain significant insights after engaging with local communities, where the work of researchers and engineers is put into practice. Currently, and despite the pivotal role of the oil and gas industry in lifting millions of people out of poverty, public perceptions regarding industry practices may not be as good as they could be, mainly because stakeholders (understood as the communities that deal with the brunt of the consequences of the actions of the industry) do not see a credible set of commitments from oil companies to tackle emissions on time.

A potential solution to this lack of communication across disciplines is using a different lens when considering the repercussions of some oil and gas operations. Lee et al. (2022) focused on the concern regarding greenhouse emissions from upstream processes and proposed a public health approach that facilitates policy discussions to effect change. Due to the potential dangers from the release of methane, hazardous air pollutants (HAPs) like benzene and toluene, sulfur dioxide, and particulate matter emanating from combustion processes, the role of health authorities in establishing accountability and regulation strategies becomes essential.

As we have observed in this section, the current economics of oil and gas processes involve more actors than ever before, all with a similar goal of obtaining maximum benefits while also taking environmentally friendlier steps. The relationships built between states, international organizations, companies, and stakeholders must be strengthened to ensure the industry's future is cleaner and safer. Also, the commitment to innovation and investment in alternative energy sources must be the staple, leading humanity to transition to renewable energies successfully.

3.6 POTENTIAL OF OIL AND GAS IN AN ENERGY TRANSITION CONTEXT

The crucial role that oil and gas resources have played in shaping the world as we know it today is hard to argue. Many economic and prosperity indicators would probably not be as positive without the widespread use of fossil fuels, both in the developed and developing world. With that said, we are at a time when humanity is considering the risks that hydrocarbon operations are exerting on the planet more seriously, as well as the seemingly irreversible repercussions of carbon processes that have contributed to the worsening of the climate. So there is a tension between the steady role of oil and gas and the pressing need to act to protect the environment.

We are currently in a time where the concept of the energy transition to renewable energy sources is not only a pipe dream but also a realistic enterprise. However, this move requires a considerable amount of time, where oil and gas processes will continue to be regarded as significant energy sources, particularly in developing countries that do not have the resources to explore more expensive sources like solar, wind, and nuclear. Thus, another tension appears between countries that utilize scarce polluting resources for the betterment of their population and those nations that can afford to "go green" only because they have already used fossil fuels to achieve a stage of economic development.

The threading of this very complex needle rests on a set of expectations, where a few of them constitute key pillars: first, the strengthening of global policies; then, the advent of more sophisticated technologies that make conventional processes cleaner and safer; and lastly, a controlled yet steady phasing out of many of those fossil fuel operations that contribute to the problem of climate change. Therefore, a sustained effort to enforce climate legislation, coupled with technologies that target the reduction of the carbon footprint of oil and gas operations, along with the continued development of safer practices of extraction, which cuts risks to workers and lowers greenhouse gas emissions, is required for the process to go on.

These dynamics leave oil and gas resources in a place where they will continue to have a significant role in the lives of millions of people in the medium term, but with a limited window of time to significantly improve its processes, or face the possibility of a phasing-out stage that could be enforced and realized by the joint work of international organizations and governments. Therefore, the push for an energy transition can be a spark plug to pursue safer and cleaner processes or the beginning of the end for oil and gas resources as we know them today. Coal resources face similar challenges and opportunities, which we will review in detail in the next section.

3.7 COAL RESOURCES: INTRODUCTION

Coal is one of the most important fossil fuels in the world and has a long track record of providing large quantities of energy for millions of people over hundreds of thousands of years. Coal is energy-dense, meaning, rocks have a high amount of carbon and hydrocarbons, a fact that many civilizations have relied upon for cooking, heating, propelling their industries, and ultimately, generating electricity. Coal rocks are black and take millions of years to form, as dead plants, animals, and organisms are turned into coal after heat and pressure are applied for centuries. Because of this long process, coal is considered a non-renewable source of energy (EIA, 2022).

Many countries have thrived thanks to the use of coal and the reliable and cheap energy it produces, which helped improve their industries and processes, develop a professional petrochemical

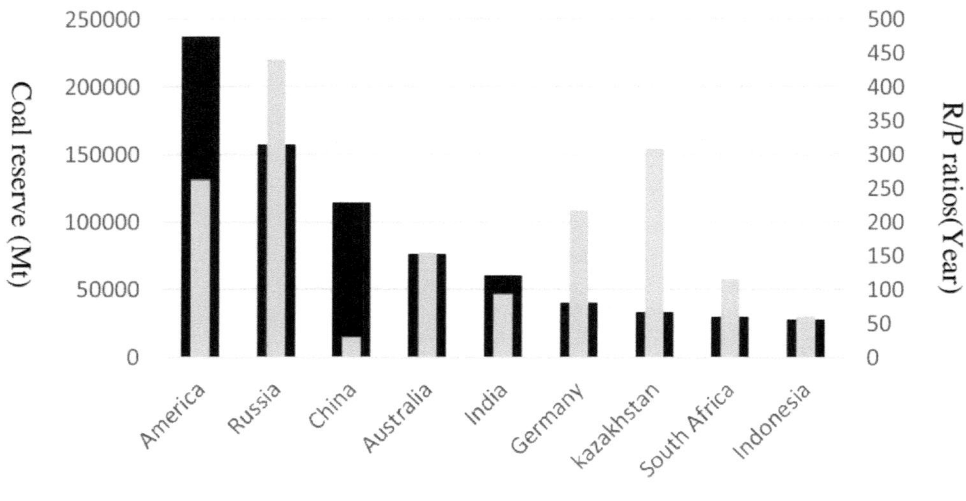

FIGURE 3.21 Coal reserves and reserve/production (R/P) ratios in 2015 for the top nine coal reserve countries.

Source: Han et al. (2018).

industry that spurred many high-paying jobs, and increase prosperity for millions of people. Moreover, significant events in human history, such as the Industrial Revolution, were mainly fueled by coal, which in turn powered the steam engine that was pivotal in the increase of trade connections and the establishment of a supply chain that provides goods and services for millions of people around the world.

Currently, coal continues to be one of the primary sources of energy for many countries. As Figure 3.21 shows, America, Russia, and China have the most significant amounts of proven coal reserves, and many other countries continue to explore and exploit reservoirs, as we may be in a time of "peak coal" (Han et al., 2018). The pivotal role of coal as an energy and jobs provider has come with the cost of being a high-emissions source, as it is one of the main culprits in the current global warming trends. This has led researchers and experts to warn against the widespread continued use of coal, which can lead to land and water pollution, the release of greenhouse gases, and the future scarcity of non-renewable energy sources.

This section delves into the development of coal resources, the exploration of sites, and the advent of new technologies for recovery. We review the strengths and weaknesses of some techniques and the different applications that coal has in our daily lives, along with the discussion regarding its potentially dangerous environmental effects. The history of coal is one of progress, but also one of change, as the oil and gas industry is today at a crossroads between continuing its path with fossil fuels as a reliable companion or transitioning to an entirely novel energy suite led by renewable energy sources. The path may be complex, but it is no less intriguing.

3.8 SCREENING, SELECTION, AND SITE ECONOMICS FOR COAL RESOURCES

The current dynamics in the oil and gas sector call for a refreshed look at site management strategies for coal resources. Given the decline in its use, partly explained by the concerns generated by its effects on the environment, selection guidelines for coal sites are now increasingly more complex, requiring a multifaceted approach concerned not only with the economic viability of sites but also with the prospects for reservoir use, along with safety mining considerations. Therefore, the prospects of successful hydrocarbon recovery and safe mining practices are only two of the parameters driving site selection, as prospects for future carbon storage and sequestration endeavors should also be included.

Over the years, screening and selection criteria for energy sites have generally centered on the geological features of a site, such as permeability and porosity, as well as on economic factors, like demand for the resource, current and expected price, availability of technologies to access the resource, coal contracts, and capital investments. Additionally, the literature assesses societal factors, like the degree of impact of construction and operations in surrounding communities, the presence or absence of protected areas, and social perceptions of the hydrocarbon and the work of the oil and gas industry.

Omitaomu et al. (2012) investigated the factors that generally contribute to the selection of sites for energy projects, with findings reported in Table 3.4. As we can see, advanced coal sites consider population, slope, stream flows, protected lands, information from the Environmental Protection Agency (EIA), and access to railroads and external pipelines. Starting from that baseline, other considerations have been added depending on, among other things, the site's location and the project's economic viability. Thus, the mining method, the infrastructure and equipment, and the prospects of dealing with wastewater and other externalities from industrial processes are other parameters that have aided researchers in choosing a site for coal processes.

Along with variables like demand, price, and the project's economic viability, current site selection methods also involve significant safety and environmental considerations. Thus, most coal projects today come with heavy constraints that deal with workplace safety and the potential reutilization of the reservoir as a storage unit for carbon dioxide or other greenhouse gases. Because of these dynamics and concerns, novel coal exploration and recovery mechanisms have been introduced in the past few years. One example is underground coal gasification (UCG). In this method, coal is not combusted completely but, rather, injected with oxidants *in situ* so that coal achieves a gasification state instead (Imran et al., 2014).

A few considerations regarding UCG procedures are related to worker safety, as one objective of these new techniques is to avoid using men underground to work the hydrocarbon, while also significantly improving the quality of life of people living near the sites. Moreover, UCG processes have shown essential noise and air pollution reductions, releasing fewer gases into the atmosphere and reducing groundwater contamination (Pei et al., 2016; Škvareková et al., 2019). All these improvements are achieved mainly due to the *in situ* treatment of coal, which simplifies the process of extracting, preparing, transporting, storing, and ultimately combusting the rock to generate energy (Imran et al., 2014).

In addition, the more novel exploration and recovery technologies have allowed engineers to explore and optimize closed or abandoned reservoirs while maintaining suitable resource conditions (Huang et al., 2023). Through UCG processes, deep coal seams and high-outburst coal seams can now realistically be recovered with a much smaller risk to workers and a low environmental impact. That said, essential risks remain when not properly vetting and screening a site. As shown in Figure 3.22, water fissures or roof cracks can occur because of the cavities created by UCG operations, all avoidable problems that could lead to site instability or reservoir discontinuity.

Due to this, necessary constraints in coal operations have mobilized the oil and gas sector to explore alternative techniques that improve site optimization. At the same time, it also secures more excellent worker safety and environmental protections through more responsible mining. Given that coal resources will continue to be a significant part of the energy suite in many countries for some decades, methods like underground coal gasification are pivotal for developing countries to achieve energy independence and open avenues for the exploration of alternative and renewable energies that can eventually drive them toward a successful energy transition process.

As a promising procedure that builds on previous coal resource methods, UCG is subject to similar site selection parameters. Exploring the conventional and novel parameters for UCG site

TABLE 3.4
Siting Criteria for Various Power Generation Options

Criteria	Data Sources	Nuclear	Advanced Coal	Concentrated Solar	CAES
Population	ORNL (LandScan)	X	X	X	X
Slope	National Geospatial-Intelligence Agency	X	X	X	X
Safe shutdown earthquake	US Geological Survey	X			
Wetlands/open water	US Geological Survey; US Department of Interior	X	X	X	X
Landslide hazards	US Geological Survey	X	X	X	X
100-year floodplain	Federal Emergency Management Agency	X	X	X	X
Stream flow	US Geological Survey; Environmental Protection Agency	X	X	X (for the wet option)	
Protected lands (a composite of 13 data layers)	National Forest Service; Bureau of Land Management; US Fish and Wildlife Service; National Wild and Scenic Rivers; National Atlas of the USA; commercial sources	X	X	X	X
Proximity to hazardous facilities (a composite of airports, military bases, and petroleum refineries)	Federal Aviation Administration; Bureau of Transportation Statistics; US Census Bureau	X			
Proximity to fault lines	US Geological Survey	X			
Salt formations	US Geological Survey				X
Aquifer	US Geological Survey			X	X
EPA non-attainment data	Environmental Protection Agency		X		
Access to railroad or barge	ORNL (TRAGIS)		X		
Avoiding piping through fault lines, protected lands, and slope	None		X		
Solar radiation	National Renewable Energy Laboratory (NREL)			X	

Source: Omitaomu et al. (2012).

selection, Mohanty (2017) found that the area's geology is an imperative factor that must be assessed at every stage, from planning to execution, and even for decommission and site restoration. Plus, impermeability with low porosity and deformation are important considerations so that the coal seam remains isolated from nearby aquifers. Meanwhile, Pei et al. (2016) reached a similar conclusion while also advancing that the potential for quality syngas production from coal processes can be added as another important parameter for UCG site screening.

Lastly, the potential for reutilizing depleted coal reservoirs or repurposing existing sites is another factor that invites a closer look at site selection parameters. Jindal and Shrimali (2022) analyzed

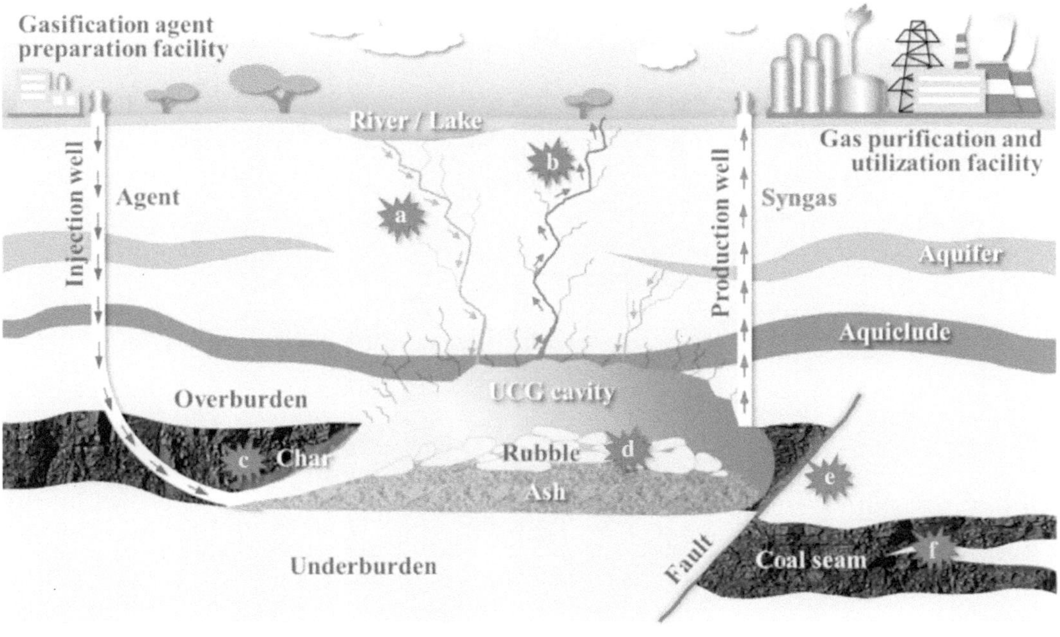

FIGURE 3.22 Potential risks caused by the improper site selection of underground coal gasification (UCG) sites: (a) forming water-flowing fissures; (b) roof cracks connecting the UCG cavity with the surface; (c) poor gasification characteristics of raw coal; (d) large-area roof collapse; (e) discontinuity of the coal seam; (f) instability of seam occurrence.

Source: Huang et al. (2023).

the economic considerations behind coal plant repurposing in developing countries, an overlooked aspect of coal resources until recently. Their work provides an in-depth look into the cost–benefit analysis of repurposing plants to allow for the development of renewable energy sources, contrasting it to two other scenarios: continuing to work the site as normal, or decommissioning the plant while renewable alternatives are built elsewhere. Findings show that the benefits of repurposing outweigh the costs of decommissioning.

To finalize, site selection for coal resources has undergone a similar transitional process as other conventional energy sources. From more established geological and economic considerations, current parameters have expanded to include social perceptions, achieving environmental commitments, improving worker safety, and plant reutilization or repurposing efforts. The reality is that coal resources will continue to be in the energy dockets of many countries for the foreseeable future, so it became necessary to advance proactive approaches to screen and choose sites for coal recovery, accounting for environmental pledges and the eventual energy transition process.

3.9 COAL APPLICATIONS AND INDUSTRY PROCESSES

As with most conventional energy sources, coal's principal use is to generate energy at reduced costs compared to sources like nuclear or natural gas and through more efficient processes than older sources like wood. Within coal resources, however, a few distinct types of matter exist, which are valued differently and provide different degrees of carbon content. In this section, we first review different types of coal matter and their properties and uses. Then, we focus on the economics

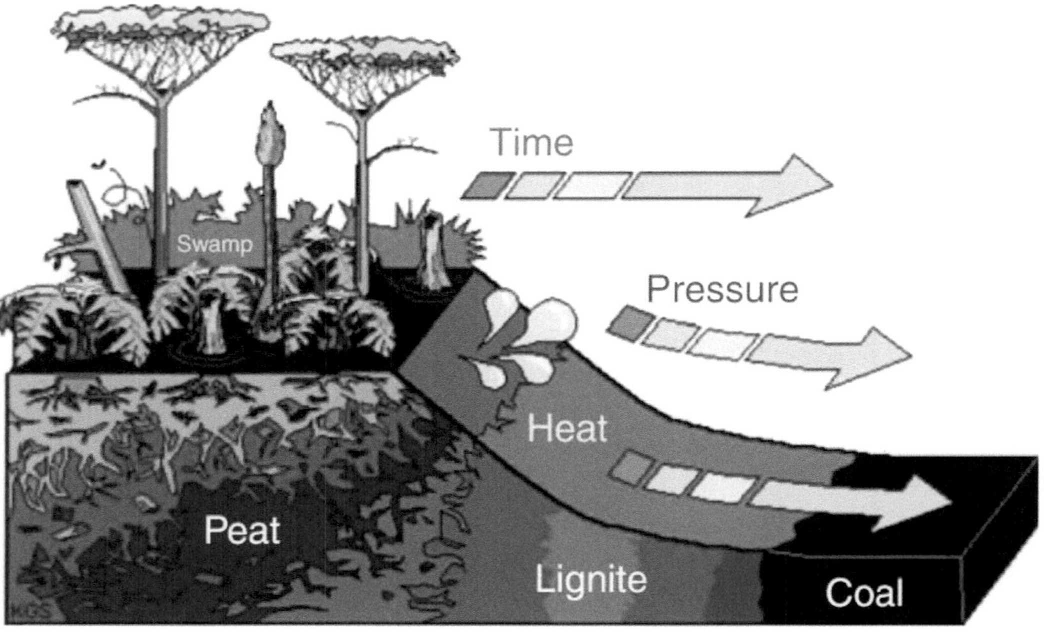

FIGURE 3.23 Principal mechanism of the formation of coal and peat.

Source: Höök (2012).

of coal resources and the best practices that have led to current applications and experiments performed in the oil and gas industry.

3.9.1 Coal Ranks and Applications

Coal types can vary in the literature, with sub-types and mid-ranks specified by some authors, but a plain hierarchy of coal quality has been established, which goes from the lowest-ranked peat to the purest form in graphite. Peat is considered almost a precursor to coal that consists of partly decayed plant and mineral matter which still needs greater heat and pressure to become what we consider as coal (USGS, 2023). Peat is the least valuable form of coal due to its low carbon content, and because of this, it is rarely used as a fuel source. With that said, some uses have been found for peat, especially in gardening, horticulture, and agriculture, where it is used as a soil strengthener and a tool to improve water circulation in the mixture.

Russia, with the current largest proven peat reserves, was also a considerable part of the energy suite in the last few decades of the Soviet Union. Currently, Russia holds almost 30% of the world's reserves of peat, while some notable countries with established peat extraction industries are Finland, Ireland, Sweden, Germany, and Belarus (Tcvetkov, 2017). Peat forms in mires are generally understood as wet places that arise due to incomplete decomposition of organic matter, where the water impedes proper oxidation of the remains. Despite their limited use in the fuel industry, Kim et al. (2014) explored an application as part of a blended fuel for commercial power plants due to the exceptional combustion efficiency of the mixture analyzed.

Then, a second stage in coal formation, or "coalification," is observed with lignite. This low-ranking matter is called "brown coal" due to its color and relatively low energy content, yet it is better regarded than peat due to being in a more advanced state in the coalification phase (Schweinfurth, 2009). As illustrated in Figure 3.23, the effects of heat and pressure over a certain period trigger different coalification phases for organic matter. Due to its low carbon content and heating value, lignite does not have significant uses other than electricity generation. However, some oil and gas players are exploring lignite gasification techniques to reduce the carbon footprint of the associated hydrogen generation process due to lignite's high sulfur and nitrogen content (Burmistrz et al., 2016).

Meanwhile, sub-bituminous coal is the next stage of coal in its coalification process. A black and dull matter, this coal has low to moderate heating values and is used mainly for electricity generation (USGS, 2023). Then, bituminous coal is a middle-ranked coal with high carbon and low sulfur content abundant in the United States, where it is also used as coking coal to produce steel (EIA, 2011). Safety concerns marked the early decades of bituminous coal exploration for miners and people living in coal towns, but with more advanced technologies, recovery is now safer and more efficient. Thus, current studies explore properties such as the pyrolysis potential or the capacity to be converted from biomass to biofuel through heating (Geng et al., 2016).

Meanwhile, anthracite is a high-ranking coal with the highest carbon content, between 86% and 97%, while boasting the lowest amount of volatile matter (EIA, 2011; Höök, 2012). Due to its low sulfur content, anthracite is one of the cleanest types of coal, mainly used as a producer of electricity and as a fuel in industrial processes. Some attempts are being made to utilize anthracite, peat, and bituminous coal as vital elements in direct carbon solid oxide fuel cells (DC-SOFC). These cells help enable a green pathway for brown coal yet remain experimental due to the negative effects of coal impurities (Wu et al., 2019a).

Lastly, graphite is a pure carbon mineral with a soft and slippery texture that turns into diamonds under tremendous pressure and temperature. Despite having limited use as a fuel source, graphite has a promising pathway in current oil and gas industry processes. Due to its versatile structure–property relationship, where many purification methods liberate a large amount of the rock flake to be exploited, graphite has found a comfortable spot in industrial processes like the production of lithium-ion batteries, fuel cells, graphene, pencils, fiber optics, electronics, electric vehicles, lubricants, and general electrical products, to name a few (Jara et al., 2019).

Countries with large reserves of graphite, like China, India, Brazil, Canada, and Russia, have a growing market to explore, highlighted by industries focusing on cleaner technologies, such as electric vehicles and fuel cells. The current trends that point to the energy transition process as the alternative to choose in the upcoming decades constitute a valuable opportunity for these countries, inviting them to continue to explore and recover coal, but doing it more cleanly and efficiently. Despite the promise, however, there continues to be a supply–demand mismatch and significant battery technology shortcomings that need to be tackled by innovation and future technologies (Jara et al., 2019).

The advent of unprecedented technological advancements, coupled with the environmental concerns raised by policymakers and international organizations, has led the industry to experiment with new ways to recover and use hydrocarbons and more in-depth exploration of the rocks and their properties. Regarding coal, it appears that the era of conventional exploration and recovery is mostly over, yet the need for coal resources will remain for many years. Because of this, significant efforts have been made to explore the potential for novel recovery methods such as gasification or for developing by-products such as nanomaterials or rare earth elements.

On the first point, the work of Hoang et al. (2018) explores the potential of coal-derived carbon nanomaterials such as nanotubes, carbon dots, graphene oxide, graphene quantum dots, and various carbon nanofibers. Applications of these materials have been performed in the energy, environmental, and biomedical sectors with great promise. Meanwhile, rare earth elements (REEs), a group of 15 lanthanide elements and 2 transition metal elements, are critical to the future green economy due to their role in constructing batteries and fuel cells. Coal reservoirs contain varying quantities of REEs, which has invited researchers to explore economically viable recovery techniques (Zhang et al., 2015).

As we can see, coal resources remain an essential part of the energy sets of many countries, and the need for coal worldwide will continue to be very significant, even amid the upcoming energy transition process. The many technological advancements around the exploration and recovery of hydrocarbons have led to safer practices and more efficient operations, while also lowering emissions through a better understanding of the physical properties of the rock in its different stages of coalification. In the following section, we review the various processing phases of coal and the latest technologies used to streamline them.

3.10 COAL PROCESSING: MINING, PROCESSING, AND TRANSPORTATION

Mining is the first step in the life cycle of a coal site. This process involves extracting coal from the ground using various mining techniques. Two main types of mining methods are used to extract coal: surface mining and underground mining. The former technique involves extracting the rock from open pits or sites where the coal is located fewer than 200 ft underground. Meanwhile, the latter adds a layer of complexity due to digging large holes in the ground to reach the coal. Also called deep mining, this extraction method requires an intricate system of tunnels and shafts that can extend for many miles underground (EIA, 2023).

Once the coal has been extracted from the ground, it must be processed and cleaned before being utilized. The process of coal preparation, also called coal washing, involves removing impurities such as ash, sulfur, dirt, and other materials that are not needed. In addition to cleaning and preparing the coal, this process increases its heating value. Then, coal is transported to its final destinations, such as power plants, where it is burned, and electricity is produced due to the steam resulting from the combustion process. Different ways of transporting coal include trains, ships, and barges for large distances, or conveyors, trucks, and trams for shorter distances (EIA, 2023).

These processes have been replicated and perfected over many generations. Still, the current environmental situation has pushed companies and governments to seek strategies to make these processes safer and cleaner while securing their efficiency and service to the end user. To achieve these goals, the literature has addressed new methods to tackle the shortcomings in the coal life cycle. For example, Burchart-Korol et al. (2016) developed a computational environmental life cycle assessment (LCA) model for coal operations that allows for incorporating variables like greenhouse gas emissions and other damage factors associated with fossil fuels, such as waste production and methane emissions.

The invaluable aid that computational methods have provided to the oil and gas industry cannot be overstated. From production and optimization models to recovery prediction based on historical records, digital technologies have revolutionized the field, assisting with techniques like neural networks powered by machine learning. The literature has developed frameworks and workflows to optimize practices using deep learning for coal processing. One example of this is the work of Yin et al. (2020), who systematized the coal preparation phase using stacked auto-encoder (SAE) methods to learn of the hidden features of the data, and bidirectional long short-term memory (BLTSM) techniques to predict the required coal quality metrics.

Similarly, Internet of Things (IoT) technologies can be applied to coal operations, developing strategies like centralizing processes and establishing clear environmental and recovery metrics milestones. In this sense, Wang et al. (2019) developed a framework to apply IoT to China's mining operations, both improving efficiency and prompting the conversion of many processes to green practices. As we see in Figure 3.24, China's Coal Industry 3.0 stage is characterized by automation and optimization of many mining processes across many operational levels, including drilling, seismic prospecting, and radar scanning information, as well as safety monitoring, personal monitoring, and other metrics that are fused through cloud computing mechanisms.

These frameworks applied to coal processing strategies are an ideal segue to develop what we know as Industry 4.0, or the digitizing of processes to centralize operations and follow global environmental guidelines more closely. The oil and gas sector is not the only one to follow big data analyses and promising machine learning techniques to perform economic assessments and prediction models.

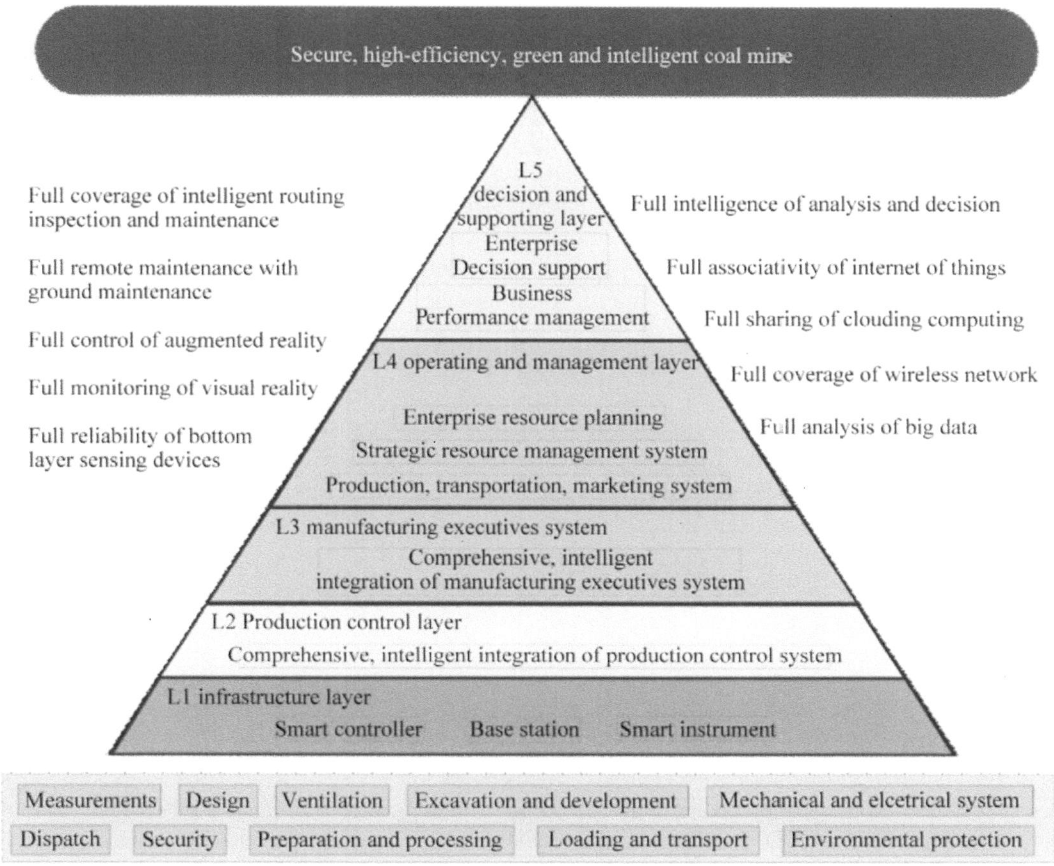

FIGURE 3.24 Internet-based multi-information fusion digital mining technology.

Source: Wang et al. (2019).

Still, it may be the industry with the most significant stake due to its pivotal role as an energy provider for billions of people. After reviewing the economics and best practices for coal resources, we touch on the scenarios shaping up for the sector amid the transition to renewable sources of energy.

3.11 ECONOMICS AND BEST PRACTICES FOR COAL RESOURCES

Historically, coal has provided cheap energy for millions of people, which has helped industrialization efforts and prosperity virtually worldwide. However, links between coal extraction and significant mining accidents, which led to greater fatalities in coal towns, were common throughout the first decades of coal recovery. Similarly, a positive correlation between living in cities with heavy coal combustion processes and various respiratory diseases has been found by some authors (Li et al., 2019b; Shekarian et al., 2021; Torres Rey et al., 2015). This, along with the global situation regarding climate change and the action taken by international players and governments, invites the coal industry to think about best practices to take moving forward.

The challenge appears when contrasting the strengths of coal resources to many countries regarding prosperity and energy independence with the shortcomings related to worker safety, public perception, and environmental impact that have stained the sector for decades. Thus, the notion that coal will remain a much-needed resource for the next few decades needs to be coupled with the concept that more actors are following the moves of processes that have created negative externalities affecting many people. This clash of concepts has spawned third-party actors to propose binding regulation that invites better research, practices, and output for all actors involved.

With that said, another consideration that oil companies must assess stakeholders is the economic viability of a project. Many optimization efforts have been made to prepare sites for optimal processes when a reservoir is operational, while also considering future site use once recovery stops being profitable. These efforts are undoubtedly commendable, but they may also hinder the ability of many actors to select and operate sites that would provide cheap energy for many people. Therefore, another challenge to the search for commonsense guidelines is the potentially too restrictive commitments, especially for small and medium energy producers, who may not afford to skip production on a site, as opposed to larger companies, who could take a loss on one site due to the profits made on many others.

Under these dynamics, not every actor will likely get what they want. However, all players have pledges to respect and general goals to achieve, so a significant degree of negotiation is essential to the new energy context. With that said, conditions in this setting are constantly changing and evolving as new technologies promise to rapidly break previously unconquered milestones. For example, many advances have been achieved in deep mining operations, where unmanned operations are a very realistic option in the short term, along with intelligent mining powered by artificial intelligence and predictive technologies (Ranjith et al., 2017).

All in all, planning, screening, operating, and ceasing operations are now essential parts of the set of parameters that ought to be considered when delineating commonsense practices for an economically viable and environmentally friendly coal project. Relatedly, a report from the Lawrence Livermore National Institute explores the best practices for the planning, approval, and operating of underground coal gasification (UCG) project, focusing on a few key areas, like the proper geological characterization of a site, the use of advanced simulation techniques used to provide insights into planning and operation, and technical issues, like the need to operate at a negative reactor pressure so that water is drawn into the reactor zone (Burton et al., 2019).

Another crucial part of improving industrial operations for coal resources is grasping each part of coal processing with intricate detail so that novel technological advancements can be applied seamlessly. In the next section, we review the different phases of coal processes that follow optimization strategies and deal with work done on the ground up to the point of utilization. These steps are summarized into three main ones: mining processes, processing schemes, and ultimate hydrocarbon transportation. Each stage has parameters for optimization and improvement and a tailored group of technologies that can move the industry forward to the next frontier.

3.12 POTENTIAL OF COAL IN AN ENERGY TRANSITION CONTEXT

In this chapter, we have reviewed the changing scenarios that fossil fuels have endured and the novel technologies that have opened the door to using computational methods and techniques that have allowed for a cleaner, safer industry. The avenues for the future of coal are myriad, as it appears that environmentally friendlier ways to recover coal have been achieved and have gone past the experimental stages. This helps coal establish itself as a potential alternative to sources like solar, wind, and nuclear once humanity transitions to cleaner energy sources.

Coal is a versatile hydrocarbon found in different states in nature, satisfying different needs. Thus, coal is limited to electricity generation and industrial processes like steel production and liquid fuels like diesel and gasoline. The widespread availability of coal, with many countries having significant reservoirs, has allowed for the development of various coal industries worldwide, which has helped these nations achieve a significant degree of energy independence. That is an important thing to have in a world marked by uncertainty and conflict.

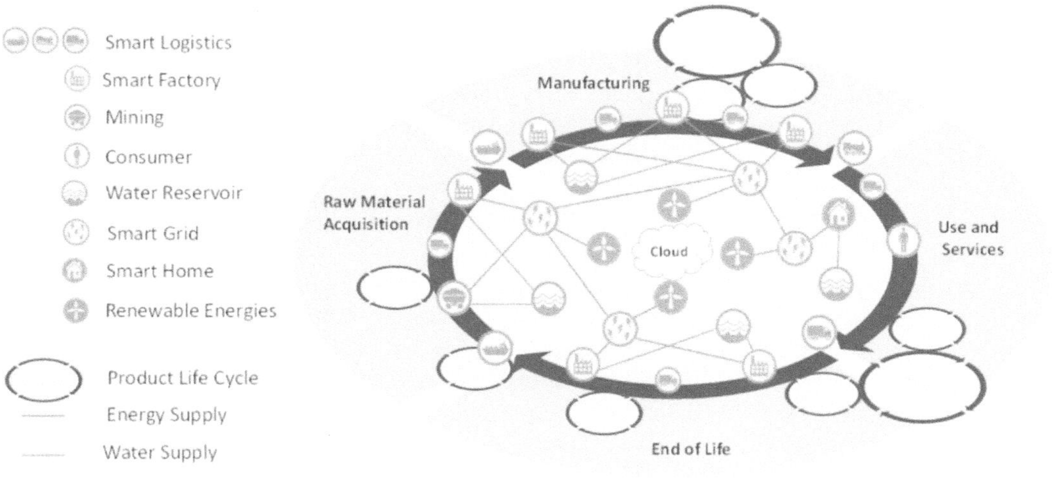

FIGURE 3.25 Macro perspective of Industry 4.0.

Source: Stock and Seliger (2016).

The current energy scenario is also one of compromises, commitments, and environmental concerns. Under changing dynamics, the role of coal has been scrutinized by environmentalists, governments, and international organizations. Historically, the coal industry has provided cheap energy and, consequently, greater prosperity to many people. However, the other side of the same coin shows that coal is a polluting element that can also have very unsafe practices for workers and communities while contributing to the warming of the climate. Today's world faces the coal sector with the challenge to improve its practices and perfect its techniques so it can continue to make people thrive.

As reviewed in this chapter, many promising technologies have helped mitigate the repercussions of coal operations observed in previous decades. From coal gasification to carbon capture and storage (CCS) projects that sequester CO_2, coal in sites can now be explored and recovered without any people going down a shaft, which increases worker safety. Similarly, injecting a catalyst down a coal reservoir can provide cleaner recovery efforts that limit the release of carbon dioxide and other dangerous pollutants into the atmosphere.

Meanwhile, the new dawn of smart technologies and advancements like cloud computing and the Internet of Things has taken us to the cusp of Industry 4.0, a concept that imitates the novelty of the first Industrial Revolution but is tailored to current seismic changes, like the energy transition to renewables. This massive undertaking can be accomplished by using the novel technologies that we have reviewed in this text, which allow for a greater variety of energy sources powering the world. Figure 3.25 provides a schematic of a macro perspective of Industry 4.0, where industrial sectors work together, big data helps tackle shortcomings, and the cloud keeps all systems synchronized and running.

The potential for a qualitative leap in how we see energy is coming together in front of our eyes very quickly. The environmental concerns driving the conversation about energy provide us with an opportunity to explore the energy transition to renewable sources, and the advent of new information technologies can help us achieve the goal of gradually phasing out fossil fuels in the next few decades. The challenge is on, but conventional sources of energy continue to be heavily relied upon in some parts of the world, where novel methods have helped develop more efficient and cleaner ways of recovery. More technologically advanced oil, gas, and coal processes have proven their worth in the medium term.

3.13 CONCLUSIONS

Conventional energy sources have contributed to lifting millions of people from poverty and starvation by constantly providing cheap energy that allows them to live their lives to their fullest potential. However, the many challenges and potential environmental dangers stemming from the indiscriminate use of fossil fuels must be addressed sooner rather than later. In this chapter, we explored the many balances and trade-offs required to continue providing the cheap energy that millions need with the environmental concern that these same processes help exacerbate.

Positive developments are seen in the many significant advances made in recent years to mitigate the environmental impacts caused by greenhouse emissions stemming from oil, gas, and coal operations. For example, methods increasing the efficiency and lowering the pollution of those procedures have been mainly a consequence of relying on computational models aided by deep learning and artificial intelligence (AI). Similarly, current site selection optimization efforts result from assessing more variables in the analysis and centralizing information in the cloud, where employees can easily and quickly access it.

Then, carbon capture and storage opportunities have opened up for depleted sites, where CO_2 from flue gases can be safely sequestered underground for decades. Likewise, *in situ* recovery techniques have shown to be cleaner and more efficient, and hybrid prediction models have shown promising results. Improved methods to repurpose abandoned fossil fuel plants have also been advanced with the help of tools like neural networks. At the same time, safer decommission strategies for old plants have been proposed when including a multi-criteria approach.

To be sure, conventional sources of energy, like coal, oil, and natural gas, will remain in the energy dockets of many countries, whether by inertia, lack of funding of research for alternative sources, or the simplicity of an already well-traveled energy generation route. However, the potential environmental dangers associated with fossil fuel operations are a direct invitation to rethink how energy serves our lives and how it will serve future generations. The energy transition to renewables can be an indelible legacy from us to them.

REFERENCES

Abas, N., A. Kalair, and N. Khan. 2015. Review of fossil fuels and future energy technologies. *Futures* 69: 31–49. doi:10.1016/j.futures.2015.03.003.

Abudu, D., and M. Williams. 2015. GIS-based optimal route selection for oil and gas pipelines in Uganda. *Advances in Computer Science: An International Journal* 4(4): 93–104.

Afari, S., K. Ling, B. Sennaoui, D. Maxey, T. Oguntade, and J. Porlles. 2022. Optimization of CO2 huff-n-puff EOR in the Bakken formation using numerical simulation and response surface methodology. *Journal of Petroleum Science and Engineering* 215 (Part A): 110552. doi:10.1016/j.petrol.2022.110552.

Al-Fatlawi, O., M. Hossain, N. Patel, and A. Kabir. 2019. Evaluation of the potentials for adapting the multi-stage hydraulic fracturing technology in tight carbonate reservoir. *SPE Middle East Oil and Gas Show and Conference*. doi:10.2118/194733-ms.

Altowilib, A., A. AlSaihati, H. Alhamood, S. Alafnan, and S. Alarifi. 2020. Reserves estimation for coalbed methane reservoirs: A review. *Sustainability* 12(24): 10621. doi:10.3390/su122410621.

Anderson, T. I., B. Vega, and A. R. Kovscek. 2020. Multimodal imaging and machine learning to enhance microscope images of shale. *Computers & Geosciences* 145: 104593. doi:10.1016/j.cageo.2020.104593.

Bach, M. 2019. The oil and gas sector: from climate laggard to climate leader? *Environmental Politics* 28(1): 87–103. doi:10.1080/09644016.2019.1521911.

Baz, K., J. Cheng, D. Xu, K. Abbas, I. Ali, H. Ali, and C. Fang. 2021. Asymmetric impact of fossil fuel and renewable energy consumption on economic growth: A nonlinear technique. *Energy* 226: 120357. doi:10.1016/j.energy.2021.120357.

BDO. 2018. *How industry is transforming the oil & gas supply chain*. Available in www.bdo.com/insights/industries/natural-resources/how-industry-4-0-is-transforming-the-oil-gas-supply-chain [accessed February 16th, 2023].

Bildirici, M. E., and T. Bakirtas. 2014. The relationship among oil, natural gas and coal consumption and economic growth in BRICTS (Brazil, Russian, India, China, Turkey and South Africa) countries. *Energy* 65: 134–144. doi:10.1016/j.energy.2013.12.006.

Braungardt, S., J. van den Bergh, and T. Dunlop. 2019. Fossil fuel divestment and climate change: Reviewing contested arguments. *Energy Research & Social Science* 50: 191–200. doi:10.1016/j.erss.2018.12.004.

Bret-Rouzaut, N. 2022. Economics of oil and gas production. In *The Palgrave handbook of international energy economics* (pp. 3–22). Cham: Springer International Publishing. doi:10.1007/978-3-030-86884-0.

Burchart-Korol, D., A. Fugiel, K. Czaplicka-Kolarz, and M. Turek. 2016. Model of environmental life cycle assessment for coal mining operations. *Science of the Total Environment* 562: 61–72. doi:10.1016/j.scitotenv.2016.03.202.

Burmistrz, P., T. Chmielniak, L. Czepirski, and M. Gazda-Grzywacz. 2016. Carbon footprint of the hydrogen production process utilizing subbituminous coal and lignite gasification. *Journal of Cleaner Production* 139: 858–865. doi:10.1016/j.jclepro.2016.08.112.

Burton, E., J. Friedmann, and R. Upadhye. 2019. *Best practices in underground coal gasification* (LLNL-TR-225331). Livermore, CA: Lawrence Livermore National Laboratory.

Chandra, D., and V. Vishal. 2021. A critical review on pore to continuum scale imaging techniques for enhanced shale gas recovery. *Earth-Science Reviews* 217: 103638. doi:10.1016/j.earscirev.2021.103638.

Chen, X., Z. Zhang, Q. Liu, L. Sun, H. Xiao, J. Gao, and H. Kang. 2023. Huff and puff by flue gas for tight oil recovery to achieve sustainable energy future. *Strategic Planning for Energy and the Environment* 42(2): 353–374. doi:10.13052/spee1048-5236.4225.

Chopra, S., L. R. Lines, D. R. Schmitt, and M. L. Batzle (eds.). 2010. *Heavy oils: Reservoir characterization and production monitoring*. Tulsa, OK: Society of Exploration Geophysicists.

Chudy-Laskowska, K., and T. Pisula. 2022. An analysis of the use of energy from conventional fossil fuels and green renewable energy in the context of the European Union's planned energy transformation. *Energies* 15: 7369. doi:10.3390/en15197369.

Davarazar, M., B. Nemati, M. Gorgich, S. Maheronnaghsh, S. A. Bayat Ghiasi, S. Zandi, and M. Mohammadi. 2020. A sustainable approach for the site selection of CO2 underground storage: Application of fuzzy-Delphi methodology. *Journal of Settlements and Spatial Planning*, Special Issue 6: 113–123. doi:10.24193/JSSPSI.2020.6.11.

Devold, H. 2009. *Oil and gas production handbook: An introduction to oil and gas production*. OSLO: ABB Oil and Gas.

Department of Energy. 2012. *Coalbed methane*. Available in www.energy.gov/sites/default/files/coalbed_methane_factcard.pdf [accessed April 27th, 2023].

Department of Energy. 2013. *How is shale gas produced?* Available in www.energy.gov/sites/prod/files/2013/04/f0/how_is_shale_gas_produced.pdf [accessed April 20th, 2023].

Department of Energy. 2023. *Fossil*. Available in www.energy.gov/fossil [accessed April 22nd, 2023].

Ding, M., Y. Wang, D. Liu, X. Wang, H. Zhao, and W. Chen. 2021. Enhancing tight oil recovery using CO2 huff and puff injection: An experimental study of the influencing factors. *Journal of Natural Gas Science and Engineering* 90: 103931.

Du, F., and B. Nojabaei. 2019. A review of gas injection in shale reservoirs: Enhanced oil/gas recovery approaches and greenhouse gas control. *Energies* 12: 2355. doi:10.3390/en12122355.

EIA. 2011. *Subbituminous and bituminous coal dominate U.S. coal production*. Available in www.eia.gov/todayinenergy/detail.php?id=2670 [accessed May 13th, 2023].

EIA. 2022. *Coal explained*. Available in www.eia.gov/energyexplained/coal/ [accessed May 2nd, 2023].

EIA. 2023. *Coal explained: Mining and transportation of coal*. Available in www.eia.gov/energyexplained/coal/mining-and-transportation.php [accessed May 5th, 2023].

Elturki, M., and A. Imqam. 2020. Application of enhanced oil recovery methods in unconventional reservoirs: A review and data analysis. In *ARMA US Rock mechanics/geomechanics symposium* (pp. ARMA–2020). ARMA.

European Commission. 2021. 2050 long-term strategy. *Climate Action*. Available in https://climate.ec.europa.eu/eu-action/climate-strategies-targets/2050-long-term-strategy_en [accessed October 4th, 2024].

Fowler, A. M., P. I. Macreadie, D. O. B. Jones, and D. J. Booth. 2014. A multi-criteria decision approach to decommissioning of offshore oil and gas infrastructure. *Ocean & Coastal Management* 87: 20–29. doi:10.1016/j.ocecoaman.2013.10.019.

Geng, C., S. Li, C. Yue, and Y. Ma. 2016. Pyrolysis characteristics of bituminous coal. *Journal of the Energy Institute* 89(4): 725–730. doi:10.1016/j.joei.2015.04.004.

Ghaithan, A. M., A. Attia, and S. O. Duffuaa. 2017. Multi-objective optimization model for a downstream oil and gas supply chain. *Applied Mathematical Modelling* 52: 689–708. doi:10.1016/j.apm.2017.08.007.

Green, J., J. Hadden, T. Hale, and P. Mahdavi. 2022. Transition, hedge, or resist? Understanding political and economic behavior toward decarbonization in the oil and gas industry. *Review of International Political Economy* 29(6): 2036–2063.

Guo, K., H. Li, and Z. Yu. 2016. In-situ heavy and extra-heavy oil recovery: A review. *Fuel* 185: 886–902. doi:10.1016/j.fuel.2016.08.047.

Han, S., H. Chen, R. Long, and X. Cui. 2018. Peak coal in China: A literature review. *Resources, Conservation and Recycling* 129: 293–306. doi:10.1016/j.resconrec.2016.08.012.

Hassanpouryouzband, A., E. Joonaki, M. V. Farahani, S. Takeya, C. Ruppel, J. Yang, . . ., and B. Tohidi. 2020. Gas hydrates in sustainable chemistry. *Chemical Society Reviews* 49(15): 5225–5309. doi:10.1039/c8cs00989a.

Hoang, V. C., M. Hassan, and V. G. Gomes. 2018. Coal derived carbon nanomaterials – recent advances in synthesis and applications. *Applied Materials Today* 12: 342–358. doi:10.1016/j.apmt.2018.06.007

Höök, M. 2012. Coal and peat: Global resources and future supply. *Fossil Energy*: 311–341. doi:10.1007/978-1-4614-5722-0_9.

Huang, W. G., S. W. Zhang, G. Z. Wang, J. Huang, X. Lu, S. L. Wu, and Z. T. Wang. 2023. Modeling methodology for site selection evaluation of underground coal gasification based on combination weighting method with game theory. *ACS Omega* 8(12): 11544–11555.

IEA. 2021. *The case for regulating downstream methane emissions from oil and gas*. Available in www.iea.org/commentaries/the-case-for-regulating-downstream-methane-emissions-from-oil-and-gas [accessed April 20th, 2023].

IEA. 2022. *World energy outlook 2022*. Paris: IEA. Available in www.iea.org/reports/world-energy-outlook-2022 [accessed February 20th, 2023].

Imran, M., D. Kumar, N. Kumar, A. Qayyum, A. Saeed, and M. S. Bhatti. 2014. Environmental concerns of underground coal gasification. *Renewable and Sustainable Energy Reviews* 31: 600–610. doi:10.1016/j.rser.2013.12.024.

Jara, A. D., A. Betemariam, G. Woldetinsae, and J. Y. Kim. 2019. Purification, application and current market trend of natural graphite: A review. *International Journal of Mining Science and Technology* 29(5): 671–689. doi:10.1016/j.ijmst.2019.04.003.

Jia, Q., D. Liu, Y. Cai, X. Fang, and L. Li. 2020. Petrophysics characteristics of coalbed methane reservoir: A comprehensive review. *Frontiers of Earth Science*. doi:10.1007/s11707-020-0833-1.

Jindal, A., and G. Shrimali. 2022. Cost-benefit analysis of coal plant repurposing in developing countries: A case study of India. *Energy Policy* 164: 112911. doi:10.1016/j.enpol.2022.112911.

Kalair, A., N. Abas, M. S. Saleem, A. R. Kalair, and N. Khan. 2020. Role of energy storage systems in energy transition from fossil fuels to renewables. *Energy Storage* 3(1): 1–27. doi:10.1002/est2.135.

Kassem, M. A., M. A. Khoiry, and N. Hamzah. 2019. Risk factors in oil and gas construction projects in developing countries: A case study. *International Journal of Energy Sector Management* 13(4): 846–861. doi:10.1108/ijesm-11-2018-0002.

Kilian, L. 2016. *The impact of the shale oil revolution on U.S. oil and gasoline prices* (Working Paper N°5723). Munich: Center for Economic Studies and IFO Institute (CESifo).

Kim, J. K., H. D. Lee, H. S. Kim, H. Y. Park, and S. C. Kim. 2014. Combustion possibility of low rank Russian peat as a blended fuel of pulverized coal fired power plant. *Journal of Industrial and Engineering Chemistry* 20(4): 1752–1760. doi:10.1016/j.jiec.2013.08.027.

Lee, J., C. Sorensen, J. Lemery, C. F. Workman, H. Lindstadt, and M. D. Bazilian. 2022. Managing upstream oil and gas emissions: A public health oriented approach. *Journal of Environmental Management* 310: 114766. doi:10.1016/j.jenvman.2022.114766.

Leporini, M., B. Marchetti, F. Corvaro, and F. Polonara. 2019. Reconversion of offshore oil and gas platforms into renewable energy sites production: Assessment of different scenarios. *Renewable Energy* 135: 1121–1132. doi:10.1016/j.renene.2018.12.073.

Li, F., Q. Yuan, T. Li, Z. Li, C. Sun, and G. Chen. 2019a. A review: Enhanced recovery of natural gas hydrate reservoirs. *Chinese Journal of Chemical Engineering* 27(9): 2062–2073. doi:10.1016/j.cjche.2018.11.007.

Li, J., J. Ran, L.-C. Chen, M. Costa, Y. Huang, X. Chen, and L. Tian. 2019b. Bituminous coal combustion and Xuan Wei lung cancer: A review of the epidemiology, intervention, carcinogens, and carcinogenesis. *Archives of Toxicology* 93: 573–583.

Li, S., B. Li, Q. Zhang, Z. Li, and D. Yang. 2018. Effect of CO2 on heavy oil recovery and physical properties in huff-n-puff processes under reservoir conditions. *Journal of Energy Resources Technology* 140(7): 072907. doi:10.1115/1.4039325.

Liang, W., J. Yan, B. Zhang, and D. Hou. 2021. Review on coal bed methane recovery theory and technology: Recent progress and perspectives. *Energy & Fuels* 35(6): 4633–4643. doi:10.1021/acs.energyfuels.0c04026.

Library of Congress. 2022. *Oil and gas industry: A research guide*. Available in https://guides.loc.gov/oil-and-gas-industry [accessed May 14th, 2023].

Liu, L., B. Ryu, Z. Sun, N. Wu, H. Cao, W. Geng, . . ., and L. Wang. 2019. Monitoring and research on environmental impacts related to marine natural gas hydrates: Review and future perspective. *Journal of Natural Gas Science and Engineering* 65: 82–107. doi:10.1016/j.jngse.2019.02.007.

Litvinenko, V. 2020. The role of hydrocarbons in the global energy agenda: The focus on liquefied natural gas. *Resources* 9: 59. doi:10.3390/resources9050059.

Lyu, Q., J. Tan, L. Li, Y. Ju, A. Busch, D. A. Wood, . . ., and R. Hu. 2021. The role of supercritical carbon dioxide for recovery of shale gas and sequestration in gas shale reservoirs. *Energy & Environmental Science* 14(8): 4203–4227. doi:10.1039/d0ee03648j.

Ma, H., Y. Yang, and Z. Chen. 2022. Numerical simulation of bitumen recovery via supercritical water injection with in-situ upgrading. *Fuel* 313: 122708. doi:10.1016/j.fuel.2021.122708.

Mahzari, P., T. M. Mitchell, A. P. Jones, E. H. Oelkers, A. Striolo, F. Iacoviello, . . ., and J. E. Juri. 2021. Novel laboratory investigation of huff-n-puff gas injection for shale oils under realistic reservoir conditions. *Fuel* 284: 118950. doi:10.1016/j.fuel.2020.118950.

McKinsey & Company. 2022. *Global energy perspective 2022*. Available in www.mckinsey.com/industries/oil-and-gas/our-insights/global-energy-perspective-2022 [accessed February 14th, 2023].

Mohamed, T., and M. Mehana. 2020. CoalBed methane characterization and modeling: Review and outlook. *Energy Sources, Part A: Recovery, Utilization, and Environmental Effects*: 1–23. doi:10.1080/15567036.2020.1845877.

Mohammadpoor, M., and F. Torabi. 2020. Big Data analytics in oil and gas industry: An emerging trend. *Petroleum* 6(4): 321–328. doi:10.1016/j.petlm.2018.11.001.

Mohanty, D. 2017. An overview of the geological controls in underground coal gasification. *IOP Conference Series: Earth and Environmental Science* 76: 012010. doi:10.1088/1755-1315/76/1/012010.

Mojarad, A. A. S., V. Atashbari, and A. Tantau. 2018. Challenges for sustainable development strategies in oil and gas industries. *Proceedings of the International Conference on Business Excellence* 12(1): 626–638. doi:10.2478/picbe-2018-0056.

Mu, Y., Y. Fan, J. Wang, and N. Fan. 2019. Numerical study on injection of flue gas as a heat carrier into coal reservoir to enhance CBM recovery. *Journal of Natural Gas Science and Engineering* 72: 103017.

Mukherjee, M., and S. Misra. 2018. A review of experimental research on enhanced coal bed methane (ECBM) recovery via CO2 sequestration. *Earth-Science Reviews* 179: 392–410. doi:10.1016/j.earscirev.2018.02.018.

Nguyen, T. V., M. Voldsund, P. Breuhaus, and B. Elmegaard. 2016. Energy efficiency measures for offshore oil and gas platforms. *Energy* 117: 325–340.

OGCI. 2022. *Leadership with impact: Annual progress report from the oil and gas climate initiative*. Available in https://info.ogci.com/hubfs/OGCI_ProgressReport_2022.pdf [accessed May 4th, 2023].

Omitaomu, O. A., B. R. Blevins, W. C. Jochem, G. T. Mays, R. Belles, S. W. Hadley, . . ., and A. N. Rose. 2012. Adapting a GIS-based multicriteria decision analysis approach for evaluating new power generating sites. *Applied Energy* 96: 292–301. doi:10.1016/j.apenergy.2011.11.087.

Ore, O. T., and F. M. Adebiyi. 2021. A review on current trends and prospects in the pyrolysis of heavy oils. *Journal of Petroleum Exploration and Production* 11: 1521–1530. doi:10.1007/s13202-021-01099-0.

Pan, Z., J. Ye, F. Zhou, Y. Tan, L. D. Connell, and J. Fan. 2017. CO2 storage in coal to enhance coalbed methane recovery: A review of field experiments in China. *International Geology Review* 60(5–6): 754–776. doi:10.1080/00206814.2017.1373607.

Patel, H., D. Prajapati, D. Mahida, and M. Shah. 2020. Transforming petroleum downstream sector through big data: A holistic review. *Journal of Petroleum Exploration and Production Technology* 10: 2601–2611. doi:10.1007/s13202-020-00889-2.

Pei, P., J. Nasah, J. Solc, S. F. Korom, D. Laudal, and K. Barse. 2016. Investigation of the feasibility of underground coal gasification in North Dakota, United States. *Energy Conversion and Management* 113: 95–103. doi:10.1016/j.enconman.2016.01.053.

Popoola, L.T., A. S. Grema, G. K. Latinwo, B. Gutti, and A. S. Balogun. 2013. Corrosion problems during oil and gas production and its mitigation. *International Journal of Industrial Chemistry* 4: 35. doi:10.1186/2228-5547-4-35.

Ranjith, P. G., J. Zhao, M. Ju, R. V. S. De Silva, T. D. Rathnaweera, and A. K. Bandara. 2017. Opportunities and challenges in deep mining: A brief review. *Engineering* 3(4): 546–551.

Reed, A., S. Ericson, M. Bazilian, J. Logan, K. Doran, and C. Nelden. 2019. Interrogating uncertainty in energy forecasts: The case of the shale gas boom. *Energy Transitions* 3: 1–11. doi:10.1007/s41825-019-00015-9.

Ruppel, C.D. 2018. Gas hydrate in nature: U.S. geological survey. *Fact Sheet 2017–3080*. doi:10.3133/fs20173080.

Santibañez-Aguilar, J. E., J. B. González-Campos, J. M. Ponce-Ortega, M. Serna-González, and M. El-Halwagi. 2014. Optimal planning and site selection for distributed multiproduct biorefineries involving economic, environmental and social objectives. *Journal of Cleaner Production* 65: 270–294. doi:10.1016/j.jclepro.2013.08.004.

Schweinfurth, S. P. 2009. *An introduction to coal quality* (U.S. Geological Survey Professional Paper 1625-F). Reston: U.S. Geological Survey.

Shafi, H., K. Velswamy, F. Ibrahim, and B. Huang. 2020. A hierarchical constrained reinforcement learning for optimization of bitumen recovery rate in a primary separation vessel. *Computers & Chemical Engineering* 140: 106939. doi:10.1016/j.compchemeng.2020.106939.

Shafiee, M., I. Animah, B. Alkali, and D. Baglee. 2018. Decision support methods and applications in the upstream oil and gas sector. *Journal of Petroleum Science and Engineering*. doi:10.1016/j.petrol.2018.10.050.

Shahriar, A., R. Sadiq, and S. Tesfamariam. 2012. Risk analysis for oil & gas pipelines: A sustainability assessment approach using fuzzy based bow-tie analysis. *Journal of Loss Prevention in the Process Industries* 25(3): 505–523.

Shekarian, Y., E. Rahimi, N. Shekarian, M. Rezaee, and P. Roghanchi. 2021. An analysis of contributing mining factors in coal workers' pneumoconiosis prevalence in the United States coal mines, 1986–2018. *International Journal of Coal Science & Technology* 8: 1227–1237. doi:10.1007/s40789-021-00464-y.

Shi, J. Q., and S. Durucan. 2005. A numerical simulation study of the Allison Unit CO2-ECBM pilot: The impact of matrix shrinkage and swelling on ECBM production and CO2 injectivity. *Proceedings of the Greenhouse Gas Control Technologies* 7: 431–439.

Škvareková, E., M. Tomašková, G. Wittenberger, and Š. Zelenák. 2019. Analysis of risk factors for underground coal gasification. *Management Systems in Production Engineering* 27(4): 227–235. doi:10.1515/mspe-2019-0036.

Stock, T., and G. Seliger. 2016. Opportunities of sustainable manufacturing in industry 4.0. *Procedia CIRP* 40: 536–541.

Tang, D., and K. Spikes. 2017. Segmentation of shale SEM images using machine learning. *SEG Technical Program Expanded Abstracts*. doi:10.1190/segam2017-17738502.1.

Tao, S., S. Chen, and Z. Pan. 2019. Current status, challenges, and policy suggestions for coalbed methane industry development in China: A review. *Energy Science & Engineering* 7(4): 1059–1074. doi:10.1002/ese3.358.

Tariq, Z., M. S. Aljawad, A. Hasan, M. Murtaza, E. Mohammed, A. El-Husseiny, . . ., and A. Abdulraheem. 2021. A systematic review of data science and machine learning applications to the oil and gas industry. *Journal of Petroleum Exploration and Production Technology* 11: 4339–4374. doi:10.1007/s13202-021-01302-2.

Taylor, S. E. 2018. Interfacial chemistry in steam-based thermal recovery of oil sands bitumen with emphasis on steam-assisted gravity drainage and the role of chemical additives. *Colloids and Interfaces* 2(2): 16. doi:10.3390/colloids2020016.

Tcvetkov, P. S. 2017. The history, present status and future prospects of the Russian fuel peat industry. *Mires and Peat* 19(14): 1–12. doi:10.19189/MaP.2016.OMB.256.

Tong, X., G. Zhang, W. Zhaoming, W. Zhixin, T. Zuoji, W. Hongjun, M. Feng, and W. Yiping. 2018. Distribution and potential of global oil and gas resources. *Petroleum Exploration and Development* 45(4): 779–789.

Torres Rey, C. H., M. Ibañez Pinilla, L. Briceño Ayala, D. M. Checa Guerrero, G. Morgan Torres, H. Groot de Restrepo, and M. Varona Uribe. 2015. Underground coal mining: relationship between coal dust levels and pneumoconiosis, in two regions of Colombia, 2014. *BioMed Research International* 2015. doi:10.1155/2015/647878.

USGS. 2023. *What are the types of coal?* Available in www.usgs.gov/faqs/what-are-types-coal [accessed May 6th, 2023].

Wang, G., Y. Hu, and H. Ren. 2019. Intelligent and ecological coal mining as well as clean utilization technology in China: Review and prospects. *International Journal of Mining Science and Technology* 29: 161–169. doi:10.1016/j.ijmst.2018.06.005.

Wang, H., F. Ma, X. Tong, Z. Liu, X. Zhang, . . ., and L. Wang. 2016. Assessment of global unconventional oil and gas resources. *Petroleum Exploration and Development* 43(6): 925–940. doi:10.1016/S1876-3804(16)30111-2.

Wang, Y., X. Lang, S. Fan, S. Wang, C. Yu, and G. Li. 2021. Review on enhanced technology of natural gas hydrate recovery by carbon dioxide replacement. *Energy & Fuels* 35(5): 3659–3674. doi:10.1021/acs.energyfuels.0c04138.

Welsby, D., J. Price, S. Pye, and P. Ekins. 2021. Unextractable fossil fuels in a 1.5° C world. *Nature* 597(7875): 230–234.

Wu, H., J. Xiao, X. Zeng, X. Li, J. Yang, Y. Zou, . . ., and J. Liu. 2019a. A high performance direct carbon solid oxide fuel cell – a green pathway for brown coal utilization. *Applied Energy* 248: 679–687. doi:10.1016/j.apenergy.2019.04.104.

Wu, Y., S. Misra, C. Sondergeld, M. Curtis, and J. Jernigen. 2019b. Machine learning for locating organic matter and pores in scanning electron microscopy images of organic-rich shales. *Fuel* 253: 662–676. doi:10.1016/j.fuel.2019.05.017.

Yin, X., Z. Niu, Z. He, Z. Li, and D.-H. Lee. 2020. Ensemble deep learning based semi-supervised soft sensor modeling method and its application on quality prediction for coal preparation process. *Advanced Engineering Informatics* 46: 101136. doi:10.1016/j.aei.2020.101136.

Zhang, G., H. Qu, G. Chen, C. Zhao, F. Zhang, . . ., and M. Ma. 2019. Giant discoveries of oil and gas fields in global deepwaters in the past 40 years and the prospect of exploration. *Journal of Natural Gas Geoscience* 4(1): 1–28. doi:10.1016/j.jnggs.2019.03.002.

Zhang, J., L. Si, J. Chen, M. Kizil, C. Wang, and Z. Chen. 2020. Stimulation techniques of coalbed methane reservoirs. *Geofluids* 2020: 1–23. doi:10.1155/2020/5152646.

Zhang, W., M. Rezaee, A. Bhagavatula, Y. Li, J. Groppo, and R. Honaker. 2015. A review of the occurrence and promising recovery methods of rare earth elements from coal and coal by-products. *International Journal of Coal Preparation and Utilization* 35(6): 295–330. doi:10.1080/19392699.2015.1033097.

Zhang, Y., M. Cui, D. Li, and G. Xin. 2022. Microscopic insights and optimization of the CH4–CO2 replacement in natural gas hydrates. *ACS Omega* 7: 47239–47250. doi:10.1021/acsomega.2c06502.

Zhao, F., Y. Liu, N. Lu, T. Xu, G. Zhu, and K. Wang. 2021. A review on upgrading and viscosity reduction of heavy oil and bitumen by underground catalytic cracking. *Energy Reports* 7: 4249–4272. doi:10.1016/j.egyr.2021.06.094.

Zhao, J., and D. Guo. 2018. Cracking mechanism of coal under high-pressure water jet and its applications for enhanced coalbed methane drainage. *Arabian Journal of Geosciences* 11(15): 427. doi:10.1007/s12517-018-3783-0.

Zhou, X., Q. Yuan, X. Peng, F. Zeng, and L. Zhang. 2018. A critical review of the CO2 huff 'n' puff process for enhanced heavy oil recovery. *Fuel* 215: 813–824. doi:10.1016/j.fuel.2017.11.092.

4 Energy Efficiency

Jorge Saldaña, Mesut Yurukcu, Narendra Boppana, Sepehr Arbabi, Jim Henry, and Sebahattin Ziyanak

4.1 INTRODUCTION

In today's world, development needs for energy efficiency and effective energy management have become an increasingly important item with a great sense of urgency. The energy industry plays a key role in our society, powering everything from businesses to homes (Benedek et al., 2023). However, it is also a contributor to greenhouse gas emissions, which has drawn attention from policymakers and environmental advocates alike. Although renewable energy technologies have shown promise, the energy landscape remains intricate. Energy efficiency and effective energy management require solutions that consider both demand and supply (Hussain et al., 2023). This session delves into the aspects of this scenario within this context. As a starting point, energy efficiency and discussing measures such as the energy use index (Wu et al., 2020) and energy intensity (Ang & Goh, 2018) as optimization methodologies for systems and designs (Zhang et al., 2016) are introduced. Next, the potential of smart grids and energy management (Fakhar et al., 2023) is examined by exploring its significance, applications, and various tools, like EMS, BMS, SCADA automation, and PLC.

Furthermore, we investigate demand-side and supply-side management strategies, from load sharing and balancing to supply chain and infrastructure considerations. Amidst these intricate components, the chapter identifies the challenges and constraints that must be navigated for effective implementation. We also delve into the crucial but often overlooked aspects of coordination between energy efficiency and demand response, including the enabling technologies and regulatory frameworks that can facilitate or hinder such coordination.

Therefore, the chapter comprehensively examines the intricacies of energy efficiency and smart grid management, aiming to offer stakeholders actionable insights to achieve a sustainable and energy-efficient future.

4.2 ENERGY EFFICIENCY

The energy use index (EUI) is a pivotal metric that provides a standardized measure of energy consumption in buildings or industrial operations (Thompson, 2023). Calculated as the total energy consumed per square foot (or per square meter) over a specific period, EUI facilitates comparing energy performance across different facilities or sectors (Thompson, 2023). In the oil and gas industry, the EUI is particularly useful for benchmarking the energy performance of different extraction or refining processes, thereby identifying areas that require optimization. Lower EUI values generally indicate better energy performance, while higher values may signal inefficiencies, requiring further investigation (Kong et al., 2023).

Energy intensity, on the other hand, is a broader metric that measures the energy required to produce a unit of output, whether it be a barrel of oil, a megajoule of natural gas, or a unit of economic value, such as gross domestic product (GDP) (Oppen et al., 2023). Figure 4.1 shows an overview of the economic and environmental impacts of sustainability assessment.

Unlike the EUI, which is often specific to a particular type of facility or operation, energy intensity can be applied more universally across different sectors, or even countries. Reducing energy intensity can have profound economic and environmental benefits in the oil and gas industry (Chen

 DOI: 10.1201/9781003282273-4

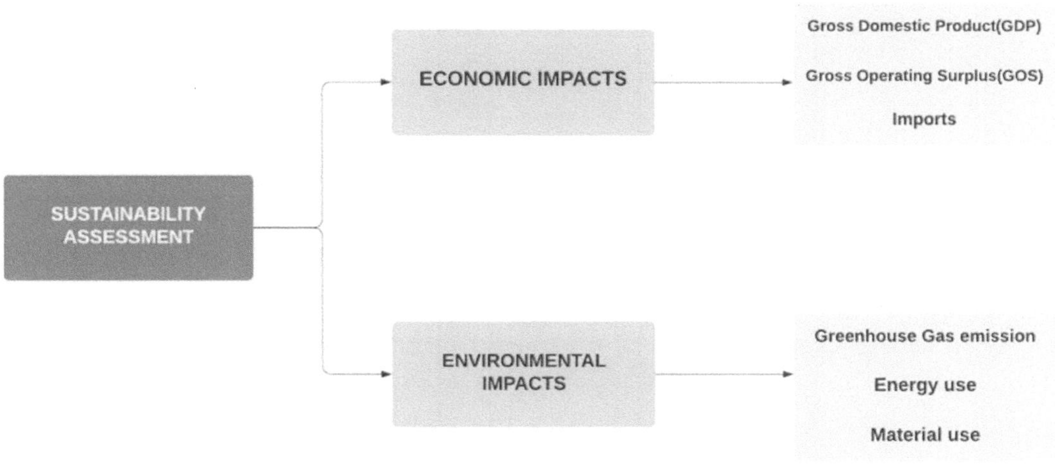

FIGURE 4.1 Overview of the economic and environmental impacts of sustainability assessment.

Source: Oppen et al. (2023).

et al., 2023). For instance, lower energy intensity can reduce operational costs and have a smaller carbon footprint, contributing to sustainability goals and bottom-line performance (Jamil et al., 2023).

The energy use index and energy intensity are invaluable tools for gauging energy consumption efficiency, but they are best utilized in conjunction. For example, while EUI might provide insights into the efficiency of individual processes or facilities, energy intensity can offer a macroscopic view of performance across an entire organization or sector (Guy et al., 2023). By employing these metrics in a complementary fashion, stakeholders in the oil and gas industry can develop a more holistic understanding of energy efficiency, thereby informing more effective strategies for optimization (Gonzalez-Torres et al., 2023). By meticulously tracking these metrics and applying them to various operational facets, organizations can comply with regulatory requirements and pioneer innovations in energy efficiency. Accurate measurement and interpretation of these metrics are the bedrock upon which sustainable and efficient energy management practices are built (Torielli et al., 2011). Figure 4.2 illustrates the matrix tools for the lean and green implementation framework for foundry.

System optimization is modifying a system to improve efficiency or performance according to specified criteria. In the oil and gas industry context, this typically involves optimizing the processes and technologies used in extraction, production, and refining to maximize output while minimizing energy consumption and waste (Barakat et al., 2020). Techniques often include real-time monitoring, predictive analytics, and advanced control strategies to adapt to system variations dynamically. The ultimate aim is to streamline operations, reduce operating costs, and diminish environmental impact (Foresti et al., 2020). Organizations can improve their energy use index and energy intensity metrics by employing system optimization methodologies, providing tangible benefits to the environment and the bottom line (Jazizadeh & Jung, 2018).

While system optimization focuses on improving existing systems, design optimization aims at enhancing the system from the ground up during the planning and development phase. By integrating energy-efficient technologies and practices into the initial design of facilities, machinery, and operational protocols, the industry can preemptively mitigate energy inefficiencies that may become problematic later (Bharany et al., 2022). It can involve selecting materials with lower thermal conductivities, designing layouts that minimize heat loss, or integrating renewable energy sources into the energy mix. Such proactive measures can significantly improve a new project's energy profile, making it more sustainable and often more financially viable in the long term (Li et al., 2017).

key lean tools enabling foundry lean and green implementation framework	Organization philosophy	Improve throughput	Energy efficiency	Innovative technologies	Community partnership
Green value stream mapping		×	×	×	×
5S and safety	×	×	×		
Statistical process improvement		×	×		
7R				×	×
Management operating system	×	×	×	×	×

FIGURE 4.2 Mapping key lean tools to a foundry lean and green implementation framework.

Source: Torielli et al. (2011).

Both system and design optimization are most effective when employed synergistically. While design optimization sets the stage for energy-efficient operations, system optimization ensures the maintenance of this efficiency over time. Advanced analytical tools, such as machine learning algorithms and simulation software, can continuously update and improve both types of optimization, creating a dynamic, adaptive system that evolves with technological advancements and changes in operational demand (Ahmad et al., 2022).

By embracing these methodologies, the oil and gas industry can significantly improve its energy efficiency, reducing its environmental impact and operating costs. Through rigorous system and design optimization, organizations can create a road map to more sustainable and economically viable operations, marking a significant step toward a cleaner, more energy-efficient future.

4.3 SMART GRID AND ENERGY MANAGEMENT

"Smart grid" refers to an intelligent, modernized electrical grid system that uses advanced technologies to monitor, control, and optimize electrical power generation, transmission, and distribution. In tandem, "energy management" refers to the strategic planning and operation of energy production and consumption units to ensure efficiency, reliability, and sustainability (Ahmad et al., 2022). As renewables are a main part of the energy mix, the need for smart grid is becoming a necessary concept for grid reliability, as shown in Figure 4.3.

Integrating smart grid and energy management systems is crucial for several reasons. Primarily, they enable better control and optimization of energy resources, leading to cost savings and reduced environmental impact. These systems also increase the resilience and reliability of the energy infrastructure, offering capabilities such as real-time monitoring and rapid response to system anomalies. Moreover, they pave the way for the seamless integration of renewable energy sources and distributed generation, which is vital for transitioning to a more sustainable energy landscape and providing jobs, as shown in Figure 4.4 (Ahmad et al., 2022).

Smart grids and energy management systems find applications across various residential, commercial, and industrial sectors (Figure 4.5). In the oil and gas industry, for instance, they can optimize the energy consumption of drilling rigs, refineries, and transportation systems. They also facilitate the integration of renewable energy into the energy mix, allowing for more sustainable operations.

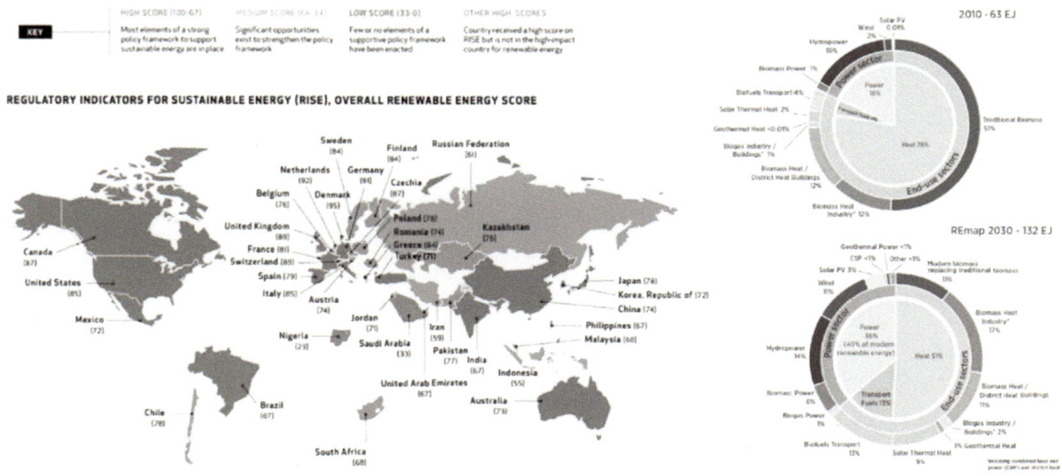

FIGURE 4.3 An overview of global renewable energy usage in 2010, and the renewable energy map 2030 projections are divided into technologies and sectors.

Source: Ahmad et al. (2022).

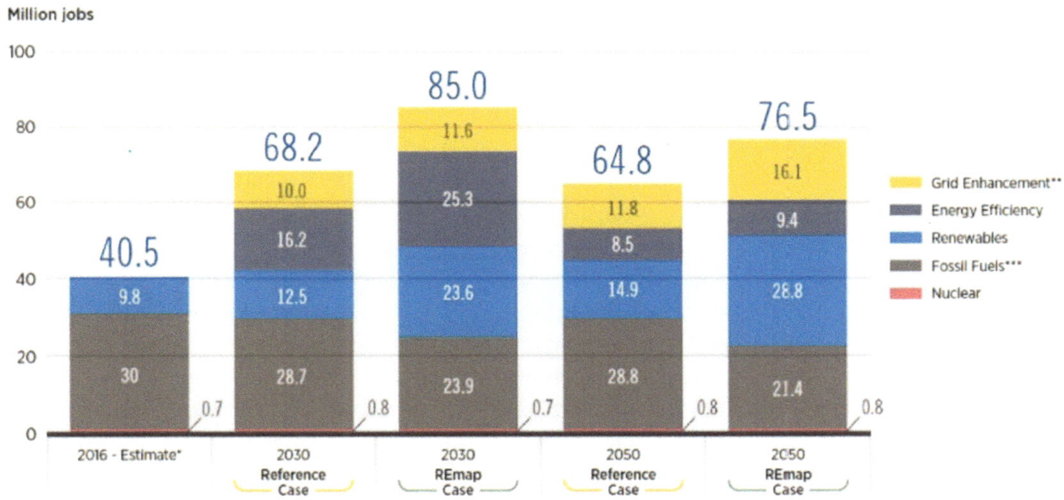

FIGURE 4.4 The energy transition process will generate an additional 11 million jobs in the energy sector by 2050.

Source: Ahmad et al. (2022).

Various tools and technologies are available to make smart grid and energy management effective, such as EMS (energy management system), BMS (building management system), SCADA (supervisory control and data acquisition), PLC (programmable logic controller), advanced automation technologies, and machine learning (ML).

The ML techniques link devices, including rooftop solar PV, local battery storage, smart meters, and home appliances, to the internet, allowing for data collection and sharing (Table 4.1) (Ahmad et al., 2022). By employing smart grid and energy management tools frcm end user to energy supplier and distributor, all stakeholders of the broader energy sector can make strides toward more sustainable, efficient, and reliable energy systems.

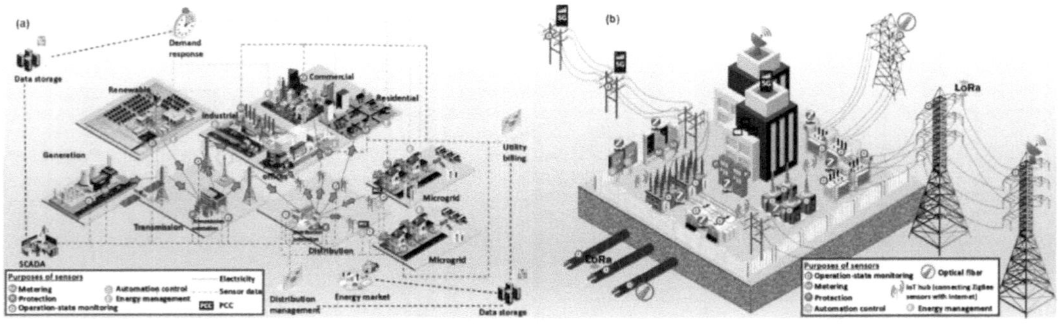

FIGURE 4.5 Pervasive sensing (a) in the power grid and (b) in the substation side.

Source: Ahmad et al. (2022).

4.4 SMART GRID AND ENERGY MANAGEMENT TYPES

4.4.1 Demand-Side Management

Smart grid represents a vision of the future power systems integrating advanced sensing technologies, control methodologies, and communication technologies at transmission and distribution levels to supply electricity in a smart and user-friendly way. According to the US Department of Energy's modern grid initiative report, the main characteristics of a smart grid are consumer friendliness, hack-proof self-healing, resistance to attack, ability to accommodate all types of generation and storage options, electricity market–based efficient operation, high power quality, and optimal assets. This modern grid is prompted by several economic, political, environmental, social, and technical factors (Logenthiran et al., 2012).

There are two components of demand-side management (DSM), as can be seen in Figure 4.6.

Energy savings. The goal of this method is to reduce energy costs, which is crucial for countries, including member states of the European Union, to achieve energy security and decrease carbon emissions. The International Energy Agency suggests that optimizing energy use could help meet carbon reduction targets by 2035. However, there is still a gap due to consumers having information and being hesitant to modify their consumption patterns based on prices (Stanelyte et al., 2022).

Demand response (DR). DR strategies involve adjusting power usage based on factors such as prices or incentives. Fixed price structures and technological barriers often limit the flexibility of demand in electricity markets (Stanelyte et al., 2022).

Demand-side management (DSM) is one of the important functions in a smart grid that allows customers to make informed decisions regarding their energy consumption and helps the energy providers reduce the peak load demand and reshape the load profile; this results in increased sustainability of the smart grid and reduced overall operational cost and carbon emission levels. Most existing demand-side management strategies, shown in Figure 4.7, in traditional energy systems employ system-specific techniques and algorithms. In addition, the existing strategies handle only a limited number of controllable loads of limited types (Logenthiran et al., 2012; Macedo et al., 2015).

Figure 4.8 shows the generation scheduling of residential, commercial, and industrial areas with the proposed demand-side management strategy.

The current trend indicates that energy demand and supply will eventually be controlled by autonomous software that optimizes decision-making and energy distribution operations. New, state-of-the-art machine learning (ML) technologies are integral in optimizing decision-making

TABLE 4.1
ML Designed to Help Record Huge Improvements across the Value Chain in the Energy Industry

Sr. #	Description	Applicable Technologies	Produce	Project	Provide	Promote
ML can produce values across the value chain.						
1	Retail		Automate the store operations and optimize the merchandising energy devices and products at micro space.	Anticipate energy demand trends, however, automating and optimizing supplier contracting and negotiation.	Tailored, enriched, and convenient customer experience.	Services and products at low prices, with the right target and the right message.
2	Power utilities		Improve electricity generation yield, optimize preventive maintenance, prevent electricity theft, and decrease energy waste.	Enhance energy supply and demand, measure the reliability of integrated energy production assets, and automate the control of energy demand-side response.	Provide accurate energy consumption insights, automate the consumer services with the use of virtual agents.	Day-of-day dynamic tariffing, as well as match producers of different consumers in real time.
3	Manufacturing devices		Reduce the error at different states, improve the process tasks, reduce the energy material delivery time, and limit the different product rework.	Enhance the device efficiency and yield, automate supplier requirements, and anticipate different parts of requirements.	Enhance the skills for pilot training and maintenance engineer.	Optimize the energy pricing, forecast maintenance services, and refine sales leads prioritization.

(Continued)

TABLE 4.1 (*Continued*)

ML Designed to Help Record Huge Improvements across the Value Chain in the Energy Industry

Sr. #	Description	Applicable Technologies	Produce	Project	Provide	Promote
The ML techniques can help capture a significant number of chains across the value chain.						
1	Power utilities	–	20% energy production increased with the use of ML models and smart sensors to optimize the energy yields; improvement in EBIT of 10–20% by applying ML to increase the predictive maintenance, automate the line fault forecasting, and enhance the capital productivity [79].	The key aim of reducing the 10% in-country energy usage is by applying the deep learning models to forecast the energy supply and demand [79].	There is a huge saving in energy bills for illustrating \$10–\$130 by applying ML models to automatically switch the energy supply deals [79].	–

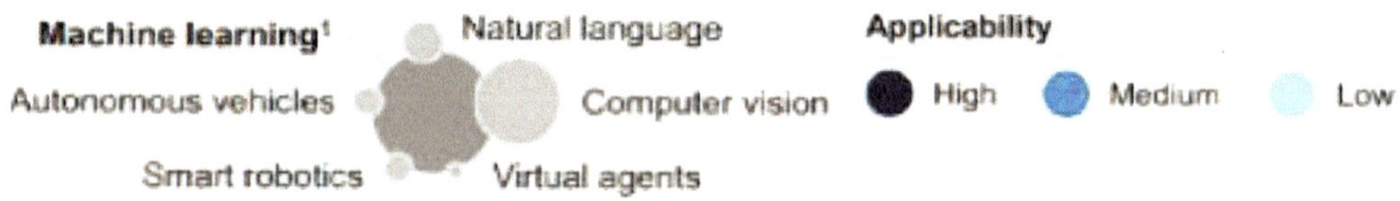

Source: Ahmad et al. (2022).

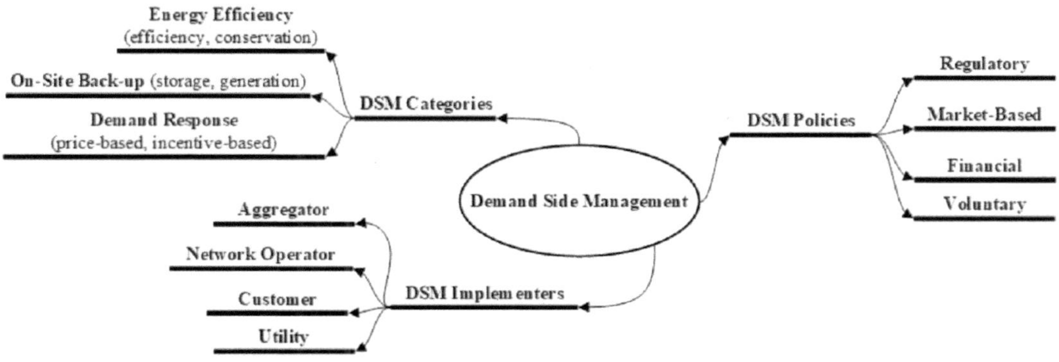

FIGURE 4.6 Defining demand-side management (DSM).

Source: Stanelyte et al. (2022).

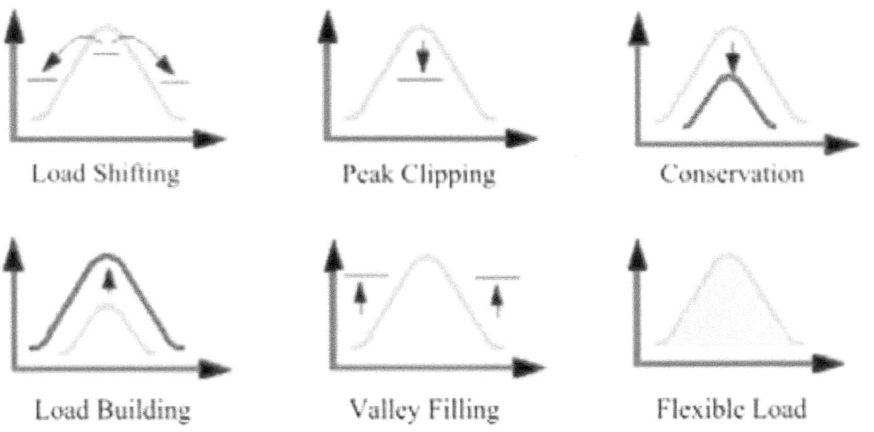

FIGURE 4.7 Demand-side management techniques.

Source: Logenthiran et al. (2012).

in energy distribution networks and systems. Strategic energy planning aims to use ML models to increase awareness, identify the overall market needs for incorporation into the basic and advanced energy infrastructure, and agree on the best integration strategies and industry practices focused on these strategies, as shown in Figure 4.9.

Paul et al. (2023) studied a smart energy management controller that modifies consumer energy usage projections using demand response programs to create an operation schedule. The suggested hybrid genetic algorithm (GA) and particle swarm optimization (PSO) techniques for intelligent load management in the smart grid have demonstrated that it greatly lowers energy costs and system efficiency. As can be seen in Figure 4.10, the suggested approach ensures the smart grid is interactive and dependable, while allowing customers to participate in the system through various demand response programs by integrating renewable resources and energy storage systems. The quantitative findings reported in this research show how the suggested strategy effectively lowers energy expenses associated with greenhouse gas emissions and electricity bills.

The advent of Internet of Things (IoT) technology has increased appliances integrated with capabilities. These smart energy appliances simplify the connection between power consumers and the energy market, especially when compared to devices (Sonsaard et al., 2023)

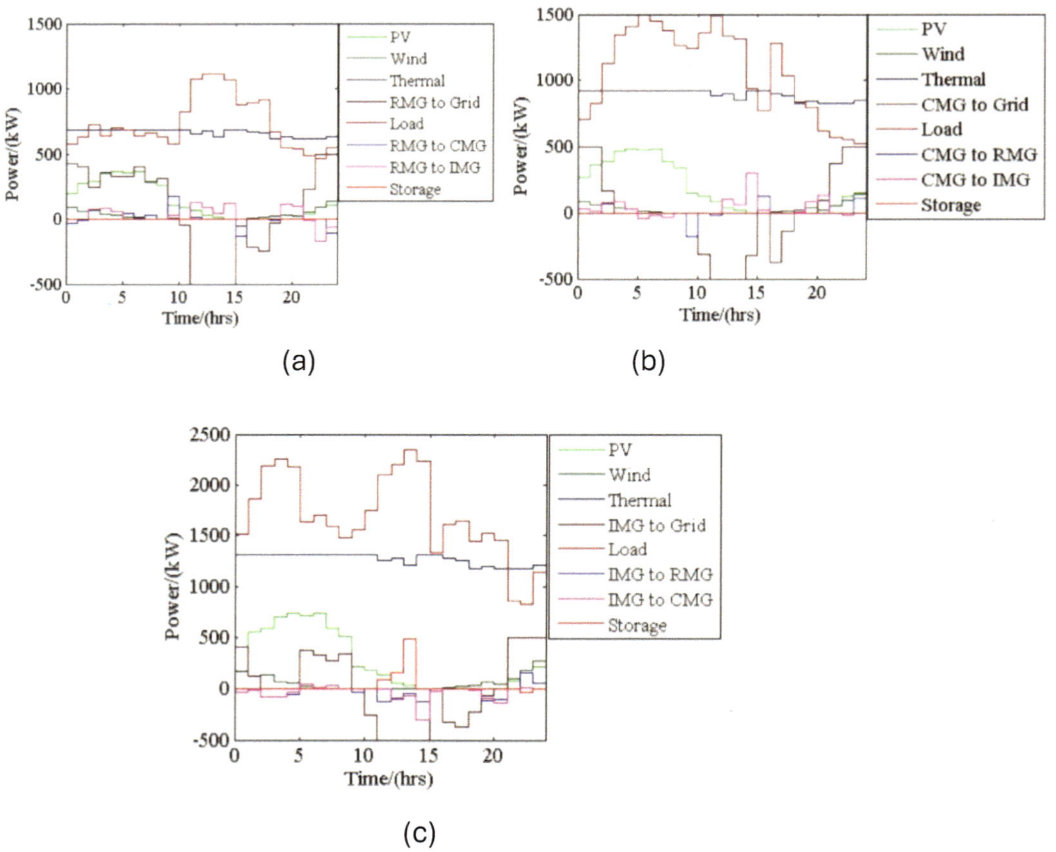

FIGURE 4.8 Areas with DSM: (a) residential, (b) commercial, and (c) industrial.

Source: Logenthiran et al. (2012).

Figure 4.11 demonstrates the synergies between developments for demand-side management integration. Internet of Things (IoT), artificial intelligence, energy storage, and smart EV charging play a key role on effective demand management and optimization.

Recently, progress has been made in achieving a balance between energy conservation (EC) and renewable energy (RE) generation by improving demand-side management programs (DSMPs) (Diaz et al., 2020) utilizing WSR-based smart grid management through a demand response program. Instead of relying on pollution emission models for microgrid management, the particle swarm optimization algorithm is used as the best solution (Bakare et al., 2023).

In another study, they introduced a way to manage electricity demand by breaking it down into three categories: baseline, burst, and regular load. Baseline loads are the essentials, like TVs, lights, and computers, that need a steady power supply. Knowing these baseline needs is crucial for figuring out capacity and ensuring a balanced grid. Burst loads are the big energy hogs, like dishwashers and dryers. Managing these effectively can help reduce energy costs (Khan et al., 2015).

To wrap up the discussion on demand-side management as shown in Figure 4.12, it is all about smart planning and efficient energy use. We can better balance the grid and save money by categorizing our power usage into essentials, like lighting and computing (baseline loads), and more variable demands, like dishwashers (burst loads). It is like creating a thoughtful budget, but for electricity.

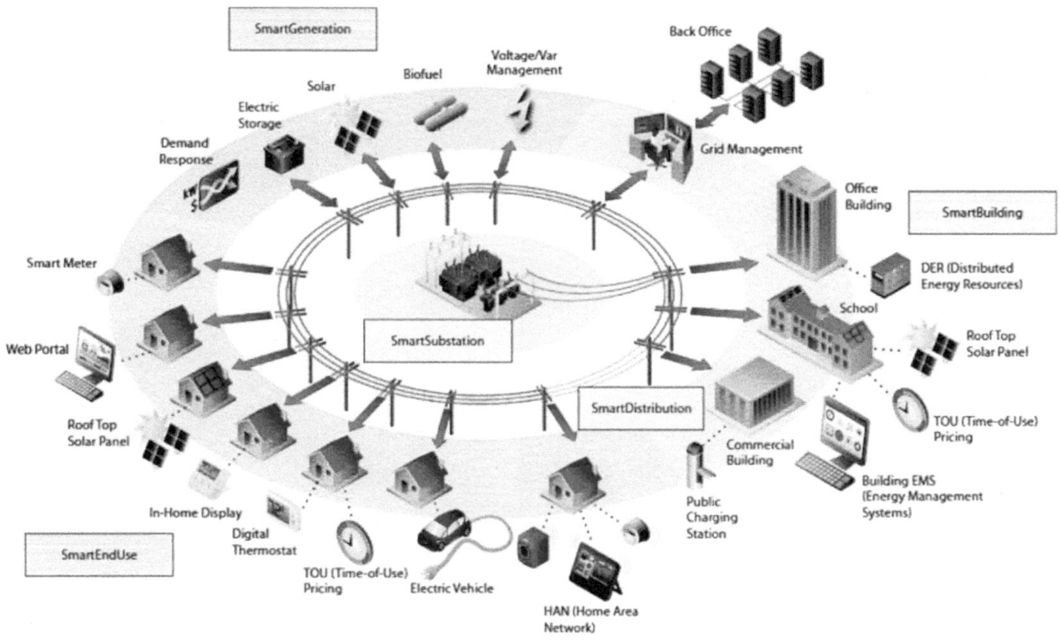

FIGURE 4.9 Smart grid energy resources and energy management.

Source: Ahmad et al. (2022).

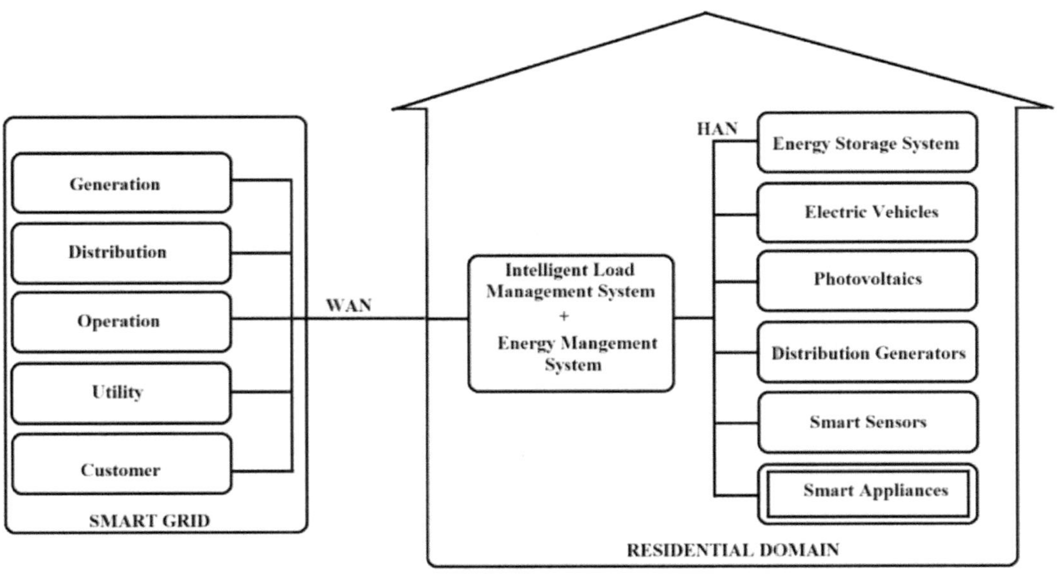

FIGURE 4.10 Architecture of proposed intelligent load management system model.

Source: Paul et al. (2023).

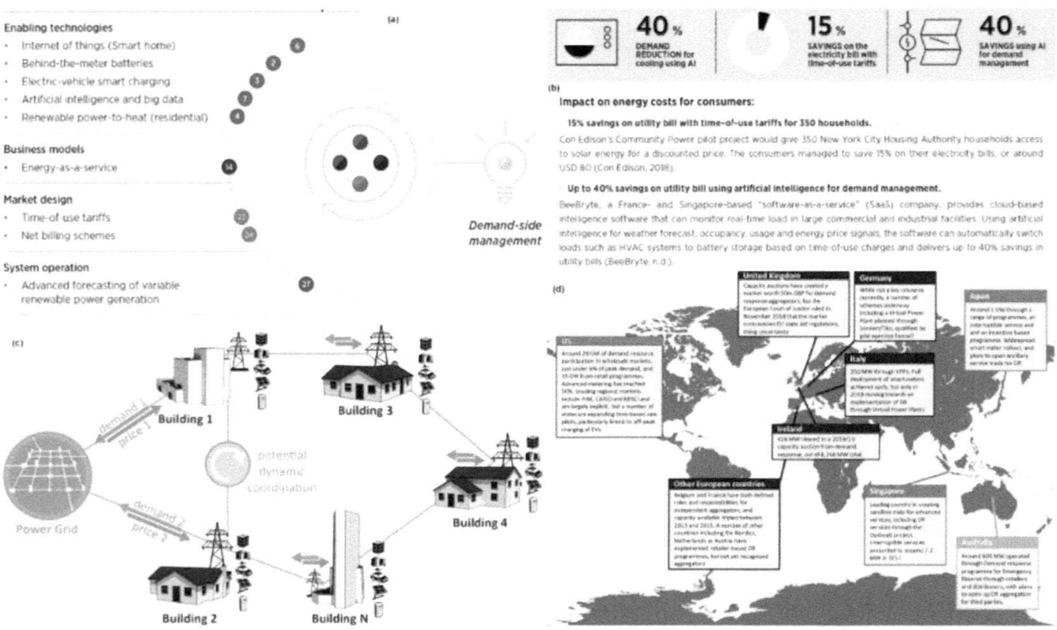

FIGURE 4.11 ML for (a) synergies between developments for demand-side management integration; (b) 40% energy demand reduction using AI; (c) multi-agent-based demand response; (d) in the sustainable development scenario, from 2018 to 2040, there is a huge potential for demand response.

Source: Ahmad et al. (2022).

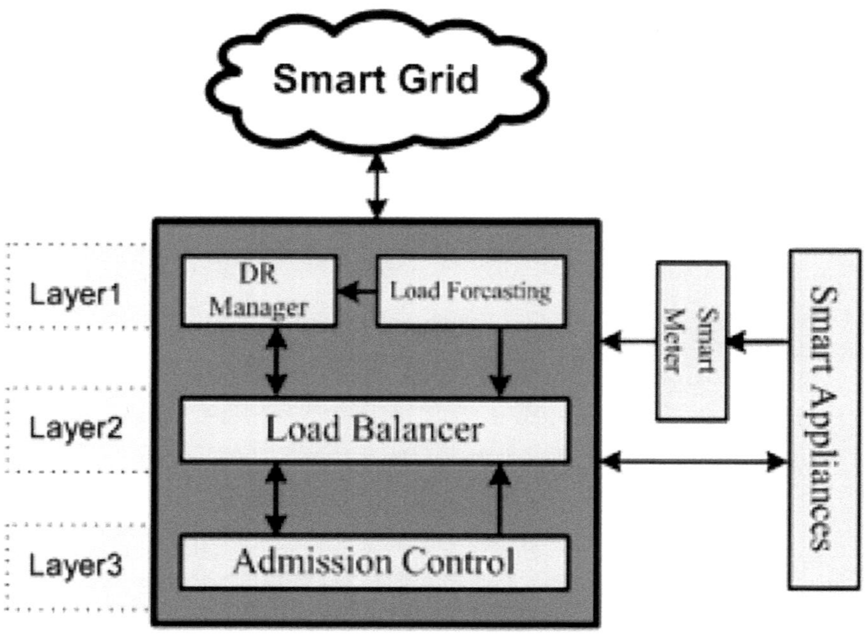

FIGURE 4.12 Demand-side load management architecture.

Source: Khan et al. (2015).

4.4.2 Supply-Side Management

In the previous section, we discussed managing energy demand, focusing on how homes use electricity. While it is crucial to understand and manage how much energy we consume, it is equally important to have a grip on how that energy is supplied. The other half of the puzzle is ensuring there is enough electricity to go around, especially during peak times. This two-sided approach ensures a more efficient and reliable power grid for everyone (Cai et al., 2019).

Cai et al. (2019) offered insights into predicting home energy use more accurately. As shown in Figure 4.13, the approach is obtaining actual demand data by performing extensive cleaning of raw

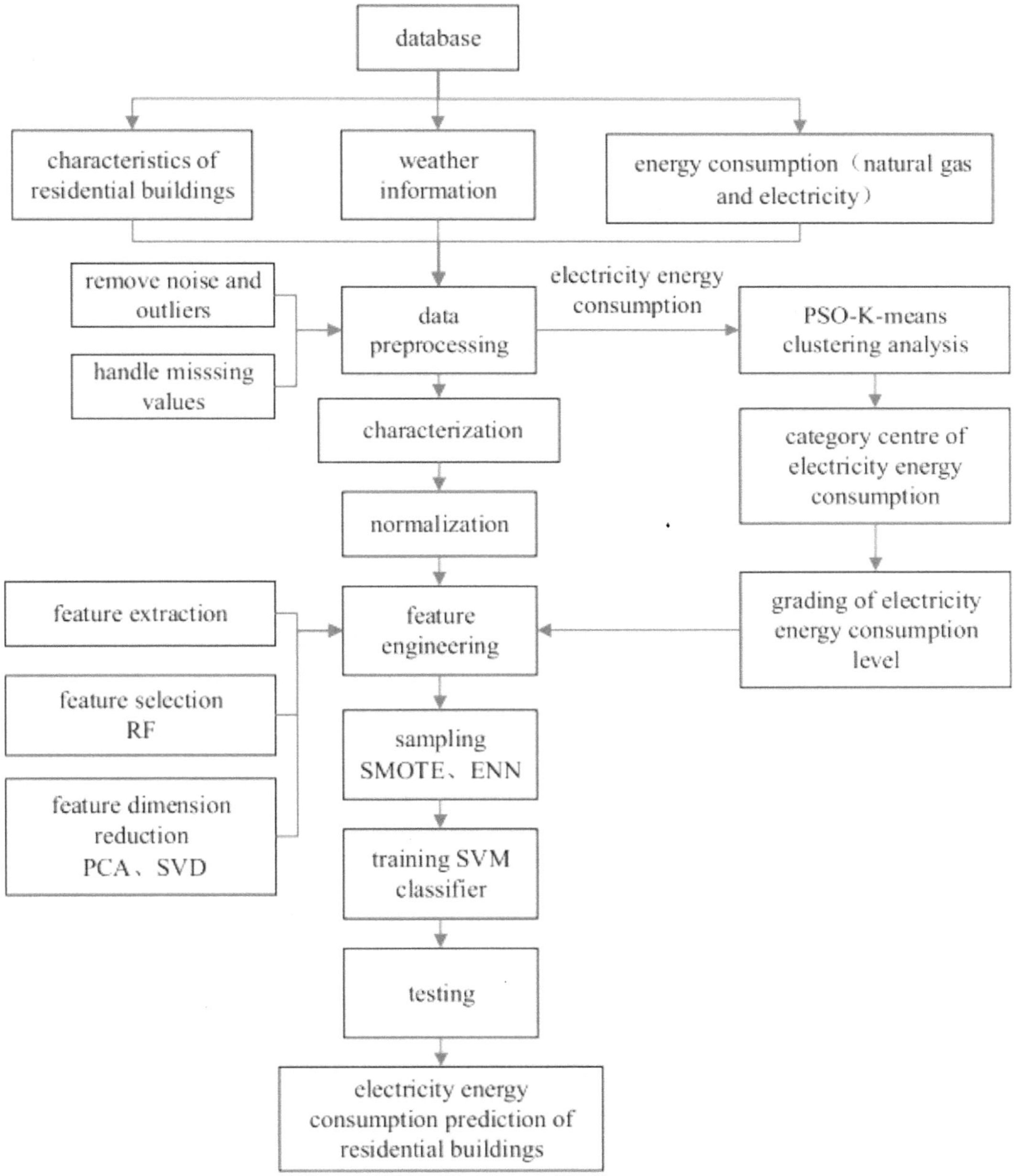

FIGURE 4.13 Technical route of the research.

Source: Cai et al. (2019).

data. They start by scrubbing the information gathered on residential buildings, weather conditions, and past energy usage to remove any inaccuracies or gaps. This way, they can make more spot-on predictions about how much energy a home will likely consume.

Goyal et al. (2023) also dived into this complex world of supply-side energy management. Their work complements, as can be seen in Figure 4.14, the existing efforts by introducing additional techniques for optimizing how we generate and use electricity. So between these studies, they were looking at a smarter, more efficient, and cleaner future for our electrical grid. With better planning and more advanced tech, they took steps to lower the financial and environmental costs of keeping the lights on.

Climate change is shaking up both how much energy we use and how much we can generate. That is why having a dynamic model that rolls with these changes is super important. Wang et al.'s (2022a) model would predict how our energy use changes with the weather, calculate the impact on power generation, and optimize the whole operation in real time. If you look at Figure 4.15, it breaks down how this all-in-one approach to energy management would work in a world affected by climate change.

In their 2022 study, Wang, X., and colleagues took a deep dive into optimizing the combined cooling, heating, and power (CCHP) system of a hospital in Shanghai. Nevertheless, this is not your run-of-the-mill study; they elevated the game by integrating the impacts of climate change. Imagine a thermostat constantly getting fidgety because the outside temperature keeps shifting – that is what climate change does to user demand and equipment efficiency, like gas turbines. They have also layered in pretty sophisticated strategies, as shown in Figure 4.16, like energy cascade utilization and tiered electricity pricing, showing how these could alter the operation playbook (Mokhtara et al., 2020).

	Operational planning	Day-ahead scheduling	Economic dispatch	Real-time dispatch
Price based DR	Time of Use (ToU) price	Real time pricing (RTP)	Critical peak pricing (CPP)	
Incentive based DR	Ancillary services market programs (ASMP)		Emergency demand response programs (EDRP)	Direct load control (DLC)
	Capacity market program (CMP)	Demand bidding	Interruptible service program (ISP)	

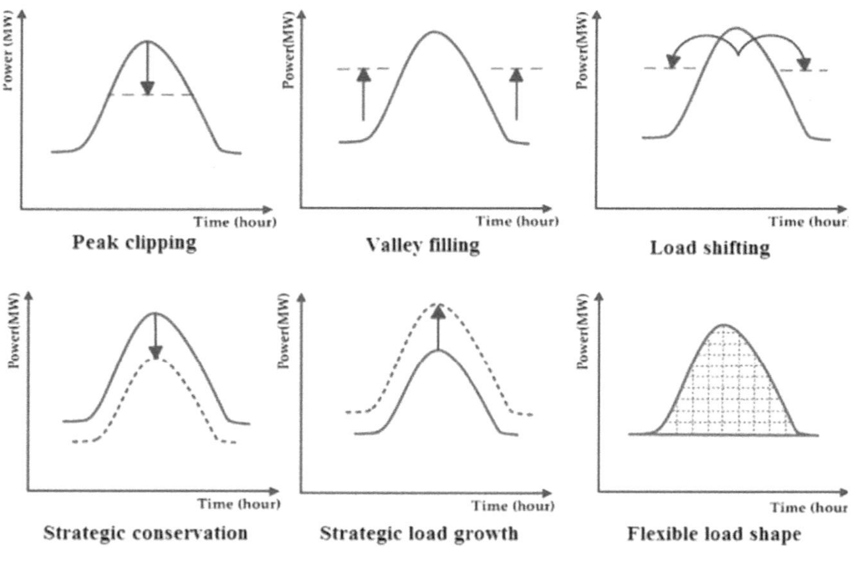

FIGURE 4.14 Forms of load management and load growth.

Source: Goyal et al. (2023).

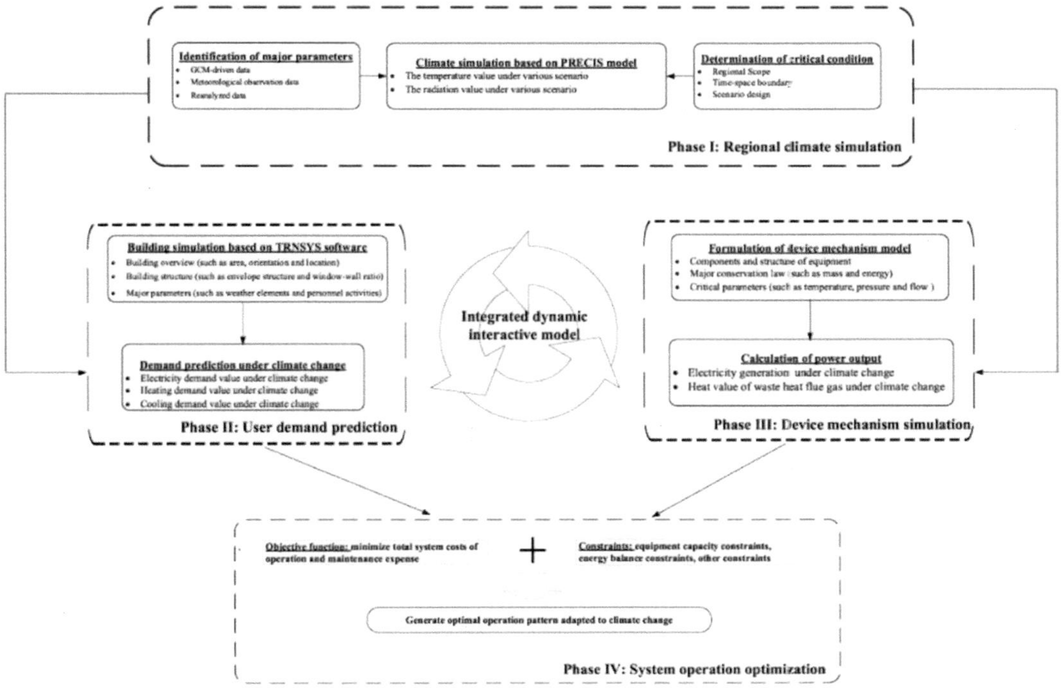

FIGURE 4.15 The general procedure of the dynamic interaction model under climate change.

Source: Wang et al. (2022a).

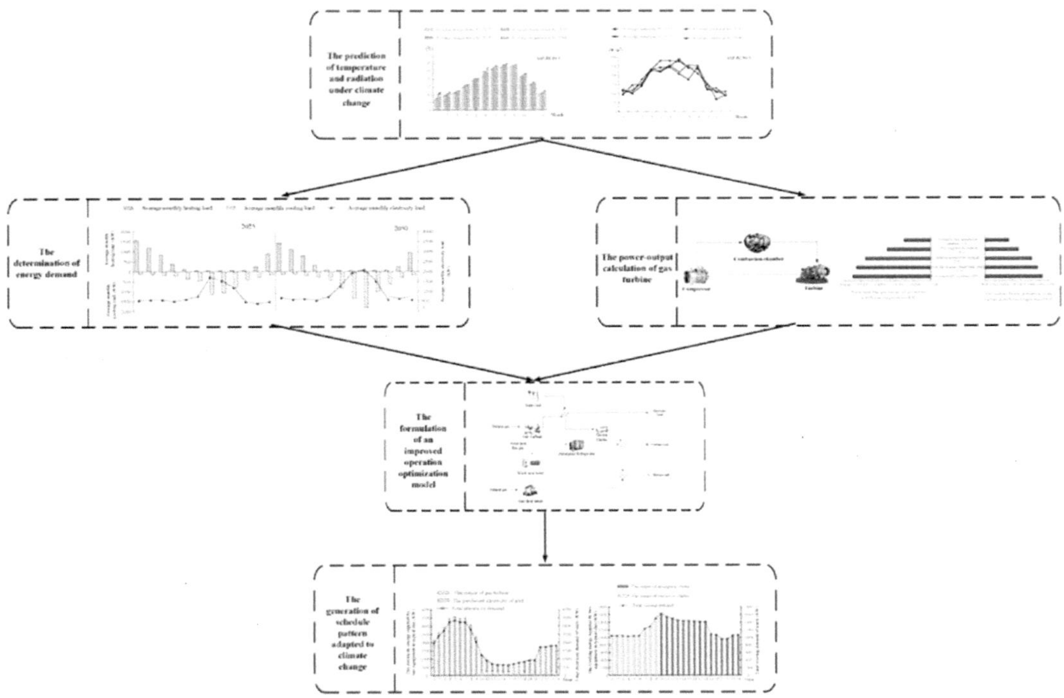

FIGURE 4.16 Framework for climate-adaptive energy system optimization.

Source: Wang et al., part 2, (2022b).

4.4.2.1 Supply Chain Management

Supply chain categories help gauge the system's performance, hinting at possible ways to tweak and improve it. We divide it into four main areas: infrastructure design, policymaking, planning, and scheduling (Emenike & Falcone, 2020). The first two are all about setting up the system before it runs, while the latter is focused on its day-to-day operations. The design and policy are about laying the groundwork and deciding the best ways to set up and manage the supply chain. Planning and scheduling, on the other hand, ensure the established network runs smoothly, adapting to any hiccups along the way (Emenike & Falcone, 2020). When designing the infrastructure, you must consider factors like varying costs based on location, the complexity of production, clear network routes, and shifting exchange rates. Using the right model helps to analyze these factors effectively.

The answer Anser et al. (2020) presented is a sophisticated green supply chain management framework intricately weaving logistics and technological advancements. Drawing upon comprehensive data from 2010 to 2018, the investigation elucidates the pivotal role of cleaner production methodologies in mitigating global carbon footprints. Moreover, it accentuates the indispensability of fuel-efficient mechanisms within logistics to bolster the caliber of services rendered. Rigorous adherence to environmental protocols, coupled with the strategic fortification of shipments via insurance and the establishment of robust financial infrastructures, can significantly augment the efficacy of green supply chain endeavors. The trajectory toward a sustainable supply chain paradigm is contingent upon a confluence of pioneering logistics, avant-garde technological innovations, and sagacious financial strategizing.

4.4.2.1.1 Energy Generation

In energy generation, envision the intricate dance of supply-side management as an orchestration of various elements, each requiring meticulous attention and expertise. Within this tapestry, energy generation – whether it is singular in the form of power or a dual output of power and steam – becomes a central theme, demanding a sophisticated understanding of its nuances.

Consider the diverse energy generation types seen in Figure 4.17, such as solar arrays, wind turbines, hydroelectric systems, nuclear facilities, and traditional fossil fuel establishments. Each can be likened to an instrument in a grand philharmonic ensemble. Every instrument's unique timbre and capabilities reflect these energy sources' distinct efficiencies and operational intricacies (Zhang et al., 2023).

In another study, Khan et al. (2015) focused on reducing costs and emissions in a smart grid with wind and solar resources. The variations in wind power and solar power were estimated using statistical tools called probability density functions. However, this study did not factor in how solar and wind power generations interact.

4.4.2.1.2 Energy Transmission

A *smart grid* is an evolved electricity network harnessing digital tech for efficient energy management. It uses advanced mechanisms to curtail costs and environmental impact while boosting system robustness and efficiency. This grid supports varied power sources, promotes consumer engagement, and showcases resilience against threats, as highlighted in Figure 4.18 (Jayachandran et al., 2021).

The smart grid (SG) idea is being integrated throughout power systems to achieve reliable, cost-efficient, and sustainable electricity networks, utilizing advanced information technologies and eco-friendly methods. This becomes especially crucial when incorporating renewable energy sources (RESs), battery storage, and EVs, demanding distinct control strategies for distribution networks. Therefore, distribution networks require entirely different control methods (Judge et al., 2022).

Many EVs demand simultaneous charging in densely populated areas, prompting researchers to propose battery swapping stations (BSS) for quicker, more efficient exchanges. Compared to traditional EV charging stations, BSS offers speed and flexibility. However, integrating a large-scale

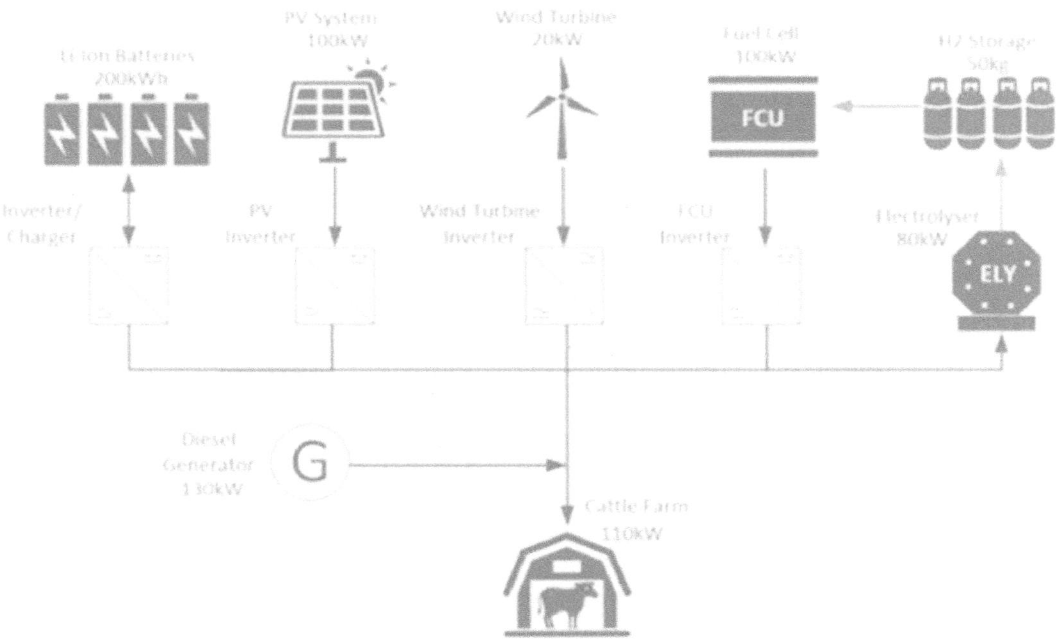

FIGURE 4.17 Remote's layout: renewable-powered cattle farm.

Source: Alonso et al. (2023).

energy management system (EMS) in the vehicle-to-grid (V2G) framework raises concerns about security, privacy, and grid resilience. The effects of different control strategies for mixed-energy EV charging on smart grids are detailed in Table 4.2 (Jayachandran et al., 2021).

The effects of different control strategies for mixed-energy EV charging on smart grids are detailed in Table 4.3 (Jayachandran et al., 2021).

4.5 SYNERGY OF ENERGY EFFICIENCY AND DEMAND RESPONSE COORDINATION

The study by Alireza Nouri and colleagues emphasizes the importance of citizen engagement in Europe's transition to renewable energy. Insights from the PANTERA project spotlight challenges in regulations and standards impeding this involvement, especially in lower-investing EU states. The article identifies demand response, renewable sources, and innovative local energy markets as key enablers. A case from Ireland illustrates efforts to adapt network codes and boost citizen participation. The PANTERA EIRIE platform's role in advancing smart grid research through knowledge sharing is also highlighted (Nouri et al., 2022).

Malik and Harish (2019) emphasized the intricate relationship between demand response and energy efficiency within the demand-side management spectrum, with automation as the central connector. They noted that utilities, despite having sufficient power, often confront inconsistencies due to fluctuating consumption and environmental variables. The automated demand response (ADR) scheme was introduced to address these challenges, encouraging consumers to modulate non-essential loads and ensuring enhanced power reliability. Furthermore, as discussed by Malik et al., the advent of OpenADR is being propelled by various pilot projects that primarily target industries and commercial entities. These endeavors offer promising economic and ecological benefits, yet transitioning to a mainstream program remains complex.

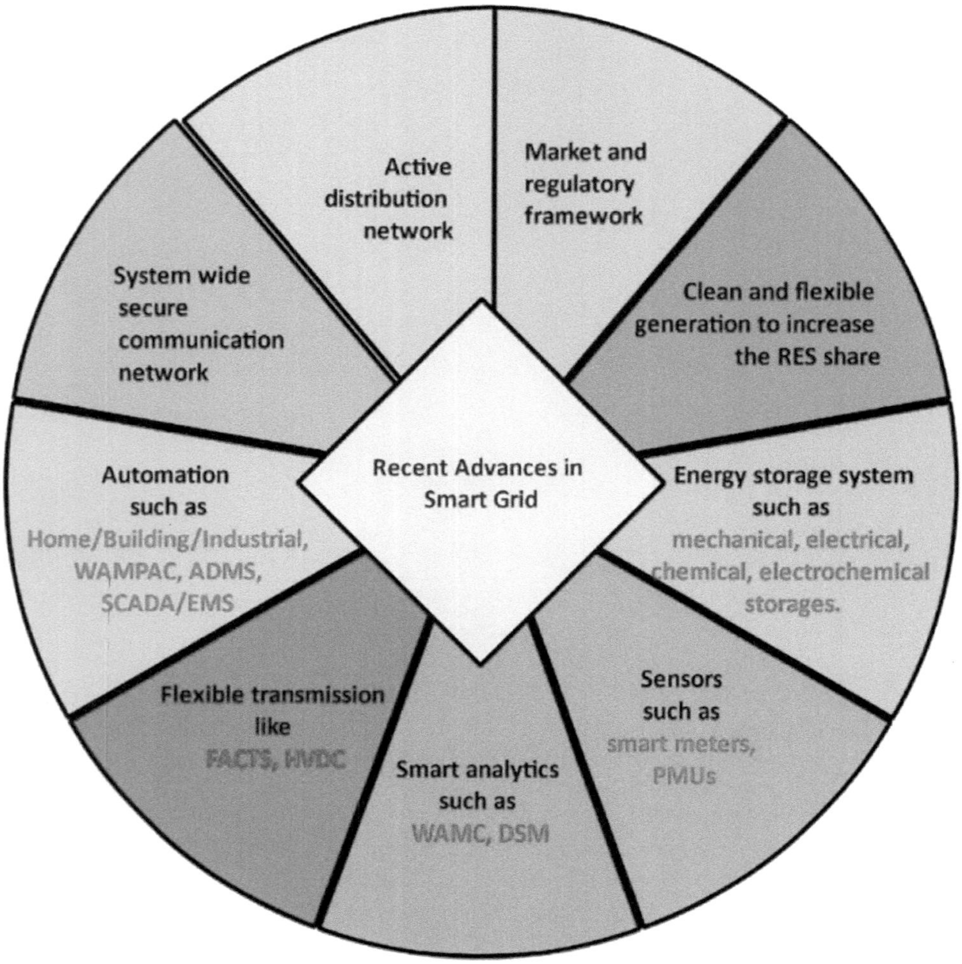

FIGURE 4.18 Recent advances in smart grid.

Source: Jayachandran et al. (2021).

According to a study conducted by P. Sonsaard in 2023, the Energy Policy and Planning Office (EPPO) plays a role in guiding the direction of the demand response (DR) market. EPPO's strategic policy initiatives are instrumental in establishing a DR market size for new participants. Energy service companies (ESCOs) are also players in shaping DR resources (DRRs), as they provide consultancy and implementation services for energy management systems in the industrial and commercial sectors, as can be seen in Figure 4.19.

In Sonsaard et al. (2023), Table 4.4 presents the outcomes of a SWOT analysis undertaken for state utilities considering entering the demand response (DR) business as load aggregators (LAs).

This is contrasted against a situation where non-utility private firms act as LAs in the DR domain. This LA business management aligns with the proposed structure of Thailand's power market, as delineated earlier and illustrated in Figure 4.20.

The Business Model Canvas (BMC) offers a unified strategic approach for developing new business models. Central to the BMC's design are considerations of the business's potential revenue streams and the value it offers. Typically, the BMC aids entrepreneurs in addressing four

TABLE 4.2
EV Charger and Charging Infrastructure in Smart Grids

Methodology	Control Method	Merits	Outcomes
Hydrogen storage for fuel cell vehicles	• Electrolysis	• Lower solar energy curtailment and storage of the energy in hydrogen for transportation	• Increased renewable penetration and negative pricing, providing clean energy transportation
In-motion charging technologies for electric vehicles	• Dynamic charging	• Solving charge-sustaining model for EV on highways	• Planning of in-motion EV charging for realizing charge-sustaining operation at minimum infrastructure cost
Electric vehicle charging management	• EV charging station	• Depleted battery either charged in EVCS with less charging time or exchanged with a charged battery in BSS	• Appropriate EVCS to be installed into the existing power network
	• Battery swapping station	• BSS offers more flexibility and convenience	• Managing the charging of the EVs at EVCS in densely populated areas
Grid-connected EV fast charging station with PV and ESS	• Modular multi-port power transformer cascaded H-bridge concept	• Possible to charge multiple and simultaneous EVs	• Planning of EVCS to maximize the contribution of EV to the power distribution grid

Source: Jayachandran et al. (2021).

TABLE 4.3
Smart Grid Stability and Control

Methodology	Control Method	Merits	Outcomes
		• VVM can regulate distribution system voltage profile and minimize reactive power flow	• Planning and deployment of monitoring system to monitor and analyze the power quality problems
Monitoring communication and computing	• VVM, AMI, GIS, SCADA, DA, and PQA	• FILR capability • PQA can investigate disturbances in power system	• Provide clean power to end user equipment with minimal disturbances
Integration of non-synchronous generators into reserve provision	• Virtual synchronous generator	• Artificially add inertia through converter controls	• Synchronism ensured in large-scale interconnected power systems
	• Fast-frequency response control	• Frequency gradient has been reduced in small island power system	• Ensure system stability

(Continued)

TABLE 4.3 (*Continued*)
Smart Grid Stability and Control

Methodology	Control Method	Merits	Outcomes
		• Increasing the inertia of the system	• Enhancing ancillary services by virtual inertia emulation to regulate frequency in active distributed network
Virtual inertia	• Voltage source converter control synchro-converter	• Enhance frequency stability profile of the system • Damp the oscillations in the power system	• Maintain power system frequency stability by providing virtual inertia, and ensure the long-term frequency stability of the system
Data analytics	• Descriptive, diagnostic, predictive, and prescriptive	• Reducing demand–supply gap • Regulate power flow in generation, distribution, and consumption	• Finding the optimal solutions for managing the load of end users in smart grid
Wide-bandgap power semiconductor devices	• AlGaN/GaN-based HEMT devices	• Improve power density and power efficiency	• HEMT-based technology meeting the needs for power distribution grid and end use utilization

Source: Jayachandran et al. (2021).

fundamental queries pivotal to conceptualizing new business models and products: WHO, WHAT, HOW, and MONEY (Sonsaard et al., 2023).

The study conducted by Sonsaard et al. (2023) explores the concept of "demand response" (DR), which refers to how industrial and commercial electricity users adjust their consumption behaviors. These consumers willingly modify their energy usage patterns in accordance with the requirements of their utility provider. The goal of this adjustment is to optimize benefits for both parties by reducing or reallocating energy consumption during periods of demand. While DR can encompass forms of energy, such as thermal and gas, this research specifically focuses on the aspect of power generation. Figure 4.21 presents a classification of DR programs categorizing them into two groups based on the incentives offered to customers.

In their study, Sonsaard et al. (2023) explored a data-focused approach to artificial intelligence (AI), moving away from symbolic methods. The research covers artificial intelligence (AI) techniques in both agent frameworks, providing a comprehensive view. Figure 4.22 presents a representation of AI approaches used in demand response (DR).

According to Stanelyte et al. (2022), the energy union introduced an energy efficiency model in 2018 to integrate energy sources (RES) and improve energy consumption. This approach combines energy savings and demand response (DR), which involves initiatives to enhance energy efficiency and adjust consumption timing, also known as load shifting (Stanelyte et al., 2022).

In the research by Stanelyte et al. (2022), the viability and feasibility of demand response (DR) are closely intertwined with the structure of electricity markets. These markets are influenced by various factors, including the interplay between generating and reserve capacities, the presence of fluctuating renewable energy sources, the requisites for offsetting these fluctuations, and the upper threshold of consumer demand amenable to load shifting, along with the reserve's temporal

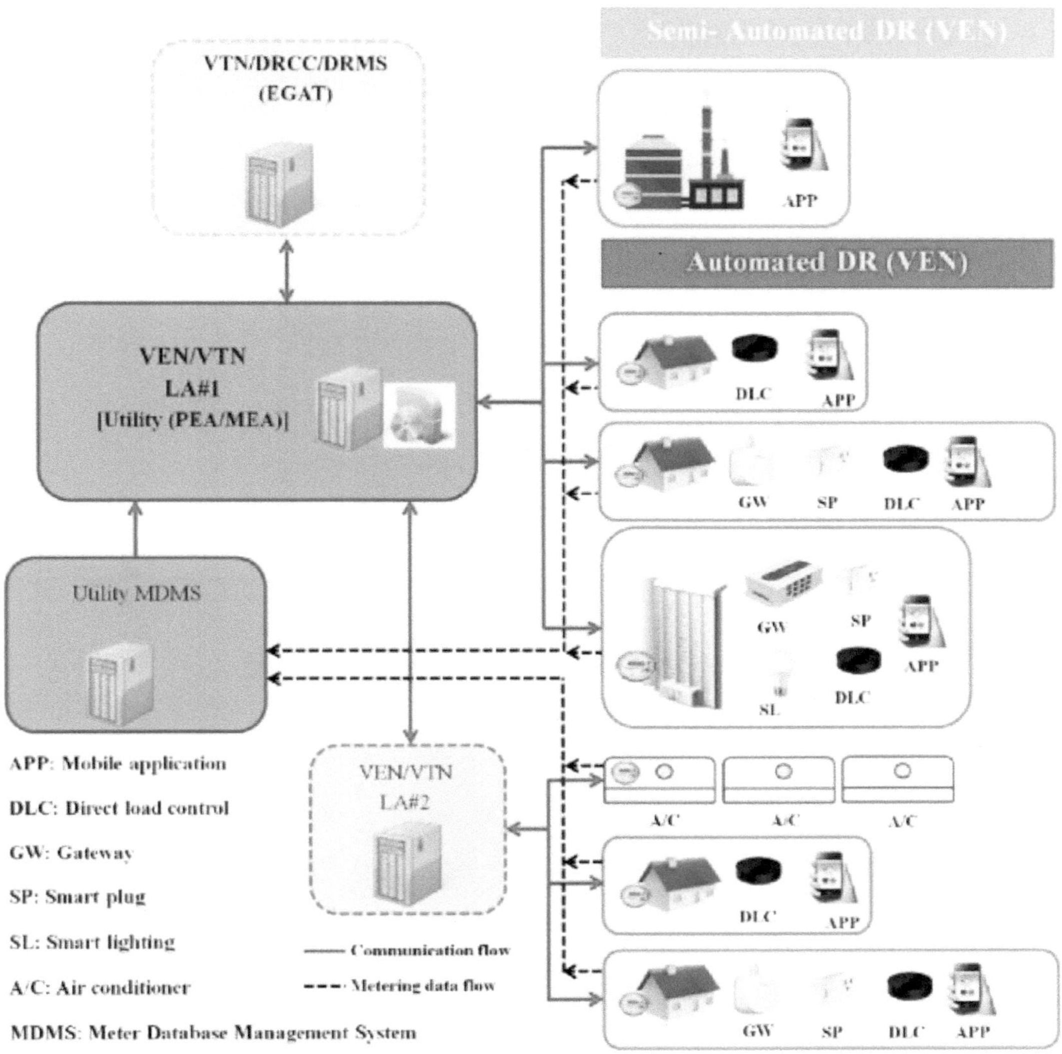

FIGURE 4.19 Proposed DR pilot project implementation plan for the utilities.

Source: Sonsaard et al. (2023).

TABLE 4.4
SWOT Analysis: State Utilities as the Lead LAs and the Ones Primarily Doing the DR Business

Strengths	Weaknesses
• Current power market structure • Group of customers with strong relationship and trust • Customer database • Strong financial status	• Government bureaucracy • High operation cost • Legal, regulatory, and policy constraints
Opportunities	**Threats**
• Seeking and meeting target customers from the utility database • Strong financial status that builds customer trust	• Emergence of LA#1 and LA#2 as private DR companies

Source: Sonsaard et al. (2023).

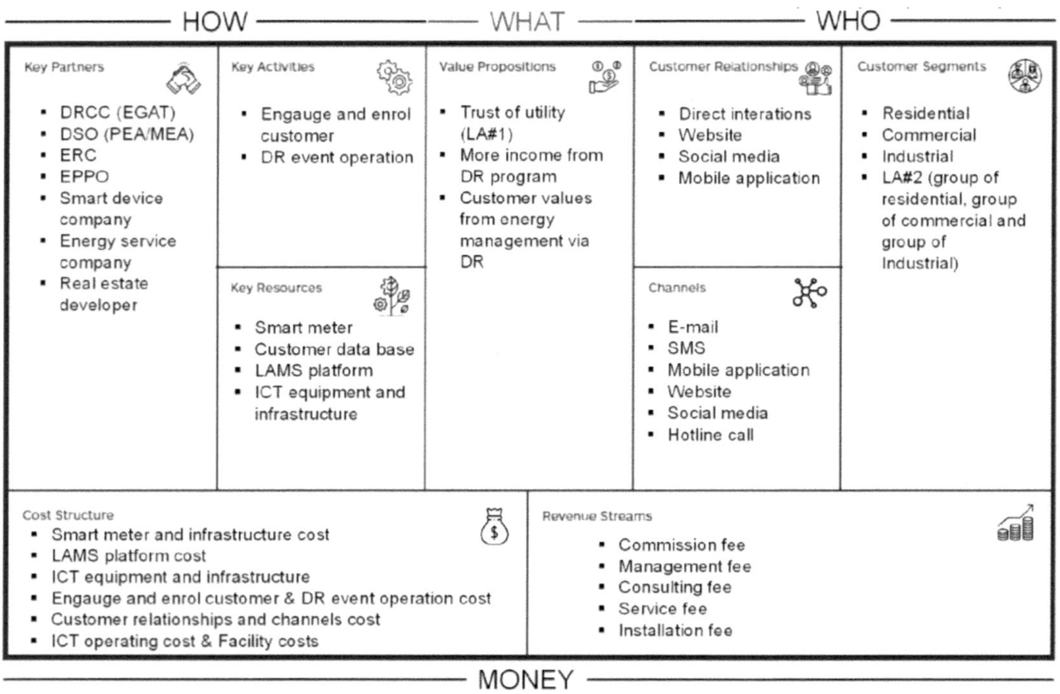

FIGURE 4.20 BMC for LA business operation for DR in Thailand power market.

Source: Sonsaard et al. (2023).

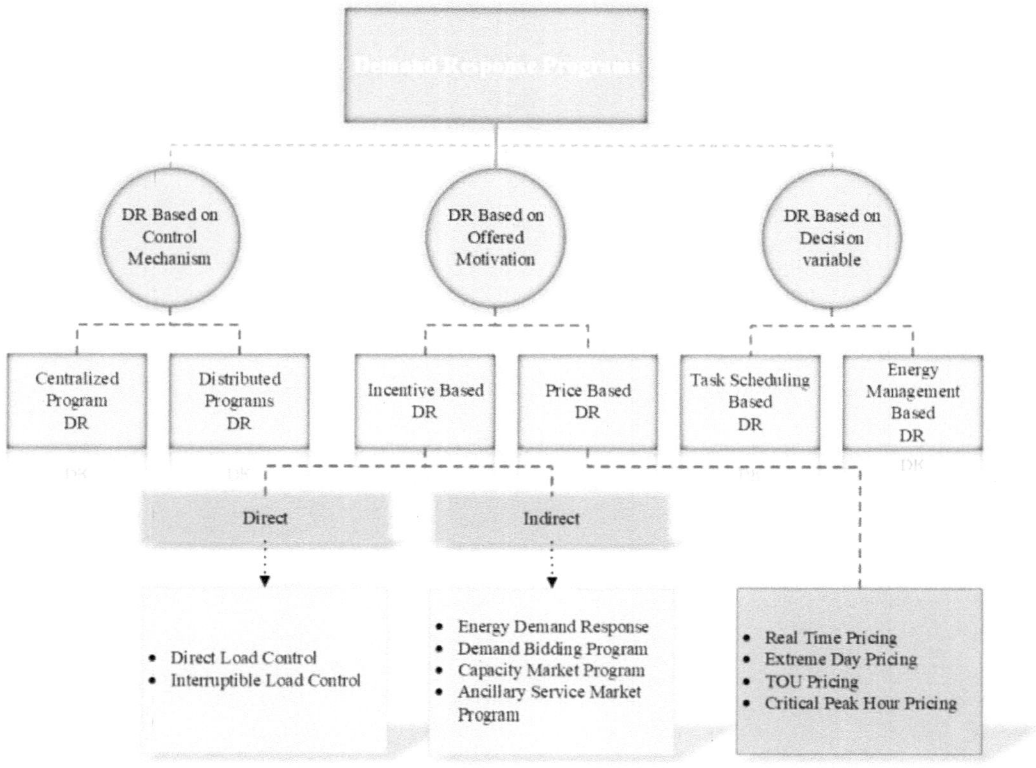

FIGURE 4.21 Classification of DR resources, operation, and structure.

Source: Sonsaard et al. (2023).

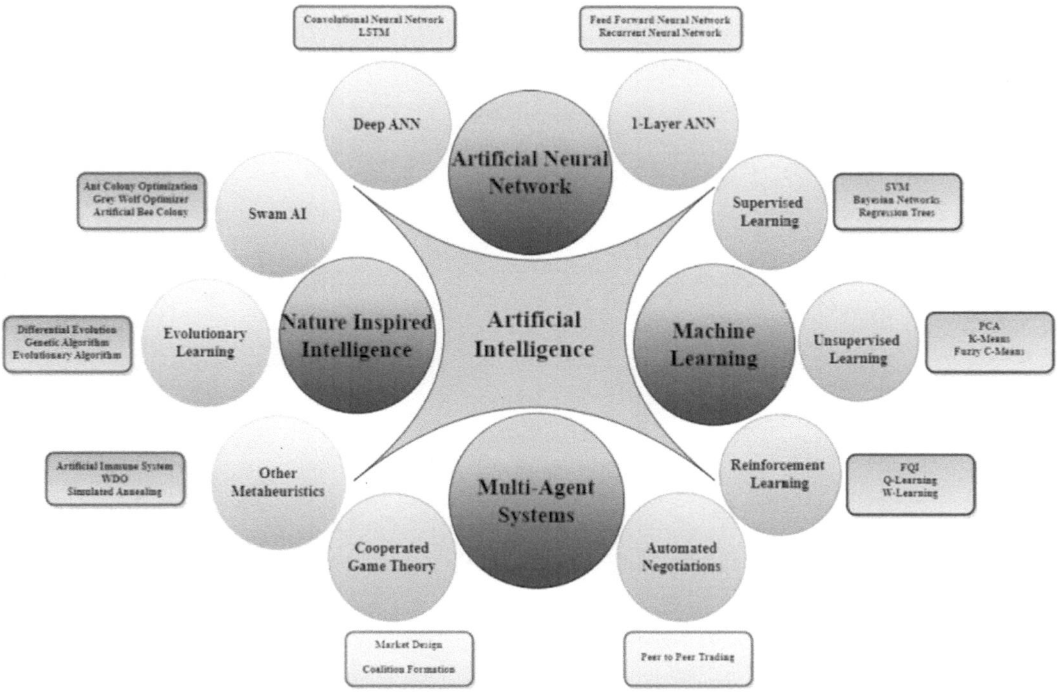

FIGURE 4.22 Different artificial intelligence techniques used for DR.

Source: Sonsaard et al. (2023).

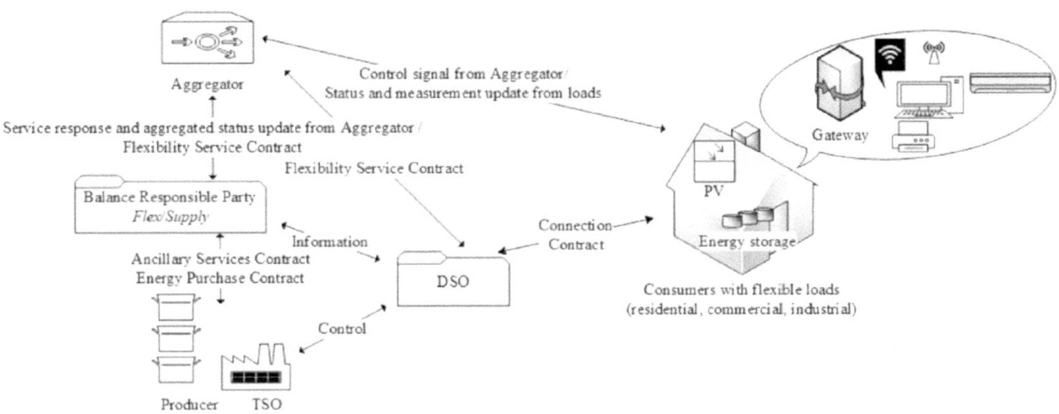

FIGURE 4.23 Major participants of a DR program.

Source: Stanelyte et al. (2022).

parameters. The electricity market is bifurcated into two main sectors: retail markets, where power retailers establish contracts with both suppliers and final consumers, and wholesale markets, which facilitate interactions among a broader range of stakeholders, including retailers, suppliers, producers, infrastructure providers, and aggregators. In the realm of DR programs, there is a collaborative dynamic among four primary stakeholder groups, a relationship depicted in Figure 4.23.

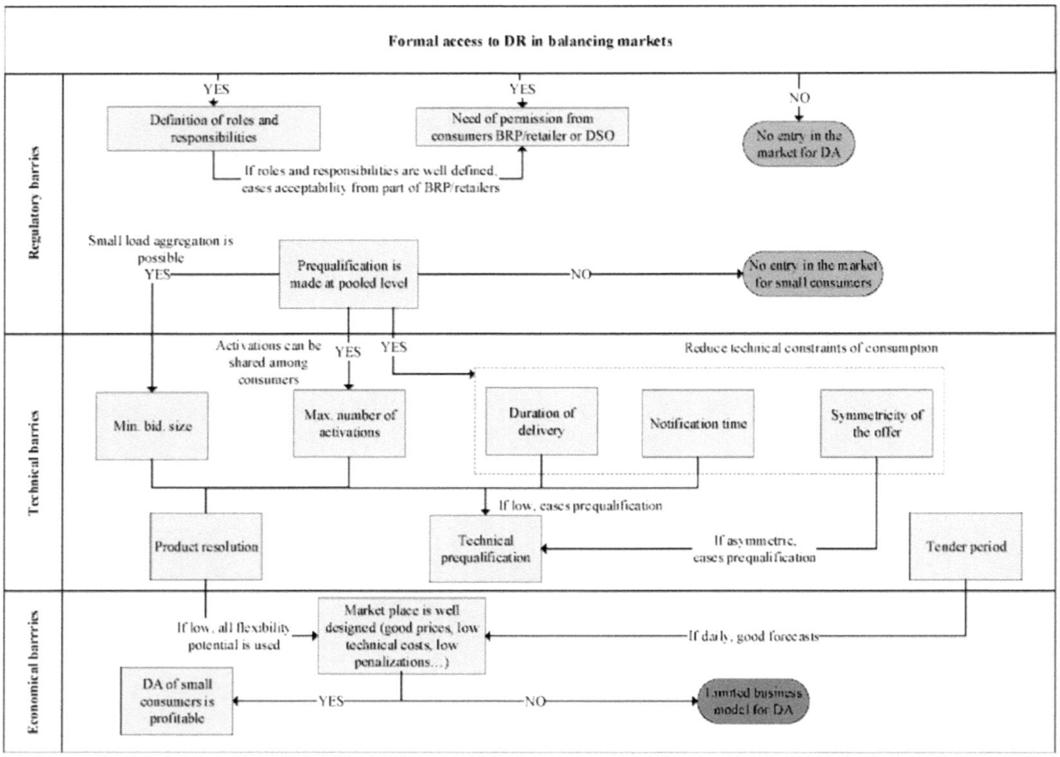

FIGURE 4.24 Relationships and challenges in the order of hierarchy and significance for aggregators in balancing markets.

Source: Stanelyte et al. (2022).

In Stanelyte et al. (2022), commercial offers align demand response (DR) with organizational goals. However, barriers remain in balancing markets for demand aggregators. Aggregators, acting as third-party intermediaries in the electricity sector, work with both industrial entities and residential consumers. Although many focus on industries, the literature emphasizes residential buildings, which constitute about 40% of total electricity use. These homes have systems like HVAC that offer demand-side flexibility. Aggregators manage these consumers either through direct agreements or via pricing incentives. The dynamics and challenges for aggregators in balancing markets are illustrated in Figure 4.24.

4.6 CONCLUSION

Demand response (DR) plays a role in managing high-demand periods and promoting competition in the energy market. While we acknowledge its benefits, there are challenges to adopting DR due to the range of users and their different needs. Some key obstacles include the lack of alignment between stakeholders, such as transmission system operators (TSOs) and distribution system operators (DSOs), the absence of guidelines, and insufficient market incentives.

Although the centralized market model allows for participation, we should also consider exploring models like blockchain that have potential advantages. However, implementing these models in the energy sector is complex. We can see promise in leveraging intelligence to enhance efficiency through energy services, but practical testing is necessary before widespread adoption. To shape strategies, we should focus on:

1. Integrating devices into retrofitting projects to improve energy efficiency
2. Providing service offerings to consumers based on their lifestyles
3. Utilizing AI for load predictions
4. Advancing DR applications to enhance energy stability and quality
5. Streamlining communication between distributed energy resources to optimize electricity quality

REFERENCES

Ahmad, T., Madonski, R., Zhang, D., Huang, C., & Mujeeb, A. (2022). Data-driven probabilistic machine learning in sustainable smart energy/smart energy systems: Key developments, challenges, and future research opportunities in the context of smart grid paradigm. *Renewable and Sustainable Energy Reviews*, *160*, 112128.

Alonso, A. M., Matute, G., Yusta, J. M., & Coosemans, T. (2023). Multi-state optimal power dispatch model for power-to-power systems in off-grid hybrid energy systems: A case study in Spain. *International Journal of Hydrogen Energy*, *52*.

Ang, B. W., & Goh, T. (2018). Bridging the gap between energy-to-GDP ratio and composite energy intensity index. *Energy Policy*, *119*, 105–112.

Anser, M. K., Khan, M. A., Awan, U., Batool, R., Zaman, K., Imran, M., . . . & Bakar, Z. A. (2020). The role of technological innovation in a dynamic model of the environmental supply chain curve: Evidence from a panel of 102 countries. *Processes*, *8*(9), 1033.

Bakare, M. S., Abdulkarim, A., Zeeshan, M., & Shuaibu, A. N. (2023). A comprehensive overview on demand-side energy management towards smart grids: Challenges, solutions, and future direction. *Energy Informatics*, *6*(1), 1–59.

Barakat, A. A., Diab, J. H., Badawi, N. S., Nader, W. S. B., & Mansour, C. J. (2020). Combined cycle gas turbine system optimization for extended range electric vehicles. *Energy Conversion and Management*, *226*, 113538.

Benedek, A., Rokicki, T., & Szeberényi, A. (2023). Bibliometric evaluation of energy efficiency in agriculture. *Energies*, *16*(16), 5942.

Bharany, S., Badotra, S., Sharma, S., Rani, S., Alazab, M., Jhaveri, R. H., & Gadekallu, T. R. (2022). Energy efficient fault tolerance techniques in green cloud computing: A systematic survey and taxonomy. *Sustainable Energy Technologies and Assessments*, *53*, 102613.

Cai, H., Shen, S., Lin, Q., Li, X., & Xiao, H. (2019). Predicting the energy consumption of residential buildings for regional electricity supply-side and demand-side management. *IEEE Access*, *7*, 30386–30397.

Chen, X., Wang, M., Wang, B., Hao, H., Shi, H., Wu, Z., . . . & Wang, B. (2023). Energy consumption reduction and sustainable development for oil & gas transport and storage engineering. *Energies*, *16*(4), 1775.

Díaz Redondo, R. P., Fernández-Vilas, A., & Fernández dos Reis, G. (2020). Security aspects in smart meters: Analysis and prevention. *Sensors*, *20*(14), 3977.

Emenike, S. N., & Falcone, G. (2020). A review on energy supply chain resilience through optimization. *Renewable and Sustainable Energy Reviews*, *134*, 110088.

Fakhar, A., Haidar, A. M., Abdullah, M. O., & Das, N. (2023). Smart grid mechanism for green energy management: A comprehensive review. *International Journal of Green Energy*, *20*(3), 284–308.

Foresti, R., Rossi, S., Magnani, M., Bianco, C. G. L., & Delmonte, N. (2020). Smart society and artificial intelligence: Big data scheduling and the global standard method applied to smart maintenance. *Engineering*, *6*(7), 835–846.

Gonzalez-Torres, M., Bertoldi, P., Castellazzi, L., & Perez-Lombard, L. (2023). Review of EU product energy efficiency policies: What have we achieved in 40 years? *Journal of Cleaner Production*, 138442.

Goyal, G. R., & Vadhera, S. (2023). Solution to objectives of supply side energy management by integrating enhanced demand response strategy. *Journal of Operation and Automation in Power Engineering*, *12*.

Guy, H., Vittoz, S., Caputo, G., & Thiery, T. (2023). Benchmarking the energy performance of European commercial buildings with a Bayesian modeling framework. *Energy and Buildings*, 113595.

Hussain, S. A., Razi, F., Hewage, K., & Sadiq, R. (2023). The perspective of energy poverty and 1st energy crisis of green transition. *Energy*, *275*, 127487.

Jamil, M., He, N., Wei, Z., Gupta, M. K., & Khan, A. M. (2023). A novel quantifiable approach of estimating energy consumption, carbon emissions and cost factors in manufacturing of bearing steel based on triple bottom-line approach. *Sustainable Materials and Technologies*, *36*, e00593.

Jayachandran, M., Reddy, C. R., Padmanaban, S., & Milyani, A. H. (2021). Operational planning steps in smart electric power delivery system. *Scientific Reports*, *11*(1), 17250.

Jazizadeh, F., & Jung, W. (2018). Personalized thermal comfort inference using RGB video images for distributed HVAC control. *Applied Energy*, *220*, 829–841.

Judge, M. A., Khan, A., Manzoor, A., & Khattak, H. A. (2022). Overview of smart grid implementation: Frameworks, impact, performance and challenges. *Journal of Energy Storage*, *49*, 104056.

Khan, A. A., Razzaq, S., Khan, A., & Khursheed, F. (2015). HEMSs and enabled demand response in electricity market: An overview. *Renewable and Sustainable Energy Reviews*, *42*, 773–785.

Kong, D., Cheshmehzangi, A., Zhang, Z., Ardakani, S. P., & Gu, T. (2023). Urban building energy modeling (UBEM): A systematic review of challenges and opportunities. *Energy Efficiency*, *16*(6), 69.

Li, Y., Rezgui, Y., & Zhu, H. (2017). District heating and cooling optimization and enhancement: Towards integration of renewables, storage and smart grid. *Renewable and Sustainable Energy Reviews*, *72*, 281–294.

Logenthiran, T., Srinivasan, D., & Shun, T. Z. (2012). Demand side management in smart grid using heuristic optimization. *IEEE Transactions on Smart Grid*, *3*(3), 1244–1252.

Macedo, M. N., Galo, J. J., De Almeida, L. A. L., & Lima, A. D. C. (2015). Demand side management using artificial neural networks in a smart grid environment. *Renewable and Sustainable Energy Reviews*, *41*, 128–133.

Malik, S., & Harish, V. S. K. V. (2019, October). Integration of automated demand response and energy efficiency to enable a smart grid infrastructure. In *2019 2nd international conference on power energy, environment and intelligent control (PEEIC)* (pp. 371–377). IEEE.

Mokhtara, C., Negrou, B., Bouferrouk, A., Yao, Y., Settou, N., & Ramadan, M. (2020). Integrated supply-demand energy management for optimal design of off-grid hybrid renewable energy systems for residential electrification in arid climates. *Energy Conversion and Management*, *221*, 113192.

Nouri, A., Khadem, S., Mutule, A., Papadimitriou, C., Stanev, R., Cabiati, M., . . . & Carroll, P. (2022). Identification of gaps and barriers in regulations, standards, and network codes to energy citizen participation in the energy transition. *Energies*, *15*(3), 856.

Oppon, E., Richter, J. S., Koh, S. L., & Nabayiga, H. (2023). Macro-level economic and environmental sustainability of negative emission technologies; Case study of crushed silicate production for enhanced weathering. *Ecological Economics*, *204*, 107636.

Paul, W. U. H., Siddiqui, A. S., & Kirmani, S. (2023). Intelligent load management system development with renewable energy for demand side management. *International Journal of Advanced Engineering and Management Research*, *8*(2).

Sonsaard, P., Ketjoy, N., & Mensin, Y. (2023). Market strategy options to implement Thailand demand response program policy. *Energy Policy*, *173*, 113388.

Stanelyte, D., Radziukyniene, N., & Radziukynas, V. (2022). Overview of demand-response services: A review. *Energies*, *15*(5), 1659.

Thompson, S. (2023). Strategic analysis of the renewable electricity transition: Power to the world without carbon emissions? *Energies*, *16*(17), 6183.

Torielli, R. M., Abrahams, R. A., Smillie, R. W., & Voigt, R. C. (2011). Using lean methodologies for economically and environmentally sustainable foundries. *China Foundry*, *8*(1), 74–88.

Wang, X., Xu, Y., Fu, Z., Guo, J., Bao, Z., Li, W., & Zhu, Y. (2022a). A dynamic interactive optimization model of CCHP system involving demand-side and supply-side impacts of climate change. *Part I: Methodology development. Energy Conversion and Management*, *252*, 115112.

Wang, X., Xu, Y., Fu, Z., Guo, J., Bao, Z., Li, W., & Zhu, Y. (2022b). A dynamic interactive optimization model of CCHP system involving demand-side and supply-side impacts of climate change. *Part II: Application to a hospital in Shanghai, China. Energy Conversion and Management*, *252*, 115139.

Wu, J., Lian, Z., Zheng, Z., & Zhang, H. (2020). A method to evaluate building energy consumption based on energy use index of different functional sectors. *Sustainable Cities and Society*, *53*, 101893.

Zhang, S., Huang, P., & Sun, Y. (2016). A multi-criterion renewable energy system design optimization for net zero energy buildings under uncertainties. *Energy*, *94*, 654–665.

Zhang, X., Liu, Z., Ding, T., Zhang, H., Siano, P., Meng, H., & Zhu, C. (2023). Multi-stage stochastic dual dynamic programming to low-carbon economic dispatch for power systems with flexible carbon capture and storage devices. *IET Generation, Transmission & Distribution*, *17*.

5 Solar Energy in the Oil and Gas Industry

Jorge Saldaña, Mesut Yurukcu, Narendra Boppana, Sepehr Arbabi, Jim Henry, and Sebahattin Ziyanak

5.1 INTRODUCTION

The sun is an inexhaustible source of free energy (Kabir et al., 2018), providing large amounts of it, as heat and light, every day. This energy is called *solar energy*. The amount of energy reaching Earth is greater than the amount produced by other sources, like water and wind. At least theoretically, solar energy could provide us with all the energy we would need, for as long as the sun exists. This means, according to NASA (2021), approximately 5 billion more years. It is important, then, to explore the possibilities to capture this energy to distribute to the population. Evidently, using the energy from the sun for human development is gaining more momentum than ever before. Because of this, researchers in all areas of knowledge are communicating and working on improving the ways to collect and distribute this energy. Figure 5.1 illustrates "Earth's energy budget," or how solar energy is distributed on the planet after arriving from the sun. Figure 5.1 shows the expected growth of renewable energy generation over the next couple of decades, which follows the improvement in technology and human ingenuity.

Considering that the current most used forms of energy are linked to problems like the warming of the Earth and climate change, making efforts to transitioning from one energy source to another becomes an important challenge. The process of energy transition can be understood as the pathway to a transformation of the entire global energy sector from fossil fuels to a sector that is based on clean, zero-carbon energy. Solar energy is a zero-carbon energy source within this transformation that will be a major part of renewables' future growth prospects, as shown in Figure 5.2, the IEA net-zero report. According to the International Energy Agency (IEA), the goal of this transition is to reach net-zero CO_2 emissions by 2050 and to limit the rise of temperature by 1.5°C.

In this chapter, an overview of global solar deployment is provided, then solar technology types and technology advancements, as well as application types reviewed in detail. Solar energy design principles, challenges, best practices, and economics are studied, which are followed by solar energy application areas and its role in energy transition within the oil and gas industry.

5.2 HISTORY OF SOLAR ENERGY

Some researchers have credited French scientist Edmond Becquerel for the invention of the solar cell because he found that light could increase electricity generation when two metal electrodes were placed into a conducting solution. This discovery, known as the "photovoltaic effect," was pivotal in subsequent developments of PV technology, mainly with selenium (Energy Sage, 2018).

Later, in 1873, Willoughby Smith realized that selenium had the potential to conduct solar energy, which led to the discovery from William Grylls Adams and Richard Evans Day that selenium can create electricity when exposed to sunlight, in 1876. Seven years later, Charles Fritts built the first solar cells containing selenium wafers, which has led other researchers to credit Fritts with their invention. With that said, current solar cells are made with silicon, an improvement produced

DOI: 10.1201/9781003282273-5

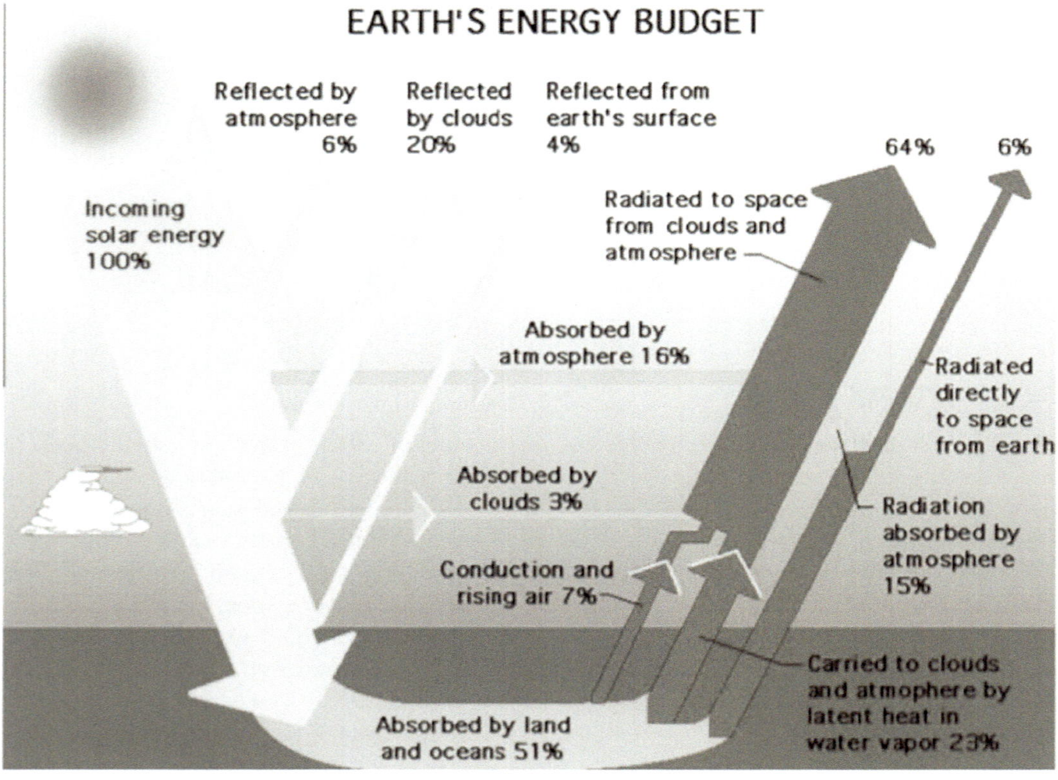

FIGURE 5.1 The Earth's energy budget.

Source: Tian and Zhao (2013).

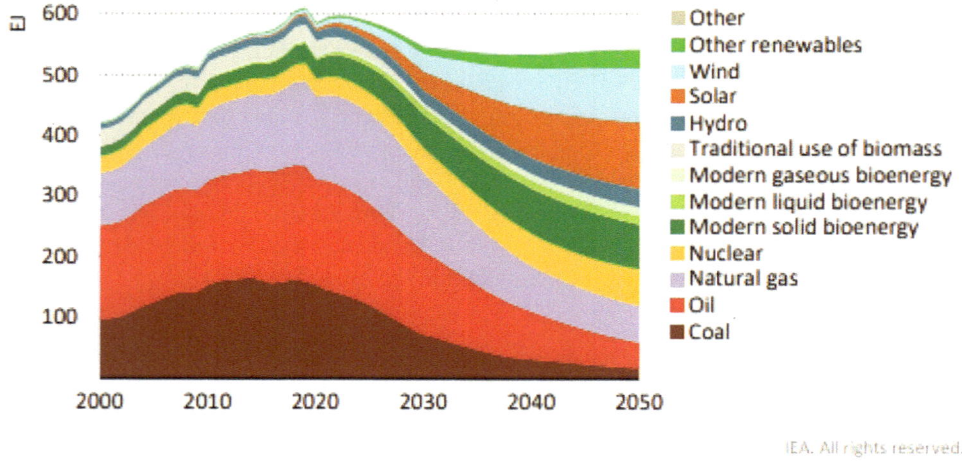

FIGURE 5.2 Projected renewable energy generation, 2000–2050.

Source: IEA (2021a), Net Zero by 2050: A Roadmap for the Global Energy Sector, p. 57.

by the development of the silicon photovoltaic (PV) cell, created at Bell Labs in 1954 by Daryl Chapin, Calvin Fuller, and Gerald Pearson. This advancement could help power an electric device for many hours, thus leading other thinkers to mark this finding as the true point for the invention of PV technology (Energy Sage, 2018).

5.3 GLOBAL SOLAR ENERGY DEVELOPMENT

This section explores the different ways that countries have coordinated their efforts in promoting and deploying the different solar applications and technologies that we have reviewed throughout this chapter. It is key that developed countries can export the results of the research performed, and that emerging democracies can adapt these programs and installations in their countries. Additionally, the capacity for deployment of solar applications and technologies can be realized faster when countries are aided by organizations like the United Nations, or by departments like the United States Department of Energy.

The focus of energy security will evolve as reliance on renewable electricity grows and the role of oil and gas diminishes. Potential vulnerabilities from the increasing importance of electricity include the variability of supply and cybersecurity risks. Governments need to create markets for investment in batteries, digital solutions, and electricity grids that reward flexibility and enable adequate and reliable supplies of electricity. The growing dependence on critical minerals required for key clean energy technologies calls for new international mechanisms to ensure both the timely availability of supplies and sustainable production. At the same time, traditional energy security concerns will not disappear, as oil production will become more concentrated (see Figure 5.3) (IEA, 2021a, *Net Zero by 2050: A Roadmap for the Global Energy Sector Report*, p. 24).

According to the IEA Renewables 2021 report, despite rising commodity prices and manufacturing costs, solar PV alone accounts for 60% (160 GW) of all renewable capacity additions (290 GW) in 2021, which is about double the rate over the previous five years. Globally, utility-scale solar PV is observed largely for adding new electricity capacity with the least costly option competing against rising natural gas and coal prices. Utility-scale solar project deployments are expected to continue 60% or more of all solar PV additions. In the meantime, China, the European Union, and India are boosting the deployment of commercial and residential PV projects by making policy changes (IEA, 2021b, *Renewables 2021 – Analysis and Forecast to 2026* report, p. 14–15).

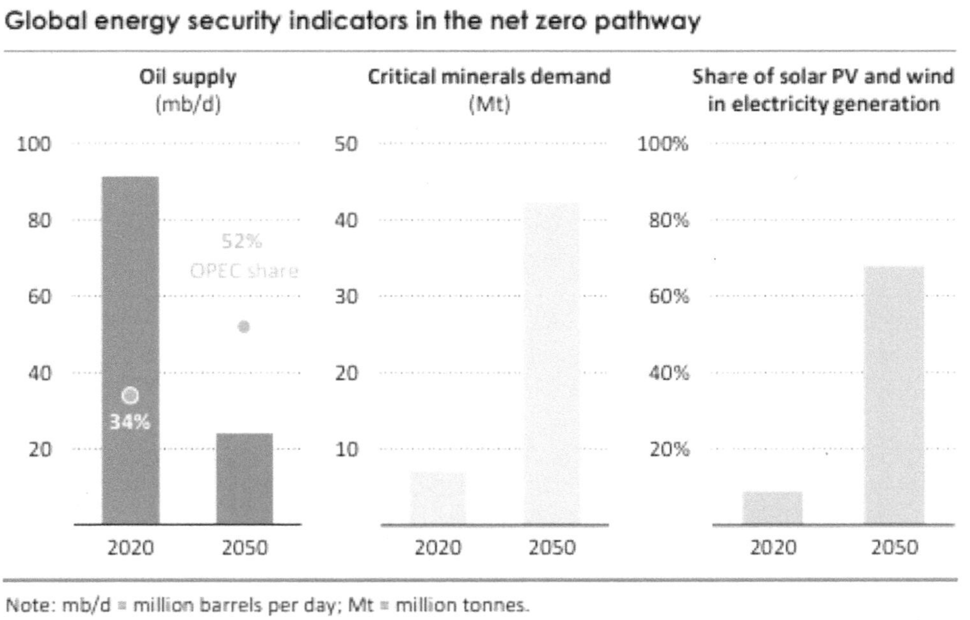

FIGURE 5.3 Global energy security indicators in the net-zero pathway.

Source: IEA (2021a), Net Zero by 2050: A Roadmap for the Global Energy Sector, p. 24.

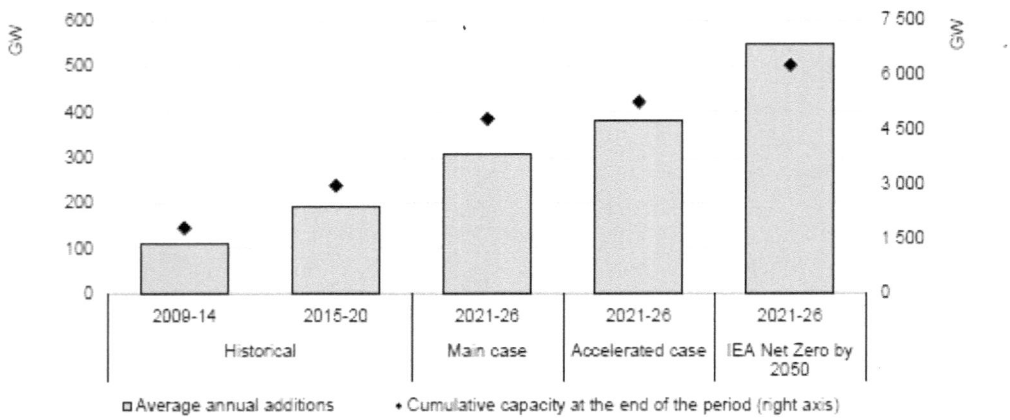

FIGURE 5.4 Average annual renewable capacity additions and cumulative installed capacity, historical, forecasts, and IEA Net Zero Scenario, 2009–2026.

Source: IEA (2021b), *Renewables 2021: Analysis and Forecast to 2026* report, p. 22.

Figure 5.4 shows both of IEA's average and cumulative average renewable capacity additions, historical and forecast, for 2021–2026.

China and the European Union (European Union, 2021) are set to overshoot their current targets, setting the stage for a more ambitious growth trajectory. China's commitment to reach carbon neutrality before 2060 has led to new nearer-term targets, such as 1,200 GW of total wind and solar PV capacity by 2030. In addition, improving competitiveness, ambitious targets, and policy support are putting renewable power on course for new highs in India and the United States. While it is expected that the expansion of renewable capacity in the United States is 65% greater than in the previous five years, within the 2021–2026 period, solar PV is anticipated to lead the way, driven by competitive auctions aimed at achieving the government's ambitious renewable power target of 500 GW by 2030 in India. Continuous policy support in more than 130 countries, ambitious net-zero goals announced by nations accounting for almost 90% of global GDP, and improving competitiveness of wind and solar PV are all driving this expansion. Nonetheless, despite this growing support, renewables face a range of policy uncertainties and implementation challenges, including those relating to financing, permitting, social acceptance, and grid integration (IEA, 2021b, *Renewables 2021: Analysis and Forecast to 2026* report, p. 14–15, 21).

Figure 5.5 shows that China, the Unites States, and India will be the major players in the economy of scale deployments, which currently account for 62% of global installed renewable capacity.

Figure 5.6 shows renewables' main and accelerated forecasts. Utility solar PV deployments will take significant share of overall renewable deployments within the next five years, as unit PV module costs are expected to be lower in the future. Figure 5.7 shows the unit PV cost projections by 2050.

In order to reach the accelerated case in Figure 5.6 illustrated by IEA's *Renewables 2021: Analysis and Forecast to 2026* report, the major markets, such as China, the United States, India, the Euro zone, and others, need to address their persistent policy challenges to be resolved. For China, the new renewable portfolio standard rules and green certificate scheme requires further clarifications for utility-scale and large-scale commercial applications, while the continuation of residential solar PV support needs extension following the phaseout of incentives in 2021. For the United States, further extension is required for investment tax credits and the monetization of credits through a direct-pay scheme. For India, improvements on the distribution companies' financial health, reduction of delays associated with signing PPAs following solar PV auctions, acceleration

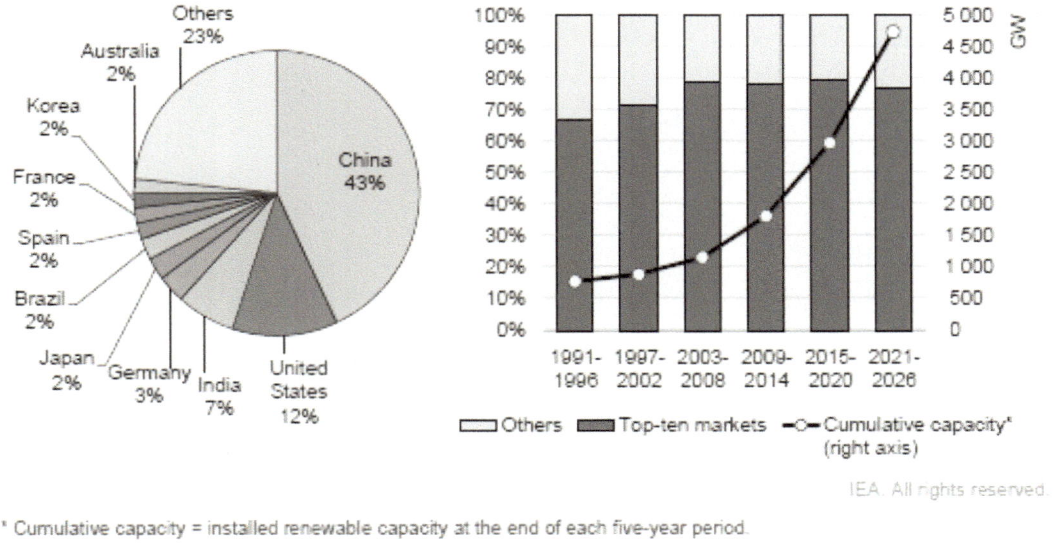

FIGURE 5.5 Top ten countries' share of total installed renewable capacity, historical and main case forecast, 1991–2026.

Source: IEA (2021b), *Renewables 2021: Analysis and Forecast to 2026* report, p. 26.

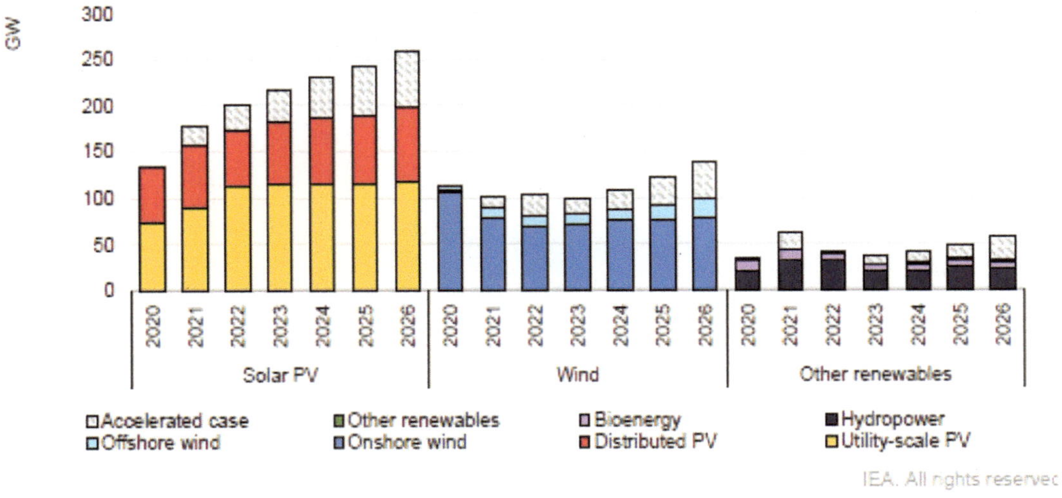

FIGURE 5.6 Annual capacity additions of solar PV, wind, and other renewables, main and accelerated forecast cases, 2020–2026.

Source: IEA (2021b), *Renewables 2021: Analysis and Forecast to 2026* report, p. 27.

of grid expansion, and further improvement on distributed PV applications are required. For the EU, an increase in auction capacity and an improvement of participation rates in utility and large commercial-scale auctions, combined with smoother permitting procedures, are needed when these are required for bidding. Further support is required on distributed solar PV by utilizing the EU resilience and recovery fund. For Japan, a smooth transition plan must be in place, from feed-in tariff (FIT) to the feed-in premium (FIP) scheme, to build up a declining project pipeline. For the MENA region, acceleration of the frequency of auction rounds and speeding up bidder selection

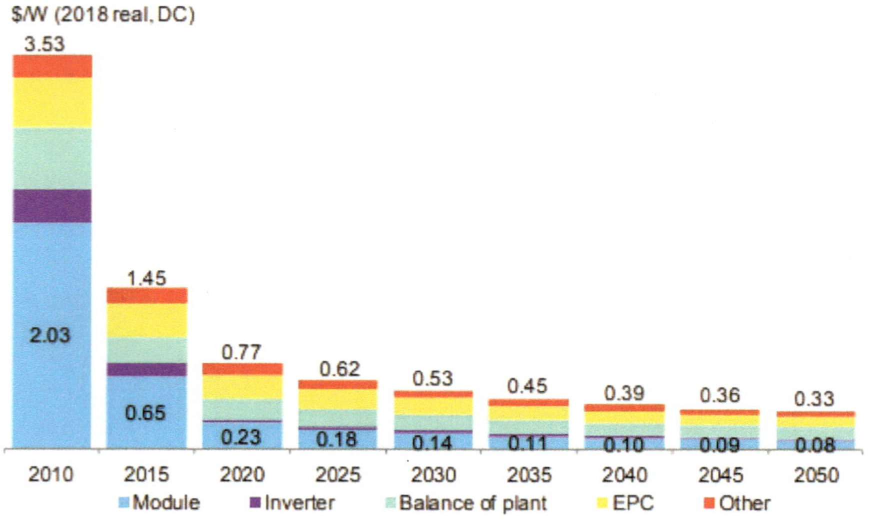

FIGURE 5.7 BNEF global benchmark capex forecast for fixed-axis utility-scale PV systems.

Source: BloombergNEF, Global Solar Investment Report (October 2019), p. 12.

TABLE 5.1
Key Deployment Milestones for Renewables Pathway

Sector	2020	2030	2050
Electricity sector			
Renewables share in generation	29%	61%	88%
Annual capacity additions (GW): Total solar PV	134	630	630
Total wind	114	390	350
- of which: Offshore wind	5	80	70
Dispatchable renewables	31	120	90
End uses sectors			
Renewable share in TFC	5%	12%	19%
Households with rooftop solar PV (million)	25	100	240
Share of solar thermal and geothermal in buildings	2%	5%	12%
Share of solar thermal and geothermal in industry final consumption	0%	1%	2%

Note: TFC = total final consumption.
Source: IEA (2021a), *Net Zero by 2050: A Roadmap for the Global Energy Sector*, p. 74.

and contractual negotiations, combined with rapid expansion of transmission and distribution grids, are needed to build up more utility-scale PV plants (IEA, 2021b, *Renewables 2021: Analysis and Forecast to 2026* report, p. 27).

Table 5.1 illustrates how the breakdown of solar PV deployment capacities in the electricity sector is expected to be two times higher than that of wind by 2050, which accounts for 64% of renewable deployments.

In buildings, renewable energy is mainly used for water and space heating. The direct use of renewable energy rises from about 10% of heating demand globally in 2020 to 40% in 2050; about three-quarters of the increase is in the form of solar thermal and geothermal. Deep retrofits and energy-related building codes are paired with renewables whenever possible: almost all buildings with available roof space and sufficient solar insolation will be equipped with solar thermal water heaters by 2050. Rooftop solar PV, which produces renewable electricity on-site, is currently installed on around 25 million rooftops worldwide; the number is expected to be 100 million rooftops by 2030, and 240 million by 2050. Solar thermal and geothermal also produce low-temperature heat for use in non-energy-intensive industries and ancillary or downstream processes in heavy industries. Bioenergy, solar thermal, and geothermal together provide about 15% of industry heat demand in 2030, roughly double their share in 2010, and this will increase to 40% in 2050. The indirect use of renewable energy via electricity adds 15% to the contribution that renewables make to total industry energy use in 2050 (IEA, 2021a, *Net Zero by 2050: A Roadmap for the Global Energy Sector* report, p. 74).

5.4 SOLAR ENERGY TECHNOLOGIES

In this section, we will review the different types of solar energy technologies that have been developed over time, the changes that they have experienced over time, and the challenges and opportunities that these bring for a conscious switch from using conventional sources of energy to renewables, particularly solar (see Figure 5.8).

After arriving at more efficient ways to collect the limitless energy of the sun, researchers started focusing on the best ways to accomplish this, expanding the use of these technologies across disciplines.

The two main technologies to actively exploit the energy of the sun are based on two different methods to dispose of the inexhaustible energy coming from the sun. First, solar photovoltaic (PV) modules contain a series of cells that convert the energy of the sun into electrical energy. Second, concentrating solar power (CSP) is a technique that relies on reflective surfaces to redirect sunrays, creating a beam that heats up a fluid that is stored inside a receiver. The steam that is produced from this heat is the energy that drives up the turbine, which, in turn, runs the generator that eventually produces the energy (Aman et al., 2015).

Additionally, solar thermal applications are a third developing method, where the energy of the sun is absorbed and gathered in a solar collector that transfers the heat into the working fluid used by the installation, which can be either air, water, or oil (Tian and Zhao, 2013). These applications

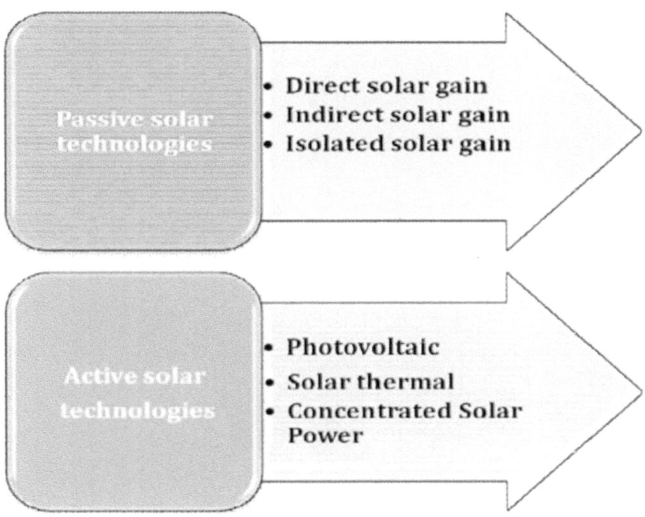

FIGURE 5.8 Classification of the present solar energy technologies.

Source: Kabir et al. (2018).

are of a smaller scale, as is the case with water or space heaters. As we can see, these methods are different ways to arrive at a similar objective, which is the transformation of the energy coming from the sun into a useful and serviceable output for human uses. In the following pages, we review these approaches in more detail.

5.4.1 Photovoltaic (PV) Devices

Solar PV is the leading solar technology today, constituting 98% of the global solar capacity. This technology is based on photovoltaic cells, which convert solar irradiance into direct current electricity passively in a scalable way.

PV technologies are classified into two categories based on manufacturing processes: wafer-based cells and thin film, as shown in Figure 5.9. Thin film PV technology requires a continuous manufacturing process which deposits the semiconducting material into an insulating substrate, such as glass or flexible plastic, whereas wafer-based technology does not require any substrate deposition and is fabricated from polysilicon into semiconducting wafers with multiple steps, as shown in Figure 5.10.

Currently, there are three major technology types of PV modules commercially used in the market. These distinct technology types are monocrystalline silicon, polycrystalline silicon, and thin film, specifically CIGS (copper indium gallium selenide) and CdTe (cadmium telluride) types.

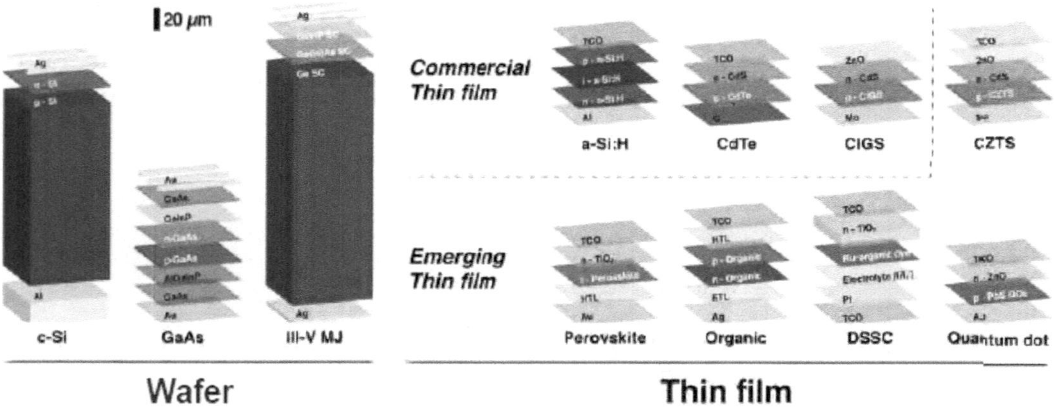

FIGURE 5.9 Solar energy conversion.

Source: MIT (2015), *The Future of Solar Energy*, p. 23.

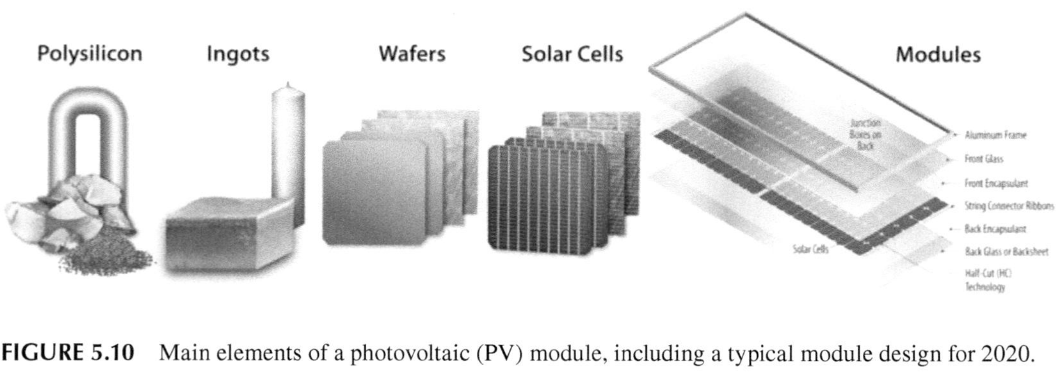

FIGURE 5.10 Main elements of a photovoltaic (PV) module, including a typical module design for 2020.

Source: United States Department of Energy (2021b).

However, GaaS (gallium arsenide) and III-V multijunction (tandem/hybrid), as well as perovskite thin film PV technologies, are currently under research and development stage, with promising prospects in the near future term. Figure 5.11 provides an overview of the PV technologies road map.

Due to learning curve, 60 years of development, the majority of commercial PV module manufacturing remains to be silicon-based, which are non-toxic, highly reliable, and efficient. Silicon-based PV technology is divided into two main categories: monocrystalline (m-Si) and multi- or polycrystalline cells (mc-Si). Monocrystalline PV cells are made from a single crystal of silicon ingot, more efficient and expensive to produce, whereas multicrystalline PV cells are a blended form of crystalized various silicon grains, cheaper to produce, with slightly lower efficiency.

A PERC (passivated emitter and rear contact) cells use an advanced silicon cell architecture. PERC cells are not much different in construction from a typical monocrystalline PV cell; however, the key improvement is the integration of a back-surface passivation layer, which is a layer of material on the back of the cells that is able to improve the cell's efficiency (Shravan and Chunduri, 2018). The efficiency gain of implementing the PERC architecture for monocrystalline cells is about 0.8% to 1% absolute, while the boost for multicrystalline cells is a little lower, at 0.4% to 0.8% (Shravan and Chunduri, 2018). According to Figure 5.12, PERC improves overall mono- and multicrystalline cell efficiencies at around 26.7% and 24.4%, respectively.

CIGS cells have achieved high efficiency levels (23.4%) comparable to commercial crystalline silicon (Fraunhofer ISE, 2022). However, manufacturing CIGS cells can be difficult due to the rarity of indium, as well as the complex stoichiometry and multiple phases to produce them, restricting large-scale production in the near term (Cherradi, 2019).

Cadmium telluride (CdTe) cells have achieved an efficiency of 21%, very similar to that of CIGS, and are characterized by good absorption and low energy losses (Fraunhofer ISE, 2022). CdTe solar cells are made through low-temperature processes, which makes their production very flexible and affordable. CdTe currently has the largest market share of all thin film technologies (Cherradi, 2019).

PV module efficiencies are lower than cell efficiencies due to losses associated with cell mismatches, parasitic absorption by the front glass and encapsulation, electrical wiring, and bypass diode at the junction box. Figure 5.13 shows commercial PV module efficiencies.

While continuous research and development are concentrated on developing new cell architecture toward higher efficiencies, PV manufacturers are focusing on new module architectures for higher energy production density. These new module architectures are half-cut cell, bifacial, and multi-busbar PV module cell architecture.

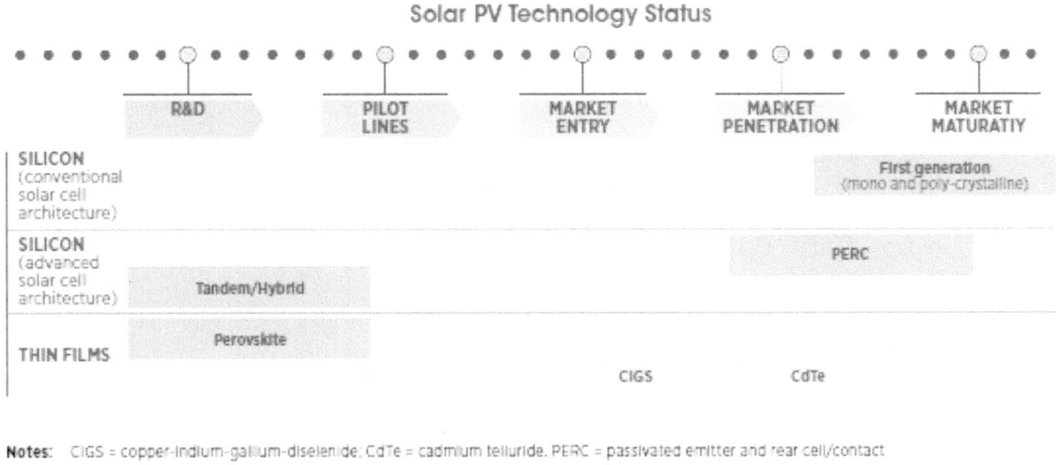

FIGURE 5.11 Solar PV technology status.

Source: IRENA Future of Solar Photovoltaic (2019), p. 41.

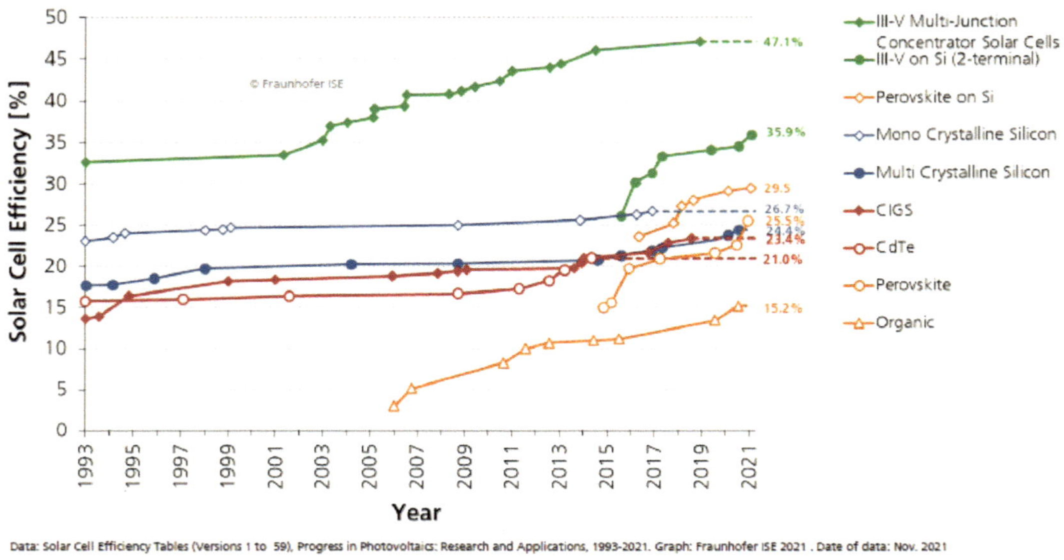

FIGURE 5.12 Solar PV technology cell efficiencies.

Source: Fraunhofer, Photovoltaics Report (2022), p. 27.

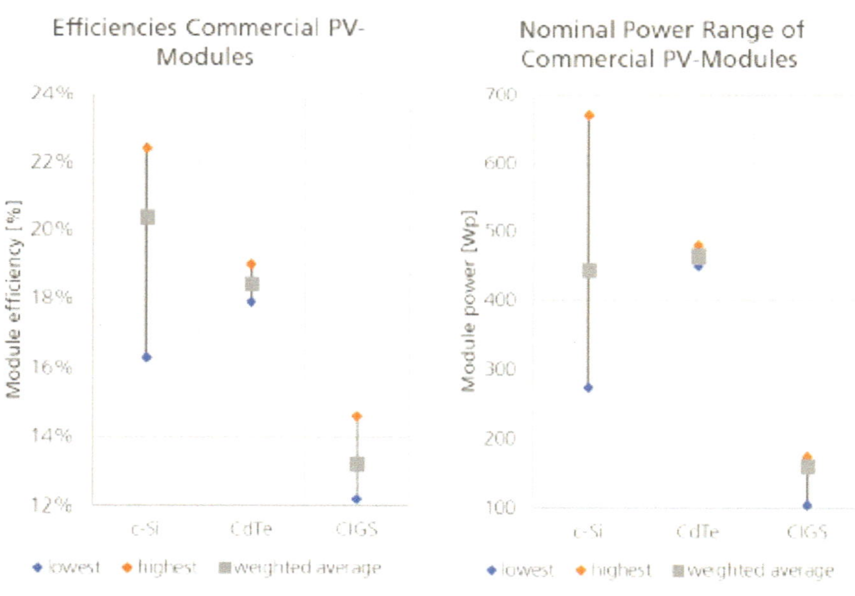

FIGURE 5.13 Commercial solar PV module efficiencies.

Source: Fraunhofer, Photovoltaics Report (2022).

A bifacial PV module cell architecture is capable of generating electricity not only from sunlight received on their front but also from reflected sunlight received on the reverse side of the cell. Energy production gain from reflectivity can be up to 6%. Bifacial modules are very suitable for utility-scale PV plants (IRENA the Future of Solar PV, 2019, p. 43).

A half-cut cell PV module architecture improves module performance and durability and can provide an instant power boost of 5–6 W (watts) (Shravan and Chunduri, 2019). PV module performance is improved by the reduction of cell interconnection wiring losses.

Silicon solar cells are metallized with thin strips printed on the front and rear of a solar cell; these are called busbars and have the purpose of conducting the electric direct current (DC) power generated by the cell. The increased number of busbars has several advantages: first is the high potential for cost saving due to a reduction in metal consumption for front-facing metallization (Braun et al., 2013); second, series resistance losses are reduced by employing thin wires instead of regular ribbon (Shravan and Chunduri, 2019); and third, optimizing the width of the busbars leads to an additional rise in efficiency. A higher number of busbars leads to higher module efficiencies because of reduced internal resistance losses; this is due to the lower distance between the busbars. Finally, a multi-busbar design is highly beneficial for bifacial technology, especially for improving bifaciality for PERC cells of 90% (Shravan and Chunduri, 2019).

A typical solar PV module is formed by combining interconnected PV cells with various cell configurations (60, 72, or 96 cells, as well as 120 or 144 half-cut cells), aluminum frame for mounting and protection, front and back laminated encapsulation layers using ethylene vinyl acetate (EVA) material for ultraviolet (UV) and moisture protection, and back sheet to provide more robustness as well as further environmental protection.

A PV plant is formed by combining one or more PV arrays, which are electrically connected PV modules, in serial and parallel configurations, integrated via a balance of system hardware components (BOS), such as inverters, combiner boxes, disconnect switches, wiring, and mounting systems. Figure 5.14 shows a complete solar PV system along with cross sections of the solar PV module and cell.

PV plant economics are measured by a metric called levelized cost of energy (LCOE), which measures the present value of lifetime energy production over the present value of lifetime plant cost (investment and operational). A typical PV plant lifetime start varies between 20 and 30 years.

With the support of the United States Department of Energy, the market for photovoltaics has blossomed. A decade ago, the estimated value of residential PV systems, measured in the levelized

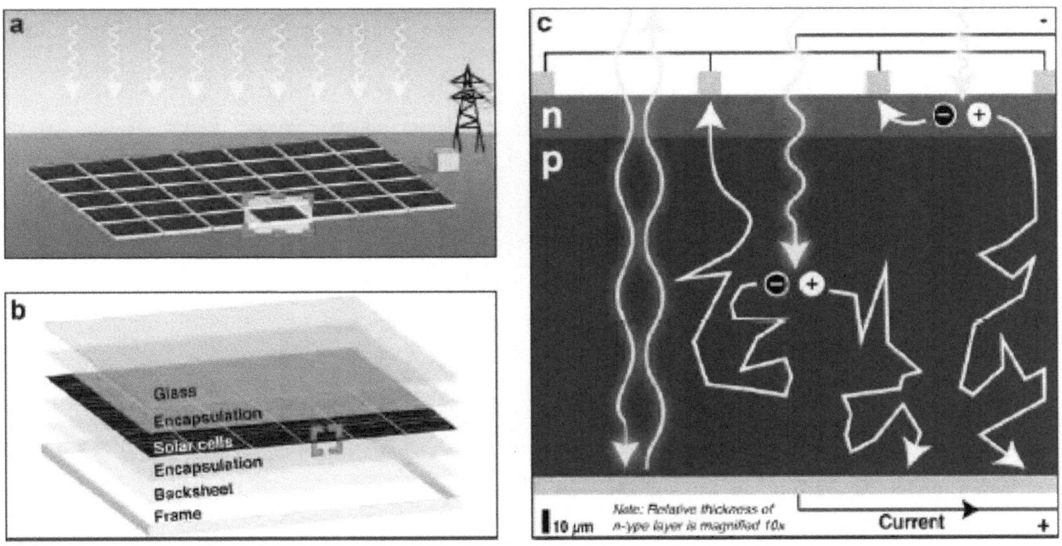

(a) Illustration of grid-connected PV system

(b) Breakout view of PV module

(c) Cross section of silicon solar cell showing PV mechanism

FIGURE 5.14 Solar PV cell structures and types.

Source: **MIT (2015),** ***The Future of Solar Energy*****, p. 22.**

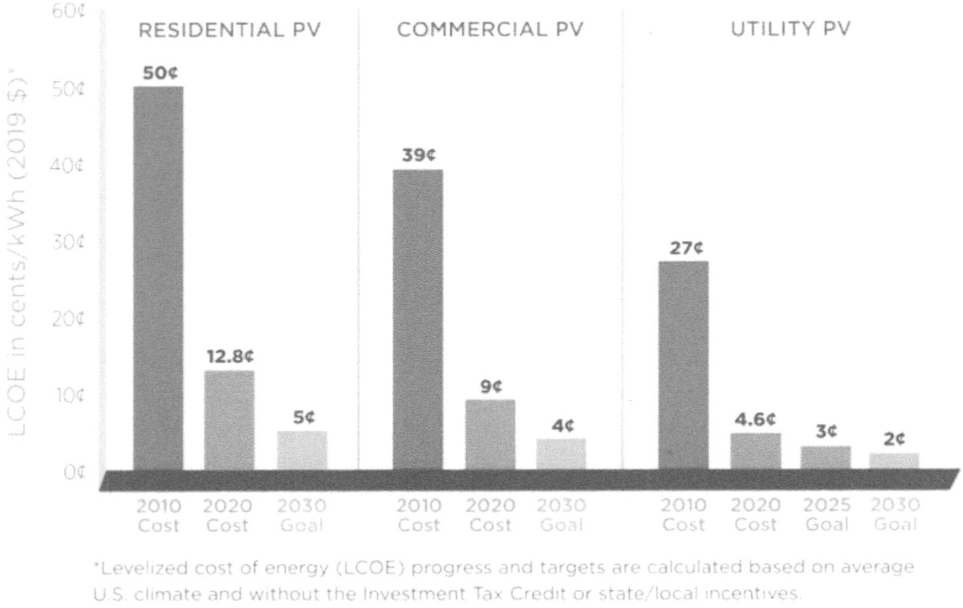

FIGURE 5.15 Photovoltaics progress and goals.

Source: United States Department of Energy (2021a).

cost of energy (LCOE) in 2019 dollars, was 50 cents per kilowatt-hour (kWh), while commercial PV systems were valued at 39 cents per kWh. Meanwhile, utility PV was estimated at 27 cents per kWh. Ten years later, these figures fell more than one-third, and the expectations of the Department are that these costs will reach comparable prices between each other, and that they remain below 5 cents per kWh (United States Department of Energy, 2021a). Figure 5.15 illustrates these trends.

Due to the price of solar panels having decreased almost 10% every year since the 1980s, the opportunity to invest in clean technologies is more enticing than ever before, and the reliability of PV technologies can be a safe bet to consider a switch to solar energy as powering our daily lives.

5.4.2 Concentrating Solar Power (CSP)

CSP technologies produce electricity by converting the rays coming from the sun to heat a medium that drives an electrical generator through a heated fluid that works like a steam turbine. The fluid can vary, and in many current designs, a heat transfer oil is circulated through the steel pipe, where it is heated up to a temperature of 400 °C (Aman et al., 2015). Molten salt thermal storage can be added for efficient heat storage, which makes CSP a "dispatchable" renewable power. CSP technology is suitable for hybrid configurations with conventional power sources, specifically combined cycle plants. Although CSP technologies offer attractive solutions, they have several limitations: First, CSP requires direct solar radiation. Unlike solar PV, intermediate cloud covers, hazy skies, humidity, sand, and dust have a great impact on system performances. Second, CSP is very sensitive to scale. Large-scale (100 MW or above) deployment is needed for maximum efficiency and lower costs. Third, CSP requires a water source for cleaning and cooling, and a flat large land.

Figure 5.16 illustrates the typical efficiency of a CSP plant. Due to losses associated with CSP's mirror array and thermal receiver, heat exchanger, etc., 42% of total solar energy can be delivered to the boiler as heat. Then, only 40% of the captured heat can be converted to electricity. As a result, overall system efficiency is around 16%. On the other hand, combined cycle conventional (fossil fuel) generation plant is around 55% consideration of similar steam turbine efficiencies.

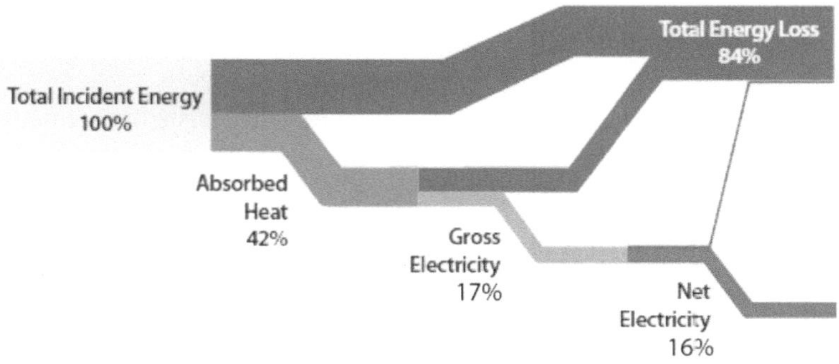

FIGURE 5.16 Efficiency of a typical CSP plant.

Source: MIT, (2015), *The Future of Solar Energy*, p. 52.

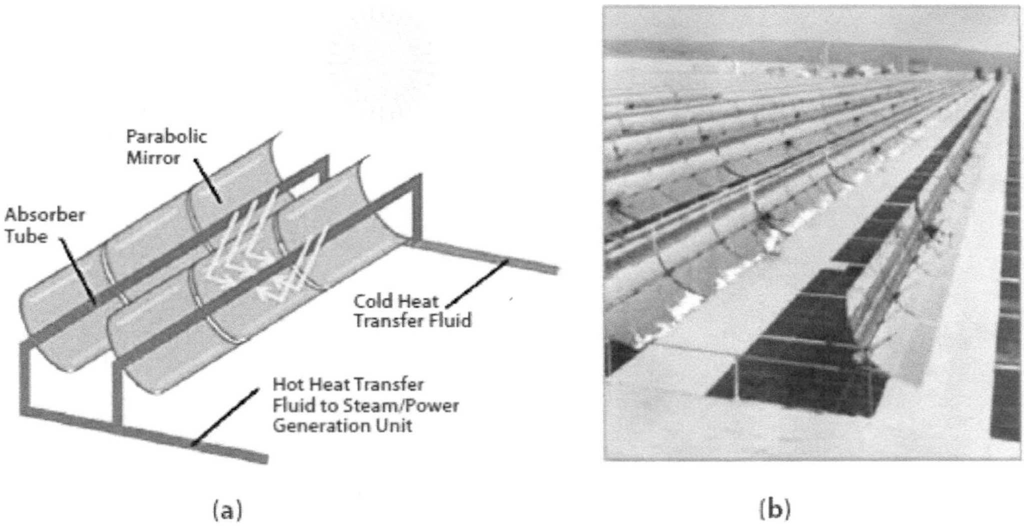

Note: (a) Schematic diagram of a parabolic trough solar field. (b) Solar Energy Generating Systems (SEGS) parabolic trough plant in the Mojave Desert at Kramer Junction, California. The tubes at the focal line of the trough carry the heat transfer fluid to the power generation system to generate electricity.

FIGURE 5.17 Parabolic trough CSP design.

Source: MIT (2015), *The Future of Solar Energy*, p. 53.

There are five CSP technology types available: parabolic trough (line-focus design), solar tower (point-focus design), linear Fresnel (low-cost and more reliable variation of line-focus design), beam down (recent low-cost variation of point-focus design), and Stirling dish.

The first approach, trough systems, uses large parabolic – or U-shaped – reflectors which have oil-filled pipes running along their center. The mirrored reflectors are tilted to face the sun, which helps heat up the oil inside to temperatures close to 750°F (400°C or less). Later, via heat exchange, the heated fluid is used to boil water, making steam that helps run conventional steam turbines and generators. The parabolic trough CSP system is equipped with a single-axis tracker to track the sun from east to west.

Although the parabolic trough CSP system (see Figure 5.17) is a mature technology, so considered as commercial technology, it requires a high capital cost associated with the many required mirrors and suffers from mechanical problems, convective heat loss, and radiation, which increase operation and maintenance costs. These systems also lead to lower thermodynamic efficiency due to the limited temperature of the heated fluid.

The second technique – also called the method of central receivers – uses multiple large mirrors which track the sun, focusing its rays onto a receiver that sits atop a tall tower, where the concentrated sunlight heats a fluid-like molten salt that can be heated up to 1,050°F (ranges between 250°C and 565°C). The hot fluid can then be used to generate steam for the generation of electricity, or stored for later use. The solar tower mirrors' function is to transfer maximum energy from the sun to the tower by dual-axis trackers. The tower height is determined based on solar field dimensions. Electricity can be generated two ways: direct and indirect steam generation. Direct steam generation occurs within the tower, while indirect steam generation requires a heat transfer approach from synthetic oil, molten salt, or air. The centralized collector increases overall efficiency.

Compared to parabolic trough systems, solar tower systems (see Figure 5.18) operate at higher-operating-temperature fluid and lower-heat-transfer surface area, which leads to higher efficiency and lower convective losses. In addition, the higher-operating-temperature fluid will help store the energy efficiently with the requirement of a smaller-size thermal storage, which reduces both cost and thermal storage heat loss. The solar tower requires dual-axis trackers for the mirrors, which leads to a higher maintenance cost associated with mirror alignment challenges. Intermittent cloud cover changes receiver fluid temperature rapidly, resulting to intermittent electricity generation as well as mechanical strain on the system. In addition, solar tower development requires careful environmental study for wildlife and rare species.

The third technique, beam-down CSP systems (see Figure 5.19), is a similar concept to solar towers, which consists of an array of tracking heliostat mirrors that reflect light to a single, centrally located mirror or a secondary heliostat atop a tower, which, in turn, reflects light down to an enclosed, secondary collector. This enclosed collection system may allow for very high-temperature working fluids and thus increased thermodynamic efficiencies. The beam-down technology has not yet been implemented at full plant-size scale. Current technical difficulties include geometric design issues, fabrication and control of the secondary heliostat, loss of light around collectors, and mirror material issues involving reflectivity and thermal strain (*The Future of Solar Energy*, MIT, 2015, p. 55).

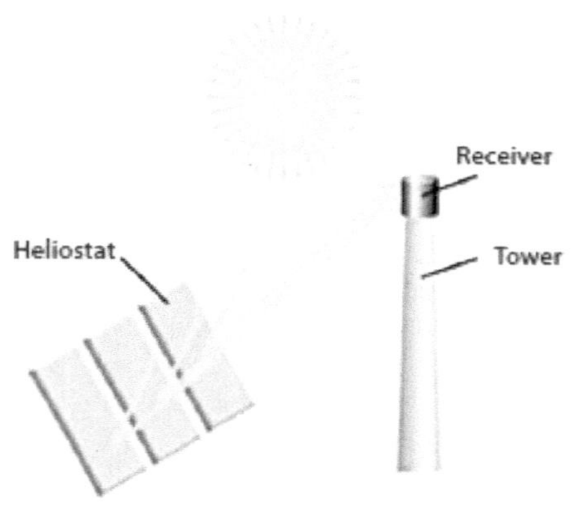

(a)

(b)

Note: (a) Schematic diagram of a solar tower design. (b) The Ivanpah SEGS plant in California's Mojave Desert is a 392-MW plant with 347,000 mirrors surrounding three 459-foot towers. Each tower is the height of a 33-story building.[6]

FIGURE 5.18 Solar tower CSP design.

Source: MIT (2015), *The Future of Solar Energy*, p. 54.

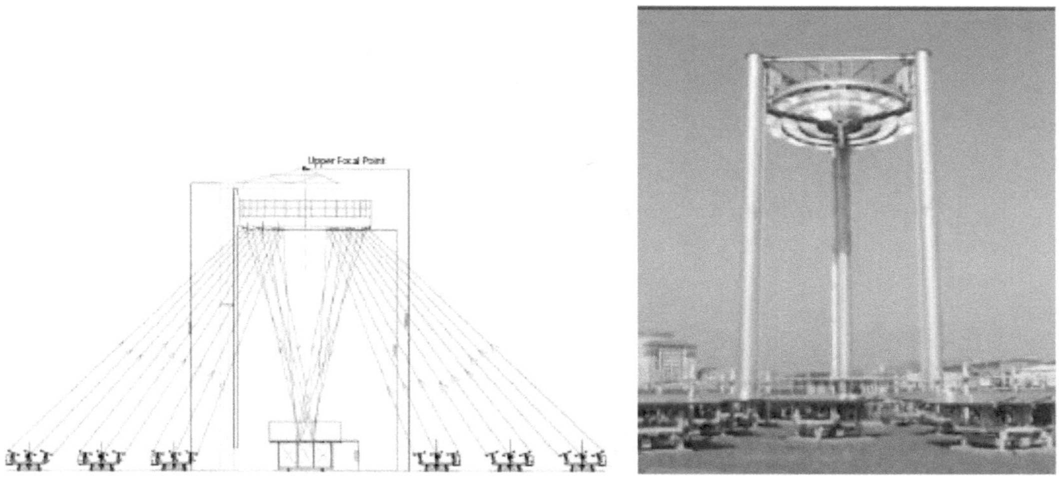

FIGURE 5.19 Beam-down CSP design.

Source: MIT (2015), *The Future of Solar Energy*, p. 56.

FIGURE 5.20 Linear Fresnel CSP design.

Source: MIT (2015), *The Future of Solar Energy*, p. 57.

The fourth technique, linear Fresnel CSP systems (see Figure 5.20), is a similar version to parabolic trough which consists of flat and/or slightly curved fixed mirrors to concentrate sunlight on a tube through which working fluid circulates. The entire system remains to be fixed, which makes the design simpler and subsequently cheaper overall. A Fresnel CSP system operates with a lower-temperature fluid, which results in lower efficiency compared to other CSP systems. A new version of the Fresnel design may operate with higher-temperature molten salt.

The last technique, Stirling engine dish systems, illustrated in Figure 5.21, uses mirrored dishes that are about ten times larger than a regular satellite dish which concentrate sunlight onto a receiver

FIGURE 5.21 Dish Stirling CSP design.

Source: MIT (2015), *The Future of Solar Energy*, p. 58.

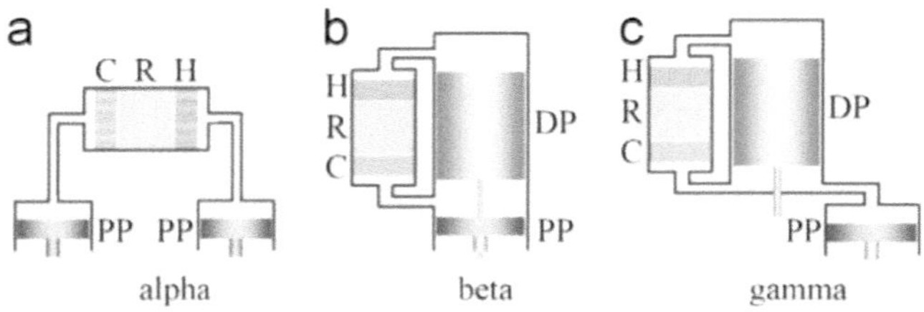

FIGURE 5.22 Three common arrangements of pistons in Stirling engines. (Note: cooler, *C*; regenerator, *R*; heater, *H*; displacer, *DP*; and power piston, *PP*).

Source: Vasu and Ismail (2018).

that is mounted at the focal point of the dish. To collect the maximum amount of solar energy (up to 750°C), the assembly of the dish tracks the sun across the sky, and the receiver is combined into a high-efficiency combustion engine (Stirling). The solar radiation that is picked up by the panel is redirected to the hot side of the Stirling engine, where there is a receiver that maximizes the amount of heat transferred to the engine while minimizing the thermal losses. Then, the engine is powered thanks to the constant flow of the working gas back and forth through the tubes of the heater (Howard and Harley, 2010), and eventually, solar energy is converted into electric power (Ahmadi et al., 2013). Altogether, the receiver, engine, and generator create a single assembly that is mounted at the focal point of the mirrored dish. Each dish has a unit rated power output typically ranging from 10 KW to 25 KW, which makes the technology modular and easily scalable. Since each dish requires a dedicated Stirling engine, these systems become capital-intensive, with high operating and maintenance (O&M) costs for larger-scale deployments.

Stirling engines are an excellent option to use for solar energy. The work of Vasu and Ismail (2018) reviews the three more common configurations of a Stirling engine, shown in Figure 5.22. The alpha design uses two pistons and two cylinders, while the beta and gamma designs each have

TABLE 5.2
Performance Data for Various Concentrating Solar Power (CSP) Technologies

	Capacity (MW)	Concentration	Peak Solar Efficiency	Annual Solar Efficiency	Thermal Cycle Efficiency	Capacity Factor (Solar)	Land Use (m^2 MWh^{-1} y^{-1})
Trough	10–200	70–80	21% (d)	10–15% (d)	30–40% ST	24% (d)	6–8
				17–18% (p)		25–70% (p)	
Fresnel	10–200	25–100	20% (p)	9–11% (d)	30–40% ST	25–70% (p)	4–6
Power tower	10–150	300–1,000	20% (d)	8–10% (d)	30–40% ST	25–70% (p)	8–12
			35% (p)	15–25% (p)	45–55% CC		
Dish Stirling	0.01–0.4	1,000–3,000	29% (d)	16–18% (d)	30–40% Stirl.	25% (p)	8–12
				18–23% (p)	20–30% GT		

Note: (d) = demonstrated; *(p)* = projected; *ST* = steam turbine; *GT* = gas turbine; *CC* = combined cycle.

$$\text{solar efficiency} = \frac{\text{net power generation}}{\text{incident beam radiation}} \qquad \text{capacity factor} = \frac{\text{solar operating hours per year}}{\text{8,760 hours per year}}$$

Source: Müller-Steinhagen and Treib (2004).

one piston; but the beta formation has one cylinder, and the gamma design has two. The authors find that the highest possible mechanical efficiency is achieved utilizing a double-acting piston, provided by the gamma configuration. Also, the working fluid should provide a high thermal conductivity, a high specific heat capacity, and a low viscosity.

Table 5.2 summarizes the performance data for various CSP technologies, according to the work of Müller-Steinhagen and Trieb (2004). Parabolic troughs, the technology with the most development, can show the greatest demonstrated numbers for efficiency. Moreover, these technologies have the potential to be "hybridized" with conventional (fossil fuel–based) combine cycle generation systems that can co-fire the plant and cover the needs of the plant during cloudy season.

Meanwhile, Table 5.3 summarizes and compares solar CSP technologies. Only dish Stirling engine systems are not suitable for large-scale deployment due to the high cost.

5.4.3 Solar Thermal

Solar thermal is an umbrella term for the different technologies that absorb the heat coming from the sun, transferring it to useful applications that are usually small- or medium-scale, like heating buildings or water (United States Environmental Protection Agency, 2021). Concentrating solar systems are a variation of a solar thermal installation, applied to larger settings. Meanwhile, small-scale PV systems can be turned into solar thermal applications when the panels are connected to water boilers, for example (see Figure 5.23).

Thus, solar thermal is understood more as an application of the first two methods to specific circumstances. A significant breakthrough was the work of Roland Winston, who developed non-imaging optics, a technology that now allows for solar thermal energy collectors to heat fluids up to temperatures of 750°C to use in varying types of buildings (Henkel, 2005). A non-imaging optical device is one that redirects light rays on a surface such that the reflection does not produce an image of the object producing the light.

The advantages of this technology to concentrate solar energy focus on the attainment of cheaper, simpler, and easier-to-maintain devices for solar concentration. Given that non-imaging

TABLE 5.3
Comparison of Solar CSP Technologies

Focal Geometry	Line-Focus Technologies		Point-Focus Technologies		
Design	Parabolic Trough	Linear Fresnel	Solar Tower	Beam-Down	Dish-Stirling Engine
Technology maturity	Most mature	Few installations	Commercial deployments	Early development	Proposed installations
Preferred scale	Large	Large	Large	Large	Small
Capital cost (relatival)	Moderate	Low	High	Moderate to high; low storage costs	High (low per unit)
Operating cost (relatival)	High	Low	High	Similar to solar tower	High (one engine per dish)
Annual solar-to-net-electricity conversion efficiency[a]	~5%[b]	~11%[b]	~17%[b]	~15%–19%[13, c]	~22%[b]
Thermal storage	Feasible	Feasible	Feasible and more efficient due to higher temperature	Feasible; very little energy lost	Not currently possible
Characteristics	• Significant construction and operational experience • High radiative and convective energy losses	• Low cost due to fewer moving parts and no tracking • Lower efficiency	• High cost due to expensive heliostat field • High-temperature HTF possible • High efficiency	• Lower efficiency than best solar tower due to added mirrors • Lower storage cost (ground storage)	• High engine efficiency • High cost due to expensive engines (one for each dish)

Notes:

[a] The efficiency figures are indicative values. Many factors affect the efficiency of CSP plants, such as plant size and location, technologies selected for plant components, efficiency of the heat-to-electricity system, etc.

[b] The calculated efficiencies are obtained from SAJV1 simulations (except for the beam-down entry). All simulated cases are assumed to be located in Daggett, California, and have 150 MW_C net capacity.

[c] We do not have detailed information on the beam-down technology reported in Ref. 13. As a result, the reported efficiency may not be on the same basis as the other technologies.

Source: The Future of Solar Energy, MIT (2015), p. 64.

techniques permit comparable levels of concentration of energy without the level of tracking that previous methods have, non-imaging technology leads to a more cost-effective solar infrastructure (O'Gallagher, 2008). As a positive externality of this technological advancement, we can now see more replicable efforts for small- and medium-scale projects in different parts of the world that governments and private actors can finance with more ease.

Briefly, we note a few notable works that deal with new solar thermal applications in small-scale settings. First, the work of Delgado-Torres (2009) develops a comprehensive review of several solar thermal engines created for water pumping. Then, Winston et al. (2015) summarize the development of non-imaging technology in the last couple of decades and apply it to photovoltaic collectors (see Figure 5.24). In turn, Widyolar et al. (2020) comment on an experimental framework for a photovoltaic thermal (PVT) ultra-low-cost solar collector that uses aluminum mini-channels and non-imaging optics.

FIGURE 5.23 Roof solar panels connected to a water heater.

Source: Greenmatch.co.uk (2021).

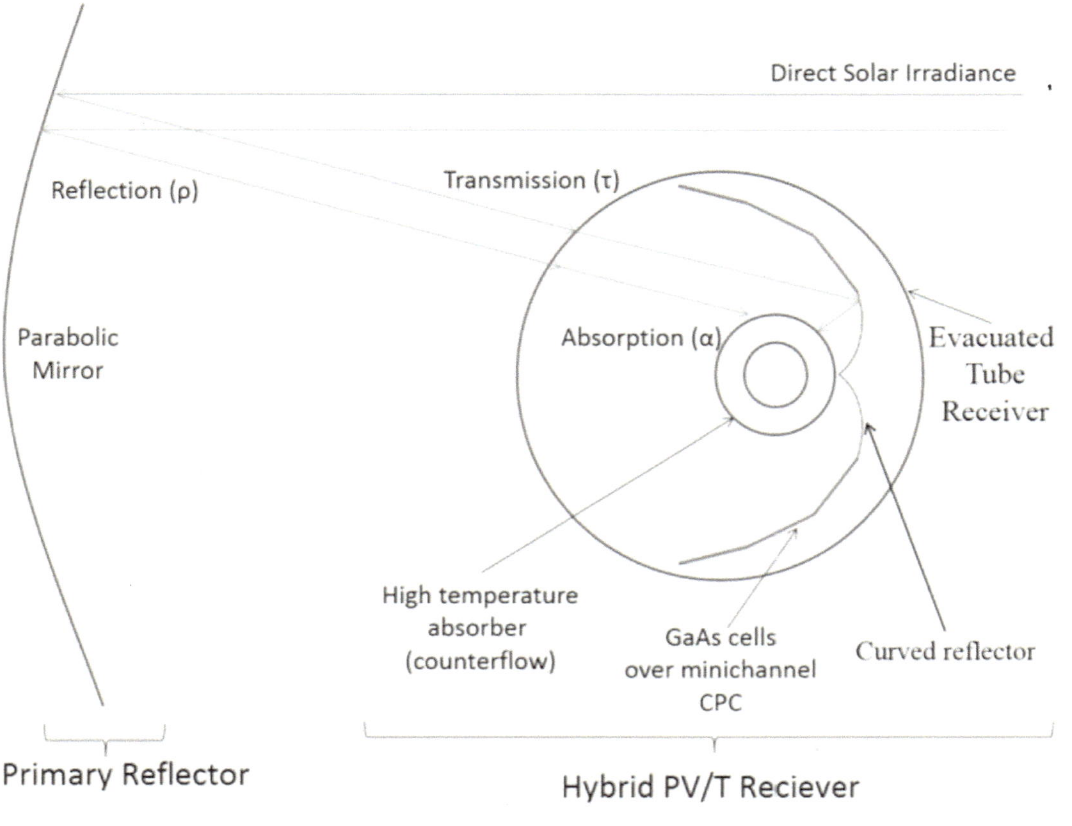

FIGURE 5.24 Photovoltaic thermal (PVT) collector components.

Source: Winston et al. (2015).

The many advances in non-imaging optics constitute a big leap forward for technologies that seek to capture the energy of the sun and use it for small- and medium-scale endeavors. The development of modern buildings includes considerations for installations that capture solar or wind energy to power some aspects of community life, and the retrofitting of old buildings is another consideration that the literature is exploring (Soutullo et al., 2020).

Relatedly, the next section centers on the many applications of solar technologies in daily life, starting with an exploration of a series of solutions offered for the residential sector. Improving the houses in which we live, making them fit to be powered by the energy from renewable sources, is a worthwhile endeavor and is a starting point for a larger switch that includes the commercial and industrial sectors. In those two settings, we can deploy larger installations that use a greater amount of energy, and we can foster the construction of solar plants that increase the share of the energy coming from the sun to power our daily activities.

5.4.4 Solar Applications

Even though an expansion in the share of solar energy to power our daily lives can help benefit the whole world, applications of the technologies explored in the previous section are not infinite. Due to resources, land, or other types of constraints, solar technologies are still in a phase where researchers must be conscious of the benefits but also the costs that installing or building a solar plant can have on a region or a population. In the following section, we review a series of applications of solar installations that have been picked up by the literature, and we analyze the potential opportunities for similar endeavors in the future.

5.4.5 Residential Applications

The most common and cost-effective forms of residential solar applications are mainly classified into two simple categories: solar water heating via solar thermal collectors, and power generation to offset electricity consumption using a grid-connected solar PV system. Adequate space and solar resource availability are the core limitations for both applications.

Grid-connected solar PV system applications are classified as distributed generation, which targets minimizing energy use from the grid. Besides space and solar resource, grid regulation must be taken into account for system capacity sizing and configuration. Figure 5.25 shows a basic grid-connected solar PV configuration. The PV array generates direct current (DC) power from solar irradiance. Then, the generated DC power is converted to alternative current (AC) power by the PV inverter, which gets connected to the main distribution panel via a disconnect switch. The generated power supply from the solar PV is consumed via on-site demand. If net metering regulations are in place, then surplus energy is utilized for nighttime usage simply by reverse-metering scheme; otherwise, battery storage must be needed to store surplus energy for nighttime usage.

Similarly, Ahmadi et al. (2020) developed a Monte Carlo simulation technique for a smart home that has two main variables of interest: outdoor temperature and solar radiation. "Smart home" is a main component of the Internet of Things (IoT) and a concept that involves all the physical devices that help the home user with electronic, sensor, software, and network connectivity. Figure 5.26 shows the proposed energy supply for the model home. Using a Monte Carlo simulation, the authors created 1,000 random scenarios for the two variables of interest, with the goal of minimizing the total energy cost of the house. For this, the authors set the thermal load, the output power of the PV, the power generation of the solar collector, and the electricity load as stochastic parameters. Results show an energy cost saving of 20% and a lowering of the peak electricity demand from 7.6 kWh to 4.2 kWh.

Researchers from Paderborn University in Germany have developed a model to deploy a residential rooftop PV in combination with batteries for short-term storage and hydrogen for long-term

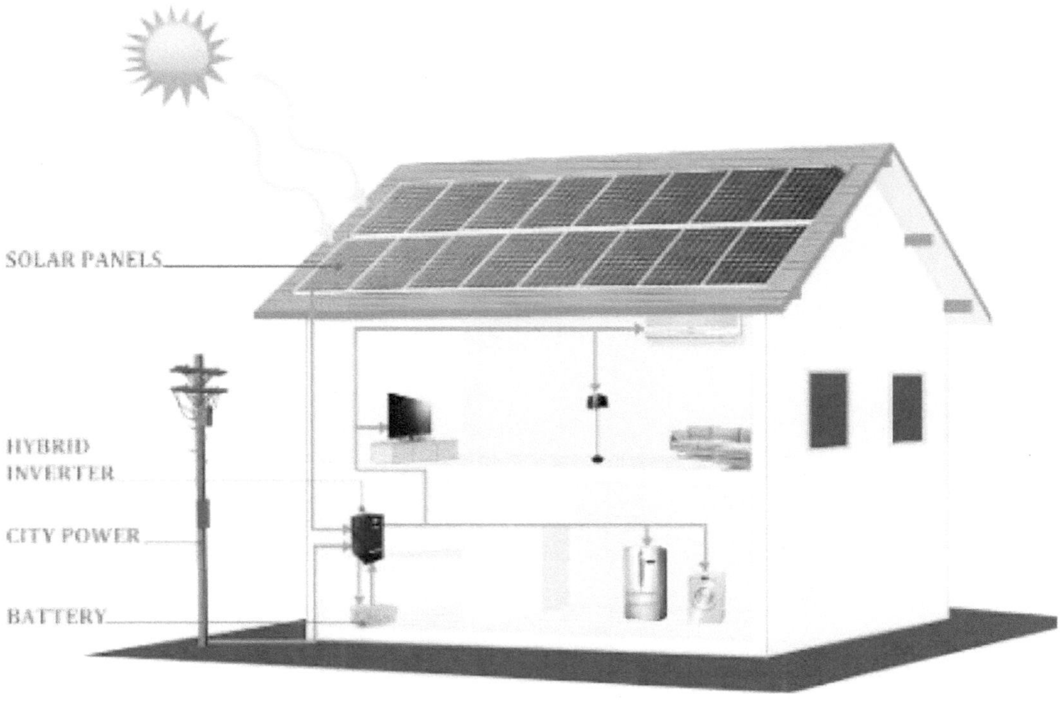

FIGURE 5.25 Residential grid-connected solar photovoltaic system.

Source: With the advancement of other technologies, hybrid systems like PV storage, PV thermal storage, and PV hydrogen storage are starting to become more popular options to cover all net-zero requirements.

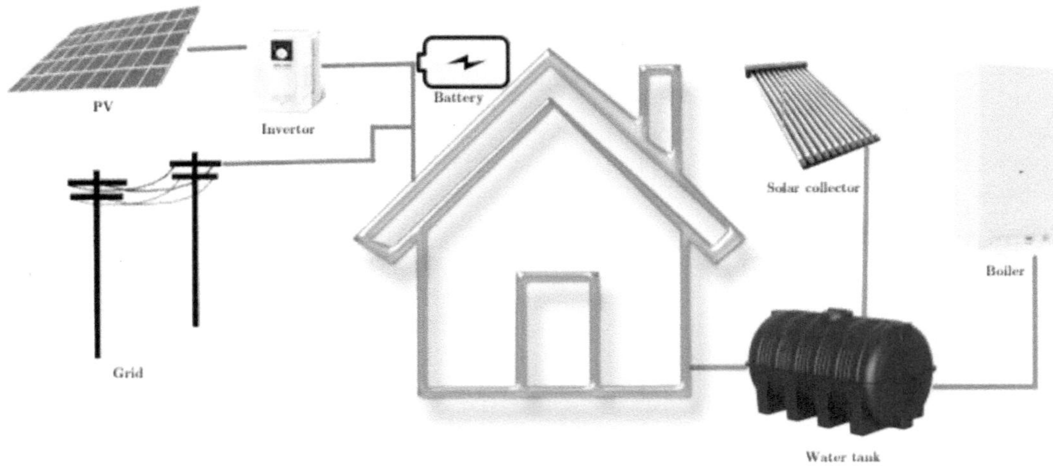

FIGURE 5.26 Proposed energy supply for the case study.

Source: Ahmadi et al. (2020).

storage, as shown in Figure 5.27. "The aim is to minimize the grid load by achieving low or even zero grid interaction, which is realized by the use of decentralized long-term energy storage," the scientists explained. The system utilizes PV array to generate power to electrical loads. Excess power is utilized to store energy as well as produce hydrogen through an electrolyzer, which is

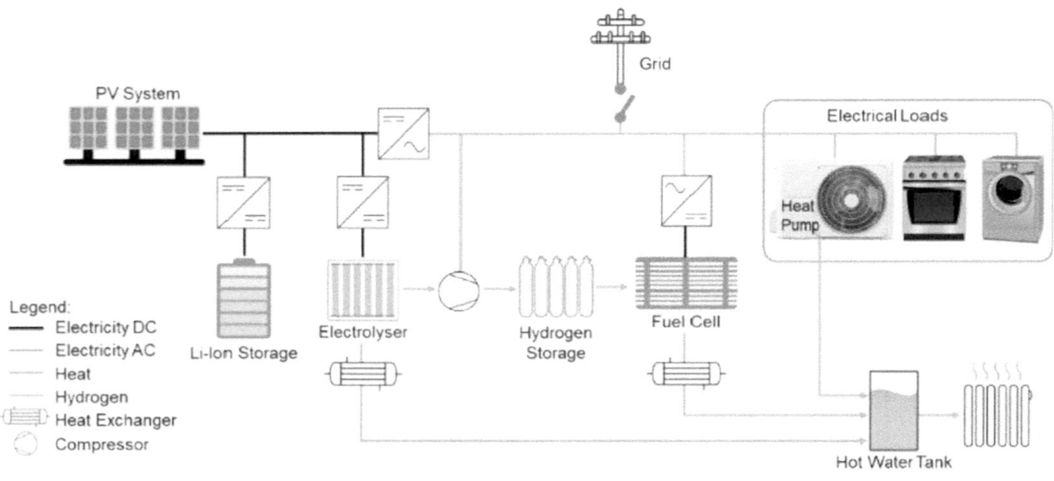

FIGURE 5.27 PV hydrogen fuel cell storage hybrid system.

Source: PV Magazine (2022), "Designs for solar+storage+hydrogen systems in buildings."

then stored in a hydrogen tank via a compressor. During high-demand usage, hydrogen fuel will be used by the fuel cell in parallel with PV to generate power. Li-ion batteries are configured for intermittent changes on the load demand, where the fuel cell will be able to compensate. The heat generated from the fuel cell and electrolyzer can be utilized via heat exchangers and a heat pump for water heating and space heating (PV Magazine, March 2022, "Designs for solar+storage+hydrogen systems in buildings").

One other application of solar energy in the residential sector consists of solar heat pump systems (SHPs), which are hybrid systems where heat pumps – mostly electrical – are used in combination with solar thermal (ST), solar photovoltaic (PV), or both types of systems, with the objective of residential heating. Compared to pumps without contributions from solar energy, these hybrid pumps hold a superior seasonal performance factor (SPF), and pumps combined with photovoltaic technology use less electricity, which results in a lower energy consumption (Poppi et al., 2018). Figure 5.28 shows the different system boundaries for SHP systems.

Solar thermal collectors are divided into two main categories: (1) covered or glazed and (2) uncovered or unglazed. At temperature levels typical for space heating (SH) and domestic hot water (DHW) preparation, covered collectors have higher efficiencies compared to uncovered collectors, thanks to their transparent cover that reduces heat loss to the environment. Covered collectors are divided into flat-plate collectors and evacuated-tube collectors. Flat-plate collectors, in parallel to heat pumps, represent the state of the art for this type of system for residential heating applications (Hadorn, 2015).

Solar photovoltaic systems can be integrated with heat pumps for (a) increasing the share of locally consumed electricity with respect to the total electricity generated by the photovoltaic array or (b) reducing electricity feed-in peaks to the electrical distribution grid (Fischer and Madani, 2017). As buying/selling electricity from/to the electrical grid has become more expensive than using electricity generated by photovoltaics, at least in the residential sector, self-consumption has become an economically viable option (Fischer and Madani, 2017).

Another residential use for solar heat pumps concerns domestic water heating. Aye et al. (2002) state the importance of these alternative heating mechanisms, especially due to the goal of discontinuing the use of R-22, or chlorodifluoromethane, a gas used as a working fluid to exchange heat between the heat source and the heat sink that possesses both a relatively high ozone depletion potential (ODP)

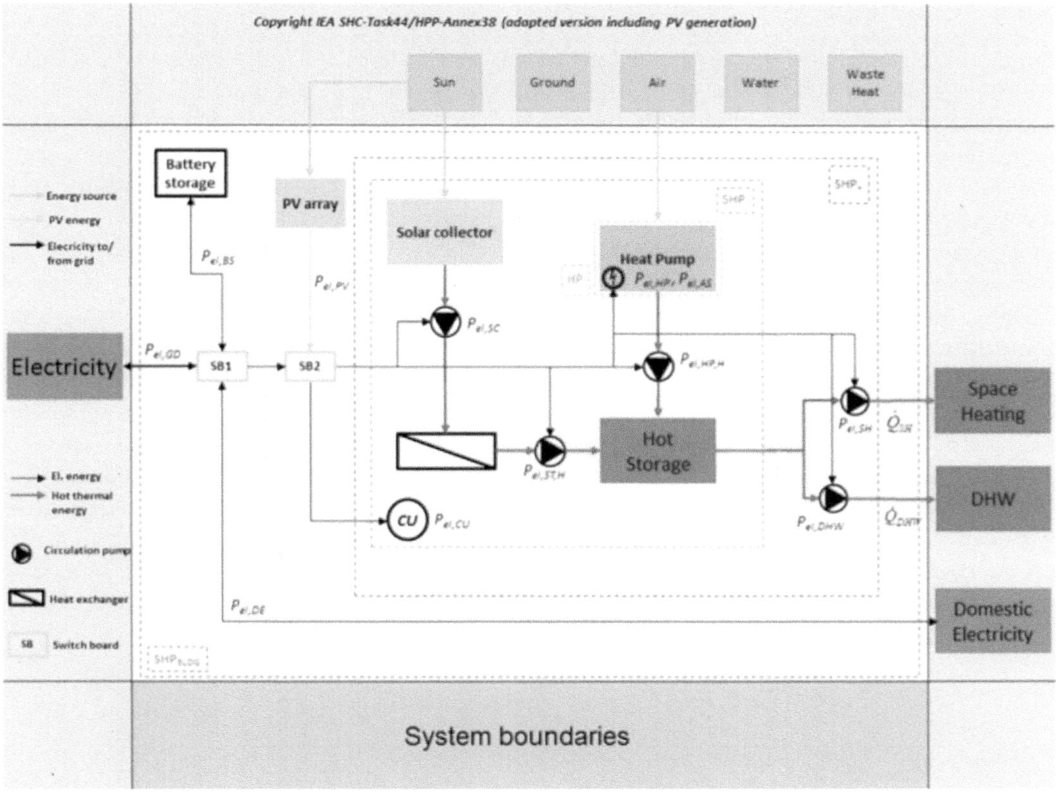

FIGURE 5.28 Different system boundaries: (a) system boundary HP includes heat pump only; (b) system boundary SHP does not include circulation pumps for space heating and hot water preparation, while (c) system boundary SHP_+ does include these pumps; (d) system boundary SHP_{PV} includes the PV array but does not include battery storage; (e) system boundary SHP_{BLDG} includes battery storage, which does not include, but is influenced by, domestic appliances.

Source: Poppi et al. (2018).

and global warming potential (GWP). For this purpose, the authors compared three different water heating systems and their performance to make recommendations for their use in Australia.

The first system analyzed was a thermosiphon water heater that relies on thermosiphon technology – or the passive exchange of heat to make hotter water less dense and ready to be delivered to the residence by a collector placed close to the top of the storage tank – to heat up the water. Then, the second system was an air source heat pump water heater, where ambient air is forced to transfer heat to the liquid working fluid – which, in most cases, is R-22 – by a fan. The third and last system is the solar heat pump water heater, which works like air source heaters but relies on solar radiation as a primary source of heat instead of air (Aye et al., 2002). In the end, location is key in the decision to move forward with one of the three water heating systems. In cities with stable solar radiation availability, the most appropriate water heating system is the one that uses thermosiphon.

Residential applications for solar installations are only one part of the switch to renewables from fossil fuels. Despite considerable acceptance to solar equipment from large groups of the population, individual efforts are insufficient to complete a more meaningful change to take advantage of the inexhaustible energy of the sun. However, grid regulations, space availability, and solar resources are the limiting factors for residential applications.

5.4.6 Commercial Applications

Similar to residential, the most common form of solar applications are grid-connected PV systems, which target for energy cost reduction. The overall system configuration is similar, but the system capacities are scaled to be larger, as shown in Figure 5.29. The installations can be either rooftop- or ground-mounted, based on the available space. Some configuration may require transformers to change voltage levels. As system capacity gets larger, monitoring PV system performance becomes more important to rectify any faults or malfunction quickly.

The number of innovative solutions for implementing PV is becoming increasingly relevant. Building-integrated photovoltaics (BIPV) (use of existing building surfaces), floating PV (use of existing water surfaces), or agrivoltaic systems (APV) (double use of land for food and energy) are some of these new examples. New conceptual designs using PV arrays to cover open farm fields have shown promising results through optimizing light availability while reducing the need for irrigation and protecting from extreme weather phenomena. APV denotes sharing the sunlight for the co-production of food and energy on the same piece of land; therefore, designs must overcome, as far as possible, physical constrains of covering crops with photovoltaic modules in order to alleviate the reduction in crop profitability (Toledo and Scognamiglio, 2021). Generated energy from APV can be used in water pumping and smart water irrigation systems to optimize water usage for on-site farming. Although APV is still in its very early development stage, it may have a significant impact on climate challenges to overcome related to food and energy security, water conservation,

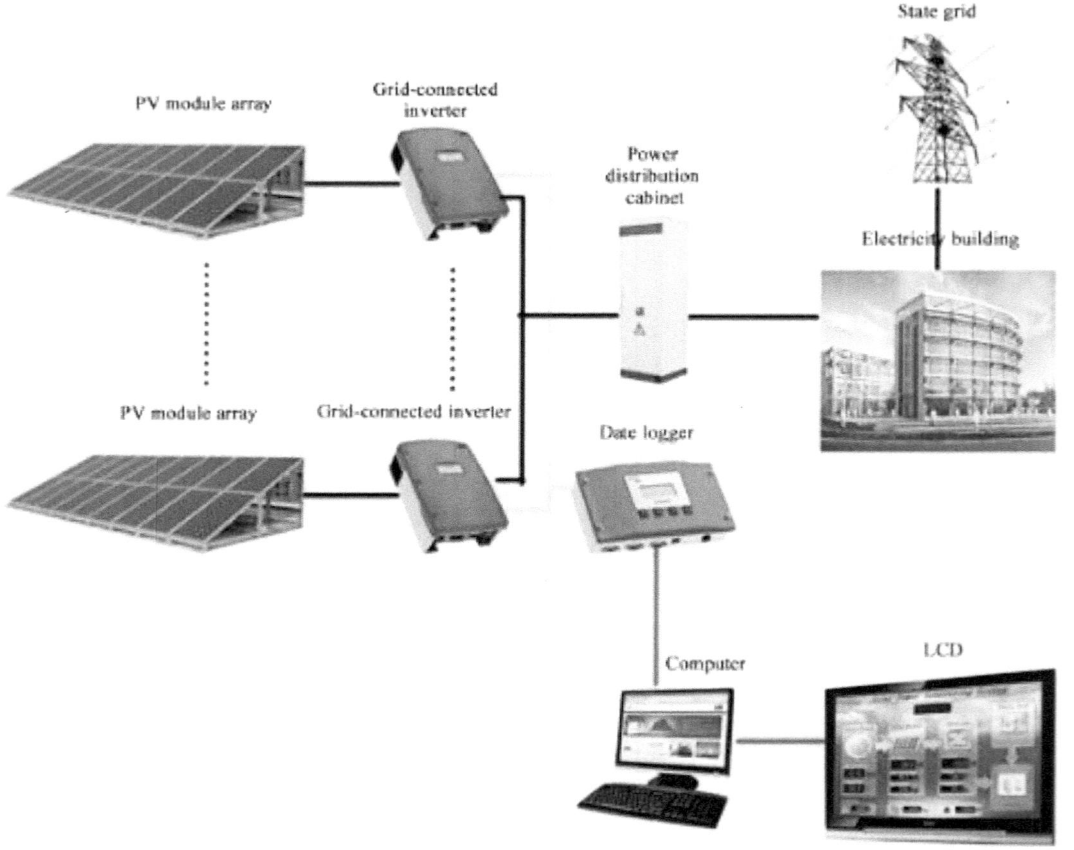

FIGURE 5.29 Commercial grid-connected solar photovoltaic system.

Source: longi.com.

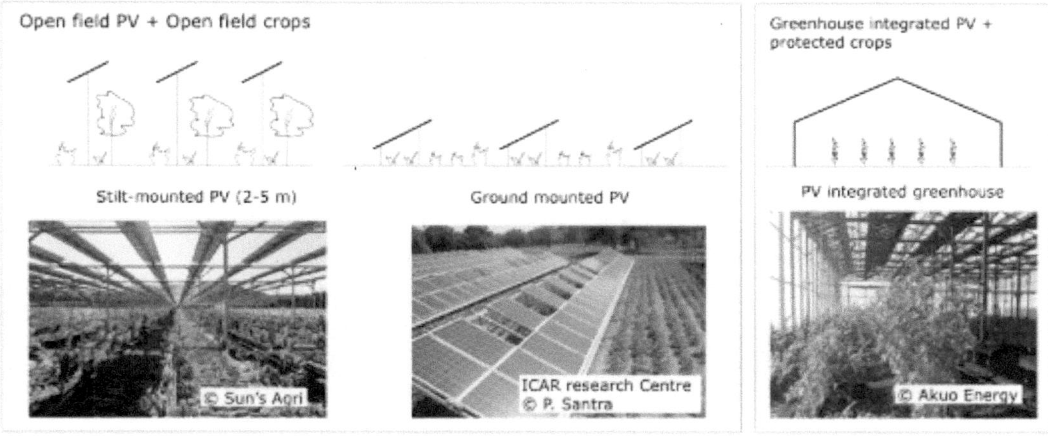

FIGURE 5.30 Commercial agrivoltaics solar photovoltaic system.

Source: Toledo and Scognamiglio (2021).

as well as agricultural land development in the future. Several types of APV installations are shown in Figure 5.30.

Ideally, PV projects ought to be replicated across the world. However, distinct conditions and constraints that affect the land within a country (though, for a "floating" PV power plant that looks to solve the land constraint that some countries have, see Sahu et al., 2016), or various legal or governmental restrictions, can hinder the potential of a solar installation. Figure 5.31 shows the typical cross section of an FPV system primarily constructed from a floating structure for PV array installation and cabling and mooring system to secure floating structure to limit free movement. Depending on distance, grid connection can be made by installing a cable on a floating structure to be transferred to the nearest ground location or by installing underwater leak- and corrosion-proof cable supported by bare steel cables. Overall system weight must be reduced and balanced. Cabling and mounting shall be carefully designed to overcome contraction and expansion movement caused by waves, earthquake, etc. All components must be made from corrosion-resistant material. FPV systems tend to produce more energy than ground or rooftop systems primarily due to the cooler water temperature, which leads to lower PV cell temperature and better power output performance and higher sunlight reflectivity from water. FPV also preserves water by preventing water evaporation.

Regarding the use of solar PV technologies in emerging sectors, there is promising evidence for the use of solar-powered unmanned aerial vehicles (UAVs) for the practice of precision agriculture, which is the use of technology and data to improve the process of agriculture and crop management (Herwitz et al., 2002). In this study, researchers analyzed the improvement in the Kauai Coffee Plantation, the largest of its type in the United States. Mapping was achieved with the use of two digital camera systems mounted on solar-powered UAVs developed by NASA and AeroVironment Inc.

Besides solar PV, solar thermal is considered the cheapest alternative among renewable energy systems (Mekhilef et al., 2011). These technologies typically work with a solar collector and concentrator to gather solar radiation, which is then stored and used for heating air or water in either domestic, industrial, or commercial settings. Figure 5.32 illustrates the basic schematic of a solar thermal conversion system. Different types of solar collectors will be better suited for different sectors, and the commercial sector appears poised to receive a good portion of experimental and novel installations. Rejected heat can be fed back to the storage tank or utilized for other needs, such as space heating.

Current solar technologies have reached a point where we can move much advancements which have shown relative success from the laboratory – or the experimental setting – to the market. The

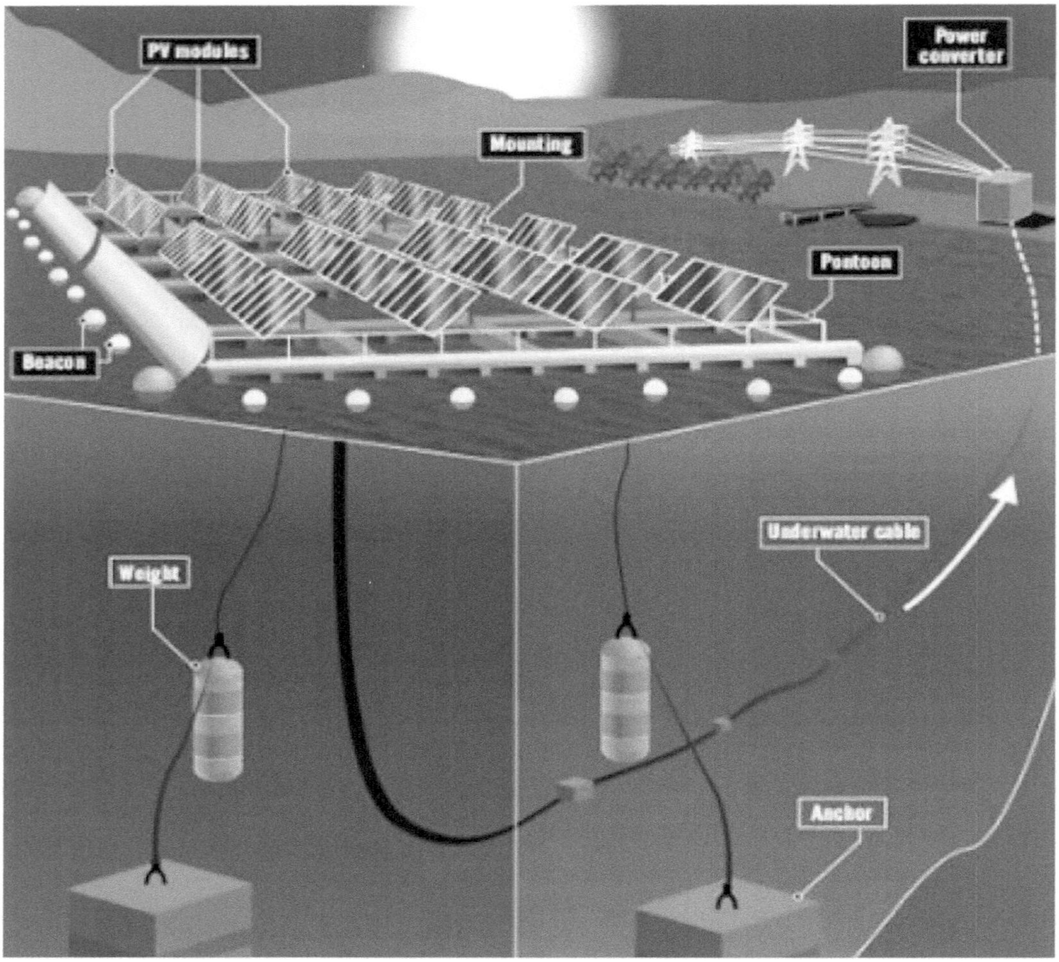

FIGURE 5.31 Commercial floating solar photovoltaic system (FPV).

Source: waaree.com.

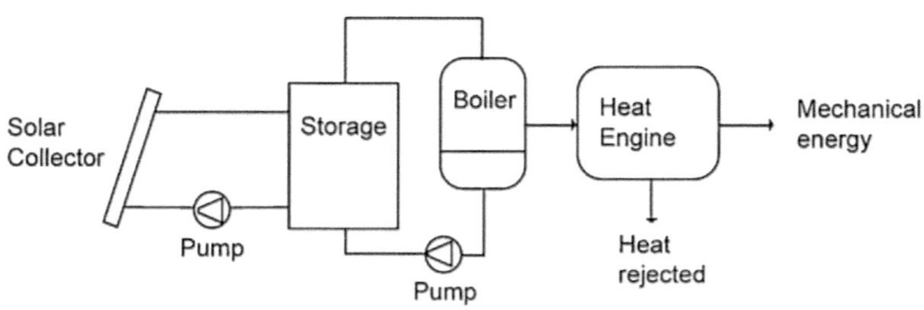

FIGURE 5.32 Schematic of a solar thermal conversion system.

Source: Mekhilef et al. (2011).

work of Qu et al. (2010) reviews the experience of building a cooling and heating system at Carnegie Mellon University that consists of several parabolic trough solar collectors. Based on their experience, the authors recommend paying close attention to the orientation of the collectors with respect to the sun, placing a storage tank in the solar collection loop, minding the length and diameter of the pipe, and certain features of the building, the economics of the energy supply, and the solar radiation profile of the geographical location (see Figure 5.33).

Similarly, the work of Ozsoy and Corumlu (2018) reviews the performance of a thermosiphon heat pipe evacuated-tube solar collector (THP-ETSC) that uses silver-water nanofluid for commercial applications. A THP is a closed passive heat transfer device that works with a fluid, and it has three basic parts: the condenser, the adiabatic, and the evaporator (see Figure 5.34). The authors tested the use of a nanofluid and compared it to pure water and found that the nanoparticles in the fluid increase the degree of internal heat transfer due to the thermal conductivity of the silver/water nanofluid being higher than that of pure water.

Another technique that is seeing positive results after real-world experiments is that of solar photocatalysis. In short, this technology consists of utilizing the energy of the sun to accelerate a process of photoreaction that is moved forward by a catalyst. Thus, a polluted resource is targeted by solar radiation and a catalyst to help separate it, eliminating the polluting elements and keeping the pure resource. Through a series of reduction and oxidation processes (Bahnemann, 2004), resources

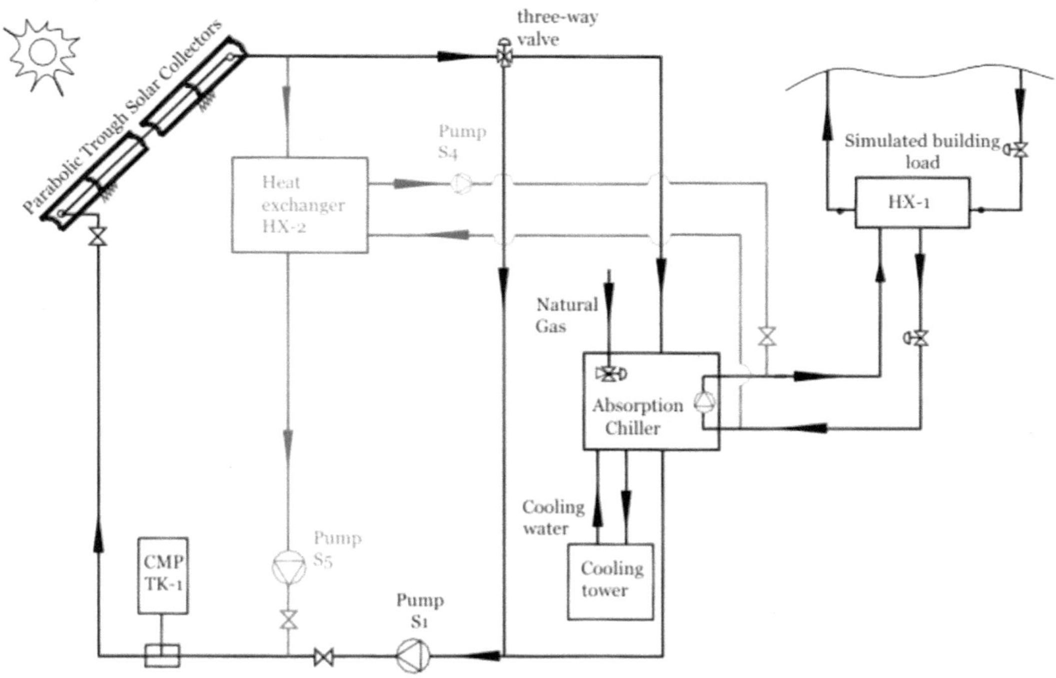

FIGURE 5.33 Process and instrumentation diagram of the solar absorption cooling and heating system.

Source: Qu et al. (2010).

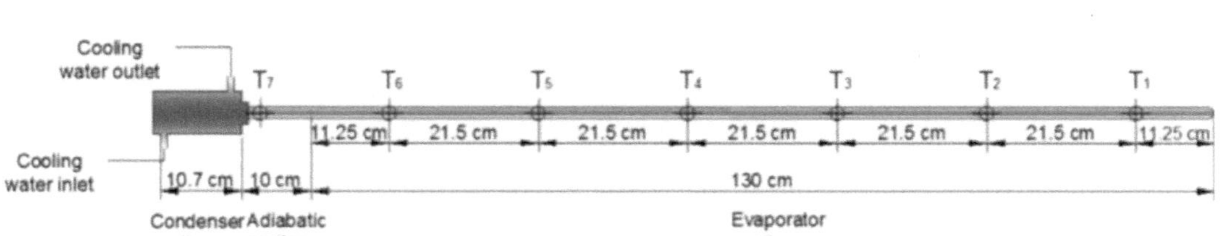

FIGURE 5.34 Schematic view of the thermosiphon heat pipe (THP).

Source: Ozsoy and Corumlu (2018).

TABLE 5.4
Main Commercial Applications of Titanium Dioxide with Sunlight

Commercial Field	Function	Use
Glass	Self-cleaning, antibacterial, and antifogging	Windows, mirrors, and vehicle glasses
Vehicles	Self-cleaning, antibacterial, antifogging, and avoiding dewdrops	Coating and painting of vehicles and side-view mirror
Textiles	Self-cleaning and antibacterial	Tent materials, storage structures, and stations and domes
Food and flowers	Antibacterial and anti-aging	Fruit and flower preservation
Buildings, hospitals, and concrete	Self-cleaning, antibacterial, and antifogging	Kitchen, bathrooms, exterior tiles and coating, and walls and roofs
Road materials	Self-cleaning and antifogging	Road pavements and mirrors, sidewalks, and lighting and signs
Water treatment	Decontamination, antibacterial, and mineralization	Small industrial and agricultural wastewater treatment plants
Industrial chemistry	Selective oxidation	Fine chemical solar productions

Source: Spasiano et al. (2015).

like water can be returned to a more natural state, safe for human consumption. These dynamics, in turn, can help develop markets and commercial ventures that take advantage of renewable energies, enabling and maintaining a virtuous circle that serves locally but has more global objectives.

As Table 5.4 illustrates, applications of photocatalysis technologies that utilize the energy of the sun to separate elements within a resource are numerous, going from textiles and vehicles to the treatment of wastewater as well as building materials that need bacterial removal. Unsurprisingly, the expansion of the opportunities has fostered a larger market for photocatalyst products, where proactive players can initiate endeavors and both profit and promote a safer environment.

In this section, we reviewed the commercial applications of solar technologies as well as experimental ones from researchers and engineers to increase the degree of commercialization of solar installations. Promising new technologies and underutilized methods can be coupled to improve the ways we use solar energy for our daily needs and to reach sustainable levels of production in commercial settings. Positive results from experiments like the ones we reviewed here allow for a greater reliance on renewable energy, progressively leaving conventional sources behind.

5.4.7 Industrial Applications

Energy intensity within industrial processes is highly concentrated around heating and cooling so that solar thermal energy is gaining significant momentum. In terms of thermal production, solar energy is divided into three main categories: (1) passive heating, (2) active heating, and (3) active cooling. Active heating is further divided into two categories: (1) solar thermal systems for low heat (<150°C) and (2) solar thermal systems for medium heat (150°C–400°C).

Passive heating, the simplest form of heating technology, refers to solar air dryers or solar air heating mainly used in the agricultural industry to prevent spoilage and contamination and lengthen shelf life, maintain or enhance product quality, and facilitate transportation of natural food products, as shown in Figure 5.35.

Figure 5.36 shows preheating water for steam boilers that will be used in industrial processes. Depending on water capacity, pressure, and temperature requirements, boilers can be designed to operate with electricity which can be supplied by solar PV. If industrial processes require steam, then a heat exchanger and thermal storage can be added as intermediate-stage heating before the boiler. In this case, thermal storage may need additional electrical power, which can be compensated by adding solar PV.

FIGURE 5.35 Direct solar drying.

Source: IRENA (2015), *Solar Heat for Industrial Applications*, p. 8.

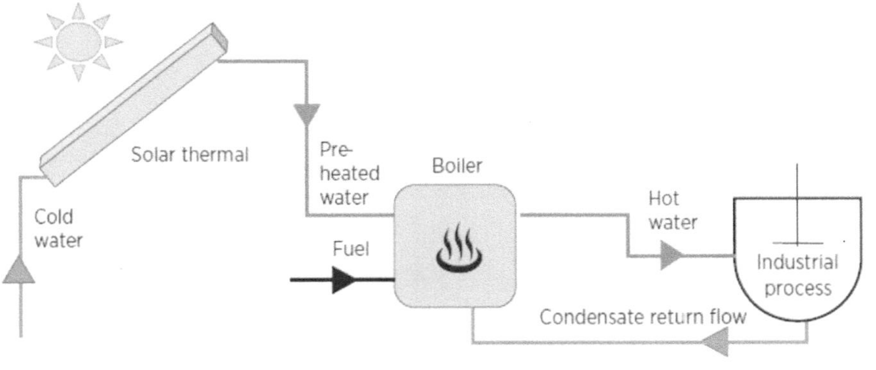

FIGURE 5.36 Active heating with solar thermal.

Source: IRENA (2015), *Solar Heat for Industrial Applications*, p. 14.

A solar thermal implementation for industrial settings is seen in the work of Allouhi et al. (2017), who studied a milk processing company in Morocco to implement an optimization procedure for a centralized solar water heating system. The process of treating fresh milk to make it fit for human consumption consists of several steps, and some of them require hot water at different temperatures. In this study, the authors developed an installation that covers the following four processes, two exclusively for milk, and two for the fruit used to make mixed milk-based beverages:

1. *Process 1, milk chilling.* To avoid spoilage and proliferation of microorganisms, milk needs to be chilled to a temperature of 5–7°C. For this, hot water at 80°C is required.
2. *Process 2, fruit cleaning.* To reduce bacteria and prevent contamination, fruit is treated at a temperature of 75°C.
3. *Process 3, milk pasteurization.* To eliminate almost all pathogenic germs, milk must be treated at a temperature of 90°C.
4. *Process 4, fruit drying.* Before mixing with the milk, fruit must be dried with water–air heat exchangers, which require water heated at a temperature of 60°C.

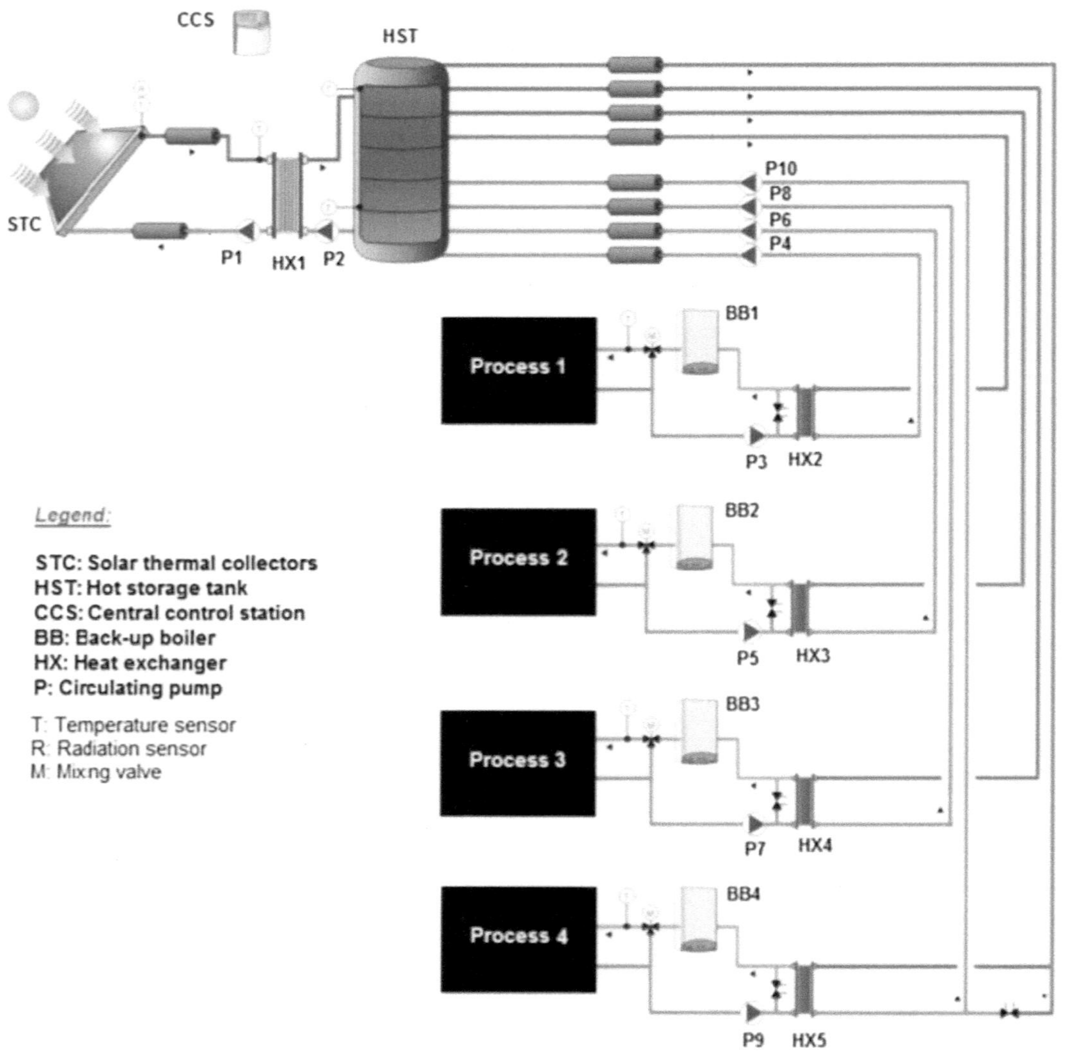

FIGURE 5.37 Schematic diagram of the proposed installation for a milk processing company.

Source: Allouhi et al. (2017).

Figure 5.37 shows the diagram for the installation that deals with these four processes, relying on a main setup that collects solar energy to power each stage. Allouhi et al. (2017) found that this setup can reduce CO_2 emissions by almost 80 tons a year. Although higher solar resource availability in Morocco provides encouraging results, many factors exist that influence the degree of success of solar technologies.

Besides heating applications, solar thermal systems can also meet cooling demands. In this case, solar thermal cooling systems can be used to replace gas-driven or electricity-driven absorption/adsorption chillers or to replace electricity-driven, vapor-compression air-conditioning systems.

Absorption and adsorption chiller systems use liquid or solid refrigerants to cool the environment. In absorption chillers (the most common system), solar energy is used to regenerate the absorber fluid, which contains the refrigerant, after it has been evaporated. Two common systems are closed absorption chiller systems with ammonia water (Figure 5.38(a)) or water lithium bromide

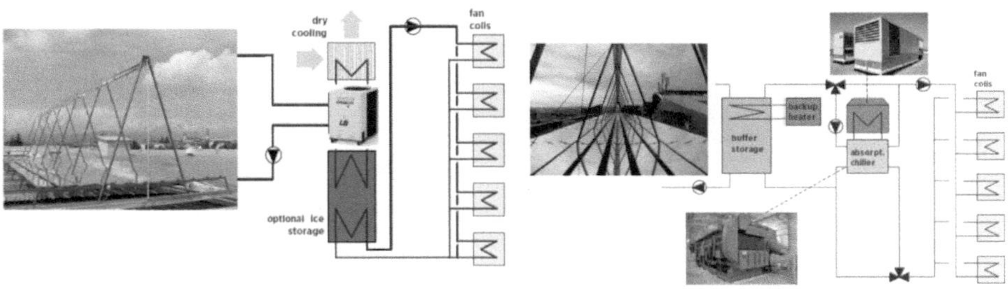

(a) Ammonia Water Chiller *(b) Water-lithium-bromide Chiller*

FIGURE 5.38 Solar cooling with absorption chiller.

Source: IRENA (2015), Solar Heat for Industrial Applications.

(Figure 5.38(b)) as refrigerant/absorber fluids. Desiccant systems are used to provide air-conditioning and use a desiccant material to absorb or adsorb warm water from the air and pass cooled air back into the building. Solar energy is used to regenerate the desiccants. Single-effect chillers, like desiccant cycles, have lower efficiency but also require lower temperatures $(70^\circ\text{C} \text{to} 100^\circ\text{C})$ to operate. Double- and triple-effect absorption chillers are only available for capacities of 100 kW or more and have higher efficiencies, but also higher heat requirements of $150-180^\circ\text{C}$ and $200-250^\circ\text{C}$, respectively (International Energy Agency (IEA), 2012).

Most thermally driven chillers used in industry have large cooling capacities $(>100\ \text{kW})$ and require high temperatures $(>100^\circ\text{C})$ and so cannot be driven by conventional solar thermal collectors. However, the market has changed, with the working temperature requirements for chillers decreasing and the temperature range from advanced solar water heaters and solar concentrators increasing (IEA-SHC, 2013). These higher temperatures from solar process heat installations and solar collectors are also needed to couple them to more efficient double- and triple-effect chillers. Furthermore, a number of companies in Austria, China, Germany, India, Italy, Ireland, and Japan now offer standard system designs to integrate and complement other systems (Augsten, 2012). Consequently, FPC and ETC can now be used to drive air-conditioning systems and slab cooling in the range of +20°C to −10°C, while solar concentrators can be used to drive refrigeration up to −20°C (Jakob, 2014).

Table 5.5 and Table 5.6 summarize the most common solar systems used in the industrial sector for heating and cooling, as well as specific industry heating requirements.

Water purification and distillation are another important area for solar applications. In a somewhat different vein, Katekar et al. (2020) explored the development of solar stills for industrial, as well as domestic, applications. These stills are water distillers that use the energy of the sun to evaporate water to purify it. This method is an alternative to large desalination plants, which are energy-intensive, are expensive, and generate a significant carbon footprint. In their review, the authors analyzed several different types of distillers and modifications done to this method of water purification.

Figure 5.39 is an illustration of one of the water stills, a tilted solar distiller with a wick. The inclination of this still provides an advantage over other types of still. Katekar et al. (2020) developed many kinds of solar distillers, both for domestic and industrial uses, and their summary is shown in Table 5.7. As we can see, the single-basin, single-slope cascade passive solar still is the one found most productive for domestic applications. Meanwhile, the tubular solar still (TSS) with a wick is the most optimal still for industrial applications.

This type of design consists of a transparent tube, cylindrical in shape, where the rectangular absorber is placed. Figure 5.40 shows a simple design of this still. The solar radiation coming in

TABLE 5.5
Solar Energy Applications, System Technologies, and Type of Systems Commonly Used in Industry

Solar Energy Application	Solar System Technology	Type of System
SWH	Thermosiphon systems	Passive
	Integrated collector storage	Passive
	Direct circulation	Active
	Indirect water heating systems	Active
	Air systems	Active
Space heating and cooling	Space heating and service hot water	Active
	Air systems	Active
	Water systems	Active
	Heat pump systems	Active
	Absorption systems	Active
	Adsorption (desiccant) cooling	Active
	Mechanical systems	Active
Solar refrigeration	Adsorption units	Active
	Absorption units	Active
Industrial heat demand process	Industrial air and water systems	Active
	Steam generation systems	Active
Solar desalination	Solar stills	Passive
	Multistage flash (MSF)	Active
	Multiple effect boiling (MEB)	Active
	Vapor compression (VC)	Active
Solar thermal power systems	Parabolic trough collector systems	Active
	Parabolic tower systems	Active
	Parabolic dish systems	Active
	Solar furnaces	Active
	Solar chemistry systems	Active

Source: Mekhilef et al. (2011).

TABLE 5.6
Industrial Processes and Temperature Requirements

Industrial Sector	Unit Operation	Temperature Range (°C)
Food	Drying	30–90
	Washing	60–90
	Pasteurizing	60–80
	Boiling	95–105
	Sterilizing	110–120
	Heat treatment	40–60
Beverages	Washing	60–80
	Sterilizing	60–90
	Pasteurizing	60–70
Paper industry	Cooking and drying	60–80
	Boiler feedwater	60–90
	Bleaching	130–150
Metal surface treatment	Treatment, electroplating, etc.	30–80

(*Continued*)

TABLE 5.6 (*Continued*)
Industrial Processes and Temperature Requirements

Industrial Sector	Unit Operation	Temperature Range (°C)
Bricks and blocks	Curing	60–140
Textile industry	Bleaching	60–100
	Dyeing	70–90
	Drying, de-greasing	100–130
	Washing	40–80
	Fixing	160–180
	Pressing	80–100
Chemical industry	Soaps	200–260
	Synthetic rubber	150–200
	Processing heat	120–180
	Preheating water	60–90
Plastic industry	Preparation	120–140
	Distillation	140–150
	Separation	200–220
	Extension	140–160
	Drying	180–200
	Blending	120–140
Flour by-products	Sterilizing	60–90
All industrial sectors	Preheating of boiler feedwater	30–100
	Industrial solar cooling	55–180
	Heating of factory buildings	30–80

Source: Kalogirou (2003).

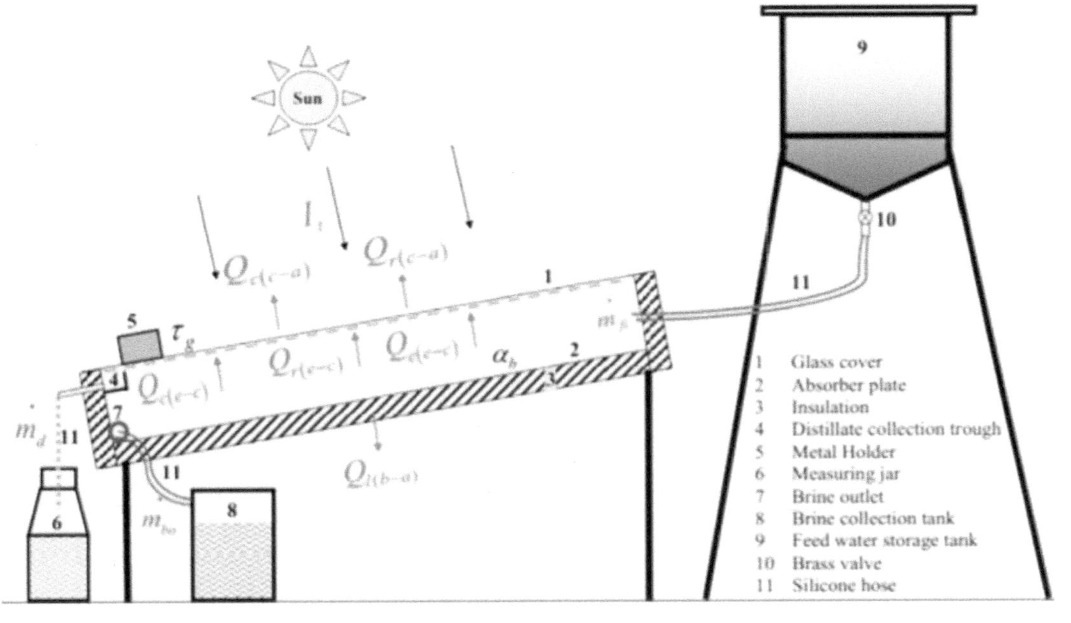

FIGURE 5.39 Tilted solar still with wick.

Source: Katekar et al. (2020).

heats up the saline water placed in the trough, evaporating the water and transferring the vapor to the inner surface of the cylinder, where it condenses and flows downward, being finally stored in the collector placed at the bottom of the trough (Kabeel et al., 2019).

Applications for this type of distiller in the industrial sector range from agricultural uses, to dispose of liquid waste (Nadal-Bach et al., 2021), to hospitals due to sterilization purposes, schools for the availability of pure water, and many other industries which require humidification and dehumidification techniques (Panchal and Mohan, 2017).

5.4.8 Utility Applications

Another practical use for solar energy is for power generation, what is referred to as "utility-scale" solar projects. This refers to the concept of using equipment or developing a plant to generate a large-enough quantity of electrical energy so that this causes a noticeable change in the running of a utility. According to the Solar Energy Industries Association (SEIA), solar ventures of this type can rely on technologies like PV or CSP to generate energy, but what differentiates them from distributed generation projects is twofold: their size, and that the electricity generated here is sold to wholesale utility buyers, not end use consumers (SEIA, 2021).

Also known as solar farms, PV power plants are large installations of photovoltaic modules that are connected to the electrical grid to provide power to the system. These projects are also called utility-scale PV power plants. Such larger-size plants come with complexities about their reliable integration into existing electrical grid, given dynamic power output variability. Therefore, real-time plant monitoring and control systems are the vital part of these plants for the grid operator to isolate and control power output remotely. Figure 5.41 shows a typical utility-scale PV plant block diagram, and the plant size varies between 100MW and 3000MW. Bhadla Solar Park in India was commissioned in 2018 with the capacity of 2700MW, which covers about a 160km^2 area. Pavagada Solar Park in India was completed in 2020, with a total capacity of 2,050 MW, which covers about a 53km^2 area. Longyangxia Dam Solar Park in China was commissioned in 2015 with the capacity of 850 MW. Tengger Desert Solar Park is the sixth-largest photovoltaic plant in the world as of December 2021. It is located in Zhongwei, Ningxia, China. It covers an area of 43 km^2. In 2018, it was the solar park with the largest peak power capacity (1,547 MW). Mohammed bin Rashid Al Maktoum Solar Park is a solar park spread over a total area of 77 km^2 in Saih Al-Dahal, about 50 km south of the city of Dubai. Besides solar farms using PV technology, the long-term project will also include concentrating solar power (CSP). The total capacity of the entire project is planned to reach 3,000 MW.

PV plants require a significantly large area, which cannot be available especially on islands (and other land-constrained countries), so that demand for floating PV (FPV) is expanding for utility power generation, because the cost of water surface is generally lower than the cost of land (GlobalData, 2018).

Floating solar is particularly well suited to Asia, where land is scarce but there are many hydroelectric dams with existing transmission infrastructure. Unsurprisingly, the world's top ten plants are located in Asia, namely, in China, Japan, and the Republic of Korea. Other Asian countries, namely, Singapore, Thailand, Vietnam, and India, are also actively pursuing floating solar project development. Europe has huge potential and demand for floating PV – especially in the Netherlands and France. The largest FPV plant, with 102.5 MW capacity, is commissioned in Korea (IRENA, *Future of Solar Photovoltaic*, 2019, p. 44).

Figure 5.42 shows active medium heating ($150^\circ C - 400^\circ C$) for utility power generation application using a solar parabolic trough thermal system. Feedwater from the deaerator is fed into the heat exchanger to generate high-pressure steam to be supplied into the waste heat recovery unit, which combines high-pressure steam from both solar and gas turbines. Both accumulated high-pressure steam is fed to the steam turbine to generate electricity. This is a very common integrated combined cycle power generation application combined with concentrated solar thermal system.

TABLE 5.7
Consolidated Report on the Literature Survey

Solar Still	Sub-Type	Best-Performing Design from Each Group	Final Investigation
Passive solar stills	• Single-basin, single-slope passive solar still	• The highest productivity augmentation of 638% was recorded for a solar still with Fresnel lens (Johnson et al., 2019).	• For domestic application, single-basin, single-slope cascade passive solar still is found more productive and economical. • For industrial application, tubular solar still with wick is found the most appropriate design.
	• Single-basin, multiple-slope passive solar still	• The highest enhancement in productivity was found as 171.43% for a single-basin, double-slope finned corrugated wick-type solar still integrated with external reflector (Gnanaraj and Ramachandran, 2022).	
	• Multiple-basin, single-slope passive solar still	• The most significant improvement in productivity was recorded as 44.57% for single-slope, double-basin solar still (Al-Hinai et al., 2002).	
	• Single-basin, multiple-slope solar still	• The maximum productivity improvement was found as 35% (slope to all the three sides, with PCM) (Sathyamurthy et al., 2022).	
	• Multiple-basin, multiple-slope passive solar still	• For double-basin, double-slope solar still, the highest productivity was recorded as 85% (Rajaseenivasan and Kalidasa Murugavel, 2013).	
Active solar stills	• Single-basin, single-slope active solar still • Multiple-slope, single-basin active solar still • Single-slope, multiple-basin active solar still	• The most substantial improvement in productivity was 307.54%, recorded for single-basin conical solar still with PCM and integrated with concentrating solar water heater (Chaichan and Kazem, 2015).	
Other miscellaneous designs	• Solar stills with greenhouse • Tubular solar still • Hemispherical solar still • Spherical solar still • Concave basin solar still • Multiple-sleeve-type solar still	• Tubular solar still with a semicircular trough filled with a black cloth and reflector was found the most productive, with 676% increase in productivity (Elashmawy, 2019).	
Hybrid solar still	• Single-basin, single-slope hybrid solar still • Single-basin, double-slope hybrid solar still	• The single-basin, multiple-slope solar still with PV/T collector was found as more productive, with 370% productivity improvement (Al-Hamadani and Yaseen, 2019).	

Source: Katekar et al. (2020).

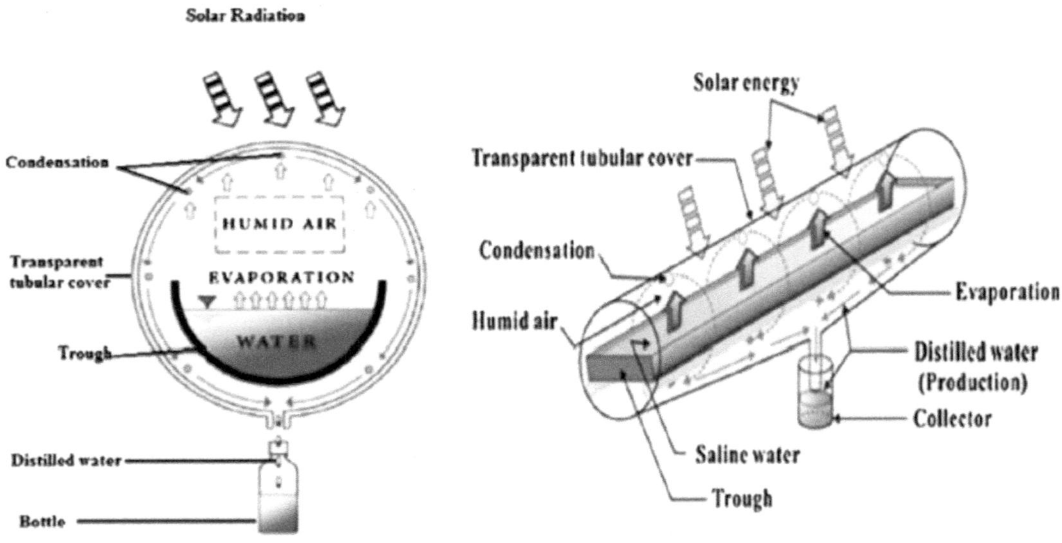

FIGURE 5.40 Layout of a single-basin tubular solar still.

Source: Kabeel et al. (2019).

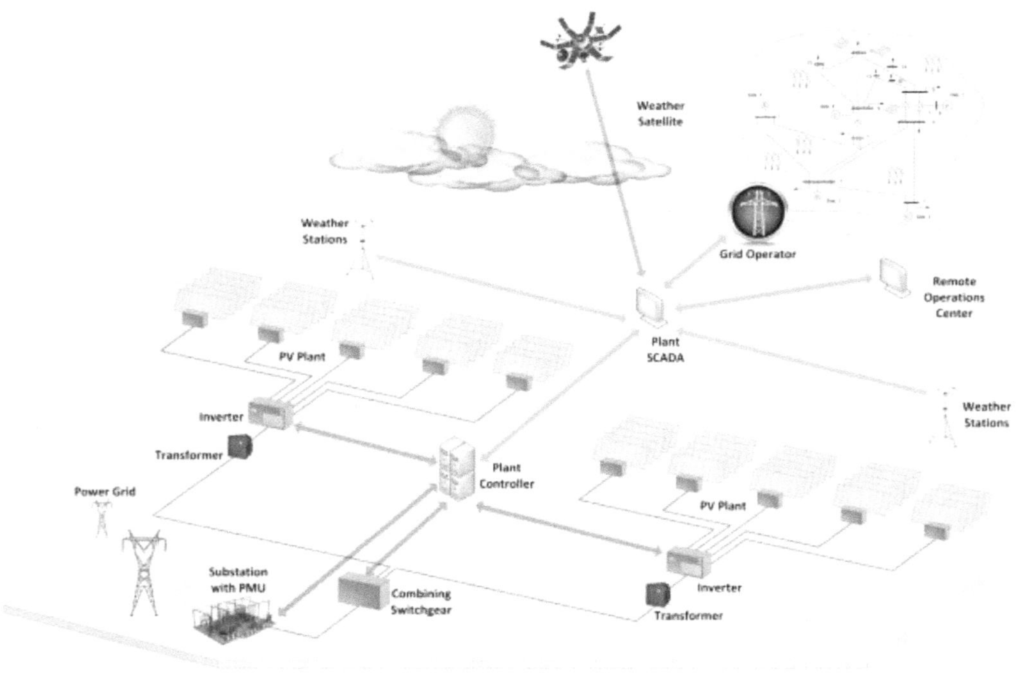

FIGURE 5.41 Typical utility-scale PV power plant.

Source: NREL (2016), Advanced Grid-Friendly Controls Demonstration Project for Utility-Scale PV Power Plants, p. 11.

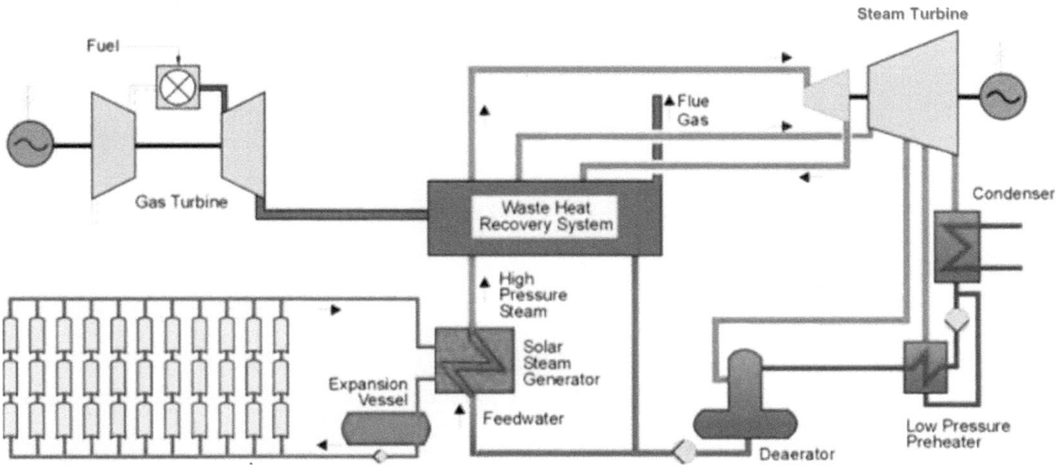

FIGURE 5.42 Active medium heating for utility power generation applications.

Source: Perel (2013), *Innovation in Power Plants*, Siemens, p. 34.

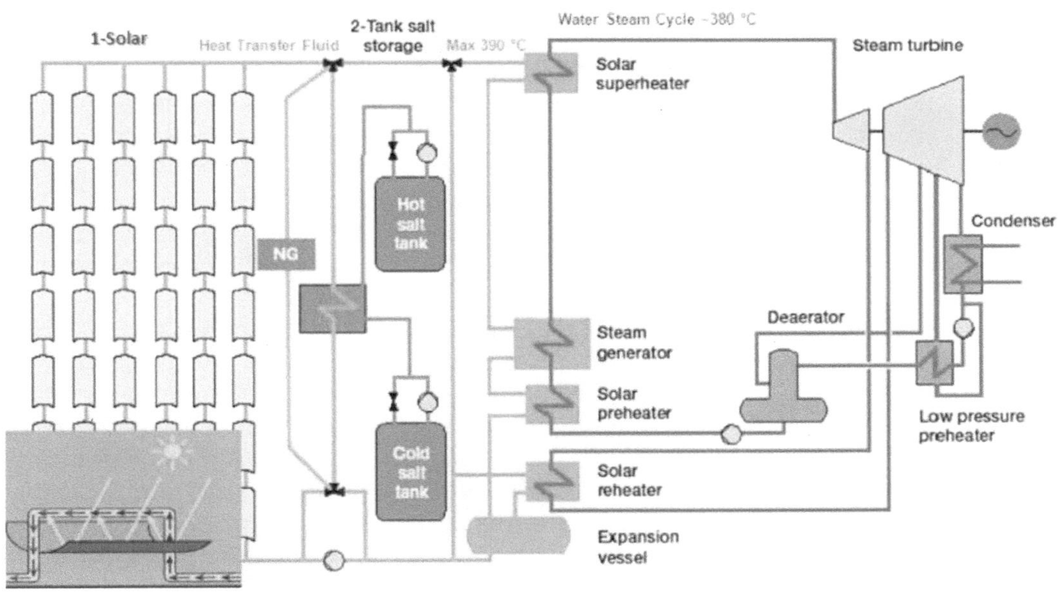

FIGURE 5.43 Active medium heating for utility power generation applications.

Source: Perel (2013), *Innovation in Power Plants*, Siemens, p. 35.

Figure 5.43 shows another form of power generation using only steam turbine with parabolic trough concentrated solar systems. Heated fluid from concentrated solar is fed into the solar superheater, and excess heat is fed into the heat exchanger to transfer heat into thermal storage. High-pressure steam from the solar superheater is fed into the steam turbine to generate power. Compared to combine cycle in Figure 5.37, its overall system efficiency is lower, with higher cost. Such system is very sensitive to solar resource availability. The boiler might be used to increase steam temperature and pressure. Thermal storage is utilized to provide stored heat for nighttime usage.

The first commercial parabolic trough plant was developed in 1984 in the Mojave Desert, California, by the now-defunct Luz Industries (SEGS, 1989). Over the years, this concentrating solar power plant has expanded, giving way to what is now known as the Solar Energy Generating Systems (SEGS) project, which currently boasts nine iterations that have a total capacity of 354 MW, feeding about 800 million kWh per year into the grid and built on more than 7 km^2 (Quaschning and Blanco Muriel, 2001).

Other solar thermal power plants around the world include the Nevada Solar One, located in the United States; the Andasol plants of Andalucía as well as PS10 in Seville, both complexes located in southern Spain (Giostri et al., 2012; Poullikkas, 2009); and the recently opened Condat in France. Despite growing initial interest, investment strategies for a wider expansion of these plants in other countries have not grown at a similar pace. Spain has a scheme of feed-in tariffs, which helped develop their plants (Solar Millennium, 2008) and has helped develop Solnova 1, PS20, Solar Tres, and Ibersol 1, but other countries have not followed suit momentarily.

5.5 SOLAR ENERGY APPLICATIONS: DESIGN, PRINCIPLES, CHALLENGES, BEST PRACTICES, AND ECONOMICS

Given the many constraints in terms of resources, land, and government regulations, along with the costs associated with solar technologies and renewable applications in general, it is neither efficient nor advisable to go ahead with every single project or to follow every encouraging result found in the literature without evaluating its merits in practice. Because of this, researchers need to develop and follow a framework for the selection of the best solar projects, after paying attention to factors like the cost of the endeavor, the long-term economic and energy-saving prospects, the replicability of the project in other contexts, and the geographical considerations explored previously. Sindhu et al. (2017) developed a strength-weakness-opportunity-challenges (SWOC) analysis, illustrated in Figure 5.44, for the prospect of deploying solar energy technologies in India which is valid for everywhere across the globe. Additionally, the key variables yielded by this analysis are later input into an analytic hierarchy process (AHP) to obtain their priority in a potential implementation of deployment strategies in the country.

The second section of Sindhu et al. (2017) consists of a hierarchical analysis of these considerations, which were presented to a panel of five experts in the area. As Table 5.8 shows, there are a series of priorities that are important to review. First, the group of weaknesses from deploying solar technologies is prioritized last by the people consulted, which reveals a degree of expectation and enthusiasm about adapting solar applications from Indian experts. Then, different considerations within each group have varying priorities, as "complex administrative procedures" are the main priority within the "challenges" group, but the fourth overall variable for respondents and the promise of a sustainable future are both the main priority within the group of "opportunities," as well as the main priority of respondents.

Analyzing regional and country-level developments reveals that several actors have embraced a switch to renewable energies, looking to deploy strategies and become proactive in this emerging market. The different strategies seen across regions and countries include the implementation of governmental schemes like special tariffs, the reduction of entry barriers, the development of infrastructure, and the capturing of promising talent. Coupled with research to improve module performance and the use of better materials to build lighter, more efficient systems, countries can work proactively to deploy solar energy technologies in a safe manner.

This degree of acceptance of solar installations in many places around the world appears ideal to move forward with the planned utility-scale solar PV projects. The top countries leading the switch to renewables have found a good balance between government incentives and reasonable energy targets, private investment, relatively ambitious bank loans, finding the best solar irradiance spots within a country, and appropriate resource allocation as well as land selections (Kumar Sahu, 2015). Using real-world data that includes climatology and legislation data and information

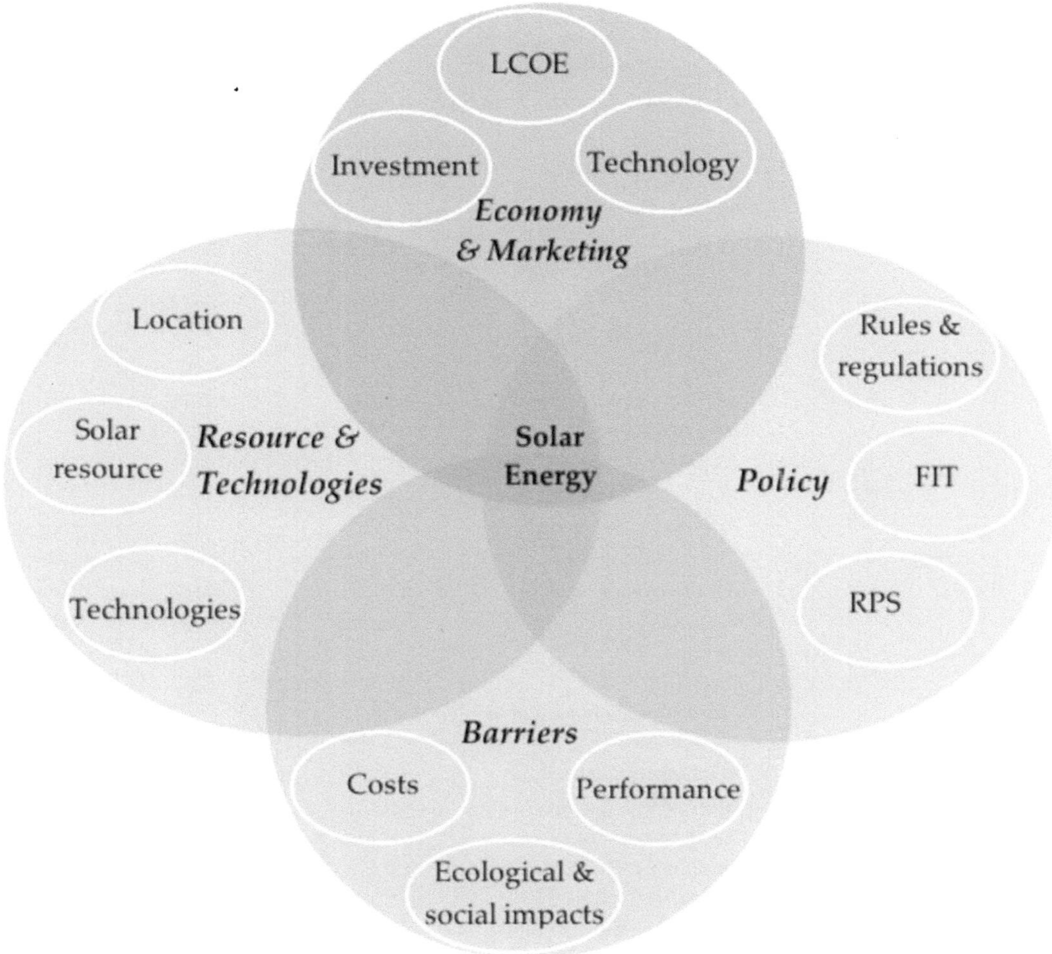

FIGURE 5.44 RPS renewable portfolio standard. (Note: *FIT*, feed-in tariff; *LCOE*, levelized cost of electricity.)

Source: Alsharif et al. (2019).

about roads, mountains, water bodies, air temperature, slope, and urban as well protected areas, Al Garni and Awasthi (2017) developed a model that assesses the strength of factors like solar irradiation, the proximity to roads, urban areas, and power lines (see Figure 5.45). Results of the analysis show that the north and northwest areas in Saudi Arabia are the most suitable for utility-scale PV projects, as solar irradiation continues to be the strongest parameter for the suitability of solar power plants.

Despite the enormous recognizable potential of solar energy, there are a series of significant challenges that need to be addressed before moving on. These considerations touch on technical aspects, as well as economic and institutional barriers that are in the way of a more seamless transition, which should be analyzed rigorously when thinking about a large-scale deployment of technologies relying on solar energy for electricity generation. According to Alsharif et al. (2019), the performance of PV panels can be affected by natural elements, such as dust, snow, or tree debris. Then, the intermittent and unpredictable nature of solar energy as well as the relatively high costs of PV modules are hurdles that may still be too high to climb for some.

TABLE 5.8
Overall Priority Scores of SWOC Variables

SWOC variables	Group priority/global priority	SWOC variables	Variables priority within the Group	Overall priority of Variables	Ranking of Variables
Strength	0.2390	High solar potential in India	0.1190	0.0284	16th
		Environment-friendly technology	0.1529	0.0365	13th
		Zero fuel cost	0.1107	0.0264	18th
		Suitability to rural and remote areas	0.1831	0.0438	8th
		Reduces carbon footprints in the environment	**0.2624**	**0.0627**	4th
		No noise pollution	0.1718	0.0411	11th
Weakness	0.1682	Intermittent source	0.1510	0.0254	21st
		Performance limitations	0.1380	0.0232	22nd
		Transmission issues	0.1529	0.0257	19th
		Manufacturing issues	**0.2478**	**0.0417**	10th
		Ecological impacts of the solar energy	0.1835	0.0309	15th
		Safety, health, and environmental impacts of the solar energy	0.1268	0.0213	23rd
Opportunity	0.3952	Employment opportunities	0.1095	0.0433	9th
		Improvement in quality of life	0.1572	0.0621	5th
		Application in vast sectors	0.2052	0.0811	2nd
		Opportunity to fulfill nation energy demands	0.1462	0.0578	6th
		Promises sustainable future	**0.2221**	**0.0878**	1st
		Reduction of dependency on conventional energy sources	0.1599	0.0631	3rd
Challenges	0.1976	Huge initial investment	0.1293	0.0255	20th
		Technical hindrances	0.1403	0.0277	17th
		Poor financing system	0.1820	0.0359	14th
		Lack of social acceptability	0.0925	0.0182	24th
		Market distortions	0.2002	0.0396	12th
		Complex administrative procedures	**0.2558**	**0.0505**	7th

Source: Sindhu et al. (2017).

All methods of energy generation and transmission affect the environment. Renewable energies offer a tremendous number of upsides, but they also bring challenges, like a greater land use for a relatively lower amount of energy produced, accidental discharges of pollutants, and the depletion of natural resources (Tsoutsos et al., 2005). Moreover, Hernandez et al. (2014) alerted on the proximate impacts on biodiversity, as some projects have caused the death of birds due to collision with site infrastructure or after being burned alive by sunrays directed at receivers (also see Visser et al., 2018). Furthermore, some projects have caused significant repatriation and translocation of some

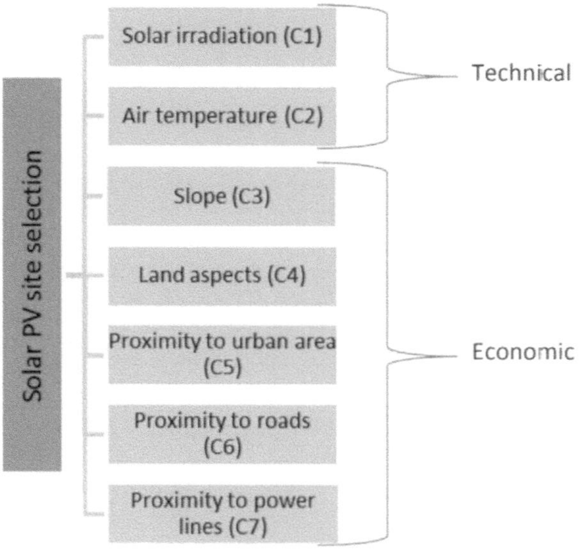

FIGURE 5.45 Decision criteria considered in solar site selection.

Source: Al Garni and Awasthi (2017).

bird communities, which themselves have a low success rate. Table 5.9 shows a summary of the impacts of solar technologies and a list of alleviation techniques.

As noted, the vast amounts of land needed for solar farms, the potential transmission system issues for energy that comes from a place that most of the time is located far from the cities that need the energy, and the disposal of dangerous materials are important issues to be considered. Due to the exponential growth in the development of solar-related technology, end-of-life (EOL) panels have become another challenge related to the expansion of solar energy. Solar panels serve the user for an important period of years; however, once their useful life comes to an end, EOL panels become a hazardous waste (Xu et al., 2018). Figure 5.46 shows a schematic for the life cycle of a typical photovoltaic solar panel.

As a response, the recovery and recycling of solar panels have drawn the attention of researchers and international organizations, like the European Union. This conglomerate has written a series of norms and rules on the management of waste from electrical and electronic equipment (WEEE), to drive a sustainable production and consumption. Regarding solar panels, their *Study on Quality Standards for the Treatment of Waste Electrical and Electronic Equipment* (2021) forbids the mixed collection of panels with construction and demolition waste, encourages the separation in the treatment of silicon-based and non-silicon-based PV panels, and demands a careful monitoring of depollution of hazardous substances like cadmium, selenium, and lead.

5.6 SOLAR ENERGY APPLICATIONS IN THE OIL AND GAS INDUSTRY

After reviewing the main solar technologies as well as their uses and applications in different contexts, we now move on to a review of the challenges and opportunities of solar energy applications in the oil and gas industry. Motivated by a similar set of constraints as the ones already explored, with the added requirement of being able to generate successful extraction and optimization strategies, the oil and gas sector is positioned as an actor with the opportunity to take noticeable steps

TABLE 5.9
Solar Energy Technology's Negative Impacts

Impacts/Burdens	Alleviation Technologies/Techniques
Solar thermal heating	
Visual impact on buildings' aesthetics	Adoption of standards and regulations for environmentally friendly design Good installation practices Improved integration of solar systems in buildings Avoiding siting of solar panels on buildings of historic interest or in conservation areas
Routine and accidental releases of chemicals	Recycling of the used chemicals Good practices – appropriate disposal
Land use	Proper siting and design
Photovoltaic power generation	
Land use: large areas are required for central systems	Use in isolated and deserted areas
Reduction of cultivable land	Avoidance of ecologically and archeologically sensitive areas Integration in large commercial buildings (facades, roofs) Use as sound isolation in highways or near hospitals
Visual intrusion – aesthetics	Careful design of systems Integration in buildings as architectural elements Use of panels in modem architecture instead of mirrors onto the facade of buildings
Impact on ecosystems (applicable to large PV schemes)	Avoidance of sensitive ecosystems and areas of natural beauty, archaeological sites
Use of toxic and flammable materials (during construction of the modules)	Avoidance of release of potentially toxic and hazardous materials with the adoption of existing safety regulations and good practice
Slight health risks from manufacture, use, and disposal	Good working practices (use of protecting gloves, sunglasses, clothing during construction)
Solar thermal electricity	
Construction activities	Good working practices Site restoration Avoidance of sensitive ecosystems and areas of natural beauty.
Visual impact – aesthetics	Proper siting (avoidance of sensitive ecosystems and areas of natural beauty, densely populated areas)
Land use	Proper siting
Effect on the ecosystem, flora, and fauna (especially birds)	Proper siting (avoidance of sensitive ecosystems)
Impact on water resources, water use (for cooling of steam plant), and possibly, water pollution due to thermal discharges or accidental discharges of chemicals used by the system	Appropriate constraints (not the excessive use of existing resources) Improved technology (use of air as heat transfer medium) Exploitation of the warm water in the nearest industry in the production stream Good operating practices and compliance with existing safety regulations Employees should be educated and familiarized with the systems
Safety issues (occupational hazards)	

Source: Tsoutsos et al. (2005).

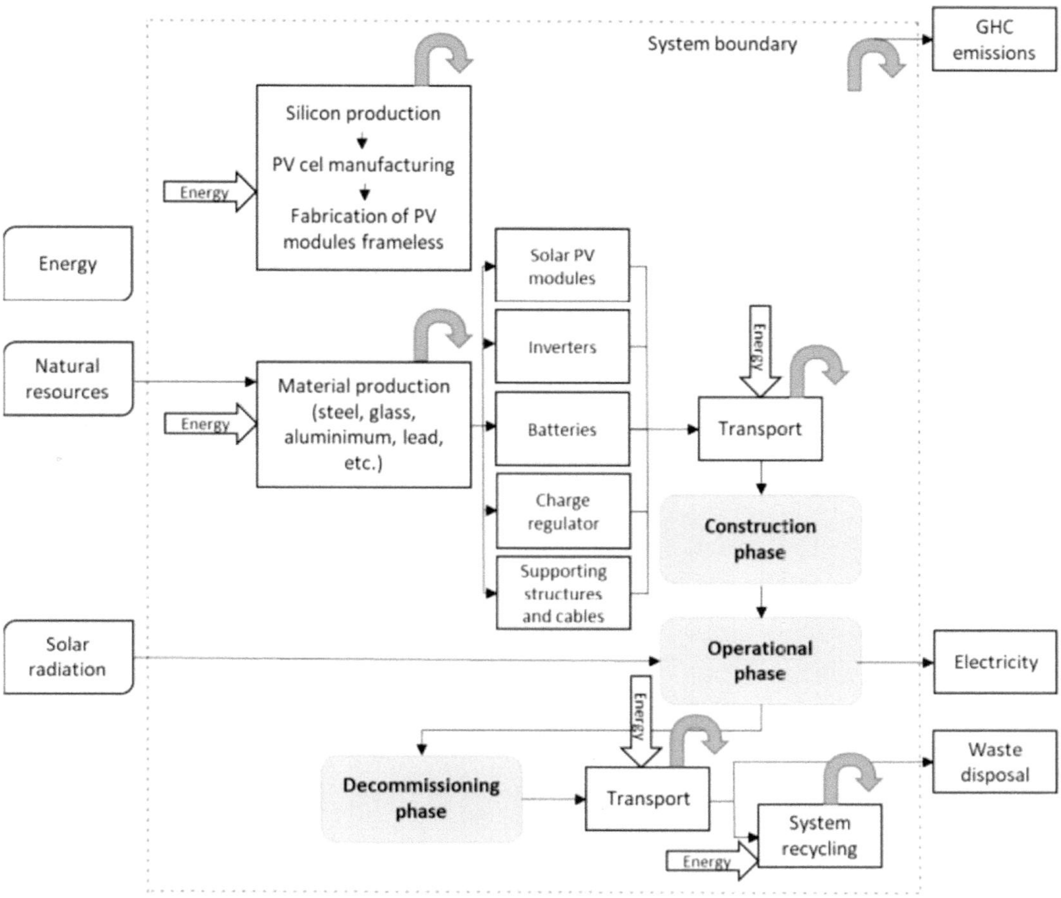

FIGURE 5.46 Life cycle assessment (LCA) boundary for a typical PV system.

Source: Bogacka et al. (2020).

forward in the implementation of technologies increasingly more relying on solar energy, which help generate a more sustainable energy sector.

Figure 5.47 shows the potential capacity for solar thermal installations, as well as photovoltaic projects for upstream production, across the world. As we can see, the North African region, the Middle East, and the United States appear poised to make a strong push to the switch to solar energy in the oil and gas industry (Wang et al., 2017). There is a golden opportunity for onshore reservoirs to take significant steps toward the implementation of solar installations in the next few years. The following section explores the key position that the oil and gas industry has in the process of the energy transition we are currently experiencing.

5.6.1 Enhanced Oil Recovery (EOR) Applications

As a technique introduced with the objective of recovering heavy oil in tertiary operations, steam flooding consists of a constant injection of vapor into a reservoir, which introduces latent heat that

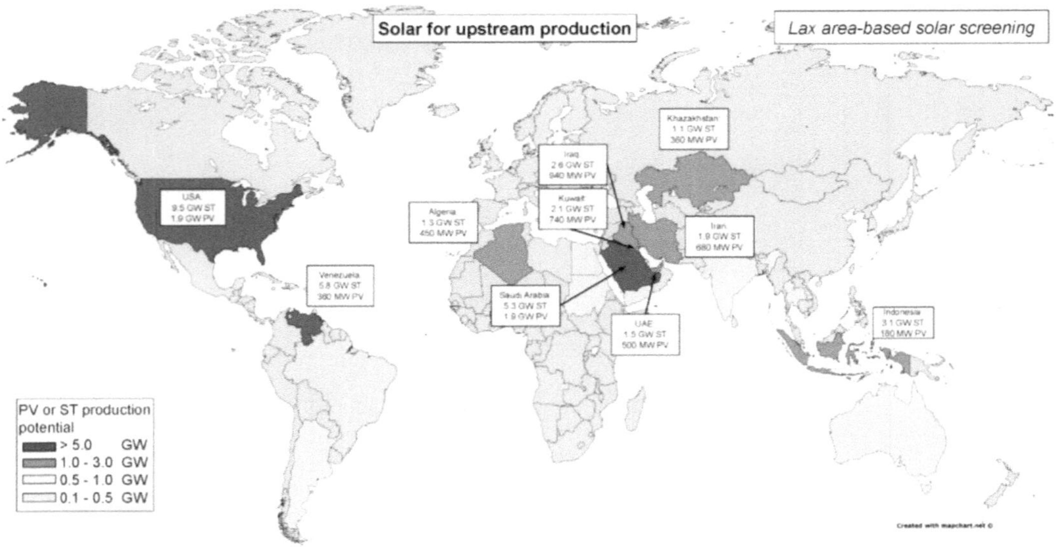

FIGURE 5.47 Photovoltaic (PV) and solar thermal (ST) potential capacity in upstream production sector under lax area–based solar resource screen.

Source: Wang et al. (2017).

helps displace oil either through vaporization, a reduction in viscosity, or a thermal expansion of rocks and fluids (Al-Saedi et al., 2018; Bagheripour Haghighi et al., 2012; Ud Din Kirmani et al., 2021). As a tool for enhanced oil recovery (EOR), solar-generated steam flooding has gained more popularity in recent years due to its economic and environmental advantages (Afsar and Akin, 2016; V. K. et al., 2019). Figure 5.48 shows the different methodologies used for enhanced oil recovery.

The oil and gas sector is full of opportunities for a fruitful application of renewable energies to help meet the energy requirements of the industry. For tertiary oil recovery, concentrating solar power (CSP) is a very good choice for steam flooding (see Figure 5.49). However, engineers need to locate these solar EOR fields in reservoirs that may not have the ideal irradiation areas. The work of Afsar and Akin (2016) shows the case of Turkey, a country with heavy oil reservoirs and with plenty of sunshine areas. The authors studied the feasibility of a solar EOR application for a heavy oil field in the southeast area of the country and concluded that solar input is not sufficient to satisfy the steam injection needs of a heavy oil well, as only over 40% of the required steam is produced by the solar field.

Generally, steam-based thermal EOR techniques can be divided into three categories: (1) cyclic steam stimulation, (2) continuous steam injection, and (3) steam-assisted gravity drainage (Gupta et al., 2017). For all these processes, EOR is generated through boilers that are powered by natural gas, a process that emits a significant amount of greenhouse gases into the atmosphere. As a response to this potential problem, the boiler can be replaced by thermal sand–based storage and solar PV plant to provide continuous steam for the EOR process. A list of oil fields which have implemented thermal EOR mechanisms is provided in Table 5.10.

Given its recent challenges of decreasing domestic natural gas production, the authors developed a techno-economic model for the feasibility of solar applications in EOR in a field located in Issaran, Egypt. Figure 5.50 shows the outcome of this model, where we see that, over time, investments made to further the use of solar technologies help save on carbon emissions. Additionally, the payback period is six years, while the internal rate of return (IRR) is 27.4%. Lastly, the levelized cost of storage (LCOS) is comparable to estimation for other reservoirs that adapt solar technologies to their EOR processes (Gupta et al., 2017).

FIGURE 5.48 Different methodologies used for oil recovery.

Source: V. K. et al. (2019).

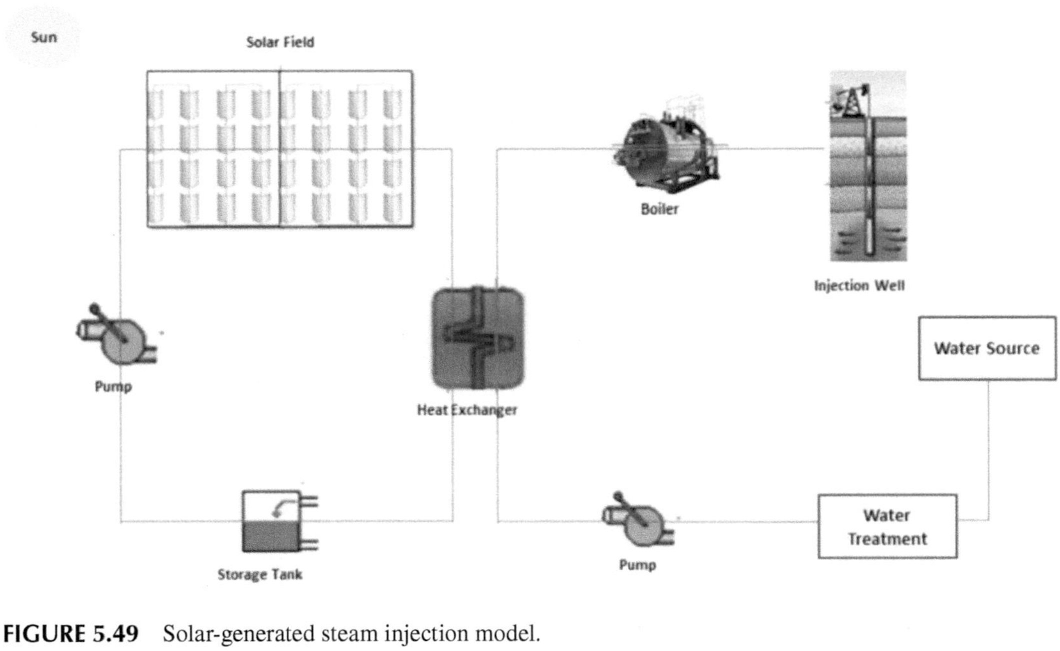

FIGURE 5.49 Solar-generated steam injection model.

Source: Afsar and Akin (2016).

TABLE 5.10
Parameters of Oil Fields with Thermal EOR

Coalinga, California USA [9]			**Steam temperature**	**229**	**[°C]**
Start of steam injection	1962	[-]	Steam quality	80	[%]
Oil production rate	27,000	[BOPD]	Steam injection method	Steam drive	
Steam injection rate	0.6–1.4	[ktons/day]	DNI	2,222	[kWh/m²/yr]
Steam pressure	406	[psi]	Topography	Flat, on land	
Cold Lake, Alberta, Canada [10]			Steam temperature	314	[°C]
Start of steam injection	1978	[-]	Steam quality	80	[%]
Oil production rate	150,000	[BOPD]	Steam injection method	CSS	
Steam injection rate	262	[tons/day]	DNI	1,738	kWh/m²/yr]
Steam pressure	1,500	[psi]	OIIP	20,000	[MMSTB]
Peace River, Alberta, Canada [11] [12]			Steam temperature	278	[°C]
Start of steam injection	1973	[-]	Steam quality	95	[%]
Oil production rate	80,000	[BOPD]	Steam injection method	SAGD	
Steam injection rate	640	[tons/day]	DNI	1,586	kWh/m²/yr]
Steam pressure	915	[psi]	OIIP	20,000	[MMSTB]
Estreito, Brazil [13]			Steam temperature	219	[°C]
Start of steam injection	1984	[-]	Steam quality	75–80	[%]
Oil production rate	28,000	[BOPD]	Steam injection method	CSS and steam drive	
Steam injection rate	9.6–22	[ktons/day]	DNI	1,413	[kWh/m²/yr]
Steam pressure	333	[psi]	OIIP	725	[MMSTB]
Carmopolis, Brazil [14]			Steam temperature	179	[°CI
Start of steam injection	1978	[-]	Steam quality	75–80	[%]
Oil production rate	18,100	[BOPD]	Steam injection method	Steam drive	
Steam injection rate (C-4 Zone)	1	[ktons/day]	DNI	1,588	[kWh/m²/yr]
Steam pressure	129	[psi]	OIIP	1,591	[MMSTB]
Tia Juana, Bolivar Coast, Venezuela [15]			Steam temperature	252	[°C]
Start of steam injection	1959	[-]	Steam quality	80	[%]
Oil production rate (M-6, del este)	20,500	[BOPD]	Steam injection method	CSS and steam drive	
Steam injection rate (M-6, del este)	6–8.5	[ktons/day]	DNI	891	[kWh/m²/yr]
Steam pressure	500	[psi]	OIIP	6,015	[MMSTB]
Bachaquero, Bolivar Coast, Venezuela [16]			Steam temperature	307	[°C]
Start of steam injection	1971	[-]	Steam quality	80	[%]
Oil production rate (Bachaquero-01)	40,000	[BOPD]	Steam injection method	CSS	
Steam injection rate	362	[tons/day]	DNI	887	[kWh/m²/yr]
Steam pressure	1,394	[psi]	OIIP	7,280	[MMSTB]
Duri, Indonesia [17]			Steam temperature	237	[°C]
Oil production rate	236,000–285,000	[BOPD]	Steam quality	70	[%]
Steam injection rate	190–206	[ktons/day]	Steam injection method	Steam drive	
Steam pressure	475	[psi]	DNI	892	[kWh/m²/yr]
Issaran, Egypt [18] [19]			Steam temperature	260	[°C]
Start of steam injection (CSS)	2004	[-]	Steam quality	80	[%]
Oil production rate 5,000–6000		[BOPD]	Steam injection method	Steam drive (earlier, CSS)	
Steam injection rate	317–476	[tons/day]	DNI	2,736	[kWh/m²/yr]
Steam pressure	680	[psi]	OIIP	500	[MMSTB]
Amal-West, Oman [20]			Steam temperature	311	[°C]

(Continued)

TABLE 5.10 (*Continued*)
Parameters of Oil Fields with Thermal EOR

Coalinga, California USA [9]			**Steam temperature**	**229**	**[°C]**
Start of steam injection	2007	[-]	Steam quality	80	[%]
Oil production rate	358	[BOPD]	Steam injection method	Steam drive	
Steam injection rate	7,500	[tons/day]	DNI	2,057	[kWh/m²/yr]
Steam pressure	1,160	[psi]	OIIP	830	[MMSTB]
Cruse E, Trinidad [21]			Steam temperature	283	[°C]
Start of steam injection	1996	[-]	Steam quality	80	[%]
Oil production rate	350	[BOPD]	Steam injection method	Steam drive	
Steam injection rate	317	[tons/day]	DNI	1,201	[kWh/m²/yr]
Steam pressure	1,000	[psi]	OIIP	31.1	[MMSTB]
Schoonebeek, Netherlands [22]			Steam temperature	225	[°C]
Start of steam injection	2008	[-]	Steam quality	80	[%]
Oil production rate	18,870	IBOPD]	Steam injection method	SAGD	
Steam injection rate	8–10	[ktons/day]	DNI	870	[kWh/m²/yr]
Steam pressure	350	[psi]	OIIP	1,069	[MMSTB]

Source: Gupta et al. (2017).

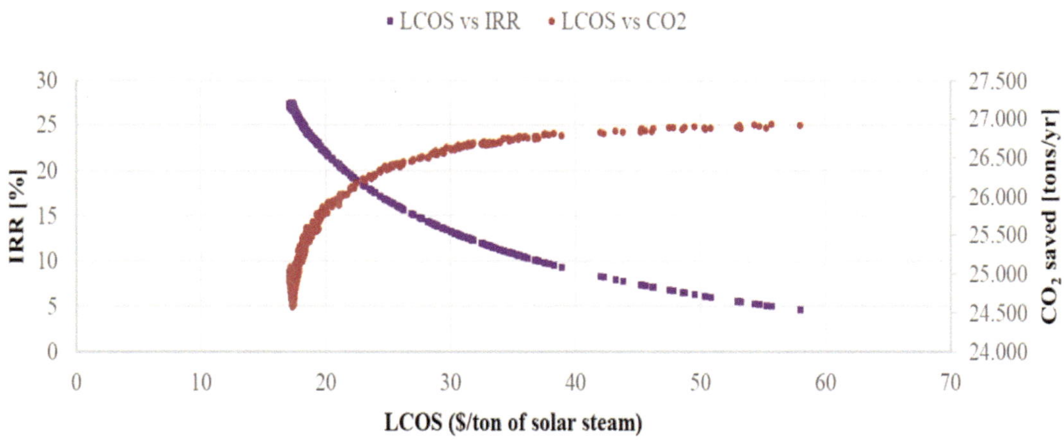

FIGURE 5.50 Optimization results for the S-EOR plant.

Source: Gupta et al. (2017).

In turn, the work of Chaar et al. (2014) explores the economics of three thermal EOR steam generation methods: (1) once-through steam generator (OTSG), (2) once-through heat recovery steam generator (OT-HRSG), and (3) solar steam generator (SSG). In what regards this chapter, the SSG method is the one that utilizes solar energy, through a concentrating solar power (CSP) field, to create the steam flooding. The set of images in Figure 5.51 illustrates each of the methods explored in this article.

When comparing the pros and cons of each of these methods, we observe that both OTSG and OT-HRSG have low capital costs per ton of steam produced, but these calculations are significantly

OTSG OT-HRSG SSG

FIGURE 5.51 Three methods of steam generation.

Source: Chaar et al. (2014).

TABLE 5.11
Summary of Solar EOR Projects

Project	Kern County 21Z	Coalinga	Amal West
Technology provider	Glass Point	BrightSource	Glass Point
Location	McKittrick Oil Field, McKittrick, California, USA	Coalinga Oil Field in Fresno County, California, USA	Amal West Oil Field, Southern Oman
Commissioning date	February 2011	October 2011	February 2013
Peak capacity	300 kW	29 MW	7 MW
CSP technology	Enclosed trough	Solar tower	Enclosed trough
Key project details	First commercial solar EOR project System spans c. 1 ac and produces c. 1 MMBTU/hr of solar heat First project to use Glass Point's enclosed trough technology in an oil field	Project spans 100 ac and consists of 3,822 mirror systems, or heliostats, each with two 10 ft (3 m) by 7 ft mirrors mounted on a 6 ft steel pole focusing light on a 327 ft solar	Middle East's first solar EOR project Produces a daily average of 50 t of steam feeding directly into existing thermal EOR operations

Source: Gregory et al. (2014).

dependent on the price of fuel. On the other hand, SSG does not consume fuel and, because of this, does not produce greenhouse emissions. Additionally, this method can extend the life of the field, given that fuel costs are eliminated. The cons of this method, however, are a high capital cost and a high degree of dependency on the weather (Chaar et al., 2014).

With that said, results of this study show that the cheapest method, at a $6/MMTBU, is SSG ($17/ton), followed by OT-HRSG ($20/ton), with the most expensive being OTSG ($27/ton). The attractiveness of solar techniques being deployed for EOR methods is also confirmed in the work of Sandler et al. (2014), who find that a 100% solar steam technique offers the best net present values (NPV) when compared to five other steam generation scenarios: direct steam generation with natural gas, all cogeneration using natural gas, "real" cogeneration using natural gas (that approximates actual oil field steam generation conditions), 50% solar fraction steam with the balance from natural gas fired direct steam generation, and 25% solar fraction steam with the balance from natural gas fired direct steam generation.

According to the work of Gregory et al. (2014), there are only three solar EOR projects operational as of 2013, but there are a few more in the works. Table 5.11 shows that two of the three endeavors are in the United States, while the third one is being performed in Oman. The Kern

County field is one of the oldest reservoirs, and the objectives of this project were to generate preheated water and steam from the solar field, minimizing the use of fossil fuels to achieve this. Then, Coalinga works with a system of heliostats which have a tracking system that concentrates the solar radiation onto the tower. The receiver is at the top of the tower, with water circulating through the receiver to produce steam. Finally, the Amal field generates steam that is injected into the oil well to heat the oil, so extraction is easier (Ravi Kumar et al., 2021).

5.6.2 Solar Photovoltaics with Electric Boilers

When considering solar thermal combined heat and power (CHP) plants, as well as electric boilers, a solution to combat the common shortcomings of wind and solar technologies, systems can integrate electrical, natural gas, and district heating (IENGDH). The work of Yu et al. (2018) advances a risk assessment method to analyze the combination of these technologies. Using two case studies, the authors developed seven risk indices which help quantify the severity of the constraints of wind and solar technologies in IENGDH systems. Figure 5.52 illustrates the steps required to develop the workflow.

Conclusions establish that the effect of electric boilers on the shortcomings of wind and solar technologies is greater than the effect of thermal energy storage. These shortcomings and intermittence problems were also addressed by Blarke (2012), who proposes a strategy where cogeneration (or when different energy sources are combined to achieve energy or heating goals) and renewable sources coexist better. The author finds that electric boilers provide an improvement in the intermittence problems between different sources of energy, but an even more efficient design is that of compression heat pumps combined with intermediate thermal storages. Besides the intermittence issues, this strategy shows more efficiency in the use of electricity in the production of heating and cooling.

Figure 5.53 illustrates the energy cogeneration issue more clearly. Given all the current constraints to renewable energies, the simultaneous concerns about greenhouse emissions, and the worsening of the environment when only using fossil fuels both for domestic and industrial applications, considering all options is imperative. In a paper soon to be released, Dimri (2021) compares four distinct combined heating and power systems (CHP) with the conventional method that relies on the national grid and a gas boiler for domestic uses. The author finds that, even though solar thermal collectors suffer from intermittence issues, when coupled with grid-driven electric boilers as backup, they provide a competitor to the conventional gas boilers.

5.6.3 Off-Grid Solar Applications

The process of exploration and production of oil consists of many phases. Both onshore and offshore operations require a considerable number of resources before operations can function properly. Given the constraints already explored related to limiting the use of fossil fuels and to making the process of extraction and transport of oil both more efficient and as well as cleaner, several avenues are explored constantly by researchers and engineers. Another one of these avenues is the application of off-grid power systems that rely on renewable energy to feed both onshore and offshore reservoirs.

Off-grid solar PV and battery systems are well suited for primary oil recovery, with natural flow well sites located in highly remote areas providing significant cost avoidance benefits against power line infrastructure cost. Similar off-grid systems are very suitable to power secondary oil recovery wells both onshore and offshore. For example, replacing conventional power sources with off-grid photovoltaic (PV) battery power supply systems with hybrid options, including small wind turbines, if applicable, to operate well site equipment, such as safety and pressure valves, chemical injection pumps, flow rate devices, communication equipment, emergency shutdown equipment, lights, cathodic protection (see Figure 5.54).

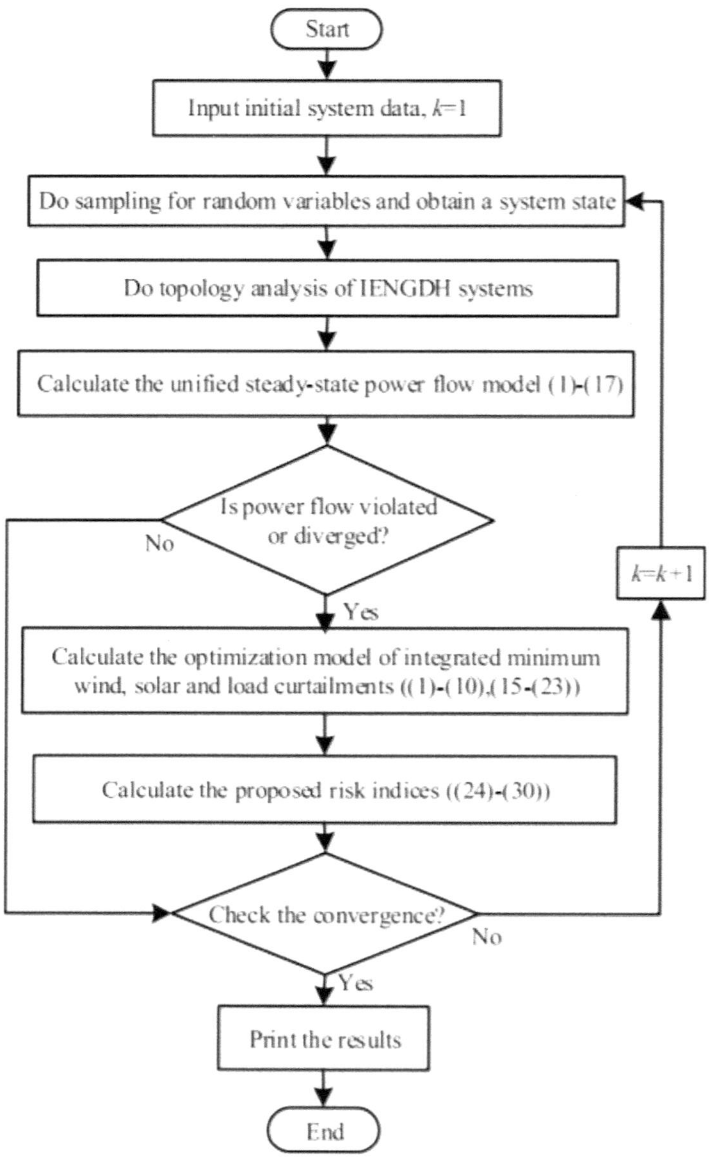

FIGURE 5.52 Risk assessment solution process for IENGDH systems using the developed method.

Source: Yu et al. (2018).

One example of the latter can be observed thanks to the work of Kalogera et al. (2014), who developed a hybrid power system with solar and wind technologies adapted to arctic climates and ready to supply offshore platforms in the Beaufort Sea. The authors find that, when counting on diesel as a backup source, a hybrid system with chemical storage featuring NiCd batteries yields the best trade-off for variables like efficiency, reliability, maturity, and cost. Recently, lithium iron phosphate (LFP) batteries are becoming more popular because they are safe, are high-temperature-resistant, have higher energy density, and are non-toxic. Meanwhile, Ahmed (2015) explored the feasibility of a hybrid power system in an unmanned offshore installation in Malaysia, and Figure 5.55 illustrates the framework of their study.

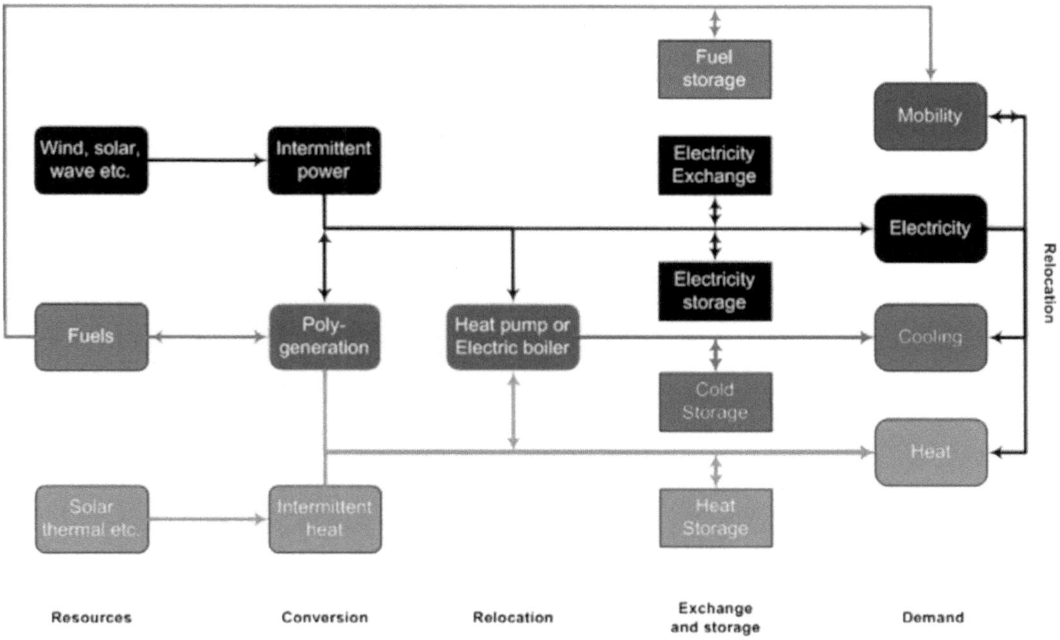

FIGURE 5.53 The intermittency-friendly energy system.

Source: Blarke (2012).

FIGURE 5.54 Wind turbines and solar panels on offshore platform.

Source: Tiong et al. (2015).

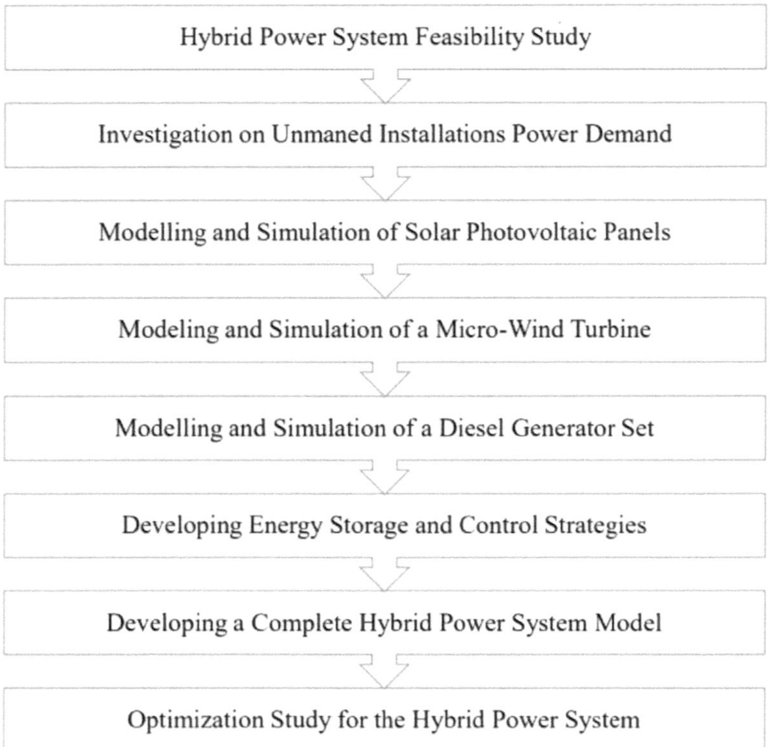

FIGURE 5.55 Research methodology for offshore installations in eastern Malaysia.

Source: Ahmed (2015).

The author found that it is possible to install a hybrid system, and that optimizing the use of solar PV panels can improve performance even more. However, given that diesel is kept as a backup source of energy, the optimization calculation is highly dependent on diesel prices. The simulation recommends the development of more solar panels, or more wind turbines, to tackle any significant problems in this dependency. In the end, the goals of reducing carbon emissions and cost cutting can be achieved, something that is a promising result for the entire industry in the region of the Southeast China Sea. A few years later, these results were replicated by the work of Abdullah et al. (2018).

On the other hand, Ohira et al. (2018) studied the implications of developing solar power plants to replace diesel power installations in onshore reservoirs, particularly those that are far from the electricity grid. The authors found that the presence of empty space around the well area improves the chances of a site to deploy a solar field to power itself, and other variables of interest are, as expected, the potential of solar intensity and the number of solar radiation days. Meanwhile, Abureden et al. (2015) looked to assist an existing oil operation in Abu Dhabi with systems of power generation that are based in renewables. The authors concluded that techniques like CSP and steam EOR are good methods to implement due to their efficacy and their potential to reduce carbon emissions.

Furthermore, the work of Okpala et al. (2020) centers on the implementation of a real-time reservoir management (RTRM) solution on an unmanned onshore reservoir in Nigeria, where part of the project consists of solar panels to meet its power requirements. Due to the increasing need to save on unnecessary trips back and forth from reservoirs, especially when most of the maintenance work can be done from a remote location, adapting solar installations to this way of production enables an energy-saving mindset while also permitting a more optimal recovery of an oil and gas site. This

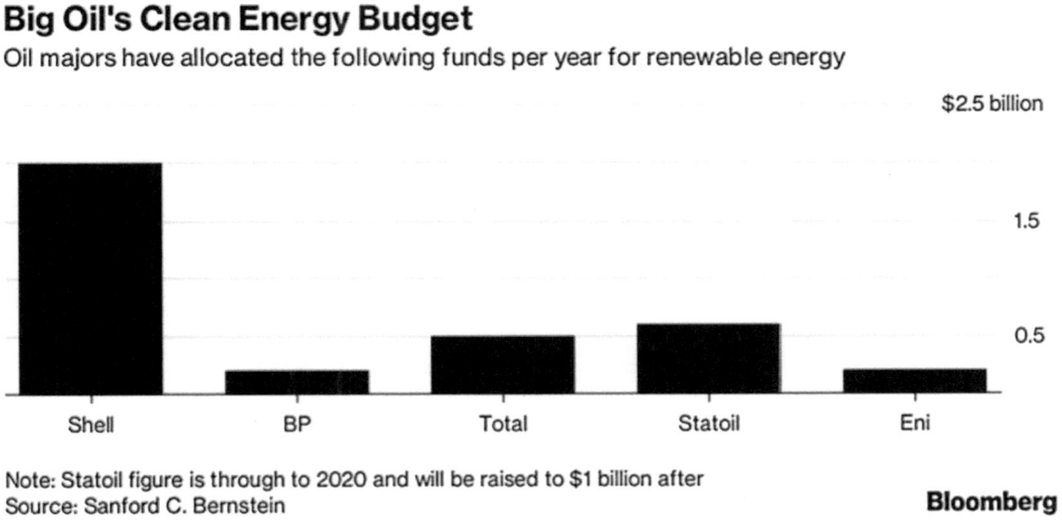

FIGURE 5.56 Oil majors' allocated funds per year in renewable energy.

Source: Hirtenstein (2018); Zhong and Bazilian (2018).

switch to remote reservoir management can be especially fruitful in sites located in what Kabir et al. (2018) call "black dot" zones, where solar irradiation is greater than the global average.

5.7 THE ROLE OF SOLAR IN ENERGY TRANSITION IN THE OIL AND GAS INDUSTRY

A process of energy transition is understood as a pathway toward the transformation of the global energy sector from the conventional way to develop and advance it into a different one. In the case of our current times, this process implies moving from industries that base their operations on fossil fuels to one that bases them on renewable energy (IRENA, 2020). With the potential problems that climate change could bring to humanity, as well as the limited availability of fossil fuels, options to begin this process of transition need to be researched and projected.

The role of oil and gas companies in this process of transition can be of manifest importance. On the one hand, new opportunities for investment and a positive return can appear with the development of strategies that involve renewable energy, particularly solar, and on the other, companies can benefit from being the "first movers" in a somewhat uncertain terrain if these technologies become established in the medium term. By virtue of this, some international oil and gas companies (IOC) have started to deploy a set of strategies to invest in cleaner technologies, with varying degrees of success (Zhong and Bazilian, 2018).

According to the authors, the role of IOCs can be a fundamental game changer in perceptions and in the way of doing business regarding technologies like wind and solar. In particular, "oil majors" can be crucial in the development and success of an energy transition process (see Figure 5.56). When climate change projections advance a change in average global temperatures and the international goals are to keep this change in the Earth's temperature below 2°C, it is perhaps more advisable to keep fossil fuel reserves in the ground. Because of this, the bottom line for many companies is also affected. So some sort of change is needed in the short term.

Over time, the public image of these companies can also be affected significantly. If the political class establishes penalties to companies for not keeping to environmental commitments, petroleum firms will be constantly judged by the population as the culprit after failed expectations and a

TABLE 5.12
Renewable Energy Targets Conceived for Saudi Arabia and Other Gulf Cooperation Council (GCC) Countries

Country	2020 Target	2030 Target
UAE	Dubai: 1% of electricity installed capacity	Dubai: 5% of electricity installed capacity
	Abu Dhabi: 7% of electricity generation	N/A
Qatar	6% of electricity generation[a]	20% of electricity generation[a]
Oman	N/A	N/A
Kuwait	5% of electricity generation	10% of electricity generation
Bahrain	N/A	5% of electricity generation[a]
Saudi Arabia	N/A	30% of electricity generation[b]

Source: Kuwait, Saudi Arabia, and Yemen from the League of Arab States (2013); RCREEE (2013a, 2013b); REN2I (2013); Dubai and Qatar (2020) from EU-GCC Clean Energy and Masdar Institute (2013); Qatar (2030) from Eversheds, and Ernst and Young (2013); Kuwait 2030 target from Al Sayegh et al. (2012).

Note:

[a] Unofficial targets.

[b] 2032 target.

Source: Salam and Khan (2017).

worsening climate. This constant measurement of companies by their ability to reduce their carbon footprint and the promotion of sustainability (Garcia et al., 2014) may look good for the political class and feel good for the population but is not a good long-term strategy for oil producers. Thus, aligning with *ex ante* sustainability regulations appears to be the best strategy moving forward (Zhong and Bazilian, 2018).

One example of the road to a transition process in a region of the world is seen in the work of Salam and Khan (2017), who detail the challenges that one of the most oil- and gas-dependent countries, Saudi Arabia, has endured to expand the role of solar energy in the country. Table 5.12 shows the list of targets that countries belonging to the Gulf Cooperation Council (GCC) have agreed to for the next couple of decades. Along with opening the door to investment in PV projects and enabling students and universities to perform research in these areas, the authors recommend less subsidization of electricity from the government and a more balanced investment strategy that includes a greater role for renewables.

In another country whose economy is heavily dependent on oil and gas, Russia, there appears to be a top-down approach to developing a local solar energy manufacturing industry. The work of Boute and Zhikharev (2019) explores the efforts from the Russian government to make the decarbonization of electricity production a main priority in their energy industry. According to the authors, a virtuous circle developed when an influential local solar energy industry, backed by major Russian industrial groups, was able to eventually overcome previous unsuccessful attempts to move the technology forward. Despite some of these players being invested heavily in the conventional energy sector, being first movers in a new environment was incentive enough to explore a switch to renewable energy.

In conclusion, when the appropriate incentives are being laid out and legislators are proactive enough to either foster new endeavors in solar energy or get out of the way when these are being explored, projects can see the light of day with less hurdles to pass. The following section focuses on the areas of change in the current dynamics and how some players can help foster a thriving renewable energy sector. Different players can set different goals for a similar objective so they can work in tandem to improve our environment.

As we can see, there are many areas that organizations are considering as a more fundamental part of their relationship with not only the environment but also society. Thus (see Figure 5.57),

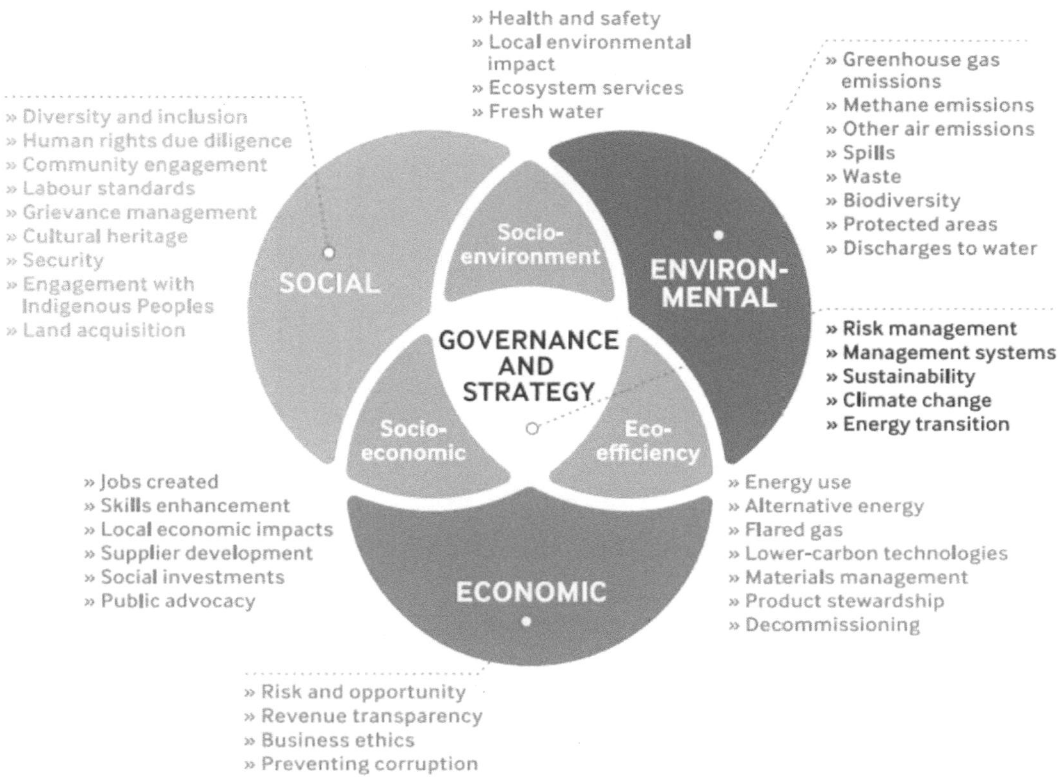

FIGURE 5.57 Sustainability issues in the oil and gas industry.

Source: IPIECA/API/IOGP (2020).

starting from what they produce as externalities of their functions, like greenhouse emissions or waste, organizations are now focusing on matters like engaging with the communities, improving the relationship with Indigenous people, and bettering labor standards. Additionally, companies are very aware of developing processes for energy transition, recognizing that hydrocarbons are expected to remain the dominant source of energy until at least 2035; renewables are to be given an increasing role (IPIECA/API/IOGP, 2020).

Similarly, in the context of a national oil company (NOCs), Krishna et al. (2012) mention the sustainability strategy of PETRONAS, Malaysia's national oil company. The corporation began their reporting in 2007, identifying key areas of development and improvement, like safety and health, climate change, and various environmental considerations. Likewise, RasGas, a Qatari firm owned by the national oil company Qatar Petroleum, is focused on improving the link between the company and its stakeholders through the betterment of sustainability issues.

Meanwhile, reporting for international oil companies (IOCs) follows a similar pattern that also relies on international organizations that offer guidelines and agreements, like the United Nations Global Compact and the Sustainable Development Goals.

Krishna et al. (2012) cite the cases of ExxonMobil and Total as two leading IOCs with superior sustainability reporting and guidelines for better practices. In general, both firms develop a healthy channel of communication between stakeholders and the heads of the company, with Total hosting a Q&A session with investors called "Ten Questions to Total" and ExxonMobil setting up an external citizenship advisory panel. Other examples of IOCs developing sustainability plans with the objective of increasing the role of renewable energy in the oil and gas sector include Equinor, previously

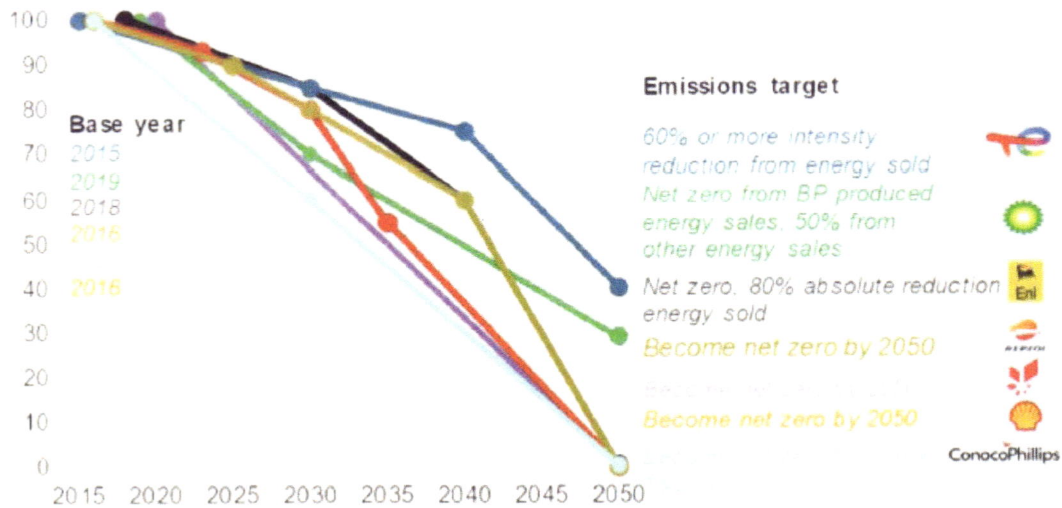

FIGURE 5.58 IOCs net-zero emissions target by 2050 Exxon-Mobil and Chevron do not have a GHG emission reduction target.

Source: BloombergNEF, Oil and Gas Divestment Trends (2021).

known as Statoil, a Norwegian company, which pledged \$200 million US dollars to invest in renewables, with themes like offshore and onshore wind, solar energy, transportation, energy storage, smart grids, and energy efficiency (Equinor, 2016). In turn, in their Energy Transition Strategy document (2021), Shell has pledged to become a net-zero company by 2050. Figure 5.58 shows a graphical view of international oil companies' (IOCs) announcements for net-zero emissions target.

In terms of renewable energy investments, European IOCs are taking the lead in transitioning themselves to become integrated energy companies. While accelerating clean energy investments in the meantime, they are divesting existing assets against mounting public and investor pressure to decarbonize themselves. The asset sales from 2015 to 2020 hit \$198 billion, specifically with more focus on upstream assets. Energy transition, maximizing portfolio value, debt reduction, and climate risk are the main influential factors of divestment (BloombergNEF Oil & Gas Divestment Trends, 2021). Figure 5.59 shows divestment and clean energy investments for leading IOCs.

In terms of clean energy investment with 5.4 GW installed renewable capacity, TotalEnergies is leading, followed by Equinor at 4.7 GW, Shell at 3.8 GW, and BP at 2.3 GW, Eneos at 1.9 GW, and ENI at 1.1 GW (BloombergNEF, Oil and Gas Transition Scores, 2022).

5.8 CONCLUSIONS AND FUTURE WORKS

This chapter took a comprehensive dive into technologies relying on solar energy and their application in the oil and gas industry. By virtue of being a limitless resource and one that will be provided to us virtually forever, researchers have started looking into more ways of taking advantage of this resource, helping power our daily activities of consumption and production. This is an especially salient topic when considering that other non-renewable resources are starting to dwindle, and that the excessive use of these resources without pivotal restrictions can lead to an increase of greenhouse emissions, and thus to a greater role in climate change.

This window of opportunity invites engineers and researchers to explore the incredible potential of renewable resources, and this chapter reviewed the most important and applied technologies that utilize the energy of the sun for our everyday use. The first section of this work centered on the three most important technologies to transform solar radiation into electricity: solar photovoltaic (PV)

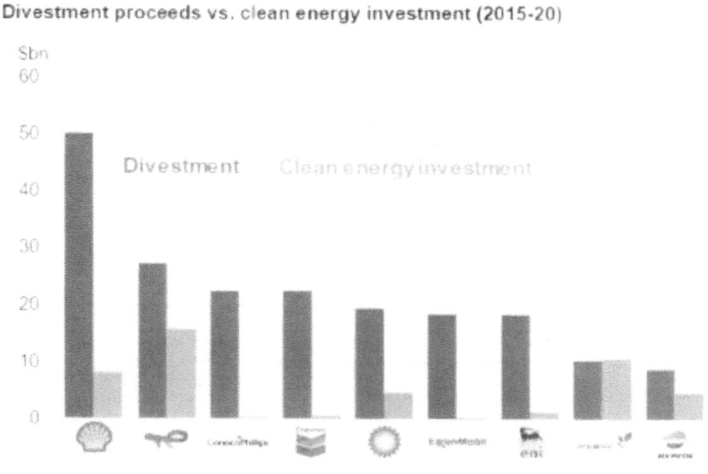

FIGURE 5.59 IOCs' divestments and clean energy investments.

Source: BloombergNEF, Oil and Gas Divestment Trends (2021).

plants, solar thermal installations, and concentrating solar power (CSP) plants. Important technical differences between these methods make them applicable to different parts of the world with varying levels of solar irradiation.

Then, we reviewed the types of uses that solar energy can benefit, going from the simple small-scale domestic to the large-scale generation of electricity to be included in the national grid. A brief mention of the costs of these technologies and a way to improve the handling and disposing of expired equipment, which contains some toxic materials, is also included.

Furthermore, we performed a review of the best practices and the different solar applications in more detail, as well as the economics of different designs for these types of solar installations. Adding more parameters for which to assess and anticipate the benefits and the challenges brought by solar projects helps fine-tune the process of arriving at the best design for the challenge at hand.

Because some hurdles to these resources are harder to jump than others, it is important to consider the role of hybrid installations. In this sense, the role of combined heat and power (CHP) plants was analyzed. With the advent of notions like the Internet of Things (IoT) and the smart home concept, the role of several sources of energy that work in combination to satisfy the needs of the end user is another topic of interest. The findings in this area suggest that solar will be a major addition to the development of these concepts, and potential challenges brought up by scarce sunlight can be overcome by complementary sources of energy.

Moving on, the role of solar energy in the advancement of the oil and gas industry was reviewed. The sector is no exception to the challenges presented at the beginning of the chapter, and we can add another one when we ponder the need to develop optimization strategies for old and current reservoirs. The process of enhanced oil recovery (EOR) is a tertiary retrieval stage, where using external techniques leads to more oil being dug up from the reservoir. This process, however, requires considerable amounts of energy. So the exploration of solar technologies to supply a larger percentage of this energy is key to improving the work of the sector in achieving a cleaner environment.

The last, NOCs' and IOCs' accelerated adoption on clean energy investments was discussed to become integrated energy companies due to public and investor pressure. European IOCs are taking the lead.

Solar technologies can be a key element in the transition from economies and businesses centering on fossil fuels to systems relying primarily on energies like the one from the sun, but also from the wind. Despite a considerable number of challenges, it remains a pending assignment to develop the necessary steps that could lead us to a future where we can generate increasing amounts of electricity and exploit significantly more quantities of oil to fulfill the needs of an expanding world but, at the same time, keeping the potential costs on the environment and on people on check. Complementing conventional sources of energy with renewables is the key to a world with more production and more energy, and a cleaner environment.

REFERENCES

Abdullah, M. F., M. Ahmed, and K. N. M. Hasan. 2018. Chapter 5: Feasibility Study on Hybrid Renewable Energy to Supply Unmanned Offshore Platform. In S. A. Sulaiman, R. Kannan, S. A. Abdul Karim, and N. Mohd Nor (eds.), *Sustainable Electrical Power Resources Through Energy Optimization and Future Engineering*. Springer.

Abureden, S., S. Taqvi, I. H. Mustafa, I. H. Alhajri, A. Alsahli, and A. Elkamel. August 2015. Case Study: Integrating Solar Energy in the Upstream Oil Supply Chain (USOSC) of Abu Dhabi, UAE. *International Conference on Science, Ecology and Technology, Vienna* I, 638–649.

Afsar, C., and S. Akin. 2016. Solar Generated Steam Injection in Heavy Oil Reservoirs: A Case Study. *Renewable Energy* 91, 83–89.

Ahmadi, E., Y. Noorollahi, B. Mohammadi-Ivatloo, and A. Anvari-Moghaddam. 2020. Stochastic Operation of a Solar-Powered Smart Home: Capturing Thermal Load Uncertainties. *Sustainability* 12, 5089.

Ahmadi, M. H., H. Sayyaadi, S. Dehghani, and H. Hosseinzade. 2013. Designing a Solar Powered Stirling Heat Engine Based on Multiple Criteria: Maximized Thermal Efficiency and Power. *Energy Conversion and Management* 75, 282–291.

Ahmed, M. 2015. *Modelling and Optimization of an Off-Grid Hybrid Power System for Supplying Unmanned Offshore Installations in Eastern Malaysia*. A project dissertation submitted to the Electrical & Electronics Engineering Programme, Universiti Teknologi PETRONAS.

Al Garni, H. Z., and A. Awasthi. 2017. Solar PV Power Plant Site Selection Using a GIS-AHP Based Approach with Application in Saudi Arabia. *Applied Energy* 206, 1225–1240.

Al-Hamadani, A., and A. Yaseen, A. 2019. Experimental Study of Multi-Effect Stages PV Panels Solar Still to Enhance the Productivity by Utilizing Water Heater and Cooling Fan. *Journal of University of Babylon for Engineering Sciences* 27, 275–286.

Al-Hinai, H., M. S. Al-Nassri, and B. A. Jubran. 2002. Parametric Investigation of a Double-Effect Solar Still in Comparison with a Single-Effect Solar Still. *Desalination* 150(1), 75–83.

Allouhi, A., Y. Agrouaz, M. Benzakour Amine, S. Rehman, M. S. Buker, T. Kousksou, A. Jamil, and A. Benbassou. 2017. Design Optimization of a Multi-Temperature Solar Thermal Heating System for an Industrial Process. *Applied Energy* 206, 382–392.

Al-Saedi, H. N., R. E. Fiori, and P. Brady. 2018. *Coupling Low Salinity Water Flooding and Steam Flooding for Sandstone Reservoirs; Low Salinity-Alternating-Steam Flooding (LSASF)*. SPE Kingdom of Saudi Arabia Annual Technical Symposium and Exhibition.

Alsharif, M. H., K. Yahya, and Z. W. Geem. 2019. Strategic Market Growth and Policy Recommendations for Sustainable Solar Energy Deployment in South Korea. *Journal of Electrical Engineering & Technology* 15.

Aman, M. M., K. H. Solangi, M. S. Hossain, A. Badarudin, G. B. Jasmon, H. Mokhlis, A. H. A. Bakar, and S. N. Kazi. 2015. A Review of Safety, Health and Environmental (SHE) Issues of Solar Energy System. *Renewable and Sustainable Energy Reviews* 41, 1190–1204.

Augsten, E. September 2012. Europe/Asia: Solar Cooling Gains Traction. *Solar Thermal World*. Available in: http://solarthermalworld.org/content/europeasia-solar-cooling-gains-traction.

Aye, L., W. W. S. Charters, and C. Chaichana. 2002. Solar Heat Pump Systems for Domestic Hot Water. *Solar Energy* 73(3), 169–175.

Bagheripour Haghighi, M., S. Ayatollahi, and M. Shabaninejad. 2012. *Comparing the Performance and Recovery Mechanisms for Steam Flooding in Heavy and Light Oil Reservoirs*. SPE Heavy Oil Conference Canada.

Bahnemann, D. 2004. Photocatalytic Water Treatment: Solar Energy Applications. *Solar Energy* 77(5), 445–459.

Blarke, M. B. 2012. Towards an Intermittency-Friendly Energy System: Comparing Electric Boilers and Heat Pumps in Distributed Cogeneration. *Applied Energy* 91(1), 349–365.

Bloomberg NEF. May 5, 2022. Oil & Gas Transition Scores 2022 Part II Result and Analysis. *Bloomberg NEF.*

Bloomberg NEF. July 27, 2021. Oil & Gas Divestment Trends Big Asset Sales by International Oil Companies. *Bloomberg NEF*.

Bloomberg NEF. October 2019. Global Solar Investment Report 12. *Bloomberg NEF*. Available in: www.climateinvestmentfunds.org/sites/cif_enc/files/knowledge-documents/bnef_2019-10-30_isa-cif_report-final_002.pdf.

Bogacka, M., M. Potempa, B. Milewicz, D. Lewandowski, K. Pikoń, K. Klejnowska, . . . and E. Misztal. 2020. PV Waste Thermal Treatment According to the Circular Economy Concept. *Sustainability* 12(24), 10562.

Boute, A., and A. Zhikharev. 2019. Vested Interests as Driver of the Clean Energy Transition: Evidence from Russia's Solar Energy Policy. *Energy Policy* 133, 110910.

Braun, S., G. Hahn, R. Nissler, C. Ponisch, and D. Habermann. 2013. The Multi-Busbar Design: An Overview. *Energy Procedia* 43, 86–92. Available in: https://doi.org/10.1016/j.egypro.2013.11.092.

Chaar, M., M. Venetos, J. Dargin, and D. Palmer. 2014. *Economics of Steam Generation for Thermal EOR*. Abu Dhabi International Petroleum Exhibition and Conference.

Cherradi, N. 2019. *Solar PV technologies what's next?* Becquerel Institute.

Delgado-Torres, A. M. 2009. Solar Thermal Heat Engines for Water Pumping: An Update. *Renewable and Sustainable Energy Reviews* 13(2), 462–472.

Dimri, N. September 2021. Thermoeconomic Optimization and Performance Analysis of Solar Combined Heating and Power Systems: A Comparative Study. *Energy Conversion and Management* 244, 114478.

Elashmawy, M. 2019. Effect of Surface Cooling and Tube Thickness on the Performance of a High Temperature Standalone Tubular Solar Still. *Applied Thermal Engineering* 156, 276–286.

Energy Sage. 2018. *What Is the History of Solar Energy and When Were Solar Panels Invented?* Available in: https://news.energysage.com/the-history-and-invention-of-solar-panel-technology/. Accessed July 18th, 2021.

Equinor. 2016. *Statoil Launches USD 200m New Energy Investment Fund*. Available in: www.equinor.com/en/news/launches-usd200m-new-energy-investment-fund.html. Accessed August 25th, 2021.

European Union. 2021. *Study on Quality Standards for the Treatment of Waste, Electrical, and Electronic Equipment (WEEE)*. Available in: https://op.europa.eu/en/publication-detail/-/publication/2004b067-726a-11eb-9ac9-01aa75ed71a1/language-en/format-PDF/source-193365602. Accessed July 30th, 2021.

Fischer, D., and H. Madani. 2017. On Heat Pumps in Smart Grids: A Review. *Renewable and Sustainable Energy Reviews* 70, 342–357.

Fraunhofer ISE. February 2022. *Photovoltaics Report*. Available in: www.ise.fraunhofer.de/content/dam/ise/de/documents/publications/studies/Photovoltaics-Report.pdf.

Garcia, R., D. Lessard, and A. Singh. 2014. *Strategic Partnering in Oil and Gas: A Capabilities Perspective*. MIT Sloan School Working Paper 5092-14. MIT.

Giostri, A., M. Binotti, M. Astolfi, P. Silva, E. Macchi, and G. Manzolini. 2012. Comparison of Different Solar Plants Based on Parabolic Trough Technology. *Solar Energy* 86(5), 1208–1221.

GlobalData. 2018. *Solar PV Module, Update 2018: Global Market Size, Competitive Landscape and Key Country Analysis to 2022*. GlobalData.

Gnanaraj, S. J. P., and S. Ramachandran. 2022. Identification of Operational Parameter Levels that Optimize the Production in Solar Stills with Plain, Corrugated, and Compartmental Basin. *Environmental Science and Pollution Research* 29, 7096–7116.

Greenmatch. 2021. *Solar Panels vs Solar Thermal: Solar Thermal or Solar Panels?* Available in: www.greenmatch.co.uk/blog/2015/04/solar-panels-vs-solar-thermal. Accessed September 5th, 2021.

Gregory, M., D. Omom, and P.-A. Greil. 2014. *Solar Enhanced Oil Recovery: An in-Country Value Assessment for Oman—January 2014*. Ernst & Young LLP.

Gupta, S., R. Guédez, and B. Laumert. 2017. Market Potential of Solar Thermal Enhanced Oil Recovery-a Techno-Economic Model for Issaran Oil Field in Egypt. *AIP Conference Proceedings* 1850, 190001.

Hadorn, J. C. 2015. *Solar and Heat Pump Systems for Residential Buildings*. John Wiley & Sons.

Henkel, E. T. 2005. New Solar Thermal Energy Applications for Commercial, Industrial, and Government Facilities. *Energy Engineering* 102(2), 39–58.

Hernandez, R. R., S. B. Easter, M. L. Murphy-Mariscal, F. T. Maestre, M. Tavassoli, E. B. Allen, C. W. Barrows, J. Belnap, R. Ochoa-Hueso, S. Ravi, and M. F. Allen. 2014. Environmental Impacts of Utility-Scale Solar Energy. *Renewable and Sustainable Energy Reviews* 29, 766–779.

Herwitz, S. R., L. F. Johnson, J. C. Arvesen, R. G. Higgins, J. G. Leung, and S. E. Dunagan. 2002. *Precision Agriculture as a Commercial Application for Solar-Powered Unmanned Aerial Vehicles*. 1st UAV Conference.

Hirtenstein, A. January 10, 2018. Big Oil Weighs Acquisitions to Boost Foothold in Green Industry. *Bloomberg News*.

Howard, D., and R. G. Harley. 2010. *Modeling of Dish-Stirling Solar Thermal Power Generation*. IEEE PES General Meeting.

International Energy Agency (IEA). 2012. *Renewables 2020*. IEA. Available in: www.iea.org/reports/renewables-2020. Accessed September 5th, 2021.

International Energy Agency (IEA). 2021a. *Net Zero by 2050: A Roadmap for the Global Energy Sector 24, 57, 74*. Available in: https://iea.blob.core.windows.net/assets/deebef5d-0c34-4539-9d0c-10b13d840027/NetZeroby2050-ARoadmapfortheGlobalEnergySector_CORR.pdf.

International Energy Agency (IEA). December 2021b. *Renewables 2021 Analysis and Forecast to 2026 Report 14, 15, 21, 26–28*. Available in: https://iea.blob.core.windows.net/assets/5ae32253-7409-4f9a-a91d1493ffb9777a/Renewables2021-Analysisandforecastto2026.pdf.

International Energy Agency (IEA-SHC). December 2013. Executive Summary: Solar Cooling Position Paper. *Task 38: Solar Air-Conditioning and Refrigeration*. www.iea-shc.org/data/sites/1/publications/IEA-SHC-Solar-Cooling-Position-Paper-Executive-Summary.pdf.

International Renewable Energy Agency (IRENA). 2015. *Solar Heat for industrial Processes, 8*. Available in: www.irena.org/publications/2015/Jan/Solar-Heat-for-Industrial-Processes#:~:text=January%202015&text=In%20developing%20countries%2C%20especially%20in,%2C%20boiling%2C%20pasteurisation%20and%20sterilisation.

International Renewable Energy Agency (IRENA). 2019. *Future of Solar Photovoltaic Report, 41, 44*. Available in: https://irena.org//media/Files/IRENA/Agency/Publication/2019/Nov/IRENA_Future_of_Solar_PV_2019.pdf.

International Renewable Energy Agency (IRENA). 2020. *Energy Transition*. Available in: www.irena.org/energytransition. Accessed August 8th, 2021.

IPIECA/API/IOGP. 2020. *Sustainability Reporting Guidance for the Oil and Gas Industry: Advancing Environmental and Social Performance Across Oil and Gas*, 4th edition. IPIECA/API/IOGP.

Jakob, U. 2014. *Technologies and Perspectives of Solar Cooling Systems*. AHK Conference. Available in: www.docstoc.com/docs/169908741/DrJakob_Green_Chiller_Assoc.pdf

Johnson, A., L. Mu, Y. H. Park, D. J. Valles, H. Wang, P. Xu, . . . and S. Kuravi. 2019. A Thermal Model for Predicting the Performance of a Solar Still with Fresnel Lens. *Water* 11, 1860.

Kabeel, A. E., K. Harby, M. Abdelgaied, and A. Eisa. 2019. A Comprehensive Review of Tubular Solar Still Designs, Performance, and Economic Analysis. *Journal of Cleaner Production* 119030.

Kabir, E., P. Kumar, S. Kumar, A. A. Adelodun, and K.-H. Kim. 2018. Solar Energy: Potential and Future Prospects. *Renewable and Sustainable Energy Reviews* 82, 894–900.

Kalogera, M., and P. Bauer. 2014. *Optimization of an Off-Grid Hybrid System for Supplying Offshore Platforms in Arctic Climates*. 2014 International Power Electronics Conference (IPEC-Hiroshima 2014—ECCE ASIA).

Kalogirou, S. 2003. The Potential of Solar Industrial Process Heat Applications. *Applied Energy* 76(4), 337–361.

Katekar, V. P., and S. S. Deshmukh. 2020. A Review on Research Trends in Solar Still Designs for Domestic and Industrial Applications. *Journal of Cleaner Production* 120544.

Krishna, P., B. Janus, C. A. Peterson, B. Doherty, R. Sherman, H. Murphy, and C Curlee. 2012. *Sustainability Reporting on the Rise Within the Petroleum Industry—A View of the Industry and Experiences from an NOC, IOC and Service Company*. International Conference on Health, Safety and Environment in Oil and Gas Exploration and Production.

Kumar Sahu, B. 2015. A Study on Global Solar PV Energy Developments and Policies with Special Focus on the Top Ten Solar PV Power Producing Countries. *Renewable and Sustainable Energy Reviews* 43, 621–634.

Mekhilef, S., R. Saidur, and A. Safari. 2011. A Review on Solar Energy Use in Industries. *Renewable and Sustainable Energy Reviews* 15(4), 1777–1790.

MIT Study. 2015. *The Future of Solar Energy 23, 52–64*. Available in: https://energy.mit.edu/wp-content/uploads/2015/05/MITEI-The-Future-of-Solar-Energy.pdf.

Müller-Steinhagen, H., and F. Trieb. 2004. Concentrating Solar Power: A Review of the Technology. *Ingenia* 1–9.

Nadal-Bach, J., J. C. Bruno, J. Farnós, and M. Rovirac. May 2021. Solar Stills and Evaporators for the Treatment of Agro-Industrial Liquid Wastes: A Review. *Renewable and Sustainable Energy Reviews* 142, 110825.

National Renewable Energy Laboratories (NREL). 2016. *Advanced Grid-Friendly Controls Demonstration Project for Utility-Scale PV Power Plants, 11*. Available in: www.nrel.gov/docs/fy16osti/65368.pdf.

NASA. 2021. *How Old Is the Sun? Space Place*. Available in: https://spaceplace.nasa.gov/sun-age/en/. Accessed July 21st, 2021.

O'Gallagher, J. J. 2008. Nonimaging Optics in Solar Energy. *Synthesis Lectures on Energy and the Environment: Technology, Science, and Society* 2(1), 1–120.

Ohira, Y., A. Indarto, and C. Hudaya. 2018. *Analysis of Solar Power as an Alternative Electrical Supply for the Oil and Gas Field Development.* 10th International Conference on Information Technology and Electrical Engineering (ICITEE).

Okpala, J., O. Adedokun, A. Jaja, O. Olubukola, E. Ogugua, J. Olayomi, I. Aitokhuehi, P. Korede, K. Awonuga, and O. Eruvbentine. 2020. *Enabling Reservoir Management Excellence Through Real Time Surveillance of an Unmanned Onshore Gas-Condensate Field Platform.* SPE Nigeria Annual International Conference and Exhibition.

Ozsoy, A., and V. Corumlu. 2018. Thermal Performance of a Thermosyphon Heat Pipe Evacuated Tube Solar Collector Using Silver-Water Nanofluid for Commercial Applications. *Renewable Energy* 122, 26–34.

Panchal, H., and I. Mohan. 2017. Various Methods Applied to Solar Still for Enhancement of Distillate Output. *Desalination* 415, 76–89.

Perel, D. October 2013. *Innovation in Power Plants-Siemens*, 34–35. Available in: www.slideserve.com/flynn/innovation-in-power-plants-dimitry-perel-e-o-ip-sales-sil-dimitry-perel-siemens-october-2013.

Poppi, S., N. Sommerfeldt, C. Bales, H. Madani, and P. Lundqvist. 2018. Techno-Economic Review of Solar Heat Pump Systems for Residential Heating Applications. *Renewable and Sustainable Energy Reviews* 81, 22–32.

Poullikkas, A. 2009. Economic Analysis of Power Generation from Parabolic Trough Solar Thermal Plants for the Mediterranean Region – A Case Study for the Island of Cyprus. *Renewable and Sustainable Energy Reviews* 13(9), 2474–2484.

PV Magazine, March 2022. Designs for Solar+Storage+Hydrogen Systems in Buildings. *PV Magazine*. Available in www.pv-magazine.com/2022/03/29/designs-for-solarstoragehydrogen-systems-in-buildings/.

Rajaseenivasan, T., and K. K. Murugavel. 2013. Theoretical and Experimental Investigation on Double Basin Double Slope Solar Still. *Desalination* 319, 25–32.

Qu, M., H. Yin, and D. H. Archer. 2010. A Solar Thermal Cooling and Heating System for a Building: Experimental and Model Based Performance Analysis and Design. *Solar Energy* 84(2), 166–182.

Quaschning, V., and M. Blanco Muriel. October 10–12, 2001. *Solar Power – Photovoltaics or Solar Thermal Power Plants?* VGB Congress Power Plants.

Ravi Kumar, K., N. V. V. Krishna Chaitanya, and N. Sendhil Kumar. February 2021. Solar Thermal Energy Technologies and Its Applications for Process Heating and Power Generation – A Review. *Journal of Cleaner Production* 282, 125296.

Sahu, A., N. Yadav, and K. Sudhakar. 2016. Floating Photovoltaic Power Plant: A Review. *Renewable and Sustainable Energy Reviews* 66, 815–824.

Salam, M. A., and S. A. Khan. 2017. Transition Towards Sustainable Energy Production – A Review of the Progress for Solar Energy in Saudi Arabia. *Energy Exploration & Exploitation* 36(1), 3–27.

Sandler, J., G. Fowler, K. Cheng, and A. R. Kovscek. 2014. Solar-Generated Steam for Oil Recovery: Reservoir Simulation, Economic Analysis, and Life Cycle Assessment. *Energy Conversion and Management* 77, 721–732.

Sathyamurthy, R., W. M. El-Maghlany, M. E. H. Attia, A. E. Kabeel, M. Abdelgaied, M. Abdel-Aziz, . . . and S. Vasanthaseelan. 2022. Hemispherical Solar Distiller Performance Utilizing Hybrid Storage Media, Paraffin Wax with Nanoparticles: An Experimental Study. *Molecules* 27(24), 8988.

Shravan, K., and K. Chunduri. 2018. PERC Solar Cell Technology- 2018 Edition. *TaiyangNews, China.*

Shravan, K., and K. Chunduri. 2019. Advanced Solar Module Technology – The Time for a New Generation of Solar Modules Has Come. *TaiyangNews, China.*

Sindhu, S., V. Nehra, and S. Luthra. 2017. Solar Energy Deployment for Sustainable Future of India: Hybrid SWOC-AHP Analysis. *Renewable and Sustainable Energy Reviews* 72, 1138–1151.

Solar Energy Industries Association. 2021. *Utility-Scale Solar.* Available in: www.seia.org/initiatives/utility-scale-solar-power.

Solar Millennium AG. 2008. *The Parabolic Trough Power Plants Andasol 1 to 3.* Available in: http://large.stanford.edu/publications/power/references/docs/Andasol1-3engl.pdf. Accessed August 18th, 2021.

Soutullo, S., E. Giancola, M. N. Sánchez, J. A. Ferrer, D. García, M. J. Súarez, J. I. Prieto, E. Antuña-Yudego, J. L. Carús, M. A. Fernández, and M. Romero. 2020. Methodology for Quantifying the Energy Saving Potentials Combining Building Retrofitting, Solar Thermal Energy and Geothermal Resources. *Energies* 13(22), 5970.

Spasiano, D., R. Marotta, S. Malato, P. Fernandez-Ibañez, and I. Di Somma. 2015. Solar Photocatalysis: Materials, Reactors, Some Commercial, and Pre-Industrialized Applications. A Comprehensive Approach. *Applied Catalysis B: Environmental* 170–171, 90–123.

Tian, Y., and C. Y. Zhao. 2013. A Review of Solar Collectors and Thermal Energy Storage in Solar Thermal Applications. *Applied Energy* 104: 538–553.

Tiong, Y. K., M. A. Zahari, S. F. Wong, and S. S. Dol. 2015. The Feasibility of Wind and Solar Energy Application for Oil and Gas Offshore Platform. *IOP Conference Series: Materials Science and Engineering* 78, 012042.

Toledo, C., and A. Scognamiglio. 2021. Agrivoltaic Systems Design and Assessment: A Critical Review, and a Descriptive Model towards a Sustainable Landscape Vision (Three-Dimensional Agrivoltaic Patterns. *Sustainability* 13(12), 6871.

Tsoutsos, T., N. Frantzeskaki, and V. Gekas. 2005. Environmental Impacts from the Solar Energy Technologies. *Energy Policy* 33(3), 289–296.

Ud Din Kirmani, F., A. Raza, R. Gholami, M. Zeshan Haidar, and C. Sarosh Fareed. 2021. Analyzing the Effect of Steam Quality and Injection Temperature on the Performance of Steam Flooding. *Energy Geoscience* 2, 83–86.

United States Department of Energy. 2021a. *Photovoltaics*. Available in: www.energy.gov/eere/solar/photovoltaics. Accessed August 28th, 2021.

United States Department of Energy. September 2021b. *Solar Futures Study*. Office of Energy Efficiency and Renewable Energy. Available in: www.energy.gov/sites/default/files/2021-09/Solar%20Futures%20Study.pdf. Accessed September 5th, 2021.

United States Environmental Protection Agency. 2021. *Solar Heating and Cooling Technologies*. Available in: www.epa.gov/rhc/solar-heating-and-cooling-technologies. Accessed September 4th, 2021.

V. K., R., V. Chintala, and S. Kumar 2019. Recent Developments, Challenges and Opportunities for Harnessing Solar Renewable Energy for Thermal Enhanced Oil Recovery (EOR). *Energy Sources, Part A: Recovery, Utilization, and Environmental Effects* 1–18.

Vasu, R., and F. B. Ismail. 2018. Design and Implementation of Solar Powered Stirling Engines: Review. *AIP Conference Proceedings* 2035, 040002.

Visser, E., V. Perold, S. Ralston-Paton, A. C. Cardenal, and P. G. Ryan. 2018. Assessing the Impacts of a Utility-Scale Photovoltaic Solar Energy Facility on Birds in the Northern Cape, South Africa. *Renewable Energy* 1–10.

Wang, J., J. O'Donnell, and A. R. Brandt. 2017. Potential Solar Energy Use in the Global Petroleum Sector. *Energy* 118, 884–892.

Widyolar, B., L. Jiang, J. Brinkley, S. K. Hota, J. Ferry, G. Diaz, and R. Winston. 2020. Experimental Performance of an Ultra-Low-Cost Solar Photovoltaic-Thermal (PVT) Collector Using Aluminum Minichannels and Nonimaging Optics. *Applied Energy* 268, 114894.

Winston, R., E. Yablonovitch, L. Jiang, B. K. Widyolar, M. Abdelhamid, G. Scranton, D. Cygan, and A. Kozlov. 2015. Hybrid Solar Collector Using Nonimaging Optics and Photovoltaic Components. *Nonimaging Optics: Efficient Design for Illumination and Solar Concentration XII*. Invited Paper.

Xu, Y., J. Li, Q. Tan, A. L. Peters, and C. Yang. 2018. Global Status of Recycling Waste Solar Panels: A Review. *Waste Management* 75, 450–458.

Yu, J., L. Guo, M. Ma, S. Kamel, W. Li, and X. Song. 2018. Risk Assessment of Integrated Electrical, Natural Gas and District Heating Systems Considering Solar Thermal CHP Plants and Electric Boilers. *International Journal of Electrical Power and Energy Systems* 103, 277–287.

Zhong, M., and M. D. Bazilian. 2018. Contours of the Energy Transition: Investment by International Oil and Gas Companies in Renewable Energy. *The Electricity Journal* 31(1), 82–91.

6 Geothermal Energy

Coruh Celikbilek and Osman Sen

6.1 GEOTHERMAL ENERGY AND USAGE AREAS

A geothermal source is hot water, steam, or gas containing much more melted matter than the normal underground and surface waters around it, which are heated by a magma focused deep in the Earth's crust and carried to the earth by fractures and cracks formed by tectonic movements in the Earth. Geothermal energy is the heat energy obtained artificially from hot water, steam, dry steam, and hot dry rocks, which are formed by storing the heat accumulated in the rocks deep in the ground by the fluids. Rocks are formed from the accumulation of various minerals or stone particles. Their most important feature is that they are a kind of document that reflects the natural environment at the time of their formation. Thus, traces of the geological development of the Earth's crust can be studied by analyzing various rocks.

A geothermal model consists of the heat source, the fluid that carries the heat from underground to the earth, and the permeable rocks that provide the circulation of this fluid. As seen in Figure 6.1, the Earth's crust has a total diameter of 6,378 km and is thinner than the mantle. The magma intrusion areas formed just below this crust constitute potential geothermal regions. These cover all areas whose boundaries have been determined by scientific and technical studies, such as the geothermal springs or the geothermal springs with natural mineral waters (Kılıç and Kılıç, 2013).

If the hot rock and underground waters at high temperatures are located in shallower parts compared to other places, this region is defined as a *geothermal area*. As shown in Figure 6.2, in places where the Earth's crust is thinner, the magma carrying high temperatures gets very close to the crust, forming geothermal fields. In addition, the warming of the groundwater at a depth of several kilometers due to weather events, and then its rise to the surface, allows this region to be defined as a geothermal area (Kılıç and Kılıç, 2013).

A geothermal system includes the feeding area, fluid, heat source, reservoir area, cover rock, and discharge area, where geothermal springs and natural mineral waters are extracted or produced, with a unique geological structure and hydrogeological and chemical properties (Kılıç and Kılıç, 2013). The permeable rock area where the heated waters are located is defined as a *geothermal reservoir*. The geothermal reservoir is in a natural balance regarding temperature and geochemistry. It consists of semi-open or closed hot water/steam production environments fed from outside differently. One of the most essential features of geothermal energy, unlike other sources produced from underground, is its dynamic structure. The heat from geothermal energy can be gained in three different ways (Figure 6.3):

1. The heat from shallow formations
2. The heat from hydrothermal systems (hot underground brine)
3. The heat from enhanced or engineered geothermal systems (EGS)

6.1.1 Hot Water System

There are 20 times more hot water–based systems than steam-based systems on Earth. In the hot water system, fluids with different chemical properties, rich in molten chemicals, under high temperature and pressure, are in the deep reservoir rock. Electrical energy is produced from the steam obtained from the hot water + steam mixture extracted from such systems by drilling.

DOI: 10.1201/9781003282273-6

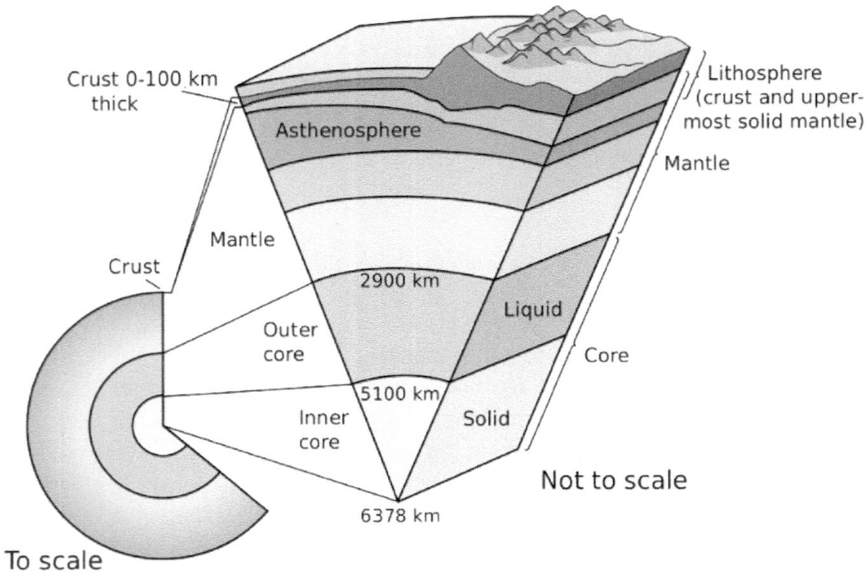

FIGURE 6.1 The Earth's internal structure, and the appearance of the Earth's crust.

Source: Kılıç and Kılıç (2013).

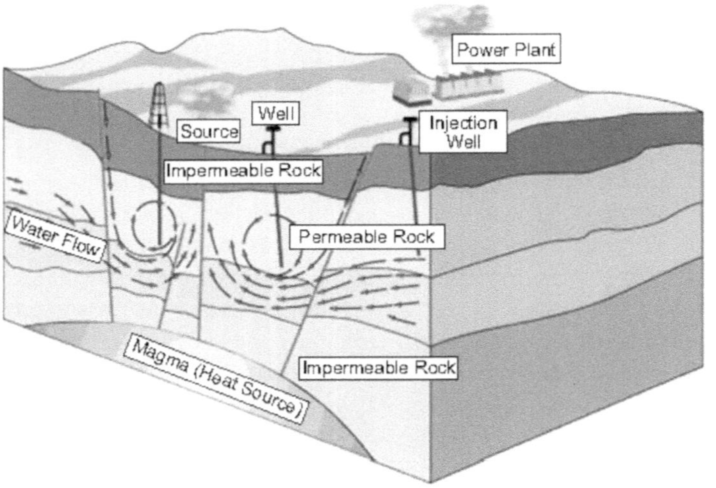

FIGURE 6.2 Geothermal system model.

Source: Günhan (2016).

The remaining steam-free hot water is used in other geothermal application areas. Wastewater is reinjected into the reservoir.

6.1.2 Dry Steam System

Unlike hot water–based systems, steam-based systems produce highly heated steam with low humidity and high temperature. This type of steam is sent directly to geothermal power plants as an energy source and converted into electrical energy. In a way, these are natural nuclear reactors formed on the Earth's crust.

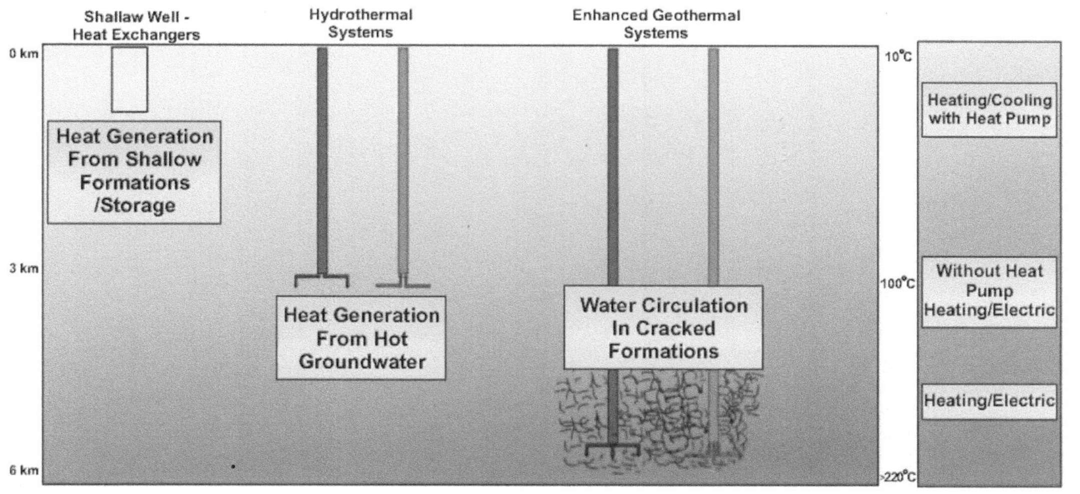

FIGURE 6.3 Types of heat production from geothermal resources.

Three types of geothermal systems have been identified so far.

6.1.3 Hot Dry Rock System

In our world, especially in young, active volcanic belts, where the geothermal gradient is very high, there are hot dry rocks with high temperatures that do not contain hot water. It is an energy source for transferring heat to the Earth by pressing cold water or suitable gas into such systems (Mertoğlu, 2000). Conventional geothermal systems require three components: a heat source, heat transfer fluid (water), and reservoir. However, only the heat source is sufficient in enhanced geothermal system (EGS) applications (Sen et al., 2015). With EGS, the target rock is descended by drilling. A reservoir is created underground by hydraulic fracturing or excitation in the area between two wells. The heat here is brought to the surface by the water or gas given from the surface. In the field of EGS, R&D studies are carried out in some countries, especially in the USA, Germany, England, France, Japan, Australia, and Turkey. For EGS applications, hot and low-porosity granite and granodiorite rocks are preferred. The most suitable rocks for creating artificial reservoirs are granite and granodiorite rocks (Sen et al., 2015).

Geothermal energy can be used in many areas, primarily in electrical energy production. We can count many different usage areas, such as city and greenhouse heating/cooling, growing tropical plants and fish, swimming pools, thermal tourism, process heat supply, drying and sterilization of foods, canning, sugar, medicine, pasteurized milk factories, dry ice, carbon dioxide, hydrogen and fertilizer production, cosmetic industries, and recovery of lithium, which is used as a battery raw material (Figure 6.4).

6.2 ELECTRICITY GENERATION WITH GEOTHERMAL ENERGY

Deep wells are drilled into underground hot water reservoirs to generate electricity from geothermal sources. There are three types of geothermal power plants: dry steam, flash, and binary. The oldest geothermal technology, the dry steam system, takes steam from cracks in the ground and uses it directly to drive a turbine. In flash power plants, the geothermal fluid is separated in the separator and used to drive the steam turbine. Flash steam is the most common technology used in geothermal power plants globally (Anderson & Rezaie, 2019). This technology uses two-phase geothermal fluids to generate electricity under high pressure and temperatures.

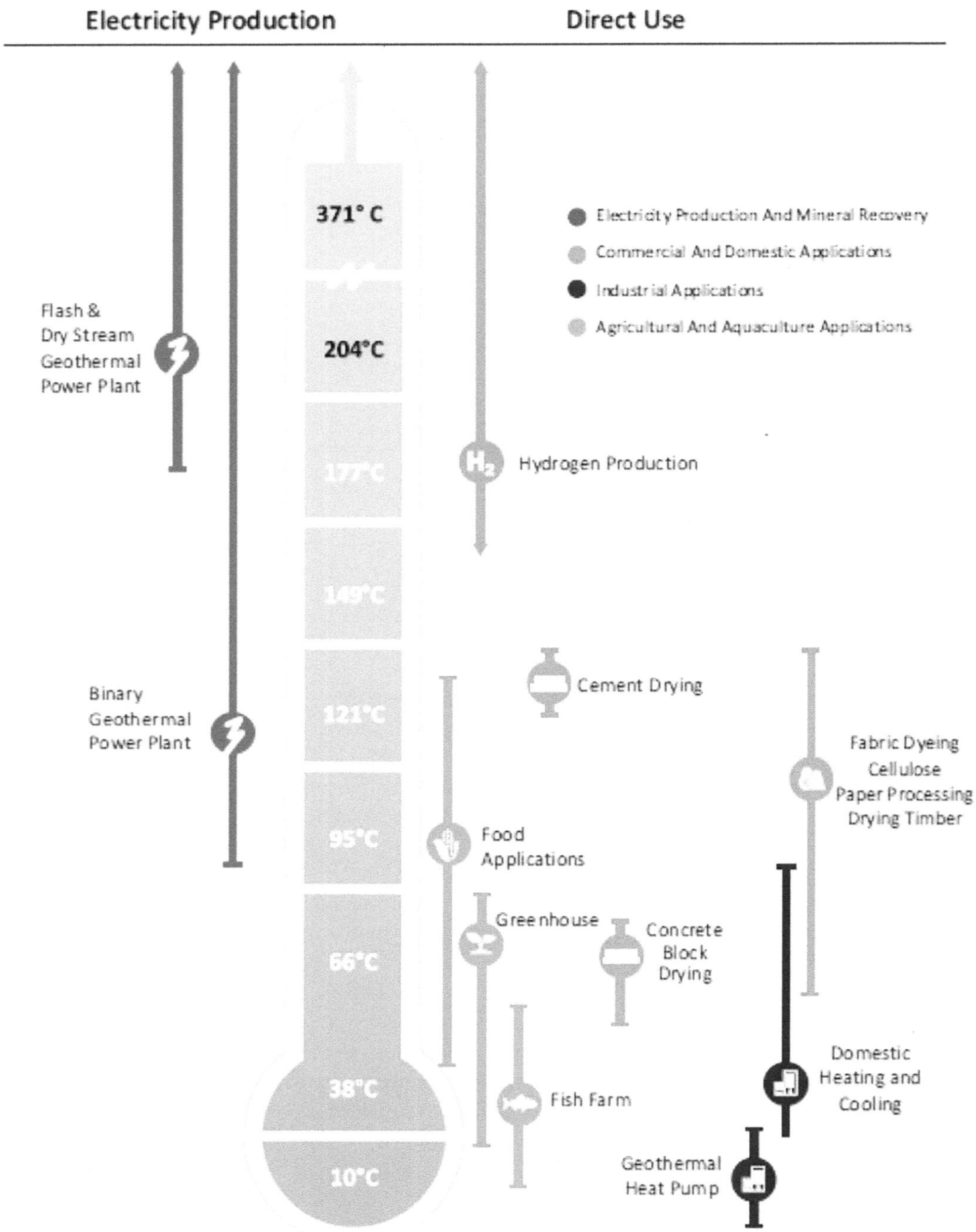

FIGURE 6.4 Usage areas of geothermal energy.

In binary plants, the geothermal fluid transfers energy to the process fluid and is reinjected into the reservoir. A process fluid with a lower boiling point is chosen. This fluid is turned into steam that drives the turbine. Binary plants use geothermal fluids, which require lower temperatures, unlike flash and dry steam.

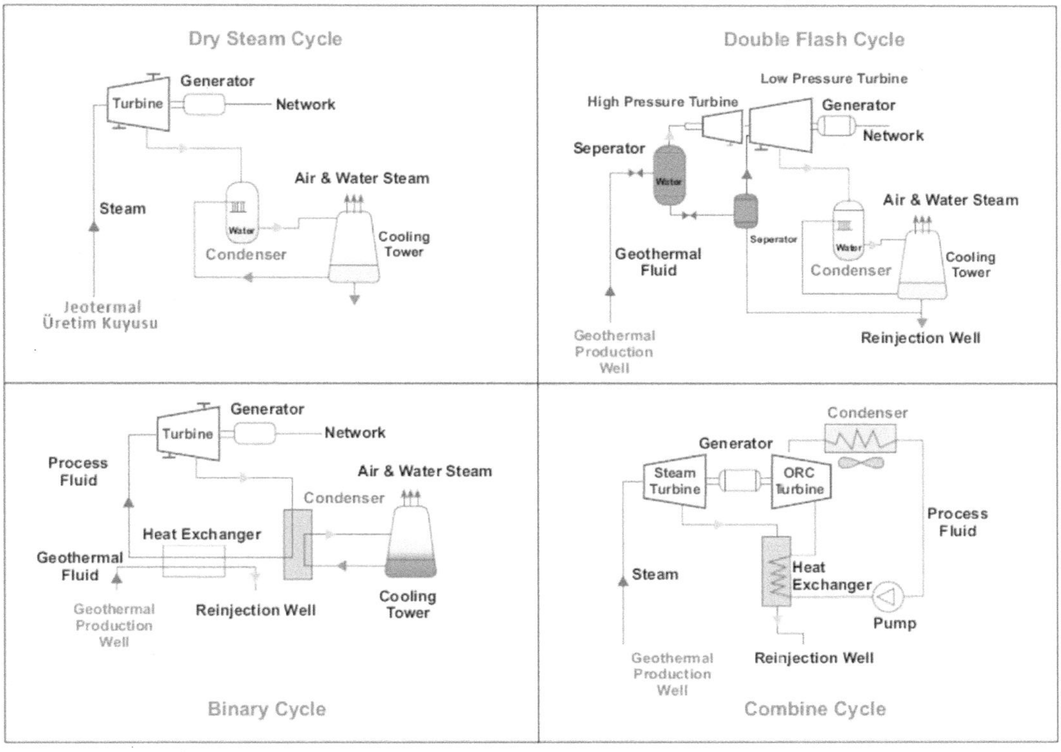

FIGURE 6.5 Electricity generation from geothermal energy.

Source: USGS (2003).

Advances in this technology, albeit on a small scale, have enabled electricity generation from geothermal sources with temperatures as low as 70–80°C. Binary plants can operate on either the organic Rankine cycle (ORC) or the Kalina cycle, depending on the type of secondary fluid (ammonia and water mixture in the Kalina cycle and butane or pentane in ORC) used (Anderson, A., et al., 2019). Binary plants can operate in a completely closed loop, where 100% of the extracted geothermal fluid is returned to the reservoir by reinjection. This allows for an emission-free operation and maximizes the sustainability of resource use. Most future geothermal power plants will be binary plants. In Figure 6.5, the studies of the systems of geothermal electricity production are given.

Electricity is produced with geothermal energy in more than 30 countries worldwide. The primary technologies used are dry steam, flash steam, and binary power plants. Production has been carried out for many years in countries producing more than 1 GWe (USA, Indonesia, Philippines, Turkey, and New Zealand). Other countries have started producing and developing energy (including Belgium, Chile, Colombia, Croatia, Honduras, and Hungary). The regions with the largest installed capacity worldwide are Asia and Oceania (5.9 GWe), North America (3.7 GWe), and Eurasia (3.5 GWe) (IRENA, 2023).

Electricity generation from geothermal energy has developed more in some countries and regions than others, regardless of the availability of suitable geothermal resources. For example, it can be seen that there are differences between the countries and regions in the Pacific Ring of Fire, which have very favorable resources in terms of geothermal development. For example, North America and Central America have more mature geothermal industries than South America and the Caribbean islands. In addition, the most successful developments in geothermal energy production in recent periods have occurred in Turkey and some European countries using low-temperature resources.

6.3 HEATING AND COOLING SYSTEMS WITH GEOTHERMAL ENERGY

Energy can be converted from one type to another. Scientists have called this principle the "first law of thermodynamics." Systems in houses heated with geothermal energy are not much different in principle from other heating systems. The most significant differences that distinguish this system from others are fluid temperature and flow rate. The geothermal resource represents the heat found underground. Systems heated by hot water, steam, and gases accumulated under the Earth's crust are economical and environmentally friendly because natural resources are used. Heat pumps are electrically powered machines by compressing steam and raising heat. Cooling is achieved by expanding the steam and reducing the temperature. This system uses basic heat pump principles with innovative, advanced, and patented designs. Thus, it enables the creation of the most advanced heating/cooling systems utilizing geothermal heat.

Place heating systems are designed assuming water enters the heat exchanger at 90°C and leaves at 70°C. In the case of using geothermal energy for this purpose, non-standard situations may occur as the operating temperatures of the system depend on the temperature of the geothermal source. Expert engineers prepare projects within the framework of the principles specified. Recently, programs that simulate the dynamic thermal behavior of the building have been widely used in calculating heat loss and gains.

Worldwide, the largest installed capacity in district heating systems is in China, Iceland, Turkey, France, and Germany (Lund & Toth, 2021). China, which accounts for 38% of the world's capacity in heating/cooling applications using geothermal energy, is the world leader in geothermal heat pumps and geothermal district heating, with an installed capacity of 40.6 GWth. Various incentives are available for local development, including planning guidance, geothermal standards, and scientific research for developing and using geothermal energy in China.

Around 475 MWth of energy is used in heating/cooling systems in Africa and the Middle East. There are common hot springs in the region. In addition, geothermal heating/cooling applications in the region include fish farming in Algeria, Egypt, Israel, Jordan, and Kenya; greenhouses in Algeria, Egypt, Israel, Kenya, the United Republic of Tanzania, and Tunisia; drying in Kenya and Malawi; milk processing in Kenya; irrigation in Tunisia; heat pumps in Algeria, Iran, and Jordan (Lund & Toth, 2021); and poultry in Tanzania (Kajugus, 2022). As you see, it can be used for many purposes.

The installed geothermal heating/cooling capacity in Latin America is estimated to be 0.8 GWth, with more than 90% contribution from Argentina, Brazil, and Mexico. About 0.8 GWth of the region accounts for less than 1% of the global installed capacity. The most common heating systems are used in hot springs. Apart from this, Argentina, Brazil, and Mexico use geothermal heat to improve greenhouse cultivation, fish farming, desalination, snow melting, and building heating/cooling applications.

By 2020, North America, with a capacity of 22.54 GWth, accounts for more than 20% of the global geothermal heating/cooling installed capacity of 107 GWth. Of this, the USA produces 20.71 GWth (91%), and 1.83 GWth (9%) is produced by Canada (Lund et al., 2022). Apart from the widely used geothermal heat pumps and hot springs, approximately 393 MWth of energy is used in the USA in various applications, such as in greenhouses, aquaculture, district heating, agri-food processing, and snowmelt.

Although high-enthalpy geothermal resources are limited in Eurasia, the use of geothermal energy for heating/cooling is widespread in the region. When the use of heat pumps is included in the installed capacity of heating/cooling systems, the leading countries of Eurasia are Germany, Sweden, and Turkey (Lund et al., 2022). When the heating/cooling installed geothermal capacities by population are compared, the top five global countries are Iceland (2,373 MWth), Sweden (6,680 MWth), Finland (2,300 MWth), Switzerland (2,197 MWth), and Norway (1,150 MWth) (Lund et al., 2022). Iceland is a world leader in the use of heat in heating/cooling applications relative to its population; more than 90% of buildings are heated using geothermal energy. Using geothermal heat pumps is also more important in Eurasia than in other continents. China, the USA, Sweden, Germany, and Finland are among the top five countries worldwide in geothermal installed capacity, including heat pumps (Lund et al., 2022). Shallow closed-loop systems are mostly used in Sweden, while deeper wells are used for heat pumps in Germany. Within Eurasia, district heating use is highest in Iceland (1,650 MWth), Turkey (1,033 MWth), France (509 MWth), and Germany (346 MWth) (IRENA, 2023).

6.4 USE OF GEOTHERMAL ENERGY IN GREENHOUSES

Geothermal energy is an energy source that can produce power continuously and generate energy from domestic sources regardless of climatic conditions. It has become an indispensable source of energy for environmentalists because it is suitable to be established and developed as small power plants with 5–10 MW capacity, not being affected by weather changes and users in the long term, being independent of price fluctuations of fossil fuels, being low enough to compete with coal-fired thermal power plants and natural gas, and having zero-emission value in closed systems. In other words, geothermal resources are the most suitable places for greenhouse cultivation with environmentally friendly production techniques. With a proper system design, although the installation cost of greenhouses is high, it can be much more economical than traditional heating systems, and more homogeneous heating can be achieved. While the stages, such as the fluid delivery to the greenhouse and the installation of the system, constitute an essential cost, the low heating cost increases the profit per unit for the future. Many consumers who think that production with hot water is healthier demand more geothermal greenhouse products (Sen & Yanar Sözal, 2022a).

Greenhouses are structures that have suitable environmental conditions for economically growing vegetables, fruits, and flowers throughout the year, regardless of climate. Greenhouse cultivation is one of the most important agricultural activities because it provides fresh and high-quality vegetables and fruits at all times, obtains high yields from the unit area, creates employment, and contributes to the development of other industries during its construction and production (Sen & Yanar Sözal, 2022b).

Technology is being used more and more in greenhouses day by day, and efficiency is increasing. The main goal is to provide a homogeneous temperature higher than the region's average temperature and to protect the plants from environmental effects, such as wind, rain, and hail. For this purpose, to grow plants successfully, it is necessary to regulate the indoor climate close to nature. While controlling the climate, high temperatures should be avoided, as they can stress the plants; fluctuation of the humidity level in the greenhouse between the maximum and minimum limits should be prevented; the carbon dioxide level should be adjusted; and the disease risks that may endanger the production should be controlled (Baille, 2001).

Greenhouse heating systems with geothermal energy are considered a collection of elements used to transport the geothermal fluid from the area where it is extracted to the areas where the consumers are located. These systems can be grouped as heating systems placed in the soil, on the soil surface, or on growing tables; air heating systems using fans and heat exchangers; and combined heating systems according to their technical characteristics. Among the heating systems, it has been determined that the most suitable system for geothermal energy applications is the heating system made from the ground or underground. Supporting this system with an ambient air heating system fed from the same source provides the best solution. The soil heating system is buried at specific depths and intervals and consists of heating pipes where hot fluid circulates. Today, heating pipes made of plastic materials are widely used because they are resistant to high temperatures and can be laid easily.

In the world, 58.8% of geothermal energy is in heat pumps, 16% in district heating, 18% in thermal springs, 3.5% in greenhouse cultivation, 1.3% in fish farms, 1.6% in industrial applications, and the rest in other fields. The use of thermal water, especially in greenhouse cultivation, has doubled globally in the last 15 years, and greenhouse areas have reached approximately 420,000 ha.

Geothermal greenhouse cultivation is highly developed in Africa, the Middle East, North America, Eurasia, and Latin America. Environmental factors such as global climate change, increasing population, decreasing energy resources, and shrinking agricultural areas increase the interest in greenhouse cultivation. Greenhouse cultivation is spreading to almost all countries worldwide, mainly thanks to the possibility of producing more controlled products per unit area (Anonymous, 2019).

It is necessary to increase the government's support to the producers engaged in geothermal greenhouse cultivation. Up to a specific upper limit, percentage support is provided to greenhouse projects using alternative energy sources (geothermal, solar energy). With the increase in government incentives, the way for geothermal greenhouse cultivation will be cleared (Sen & Yanar Sözal, 2022c).

Suppose we want to evaluate the relationship between greenhouses and geothermal power plants, for example. In that case, a geothermal power plant with a power of 10 MW can produce 80.000.000 kWh of electrical energy if it operates for 8,000 hr per year. If the same energy is used in geothermal greenhouses, approximately 500 daa (decares) of greenhouse can be heated. If the same fluid is used for greenhouse heating after electricity is produced in a 10 MW power plant, approximately 200 daa of geothermal greenhouse can be heated (Sen & Yanar Sözal, 2022d). In this case, the extra gain provided by using enough geothermal fluid for a 10 MW power plant directly in the greenhouse without generating electricity is equivalent to heating a 300 daa greenhouse. The annual gain of a 10 MW power plant using the same energy is lower than the current prices in many countries.

According to this result, geothermal-heated greenhouses are more profitable today than those using electricity. However, considering the long term, the ever-increasing energy deficit, and the rising costs worldwide, it is confirmed that we have to generate electricity from our domestic resources. For this reason, the most accurate and most profitable method is to use geothermal fluid in greenhouse heating after electricity generation as an integrated approach. Thus, it will be possible to produce electrical energy in a geothermal fluid, an efficient and environmentally friendly method, and secondly, by employing a large number of people, it will be possible to make a significant contribution to employment in greenhouses and to grow healthier products at the same time (Sen & Yanar Sözal, 2022e).

6.5 ENHANCED GEOTHERMAL SYSTEMS (EGS)

The conventional method can be utilized where a reservoir and other necessary conditions are present. However, energy generation via the injection of water through wells to develop a reservoir where geothermal fluid is not available has emerged as another energy generation technique. This method where underground heat is utilized as a source is defined as **enhanced geothermal systems (EGS)** or **hot dry rock (HDR)**. Work on this technology started in the 1970s.

Trial studies started as the Fenton Hill Project in the USA for the first time in the 1970s. They continued in many countries, especially Australia, France, Germany, England, and Japan. According to Massachusetts Institute of Technology (MIT) reports, approximately 100,000 MW of electricity is expected to be generated via the EGS method until 2050. The International Geothermal Association gives this value as 80,000 MWe.

Governments have been increasingly funding EGS R&D studies and experiments in recent years. These demonstrate the importance given to EGS studies and applications in developed countries. In recent conferences/seminars on geothermal energy, it has been observed that studies with EGS are getting more intense and interest in this subject is increasing.

Positive results have been obtained from the studies and experiments on this subject, and progress has been made in overcoming the technical problems encountered. In many countries, trial works in the fields continue rapidly. In the EGS method, unlike traditional geothermal methods, the source rock (reservoir) to be used for electricity generation is defined as a geological unit that is spread over large areas that can be encountered at different depths and has high temperatures but does not contain water due to the poor development of crack systems or other geological conditions

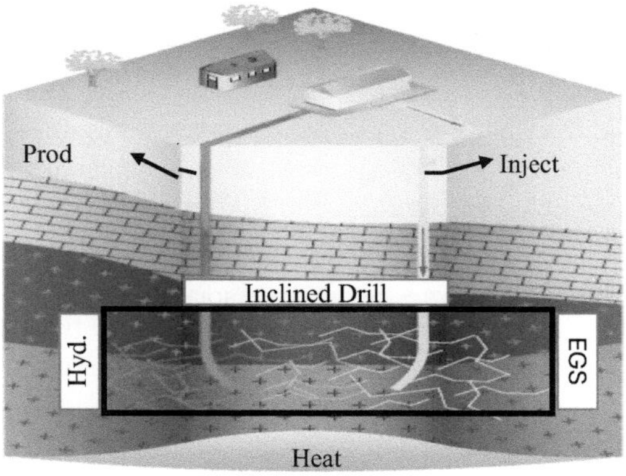

FIGURE 6.6 Schematic diagram for EGS application in general.

Designed or enhanced geothermal systems (EGS) improve the permeability of geothermal systems through hydraulic, chemical, and thermal stimulation. This stimulation can be done in some geological environments with high subsurface temperatures but where fluid volumes and/or rock permeability is insufficient to allow economic extraction using current techniques. In such cases, permeability can be increased by pumping water (or other liquids such as carbon dioxide [CO_2]) to break up the rock, stimulating the reservoir by creating an artificial reservoir (Sen et al., 2015).

in the region. Regarding EGS applications, the fact that the region has a high thermal gradient is not the only criterion in determining the economic applicability of the EGS method. Still, it is an important and decisive parameter (Ledru et al. 2006).

The conceptual basis of EGS projects. The aim is to descend into the target rock with drillings to ensure that the heat hidden in this rock passes from the surface to a given liquid or gas and that this fluid rises to the surface with its new (desired) temperature (Figure 6.6). Water is usually used as fluid because it is the cheapest and most trouble-free option. Various gases are also used in some applications. Drillings and hydraulic fracturing, or excitation processes, are intended to increase the passage of this water through the reservoir rock and the heat transfer efficiency from the rock to the water. The fluid, sent from the ground surface with the injection well, heats up as it moves toward the production well in the rock and is transferred back to the surface. This is the basic principle in the implementation of the EGS project (Tester et al., 1977). The EGS reservoir is a deep geological unit containing heat but no water. It does not contain water because the crack systems in the unit are not well-developed, and these cracks are not connected to each other (Sen et al., 2015). For this reason, hydraulic fracturing creates cracks within the unit in the area or connects existing cracks by increasing their permeability (Evans et al., 1999).

Since the 1970s, various technologies have been used worldwide to improve the permeability of hot dry rock through hydraulic, chemical, and thermal stimulation. Most EGS projects are in Australia, Europe, Japan, the Philippines, and the USA (Pollack et al., 2021).

The first studies to use geothermal heat from hot rocks with low permeability began in 1974 at the Fenton Hill pilot-scale site, conducted by the Los Alamos National Laboratory in New Mexico. The locations where these projects took place are Desert Peak and Brady (Nevada), Raft River (Idaho), and the Geyser and Bottle Rock (California). The Desert Peak, Raft River, and Geysers projects have achieved commercial results. A research project for the EGS initiative is underway at Bottle Rock (NREL, 2021).

In recent years, in the USA, the Frontier Observatory for Geothermal Energy Research (FORGE) and its efforts to support the technological development of EGS have intensified. Various funding opportunities are provided for the EGS collaboration project and are dedicated to EGS development, such as drilling efficiency, zone isolation, waterless stimulation, and utilization. FORGE began in

2015 as a dedicated underground field laboratory near Milford (Utah) to accelerate research and technological advances to exploit EGS resources. The first high-deflection (65° from vertical) deep well and two vertical monitoring wells were completed in 2021 with significant improvements to the drilling program. Hydraulic fracturing research and tests are continuing (UTAH FORGE, 2022).

Canada also has significant EGS potential, with promising data along parts of the Canadian Cordillera (southern British Columbia and southern Yukon), the Mackenzie Basin (northeast British Columbia), northwest and central Alberta, and Saskatchewan (Grasby et al., 2012).

In 2010, the DESTRESS hot dry rock (EGS) demo project with a 1 MWe production capacity and a binary system was started in Pohang, Korea. Two deep boreholes with a depth of 4 km were drilled, and hydraulic warnings were made in 2016–2017. However, in November 2017, two months after ten hydraulic stimulations, an earthquake with a magnitude of 5.5 occurred near the field (Song et al., 2021). The earthquake led to suspending all research activities in Pohang, and many studies determined the earthquake resulted from geothermal hydraulic stimulation (Lee et al., 2019). As of 2021, all deep geothermal explorations across the country have been suspended due to public concerns about seismic hazards and potential damage from EGS activities (Song et al., 2021). The Pohang earthquake is the largest earthquake caused by EGS, and this experience highlights the difficulty of managing seismicity at EGS projects.

There is significant potential for developing active volcanism, high-temperature geothermal systems, EGSs, and supercritical geothermal resources along Latin America's Pacific border and the eastern Caribbean islands. EGS had been explored for the Berlin geothermal field of El Salvador in the early 2000*s* (Majer et al., 2007). More recently (2016–2021), supercritical fluids and potential EGS applications have been investigated in the GEMex Project, a Euro-Mexico collaboration initiative, in Mexico's Los Humeros geothermal field and the Acoculco geothermal research area (GEMex, 2023).

EGS studies in Turkey started in 2011, especially in the Dikili district of İzmir. It is thought that with hot dry rock studies, significant contributions will be made to achieving many important socioeconomic goals, such as reducing greenhouse gas emissions and the ever-increasing need for electrical energy. While it is estimated that the total energy produced by traditional geothermal methods in the currently known and currently used hydrothermal fields worldwide is around 72 GWe, this capacity is predicted to reach 138 GWe with the hydraulic fracturing and/or excitation method (Sen et al., 2015).

6.6 RECOVERY OF MINERALS FROM GEOTHERMAL WATERS

When the chemical compositions of geothermal fluids are examined, it is seen that they are rich in many minerals and minerals included in the "technologically critical elements" classification (European Commission, 2023). The first known application of recovering valuable minerals from geothermal resources was boric acid production from thermal pools in Larderello, Italy, in 1818 (Cetiner et al., 2017).

Research on the recovery of minerals from geothermal waters has increased in recent years. Successful projects in France, Germany, Iceland, New Zealand, the UK, and the USA have shown that it is technically possible to recover lithium and silica. Large-scale commercial projects need further developments to make the process economically and environmentally sustainable. Countries with high mineral structures, such as Argentina, Bolivia, Chile, and Italy, have more mineral recovery potential. For example, a license was granted to the Cesano region of Italy in 2022 for lithium recovery (ThinkGeoEnergy, 2022a). It is also called the "lithium triangle" because it contains concentrated lithium in the salt lakes of Argentina's Andean region, Southern Bolivia, and Northern Chile. There are many existing projects for mineral recovery from geothermal waters worldwide.

Lithium concentrations of deep geothermal waters in the Upper Rhine Graben from Strasbourg (France) to Mannheim (Germany) are about 200 mg/L. In 2021, a pilot lithium recovery system was successfully founded at the Rittershoffen geothermal power plant in northern Alsace (France).

Following this study, lithium carbonate, which has a purity enough to produce batteries, was produced in Soultz-sous-Forets (Bas-Rhin, France) in 2021 (Eramet, 2021). In addition, with the "Zero Carbon Lithium" project, commercial lithium production is aimed until 2024 from the geothermal fluid in the German part of the Upper Rhine Graben (ThinkGeoEnergy, 2022b).

In Iceland, silica is recovered from the geothermal water used for electricity generation at Hellisheidi and Svartsengi geothermal fields, to be used to produce silica dietary supplements and natural skin care products, respectively (ThinkGeoEnergy, 2020).

New Zealand has developed technology to recover minerals such as silica, lithium, boron, rubidium, and cesium from geothermal waters. From 2013 to 2015, studies were carried out on the commercial potential of mineral recovery and processing technology with the project "From Waste to Wealth," and research conducted by government-funded scientific institutions carried out this project (Climo et al., 2016). The recovery of silica from the geothermal fluid during reinjection reduces the calcification potential and the need for chemical treatment of the fluid, operation, and maintenance costs, reduces well labor, simplifies the operating processes, and ensures the efficient use of geothermal fluids. In 2021, the world's first sustainable large-scale commercial silica recovery plant was commissioned at the Ohaaki geothermal field in New Zealand (Geo40, 2023). In addition, tests have been carried out for lithium recovery from geothermal fluid in this region, and investment plans have been made to increase lithium production with a pilot project (ThinkGeoEnergy, 2022b).

Lithium-enriched geothermal fluids were first discovered in England in 1864 in mines in Cornwall. The tests carried out in 2021 showed that lithium is in concentrations above 250 mg/L in geothermal waters (ThinkGeoEnergy, 2022b). Various projects have been developed to extract lithium from geothermal fluids at different depths (Cornish Lithium, 2023).

In the USA, geothermal wells in the Salton Sea region (Southern California) contain geothermal fluids rich in lithium mineral at concentrations up to $440\,\text{mg}/\text{L}$ (NREL, 2021). A new commercial geothermal power plant, which will generate electricity and extract lithium, is being planned in this geothermal field. The project to be implemented here is planned to produce 49.9 MWe and 20,000 tons of lithium carbonate in 2024, increase this capacity to approximately 260 MWe in 2026, and produce 80,000 tons of lithium carbonate (ThinkGeoEnergy, 2023).

Vast lithium deposits are present in salt waters in northern Argentina, Chile's Andean region, and southern Bolivia but have not been utilized. Geothermal development and operation may be more operative when lithium recovery is possible. Exploration of mineral extraction potential continues at Cerro Pabellon and other geothermal fields in northern Chile. A German Chilean research project, "Brine Mine," investigates the selective separation of lithium, other minerals, and fresh water from geothermal waters (KIT, 2021).

The recovery of minerals from geothermal fluids is also under study in Turkey. Since 76% of the global boron mineral is found in Turkey, there is a high amount of boron in geothermal waters. Apart from these, the most essential mineral is lithium. Trial productions are carried out with many pilot-scale facilities, and successful results have been obtained. Very successful results have been obtained with high recovery rates in experimental lithium, boron, and arsenic recovery projects (Sen & Yanar Sözal, 2018).

6.7 GREEN HYDROGEN PRODUCTION

Hydrogen production from renewable energy sources is defined as *green hydrogen*. Hydrogen was discovered in the 1500s, its flammability was realized in the 1700s, and only water vapor was produced as a result of its combustion with oxygen. A temperature of approximately 2,600°C is released (Ozdemir et al., 2019). Hydrogen energy is the energy carrier produced using various energy sources and has the highest energy content per unit mass among known fuels, with an energy content three times higher than fossil fuels thermodynamically (Sen & Yanar Sözal, 2023).

The most important production method of hydrogen, which is not directly found in nature and can be produced by many methods, is the electrolysis of water, which offers an essential solution to the problem of hydrogen production. The energy required for electrolysis can be obtained from fossil fuels, hydroelectric resources, nuclear power, and geothermal, solar, and wind energies. One of the important benefits of water electrolysis over other hydrogen production technologies is that it is compatible with renewable energy technologies, such as solar, wind, wave, and geothermal. In this way, environmentally friendly green hydrogen production can be achieved (Sen & Yanar Sözal, 2023).

The hydrogen energy production initiative is a part of the "sustainable development" goals put forward by world organizations such as the United Nations much earlier. There are many energy goals, and the aim is "to provide access to affordable, reliable, sustainable, and modern energy for all." In this context, the proposed targets to be reached by 2030 are as follows:

- Ensuring universal access to energy
- Significantly increasing the share of renewable energy in global life
- Doubling energy efficiency
- Developing international cooperation in all systems related to energy supply
- Developing infrastructure and technology to provide sustainable energy services for all developing countries

(Gondal et al., 2018)

Geothermal energy has great potential in green hydrogen production, providing stable electrical energy, thermal power, and efficiency. The world's first geothermal green hydrogen production facility was established in Iceland. The plants that produce green hydrogen using geothermal energy, under construction in Japan and operating in New Zealand, are leading in other countries worldwide. International cooperation and partnerships between companies of both countries benefit from the latest hydrogen technology, geothermal power plant operations, environmental protection, and sustainability vision. Depending on these, green hydrogen promotion facilities have been established in Kyushu, Japan, and Taupo, New Zealand (IRENA, 2023).

Global hydrogen investments are approaching 1.1 billion dollars. These investments have been realized in the USA at 44%, and in Asia Pacific countries at 33%. China has invested 2.5 times more than the USA in electrolysis technology, essential for producing green hydrogen. In addition, the electrolysis system can be installed in China at 80% less than in other regions. In some projects to be realized in the Middle East and Europe, investments cannot be made due to unclear hydrogen standards. However, more than 60 countries are developing a hydrogen strategy. While it is calculated that the electrolysis capacity to produce green hydrogen worldwide will reach 115 GW (gigawatts) in 2030, it is predicted that 73% of this will occur in the European region (IRENA, 2023).

A new investor has purchased a defunct geothermal project at Canada's Meager Creek geothermal field, and it has been planned to be revived by 2025 with a focus on green hydrogen production (ThinkGeoEnergy, 2023).

Also, interest in green hydrogen production from geothermal sources is increasing in Central and South America and the Caribbean islands. Directives and specific road maps have been issued recently to promote green hydrogen in Argentina, Brazil, Chile, Colombia, Costa Rica, and El Salvador and to facilitate private investment in this new energy sector (Nunes et al., 2021).

Green hydrogen is considered an option to improve the economies of local geothermal projects in the Caribbean islands. An industrial facility in Trinidad and Tobago already has infrastructure for processing and exporting hydrogen, methanol, and ammonia. The government of Dominica has dealt with a private company to develop geothermal resources associated with the green hydrogen production potential (ThinkGeoEnergy, 2023). Another opportunity being considered (in Colombia and Mexico) is generating geothermal energy using existing wells and related infrastructure in oil fields and combining it with green hydrogen production.

R&D studies on hydrogen have been carried out before in Turkey. However, until 2020, hydrogen has remained in the background in Turkey's energy policies. It is required to take place in the future of energy strategies, and hydrogen studies started again in 2020. Hydrogen is desired to be obtained through four main benefits: to be included in more renewable energy systems, to make the heat sector carbon-emission-free, to produce hydrogen from domestic coal, and to increase the use of boron as hydrogen storage and retainer. Therefore, Turkey aims to mix hydrogen and natural gas in the near future. In the latest combustion tests, natural gas import costs equivalent to approximately \$600 million can be eliminated by injecting 5% hydrogen into natural gas. It is estimated that Turkey has a green hydrogen potential of about 4.6 MTEP per year. This potential is predicted to be met with an electrolyzer capacity of 12.1 GW. Hydrogen is expected to have a share of 1% of Turkey's total energy consumption by 2030 (Sen & Yanar Sözal, 2023).

6.8 ENVIRONMENTAL IMPACTS OF GEOTHERMAL APPLICATIONS

Geothermal energy is important in preventing environmental problems due to global warming and climate change caused by the consumption and use of fossil fuels. This situation is primarily due to the natural advantages of geothermal energy compared to other types of energy in terms of the environment. On the other hand, significant developments have been made recently in the solution of environmental problems related to the use of geothermal energy (Sen & Yanar Sözal, 2022a).

The energy produced by humans or converted into another form will inevitably have environmental effects. As a result of the research, it has been determined that 17% of the world population's electricity needs can be met using geothermal energy. Using geothermal energy, one of the renewable energy sources, instead of fossil fuels reduces greenhouse gas emissions (Dickson et al., 1990). Therefore, deep geothermal waters, used for electricity generation or other reasons, greatly impact the environment. These effects can range from subsidence on the surface to the destruction of natural beauties (such as the Pamukkale travertines) caused by the geothermal fluid (Ozkaya et al., 2008).

Modern geothermal power plants use no fuel, so nitrogen emission does not occur and sulfur dioxide emission is also very low. Especially in district heating systems, emissions are zero (Memis et al., 2015). Rich minerals such as boron, mercury, arsenic, lead, ammonia, antimony, lithium, carbon dioxide, hydrogen sulfide, and salt, which are found in high amounts in the geothermal fluid, can pollute the soil, air, surface waters, and underground waters if correct applications and precautions are not taken. It can cause adverse effects on human life, domestic animals, agricultural products, and wildlife (Sen & Yanar Sözal, 2022b). In binary power plants and geothermal heating systems, the geothermal fluid is pressed back into the reservoir with the non-condensable gases in its structure, so there is no negative impact on the environment since nothing is thrown out.

The interest in clean and renewable energy sources in the world, the transformation to the use of local resources, and therefore, the studies on the more efficient use of geothermal resources have been accelerating. According to the developing technology, some chemicals in geothermal wastewater can be separated, and the fluid can become harmless (Sen & Yanar Sözal, 2022c). In addition, wastewater can be purified by precipitating some components in the settling pools. In some geothermal areas close to the sea, the fluid may have the same chemical characteristics as seawater; in such cases, discharging the wastewater to the sea is not a problem (Verel, 2003). Underground reinjection of wastewater is extremely important in protecting the environment and the reservoir parameters. Therefore, this method is used in many geothermal fields (Sen & Yanar Sözal, 2022d).

It is important to examine the environmental effects of geothermal energy applications, an interdisciplinary field of study, and to raise awareness of environmental issues related to society (Sen & Yanar Sözal, 2022e). In addition to geothermal power plants, the problems that arise when using district heating systems are directly reflected to the users since the wells in the living areas, distribution system, heat center, residential heating systems, and project application areas are intertwined with the users. For this reason, public awareness of environmental impacts should be raised in addition to the importance of resource use, necessary social consensus, and technical issues (Kutluca et al., 2005).

6.8.1 Physical Pollution

The first visible impact on the environment occurs during drilling. To establish the necessary equipment for drilling, the road and the surroundings of the well should be arranged to provide access to the wellhead. These operations can alter the surface morphology of the field and harm local vegetation and wildlife. However, when the drilling process is completed, the negativities outside of an area as small as the drilling field should be eliminated in the most compatible way with nature (Sen & Yanar Sözal, 2022a).

Undesired gases can be released into the atmosphere during drilling and tests, and these processes also cause noise pollution. In geothermal fluid exploration, chemical mud is used, but after drilling is completed, it should be cleaned and separated from the fluid. In addition, solid materials extracted by drilling should be stored in special waste tanks or pools (Sen & Yanar Sözal, 2022b). If these processes are carried out, the environmental damage will decrease to its minimum level when the drilling is completed.

The next step after drilling is the establishment of the pipeline through which the geothermal fluid is transported. Meanwhile, the surface morphology of plant and animal life is affected. Also, the view of pipelines distorts the panorama. Therefore, it should be made as environmentally friendly as possible. The problems encountered in power plant operation are given in the following.

- Since the wastewater of geothermal power plants has a higher temperature than natural water sources, if discharged to surface waters without cooling, it causes thermal pollution in the surface waters and destruction of the habitat in the discharged area. However, if these fluids do not harm the environment in terms of chemical content, they can be thrown into surface waters after cooling.
- Large amounts of fluid extraction from shallow reservoirs for investments, such as thermal tourism and heating, can reduce the groundwater level and adversely affect the surrounding geothermal elements. It can cause gradual precipitation on the soil surface due to hydrothermal eruptions, steam zone increase, salinity increase, or morphology. Therefore, production and reinjection wells should be planned and constructed very carefully. If there is a sufficiently thick, impermeable cover layer on the geothermal reservoirs, these problems do not occur.
- Cooling waters used in power plants are generally used to condense process water. Some waste heat is generated here. Thanks to the integrated use of geothermal energy, this heat can be used in fields that need energy, such as drying, fish farming, greenhouses, and heating/cooling.
- In the operation of geothermal power plants, there is a noise originating from carbon dioxide gas. If the CO_2 gas in the fluid is removed, the noise will be minimized and no longer be a problem. The noise generated in district heating applications remains negligible in the heat generation center.

6.8.2 Chemical Pollution

Geothermal fluids contain non-condensable gases and solid particles. Non-condensable gases consist primarily of carbon dioxide (CO_2), hydrogen sulfide (H_2S), ammonia (NH3), nitrogen (N_2), hydrogen (H_2), mercury (Hg), boron (B), radon (Rn), and hydrocarbons, such as methane (CH_4). Geothermal processes' most important gas emission points are chimneys, cooling towers, mufflers, and steam pipes. In addition, chemical mud is used in drilling; toxic and corrosive chemicals, such as caustic soda and sulfuric acid, are used to prevent bacterial formation or crusting in wells; and solid and gaseous wastes are produced during tests and reinjection (Sen & Yanar Sözal, 2022a).

The chemical emissions released by geothermal energy production increase the concerns about soil, water, and air pollution day by day. However, geothermal power plants are considered

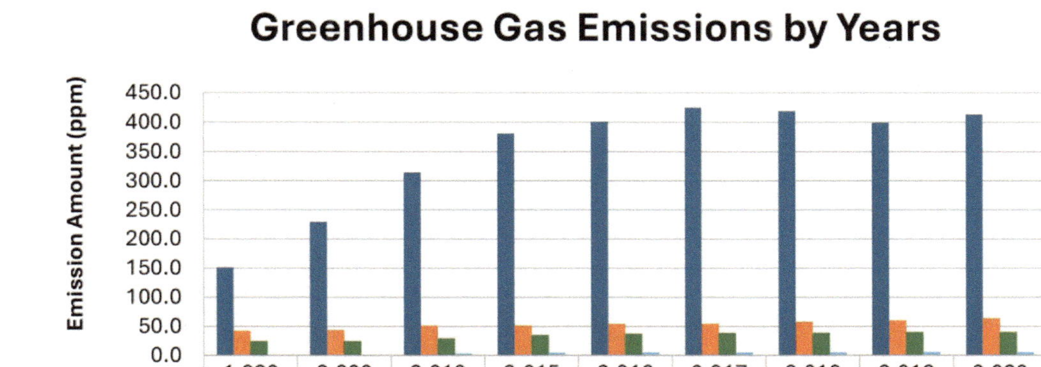

	1,990	2,000	2,010	2,015	2,016	2,017	2,018	2,019	2,020
CO_2 (ppm)	151.5	229.8	314.4	381.3	401.2	425.3	419.4	399.3	413.4
CH_4 (ppm)	42.5	43.7	51.4	51.6	54.5	54.8	58.1	60.3	64.0
N_2O (ppm)	25.0	24.8	29.8	35.4	37.7	39.1	39.3	40.2	40.3
F-gas (ppm)	0.6	0.7	3.6	5.0	5.5	5.7	5.7	6.2	5.8

FIGURE 6.7 Greenhouse gas emissions.

Source: TUIK (2022).

an "environmental" alternative to fossil fuels and nuclear power plants. CO_2 emission values are $900 - 1300\,g / k\,Wh$ in coal, $500 - 1250\,g / k\,Wh$ in natural gas, $20 - 250\,g / kWh$ in solar energy, $20 - 50\,g / k\,Wh$ in wind energy, and $20 - 35\,g / k\,Wh$ in geothermal energy (Hepbasli et al., 2003). The main cause of global carbon emissions is the use in the energy sector. Greenhouse gases, mostly originating from fossil fuels, can cause global warming. Figure 6.7 shows greenhouse gas emissions year by year. According to studies, using renewable energy sources instead of fossil fuels decreased CO_2 emissions by 2% worldwide in 2019. It was the most significant drop from 1990.

Since no fuel is burned in geothermal power plants, N_2 emission does not occur, and H_2S emission is very low. There is no gas emission, thanks to binary geothermal power plants. No fluid is discharged to the surface with binary geothermal power plants.

To take chemical pollution precautions, it is necessary to constantly monitor the gases in the air, especially in the power plants, by keeping $CO_2, H_2S, Hg, CH_4, NH_3, N_2$ and Rn gas measuring devices. Many of these gases can be measured with a single instrument. Companies should take additional precautions if these gases are in an amount that will harm the environment. It would benefit state authorities with control and inspection authority to control the facilities by having gas measuring devices.

Even if power plants and district heating systems are entirely closed-loop, they can cause minor problems, but this is negligible. These wastes can also be minimized or eliminated with careful management.

Many chemical wastes occur during geothermal energy production and power plant installation. In addition, the remaining fluids after the use of geothermal water contain large amounts of carbonate, silica, sulfate, and chloride, along with chemical pollutants, such as lithium, boron, arsenic, fluoride, hydrogen sulfide, mercury, lead, zinc, and ammonia.

6.8.3 Thermal Pollution

After energy production from geothermal systems, much waste heat is generated. It must be distributed without harming the environment. With their high temperature and chemical composition, geothermal fluids are important pollutants in groundwater, and their interactions should be strictly avoided. When this hot fluid is discharged into the atmosphere, it affects the regional climate and animal and plant life. For the last five years, many geothermal energy production companies have been discharging waste geothermal fluid to reservoirs by reinjection. However, discharging the waste fluid by reinjection after the heat is taken provides gains in terms of environment and economy. To use geothermal energy's full potential and thus become a much more economical and environmentally friendly energy type, its integrated use must be fully adopted and implemented. The integrated use of geothermal energy can be explained as shown in Figure 6.8. The energy of the high-temperature fluid extracted by drilling is transferred to the fluid used in the process by heat exchangers, generating electricity. Reusing high-temperature fluid energy, which remains after electricity generation, in fields such as greenhouse cultivation, district heating/cooling, fisheries, and drying vegetables and fruits means its integrated use. The remaining low-temperature waste fluid is reinjected into another geothermal well. Thus, the continuity and sustainability of the geothermal fluid are ensured by maintaining the fluid amount and pressure in the reservoir.

Although geothermal energy does not have adverse effects when the right technologies are used, it can cause environmental pollution if the right technologies are not used.

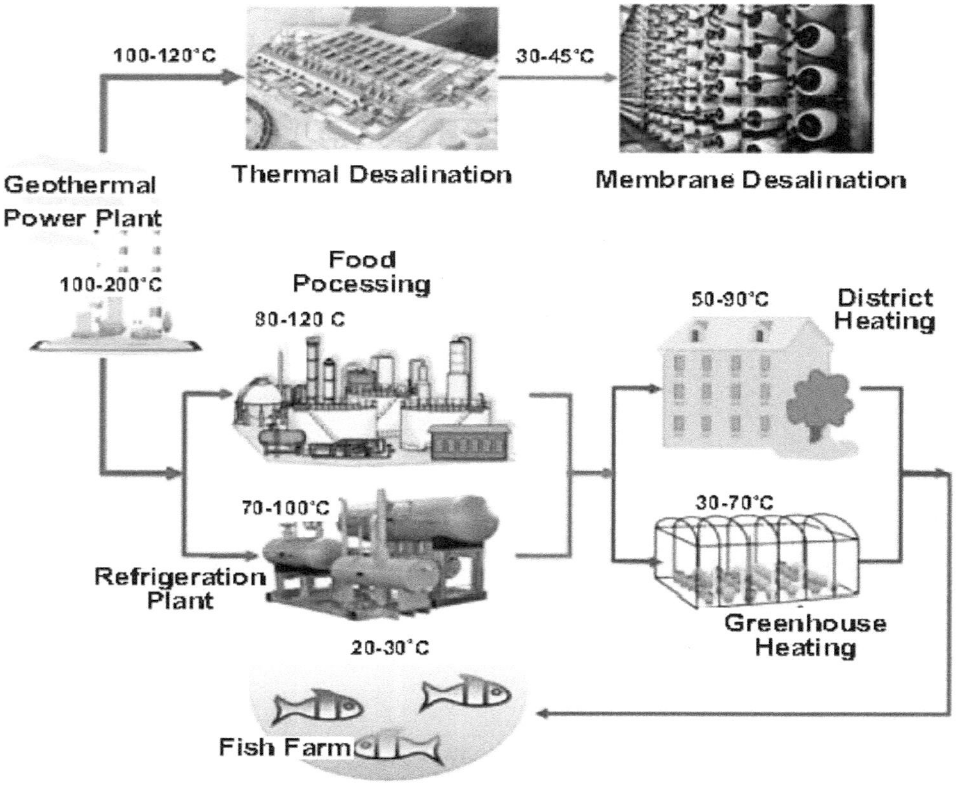

FIGURE 6.8 Integrated usage of geothermal energy.

Source: Bundschuh et al. (2017).

6.9 GEOTHERMAL ENERGY IN THE WORLD

Globally, electricity has been produced in more than 30 countries by using geothermal energy. The technologies mainly used are dry steam, flash steam, and binary power plants. The USA, Indonesia, the Philippines, Turkey, and New Zealand are the countries that have been producing geothermal energy for many years. Generally, their production capacity is more than 1 GWe. Countries with lower capacity generally have just started to produce and are developing countries, such as Belgium, Chile, Colombia, Croatia, Honduras, and Hungary. The regions with the largest installed capacity worldwide are Asia and Oceania (5.9 GWe), North America (3.7 GWe), and Eurasia (3.5 GWe). Worldwide geothermal electricity installed capacity distribution is given in Figure 6.9.

As of 2020, the installed capacity of heating/cooling systems using geothermal energy worldwide is 107.4 GWth. While 72% of this capacity consists of geothermal heat pumps, the remaining 28% comprises heating/cooling systems directly using geothermal fluids. Heating/cooling applications with geothermal energy are concentrated in three regions. The Asia and Oceania region is the world leader, with 43% of the part and an installed capacity of 45.8 GWth. Eurasia follows it with 38%, and North America with 21%. Latin America, the Caribbean, Africa, and the Middle East contribute 1% or less. Worldwide geothermal heating/cooling installed capacity distribution is also given in Figure 6.10.

The USA is the world's largest producer of geothermal power, and the largest geothermal development in the world is at Geysers north of San Francisco in California. In the last decade, significant developments have been seen in geothermal installed electricity capacity in many countries (mainly including Indonesia, Kenya, New Zealand, and Turkey) and in geothermal heating/cooling in some countries (including China and Turkey), and geothermal energy targets have been determined by these countries. For example, the USA plans to increase geothermal electricity generation to 60 GWe installed capacity by 2050 and significantly increase its EGS and geothermal heating/cooling potential. In 2017, China prepared the "13th Five-Year Geothermal Energy Development and Utilization Plan," which supports the development of geothermal energy to reduce air pollution and provide electricity and heat. In 2022, the 14th Five-Year Plan for Renewable Energy plans to develop and use geothermal energy, focusing on optimizing geothermal heating/cooling distribution (Nextrends Asia, 2021). In Europe, the International Energy Agency (IEA) and the French Environmental and Energy Management Agency (ADEME) have determined that geothermal energy is the most cost-effective solution for district heating (EU Reporter, 2022). The European Geothermal Energy

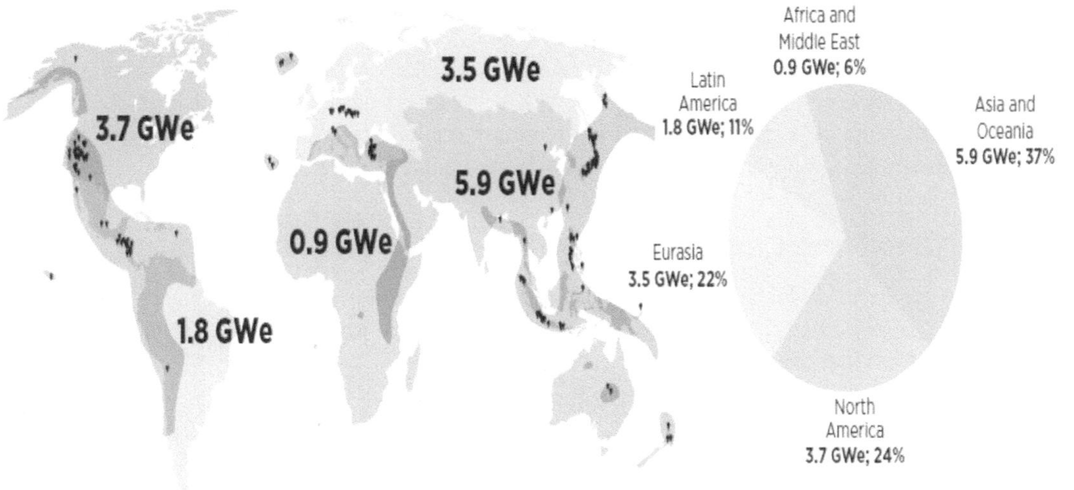

FIGURE 6.9 Worldwide installed capacity of electricity generation from geothermal energy (2021)

Source: IRENA (2023).

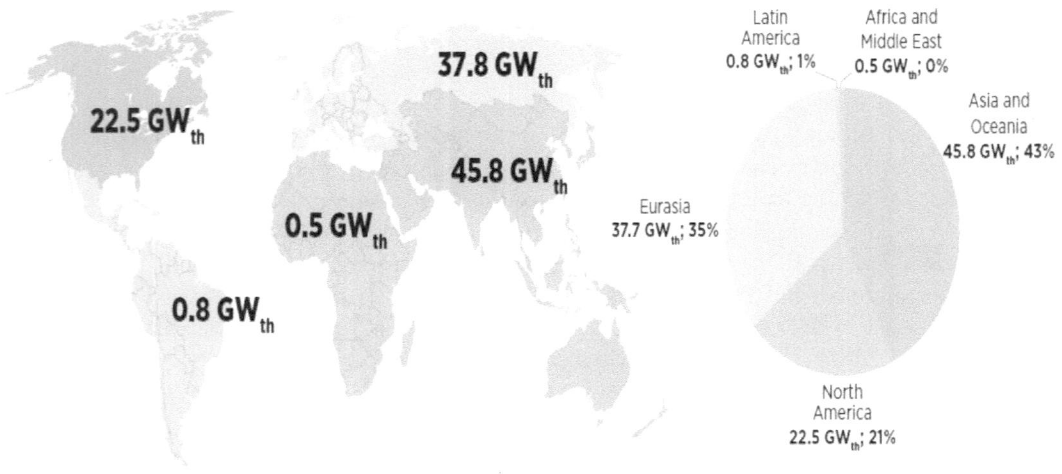

FIGURE 6.10 Worldwide geothermal energy heating/cooling installed capacity (2020).

Source: IRENA (2023).

Council (EGEC) encourages using geothermal energy resources not only for heating/cooling and electricity generation but also for recovering lithium and other essential minerals. Kenya plans to produce 1.6 GWe of energy in Africa by approximately doubling its geothermal electricity capacity by 2030 (Burkardt et al., 2021).

6.10 TOP 10 GEOTHERMAL POWER PLANTS WORLDWIDE

6.10.1 Geysers Geothermal Complex, USA

The world's largest geothermal power plant, located approximately 100 km north of San Francisco, California, consists of 18 power plants with 900 MW of active generation capacity. It is founded on approximately 117 km^2 and generates energy for almost 1 million homes (Dowsclimatecare.com, 2023).

6.10.2 Larderello Geothermal Complex, Italy

It is the second largest facility in the world and is a huge field consisting of 34 power plants. The net capacity of this geothermal plant is 769 MW. It is the world's first and oldest power plant, established in 1913. This field supplies 10% of all geothermal energy produced worldwide and approximately 27% of the region's energy demand (Dowsclimatecare.com, 2023).

6.10.3 Cerro Prieto Geothermal Power Station, Mexico

Located in northern Mexico, the power plant is the world's third largest, with a capacity of 720 MW. This facility is owned by the Comisión Federal de Electricidad and is built in a unique geological location (Dowsclimatecare.com, 2023).

6.10.4 Mak–Ban Geothermal Complex, Philippines

The fourth largest geothermal facility with a capacity of 458 MW is Mak–Ban Geothermal Energy Complex. There are six power plants in this field located in Laguna and Batangas. The Philippines has an excellent geothermal infrastructure but needs to increase production to ensure 100% sustainability (Dowsclimatecare.com, 2023).

6.10.5 CalEnergy Generation's Salton Sea Geothermal Plants, USA

The facility comprises ten geothermal power plants clustered together and located in Southern California. With a total production of 340 MW, this series of power plants is the fifth largest geothermal power producer in the world. Experts report that this region has more geothermal potential than any other power plant in the United States, and the plant is helping combat soil erosion and the spread of toxic salt levels across the region (Dowsclimatecare.com, 2023).

6.10.6 Hellisheidi Geothermal Power Plant, Iceland

The sixth largest geothermal power plant, with 400 MW of thermal energy and 303 MW of electrical power generation capacity, is located in Iceland's Hengill volcano. Also, Iceland is a country full of hot springs (Dowsclimatecare.com, 2023).

6.10.7 Tiwi Geothermal Complex, Philippines

More than 1/4 of the Philippines' energy power is met by geothermal power plants. The seventh largest geothermal power plant is located in Albay, Philippines. It consists of three power plants with a capacity of 289 MW, each composed of two units. This power plant has operated since 1979 and is among the world's oldest operational geothermal power plants (Dowsclimatecare.com, 2023).

6.10.8 Malitbog Geothermal Power Station, Philippines

This power plant in the Philippines is the eighth largest in the world. It has 232.5 MW capacity. The Energy Development Corporation owns this facility, and it supplies power to the island of Luzon. It is also the world's largest single-roof geothermal power plant (Dowsclimatecare.com, 2023).

6.10.9 Wayang Windu Geothermal Power Plant, Indonesia

The world's ninth largest geothermal power plant is in West Java, Indonesia. It has an energy capacity of 227 MW, and the geothermal field is 40 km^2 in size (Dowsclimatecare.com, 2023).

6.10.10 Darajat Geothermal Power Plant, Indonesia

The world's tenth largest geothermal power plant is located in the Pasirwangi district of Garut, Indonesia. It consists of three facilities that were established separately between 1994 and 2007. This plant has a capacity of 259 MW. Since Indonesia has a very active volcanic geology, geothermal energy is becoming increasingly popular in every corner of the country (Dowsclimatecare.com, 2023).

6.11 COUNTRIES WITH THE MOST GEOTHERMAL ENERGY GENERATION CAPACITY WORLDWIDE

Worldwide, the total installed geothermal electricity generation capacity at the end of 2022 was calculated as 16,127 MW. This value increased by 286 MW compared with 2021. Newly added power plants lead with Kenya (83 MW), Indonesia (80 MW), USA (72 MW), and Nicaragua (10.4 MW), and other small-scale power plants have been found in China, the Philippines, and Japan. Worldwide geothermal energy production is foreseen to be 19.10 GW by the end of 2025, 30 GW by the end of 2030, and 70 GW by the end of 2050. Following are the countries with the highest geothermal energy production in 2022 (Table 6.1).

TABLE 6.1
10 Countries with the Highest Geothermal Energy Production in 2022

No.	Country	Capacity (Mwe)
1	USA	3,794
2	Indonesia	2,356
3	Philippines	1,935
4	Turkey	1,682
5	New Zealand	1,037
6	Mexico	963
7	Kenya	944
8	Italy	944
9	Iceland	754
10	Japan	621

Source: ThinkGeoEnergy (2021).

6.11.1 United States (USA) – 3,794 MW

Electricity generation from geothermal sources in the USA started in the 1950s. Today, there are 93 geothermal power plants and 32 geothermal fields, all operating in the USA (NREL, 2021). In 1960, the first commercial geothermal power plant in which geothermal steam was used for electricity generation was established in Geysers, California, with a capacity of 11 MWe (Cataldi et al., 2016). At the end of 2021, the USA will constitute 23% of the world's installed capacity with its geothermal electricity capacity of 3.7 GWe.

Most of this capacity is in California (2.6 GWe; 72%) and Nevada (0.8 GWe; 22%). However, these two states account for more than 90% of the geothermal electricity produced in the USA (North America). The remaining 10% comes from other Western states (Idaho, New Mexico, Oregon, Utah) and Hawaii, with installed generation capacities ranging from 15 MWe in New Mexico to 90 MWe in Utah. A small geothermal power plant in the Chena Hot Springs area of Alaska is also operating with a capacity of 400 KWe in 2006 and increased to 680 KWe in 2008.

The overall geothermal generation capacity in the USA accounts for approximately 0.4% of the national electricity production (Hutter, 2021). During the last 30 years, geothermal electricity capacity grew at an average annual rate of 1.4%. The growth rate has ranged between 1.6% in the last ten years and 1.1% in the last five years (IRENA, 2023). As of 2020, with a capacity of 20.71 GWth, the USA corresponds to more than 20% of the global installed geothermal heating/cooling capacity of 107 GWth (Lund et al., 2021). The region's geothermal heating/cooling industry consists mainly of geothermal heat pump (GHP) applications, which are widely used and have grown steadily over the past decade. GHP corresponds to 98% of the USA's geothermal heating/cooling capacity. In the United States, GHP installations stand at 20.23 GWth. Hot springs are common in the region, with an installed capacity of approximately 89.85 MWth (Thompson et al., 2021).

Apart from the widely used GHP and spas in the USA, several heating and cooling applications have been developed in the USA, such as greenhouses (80 MWth), aquaculture (122 MWth), individual space heating (89 MWth), district heating (90 MWth), and agro-food processing, snow melting, and other applications (12 MWth), a total of approximately 393 MWth, and mostly located in Idaho, California, and Oregon.

Geothermal heating technology was started in the USA more than a century ago. The first district heating system was developed in Boise (Idaho) in the early 1890s. Later, similar applications were made in Klamath Falls (Oregon) in 1900, and greenhouse heating systems were successively developed in the late 1920s (Stober et al., 2013). Currently, 23 geothermal district heating systems have been operating in the USA, with capacities ranging from 0.1 MWth to 20 MWth. While most of these systems were installed in the 1980s, only four have been built since 2000. Conventional heating/cooling use of geothermal resources (excluding GHP) in the USA has stopped in the last 30–40 years. Many facilities have been closed in the last ten years. Of the 550 historically recorded heating/cooling applications in the USA, more than 100, or about 19%, have been abandoned for various reasons, including high operating costs, lack of specialized personnel for operation and maintenance, and low government-level attention (Lund et al., 2021). Geothermal development in the region has been managed by private companies. Except for the steadily growing GHP industry, the geothermal market has seen little growth in electricity or heating/cooling systems in recent years. In the USA, which represents the largest segment of the regional market for GHP, even GHP installation slowed in recent years, from a growth rate of 8% in 2010–2015 to 3.7% in 2020 (NREL, 2021).

Enhanced geothermal systems (EGS) resources are predicted to increase the distribution of geothermal electricity through commercial distribution. According to the foreseen scenarios, by 2050, the geothermal generation capacity could reach 6 GWe. If permits are obtained, it can reach 13 GWe. In addition, technological developments can succeed in reducing project costs, and if consensus can be reached on the deployment of deep EGSs, which are considered to have the highest potential to accelerate growth in the geothermal electricity sector, the capacity can reach 60 GWe.

In 2020, a new Energy Act improved access to federal land for renewable energy investors, including geothermal, and set new targets for renewable electricity generation on federal land. In addition, the 2020 Energy Act has created an annual budget of US$170 million for geothermal research, development, and promotion projects. In California, Hawaii, Nevada, New Mexico, and Washington, there has been state support for the improvement of geothermal permits and regulations, and it was thought that geothermal would make a significant contribution to the achievement of decarbonization targets (US DoE, 2020).

The GHP sector is expected to grow in residential and commercial applications, offering reliable and energy-efficient heating/cooling solutions. In the USA, GHPs in the residential sector alone have a market potential of 235 GWth, or 28 million households, representing 23% of the projected residential heating and cooling demand by 2050 (IRENA, 2023). It is estimated that regional geothermal heating has a growth potential of 17,500 new household installations and 45 million households, with the capacity to provide reliable, energy-efficient heating and cooling (US DoE, 2019).

In the last decade, geothermal operators in the USA have begun to develop commercial or pilot-scale geothermal–solar hybrid systems. Stillwater, Patua, and Tungsten Mountain geothermal power plants in Nevada have been combined with solar panels. In Stillwater, a solar power system was installed to preheat the fluid used in the 33 MWe binary geothermal plant, with a resulting increase in electricity generation of 3.6% compared with production from geothermal alone (Enel Green Power, 2021).

In addition to geothermal heat pumps widely used for district heating/cooling, various cooling applications are also developing, such as geothermal-powered absorption chillers, district cooling using medium-temperature geothermal resources, ice production, and industrial cooling. In the early 1980s, the Oregon Institute of Technology campus commissioned an absorption cooling system powered by geothermal heat. It worked until 1999, when it was taken out of service due to its high water usage and low cooling efficiency (NREL, 2021). An absorption cooling system with geothermal heat extracted from shallow wells at 74°C was installed in Chena in 2006. At the same time, this spa has been heated gradually, and electricity has been produced with a capacity of 680 KWe.

Low-temperature electricity generation, geothermal microgrids, and small-scale geothermal power plants operate successfully in many countries. In recent years, the ORC system has become widespread, which enables the modularization of small power plants optimized to operate with low-temperature liquids. The USA's Wabuska geothermal power plant was the first to operate at 107°C with low-temperature fluids, with the first binary unit commissioned in 1984 and 1987 (0.6 MWe). The installed equipment has been successively renewed and expanded to its current production capacity of 4.4 MWe (NBMG, 2012). Another exciting example of low-temperature electricity generation was a micro geothermal power plant co-produced with the Florida Canyon gold mine using geothermal water at 110°C. The first experimental low-temperature ORC unit with a capacity of 50 KWe was installed in 2009 and increased to 75 KWe in 2012 (IRENA, 2023).

6.11.2 Indonesia – 2,356 MW

In Indonesia, the Kamojang geothermal field was discovered in the 1980s with the help of the New Zealand government, and significant geothermal developments have been made. Indonesia leads with 2,356 MWe, the region's largest electricity generation installed power. It ranks second in the world. In Indonesia, with geothermal energy, 6% of the country's electricity needs are met, and new policies and financial incentives have been initiated to generate energy from renewable energy sources. With various geothermal laws and regulations that have come into force since 2020, permit and license procedures have been facilitated. In November 2022, trainings were organized by the Indonesian government for project developers and investors to receive technical guidance on applying for geothermal heating/cooling permits (EBTKE, 2022). As Indonesia moves toward its energy transition goals, it focuses on geothermal heating and cooling and aims to reach 3.3 GWe of installed geothermal capacity by 2030.

6.11.3 Philippines – 1,935 MW

In the 1980s, the Palinpinon/South Negros geothermal field was discovered in the Philippines. With an installed capacity of 1,935 MWe, it has the third largest installed capacity of electricity generation in the world and meets 12% of the country's electricity needs.

In the Philippines, the ORC binary power plant with a capacity of 3.6 MWe, Mindanao-3, built during the COVID-19 pandemic, was commissioned in April 2022. This power plant has generated energy using the waste heat from the Mindanao-1 and Mindanao-2 geothermal power plants (Exergy, 2022). Thus, in addition to the energy produced from two flash plants, total electricity production is increased with 1 ORC plant. In this system, the chemical composition of the geothermal fluid and the calcification may cause restrictions in energy production.

6.11.4 Turkey – 1,682 MW

The geothermal energy sector in Turkey is in continuous development. However, although it has a very high geothermal potential and is among the top ten in the world (JED, 2022), its primary energy supply is still provided by thermal power plants. Turkey's advantageous geological conditions, technological developments, and incentives for developing geothermal energy laws and regulations have made geothermal energy economically and technically feasible (Eker et al., 2017). Installed power in 2008 was 30 MWe (IGA, 2017). It had been calculated as 1,549 MWe in 2019, and 1,682 MWe in 2022 (TEIAS, 2023). Turkey is the leader in its region in geothermal heating/cooling systems, with an installed power of 3,488 MWth (NREL, 2021). This energy is mainly used for spas, swimming, central heating, greenhouses, and district heating. Turkey has a capacity of 1,033 MWth in district heating systems using geothermal heat pumps.

Turkey has been among the fastest-growing geothermal markets in electricity generation in the last 15 years. Thus, it has risen to the first place in Europe and the fourth place in the world. Turkey is working intensely to develop further geothermal energy. The financial support of the World Bank within the scope of the Global Geothermal Development Plan plays a vital role in the rapid growth of the Turkish geothermal sector. The World Bank has a risk-sharing mechanism. It offers loans to increase private sector investments in geothermal energy development and to develop resources in Turkey. Turkey's 2030 target in geothermal electricity production is 3,000 MWe, and the target for 2050 is 5,000 MWe (Sen & Yanar Sözal, 2022d).

6.11.5 New Zealand – 1,037 MW

In the 1950s, the Wairakei and Kawarau geothermal fields were first discovered in New Zealand, and significant developments were made in the geothermal sector. With an installed capacity of 1,037 MWe, New Zealand has the world's fifth-largest installed capacity for electricity generation. With geothermal energy, 22% of the country's electricity needs have been met. In addition to electricity generation, the Kawarau geothermal field has multiple heating/cooling applications (paper and pulp mill, wood drying, dairy processing, and greenhouses), and new initiatives have been built. In addition, studies are carried out on technologies used to recover minerals such as silica, lithium, boron, rubidium, and cesium from geothermal waters in New Zealand. In light of these studies, it is seen that the system should be economical to develop successful projects. This depends on the technology used, the material, the commercial risks to be taken, and the laws. Regarding the recovery of minerals from geothermal water, silica has been studied mostly because it precipitates during reinjection. Thus, the calcification potential in the wells, the need for chemical treatment of the fluid, the operating and maintenance costs, the reduction of well labor, and the efficient use of geothermal fluids have been ensured. In 2021, the world's first sustainable large-scale commercial silica recovery plant was commissioned at the Ohaaki geothermal field in New Zealand (Geo40, 2023). In addition, lithium recovery tests from geothermal water have been carried out in this region, and a pilot project has been planned.

6.11.6 Mexico – 963 MWe

In the early 1970s, the first large-scale geothermal electricity generating plants started operating in Mexico. With an installed capacity of 963 MWe, it ranks sixth among the top ten geothermal electricity–producing countries worldwide. Geothermal power plants have been operating uninterruptedly for over 40 years. Most of the installed capacity used at high-temperature volcanoes hosted hydrothermal resources. In Mexico, the discovery and development of geothermal fields was provided through the CFE Company. Geothermal heating/cooling installed capacity has been estimated at around 0.8 GWth in Argentina, Brazil, and Mexico combined, and it represents less than 1% of the 108 GWth global installed capacities in 2020. Besides electricity generation in Mexico, geothermal heat is used to develop applications such as in the agro-industry, fish farms, desalination, snow melting, and heating/cooling in buildings. The use of shallow geothermal resources with GHPs is highly developed, especially in hotels, schools, hospitals, and universities.

For the last ten years, studies have been carried out on ORC power plants, which use the remaining heat from existing power plants to increase the production capacity of an operating field without additional production wells. In 1993, the first trial was conducted by commissioning 2×1.5 MWe ORC units at the Los Azufres geothermal field in Mexico, and it has been deactivated currently. In Mexico, the state-owned power company CFE and the French agency FASEP have recently identified significant potential for increasing generation capacity with ORC technology (ThinkGeoEnergy, 2023).

6.11.7 Kenya – 970 MWe

Kenya ranks seventh among the top ten countries producing geothermal electricity worldwide (ThinkGeoEnergy, 2021). In 2019, 29% of the national electricity installed capacity and 47% of the consumed electricity were produced with geothermal energy (Omenda et al., 2023). Kenya's geothermal market is one of the fastest-growing markets worldwide. In the last ten years, Kenya has achieved a growth rate of 6–13% per year, well above the world average of 2–4.6%. Until the late 1990s in Kenya, the state-owned utility/generating company developed geothermal projects until the late 1990s, when regulatory reforms allowed independent power producers to generate geothermal electricity (IRENA, 2023). Thus, 472 MWe were commissioned between 2000 and 2014, and

218 MWe between 2014 and 2019 (Omenda et al., 2023). In addition to these, another 86.6 MWe was commissioned in 2022. Kenya's rapid growth has been attributed to substantial public funding and favorable government policy supporting private sector investment and incentives. A private investor in Kenya established a power plant with an installed capacity of 171 MWe. This state-controlled electricity generator can produce 799 MWe of the country's total installed geothermal electricity capacity of 970 MWe. The government-controlled electricity generator in Kenya has been in the process of establishing a public–private partnership for the construction and operation of a 140 MWe geothermal power plant. In addition, exploration drilling has been carried out in the Baringo-Silali geothermal block in Kenya, and the work is planned to continue by expanding the Olkaria geothermal field. In addition to electricity generation, geothermal energy is used for many purposes, such as in spas, district heating/cooling applications, fish farming, dairy management, and greenhouse cultivation.

6.11.8 Italy – 944 MW

The first experiments to generate electricity using geothermal steam were conducted in Larderello, Italy, in 1904. In 1913, a commercial geothermal power station was established in Tuscany, where most of its geothermal heat originates from a magmatic depression. More than 40% of the region's annual electricity needs have been provided by geothermal energy. Geothermal energy facilities have made up 7% of Italy's national income, met the region's 16,000 GWh electricity needs, and provided employment to 3,000 people. The use of geothermal energy in Italy has increased rapidly in the 20th century, following the developments that occur every day. In 2020, 916 MWe (Hutter, 2021) was produced from more than 30 geothermal power plants in Italy, which increased to 944 MWe in 2021 (ThinkGeoEnergy, 2021). These power plants have produced approximately 6 TWh (terawatts) of electricity annually and met the electricity needs of 2.5 million households. In addition to 30 ha of greenhouses, there are more than 16,000 residential users and facilities in geothermal regions heated by geothermal energy.

There has been a lack of incentives and regulation in Italy. Geothermal energy has been excluded from the support system decree for financing renewable energy resources and technologies. Therefore, support for new geothermal projects has been stopped. Also, the authorization process for geothermal projects in Italy is complex and takes a long time.

6.11.9 Iceland – 754 MW

Iceland has significant geothermal potential due to its location as a volcanically active region. Iceland's geothermal electricity sector has grown significantly in recent years. The first geothermal power plant was put into operation in 1969; eight power plants are currently operating in the region, and the total installed power is 754 MWe (Ragnarsson et al., 2018). Iceland is a world leader in heating/cooling applications, with a total capacity of 2,373 MWth. More than 90% of buildings are heated using geothermal energy. Iceland has a capacity of 1,650 MWth in district heating with a geothermal heat pump.

In Iceland, the National Energy Fund has greatly affected the geothermal sector since the 1960s by promoting geothermal developments through loans and risk reduction (Atlantic Council, 2018). Loans up to 60% of total drilling costs for drilling geothermal wells turn into grants if the drilling phase fails.

In Iceland, with the Carbfix initiative to reinject carbon dioxide and store it via natural underground mineralization, which was developed in collaboration with the geothermal operator of the Hellisheidi geothermal field, the country started storing captured non-condensable gases from the geothermal electric plant and then expanded to apply the underground storage technology to carbon dioxide directly captured from the air. This technology is currently rolling out to multiple sites, and funding has been received from the European Commission for the Silverstone project (IRENA, 2023).

6.11.10 Japan – 621 MW

In the 1960s, the Matsukawa and Otake geothermal fields were discovered in Japan, and studies on geothermal energy production were started. Japan ranks fourth in its region and tenth in the world, with an installed capacity of 621 MWe. After the Fukushima nuclear disaster in Japan in 2011, developments in geothermal energy have increased with the small-capacity binary power plants built with government incentives. Sixty geothermal power plants of less than 2 MWe each have been installed in 45 geothermal fields in Japan. The use of GHPs (geothermal heat pumps) in Japan is minimal (163 MWth) but is expected to increase rapidly thanks to government incentives (Yasukawa et al., 2021).

6.12 GEOTHERMAL RESOURCE RESEARCH AND DETECTION METHODS

The primary purpose of geothermal resource exploration is to find a suitable geothermal system and to ensure that energy is produced economically. Generally, geothermal fields are pretty large. The main thing to do here is to narrow the search area as the data is obtained and to continue the search activities until the source is found. For this, first of all, it is necessary to determine the location of the geothermal field and estimate the fluid potential in the field, that is, whether water or steam dominates the reservoir. The area, depth, and temperature range should be determined as accurately as possible in the area found, and the magnitude of the heat potential and the capacity of the entire geothermal resource should be estimated with this data. Rapid and low-cost exploration techniques are applied in the initial stages, where research areas are large (Serpen, 2023). As data is collected and search spaces narrow, reliability increases and more expensive techniques become available. In geothermal exploration, this approach continues until the most expensive technique, "exploration well," is drilled and the possible field tested. Before the geothermal exploration program is planned, the data from the water, mine, and oil exploration studies carried out in the research area and nearby fields should be collected. This obtained data plays a significant role in planning geothermal exploration and significantly affects cost reduction.

Geothermal springs can be seen in places with hot fluid discharges, such as volcanic activity, geysers, fumaroles, and hot springs, generally extensions of the Earth. However, high-temperature fluid may not always be found where these signals are present. Sometimes, deep geothermal reservoirs can be encountered without any indication on the surface, and these resources can only be seen by solving the surface geology. To carry out the geological research and detection methods with all these assumptions correctly, data analysis must be done by experts (geologists, hydrogeologists, geochemists, and geophysicists) first. Data such as topography, meteorology, geology, hydrogeology, observation of hot spring waters, and geophysical and geochemical measurements should be collected and reviewed (Serpen, 2023). The methods used in geothermal exploration are given in Figure 6.11.

- With geological studies, lithological, stratigraphic, petrographic, and geological structures of the formations
- With geophysical studies, underground structural geology
- With photogeological studies, geological, tectonic, and lithological features
- With geochemical studies, water chemistry and water–rock relationships

Underground geology, hydrothermal alteration zones, reservoir and cover rocks, and lithological-stratigraphic and hydraulic properties of the formations should be investigated with drilling (Serpen, 2023). The geothermal resource exploration and detection method is given in Figure 6.12.

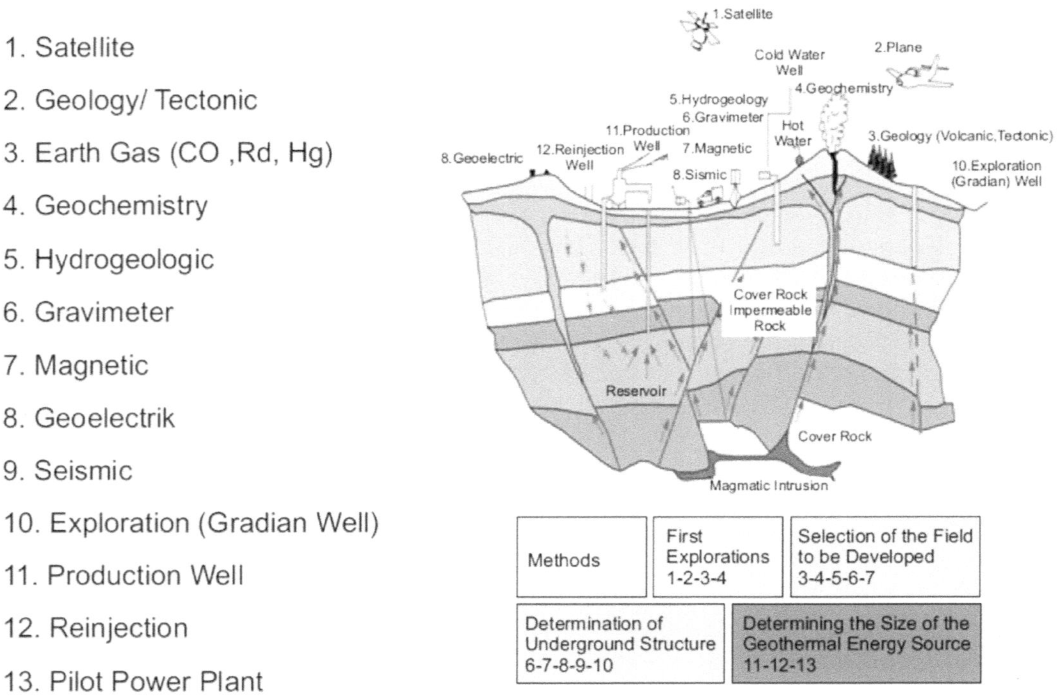

FIGURE 6.11 Methods used in geothermal explorations.

Information on the detailed geological, geophysical, geochemical, and drilling studies of the area where the geothermal energy potential will be investigated is given in the following.

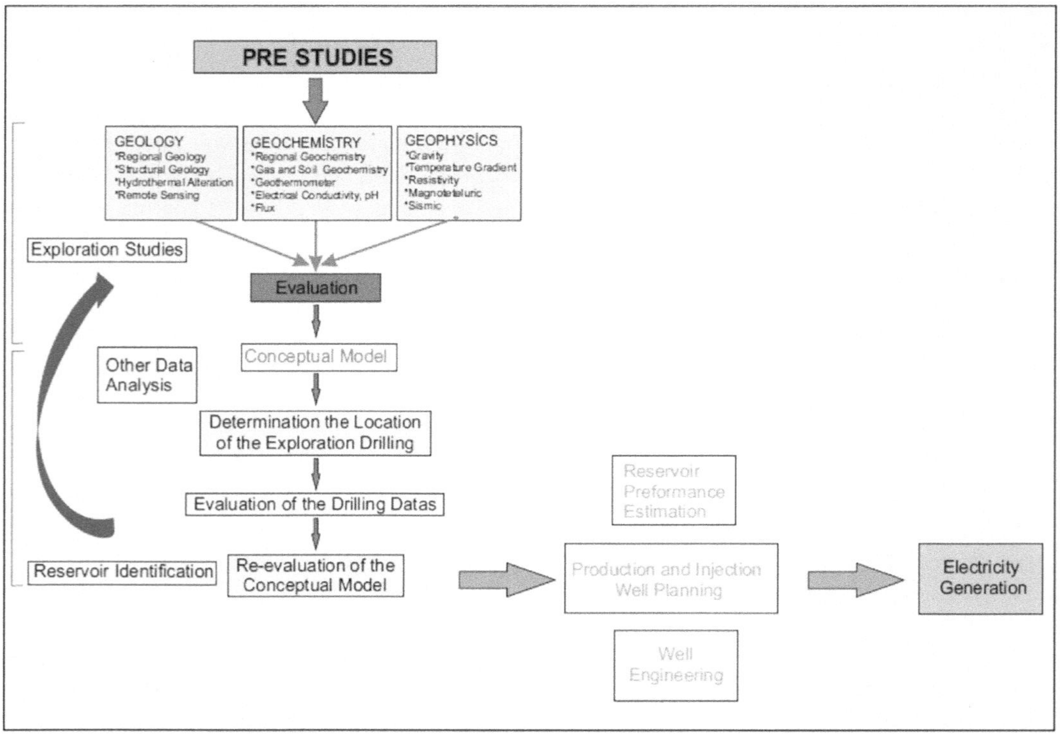

FIGURE 6.12 Geothermal resource exploration and detection method.

6.12.1 Geological Exploration Techniques

Search programs start with collecting all data by examining the studies of government institutions, universities, and published books, journals, and articles. The literature search is the cheapest search technique. In the next step, remote sensing studies are carried out. With the interpretation of aerial photographs, structural analysis, and geological maps, geophysical measurements can be made in areas with challenging topography. If the regional and stratigraphic framework is unknown, aerial magnetic measurements can provide important information. In addition, hot regions can be detected with infrared ray photographs taken from satellites such as Landsat from the air.

On the 1/25,000 scaled geological map in the region belonging to the geothermal field, contact relationships of rock units, strike and slope, pore and permeability, thickness, structural elements, hot/cold water points, magmatic dykes and intrusions, volcanic centers, hydrothermal alteration zones, separation and age determination of sedimentary formations and volcanic rocks of different lithology, tectonism, magmatism-volcanism, and possible reservoir rock cover rock and heat source to be provided with geothermal energy should be highlighted and marked (Laughlin, 1982).

In addition, when occurring on the surface of the map, it is important to mark all surface activities in the form of diffuse outlets, direct discharge, geysers, and concealed discharge. Samples should be taken from the fluids to measure the flow rates and temperatures of the outgoing fluids and determine their composition. Based on these data, the heat discharge amount of the geothermal system should be determined; the locations of these activities on the geological map, the rough structure of the system, the tectonism, and the hydrological model should be reinterpreted; and the most intense of the activities should be reinterpreted to be accepted as the center. Geophysical surveys should be carried out within a 10–15 km radius, with the most severe of these activities being the center.

In the next stage of the exploration program, geochemical studies are evaluated. For this, samples from various geothermal sources in the field, such as rocks, wells, geysers, fumaroles, and hot or cold water sources, should be taken, and physical and chemical analyses should be made. Analyses should be evaluated by a qualified geochemist. Thus, in the geothermal system, information such as hot fluid composition, original temperature, formation type, geothermal fluid–rock interactions, origin of recharge, flow direction in the region, and duration of underground residence can be obtained first. Of the geothermal fluid, class, group, phases in the system and their changes, hot–cold water mixing, mineral precipitation potential, direction of rise and flow of geothermal fluid to the Earth, duration of undergrounding, possible environmental impact, estimated minimum natural heat flux and potential of the region, analysis of economically essential elements, and reinjection of the fluid are determined (Serpen, 2023). Moreover, the determination of reservoir rock temperatures, origin of fluids, determination of recharge area, production capacity estimation, flow and break times of hot fluids, definition of hot–cold water mixtures and boiling processes, definition of hydrological circuit, fluid and rock balance, and definition of water–rock reactions can be made by using natural stable isotopes in the fluid. In addition, active fractures and faults in the geothermal area can be determined from the amount of mercury, radon, and/or helium gases in the soil.

The physical properties of the Earth are examined with geophysical exploration methods. These studies are carried out to determine temperature, mineralization, gas and fluid movement, sudden changes in the thickness of some lithostratigraphic units, and structural features of shallow and deep reservoirs.

Thermal exploration techniques are very important in estimating the potential of a geothermal system. Therefore, techniques such as surface and shallow temperature measurements, geothermal gradient measurements up to 100 m deep, and heat flow determinations at depths of more than 100 m are used. Another technique, gravity studies, is gravity measurements made in suitable environments, which can indicate local positive anomalies that may be associated with important structural formations and geothermal systems. Regional structural uplift and buried volcanic or intrusive structures may create positive anomalies (Laughlin, 1982). The gravity method should be used to determine the magmatic intrusions underground and their size, which are the heat source, the silica sedimentation structures created by the geothermal fluids blocking itself during the uplift, and the important faults controlling hydrothermal systems dominated by water (Serpen, 2023).

SP studies, which are correct to be applied on an area of 50–100 km^2, are among the important exploration techniques for giving meaningful results to determine the geothermal gradient and fluid motion. This method measures the natural voltage changes arising from the interaction of groundwater with conductive rocks, high geothermal gradients, and moving fluids and is widely used in geothermal exploration (Laughlin, 1982). In the case of volcanoes, magnetic studies are another exploration technique used to determine the extent of the geothermal anomaly. This method should be used especially in the case of regional structural analysis. If there are magnetic measurements from the air, it is possible to calculate the average geothermal gradient and heat flux from the Curie point depths in the anomaly region (Laughlin, 1982). Telluric studies are based on vector-telluric and telluric profiling methods measuring frequencies below 20 Hz. Since this method is sensitive to lateral changes in resistivity, it can be used in geothermal exploration. Especially if the direction of the geological contacts is known, measuring lines can be placed perpendicular to this direction, making it very useful in the search. Magnetotelluric studies are applied to identify deep conductive semi-molten zones that act as heaters under possible geothermal fields. The AMT method can be performed at depths of up to 2 km (frequency range 8–18,000 Hz) to identify shallow conductive zones (Laughlin, 1982). The dimensions of the buried hydrothermal structure are determined by the resistivity method, and the Schlumberger or Wenner electrode array is used to correlate it with the thermal and hydrogeological systems related to geothermal reservoirs. The actual resistivity structure of the field can be interpreted by evaluating all the measurements one by one and by evaluating the electrical structure sections prepared in two dimensions, on a field basis, in three dimensions. The seismic methods used can be applied as active and passive. Active seismic is a rarely used method in geothermal exploration in terms of both its cost and benefit. Active seismic methods can be used to detect the condition of the basement rock and the aquifer structure. Microearthquake method, one of the passive seismic methods, can also be applied to identify active faults. Passive methods are more commonly used in geothermal exploration (Serpen, 2023).

6.12.2 Exploration Drills

Exploration drilling fields should be determined by considering geological, geophysical, and geochemical criteria. The first well location should provide as much data as possible about the deep system and, at the same time, generate fluid. From a geological point of view, geological evidence of permeability and porosity at depths should be sought, and it should foresee faults close to vertical cuts at the depth where safe reservoir operation can be made. From a geophysical point of view, fields with significant discontinuities and fault zones are predicted, and the lowest resistivity values at the investigated depth should be selected. Regarding geochemicals, selection should be made according to the type of surface activity. Places with high temperatures and low groundwater mixture should be chosen in hot water systems (Serpen, 2023).

6.13 GEOTHERMAL SOURCE PRODUCTION AND REINJECTION WELLS

The constants for drilling are knowledgeable, experienced people and their equipment. In a geothermal project, drilling is used from the beginning to the final stage and throughout the operation. In this type of project, drilling starts with information, followed by information and production; after that, it undertakes strategic and indispensable functions, such as information and reinjection of the produced fluid.

6.13.1 Production Well

Well design is very important within the scope of geothermal drilling works. The issues to be considered in the design of the well are the lithological structure of the field, the reservoir depth, the diameter of the production pipe, the reservoir fluid type and properties, and the geological and tectonic structure. In drilling, in addition to selecting machines and pumps suitable for the depth of the geothermal reservoir and the diameter of the well, selecting the appropriate equipment for the capacity of the machine and pump is also very important (Toka, 2023a).

There are two important stages for drilling: well planning and well design. *Well planning* for drilling means defining, scheduling, and budgeting all activities required in drilling. *Well design* is to define all physical parameters (depth, diameter, etc.) of the well (Kutluca et al., 2005). Careful planning is critical for the drilling process. Not only will it minimize the cost, but it will also reduce the risk of unexpected events, injuries, or property damage accidents. To complete the well, a drilling plan should be listed and describe all activities, their time, and the costs.

For problems that may be encountered in the follow-up of the well at the drilling stage (blow-out risk, if more than one production zone is expected; material filling from the production zones to the well; clay zone problems; selection of drill diameter and type according to lithology; cutting the production zone without lowering the intermediate casing), the following information is required to take countermeasures:

- Study reports of the field's geology and geothermal purposes (geological, geophysical, geochemical data)
- Determination of all parts from the bottom of the well by examining the possible log
- Hole size and recommended drill bit
- Estimated penetration and expected use of the drill
- Provision of instructions if directional drilling is to be done
- Type of fluid to be drilled, and flow rate
- Required log before the casing is lowered or during drilling
- Necessary tests after cementing or at the end of the well
- Weight, type, and connection of casing pipes
- Recommended cementing schedule
- Measures to be taken against possible problems

(Kutluca et al., 2005)

Well design defines all the well's physical parameters (depth, diameter, etc.). Although a lot of information is needed for well design, all this information may not be available. While the following information is necessary, it should not be limited to this information for the design of a well:

- Purpose of drilling
- Surface or shallow hole conditions
- Reservoir conditions
- Logistics conditions
- Possible problems in drilling
- Casing pipe requirements

(Kutluca et al., 2005)

When the cover layers (weak, cracked, loose, and swelling formations) on the reservoir are not controlled during the drilling operation and underground fluid production, it causes problems with wells. Such problematic formations should be controlled with casing pipes so the drilling is at the desired diameter and depth for smooth fluid production. While large-diameter casing pipes are used at the beginning of the well, the diameters of the casing pipes gradually decrease as the well gets deeper. After the closed protection pipes are lowered to the determined depth, the back of the pipe is cemented from the depth where the pipe is lowered to the surface (Toka, 2023a). In geothermal wells, casing pipes are lowered into the well for the following reasons:

- To protect the surface aquifer (regulations required to prevent contamination of groundwater by drilling fluids)
- To close weak and cracked levels to get better efficiency from the well

- To prevent fluid transfer between formations
- To prevent unwanted liquids from entering the well during drilling
- To control the fluid pressure
- To protect the well age and production zone
- To prevent sudden arrivals during drilling

(Kutluca et al., 2005)

Preferred protection pipe diameters in geothermal wells vary from field to field, depending on the well's depth and the fluid's physical properties. A typical geothermal well design is given in Figure 6.13. This design is especially ideal for wells with more than 3,500 m depth. In shallower wells, it is possible to start with a 26" diameter well instead of the 36" diameter well section. In wells shallower than 1,000 m, starting with a 17 Ω" well is more appropriate. Well and casing diameters and cementing depths should be determined according to the region's structural, geological, and lithological conditions.

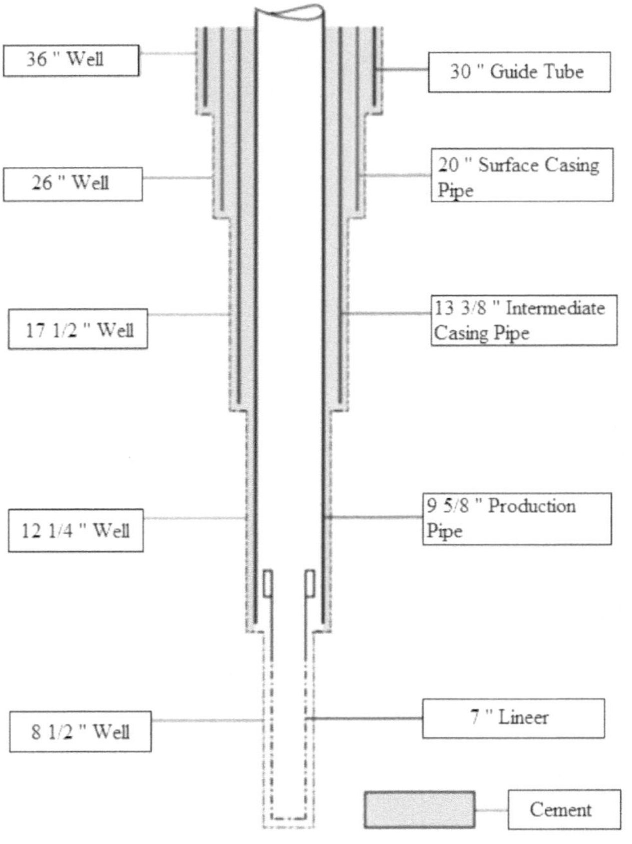

FIGURE 6.13 Typical geothermal well design.

In the *protection pipe design* of geothermal wells, the outermost first pipe used at the beginning of the well is called the guide pipe. This pipe prevents unstable geological layer formations (such as alluvium, sandstone, and pebble) from flowing into the well, spilling, and falling pieces during progress (Rabia, 2001). After the guide pipe is equipped and the back is cemented, the guide pipe is kept open in case of uncontrolled fluid flow during advancement.

The *surface casing pipe* is used after the guide pipe. Its main task is to prevent soft formations, such as clay and marl under the alluvium, to prevent the formations from swelling and causing problems during progress, by preventing their contact with the deep fluid, and to support the installation of the wellhead safety valve (BOP, blow-out preventer) equipment. The surface casing pipe is lowered into the harder formations under the soft formations, and the back of the pipe is cemented up to the surface (Heriot-Watt University).

The *intermediate casing pipe* is used to separate the problematic formations between the surface pipe and the production pipe. The most important purpose of using this pipe is to control the problematic formations encountered to reach the desired depth (Heriot-Watt University).

The *production pipe* is lowered to the upper part of the production zone, and the back of the pipe is cemented. They are generally designed to withstand reservoir pressure, and pressure test must be done. Its main purpose is to separate the production range from other formations and/or act as a channel for production (Heriot-Watt University).

The *liner pipe* does not rise to the surface. It is left at the bottom of the well to rise 20–30 m in deep drillings in the production pipe. This pipe is used if the fluid production zone can cause debris. If the production zone is intact, the pipe is not lowered into the production zone, and the well is left bare (Toka, 2023b).

In a geothermal well, the first data about the reservoir begins to be obtained during drilling. The inlet and outlet temperatures of the drilling mud give us preliminary information about the reservoir temperature. Mud leaks and the number of leaks that occur during drilling are important in terms of giving us first information about the production zone. Following the completion of the drilling works, the first information about the well is obtained by performing well completion tests. Mining and petroleum engineers also carry out well tests to ensure a more efficient use of geothermal resources after the well's planning, drilling, and completion. With these tests, reservoir parameters are learned in a healthy way, and these parameters are constantly monitored during operation. Thus, the field's potential is revealed correctly, the reservoir behaviors are predicted, and the data that will enable the necessary measures to be taken in a timely manner are reached. In addition to measuring the physical properties of the well with these tests, it is also possible to obtain information about the reservoir by observing the behavior of the well. The most important aim of these tests is to create a reservoir model in light of the information obtained from the well and to predict the future behavior of the reservoir through this model (Toka, 2023b).

The tests performed in geothermal wells can be grouped into five main groups: pressure tests, temperature tests, production tests, gas measurements, and tracer tests. Before starting the studies on geothermal energy, as much information as possible from the thermal and natural formations on the surface should be recorded to control the changes that will occur in the future. From the data obtained during these research studies, information about the hot fluid in the depth can be obtained, and these data can also be used as basic comparison data for future use.

6.13.2 Reinjection Well

After the fluid obtained from geothermal resources is used, it is pumped back into the geothermal reservoir. It is called *reinjection*, which provides significant benefits. A part of the reservoir volume emptied due to production can be filled with natural nutrition, but it may not correspond to all the water used. In these cases, the reservoir pressure or downhole level decreases. When reinjection is made, the reservoir pressure is maintained, the continuity of the geothermal resource is ensured, the wastewater is prevented from harming the environment, and the surface collapses that occur on the earth due to production is minimized. The reinjection process has three important purposes:

- To maintain the reservoir pressure
- To provide more heat generation from the reservoir
- To dispose of the remaining wastewater after the production of hot water

Detection of reinjection well locations in geothermal fields is more difficult than determining production well locations. Various problems arising from reinjection may be encountered, such as not finding a suitable injection field, the injected water reaching the production zone very quickly and cooling the production zone, excessive injection pressure, contamination of underground freshwater aquifers, interruption of production in steam wells, short-circuiting of the injected fluid to the surface, and change in the chemistry of the produced fluid. For this reason, it is necessary to carry out test and observation studies in detail and to reach the 3D model of the system by interpreting it well (Korkmaz Basel et al., 2009).

If the flow rate of the pumped water is known, the most important issue that needs to be examined in the application of reinjection is how far from a well or wells the water with a lower temperature than that of the reservoir should be pumped from the production area in order to prevent thermal pollution. When choosing the locations of the injection wells, the cold water injection should not immediately affect the hot reservoir water produced.

Pumped water flow in highly permeable flow channels and early arrival at production wells should be prevented. Generally, the flow is not expected to progress symmetrically around the injection well in fractured rocks. It should be considered that the fluid will move faster in some directions. Therefore, there must be a reliable gap between the injection and production wells. These ranges can only be determined after proper pressure interference, and tracer tests, especially, have defined the flow paths in the reservoir. In Figure 6.14, the design of the reinjection well is given (Korkmaz Basel et al., 2009).

6.13.2.1 Factors in Determining the Distance between the Production Well and the Reinjection Well

1. Capacities of the production wells
2. Temperature and steam ratio
3. Presence of zones with high permeability
4. The temperature of the injected water
5. Amount of injected fluid

During reinjection, a zone is formed around the injection well due to the reinjected water. The temperature of this zone is lower than the original reservoir temperature and grows over time and reaches the production well. As the water in contact with the rock heats up, the rock cools. After a certain period of time has passed from the arrival of the cold water to the production wells, the temperature of the produced water decreases. Therefore, estimating the water temperature in the production wells depends on the discharged water flow, time, and the interval between injection and production wells.

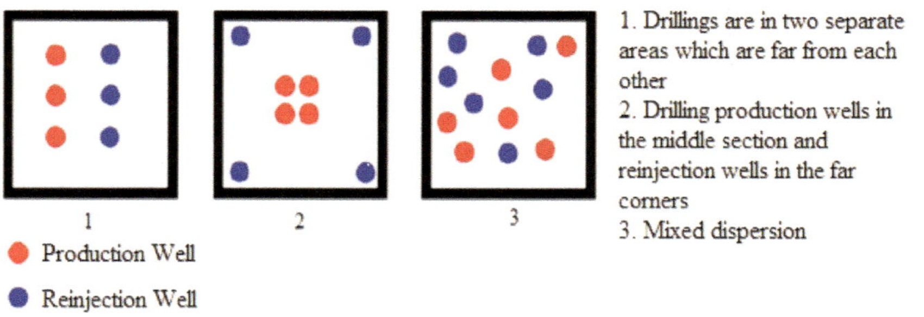

FIGURE 6.14 Reinjection well location design.

Source: Korkmaz Basel et al. (2009).

If any well is wanted to be used as an injection well in a geothermal field, these three important features of the well should be evaluated.

6.13.2.2 Suitability of the Wellhead and the Downhole Equipment for Injection

Production wells can be converted to injection wells. However, this conversion process requires examining, designing, and preparing the well. The technical and economic suitability of the well for reinjection should be investigated. If the well is used as an injection well, the first thing to consider is that the wellhead equipment is resistant to the high pressure required for injection. Pickling, perforation, and hydraulic fracturing processes should be designed and applied when necessary (Korkmaz Basel et al., 2009).

6.13.2.3 Ability to Provide the Desired Injection Flow Continuously

The conversion of wells with low manufacturability into injection wells is one of the first approaches that come to mind. However, the most important feature that should not be forgotten is that low manufacturability may be due to a possible low permeability of the formation and low net thickness; therefore, these parameters can also negatively affect injectivity in the same way. In the case of injection, two important parameters to be controlled are the injection flow rate and the required wellhead injection pressure. Depending on the reservoir conditions, the injection flow may increase, decrease, or remain constant over time. If the flow rate decreases over time, the field behaves as a closed system, or the pumped fluid contaminates the formation around the well and reduces the permeability; if the flow rate increases over time, it can be interpreted that the injection conditions positively affect the formation permeability around the well. In addition, the fact that the injection flow rate may be affected by the production of the production wells in the region close to the injection well should not be neglected (Korkmaz Basel et al., 2009). To achieve the desired injection flow rate, it is important that the well is placed in areas with the appropriate permeability required for the flow rate.

6.13.2.4 Location of the Injection Well in the Field Relative to the Production Wells

The most important feature to be considered when choosing the location of the injection well is the effect of the pumped fluid on the production performance of the area. During injection, it is aimed to increase the presence of fluid in the reservoir, increase the temperature of the pumped fluid during flow, and increase the heat production from the field by producing it from production wells over time. However, the thing to be avoided is that the fluid with a low temperature reaches the production well before it heats up sufficiently and negatively affects production performance.

6.14 MODELING OF GEOTHERMAL RESERVOIRS

The main purpose of geothermal reservoir modeling is to accurately and realistically predict the long-term geothermal reservoir potential. For this purpose, many methods have been developed to model the behavior of geothermal reservoirs.

These methods vary in their complexity and cost. The selection of the appropriate model should consider the reliability and amount of data available, as well as the purpose of the modeling. Geothermal systems are often very complex due to faults, cracks, phase changes, chemical reactions, and thermal effects (Akan, 2013). While developing the geothermal system model, the existing field data should be carefully evaluated, and the system's physical structure should be determined. After creating a conceptual model in this way, the future performance of the geothermal reservoir is predicted by using the model that best represents the behavior of the geothermal reservoir.

There are various methods in the literature that model geothermal reservoir behaviors. These can be grouped under three main headings: production flow reduction method, dimensionless reservoir modeling method, and 3D numerical modeling method. The most used method in recent years is the numerical modeling method (Akan, 2013).

In the production flow reduction analysis method, the production data of the wells and the reservoir are used in appropriate mathematical equations, and future production flows are estimated. With the estimation of the production flow rates, the number of new wells required can be determined by adding them to the system in the future. However, sufficient production data must be available for the modeling. The most important problem with this method is that it does not consider changes in field operation (development drillings, additional wells, reinjection wells, etc.) (Karahanoğlu, 2019).

In dimensionless reservoir modeling, the reservoir is taken as a whole. It is a modeling method that aims to determine the behavior of reservoir pressure and temperature according to time or production from the reservoir, considering what enters and exits the reservoir and using fluid/rock properties. The main advantages of this method are its simplicity of use and the absence of the need for extensive computer capacities (Akan, 2013).

In the digital models where the reservoir is taken as 3D, all kinds of fluid and well properties of the reservoir are entered into the model, and the reservoir performance is examined by considering the distribution of all parameters valid locally, regionally, and throughout the reservoir (Akan, 2013). The production of geothermal reservoirs activates the hydrodynamic and hydrothermal mechanisms in natural equilibrium, and as a result, fluid movement and heat dissipation occur within the reservoir. Numerical modeling studies, which start with the mathematical description of these mechanisms, can investigate the effect of different operating conditions on the reservoir and thus enable the sustainable production of geothermal systems. Numerical models previously prepared for two-dimensional and single-phase fluids have become able to model the movement of multi-phase (liquid, gas) fluids (hot water, steam, air, CO_2) in three-dimensional reservoirs over time and beyond fluid motion and energy transport mechanisms, surface collapse, chemical reactions, CO_2 storage, various geomechanical processes, biodegeneration, and similar simulations have been developed (Karahanoglu, 2019). However, the accuracy of such models is related to the excess of the entered data and the accuracy. Generally, such numerical modeling studies are preferred at the stages where the data about the reservoir is sufficient, and there is a certain production and reservoir performance history. The main advantage of this method is that it stores all the mathematical equations in a computer code and allows the user to model by deciding how detailed the model and what physical operations are required. The most important disadvantages are that much information about fluids and reservoirs is used as data in the model, and large computer capacities and experienced users are needed (Karahanoglu, 2019). The negative aspects of numerical modeling are the need for significant geological, geophysical, geochemical, and hydrological data in 3D reservoir models, the long time it takes to obtain the necessary data, and the high cost. There is not enough data available for newly discovered geothermal fields. When production data about the field is obtained, 3D reservoir modeling can be the most appropriate and realistic approach.

With the application of the developed numerical models, the numerical modeling of the geothermal fields in different regions of the world is made. These fields are developed most appropriately, and their optimum and sustainable operation is ensured (Satman et al., 2023). The complexity of geothermal systems, which are formed under various factors in the natural environment and over many years; the structural elements and all physical properties of geothermal systems; and the transfer of these physical parameters to numerical models with the help of conceptual models are realized by representing the reservoir environment in the best way (Akan, 2013). Similarly, determining the boundary and initial conditions of geothermal systems to represent the terrain conditions affects the success of the modeling studies at the highest level. In addition to being able to model almost all kinds of complex mechanisms, numerical models can be created in which very different geological structures can be defined as conceptual models.

Thanks to the modeling of the reservoirs, it is possible to provide the most suitable, efficient, and sustainable production. Since geothermal reservoirs are constantly dynamic, they require serious scientific follow-up and data collection as long as production continues (Satman et al., 2023). Otherwise, it is possible to encounter irreparable results in the case of correct modeling and, thus, correct production and injection in production reservoirs.

6.15 CONCLUSIONS AND RECOMMENDATIONS

- There are high first costs and resource risks in the early stages of geothermal energy investments. To encourage both electricity generation and heating/cooling from geothermal sources, new risk mitigation plans should be established and adapted to market conditions.
- To use renewable energy sources more effectively, there is a need for new legal regulations supporting this process in many countries. Considering the high investment costs of renewable power plants, especially during the installation phase, current legal regulations are not enough to encourage investors. More support should be provided for new renewable energy investments, and societies should be made aware.
    - Newly constructed buildings should be built with renewable energy.
    - More funds and time should be allocated for R&D studies on renewable energy.
    - Appropriate land use opportunities for renewable energy should be created.
    - More courses on renewable energy should be taught in some undergraduate and graduate programs at universities.
    - Appropriate conditions should be provided for the transmission and storage of energy produced with renewable energy.
    - Efficiency and energy saving should be given importance in energy consumption.
    - Public awareness should be raised, and the environmental, financial, and political benefits of these energy sources and the disadvantages of nuclear and fossil fuels should be explained in detail by experts in panels organized in the media, universities, and society.
- Geothermal energy is not infinite or trouble-free energy. In the use of geothermal energy, which is energy that can be maintained under suitable conditions, recovery must be ensured. Due to the chemical structure of geothermal water, reinjection wells should be opened to protect the environment and ensure the geothermal resource's continuity.
- Geothermal energy sources have enormous environmental benefits when compared to other energy sources. The most significant advantages are the absence of air emissions and waste products or their negligible level. The use of these energy sources helps maintain air, soil, and water quality and natural balance.
- When planning production or reinjection wells, wells should be equipped so as not to damage groundwater. Wells produced with drill pipes that do not comply with the technical conditions or are corroded should be closed immediately. The damage they cause to the groundwater should be terminated immediately.
- After the research is carried out using geological, geochemical, and especially geophysical methods in geothermal fields, by making aquifer modeling, boreholes should be drilled at appropriate points according to the technique to protect both the aquifer and groundwater. Geothermal wells to be drilled to protect cold water aquifers or hot aquifers must be controlled by the relevant geology, geophysics, petroleum, and/or mining engineers.
- While the well equipment is being made, it should be planned to prevent the contamination of cold water aquifers. Using high-temperature sources for electricity generation from geothermal energy is common, but there is significant potential in low- and medium-temperature sources. By using binary geothermal power plants more widely, the damage to the environment can be minimized and maximum efficiency can be obtained from the geothermal fluid using ORC technology. Various studies should be carried out to increase efficiency.
- The environment is affected during the production and transportation of every energy used. Although geothermal energy does not have a negative effect when the right technologies are used, it can cause environmental pollution if the right technologies are not used.
- Geothermal fluids naturally contain high concentrations of molten minerals. For this reason, protective or preventive measures should be taken by considering harmful elements. It is necessary to determine the possible pollution effects of chemicals given to the

environment during geothermal research and applications and to protect them from these harmful effects.

- To prevent groundwater contamination, it is necessary to protect (casing) wells and prevent the uncontrolled flow of drilling fluids and mud throughout the groundwater region.
- Abandoned mines or oil gas wells can be used in geothermal projects to reduce drilling costs and risk.
- Energy efficiency can be increased, and the widespread use of heat pumps in geothermal energy and district heating/cooling systems can reduce costs.
- Recovering precious elements such as calcium silicate, colloidal silica, arsenic, lithium, gold, and silver from geothermal fluids is possible. Various successful studies have been carried out on these issues.
- Lithium, one of the rarest elements on Earth, is found in inorganic environments. Its high energy density, low weight, and environmental and performance properties enable lithium to be used in many areas, from portable electronic devices to power supplies for vehicles in the battery sector. Therefore, lithium recovery studies should be given importance, considering the use of technology in the world and potential technological and economic development.
- Green hydrogen is expected to play an important role in the world's energy future as it is a clean fuel and energy carrier that does not harm the environment. Therefore, it is important to use alternative renewable energy sources, such as geothermal energy sources, and to encourage them technically and economically to solve the technological, practical, and economic problems related to hydrogen production, transportation, and use. For the energy sector to be decarbonized, more renewable energy should be integrated into power generation, and its capacity should be maximized.
- Necessary support should be provided to investors to expand the usage areas of renewable energy resources and make more investments in this area. It is foreseen that different studies can be developed for using geothermal energy in hydrogen production and can be successfully applied in terms of thermodynamics.
- The development and use of geothermal energy are intertwined with various sectors, such as carbon sequestration and storage, hydrogen production, the mining industry, other renewable energies, such as wind/solar, and end use sectors, including the industrial and agri-food sectors. Improvements are needed to promote synergy between geothermal and other industries and to ensure inter-sectoral cooperation.
- Regulations should be developed to encourage the sharing of underground data among the oil, gas, and mining sectors with the relevant data repository on the underground distribution of temperatures and geological formations.
- Hybrid geothermal electricity generation with other renewable sources, such as concentrated solar power, solar photovoltaic, and biomass, should be encouraged to maximize the efficiency of electricity generation.
- Pilot projects should be implemented to demonstrate commercial project viability and gain the trust of communities, politicians, and other partners.
- A regional or global approach should address common issues hindering geothermal development.
- Studies on geothermal energy should be carried out on international platforms with various investors from energy and other related sectors, for example, during meetings on global warming and climate change.
- With new technologies and conscious use, the environmental impact of geothermal energy will be minimized and be almost nonexistent. In binary systems, "zero-waste and 100% environmentally friendly" applications can be developed. No waste fluid will be generated if the geothermal fluid can be circulated and reinjected at high pressure in the system starting from the well. Thus, the environment will not be harmed.

REFERENCES

Akan, B. (2013). *Jeotermal sistemlerde modelleme yaklaşımı. Maden Tetkik ve Arama Genel Müdürlüğü, Enerji Hammadde Etüt ve Arama Dairesi*. Retrieved April 7, 2023, from www.mta.gov.tr/v3.0/sayfalar/hizmetler/kutuphane/ekonomi-bultenleri/2013_15/149.pdf

Anderson, A., & Rezaie, B. (2019). Geothermal technology: Trends and potential role in a sustainable future. *Applied Energy*, 248(August), 18–34.

Anonymous. (2019). *Eğitim, danışmanlık ve yurtdışı faaliyet ihtiyaç analizi özeti*. Zobu Consulting.

Atlantic Council. (2018). *A geothermal leader: The case of Iceland*. Atlantic Council. Retrieved March 20, 2023, from www.atlanticcouncil.org/blogs/energysource/a-geothermal-leader-the-case-of-iceland/

Baille, A. (2001). Trends in greenhouse technology for improved climate control in mild winter climates. *Acta Horticulturae*, 559, 161–168.

Bundschuh, J., Chen, G., Chandrasekharam, D., & Piechocki, J. (2017). *Solar, wind and geothermal energy applications in agriculture and aquaculture*. Sustainable Energy Developments Series No. 13 (Chapter 1).

Burkardt, P., & Herbling, D. (2021). The world's no. 1 in geothermal electricity, Kenya aims to export its know-how. *Bloomberg*. Retrieved March 1, 2023, from www.bloomberg.com/news/features/2021-07-19/kenya-aims-to-double-geothermal-capacity-by-2030

Cataldi, R., & Súarez Arriaga, M. C. (2016). *Our geothermal legacy: A historic overview*. IGA News, Newsletter of the International Geothermal Association, Special Number on the Anniversary of the IGA. Retrieved from www.lovegeothermal.org//IGAnews_archive/IGANews-Special20Number-2.pdf

Çetiner, Z. S. (2017). Ayvacık-Tuzla bölgesindeki jeotermal kaynaklardan metal/mineral kazanım olanakları. *Ömer Halisdemir University Journal of Engineering Sciences*, 7(1).

Climo, M., Carey, B., & Mroczek, M. (2016). *Mineral extraction from geothermal brines in New Zealand*. GNS Science.

Cornish Lithium. (2023). *Lithium in geothermal waters*. Retrieved from https://cornishlithium.com/projects/lithium-in-geothermal-waters/

Dickson, M. H., & Fanelli, M. (Eds.). (1990). *Small geothermal resources: A guide to development and utilization*. UNITAR/UNDP Centre on Small Energy Resources.

Directorate General of New Renewable Energy and Energy Conservation. (2022). *To support the energy transition the government encourages direct utilization of geothermal energy*. EBTKE.

Eker, A. M., Yal, G. P., Cambazoğlu, S., Şen, O., & Dünya, H. (2017). *Enhancement of geothermal reservoir through varying acidizing operation procedures in Aydın Kuyucak geothermal field*. Proceedings of the 42nd Workshop on Geothermal Reservoir Engineering, Stanford University.

Enel Green Power. (2021). *Stillwater triple hybrid plant, USA*. Enel Green Power. Retrieved March 15, 2023, from www.enelgreenpower.com/our-projects/operating/stillwater-hybrid-plant

Eramet. (2021). *Eramet and Électricité de Strasbourg announce the success of the first pilot test to extract lithium from geothermal brine in Alsace, France*. Retrieved from www.eramet.com/en/eramet-and-electricite-de-strasbourg-announce-success-first-pilot-test-extract-lithium-geothermal

EU Reporter. (2022). *150 companies call on President Von Der Leyen to prepare a European strategy to unlock geothermal*. EU Reporter. Retrieved March 1, 2023, from www.eureporter.co/energy/2022/04/13/150-companies-call-on-president-von-der-leyen-to-prepare-a-european-strategy-to-unlock-geothermal/

European Commission. (2023). *Critical raw materials*. Retrieved from http://ec.europa.eu/growth/sectors/raw-materials/specific-interest/critical/

Evans, K., Ito, T., Kawai, K., & Hayashi, K. (1999). Hydraulic fracture reopening pressure and the estimation of maximum horizontal stress. *International Journal of Rock Mechanics and Mining Sciences*, 36(6), 811–826.

Exergy. (2022). *EDC's first geothermal brine recovery plant with Exergy technology is now operational in the Philippines*. Exergy. Retrieved March 15, 2023, from www.exergyorc.com/edcsfirstgeothermal brinerecoveryplantwithexergytechnologyisnowoperationalinthephilippines

GEMex. (2023). *GEMex final conference*. Retrieved from www.gemex-h2020.eu/

Geo40. (2023). *Geothermal silica*. Retrieved from https://geo40.com/geothermalsilica/

Gondal, I. A., Masood, S. A., & Khan, R. (2018). Green hydrogen production potential for developing a hydrogen economy in Pakistan. *International Journal of Hydrogen*, 1–29.

Grasby, S. E., Allen, D. M., Bell, S., Chen, Z., Ferguson, G., Jessop, A., Kelman, M., Ko, M., Majorowicz, J., Moore, M., Raymond, J., & Therrien, R. (2012). Geothermal energy resource potential of Canada. *Geological Survey of Canada*, Open File 6914.

Günhan, Ö. (2016). *Türkiye'de jeotermal aktivitelerden kaynaklanan su kirliliği*. General Directorate of Water Management, Department of Water Quality Management.

Hepbaşlı, A., & Çanakçı, C. (2003). Geothermal district heating applications in Turkey: A case study of İzmir-Balcova. *Energy Conversion and Management*, 44(8), 128.

Heriot-Watt University Department of Petroleum Engineering, Drilling Engineering. Retrieved from https://www.academia.edu/39642604/Heriot_Watt_University_Drilling_Engineering

Hutter, G. W. (2021). *Geothermal power generation in the world 2015–2020 update report*. Proceedings of the World Geothermal Congress 2020+1. Retrieved from www.geothermal-energy.org/pdf/IGAstandard/WGC/2020/01017.pdf

International Geothermal Association. (2017). *Global geothermal development: An overview* [Presentation]. Global Geothermal Alliance Meeting, IRENA.

International Renewable Energy Agency (IRENA). (2021). *Geothermal: The solution underneath*. IRENA.

IRENA. (2023). *Global geothermal market and technology assessment*. IRENA.

JED. (2022). *Clean, safe, renewable energy*. Jeotermal Enerji Derneği. Retrieved March 15, 2023, from https://jeotermalenerjidernegi.org.tr/en

Kajugus, S. I. (2022). *Geothermal innovations in Tanzania: Geo-hatchery project*. Retrieved from https://grmf-eastafrica.org/wp-content/uploads/2022/05/01Geo-hatcheryMay.2022.pdf

Karahanoğlu, N. (2019). Jeotermal rezervuarlarla ilgili sayısal modelleme/benzeşim çalışmaları. *Jeoloji Mühendisliği Dergisi*, 43, 99–130. Retrieved April 7, 2023, from https://dergipark.org.tr/tr/download/article-file/728101

Kılıç, F. Ç., & Kılıç, M. K. (2013). Jeotermal enerji ve Türkiye. *Engineer and Machinery*, 54(639), 45–56.

KIT. (2021). *Brine mine: A BMBF project of Chilean and German partners to develop a geothermal system for mineral extraction, provision of drinking water and heat generation*. Karlsruhe Institute of Technology. Retrieved from https://geothermics.agw.kit.edu/english/brinemine.php

Korkmaz Basel, E. D., Serpen, U., & Satman, A. (2009, February 9–11). *Assessment of Turkey geothermal resources*. Paper presented at the Thirty-Fourth Workshop on Geothermal Reservoir Engineering, Stanford University, Stanford, CA.

Kutluca, A. K., & Gökçen, G. (2005). *Jeotermal elektrik üretiminin sosyo-ekonomik etkileri: Kızıldere jeotermal sahası*. Geothermal Energy Seminar Book (MMO No: E/2005/393-2).

Laughlin, A. W. (1982). Exploration for geothermal energy. In *Handbook of geothermal energy*. Edited by Edwards, L. M., Chilingar, G. V., Rieke, H. H., & Fertl, W. H. Gulf Publishing Company, Houston.

Ledru, P., Calcagno, P., Genter, A., Huenges, E., Kaltschmitt, M., Karytsas, C., Kohl, T., Lokhorst, A., Manzella, A., & Thorhallsson, S. (2006). Enhanced geothermal innovative network for Europe. *GRC Transactions*, 30.

Lee, K. K., Ellsworth, W. L., Giardini, D., Townend, J., Ge, S., Shimamoto, T., Yea, I. W., Kang, T., Rhie, J., Sheen, D. H., Chang, C., Woo, J. U., & Langenbruch, C. (2019). Managing injection-induced seismic risks. *Science*, 364(6442), 730–732.

Lund, J. W., Huttrer, G. W., & Toth, A. N. (2022). Characteristics and trends in geothermal development and use, 1995 to 2020. *Geothermics*, 105, 102522.

Lund, J. W., Sifford, A., Hamm, S. G., & Anderson, A. (2021). *The United States of America direct utilization update 2019*. Proceedings of the World Geothermal Congress 2020+1. Retrieved from www.geothermal-energy.org/pdf/IGAstandard/WGC/2020/01011.pdf

Lund, J. W., & Toth, A. N. (2021). Direct utilization of geothermal energy: 2020 worldwide review. *Geothermics*, 90. https://doi.org/101016/j.geothermics.2020.10915

Majer, E. L., Baria, R., Stark, M., Oates, S., Bommer, J., Smith, B., & Asanuma, H. (2007). Induced seismicity associated with enhanced geothermal systems. *Geothermics*, 36(3), 185–222.

Memiş, Ü., & Demer, S. (2015). Afyonkarahisar jeotermal sahalarının hidrojeokimyası ve jeotermometre uygulamaları. *Süleyman Demirel Üniversitesi Fen Bilimleri Enstitüsü Dergisi*, 19(2), 66–77.

Mertoğlu, O. (2000). Türkiye'deki jeotermal enerji uygulamalarının çok yönlü önemi ve dünyadaki yeri. *Engineer and Machinery*, 485, 1.

National Renewable Energy Laboratory (NREL). (2021). *U.S. geothermal power production and district heating market report*. Retrieved from www.nrel.gov/doc/fy21osti/78291.pdf

Nevada Bureau of Mines and Geology. (2012). *Mason Valley site description*. NBMG.

Nextrends Asia. (2021). *Energy in China's 14th five year plan*. Nextrends Asia. Retrieved March 1, 2023, from https://nextrendsasia.org/energy-in-chinas-14th-five-year-plan/

Nunes, J. C., Coelho, L., Carvalho, J. M., do Rosário Carvalho, M., & Garcia, J. (2021). *Portugal country update 2020*. Proceedings of the World Geothermal Congress 2020+1.

Omenda, P., Mangi, P., Ofwona, C., & Mwangi, M. (2023). *Country update report for Kenya 2015–2019*. Proceedings of the World Geothermal Congress 2020+1.

Özdemir, Z. Ö., & Mutlubaş, H. (2019). *Enerji taşıyıcısı olarak hidrojen ve hidrojen üretim yöntemleri*. e-ISSN: 2667-5048, 2(1), 16–34. Retrieved from https://dergipark.org.tr/tr/download/article-file/771874

Özkaya, M. G., Variyenli, H. İ., & Yonar, G. (2008). Jeotermal enerji ile ısıtılan Kütahya ili Simav ilçesindeki ısıtma sisteminin çevresel etkilerinin değerlendirilmesi ve uygulanması gereken yenilikler. *Journal of Science*, 29(2).

Pollack, A., Horne, R., & Mukerji, T. (2021). *What are the challenges in developing enhanced geothermal systems (EGS)? Observations from 64 EGS sites*. Proceedings of the World Geothermal Congress.

Rabia, H. (2001). *Well engineering and construction*. Entrac Consulting.

Ragnarsson, A., Steingrímsson, B., & Thorhallsson, S. (2018). *Geothermal country update for Iceland*. Proceedings of the 7th African Rift Geothermal Conference.

Rogner, H. H. (2012). Energy resources. *Environment & Policy*, 54.

Satman, A., Onur, M., & Sarak, M. (n.d.). Jeotermal rezervuarların modellenmesi. *Jeotermal Enerji Semineri*. Retrieved April 7, 2023, from http://www1.mmo.org.tr/resimler/dosya_ekler/476806bbbdae84d_ek.pdf

Şen, O., Eker, M. A., Cambazoğlu, S., Yal, G. P., & Hou, M. Z. (2015). *Türkiye'nin ilk geliştirilmiş jeotermal sistemler (EGS) projesi: Dikili-İzmir*. Retrieved from https://www.jmo.org.tr/resimler/ekler/a2bcf34ff692a9a_ek.pdf

Sen, O., & Yanar Sözal, P. (2018). Jeotermal sulardan bor, lityum ve arsenik elementlerinin seçimli ayrılması. *Jesdergi*, 32–39.

Sen, O., & Yanar Sözal, P. (2022a). *Jeotermal uygulamaların çevresel etkileri* (2nd ed.). HEZ Energy.

Sen, O., & Yanar Sözal, P. (2022b). *Su kalitesi ve reenjeksiyonun önemi* (2nd ed.). HEZ Energy.

Sen, O., & Yanar Sözal, P. (2022c). *Jeotermal akışkan kullanımında oluşan hava emisyonu ve diğer enerji türleriyle kıyaslanması* (2nd ed.). HEZ Energy.

Sen O., & Yanar Sözal, P. (2022d, April). *Yenilenebilir Enerji Kaynaklarının Önemi ve Türkiye'de Yapılabilecek En Uygun Yenilenebilir Enerji Yatırımı*. HEZ Energy.

Sen, O., & Yanar Sözal, P. (2022e, February). *Jeotermal enerjinin seracılıktaki önemi, uygulamalar ve yeni gelişmeler* (2nd ed.). HEZ Energy.

Sen, O., & Yanar Sözal, P. (2023, January). *Hidrojen enerjisi*. HEZ Energy.

Serpen, Ü. (2023). *Jeotermal enerji arama teknikleri*. Jeotermal Enerji Semineri.

Song, Y., & Lee, T. J. (2021). *Geothermal development in the Republic of Korea: Country update 2015–2019*. Proceedings of the World Geothermal Congress. Retrieved from www.geothermal-energy.org/pdf/IGAstandard/WGC/2020/01012.pdf

Stober, I., & Bucher, K. (2013). History of geothermal energy use. In *Geothermal energy* (pp. 15–24). Springer.

TEİAŞ. (2023). *Aralık Ayı Kurulu Güç Raporu*. TEİAŞ. Retrieved from www.teias.gov.tr/kurulu-guc-raporlari

Tester, J., & Smith, M. (1977). Energy extraction characteristics of hot dry rock geothermal systems. In *Geology*.

ThinkGeoEnergy. (2020). *Innovation in geothermal energy utilization – The Icelandic story*.

ThinkGeoEnergy. (2021). *ThinkGeoEnergy's top 10 geothermal countries 2021 – installed power generation capacity (MW)*. ThinkGeoEnergy. Retrieved March 15, 2023, from www.thinkgeoenergy.com/thinkgeoenergys-top-10-geothermal-countries-2021-installed-power-generation-capacity-mwe/

ThinkGeoEnergy. (2022a). Vulcan and Enel to jointly develop Cesano geothermal lithium project in Italy. *ThinkGeoEnergy*.

ThinkGeoEnergy. (2022b). NZ government to fund geothermal lithium extraction technologies.

ThinkGeoEnergy. (2023). Pilot project in Nevada aims to convert oil wells into geothermal producers. *ThinkGeoEnergy*. Retrieved from www.ThinkGeoEnergy.com/pilot-project-in-nevada-aims-to-convert-oil-wells-into-geothermal-producers/

Thompson, A., Harmer, Z., & Fong, W. (2021). *Geothermal industry development in Canada 2020 country update*. Proceedings of the World Geothermal Congress 2020+1. Retrieved from www.geothermal-energy.org/pdf/IGAstandard/WGC/2020/01026.pdf

Toka, B. (2023a). *Derin jeotermal sondajlarda kuyu tasarımı ve sondaj makineleri için takım dizisi seçimi*. MTA Sondaj Dairesi.

Toka, B. (2023b). *Jeotermal enerji*. Chamber of Mining Engineers.

Toksoy, M. (2001). *Jeotermal enerji doğrudan ısıtma sistemleri*. Fundamentals and Design Seminar Book (MMO Public No: MMO/2001/270).

TUIK. (2022). *Sera gazı emisyon istatistikleri, 1990–2020* (No: 45862).

U.S. Department of Energy. (2019). *GeoVision: Harnessing the heat beneath our feet*. U.S. Department of Energy. Retrieved March 1, 2023, from www.energy.gov/eere/geothermal/downloads/geovision-harnessing-heat-beneath-our-feet

U.S. Department of Energy. (2020). *Guadeloupe: Energy snapshot*. Energy Transition Initiative, U.S. Department of Energy. Retrieved from www.energy.gov/sites/default/files/2020/11/f80/ETI-Energy-Snapshot-Guadeloupe-FY21.pdf

USGS. (2003). *Geothermal energy: Clean power from the Earth's heat*. US Geological Survey. Retrieved from https://pubs.usgs.gov/circ/2004/c1249/c1249.pdf

Utah FORGE. (2022). *Frontier observatory for research in geothermal energy – FORGE*. Retrieved from https://utahforge.com

Verel, A. R. (2003). Jeotermal enerjinin tanıtımı ve kullanımı. *Journal of Electrical Engineering*, 418, 58.

The World's 10 Biggest Geothermal Energy Plants. (n.d.). *Dow's climate care*. Retrieved March 1, 2023, from www.dowsclimatecare.com/blog/theworlds-10-biggest-geothermal-energyplants/

Yasukawa, K., Nishikawa, N., Sasada, M., & Okumura, T. (2021). *Country update of Japan*. Proceedings of the World Geothermal Congress 2020+1. Retrieved from www.geothermal-energy.org/pdf/IGAstandard/WGC/2020/01037.pdf.

7 Wind Energy in the Context of the Energy Transition

Jorge Saldaña, Mesut Yurukcu, Narendra Boppana, Sepehr Arbabi, Jim Henry, and Sebahattin Ziyanak

7.1 INTRODUCTION

Despite the label as a "newer" form of energy, particularly in a world that is now looking for cleaner, more sustainable sources of power, wind has a rich history that dates to several centuries. This has established the power of wind as one of the more mature forms of energy, and one of the most promising primary energy sources for a future that is looking to transition away from fossil fuels (Leung and Yang, 2012). With ever-increasing demands from international organizations like the United Nations to curb the trajectory of fossil fuels, linked with the increase in global temperatures in the last few decades, the oil and gas industry must work closely with governments and stakeholders to develop new technologies that harness the power of renewables, like solar and wind. Climate change is a real threat, but with that said, humanity has both the tools and the ingenuity to develop the necessary response that can withstand the many current challenges, while also continuing to fulfill the energy needs of billions of people around the world.

Wind energy has served humans as a valuable source of power for centuries. Simple wind devices like vertical-axis windmills dating back to 200 BC have been found at the Persian Afghan borders, while almost 2,000 years later, in what is now the Netherlands, horizontal-axis windmills were used extensively to harness the energy coming from the wind (Kaldellis and Zafirakis, 2011). The first mechanisms exploiting the power of the wind were based on aerodynamic drag forces, while the more sophisticated current devices follow advanced aerodynamic lift forces (Allamehzadeh, 2016).

Figure 7.1 shows a series of wind devices that have been used throughout history for the betterment of the human experience. From the Persian Sistan windmills with a panemone design made to pump water, to more sophisticated vertical-axis windmills built to grind grains or for other general purposes, to the famous Dutch windmills which have become a UNESCO World Heritage Site, humans have explored the capabilities of the power of the wind to meet their needs, with largely satisfactory results. Current times call for a return to the reliance of humans on wind, but this time, needs are larger than before.

With the threat of climate change affecting life on Earth through a warmer climate, achieved partly due to our overreliance on fossil fuels, we need to take a harder look at the sources used to generate power to meet our needs. Going back to cleaner, renewable sources of energy, increasing their share while promoting a phaseout of older, more polluting sources is part of this new energy puzzle that humanity needs to complete rather quickly. However, this challenge is replete with significant hurdles on many areas, which go from infrastructure woes to government passivity, as well as limitations of these sources of energy, the potential of other alternatives as complements, and the role of other industries in this entire revolution.

The first section of this chapter provides an in-depth look at wind energy, its applications, its strengths and weaknesses, its value and potential when compared to other sources of energy, and the possibilities of counting on it as a primary source to power the energy needs of millions of people. The main sections cover current technologies, like wind turbines and onshore/offshore installations,

DOI: 10.1201/9781003282273-7

FIGURE 7.1 From the early stages of wind energy exploitation to the outbreak of California.

Source: Kaldellis and Zafirakis (2011).

as well as the plans to expand their reach across countries. Meanwhile, later sections deal with the main challenges for the oil and gas industry vis-à-vis the energy transition process and how it can help raise awareness on the current decarbonization efforts from most of its major players.

The later sections of the chapter deal with the new challenges and opportunities brought up by alliances between wind companies, oil companies, governments, and supranational organizations. Likewise, new technologies that apply the same concepts of wind turbines but optimize land or materials shortages are explored, with the expectation that innovations in these areas can help export these methods to more regions of the world. Finally, conclusions raise awareness for the need to move further on the joint goal for a cleaner environment for future generations and the realization that a new energy paradigm's best weapon may come from a familiar source: the power of the wind.

7.2 PHYSICS AND PRINCIPLES OF SCREENING AND SELECTION OF WIND ENERGY

The potential of wind energy is, at least on paper, limitless. This source of energy is inexhaustible, which means that, as long as the Earth continues to behave as it has for most of its existence, wind will continue to blow. The challenge for researchers is, however, to find improved ways to harvest and to utilize this energy efficiently. Wind and solar have the advantage of being clean and renewable sources of energy, but there are still many challenges ahead, like climatic changes impacting alterations over time in wind resources.

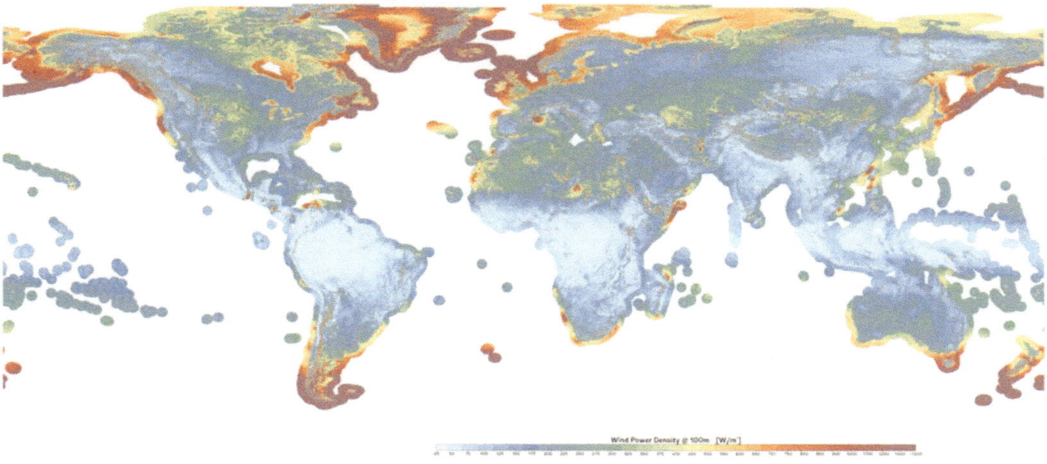

FIGURE 7.2 Wind power density potential.

Source: The World Bank (2019).

One of the first questions concerns the screening of potential locations for wind resources. Not every place in the world has an equal exposure to wind resources; therefore, the placement of facilities is a task that requires research and planning and a degree of involvement from all stakeholders. This level of preparedness is done to maximize the degree of efficiency, because the location of a wind facility by itself can vary greatly. Because of this, one of the first considerations to have in mind is the degree of wind power density, a measure that the World Bank and other organizations have measured in the past few decades.

Figure 7.2 shows a map of the world and its potential for wind power density, or the quantity of wind available at a given location (Wang et al., 2021). As the image shows, the areas close to the poles generally have a greater wind density, and there is some potential in monsoon climate zones as well. Similarly, some areas appear perfectly suited for offshore installations. To calculate the wind power density of a location, wind speed and mean wind power are required metrics which can be obtained from 12 months of clean wind data collected from a wind mast, which is called a measurement tower or a measurement mast, also known as meteorological tower or meteorological mast. The number of required wind mast depends on site terrain. A flatter terrain requires fewer number of wind mast. To predict long-term wind power density, various statistical computational techniques can be utilized. For example, Mohammadi et al. (2015) relied on extreme learning machine (ELM) techniques that build on those two metrics and can predict wind power density accurately.

Relatedly, Jeon and Taylor (2012) employed conditional kernel density (CKD) estimations, which enable the density estimation of a variable conditional to the value of one or more explanatory variables. Because the dependency of wind power on wind speed is non-linear and stochastic (thus distributed mostly randomly), CKD adapts smoothly to estimations of wind power densities that are conditional on wind speed values, but also on contextual variables like temperature, air pressure, and others, as well as on other random and uncertain elements which sometimes are not being measured. These latent variables may become known factors in the future, but it is key that current models include space for unknowns that could be correlated to geography, temperature, or other natural aspects.

Then, screening processes can be performed with the help of hierarchical models that include a large set of variables. Wang et al. (2021) examined the case of Vietnam and developed a two-stage fuzzy numbers process based on data envelopment analysis (DEA), fuzzy analytic hierarchy process (FAHP), and fuzzy weighted aggregated sum-product assessment (FWASPAS) techniques. The

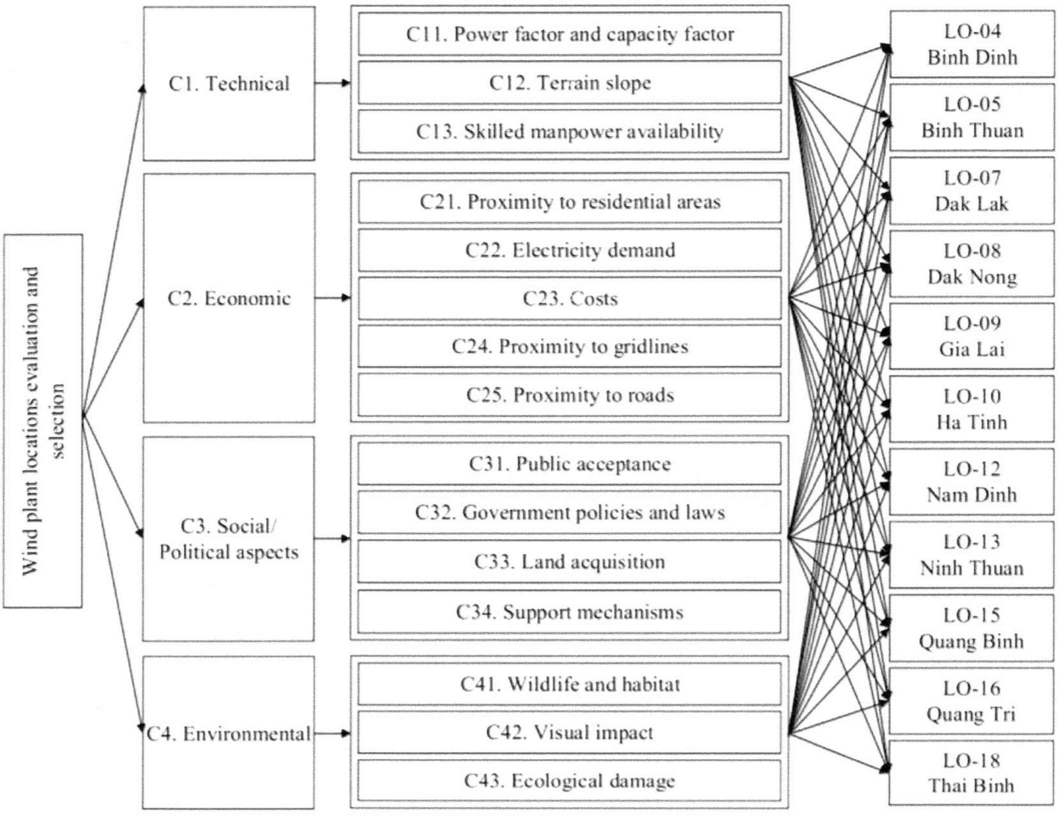

FIGURE 7.3 The decision hierarchy tree for selecting wind plant locations.

Source: Wang et al. (2021).

authors found that, out of the 20 potential provinces in Vietnam, the Binh Thuan province obtained the highest rank in their hierarchical modeling for sustainability in placement of wind plants.

Figure 7.3 provides an illustration of the decision hierarchy tree used to rank the last 11 provinces that made it to the second stage of the selection process. Each of those categories produced a coefficient for each province, and Bihn Thuan was placed first, especially after considering the criterion of natural disaster occurrences, a somewhat novel factor in the literature.

Likewise, the estimation for wind energy potential has been aided by different types of software. Mentis et al. (2015) developed a framework to quantitatively determine the wind power potential in Africa, using a geographic information systems (GIS) software. This continent not only requires the power of wind to complement their energy grid, but some areas also need to harness this power to build their own systems and achieve energy independence. Figure 7.4 shows the authors' framework, which consists of a few main parameters, like annual and daily wind speeds, elevation and slope, and distribution of areas, like their protected status, agricultural zones, or the presence of a population center. The results show that the largest potential for wind plants is in both extremes of the continent, in the north and in the south of Africa. Particularly, Algeria, Egypt, Libya, Niger, South Africa, and Sudan are the countries with a wind potential of over 5,000 TWh/year.

Meanwhile, a similar framework was applied by Atici et al. (2015) in the Balıkesir and Çanakkale areas of western Turkey. Integrating GIS-powered techniques with multiple-criteria decision

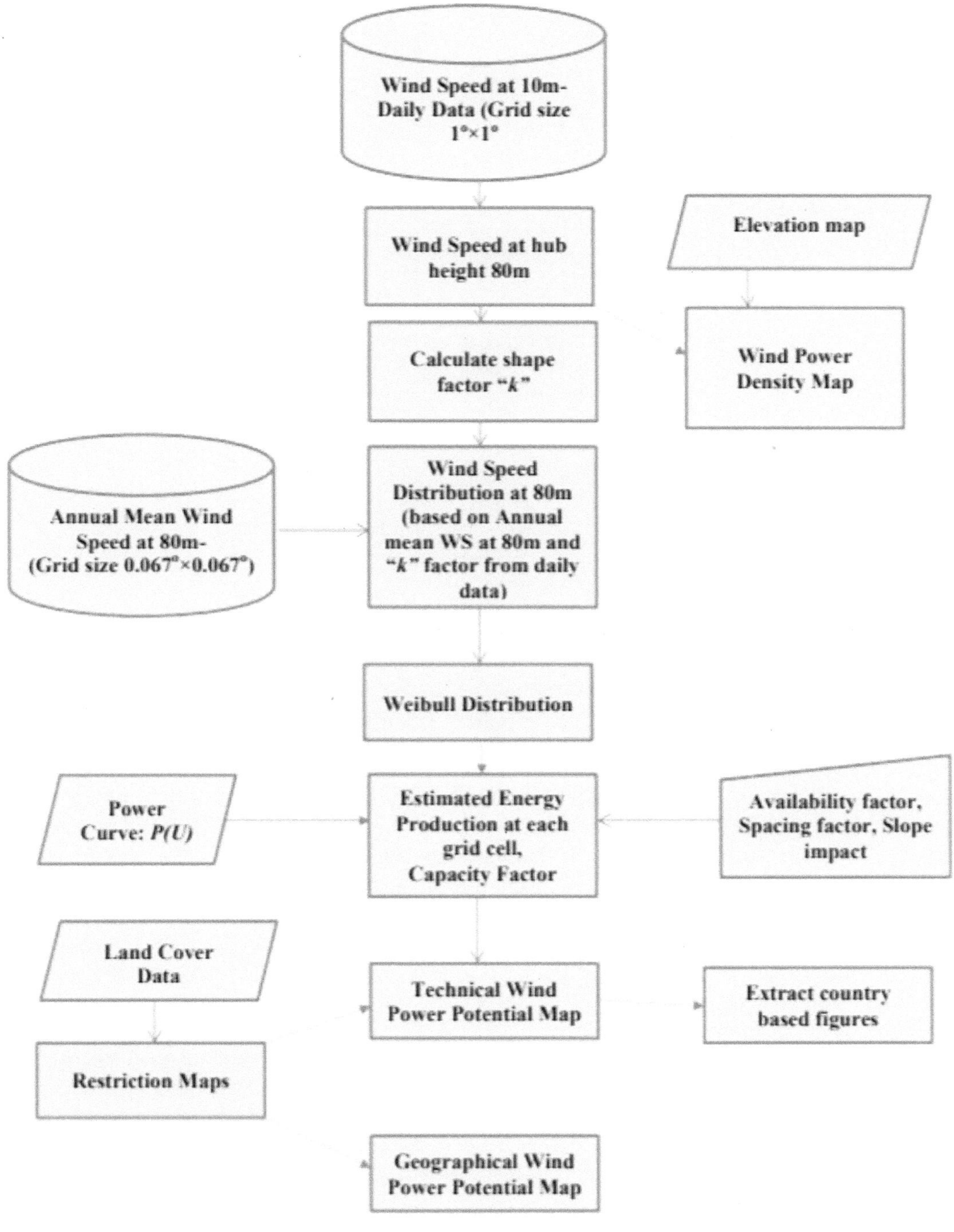

FIGURE 7.4 Flowchart methodology for a GIS-based approach to determining wind energy potential in Africa. (Color convention: light orange is "data," light blue is "process/analysis," and green is "results/products of the assessment.)

Source: Mentis et al. (2015).

FIGURE 7.5 Alternative fields for wind sites in western Turkey.

Source: Atici et al. (2015).

analysis allowed them to find the best areas for potential wind sites, seen in Figure 7.5. These areas are suggested by the analysis after eliminating infeasible and greatly populated areas. In the same setting, Derse and Yılmaz (2021) utilized a Pareto diagram that yields a list with the set of geographical locations with the best potential for wind energy. This technique is based on Vilfredo Pareto's 80/20 principle, which suggests that for many outcomes, 80% of the output is generated from 20% of the inputs. Thus, the authors determined which of the 13 Turkish provinces account for close to 80% of the wind potential for the country. Later, an autoregressive integrated moving average (ARIMA) model and a time series analysis were performed. Results show that, given the installation decisions, the provinces of Balıkesir, Çanakkale, and İzmir regions are the ones with the highest wind potential.

Moreover, in what concerns the decision-making process of offshore locations, Abdel-Basset et al. (2020) offered light into the subject. The success of an offshore installation depends almost entirely on the selection of the optimal offshore wind power station (OWPS), a process that is contingent upon a process of multi-criteria decision-making (MCDM). However, this process has its own limitations, like loss of data and incomplete use of the collected information, which the authors try to alleviate using a hybrid methodology that combines an analytical hierarchy process (AHP) method and a preference ranking organization method for enrichment evaluations (PROMETHEE) II technique, which includes the input from specialists to cover that information loss, and to ground the quantitative results better.

Another method for the estimation of sites for wind installations is provided by Onar et al. (2015). The authors followed the fuzzy framework and developed an interval-valued intuitionistic fuzzy (IVIF) approach, which can analyze large numbers of infrastructure and technology criteria brought by multi-expert evaluations. These assessments are put in pairs and distributed in matrices that yield a ranked preference for one site or another based on the positive or negative rankings for

TABLE 7.1
Wind Energy Technology Selection Criteria

Criteria
Reliability
Supplier location
Warranty issues
Domesticity
Production statistics and track record
Suitability for environmental conditions
Production volume
Technical characteristics
Hub height
Rotor diameter
Cut-out speed
Cut-in speed
Wind turbine performance
Power curve (power produced by the turbine at the given pole height)
Unit production price
Cost factors
Initial investment (price of the wind turbine)
Installation costs
Operating costs
Connection costs
Availability factors
Availability of desired turbine type
Spare part availability
On-time delivery of the turbines
Maintenance service
Cooperation capability

Source: Onar et al. (2015).

each initial criterion. A tentative list of the many criteria for wind energy technology selection can be seen in Table 7.1.

7.3 TECHNO-ECONOMIC FEASIBILITY OF WIND ENERGY

As noted, a calculation for the feasibility of wind power plants should include a multi-factorial analysis, where the potential for wind energy is one of the main geographical explaining variables, but other distinctions regarding infrastructure and technical issues must be considered as well. Thus, grid network capacity, availability of materials, installation and operational costs, maintenance service, general outputs, and even the location of the suppliers can become relevant difference-makers in the process of technology selection for wind power installations. All this opens a second branch of variables that may affect the decisions from governments and stakeholders to decide to invest on wind energy. For example, one country with a strong wind potential profile could perform analyses of this type and conclude that the investment will not be as attractive life cycle cost as another country with a similar profile does move forward with investment on wind energy. In the end, a multiplicity of factors is what determines these decisions, so outcomes should be weighted under those merits as well.

In another branch of the decision-making process for wind technologies, the design of parts and systems to place in wind farms and plants is considered. Environmental aerodynamics has a vital role in the improvement and efficient delivery of products designed to contribute to the harnessing of the power of the wind. Schaumann et al. (2011) reviewed the design of wind energy converters in detail, analyzing several aspects in the development and construction phase. The authors considered the distinct features of onshore wind structures, such as the diameter of the tower tubes, the potential for fatigue of the materials, the degree of functionality of the structure, and even the type and number of bolts placed in a row inside the tubular structures.

All these design and efficiency features are closely linked to the next key factor in the development of wind plants and the proliferation of wind technologies in a jurisdiction, which is the economic dimension. This aspect considers investments, costs, placement, and outputs, among other features. Blanco (2009) developed a comprehensive list of the key parameters that rule wind power costs, mentioning *capital costs*, which include wind turbines, foundations, road construction, and grid connection. Then, the author mentioned *variable costs*, where operation and maintenance are the key components, including service and spare parts, insurance, and land rent, among others. Later, the *discount rate* and the *economic lifetime of the investment*, which reflect the perceived risk of the project, are also a worthwhile consideration. Finally, the *capacity factor* expresses the percentage of time that a wind energy facility produces electricity each year.

Figure 7.6 shows an illustration of the many components of a wind turbine, demonstrating why the capital costs can rise to 80% of the total costs of a wind energy endeavor. As depicted, only the tower and the rotor blades can constitute half of the entire cost of a wind turbine, with the power

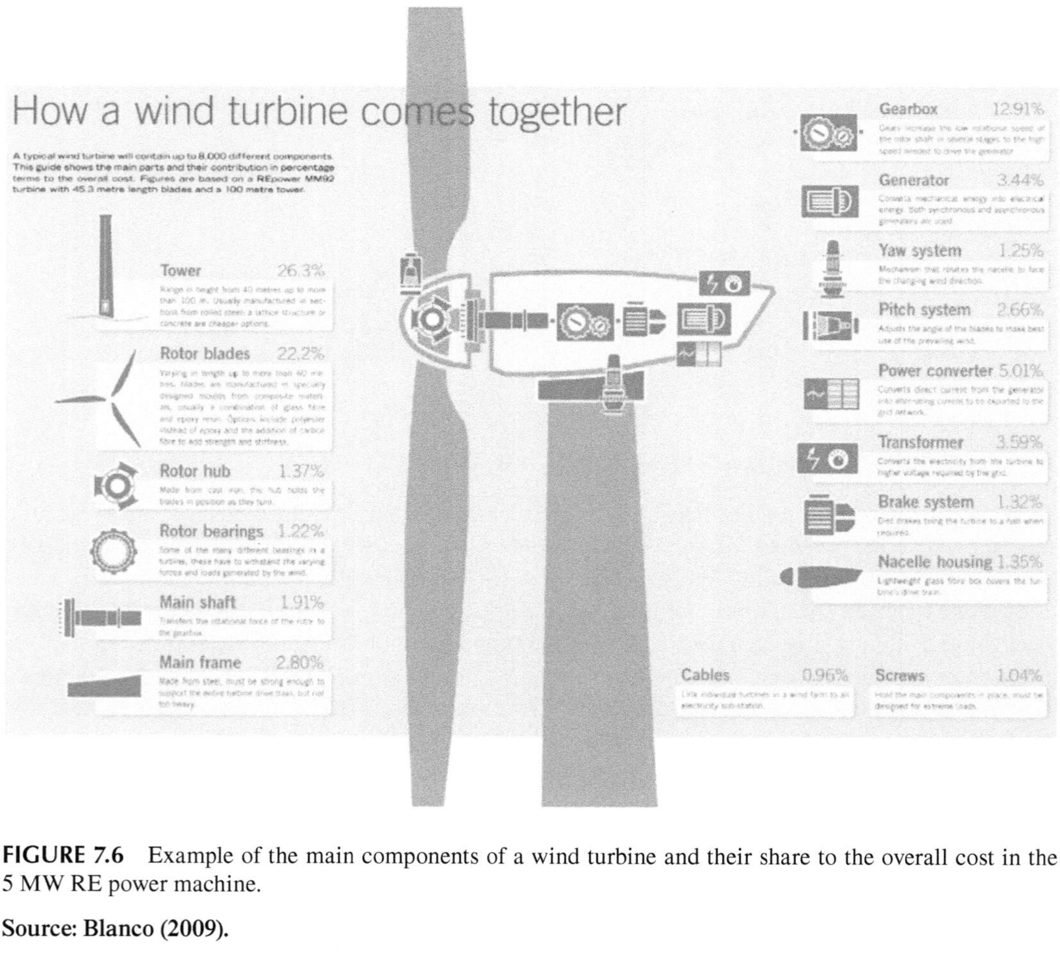

FIGURE 7.6 Example of the main components of a wind turbine and their share to the overall cost in the 5 MW RE power machine.

Source: Blanco (2009).

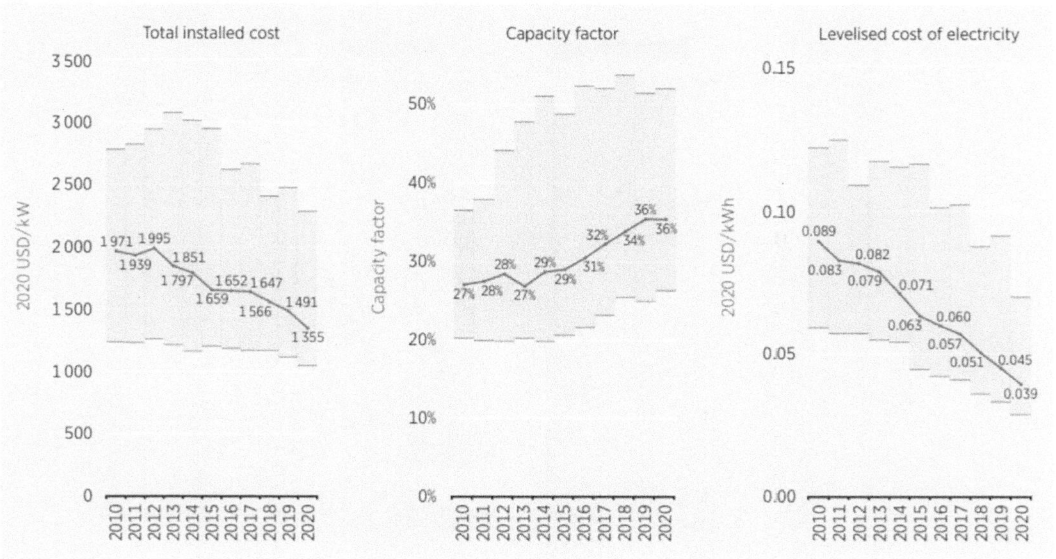

FIGURE 7.7 Global weighted average of total installed costs, capacity factors, and LCOE for onshore wind, 2010–2020.

Source: IRENA (2021).

converter and the transformer being the two next more costly parts (Blanco, 2009). Additionally, the author notes that the generation costs of wind energy have increased significantly, something caused in part by rising prices of important raw materials and a surge in the demand for wind turbines that followed the approval of policies in favor of wind energy in countries like the United States, China, and European Union member states. For offshore development, support structures cost may change the cost breakdown ratio.

Figure 7.7 shows these trends in more detail, where we can note that the past decade has brought a steady decrease in both total installed costs and levelized cost of electricity (LCOE), with a parallel increase in the actual energy output from wind plants when contrasted to their maximum possible electrical output (IRENA, 2021).

To conclude, both the development and as well as the success of wind methods and technologies depend on many factors, which touch on technical aspects, as well as on varied infrastructure, location, and economic factors. From the very basic consideration of building a wind power plant in a place where more wind blows so that efficiency and performance improve, to the more complex set of rules and procedures that govern emerging markets, which help explain price variations for materials and incentives for the varying degrees of investment from private actors or governments, a large set of variables helps determine the decisions made regarding the selection, screening, and design of wind-based technologies.

7.4 WIND ENERGY APPLICATIONS, DESIGN PRINCIPLES, AND BEST PRACTICES

This section explores some examples of wind energy facilities and their outcomes in terms of economics and the many lessons learned from varying applications across the world. We analyze how practical differences can influence efficiency or performance of a wind installation, and which are the steps to replicate and which ones to avoid.

FIGURE 7.8 Two main types of wind turbines: horizontal-axis (left) and vertical-axis (right).

Source: Office of Energy Efficiency and Renewable Energy (2021).

7.4.1 Wind Turbine Design Principles, Types, and Capacities

A *wind turbine* is a structure that converts the kinetic energy of the wind into electrical energy (Khudri Johari et al., 2018). These turbines work when the wind turns its blades around a rotor, which spins the generator that creates electricity (Office of Energy Efficiency & Renewable Energy, 2021). The preferred kind of turbine is determined due to its greater degree of efficiency when extracting energy, particularly because of a larger energy generation after a full rotation of the blades. However, their main disadvantage is that they need to be facing the wind to be most efficient, something which is not always achieved.

In a broad sense, most of the wind turbines fall into two basic types: horizontal-axis wind turbines (HAWT) or vertical-axis wind turbines (VAWT). Vertical-axis wind turbines (VAWT) are those with a rotating surface that is perpendicular to the ground, which have two main types, *Savonius* and *Darrieus*. The former uses drag forces, producing electricity in the same way that a waterwheel works, while the former uses aerodynamic blades to generate lift that turns the turbine. As noted, the main disadvantage of HAWTs is one of the VAWTs' greatest strengths. Thus, omnidirectionality helps VAWTs generate electricity in slow and more turbulent wind environments. With that said, disadvantages are their lower degree of efficiency and their vulnerability to backtracking due to the blades moving in the same direction of the wind (Khudri Johari et al., 2018). Figure 7.8 shows an illustration of both horizontal-axis and vertical-axis turbines.

Meanwhile, a more detailed look at the distinct types of wind turbines is performed by Hansen et al. (2004), who classify these devices according to their speed control and power control, generating a more complex grid of varying wind turbines with different specifications. Along the first dimension, there are fixed-speed and variable-speed wind turbines, while for the second we see stall (or passive) control, pitch control, or active control wind turbines. Different combinations of specifications can be used to build different wind turbines, which will also show different results over time in different regions of the world.

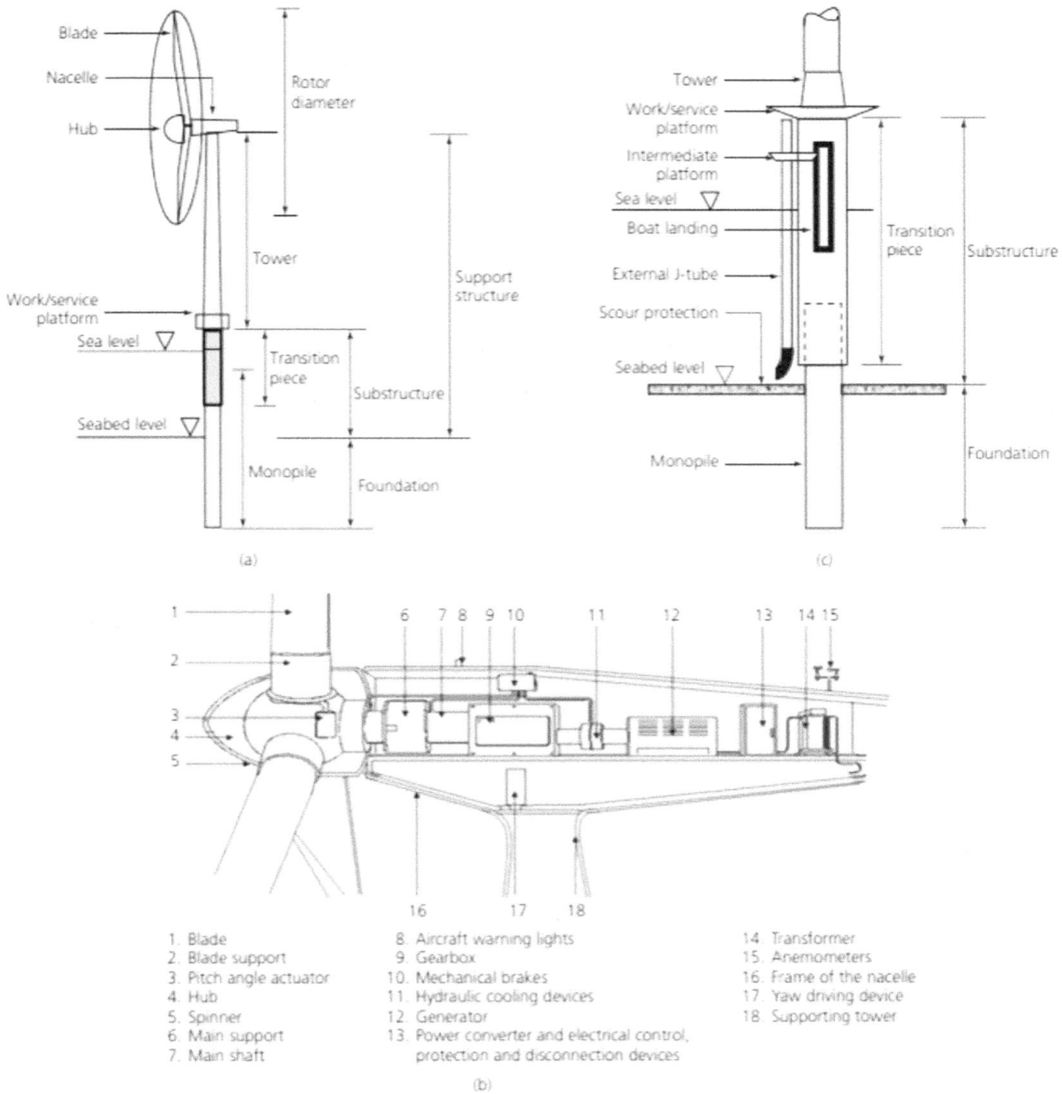

FIGURE 7.9 Major components of offshore wind turbine (OWT) systems: (a) wind turbine system; (b) electromechanical parts; (c) details of monopile and transition piece.

Source: Arshad and O'Kelly (2013).

In order to achieve high-power performance from wind towers, Haraguchi et al. (2009) developed four different rotor structures for large-scale wind power generators. The difference in torque, power, and efficiency of these structures is another key factor that researchers and engineers ought to consider when choosing the designs for their wind energy facilities. On that note, the work of Arshad and O'Kelly (2013) investigates the potential for offshore wind turbines and structures. Figure 7.9 shows the most important components of an offshore wind turbine (OWT), where we see the many electromechanical parts as well as the tower, rotor blades, and nacelle (or hub).

On a similar vein, Zhang and Ula (2008) analyzed three different types of wind turbines (WT) in terms of their performance, cost, and efficiency. Considering the main types of WT utilized in the two decades prior to their research, these three categories are (a) wind turbines with induction

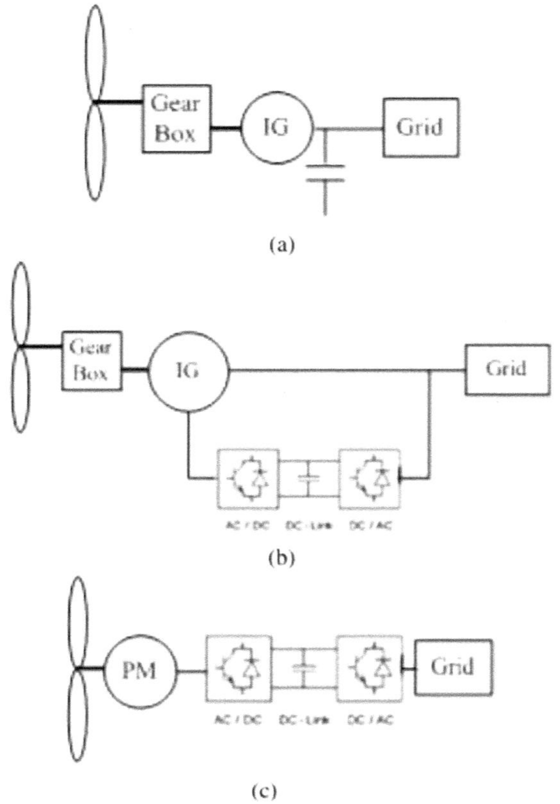

FIGURE 7.10 General structures of three different types of wind turbines (WT): (a) fixed-speed WT with induction generator (IG); (b) variable-speed WT with a doubly fed IG; (c) variable-speed WT with a permanent magnet synchronous generator (PM).

Source: Zhang and Ula (2008).

generators, (b) wind turbines with double-fed induction generators, and (c) wind turbines with direct drive. The first kind of turbine is used in small wind farms, while the second type is the more widespread due to their lower initial capital, despite the third type being the wind turbine that can support the grid voltage better than all of them. Meanwhile, maintenance costs are relatively similar, but turbines that have a gearbox (and thus more smaller moving parts) are expected to incur higher costs. Figure 7.10 shows a schematic of the three types of WT analyzed.

Meanwhile, Pfaffer et al. (2017) developed a comprehensive analysis of the performance and reliability of wind turbines according to a series of features that differentiate them. Using six different WT characteristics, the authors reviewed a series of wind initiatives where information from different countries and wind projects was collected, to compare these features and the performance provided by varying combinations of them. The six features are the capacity factor, time-based availability, technical ability, energetic availability, failure rate, and mean downtime of wind turbines.

The potential for a comparative work and for strong recommendations for certain features over others is very limited due to the difficulty that the geographical context adds to the whole endeavor. For example, Lu et al. (2002) found that the hub height of a wind turbine is a significant factor in its power generation. Then, Hahn et al. (2007) highlighted the importance of service teams that respond quickly to turbine failures as another factor that can improve the reliability of

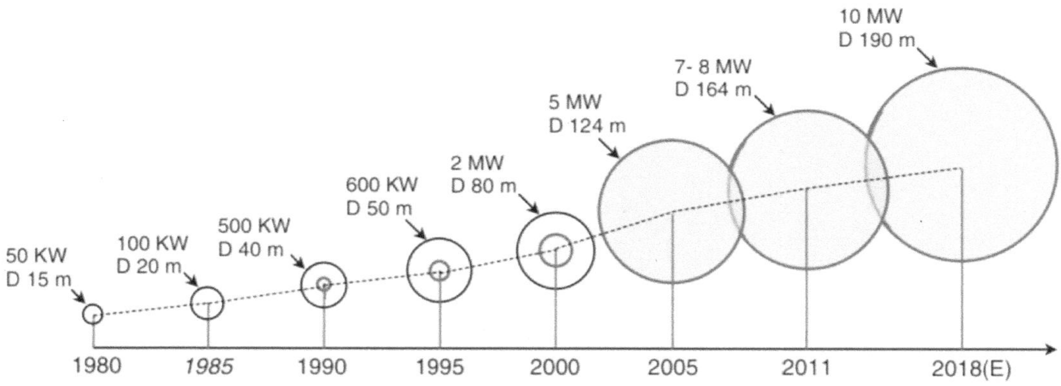

FIGURE 7.11 Evolutions of wind turbine dimension and the corresponding capacity coverage by power electronics converters seen from 1980 to 2018 (estimated).

Source: Errami et al. (2014).

WTs. Furthermore, Buckney et al. (2013) focused on the blades of a turbine to assess which factors could help in their improvement, concluding that alternative configurations, like a different pitch angle of a blade, could help diminish the weight and improve their design, which, in turn, reduces the cost of wind energy by requiring less materials (for a similar study, also see Schubel and Crossley, 2012).

Figure 7.11 explores the evolution in the dimension of wind turbines, as well as the corresponding capacity coverage by evolving converters. At the beginning of the 1980s, the diameter of a wind turbine could reach 15 m, while the corresponding wind energy conversion system (WECS) could be up to 50 kW. Thirty years later, advancements have allowed turbines to have diameters of 164 m, which translate into a tenfold increase, while having converters with a capacity of 7–8 mW. For example, rotor diameter has increased from less than 50 m in 1985 to almost 250 m in 2020, with a rated power capacity (RPD) that is expected to grow by almost 20 times, with estimations placing it at 20 MW (Arshad and O'Kelly, 2013).

This change is more than a hundredfold in the potential power for a power plant to produce electricity (Errami et al., 2014) and shows how maximizing the structural efficiency of wind turbine blades constitutes another dimension in the success or failure of the design of these structures. The main driver for large-capacity wind turbines is to generate the lowest cost of energy by reducing installation cost.

Meanwhile, in addition to being less easily accessible, offshore converters must account for harsher weather conditions, such as strong wave loads, salt, humidity, and irregular sea states. Thus, offshore wind energy converters (OWECs) need materials that can withstand these phenomena, providing a higher structural safety and reliability. Of particular importance is the protection of materials from corrosion, where surface preparation and multiple coatings are two important criteria (Schaumann et al., 2011). Moreover, substructures must be added to attach the towers to steady ground. The most common support structure is the monopile, a steel pile that is drilled up to 40 m into the seabed.

Figure 7.12 shows other widespread types of offshore support structures. The design of both onshore and offshore structures is subject to a continuous scrutiny by the conditions to which they will be subject, as well as by the performance that is expected from them. For example, Makarios and Baniotopoulos (2012) investigated methods and models that are applied to circumstances where wind towers are subject to strong dynamic loading (such as wind or other similar conditions) while being affected by a strong seismic motion.

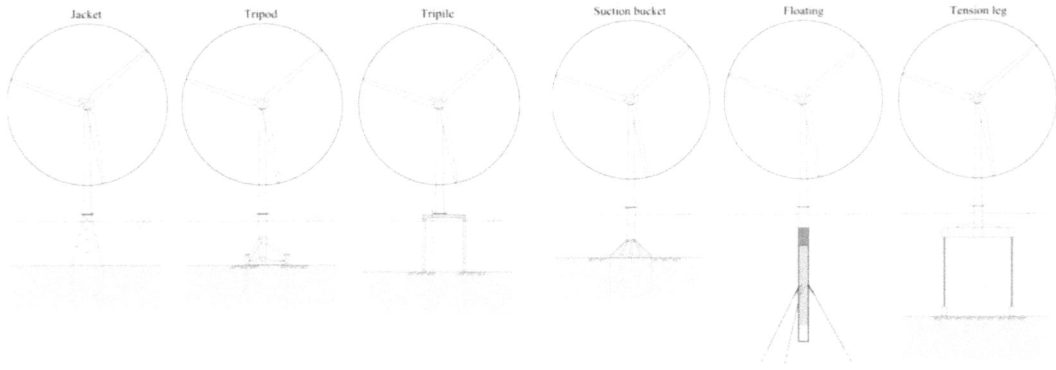

FIGURE 7.12 Support structures of offshore wind energy converters.

Source: Schaumann et al. (2011).

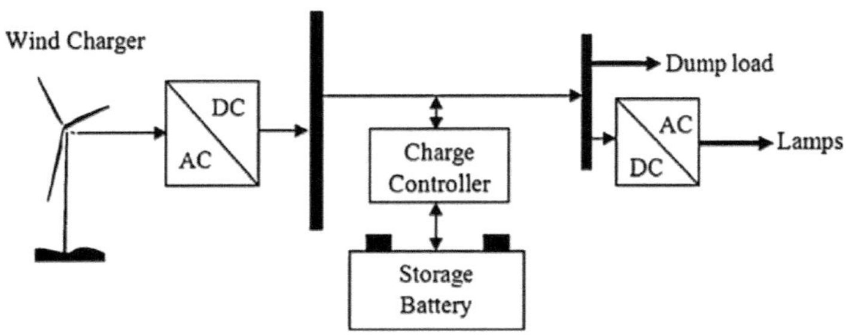

FIGURE 7.13 Typical wind charger components.

Source: Khatib et al. (2012).

7.4.2 Wind Chargers

A smaller subset of wind installations, *wind chargers* are structures that can simplify a power system while delivering continuous electricity with a greater degree of reliance (Chen et al., 2014). These chargers can help remote locations without access to the electricity grid with small tasks like night lighting or house electrification (Khatib et al., 2012) or aid with the storage of excess energy for later use. Additionally, the efficiency of the energy capture from a wind turbine can be improved with a battery charging system (Lo et al., 2011), while also overcoming high voltage transfer ratio problems.

Figure 7.13 shows a schematic of a typical wind charger. These consist of a wind generator, a storage battery, converters (both DC–DC and AC–DC), and loads like batteries. Therefore, chargers are basically an extension to a wind power structure that is used to store, convert, and distribute energy at different moments, either when wind is not present or when energy demand is greater than what plants can generate at one moment.

In some cases, grid-independent wind facilities allow for the possibility of a hybrid power generation system, shared with solar photovoltaic (PV) technologies and with fuel cell technologies, to increase the reliability of these installations and to achieve the real penetration of renewable energy into the daily lives of millions of people (Ekren and Ekren, 2010; Kaabeche et al., 2011).

Figure 7.14 shows a system where solar and wind methods share the task to power a small endeavor. But not only rural or remote locations can benefit from these installations, as pivotal systems can work with this hybrid stand-alone technologies, like water pumping and telecommunication. Another example is seen in the work of Bendary and Ismail (2019), who simulated a hybrid

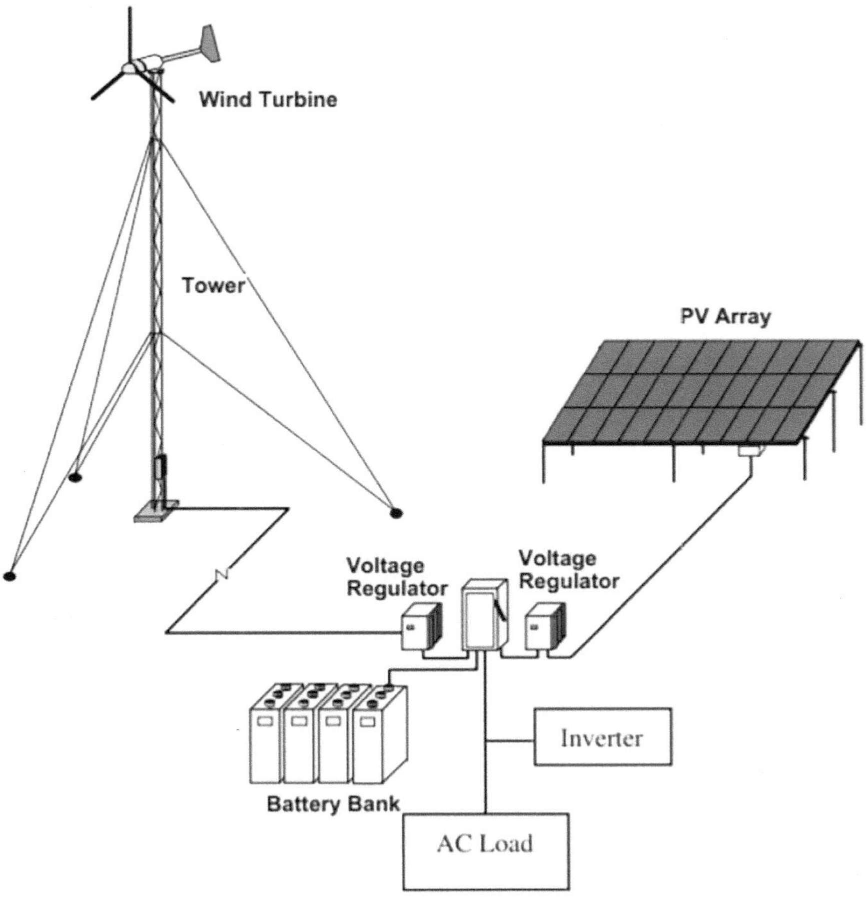

FIGURE 7.14 Schematic diagram of a basic hybrid PV/wind energy system.

Source: Ekren and Ekren (2010).

system that includes PV, wind, and fuel cell technologies and relies on artificial intelligence (AI) techniques to provide an enhanced battery charging performance.

7.4.3 Offshore and Onshore Wind Farms

Part of this discrepancy between developed and developing countries can be pointed at the cost of building and maintaining offshore wind plants. Currently, offshore technology is further from the reach of emerging countries due to its prohibitive costs compared to onshore installations. While electricity from onshore wind plants can very closely compete with the cheapest fossil fuel–based electricity production, offshore facilities still need a significant reduction in its costs to be competitive (Junginger et al., 2004). The differences in cost are attributed to greater investment costs for offshore plants, mainly in the foundation of the turbines, the maintenance of electrical connections, and the installation and operation expenditures (Klinge Jacobsen et al., 2019). Table 7.2 offers a list of relevant factors behind potential cost reductions for offshore wind farms.

On the other hand, offshore wind farms offer the greatest potential of electricity generation when compared to onshore locations. Due to the more constant wind speeds at sea and less resistance from the public to their construction, the potential for offshore wind plants is much greater than for

TABLE 7.2
Overview of Relevant Factors behind Cost Reductions of Offshore Wind Farms

	Specific Offshore Wind Developments	Exogenous Developments
Wind turbine	Upscaling Improved design Standardization Economies of scale	Further development of onshore turbines Steel price
Grid connection	Standardizing the design of HVDC cables Applicability of XLPE insulation to HVDC cables Advances in valve technology and power electronics	Further development and diffusion of submarine HVDC interconnectors
Foundations	Standardization Economies of scale Design regarding dynamic loads	Steel prices
Installation	Learning-by-doing Development and structural purpose-built ships Standardization of turbines and equipment	(Oil prices)

Source: Junginger et al. (2004).

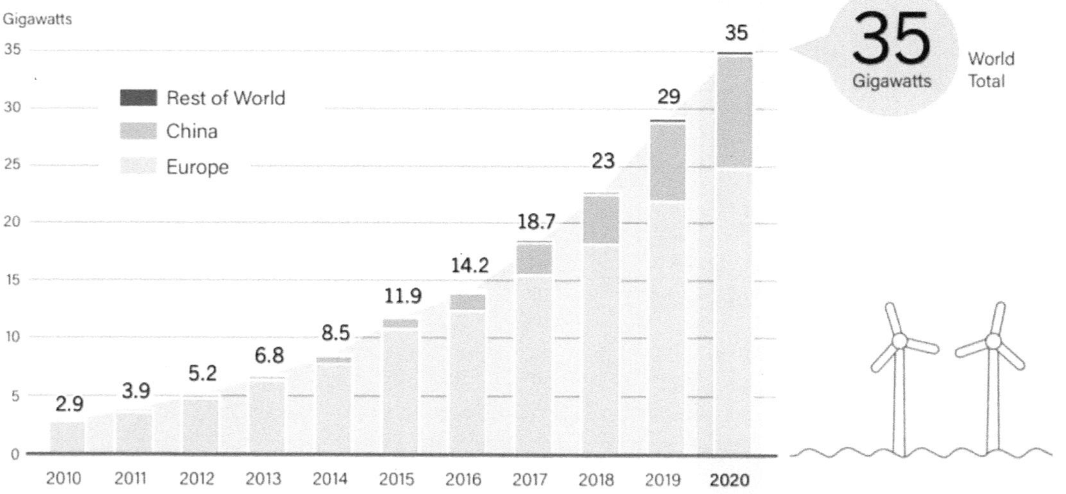

FIGURE 7.15 Wind power offshore global capacity by region, 2010–2020.

Source: REN21 (2021).

onshore farms (Leung and Yang, 2012). According to the location of the field, wind farms can be classified into one of these two categories. In their 2021 Global Status Report, the think tank REN21 estimated the global wind capacity as more than 740 GW for 2020, with its biggest jump from one year to the next experienced that year, moving from 650 GW in 2019. Of those, only 35 GW are produced in offshore facilities, with Europe producing most of it (see Figure 7.15).

The world's first offshore wind farm was the Vindeby plant in Denmark, built in 1991 and decommissioned in 2017, after 25 years of service (Feng and Shen, 2017). According to Díaz and Guedes Soares (2020), there are currently 112 offshore wind farms in operation, while there are 712 projects in different development stages, and 53 more are under construction or close to start. Figure 7.16 shows the global distribution of offshore wind farm by country, and it is still glaringly evident

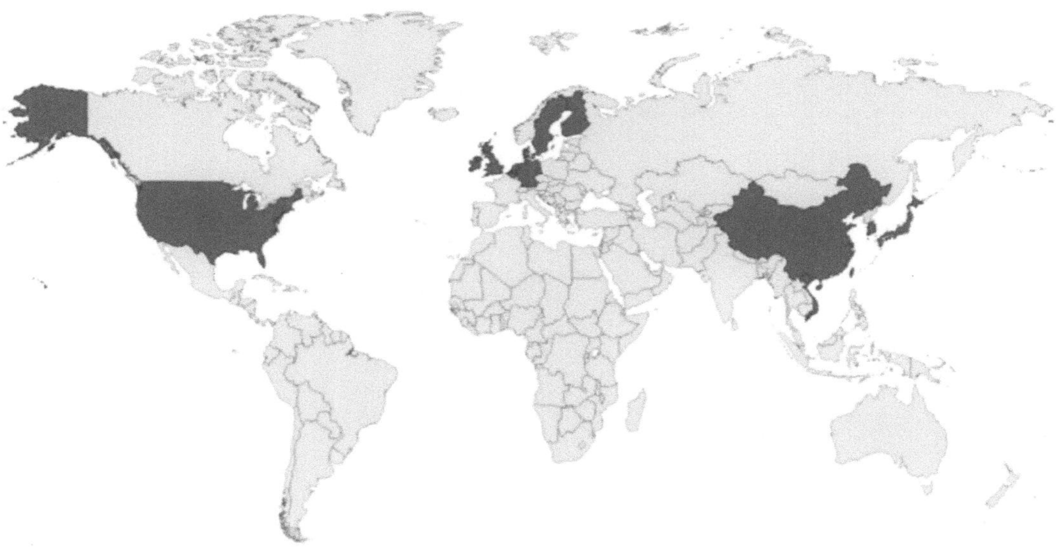

FIGURE 7.16 Global distribution of offshore wind farms by country.

Source: Díaz and Guedes Soares (2020).

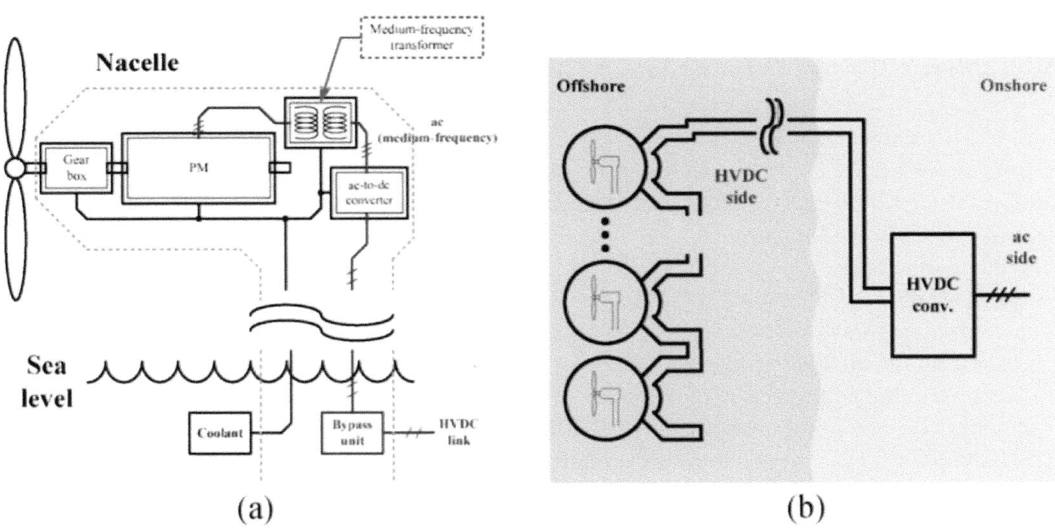

FIGURE 7.17 Proposed wind farm topology with (a) a windmill unit with a permanent magnet (PM) generator and a medium-frequency (MF) transformer and (b) an offshore wind farm using a series of shunt connected turbines.

Source: Prasai et al. (2008).

that developed countries hold most offshore projects, with the United States, United Kingdom, Germany, and China leading the way.

To deal with cost and efficiency issues, some more novel forms of offshore installations have been proposed by different authors. For example, Prasai et al. (2008) proposed a system that uses permanent magnet machines with a high pole number as the primary generator of energy, illustrated in Figure 7.17. These designs tend to feature a high power density while being more reliable under more pressure to perform. Meanwhile, Feng and Shen (2017) experimented with different sizes

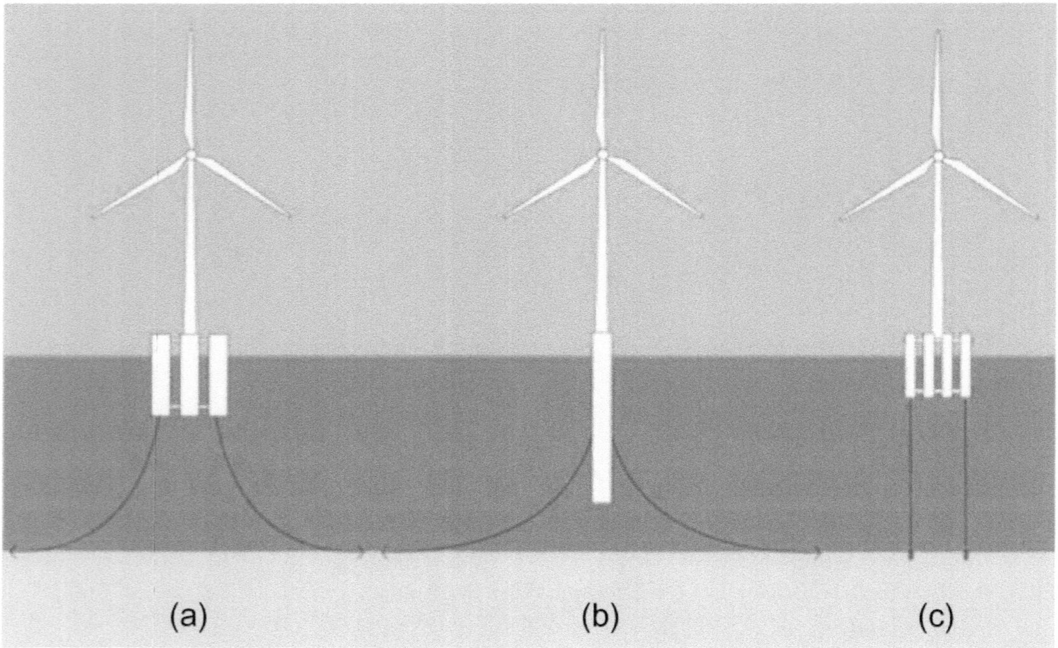

FIGURE 7.18 Three floating wind turbine platform concepts: (a) semi-submersible, (b) spar buoy, (c) tension leg (TLP).

Source: Ferreño González and Diaz-Casas, in Castro-Santos and Diaz-Casas (2016).

of offshore wind farms, trying to find an optimal configuration for distinct settings. Their non-uniform design framework includes considerations like wind speed, wake effects (which translate into spacing between turbines and farms), and capital to delineate offshore projects with the greatest likelihood of success.

Then, floating offshore wind structure development has gained momentum in the past few years. These structures have been developed to satisfy distinct efficiency needs along with technology transfer from the oil and gas industry and the use and exploitation of deepwater soil. In fact, a contribution from this industry is seen in the use of component technologies like anchors and mooring.

Figure 7.18 shows an illustration of three different concepts for floating offshore wind turbines. The first concept has a semi-submersible platform with a distributed buoyancy that is anchored to the seabed with catenary (or a free-hanging line) mooring lines. The second concept consists of a cylindrical ballast-stabilized structure which obtains its stability for having its center of gravity lower in the water. Also known as spar buoy platforms, these structures are easier to build but may have difficulties when transporting and in more shallow waters. Lastly, tension leg platforms (TLP) consist of a straight mooring line which allows for a smaller and lighter structure (Ferreño González and Diaz-Casas, in Castro-Santos and Diaz-Casas, 2016).

Moreover, Table 7.3 shows the main advantages and disadvantages of these three types of floating wind platforms. As expected, these three designs adapt better to different weather and depth conditions, placing important trade-offs to the production of electricity from the energy of the wind. With that said, floating structures have become yet another promising avenue for the generation of large quantities of low-carbon electricity (Bento and Fontes, 2019).

In another set of innovations, the work of Klinge Jacobsen et al. (2019) explores the development of an option in between onshore and offshore: nearshore wind farms. These are defined by the authors as turbines that are up to 15 km off the coast, a very attractive option in Denmark, the

TABLE 7.3
Advantages and Disadvantages of the Three Types of Floating Wind Platforms

Type	Advantages	Disadvantages
Spar buoy, ballast-stabilized	• Simple design is amenable to serial fabrication processes • Few moving parts (no active ballast required) • Excellent stability	• Constrained to deepwater locations • Offshore turbine assembly requires dynamic positioning vessels and heavy lift cranes • Large draft limits ability to tow the structure back to port for major repairs
Tension leg platform (TLP), mooring line–stabilized	• Low structural mass • Onshore turbine assembly • Few moving parts (no active ballast required) • Excellent stability	• High loads on the mooring-and-anchoring system • Challenging installation process • Bespoke installation barge often required
Semi-submersible platform, buoyancy-stabilized	• Flexible application due to the ability to operate in shallow water depths • Low vessel requirement – only basic tugboats required • Onshore turbine assembly • Amenable to port-side major repairs	• High structural mass to provide sufficient buoyancy and stability • Complex steel structures with many welded joints can be difficult to fabricate • Potentially costly active ballast systems

Source: Ferreño González and Diaz-Casas, in Castro-Santos and Diaz-Casas (2016).

TABLE 7.4
Recommendations for the Analysis of the Potential for Development of Nearshore Wind in Denmark

Category of Issues/ Cost Drivers	Overall Potential Screening	Evaluation of Cost Advantage versus far Offshore	Cost Ranking of Nearshore Sites	Studies Recommended
Nearshore site characteristics (depth, seabed conditions)	Identify if nearshore sites with low water depth (5–15 m) are available. Check that challenging seabed conditions are not present.	Depth dropping to levels below 25 m already at a distance of 15 km from shore indicates cost advantage. Clarify that no offshore banks with low water depth are available.	Rank nearshore sites based on basic cost criteria.	Combine with preference study results.
Connection cost and transmission reinforcements	Establish connection points and direct cost for potential nearshore.	Estimate if large offshore requires domestic transmission grid reinforcements to a larger degree than nearshore.	Rank connection cost for the nearshore sites, and combine with other costs.	Possible study of offshore versus nearshore impact on the national grid.
Planning/alternative use barriers for nearshore	Avoid recreational high-value area (tourist beaches, summer residences). Consider harbor, shipping routes, and military areas restriction.		Possibly combine ranking on basic costs with alternative value use (cost of tourist impact).	Estimate alternative use value or loss of revenue from the impact on existing activities.

(Continued)

TABLE 7.4 (*Continued*)
Recommendations for the Analysis of the Potential for Development of Nearshore Wind in Denmark

Category of Issues/ Cost Drivers	Overall Potential Screening	Evaluation of Cost Advantage versus far Offshore	Cost Ranking of Nearshore Sites	Studies Recommended
Impact on population	Identify low-population-density areas – or areas with existing infrastructure already with visual impact (no additional impact from turbines).	Evaluate if local population generally positive to wind development in the area.	Rank nearshore sites based on the impact on local population (people affected).	Carry out site-specific preference studies/ willingness-to-pay studies for the highest-ranked (cost-attractive) nearshore sites. Combine with other costs.
National policy priorities, for example, domestic developer concerns	Clarify what national policy priorities exist and the expected impact on priorities of nearshore development in general.	Clarify if domestic developers will have easier access to bidding for smaller-scale nearshore wind compared to larger offshore farms. Evaluate the relative potential local employment effect.	No effect on the ranking of nearshore sites expected.	Possible study of domestic developer capability to bid and develop nearshore and preference for these projects.

Source: Klinge Jacobsen et al. (2019).

country where this study was performed. Nearshore wind installations mitigate some of the costs of onshore facilities and can also reap some of the benefits of offshore plants. For example, nearshore farms have a lower visual impact than onshore fields, a significant concern in the public opinion of some countries (Ladenburg, 2009; Lee et al., 2020). Conversely, the fact that these fields are closer to the shore may decrease installation costs due to a lower water depth. Ideally, if these costs are lower, more of these fields can be built, which can create a virtuous circle of electricity generation via renewables and simultaneous decreasing costs. Table 7.4 shows a list of suggestions for the development of this middle ground type of wind farm.

Related to this high-stakes process of selecting the appropriate ways to proceed with a wind farm project is the concept of wind farm design and optimization (WFDO). This concept captures all the decisions and planning actions needed to optimize the performance of a wind farm, whether it is formed by a few turbines or by thousands of them (Herbert-Acero et al., 2014). Generally, optimizing a wind farm project involves familiarizing oneself with the many trade-offs and conflicting objectives that they entail, with all the potential benefits weighed along with the costs. For example, Kusiak and Song (2010) touched on the wake loss that placing turbines in the wind trajectory of other turbines could entail on overall electricity generation. Figure 7.19 shows the potential issues of this variable when the wind flows in one direction. As noted in the figure, if the wind blows from west to east, the wake of Turbine 1 affects Turbine 2, while Turbine 4's wake affects Turbine 3. Meanwhile, Turbine 5 remains unaffected. If the wind blew in a different direction, we would see a different pattern where potential losses for other turbines could be discovered. This shows that the challenge is dynamic, and an optimization needs to account for changing conditions and scenarios.

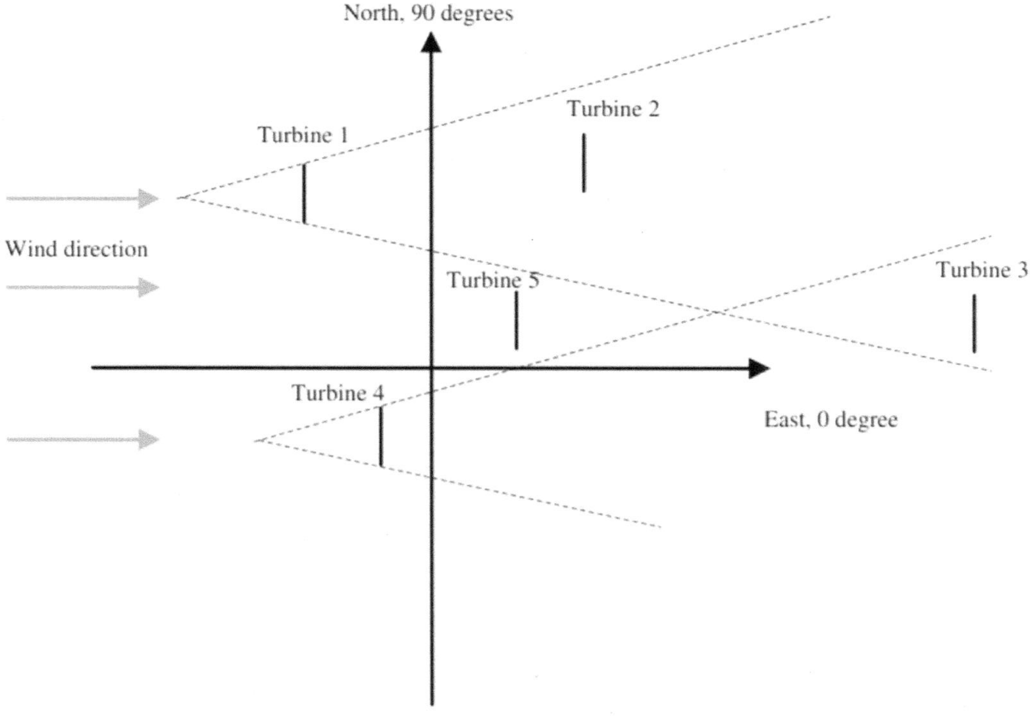

FIGURE 7.19 Turbines affected by the wake of other turbines.

Source: Kusiak and Song (2010).

TABLE 7.5
List of Known and Potential Impacts of Utility-Scale Wind Energy Development, Operation, Maintenance, and Decommissioning on Wildlife

Impacts Due to Facility Construction and Decommissioning	Impacts Due to Facility Presence, Operation, and Maintenance
Direct mortality	Habitat fragmentation
Destruction and modification of habitat	Noise
Impacts of roads	Vibration and flicker effects
Off-site impacts	Electromagnetic field generation
	Macro- and micro-climate change
	Predator attraction
	Fire

Source: Lovich and Ennen (2013).

Another important factor to consider is the potential wildlife and biodiversity effects from wind farms. Especially during the construction phase, the development of a wind farm can have a series of effects, like the generation of micro-climate changes that affect vegetation and the local wildlife. These problems can be remedied by friendly designs with tailored solutions that include, among other actions, building the power lines underground and placing turbines far enough from migration spots (Herbert-Acero et al., 2014). This way, birds, bats, and other small animals are not disturbed and confused by the heavy machinery (Davy et al., 2020). Further, staying away from protected areas has been another regular consideration in most of the literature on the subject.

Table 7.5 shows two main categories of potential impacts of wind facilities, the first related to problems when building and decommissioning wind plants, and the second related to their presence, operation, and maintenance. The first list of problems includes potential human casualties when erecting and installing wind turbines, impacts on roads, and the significant modification of a habitat that could disrupt animals as well as people. Then, another potential impact that is not felt in the location of the wind plant, but rather in poorer countries, can occur when mining materials in excess to produce a large amount of turbine parts, or when dumping used parts after their life cycle has ended.

Meanwhile, potential impacts of functional plants include turbines generating low-frequency noise that could disturb sleep, particularly in children (Hanning and Evans, 2012). This noise, normally in the range of 20 to 200 Hz, has been associated with increased risk of epilepsy, cardiovascular effects, and even coronary artery disease (Chou et al., 2021). Moreover, other types of vibrations from these large structures could generate other issues, like the potential creation of an electromagnetic field that could cause localized changes in the climate of a region or increase the hazards of fires.

Relatedly, Diógenes et al. (2020) reviewed the main barriers to the implementation of renewable energy technologies, with an emphasis on onshore wind plants. Table 7.6 shows a summary of these limitations that focus on economic, institutional, social, and technical shortcomings. Throughout this chapter, we have reviewed a series of design improvements that can be performed to obtain a better efficiency from wind turbines, as well as the analysis of the optimal geographical locations for potential plants and the social perception of a wind plant. However, the economic viability and the lack of skilled personnel are two important factors that also need to be considered and could make or break a wind farm project in a country.

TABLE 7.6
Barriers to Renewable Energy Implementation

Barrier category	Barriers
Market failures (MF)	(1) Highly controlled energy sector; (2) lack of awareness and information; (3) restricted access to technology; (4) lack of competition; (5) high transaction costs; (6) poor market infrastructure; (7) high investment requirements
Market distortions (MD)	(8) Favorable treatment to conventional energy; (9) taxes on RES; (10) non-consideration of externalities; (11) trade barriers
Economic and financial (EF)	(12) Economically not viable; (13) high discount rates; (14) high payback period; (15) small market size; (16) high cost of capital; (17) lacking/inadequate access to capital; (18) lack of access to credit to consumers; (19) high up-front capital costs; (20) lack of financial institutions and instruments
Institutional (I)	(21) Lack of institutions and mechanisms; (22) lack of a legal/regulatory framework; (23) problems in realizing financial incentives; (24) unstable macroeconomic environment; (25) lack of involvement of stakeholders in decision-making; (26) clash of interests; (27) lack of R&D culture; (28) lack of private sector participation; (29) lack of professional institutions
Technical (T)	(30) Lack of standards, codes, and certification; (31) lack of skilled personnel; (32) lack of O&M facilities; (33) lack of entrepreneurs; (34) system constraints; (35) unreliable product
Social (S)	(36) Lack of consumer acceptance of the product; (37) lack of social acceptance for some RES
Other barriers (OB)	(38) Uncertain and unsupportive governmental policies; (39) environmental; (40) high-risk perception for RES; (41) lack of infrastructure

Source: Diógenes et al. (2020).

In conclusion, the correct functioning of wind farms requires a sequential approach that involves many steps that need to work properly to have a satisfactory result. Thus, it is not only necessary to be in a zone with a large wind potential, have the necessary approval from the neighborhood or residential zone where these will be built, and have up-to-date designs and technology. Therefore, a successful major transition to wind power as a primary form of energy will take significant effort from all actors involved. One of these actors is the oil and gas industry, whose role we will review in more detail in the next section.

7.5 WIND ENERGY APPLICATIONS IN THE OIL AND GAS INDUSTRY

The oil and gas sector is exposed to a constant evaluation for the optimization and modernization of its methods to reduce their carbon emissions. In this section, we perform a detailed review of a series of steps taken by the industry to improve their carbon footprint across many processes that are related to the extraction and transportation of raw materials. Additionally, we review current applications, as well as challenges and opportunities for the industry in the future.

The oil and gas sector has over 70 years of experience in the development of onshore and offshore structures, with researchers working on the latter providing many examples of foundation concepts developed in marine environments (Schneider and Senders, 2010). This way, a close interrelation between the renewable energy sector and the oil and gas sector can prove beneficial for both industries, as successful techniques with decades of work behind are applied to a newer endeavor that is looking to achieve a clean energy revolution to effectively confront the climate challenges of the 21st century.

The petroleum sector has many decades of experience in building structures that optimize resources to exploit reservoirs, both onshore and offshore. Meanwhile, the wind energy sector, despite utilizing notions and techniques that span thousands of years, can benefit from the technological advancements brought by engineers.

This interconnectedness between previously competing industries is what Ingstrup and Menzel (2019) refer to as “relatedness” between industries. Seemingly unrelated at first glance, offshore oil and gas and offshore wind energy have benefitted from knowledge transfers that help simplify

FIGURE 7.20 Multiple units of wind turbines and solar panels installed on offshore platform.

Source: Tiong et al. (2015).

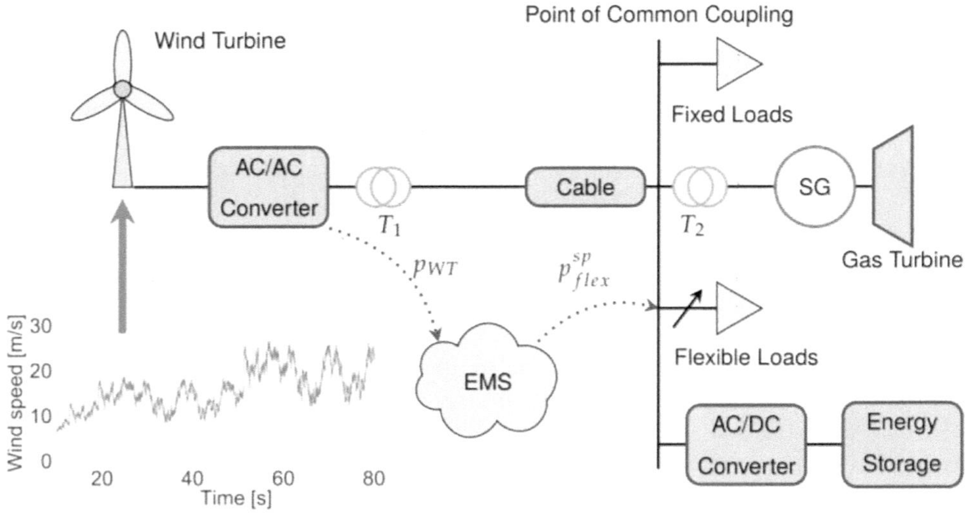

FIGURE 7.21 Main elements of the electric power system for a wind-powered offshore oil and gas platform (OOGP).

Source: Alves et al. (2019).

processes and expedite solutions which benefit not only the actors within these industries but also the populations affected by the reach of them.

One common example of this relatedness between industries is seen in offshore oil and gas installations that have optimized their processes to include wind turbines and solar panels to provide them with electricity, thus phasing out the use of polluting gas turbines (Korpås et al., 2012). Depending on the location of the platform, more solar exposure or wind speeds will lead researchers to choose which renewable energy is more convenient to exploit, while at times it is convenient to add both turbines and panels, as the model illustrated in Figure 7.20 shows. As noted previously, offshore wind farms are the most effective at producing electricity, which means that they are an almost-seamless addition to aid offshore rigs that are looking to minimize carbon-producing activities to comply with more strict environmental agreements.

A well-known shortcoming of renewable energies is their intermittence, as wind installations work best when there are high wind speeds and solar farms collect plenty of energy when the sun is shining bright. However, a night with no wind speeds will render installations almost useless, unless they have batteries which distribute energy collected earlier to those who need it.

The work of Alves et al. (2019) seeks to solve this problem for offshore oil and gas platforms (OOGPs), proposing a smart load management scheme along with an energy storage system (ESS) led by an algorithm to assess power quality. Figure 7.21 illustrates the proposed framework, where wind turbines collect energy and any excess generated due to high wind speeds is kept in the storage unit for later use. Meanwhile, the energy management system (EMS) can react to sudden changes in the quality of power, allowing for a smart load management that keeps the installation running despite unexpected changes in the input from the turbines, relying on the stored energy to back up electricity needs from the OOGP.

7.6 THE ROLE OF WIND IN THE ENERGY TRANSITION

As major oil and gas players in the current energy scenario, they also have incentives to modernize some of their processes. For example, many procedures that involve the extraction and processing of

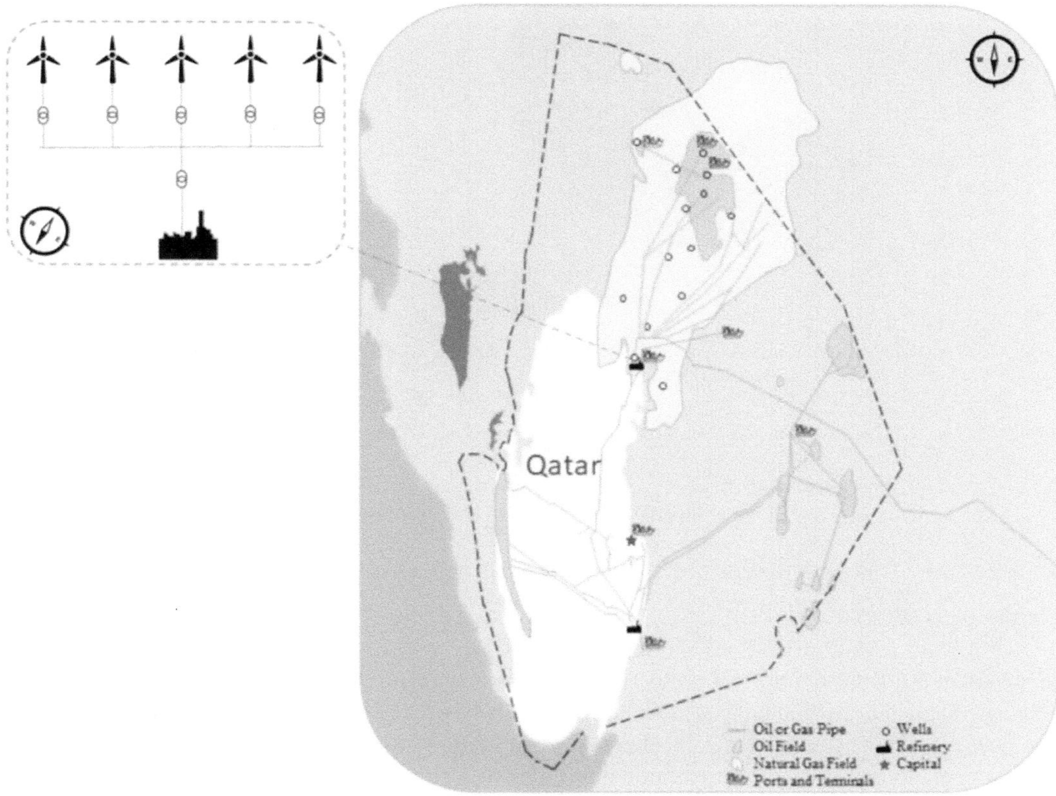

FIGURE 7.22 Map of Qatar with oil and gas field, refinery, and the proposed wind farm.

Source: Méndez and Bicer (2019).

raw materials have the potential to be converted into clean energy endeavors. The work of Méndez and Bicer (2019) exemplifies this effort in Qatar, where the possibility to use the energy of the wind to power the natural gas processing industry was analyzed. Figure 7.22 shows the proposal, which serves as a potential blueprint for other projects across the world.

A process of energy transition is understood as the move away from a fossil fuel–dominated economy to a more diversified sector, where renewable sources of energy, like solar and wind, have the predominant role. As the threat of a climate changed by human excesses in burning polluting sources of energy becomes more real, a transition process has become a necessary step to mitigate most of the potentially negative impacts of these practices.

Many legacy players in the oil and gas industry have started investing and lobbying for a greater space for renewable energy, both for economic and for environmental reasons. In what follows, we will review some oil companies, both national and international, that have developed long-term strategies to include wind power into their arsenal of resources for the new century.

With that said, the inclusion of oil and gas players and corporations can be a net positive in this process of energy transition. Given their level of expertise in technological advancements developed over the years, as well as the institutionalization of practices related to obtaining investment and funding from various government offices, the wind energy industry can benefit from this "insider" feature of oil and gas actors.

For example, in 2016, Equinor (then known as Statoil) established a capital fund of 200 million dollars to invest in renewable energy projects. A project of significant relevance at that time was the investment of 3 million dollars in United Wind, an American wind energy developer that has

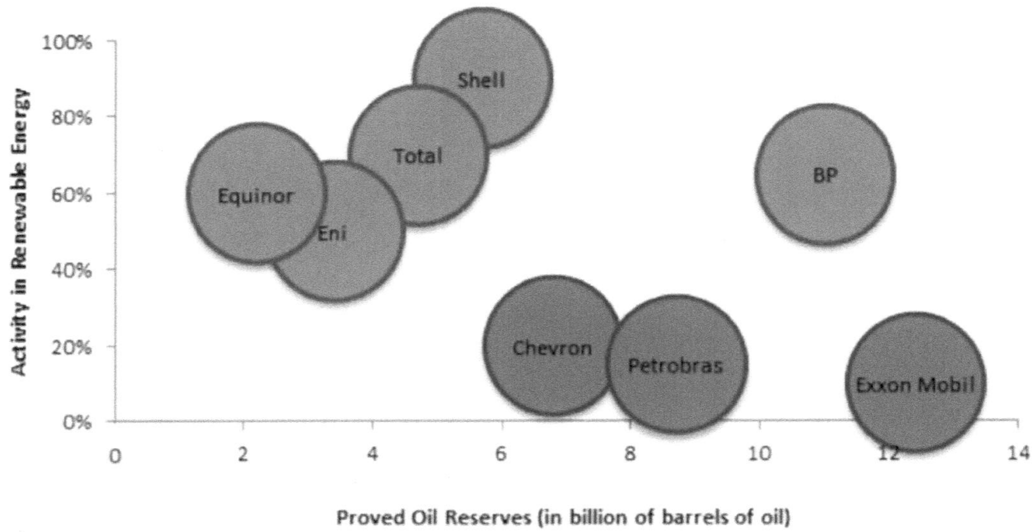

FIGURE 7.23 Linkage of renewable energy activity of oil majors with their proved oil reserves.

Source: Pickl (2019).

helped install turbines in several states across the country (Equinor, 2016; Windpower Engineering and Development, 2019; Zhong and Bazilian, 2018). Steps like those taken by private companies can have a domino effect that makes other companies follow suit, later encouraging other actors like governments to invest in cleaner technologies.

Figure 7.23 illustrates the activity of some of the big players in the oil industry, or "oil majors," regarding renewable energy investments, relative to their proved reserves of oil. As we can see, there appears to be a close negative relationship between the reserves of oil from a company and their degree of investment on renewable energy technologies. Therefore, the fewer oil reserves one of these big players possesses, the larger the investment in renewable energy methods, with the exceptional case being provided by BP, which still has a large percentage of activity in the renewable industry despite counting with the second largest proven oil reserves after only ExxonMobil.

It appears that a strong incentive to invest in renewables is a stagnating or decreasing reserve of oil. But additionally, the public perception that an oil company is investing in newer methods to provide electricity through cleaner sources is also another very strong incentive, especially considering that there is significant pressure from environmental groups and activists, as well as from shareholders, to reduce carbon emissions with the expectation of limiting the extent of the warming of the climate that can be traced back to these sources of energy (Pickl, 2019). Thus, despite a plausible alternative for easy oil and gas in unconventional source rocks, shale resource exploitation technology is still in its infancy, uncertainties are high, and social acceptance for these types of resources continues to be low (Weijermars et al., 2014).

Figure 7.24 represents the different levels of investment from oil majors on low-carbon technologies in the period 2015–2020, according to the Global Wind Energy Council's 2021 Global Wind Report. Of all renewable sources, wind was the preferred target for the oil and gas sector in cumulative terms, with Total and Equinor being large investors in this technology and the latter investing in wind more than any other technology. Coming from these large actors, pledges like these also help raise awareness about the environmental challenges of the current century while benefitting the image of a sector that may not be seen with the best eyes by some people.

This is particularly relevant in the era of social media and global communications. The concept of corporate social responsibility (CSR) has evolved in the last decade to include considerations of sustainability, community outreach, and more friendly business practices (Csomós, 2014; Frynas,

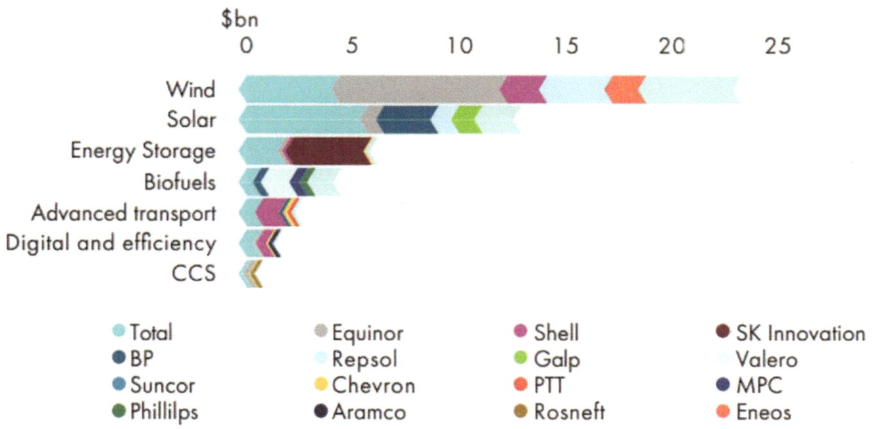

FIGURE 7.24 Oil and gas sector investment in low-carbon technologies, 2015–2020.

Source: GWEC (2021).

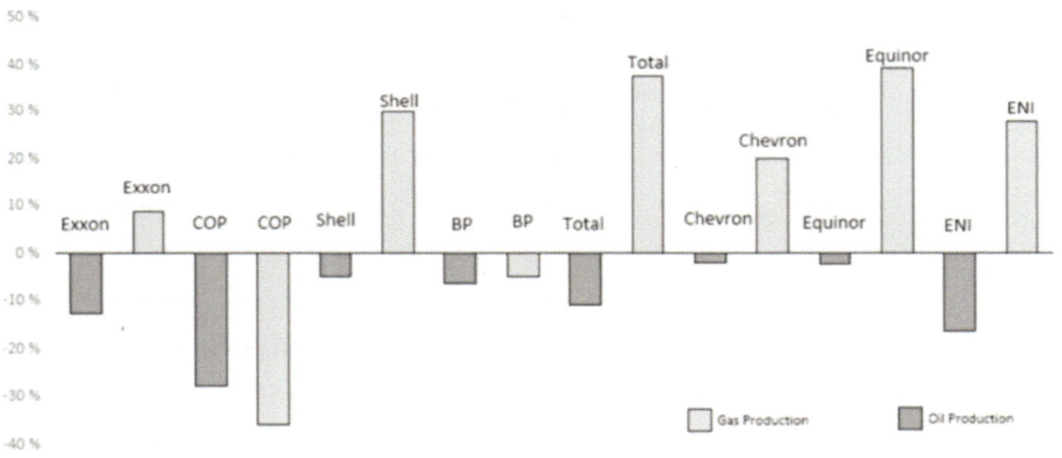

FIGURE 7.25 Percentage change in oil and gas production for selected oil majors, 2007–2017.

Source: Vormedal et al. (2020).

2009; Yousefi and Mataix Aldeanueva, 2013). In the case oil majors, CSR can be a key aspect to help them move the needle toward a more complete view that also highlights the role of these companies as key in the improvement of the quality of life of millions of people across the world. This way, a search to maintain these standards through strategic alliances with wind and solar endeavors is a very attractive way to change the previous public perceptions of oil majors.

Additionally, a move to a greater degree of investment in renewable energies is intended to help the bottom line of these companies in the long term, due to newer sustainability regulations placed by international organizations, many of them being supported by these same companies. In their analysis of "big oil" players, Vormedal et al. (2020) investigated why oil majors make moves that supposedly go against their own interests, concluding that these actors have thought about moving from coal to gas in advance, and supporting limitations to the more polluting sources of energy places them on the driving seat to move from big oil to "big gas" while championing cleaner sources of energy (which improves their image toward the public) and still profiting off these moves.

Figure 7.25 shows the percentage change in oil and gas production in a decade for a selection of oil majors. Most of the players shown in the figure have decreased their production of oil while increasing their production of gas significantly, with only BP and ConocoPhillips divesting in both sources of energy in the included time frame, with the former divesting after the Deepwater Horizon accident of 2010, and the latter reducing debt and selling noncore assets. However, the consistent move by all players is to rely more heavily on gas, both due to environmental pressures on more polluting forms of energy and because a coal-to-gas switch is still a very attractive opportunity for these actors to remain as powerful players in a changing industry that will be relevant for many years to come (Vormedal et al., 2020).

As we can witness, the current standing of the industry shows some signs of opening to a large inclusion of renewables, but also to a big investment on natural gas as a "bridge fuel" and to overcome the potential intermittence problems of renewables like solar and wind (Delborne et al., 2020). This opens the door to some conflicts between actors who wish that the energy transition process occurred faster and those that still see renewable energies as incapable of providing what current sources of energy provide. It is in this dynamic scenario that we see the interaction of international and national oil companies with governments and international organizations. In the rest of this section, we will review some of these dynamics at play and how companies have been able to increase the role of wind technologies over the years.

7.6.1 International Oil Companies (IOCs) and National Oil Companies (NOCs)

Oil companies can become leaders in this new scenario of energy transition. Despite working against some of their interests, the world has shifted to a position where climate considerations cannot be ignored for much longer, while governments have placed their attention and most of its resources on projects that prioritize renewable sources of energy (Shojaeddini et al., 2019). This has created the incentive for oil companies to branch out and seek for alliances and partnerships with renewable companies, as well as with governments who seek to make these sources of energy the main providers of

TABLE 7.7
Oil and Gas Companies' Net-Zero Commitments and Renewable Targets

Selected Companies	Climate Change Targets	Renewable Investment Targets
BP	• Net zero by 2050 • Carbon intensity 50% lower in 2050 compared to 2019 levels	• 50 GW renewable installations by 2030
ENI	• Carbon intensity 50% lower in 2050 compared to 2018 levels	• 4 GW renewable installations by 2024, 15 GW by 2030, and 60 GW by 2050
Equinor	• Net zero by 2050 • Reduce net carbon intensity to zero by 2050	• 4–6 GW renewable installations by 2026, and 12–16 GW renewable installations by 2035
Repsol	• First oil and gas company to target net-zero by 2050 • Carbon intensity 20% lower in 2030 and 40% lower in 2040 compared to 2016 levels	7.5 GW low carbon power capacity by 2025, and 15 GW by 2030
Shell	• Net-zero by 2050 • Reduce carbon intensity of energy products by 100% by 2050 compared to 2016 levels	Investing $2–3bn a year in renewables and energy solutions, including hydrogen, and doubling electricity sales by 2030 from current levels
Total	• Net-zero by 2050 • Carbon intensity 60% lower in 2050	35 GW renewable installations by 2025, and 100 GW of renewables capacity by 2030
Petronas	• Net zero by 2050	3 GW renewable installations by 2024
CNOOC	• Emissions peaking and carbon neutrality plan being drafted	Clean energy accounts for 60% of its energy mix by 2025

Source: GWEC (2021).

TABLE 7.8
Overview of Oil Companies' Engagement with Low-Carbon Technologies

Companies	Expanding beyond Oil Production to Clean Energy Technologies	Integrating Low-Carbon Technologies in Oil Production	Aiming to Lower Operational Emissions	Renewable Energy Technology Main Investments	Investments in Downstream Electricity	Renewable Energy Targets	Other Engagements in Low-Carbon Initiatives (i.e., Joint Ventures or Funds)
BP plc	✓	✓		Onshore wind, solar, biofuels, EVs infrastructure, batteries		50 (GW) by 2030	Joint ventures with renewable companies
Chevron Corporation		✓	✓	N/A		N/A	Future energy fund to invest in breakthrough low-carbon-emission technologies
Eni S.p.A.	✓	✓		Solar, wind, hydrogen, EVs batteries and chargers, biofuels	✓	15 GW by 2030, and 55 GW by 2050	Venture capital fund for R&D in renewables with universities and research centers
Equinor ASA	✓	✓		Solar, offshore wind, hydrogen, EVs	✓	4–6 GW by 2026, and 12–16 GW by 2035	Joint ventures with renewable companies
ExxonMobil Corporation		✓	✓	N/A		N/A	N/A
Royal Dutch Shell plc	✓	✓		Offshore wind, hydrogen biofuels, EVs	✓	Invest USD 3 billion in renewable energy (including hydrogen) per year by 2030	Investments in renewable start-ups and innovation hubs
Total SE	✓	✓		Solar, wind, hydrogen, biofuels	✓	35 GW of renewable electricity by 2025	Joint ventures with renewable companies

Source: Asmelash and Gorini (2021).

electricity for millions of people. Thus, the goal for some of these big oil companies is now to become significant players in a different industry, rebranding themselves as energy companies (Pickl, 2019).

Likewise, most oil majors have pledged to different strategies related to achieving net-zero. Table 7.7 shows that BP, Equinor, Shell, Total, and Petronas have all committed to achieving this goal by the year 2050, while most of them have set targets for investment in renewables for the next decade. With different levels of investments in renewable installations, these companies are only one part of the puzzle regarding the fulfillment of carbon reduction strategies.

Table 7.8 shows a summary of the steps taken by oil companies to engage with low-carbon technologies. According to the authors, BP plc was the first company to diversify to renewable energy and has committed to becoming a net-zero company by 2030. Then, Eni has struck deals with General Electric and Equinor to work on offshore wind installations, and in late 2021, the company acquired a 20% stake in the United Kingdom's Dogger Bank C project, the world's largest wind farm currently under construction (Eni, 2021). Meanwhile, Equinor developed the world's largest floating offshore wind farm, the Hywind Tampen project (Asmelash and Gorini, 2021).

Moreover, Shell has rebranded itself from an oil and gas company into an "energy transition" company and launched a New Energies division, where wind power is explored in combination with solar, to develop integrated energy solutions. Also, Total has followed a similar strategy as Shell, acquiring electricity supplying companies and investing in businesses like United Wind. Finally, while Chevron and ExxonMobil have not undertaken direct attempts at wind projects, both companies have committed to addressing climate change and coming up with strategic partnerships or investment schemes in breakthrough technologies like electric and autonomous vehicles, carbon capture and storage (CCS), or biofuels research (Asmelash and Gorini, 2021).

The United States serve as a microcosm of the situation in the rest of the world. With the country being the largest producer of oil in the planet, every oil company that is undertaking significant investments in wind technologies has looked at the United States to advance and develop their wind projects. Figure 7.26 shows the wind projects associated to BP in the United States, which span from north to south and from east and west of the territory. The two largest projects are Flat Ridge 2 and

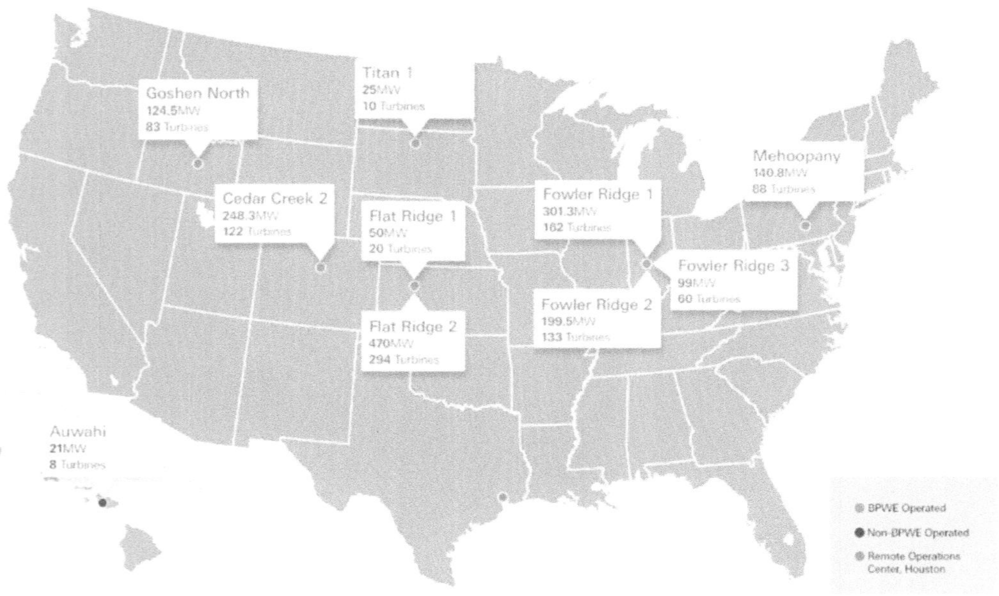

FIGURE 7.26 Wind energy sites associated with BP in the United States.

Source: BP (2022).

Fowler Ridge 1, both endeavors operated by BP's Wind Energy, which are in the states of Kansas and Indiana, respectively. These two projects alone utilize over 450 turbines and have a total capacity of 750 MW (BP, 2022).

Another practical example in a different part of the world is demonstrated in the work of Backer (2009), who follows the close interaction between Shell and Vestas, the largest wind turbine company. Their shared objectives of being relevant actors in the energy transition process and the green sensemaking processes that drive the current environment-protecting policies and regulations aligned closely enough for the two actors to develop a business relationship. Additionally, this close alliance helped with another stated goal of these actors, which is a positive return on their investments. In the end, thanks to a process of deliberation and strategic organizational improvements, both companies capitalize on the strength of the other while contributing to a shared societal goal that ranks high on people's minds.

As we can see, international oil companies (IOCs) have invested significant resources to expand their reach and become sustainable businesses that are aware of the climate challenges of this century while they remain competitive in this new effort. Despite having somewhat different motivations and operational directives than IOCs (Hartley and Medlock III, 2008), national oil companies (NOCs) are faced with similar trials and opportunities to transform renewables like wind and solar into the main sources of energy for the 21st century. In what follows, we will review some examples of NOCs that have decided to embrace wind technologies, the paths that they followed to achieve their plans, and the outcomes that these strategies have registered.

To start, Mäkitie et al. (2019) analyzed the case of Norwegian oil and gas companies, which assessed the changing environment both in terms of the economic and market conditions, as well as from a socio-political perspective, and moved forward on a strategy to develop offshore wind projects. In Figure 7.27, the authors summarize the rationale likely followed by most oil and gas firms which saw the changing environment and adjusted their priorities to reflect the saliency of renewable energy markets. These organizational dynamics have also been observed in the case of Chinese NOCs who developed an aggressive strategy of going abroad and investing in emerging technologies, with commercial incentives as the main driver (Jiang and Sinton, 2011).

Moreover, China is today the largest wind energy market in the world and is home to many global leaders in turbine manufacturing (Klagge et al., 2012). This has been achieved through a combination

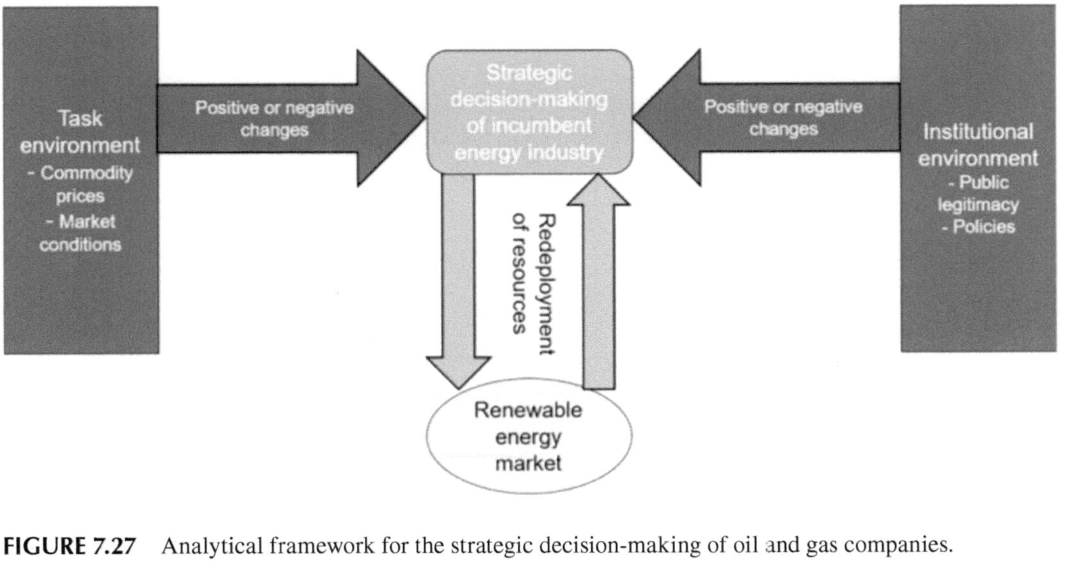

FIGURE 7.27 Analytical framework for the strategic decision-making of oil and gas companies.

Source: Mäkitie et al. (2019).

TABLE 7.9
Current Wind Energy Projects in the GCC

Country	Project	Size (MW)	Completion Year
Bahrain	A1 Dur (solar–wind hybrid)	5	N/A
Kuwait	Shagaya	10	N/A
Oman	Dhofar phase I	50	2019
	Dhofar phase II	150	2023
Saudi Arabia	Dumat A1 Jandal	400	2022

Source: Connelly and Xydis (2021).

of, among other things, technological advancement, organizational adaptation, coordinated governmental involvement, and a proper understanding of the long-term economics of wind energy. These factors may not be sufficient to generate a successful local wind power industry; however, they are necessary to be replicated if other developed countries want to follow a similar path (Martin and Ramsey, 2009). As noted in the case of Norway, reorientations of national oil companies toward sustainability followed market downturns in the oil and gas sector, which were later galvanized by more long-term environmental policies seeking to decarbonize the global energy sector (Mäkitie et al., 2019).

With that said, as entities depending on governments, NOCs are much more driven by policy than motivated by efficiency. Because of this, national companies tend to delegate the renewable energy ownership to actors in their respective electricity sectors, and the reason we observe countries like Japan deciding (through the Japan National Oil Corporation) against a widespread wind energy policy (Valentine, 2011). However, other national companies have helped launch renewable projects, like the case with the Kuwait Petroleum Corporation (KPC) and the Waha Al-Subaihiya project (Shojaeddini et al., 2019).

Relatedly, countries in the Gulf Cooperation Council (GCC) have developed a common energy agenda and included wind power into the mix, despite the very strong competitive advantage of solar power in the region. Table 7.8 shows a list of current wind energy projects in GCC countries, which would total over 600 MW of capacity in the next few years. These projects offer an optimistic view on the development of wind energy ventures worldwide, because the GCC has been considered a historically oil-heavy region (Al-Salem et al., 2018).

The following section explores the role of service and utility companies which provide operations and maintenance to wind installations and electricity for consumers, respectively. Actors in these two sectors have their own set of incentives to explore the market in search of economic opportunities as well and can be a pivotal element to the increase in awareness of the use of renewable energy to power the world.

7.6.2 Service Companies

Service companies stand between a renewable energy installation and the electricity provided to the final energy consumer. They oversee the maintenance of operations through providing technical assistance and materials to energy companies so that they keep their efficiency levels to a maximum. According to Bertoldi et al. (2006), the term *energy services* is sort of a catchall concept that includes several activities like audits and energy analysis, project design and implementation, maintenance and operation, energy equipment and supply, among others. Thus, the companies that provide these services, known as ESCOs, cover many issues related to the improvement of energy generation, production, and distribution.

Given the position of ESCOs in the overall renewable energy industry, they have incentives to improve the reach of these technologies to all types of users. Thus, their span may include from

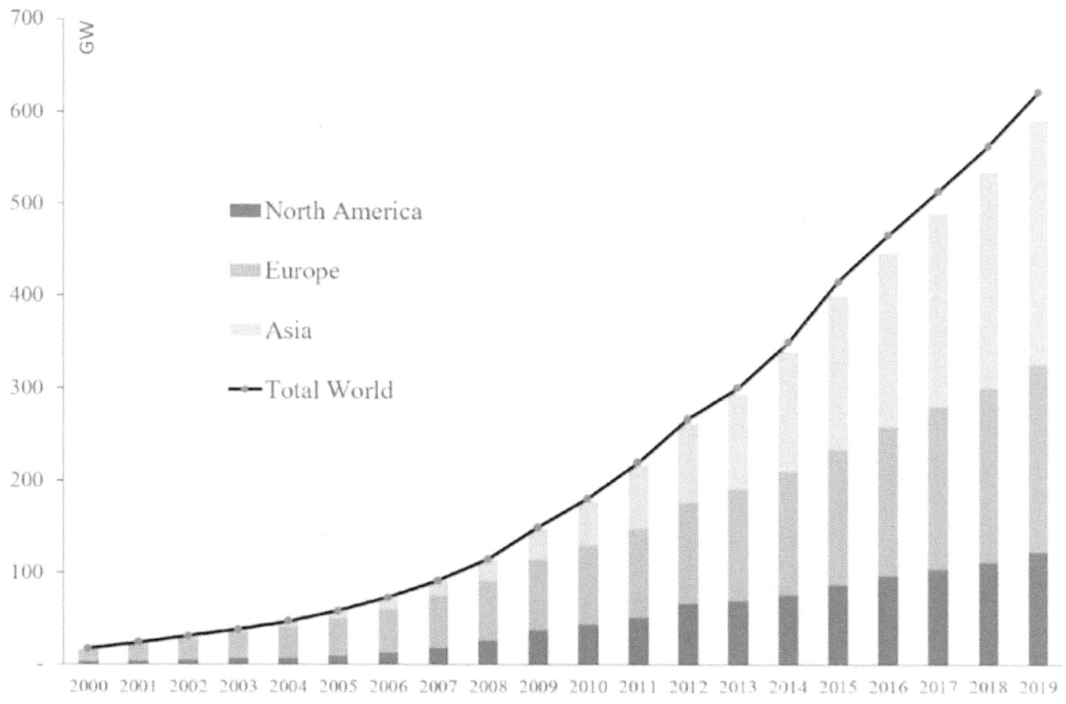

FIGURE 7.28 Global wind capacity, 2000–2019.

Source: Ren et al. (2021).

governments to small-scale users. Depending on their role, ESCOs can help mitigate CO_2 emissions when working in tandem with government organizations in a smart energy market (Das and Balakrishnan, 2010), or they can help optimize the series of steps that an energy company is encouraged to take, so that they can later achieve those environmental goals by themselves (Ren et al., 2021).

The role of service companies in the operation and maintenance of wind installations is important, with at least two aspects: (1) the development of more wind facilities in the future, which has direct consequences on (2) the future pricing of electricity for final energy users.

Figure 7.28 shows the marked increase in wind energy capacity in GW, which has created a strong market for the offering of services associated with the maintenance and improvement of wind installations. Because an increasing number of the wind installations built in the past few years are located offshore, service companies have had to adapt to this, acquiring equipment that allows them to serve their clients on time and distinguish themselves from other competition.

The improvement in the work of service providers can send positive shock waves across the entire renewable energy sector. Figure 7.29 shows the change in the development of offshore installations, measured in megawatts. In 2019 and 2020, the increase in megawatts coming from offshore installations surpassed 6,000 for each year, while the compound annual growth rate (CAGR) for electricity coming from offshore wind in the five years analyzed reached 28.5%. The largest contributors of this increase in 2020 and 2021 were China, Germany, the United Kingdom, and several other European countries.

Other functions that service companies provide relate to operational aspects, like replacing or repairing entire wind turbines, parts of them (like blades), or generators and gearboxes. Figure 7.30 provides an example of a lifting process for a wind turbine, a service provided by an ESCO. Operations like these have their own challenges, like weather conditions, their own technical failures, and even the skill barriers from technicians and engineers working for the companies. Other identified challenges relate to the waste generated by these new processes and the understudied effects that these operations may have on marine wildlife (Ren et al., 2021).

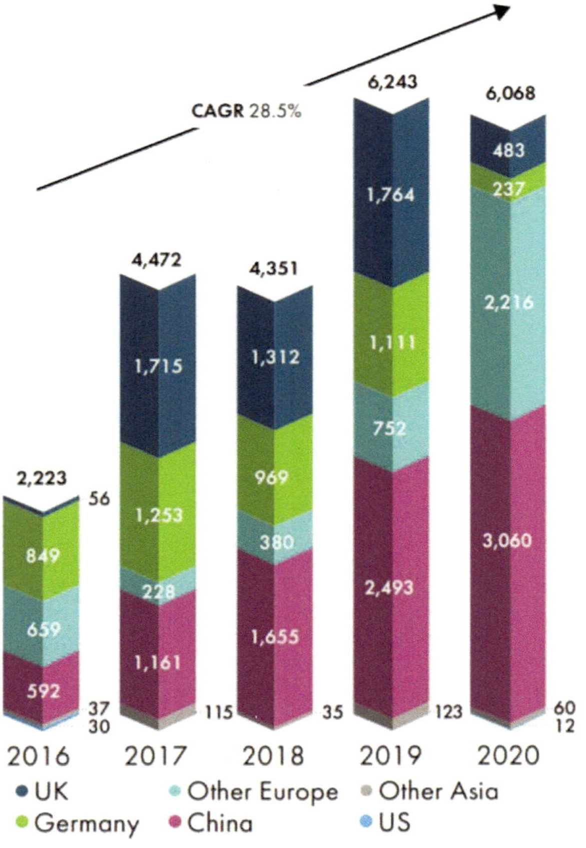

FIGURE 7.29 New offshore installations in MW.

Source: GWEC (2021).

Additionally, technologies to improve the service industry have been developed to simplify many processes in the field, particularly when wind conditions act against equipment designed to capture its energy. For example, different maintenance strategies powered by artificial intelligence (AI) have been tried and executed, which relate to a more predictive way of maintenance that could anticipate any potential failures before they occur (Yang et al., 2020).

Figure 7.31 shows the evolution of artificial intelligence (AI) use in the wind energy. One of those techniques is known as condition-based monitoring (CBM), which is based on prediction analyses for areas like fault detection and prognosis, establishing systems for early detection so that wind turbines are kept in continued operation, thus lowering efficiency loss (Bakir et al., 2021; Chatterjee and Dethlefs, 2021).

Lastly, a couple of other factors to consider when thinking about the optimization of the service industry, particularly for offshore wind installations, are linked to the availability of maintenance crews, the efficient routing of vessels, as well as their own upkeep strategies (Seyr and Muskulus, 2019). As the operation and maintenance sector seeks to optimize, more moving parts are added to the equation that need to be analyzed and optimized. For example, stochastic elements need to be studied and predicted, to the best of the researchers' abilities, after thorough processes of data collection, analysis, and model creation.

Besides service companies, utility companies can contribute to this process of decarbonization through facilitating access to renewable energy to final consumers, expanding the reach for the use

FIGURE 7.30 Lifting operation for an offshore wind turbine.

Source: Ren et al. (2021).

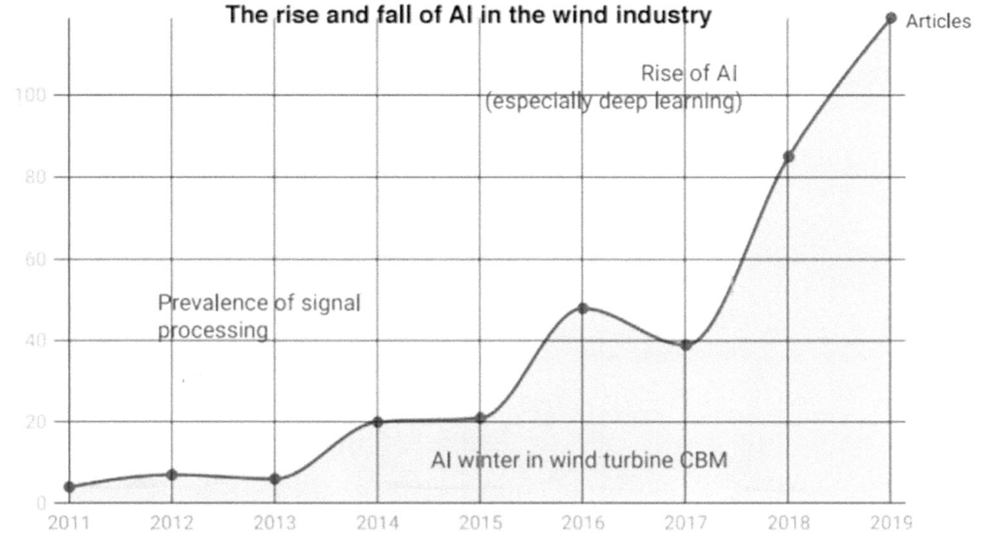

FIGURE 7.31 Evolution of artificial intelligence (AI) in the wind industry literature.

Source: Chatterjee and Dethlefs (2021).

of these sources in more remote areas, and investing in the proliferation of wind farms, both onshore and offshore. The following segment deals with the role of utility companies in the energy transition process of the 21st century.

7.6.3 Utility Companies

The nature of the collaborative process of decarbonization and fulfillment of international treaties involves the participation of many actors. One of the last players in this chain are utility companies, which provide electricity to the final user, while meeting the objectives of decarbonization and energy transition.

As Figure 7.32 shows, wind energy can be integrated into the electrical grid through various ways, which go from the large offshore plant to the onshore installation for medium-scale use, all the way to the local small turbine tailored for small residential uses. The more widespread this integration becomes, the easier it will be to phase out many of the current polluting power plants that are still required because of the current shortcomings of renewables and the limitations on energy storage and battery technology for large quantities of energy that still hinder the objectives for a faster energy transition process.

This new structuring of the electrical grid has multiple objectives. Along with the inclusion of cleaner sources of energy, decentralization of the market to decrease dependence on one country or region can be another helpful goal. This topic became especially salient once more in February 2022, after the conflict between Russia and Ukraine highlighted the dependence of many European Union on Russian oil and gas and the perilous state that some of these countries could be left in if

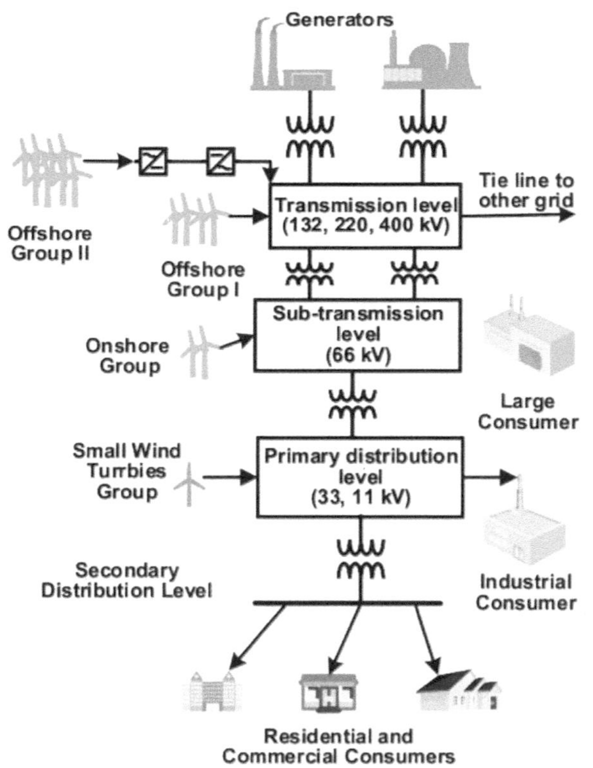

FIGURE 7.32 Power system grid with wind integration.

Source: Ahmed et al. (2020).

they are not proactive enough about integrating renewable sources into their grids, whether through large-scale or utility-scale projects.

In the work of Aman et al. (2013), the authors studied the potential incorporation of small windmills for residential use in Karachi, one of the largest cities in Pakistan, but with increasing needs for more electricity generation from a wider set of energy sources. The Karachi Electricity Supply Company (KESC) oversees the provision of electricity to most of the city, and the authors proposed that the utility company introduce a scheme where some residential users in high-wind sectors are provided with small windmills to power some low- to middle-energy needs. This move could save close to 1,700 MWs of energy while paving the way to the development of a smart grid that relies on several sources of power to satisfy users.

Figure 7.33 shows a map of Pakistan and its wind speed profile. At the lower part of the map, Karachi shows a moderate potential for wind speeds that could make small wind farms contribute significantly to the electrical grid of the city. This smart integration of wind in the grid is possible due to the improvement in the reliability for this energy, through infrastructure improvements that themselves are a result of increased funding and better research. The experiences from Europe and China, where wind has become part of the suite of energy sources, can help researchers with challenges related to interconnection, integration, while making wind a sustainable energy source not only from an environmental standpoint but also from an economic perspective (Ahmed et al., 2020; Smith et al., 2007; Welch and Venkateswaran, 2009).

Figure 7.34 shows a series of changes that have occurred over the years which signify a more active role from the populace in the development of a smart grid, as well as the transition to an

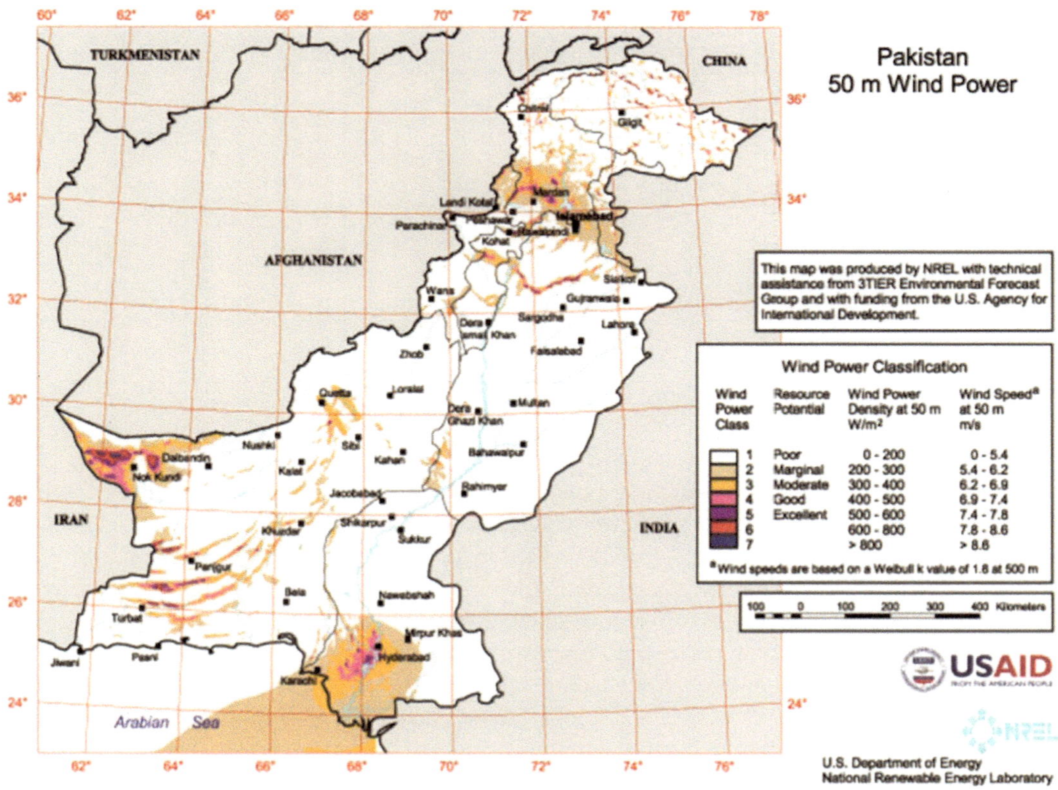

FIGURE 7.33 Pakistan average wind speed classification.

Source: Aman et al. (2013).

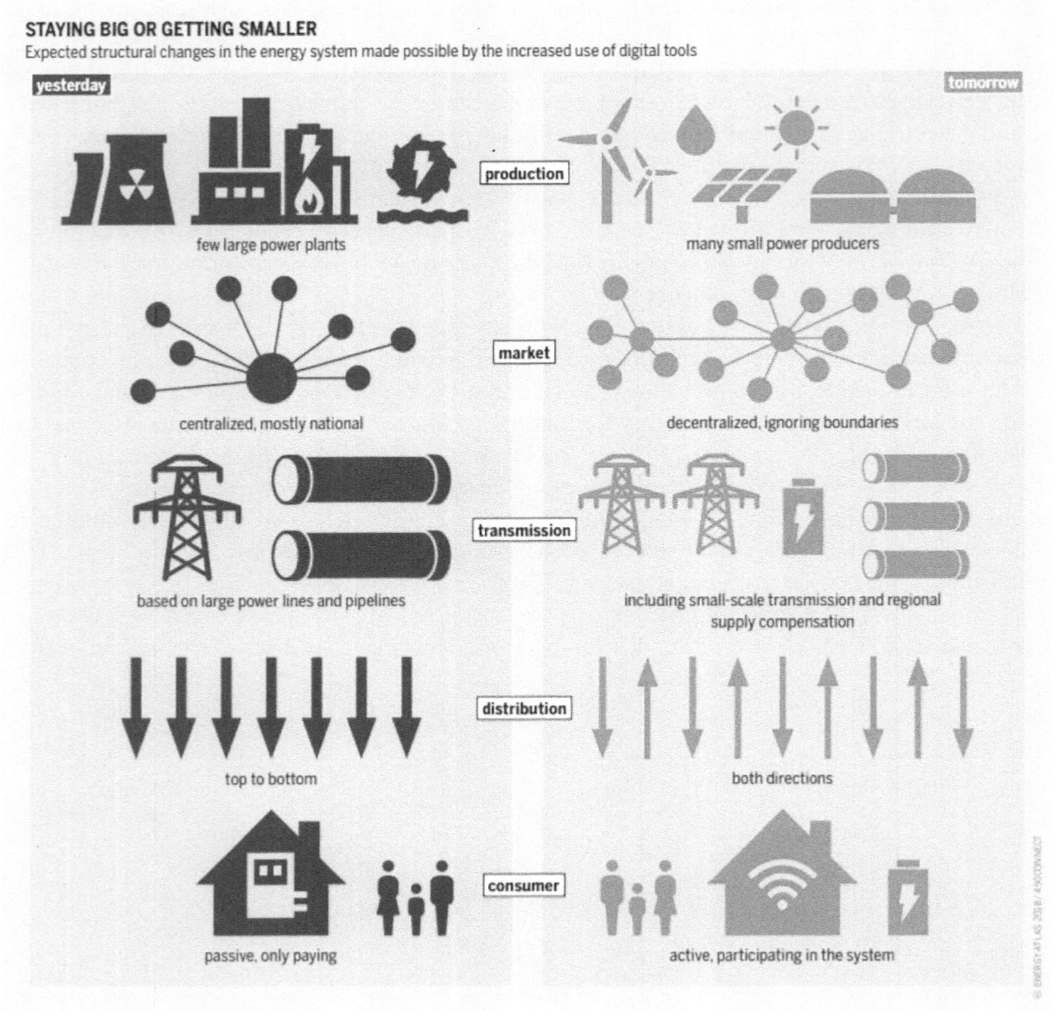

FIGURE 7.34 Comparison between older grid system and the smart grid.

Source: GEF (2018).

energy sector that is led by the power of renewables. Ideally, utility companies will be essential actors in the implementation of many of these changes, particularly those that relate to transmission, distribution, and consumer access. A grid that is fed by multiple sources and is more decentralized than current arrangements has the advantage of allowing smaller actors that produce electricity for local or remote areas to contribute directly (Richter, 2013). Utility companies can place themselves in the role of serving as a bridge between larger companies and smaller companies, but also in the position of providers and promoters of technologies like energy storage and at-home solar and wind installations.

With advancements like smart controls for wind plants and the automation of turbines, advanced wind energy technology fits right in the middle of this revolution (Glinkowski et al., 2011). Meanwhile, utility companies can help in achieving a successful interconnection through improving the quality of power that is provided by different actors, which is sometimes hindered by intermittence and unreliability problems (Völler et al., 2008).

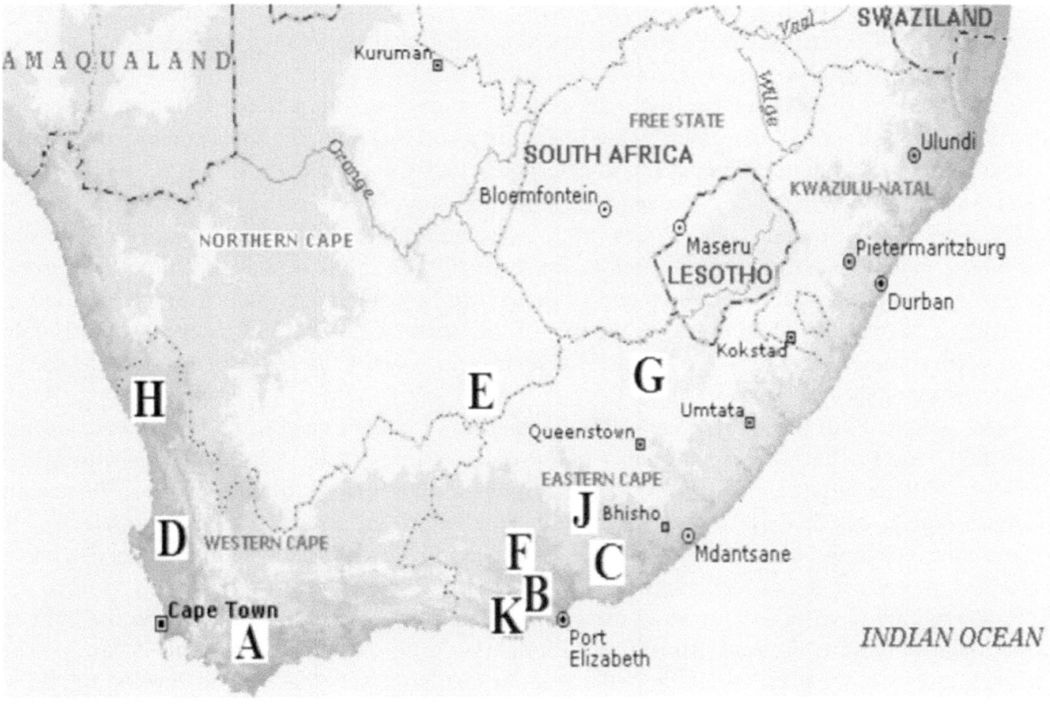

FIGURE 7.35 Location of wind farms in South Africa.

Source: Nobela et al. (2019).

The case of integration of wind energy in South Africa in the last few years can provide some valuable pointers. The work of Nobela et al. (2019) shows the potential challenges of renewable energy integration, particularly when independent power producers (IPPs) are encouraged to inject power into the grid and some IPP farms produce voltage fluctuations, flicker emissions, and harmonic distortions, which decrease the quality of the power being delivered. A solution to this problem is the installation of power quality recorders that assess the compatibility of wind energy conversion systems (WECS) with the utility grid.

As Figure 7.35 shows, some regions of South Africa are friendly to the installation of wind farms due to higher wind speed averages, something that has opened the door to the proliferation of these projects in the country. However, an immediate potential shortcoming comes when considering that the power quality (PQ) is bound to suffer, at least at the beginning of the integration process (Mahela et al., 2018, 2020). Utility companies are the right actor in this dynamic to tackle issues of supply, through steps like assisting small-scale energy providers, improving the power compatibility of WECS with the national grid, and deploying a strategy for continuous monitoring and data collection of these dynamics (Nobela et al., 2019).

Likewise, the work of Atwa and El-Saadany (2010) shows similar findings when a certain level of penetration of wind energy is achieved in the power grid. The authors advance that after a 20–30% level of dissemination of renewable energy in a city or country, there is a greater likelihood of significant intermittence problems. This problem with wind energy integration has been a common topic of study in the literature (see Ahlstrom et al., 2005). Using Ontario, Canada, as their case study, their proposed solution consists of an optimal allocation of energy storage systems (ESSs) so that spilled energy is accommodated and the annual electricity costs are minimized to the fullest.

For the case of wind energy optimization, ESSs can help with improving the grid-friendliness of wind power so that this energy can be controlled just like conventional sources coming from power plants. Similarly, grid variability and uncertainty problems can be curtailed when relying on stored energy in a properly selected storage system, which can also contain hybrid elements of those reviewed and analyzed previously (Zhao et al., 2015).

In this section, we reviewed the importance of oil companies, both national and international, as well as service and utility companies, in the search for a successful energy transition process. As we have come to understand throughout this chapter, no actor by itself may be able to achieve the proposed goals of decarbonization and reliance on renewable sources of energy for a cleaner future.

The work of these companies has been key to overcome challenges that the wind sector by itself could have taken much more time to tackle. Currently, we are in the middle of a process of interconnection and dialogue between companies thought to be competitors, which are now allied in the struggle to overcome the climate challenges of this century. Success in this endeavor will only occur when some of these steps are taken: first, smart grid network management through the use of advance technologies.

Then, compatibility issues between different sources of energy need to be minimized, so that renewable energies start erasing the label of "unreliable," moving to more stable structures and frameworks. This can be achieved through the improvement of energy storage systems. This chain of improvements can be optimized when a set of actors which are strong in their own areas is able to create bridges to share knowledge and create a decentralized system where innovation can come from the top or the bottom of the hierarchy.

In the following section, we review the efforts made by companies and actors on the goal of carbon neutrality and "net-zero" strategies. From entire sectors that have contributed more to the carbon footprint of the planet throughout the years to smaller players that want to develop and grow using renewable energy, these pledges have become an important part of the current energy transition process. In the quest for the fulfillment of these commitments, wind energy has turned into a central actor that, along with solar and other cleaner energies, can completely change the equation and lead us to a successful energy transition in the next few decades.

7.7 CONCLUSIONS

This chapter focused on the development and propagation of wind energy technologies and the role that the oil and gas industry has provided in the energy transition toward renewable sources of energy. From one of the oldest ways to generate electricity in small-scale settings to a leading source of energy for large-scale endeavors in the 21st century, wind stands out as a competitive form of energy to power the world. The collaboration with a powerful and resourceful sector like oil and gas stands to reap positive benefits for everyone involved, as goals for decarbonization can be achieved more rapidly.

First, we reviewed the different criteria for selection and design of wind installations. Perhaps unsurprisingly, the most important factor to select the appropriate location for the construction of a wind farm is the potential for high wind speeds that keep wind turbines running, constantly generating electricity. Then, we analyzed the latest developments in the design for structures and the criteria for wind technology selection, which go from turbine types and performance, availability of spare parts, medium- and long-term investment opportunities, to the quality of the assistance provided by service companies.

Later, we checked on the status of onshore and offshore wind facilities and the advantages and challenges of each type of installation for the efficiency in electricity production, for the public and the landscape, and for the prospects of achieving the goals of cutting on carbon emissions. Due to a better profile for benefits against costs, offshore wind plants appear to be the better option for large-scale wind projects, which also take advantage of higher wind speeds on average, while also disturbing the public less than with onshore projects that can contribute visual contamination and alter some birds' migration patterns.

In this search for the improvement of wind technologies, the oil and gas industry can offer many valuable tools and procedures. For example, the many decades of expertise applied to offshore rigs can be translated quickly to offshore wind installations, particularly the design of bases for turbines, whether they are attached to the seabed or are floating turbines. On the latter design, floating turbines can help the oil and gas sector through not interfering with potential sites for future exploration of oil sites in unfamiliar territories. Similarly, advancing on storage technologies for wind energy is a much-needed avenue of exploration.

These joint efforts bring marked benefits for both industries. The petroleum sector receives the many lessons on the decarbonization efforts from a growing industry that, in turn, needs the professionalization and institutionalization of the many processes that have led the oil and gas sector into being a leading energy industry. Together, the two areas are bound to tackle the challenges of a future with net-zero commitments from big companies and leading businesses, and with an international community that is more vigilant about climate considerations than ever before.

Part of the global objectives stated by international organizations is to replicate many of the low-carbon projects that have showed moderate success so that the process of energy transition advances swiftly across many countries. To achieve this, the role that international and national oil companies have can make or break a particular endeavor in renewable energies.

Then, the contribution that service companies offer has helped the sector grow exponentially in the last few years. ESCOs assist wind energy companies in many different areas, like technical assistance, operation and maintenance issues, installation and failure prediction, and distribution challenges. The degree of professionalization reached by service companies in the past couple of decades reflects the steps forward that wind technologies have taken, and the specialization from these businesses sends the signal that renewables are now an important part of the suite of sources of energy for human flourishing.

Moreover, wind energy aligns with the decarbonization efforts made by governments and some very important private companies, like airlines and multinational corporations. In the same spirit, the oil and gas industry has become a leader in these efforts through a series of pledges made by its most important companies, like BP, Equinor, Shell, and Total, to become "net-zero" companies by 2050. Also, think tanks and international organizations have developed strategies for industries and for people to start their own awareness on the carbon footprint they generate and how to lower it in the next few decades.

The potential of wind energy is effectively limitless, provided that the technology advances at a pace that can progressively capture, store, and effectively distribute more of the energy that we get every day. Currently, it is not possible to materialize a full energy transition to renewable energies, but the seeds of a transformation are undoubtedly planted. We can only hope that future research targets the correct issues to improve, and that wind energy becomes one of the cornerstones of the switch away from fossil fuels that is much needed for the decarbonization of most industrial, commercial, and residential energy processes.

REFERENCES

Abdel-Basset, M., A. Gamal, R. K. Chakrabortty, and M. Ryan. 2020. A New Hybrid Multi-Criteria Decision-Making Approach for Location Selection of Sustainable Offshore Wind Energy Stations: A Case Study. *Journal of Cleaner Production*: 124462.

Ahlstrom, M. L. Jones, R. Zavadil, and W. Grant. 2005. The Future of Wind Forecasting and Utility Operations. *IEEE Power and Energy Magazine* 3(6): 57–64.

Ahmed, S. D., F. S. M. Al-Ismail, M. Shafiullah, F. A. Al-Suliman, and I. M. El-Amin. 2020. Grid Integration Challenges of Wind Energy: A Review. *IEEE Access*: 1–23.

Allamehzadeh, H. 2016. *Wind Energy History, Technology and Control*. 2016 IEEE Conference on Technologies for Sustainability (SusTech).

Al-Salem, K., S. Neelamani, and W. Al-Nassar. 2018. Wind Energy Map of Arabian Gulf. *Natural Resources* 9: 212–228.

Alves, E., S. Sanchez, D. Brandao, and E. Tedeschi. 2019. Smart Load Management with Energy Storage for Power Quality Enhancement in Wind-Powered Oil and Gas Applications. *Energies* 12(15): 2985.

Aman, M. M., G. B. Jasmon, A. Ghufran, A. H. A. Bakar, and H. Mokhlis. 2013. Investigating Possible Wind Energy Potential to Meet the Power Shortage in Karachi. *Renewable and Sustainable Energy Reviews* 18: 528–542.

Arshad, M., and B. C. O'Kelly. 2013. Offshore Wind-Turbine Structures: A Review. *Proceedings of the Institution of Civil Engineers—Energy* 166(4): 139–152.

Asmelash, E. E., and R. Gorini. 2021. *International Oil Companies and the Energy Transition.* International Renewable Energy Agency (IRENA).

Atici, K. B., A. B. Simsek, A. Ulucan, and M. U. Tosun. 2015. A GIS-based Multiple Criteria Decision Analysis Approach for Wind Power Plant Site Selection. *Utilities Policy* 37: 86–96.

Atwa, Y. M., and E. F. El-Saadany. 2010. Optimal Allocation of ESS in Distribution Systems with a High Penetration of Wind Energy. *IEEE Transactions on Power Systems* 25(4): 1815–1822.

Backer, L. 2009. When Oil and Wind Turbine Companies Make Green Sense Together. *Business Strategy and the Environment* 18: 43–52.

Bakir, I., M. Yildirim, and E. Ursavas. 2021. An Integrated Optimization Framework for Multi-Component Predictive Analytics in Wind Farm Operations & Maintenance. *Renewable and Sustainable Energy Reviews* 138: 110639.

Bendary, A. F., and M. M. Ismail. 2019. Battery Charge Management for Hybrid PV/Wind/Fuel Cell with Storage Battery. *Energy Procedia* 162: 107–116.

Bento, N., and M. Fontes. 2019. Emergence of Floating Offshore Wind Energy: Technology and Industry. *Renewable and Sustainable Energy Reviews* 99: 66–82.

Bertoldi, P., S. Rezessy, and E. Vine. 2006. Energy Service Companies in European Countries: Current Status and a Strategy to Foster Their Development. *Energy Policy* 34: 1818–1832.

Blanco, M. I. 2009. The Economics of Wind Energy. *Renewable and Sustainable Energy Reviews* 13: 1372–1382.

BP. 2022. *Wind Energy Sites.* Available at: www.bp.com/en_us/united-states/home/where-we-operate/wind-energy-sites.html [Accessed February 8th, 2022].

Buckney, N., A. Pirrera, S. D. Green, and P. M. Weaver. 2013. Structural Efficiency of a Wind Turbine Blade. *Thin-Walled Structures* 67: 144–154.

Castro-Santos, L., and V. Diaz-Casas (eds.). 2016. *Floating Offshore Wind Farms.* Green Energy and Technology, Springer.

Chatterjee, J., and N. Dethlefs. 2021. Scientometric Review of Artificial Intelligence for Operations & Maintenance of Wind Turbines: The Past, Present and Future. *Renewable and Sustainable Energy Reviews* 144: 111051.

Chen, C.-W., Chen, K.-H., and Chen, Y.-M. 2014. *Modeling and Controller Design for the Multi-Input PV/ Wind Charger.* 16th European Conference on Power Electronics and Applications.

Chou, T., V. Kosmas, M. Acciaro, and K. Renken. 2021. A Comeback of Wind Power in Shipping: An Economic and Operational Review on the Wind-Assisted Ship Propulsion Technology. *Sustainability* 13(4): 1880.

Connelly, C., and G. Xydis. 2021. Wind Energy in the Gulf Cooperation Council Region: Progress, Challenges and Strategies for Development. *Review of Economics and Political Science* 6(4): 278–291.

Csomós, G. 2014. Relationship Between Large Oil Companies and the Renewable Energy Sector. *Environmental Engineering and Management Journal* 13(11, November): 2781–2787.

Das, A., and V. Balakrishnan. 2010. *Energy Service Companies (ESCOs) to Optimize Power in Peak Demand Period in Hybrid Energy System: An Impact on Climate Change.* IEEE Green Technologies Conference.

Davy, C. M., K. Squires, and J. R. Zimmerling. 2020. Estimation of Spatiotemporal Trends in Bat Abundance from Mortality Data Collected at Wind Turbines. *Conservation Biology* 35(1): 227–238.

Delborne, J. A., D. Hasala, A. Wigner, and A. Kinchy. 2020. Dueling Metaphors, Fueling Futures: "Bridge Fuel" Visions of Coal and Natural Gas in the United States. *Energy Research & Social Science* 61: 101350.

Derse, O., and E. Yılmaz. 2021. Optimal Site Selection for Wind Energy: A Case Study. *Energy Sources, Part A: Recovery, Utilization, and Environmental Effects*: 1–18.

Díaz, H., and C. Guedes Soares. 2020. Review of the Current Status, Technology and Future Trends of Offshore Wind Farms. *Ocean Engineering* 209: 107381.

Diógenes, J. R. F., J. Claro, J. C. Rodrigues, and M. V. Loureiro. 2020. Barriers to Onshore Wind Energy Implementation: A Systematic Review. *Energy Research & Social Science* 60: 101337.

Ekren, O., and B. Y. Ekren. 2010. Size Optimization of a PV/Wind Hybrid Energy Conversion System with Battery Storage Using Simulated Annealing. *Applied Energy* 87(2): 592–598.

Eni. 2021. *Eni Strengthens Its Presence in the UK Offshore Wind Market.* Available at: www.eni.com/en-IT/media/press-release/2021/11/eni-strengthens-its-presence-in-the-uk-offshore-wind-market.html [Accessed February 8th, 2022].

Equinor. 2016. *Statoil Launches USD 200m New Energy Investment Fund*. Available at: www.equinor.com/en/news/launches-usd200m-new-energy-investment-fund.html [Accessed January 25th, 2022].

Errami, Y., M. Ouassaid, M. Cherkaoui, and M. Maaroufi. 2014. Sliding Mode Control Scheme of Variable Speed Wind Energy Conversion System Based on the PMSG for Utility Network Connection. *Studies in Computational Intelligence*: 167–200.

Feng, J., and W. Z. Shen. 2017. Design Optimization of Offshore Wind Farms with Multiple Types of Wind Turbines. *Applied Energy* 205: 1283–1297.

Frynas, J. G. (2009). Corporate Social Responsibility in the Oil and Gas Sector. *The Journal of World Energy Law & Business* 2(3): 178–195.

GEF. 2018. *Energy Atlas 2018: Facts and Figures About Renewables in Europe*. Green European Foundation.

Glinkowski, M., J. Hou, and G. Rackliffe. 2011. Advances in Wind Energy Technologies in the Context of Smart Grid. *Proceedings of the IEEE* 99(6): 1083–1097.

GWEC. 2021. *Global Wind Report 2021*. Global Wind Energy Council.

Hahn, B., M. Durstewitz, and K. Rohrig. 2007. *Reliability of Wind Turbines: Experiences of 15 Years with 1,500 WTs*. Institut für Solare Energieversorgungstechnik (ISET). Universität Kassel e.V.

Hanning, C. D., and A. Evans. 2012. Wind Turbine Noise. *British Medical Journal* 344(7853): 1–2.

Hansen, A. D., F. Iov, F. Blaabjerg, and L. H. Hansen. 2004. Review of Contemporary Wind Turbine Concepts and Their Market Penetration. *Wind Engineering* 28(3): 247–263.

Haraguchi, H., S. Morimoto, and M. Sanada. 2009. *Suitable Design of a PMSG for a Large-Scale Wind Power Generator*. IEEE Energy Conversion Congress and Exposition.

Hartley, P., and K. B. Medlock III. 2008. A Model of the Operation and Development of a National Oil Company. *Energy Economics* 30(5): 2459–2485.

Herbert-Acero, J. F., O. Probst, P.-E. Réthoré, G. C. Larsen, and K. K. Castillo-Villar. 2014. A Review of Methodological Approaches for the Design and Optimization of Wind Farms. *Energies* 7(11): 6930–7016.

Ingstrup, M. B., and M.-P. Menzel. 2019. *The Emergence of Relatedness Between Industries: The Example of Offshore Oil and Gas and Offshore Wind Energy in Esbjerg, Denmark*. Papers in Evolutionary Economic Geography #19.29. Utrecht University.

IRENA. 2021. *Renewable Power Generation Costs in 2020*. International Renewable Energy Agency.

Jeon, J., and J. W. Taylor. 2012. Using Conditional Kernel Density Estimation for Wind Power Density Forecasting. *Journal of the American Statistical Association* 107(497): 66–79.

Jiang, J., and J. Sinton. 2011. *Overseas Investments by Chinese National Oil Companies*. IEA Information Paper.

Junginger, M., A. Faaij, and W. C. Turkenburg. 2004. Cost Reduction Prospects for Offshore Wind Farms. *Wind Engineering* 28(1): 97–118.

Kaabeche, A., M. Belhamel, and R. Ibtiouen. 2011. Sizing Optimization of Grid-Independent Hybrid Photovoltaic/Wind Power Generation System. *Energy* 36(2): 1214–1222.

Kaldellis, J. K., and D. Zafirakis. 2011. The Wind Energy (R)evolution: A Short Review of a Long History. *Renewable Energy* 36(7): 1887–1901.

Khatib, T., K. Sopian, A. Mohamed, and M. Z. Ibrahim. 2012. Sizing of a Wind Charger at Minimum Cost for Remote Housing Electrification: A Case Study for Nine Coastal Sites in Malaysia. *Energy and Buildings* 51: 185–190.

Khudri Johari, M., M. A. A. Jalil, and M. F. M. Shariff. 2018. Comparison of Horizontal Axis Wind Turbine (HAWT) and Vertical Axis Wind Turbine (VAWT). *International Journal of Engineering & Technology* 7 (4.13): 74–80.

Klagge, B., Z. Liu, and P. Campos Silva. 2012. Constructing China's Wind Energy Innovation System. *Energy Policy* 50: 370–382.

Klinge Jacobsen, H., P. Hevia-Koch, and C. Wolter. 2019. Nearshore and Offshore Wind Development: Costs and Competitive Advantage Exemplified by Nearshore Wind in Denmark. *Energy for Sustainable Development* 50: 91–100.

Korpås, M., Warland, L., He, W., and J. O. G. Tande. 2012. A Case-Study on Offshore Wind Power Supply to Oil and Gas Rigs. *Energy Procedia* 24: 18–26.

Kusiak, A., and Z. Song. 2010. Design of Wind Farm Layout for Maximum Wind Energy Capture. *Renewable Energy* 35(3): 685–694.

Ladenburg, J. 2009. Visual Impact Assessment of Offshore Wind Farms and Prior Experience. *Applied Energy* 86(3): 380–387.

Lee, H.-J., S.-H. Yoo, and S.-Y. Huh. 2020. Public Perspectives on Reducing the Environmental Impact of Onshore Wind Farms: A Discrete Choice Experiment in South Korea. *Environmental Science and Pollution Research* 27(20): 25582–25599.

Leung, D. Y. C., and Y. Yang. 2012. Wind Energy Development and Its Environmental Impact: A Review. *Renewable and Sustainable Energy Reviews* 16(1): 1031–1039.

Lo, K.-Y., Y.-M. Chen, and Y.-R. Chang. 2011. MPPT Battery Charger for Stand-Alone Wind Power System. *IEEE Transactions on Power Electronics* 26(6): 1631–1638.

Lovich, J. E., and J. R. Ennen. 2013. Assessing the State of Knowledge of Utility-Scale Wind Energy Development and Operation on Non-Volant Terrestrial and Marine Wildlife. *Applied Energy* 103: 52–60.

Lu, L., H. Yang, and J. Burnett. 2002. Investigation on Wind Power Potential on Hong Kong Islands—an Analysis of Wind Power and Wind Turbine Characteristics. *Renewable Energy* 27(1): 1–12.

Mahela, O. P., K. Dev Kansal, and S. Agarwal. 2018. *Detection of Power Quality Disturbances in Utility Grid with Wind Energy Penetration*. 8th IEEE India International Conference on Power Electronics (IICPE).

Mahela, O. P., B. Khan, H. Haes Alhelou, and S. Tanwar. 2020. Assessment of Power Quality in the Utility Grid Integrated with Wind Energy Generation. *IET Power Electronics* 13(13): 2917–2925.

Makarios, T. K., and C. C. Baniotopoulos. 2012. Wind Energy Structures: Modal Analysis by the Continuous Model Approach. *Journal of Vibration and Control* 20(3): 395–405.

Mäkitie, T., H. E. Normann, T. M. Thune, and J. Sraml Gonzalez. 2019. The Green Flings: Norwegian Oil and Gas Industry's Engagement in Offshore Wind Power. *Energy Policy* 127: 269–279.

Martin, J., and D. Ramsey. 2009. The Economics of Wind Energy. *Journal of Applied Corporate Finance* 21(2): 100–109.

Méndez, C., and Y. Bicer. 2019. Qatar's Wind Energy Potential with Associated Financial and Environmental Benefits for the Natural Gas Industry. *Energies* 12(17): 3329.

Mentis, D., S. Hermann, M. Howells, M. Welsch, and S. H. Siyal. 2015. Assessing the Technical Wind Energy Potential in Africa a GIS-Based Approach. *Renewable Energy* 83: 110–125.

Mohammadi, K., S. Shamshirband, P. L. Yee, D. Petković, M. Zamani, and S. Ch. 2015. Predicting the Wind Power Density Based Upon Extreme Learning Machine. *Energy* 86: 232–239.

Nobela, O. N., R. C. Bansal, and J. J. Justo. 2019. A Review of Power Quality Compatibility of Wind Energy Conversion Systems with the South African Utility Grid. *Renewable Energy Focus* 31: 63–72.

Office of Energy Efficiency & Renewable Energy. 2021. *How Do Wind Turbines Work?* Available at: www.energy.gov/eere/wind/how-do-wind-turbines-work [Accessed January 4th, 2022].

Onar, S. C., B. Oztaysi, İ. Otay, and C. Kahraman. 2015. Multi-Expert Wind Energy Technology Selection Using Interval-Valued Intuitionistic Fuzzy Sets. *Energy* 90: 274–285.

Pfaffer, S., S. Faulstich, and K. Rohrig. 2017. Performance and Reliability of Wind Turbines: A Review. *Energies* 10: 1904.

Pickl, M. J. 2019. The Renewable Energy Strategies of Oil Majors – from Oil to Energy? *Energy Strategy Reviews* 26: 100370.

Prasai, A., J.-S. Yim, D. Divan, A. Bendre, and S.-K. Sul. 2008. A New Architecture for Offshore Wind Farms. *IEEE Transactions on Power Electronics* 23(3): 1198–1204.

Ren, Z., A. S. Verma, Y. Li, J. J. Teuwen, and Z. Jiang. 2021. Offshore Wind Turbine Operations and Maintenance: A State-of-the-Art Review. *Renewable and Sustainable Energy Reviews* 144: 110886.

REN21. 2021. *Renewables 2021 Global Status Report*. REN21 Secretariat.

Richter, M. 2013. Business Model Innovation for Sustainable Energy: How German Municipal Utilities Invest in Offshore Wind Energy. *International Journal of Technology Management* 63(1–2): 24–50.

Schaumann, P., C. Böker, A. Bechtel, and S. Lochte-Holtgreven. 2011. Support Structures of Wind Energy Converters. In Baniotopoulos, C. C., Borri, C., and Stathopoulos, T. (eds.), *Environmental Wind Engineering and Design of Wind Energy Structures*. Springer.

Schneider, J. A., and M. Senders. 2010. Foundation Design – a Comparison of Oil and Gas Platforms with Offshore Wind Turbines. *Journal of the Marine Technology Society* 44(1): 32–51.

Schubel, P. J., and R. J. Crossley. 2012. Wind Turbine Blade Design. *Energies* 5(9): 3425–3449.

Seyr, H., M. Muskulus. 2019. Decision Support Models for Operations and Maintenance for Offshore Wind Farms: A Review. *Applied Sciences* 9(2): 278.

Shojaeddini, E., S. Naimoli, S. Ladislaw, and M. Bazilian. 2019. Oil and Gas Company Strategies Regarding the Energy Transition. *Progress in Energy* 1(1): 012001.

Smith, J. C., M. R. Milligan, E. A. DeMeo, and B. Parsons. 2007. Utility Wind Integration and Operating Impact State of the Art. *IEEE Transactions on Power Systems* 22(3): 900–908.

Tiong, Y. K., M. A. Zahari, S. F. Wong, and S. S. Dol. 2015. The Feasibility of Wind and Solar Energy Application for Oil and Gas Offshore Platform. *IOP Conference Series: Materials Science and Engineering* 78: 012042.

Valentine, S. V. 2011. Japanese Wind Energy Development Policy: Grand Plan or Group Think? *Energy Policy* 39: 6842–6854.

Völler, S., A.-R. Al-Awaad, and J. F. Verstege. 2008. *Benefits of Energy Storages for Wind Power Trading*. 2008 IEEE International Conference on Sustainable Energy Technologies.

Vormedal, I., L. H. Gulbrandsen, and J. B. Skjærseth. 2020. Big Oil and Climate Regulation: Business as Usual or a Changing Business? *Global Environmental Politics* 20(4): 143–166.

Wang, C.-N., T.-T. Dang, and N.A.T. Nguyen. 2021. Location Optimization of Wind Plants Using DEA and Fuzzy Multi-Criteria Decision Making: A Case Study in Vietnam. *IEEE Access* 9: 116265–116285.

Weijermars, R., O. Clint, and I. Pyle. 2014. Competing and Partnering for Resources and Profits: Strategic Shifts of Oil Majors During the Past Quarter of a Century. *Energy Strategy Reviews* 3: 72–87.

Welch, J. B., and A. Venkateswaran. 2009. The Dual Sustainability of Wind Energy. *Renewable and Sustainable Energy Reviews* 13(5): 1121–1126.

Windpower Engineering and Development. 2019. *United Wind to Install More Than 50 Distributed Wind Turbines to Power Smithfield Farms in Colorado.* Available at: www.windpowerengineering.com/united-wind-to-install-more-than-50-distributed-wind-turbines-to-power-smithfield-farms-in-colorado/ [accessed January 23rd, 2022].

The World Bank. 2019. *Wind Power Density Potential.* Available at: https://s3-eu-west-1.amazonaws.com/globalwindatlas3/HR_posters/pd_World.pdf [Accessed January 3rd, 2022].

Yang, L., G. Li, Z. Zhang, X. Ma, and Y. Zhao. 2020. Operations & Maintenance Optimization of Wind Turbines Integrating Wind and Aging Information. *IEEE Transactions on Sustainable Energy*: 1–10.

Yousefi, A., and C. Mataix Aldeanueva. 2013. *Social Responsibility—A Comparative Study of Oil Majors—Who Is the Best?* SPE Latin-American and Caribbean Health, Safety, Environment and Social Responsibility Conference.

Zhang, Y., and S. Ula. 2008. Comparison and Evaluation of Three Main Types of Wind Turbines. *2008 IEEE/PES Transmission and Distribution Conference and Exposition* (pp. 1–6). IEEE.

Zhao, H., Q. Wu, S. Hu, H. Xu, and C. N. Rasmussen. 2015. Review of Energy Storage System for Wind Power Integration Support. *Applied Energy* 137: 545–553.

Zhong, M., and M. D. Bazilian. 2018. Contours of the Energy Transition: Investment by International Oil and Gas Companies in Renewable Energy. *The Electricity Journal* 31(1): 82–91.

8 Hydropower

Jorge Saldaña, Mesut Yurukcu, Narendra Boppana, Sepehr Arbabi, Jim Henry, and Sebahattin Ziyanak

8.1 INTRODUCTION

The energy transition process has accelerated the search for various clean sources that can replace fossil fuels in the medium to long term, with hydropower emerging as one of the frontrunners. Similar to ocean and tidal energy projects, hydropower endeavors to harness the power of water to generate electricity. However, a significant difference for the latter form of energy has involved building big installations such as dams to control large flows of water that capture potential and kinetic energy to produce electrical energy. With time, new technologies have allowed for the advancement of smaller-scale projects that can more optimally adapt to the locations they look to serve.

Efforts to harness the untapped energy of water are by no means a new development. Streams and currents have been explored and exploited for many decades, with one early highlight of the professionalization of this exploration unveiled in the 1930s with the opening of the Hoover Dam in the state of Nevada. Generally, hydropower technologies rely on the water cycle, as precipitation that drains from rivers can be fed to dams to produce significant quantities of energy. Thus, drought periods can significantly affect the energy production from these types of energy installations (U.S. Energy Information Administration, 2023a).

In this chapter, we perform a detailed review of the methods used to capture energy from the controlled flows of water, the strengths and challenges of the technologies, and the advancements that have led to novel techniques being explored in the field with promising results. Also, we analyze the relationship that hydropower has had with the oil and gas sector and the steps taken by the industry to increase the share of hydropower installations around the world. In the context of the energy transition, we also assess the opportunities for hydropower to assist and eventually replace fossil fuels, serving as a potential leader in the renewable energy sector of the 21st century.

The primary mechanism of energy production from hydropower technologies is the use of the elevation difference caused by a dam or diversion structure or water flowing from high to low heights. Due to water being moved as the only element used in electricity generation, hydropower projects cost less than others. This explains the lower bills found in states with a significant share of hydropower in their energy suite. Moreover, despite being recognized by large projects like the Hoover Dam, more novel technologies allow for the construction of much smaller plants with multiple uses (U.S. Department of Energy, 2023b).

Due to their source being water provided by natural cycles, hydropower is a vital renewable source of energy, key in the energy transition process away from fossil fuels. Additionally, over a hundred years, incremental improvements to these methods have placed hydropower as a mature technology that produces reliable, low-carbon electricity (IRENA, 2023). While the main goal of a hydropower project is to generate electricity, these installations can serve other uses, such as irrigation, flood control, and water supply to communities. In this sense, hydropower endeavors can support the economic growth of geographical locations as they also tackle critical social and environmental challenges.

The first models using water to generate energy were waterwheels placed next to properties to help with industrial activities, such as grinding grain into flour, sawing wood into lumber, or serving as power textile mills. These technologies were advanced in the United States in the 19th century

 DOI: 10.1201/9781003282273-8

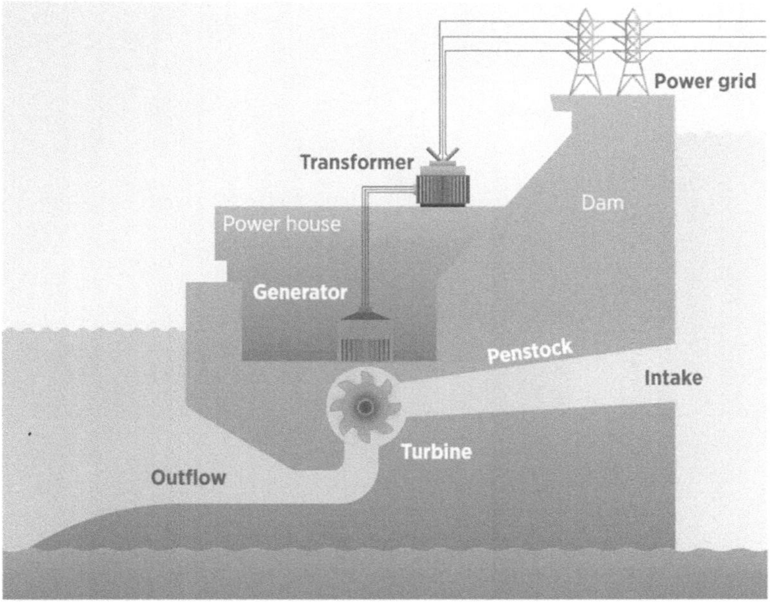

FIGURE 8.1 Generic reservoir hydropower plant.

Source: IRENA (2023).

thanks to the development of the electric generator by Michael Faraday in 1831 and of the hydroturbine by James Francis in 1849. With that knowledge, the first hydropower plant worldwide opened in Wisconsin in 1882, cementing the big role of hydropower in the United States, where it is currently the second renewable electricity source after wind (US Department of Energy, 2016; U.S. Energy Information Administration, 2023b).

A typical modern hydropower plant setup is illustrated in Figure 8.1, featuring the already-familiar notions of a water current used to operate a turbine, which in turn powers a transformer that produces electricity to be added to the local or regional grid. Different types of hydropower plants have been devised from that basic framework over the years, with three types standing out as the most common. The first two, which belong to the category of conventional methods, are impoundment of reservoir and run-of-river. Meanwhile, a third method that has gained momentum in the past few decades is pumped storage hydropower, or PSH.

When considering these main types of hydropower technologies, the current installed capacity around the world exceeds 1,500 GW (gigawatts) between PSH and conventional methods, with the latter still leading the former by a factor of almost 10 to 1. Figure 8.2 shows that conventional technologies reached over 1,200 GW in 2021, while PSH features were close to 140 GW. Conventional methods have been around for longer and vary more in their power capacity; thus, they can be better adapted to different contexts, satisfying different levels of energy demands around the world.

A promising role of pumped storage hydropower configurations is their capacity to collect excess energy in periods of low demand, releasing it during peak demand. In this sense, PSH facilities can act as large-scale batteries and assist intermittent sources like solar and wind when these energy sources present variations that could compromise the grid (Yang and Yang, 2019). Likewise, hydropower projects in rich water environments provide consistent electricity, ensuring grid stability and meeting baseload power expectations. All in all, hydropower can be seen as a mature and reliable renewable energy source that can be the backbone of the energy transition of the 21st century.

The following sections analyze the multiple criteria for site selection and the general economics behind hydro projects. We also more deeply assess the different types of hydropower technologies,

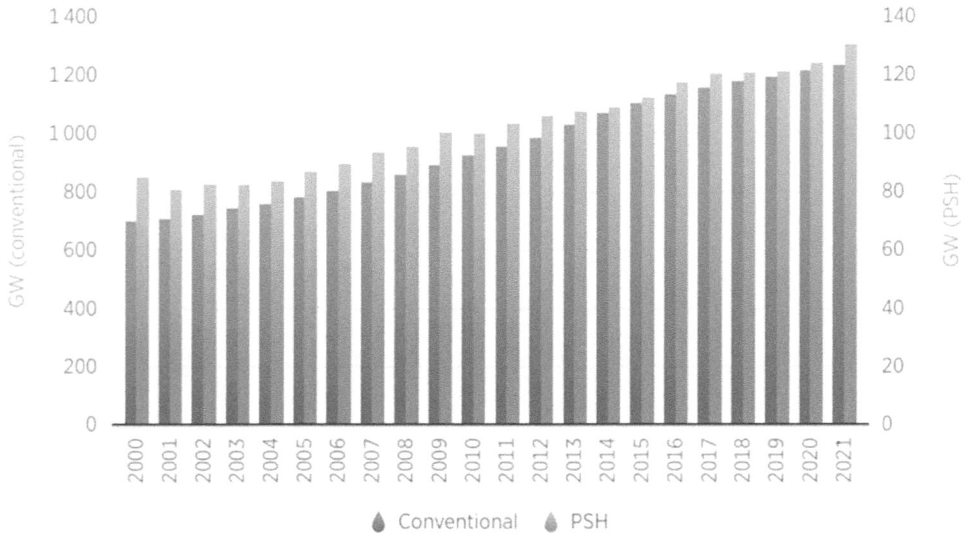

FIGURE 8.2 Hydropower installed capacity by year 2000–2021. (*GW* = gigawatt; *PSH* = pumped storage hydropower.)

Source: IRENA (2023).

evaluating their advantages, drawbacks, and opportunities in varying contexts. Additionally, we analyze the role that the oil and gas industry has had in the development of hydropower endeavors, the latest technologies spurred by the joint work between the oil and gas sector and the renewable sector, and we review the collaboration efforts that are required to fund more of these projects and the participation that governments, entrepreneurs, and international organizations have in the advancement of these methods.

8.2 SCREENING AND SITE SELECTION FOR HYDROPOWER PROJECTS

The parameters to screen and select a site for hydropower purposes depend on many factors, such as the maturity of the technology and the scope of the specific project being examined. In recent years, the role of geographic information systems (GIS) and artificial intelligence (AI) has simplified the process of tailoring these parameters to a particular location to achieve optimal production and operation periods. This section reviews the latest advances in choosing a site for developing a hydropower endeavor and the implications for the renewable sector after more of these projects are successfully greenlit in the upcoming years.

The first and most rudimentary consideration for site selection for a hydropower project is the availability and flow of water. Thus, proximity to a body of water is the baseline initial factor, but a few more are required to tip the scale toward deciding to green-light a project near water. Thus, parameters like river discharge, rainfall patterns, snowmelt, and water storage capacity are pivotal to choosing a location. For example, one piece of data that shows a strong correlation with hydropower generation estimations is regional runoff data, or the forecast of excess water flow that helps operate the turbine of a hydropower plant. For the United States, Kao et al. (2015) refer to the US Geological Survey WaterWatch runoff data to better predict hydropower production at a specific site.

Similarly, the water energy potential of the body of water, also known as its reach, is another valuable computation pivotal to the ulterior selection of a site to host a hydropower plant. In the case of small hydropower (SHP) projects, the reach of a river can be expressed in terms of average annual output (power in kilowatts) or the average annual energy (kWh) (UNIDO, 2019). Relatedly, the importance of collecting these data is even greater when considering that SHP structures generally lack a storage component, which makes them utterly dependent on the gross head (or the calculation of the height of standing water) and the available flow, both of which vary spatially (Fang and Karki, 2018; Moiz et al., 2018).

Then, the topography and geology of a site greatly influence the development and construction of hydropower projects. Hence, evaluating factors such as elevation, slope, and geological stability is essential to determine the feasibility of dam construction, reservoir formation, and turbine installation. For instance, steep slopes and stable geological formations offer reduced construction costs and improved project longevity. The various environmental considerations for hydropower projects are closely related to this. Here, factors like biodiversity, fish migration routes, and habitat disturbance are essential to determine the potential adverse impacts of hydropower endeavors.

Further, the economic viability of a project can be significantly affected by the proximity of the site to existing infrastructure and the availability of grid connections. In this sense, construction and operation costs can be lowered substantially if a location is already close to transportation networks, roads, and transmission lines. Similarly, a significant distance from urban centers can trigger losses on transportation and other unforeseen expenses, which could affect the construction stage and even the future efficiency of the plant.

In other considerations, social factors like the potential of adding jobs to a community or stimulating economic growth, increasing the notoriety and the tourism of a specific area, and the presence of heritage or Indigenous population sites can significantly alter site selection plans. In this sense, considering the nexus between ecosystems and water energy systems is key (Temel et al., 2023). So locations that allow for minimal displacements have limited social conflicts and present enticing opportunities, for benefitting local communities and economies is to be preferred when developing a hydropower project, particularly when the feasibility and costs of other projects are prohibitive (Wu et al., 2019).

As we can see, the difficulty of finding multiple trade-offs when deciding is apparent. For example, prioritizing biodiversity could mean compromising on the size of the installation, choosing a small hydropower project instead, and giving in on the infrastructure in place (Dorber et al., 2020). Concurrently, Chala et al. (2019) observed that large plants require significant deforestation, affecting biodiversity and potentially affecting local communities and their habitat. This stresses the importance of placing the features of each individual project in the center of the task, rank-ordering the different parameters according to what the specific location being assessed requires a holistic approach (Kwartnik-Pruc and Mączyńska, 2022; Romanelli et al., 2018).

Likewise, the legal and policy frameworks of the location in question may become a significant factor in choosing a site for a hydro energy project. In this sense, Tomczyk and Wiatkowski (2020) explored the country-level variations in European countries bounded by European Union renewable energy policy, smaller regional commitments, or both. Studying member states and countries outside the Union, the authors found that the framework nudges actors to comply. However, local regulations and their geographical constraints may play a part in some countries not reaching the stated goals. Conversely, actors outside the Union, like Albania, have taken advantage of their natural advantages and made hydropower their first energy source.

Table 8.1 shows a series of criteria for site selection of small hydropower projects that produce below 50 MW of electricity. However, these values may vary across authors and regions of the world, as Manzano-Agugliaro et al. (2017) point out. The seven main areas seen in the table encompass environmental considerations, geological and economic criteria, and various economic and legal factors, some of which may contradict or work against others (Moiz et al., 2018). Presented with these parameters, researchers have developed more sophisticated tools to rank them, including multi-criteria analyses, geographic information systems (GIS), and machine learning (ML).

In their work selecting a site for a PSH plant, Wu et al. (2019) developed a multi-attribute decision-making (MADM) framework to analyze the Chinese reality, where an evaluation criteria system,

TABLE 8.1
Multi-Criteria Analysis of Essential Factors to Select Locations for the Construction of Small Hydroelectric Power Plants

Environmental/Ecological Criteria in the Context of Minimizing the Impact of Building Structures on the Surrounding Environment	Economic Criteria in the Context of Increasing the Efficiency of Small Hydropower Plants	Social Criteria in the Context of the Profits from the Construction of the Power Plant for the Local Society	Topographic and Geological Criteria	Technical (Engineering) Criteria in the Context of Increasing the Efficiency of Small Hydropower Plants	Hydrological Criteria	Legal and Administrative Criteria in the Context of Constraints Related to the Possibility of Building and Locating Structures
• Maintaining minimum flow of the watercourse • Minimization of flooded area • Maintaining stability of slopes • Low movement of Earth masses • The least possible disturbance of the watercourse ecosystem • Maintaining direction of watercourse flow • Maintaining water quality in watercourse	• Valuation of investment costs, costs of maintenance, and operation of a water plant in a given period of time • The assessment should take the • following costs into account: • Direct costs: • Preparatory work • Environment protection • Construction work • Hydraulic equipment • Electrical equipment • Transmission lines • Modernization costs • Indirect costs: • Administrative and engineering costs • Unexpected expenses	• Increased employment • Local development of technical infrastructure (road, electricity, telecommunications networks) • Extension of public utility buildings • Economic development of local market • Tourism development	• Terrain height • Slope • Land cover (use) • Slope exposure • Size of the area to be flooded • Watercourse length • Distance from the river • Geological composition of rock formations in the area (susceptibility to landslides)	• Maximum power of a hydropower plant (maximum water flow rate) • Maximum annual electricity production • Minimum length of the transmission line (the longer the energy flow, the greater the energy losses) • Maximum ratio of annual electricity capacity (maximum land slope)	• Direction of watercourse flow • Watercourse flow rate • Energy potential (calculated from flow) • Tank size	• Legal and administrative restrictions depending on legal provisions in a given area • Restrictions related to environmental • protection zones

Source: Kwartnik-Pruc and Mączyńska (2022).

weights, and a ranking workflow are established. The authors decide on four prominent categories: the first is related to geography and engineering, the second considers the economy, the third section concerns environmental factors, and the last section groups social considerations. Within those categories, factors like reservoir volume and capacity, investment and rates of return, average rainfall and seismic activity, and residents' behavior and employment prospects are included, respectively.

Then, Chien et al. (2020) developed a similar framework for the emerging Vietnamese hydropower sector, adding many qualitative parameters that needed to be resolved in a fuzzy environment. Thus, the weight of these criteria was calculated using a fuzzy analytical network process (FANP). At the same time, the last stage consisted of a technique for order of preference by similarity to the ideal solution (TOPSIS) to rank-order all the potential plant locations. After running their model and assessing parameters like fauna protection, water quality, fish population, landscape quality, and flow regime, the authors find that the location of Nhge An is the closest to the positive ideal solution, given all these constraints.

For the case of a pumped storage hydropower installation, Jiménez Capilla et al. (2016) lay out their ranked criteria, obtained using a multi-criteria analysis coupled with other statistical tools like GIS and analytical hierarchical processes (AHP). As observed in Figure 8.3, their criteria relate to the geological features of the location, the elevation difference, the piping length, the excavation requirements, the existence of a lower reservoir, the allocation of a power station, the easiness of a grid connection, and the feasibility of low-cost electricity. Their framework also allows for a visual representation of the solution and the potential to analyze large areas of land and obtain different results, depending on the weights placed on each factor.

Continuing with frameworks assisted by GIS, the work of Romanelli et al. (2018) combines AHP with geographical information to apply selection criteria for small hydropower plants to be developed in the Brazilian state of Minas Gerais. Along with advancing these more sophisticated methods, the authors encourage the simultaneous use of environmental impact assessments (EIA) to gather more detailed information about social and environmental constraints. Meanwhile, a GIS-focused framework to choose sites for run-of-river installations is advanced by Jafari et al. (2021) for use in the Gilan province of Iran. The authors relied on the best-worst method (BWM), a multi-criteria decision-making model that seeks the optimal weights via a simple linear optimization through an expert questionnaire.

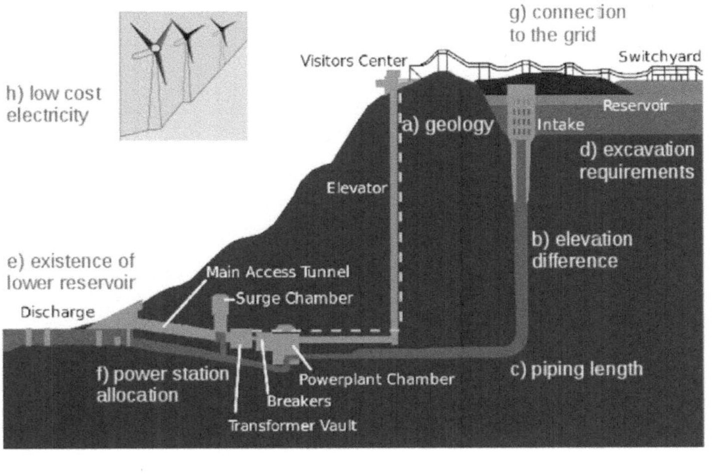

FIGURE 8.3 Illustration of a pumped storage hydroelectric plant pointing to the most important factors that affect its location.

Source: Jiménez Capilla et al. (2016).

Following this investigation, a sensitivity analysis was performed to determine the change in the target variables given changes in the input variables. A shortcoming of previous studies of GIS-AHP is that a small degree of sensitivity is observed in the rankings for the top potential options. This change is mainly due to the increased importance of parameters such as proximity to railroads, power stations, floodplains, and high-density population zones. However, this variation does not translate into significant changes for the most suitable areas (Jafari et al., 2021).

Then, in a review of selection criteria for different types of energy facilities, Khanlari and Alhuyi Nazari (2021) highlight the use of GIS-AHP and summarize the categories and subcategories of the most common criteria for site selection of hydropower plants as follows. First, an electricity generation dimension considers sub-factors like site capacity, load demand supply, and length of the transmission line. Then, an engineering and economics section has four subsections: project costs, difficulty in materials transportation, accessibility to the project site, and site obstruction. Lastly, the third main section contains socio-economic issues, where subdivisions are socially conflicted and have problems with water resources.

Shimray et al. (2017a, 2017b) get us closer to the ML frameworks by combining fuzzy approaches with neural networks in a related strategy of combining different methods to combat ambiguities and uncertainties. In this multi-layer perceptron (MLP) neural network trained with a genetic algorithm (GA), the framework separates the different attributes into different classes in an initial stage. Later, fuzzy reasoning is applied to prioritize these attributes so that policymakers can move along with selecting a site with an optimal road map for developing a hydropower plant. Figure 8.4 shows a hierarchy for these attributes to be used in different regions of India.

Meanwhile, a combined use of GIS with a MATLAB-based algorithm is seen in the work of Serpoush et al. (2017), who applied it in a large area of Iran to detect the most suitable areas for developing a hydropower project. The authors find that relying on these tools decreases the time to analyze large areas, which optimizes the selection process. Additionally, the output is as complete as a more prolonged study would yield, which means that alternatives have been identified and

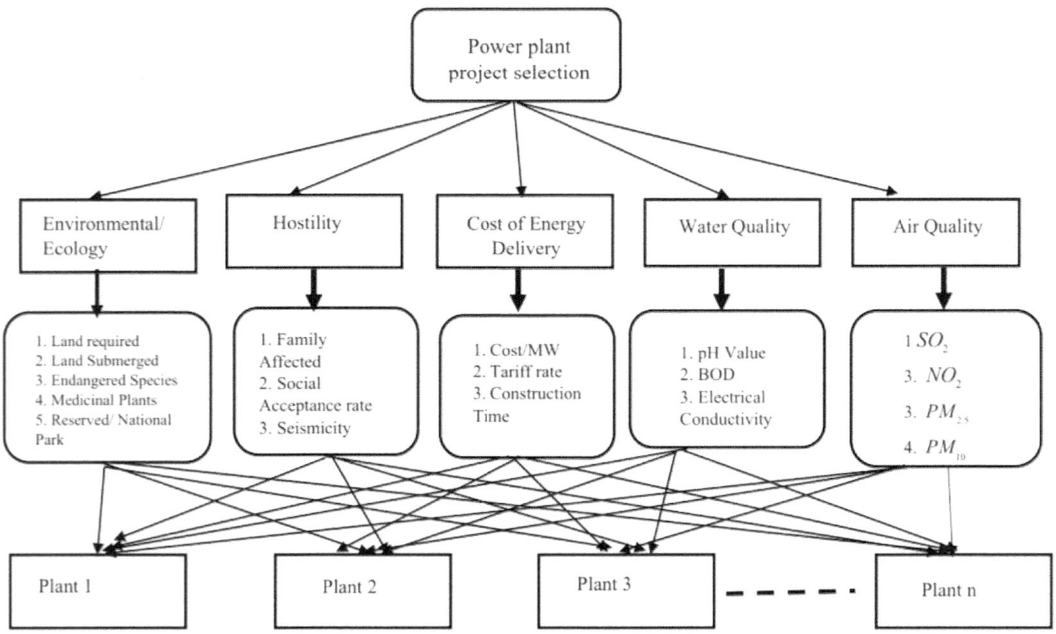

FIGURE 8.4 Hierarchical criteria structure and sub-criteria considered for hydropower site selection.

Source: Shimray et al. (2017a).

evaluated. Also, using this tool helps reduce human error, given that the mechanism consists of a taught algorithm using geographical data, which eliminates the need to rely on varying experts' opinions and surveys.

Also, Moiz et al. (2018) developed a support tool based on GIS that looks to optimize the site selection for run-of-river small hydropower setups. In this approach, the authors assess the role of topographical factors, with and without hydrological considerations, to obtain an optimized site selection. Using a case study in Pakistan, an improved water-and-energy-budget-based distributed hydrological model (WEB-DHM-S) is applied to simulate the long-term continuous spatial distribution of hydrological processes, along with snow variables added after this improvement. A significant strength of this framework is its ability to derive environmental flaws overlooked or parameterized by GIS configurations.

As Figure 8.5 shows, after deploying the WEB-DHM-S model for the target basin to simulate the basin hydrology and to generate the flow duration curves (FDCs) in each hydrological unit, the system analyzes all feasible hydropower configurations, yielding the most efficient non-conflicting alternatives. This step is based on PROMETHEE II, a decision-making software that applies weights to alternatives within a multi-criteria environment. Results show that following only topographical factors, the system provided 36 optimal site alternatives with a potential of 235 MW at 70% flow exceedance. Meanwhile, when applying topographic and hydrological factors, 26 sites were advanced with the same parameters of production (Moiz et al., 2018).

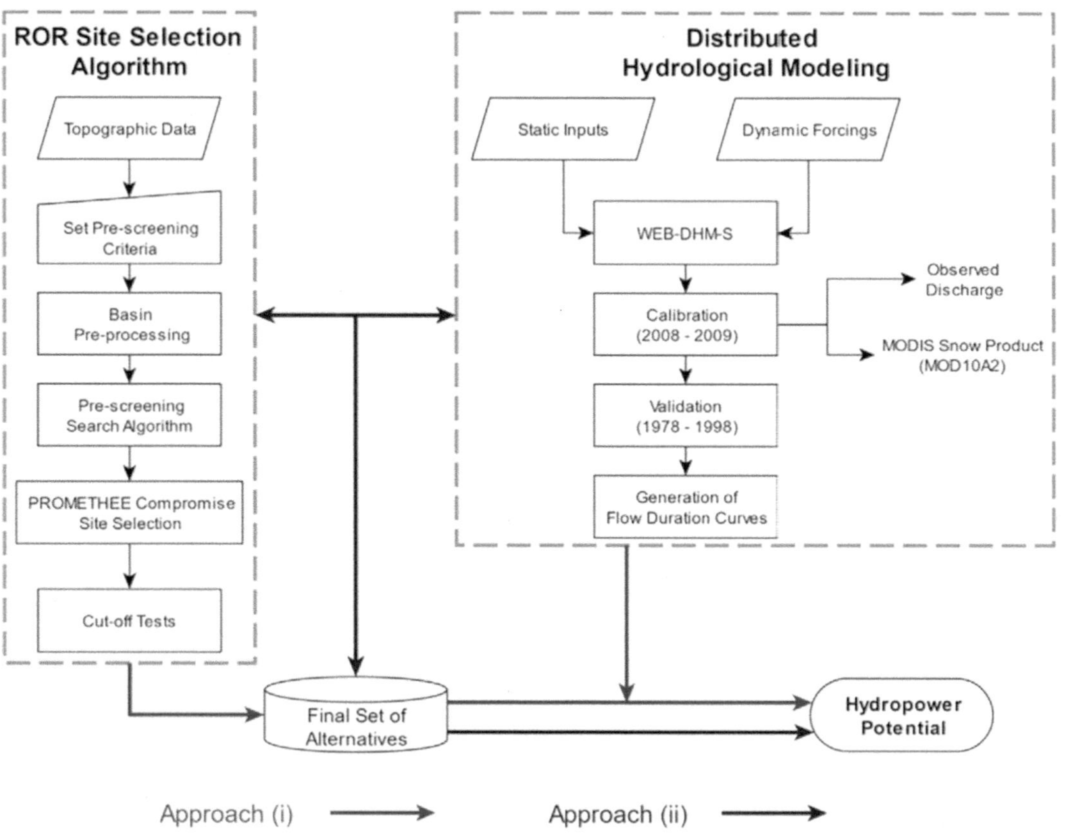

FIGURE 8.5 Analytical framework of the integrated tool.

Source: Moiz et al. (2018).

This section reviewed the main criteria to consider when selecting a site for a future hydropower project. As expected, water levels and flow considerations are first on the list, yet many other variables could promote or relegate a potential site from the list. The literature shows that geological and topographical factors are a close second. At the same time, an economic dimension is also pivotal through parameters like proximity to transmission lines and prior infrastructure in place. Moreover, environmental and social factors like the presence of heritage or native sites, public opinion and perception, as well as the legal frameworks in place are critical contributors to site selection.

Adding a significant challenge to choosing a site for optimal electricity production, many of these parameters are at times against each other, so it becomes absolutely relevant to consider a site's individual features when making the decision. One of these factors could be a project's economic feasibility, which could be higher in one location than another, keeping all else constant. In the following section, we will analyze the economics of hydropower methods and projects in more detail so that engineers and researchers have more tools at their disposal when choosing a site for future exploration.

8.3 ECONOMICS AND DESIGN FEATURES OF HYDROPOWER TECHNOLOGIES

As noted earlier in this chapter, hydropower projects are reportedly more cost-effective than similar endeavors, making them an attractive alternative to fossil fuel installations like coal or gas. However, plenty of nuances and gray areas need further exploration, such as the trade-offs observed between environmental gains and economic feasibility, particularly when adding design characteristics to the equation. This section explores both the design constraints for hydropower projects and the economics of these endeavors in more detail, providing a clearer picture to the reader as to the opportunities, challenges, and paths forward for this technology in the context of the energy transition to renewables.

Generally, hydropower endeavors require significant initial investment for construction and operation purposes. According to IRENA (2023), almost half of the total costs of a hydropower project are related to civil works or the construction of dams, tunnels, and any other infrastructure required to access the project. Then, almost a third of the costs are procurement expenditures, like the electromechanical equipment to operate a plant. Depending on the size of the project, these costs can be prohibitive for some actors. In the first decades of hydropower technologies, deploying these projects in less-developed countries was more challenging. However, with the advent of novel methods, efficiency in production and economics can be obtained with smaller projects.

That initial economic hurdle can be significant for many actors. Nevertheless, once the installation operates and benefits people, hydropower is one of the cheapest forms of electricity, so fewer future costs are passed on to the consumer compared to other energy sources. This can be further optimized with an installation design scheme that is tailored to the specific needs of the location that is being served and aided by technologies like machine learning and AI, which would optimize parameters like the number and size of turbines, the distance to transmission lines, or various generator specifications, like its speed and various auxiliary cooling systems (Nasir, 2021).

Another way to reduce costs involves deciding against an external energy storage unit. This may be an advantageous strategy at locations where costs are already prohibitive. However, the trade-off in this scenario is that lacking a storage unit can leave a location subject to intermittency issues caused by the variable water flow. This gamble could pay off in locations with high predicted water flows throughout the year but could provide a challenging test for sites with occasional droughts. This has motivated some players to develop pumped storage hydropower (PSH) installations, which essentially act as batteries, due to their ability to store excess power and release it when needed (U.S. Department of Energy, 2023d).

With that said, some authors recommend against building storage units for some plants, given the space they occupy being a potential hindrance to the environment around them. For example, the work of Chaudhari et al. (2021) explores hydropower development in the Amazon region, promoting in-stream turbines that only use a fraction of the river stretch that a larger project with dams and storage units would use. In-stream turbines are smaller mechanisms placed into the flowing body of water, neither blocking the passage as dams do nor increasing the water level significantly, thus allowing high connectivity and assuring a higher ecological safety (Brown et al., 2023).

As a key element in the optimization of hydropower installation designs, one that is also strongly related to cost management, turbines have gone through their process of development and improvement, which we will review next. In general terms, depending on the type of hydropower plant, engineers rely on different types of turbines and on a few or multiple of them to achieve the energy goals of the installation. The first hydro structures featured large hydrokinetic turbines, evolving directly from old waterwheels. However, more novel technologies have improved their size, weight, and level of interference in the environment around them (van Els and Junior, 2015).

Hydrokinetic turbines (HKTs) were developed to capture the kinetic energy of water that flows in rivers, canals, and irrigation systems without constructing large structures. Original designs featured large wheels that move with water flow, with arrangements borrowing from wind turbine technologies. However, a few problems surface when trying to scale these down, as capital costs are high while the capacity factor of the structure remains low. This means that the ratio between the observed and the potential electrical energy that would have been produced with the structure at full capacity remains inadequate to deploy these projects in many different locations. Because of this, remote locations rely on cheaper options, such as diesel, to generate much more energy for the same price (Kirke, 2020).

In addition to these economic and efficiency problems, large structures like dams have fallen out of favor in recent decades, given their significant environmental and biodiversity trade-offs. For example, Moran et al. (2018) have observed that river ecology, deforestation, loss of terrestrial and aquatic biodiversity, and life-altering population displacements are significant unintended consequences of constructing and operating large hydropower structures. Moreover, greenhouse gases emitted while building and operating these plants are becoming a real preoccupation when considering the various climate change–related pledges many governments have committed.

These significant constraints have invited engineers and researchers to rethink hydrokinetic turbine technologies to adapt their new developments to a world where sustainability and environmental protection are key considerations for viable projects. Obviating the need for large structures, novel hydrokinetic turbines follow two main principles depending on the positioning of the rotor axis, being classified as axial-flow/horizontal-axis structures if these are parallel to the water flow and use propeller-type rotors. Meanwhile, turbines with rotor axes perpendicular to the water flow are classified as cross-flow/vertical-axis structures (Kumar and Saini, 2016). Figure 8.6 shows different types of turbines developed for harnessing water energy following those two main criteria.

The main advantages of horizontal axis turbines are their high efficiency, as well as their self-starting capabilities. However, they demand higher manufacturing, transportation, and maintenance costs, can easily clog with debris floating on the water, and some of their setups require components that must remain water-sealed, like generators. For these reasons, vertical-axis turbines appear better suited for energy extraction from free-flow streams (Kumar and Saini, 2016). A handful of authors have explored the potential of Savonius vertical turbines as the structures better suited for low-speed channels and rivers, assessing parameters like size, arrangement, and blade rotor shape for their S-shaped turbines (Prabowoputra et al., 2020).

Moreover, different ways to classify turbines have been advanced in the literature over the years. One of these distinctions is drawn by Jawahar and Michael (2017), who separated the use of turbines according to the head of the particular installation, which is understood as the vertical height between the hydro intake level and the water level at the point of discharge. Under this classification, turbines are seen as having low head (up to 40 m), medium head (between 40 and 100 m), and high head (more than 100 m). Following these distinctions, propellers and Kaplan turbines fit low-head structures; Francis, Pelton wheels, pump as turbine, and cross-flow are considered medium-head; and Turgo and Pelton wheels can fit into the high-head classification.

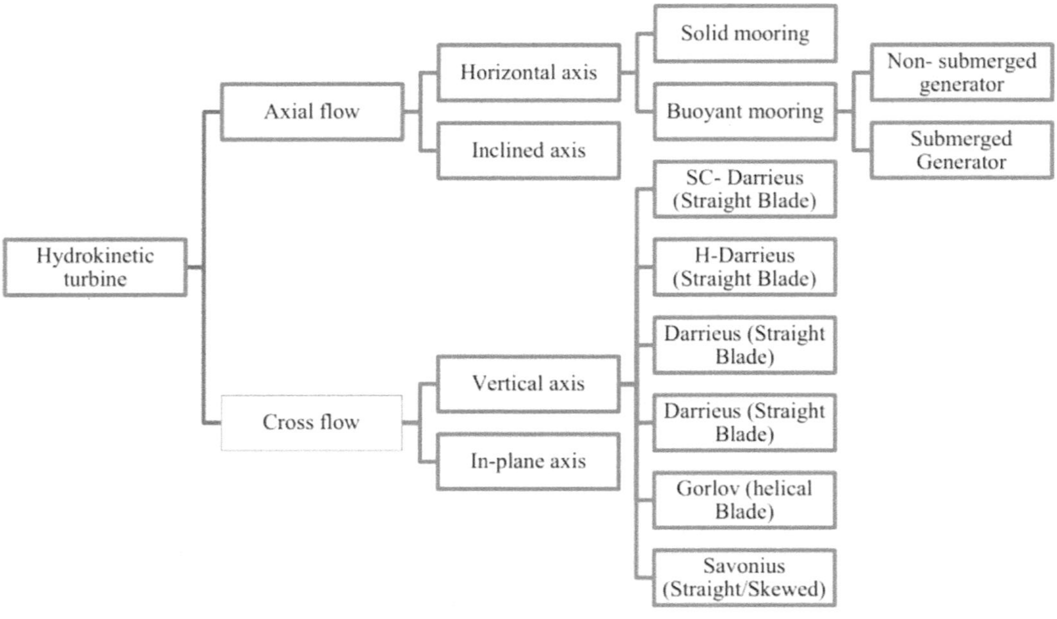

FIGURE 8.6 Classification of hydrokinetic turbines.

Source: Kumar and Saini (2016).

Figure 8.7 shows four of these types of turbines, all with notable differences that align better with particular contexts. For example, the Pelton turbine features one or more free jets that discharge water into an aerated space that impinges on the buckets of a runner. These types of turbines are ideal for high heads and low flows. Meanwhile, cross-flow turbines feature a circular runner placed against a rectangular section placed against the vanes of the runner. In this setup, water flows through the blades twice, with the first pass going from the outside of the blades to the interior and the second pass going from inside to the exterior (U.S. Department of Energy, 2023e).

Then, Turgo structures are commonly used in medium- to high-head setups and are able to handle higher flow rates, providing a high efficiency by using more water with less head. This efficiency is achieved thanks to the runner speed, which is double the speed of Pelton turbines (Elbatran et al., 2015b; Nasir, 2021). Lastly, Kaplan turbines feature adjustable blades and wicket gates, which allow for a broader range of operations. These latter structures are classified as reaction turbines, which generate power from the combined forces of constant pressure and moving water and are better suited for sites with lower head and higher flows (U.S. Department of Energy, 2023e).

Achieving a higher coefficient of power for the performance of turbines allows for higher efficiency, which translates into economic benefits for producers and consumers. The use of different types of turbines also depends on the context, as larger structures will be more appropriate for bigger projects. With that said, the emphasis in this past decade has been placed chiefly on smaller projects, where micro-hydropower endeavors have stood out as the most promising alternative, potentially being applied in the largest set of contexts. Micro-hydropower systems commonly generate up to 100 KW of energy, requiring either a turbine, a pump, or a waterwheel to transform the energy of flowing water into rotational energy, later converted to electricity (U.S. Department of Energy, 2023c).

Following this classification, *impulse turbines* are defined as structures that work with the velocity of water to move a runner and are better suited for high-head, low-flow settings. Figure 8.8 shows that most of the turbines analyzed are impulse structures, while one of the most well-known turbine setups, the Francis turbine, is categorized as a reaction turbine. Francis turbines feature a spiral casing and controllable guide vanes, where water is introduced just above the runner, falling through

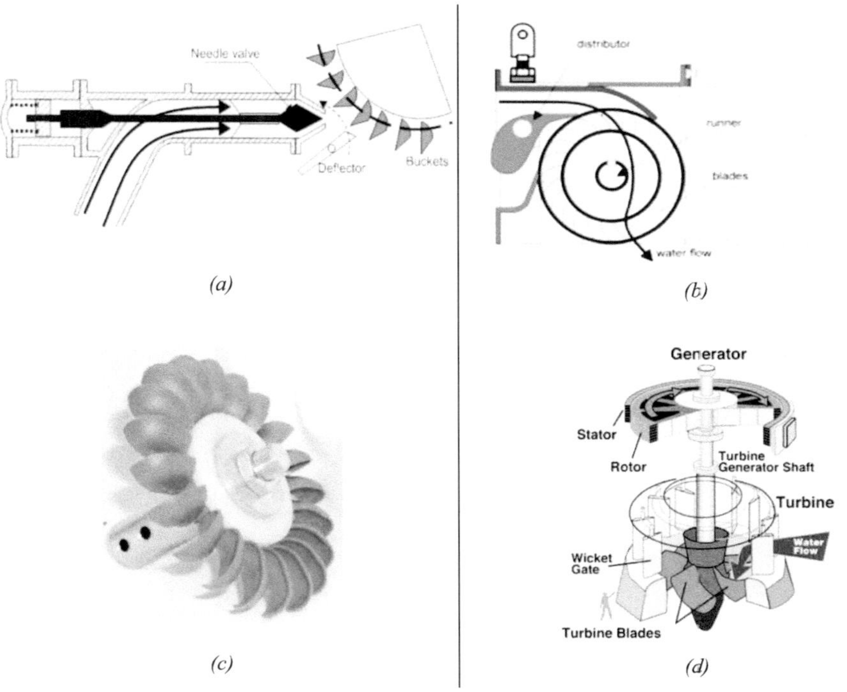

FIGURE 8.7 Illustrations of different impulse and reaction turbine types: (a) Pelton turbine; (b) cross-flow turbine; (c) Turgo turbine; (d) Kaplan turbine.

Source: Elbatran et al. (2015b); Nasir (2021).

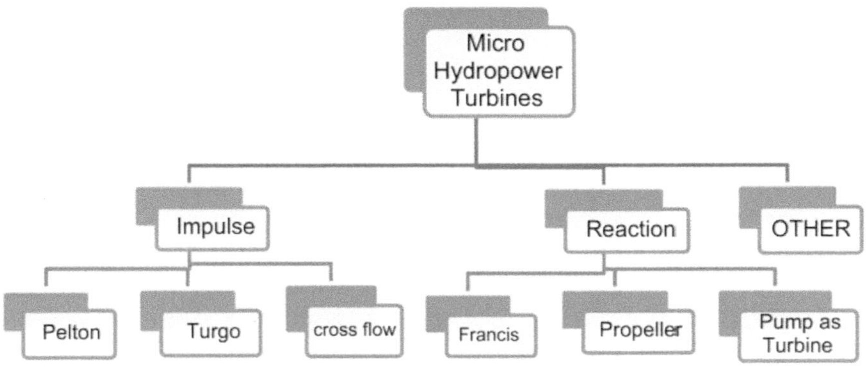

FIGURE 8.8 Micro-hydropower turbines classification.

Source: Elbatran et al. (2015a).

and spinning the runner. These structures can work both in horizontal and vertical arrangements, with Figure 8.9 illustrating a cross section of a horizontal setup (Elbatran et al., 2015b; Nasir, 2021; U.S. Department of Energy, 2023a).

Meanwhile, pump-as-turbine (PAT) systems are a more novel technology perfected to suit smaller contexts and remote villages or serve as irrigation mechanisms. PATs, also referred to as reversible pump turbines, work when a pump rotates in the inverse direction, with water entering the pump under very high pressure, moving through the impeller blades, which releases pressure

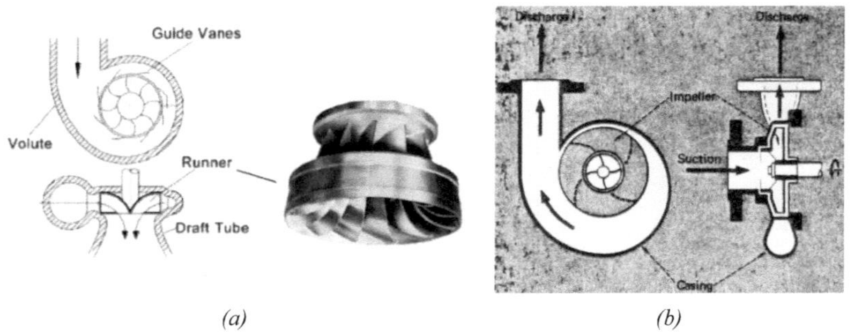

FIGURE 8.9 Illustrations of (a) Francis turbine and (b) pump as turbine (PAT) reaction structures.

Source: Elbatran et al. (2015b).

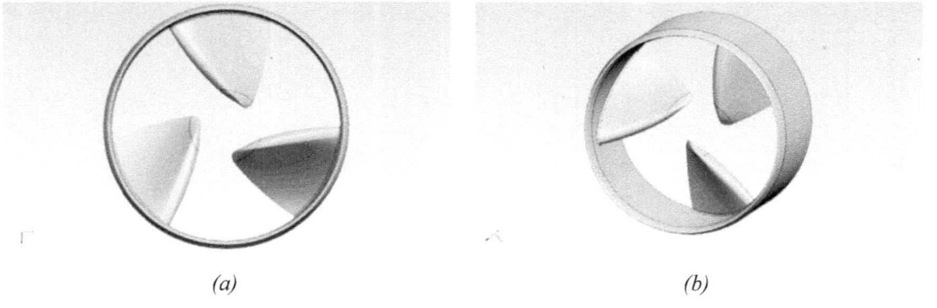

FIGURE 8.10 Illustrations of ducted rotor in-stream turbines: (a) flow-direction view; (b) oblique view.

Source: Brown et al. (2023).

and kinetic energy (Binama et al., 2017; Elbatran et al., 2015a). Despite their lower costs, shortcomings are related to the uncertainty of pump performance when used as turbines, their low part–load efficiency, or their capacity to operate optimally at off-design times (Novara and McNabola, 2018).

Then, mentioned briefly at the beginning of this section, in-stream turbines are minimal structures that seek to maintain the positive economic aspects of hydropower projects while combating the potentially negative socioenvironmental consequences of larger endeavors (Chaudhari et al., 2021). As Figure 8.10 shows, in-stream turbines are negligibly invasive and improve many of the mechanisms for injury or mortality of the fish and other marine species that come in close contact with these structures, such as pressure change, physical strike, turbulence, and shear. Additionally, in-stream turbines improve general conditions for marine life, helping to decrease the sound emissions from these devices (Brown et al., 2023).

As we can see, the design of a hydropower energy project is intrinsically linked with the economic considerations of the site. In this sense, ventures need to ponder their potential to produce vast quantities of energy and provide consistent baseload power. With that said, today's world also demands energy actors to assess variables like sustainability and the protection of marine life, which are now as important as revenue generation plans. These new variables have made the selection of design elements a pivotal consideration in determining the success or failure of hydropower projects worldwide.

Exploring this latter dimension in more detail, Quaranta et al. (2021) advanced the use of environmentally enhanced turbines (EETs) to improve water quality while reducing marine life injury. Focusing on the Alden and the minimum gap runner (MGR) turbines, the authors show that these structures are cost-effective, efficient in energy generation, and environmentally friendly. The Alden consists of an evolved design of the Francis structure, with three blades wrapped around the shaft to reduce fish injury. Meanwhile, MGRs are adapted from Kaplan bulb turbines, with reduced gaps between the runner blades and the hub that seeks to minimize marine life injury and death.

With the improvement in the selection of hydropower installation components in mind, some authors have developed statistical tools to simplify these processes, providing specific main characteristics of a hydropower venture. For example, the work of Novara and McNabola (2018) advances the use of a software program to assist the designer in selecting the most optimal plant layout, emphasizing the uncertainty observed in pump deployment as turbine projects. Their analysis shows that the design of a PAT venture is subject to a series of trade-offs, and its performance will be tied to factors like construction type, size, and flow conditions under which the machine is expected to operate.

Similarly, Binama et al. (2017) advanced a framework for selecting turbines in micro-hydropower projects that considers site parameters such as the head and flow, an estimation of turbine speed, as well as their approximate output power, rotational speed, and other pertinent factors. Figure 8.11 illustrates this workflow, which seeks to simplify an increasingly sought-after procedure centering on established criteria affecting the operation of hydro plants. In this respect, Singh and Singal (2017) assess the main parameters for hydropower projects to achieve minimal environmental effects while generating maximum energy, arriving at a list highlighting head, discharge, turbine, and generator as the most salient variables to consider.

The improvement in numerical and statistical methods to estimate ideal scenarios to deploy a hydropower project is perhaps seen best in developing disciplines like computational fluid dynamics (CFD). These studies help simulate various hydrodynamic phenomena occurring while turbines operate, which helps researchers understand the strengths and limitations of different designs, pointing them at optimal alternatives. Likewise, CFD techniques have assisted in understanding intricate details of flow features, such as rotor–stator interactions (RSIs), fluid–structure interactions (FSIs), and processing vortex core (PVC) calculations, all key to the design improvement of components like distributors, runners, and draft tubes (Tiwari et al., 2020).

In more detail, some authors have utilized CFD techniques to simulate flow conditions under varying hydraulic behaviors of a penstock, which is the name of the pipe that drives the water intake from the source to the turbine for electricity generation. Studying a refurbished Czech plant, Souček et al. (2023) found that measuring three flow measurement components, current meter, pressure time, and ultrasonic transit time, can help in reducing the uncertainty of an operating regime. The use of CFD models populated with historical data of a particular installation can help in the estimation of the most optimal characteristics of certain components, such as the penstock, or in predicting the behavior of a given section of a plant given an uncertain event.

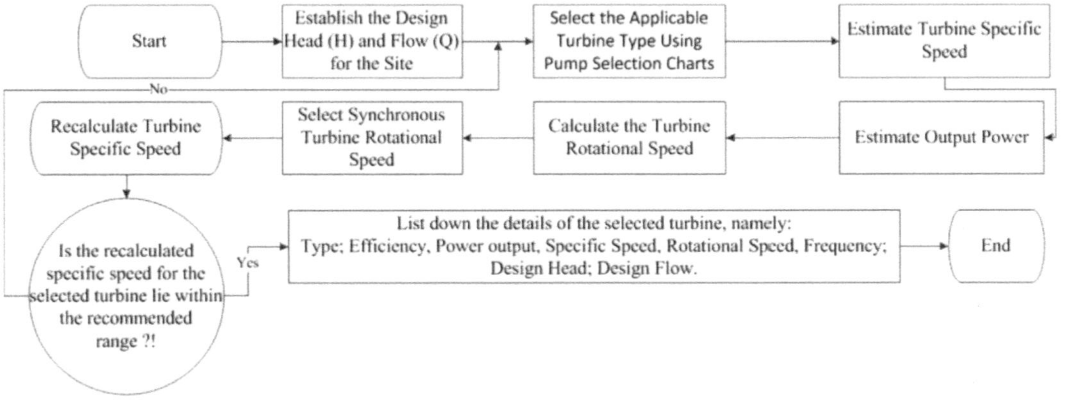

FIGURE 8.11 Turbine selection flowchart for micro-hydropower schemes.

Source: Binama et al. (2017).

In this sense, Mandair et al. (2021) explored a potentially threatening situation observed when a hydropower plant goes under a full load rejection, occurring when a sudden loss in the system is observed, which could lead to structural damage if unresolved quickly. Using a 3D CFD model coupled with a one-dimensional method of characteristics (MOC) solver, the authors compared high-level data, like head, flow, torque, and rotational speed, to estimate turbine behavior under this uncertain scenario. Moreover, the source for a potential load rejection is found to be unstable vortices inside the runner.

Then, Yildiz and Vrugt (2019) explore advanced numerical models to simulate and optimize site features in small run-of-river (RoR) plant designs, with the ultimate goal of predicting energy production with cost minimization. The authors introduced a simulation model called HYPER, coded in MATLAB®, that estimates the daily performance, investment, operation, and maintenance (O&M) costs and the potential economic profit of a RoR hydropower plant. These predictions are made following a set of project-specific variables, such as generator efficiency, penstock cost per ton, runner position above the tailrace, project lifetime, and various O&M cost coefficients.

Relatedly, thanks to the development of machine learning (ML) techniques, advanced numerical analyses like flow dynamic estimations can be improved by these simulation methods. For example, the work of Tapia et al. (2021) explores the use of a genetic algorithm (GA) to optimize the design of penstock layouts for micro-hydropower installations. Starting with a topographic survey that determines a series of optimal points, the proposed ML framework later estimates parameters like penstock length, number of elbows, and pipe diameter to prescribe the most optimal values for power generation capacity while reducing costs.

Observing the increase in the use of ML techniques for hydropower processes, Kumar and Saini (2021) explored the different areas where these methods are used. The authors find that performance optimization, forecasting plant parameters, monitoring and control of installations, policy and feature selection, feasibility studies, and both accuracy evaluation and capability assessment reviews are the areas most benefitted by ML. This development is particularly important for hydropower economics because the initial costs required to advance a project are very high, and improved simulations of energy generation, demand, supply, head, and load will help lower selection, operation, and maintenance costs.

Meanwhile, a systematic review of the penetration of AI-assisted techniques in hydropower studies performed by Bernardes Jr. et al. (2022) finds that machine learning methods are utilized across all aspects of hydropower operations, applied in pre-operation, real-time, and post-operation phases. The preferred ML technique is supervised learning, which is applied mainly in long-term scheduling horizons, while reinforcement learning and unsupervised learning are two other techniques with widespread use. Also, the hydropower areas where ML methods are most applied currently are river flow forecast and the optimization of reservoir operations.

Hydropower has gone through a period of intense technical advancement. At the same time, its large setups have been increasingly questioned due to the potentially harmful effects on the biodiversity of the location where plants are built, along with the negligible effect on the quality of life of the people in the communities where an installation is built. For example, the work of de Faria et al. (2017) suggests that the justification to build large hydropower plants due to long-term economic benefits is not supported by the evidence, which shows that social indicators like life expectancy, access to clean water and public electricity, and average income are not improved significantly in locations where hydropower projects are conducted.

These findings invite researchers to explore smaller-scale solutions featuring a solid environmental aspect, which sometimes becomes challenging for profit-seeking actors. That said, the series of trade-offs that appear under these conditions can be better confronted with a combination of the

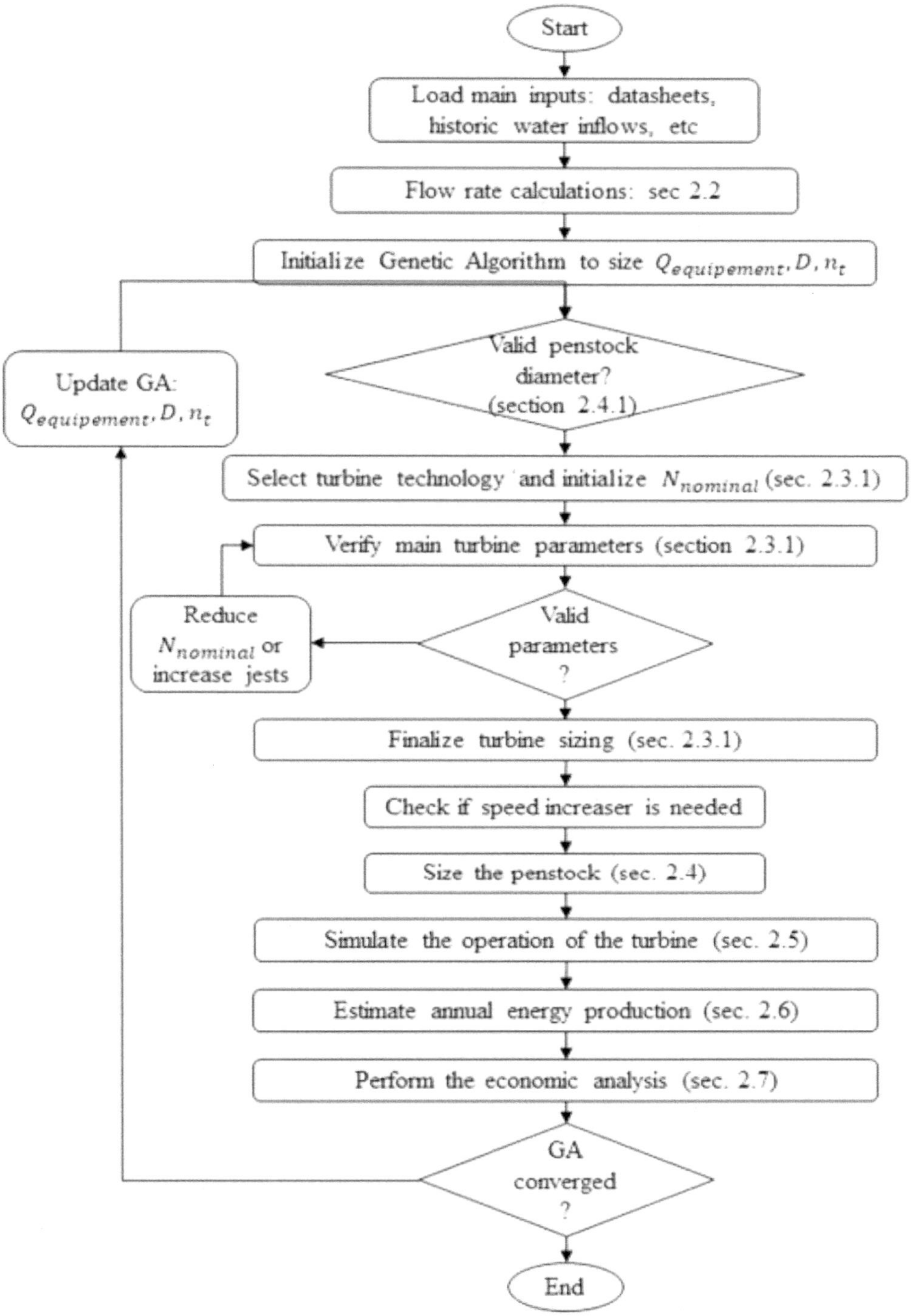

FIGURE 8.12 Optimization process minimizing levelized cost of energy (LCOE) for hydropower projects.

Source: Amougou et al. (2022).

novel techniques we have reviewed thus far. An example of this is seen in the work of Amougou et al. (2022), who developed a genetic algorithm to optimize plant characteristics that yield an optimal levelized cost of energy (LCOE) for small run-of-river hydropower plants. The model is visualized in Figure 8.12 and helps estimate variables like penstock thickness, selection and location of a turbine, energy production, and various costs for a small plant when connected to an existing electricity network.

This section dissected the various infrastructure and economic challenges confronted by engineers seeking to advance hydropower endeavors. As we saw, turbine design is a critical element in optimizing a project, which can signify the achievement of economically feasible projects in diverse parts of the world. Advanced numerical models powered by machine learning have gained popularity and are being perfected as we speak to aid in these decisions. Machine learning models can now take an increasingly large number of variables so that uncertainty is reduced and reliable estimations of energy production and its costs are reached.

In the next section, we review the different types of hydropower applications, those historically preferred and those that have gained popularity in the past couple of decades, given the new set of environmental constraints related to climate change and marine life protection. Additionally, we explore the dynamics that have led to these selections, along with the technologies that are behind the expansion of hydropower around the world. Later, we analyze the relationship between the oil and gas sector and the hydropower industry and how different actors like oil majors, governments, and international organizations have helped shape the current policy landscape.

8.4 HYDROPOWER APPLICATIONS AND CURRENT TECHNOLOGIES

The force of water is a tested method to provide energy in many parts of the world. What started as a source of power in small to medium households through waterwheels evolved into the development of large installations that operate big turbines cleanly and safely. However, some of these advances are now challenged by notions of preserving marine life and sustainable operation. In this segment, we review the different applications and their strengths and weaknesses, as well as the opportunities and the paths forward brought by novel methods and techniques to develop hydropower projects around the globe, benefitting millions of people.

Figure 8.13 shows the current hydropower capacity installed across continents, with Asia boasting the largest share of this capacity, concentrated in countries like China, India, Japan, Vietnam, and Pakistan. This capacity also features over 70 GW of energy delivered by pumped hydro storage plants, more than doubling the capacity provided by Europe and North America (Andritz, 2023; IRENA, 2023). Given the current placement of incentives and the notions of sustainability and environmental protection, we can only expect these numbers to continue growing, particularly in the BRICS countries and developing nations with enough funding to develop micro-hydropower endeavors (Zeng et al., 2017).

Hydropower project classification varies depending on their capacity and the primary technology they utilize. As shown in Figure 8.14, size considerations separate endeavors into four major categories: large-, small-, micro-, and pico-hydropower. Following the distinction made by Fortaleza et al. (2018), large projects provide over 100 MW, medium setups reach between 15 and 100 MW, and small-hydro (which includes their own class of mini-hydro) provides between 100 kW and 15 MW. Micro-hydro includes projects up to 100 kW, while pico-hydro features under 5 kW of energy generation.

Meanwhile, when talking about the technology featured in these projects, there are also a few essential distinctions. First, dammed reservoir projects, or impoundment facilities, are large projects that require the construction of a water reservoir and dams to seclude the water. Then, run-of-river (RoR), or diversion setups, are smaller stations that take advantage of the natural hydrological variations to produce electricity, meaning, no artificial reservoir is required. Further, pumped storage hydropower (PSH) consists of a facility that stores energy from other sources for later use, pumping water from a lower reservoir to one located at a higher altitude. When demand is high, the

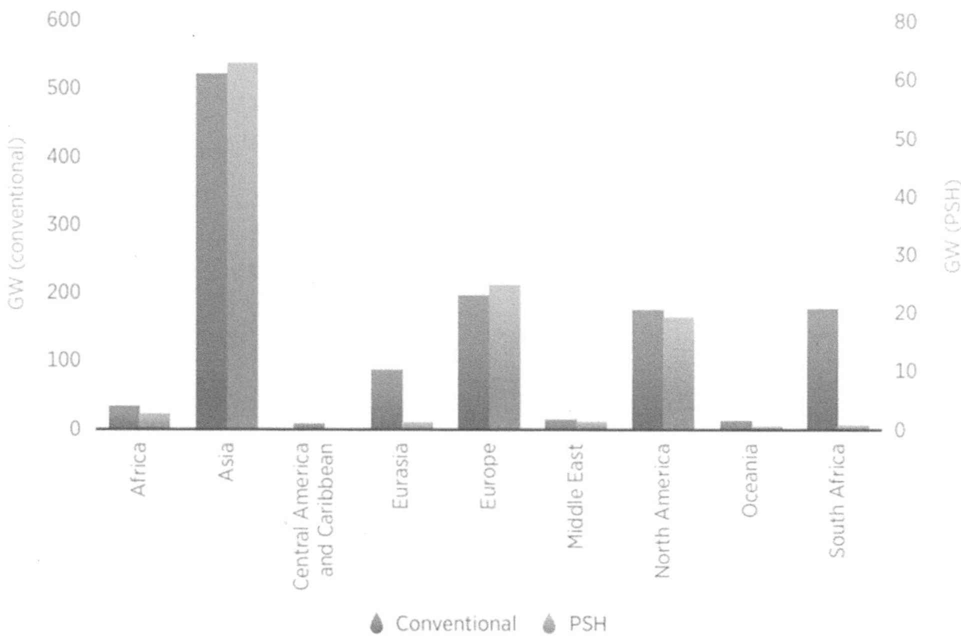

FIGURE 8.13 Hydropower installed capacity by region, 2021. (*GW* = gigawatt; *PSH* = pumped storage hydropower.)

Source: IRENA (2023).

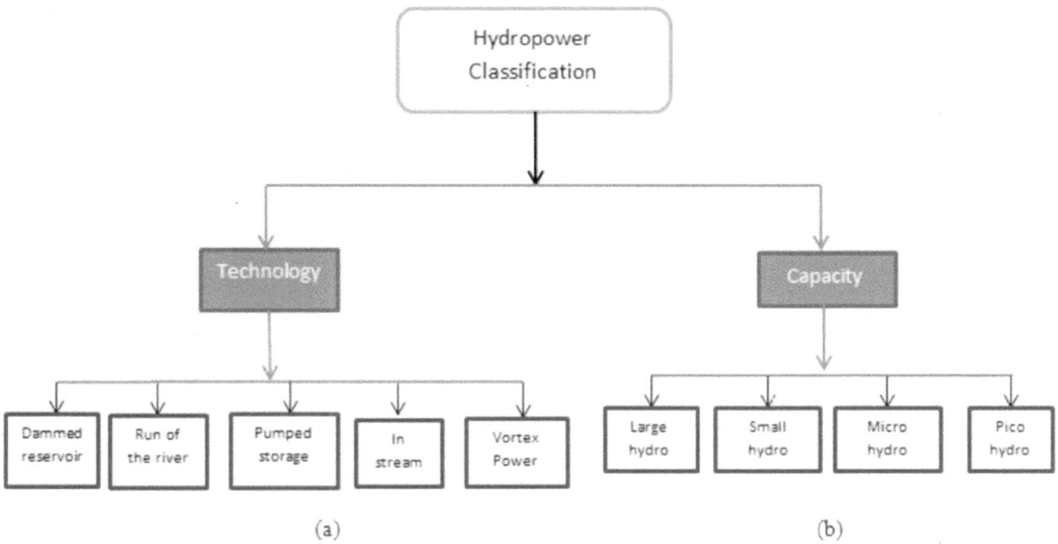

FIGURE 8.14 Hydropower categories by (a) technology and by (b) power capacity.

Source: Elbatran et al. (2015b).

water is released back down, operating a turbine that produces electricity (Elbatran et al., 2015b; U.S. Department of Energy, 2023e).

The last two types of technologies consist of in-stream interventions, which also take advantage of the natural hydrological situation of the location, placing turbines in free-flowing rivers or engineered waterways to produce energy. Then, vortex power technologies generate electrical power by creating artificial whirlwinds without propellers. Still, in a developmental stage, vortex technologies are promising alternatives that use the power of water, incurring fewer infrastructure costs, thus lowering the potential environmental impacts that plague larger projects (Elbatran et al., 2015b). In the rest of this section, we will review the leading hydropower technologies in greater detail.

8.4.1 Conventional Hydropower Plants

Conventional hydropower consists of the classic setups where a reservoir is formed, and large dams are constructed to create the electricity generation mechanisms. Power generation is achieved thanks to large quantities of stored water being released through turbines, turning a generator that produces power. Figure 8.15 shows the typical layout of these installations, with all the infrastructure required to generate large quantities of electricity that go straight to the grid. In these plants, the water can be saved and used for other purposes, such as irrigation, flood control, and even recreational uses, like fishing and boating (U.S. Department of Energy, 2023e).

The strengths of conventional hydropower applications are centering on their high energy efficiency. They can produce large quantities of electricity from a renewable and potentially inexhaustible source. Moreover, when built in locations with favorable water flows, these plants provide a reliable and consistent power supply that rivals those coming from fossil fuels, which could make hydro the main player in the suite of renewable alternatives leading the energy transition. Also, hydropower is flexible, quickly going from zero to maximum output, aiding the grid in times of significant disruptions (U.S. Department of Energy, 2023a).

Moreover, as mentioned before, large hydropower installations can have a few other uses for the water, such as irrigation, flood control, clean drinking water for different communities, and recreational uses, such as swimming, fishing, and navigation. Last but not the least, another important

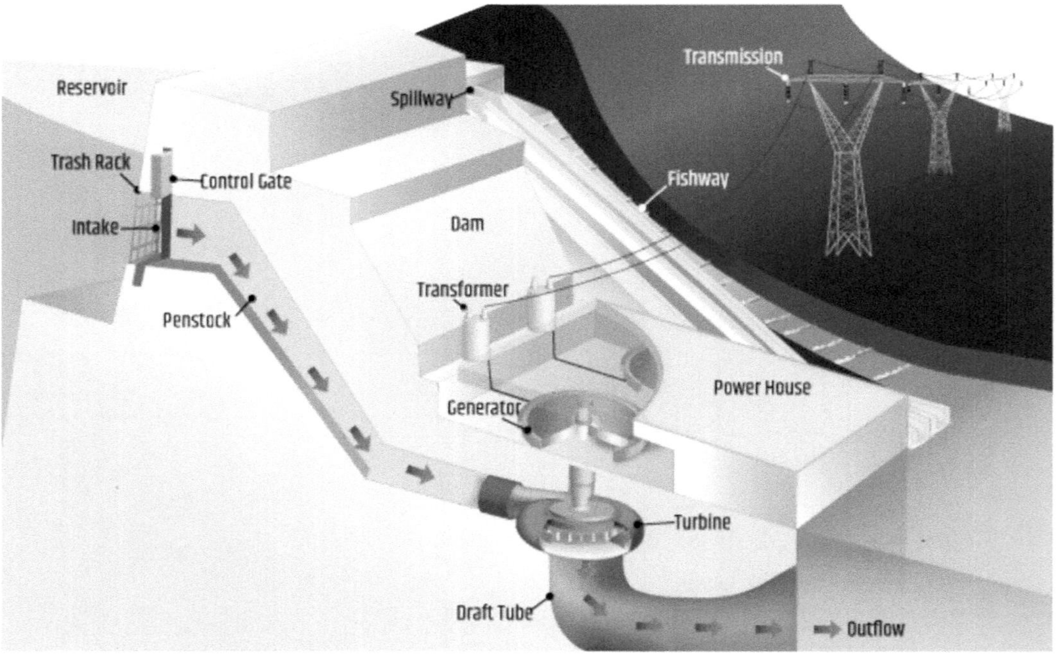

FIGURE 8.15 Impoundment, or conventional, hydropower installation.

Source: U.S. Department of Energy (2023e).

aspect of these installations is the large amount of jobs the sector has created over the decades, offering competitive wages. According to the U.S. Department of Energy (2023a), the sector employs over 65,000 people a year, potentially growing almost threefold by 2050, particularly benefitting rural locations and boosting local economies.

On the other side of this coin, there are marked environmental considerations that these large projects have not appropriately addressed for decades. As mentioned earlier, important habitat disruptions and alterations are observed on large hydropower sites, including dangers of fish and marine life injury or death, due to risky drainage operations and multiple hydraulic structures (Cox et al., 2023). Also, some communities may see their way of life threatened due to potential displacement and erosion of their cultural heritage. Additionally, sites selected for conventional hydro projects require long periods of construction and significant investments that take considerable time to repay, sometimes extending into decades (Bagher et al., 2015).

On this latter point, we are observing several plants reaching the end of their expected life, so aging infrastructure has become another challenge for conventional hydropower. As Table 8.2 shows, most components in a hydropower plant have economic and technical lifetimes of between 20 and 50 years, which places many current plants in a situation where they need replacement, refurbishment, or perhaps complete retirement (IRENA, 2023). These costs have had the added consequence of making some investment partners stay away from large projects, focusing instead on smaller, easier-to-manage hydropower endeavors.

These dynamics have relegated large hydropower projects to a diminished role in the energy transition, with relevant actors promoting smaller projects with a more sustainable angle. With that said, this technology has many paths forward, provided that these drawbacks are assessed and improved. A few avenues involve optimized site and infrastructure selection, adopting more environmentally friendly turbine systems, and various methods to diminish disruption of communities and marine life while building and operating plants. These improvements can be obtained

TABLE 8.2
Lifetimes of Selected Hydropower Plant Components

	Economic Lifetime (Years)	Technical Lifetime (Years)
Electrical Installations		
Generators, transformers	25–40	30–60
High-voltage switchgear, auxiliary electrical equipment, control equipment	20–25	30–40
Batteries, direct current equipment	10–20	20–30
Mechanical Installations		
Turbines		
Kaplan and Francis turbines	30–40	30–60
Pelton turbine	40–50	40–70
Pump turbine and storage pumps	25–33	25–50
Gates, butterfly valves, special valves, cranes, auxiliary mechanical	25–40	25–50
Civil Works		
Dams, canals, tunnels, caverns, reservoirs, surge chambers	60–80	80–150
Powerhouse structures, water catchment, spillway, sand traps, penstocks, steel linings, roads, bridges	40–50	50–80

Source: IRENA (2023).

with thorough environmental impact assessments and proactive engagement from stakeholders and governments.

Despite a significant loss in favorability over the past couple of decades, there are many contexts where an extensive hydropower system can still be very beneficial to a given community in need of reliable power. Improved techniques for site selection, component design, and the optimization of operation and management of finished plants continue to place conventional hydropower in an essential position as a technology that uses a renewable source to produce large quantities of energy. Therefore, as long as conventional hydropower provides a viable alternative to fossil fuels, transitioning to cleaner sources requires not their elimination but a better balance between the promising hydropower alternatives that have been advanced in the last few decades.

8.4.2 Run-of-River Hydropower Plants

One of the alternatives to larger, invasive projects of conventional hydropower has been to develop run-of-river (RoR) technologies, which consist of diverting a portion of the river's flow through a channel, which is eventually passed through turbines to produce electricity. After this process, the water rejoins the main body downstream with little to no intervention. As Figure 8.16 shows, the water borrowed from the river is passed through a channel and ultimately driven through a penstock built along the natural slope produced by the change in elevation, generating most of the energy via the waterdrop (U.S. Department of Energy, 2023a).

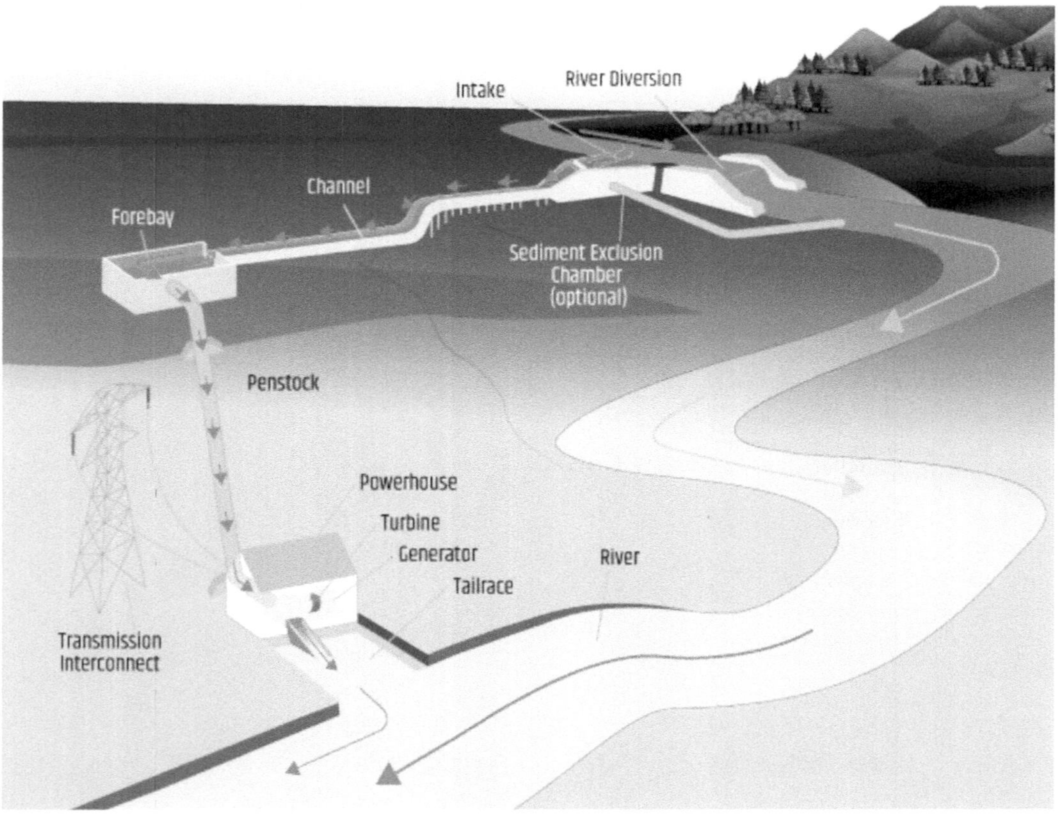

FIGURE 8.16 Illustration of a run-of-river, or diversion, hydropower facility.

Source: U.S. Department of Energy (2023e).

Minor variations within the RoR technology feature dam-toe schemes, which do not require diversion but rather a structure built on top of the flowing body of water that retains water for a negligible time to generate energy. These projects are favored in ultra-low-head settings. Meanwhile, diversion setups vary in storage capabilities, with diversion weir schemes lacking storage and being conceived to satisfy short-term electricity demands. On the other hand, pondage technologies include storage that allows water to be stored in off-peak times and released during peak electricity demand. Both of these methods are applied in medium or highland rivers with considerable gradients that create significant hydraulic heads for substantial electricity generation (Kuriqi et al., 2021).

The main advantages of these technologies center on their lower environmental impact compared to conventional hydropower methods. Their lower environmental impact is because there is minimal building and water storage, with plants primarily relying on weirs to raise the water level upstream, thus avoiding constructing larger dams. Due to this, there is much less intervention in the locations selected for plant development, which allows for preserving natural river flows and minimizing habitat and community disruption. Naturally, because costs are also reduced, developing countries and small areas can develop their own RoR endeavors.

Additionally, in the energy transition context, RoR schemes can provide a resilient alternative for energy generation in perennial rivers. The potentially damaging effects of the consequences of climate change can place heavy burdens on certain localities that rely on harnessing the power of water to generate electricity. However, given that run-of-river methods can produce power even in low-flow conditions, these small hydropower projects appear to be a key, low-budget alternative for many actors looking to diversify the grid and steadily move away from fossil fuels (Skoulikaris, 2021).

Having stated that, recurring dry seasons can still significantly limit the power generation capacity of RoR projects, which makes this technology particularly vulnerable to impacts from climate variations. Consequently, grid integration can prove challenging due to the intermittence that would materialize in those scenarios. Moreover, Kuriqi et al. (2021) highlight the environmental impacts of these setups, challenging the notion of negligible effects. In their work, they highlight how dam-toe, diversion weir, and pondage projects significantly affect river flow, fluvial habitat, and consequences on the flora and fauna of a location.

First, dam-toe projects can alter the physical and chemical parameters of water and the depth upstream and downstream of the weir. Also, they can fragment the habitat of many species, making it unsuitable for some of them. In the case of both diversion weir and pondage schemes, temperature variations across seasons, a lower quantity and quality of flow in the portion of the river where water was extracted, and pollutant retention upstream of the weir can occur (see Table 8.3 for a comprehensive list). As a consequence of these phenomena, aquatic animals could show behavioral and physical changes, reduce the abundance and richness of fish, and increase injuries to fish (Kuriqi et al., 2021).

As observed, there are a few issues where run-of-river projects have significant room for improvement. In places without an immediate need for electricity, advancements in energy storage techniques can provide reliability to the grid, tackling any problems with intermittence that could result from droughts. As an alternative, pumped storage systems have been advanced, a promising set of technologies we will review momentarily. Then, to avoid significant changes in flow after deploying a hydropower plant, comprehensive long-term hydrological studies are now an essential element of an improved site selection.

In addition, novel techniques that rely on advanced numerical and statistical methods can aid in improving RoR technologies. As the review performed by Bernardes Jr. et al. (2022) shows, machine learning methods like artificial neural networks (ANNs) have been used to forecast production based on input data, including temperature, precipitation, and historical water inflow. We are still at an early stage in developing these technologies, yet RoR hydropower applications appear well suited for an explosion in the next couple of decades. Provided that they follow strict environmental assessments for their development, we are looking at one of the two hydropower technologies of the future.

TABLE 8.3
Alteration of Upstream–Downstream Biophysical Parameters Near RoR Sites

Parameters	Dam-Toe RoR Scheme			Diversion Weir			Pondage		
	↑	↓	↔	↑	↓	↔	↑	↓	↔
Water temperature	-	-	•	$\bullet^{d}$	-	-	$\bullet^{d}$	o^{u}	-
pH	o^{u}	o^{d}	•	o^{u}	$\bullet^{d}$	-	$\bullet^{u}$	$\bullet^{d}$	-
Conductivity	-	-	•	o^{u}	o^{d}	-	$\bullet^{u}$	$\bullet^{d}$	-
Turbidity	o^{u}	o^{d}	•	o^{u}	o^{d}	•	o^{u}	$\bullet^{d}$	-
Dissolved oxygen (DO)	-	-	•	o^{u}	$\bullet^{d}$	-	o^{u}	$\bullet^{d}$	-
Water depth	o^{u}	o^{d}	•	o^{u}	$\bullet^{d}$	-	$\bullet^{u}$	$\bullet^{d}$	-
Flow velocity	-	o^{u}	•	-	$\bullet^{u}$	-	-	$\bullet^{u}$	-
Wetted perimeter	-	o^{d}	•	o^{u}	$\bullet^{d}$	-	$\bullet^{u}$	$\bullet^{d}$	-
Sediments retention	-	-	•	o^{u}	-	o	$\bullet^{u}$	-	-
Chlorophyll-*a*	-	-	•	o^{u}	o^{d}	o	$\bullet^{u}$	$\bullet^{d}$	-
Pollutants retention/concentration	-	-	•	o^{u}	o^{d}	o	$\bullet^{u}$	$\bullet^{d}$	-
Connective loss/migration	-	$\bullet^{u\text{-}d}$	-	-	$\bullet^{u\text{-}d}$	-	-	$\bullet^{u\text{-}d}$	-
Fishes injuries	$\bullet^{u\text{-}d}$	-	-	$\bullet^{u\text{-}d}$	-	-	$\bullet^{u\text{-}d}$	-	-
Fish abundance	-	$o^{u\text{-}d}$	•	-	$\bullet^{u\text{-}d}$	o	-	$\bullet^{u\text{-}d}$	-
Fish richness	-	$o^{u\text{-}d}$	•	-	$\bullet^{u\text{-}d}$	o	-	$\bullet^{d}$	-
Macroinvertebrates abundance	-	$o^{u\text{-}d}$	•	-	$\bullet^{u\text{-}d}$	o	-	$\bullet^{u\text{-}d}$	-
Macroinvertebrates richness	-	$o^{u\text{-}d}$	•	-	$\bullet^{u\text{-}d}$	o	-	$\bullet^{u\text{-}d}$	-
Riparian vegetation	-	$o^{u\text{-}d}$	•	-	$\bullet^{u\text{-}d}$	o	-	$\bullet^{u\text{-}d}$	
Macrophytes/other aquatic plants	-	$o^{u\text{-}d}$	•	o^{d}	o^{u}	•	o^{d}	$\bullet^{u}$	o

Note: Highlighted reflects most predominant and extensively reported impacts. Symbols used in this table stand for increase (↑), decrease (↓), no changes (↔), no study reported (-), scarcely reported (o) (for a number of studies ≤20), extensively reported (•) (for a number of studies >20), upstream (u), downstream (d), and upstream–downstream (u-d).
Source: Kuriqi et al. (2021).

8.4.3 Pumped Storage Hydropower Plants and Vortex Systems

Speaking of promising hydropower applications, pumped storage hydropower (PSH) applications check many boxes in the renewable energy and sustainability segments, which has helped them be the top choice for energy storage systems worldwide. First, they include a storage component that serves as a repository of energy collected during off-peak hours, which can come from different sources, such as solar and wind. Then, these installations incur fewer costs to build and operate when compared to fossil fuels like natural gas or diesel plants. Lastly, PSH installations function almost exclusively with water, a clean and virtually inexhaustible resource (Alvarez, 2020).

The way PSH installations work can be understood primarily as a giant battery, where energy from various sources is stored during low-demand hours, while some energy is also used to move water to higher-elevation reservoirs to prepare for an eventual electricity generation stage. This moment comes when, for example, the sun stops shining and the wind stops blowing and peak hours demand a share of power to fill the gaps. At this moment, water is released back to the lower reservoirs, where turbines operate a generator that produces energy (U.S. Department of Energy, 2023d).

Figure 8.17 shows a basic illustration of a typical pumped storage design continuously connected to the naturally flowing water source. These setups are also known as open-loop PSH, contrasting

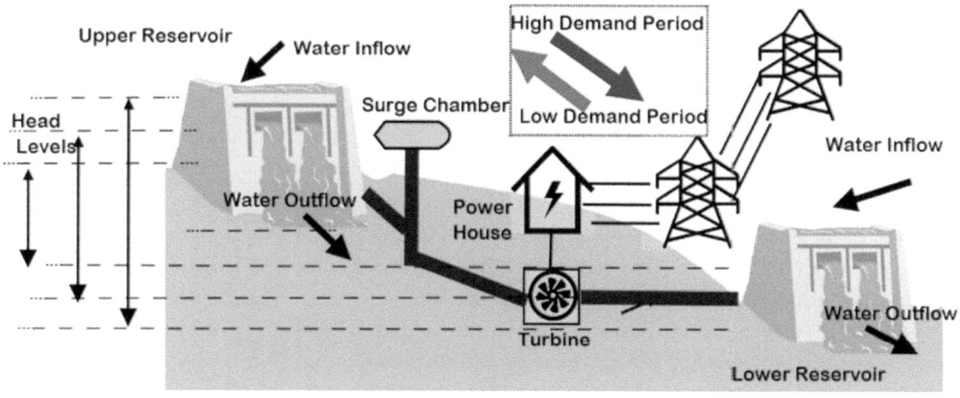

FIGURE 8.17 Basic representation of a pumped storage hydropower operation.

Source: Alvarez (2020).

with those plants without a constant plug to a water source, or closed-loop PSH projects. There are differences between these two groups, as the environmental effects of closed-loop projects are perceived to be lower due to these setups being placed "off-stream," which minimizes potential aquatic and terrestrial impacts and has greater siting flexibility. However, some closed-loop systems can be more impactful than open-loop schemes if, for example, their higher and lower reservoirs need to be built aboveground (Saulsbury, 2020).

A third common type of PSH is the pump-back storage installation, where pumps are placed in nearby reservoirs at different elevations, allowing water to flow back and forth between the two sites. The pump-back storage system provides better flexibility and operational range because installations can be used as pumped storage structures throughout the year, maintaining operations in drought seasons, when hydropower may see its production significantly reduced. In locations with substantial precipitation going downstream in river flows, some of this rainfall can be pumped back upstream, stored for a while, and used to generate electricity at a later time (Hunt et al., 2020).

PSH technologies have enticing strengths, such as their capacity for large-scale energy storage, which enables the joint contribution of multiple renewable sources to fill intermittence gaps, particularly when the energy coming from solar and wind is insufficient to satisfy the needs of final users (Alvarez, 2020). Relatedly, due to their primary role as a large battery, PSH setups provide grid stability thanks to their quick response to changes in electricity demand, making them ideal structures for peak load management. Plus, along with the high degree of efficiency in electricity generation just described, these projects also feature substantial efficiency in storage capabilities.

As alternatives to large hydropower endeavors, PSH schemes offer a lower environmental impact, placing them as one of the lowest emitters and, by extension, having significant environmental advantages over fossil fuel installations that produce comparable energy levels. Then, as a more accessible technology for a larger number of countries, PSH applications have the potential for significant economic benefits in more parts of the world, providing employment in a blossoming area that is a centerpiece of the transition to renewable energies.

Despite the ample promise, PSH projects have a set of challenges that continue to constrain their more widespread application across more contexts. First, site requirements continue to place severe burdens on many locations, which will not be suited to host an installation of this type. Thus, projects in locations lacking the necessary elevation difference may be discarded entirely. Likewise, initial capital costs are prohibitive for many actors, as they are relatively high compared with other technologies (Zvirgzdins and Linkevics, 2020). The initial capital cost means that profits are achievable almost exclusively with large plants, which affects the prospects of smaller projects to see the light of day.

These endeavors may also suffer from energy and water losses, first because the resource needs to be pumped back upstream, consuming some of the same electrical energy for which these installations are built. Furthermore, some water can be exhausted due to evaporation, leakages, misuse, or theft, which translates into some of the resources being lost and unable to be converted into electricity. This can be exacerbated in extreme weather or drought, causing variations that could jeopardize an operating plant or indefinitely pause a project in the planning phase.

Lastly, as mentioned previously, with larger hydropower projects, PSH setups can significantly alter marine life, interfere with migration patterns, force complete relocation, and cause injury or death to fish and other species (Ansorena Ruiz et al., 2022). Moreover, larger projects can incur hefty costs for the communities where these installations are being built, producing noise and polluting the air in the short term. Further, some people would be forced to leave the area entirely, given the construction troubles and the land requirements, which can demand up to 12 ha for an average setup (Blakers et al., 2021).

The avenues that pumped storage hydropower installations have opened are very enticing, as we have been presented with a renewable source that combats intermittence problems from other clean sources, like solar and wind, while also serving as a very valuable battery structure for energy coming from different places. We are now at the ideal time to explore these installations to their fullest extent, provided that we also perform adequate environmental assessments that consider reductions to habitat disruption, both for animals and humans.

At the close of this section, we also offer a summary of water vortex hydropower technologies. These setups continue to be at a very early stage. Nevertheless, they offer significant reductions in cost and environmental impacts, thus making them applicable for small contexts where funding is scarce and interventions on natural habitats are expected to be minimal. Water vortex technology works by harnessing hydroelectric power from the high angular velocity found at the core of a whirlpool that is generated in a vortex chamber (Timilsina et al., 2018). Figure 8.18 shows the schematics of a setup that creates an artificial free-surface vortex flow to generate energy.

This structure, also known as a gravitational water vortex hydropower plant (GWVHP), does not differ much from a classic Kaplan or Francis turbine, except in the system's exposure to atmospheric conditions. Thus, GWVHPs are left to operate under normal conditions, unlike other turbines that require different pressures to maximize their function (Timilsina et al., 2018). These are straightforward structures, including inlet and outlet channels, a basin, and a turbine with blades attached. These systems work after water flows in through the inlet channel, converting the flow into mechanical energy by the turbine blades and electrical energy by the generator (Maika et al., 2023).

These vortex structures are powerful for their size and requirements, producing outputs between 100 W and 20 kW at flow rates that go between 0.1 and 2 m^3, relying on low heads with ranges of around 0.7–2 m. Also, they are environmentally friendly and feature a compact design that holds

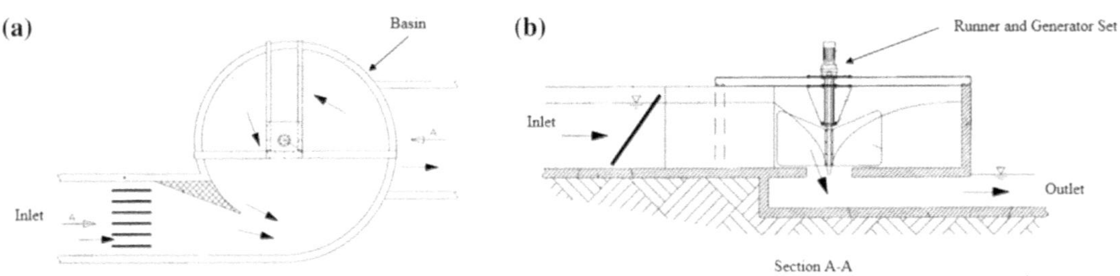

FIGURE 8.18 (a) Plan and (b) elevation of Zotlöterer's novel vortex diagrams.

Source: Timilsina et al. (2018).

great promise in hard-to-access, remote, and rural areas with little or no access to the main electricity grid (Maika et al., 2023). One example of a remote application is seen in mountainous and rural Nepal, where Dhakal et al. (2015) studied the conditions of many perennial free-flowing rivers and over half of households lacking access to electricity to advance vortex technologies as a promising solution.

Along with PSH endeavors, water vortex applications have benefitted from the advent of methods like computational flow dynamics (CFD), which allow researchers to estimate the dynamics that will govern the behavior of these structures depending on the contexts where they will be applied. For example, the work of Domfeh et al. (2020) explores CFD as an optimization tool, where its simulation instruments can be used to estimate the most optimal structural designs, as well as the best turbine designs, to create the free surface vortices in bodies of water that can offer the maximum potential for energy generation, in as many locations in the world as possible.

As we have seen in this section, different contexts call for tailored applications of hydropower technologies, each featuring its own challenges and advantages. The energy transition process invites actors to explore a constant but responsible process of developing hydropower technologies that can include as many locations and players as possible, lowering barriers of entry like construction and maintenance costs, as well as securing access to the latest technologies. To achieve these incentives, joint efforts from governments, researchers, and international organizations are required, which ought to follow standard practices that benefit all participants. In the following section, we summarize a few of these recommendations for improved hydropower endeavors.

8.5 BEST PRACTICES IN HYDROPOWER APPLICATIONS

In this chapter, we have gone over site screening and selection mechanisms and the optimization of many aspects of a new hydropower installation. Now, we will summarize a series of practices we have identified as the most common in optimizing sites and technologies that researchers and engineers could find helpful to apply in their own efforts. Hence, this section contains guidelines for the improved use of numerical and statistical methods, the increasing role of machine learning and Internet of Things (IoT) frameworks, and the inclusion of environmental considerations that align with international pledges and commitments set by international organizations.

From planning to site selection, including operation, maintenance, and even decommission, the improvement of methods for dealing with various aspects of hydropower plants has been exponential in the last few decades. Today, it is expected to delay in-field operations until comprehensive simulations are completed, and studies performed in laboratory settings provide a series of guidelines for the expected costs and returns on investment, the years a plant is expected to last, their environmental impact, and any potential structural and human errors that could be encountered along their operation.

However, for this widespread digitalization of hydropower operations to occur, large quantities of data need to be collected so that real-world information on things like turbine operation is kept for comparison and estimation purposes. Then, a process of updating older plants and infrastructure is required to connect them to this new, emerging network. According to Kougias et al. (2019), this modernization not only can prolong the lifetime of installations and prepare them for cybersecurity risks but can also increase annual energy production from a low-emission renewable source by up to 42 TWh, which equals USD$5 billion in annual operational savings.

Therefore, as a first set of practices, we invite researchers and people in the field to explore the possibilities that digitizing processes in the hydropower sector could bring. As a set of technologies over 4 decades old, with a current wave of rethinking their size and assessing their environmental downfalls, a first calculated move could be to explore downsizing large hydropower dams while collecting data on this shift to optimize future projects. Concurrently, stakeholders and engineers should use this information to replicate as many small, micro-, and pico-projects as possible, in as many locations as possible, with robust environmental and sustainability assessments.

In this respect, the work of Nautiyal and Goel (2020) focuses on improving the sustainability of hydropower projects by expanding the set of parameters usually associated with these systems. Their considerations include a series of potential biophysical impacts, such as land use, acidification, reservoir formation, and logistical and construction-related factors like fossil fuels and explosives, the installed equipment and machinery, and future demolition and recycling expectations. The conclusions of this work back up the notion that until more factors are considered for a holistic analysis, these decisions will lack the required depth to determine their long-term sustainability.

Another factor common in the literature when exploring best practices is the role of policymaking and governance. Authors have explored the energy transition as a process that needs to be governed or legislated, being more proactive in passing laws and funding small hydropower (SHP) schemes so that the appropriate incentives exist for actors to explore these projects (Kelly-Richards et al., 2017). An encouraging sign is that improved proactive governance, along with democratic indicators like political stability and government effectiveness, has been associated with the growth of the hydropower industry in European Union member states (Alsaleh and Abdur-Rahim, 2021).

These efforts are not only expected from the top to the bottom, thus directed from supranational organizations to the population, but can also appear at the local level and percolate to the regional or national legislation over time. For example, communities and local organizations that transcend country boundaries because of the link they have via obtaining resources from their local rivers can coordinate to generate a network of voluntary negotiations that predate any national legislation. Borrowing from the water-energy-food (WEF) nexus literature, Dombrowsky and Hensengerth (2018) highlighted the role of international river basin organizations (IRBOs) in applying environmental and social safeguards that national actors can reproduce in the future.

Therefore, the interconnection between regional actors and legislation that follows these voluntary dynamics can be a crucial contributor to the expansion of hydropower technologies around the world. Given that a set of trade-offs and balancing acts is required for legislation to be passed, where not all actors get what they necessarily want, transparent coordination efforts between various players should be encouraged. Facilitating this environment can help foster the advancement of novel projects in different settings through the commitment of some actors to exploring new technologies. In contrast, others offer funding schemes for their implementation.

An example of this joint work has been the recent exploration of underground reservoirs for deploying pumped storage hydropower facilities (UPSHs). Unlike regular hydropower projects, the potential advantages of these installations relate to their low environmental impacts, along with topography not being a limitation for their development. Analyzing the prospects for setting up underground plants, Lyu et al. (2022) studied the prospects of utilizing abandoned mine shafts to convert intermittent energy into useful energy, concluding that achieving success in this area requires a co-development of the industrial, education, and research sectors through a process of integration, completion, and coordination.

Without collaboration from other actors to absorb some initial costs, efforts to refurbish or retrofit underground structures may not see the light of day at all. This can be inferred from the work of Menéndez et al. (2020), who assessed the economic feasibility of UPSH under different scenarios, such as excavating new tunnels versus relying on existing underground infrastructure. For this circumstance, they found little to no economic feasibility when new infrastructure has to be built, which invites a degree of collaboration from actors in other sectors or governments looking to convert depleted or abandoned installations into the hydropower facilities of the future.

When talking about partnerships, the role of the oil and gas industry can also be critical to the development of the hydropower sector. We will review this in greater detail in the following section. However, it is relevant to highlight that joint efforts that combine the *know-how* of the fossil fuel sector and its expertise in planning and deploying energy installations, with efforts from engineers and up-and-coming entrepreneurs in renewable energies, can help materialize the energy transition to clean sources much more rapidly. Following strategies like environmental social governance (ESG), the oil and gas sector can strongly support a new, low-carbon energy period (Atlantic Council, 2020).

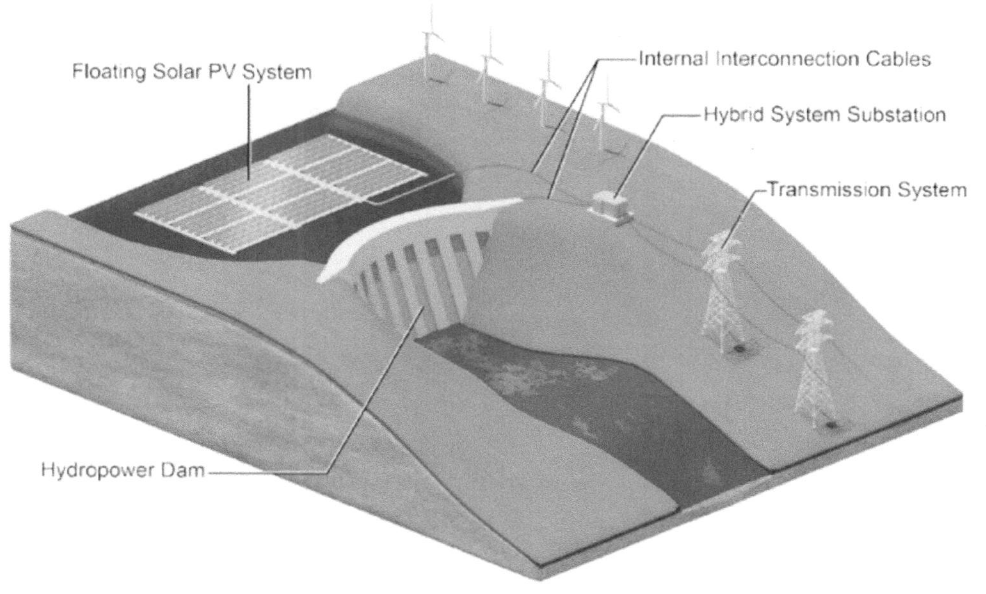

FIGURE 8.19 Illustration of a hybrid floating solar photovoltaic (FPV) and hydropower system.

Source: Lee et al. (2020).

Then, other ways to achieve productive outcomes center on exploring the possibility of mixing projects under one installation, thus producing energy from different renewable sources in the same facility. This was assessed by Xu et al. (2019), who explored a hybrid photovoltaic-wind-hydropower with a pumped storage system in the Chinese province of Xiaojin, Sichuan. This stand-alone project includes PV arrays, wind turbines, upper and lower reservoirs, a controller, an inverter, a variable-speed pump, a water turbine, and accessories like pipes and cables.

The hybrid plant proposed, similar to the illustration provided in Figure 8.19, works as follows: When the energy generated from the PV arrays and wind turbines is greater than the local load, energy is used to pump water from the lower to the higher reservoir. Then, as the power from solar and wind sources is insufficient to supply the local load, the turbine starts working, releasing water to generate more power (Xu et al., 2019). These setups not only increase energy sources, thus potentially increasing the energy output from the installations, but also reduce construction costs by grouping multiple objectives in the same project, making them more environmentally friendly (Agarkar and Barve, 2016).

Using meteorological and national grid data, the hybrid system advanced by Xu et al. (2019) yields a set of values for the relevant parameters when building a plant of these characteristics, as shown in Table 8.4. Thus, given an assumed O&M cost of 1% of the initial capital cost, excluding expenditures on wind turbines, specifications for some of the components related to costs and lifetime for solar, wind, and hydropower technologies are listed as indicators to maintain an economically viable project that provides reliable and clean energy for a remote area.

Furthermore, some hybrid installations can be built offshore, provided that the energy potential is larger at sea than onshore. With that said, this decision is also constrained by the higher costs of building offshore structures. In this sense, Lee et al. (2020) found that benefits can be maximized through hybrid projects that combine floating PV arrays with hydropower techniques. Some of the strengths of hybridization are, for example, an improved system operation, increased energy storage opportunities, transmission system benefits, resource conservation, and reduced intermittence from solar power, offset by electricity generation from hydropower.

TABLE 8.4
Costs and Lifetime Considerations for Hybrid System Components

	Initial Capital Cost	Operating and Maintenance Cost	Replacement Cost	Lifetime (Year)
PV array (200 W)	\$400/unit	\$4/unit	\$340/unit	15
Wind turbine (250 kW)	\$375,000/unit	\$7,500/unit	\$265,500/unit	20
Upper reservoir	\$170/m^2	\$1.7/m^3	–	35
Variable-speed pump	\$1,075/kW	\$10.75/MW	\$1,075/kW	10
Water turbine	\$1,000/kW	\$10/MW	\$1,000/kW	10

Source: Xu et al. (2019).

In the search to optimize hydropower endeavors, including cost reductions that allow for the expansion of these projects to more locations around the world, connectivity technologies like the Internet of Things (IoT) are being proposed by some authors. For example, Fortaleza et al. (2018) advanced a system of sensors and auxiliary software to operate a pico-hydro project in two Philippine provinces, Ifugao and Aurora. This setup relies on a Pelton turbine suited for high head and low water flow rates, a generator, and an IoT structure with a GSM module, sensors, and an ASIC microcontroller. While the sensors detect water levels and trigger the turbine to operate, the microcontroller sends the data for monitoring purposes, and the GSM module serves as a structure for the process to work seamlessly.

This move toward adding machine learning elements with different levels of sophistication is perhaps one of the most promising advancements we have witnessed for hydropower technologies in the past few years. From an optimized connectivity achieved by software programs and sensors to learning algorithms that analyze large quantities of historical data to predict future behavior from machinery and personnel, the transformation of the sector has allowed for a steady rise in hydropower installations as preparation for the energy transition process to cleaner sources of energy.

As an example of these innovations, the work of Sapitang et al. (2020) explores four different ML frameworks to accurately forecast the changes in water levels of a Malaysian reservoir under two different scenarios: one including rainfall and water level as inputs, and a second that features both mentioned inputs plus sent-out indicators. The methods proposed for this optimization are boosted decision tree regression (BDTR), decision forest regression (DFR), Bayesian linear regression (BLR), and neural network regression (NNR), with the authors finding that the BLR method performed best in the first scenario, while BDTR showed a better fit for the second proposed situation.

The potential for these technologies to aid the hydropower sector is as powerful as the potential that harnessing the force of water has to produce clean, reliable energy for billions of people. The energy transition process is the ideal time to expand on the reach of all these novel techniques, and coordinating with other players is crucial for this enhancement's continued success. In this sense, the oil and gas industry can provide valuable experience and skills as the energy players lead the way for the second half of the 20th century. In the following section, we review the prospects for the oil and gas industry to be a significant contributor to the growth of the hydropower sector.

8.6 HYDROPOWER AND THE OIL AND GAS INDUSTRY IN THE ENERGY TRANSITION PROCESS

As we have reviewed in other chapters of this book, the oil and gas industry has had varied strategies when dealing with emerging technologies that rely on clean energy sources. In the case of wind, solar, and hydrogen, we have seen a marked interest in investing and promoting new facilities and applications. Conversely, the oil and gas sector has been less animated and explicit for techniques that harness the energy of water, such as currents, tides, and hydropower. However, despite this apparent disinterest, an agenda still includes expansion to renewables, like hydropower and ocean technologies, as important elements of the energy transition.

As a powerful player in energy generation, a move which has also led to prosperity for billions of people, the oil and gas sector has obtained a significant expertise that continues to generate new knowledge and alternative mechanisms for optimizing fossil fuel utilization. Thus, within their own area, the industry is making improvements toward cleaning their act, with initiatives such as carbon capture and storage (CCS), hybridization of installations to produce electricity from cleaner sources, and switching some processes inside the plants to be powered by renewables.

Additionally, the sector has seen the opportunity to invest in a set of renewable technologies like solar and wind, whether as a result of nudging strategies performed by governments and international organizations, because some applications have shown to be economically viable in advantageous geographical contexts, or because they genuinely want to decarbonize their processes. Whatever the reason, the sector now has a window of opportunity to move from the "oil and gas" label to the "energy" label, where multiple sources of energy, both from fossil fuels and as well as from renewables, can mark the beginning of the transformation of the energy sector in the 21st century (IEA, 2020).

As advanced previously, new energy players also have incentives to collaborate with oil majors, mainly due to the latter's position in the energy markets and the expertise in their processes. The business side of these dynamics reveals that the emerging clean sector is competing for the same market as the oil and gas industry. With that said, collaboration can be the most optimal strategy for everyone involved. Therefore, this joint work can result in faster and cleaner processes, a professionalized energy industry, and more environmentally friendly energy production procedures.

For example, given that hydropower projects are significantly capital-intensive, the oil and gas industry can provide critical funding and know-how for some of these endeavors to mature and thrive without as many risks of shutting down as they would typically exist (IEA, 2020). Additionally, the refurbishment or retrofitting of installations previously used for industrial use, particularly those utilized for oil and gas processes, can be another initiative pursued by a collaboration between the sectors. Also known as brownfield development, using previously constrained land for cleaner uses is a promising development that demands fewer costs when compared to greenfield projects (IEA, 2021).

Moreover, some considerations that have gained notoriety in the past few years are energy independence and geopolitical uncertainties. Due to global events that have disturbed the previous sense of normalcy, oil and gas players and governments have assessed the need to become more localized and flexible in their energy production processes. In this sense, a collaboration between fossil fuel players and governments has risen, with the goal of promoting and funding green initiatives that initially serve as alternatives to hydrocarbons, but with the hope that, eventually, they become the primary source of power for their populations (Redko et al., 2022).

Inextricably related to the previous notion, global geopolitical concerns have heightened considerably due to the conflict between Russia and Ukraine. Some countries have explored technologies like fracking and hydrogen to remedy potential difficulties and reduce their dependence on hydrocarbons, like Russian natural gas. In contrast, others have decided to pursue applications like small hydropower and hybrid projects mixing wind and solar in offshore facilities. While starting as solutions to energy scarcity, these strategies could also provide a considerable degree of power to new players that produce surplus energy and begin to influence emerging markets (Sica and Huber, 2017).

In the middle of these dynamics, the oil and gas sector is also involved in its own search to become a more sustainable industry. According to Morgunova and Shaton (2022), companies inside the industry have adapted their outlooks for the next few decades, pledging varying degrees of commitment to net-zero pledges drafted and enforced by international organizations. These initiatives

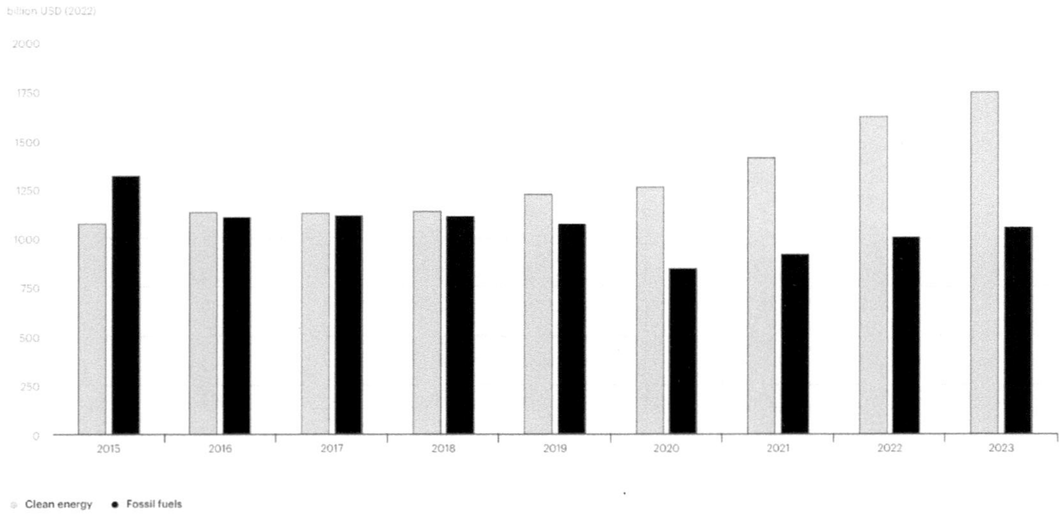

FIGURE 8.20 Global energy investment in clean energy and fossil fuels, 2015–2023.

Source: IEA (2023).

include, but are not limited to, decarbonization efforts within fossil fuel processes, carbon pricing, reductions in oil and natural gas share, and an active contribution to a sustainable energy transition.

With that said, these steps can be much more streamlined around the world with legislation and regulations that pave the way for a more seamless integration of the oil and gas sector with the renewable industry. Looking at Figure 8.20, we can see that the current global context has partly incentivized a larger investment in cleaner energy sources in the past couple of years, with an apparent increase since 2020, the first year of the COVID-19 pandemic. This seems to suggest that we are in an exciting moment to explore the galvanization of the motion toward decarbonization through encompassing legislation, especially for a technology like hydropower, which seems to be trending in the wrong direction (IEA, 2023).

A few examples that policymakers around the world can imitate are the European Union's Hydropower Alliance, where leading players in the hydropower sector joined forces in May of 2023 to aid in the fulfillment of the region's pledge of climate neutrality by 2050. The group is asking the European Commission to center current and future legislation on the repowering, refurbishment, and extension of current installations and to emphasize new pumped storage installations. Efforts like this have followed years of policies to expand hydropower that go even beyond Europe's borders, where the EU–China road map on energy cooperation is a clear example (Wagner et al., 2019).

Then, another model for oil and gas and hydropower policymaking can be seen in Norway's regulatory framework, known more informally as the "10 oil commandments," written in the 1970s. According to Gociu (2021), the legal boundaries set by the Norwegian government for the exploration and exploitation of newly found oil and gas reserves were very tight, setting a type of oil and gas constitution that gave preference to hydropower, the primary energy source of the country at that time. Thus, the role of fossil fuels was to be regulated by a national oil company, pursuing goals that satisfied national oil independence without disregarding the environment while also minding any socio-political development.

As can be inferred, the relationship that the oil and gas sector can have with the hydropower industry, emerging in some parts of the globe but very established in others, is sometimes mediated by the legal framework of the locations being analyzed. Additionally, private oil companies have their own set of strategies that sometimes align with those of the emerging hydropower players yet can interfere with them at other times. Because of this, we observe an initial degree of competition between the two energy sectors that has been mitigated by the enticing opportunities that both industries have to become leading actors in the energy transition of the 21st century.

Consequently, we also observe a series of cooperation schemes to move forward with projects that include the expertise of the oil and gas sector with hydropower applications that have benefitted millions of people. For example, hybrid and retrofitting projects that rely on a hydropower baseload are growing, with decreasing reliance on fossil fuels only as backup sources when the basic load needs cannot be covered (US Department of Energy, 2021). Moreover, many high-energy processes within oil and gas operations, such as transportation, shaft work in offshore/onshore production, and enhanced recovery, have been hybridized so that sources like hydropower can fuel them (Rafiee and Khalilpour, 2018).

Additionally, the oil and gas sector can prescribe similar decarbonizing techniques used in its own processes to help those hydropower applications in need of greening schemes. Despite relying on a clean energy source, some hydro applications can still have significant environmental effects, such as extensive deforestation, marine life injury, and various social disruptions. Shiji et al. (2021) explored the mechanisms of green small hydropower endeavors, creating an evaluation system around ecological environment indicators like hydrological condition and emission reduction, as well as various social, management, and economic indicators, such as benefit sharing, technological progress, and the contribution to the regional economy.

Meanwhile, a successful energy transition process requires the vital work that plants like pumped storage hydropower can deliver as big batteries, along with their electricity production. These types of installations can be expanded through private and public funding schemes, and oil companies have begun to provide funds for some of these endeavors to complement and assist the grid in times of scarce power. Using technology from applications such as compressed air energy storage (CAES) to balance loads in fossil fuel installations, the oil and gas sector can provide critical tools to enhance the role of hydropower in the transition to renewables.

As we have seen, despite a more diffuse role in developing hydropower as a reliable and clean energy source when compared to solar and wind, the oil and gas sector has an important role in the future of hydro technologies. For example, the human capital fostered by the hydrocarbon industry continues to provide significant advancements that have benefitted the renewables sector, allowing for different applications to be deployed worldwide. Thus, geographical locations can now be matched to particular projects with a greater degree of accuracy and maximize the likelihood of both an optimal degree of energy production and an improved well-being for the related population.

Meanwhile, funding strategies and joint collaborations are two other aspects driving the push toward greater use of hydropower, aided by the oil and gas sector. The potential of small, micro-, and pico-hydro projects in different parts of the world appears too enticing to ignore, and hydrocarbon players who are looking to diversify their grids have invested in hybrid setups or hydropower-only installations. Also, due to its maturity and strong storage capabilities, the oil and gas sector should help hydropower become the backbone of the transition to renewable energy sources.

8.7 CONCLUSIONS: THE FUTURE OF HYDROPOWER TECHNOLOGIES IN THE CONTEXT OF THE ENERGY TRANSITION

In this chapter, we have reviewed the different applications of hydropower throughout its history, their strengths and weaknesses, and the opportunities that new advancements have provided to expand the reach of hydro projects around the world. After this comprehensive analysis, we tend to concur with the assessment in IEA's 2021 report, where they characterize *hydropower* as "the forgotten giant of low-carbon electricity." Perhaps due to a rethinking of some of its applications, hydropower is currently not as popular as wind or solar. However, it appears poised for a comeback just in time for the energy transition.

The biggest point of contention for this apparent stagnation of hydropower is the environmental stress that is exerted by large installations, which require significant deforestation, land use, and potential marine life disruptions, albeit in exchange for reliable power generation. With that said, the advancement of new technologies has led to various novel applications with a lower environmental impact. It can provide an alternative to fossil fuel plants while serving as a storage facility for energy generated from other renewables, like solar and wind.

As we found in our review, significant developments in turbine technology and plant design have increased the potential of smaller installations being deployed across many locations worldwide, providing remote communities with electricity. This has eventually opened the door for more places to host hydropower installations, yet this comes with its own set of challenges. First, these projects are capital-intensive, thus requiring funding from various sources like governments or oil and gas companies to be greenlit. Then, improved environmental assessments are required so that problems like the ones encountered with large projects are not repeated in these new iterations.

These dynamics reflect one crucial feature of the new push for hydropower in an energy transition context: collaboration. Whether between governments and hydropower players or these actors and the oil and gas industry, significant joint efforts will be needed to achieve the goals necessary for this process. Additionally, policymakers' and regulators' roles become important here because friendly legislation can simplify the road toward implementing multiple hydropower facilities in a given region or country. As we see in the case of the European Union, pledges are insufficient without a legal framework that allows the right conditions for hydropower to flourish.

Additionally, efforts to expand hydropower endeavors should not only come from the top down but also from regional and local authorities and communities, who have the most to gain from these initiatives. Therefore, it is crucial that local organizations, which sometimes cross borders between countries, get together and promote the use of responsible and sustainable hydropower installations along their rivers. Furthermore, other stakeholders may provide valuable insights on new developments, while others should take more responsibility to help make hydropower projects come to fruition in locations with natural advantages, which national authorities may overlook.

Despite this multiplicity of actors in the fold, there is still ample space for a big player like the oil and gas industry to contribute to enhancing hydropower technologies, taking a commanding role in the move toward the energy transition. Understanding that part of the objectives of this process is the attainment of a grid that increasingly phases fossil fuels out, the oil and gas sector is presented with an opportunity to move away from notions of competition, shifting from a somewhat tarnished label to one that sees them become relevant energy players in the 21st century.

Even though most of their work is observed in the solar, wind, and hydrogen sectors, there are signs that the oil and gas industry is also committed to expanding the role of hydropower technologies. For example, funding schemes and technology sharing have been two strategies used by the sector, with retrofitting and refurbishing opportunities serving as an ideal setting to convert fossil fuel installations into hydropower or hybrid facilities. Moves like this take advantage of the human capital fostered by the hydrocarbon sector while also developing projects that incur fewer costs when compared to ventures that require building entirely new facilities.

Likewise, novel technological advancements in machine learning have yielded improved methods to screen and select sites for future hydropower locations and enhance everyday operation and maintenance tasks. Thus, using learning methods backed by historical data from similar projects deployed in the past can help manage new developments and estimate relevant parameters that are linked to productive facilities. In this sense, the energy transition will be fueled not only by a larger set of environmentally friendly projects but also by optimized methods to operate them, along with improved techniques to oversee existing ones.

To conclude, the energy transition requires many different actors with different strengths to come together at the table and provide their input. Governments and international organizations can set the improved rules of the game. At the same time, joint efforts from the hydropower and the oil and gas sectors will generate a more diverse and reliable energy landscape built on renewables. Focusing on sustainability, environmental protection, and efficient life cycles will help hydropower regain its status as the backbone of the clean energy movement around the world.

REFERENCES

Agarkar, B. D., and S. B. Barve. 2016. A Review on Hybrid Solar/Wind/Hydro Power Generation System. *International Journal of Current Engineering and Technology* 4(4), 188–191.

Alsaleh, M., and A. S. Abdul-Rahim. 2021. The Nexus Between Worldwide Governance Indicators and Hydropower Sustainable Growth in EU 28 Region. *International Journal of Environmental Research*. doi:10.1007/s41742-021-00366-6.

Alvarez, G. E. 2020. Operation of Pumped Storage Hydropower Plants through Optimization for Power Systems. *Energy* 202, 117797.

Amougou, C. B., D. Tsuanyo, D. Fioriti, J. Kenfack, A. Aziz, and P. Elé Abiama. 2022. LCOE-Based Optimization for the Design of Small Run-of-River Hydropower Plants. *Energies* 15, 7507. doi:10.3390/en15207507.

Andritz. 2023. *Hydropower in Asia*. Available in www.andritz.com/hydro-en/hydronews/hydro-news-asia/facts-and-figures [Accessed on July 25th, 2023].

Ansorena Ruiz, R. L. H. De Vilder, E. B. Prasasti, M. Aouad, A. De Luca, B. Geisseler, . . ., and N. Goseberg. 2022. Low-Head Pumped Hydro Storage: A Review on Civil Structure Designs, Legal and Environmental Aspects to Make Its Realization Feasible in Seawater. *Renewable and Sustainable Energy Reviews* 160, 112281. doi:10.1016/j.rser.2022.112281.

Atlantic Council. 2020. *The Role of Oil and Gas Companies in the Energy Transition*. Available in www.atlanticcouncil.org/wp-content/uploads/2020/07/OGT-final-web-version.pdf [Accessed on July 25th, 2023].

Bagher, A. M., M. Vahid, M. Mohsen, and D. Parvin. 2015. Hydroelectric Energy Advantages and Disadvantages. *American Journal of Energy Science* 2(2), 17–20.

Bernardes Jr., J., M. Santos, T. Abreu, L. Prado Jr., D. Miranda, R. Julio, . . ., and G. S. Bastos. 2022. Hydropower Operation Optimization Using Machine Learning: A Systematic Review. *AI* 3, 78–99. doi:10.3390/ai3010006.

Binama, M., W.-T. Su, X.-B. Li, F.-C. Li, X.-Z. Wei, and S. An. 2017. Investigation on Pump as Turbine (PAT) Technical Aspects for Micro Hydropower Schemes: A State-of-the-Art Review. *Renewable and Sustainable Energy Reviews* 79, 148–179. doi:10.1016/j.rser.2017.04.071.

Blakers, A., M. Stocks, B. Lu, and C. Cheng. 2021. A Review of Pumped Hydro Energy Storage. *Progress in Energy* 3, 022003. doi:10.1088/2516-1083/abeb5b.

Brown, E., S. Sulaeman, R. Quispe-Abad, N. Müller, and E. Moran. 2023. Safe Passage for Fish: The Case for In-Stream Turbines. *Renewable and Sustainable Energy Reviews* 173, 113034. doi:10.1016/j.rser.2022.113034.

Chala, G. T., M. I. N. Ma'Arof, and R. Sharma. 2019. Trends in an Increased Dependence Towards Hydropower Energy Utilization—a Short Review. *Cogent Engineering* 6, 1631541. doi:10.1080/23311916.2019.1631541.

Chaudhari, S., E. Brown, R. Quispe-Abad, E. Moran, N. Müller, and Y. Pokhrel. 2021. In-Stream Turbines for Rethinking Hydropower Development in the Amazon Basin. *Nature Sustainability* 4(8), 680–687. doi:10.1038/s41893-021-00712-8.

Chien, F., C.-N. Wang, V. T. Nguyen, V. T. Nguyen, and K. Y. Chau. 2020. An Evaluation Model of Quantitative and Qualitative Fuzzy Multi-Criteria Decision-Making Approach for Hydroelectric Plant Location Selection. *Energies* 13(11), 2783. doi:10.3390/en13112783.

Cox, R. X., R. T. Kingsford, I. Suthers, and S. Felder. 2023. Fish Injury from Movements Across Hydraulic Structures: A Review. *Water* 15, 1888. doi:10.3390/w15101888.

de Faria, F. A. M., A. Davis, E. Severnini, and P. Jaramillo. 2017. The Local Socio-Economic Impacts of Large Hydropower Plant Development in a Developing Country. *Energy Economics* 67, 533–544. doi:10.1016/j.eneco.2017.08.025.

Dhakal, S., A. B. Timilsina, R. Dhakal, D. Fuyal, T. R. Bajracharya, . . ., and A. M. Nakarmi. 2015. Comparison of Cylindrical and Conical Basins with Optimum Position of Runner: Gravitational Water Vortex Power Plant. *Renewable and Sustainable Energy Reviews* 48, 662–669. doi:10.1016/j.rser.2015.04.030.

Dombrowsky, I., and O. Hensengerth. 2018. Governing the Water-Energy-Food Nexus Related to Hydropower on Shared Rivers—The Role of Regional Organizations. *Frontiers in Environmental Science* 6, 153. doi:10.3389/fenvs.2018.00153.

Domfeh, M. K., S. Gyamfi, M. Amo-Boateng, R. Andoh, E. A. Ofosu, and G. Tabor. 2020. Free Surface Vortices at Hydropower Intakes: A State-of-the-Art Review. *Scientific African* 8, e00355. doi:10.1016/j.sciaf.2020.e00355.

Dorber, M., A. Arvesen, D. Gernaat, and F. Verones. 2020. Controlling Biodiversity Impacts of Future Global Hydropower Reservoirs by Strategic Site Selection. *Scientific Reports* 10, 21777. doi:10.1038/s41598-020-78444-6.

Elbatran, A. H., O. B. Yaakob, Y. M. Ahmed, and H. M. Shabara. 2015a. Operation, Performance and Economic Analysis of Low Head Micro-Hydropower Turbines for Rural and Remote Areas: A Review. *Renewable and Sustainable Energy Reviews* 43, 40–50. doi:10.1016/j.rser.2014.11.045.

Elbatran, A. H., M. W. Abdel-Hamed, O. B. Yaakob, Y. M. Ahmed, and M. A. Ismail. 2015b. Hydro Power and Turbine Systems Reviews. *Jurnal Teknologi (Sciences & Engineering)* 74(5), 83–90.

Fang, F., and R. Karki. 2018. Reliability Implications of Riverflow Variations in Planning Hydropower Systems. *IEEE Conference on Technologies for Sustainability (SusTech)*. doi:10.1109/sustech.2018.8671365.

Fortaleza, B. N., R. O. Serfa Juan, and L. K. S. Tolentino. 2018. IoT-Based Pico-Hydro Power Generation System Using Pelton Turbine. *Journal of Telecommunication, Electronic and Computer Engineering (JTEC)* 10(1–4), 189–192.

Gociu, A. 2021. The Norwegian Petroleum Regulatory Framework and the Transition to Green Energy. *QMLJ* 33. doi:10.26494/QMLJ72468.

Hunt, J. D., B. Zakeri, R. Lopes, P. S. F. Barbosa, A. Nascimento, N. J. de Castro, . . ., and Y. Wada. 2020. Existing and New Arrangements of Pumped-Hydro Storage Plants. *Renewable and Sustainable Energy Reviews* 129, 109914. doi:10.1016/j.rser.2020.109914.

IEA. 2020. *The Oil and Gas Industry in Energy Transitions: Insights from IEA Analysis*. Available in https://iea.blob.core.windows.net/assets/4315f4ed-5cb2-4264-b0ee-2054fd34c118/The_Oil_and_Gas_Industry_in_Energy_Transitions.pdf [Accessed on July 28th, 2023].

IEA. 2021. *Hydropower Special Market Report: Analysis and Forecast to 2030*. Available in https://iea.blob.core.windows.net/assets/4d2d4365-08c6-4171-9ea2-8549fabd1c8d/HydropowerSpecialMarketReport_corr.pdf [Accessed on July 28th, 2023].

IEA. 2023. *World Energy Investment 2023: Overview and Key Findings*. Available in www.iea.org/reports/world-energy-investment-2023/overview-and-key-findings [Accessed on July 28th, 2023].

IRENA. 2023. *The Changing Role of Hydropower: Challenges and Opportunities*. International Renewable Energy Agency.

Jafari, M., R. Fazloula, M. Effati, and A. Jamali. 2021. Providing a GIS-Based Framework for Run-of-River Hydropower Site Selection: A Model Based on Sustainable Development Energy Approach. *Civil Engineering and Environmental Systems* 38(2), 102–126. doi:10.1080/10286608.2021.1893310.

Jawahar, C. P., and P. A. Michael. 2017. A Review on Turbines for Micro Hydro Power Plant. *Renewable and Sustainable Energy Reviews* 72, 882–887. doi:10.1016/j.rser.2017.01.133.

Jiménez Capilla, J. A., J. A. Carrión, and E. Alameda-Hernandez. 2016. Optimal Site Selection for Upper Reservoirs in Pump-Back Systems, Using Geographical Information Systems and Multicriteria Analysis. *Renewable Energy* 86, 429–440. doi:10.1016/j.renene.2015.08.035.

Kao, S.-C., M. J. Sale, M. Ashfaq, R. Uria Martinez, D. P. Kaiser, Y. Wei, and N. S. Diffenbaugh. 2015. Projecting Changes in Annual Hydropower Generation Using Regional Runoff Data: An Assessment of the United States Federal Hydropower Plants. *Energy* 80, 239–250. doi:10.1016/j.energy.2014.11.066.

Kelly-Richards, S., N. Silber-Coats, A. Crootof, D. Tecklin, and C. Bauer. 2017. Governing the Transition to Renewable Energy: A Review of Impacts and Policy Issues in the Small Hydropower Boom. *Energy Policy* 101, 251–264. doi:10.1016/j.enpol.2016.11.035.

Khanlari, A., and M. Alhuyi Nazari. 2021. A Review on the Applications of Multi-Criteria Decision-Making Approaches for Power Plant Site Selection. *Journal of Thermal Analysis and Calorimetry*. doi:10.1007/s10973-021-10877-1.

Kirke, B. 2020. Hydrokinetic Turbines for Moderate Sized Rivers. *Energy for Sustainable Development* 58, 182–195. doi:10.1016/j.esd.2020.08.003.

Kougias, I., G. Aggidis, F. Avellan, S. Deniz, U. Lundin, A. Moro, . . ., and N. Theodossiou. 2019. Analysis of Emerging Technologies in the Hydropower Sector. *Renewable and Sustainable Energy Reviews* 113, 109257. doi:10.1016/j.rser.2019.109257.

Kumar, A., and R. P. Saini. 2016. Performance Parameters of Savonius Type Hydrokinetic Turbine – A Review. *Renewable and Sustainable Energy Reviews* 64, 289–310. doi:10.1016/j.rser.2016.06.005.

Kumar, K., and R. P. Saini. 2021. Application of Artificial Intelligence for the Optimization of Hydropower Energy Generation. *EAI Endorsed Transactions on Industrial Networks and Intelligent Systems* 8(28). doi:10.4108/eai.6-8-2021.170560.

Kuriqi, A., A. N. Pinheiro, A. Sordo-Ward, M. D. Bejarano, and L. Garrote. 2021. Ecological Impacts of Run-of-River Hydropower Plants—Current Status and Future Prospects on the Brink of Energy Transition. *Renewable and Sustainable Energy Reviews* 142, 110833. doi:10.1016/j.rser.2021.110833.

Kwartnik-Pruc, A., and A. Mączyńska. 2022. Assessing Validity of Employing Surveying Methods to Capture Data on Topography to Determine Hydrological and Topographic Parameters Essential for Selecting Locations for the Construction of Small Hydropower Plants. *Energies* 15, 1527. doi:10.3390/en15041527.

Lee, N., U. Grunwald, E. Rosenlieb, H. Mirletz, A. Aznar, R. Spencer, and S. Cox. 2020. Hybrid Floating Solar Photovoltaics-Hydropower Systems: Benefits and Global Assessment of Technical Potential. *Renewable Energy* 162, 1415–1427. doi:10.1016/j.renene.2020.08.080.

Lyu, X., T. Zhang, L. Yuan, K. Yang, J. Fang, S. Li, and S. Liu. 2022. Pumped Storage Hydropower in Abandoned Mine Shafts: Key Concerns and Research Directions. *Sustainability* 14, 16012. doi:10.3390/su142316012.

Maika, N., W. Lin, and M. Khatamifar. 2023. A Review of Gravitational Water Vortex Hydro Turbine Systems for Hydropower Generation. *Energies* 16, 5394. doi:10.3390/en16145394.

Mandair, S., J. F. Morissette, R. Magnan, and B. Karney. 2021. MOC-CFD Coupled Model of Load Rejection in Hydropower Station. *IOP Conference Series: Earth and Environmental Science* 774(1), 012021. doi:10.1088/1755-1315/774/1/012021.

Manzano-Agugliaro, M. Taher, A. Zapata-Sierra, A. Juaidi, and F. G. Montoya. 2017. An Overview of Research and Energy Evolution for Small Hydropower in Europe. *Renewable and Sustainable Energy Reviews* 75, 476–489. doi:10.1016/j.rser.2016.11.013.

Menéndez, J., J. M. Fernández-Oro, and J. Loredo. 2020. Economic Feasibility of Underground Pumped Storage Hydropower Plants Providing Ancillary Services. *Applied Sciences* 10, 3947. doi:10.3390/app10113947.

Moiz, A., A. Kawasaki, T. Koike, and M. Shrestha. 2018. A Systematic Decision Support Tool for Robust Hydropower Site Selection in Poorly Gauged Basins. *Applied Energy* 224, 309–321. doi:10.1016/j.apenergy.2018.04.070.

Moran, E. F., M. C. Lopez, N. Moore, N. Müller, and D. W. Hyman. 2018. Sustainable Hydropower in the 21st Century. *Proceedings of the National Academy of Sciences* 115(47), 11891–11898. doi:10.1073/pnas.1809426115.

Morgunova, M., and K. Shaton. 2022. The Role of Incumbents in Energy Transitions: Investigating the Perceptions and Strategies of the Oil and Gas Industry. *Energy Research & Social Science* 89, 102573. doi:10.1016/j.erss.2022.102573.

Nasir, B. A. 2021. Design Aspects of Small-Hydropower Plant. *Journal of Energy Research and Review* 9(2), 1–9. doi:10.9734/JENRR/2021/v9i230226.

Nautiyal, H., and V. Goel. 2020. Sustainability Assessment of Hydropower Projects. *Journal of Cleaner Production* 265, 121661. doi:10.1016/j.jclepro.2020.121661.

Novara, D., and A. McNabola. 2018. The Development of a Decision Support Software for the Design of Micro-Hydropower Schemes Utilizing a Pump as Turbine. *Proceedings* 2(11), 678. doi:10.3390/proceedings2110678.

Prabowoputra, D. M., S. Hadi, A. R. Prabowo, and J. M. Sohn. 2020. Performance Investigation of the Savonius Horizontal Water Turbine Accounting for Stage Rotor Design. *International Journal of Mechanical Engineering and Robotics Research* 9(2), 184–189. doi:10.18178/ijmerr.9.2.184-189.

Quaranta, E., J. I. Pérez-Díaz, P. Romero–Gomez, and A. Pistocchi. 2021. Environmentally Enhanced Turbines for Hydropower Plants: Current Technology and Future Perspective. *Frontiers in Energy Research* 9, 703106. doi:10.3389/fenrg.2021.703106.

Rafiee, A., and K. R. Khalilpour. 2018. Renewable Hybridization of Oil and Gas Supply Chains. In Khalilpour, K. R. (ed.). *Polygeneration with Polystorage for Chemical and Energy Hubs*. Elsevier. doi:10.1016/b978-0-12-813306-4.00011-2

Redko, K., L. Denyshchenko, O. Dobrovolska, N. Lukyanenko, and Y. Kyryllova. 2022. Development of Green Energy as a Path to Energy Independence of the National Economy. *Futurity Economics & Law* 2(4): 36–42. doi:10.57125/FEL.2022.12.25.05.

Romanelli, J. P., L. G. M. Silva, A. Horta, and R. A. Picoli. 2018. Site Selection for Hydropower Development: A GIS-Based Framework to Improve Planning in Brazil. *Journal of Environmental Engineering* 144(7). doi:10.1061/(ASCE)EE.1943-7870.0001381.

Sapitang, M., W. M. Ridwan, K. F. Kushiar, A. N. Ahmed, and A. El-Shafie. 2020. Machine Learning Application in Reservoir Water Level Forecasting for Sustainable Hydropower Generation Strategy. *Sustainability* 12(15), 6121. doi:10.3390/su12156121.

Saulsbury, J. W. 2020. *A Comparison of the Environmental Effects of Open-Loop and Closed-Loop Pumped Storage Hydropower* (No. PNNL-29157). Pacific Northwest National Lab (PNNL).

Serpoush, B., M. Khanian, and A. Shamsai. 2017. Hydropower Plant Site Spotting Using Geographic Information System and a MATLAB Based Algorithm. *Journal of Cleaner Production* 152, 7–16. doi:10.1016/j.jclepro.2017.03.095.

Shiji, C., S. Dhakal, and C. Ou. 2021. Greening Small Hydropower: A Brief Review. *Energy Strategy Reviews* 36, 100676.

Shimray, B. A., K. M. Singh, T. Khelchandra, and R. K. Mehta. 2017a. A New MLP–GA–Fuzzy Decision Support System for Hydro Power Plant Site Selection. *Arabian Journal for Science and Engineering* 43, 6823–6835. doi:10.1007/s13369-017-2885-4.

Shimray, B. A., K. M. Singh, T. Khelchandra, and R. K. Mehta. 2017b. Ranking of Sites for Installation of Hydropower Plant Using MLP Neural Network Trained with GA: A MADM Approach. *Computational Intelligence and Neuroscience*, 4152140. doi:10.1155/2017/4152140.

Sica, C. E., and M. Huber. 2017. "We Can't Be Dependent on Anybody": The Rhetoric of "Energy Independence" and the Legitimation of Fracking in Pennsylvania. *The Extractive Industries and Society* 4(2), 337–343. doi:10.1016/j.exis.2017.02.003.

Singh, V. K., and S. K. Singal. 2017. Operation of Hydro Power Plants: A Review. *Renewable and Sustainable Energy Reviews* 69, 610–619. doi:10.1016/j.rser.2016.11.169.

Skoulikaris, C. 2021. Run-Of-River Small Hydropower Plants as Hydro-Resilience Assets Against Climate Change. *Sustainability* 13, 14001. doi:10.3390/su132414001.

Souček, J., P. Nowak, M. Kantor, and D. Veselý. 2023. CFD as a Decision Tool for Pumped Storage Hydropower Plant Flow Measurement Method. *Water* 15, 779. doi:10.3390/w15040779.

Tapia, A., A. Rodríguez del Nozal, D. G. Reina, and P. Millán. 2021. Three-Dimensional Optimization of Penstock Layouts for Micro-Hydropower Plants Using Genetic Algorithms. *Applied Energy* 301, 117499. doi:10.1016/j.apenergy.2021.117499.

Temel, P., E. Kentel, and E. Alp. 2023. Development of a Site Selection Methodology for Run-of-River Hydroelectric Power Plants Within the Water-Energy-Ecosystem Nexus. *Science of the Total Environment* 856(2), 159152. doi:10.1016/j.scitotenv.2022.159152.

Timilsina, A. B., S. Mulligan, T. R. Bajracharya. 2018. Water Vortex Hydropower Technology: A State-of-the-Art Review of Developmental Trends. *Clean Technologies and Environmental Policy*. doi:10.1007/s10098-018-1589-0.

Tiwari, G., J. Kumar, V. Prasad, and V. K. Patel. 2020. Utility of CFD in the Design and Performance Analysis of Hydraulic Turbines—A Review. *Energy Reports* 6, 2410–2429. doi:10.1016/j.egyr.2020.09.004.

Tomczyk, P., and M. Wiatkowski. 2020. Challenges in the Development of Hydropower in Selected European Countries. *Water* 12(12), 3542. doi:10.3390/w12123542.

UNIDO. 2019. *Technical Guidelines for the Development of Small Hydropower Plants – Design*. United Nations Industrial Development Association. Available in www.unido.org/sites/default/files/files/2020-01/P1_DES_002-1.pdf [Accessed on July 5th, 2023].

U.S. Department of Energy. 2016. *Hydropower Vision: A New Chapter for America's 1st Renewable Electricity Source*. Available in www.energy.gov/sites/default/files/2018/02/f49/Hydropower-Vision-021518.pdf [Accessed on July 5th, 2023].

U.S. Department of Energy. 2021. *Hybrid Energy Systems: Opportunities for Coordinated Research*. National Renewable Energy Laboratory. DOE/GO-102021-5447. Available in www.nrel.gov/docs/fy21osti/77503.pdf [Accessed on July 25th, 2023].

U.S. Department of Energy. 2023a. *Benefits of Hydropower*. Available in www.energy.gov/eere/water/benefits-hydropower [Accessed on July 26th, 2023].

U.S. Department of Energy. 2023b. *Hydropower Basics*. Available in www.energy.gov/eere/water/hydropower-basics [Accessed on July 6th, 2023].

U.S. Department of Energy. 2023c. *Microhydropower Systems*. Available in www.energy.gov/energysaver/microhydropower-systems [Accessed on July 16th, 2023].

U.S. Department of Energy. 2023d. *Pumped Storage Hydropower*. Available in www.energy.gov/eere/water/pumped-storage-hydropower [Accessed on July 8th, 2023].

U.S. Department of Energy. 2023e. *Types of Hydropower Turbines*. Available in www.energy.gov/eere/water/types-hydropower-turbines [Accessed on July 18th, 2023].

U.S. Energy Information Administration. 2023a. *Electricity Explained: Electricity in the United States*. Available in www.eia.gov/energyexplained/electricity/electricity-in-the-us.php [Accessed on June 30th, 2023].

U.S. Energy Information Administration. 2023b. *Hydropower Explained*. Available in www.eia.gov/energyexplained/hydropower/ [Accessed on June 30th, 2023].

van Els, R. H., and A. C. P. B. Junior. 2015. The Brazilian Experience with Hydrokinetic Turbines. *Energy Procedia* 75, 259–264. doi:10.1016/j.egypro.2015.07.328.

Wagner, B., C. Hauer, and H. Habersack. 2019. Current Hydropower Developments in Europe. *Current Opinion in Environmental Sustainability* 37, 41–49. doi:10.1016/j.cosust.2019.06.002.

Wu, Y., T. Zhang, C. Xu, X. Zhang, Y. Ke, H. Chu, and R. Xu. 2019. Location Selection of Seawater Pumped Hydro Storage Station in China Based on Multi-Attribute Decision Making. *Renewable Energy* 139, 410–425. doi:10.1016/j.renene.2019.02.091.

Xu, X., W. Hu, D. Cao, Q. Huang, C. Chen, and Z. Chen. 2019. Optimized Sizing of a Standalone PV-Wind-Hydropower Station with Pumped-Storage Installation Hybrid Energy System. *Renewable Energy* 147, 1418–1431. doi:10.1016/j.renene.2019.09.099.

Yang, W., and J. Yang. 2019. Advantage of Variable-Speed Pumped Storage Plants for Mitigating Wind Power Variations: Integrated Modelling and Performance Assessment. *Applied Energy* 237, 720–732. doi:10.1016/j.apenergy.2018.12.090.

Yildiz, V., and J. A. Vrugt. 2019. A Toolbox for the Optimal Design of Run-of-River Hydropower Plants. *Environmental Modelling & Software* 111, 134–152. doi:10.1016/j.envsoft.2018.08.018.

Zeng, S., Y. Liu, C. Liu, and X. Nan. 2017. A Review of Renewable Energy Investment in the BRICS Countries: History, Models, Problems and Solutions. *Renewable and Sustainable Energy Reviews* 74, 860–872. doi:10.1016/j.rser.2017.03.016.

Zvirgzdins, J., and O. Linkevics. 2020. Pumped-Storage Hydropower Plants as Enablers for Transition to Circular Economy in Energy Sector: A Case of Latvia. *Latvian Journal of Physics and Technical Sciences* 3. doi:10.2478/lpts-2020-0012.

9 Fuel Cells

Mesut Yurukcu, Jorge Saldaña, Narendra Boppana, Sepehr Arbabi, Jim Henry, and Sebahattin Ziyanak

9.1 INTRODUCTION

Fuel cells stand at the forefront of the movement toward renewable energy sources, playing a vital role in the global shift from reliance on fossil fuels to embracing renewable energy (Figure 9.1). These devices are known for efficiently converting chemical energy from fuels into energy through a process, thus avoiding the inefficiencies associated with traditional combustion processes (Carrette et al., 2000).

It is widely used in aerospace, electric vehicles, and scale stationary engines. Fuel cells with complex involved processes, such as mass and heat transfer, electrochemical reactions, and ion/electron conduction, are essential for efficiency and durability (Wilberforce et al., 2023; Ming et al., 2023). In the literature, for energy solutions, fuel cells have become a technology in reshaping the global energy scene. These devices use chemical reactions to convert fuel into electricity, sidestepping combustion-based power generation inefficiencies (Skaloumpakas, P., et al., 2024). Unlike methods that rely on combustion, fuel cells produce electricity by harnessing electron flow through reactions, leading to improved efficiency and reduced environmental impact (Abdin, 2024). A fuel cell operation comprises three elements: an anode, a cathode, and an electrolyte (Khudhayer et al., 2015; Kariuki et al., 2016; Yurtsever et al., 2018; Yurukcu, 2018). These components collaborate to enable ion movement and electron flow, which is essential for generating power. Fuel cells (Figure 9.2) are categorized based on their type, which dictates their features, suitable uses, and operating temperatures (Wilberforce et al., 2024).

Different kinds of fuel cells are available, such as proton exchange membrane fuel cells (PEMFCs) (Farghali et al., 2023; Laimon & Yusaf, 2024), which are valued for their start-up times and suitability for different uses, and solid oxide fuel cells (SOFCs) are known for operating at high temperatures and being perfect for stationary power generation (Abouemara et al., 2024). Molten carbonate fuel cells (MCFCs) find uses in industrial settings due to their ability to use various fuels and capture carbon dioxide (Shuhayeu et al., 2024). Additionally, there are phosphoric acid fuel cells (PAFCs) (Sammes et al., 2004; Oh et al., 2023) and alkaline fuel cells (AFCs) (Hamada et al., 2023), each with benefits tailored to different environments. Alkaline fuel cells (AFCs) (seen in Figure 9.3) become prominent with their potential for power output and extended lifespan compared to proton exchange membrane fuel cells (PEMFC). AFCs benefit from corrosive electrolytes, enabling the use of various catalyst materials apart from platinum, leading to cost savings and easier mass production. Hamada et al. (2023) covered the principles of AFCs' recent advancements in catalyst technology, component, and stack designs, including performance implications under different operational conditions (i.e., electrolyte concentration and temperatures). Increasing the concentration has shown an increase in AFC peak power density. Hamada and colleagues addressed challenges such as AFC's sensitivity to carbon dioxide, which explores solutions aimed at overcoming these obstacles to stimulate further progress in AFC technology development (Hamada et al., 2023).

Another research study, by Oh, S., and colleagues in 2023, examined how combining a phosphoric acid fuel cell (PAFC) with a Rankine cycle (ORC) can boost the electrical efficiency of the PAFC system (seen in Figure 9.4). This innovative PAFC–ORC hybrid system harnesses the waste heat from the PAFC to warm up working fluids in the ORC, generating electricity. Oh et al. (2023) developed a model of the PAFC system using Aspen Plus® software and validated it against

DOI: 10.1201/9781003282273-9

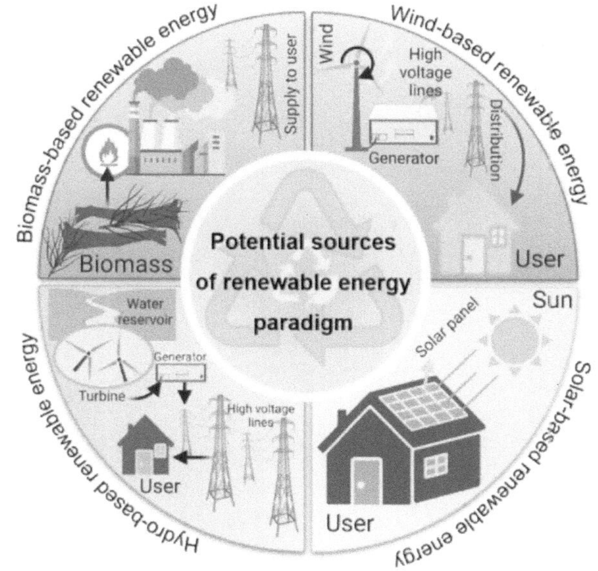

FIGURE 9.1 Potential renewable energy sources.

Source: Nazir et al. (2020).

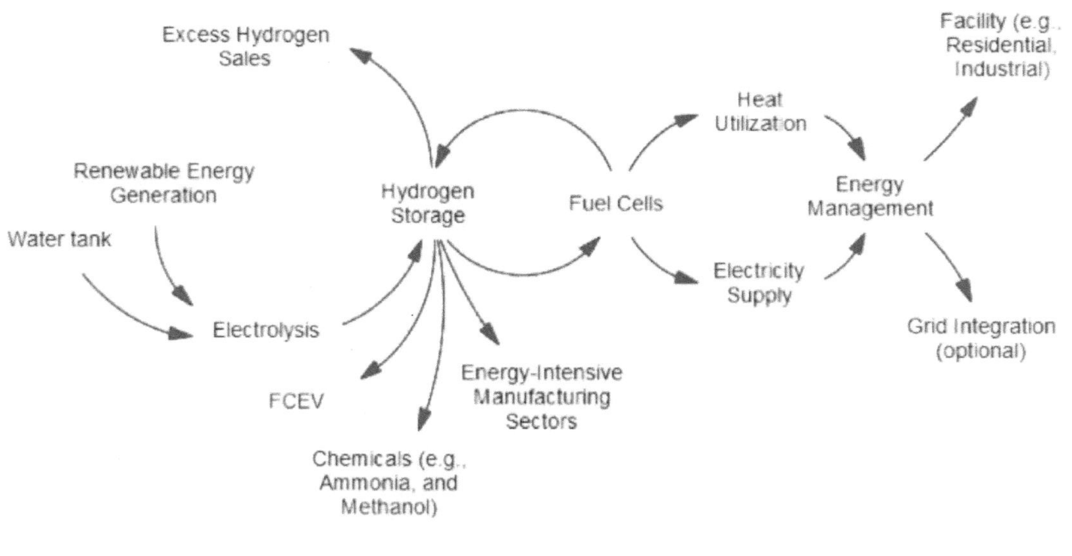

FIGURE 9.2 Interconnected loop for energy independence and sustainability.

Source: Laimon and Yusaf (2024).

real-world data obtained from the current–voltage polarization curve of the PAFC. Using different working fluids, the team conducted exergetic analyses to assess the efficiency and energy quality at operational temperatures. Among these options, methanol proved to be the most efficient, achieving an efficiency of 60.49% when the PAFC was operating at 190°C. Methanol also caused minimal exergy loss within the ORC system (Oh et al., 2023).

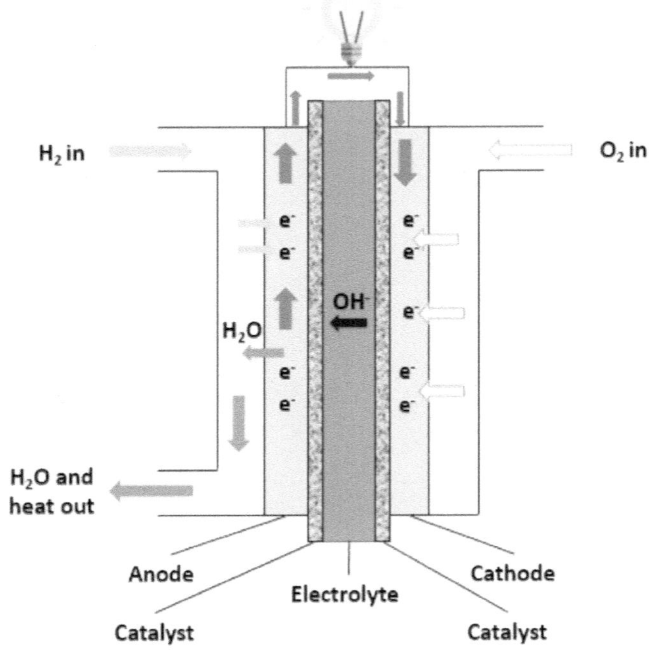

FIGURE 9.3 An AFC's working principle and constituent components.

Source: Hamada et al. (2023).

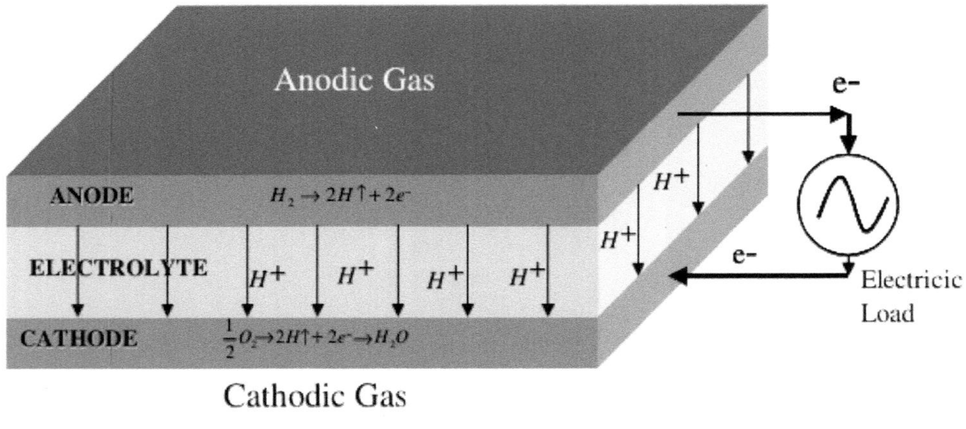

FIGURE 9.4 Schematic representation of a phosphoric acid fuel cell.

Source: Sammes et al. (2004).

On the other hand, the preheater and condenser exhibited the exergoeconomic factor at 5%, pointing toward areas where economic efficiency could be boosted. The research underscores the benefits of integrating a PAFC with an ORC using methanol to enhance both energy efficiency and economic performance of fuel cell systems (Sammes et al., 2004; Oh et al., 2023; Cheekatamarla, 2024).

The capacity of fuel cells can vary significantly, ranging from devices producing a few watts to large systems capable of generating megawatts of power. When it comes to efficiency, fuel cells typically achieve over 60% efficiency, which can increase to above 80% when utilized in cogeneration systems that harvest both electricity and heat (Wang & Gao, 2023; Hassan et al., 2024)

This chapter will explore the details of each type of fuel cell, looking into the benefits, drawbacks, and technological progress accelerating their evolution. By analyzing these aspects, this chapter aims to provide a basis for grasping the significance of fuel cells in promoting sustainable energy practices and overcoming obstacles hindering their widespread use. Therefore, this introduction not only gives information but also encourages creativity and implementation of fuel cell technology across different industries.

In the following sections, there will be an examination of how fuel cells work, a thorough exploration of the different types available, and an overview of their benefits and applications. This foundational knowledge will pave the way for selecting strategies that consider design elements and analyze the economic viability of fuel cell technologies. These steps are crucial for leveraging their capabilities within the evolving energy landscape. This in-depth understanding will shed light on aspects and the financial and environmental impacts associated with integrating fuel cells into a changing energy sector by using hydrogen, as shown in Figure 9.5 (Laimon & Yusaf, 2024).

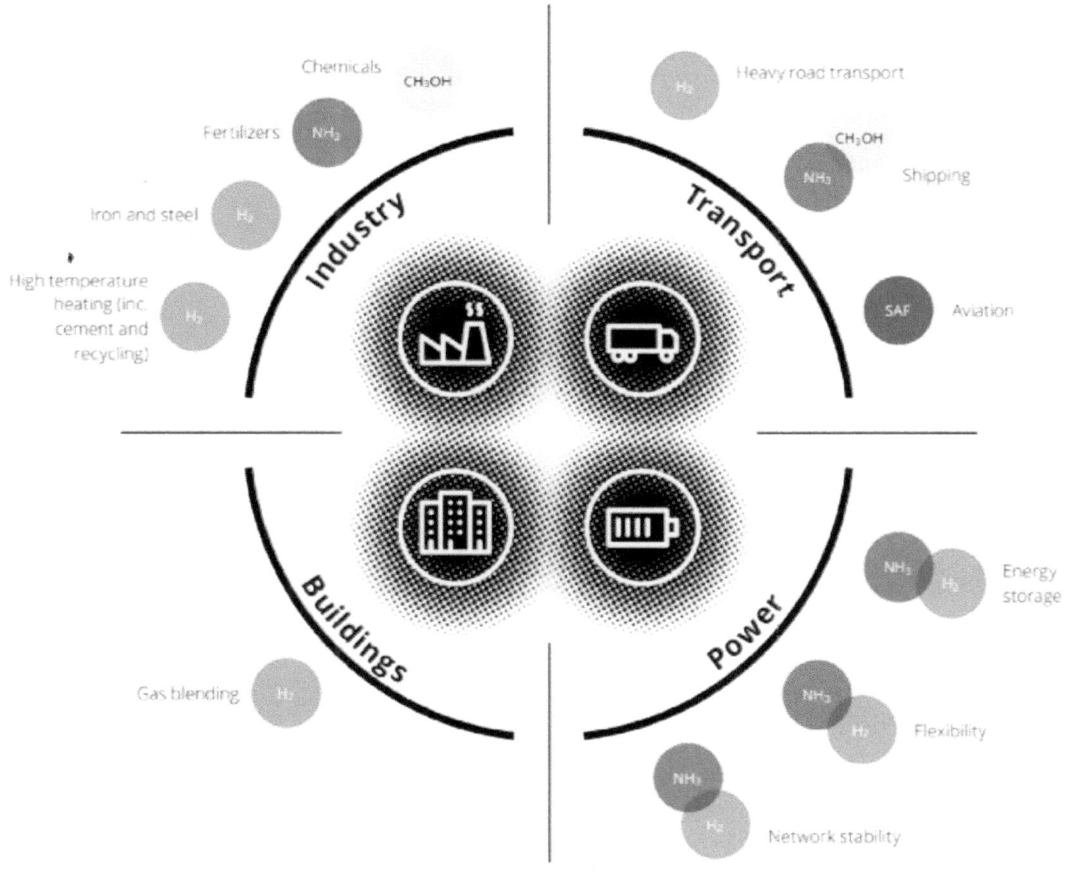

FIGURE 9.5 Main uses of hydrogen across different sectors.

Source: Laimon and Yusaf (2024).

9.2 SCREENING, SELECTION, DESIGN, AND ECONOMICS OF FUEL CELLS

The utilization of fuel cell technology requires a procedure encompassing screening, selection, design, and economic assessment to guarantee functionality, dependability, and affordability (Zhang et al., 2024). This thorough method supports incorporating fuel cells into different uses and boosts their business feasibility. When choosing fuel cells, the screening and selection process prioritizes finding the type that fits the needs of a given application (Gòdia Sangrà, 2024).

One of the applications studied by Ming and colleagues examines how machine learning, encompassing both ML and deep learning (DL), is utilized to improve the assessment of fuel cell performance, predict durability, and monitor applications. It explores the differences among ML techniques, combines them with physics simulations, and suggests avenues for future research in ML applications for fuel cells (Ming et al., 2023).

Another application is studied by Bello and colleagues (2023), who investigate the hurdles and advancements in creating materials for protonic ceramic fuel cells (PCFCs). The PCFCs are essential for achieving efficiency and minimal emissions in power production. The conventional trial and error with machine learning (ML) techniques for developing cathode materials in PCFCs face drawbacks, such as expenses and reliance on data. It impedes the discovery of optimal materials. As seen in Figure 9.6, the authors suggested using the design paradigm (EDP), which systematically employs controlled experimental planning and analysis to pinpoint and enhance new cathode materials. This strategy showed through the screening of the BaCoαCeβFeγYζO3-δ (BCCFY) family exhibits performance in reducing area-specific resistance and improving electrochemical properties, surpassing existing methods, in both effectiveness and cost-effectiveness (Bello et al., 2023).

Wilberforce and colleagues (2023) explored the role of fuel cells in meeting the growing energy needs and combating the escalating emissions from fossil fuel dependence. Their in-depth study enhances proton exchange membrane fuel cell (PEMFC) designs by innovating plate geometries to improve water management and overall cell performance. The research utilizes a mix of numerical techniques to evaluate designs, such as serpentine, parallel, and interdigitated flow channels, while also introducing a unique method leveraging machine learning to forecast the efficacy of innovative designs. (Wilberforce et al., 2023).

In addition, Alvi and Sandberg (2024) explored ways to improve the sustainability and cost-efficiency of fuel cells (PCFCs) by utilizing affordable and eco-friendly electrode materials. For platinum cathodes, the research suggests using alternative substances, like PEDOT and PSS, a polymer that proves to be an eco-conscious choice. The research highlights the creation of zinc oxide nanorods on carbon fiber surfaces through a temperature chemical growth technique, which is economical and adaptable for industrial use (Figure 9.7). This dual-use strategy aligns with optimizing resources and presents a pathway for developing economically feasible solutions for eco-friendly wastewater treatment and energy production (Alvi & Sandberg, 2024).

Different applications have different needs regarding power output, durability, and operating conditions (Figure 9.8). For example, portable devices and backup power systems require fuel cells that are lightweight and compact, such as proton exchange membrane fuel cells (PEMFCs). On the other hand, solid oxide fuel cells (SOFCs) are more suitable for power generation due to their higher efficiency on a larger scale (Adebisi et al., 2023).

The availability of fuel plays a significant role in determining the type of fuel cell to be used. For instance, direct methanol fuel cells are preferred in situations where methanol is easily accessible. In choosing the fuel cell technology, it is essential to consider its environmental impact. Alkaline fuel cells (AFCs), which only emit water during operation, are favored for applications where environmental concerns are critical (Ming et al., 2023).

As discussed in the mentioned research, the process of selecting fuel cells involves utilizing machine learning (ML) techniques to improve performance assessment, forecast longevity, and oversee fuel cell applications. This comprehensive examination highlights ML and deep learning (DL) approaches that are effectively utilized in fuel cell technology. Fuel cells replace renewable

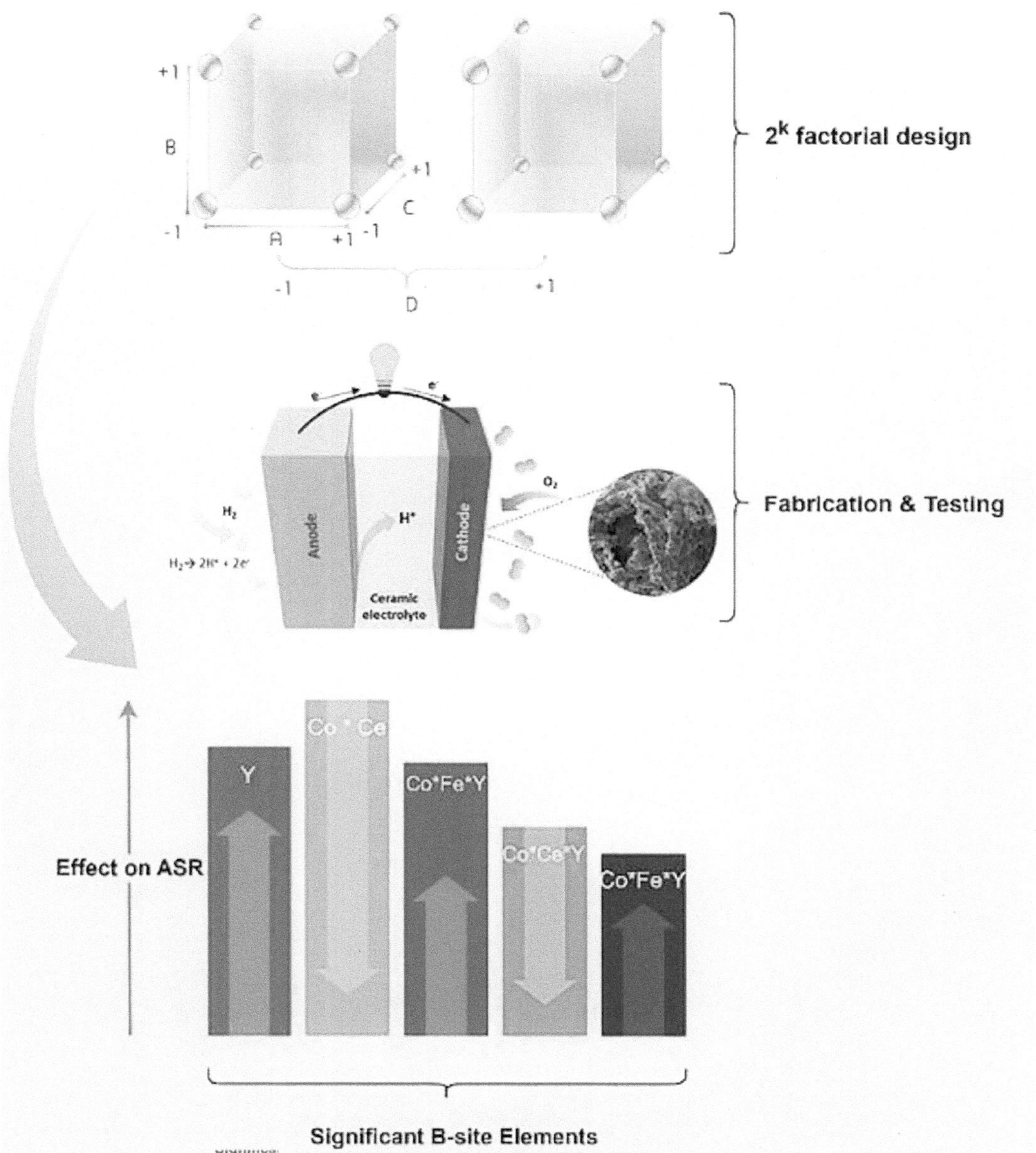

FIGURE 9.6 A novel experimental design paradigm (EDP) was introduced to design cathode materials for ceramic fuel cells.

Source: Bello et al. (2023).

energy sources due to their enhanced energy efficiency and reduced emissions. They directly convert chemical energy into electrical energy. Noteworthy types of fuel cells include the polymer membrane (PEM) and solid oxide fuel cells (SOFCs) (Ming et al., 2023).

The text discusses the hurdles in meeting cost and durability targets for achieving feasibility as indicated by benchmarks established by the US Department of Energy (DOE). It underscores the significance of ML in advancing fuel cell technologies through optimizing material choices, simulating

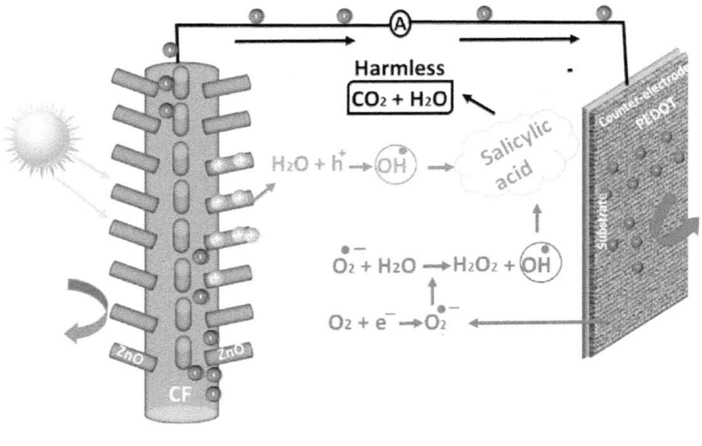

FIGURE 9.7 Schematic illustration of charge generation, transfer, and pollutant degradation at ZnO/carbon fiber anode and PEDOT: PSS cathode under the sunlight.

Source: Alvi and Sandberg (2024).

Metric	AFC	PEMFC	DMFC	PAFC	MCFC	SOFC
Electrolyte	KOH Solution	Proton-conducting polymer membrane	Proton-conducting polymer membrane	Phosphoric acid	Li2CO3; KCO3 melted in LiALO2	ZrO2 and Y2O3
Ions in the electrolyte	OH^-	H^+	H^+	H^+	$CO^{2-}{}_3$	O^{2-}
Temperature	60–80°°C	60–100°C	60–100°C	180–220°C	600–660°C	700–1,000°C
Fuel	H_2	H2 (pure or reformed)	Methanol	H2 (pure or reformed)	H2 (pure or reformed)	H2 (pure or reformed)
Application domains	Spatial	Automobile, portable, cogeneration and maritime	Portable	Cogeneration	Cogeneration centralized electricity production, maritime	Cogeneration centralized electricity production, maritime

FIGURE 9.8 Comparison of the available FC technologies for hydrogen production.

Source: Adebisi et al. (2023).

chemical reactions, and interpreting polarization curves. Predicting durability is also essential, as well as monitoring health status, diagnosing faults, and estimating the remaining lifespan of the fuel cells. After comparing machine learning methods, it is evident that deep learning can handle intricate datasets and extract features without requiring in-depth domain expertise. This advancement dramatically improves the capabilities and operational monitoring of fuel cells. The analysis also outlines research pathways highlighting the ongoing development of machine learning applications in fuel cells to address growing efficiency and durability requirements (Peters et al., 2024).

Milewski and colleagues investigated the use of boron-containing compounds borates as molten electrolytes in a unique type of fuel cell known as molten borates fuel cells (MBFCs). Through the application of models, they projected these fuel cells' performance and compared them to traditional molten carbonate fuel cells (MCFCs), emphasizing their ability to function effectively at moderate temperatures. By incorporating data on how ceramic matrices interact with electrolytes into their models, the researchers were able to identify promising compounds for experimental testing (Milewski et al., 2024).

Opakhai et al.'s (2023) article examines the progress and obstacles in metal-supported solid oxide fuel cells (MS SOFCs). These fuel cells are selected to take advantage of metal supports for improved strength, thermal shock resistance, and cost-efficiency compared to ceramic-supported SOFCs. Metal substrates like steel are preferred due to their costs, quick start-up capabilities, and simpler production processes. It also addresses challenges like electrode material degradation and thermal expansion differences between metal supports and electrolytes that can affect the long-term stability and efficiency of MS-SOFCs. To overcome these challenges in the future, the article proposes focusing on enhancing machine learning methods for performance prediction and operational monitoring of fuel cells (Opakhai & Kuterbekov, 2023; Hong et al., 2024)

When creating a fuel cell system, it is essential to integrate operational approaches to achieve the best performance and durability. The fuel cell stack configuration is critical as it determines the voltage and current output by deciding the number and arrangement of cells. Managing temperature and water levels is also crucial in PEMFCs, where precise moisture control is vital for maintaining the membrane's proton conductivity and the cell's overall efficiency. The choice of materials significantly impacts how lasting and cost-effective the system will be. Incorporating materials like reinforced composites not only improves the structural strength of components but also reduces wear and tear, ensuring that the fuel cell system remains efficient and economical in the long run (Peters et al., 2024; Kazda & Sedlarikova, 2024)

It is crucial to examine the up-front expenses, ongoing costs, and potential returns on investment to assess the aspects of fuel cell systems. The initial costs mainly involve purchasing and setting up the system, while continuous expenses cover maintenance, fuel, and component replacements. Finding ways to lower these costs through economies of scale and leveraging technological advancements is vital (Dezhdar et al., 2024; Mariscal et al., 2024). Moreover, the economic viability of fuel cells heavily depends on energy prices and government incentives that support energy technologies. Fuel cells have become feasible choices in areas with high electricity rates or generous incentives. Conducting a cost–benefit analysis and evaluating the payback period is essential for determining the sustainability of fuel cell projects in the long run (Yamate & Otomo, 2024; Satyanarayana & Naidu, 2024). This assessment considers life cycle costs and benefits, including improvements in energy security. Additionally, employing exergoeconomic analysis aids in optimizing fuel cell system designs by pinpointing and reducing exergy losses to enhance energy efficiency and operational performance (Dorregaray-Oyaregui et al., 2024; Hassouna & Shin, 2024).

In 2024, Peters, R., and colleagues explored the aspects of transitioning from fossil fuels to hydrogen and synthetic fuels by leveraging advancements in fuel cell technology (seen in Figure 9.9). They designed hydrogen production for fuels and are now repurposed to create fuels, which is viewed as a move toward establishing an environmentally friendly economy. Their study promotes an integrated approach that enhances fuel cell efficiency and synthetic fuel production processes to enhance energy system efficiency and sustainability (Peters et al., 2024).

Agnolucci (2007) investigated the economics of fuel cells, underscoring the market adoption despite earlier optimistic predictions. The author pointed out the challenges in achieving cost competitiveness with established energy sources like batteries. While offering benefits (i.e., power output and efficiency), fuel cells face obstacles due to their initial higher costs. Agnolucci and friends

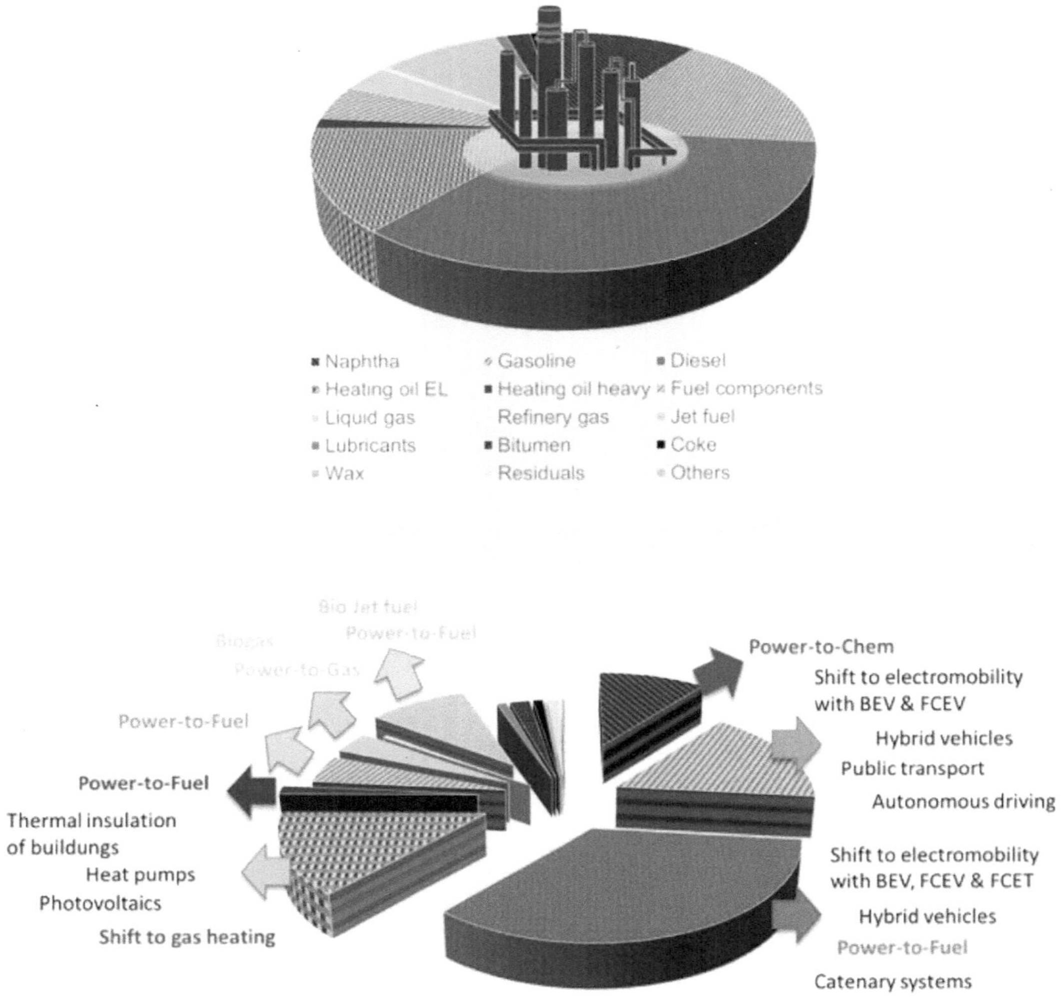

FIGURE 9.9 Refinery products in Germany.

Source: Peters et al. (2024).

highlighted that for fuel cells to become widely accepted, they must not only deliver performance but also be cost-competitive with or cheaper than traditional power sources (Figure 9.10). Their analysis indicates that unless these economic hurdles are overcome, fuel cells will likely remain confined to niche markets where their specific advantages outweigh the added expense. This examination of the feasibility of fuel cells emphasizes the importance of innovation and cost reduction in advancing their market viability (Agnolucci, 2007).

In a 2022 publication, Mayyas, A., and colleagues thoroughly investigated the feasibility of lithium-ion batteries and reversible fuel cells (Figure 9.11) for storing energy, especially in the context of fluctuating electricity prices. The research highlights how these technologies prove to be cost-effective when integrated into power grids that handle varying energy outputs. By using the levelized cost of energy storage (LCOS) as a metric for evaluation, the study demonstrates that both technologies can store electricity at a higher price. Reversible fuel cells stand out for their flexibility and ability to bolster grid resilience during prolonged power disruptions (Mayyas et al., 2022; Chadly et al., 2022).

Adopting new technologies can improve the efficiency and sustainability of energy storage systems (Table 9.1) within the grid, supporting goals of integrating renewables and reducing carbon emissions.

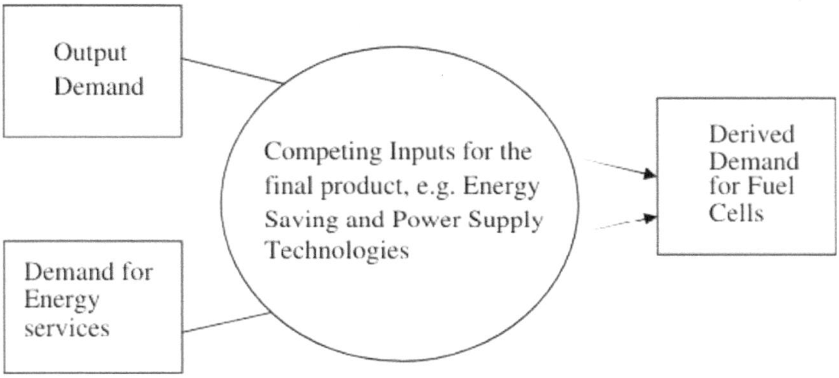

FIGURE 9.10 Schematic representation of the twice-derived demand for fuel cells.

Source: Agnolucci (2007).

The economic analysis shows the way affordable energy storage solutions align with the industry's transition toward sustainable practices (Wang et al., 2018; Mayyas et al., 2022; Metzger & Li, 2022).

To sum up, the processes of screening, selecting, designing, and economically evaluating fuel cells are crucial in determining the viability and success of fuel cell technology across applications. With advancements and a growing emphasis on sustainability, fuel cells are poised to impact the future energy landscape as long as these aspects are managed with care.

9.3 FUEL CELLS APPLICATIONS, PRINCIPLES, BEST PRACTICES, AND ECONOMICS

Fuel cells find various uses due to their ability to convert energy efficiently while keeping it clean. In the realm of transportation, they find uses in a variety of vehicles, such as cars, buses, and handling equipment, enabling longer travel distances and faster refueling times compared to electric vehicles (Alzahrani et al., 2022; Mohammadi et al., 2024). In terms of power supply, fuel cells offer electricity for homes, businesses, and industries, proving particularly useful for backup power systems and remote locations with limited access to the grid. Additionally, for applications (Figure 9.12) like military missions, outdoor activities, and charging electronic devices, fuel cells provide enhanced durability and energy storage capacity compared to traditional batteries (Nagao, 2024; Singh et al., 2024).

Fuel cells convert chemical energy into electrical energy using a process that generally involves hydrogen and oxygen. This process occurs in stages; hydrogen is split into protons and electrons at the anode, with the protons moving to the cathode through an electrolyte and electrons flowing through a circuit to produce electricity (Vignesh et al., 2024; Lepikash et al., 2024). Various kinds of fuel cells, including polymer electrolyte membrane (PEM), solid oxide fuel cells (SOFC), and molten carbonate fuel cells (MCFC), are tailored for uses based on factors like temperature, efficiency, and fuel needs. The selection of catalysts, often platinum, is crucial in improving reaction rates and overall effectiveness (Seyam et al., 2024)

Following certain practices is essential to ensure fuel cell systems (Table 9.2 and Figure 9.13) work efficiently and last long. One key aspect is to install them in ventilated areas to avoid overheating and make maintenance easier. Regular upkeep is crucial for checking and replacing worn-out parts, like electrolytes and catalysts, which can wear down over time (Giorgi & Leccese, 2013; MTU Solutions (n.d.), 2024). Also, using high-quality hydrogen fuel is vital, and impurities can harm the catalyst, reducing the efficiency and lifespan of the system. By sticking to these guidelines, reliability and performance of fuel cells can be improved significantly (Chaudhary et al., 2024; Habib, 2024; Reddy et al., 2024; Wang et al., 2024).

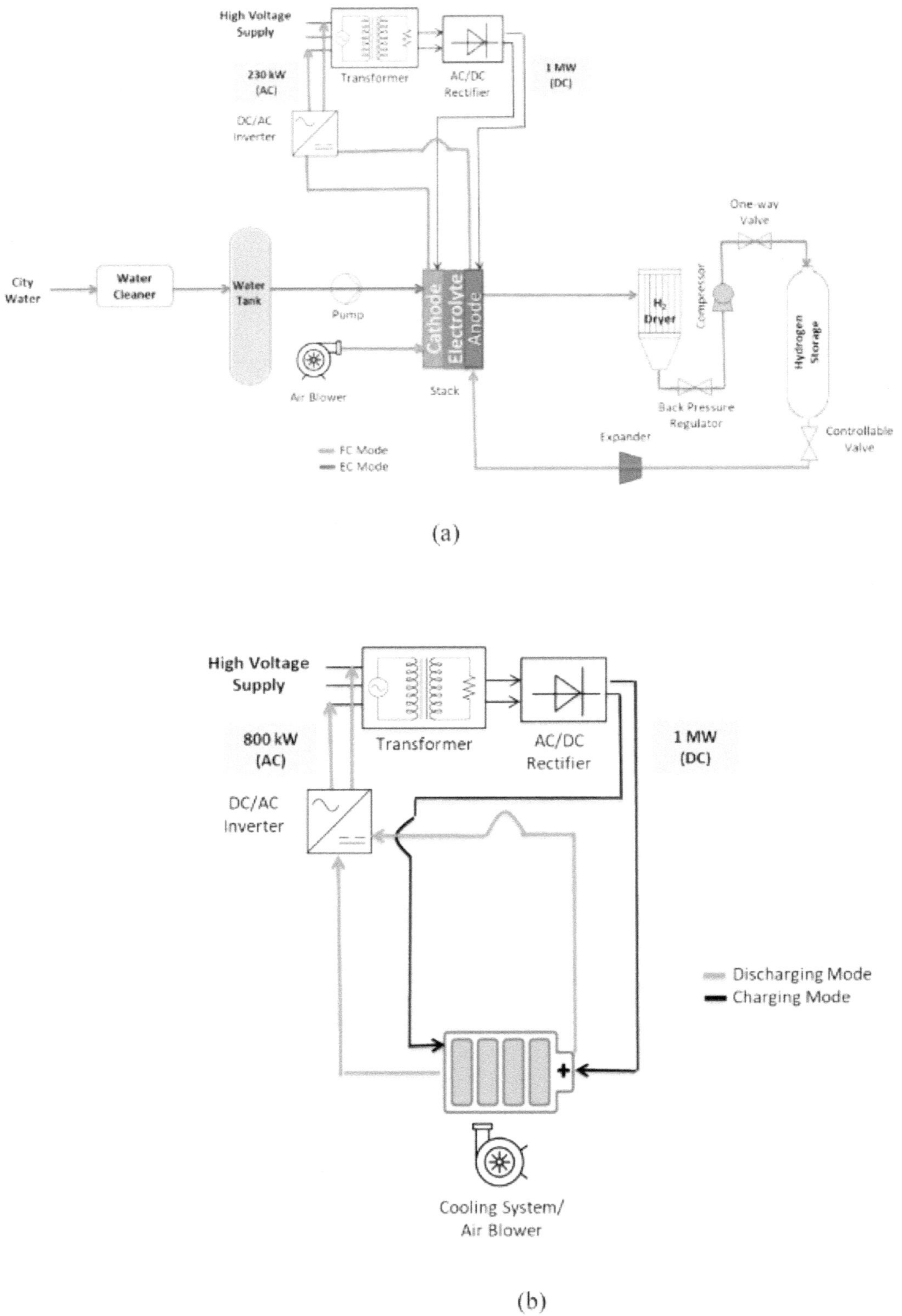

FIGURE 9.11 Schematic of the energy storage system using (a) reversible fuel cells and (b) lithium-ion batteries.

Source: Mayyas et al. (2022).

TABLE 9.1
Energy Storage System (ESS) Design Parameters and Cost Model Assumptions

Parameter	RFC	LIB	Units
System Design Parameters			
Nominal system size	1,000 (electrolyzer) 470 (fuel cell)	1,000 (nominal size)	kW for RFC, kWh for LIB
Discharge rate (kWh/h)	230	500	kW
Lowest depth of discharge (DoD)	0	250	kWh
Hydrogen storage cost	Variable (as low as $20 for underground storage; salt caverns and up to 750 per kg-H_2 for ground storage tanks)[a]	n/a	
Round trip efficiency	30%	80%	
Economic Analysis Parameters			
Discount rate	8%	8%	
Stack lifetime	3 years	3,000 cycles	
System lifetime	20	20	
Price of electricity	Variable	Variable	$/kWh
O&M cost	0.02	0.02	$/kWh
System capital cost	600	350	$
Installation factor	25%	25%	$/kW
Installed system cost (TIC)[b]	750,000	437,500	
Cycles per year	Variable	Variable	

Source: Mayyas et al. (2022).

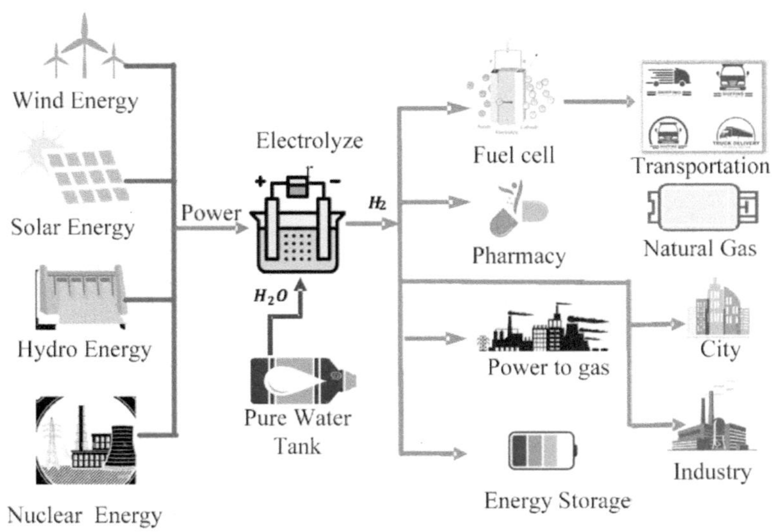

FIGURE 9.12 Applications of hydrogen energy.

Source: Alzahrani et al. (2022).

TABLE 9.2
The Different Fuel Cells That Have Been Realized and Are Currently in Use and Development

	AFC Alkaline	PEMFC Polymer Electrolyte Membrane	DMFC Direct Methanol	PAFC Phosphoric Acid	MCFC Mullen Carbonate	SOFC Solid Oxide
Operating temp. (°C)	<100	60–120	60–120	160–220	600–800	800–1,000 Low temperature (500–600) possible
Electrolyte	KOH	Perfluorosulfonic acid (Nafion membrane)	Perfluorosulfonic acid (Nafion membrane)	H_3PO_4 immobilized in SiC matrix	Li_2CO_3-K_2CO_3 eutectic mixture immobilized in γ-$LiAlO_2$	YSZ (yttria-stabilized zirconia)
Charge carrier in the electronic	Oil	H^+	H^+	H^+	**CO32–**	O^{2-}
Anode reaction	$H_2 + 2OH^+ \rightarrow 2H_2O + 2e^-$	$H_2 \rightarrow 2H^+ + 2e^-$	$CH_3OH + H_2O \rightarrow CO_2 + 6H^+ + 6e^-$	$H_2 \rightarrow 2H^+ + 2e–$	$H_2 + CO_3^{2}– \rightarrow H_2O + CO_2 + 2e^-$	$H^2 + 0^{2-} \rightarrow H_2O + 2e^-$
Cathode reaction	$1/2\ O_2 + H_2O + 2e^- \rightarrow 2OH^-$	$½O_2 + 2H^+ + 2e^- \rightarrow H_2O$	$3/2\ O_2 + 6H^+ + 6e^- \rightarrow 3H_2O$	$1/2\ O_2 + 2H^+ + 2e– \rightarrow H_2O$	$1/2\ O_2 + CO_2 + 2e– \rightarrow CO_3^{2-}$	$1/2\ O_2 + 2e^- \rightarrow O^{2-}$
Electrode materials	Anode: Ni Cathode: Ag	Anode: Pt, PtRu Cathode: Pt	Anode: Pt, PtRu Cathode: Pt	Anode: Pt, PtRu Cathode: Pt	Anode: Ni-5Cr Cathode: NiO(Li)	Anode: Ni-YSZ Cathode: lanthanum strontium manganite (LSM)
Applications	Transportation Space, military Energy storage systems			Combined heat and power for decentralized stationary power systems	Combined heat and power for stationary decentralized systems and for transportation (trains, boats, . . .)	
Realized power	Small plants: 5–150 kW modular	Small plants: 5–250 kW modular	Small plants: <5 kW	Small to medium-sized plants: 50 kW–11 MW	Small power plants: 100 kW–2 MW	Small power plants: 100–250 kW
Main producers	AFC Energy (UK) UTC Power (USA) Acta Power (Italy)	Ballard (Canada) Heliocentris (Germany)	SFC Energy (Germany)	UTC Power (USA) Fuji Electric (Japan)	Fuel Cell Energy (USA)	Ceramic Fuel Cells Limited (Australia) Hexis & Vaillant (Germany) SOFC Power (Italy) Bloom Energy (USA)
*Lifetime**	Not available	2,000–3,000 h	1,000 h	>50,000 h	7,000–8,000 h	1,000 h

Note: *Estimated values from literature data.
Source: Giorgi and Leccese, 2013].

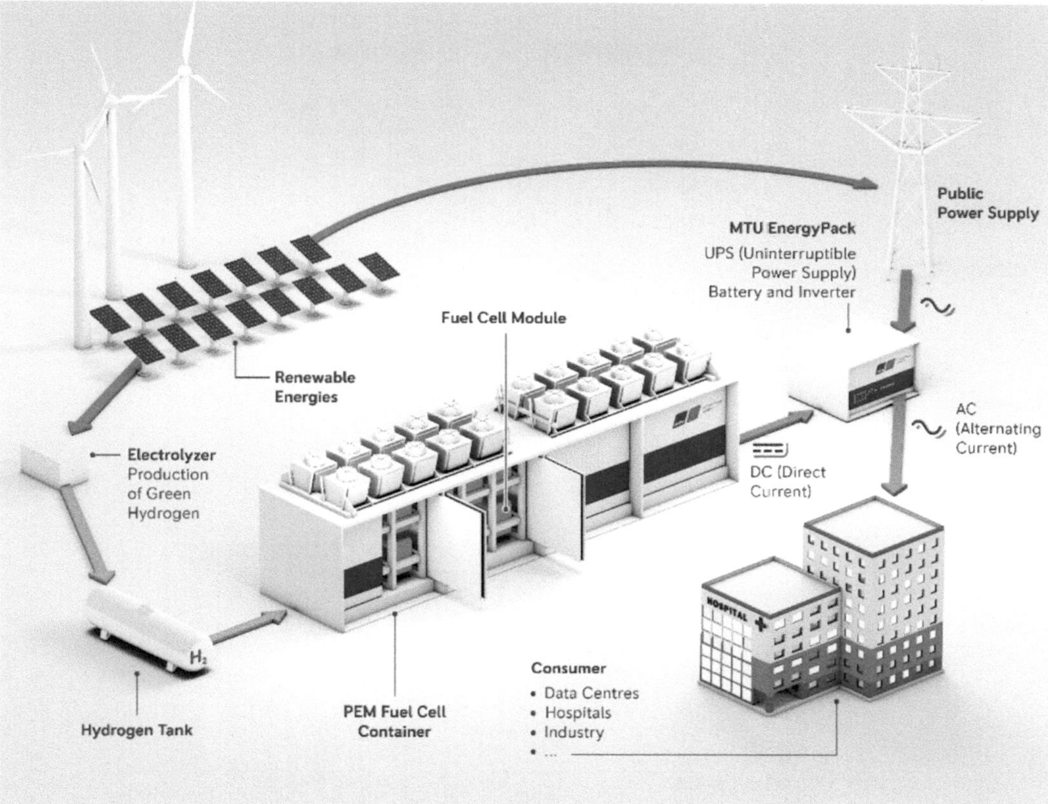

FIGURE 9.13 The fuel cell is destined to be part of a fully integrated standby power solution comprising the fuel cell system, a UPS system, batteries, and the hydrogen infrastructure.

Source: MTU Solutions (n.d.; 2024).

The economic evaluation of fuel cells aims to assess their cost-effectiveness and sustainability over time. While the up-front expenses for fuel cell systems are typically high because of the need for materials like platinum, their operational costs can be low thanks to their efficiency and potentially long lifespan (Ghimire et al., 2024; Iqbal et al., 2024). Government incentives and subsidies can significantly boost the feasibility of fuel cells by offsetting costs and promoting wider adoption. It is crucial to conduct a cost–benefit analysis to weigh the financial gains along with indirect benefits, such as environmental impact and energy security, which are key factors in determining the overall economic feasibility of implementing fuel cell technology (Hou et al., 2024; Yan et al., 2024)

9.4 POWER PRODUCTION APPLICATIONS

In the 2024 article authored by Yalcinkaya and colleagues, they explored the exploration of using microbial fuel cells (MFCs) for innovatively generating power (Figure 9.14) (Abouemara et al., 2024; Singh et al., 2024; Yalcinkaya et al., 2024).

In their approach, nanofiber technologies (Figure 9.15) are integrated into the system to improve energy output. MFCs, known for producing electricity from compounds through microbial processes, have traditionally struggled with lower power density when compared to chemical fuel cells (Abouemara et al., 2024; Singh et al., 2024; Yalcinkaya et al., 2024).

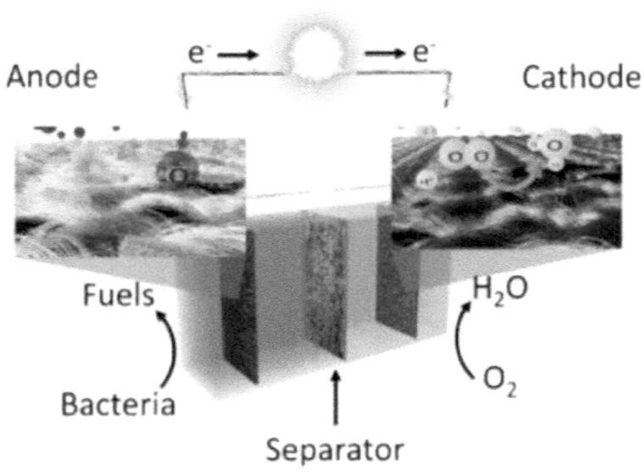

FIGURE 9.14 The schematic and fundamental principles of a conventional microbial fuel cell (MFC).

Source: Yalcinkaya et al. (2024).

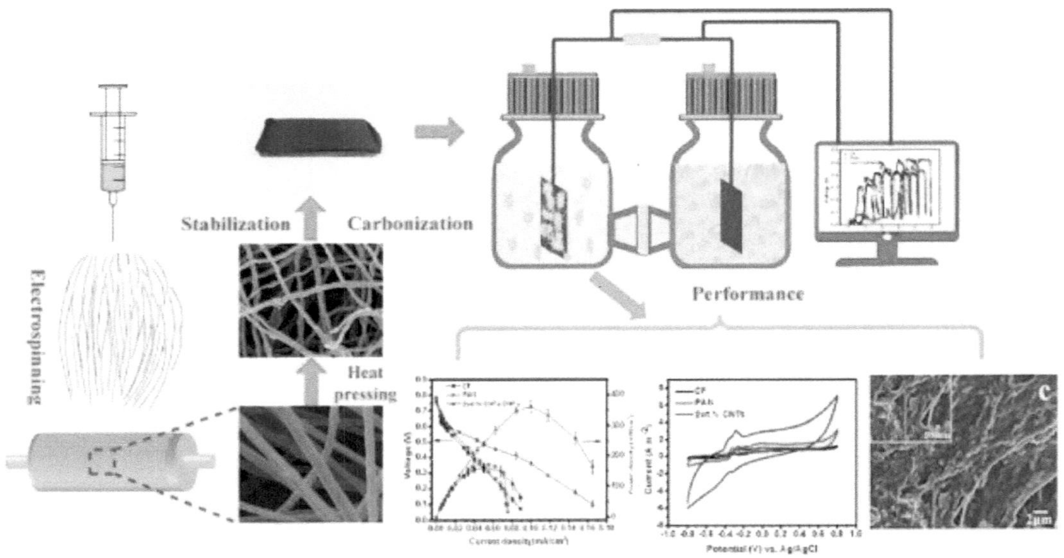

FIGURE 9.15 Composite anode of electrospun carbon nanofibers and hybrid carbon nanotubes facilitate microbial attachment and electron transfer and exhibit superior conductivity and biocompatibility compared to commercial carbon felt.

Source: Yalcinkaya et al. (2024).

They studied the incorporation of nanofiber technologies (Figure 9.16) and carbon nanofibers to enhance the performance of MFCs significantly. These nanofibers help in the attachment of microbes, increase the surface area for electron transfer, and improve the system's durability. The findings suggest an increase in power density, signaling a promising future where nanofibers could transform MFC efficiency and effectiveness (Abouemara et al., 2024; Singh et al., 2024; Yalcinkaya et al., 2024).

The 2024 study conducted by Syed, M. A., and colleagues explored the application of direct ammonia fuel cells (DAFC) (can be seen in Figure 9.17) to address the challenges associated with the variable power output from renewable sources like solar and wind (Syed et al., 2024).

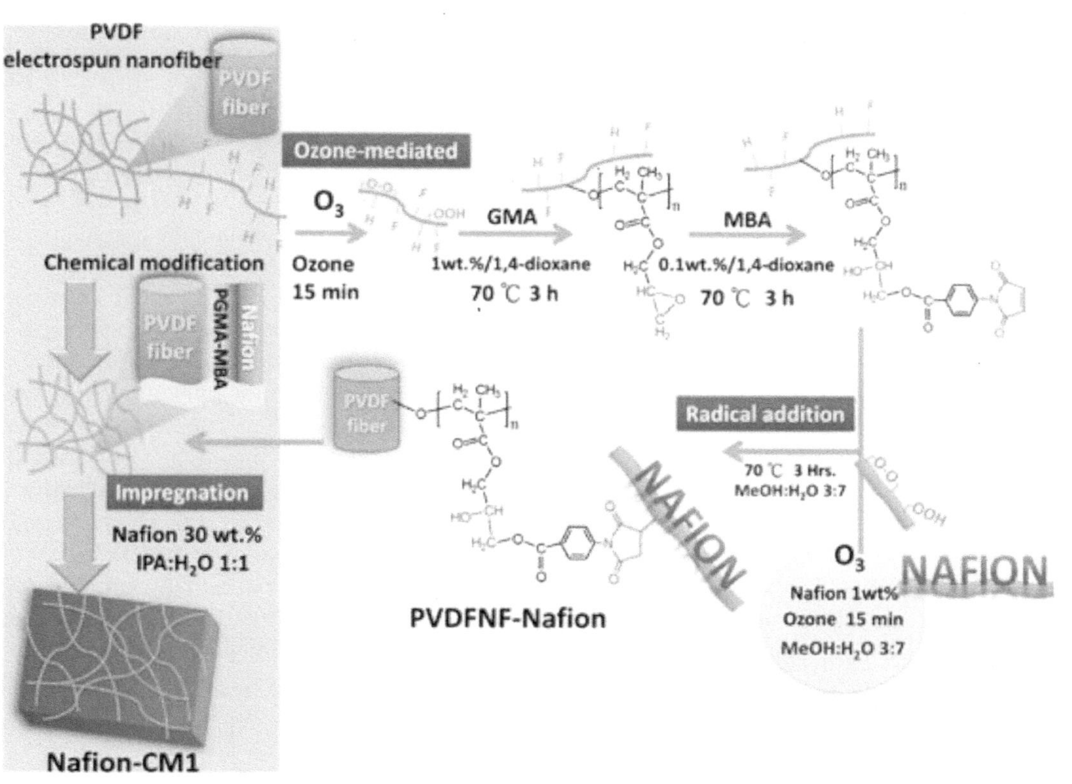

FIGURE 9.16 Synthetic route of Nafion-functionalized PVDF nanofibers.

Source: Yalcinkaya et al. (2024).

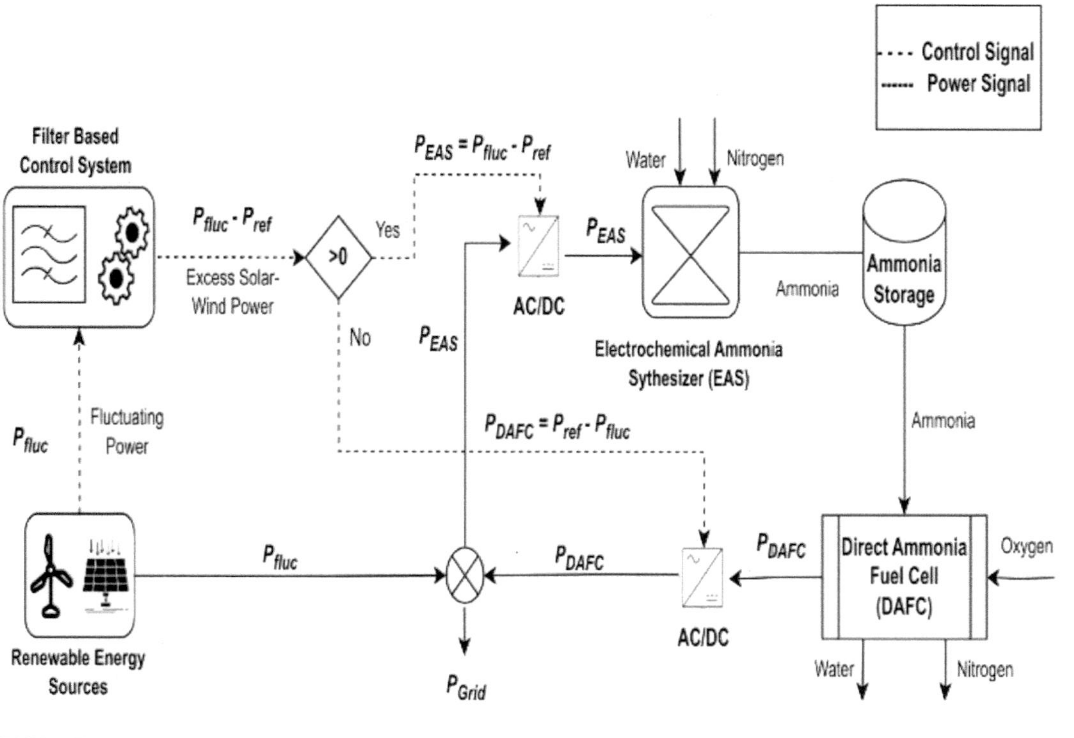

FIGURE 9.17 The proposed solar and wind power control strategy via electrochemical ammonia synthesis and direct ammonia fuel cell utilization.

Source: Syed et al. (2024).

Their research emphasized the potential of DAFCs in grid applications by offering a solution to the issues arising from the unpredictable nature of renewable energy sources, which can impact grid stability. Their methods improved the efficiency and dependability of renewable energy systems. Their results indicated that this technology could be influential in shaping the energy landscape, facilitating a more secure and sustainable integration of renewable resources into the grid (Ghimire et al., 2024; Syed et al., 2024).

In a study conducted in 2024 by Ghimire and colleagues, they extensively explored the pressing issue of power outages in Nepal and their impact on health-care services. Despite Nepal's potential for hydropower, the outdated electricity infrastructure and challenging geographical and seismic conditions often result in prolonged power interruptions that significantly affect hospitals and other essential facilities. The research focuses on Dhulikhel Hospital, a community-centered facility that heavily relies on power supply for its medical and surgical operations. The hospital has implemented a power backup system that combines panels with diesel generators to address power disruptions. However, dependence on diesel comes with economic and environmental drawbacks, such as high fuel costs and maintenance requirements, as well as notable levels of noise and pollution. Their study suggests adopting fuel cell technology, particularly emphasizing proton exchange membrane fuel cells (PEMFC), as a quieter and more efficient alternative for uninterrupted power supply in health-care settings (Figure 9.18). This study highlights the applications of fuel cells in critical health-care scenarios, where a consistent energy supply is essential (Ghimire et al., 2024).

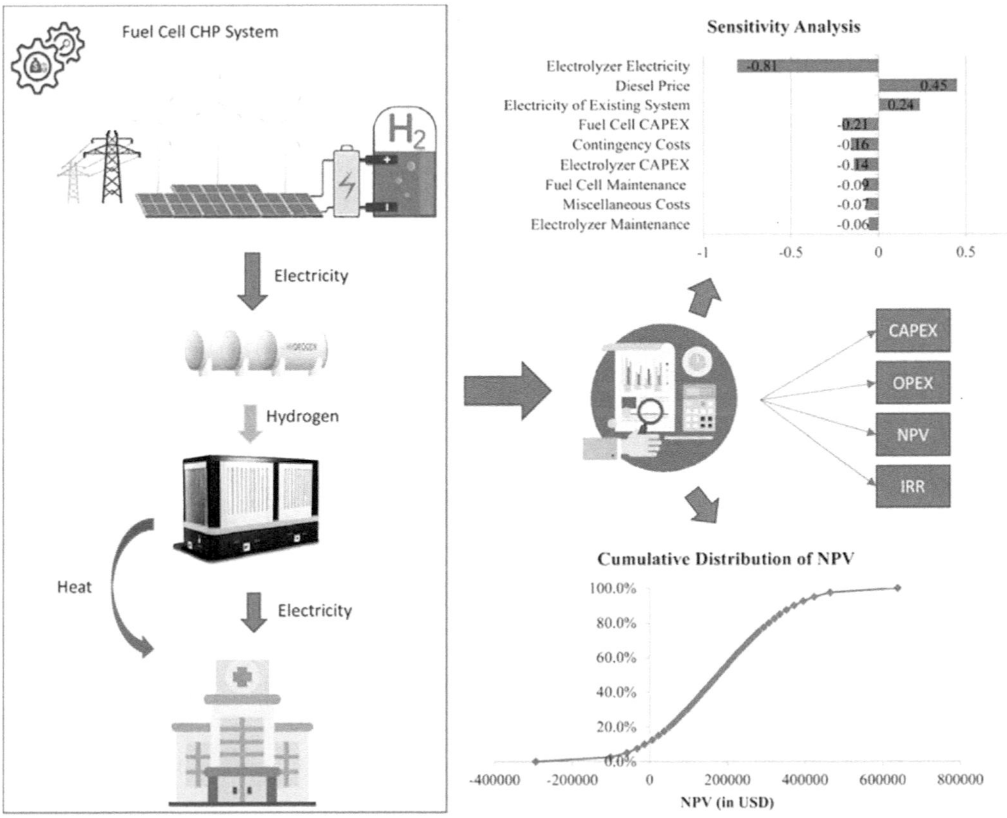

FIGURE 9.18 An analysis of fuel cells combined heat and power systems, including capital expenditures (CAPEX), operating expenses (OPEX), net present value (NPV), and internal rate of return (IRR).

Source: Ghimire et al. (2024).

Fuel cells offer a groundbreaking approach to generating power with advantages over energy sources thanks to their high efficiency and positive environmental impact. As energy options, they deliver reliable and uninterrupted power and support global sustainability efforts by reducing greenhouse gas emissions (Fazal et al., 2024). The unique ability of fuel cells to cater to energy demands – from residential and commercial backup systems to powering facilities like hospitals – positions them as a critical technology in transitioning toward a more eco-friendly energy landscape. These characteristics make fuel cells a top choice for integrating into energy systems, where they can play a significant role in load balancing and storing energy efficiently (Jaiswal et al., 2022). Moving ahead, the ongoing progress and expansion of fuel cell technologies will be crucial in addressing challenges like high initial costs and infrastructure requirements, covering the way for broader adoption and a sustainable future for energy usage.

9.5 INDUSTRIAL APPLICATIONS

Fuel cells, the proton exchange membrane (PEM) variety, are becoming increasingly popular across industries because of their efficiency and eco-friendliness. However, when it comes to using them in settings, there are challenges to overcome, mainly related to meeting the high-voltage needs of heavy machinery and large-scale operations.

PEM fuel cells can be integrated into applications to boost industrial voltage use. While PEM fuel cells are effective and environmentally friendly, they typically generate a voltage that may not meet the requirements of industrial processes (Table 9.3), and this is where advanced DC–DC boost converters come into play to increase voltage levels. The study introduces a switch-cascaded DC–DC boost converter that provides higher voltage gains and efficiency compared to traditional converters. PEM fuel cells, employed in various settings, rely on the operation of a fuzzy-controlled single-switch cascaded converter for their efficiency and reliability. This achievement represents a positive step in the adoption of clean energy technologies in sectors with substantial power demands (Gupta et al., 2024; Hu et al., 2024; Kart et al., 2024; Nagao, 2024; Zhang et al., 2024).

To address the significant industrial environmental impact, researchers have been focusing on the role of fuel cells as a sustainable option. One of the studies investigated the developments and potential of fuel cell technology in vehicles, emphasizing how they offer an alternative to traditional fuels, helping reduce emissions and improve energy efficiency. The authors highlighted the advancements that have made integrating fuel cells practical, such as enhancements in catalytic processes and innovative designs that enhance the performance and durability of the fuel cells. They point out that the technology could revolutionize global energy practices, and they advocate strongly for their wider adoption in the manufacturing processes of hydrogen (Figure 9.19) (Giorgi & Leccese, 2013; Osman et al., 2022; Olabi et al., 2021).

The 2013 paper by Giorgi and Leccese explores how fuel cells are being used in industries, highlighting their potential to bring about significant changes across various sectors. As businesses look for eco-effective energy options, fuel cells are considered valid options for producing clean energy. Giorgi and Leccese point out the suitability of types of fuel cells, such as polymer electrolyte membrane (PEMFC) and solid oxide fuel cells (SOFC) (Table 9.4), for a range of purposes due to their unique operational features. The research delves into how fuel cells are being incorporated into areas like power generation and transportation, emphasizing the need for advancements in fuel cell technologies to improve economic feasibility and operational effectiveness (Giorgi & Leccese, 2013; Van Veldhuizen et al., 2023).

In their study published in 2023, Van Veldhuizen, B., and colleagues dug into the usage of solid oxide fuel cells (SOFCs) in environments to discuss the advantages and challenges of this technology. With the naval sector aiming to comply with emission standards, SOFCs are seen as a solution for reducing greenhouse gas emissions and improving ship energy efficiency (Figure 9.20). The

TABLE 9.3
Summary of the Required Properties and Issues of Proton-Conducting Polymers on Each Application

Application	Required Properties	Present Issues
High-temperature PEMFCs Membranes	• High proton conductivity above 120°C, and low-humidity conditions • Good mechanical strength, especially in thinner membranes (essential for maintaining structural integrity and improving performance) • Adequate thermal and chemical stability • Low gas permeability for membranes.	• Managing the issue of acid and ionic liquid leakage from the host polymers • Balancing proton conductivity and mechanical properties • Long-term use • Environmental profiles due to being fluorine-free • Cost reduction
Ionomers	• High proton conductivity, thermal and chemical stability • Higher oxygen permeability for ionomers • Compatibility between membranes and ionomers	• Balancing high oxygen permeability with suitable ionomer IEC • Pt dissolution into the ionomer, and redeposition of Pt particles • Ionomer–catalyst interface resistance
Electrolyzers	• High proton conductivity, chemical stability, and sufficient mechanical strength • Low hydrogen permeability for membranes • Compatibility between membranes and ionomers	• Balancing high proton conductivity with low hydrogen permeability and high mechanical stability • Maintaining performance, durability, and cost reduction • Balancing the membrane's thickness to minimize ohmic losses, hydrogen crossover, and safety • Over time, powdered materials detach from the membrane • Swelling behavior leading to mechanical strain and delamination • Optimizing ionomer content in the anode and cathode • Environmental profiles due to being fluorine-free • Scalability
Photoelectrochemical water splitting	• Capturing water from ambient air and enabling proton conduction	• Optimizing the ionomer loading and composition to balance water vapor capture and proton conduction without imposing mass transport limitations or excessive thickness that could impede efficiency • Minimizing potential drawbacks like reduced light transmittance and long-term degradation under UV irradiation
Secondary batteries	• High proton conductivity and higher gas barrier properties • High anhydrous proton conductivity for sealed solid-state batteries • Good processability, mechanical softness, and chemical stability • Broad electrochemical potential window • Interfacial compatibility with electrodes	• Low proton conductivity due to insufficiently humidified environment • Poor interfacial compatibility with electrodes • Cyclic stability • Control of grain boundaries if they exist

TABLE 9.3 (*Continued*)

Summary of the Required Properties and Issues of Proton-Conducting Polymers on Each Application

Application	Required Properties	Present Issues
Supercapacitors	• Broad electrochemical potential window • Excellent film-forming capabilities • High ionic conductivity and mechanical flexibility • High thermal stability • Interfacial compatibility with electrodes • Proton conductive binders in electrodes	• Balancing high energy density and power density • Cyclic stability and durability • Environmental profiles and cost reduction by biopolymer electrolytes
Flow batteries	• High ion selectivity • Low permeability of active species • High proton conductivity, chemical stability, mechanical strength, and thermal stability • Low electronic conductivity • Compatibility with electrolytes to avoid unexpected reaction	• Balancing conductivity, selectivity, and stability • Active species crossover • Lifespan and cycle stability • Scalability • Environmental impact and cost reduction
• Actuators	• Flexibility • High proton conductivity, chemical stability, and mechanical strength • Lower-voltage operation • Low swelling	• Balancing proton conductivity and mechanical properties • Durability under various operational conditions (humidity, temperature, long-term use, etc.) • Minimizing acid or ionic liquid leakage • Water management for proper hydration levels • Environmental impact and cost reduction
• Sensors	• High sensitivity and/or selectivity • High proton conductivity, chemical stability, and mechanical durability • Wide temperature range operation • Various environmental stability	• Advancing sensitivity and/or selectivity in complex matrices • Long-term stability • Integration with other components • Balancing proton conductivity with mechanical properties • Cost reduction and scalability • Environmental impact during production and disposal

Source: Nagao (2024).

research investigates how SOFC systems can be integrated into types of vessels and explore their potential to work alongside alternative fuel sources. It also emphasizes the environmental benefits of using SOFCs, such as lowering NOx, SOx, particulate matter, and noise levels. By bridging power generation practices with requirements, this study underscores the importance of a comprehensive approach to implementing SOFC technology in marine settings (Van Veldhuizen et al., 2023).

Expanding on the advantages Van Veldhuizen and colleagues (2023) discussed regarding the use of solid oxide fuel cells, it is essential to investigate how this innovation could be tailored for land-based transportation purposes. In the next part, we will examine the obstacles and possibilities associated with incorporating fuel cell technology in cars and public transportation systems, showcasing their capacity to transform transportation through sustainability and enhanced performance.

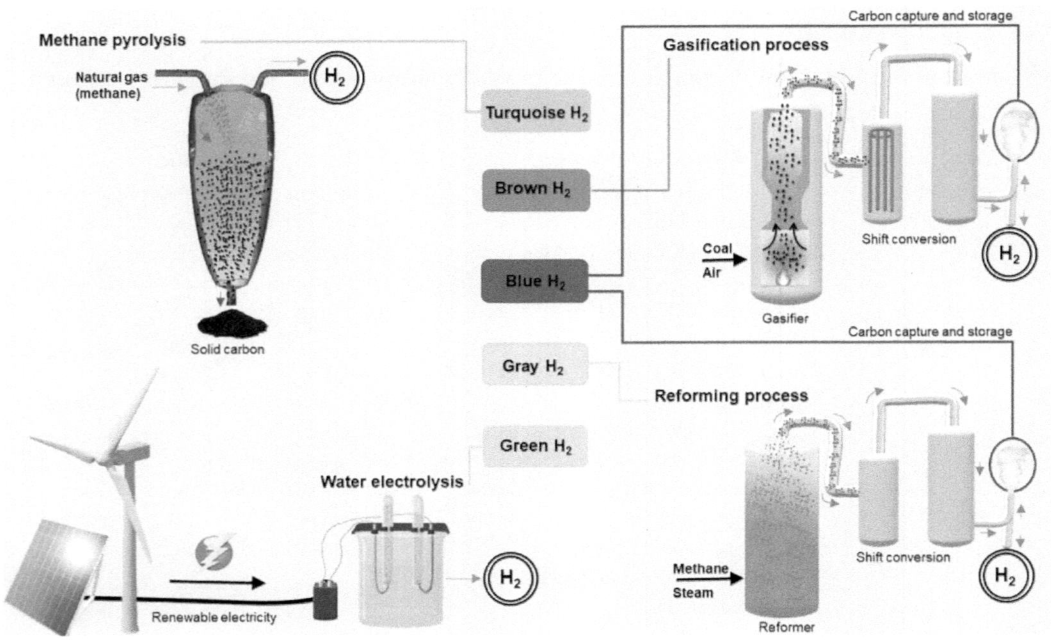

FIGURE 9.19 Hydrogen color coding for various manufacturing processes. Green hydrogen is produced using renewable energy sources, such as solar or wind energy, followed by water electrolysis. Gray and brown hydrogen are produced by methane steam reforming and coal gasification, respectively, and when combined with carbon capture and storage, blue hydrogen is produced. Turquoise hydrogen is made through the pyrolysis of methane, with solid carbon as a by-product.

Source: Osman et al. (2022).

TABLE 9.4
Performance Comparison between Off-the-Shelf SOFC Systems

Criteria	Unit	SOFC	DG
Electrical efficiency	%	43%–65%	30%–45%
Vol. power density	kW/m³	2–28	30–60
Grav. power density	kW/ton	5–30	45–75
Start-up time (cold start)	h	12–24	0.2–0.3
Start-up time (hot start)	h	0.1–0.2	0.02–0.04
Load change rate	%/min	2%–10%	10%–20%
Noise	dB (A)	40–70	80–110
Current system CAPEX (2021)	€/kW	1,500–7,000	250–400
Expected system CAPEX (2030)	€/kW	500–2,000	300–500
System lifetime	1,000 h	100–150	120–200
Major maintenance interval	1,000 h	30–90 (stack replacement)	40–60 (engine overhaul)

Source: Van Veldhuizen et al. (2023).

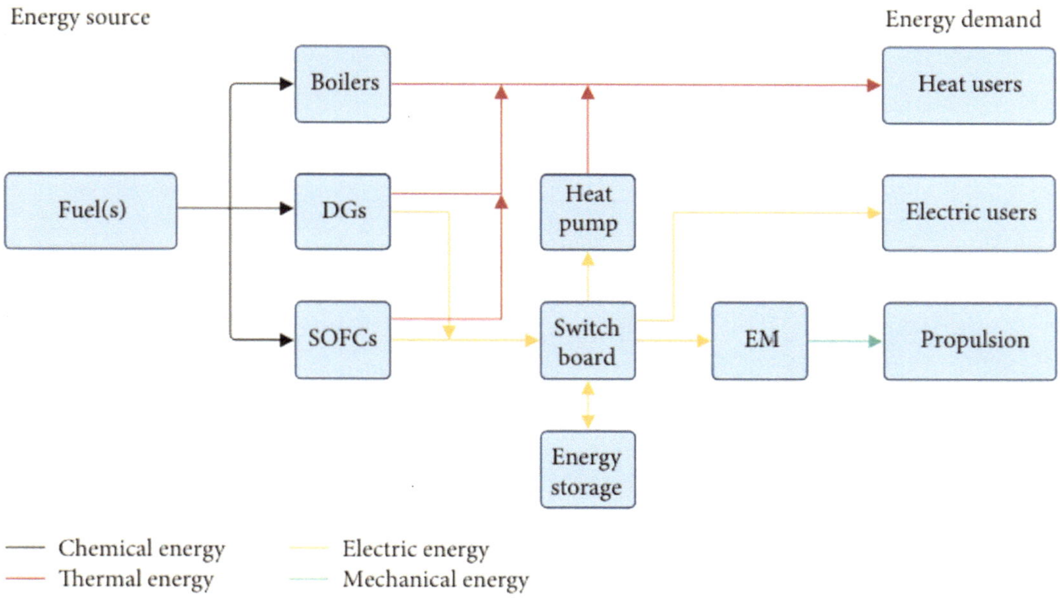

FIGURE 9.20 Generic hybrid SOFC power plant for marine applications.

Source: Van Veldhuizen et al. (2023).

9.6 TRANSPORTATION APPLICATIONS

While fuel cells have generally been praised for their emissions and efficient energy utilization, the challenges in structural design hinder their widespread use in commercial airplanes (Das et al., 2023; Dirkes et al., 2023). Nevertheless, fuel cells present benefits for land-based transport, like in cars, buses, and trains, including shorter refueling times and lower emissions compared to electric vehicles powered by batteries. The thorough assessment by Fakhreddine et al. highlights how fuel cells have the potential to revolutionize transportation by making it greener and economically feasible, positioning them as a technology for many modes of transport (Figure 9.21) (Fakhreddine et al., 2023; Pramuanjaroenkij et al., 2023).

According to a study by Dirkes and colleagues (2023), the transportation industry, which serves as a primary link and heavily relies on fossil fuels, has undergone significant changes since the start of the COVID-19 pandemic. Hydrogen fuel cell vehicles (HFCVs) (Figure 9.22) have emerged as a solution due to their ability to function without emitting greenhouse gases or involving direct combustion. Unlike batteries that have limited energy storage capabilities, fuel cells can provide power as long as there is a steady supply of fuel, making them crucial for the advancement of environmentally sustainable transportation technologies (Dirkes et al., 2023; Olabi et al., 2023).

In summary, looking ahead, the transition from fuels to hydrogen fuel cells and electric vehicles not only addresses environmental concerns but also opens up new economic possibilities (Table 9.5). With hydrogen fuel cells and electric vehicles expected to outperform petrol vehicles by 2050, it is evident that the momentum toward more efficient and sustainable transportation choices is gaining traction (Korner et al., 2015; Dirkes et al., 2023).

Embracing these technologies is crucial for curbing carbon emissions and promoting a future, underscoring the need for ongoing investments and research in these innovative transportation solutions.

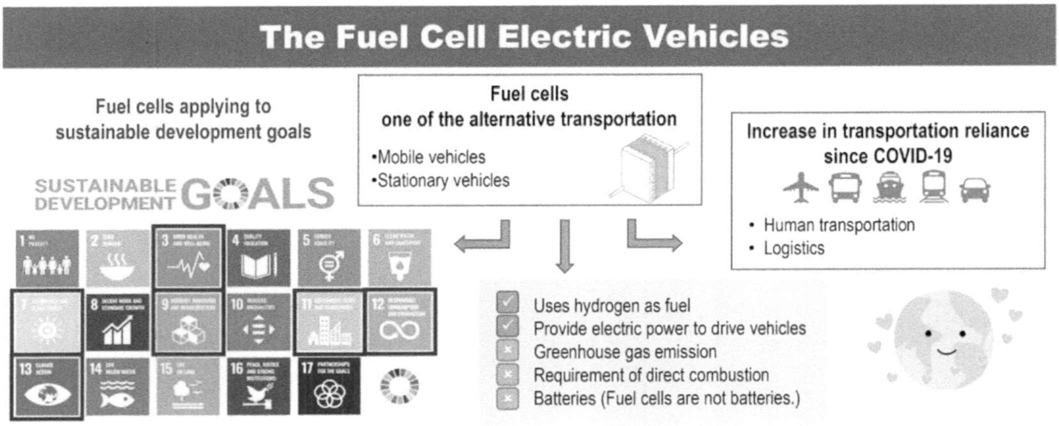

FIGURE 9.21 The fuel cell electric vehicles' sustainability development goals.

Source: Pramuanjaroenkij et al. (2023).

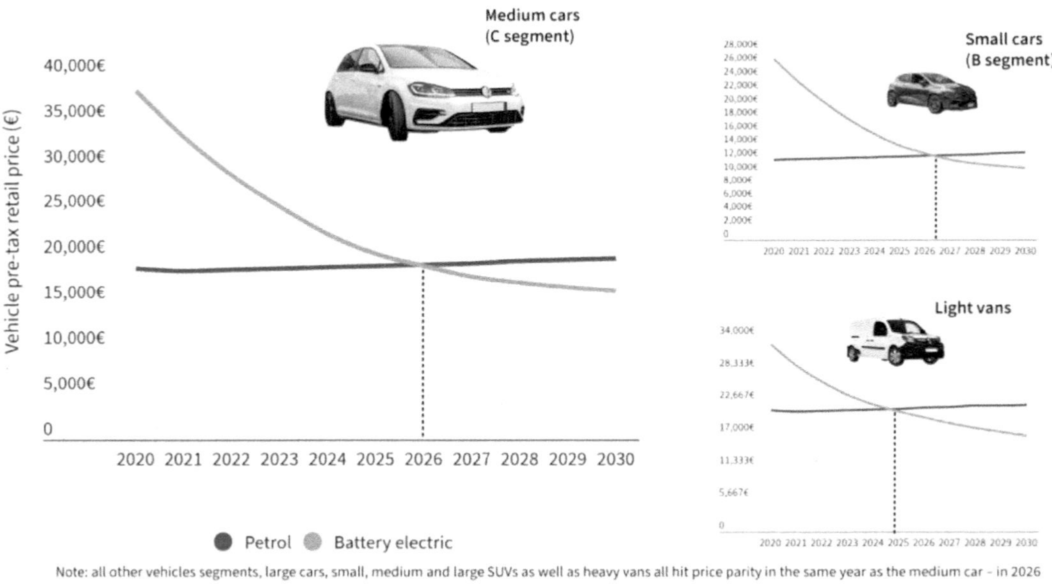

FIGURE 9.22 EVs will be cheaper than petrol cars in all segments by 2027.

Source: Transport & Environment (n.d.; 2024).

9.7 FUEL CELL APPLICATIONS IN THE OIL AND GAS INDUSTRY

The oil and gas sector, a part of the energy industry, is increasingly looking toward fuel cell technology to improve efficiency and reduce environmental impact. When researching how fuel cells are being used in this industry, their deployment, challenges, workflows, and economic aspects, it becomes clear that they offer potential alternatives to traditional power sources (Figure 9.23) (Busby, 2005; Speight, 2011; Sullivan et al., 2016; Al-Khori et al., 2020; Abdelkareem et al., 2021).

These systems not only deliver reliable and eco-friendly energy but also align with the sector's growing focus on sustainability. Yet integrating fuel cell technology into oil and gas operations

TABLE 9.5
Cost of PLDVs by Technology as Computed in the Model

		Today	2030	2050	Unit
United States	Conventional ICE gasoline	28,600	30,900	32,300	USD
	Conventional ICE diesel	29,300	31,700	33,100	USD
	Hybrid gasoline	30,000	31,800	33,200	USD
	Plug-in hybrid gasoline	32,400	33,200	34,400	USD
	BEV	35,400	32,800	34,000	USD
	FCEV	**60,000**	**33,600**	**33,400**	**USD**
EU 4	Conventional ICE gasoline	26,000	28,100	29,400	USD
	Conventional ICE diesel	26,600	28,800	30,100	USD
	Hybrid gasoline	27,200	28,900	30,900	USD
	Plug-in hybrid gasoline	29,400	30,200	31,300	USD
	BEV	32,200	29,800	30,900	USD
	FCEV	**60,000**	**30,600**	**30,300**	**USD**
Japan	Conventional ICE gasoline	20,500	23,000	25,200	USD
	Conventional ICE diesel	21,000	23,600	25,800	USD
	Hybrid gasoline	21,500	23,600	25,900	USD
	Plug-in hybrid gasoline	23,200	24,700	26,800	USD
	BEV	25,400	24,400	26,500	USD
	FCEV	**60,000**	**25,000**	**26,000**	**USD**

Source: Korner et al. (2015).

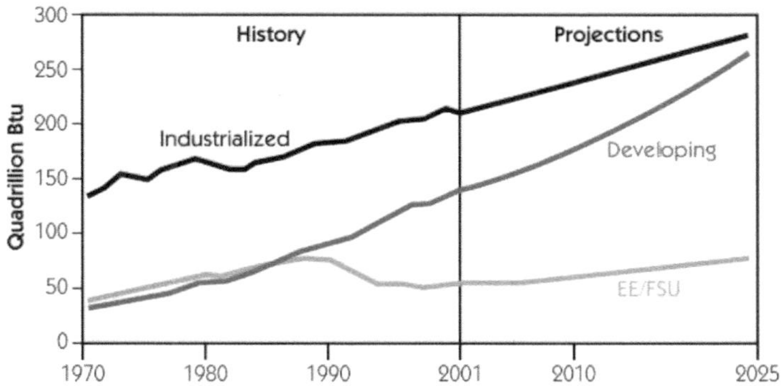

FIGURE 9.23 Energy use in developing nations. Robust economic growth in China, Asia, and other developing regions will greatly increase their demand for energy.

Source: Al-Khori et al. (2020).

poses hurdles like high initial expenses and the need for customized workflow adjustments (Busby, 2005; Speight, 2011; Al-Khori et al., 2020; Abdelkareem et al., 2021). This section will discuss the applications of fuel cells in the oil and gas industry, the existing systems in place for implementing this technology, the obstacles hindering widespread adoption, and the economic factors influencing their feasibility. By exploring these areas, we can better understand how fuel cells are shaping the future of energy in the oil and gas domain.

The use of solid oxide fuel cells (SOFC) in the oil and gas sector represents a step toward adopting eco-energy practices, especially in improving efficiency and reducing environmental harm.

Al Khori et al. (2020) investigated the role of SOFCs in shifting from energy systems to cleaner, more effective alternatives. Known for its energy conversion efficiency, this technology directly transforms chemical energy into electric power, leading to a notable decrease in greenhouse gas emissions from oil and gas activities. The study highlights how SOFC systems can easily integrate with existing energy infrastructure, such as flare and steam systems, such as photovoltaic arrays, ensuring a continuous energy supply (Busby, 2005; Speight, 2011; Al-Khori et al., 2020; Abdelkareem et al., 2021).

Moreover, Al Khori and team highlight the performance challenges hindering the adoption of SOFCs in the industry. Despite these barriers, ongoing research and development show promise for transitioning toward economically feasible energy solutions within the oil and gas field. This comprehensive review emphasizes the pressing need to reduce reliance on fuels while positioning SOFC technology and shaping future energy transformations within the sector. This need for transition sets the stage for decreasing carbon emissions and moving toward an energy model (Al-Khori et al., 2020).

In a study conducted in 2016, Sullivan and colleagues delved into the use of solid oxide fuel cell (SOFC) technology within the oil and gas industry, specifically concentrating on oil and gas processing. The research introduces the geothermic fuel cell (GFC), a system that produces electricity and utilizes the generated heat to refine oil shale into quality crude oil and natural gas. The nine-stack SOFC system, coupled with components like a natural gas fuel converter and air compressor, showcases a combined heat and power (CHP) strategy. Embedded underground, these systems transfer heat to the surrounding formations, effectively transforming kerogen into usable energy sources. A thermodynamic evaluation revealed a 55% efficiency in combined heat and power generation. This inventive application of fuel cells underscores their potential to transform energy efficiency and mitigate greenhouse gas emissions in the realm of oil and gas, representing a stride toward sustainable and eco-conscious extraction and processing techniques (Sullivan et al., 2016; Al-Khori et al., 2021).

They focused on how these cells can boost energy efficiency and reduce harm. One of the studies explores the use of oxide fuel cells (SOFCs) in oil and gas activities, which shows their ability to operate efficiently while significantly reducing greenhouse gas emissions. Stambouli et al. cast light on how these fuel cells contribute to growth and align with global energy security and environmental conservation objectives. By examining uses and potential improvements in detail, the review argues for broader adoption of fuel cell technologies to address rising energy needs and environmental sustainability challenges in this sector (Stambouli & Traversa, 2002).

In wrapping up the conversation about implementing fuel cell technology in the oil and gas sector, it is clear that solid oxide fuel cells (SOFCs) and various other fuel cell types bring advantages in improving efficiency and lessening impacts. The research studies conducted by Sullivan et al. (2016), Al Khori et al. (2020), and Stambouli et al. (2002) offer perspectives on how these technologies are being incorporated into traditional energy systems to promote more sustainable and effective operations. While challenges like expenses and integration complexities exist, ongoing research efforts are leading to more financially feasible solutions. Involving fuel cell technologies in the oil and gas field not only supports sustainability objectives but also establishes a model for future energy generation practices. This transition lays the groundwork for exploring how similar progress can be replicated across industries, potentially revolutionizing the energy sector.

9.8 FUEL CELLS IN ENERGY TRANSITION IN THE OIL AND GAS INDUSTRY

The oil and gas sector, reliant on fuels, is now looking toward fuel cells as a critical technology to help transition to cleaner energy. Solid oxide fuel cells (SOFCs), in particular, are playing a pivotal role in reshaping how energy is produced in this industry, by offering more efficient power generation solutions (Zhdaneev et al., 2024).

Fuel cells bring several advantages to the oil and gas field. Firstly, they can significantly reduce greenhouse gas emissions by using hydrogen, a fuel that only produces water when consumed in a fuel cell (Kaland, 2024, April). This is a comparison to the emissions released from fossil fuel burning processes, which emit elevated levels of carbon dioxide, nitrogen oxides, and other harmful pollutants (Wang & Azam, 2024). Secondly, SOFCs and different types of fuel cells can improve efficiencies by achieving energy conversion rates compared to conventional combustion systems that often experience energy losses due to inefficiencies like heat emission (Sarlak et al., 2024). Incorporating fuel cells into the oil and gas industry aids in transitioning from energy systems to alternatives. The change is vital not just for meeting carbon reduction goals but also for aligning with investors' and the public's expectations for more sustainable practices that are environmentally friendly. Moreover, offering power in remote oil and gas operation areas enhances energy security. It helps reduce reliance on grid power and strengthens the resilience of energy systems (Usman et al., 2024).

However, obstacles have been encountered in the integration of fuel cells in the oil and gas sector. Key challenges include the costs of fuel cell systems that need extensive infrastructure to support hydrogen operations and the technological complexities involved in scaling up fuel cell systems to meet industrial energy needs (Lahrichi et al., 2024). Additionally, there are safety issues associated with using hydrogen that require attention. Despite these problems, the potential for fuel cells to facilitate the energy transition within the oil and gas industry is significant. Continuous advancements in fuel cell technology, cost reductions, and improvements in hydrogen infrastructure are expected to make fuel cells more viable (Singh et al., 2024).

In summary, fuel cells offer a pathway for the oil and gas industry to move toward sustainable energy practices. In light of the changing landscape shaped by concerns, regulations, and market dynamics, fuel cells are poised to have an impact on facilitating the industry's ability to meet forthcoming energy needs and diminish its ecological impact. This pivotal transformation serves as both a reaction to the issues posed by climate change and a chance for the sector to drive innovation and take charge of the worldwide shift toward sustainable energy sources.

9.9 SIGNIFICANCE AND FUTURE OF FUEL CELLS IN ENERGY TRANSITION IN THE OIL AND GAS INDUSTRY

The importance of fuel cells in the oil and gas industry's transition is gaining momentum as companies strive to reduce their impact and boost efficiency. Fuel cells offer an avenue for delivering reliable energy that is essential for the industry's move toward more sustainable practices (Kumar et al., 2024). Fuel cells, including solid oxide fuel cells (SOFCs) and proton exchange membrane fuel cells (PEMFCs), are essential in cutting down greenhouse gas emissions in one of the energy-intensive sectors worldwide (Gupta et al., 2024).

Furthermore, fuel cells can significantly improve the efficiency of energy systems by generating power on-site, reducing energy losses during long-distance electricity transmission from power plants (Risco-Bravo et al., 2024). Likewise, fuel cells can use waste products from the industry, like gas, as a fuel source, transforming waste into energy and contributing to resource utilization. Examining fuel cells is expected to show an important in energy transition strategies within the oil and gas sector (Abouemara et al., 2024). With pressure on the industry to shift toward low-carbon energy sources, fuel cells will likely emerge as an important energy portfolio element. Their scalability and versatility make them suitable for applications across the industry – from powering drilling sites and offshore platforms to providing reliable energy for pipeline operations (Sharma et al., 2024).

Establishing hydrogen infrastructure is also crucial for the future of fuel cells. As the industry focuses on hydrogen as an energy carrier, fuel cells will be instrumental in utilizing this resource. Innovations in hydrogen production methods, such as electrolysis powered by energy sources, will facilitate the integration of fuel cells (Sikiru et al., 2024). Additionally, advancements in fuel cell technology (including enhancements in durability, cost-efficiency, and operational adaptability) will improve their feasibility and widespread adoption (Roslan et al., 2024). Ongoing exploration and innovation backed by measures and collaborations with industry will be crucial in addressing current obstacles and tapping into the possibilities presented by fuel cells (Farinloye et al., 2024).

To sum up, fuel cells are set to impact revolutionizing the oil and gas sector by offering eco-energy solutions that align with global environmental objectives. As the industry moves toward an effective future, fuel cells will drive the shift toward cleaner, more environmentally friendly energy practices.

9.10 FUEL CELL DEMAND, SUPPLY

The energy industry worldwide is experiencing a transformation, with fuel cells becoming increasingly crucial in the move toward energy solutions (Hassan et al., 2024). This section explores fuel cells' demand-and-supply dynamics, looking at their past, present, and potential future. We will discuss the progress that has propelled fuel cells to the forefront of energy discussions and the obstacles hindering their widespread use.

From advancements in materials and processes that boost the efficiency and longevity of fuel cells to efforts aimed at establishing hydrogen economies, we will address several factors influencing the market (Singh et al., 2024). Additionally, we will shine a light on innovations in the field that are paving the way for an uptick in demand for fuel cells across industries globally (Shaker et al., 2024). By grasping these aspects, we can better understand how technology, policies, and market dynamics interact to shape future fuel cell energy systems scenarios.

Fan et al. (2021) explored the increasing importance of hydrogen as an energy source to address sustainability and energy security issues (Figure 9.24). Countries like Japan, Korea, and China are at the forefront of advancing hydrogen technology and infrastructure to meet rising energy needs while decreasing reliance on fossil fuels. The focus is on proton exchange membrane fuel cells (PEMFCs) for their applications in vehicles like buses and general logistics that require minimal hydrogen infrastructure (Fan et al., 2021).

The article investigates the components of PEMFCs, including their water management systems and the strategic national approaches to embracing hydrogen, reflecting a global shift toward a hydrogen-centric economy (Figure 9.25). This signals a growing demand for fuel cells fueled by policies and the incorporation of PEMFCs in sectors seeking reliable and effective energy solutions (Wang et al., 2018).

9.10.1 Past, Present, and Demand and Supply

It is important to consider fuel taxes when considering the cost of driving for consumers. Fuel taxes match around 100% of petroleum fuel costs in places like the EU and Japan. This trend is expected to continue. In the United States, it is projected that a future tax rate of 30% will be in place. Exempting hydrogen from fuel taxes could reduce the total cost of driving for FCEVs, possibly leading to cost-competitiveness with gasoline vehicles by 2035 in the EU and Japan. Fluctuations of around 30% in hydrogen prices could impact this timeline by five years. Implementing hydrogen taxes at a reduced rate once FCEVs reach cost parity could help them stay competitively priced compared to vehicles in the market (Figure 9.26) (Fan et al., 2021).

In this part, we explore fuel cell possibilities, the technology behind them, the obstacles in adoption, and the latest advancements globally. The potential of fuel cells is immense due to their efficiency and versatility in using fuels like hydrogen, natural gas, and biogas, which offer an alternative compared to combustion-based power generation methods by emitting minimal to no pollutants based on the fuel used (Goria et al., 2024). In transportation, fuel cells are essential in reducing vehicle emissions, supporting the transition toward zero-emission transportation options (Ledna et al., 2024). In industrial applications, fuel cells present decentralized power generation solutions that can work alongside intermittent renewable sources, such as solar and wind power. It stabilizes the grid while providing power during emergencies.

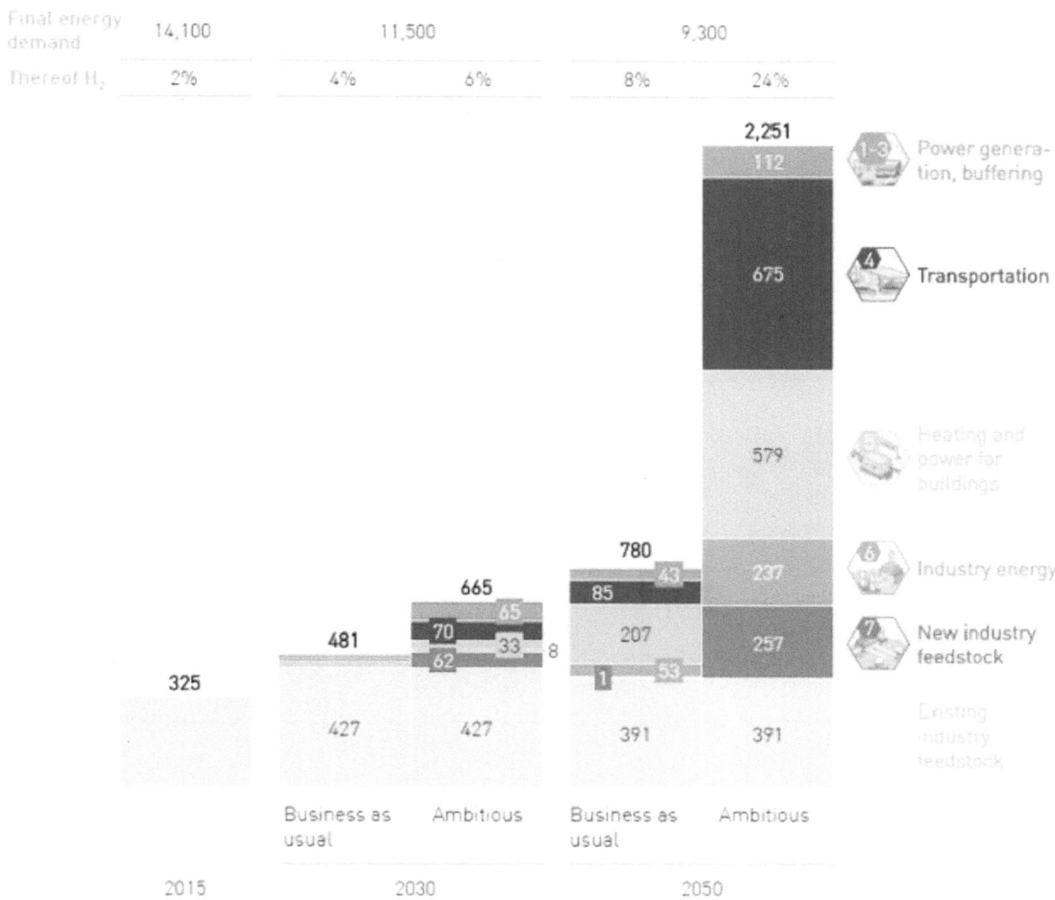

FIGURE 9.24 An ambitious scenario for energy demand and hydrogen deployment for Europe from 2015 to 2050.

Source: Fan et al. (2021).

It should be mentioned that establishing infrastructure networks poses challenges for hydrogen fuel cells, which are needed for producing, storing, and distributing hydrogen. Another significant obstacle is improving the durability and reliability of fuel cells to match or exceed engines in long-term performance. Lastly, addressing safety concerns is crucial since handling and storing hydrogen come with risks that require careful management to encourage broader adoption (Laimon & Yusaf, 2024; Rani et al., 2024; Cho et al., 2024).

Technological advancements and ongoing research efforts are actively tackling the challenges associated with the adoption of fuel cells. Progress in material sciences has notably reduced platinum usage in proton exchange membrane fuel cells (PEMFCs) and created materials for solid oxide fuel cells (SOFCs) that can function at lower temperatures. These developments contribute to cost reduction and increased durability. In manufacturing, progress like in 3D printing and automated assembly techniques is transforming fuel cell production by lowering costs and simplifying processes. Furthermore, the infrastructure expansion in Europe and Asia is where significant investments are being made in hydrogen. Projects like the Hydrogen Highways in California and Hydrogen Cities in Japan exemplify approaches to implementing fuel cell technologies. Governmental support is also robust, and incentives are offered globally to encourage research and implementation of fuel cells. Initiatives include subsidies for fuel cell vehicles, investments in hydrogen production, and grants

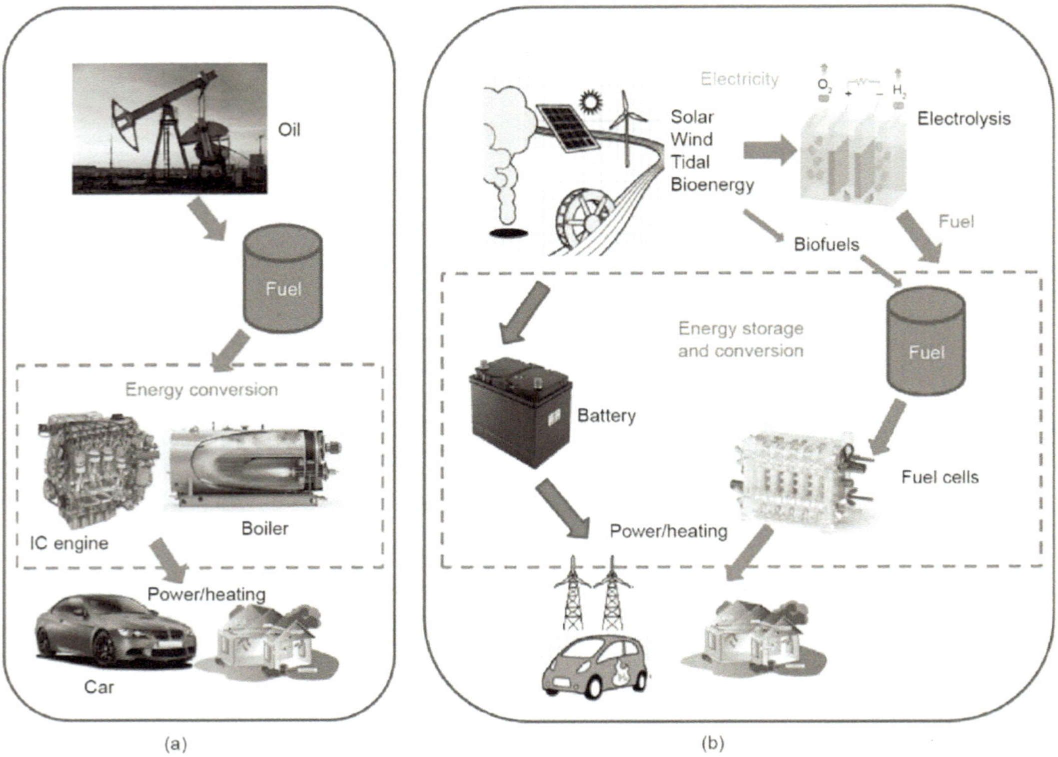

FIGURE 9.25 The energy economy: (a) fossil fuels energy economy; (b) renewable energy economy. (*IC*: internal combustion.)

Source: Wang et al. (2018).

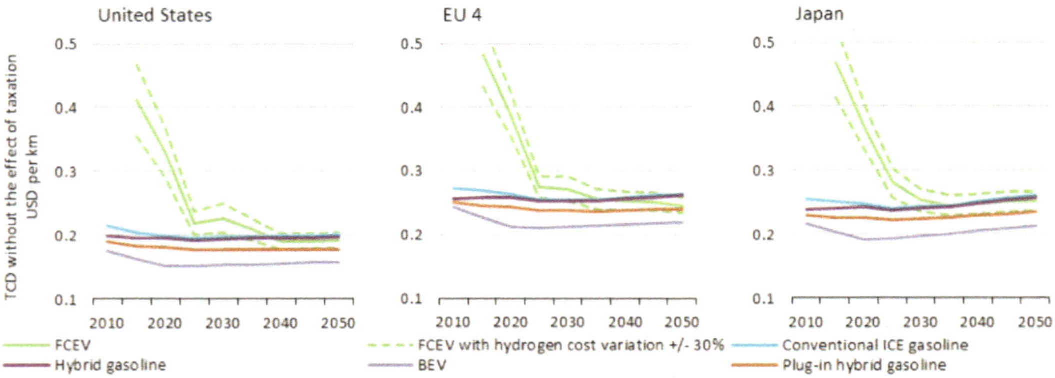

FIGURE 9.26 Total driving costs with petroleum fuel taxes and fuel tax exemption of hydrogen for EU 4, Japan, and the United States for the 2DS high H_2.

Source: Fan et al. (2021).

for fuel cell research to drive the development and adoption of this friendly technology (Baharuddin et al., 2021).

In summary, the outlook for fuel cells is positive. Like many innovative approaches, the future of fuel cells hinges on technological advancements, supportive policies, and collaborative efforts involving governments, industry stakeholders, and academic institutions. With the coming of these factors, fuel cells are positioned to have an impact on the worldwide shift toward sustainable energy, aiding in the creation of a greener and cleaner future.

This section has explored fuel cells' role in the energy sector, showcasing their principles, practical uses, and growing importance as the world moves toward sustainable energy practices, especially within the oil and gas industry. We have examined their inner workings and the advancements in their effectiveness and feasibility, and we see that fuel cells have proven to be critical players in propelling the energy transition forward. We have explored how fuel cells are not only reshaping power generation landscapes but also gaining ground in transportation and industrial sectors by offering efficient and dependable energy solutions that support the rising environmental sustainability objectives.

In the oil and gas industry, fuel cells are emerging as tools for reducing carbon emissions and boosting efficiency in traditional energy setups, aiding the industry's shift toward achieving net-zero emissions. Examining their existing applications and future possibilities underscores this technology's nature as a cornerstone of energy solutions. Furthermore, when considering the impacts of fuel cell technology on both global scales, it becomes evident that it has the potential to influence financial markets and energy policies, significantly sparking increased investments in clean energy technologies. This chapter examined the significance of fuel cells in the energy transition and emphasized their ability to reshape future energy approaches for a sustainable, effective, and environmentally friendly tomorrow.

REFERENCES

Abdelkareem, M. A., Elsaid, K., Wilberforce, T., Kamil, M., Sayed, E. T., & Olabi, A. (2021). Environmental aspects of fuel cells: A review. *Science of the Total Environment*, *752*, 141803.

Abdin, Z. (2024). Shaping the stationary energy storage landscape with reversible fuel cells. *Journal of Energy Storage*, *86*, 111354.

Abouemara, K., Shahbaz, M., Mckay, G., & Al-Ansari, T. (2024). The review of power generation from integrated biomass gasification and solid oxide fuel cells: Current status and future directions. *Fuel*, *360*, 130511.

Adebisi, J. A., Denwigwe, I. H., & Babatunde, O. M. (2023). Hydrogen storage for micro grid application: A framework for ranking fuel cell technologies based on technical parameters. *International Journal of Electrical and Computer Engineering*, 13(2), 1221–1320.

Agnolucci, P. (2007). Economics and market prospects of portable fuel cells. *International Journal of Hydrogen Energy*, *32*(17), 4319–4328.

Al-Khori, K., Bicer, Y., & Koc, M. (2020). Integration of solid oxide fuel cells into oil and gas operations: Needs, opportunities, and challenges. *Journal of Cleaner Production*, *245*, 118924.

Al-Khori, K., Al-Ghamdi, S. G., Boulfrad, S., & Koç, M. (2021). Life cycle assessment for integration of solid oxide fuel cells into gas processing operations. *Energies*, *14*(15), 4668.

Alvi, N. U. H., & Sandberg, M. (2024). Sustainable and low-cost electrodes for photocatalytic fuel cells. *Nanomaterials*, *14*(7), 636.

Alzahrani, A., Ramu, S. K., Devarajan, G., Vairavasundaram, I., & Vairavasundaram, S. (2022). A review on hydrogen-based hybrid microgrid system: Topologies for hydrogen energy storage, integration, and energy management with solar and wind energy. *Energies*, *15*(21), 7979.

Baharuddin, N. A., Yusoff, W. W., Abd Aziz, A. J., & Tahir, N. M. (2021, February). Hydrogen fuel cells for sustainable energy: Development and progress in selected developed countries. In *IOP conference series: Materials science and engineering* (Vol. 1078, No. 1, p. 012011). IOP Publishing.

Bello, I. T., Guan, D., Yu, N., Li, Z., Song, Y., Chen, X., . . . & Ni, M. (2023). Revolutionizing material design for protonic ceramic fuel cells: Bridging the limitations of conventional experimental screening and machine learning methods. *Chemical Engineering Journal*, *477*, 147098.

Busby, R. L. (2005). *Hydrogen and fuel cells*. Pennwell Corporation.

Carrette, L., Friedrich, K. A., & Stimming, U. (2000). Fuel cells: Principles, types, fuels, and applications. *ChemPhysChem*, *1*(4), 162–193.

Chadly, A., Azar, E., Maalouf, M., & Mayyas, A. (2022). Techno-economic analysis of energy storage systems using reversible fuel cells and rechargeable batteries in green buildings. *Energy*, *247*, 123466.

Chaudhary, K., Bhardvaj, K., & Chaudhary, A. (2024). A qualitative assessment of hydrogen generation techniques for fuel cell applications. *Fuel*, *358*, 130090.

Cheekatamarla, P. (2024). Hydrogen and the global energy transition—path to sustainability and adoption across all economic sectors. *Energies*, *17*(4), 807.

Cho, A., Kim, H., & Park, S. (2024). Resurgence of the hydrogen energy in South Korea's government strategies from 2005 to 2019. *International Journal of Hydrogen Energy*, *65*, 844–854.

Das, P. K., Jiao, K., Wang, Y., Barbir, F., & Li, X. (Eds.). (2023). *Fuel cells for transportation: Fundamental principles and applications*. Elsevier.

Dezhdar, A., Assareh, E., Agarwal, N., Baheri, A., Ahmadinejad, M., Zadsar, N., . . . & Lee, M. (2024). Modeling, optimization, and economic analysis of a comprehensive CCHP system with fuel cells, reverse osmosis, batteries, and hydrogen storage subsystems Powered by renewable energy sources. *Renewable Energy*, *220*, 119695.

Dirkes, S., Leidig, J., Fisch, P., & Pischinger, S. (2023). Prescriptive Lifetime Management for PEM fuel cell systems in transportation applications, Part I: State of the art and conceptual design. *Energy Conversion and Management*, *277*, 116598.

Dorregaray-Oyaregui, S., Martín-Gómez, C., & Aguado, M. (2024). Installation of fuel cells in building in use: Technical, regulatory, and economic feasibility. *Energy Conversion and Management: X*, *22*, 100536.

Fakhreddine, O., Gharbia, Y., Derakhshandeh, J. F., & Amer, A. M. (2023). Challenges and solutions of hydrogen fuel cells in transportation systems: A review and prospects. *World Electric Vehicle Journal*, *14*(6), 156.

Fan, L., Tu, Z., & Chan, S. H. (2021). Recent development of hydrogen and fuel cell technologies: A review. *Energy Reports*, *7*, 8421–8446.

Farghali, M., Osman, A. I., Mohamed, I. M., Chen, Z., Chen, L., Ihara, I., . . . & Rooney, D. W. (2023). Strategies to save energy in the context of the energy crisis: A review. *Environmental Chemistry Letters*, *21*(4), 2003–2039.

Farinloye, T., Oluwatobi, O., Ugboma, O., Dickson, O. F., Uzondu, C., & Mogaji, E. (2024). Driving the electric vehicle agenda in Nigeria: The challenges, prospects and opportunities. *Transportation Research Part D: Transport and Environment*, *130*, 104182.

Fazal, M., Janjua, A. K., Anwar, M., Hassan, M., Shakir, S., & Bahadar, A. (2024). Energy, economic and environmental (3E) analyses of solar photovoltaic and fuel cell based hybrid energy system for a developing country: A worldwide approach to sustainable energy. *Environment, Development and Sustainability*, 1–28.

Ghimire, R., Niroula, S., Pandey, B., Subedi, A., & Thapa, B. S. (2024). Techno-economic assessment of fuel cell-based power backup system as an alternative to diesel generators in Nepal: A case study for hospital applications. *International Journal of Hydrogen Energy*, *56*, 289–301.

Giorgi, L., & Leccese, F. (2013). Fuel cells: Technologies and applications. *The Open Fuel Cells Journal*, *6*(1).

Gòdia Sangrà, A. (2024). *Study of a power transmission system for a formula student vehicle*. https://upcommons.upc.edu/handle/2117/404448.

Goria, K., Singh, H. M., Singh, A., Kothari, R., & Tyagi, V. V. (2024). Insights into biohydrogen production from algal biomass: Challenges, recent advancements and future directions. *International Journal of Hydrogen Energy*, *52*, 127–151.

Gupta, P., Toksha, B., & Rahaman, M. (2024). A critical review on hydrogen based fuel cell technology and applications. *The Chemical Record*, *24*(1), e202300295.

Habib, A. R. R. (2024). A comparative study of the machine learning-based energy management system for hydrogen fuel cell electric vehicles. *Future Technology*, *3*(1), 13–24.

Hamada, A. T., Orhan, M. F., & Kannan, A. M. (2023). Alkaline fuel cells: Status and prospects. *Energy Reports*, *9*, 6396–6418.

Hassan, Q., Viktor, P., Al-Musawi, T. J., Ali, B. M., Algburi, S., Alzoubi, H. M., . . . & Jaszczur, M. (2024). The renewable energy role in the global energy transformations. *Renewable Energy Focus*, *48*, 100545.

Hassouna, F. M., & Shin, K. (2024). Economic prospects of taxis powered by hydrogen fuel cells in Palestine. *World Electric Vehicle Journal*, *15*(2), 50.

Hong, J., Yang, H., Liang, F., Li, K., Zhang, X., Zhang, H., . . . & Wang, J. (2024). State of health prediction for proton exchange membrane fuel cells combining semi-empirical model and machine learning. *Energy*, *291*, 130364.

Hou, X., Sun, R., Huang, J., Geng, W., Li, X., Wang, L., & Zhang, X. (2024). Energy, economic, and environmental analysis: A study of operational strategies for combined heat and power system based on PEM fuel cell in the East China region. *Renewable Energy*, *223*, 120023.

Hu, C., Kang, N. Y., Kang, H. W., Lee, J. Y., Zhang, X., Lee, Y. J., . . . & Lee, Y. M. (2024). Triptycene branched poly (aryl-co-aryl piperidinium) electrolytes for alkaline anion exchange membrane fuel cells and water electrolyzers. *Angewandte Chemie*, *136*(3), e202316697.

Iqbal, R., Liu, Y., Zeng, Y., Zhang, Q., & Zeeshan, M. (2024). Comparative study based on techno-economics analysis of different shipboard microgrid systems comprising PV/wind/fuel cell/battery/diesel generator with two battery technologies: A step toward green maritime transportation. *Renewable Energy*, *221*, 119670.

Jaiswal, K. K., Chowdhury, C. R., Yadav, D., Verma, R., Dutta, S., Jaiswal, K. S., & Karuppasamy, K. S. K. (2022). Renewable and sustainable clean energy development and impact on social, economic, and environmental health. *Energy Nexus*, *7*, 100118.

Kaland, T. (2024, April). *Industry-academia-contribution to energy transition*. SPE Norway Subsurface Conference? (p. D011S012R005). SPE.

Kart, S., Demir, F., Kocaarslan, İ., & Genc, N. (2024). Increasing PEM fuel cell performance via fuzzy-logic controlled cascaded DC-DC boost converter. *International Journal of Hydrogen Energy*, *54*, 84–95.

Kariuki, N. N., Cansizoglu, M. F., Begum, M., Yurukcu, M., Yurtsever, F. M., Karabacak, T., & Myers, D. J. (2016). SAD–GLAD Pt–Ni@ Ni nanorods as highly active oxygen reduction reaction electrocatalysts. *ACS Catalysis*, *6*(6), 3478–3485.

Kazda, T., & Sedlarikova, M. (2024). Selected contributions in advanced batteries, accumulators, and fuel cells. *Monatshefte für Chemie-Chemical Monthly*, *1*.

Khudhayer, W. J., Begum, M., Nasini, U. B., Cansizoglu, M. F., Yurukcu, M., Shaikh, A. U., & Karabacak, T. (2015). Oxygen reduction reaction electrocatalytic activity of tilted pt nanorod arrays fabricated by glancing angle deposition. *Journal of Applied Electrochemistry*, *45*, 1113–1121.

Körner, A., Tam, C., Bennett, S., & Gagné, J. (2015). *Technology roadmap-hydrogen and fuel cells* (pp. 1–29). International Energy Agency (IEA).

Kumar, A., Jayeoye, T. J., Mohite, P., Singh, S., Rajput, T., Munde, S., . . . & Parihar, A. (2024). Sustainable and consumer-centric nanotechnology-based materials: An update on the multifaceted applications, risks and tremendous opportunities. *Nano-Structures & Nano-Objects*, *38*, 101148.

Lahrichi, A., El Issmaeli, Y., Kalanur, S. S., & Pollet, B. G. (2024). Advancements, strategies, and prospects of solid oxide electrolysis cells (SOECs): Towards enhanced performance and large-scale sustainable hydrogen production. *Journal of Energy Chemistry*, *94*, 688–715.

Laimon, M., & Yusaf, T. (2024). Towards energy freedom: Exploring sustainable solutions for energy independence and self-sufficiency using integrated renewable energy-driven hydrogen system. *Renewable Energy*, 222, 119948.

Ledna, C., Muratori, M., Yip, A., Jadun, P., Hoehne, C., & Podkaminer, K. (2024). Assessing total cost of driving competitiveness of zero-emission trucks. *iScience*, *27*.

Lepikash, R., Lavrova, D., Stom, D., Meshalkin, V., Ponamoreva, O., & Alferov, S. (2024). State of the art and environmental aspects of plant microbial fuel cells' application. *Energies*, *17*(3), 752.

Mariscal, G., Depcik, C., Chao, H., Wu, G., & Li, X. (2024). Technical and economic feasibility of applying fuel cells as the power source of unmanned aerial vehicles. *Energy Conversion and Management*, *301*, 118005.

Mayyas, A., Chadly, A., Amer, S. T., & Azar, E. (2022). Economics of the Li-ion batteries and reversible fuel cells as energy storage systems when coupled with dynamic electricity pricing schemes. *Energy*, *239*, 121941.

Metzger, N., & Li, X. (2022). Technical and economic analysis of fuel cells for forklift applications. *ACS Omega*, *7*(22), 18267–18275.

Ming, W., Sun, P., Zhang, Z., Qiu, W., Du, J., Li, X., . . . & Guo, X. (2023). A systematic review of machine learning methods applied to fuel cells in performance evaluation, durability prediction, and application monitoring. *International Journal of Hydrogen Energy*, *48*(13), 5197–5228.

Milewski, J., Szczęśniak, A., Szabłowski, Ł., Martsinchyk, A., Siekierski, M., Dybinski, O., & Świrski, K. (2024). Molten borates fuel cells—mathematical modeling and identification of performances. *Renewable and Sustainable Energy Reviews*, *190*, 113949.

Mohammadi, T., Hosseini, M. G., Pastor, E., & Ashassi-Sorkhabi, H. (2024). One-step growth of RuNi-MOF nanoarrays on carbon felt host as a high-performance binder-free electrode for dual application: Ethanol fuel cell and supercapacitor. *Journal of Energy Storage*, *79*, 110146.

MTU Solutions. (n.d.). *Fuel cells and the quest for green power*. Retrieved April 22, 2024, from www.mtu-solutions.com/na/en/stories/technology/research-development/fuel-cells-and-the-quest-for-green-power.html

Nagao, Y. (2024). Proton-conducting polymers: key to next-generation fuel cells, electrolyzers, batteries, actuators, and sensors. *ChemElectroChem*, e202300846.

Nazir, M. S., Ali, Z. M., Bilal, M., Sohail, H. M., & Iqbal, H. M. (2020). Environmental impacts and risk factors of renewable energy paradigm—a review. *Environmental Science and Pollution Research*, *27*, 33516–33526.

Oh, S., Kim, T., Kim, S., & Kang, S. (2023). Energetic, exergetic, economic, and exergoeconomic analysis of a phosphoric acid fuel cell-organic Rankine cycle hybrid system. *Energy Conversion and Management*, *284*, 116993.

Olabi, A. G., & Sayed, E. T. (2023). Developments in hydrogen fuel cells. *Energies*, *16*(5), 2431.

Olabi, A. G., Wilberforce, T., & Abdelkareem, M. A. (2021). Fuel cell application in the automotive industry and future perspective. *Energy*, *214*, 118955.

Opakhai, S., & Kuterbekov, K. (2023). Metal-supported solid oxide fuel cells: A review of recent developments and problems. *Energies*, *16*(12), 4700.

Osman, A. I., Mehta, N., Elgarahy, A. M., Hefny, M., Al-Hinai, A., Al-Muhtaseb, A. A. H., &Rooney, D. W. (2022). Hydrogen production, storage, utilisation and environmental impacts: A review. *Environmental Chemistry Letters*, 1–36.

Peters, R., Pasel, J., & Samsun, R. C. (2024). Technology transfer from fuel processing for fuel cells to fuel synthesis from hydrogen and carbon dioxide. *International Journal of Hydrogen Energy*, *49*, 943–961.

Pramuanjaroenkij, A., & Kakaç, S. (2023). The fuel cell electric vehicles: The highlight review. *International Journal of Hydrogen Energy*, *48*(25), 9401–9425.

Rani, P., Parkash, V., & Sharma, N. K. (2024). Technological aspects, utilization and impact on power system for distributed generation: A comprehensive survey. *Renewable and Sustainable Energy Reviews*, *192*, 114257.

Reddy, V. J., Hariram, N. P., Maity, R., Ghazali, M. F., & Kumarasamy, S. (2024). Sustainable vehicles for decarbonizing the transport sector: A comparison of biofuel, electric, fuel cell and solar-powered vehicles. *World Electric Vehicle Journal*, *15*(3), 93.

Risco-Bravo, A., Varela, C., Bartels, J., & Zondervan, E. (2024). From green hydrogen to electricity: A review on recent advances, challenges, and opportunities on power-to-hydrogen-to-power systems. *Renewable and Sustainable Energy Reviews*, *189*, 113930.

Roslan, M. F., Abd Rahman, H., Abd Karim, R., & Osman, S. H. (2024). Advancements in fuel cell technology for electric vehicles. *Malaysia Journal of Invention and Innovation*, *3*(2), 11–20.

Sammes, N., Bove, R., & Stahl, K. (2004). Phosphoric acid fuel cells: Fundamentals and applications. *Current Opinion in Solid State and Materials Science*, *8*(5), 372–378.

Sarlak, M., Pirouzfar, V., Sakhaeinia, H., & Alihosseini, A. (2024). Combustion of gasoline with oxygen-containing and nano-additives: An experimental study, modeling, optimization, and analysis survey. *Journal of the Taiwan Institute of Chemical Engineers*, *159*, 105452.

Satyanarayana, A., & Naidu, I. E. S. (2024). Modeling and analysis of fuel cell power generation system using proportional integral speed controller. *Advances in Science and Technology Research Journal*, *18*(2), 187–195.

Seyam, S., Dincer, I., & Agelin-Chaab, M. (2024). Multi-objective optimization of locomotive engines hybridized with fuel cells using selected fuel blends. *International Journal of Hydrogen Energy*, *60*, 1481–1499.

Shaker, L. M., Al-Amiery, A. A., & Al-Azzawi, W. K. (2024). Nanomaterials: paving the way for the hydrogen energy frontier. *Discover Nano*, *19*(1), 3.

Sharma, S. K., Rani, A., Bakhariya, H., Kumar, R., Tomar, D., & Ghosh, S. (2024). The role of IoT in optimizing operations in the oil and gas sector: A review. *Transactions of the Indian National Academy of Engineering*, 1–20.

Shuhayeu, P., Martsinchyk, A., Martsinchyk, K., Szczęśniak, A., Szabłowski, Ł., Dybiński, O., & Milewski, J. (2024). Model-based quantitative characterization of anode microstructure and its effect on the performance of molten carbonate fuel cell. *International Journal of Hydrogen Energy*, *52*, 902–915.

Sikiru, S., Oladosu, T. L., Amosa, T. I., Olutoki, J. O., Ansari, M. N. M., Abioye, K. J., . . . & Soleimani, H. (2024). Hydrogen-powered horizons: Transformative technologies in clean energy generation, distribution, and storage for sustainable innovation. *International Journal of Hydrogen Energy*, *56*, 1152–1182.

Singh, M., Singla, M. K., Beryozkina, S., Gupta, J., & Safaraliev, M. (2024). Hydrogen vehicles and hydrogen as a fuel for vehicles: A state-of-the-art review. *International Journal of Hydrogen Energy*, *64*, 1001–1010.

Singh, P., Arora, K., & Rathore, U. C. (2024). Design and applications of fuel cells. *Electric Vehicle Design: Design, Simulation, and Applications*, *19*.

Skaloumpakas, P., Sarmas, E., Rachmanidis, M., & Marinakis, V. (2024). Reshaping the energy landscape of Crete through renewable energy valleys. *Scientific Reports*, *14*(1), 8038.

Speight, J. G. (2011). Fuels for fuel cells. In *Fuel cells: Technologies for fuel processing* (pp. 29–48). Elsevier.

Stambouli, A. B., & Traversa, E. (2002). Fuel cells, an alternative to standard sources of energy. *Renewable and Sustainable Energy Reviews*, *6*(3), 295–304.

Sullivan, N., Anyenya, G., Haun, B., Daubenspeck, M., Bonadies, J., Kerr, R., . . . & Savage, M. (2016). In-ground operation of geothermic fuel cells for unconventional oil and gas recovery. *Journal of Power Sources*, *302*, 402–409.

Syed, M. A., Siddiqui, O., Kazerani, M., & Khalid, M. (2024). Analysis and modeling of direct ammonia fuel cells for solar and wind power leveling in smart grid applications. *IEEE Access*, *12*, 46512–46523.

Transport & Environment. (n.d.). EVs will be cheaper than petrol cars in all segments by 2027, BNEF analysis finds. *Transport & Environment.* Retrieved April 22, 2024, from www.transportenvironment.org/discover/evs-will-be-cheaper-than-petrol-cars-in-all-segments-by-2027-bnef-analysis-finds/

Usman, O., Iorember, P. T., Ozkan, O., & Alola, A. A. (2024). Dampening energy security-related uncertainties in the United States: The role of green energy-technology investment and operation of transnational corporations. *Energy*, *289*, 130006.

Van Veldhuizen, B., Van Biert, L., Aravind, P. V., & Visser, K. (2023). Solid oxide fuel cells for marine applications. *International Journal of Energy Research*, *2023*.

Vignesh, D., Gupta, M. K., Mittal, R., & Rout, E. (2024). Proton-polaron and thermionic identity of BaCeO3 polymorph for intermediate temperature fuel cell technology: A first principles and molecular dynamics approach. *International Journal of Hydrogen Energy*, *57*, 394–407.

Wang, J., & Azam, W. (2024). Natural resource scarcity, fossil fuel energy consumption, and total greenhouse gas emissions in top emitting countries. *Geoscience Frontiers*, *15*(2), 101757.

Wang, J., Wang, H., & Fan, Y. (2018). Techno-economic challenges of fuel cell commercialization. *Engineering*, *4*(3), 352–360.

Wang, Y., & Gao, W. (2023). Why do we need the distributed energy solution for low-carbon society? In *Distributed energy resources: Solutions for a low carbon society* (pp. 1–23). Springer International Publishing.

Wang, Z., Dong, B., Wang, Y., Li, M., Liu, H., & Han, F. (2024). Analysis and evaluation of fuel cell technologies for sustainable ship power: Energy efficiency and environmental impact. *Energy Conversion and Management: X*, *21*, 100482.

Wilberforce, T., Olabi, A. G., Muhammad, I., Alaswad, A., Sayed, E. T., Abo-Khalil, A. G., . . . & Abdelkareem, M. A. (2024). Recovery of waste heat from proton exchange membrane fuel cells–a review. *International Journal of Hydrogen Energy*, *52*, 933–972.

Wilberforce, T., Olabi, A. G., Pritchard, D., Abdelkareem, M. A., & Sayed, E. T. (2023). Development of proton exchange membrane fuel cell flow plate geometry design. *Energy*, *283*, 128854.

Yalcinkaya, F., Torres-Mendieta, R., Hruza, J., Vávrová, A., Svobodová, L., Pietrelli, A., & Ieropoulos, I. (2024). Nanofiber applications in microbial fuel cells for enhanced energy generation: a mini review. *RSC Advances*, *14*(13), 9122–9136.

Yamate, S., & Otomo, J. (2024). Design of cost-effective and highly efficient systems for protonic ceramic fuel cells based on techno-economic analysis. *Energy Conversion and Management*, *301*, 118016.

Yan, J., Jing, J., & Li, Y. (2024). Hydrogen fuel cell commercial vehicles in China: Evaluation of carbon emission reduction and its economic value. *International Journal of Hydrogen Energy*, *52*, 734–749.

Yurtsever, F. M., Yurukcu, M., Begum, M., Watanabe, F., & Karabacak, T. (2018). Stacked and Core-Shell Pt: Ni/WC nanorod array electrocatalyst for enhanced oxygen reduction reaction in polymer electrolyte membrane fuel cells. *ACS Applied Energy Materials*, *1*(11), 6115–6122.

Yurukcu, M. (2018). *An investigation of conformality of the core-shell polymer electrolyte membrane (PEM) fuel cell catalysts by sputter deposition* (Doctoral dissertation, University of Arkansas at Little Rock).

Zhang, M., Lyu, H., Bian, H., & Ghadimi, N. (2024). Improved chaos grasshopper optimizer and its application to HRES techno-economic evaluation. *Heliyon*, *10*(2).

Zhang, Y., Liu, F., Shan, J., Zhang, X., Wu, X., & Hou, Z. (2024). Advances of high-performance and stable finely-ordered patterned membrane for proton exchange membrane fuel cell applications. *Journal of Power Sources*, *595*, 234053.

Zhdaneev, O. V., & Frolov, K. N. (2024). Technological and institutional priorities of the oil and gas complex of the Russian federation in the term of the world energy transition. *International Journal of Hydrogen Energy*, *58*, 1418–1428.

10 Hydrogen Energy and Technologies

Jorge Saldaña, Mesut Yurukcu, Narendra Boppana, Sepehr Arbabi, Jim Henry, and Sebahattin Ziyanak

10.1 INTRODUCTION

The potentially risky environmental scenarios that appear when humanity continues to stay on similar and the same fossil fuel–dependence trajectories have generated a growing interest in the development of alternative energy technologies that are cleaner and more sustainable. Another one of these sources is based on hydrogen, the most abundant element in the universe and one of the simplest chemical elements due to each atom of hydrogen having only one proton (EIA, 2022). Like electricity, petroleum, or coal, hydrogen is an energy carrier. Therefore, it is not a source of energy but rather a secondary form generated from natural and bioresources (Sazali, 2020) and requires a conversion process for the energy to be released.

First thought to be a result of the inflammation of metals, hydrogen was recognized by Antoine Lavoisier as a gas and gave it its name. Hydrogen is also the lightest of the known elements while having the maximum energy content of conventional fuels per unit of weight, three times that of gasoline (Abdin et al., 2020). Table 10.1 summarizes the energy contents of different types of fuels, where we see that hydrogen has the most significant amount of megajoules per kilogram, both in gaseous and liquid form, almost tripling the energy contents of natural gas, while holding six times more energy than ethanol (Abe et al., 2019).

Hydrogen is highly flammable, and the Hindenburg disaster in 1937 is partly attributed to people's failure to manage hydrogen effectively (Dawood et al., 2020). With that said, the more trials are performed with hydrogen, and the more technology advances, the safer it has become to work with it. The deepening of the need for cleaner energy sources to overtake fossil fuel–based sources has invited the research and development of new and refined uses of hydrogen. In this sense, hydrogen production can be divided into four main categories: electrolysis, photolysis, biolysis, and thermolysis. Each of these categorizations will be explored in more detail throughout this chapter.

First, hydrogen production through electrolysis occurs after water is split into hydrogen and oxygen through electricity generated in an electrolyzer unit. Because the only by-product of this process is pure oxygen, and because electrolysis can use DC power from sustainable energy resources like solar and wind, this technique can be placed at the center of the efforts to transition to cleaner energy sources. Different types of devices include alkaline, solid oxide, microbial, and polymer electrolyte membrane (PEM) electrolyzers (Shiva Kumar and Himabindu, 2019; US Department of Energy, 2022a).

Meanwhile, photolysis consists of hydrogen generation from water through the action of microorganisms like cyanobacteria and green algae combined with sunlight. The photons of light that are captured from light energy split water molecules into protons (H^+) and oxygen ions (Aklaghi and Najafpour-Darzi, 2020; US Department of Energy, 2022b). Then, biolysis processes include dark fermentation, a biological process that occurs in the absence of sunlight and where hydrogen is generated by anaerobic bacteria that are grown in the dark, and aqueous phase reforming (APR), where hydrogen is produced from biomass-derived compounds like glycerol, sugars, and sugar alcohols or polyols (Fasolini et al., 2019; Łukajtis et al., 2018).

 DOI: 10.1201/9781003282273-10

TABLE 10.1
Comparison of Some Selected Energy Contents of Fuels

Fuel	Lower Heating Value	Higher Heating value
Gaseous hydrogen	119.96	141.88
Liquid hydrogen	120.04	141.77
Natural gas	47.13	52.21
Liquified natural gas (LNG)	48.62	55.19
Still gas (in refineries)	46.89	50.94
Crude oil	42.68	45.53
Liquefied petroleum gas (LPG)	46.60	50.14
Conventional gasoline	43.44	46.52
Reformulated or low-sulfur gasoline (RFG)	42.35	45.42
Conventional diesel	42.78	45.76
Low-sulfur diesel	42.60	45.56
Coal (wet basis)	22.73	23.96
Bituminous coal (wet basis)	26.12	27.26
Coking coal (wet basis)	28.60	29.86
Methanol	20.09	22.88
Ethanol	26.95	29.84

Source: Abe et al. (2019).

Lastly, thermochemical processes of water splitting use high temperatures and chemical reactions to produce hydrogen and oxygen. These processes can involve thermal, chemical, or hybrid production cycles, which include two-step, three-step, four-step, and hybrid types of hydrogen production (Safari and Dincer, 2020; US Department of Energy, 2022c). The development of these technologies has led researchers to advance the possibilities for a *hydrogen economy* that is based on a revamped supply chain that revolves around renewable energies, with infrastructure that allows for the steadfast production, transport, and utilization of hydrogen to power everyday needs, as well as commercial and industrial processes (see Figure 10.1).

This chapter includes a detailed review of one of the most promising clean technologies of the 21st century, hydrogen. In the following sections, we will assess the potential for widespread implementation, the criteria for screening and building hydrogen generation sites, the strengths and limitations of these techniques, and the role that the oil and gas industry has had and will have in the development of hydrogen technologies in the future. The bridging of the gap between theory and practice is strongly related to the prominent role that oil majors can have, the collaboration between governments and international organizations, and the commitments made by countries to advance these new technologies.

10.2 SCREENING, SELECTION, DESIGN, AND ECONOMICS OF HYDROGEN ENERGY

The fruitful development of hydrogen technologies depends upon a robust framework of incentives that positions the energy transition as a primary goal while including the active participation of players from governments to private oil companies and their shareholders. Theoretically speaking, and somewhat of a contrast to solar and wind, hydrogen technologies have fewer geographical restrictions to further their development. Therefore, because hydrogen production is not as tightly subject

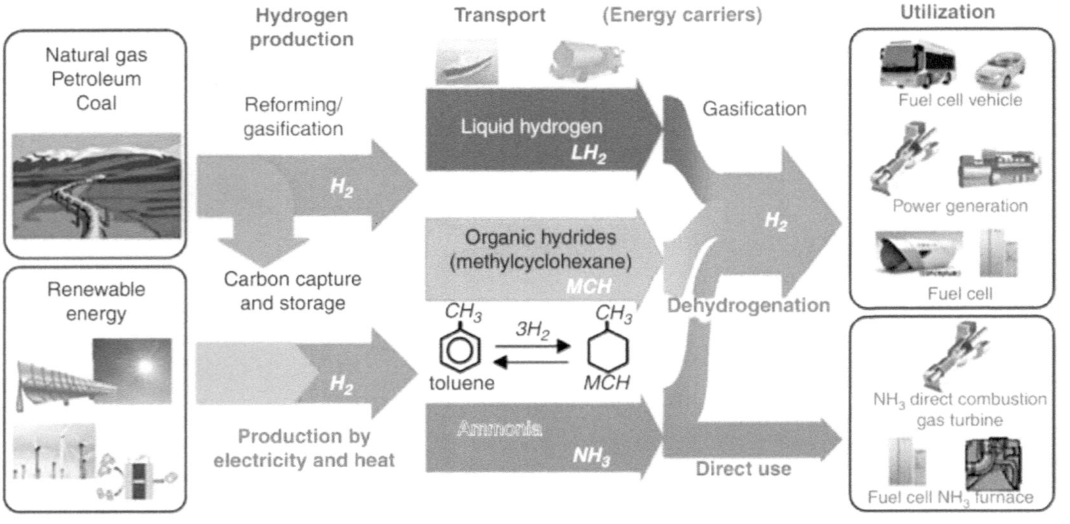

FIGURE 10.1 The hydrogen supply chain concept.

Source: Iida and Sakata (2019).

to environmental conditions of ideal wind speeds or solar radiation exposure, initial considerations for developing hydrogen sites are held to one less constraint than those of renewable energy sources.

That said, the hydrogen economy still depends mainly on energy from renewable sources for production (Lewandowska-Śmierzchalska et al., 2018). Thus, it follows that the advancement of the hydrogen economy has a larger set of incentives in countries and locations already heavily invested in renewables. And from this, we can advance, as a first approximation, that countries with a larger share of funding initiatives destined for renewable projects also have a significant interest in developing hydrogen offshoot technologies. We can expect something similar in regions with greater interest from private actors. In this section, we explore this notion and analyze other relevant criteria for the screening, selection, and design of hydrogen sites, as well as the primary economic considerations of the entire process.

Given the symbiosis between renewable energy sources and the generation of clean hydrogen, an energy vector, many projects contemplate integrated energy systems that include solar or wind generators and collectors, along with an electrolysis unit for hydrogen production. As an example, the work of Habibollahzade et al. (2018) develops an optimization project that introduces an energy system that is formed by a parabolic trough solar collector (PTSC) field, a thermoelectric generator (TEG), a Rankine cycle, and a proton exchange membrane (PEM), featured in Figure 10.2. Initiatives of this magnitude place all the pieces of the puzzle we have reviewed in this book together, further embracing the project of a medium-term transition to renewable sources of energy.

With parameters like the degree of information about the customers' acceptance of hydrogen technologies, the general characteristics of a given region, and a cost–benefit analysis of market entry and competition in that zone, the authors feel a greater degree of confidence that the deployment of one hydrogen application over another will be more optimal. Therefore, the selection of one strategy over another implies a hierarchy of decisions that, in a more general sense, may include other variables, like the emphasis on renewables and funding schemes from the government, the degree of investment from private companies, the human capital within a country, the infrastructure already in place, and the research and development of hydrogen techniques, among others.

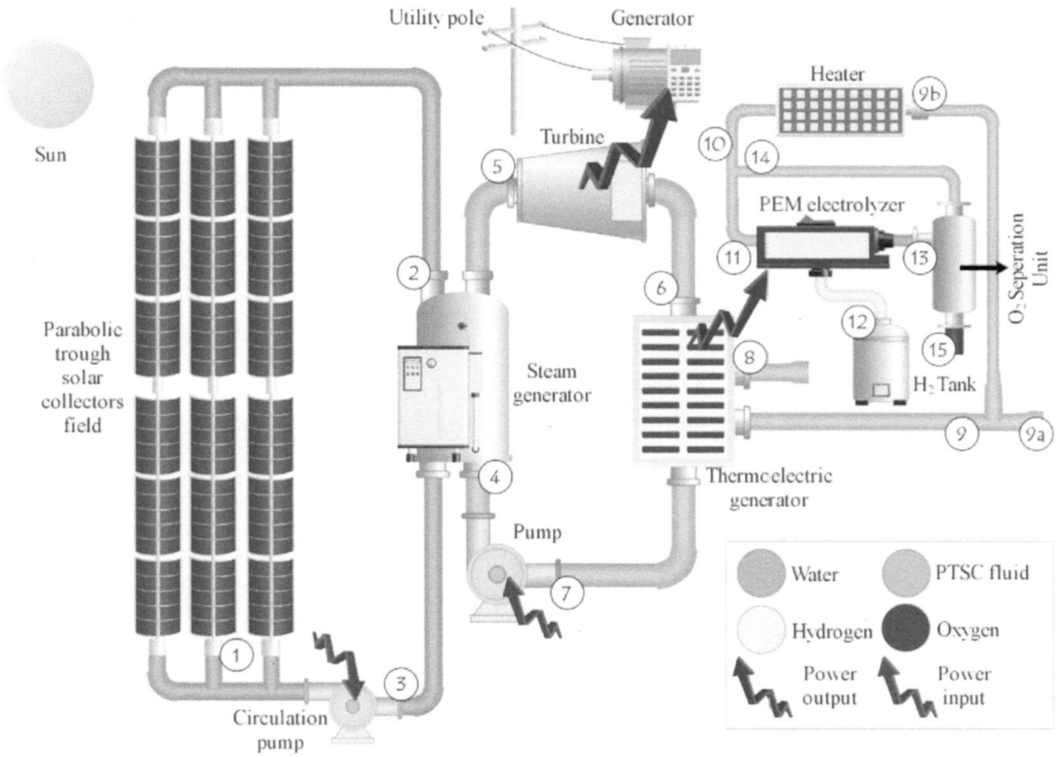

FIGURE 10.2 Schematic diagram of the proposed integrated energy system: PTSC, Rankine cycle, TEG, and PEM electrolyzer.

Source: Habibollahzade et al. (2018).

Perhaps more than with other types of technologies analyzed in this volume, hydrogen has the potential to be involved in many aspects of the daily lives of millions of people. Because of this, it is crucial to understand the public's degree of acceptability of some of the changes required to incorporate hydrogen into their lives and to know how well the end user understands hydrogen energy systems. To study the response to these novel systems from individuals, Martin et al. (2020) collected information on the literature's handling of the issue of users. They warned about the disconnection that could be occurring between an excessively techno-centric approach, with the public's acceptance of these technologies only seen as an afterthought.

Likewise, implementation effects are another essential piece related to deploying an energy system in a region. Shimizu et al. (2020) explored the ramifications of these decisions across different places, and the necessity of implementing novel energy systems with a proper screening procedure of the main characteristics of a given region appears clear. Using the case of Japan, the authors find that the development of hydrogen energy systems can have different impacts on air pollution in different parts of the country. Emissions of CO_2, nitrogen oxides, sulfur oxides, and others can affect some areas more than others, but fuel cell vehicles (FCVs) can help offset these changes more in some places than others.

Similarly, through an analytical hierarchy process (AHP) that involves the use of a fuzzy technique for order of preference by similarity to ideal solution (TOPSIS) method, Tian et al. (2019) arrived at two main selection criteria to deploy different hydrogen projects across geographical regions. These two parameters are acceptance of hydrogen applications – both from a market and a social perspective – and the strength of the energy industrial chain or how well-prepared the sector

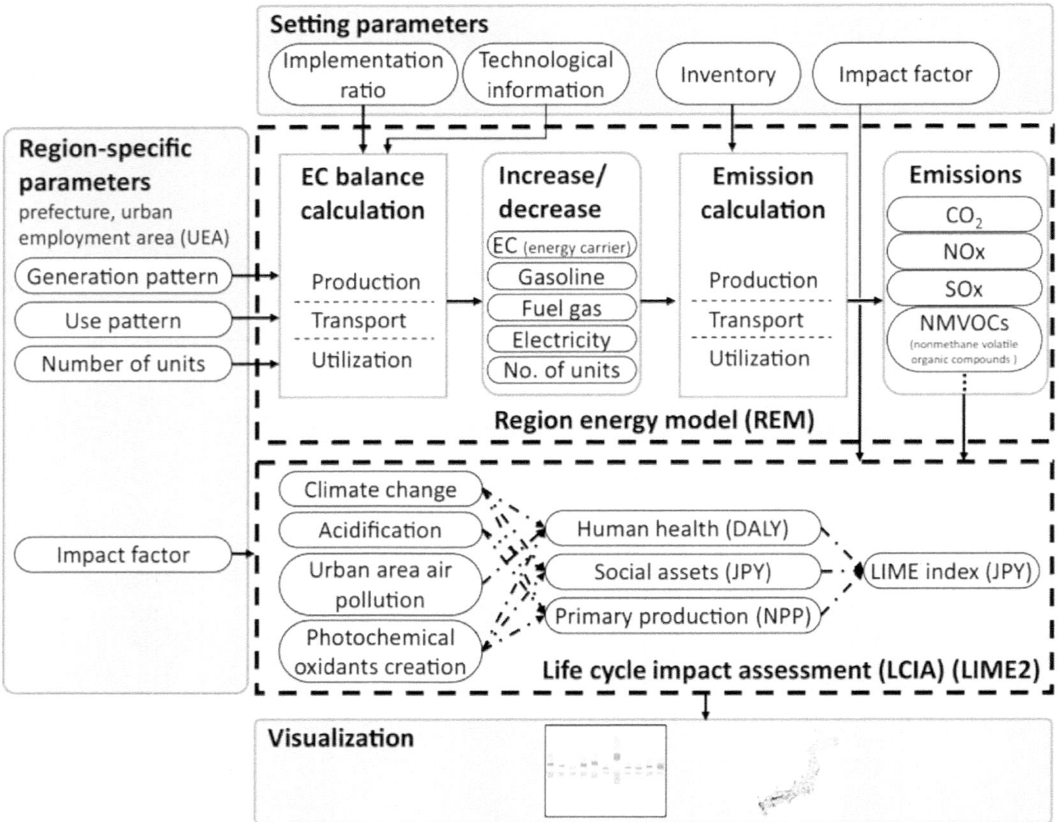

FIGURE 10.3 Region-specific analysis approach combining region energy model (REM) and life cycle assessment (LCA). (Abbreviations: *DALY*, disability-adjusted life year; *JPY*, Japanese yen; *LIME*, life cycle impact assessment method based on endpoint modeling; *LIME2*, life cycle impact assessment method based on endpoint modeling for Japan; *NOx*, nitrogen oxides; *NPP*, net primary production; *SOx*, sulfur oxides.

Source: Shimizu et al. (2020).

is in its production, distribution, and marketing stages. Hence, different degrees of market preparedness and social acceptance are essential to hydrogen energy projects getting the green light to be developed in regions with distinct features.

Therefore, the potential benefits of newer energy technologies must be weighted along with the intrinsic features of the regions where proponents want advancements to be implemented. This could mean that some proposed projects may end up being delayed or outright canceled because of time and resources spent collecting data and performing comparative assessments to improve the fit between a project and its site, which could be seen as a problem at first. However, when a proper optimization process includes a more extensive set of technical parameters while also considering public views to try to bridge any potential gaps in knowledge between the average citizen and the class of experts, trust can grow, and a hydrogen society can be promoted through education and recognizability (Kovač et al., 2021).

This degree of familiarity can also benefit potential investors that are just entering the energy arena. As with similar breakthrough technological advancements, there is an initial problem with "early adopters," or the lack thereof. Despite the variety of applications that hydrogen can offer (see Figure 10.4), making significant modifications in infrastructure and, in some cases, even people's behavior can be a

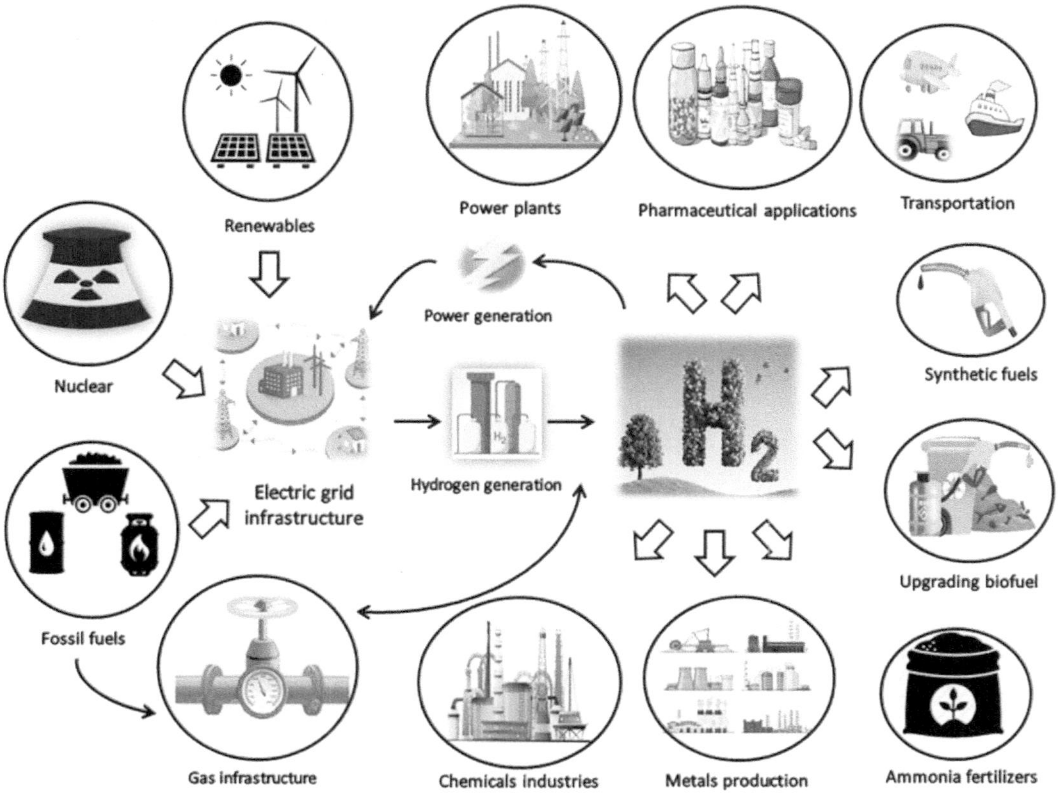

FIGURE 10.4 Hydrogen production routes, including renewables, fossil fuels, and nuclear, with hydrogen being produced in power plants, pharmaceutical applications, synthetic fuels or their upgrades in transportation, ammonia synthesis, metal production, or chemical industry applications.

Source: Osman et al. (2021).

bridge too far for many potential consumers or investors to adapt to the use of hydrogen. In this sense, until other actors jump in or other, more risk-taking entrepreneurs incite a major shift, a significant portion of investors will not be interested in making a serious shift to hydrogen technologies. However, this problem can hinder the quicker development of hydrogen as a valuable energy alternative.

These "chicken-and-egg" challenges are significant potential roadblocks to faster development of hydrogen technologies. However, these can constitute opportunities for actors who are more risk-taking in their approach. A possible advantage for actors that move first, exploring these technologies and investing in sectors where they could see future benefits, is that any coming rewards will come first to them. Further, given that, in the past few years, regulatory frameworks have been developed to encourage the use of hydrogen technologies and a more decisive move toward cleaner energy sources, those players who moved first now have a stronger playing field of investing in and reaping the rewards. So if the environment around hydrogen technologies becomes more attractive, other actors will eventually appear.

Relatedly, and regarding the economy of hydrogen technologies, it is essential to identify the conditions where these energy systems become financially attractive. One key consideration is to tackle cost issues by scaling production accordingly, as well as standardization of products, and welcome actors that can develop regulatory frameworks which allow other players to participate and propose their low-carbon applications (Parra et al., 2019). Thus, it becomes necessary to leverage many economic, social, and technical considerations to reduce the list of potential hydrogen energy systems to a shortlist that best fits the selected region.

TABLE 10.2
Cost of Hydrogen Production Methods

Process	Cost of H_2 ($USDkg^{-1}$)
Natural gas reforming	1.03
Natural gas reforming + CCS	1.22
Natural gas reforming + PSA + CCS[a]	1.56
Coal gasification	0.96
Coal gasification + CCS	1.03
Wind electrolysis[b]	6.64
Biomass gasification	4.63
Biomass pyrolysis	3.80
Nuclear thermal splitting of water	1.63
Gasoline (for comparison purposes)	0.93

Note: [a] Current central H_2 production from natural gas with pressure swing adsorption (PSA) used for H_2 purification up to 99.6% [41].
[b] Electrolysis using electricity generated by wind turbines.

Source: David et al. (2019).

Moreover, Table 10.2 compares the cost of different hydrogen production methods, with gasoline costs added as a baseline. The closest hydrogen production technique in prices stems from fossil fuels, while adding CCS technologies adds more expenditures. Whether due to economies of scale obtained after years of market growth or consumer adaptation, fossil fuel–led hydrogen production appears to be the dominant form of generating this gas. However, as cleaner hydrogen production processes develop and markets mature and adapt to these new technologies, their costs will be lowered to competitive levels.

An example of the complex equilibrium of distinct factors to choose the appropriate energy system option is illustrated in Figure 10.5, where we see a list of different criteria to be considered when selecting the most sustainable hydrogen production scheme. In their work, Acar et al. (2018) propose the following alternatives for hydrogen projects: grid electrolysis, wind electrolysis, PV electrolysis, nuclear thermochemical water splitting cycles, solar thermochemical water splitting cycles, and photoelectrochemical cells. The authors surveyed experts from the industry to obtain weights for each of the leading and sub-criteria, with *technical performance* placing first while *environmental performance* is a close second for the selection process.

Using hesitant fuzzy AHP methods, the authors evaluated the potential for each hydrogen project option following the series of variables listed earlier. When considering all factors, grid electrolysis appears as the most sustainable option, followed by solar thermochemical water splitting cycles and wind electrolysis (Acar et al., 2018). Moreover, this list of considerations may still be improved with geographical features or the potential for a hydrogen system to provide energy independence. On this, Groppi et al. (2018) provided an analysis of the small Italian islands located in a high-solar-radiation area and proposed that the development of hydrogen energy systems can help these remote locations not only in decarbonizing but also in achieving energy independence.

One of the critical aspects of the energy transition is the diffusion of alternatives, which allows more actors to contribute to the grid through different sources. Thus, a region that previously only relied on one coal plant for its energy can now develop smaller renewable facilities. At the same time, they work on phasing out the more polluting facility. Further, this diffusion can involve local actors developing their energy systems in their backyards or residential buildings having their energy systems. These alternatives are not only tailored to the needs of the people there but also provide some relief on the grid. Ultimately, the new rubric opens the door to system design features that end users can independently assess.

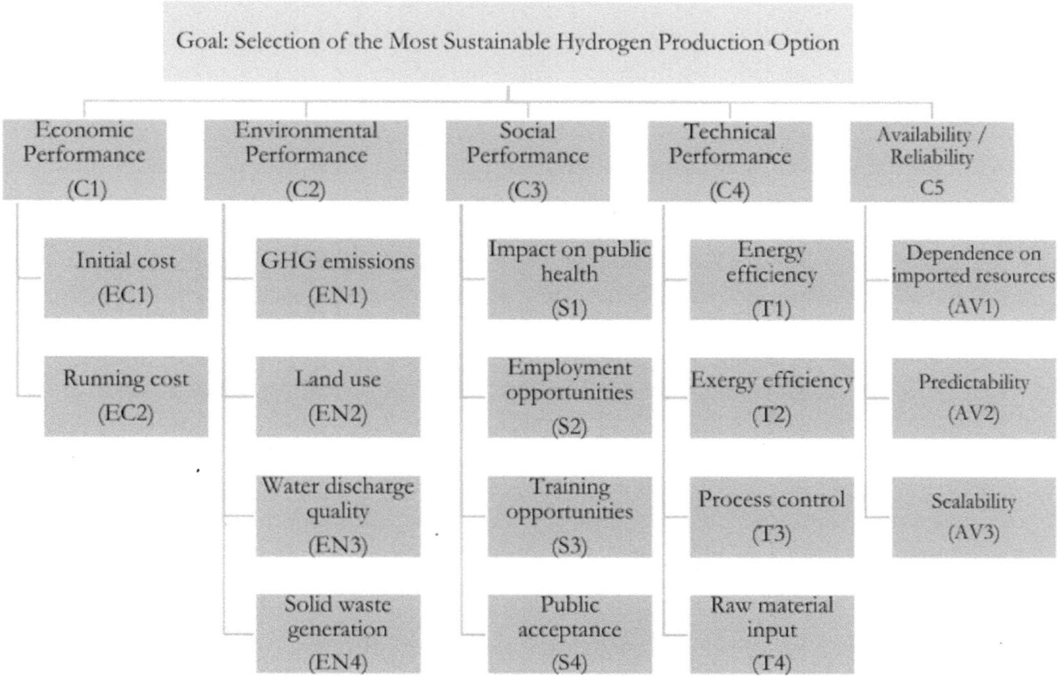

FIGURE 10.5 Sustainable hydrogen production criteria.

Source: Acar et al. (2018).

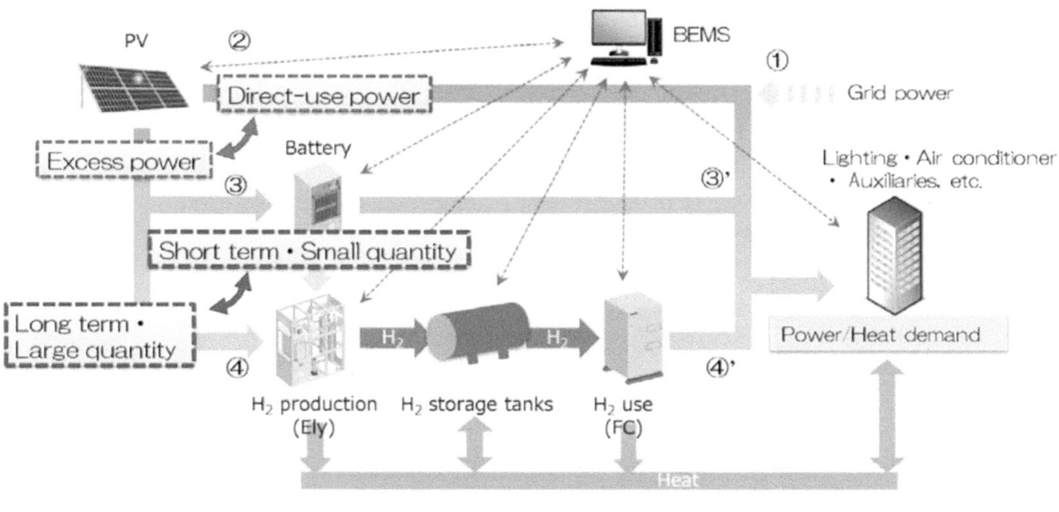

FIGURE 10.6 Energy flow of the hydrogen energy utilization system with photovoltaic (PV) energy, controlled by a building energy management system (BEMS).

Source: Endo et al. (2019).

For example, Figure 10.6 shows the framework of a hydrogen-based energy system used for a proposed zero-emissions building (ZEB) in Japan. This arrangement was designed by simulations performed in MATLAB/Simulink and included an electrolyzer, a storage tank, a fuel cell, and a battery for short-term power storage. For this example, considerations of design assessed first the size of the building, settling for a low-rise to achieve the ZEB requirement. Then, the design of the installation included 20 kW PV panels, a 36 Nm^3 (normal cubic meter) hydrogen storage tank, a 4.2 kW fuel cell, a 10 kW battery, and a hydrogen-generating electrolyzer ranging between 2 and 3 Nm^3 (Endo et al., 2019).

Meanwhile, a systematic literature review found that projects are being considered at different scales, like the household, neighborhood, university, or power plant unit. The sources of energy relied upon center on sun and wind, and the hydrogen component of the studies leans heavily on electrolyzers and fuel cells. However, researchers have looked closely at gasification and methanation processes (Fonseca et al., 2019). Another vital consideration about system design is the capability of installations to tackle intermittence shortcomings, optimally balancing energy supply and demand within a renewable-based system (Wang et al., 2018a).

Hydrogen systems are subject to similar limitations in their design, site selection, and economics. First, suppose these schemes are built within the infrastructures governments and private actors have developed to advance renewable energy sources. In that case, they suffer from a self-selection issue which translates into a larger presence of these installations in regions with high solar radiation or high winds, where renewable projects are already more likely to be developed. Under the framework of an energy transition process, this is a significant limitation that has been tackled by a few projects that focus on oil and gas resources – like coal and methane – to generate hydrogen from non-renewable sources (Kolawole Muritala et al., 2020; Mehrjerdi, 2019; Tian et al., 2018).

In the second place, other constraints relate to the energy project's setting and scope. Given that these sorts of endeavors can adapt to the end user more easily than do fossil fuel facilities, there is a greater leeway to develop smaller projects adapted to the neighborhood level, the residential building, or even the household. However, this top-down independence is met with the economic limitation of a search for the standardization of equipment and infrastructure, which helps service companies develop profitable businesses while supporting and strengthening the renewable energy market.

Lastly, a more robust market can drive more innovation, and along with a greater public awareness of the potential benefits of hydrogen endeavors, both factors can help create a fertile soil for new actors to appear and plant their seeds of innovation and entrepreneurship. In the end, understanding the determinants of success for hydrogen energy initiatives, what regions of the world would be optimal to host these projects, what incentives should the renewables market receive before or while hydrogen projects are being developed, and how governments and private actors can contribute within their realms is critical to the prosperous advancement of hydrogen energy systems.

10.3 HYDROGEN ENERGY APPLICATIONS, PRINCIPLES, BEST PRACTICES, AND ECONOMICS

The production of hydrogen to utilize it as the primary energy carrier in a post–fossil fuel world is one promising avenue for the energy transition process of the 21st century. Depending on the setting, infrastructure, and technology, hydrogen applications can vary significantly. As Figure 10.7 shows, hydrogen production methods can be derived from both non-renewable and renewable energy sources, with the latter procedures being divided mainly into biological/thermochemical and water splitting techniques (Shiva Kumar and Himabindu, 2019). Besides more novel techniques that are being explored in more detail by the literature, some of which will be reviewed in this chapter, most of the hydrogen production is achieved through water electrolysis, steam methane reforming, methane pyrolysis, and coal gasification (Osman et al., 2021).

Moreover, these hydrogen generation procedures have received different color schemes depending on the source. Thus, "green hydrogen" is obtained from renewable energy sources like solar or wind, following a clean water electrolysis process. Then, "gray hydrogen" is obtained after methane steam reforming, while "brown hydrogen" is produced through coal gasification. Instead, these

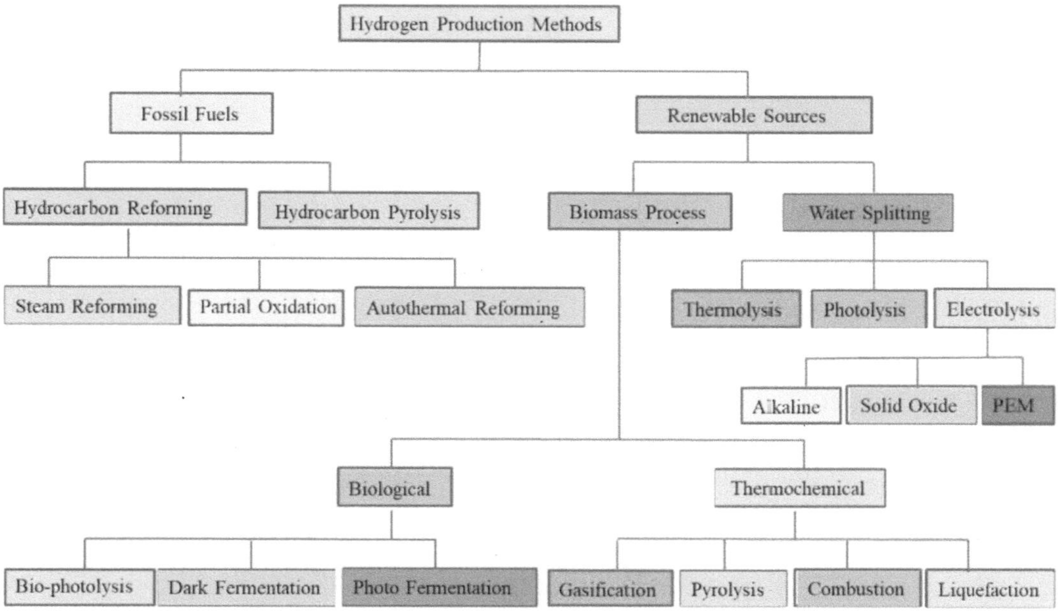

FIGURE 10.7 Various hydrogen production methods.

Source: Shiva Kumar and Himabindu (2019).

two processes can be classified as producers of "blue hydrogen" when combined with carbon capture and storage (CCS) methods. Lastly, "turquoise hydrogen" is obtained after methane pyrolysis, which leaves solid carbon as a by-product. Figure 10.8 shows an illustration of this color coding.

The following section performs a comprehensive review of these hydrogen applications, contrasting their advantages and shortcomings, their level of development, the real-life projects and applications currently in progress, and the degree to which these technologies can become a simple energy tool in the arsenal of clean alternatives. We will review different procedures using varied types of feedstocks, from fossil fuel–based to renewables, to assess the potential of hydrogen systems. The promise of these methods is only beginning to be unlocked, so more significant research and investment can make a difference in successfully disseminating these clean energy devices across several regions worldwide.

10.3.1 Chemical Production of Hydrogen

The chemical processes of hydrogen production consist of feedstock purification at high temperatures, separating the original input material into different elements, with hydrogen being those products. One of the procedures with the highest degree of maturity and the contributor to the largest share of the world's hydrogen production is steam methane reforming (SMR). Despite increased global warming potential, it is one of the cheapest methods to produce hydrogen, next to coal gasification (Safari and Dincer, 2020). Figure 10.9 shows a simplified process for SMR techniques, where synthetic gas (syngas) is produced along with a mixture of carbon monoxide and hydrogen (Luo et al., 2018).

When these processes are performed from non-renewable sources of energy, like natural gas, the hydrogen generated is referred to as "gray hydrogen" if the CO_2 from the separation is released into the atmosphere, and "blue hydrogen" if the carbon dioxide undergoes a process of storage or capture, thus somewhat mitigating its release. Meanwhile, "green hydrogen" results from a renewable

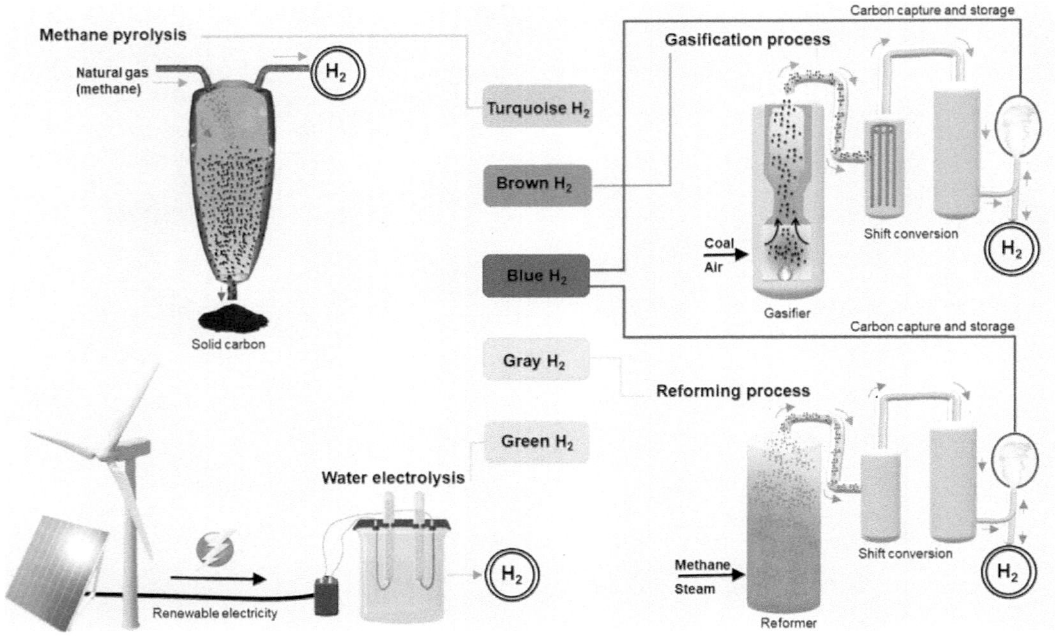

FIGURE 10.8 Hydrogen color coding for various manufacturing processes.

Source: Osman et al. (2021).

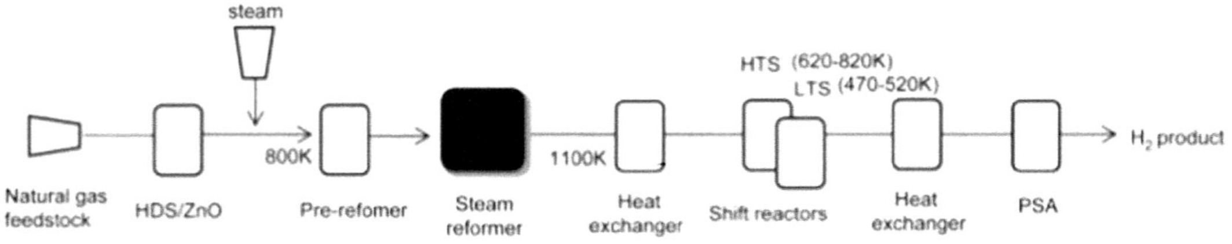

FIGURE 10.9 Simplified flow diagram of conventional steam methane reforming (SMR) process.

Source: Luo et al. (2018); Sazali (2020).

energy source process. Almost all hydrogen demand is met through SMR processes, which primarily generate gray hydrogen (Ali Khan et al., 2021). Over time, however, techniques for carbon capture and storage (CCS) have progressed, making SMR and other chemical processes cleaner, less expensive, more efficient, and more accessible.

Two other similar chemical processes that have benefitted from advances in CCS technologies are autothermal reforming (ATR) and natural gas decomposition (NGD). These methods consist of a structure with a reactor that works with high temperatures, separating hydrogen from CO_2, a syngas unit where remaining gases are recycled, a hydrogen compression and storage unit, and a carbon sequestration section. Figure 10.10 shows a simplified schematic of these processes, while Table 10.3 shows the greenhouse emissions (GHG) estimated for each of them. As noted, NGD is the least-polluting hydrogen method, while SMR is the least clean of the group. Conversely, NGD techniques have the most significant capital and operating costs (Oni et al., 2022).

To tackle the potential environmental challenge of having a technology poised to be significantly utilized to produce much-needed hydrogen but doing so at a significant polluting cost, an emerging

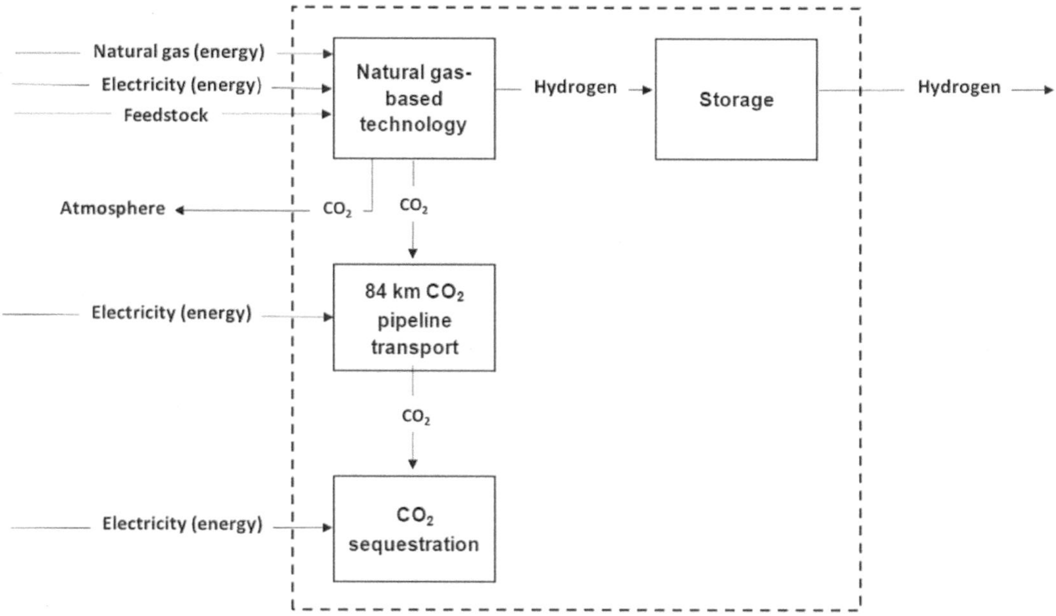

FIGURE 10.10 An overview of hydrogen-based technology.
Source: Oni et al. (2022).

TABLE 10.3
Greenhouse Emissions (GHG) Results for Blue and Gray Hydrogen Production

	Gray Hydrogen			Blue Hydrogen			
Parameter	**SMR**	**ATR**	**NGD**	**SMR 52%**	**SMR 85%**	**ATR-CCS**	**NGD-CCS**
On-site emission (kg CO_2 eq./kg H_2)	9.17	S.39	1.84	5.52	1.98	0.62	0.9
Electricity emission (kg CO_2 eq./kg H_2)	0.52	1.28	1.21	0.72	2.4	1.95	1.74
Upstream NG emission (kg CO_2 eq./kg H_2)	1.65	1.34	1.84	1.96	2.28	1.34	1.9
Total GHG emissions (kg CO_2 eq./kg H_2)	11.35	11.01	4.89	8.20	6.66	3.91	4.54

Source: Oni et al. (2022).

technology of low-carbon SMR has been advanced. Sorption-enhanced steam methane reforming (SE-SMR) combines carbon capture and hydrogen production, adding a CO_2 sorbent into either the water–gas shift reactor or the reformer and a catalyst such as a noble metal or a nickel-based compound. Trusted carbon sorbents are materials based on calcium oxide (CaO), lithium zirconate, sodium zirconate, and lithium silicate (Masoudi Soltani et al., 2021).

These optimized processes have been applied with another known set of chemical methods, chemical looping technologies. The most used procedure, chemical looping combustion (CLC), utilizes two reactors to constantly oxidize and reduce oxygen carriers (like iron oxide). Because of this process, the flue gas that leaves the fuel reactor is nitrogen-free, making it almost pure carbon that can be more easily sequestered or captured (see Figure 10.11). A hydrogen-producing chemical

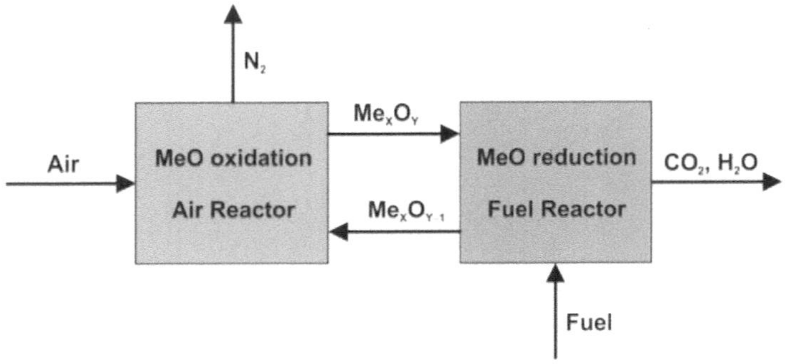

FIGURE 10.11 Chemical looping combustion concept.

Source: Czakiert et al. (2022).

looping process is *chemical looping reforming* (CLR). It consists of a similar process as CLC, but the fuel reactor is supplied with CO_2 (for dry CLR) or H_2O (for steam CLR), which react and produce hydrogen (Czakiert et al., 2022; Yang et al., 2021b).

Moreover, a chemical looping gasification (CLG) method has been advanced, which converts biomass feedstock from renewable sources into hydrogen-enriched syngas. The work of Hu et al. (2020) uses iron oxide (Fe_2O_3) and calcium oxide (CaO) in different ratios as oxygen carriers. Figure 10.12 shows that the structure consists of a vertical electric furnace used as a gasification system. This furnace is heated to a stable target temperature of close to 900°C, and rice straw, which serves as raw feedstock, is added along with a mixture of the oxygen carriers. Then, water was added to generate steam to help a chemical reaction, which is optimal at temperatures over 800°C.

Other ways to produce hydrogen from chemical processes have been explored in the literature. One such way consists of using food waste, which, through a method of anaerobic digestion (meaning, that oxygen is not a part of the biodegradation) and biogas reforming, can produce 84 L of biogas from 1 kg of food waste (Park et al., 2020a). Biogases stemming from natural processes of food waste degradation are methane (CH_4) emitters, which can contribute to climate warming. The richness of the combustible component of methane makes this gas a promising area of research for purification and optimization (Gao et al., 2018).

Nurdiawati et al. (2019) proposed a system that uses *Acutodesmus obliquus* algae as feedstock in a supercritical water gasification (SCWG) process that converts wet biomass without the need for pre-drying the feedstock. When the water reaches the supercritical point (hence, where water has gas-like viscosity but liquid-like density) through rising temperatures over 374.15°C and pressures over 22.1 MPa (megapascals), the syngas generated has a higher proportion of hydrogen, CO, CO_2, and CH_4. As Figure 10.13 shows, this syngas is later separated thanks to a chemical looping process, and the hydrogen can be combusted to produce energy for a power generation system.

Chemical technologies have advanced significantly in the last decade. However, despite their significant potential to generate hydrogen through clean processes, some shortcomings keep this technology from being deployed at a larger scale across more countries. First, capital and operational expenditure (CAPEX and OPEX, respectively) curves may still be too steep for some players, so adding more processes like producing methanol from food waste, finding better oxygen carriers, or increasing the temperature of a process to improve the purity of CO_2 may be a step that puts firms on a negative economic path (Kaplan and Kopacz, 2020).

Additionally, some questions remain about the degree of integration of hydrogen methods in power systems and the duration of this incorporation process. Because some techniques that have been

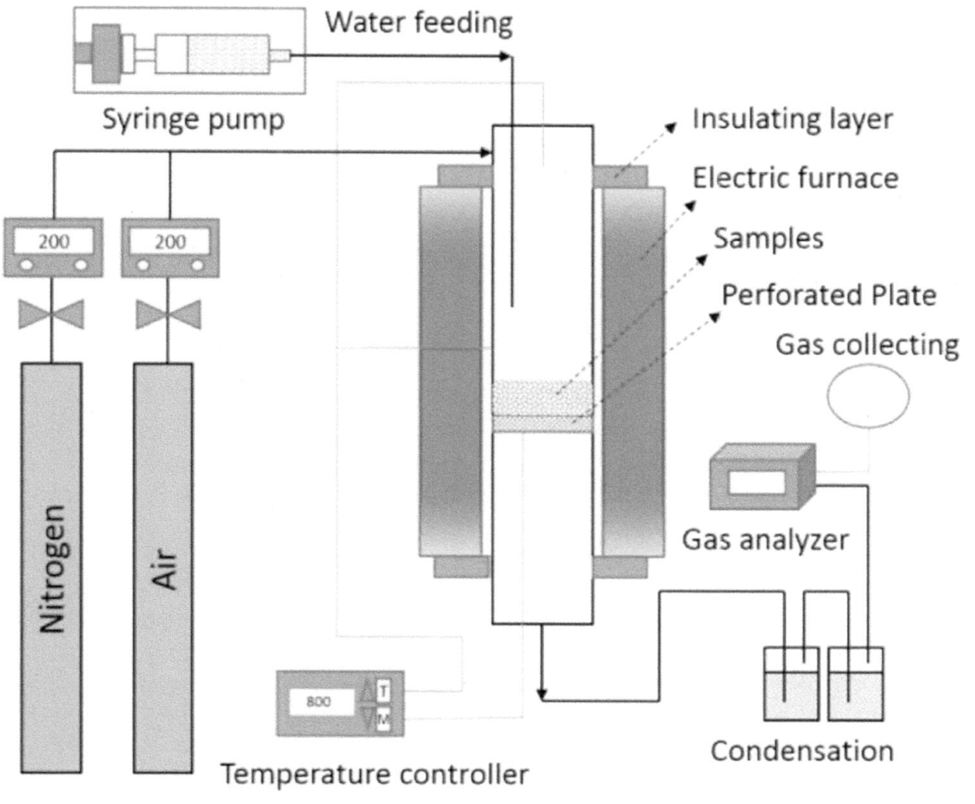

FIGURE 10.12 Schematic of the chemical looping gasification system.

Source: Hu et al. (2020).

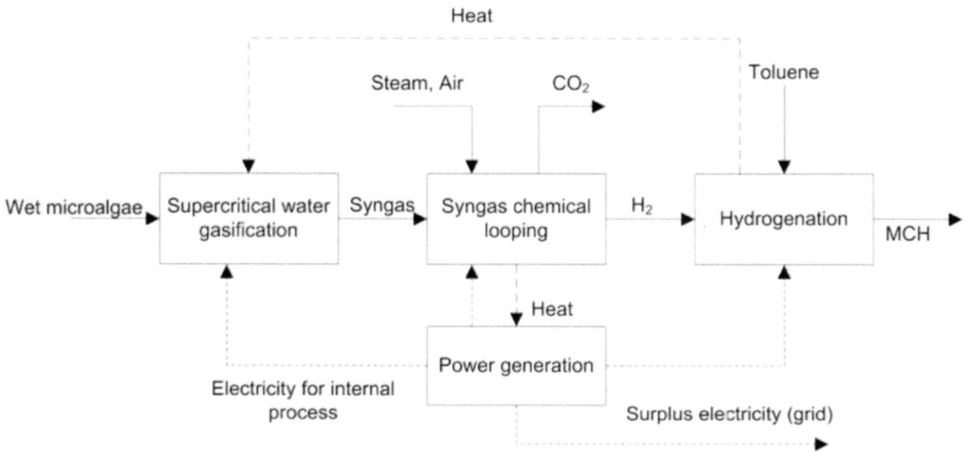

FIGURE 10.13 Schematic of proposed chemical hydrogen production system based on microalgae.

Source: Nurdiawati et al. (2019).

researched and explored for many years are still in their infancy regarding application and integration, the hydrogen economy is a project that will require a few decades to be optimally implemented (Safari and Dincer, 2020). Likewise, chemical methods may require a few more iterations of research and experimentation and a breakthrough discovery that lowers costs and increases efficiency, so we may still be a few decades away from a massive implementation of new hydrogen systems worldwide.

10.3.2 Electrolytic Hydrogen

A second set of hydrogen generation methods is those based on electrolytic processes. These are methods of obtaining hydrogen by splitting water (H_2O) into its basic components, using a separator that works with an electric current (see Figure 10.14). A structure called an electrolyzer separates water into oxygen and hydrogen by attracting oxygen molecules to its anode and hydrogen molecules to its cathode. This method is relatively cost-effective and has the significant advantage of only generating pure oxygen as a by-product (US Department of Energy, 2022a; Shiva Kumar and Himabindu, 2019). Water electrolysis methods vary in complexity so that they can be divided into three main types: alkaline, proton exchange membrane (PEM), and solid oxide electrolyzers (SOE).

Due to most of the hydrogen production coming from non-renewable energy sources, water electrolysis needs to be developed urgently. Electrolytic processes account for no more than 4% of hydrogen production. This limited production calls for improving the hydrogen evolution reaction (HER) or the cathodic reaction in various water splitting methods (Dubouis and Grimaud, 2019). In this sense, alkaline solutions used in electrolyzers offer one avenue for a larger share of hydrogen production. Despite being hampered by inferior kinetics, a better understanding of catalyst solutions can accelerate alkaline HER techniques and position this method as a leading hydrogen production procedure (Chen et al., 2019; Mahmood et al., 2018).

Alkaline water electrolysis is a process that operates at temperatures between 60 and 80°C. It uses an aqueous solution that, in most cases, consists of a 30% wt potassium hydroxide (KOH) or a 25% wt sodium hydroxide (NaOH) as the electrolyte, asbestos as the permeable membrane to separate the elements, and porous nickel materials as components of the electrode (Chi and Yu, 2018; Yu et al., 2021). These structures work after making the alkaline electrolyte enter both membrane regions, touching the anode and cathode, and applying an electrical current that generates reactions on each side. In the cathode region, water molecules combine with electrons to form hydrogen and

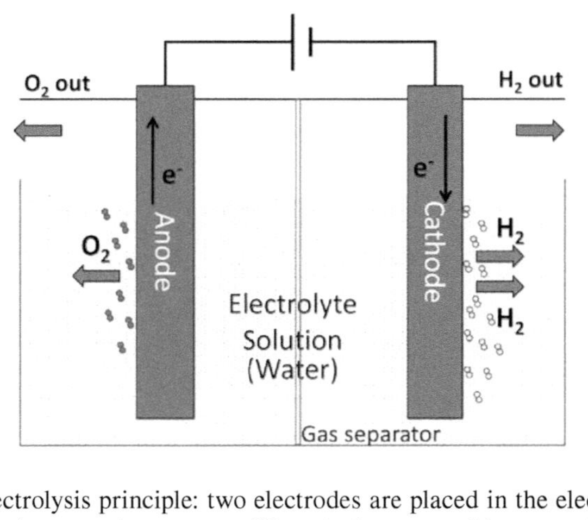

FIGURE 10.14 Water electrolysis principle: two electrodes are placed in the electrolyte solution, which is connected to the power supply to conduct current. Water is decomposed into pure hydrogen and oxygen gas, appearing at the cathode and the anode, respectively.

Source: Yue et al. (2021).

hydroxide ions (OH^-), while in the anode section, these hydroxide ions lose electrons, generating oxygen and water (US Department of Energy, 2022a; Guo et al., 2019).

Meanwhile, proton exchange membrane (PEM) electrolysis works similarly to an alkaline electrolyzer. The main difference is that the former uses a thin film electrode assembly, or a membrane electrode that replaces asbestos, allowing for the exchange of protons (H^+) (David et al., 2019). Also, the electrolyzed raw material is deionized water, which only needs to enter the anode side of the electrolyzer to generate hydrogen ions that reach the cathode side in a hydrated state ($H^+ \times H_2O$) to develop hydrogen gas molecules. Lastly, due to the catalyst using a noble metal, manufacturing a PEM electrolyzer costs more than an alkaline (Guo et al., 2019). Figure 10.15 illustrates both processes in more detail.

Lastly, solid oxide electrolyzers (SOE) use a solid ceramic material as an electrolyte and are based on high-temperature steam processes that are more efficient and less costly than the other two procedures. Figure 10.16 shows the working principle of a solid oxide electrolysis cell (SOEC), where water steam reduces hydrogen electrodes to hydrogen and oxygen ions. This process is the

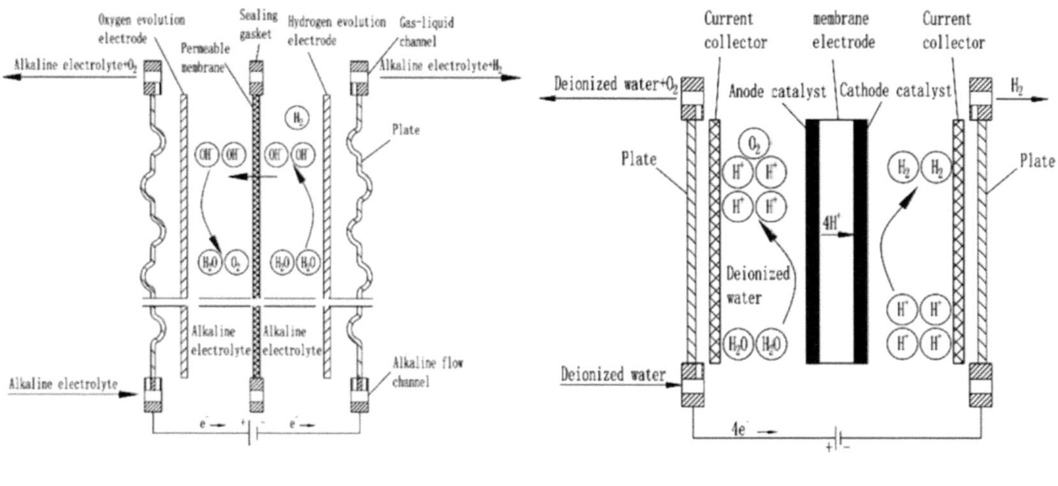

FIGURE 10.15 Schematics for an alkaline electrolyzer and a proton exchange membrane (PEM) electrolyzer reaction.

Source: Guo et al. (2019).

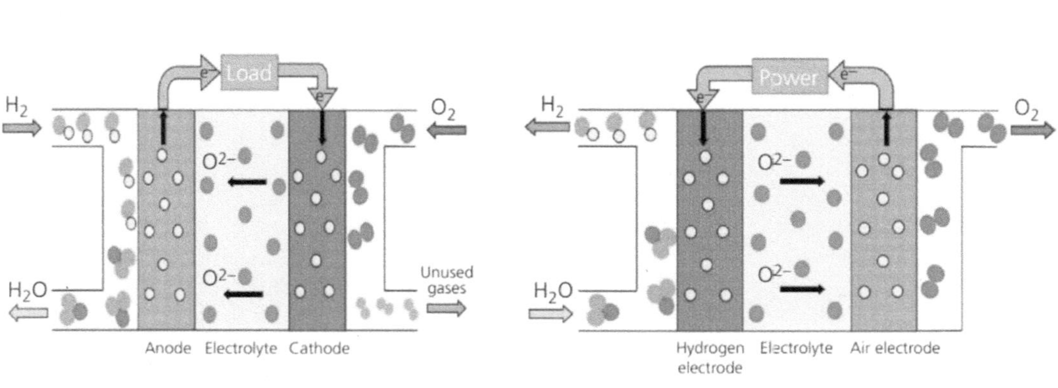

FIGURE 10.16 Working principle of a solid oxide fuel cell (SOFC) and a solid oxide electrolysis cell (SOEC).

Source: Pandiyan et al. (2019).

reverse of a solid oxide fuel cell (SOFC) method, which converts chemical energy to electrical energy (Pandiyan et al., 2019). Further, there are two distinct methods based on this technology, oxygen SOEC, and proton SOEC, with the first producing hydrogen from the water electrolysis process. At the same time, the second generates hydrogen from the proton reactions with electrons at the electrode (Kim et al., 2018).

These techniques show essential differences that affect their performance, efficiency, and cost. Table 10.4 lists the typical specifications for these three hydrogen production methods. What is perhaps most important from these methods, however, is their ability to produce hydrogen without any greenhouse gas emissions, depending on the source of electricity used for the current to operate the electrolyzer. Similarly, the potential for integration with other energy sources provides many opportunities for synergy and the ability to tackle intermittence problems anticipated from renewables like solar and wind (US Department of Energy, 2022a).

The realization of this tremendous potential, however, relies on the steady development of a strong infrastructure that is capable of transporting and storing excess electricity, some in the form of hydrogen, which can later be sold as fuel for electric vehicles or converted back to electricity when demand or price require (Mayyas et al., 2020). The market is currently developing for big industrial consumers and retailers, but the challenge is to continue this advancement toward residential and commercial settings, where users are more sensitive to short-term price fluctuations and may not have the time or resources to invest in immature technologies.

Therefore, significant challenges remain, particularly reducing the cost of producing 1 kg of hydrogen to US$1, increasing carbon capture and storage technologies' capabilities, and lowering electrolyzers' capital costs (Safari and Dincer, 2020). These advances can also be complemented by improving newer technologies with potential applications in other areas. The following sections will review other techniques of hydrogen generation in more detail, with the goal of highlighting their strengths and diverse application areas.

TABLE 10.4
Typical Specifications of Alkaline, PEM, and SOE Electrolysis Water Splitting Methods

Specification	Alkaline	PEM	SOE
Technology Maturity	**State of the Art**	**Demonstration**	**R&D**
Cell temperature, °C	60–80	50–80	900–1,000
Cell pressure, bar	<30	<30	<30
Current density, A/cm^2	0.2–0.4	0.6–2.0	0.3–1.0
Cell voltage, V	1.8–2.4	1.8–2.2	0.95–1.3
Power density, W/cm^2	Up to 1.0	Up to 4.4	–
Voltage efficiency, %	62–82	67–82	81–86
Specific system energy consumption, kWh/Nm^2	4.5–7.0	4.5–7.5	2.5–3.5
Partial load range, %	20–40	0–10	–
Cell area, m^2	<4	<300	–
Hydrogen production, Nm^2/hr	<760	<30	–
Stack lifetime, hr	<90,000	<20,000	<40,000
System lifetime, yr	20–30	10–20	–
Hydrogen purity, %	>99.8	99.999	–
Cold start-up time, min	15	<15	>60

Source: El-Shafie et al. (2019).

10.3.3 Thermolytic Hydrogen

Thermolysis is another water-based hydrogen production method, but instead of an electrical current, these techniques work through thermal decomposition, or the separation of water caused by heat. The principle is to heat water to a temperature of at least 2,500 K, where water molecules decompose into H_2 and O_2. This process can adapt better to energy transition goals using solar energy to dissociate water molecules (Hosseini and Wahid, 2020; Wang et al., 2019). Figure 10.17(a) illustrates the process of capturing solar energy for a thermolysis process. With that said, due to the problems of finding materials that can withstand those high temperatures consistently, thermolysis may not be a convenient method of hydrogen production for widespread implementation.

This second process reduces the temperature for molecule separation to around 1,200 K, which makes it easier for materials and expands its use. Thermochemical water splitting is a two-step method that relies on metal oxide redox reactions, where the first one is a solar endothermic dissociation process and the second is a non-solar metal hydrolysis occurrence that forms hydrogen and metal oxide in the way that Figure 10.17(b) illustrates (Hosseini and Wahid, 2020). Thermochemical processes are part of a set of long-term techniques with sustainable fuel production as their primary goal, constituting a step forward from decarbonization techniques that seek to convert energy from fossil fuel–based feedstock (see Figure 10.18).

The group of techniques that produce hydrogen from clean energy sources is still a relatively novel set of methods, and new research has provided more avenues for their widespread implementation. For example, the work of Chen et al. (2021) proposes a high-temperature electrolysis process integrated with an ammonia-based thermochemical energy storage unit and an s CO_2 Brayton cycle. The use of preheated ammonia, coming originally from the storage tank, in the endothermic reactor has the goal of mixing with nitrogen and hydrogen, and the Brayton cycle is added to receive high-temperature supercritical CO_2 that drives the turbine blades, which generate the shaft work.

Meanwhile, other authors have explored the two-step process, which allows for the lowering of temperature when splitting water molecules to around $900-1,500°C$, in more detail. For this change to occur, however, a thermochemical cycle that uses metal oxide redox reactions must be performed. Figure 10.19 shows the two steps of this process, with the first consisting of an endothermic phase, known as activation (which can be assisted by solar technologies for the heat source), occurring after a metal oxide of high valance is reduced, causing oxygen to be released. The second exothermic phase of this process, known as hydrolysis, occurs after the oxygen obtained from the

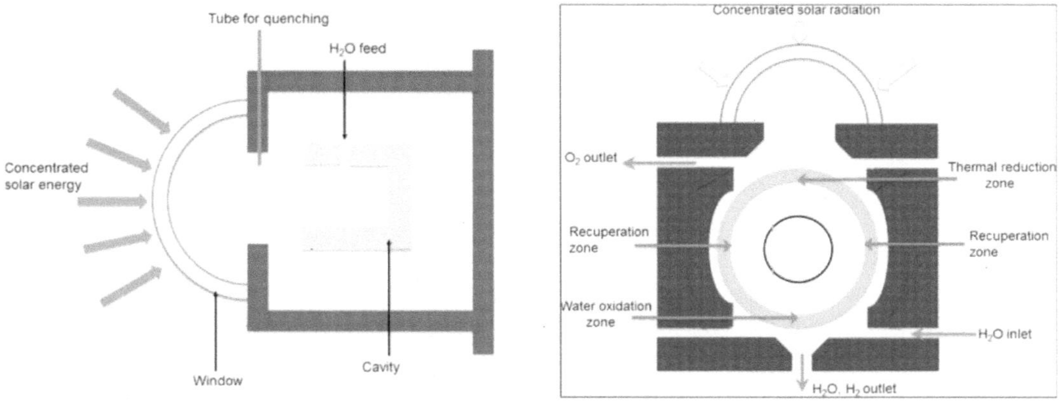

FIGURE 10.17 (a) Schematic of hydrogen production using a solar reactor for the thermolysis process; (b) schematic of hydrogen production using a thermochemical solar reactor.

Source: Hosseini and Wahid (2020).

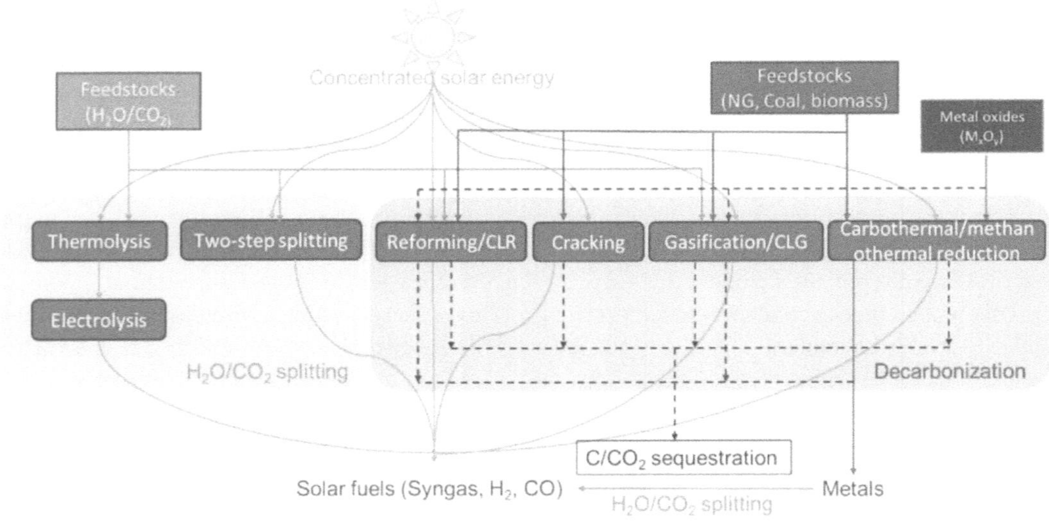

FIGURE 10.18 Thermochemical routes for solar fuel production using concentrated solar energy.

Source: Chuayboon and Abanades (2020).

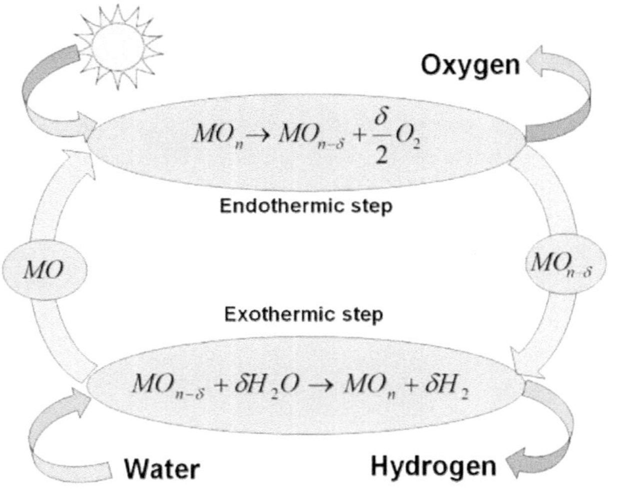

FIGURE 10.19 Schematic diagram for a two-step metal oxide thermochemical cycle.

Source: Safari and Dincer (2020).

first stage reacts with water to produce hydrogen and pristine oxide, which is then recycled to the first step (Safari and Dincer, 2020).

Part of a successful two-step process depends on the redox materials. Oudejans et al. (2022) compared a set of suitable redox materials under several criteria, finding that ferrite supported on yttrium-stabilized zirconia, ceria doped with zirconia, and samarium and ferrite doped with nickel as its core and an yttrium oxide shell appear to have the most promise. Moreover, temperature reduction caused by the redox material needs an efficient oxygen removal process. Otherwise, the operation runs the risk of significant energy penalties. To solve this, Brendelberger et al. (2019) found that perovskite-based redox materials increase the efficiency of thermochemical pumps by recovering more heat released during the oxidation of the pumping material.

10.3.4 Photobiologic/Photolytic Hydrogen

Biological processes for hydrogen production have increased in the past few years. These processes can involve anaerobic bacteria, green algae, dark fermentation, and photo-fermentative methods (El-Shafie et al., 2019). Compared to the previous approaches for hydrogen production, bio-hydrogen has a broader range of raw materials, has lower consumption, and generates virtually no pollution. According to Wang et al. (2018b), feedstock for gas production can come from five different categories: plant straw materials (like rice or corn straw), animal feces (like cow, pig, and chicken manure), food industry residue (like apple pomace and cottonseed cakes), garbage (like kitchen waste and landfill percolated liquid), and other materials (like sewage, sediment, and microalgae).

Predictably, varying types of raw materials will have different degrees of expected efficiencies to generate hydrogen. Coupled with distinct production methods, we witness potentially different production levels. Additionally, factors like temperature, pH, voltage, hydraulic retention time, light intensity, chemical oxygen demand (COD) concentration, and saline concentration will influence the reaction. At the same time, bio-hydrogen production is performed (Wang et al., 2018b). For example, Fu et al. (2021) explored the pathways for hydrogen production from waste-activated sludge (WAS). Figure 10.20 illustrates the three best-suited methods to work with WAS, a main by-product of municipal wastewater treatment plants.

Dark fermentation is one of the more mature technologies on this list, consisting of hydrogen generation from various sources of waste materials or wastewater using anaerobic organisms, which means that they do not require oxygen for growth. Akhlaghi and Najafpour-Darzi (2020) identified *Enterobacter* sp., *Bacillus* sp., and *Clostridium* sp. as producers of hydrogen through these

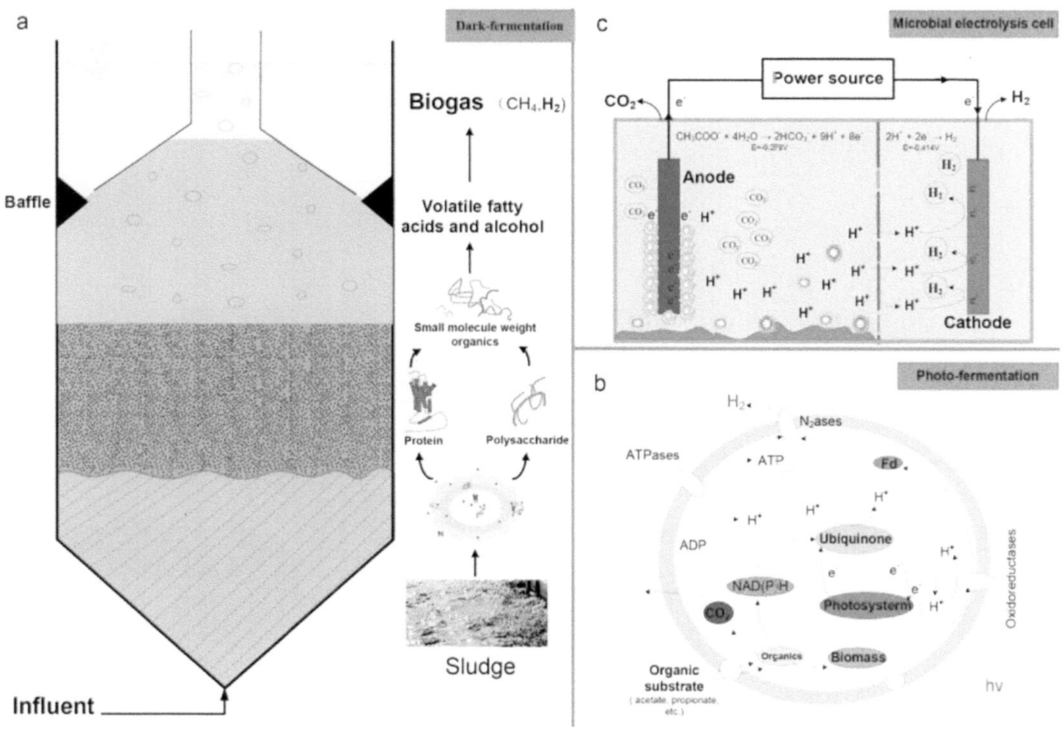

FIGURE 10.20 Schematic diagram of hydrogen production from (a) dark fermentation, (b) photo-fermentation, and (c) microbial electrolysis cells.

Source: Fu et al. (2021).

processes. Meanwhile, Pachiega et al. (2019) found that sugars such as fructose, glucose, sucrose, and xylose, present in different waste products, are rich in carbohydrates and can thus offer potential resources for dark fermentation processes. Using brewery waste, the authors found potential in fructose as a promising feedstock for bio-hydrogen generation.

A similar hydrogen production process, photo-fermentation, occurs when photosynthetic bacteria use light as energy, reducing sugars and organic acids to extract hydrogen (Zhang et al., 2020). In this technique, hydrogen production is mediated by purple non-sulfur (PNS) bacteria, which are gram-negative prokaryotes. These bacteria include *Rhodobacter sphaeroides*, *Rhodobacter capsulatus*, and *Rhodobacter sulfidophilus*. Figure 10.21 shows this process, where carbon substrate is oxidized to CO_2 in the TCA cycle and H^+ electrons are transferred to hydrogenase, which reduces the protons to hydrogen (Akhlaghi and Najafpour-Darzi, 2020).

Then, microbial electrolysis is a process where organic carbons or CO_2 are transformed into valuable chemicals. To obtain these products, researchers have used microbial electrolysis cells (MECs), which require less external energy input while generating a higher hydrogen yield than other biological technologies. As noted in Figure 10.20 earlier, a MEC works similarly to a water electrolysis device but can rely on organic wastewater as a substrate. With that said, shortcomings of these techniques include a very slow hydrogen evolution reaction (HER) on plain cathode materials, which raises the cost of the method by needing platinum as a catalyst (Yang et al., 2021a).

In a joint effort to combine the best of two technologies, Ban et al. (2018) advanced the use of bacteria to improve algal photolysis hydrogen production. Their work concludes that the bacterium *Pseudomonas* sp. strain D was pivotal in assisting *Chlamydomonas reinhardtii* in this improvement. It is a good partner for promoting H_2 by the green algae *Chlorella* and *Scenedesmus*. The mechanism was initiated by the bacterium growing, which consumed the oxygen in the environment, resulting in an entirely anaerobic setting that allowed for the activation of algal hydrogenase (Wang et al., 2020b). Moreover, there is potential for microbial enhancement through genetically modified strains that would tolerate more oxygen, translating into greater hydrogen production (Akhlaghi and Najafpour-Darzi, 2020).

The stage is set for the development and improvement of many of these hydrogen production technologies, with many of them still in their infancy. Table 10.5 shows a summary of the main

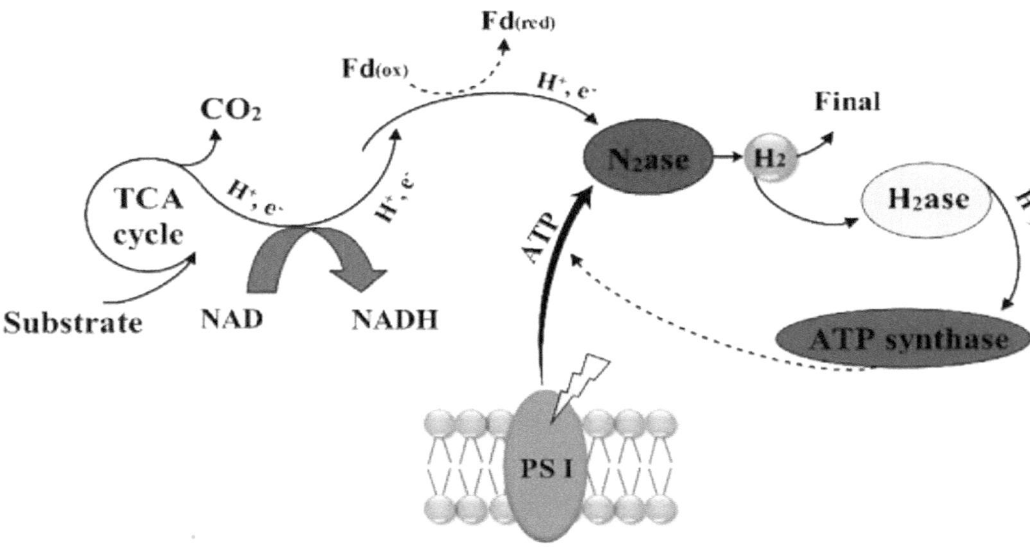

FIGURE 10.21 Photo-fermentation pathway in PNS bacteria for hydrogen production.

Source: Akhlaghi and Najafpour-Darzi (2020).

TABLE 10.5
Various Hydrogen Production Methods Along with Their Advantages, Disadvantages, Efficiency, and Cost

Hydrogen Production Method	Advantages	Disadvantages	Efficiency	Cost ($/kg)
Steam reforming	Developed technology and existing infrastructure	Produced CO, CO_2 unstable supply	74–85	2.27
Partial oxidation	Established technology	Along with H_2 production, produced heavy oils and petroleum coke	60–75	1.48
Autothermal reforming	Well-established technology and existing infrastructure	Produced CO_2 as a by-product, use of fossil fuels	60–75	1.48
Biophotolysis	Consumed CO_2, produced O_2 as a by-product, working under mild conditions	Low yields of H_2, sunlight needed, large reactor required, O_2 sensitivity, high cost of material	10–11	2.13
Dark fermentation	Simple method, H_2 produced without light, no limitation O_2, CO_2-neutral, involves waste recycling	Fatty acids elimination, low yields of H_2, low efficiency, necessity of huge volume of reactor	60–80	2.57
Photo-fermentation	Involves wastewater recycling, uses different organic wastewaters, CO_2-neutral	Low efficiency, low H_2 production rate, sunlight required, necessity of huge volume of reactor, O_2-sensitivity	0.1	2.83
Gasification	Abundant, cheap feedstock, and neutral CO_2	Fluctuating H_2 yields because of feedstock impurities, seasonal availability and formation of tar	30–40	1.77–2.05
Pyrolysis	Abundant, cheap feedstock, and CO_2-neutral	Tar formation, fluctuating H_2 amount because of feedstock impurities and seasonal availability	35–50	1.59–1.70
Thermolysis	Clean and sustainable, O_2 by-product, copious feedstock	High capital costs, elements toxicity, corrosion problems	20–45	7.98–8.40
Photolysis	O_2 as by-product, abundant feedstock, no emissions	Low efficiency, non-effective photocatalytic material, requires sunlight	0.06	8–10
Electrolysis	Established technology, zero-emission, existing infrastructure, O_2 as by-product	Storage and transportation problem	60–80	10.30

Source: Shiva Kumar and Himabindu (2019).

hydrogen production methods, their strengths, shortcomings, as well as their efficiency and cost. As we can see, there is significant variation between these techniques, which invites some players to consider some for some instances, and different ones for other contexts. For example, a country that is advanced in its energy transition can find it easier to rely on technologies like water electrolysis based on solar energy. In contrast, a less-advanced country can allocate resources to techniques like dark fermentation.

Some of these methods have added benefits, like treating wastewater or reforming carbon dioxide, previously destined to be released into the atmosphere, which would contribute to climate change. These hydrogen production methods are essential parts of a puzzle that will only be completed with all the pieces. Figure 10.22 shows an estimation for a timeline of the development of these hydrogen technologies, with a space for research advancement and lowering costs. With the start of established industrial methods, hydrogen can be produced first from fossil fuel–based sources in processes like steam methane reforming (SMR) and coal gasification within an industry that already has developed economies of scale and optimized procedures.

However, as renewable energy sources develop and costs are lowered in the medium term, thanks to the widespread implementation of clean techniques, processes like anaerobic digestion and biomass gasification can be deployed more regularly across more places worldwide. Finally, due to their prohibitive costs and the greater reliance on sources of energy that still have important intermittence challenges, most of the hydrogen production projects reviewed in the last few pages are expected to be executed at a larger scale in a longer time frame. Thus, electrolysis based on wind or solar power and photobiological and photoelectrochemical hydrogen generation from water splitting require more time and academic maturity.

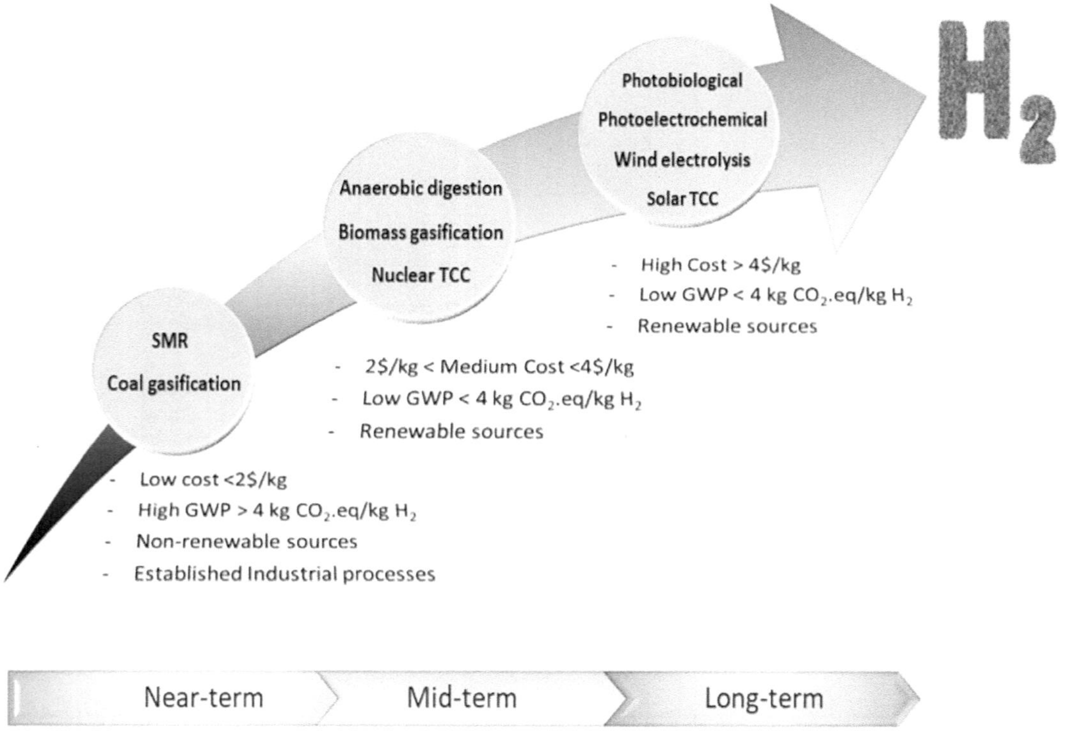

FIGURE 10.22 Pathway toward green and sustainable hydrogen production.

Source: Safari and Dincer (2020).

To be realized fully, the energy transition process requires a parallel structure of research, methods, and optimized procedures that are tested and ready to perform once our reliance on fossil fuels is phased out. The hydrogen economy is a promising avenue that can start operating concurrently, helping clean some established polluting energy processes and reutilizing natural waste and sewer water to generate hydrogen for industrial, commercial, and residential uses. Still, this alternative requires its own set of improvements and advancements, like in the storage and transportation capabilities and the optimization and efficiency departments. The following section reviews the latest advances while assessing the oil and gas industry's role in this shift.

10.4 HYDROGEN ENERGY STORAGE, TRANSPORTATION, AND EFFICIENCY

The production of hydrogen is only one part of the picture. The new energy paradigm requires an infrastructure that considers storage of the energy carrier and transportation best practices for its implementation in society, with the most optimal degree of efficiency possible. Hydrogen is more costly to store than direct energy sources, requiring both mobile and stationary systems to prosper (Abe et al., 2019). A salient example of the former is onboard automotive hydrogen storage systems, which allow cars to be driven without needing oil. Meanwhile, the latter can be exemplified in fuel cell systems that support power grids (Colbertaldo et al., 2019).

Additionally, storage infrastructure varies depending on the state of the energy carrier. Therefore, hydrogen can be stored physically as a gas or a liquid, where the former requires high-pressure tanks, while the latter needs cryogenic temperatures. And as Figure 10.23 shows, hydrogen can also be stored on the surfaces of solids, through adsorption processes, or within certain solids, through absorption (US Department of Energy, 2022d). In this section, we will review these storage methods in more detail, highlighting the advantages and challenges for each while advancing a path forward for a successful energy transition process aided by the knowledge and experience of the oil and gas sector.

10.4.1 Storage Systems

Energy storage is a key aspect of the energy supply chain because it improves the stability of grids and systems while helping increase the share of renewable energies utilized across more sectors, which reduces the environmental impact of fossil fuel–produced energy (Lewandowska-Śmierzchalska et al., 2018). Large stationary storage facilities serve for heat and power cogeneration (CHP), while mobile systems are smaller. The better-known example of these technologies are fuel cell electric

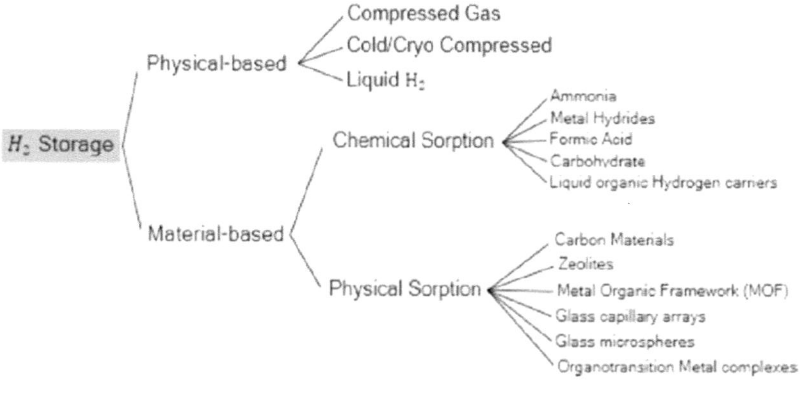

FIGURE 10.23 Hydrogen storage methods.

Source: Kovač et al. (2021).

vehicles (FCEVs), which have the highest average performance compared to conventional, hybrid, and biofuel vehicles (Kovač et al., 2021).

A key consideration for hydrogen storage systems is their ratio between gravimetric and volumetric capacity. The former concept is understood as the amount of H_2 adsorbed per unit mass, expressed as a weight percentage (wt%), while the latter refers to the volume of the tank to store hydrogen or the amount of H_2 adsorbed per unit volume (Broom et al., 2019). Ideally, better technologies will increase both their weight capacity to store more hydrogen in lighter containers and as well as the efficiency of the adsorbent to store more in the same space. However, this can prove challenging. For instance, a vehicle's hydrogen tank should not be too heavy, while offering a convenient driving range that a heavier tank can only attain.

Figure 10.24 shows the state of the art for current hydrogen storage technologies. More mature technologies like compressed, cryo-compressed, and liquefied hydrogen and chemical and complex hydrides have values between 1.5 and 6 wt% on the gravimetric scale and around 20 and 40 g/L in the volumetric capacity measure. Despite the ambitious targets set out by some players for the next couple of decades, moving forward with methods that reach greater values in either of these scales is challenging due to their various trade-offs. For example, liquid hydrogen yields sufficient volumetric capacities but shows excessive penalties for their applicability, such as long-term storage losses and cryogenic liquefaction energy requirements (Boateng and Chen, 2020).

In this section, we review the research around hydrogen storage methods, their levels of maturity and exploration, and their potential to be deployed across multiple projects. As expected, the level of development for some of these methods is not currently at an ideal stage. Thus, expectations need to be tempered as the role that hydrogen technologies could have in the short to medium term. With that said, there is great promise for the advancement of clean methods that complement fossil

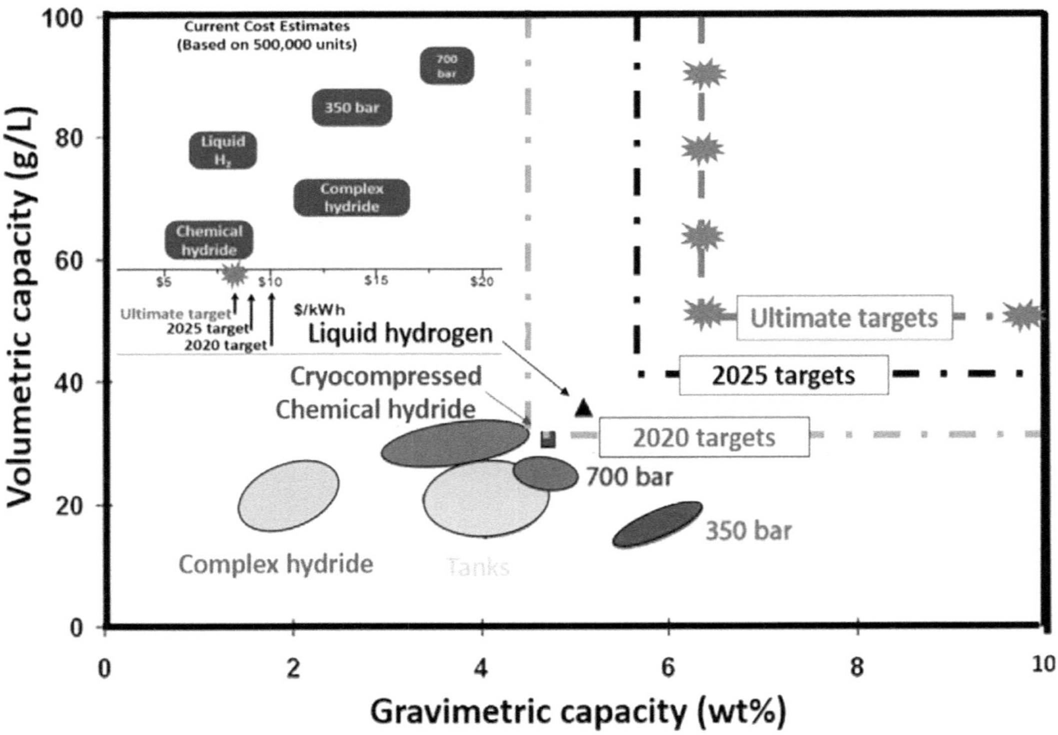

FIGURE 10.24 Status of current hydrogen storage systems.

Source: Boateng and Chen (2020).

fuel–based technologies in the next few decades, and we ought to be encouraged by the many potential opportunities for a complete replacement of devices and technologies that rely on more polluting sources of energy, even within this century.

10.4.1.1 Compressed Gas

One of the most mature and widespread technologies, the physical storage of hydrogen as a compressed gas is done for both stationary and mobile systems. For instance, the Toyota Mirai fuel cell vehicle stores hydrogen gas at a 700 bar pressure. This compression procedure is performed to achieve a lower volume, but it is an energy-demanding method due to hydrogen's lower relative density than other traditional fuels (Egeland-Eriksen et al., 2021; Elberry et al., 2021). Figure 10.25 illustrates where hydrogen is in a grid that displays storage capacity and discharge time, and its position demonstrates its promise as a long-term energy storage option.

Once compressed, hydrogen is placed inside vessels, also called tanks. Materials to make these tanks vary due to the potential of some metals for embrittlement (or the loss of material ductility to deform before it starts to tear). Therefore, recommended resources are austenitic stainless steel, aluminum, and copper alloys, which can withstand the effects of hydrogen at ambient temperatures (Elberry et al., 2021). Figure 10.26 illustrates a tank used to pressurize hydrogen, consisting of layers to insulate the compressed gas from the environment and exterior protective domes to minimize consumer risks in case of a leak.

Moreover, different types of stationary storage vessel systems include, among others, seamless vessels, multi-functional steel layered (MSLV), steel–concrete composite, wind turbine towers, and those using natural gas storage facilities. The first type is made from high-strength seamless tubes, mainly used in hydrogen fuel stations. Then, to tackle cost and embrittlement issues, MSLVs combine more materials that withstand prolonged interactions with the gas, while steel–concrete composite vessels help lower the thickness of shell layers. Lastly, crossover pressures generated by wind turbines can help store hydrogen in their tubular towers, while reusing metal pressure vessels manufactured for natural gas, such as pipe storage facilities, gas holders, and spherical pressure vessels, has proved to be a fruitful endeavor (Elberry et al., 2021).

Another way to store hydrogen as a compressed gas involves using geological sites, such as aquifers, salt or rock caverns, and depleted reservoirs or abandoned mines. According to Tarkowski

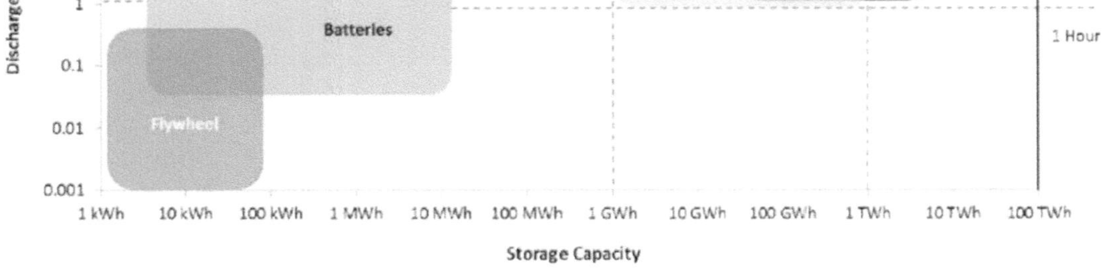

FIGURE 10.25 Comparison of storage capacity and discharge time for various energy storage technologies. **Source: Egeland-Eriksen et al. (2021).**

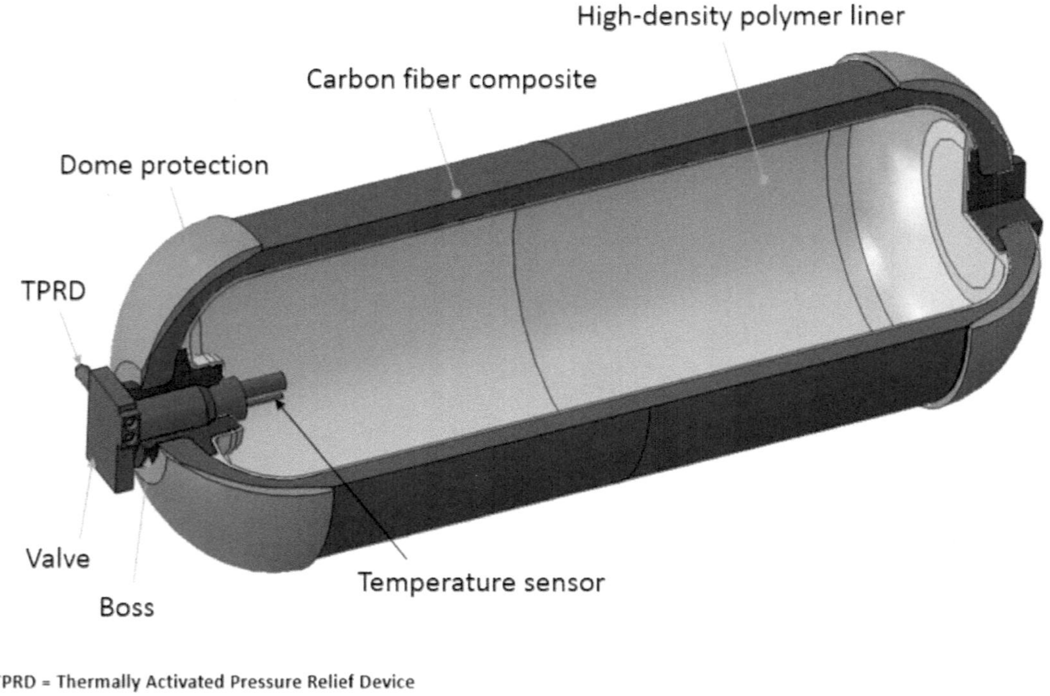

FIGURE 10.26 Components of a pressurized hydrogen storage tank.

Source: US Department of Energy (2022e).

(2019), an underground hydrogen facility works similarly to a natural gas installation, storing a gas several hundred meters below surface level. The stored gas consists of the working gas, injected and later withdrawn for its use, and the cushion gas, stored to be kept for the duration of the facility's life. The role of this cushion gas is to exercise a minimum amount of pressure needed to avoid the intrusion of water and provide the best conditions for injection.

Despite the promise of underground storage techniques, these methods still have a few drawbacks, like the technical limitations brought by a lack of equipment to deal exclusively with hydrogen technologies, as well as the potential problems of adaptation of existing boreholes, which continue to be an underexplored factor (Yue et al., 2021). However, this opens the door for exploring other physical hydrogen storage technologies in more depth. One of them is the storage of liquid forms of the element under extremely low temperatures, in what is known as cryogenic hydrogen.

10.4.1.2 Cryogenic Hydrogen

With volumetric values slightly superior to solid hydrogen, storage of this element in liquid form can be done at very low temperatures, around 20 K (or below −250°C) and ambient pressure (Moradi and Groth, 2019; Yue et al., 2021). However, these methods have the drawback of high combustion of energy stored in the hydrogen, going from 30 to 40% energy loss (Yanxing et al., 2019), while also being more time-consuming than solid hydrogen due to the ordeals of the liquefying process. These drawbacks have narrowed the field of utilization of liquid hydrogen to large-scale endeavors like hydrogen shipping projects and space exploration, where low-temperature rockets have begun to use liquid hydrogen as a propellant (Qiu et al., 2021).

Moreover, a subset of these methods is known as cryo-compression, occurring when hydrogen is stored in a vessel that can be pressurized (generally to 250–350 atm, or 25–35 MPa) and hydrogen is in a supercritical state. Figure 10.27 shows a schematic of the cryo-compressed hydrogen process, where

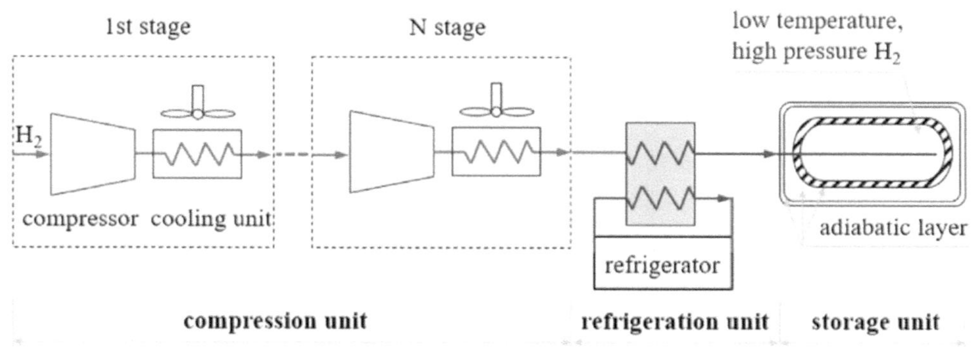

FIGURE 10.27 Schematics of the cryo-compressed hydrogen storage.

Source: Yanxing et al. (2019).

the element is compressed multiple times, refrigerated, and ultimately stored in specially designed containers. In the compression stage, the hydrogen produced is highly pressurized yet still near room temperature, but due to hydrogen's boiling point at 1 atm pressure being –253°C, a refrigerant like liquid nitrogen is used to cool the hydrogen so it liquefies (Osman et al., 2021; Yanxing et al., 2019).

Because of their support for space exploration, these liquid hydrogen technologies have also been advanced to aid the transportation sector. Thus, research has been done into developing fuel cells that use liquid nitrogen to power vehicles like trucks, buses, or smaller cars. Auto makers like Honda, Toyota, and Hyundai have started investing in and developing fuel cell vehicles (FCVs), making their new models available in mature markets like North America, Asia, and Europe. Compared to battery electric vehicles (BEVs), FCVs have a refueling style that mirrors that of internal combustion engine (ICE) vehicles, where the user refuels the tank and is ready to go. In contrast, electric batteries must be recharged for at least several minutes (Manoharan et al., 2019). Meanwhile, liquid hydrogen can be a very appealing choice for heavy-duty fuel cell trucks (Wang et al., 2020a).

We will further explore hydrogen fuel cells and vehicles momentarily, but it is important to note that the high promise of these technologies continues to be counterbalanced by increased costs. First, since it is an artificially manufactured fuel, it costs about three times more than petroleum refining. Then, because of hydrogen's low density, it cannot be stored as easily as fossil fuels can (Manoharan et al., 2019). Third, and as mentioned previously, cryogenic losses from production, storage, transportation, and handling are hefty, particularly in the case of liquefied hydrogen (Al Ghafri et al., 2022). And lastly, infrastructure is still behind more optimized and established methods. Figure 10.28 shows a diagram for a liquid hydrogen (LH_2) storage system used in a heavy-duty truck, which still needs significant modifications and optimizations to be released to the market.

The potential of cryogenic hydrogen to overcome the current dominance of fossil fuel–based devices is encouraging. It should not surprise anybody when we are surrounded by hydrogen-run vehicles in a couple of generations and hydrogen production and storage plants provide several hundreds of thousands of jobs across many countries. With that said, research needs to improve on storage techniques and devices. Currently, project development is limited to certain areas, and testing of products remains an early adapter's game and an adventure reserved for more developed countries. The next couple of decades will be crucial to begin closing those gaps and building those bridges to export these technologies to many more countries so that a real difference in the reduction of global temperatures can be achieved.

10.4.1.3 Material-Based Storage

Meanwhile, the handling of solid hydrogen appears to be the more suitable option to deal with the current challenges related to storage and transportation, despite trailing compressed gas and LH_2 in

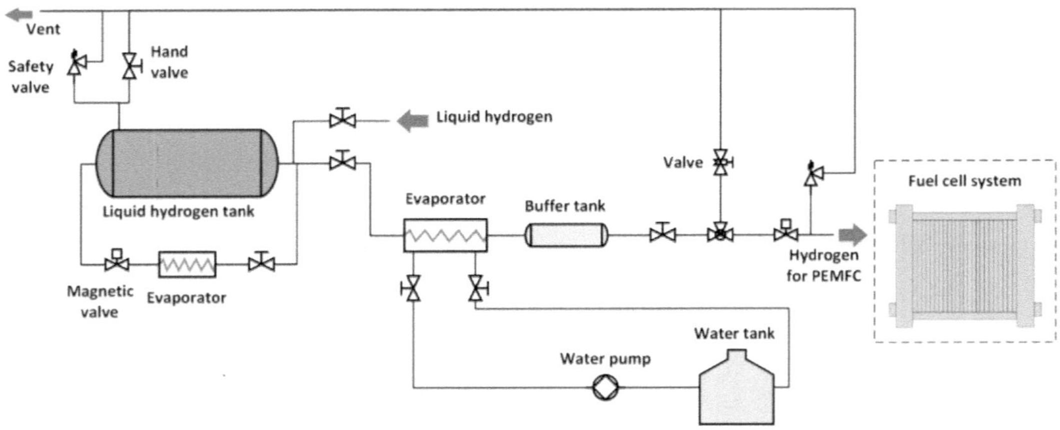

FIGURE 10.28 Diagram of liquid hydrogen system.

Source: Wang et al. (2020a).

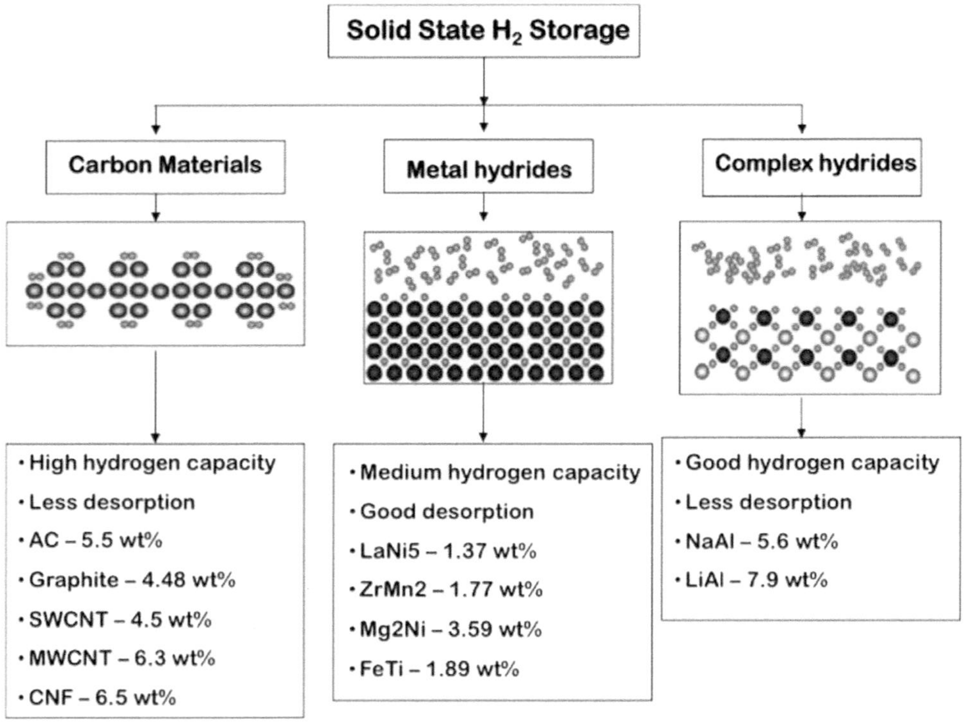

FIGURE 10.29 Different solid-state hydrogen storage options.

Source: Mohan et al. (2019).

research maturity. A series of nanomaterials, like complex chemical hydrides, clathrates, amides, and zeolites, can serve as suitable surfaces for the physical storage of hydrogen in the solid state. Meanwhile, hydrogen atoms and molecules can attach to metal hydrides in what is known as chemical sorption techniques (Boateng and Chen, 2020; Mohan et al., 2019). Figure 10.29 shows different methods of sorption based on the material used to store hydrogen molecules, while Table 10.6 compares the storage capacities of the various technologies we have reviewed thus far.

TABLE 10.6
Storage Capacities for Different Hydrogen Storage Materials

Storage Method	Materials	Storage Capacity; p_m (wt%)	P_{eq}, T
Compressed gas	Energy for compression: ≈4 kcal mol^{-1}	13	140 bar, 298 K
Cryogenic liquid	Energy for liquefaction: ≈7 kcal mol^{-1}	Size-dependent (e.g., ≈5 wt% for a can tank)	1 bar, 21 K
Adsorption	Activated carbon	5.5	80 bar, 298 K
	Graphite	4.48	100 bar, 298 K
	SWCNT	4.5	4 bar, 298 K
	MWCNT	6.3	148 bar, 298 K
	CNF	6.5	120 bar, 300 K
Absorption	AB_5 type: $LaNi_5$	1.37	2 bar, 298 K
	AB_2 type: $ZrMn_2$	1.77	0.001 bar, 298 K
	A_2B type: Mg_2Ni	3.59	1 bar, 555 K
	AB type: FeTi	1.89	5 bar, 303 K
	NaAl	5.6	493 K
	LiAl	7.9	453 K

Abbreviations: CNF, carbon nanofiber; *MWCNT*, multi-walled carbon nanotubes; *SWCNT*, single-walled carbon nanotubes.
Source: Mohan et al. (2019).

In material-based hydrogen storage, the two main techniques used are physisorption and chemisorption. The former method occurs when hydrogen molecules are attached to the surface of nanomaterials, such as carbon nanotubes and metal–organic frameworks (MOFs). Meanwhile, chemisorption happens after hydrogen molecule absorption into the host material's lattice, including metal hydrides and liquid organic carriers (Singh et al., 2020; Yanxing et al., 2019). A promising alternative that has gained momentum in the past few years is the process of adsorption on activated carbons (ACs). Ramirez-Vidal et al. (2021) studied six different commercial ACs with varying pore volumes, concluding that those with higher pore volumes and narrower micropores should be chosen for storage, while those with wider micropores and lower mesopore volumes should be used for hydrogen compression.

In turn, absorption relies on complex and chemical hydrides to store hydrogen. Some examples of the group of complex metal hydrides are alanates and borates of lithium, sodium, and potassium, like Mg_2NiH_4, $LiAlH_4$, and $NaBH_4$. On the other hand, examples of chemical hydrides are LiH, NaH, and CaH_2. Absorption methods were advanced through research on palladium (Pd), which can absorb 900 times its volume of hydrogen at room temperature and atmospheric pressure (Mohan et al., 2019; Yue et al., 2021). Moreover, contemporaneous research strongly supports magnesium-absorbent materials (Jalil et al., 2018) and an MH alloy activated with $La_{0.9}Ce_{0.1}Ni_5$ (Jenne et al., 2020).

These material-based technologies are still behind those using compressed gas and cryogenic liquid hydrogen in research and development. For example, advancements in the use of carbon nanotubes (CNTs) for hydrogen adsorption have shown inconsistent results over time, and the work of Lyu et al. (2020) indicates that there are still too many imperfect measurement methods that generate significant discrepancies between theoretical and experimental results. These differences currently do not allow for a full-scale implementation of adsorption and absorption technologies, especially considering that their costs are already prohibitive for many actors that will prefer to fall back on cheaper methods.

The challenge is thus established, and the advent of improved computer simulations and *in silico* research will open the door to more optimized processes and cheaper routes to execute them. As Figure 10.30 summarizes, improved hydrogen storage methods have an ever-increasing array of

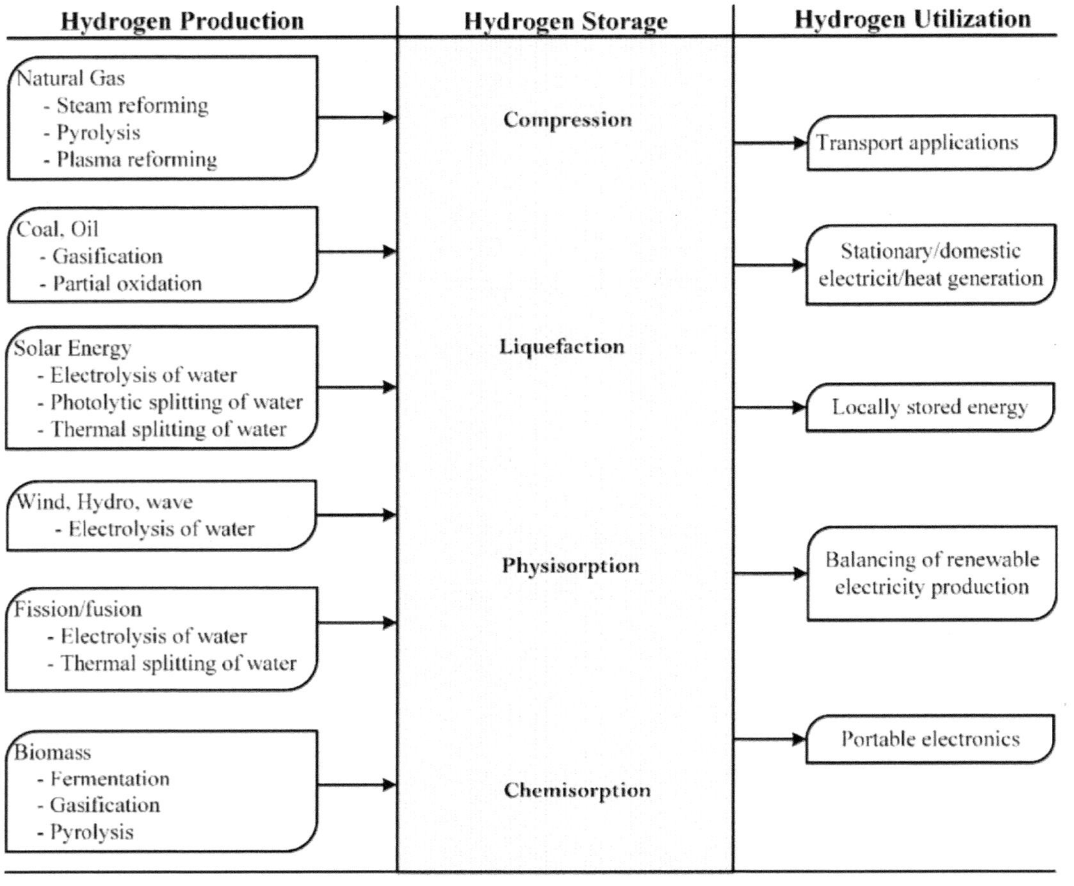

FIGURE 10.30 Hydrogen vector: sources, generation options, storage options, and end uses.

Source: Abdin et al. (2020).

areas of application in the daily lives of millions of people, from local electricity generation to portable devices, to mass transportation initiatives, to a potential contribution to the national electricity grid within a country. So like most of the applications and methods reviewed throughout this work, their significant contributions can first complement a fossil fuel–based economy, with the express desire to eventually transition to these cleaner forms of energy as the dominant form of powering the world.

The following section reviews how these technological advancements have yielded positive results in similar areas, particularly with compression and hydrogen transportation and the chemical reactions needed to operate optimally. Further, we explore the role of the oil and gas industry in developing and funding these technologies and how the sector can position itself as a protagonist in implementing a hydrogen economy. The destination we are headed toward appears promising, but engineers and researchers need to stop at every stop so that the stay is safe and, equally important, permanent.

10.4.2 Hydride Hydrogen Compressors

In the quest to advance compression technologies for hydrogen storage, hydride compressors aim to increase the density of hydrogen energy storage. They are based on the reversible interaction of metals of alloys with hydrogen gas, first absorbing hydrogen at low pressures and temperatures (lower than 10 bar and at most 30°C), to later desorb it at a higher pressure after raising the temperature of the tank through an external source of heat. Figure 10.31 illustrates this two-step process that displays the operating principle of these devices, which is the thermal hydrogen compression system (Stamatakis et al., 2018; Tarasov et al., 2020).

The advantages of metal hydride (MH) compressors are their simpler design than mechanical compressors, potentially lower costs, and a lower likelihood of leakage and contamination (Tarasov et al., 2020). Moreover, Gray (2021) reviewed the application of interstitial MHs, which can absorb and desorb hydrogen without many variations in pressure, which can help with the efficiency of the design. Lastly, Galvis et al. (2018) experimented with a three-stage compression mechanism that simulates thermodynamic features of hydride formation and decomposition with promising results.

10.4.3 Hydride Heat Pumps

Meanwhile, heat pumps have a similar mechanism of absorption and desorption in metal hydrides, with the difference that these devices rely on exothermic and endothermic reactions to operate. Additionally, they use waste heat as their energy source and have a lower maintenance cost than vapor compression heat pumps (Shafiee, 2020). Figure 10.32 shows a schematic of a hydrogen compression system that uses a metal alloy to absorb hydrogen into its structure at low temperatures and pressure levels. It also includes a cooler to avoid problems with high temperatures in the tanks.

These devices are promising avenues to increase the use of hydrogen and are also closer to everyday users. Heat pumps' goal is to find a cheap source of heat and cold that can benefit the residential and commercial sectors. Additionally, metal hydride heat pumps operate on many different energy resources and lack any large mechanical components, which translates into less noise and vibrations, potentially leading to fewer maintenance problems (Shafiee, 2020). However, we still have to deal with the potentially prohibitive costs of metal hydrides (Krane et al., 2022). And finally,

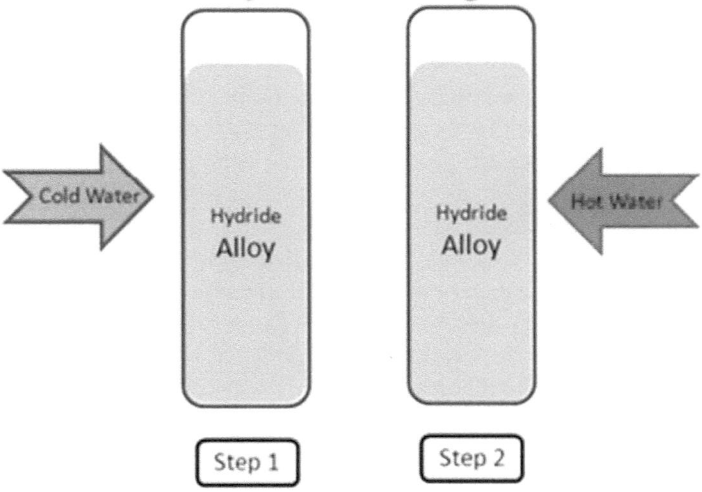

FIGURE 10.31 Metal hydrides–based compression principle.

Source: Stamatakis et al. (2018).

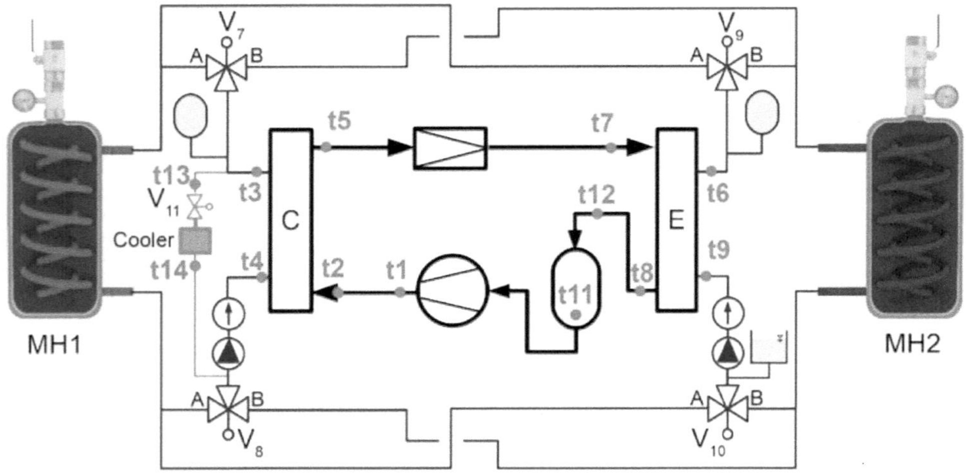

FIGURE 10.32 Diagram of a heat pump with a water circuit.

Source: Brestovič et al. (2022).

simulations performed in the literature have improved markedly over the years, now closely matching numerical results obtained by estimations and modeling (Aadhithiyan et al., 2022).

10.4.4 Transportation

Along with storage, transportation is another essential part of the hydrogen chain. If the energy supply is interrupted frequently, fewer actors will look forward to transitioning away from the sources that have kept a relatively stable and uninterrupted energy system for decades. Hydrogen-based energy must still deal with challenges related to its gravimetric density, which requires compression to be economically viable, and the inability of batteries to store electricity for extended periods (Aakko-Saksa et al., 2018). Moreover, transport and storage may be exposed to risks, like a higher-than-desired flammability hazard and low ignition energy (Wan et al., 2021).

One alternative for storage and transportation that has gained increased attention in the past few years is that of liquid organic hydrogen carriers (LOHC). Due to the higher costs of pressurized hydrogen and its safety concerns, the preferred method of transportation is through pipelines, another costly and cumbersome method. To tackle this, LOHC consists of a two-step process where hydrogen is loaded onto a carrier in a catalytic hydrogenation reaction and later unloaded in a catalytic dehydrogenation reaction that produces gaseous hydrogen (Southall and Lukashuk, 2022). This method of storage and transport allows for long-term handling without bio-off or other hydrogen losses we have touched on earlier (Niermann et al., 2019).

Figure 10.33 shows the LOHC concept in more detail, including the two steps performed between the point of production and destination or the point of utilization of the hydrogen. In the first step in the process, the volumetric energy density of the hydrogen is increased significantly, which in turn allows for safer handling of the carrier, now being compared to the handling of any other liquid already known in the industry (Niermann et al., 2021). Meanwhile, the second step of dehydrogenation allows for the release of hydrogen to be used in gaseous form in multiple cycles. Just as hydrogenation, this second process is also facilitated by a catalyst that most commonly is one from the platinum group metal (PGM), like platinum, palladium, or a ruthenium-based substance (Southall and Lukashuk, 2022).

Moreover, because of its high gravimetric density, ammonia can facilitate the transport and storage of hydrogen, contributing significantly to a hydrogen-based energy system (Makepeace

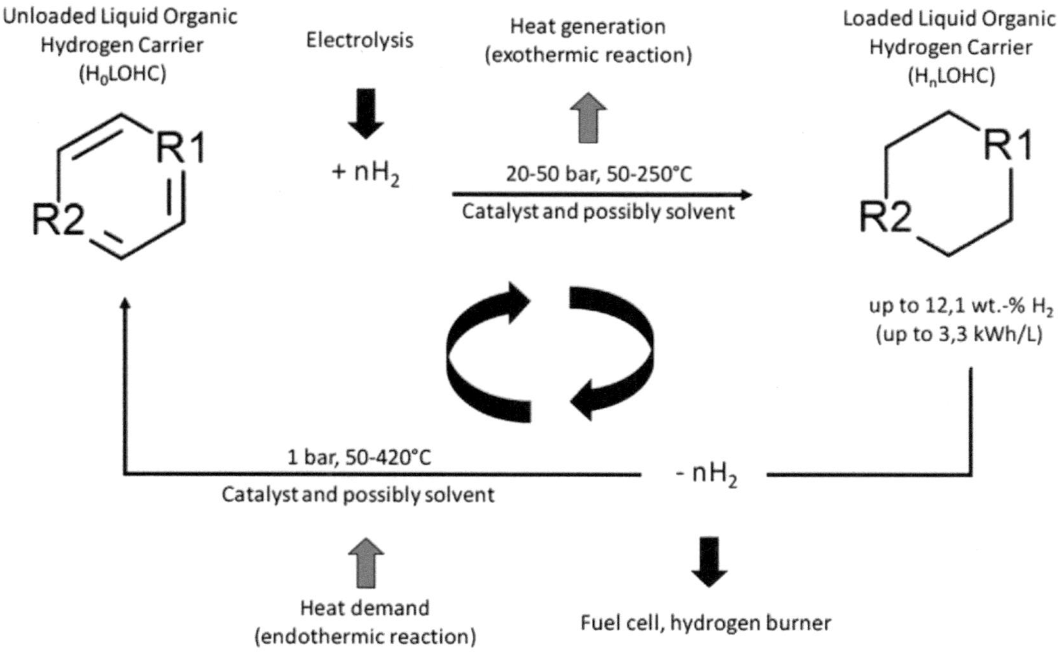

FIGURE 10.33 Concept of the liquid organic hydrogen carrier (LOHC) storage.

Source: Niermann et al. (2019).

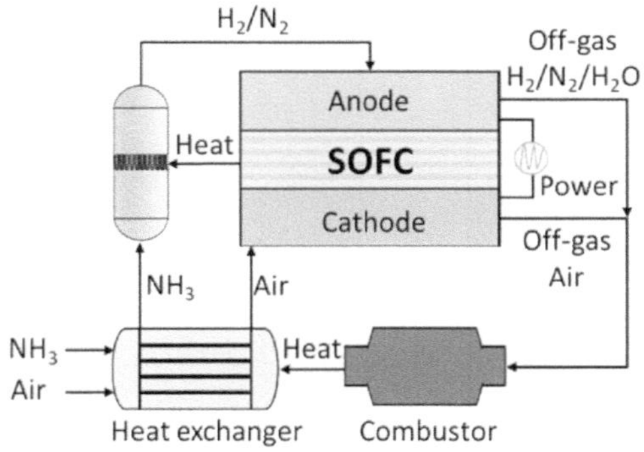

FIGURE 10.34 Schematic of ammonia decomposition reactor integrated with SOFCs with combined heat and power.

Source: Wan et al. (2021).

et al., 2019). With around 18 wt% of hydrogen and no carbon, this synthetic chemical can later be decomposed into a gas mixture of 75% H_2 and 25% N_2 that generates zero carbon emissions (see Figure 10.34). Additionally, ammonia has a boiling point of –33°C at standard pressure, much lower than the –253°C of hydrogen and the –160°C of methane, and it can also be compressed to liquid at a pressure of 10 bar. Lastly, ammonia is non-flammable, and it has general risk levels comparable to those of gasoline, liquefied petroleum gas (LPG), compressed natural gas (CNG), and methanol (Wan et al., 2021).

Along with these developments, pipelines and distribution logistics must be improved to lower costs and optimize the necessary processes to deliver hydrogen quickly and efficiently to increasing numbers of people worldwide. For this, the technical brief performed by Khan et al. (2021) presents the challenges that building a hydrogen pipeline infrastructure will bring. For example, a main consideration is repurposing pipelines with lower-strength steel, which is less prone to embrittlement. Then, safety risks are more significant than those of natural gas due to hydrogen's extensive flammability range in air, the small amount of energy required for ignition, and the invisibility of the flame in case it ignites. The following section explores these challenges in more detail and details the latest advances and expectations for the next few years and decades.

10.4.5 Efficiency, Challenges, Latest Advances

We are currently in a privileged position to witness the advancement, development, and perfecting of many methods and technologies based on hydrogen through the path initiated by the energy transition process of the 21st century. The current climate situation does give this entire push some urgency, so hydrogen technologies can start contributing quickly as complements within a fossil fuel–based energy system. One example is the potential rethinking of the role of natural gas (NG) infrastructure, eventually repurposing for hydrogen use. However, the review performed by Ogden et al. (2018) does not find this option – or an alternative, where H_2 is blended into the NG system – very attractive.

As Figure 10.35 shows, several potential avenues for the implementation of hydrogen are explored, which include scenarios where H_2 production is done on-site to avoid transportation issues, liquid hydrogen trucks move from the plant to the dispenser site, and where a blend between hydrogen and natural gas is offered to the final user. The technology has advanced to a point where hydrogen transportation has been optimized. Yet current challenges for net-zero and fuel cell vehicles require considerable investment in exclusive and dedicated energy systems that put hydrogen in the center and not as an accessory.

Over time, this evolution can lead us to a system where hydrogen leads the way, turning itself into the core of a smart energy production center, being relied upon for various energy needs. This

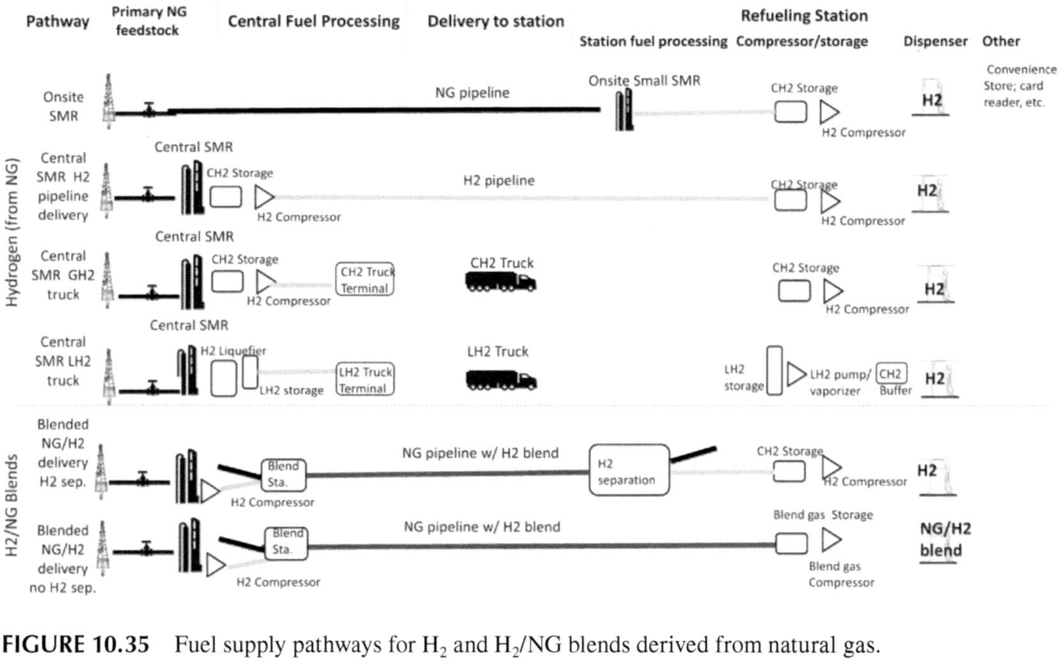

FIGURE 10.35 Fuel supply pathways for H_2 and H_2/NG blends derived from natural gas.

Source: Ogden et al. (2018).

dedicated system requires significant investment to develop an infrastructure ready to quickly convert hydrogen to tailor it to the needs of the sector or population which requires it. Figure 10.36 illustrates this poly-generation system, where renewable energies that are not immediately used for electricity or other urgent needs are collected, converted, and stored as hydrogen, which is later dispensed on demand to serve the needs of industry, commerce, residential power or heat, and transportation (Genovese and Fragiacomo, 2022).

Further, while the leading technologies and methods continue their optimization and adaptation processes, new hydrogen techniques have been developed. One of them is seen in the work of David et al. (2019), who explored advances in alkaline water electrolyzers, particularly the progress of a cell using anion exchange membrane (AEM) technology. These cells work like proton exchange membrane (PEM) electrolyzers, but instead of H^+ protons, OH^- anions are transported along the membrane. AEM methods are generally cheaper than PEM cells, while the purity of gases produced at high pressure also appears to be improved.

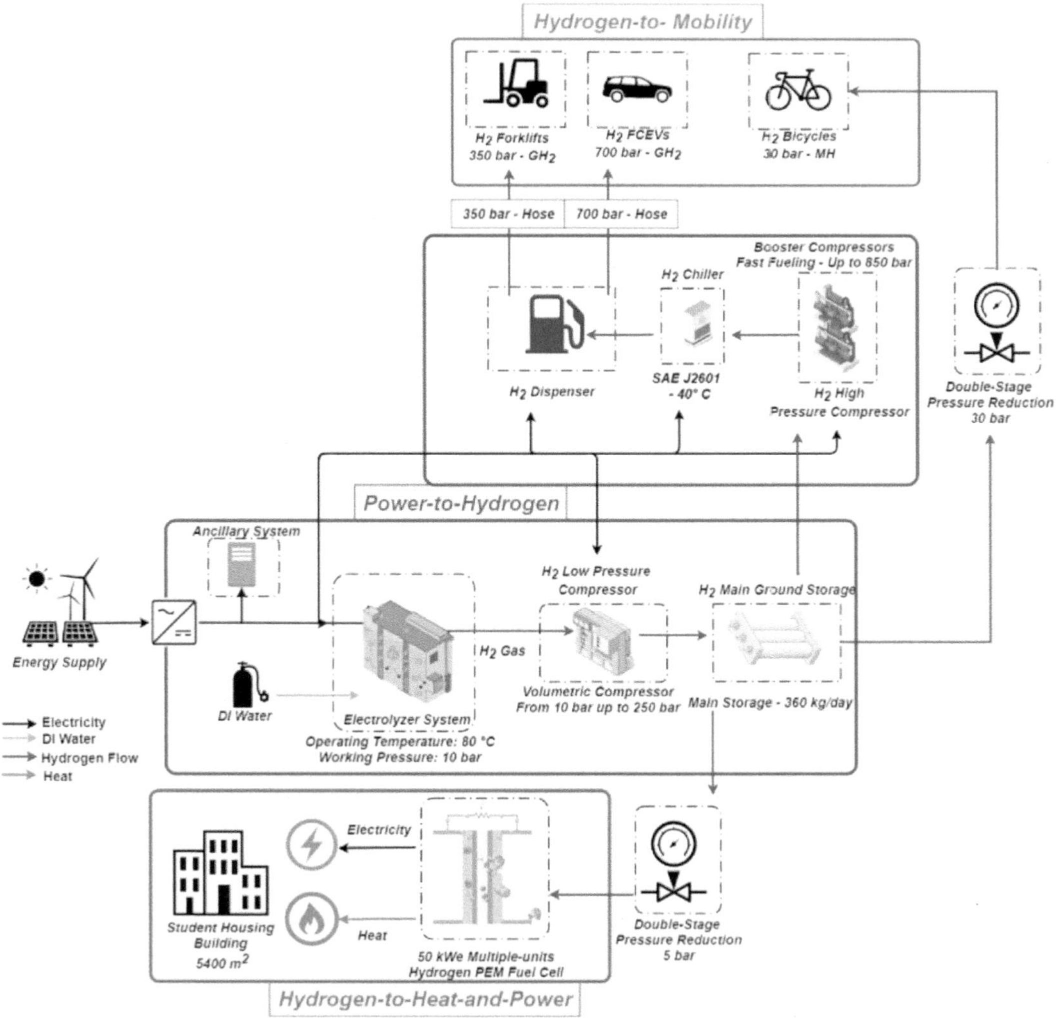

FIGURE 10.36 Hydrogen station layout for a poly-generation system.

Source: Genovese and Fragiacomo (2022).

Moreover, biomass is a productive source to produce and convert hydrogen, and we reviewed most of the techniques in these pages. However, Pandey et al. (2019) identified a few other ways of utilizing biomass, as reflected in Figure 10.37. Generally, methods have evolved because they offer advantages over techniques that have become obsolete or only show a limited set of benefits. The most common challenges for these hydrogen technologies are their production efficiency, potential pitfalls as a zero-emission method, and cost. Current technologies need to face significant trade-offs between these challenges, which means that some more efficient methods will perhaps be more polluting.

The state of the art for hydrogen technologies is in constant flux and evolution, one that has now enabled a close relationship with the organizations and institutions that fund these endeavors. In this regard, the role that oil and gas companies play in the advancement of these technologies, as well as in the raising of awareness with the public, is vital. The following section centers on the predominant role that oil majors and other prominent actors, like international organizations and governments, have played in the flourishing of research and promising technologies. As advanced previously, a close and fruitful relationship between private and public actors is necessary for the energy transition to succeed.

10.5 HYDROGEN ENERGY AND FUEL CELLS APPLICATIONS IN THE OIL AND GAS INDUSTRY

In the quest for mitigating the potential climate damage caused by constant energy processes that rely on fossil fuels, these decades have developed a proactive symbiosis between actors searching for cleaner ways to maintain energy independence and prosperity. Sharing similar goals, governments have sought help from international organizations that set global standards

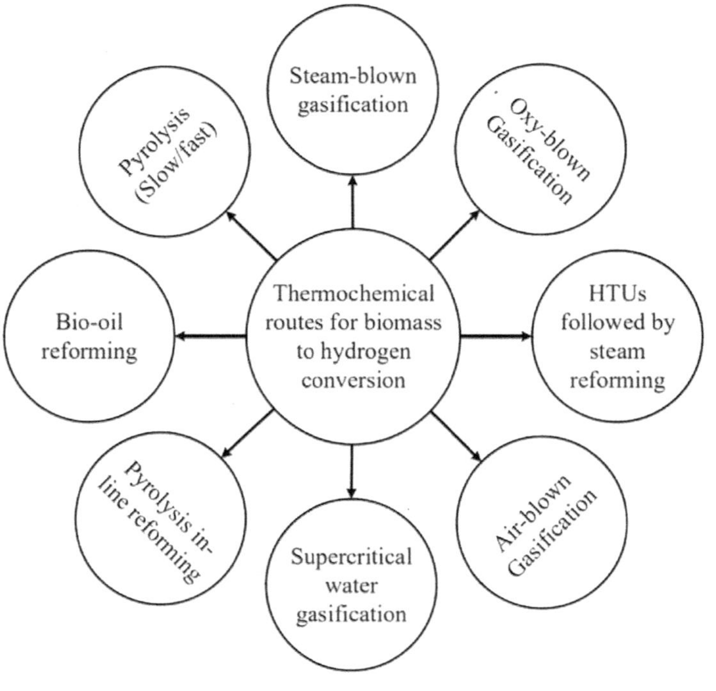

FIGURE 10.37 Possible biomass conversion processes.

Source: Pandey et al. (2019).

for pollution and mitigation while relying on private actors that fund and facilitate research on cleaner energy sources. This way, many more sectors, and industries, today use alternatives to fossil fuels. As Table 10.7 shows, hydrogen technologies now support many industries, like electronic and plastics production and the development of chemicals, metals, glass, and even food products.

Of particular interest is the penetration of hydrogen methods in the oil industry, where these techniques have helped improve many current processes that rely on fossil fuels. This switch has changed the energy source for these procedures, lowering some of the environmental costs associated with them while maintaining the benefits of electricity, heat, and cold for millions of people worldwide. Thus, an encouraging first step toward the energy transition to renewables is the successful integration of newer, cleaner hydrogen technologies into current procedures within the oil and gas industry.

Al-Khori et al. (2019) explored this integration by developing solid oxide fuel cells (SOFCs) added to oil and gas operations to improve system efficiencies and reduce their emissions. These fuel cells are not only relegated to industrial uses, however, as they can also be deployed across many other areas, satisfying residential energy needs and expanding the market in the transportation sector. In this section, we review these developments and their applications, with a particular interest in the advancement of fuel cells for vehicles and the role of the oil and gas industry in the widespread implementation of these technologies. Hydrogen has a bright future, and the strategic collaboration of actors from all persuasions will help clear the hurdles to establish clean technologies as the standard moving forward.

TABLE 10.7
Uses of Hydrogen in Different Industries

Types of Industry	Uses
Oil	• Removal of sulfur and other impurities • Hydrocracking of large hydrocarbons to fuel distillates
Chemical	• Syntheses of ammonia, methanol, etc.
Food	• Conversion of sugars to polyols • Conversion of edible oils to fats • Conversion of tallow and grease to animal feed
Plastics	• Syntheses of nylons, polyurethanes, polyesters, polyolefin • Cracking used plastics to produce lighter molecules that can be recycled
Metals	• O_2 scavenger • Reductive atmosphere for production of iron, magnesium, molybdenum, etc. • Welding torches • Heat treatment to improve ductility and machining quality, to relieve stress, to harden, to increase tensile strength, changing magnetic or electrical characteristics
Electronics	• "Epitaxial" growth of polysilicon • Manufacture of vacuum tubes • Heat bonding materials
Glass	• High-temperature cutting torches • Glass polishing • Heat treatment of optical fibers • Reductive atmosphere for float glass process
Electric power	• Coolant for large generators of motors • Nuclear fuel processing

Source: **Abdalla et al. (2018).**

10.5.1 Current Applications and Future Prospects

Oil majors are confronted with a significant conundrum. On the one hand, their technologies have helped power the needs of increasing numbers of people, but on the other, relying exclusively on fossil fuels to develop these technologies has contributed to the warming of the climate. Because of this, companies are met with the challenge of promoting methods that could potentially take them out of business. However, the potential of these technologies to reshape the world energy scenario is an enticing opportunity that oil majors have explored through investment, research, and innovation. Thus, leading the way in this endeavor will not take them out of business but make them pioneers in the quest for a cleaner energy world.

The current interconnection of global actors in climate cooperation issues has led to a standard set of rules and procedures, understood as international energy governance (Van de Graaf et al., 2020). Through the joint work of governments, stakeholders, and international organizations, efforts are being made to achieve the goals and pledges made in accords like the Paris Agreement or the recently held COP26. Here, corporations like big oil and gas players can help move the needle toward implementing national and regional rules that allow for the deployment of hydrogen technologies on a massive scale through funding, research, and development.

This expansion toward a hydrogen economy has many points of ignition, and one of them has been the promising advance in fuel cell technologies, particularly for vehicles like trucks, cars, and mass transportation buses. Manoharan et al. (2019) have explored the trade-offs in the fuel cell vehicles (FCVs) sector, where selecting an onboard tank is subject to significant considerations. For example, a storage tank for pressurized hydrogen requires impact-resistant materials, mostly made from cylinders wrapped in carbon fiber. In turn, liquid hydrogen requires high refrigeration and risks losing storage volume due to boiling. Thus, solving issues in these areas potentially increases the weight and size of the tank, which then affects efficiency, mileage, and other measures of competitiveness compared to conventional vehicles.

So a vehicle prototype like the one featured in Figure 10.38 has gone through extensive research and trial and error. From the basic structure of a fuel cell–based vehicle (seen in Figure 10.39), engineers need to consider size and weight limitations, which then affect the vehicle's mileage. Vessel

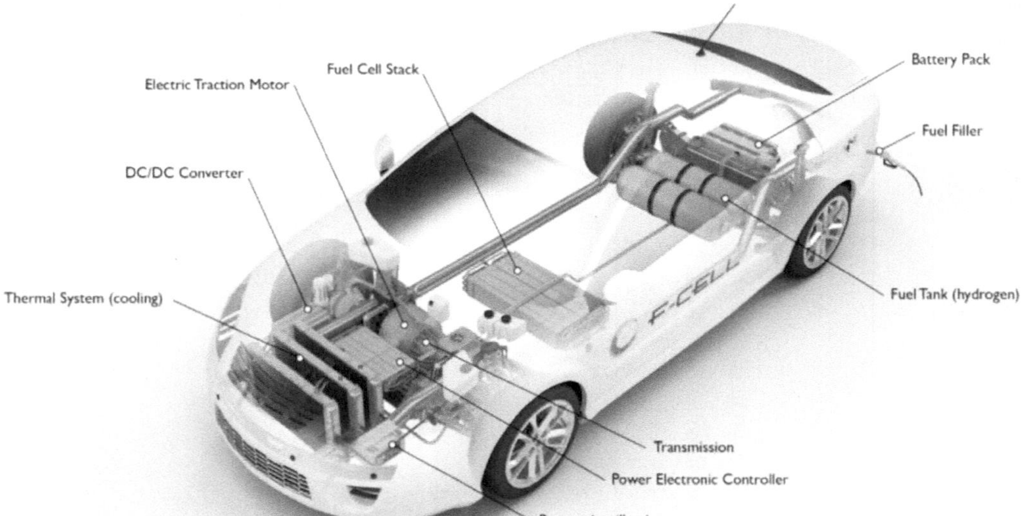

FIGURE 10.38 Fuel cell vehicle with onboard storage.

Source: Manoharan et al. (2019).

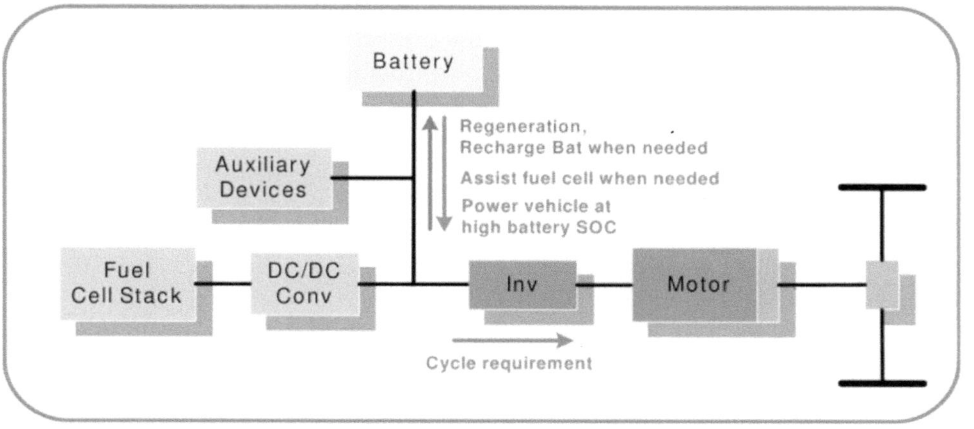

FIGURE 10.39 Basic structure of a fuel cell–based electric vehicle.

Source: Manoharan et al. (2019).

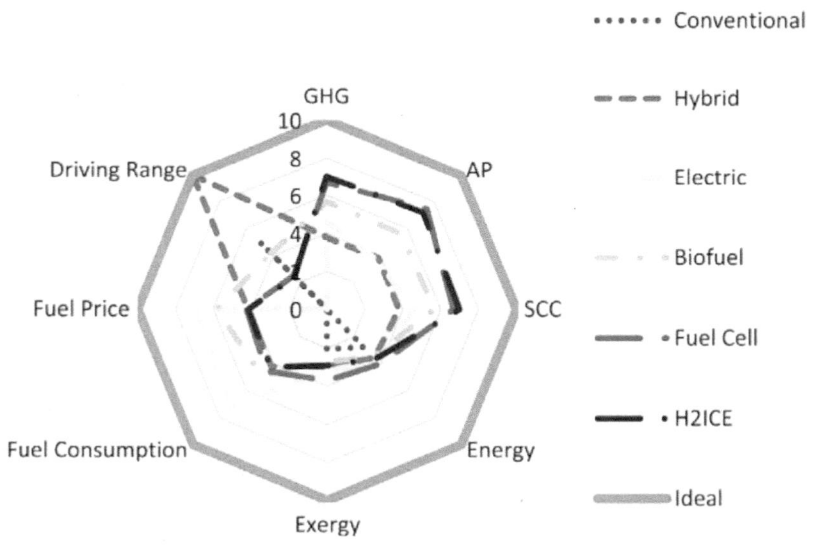

FIGURE 10.40 Normalized rankings for different types of vehicles.

Source: Acar and Dincer (2020).

features are a critical variable in the development of a competitive vehicle. Research has found that other success variables, particularly for cryo-compressed hydrogen vehicles, are the liquid hydrogen (LH_2) pump performance, the heat transfer from the environment, the conversion between the spin isomers para-H_2 and ortho-H_2, and even the users' driving habits (Moreno-Blanco et al., 2019).

These significant challenges could help explain the delay in deploying hydrogen technologies on a larger scale. Additionally, high costs in retrofitting and developing a dedicated infrastructure are partly why some countries have yet to commit to developing policies to deploy hydrogen technologies. However, an expansion of the automotive market may be a very promising first step toward the massification of these methods. When comparing hydrogen-based with conventional vehicles, Acar and Dincer (2020) found that hydrogen fuel cells and hydrogen internal combustion engine (H2ICE) vehicles behave in very competitive ways to gas-powered, hybrid, and electric models across many dimensions (see Figure 10.40).

Also, fuel cell–powered vehicles have the potential to become cheaper than fuel-powered automobiles after a critical mass starts using them, which would benefit more people on a larger time horizon, thanks to economies of scale (Tanç et al., 2019). As fuel cells are devices that directly convert chemical energy to electrical energy without carbon emissions, they are considered vital for the energy transition. However, as Abdelkareem et al. (2021) warn us, their environmental impact may not be seen at the tailpipe but at these technologies' construction and manufacturing stage. Like solar panels and wind turbines, using some precious metals like platinum makes these technologies less clean than advertised and cost-prohibitive for prospective adapters. Refueling stations and the necessary infrastructure to seamlessly include hydrogen vehicles are another high up-front cost we will discuss shortly.

Recognizing the seriousness of the environmental situation while attempting to strike a balance between building entirely new infrastructure and sustaining the present status quo, the oil and gas industry has an almost clear path toward innovation and lowering emissions. One such path concerns the incorporation of renewable energies into their production methods, particularly of the hydrogen required for crude oil upgrader facilities. Likkasit et al. (2018) explored three different methods of solar-aided hydrogen production and compared them to the conventional steam methane reforming (CSMR) method from natural gas. Findings open the door to cleaner and more efficient hydrogen production processes (see Figure 10.41).

Due to factors like market adaptation, conventional hydrogen production processes based on fossil fuel combustion are currently the least-expensive. However, their greenhouse gas (GHG)

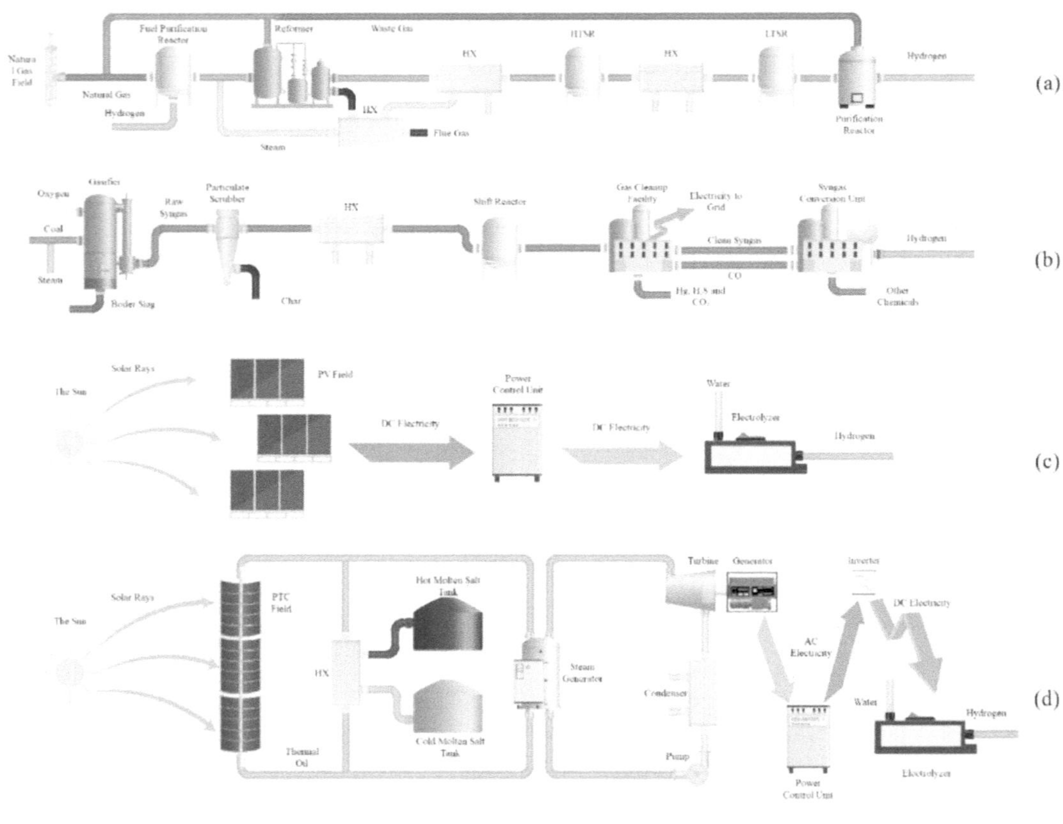

FIGURE 10.41 Various hydrogen production pathways: (a) steam methane reforming; (b) coal gasification; (c) photovoltaic water electrolysis; (d) solar thermal water electrolysis.

Source: Sadeghi et al. (2020).

emissions are relatively high compared to newer renewable energy techniques. By switching to cleaner processes for ordinary tasks, the oil and gas industry can set an example as a proactive sector that encourages investment and innovation. This can be turned into a dynamic, virtuous circle when governments develop funding schemes and international organizations keep countries to their environmental pledges.

The following section reviews the close collaboration required to advance the energy transition process, which needs to occur across sectors and levels. First, governments signal through funding and legislation that cleaner energy sources will become a priority, which is later followed by academia and researchers, who test new hypotheses and generate knowledge about these new technologies. Then, the findings obtained in academia are applied on a larger scale by actors who have the incentives and the resources to deploy them, like oil majors and other private actors, who also hire engineers and researchers to perform analyses of cost and the potential for different methods.

10.6 ENERGY TRANSITION AND THE OIL AND GAS INDUSTRY

As a groundbreaking endeavor, the energy transition has ambitious goals for the energy future of the planet. With the main objective of reducing carbon emissions by switching to renewable energy as the main power source, this transition requires the collaboration of a set of players, including governments, international organizations, and oil and gas players. Hydrogen technologies can be seen as a very accurate sample of the whole transition process, as some of these techniques can immediately help processes based on fossil fuels, but an economy built around it can constitute a complete break from the old paradigms.

In this section, we review the main drivers and incentives for major players, as well as the investments made by both international and national oil companies, the close ties (or lack thereof) that these latter entities have with their governments, and the role that international governance strategies have played on the development of regional and global environmental policies. From individual investment done by oil companies in some parts of the world to a coordinated, dedicated hydrogen pipeline in some European countries, the stage is set for the growth and widespread implementation of hydrogen technologies, which hopefully encourages other actors with fewer resources, or without many policies to facilitate this switch, to follow suit.

10.6.1 Drivers and Current Applications

The environmental issues that current generations are facing and the opportunities brought about by research and innovation are what will ultimately lead to the establishment of a hydrogen energy system. The search for the decarbonization of multiple current processes has spurred the development of several techniques, including carbon capture and storage (CCS), the advancement of technologies like fuel cells, and the increasing reliance on renewable energies to handle most energy needs across countries. With that said, depending on the level of development and prosperity within each country, we can identify which locations will consider hydrogen technologies.

In a similar way as deploying more costly technologies like CCS shows, the implementation of newer hydrogen techniques requires either a minimum level of prosperity for a country or a strong incentive structure that injects resources from abroad due to a given location's natural or geographical advantages. Figure 10.42 shows the current state of global hydrogen projects, where we see that most large-scale ventures have been implemented in developed countries in Europe, as well as in the United States, Australia, and Canada. Only a handful of "giga-scale" projects are found outside of these locations, with China as the essential player featuring several hydrogen endeavors which their government mostly funds.

Therefore, unless environmental policies favor a change, public awareness is raised, and funding opportunities reach these locations, newer technologies are still prohibitive for countries with lower levels of economic development. Regardless of the investment pledged by private actors, the push

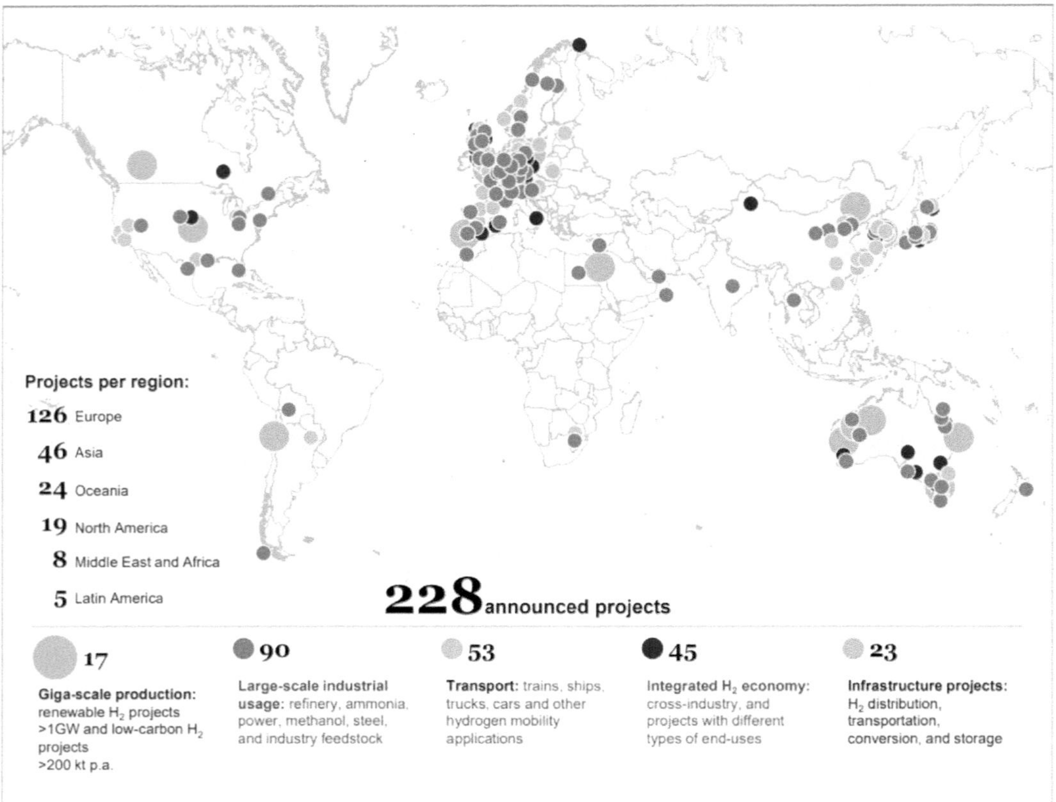

FIGURE 10.42 Global hydrogen projects across the value chain.

Source: McKinsey & Company (2021).

for the energy transition within a country starts with government regulation. So the relationship between public and private actors can truly flourish once officials develop an adequate regulatory regime that focuses on critical issues like the sustained reduction of carbon emissions, the promotion of cleaner technologies like H_2, and the provision of grants and funding for large-scale facilities (Hegnsholt et al., 2019).

For big oil and gas players, the opportunity is there to invest in new regions and promote clean technologies. Shojaeddini et al. (2019) explored the different strategies that international oil companies (IOCs) like BP, Chevron, and ExxonMobil have followed in the past few years to embrace the energy transition process, with an emphasis on newer technologies like CCS and hydrogen fuel cells. A couple of initiatives that oil companies have undertaken are the joint effort from Equinor, Shell, and Total to join the Hydrogen Council and the development of the H_2 Mobility venture, where Shell joined forces with Daimler and Total to install hydrogen vehicle refueling infrastructure in Germany, which has been replicated in the United Kingdom and California by other actors like ITM Power and Honda and Toyota, respectively.

These initiatives break the mold of a top-down environmental strategy, thus inviting more actors to join environmental partnerships that can be more horizontal. This way, we can visualize different pathways toward the common goal of decarbonization, which can be tailored differently for other countries at different times. Figure 10.43 illustrates the different roads toward hydrogen implementation in the next few decades, differentiating between global strategies and those that can be applied regionally or even locally. Consequently, a country with more limited resources can utilize its already-built natural gas infrastructure to introduce hydrogen processes and a set of CCS

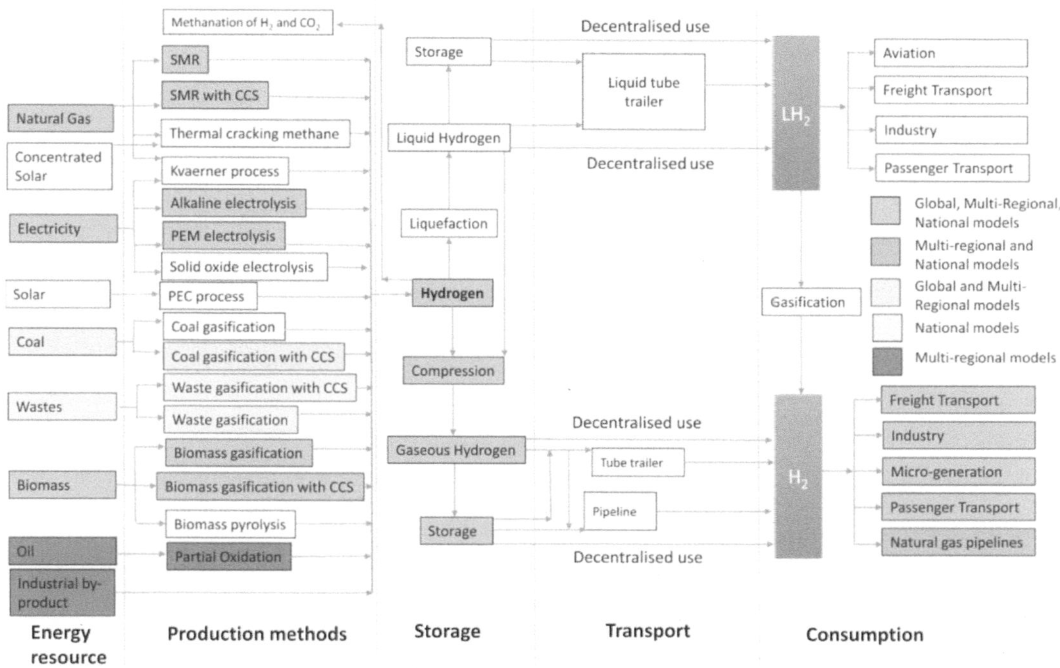

FIGURE 10.43 Global, multi-regional, and national hydrogen pathways toward 2050.

Source: Hanley et al. (2018).

technologies to lower emissions, sticking with that plan for a few years while officials secure funding for more ambitious projects in the future.

Alternatively, actors can coordinate to expand these technologies' reach across regions, bringing them closer to locations that may not be able to experience them through acting alone. One example of a regional collaboration can be seen in the European Hydrogen Backbone report (2020), where a prototype for a set of hydrogen pipelines, storage sites like aquifers and salt caverns, and industrial clusters is proposed for ten countries, with a potential for future expansion after 2030. Figure 10.44 shows a map of this proposal, with dotted lines signifying potential extension routes. Costs for these projects may not be attainable by one country, but international organizations like the European Union can create conditions that attract capital and secure long-term financing options.

Another example, this time for a country that recently broke with a supranational organization, can be seen in the HyNet project, a hydrogen production and supply facility of 100 $kNm^3\ h^{-1}$ to be constructed in the northwest of England. This project aims to deliver low-cost, low-carbon hydrogen to industries alongside the cluster while revamping the infrastructure to provide fueling stations for vehicles, hydrogen-powered trains, and blending to residential sectors (see Figure 10.45). Another case study to highlight is the Acorn Hydrogen Project in the northeast of Scotland, an H_2 production plant based around natural gas incorporating CCS technologies to mitigate emissions (French, 2020).

The success of these projects, either from IOCs or NOCs, goes hand in hand with proactive policymaking that allows for developing these projects, understanding that they are long-term commitments. According to the International Energy Agency (2019), five types of policy need to work together for hydrogen projects to flourish: (1) targets or long-term signals need to be clear and expected; (2) support for the creation of demand is manifest; (3) mitigate investment risks is

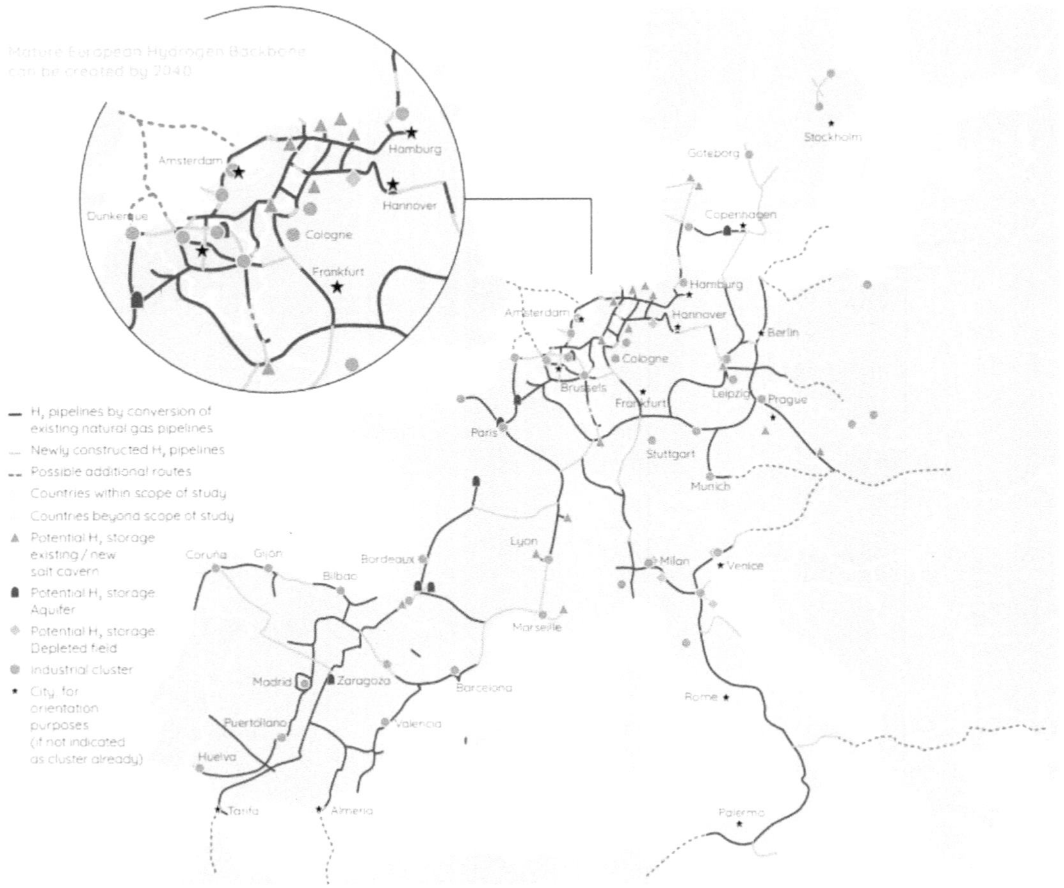

FIGURE 10.44 Potential European Hydrogen Backbone.

Source: Wang et al. (2020a).

attractive for prospective players; (4) promote research and development (R&D) strategies, as well as knowledge sharing; and (5) harmonize standards and remove barriers. Since there is no one-size-fits-all hydrogen policy, actors must proactively adapt to regional conditions.

Moreover, and returning to applications for alternatively fueled vehicles (AFV), refueling stations are their endeavor that requires individual assessments of many variables, such as distance and accessibility, optimization of supply, the economic performance of the vehicles, and environmental benefits of production and construction of these stations, among others. The work of Park et al. (2020b) explores these and other variables through a survey, concluding that user features are also vital in developing these projects, which, in the end, are social improvements. Rose and Neumann (2020) also found that investing in hydrogen refueling stations should follow a symbiotic relationship with the rest of the power system due to potential cost savings of several millions of euros for their German heavy-duty traffic case study.

As we can see, despite these highly technical discussions and assessments we have reviewed, researchers and big players should never forget that their work consists of striving to improve the lives of millions of people who are not looking to pollute indiscriminately but rather to live their lives with the best tools at their disposal, just as they see their leaders do so frequently. Therefore, these projects must include the feedback of ordinary people and be deployed at a human scale that minds the different paths people follow to succeed. The next section describes the role of the oil and

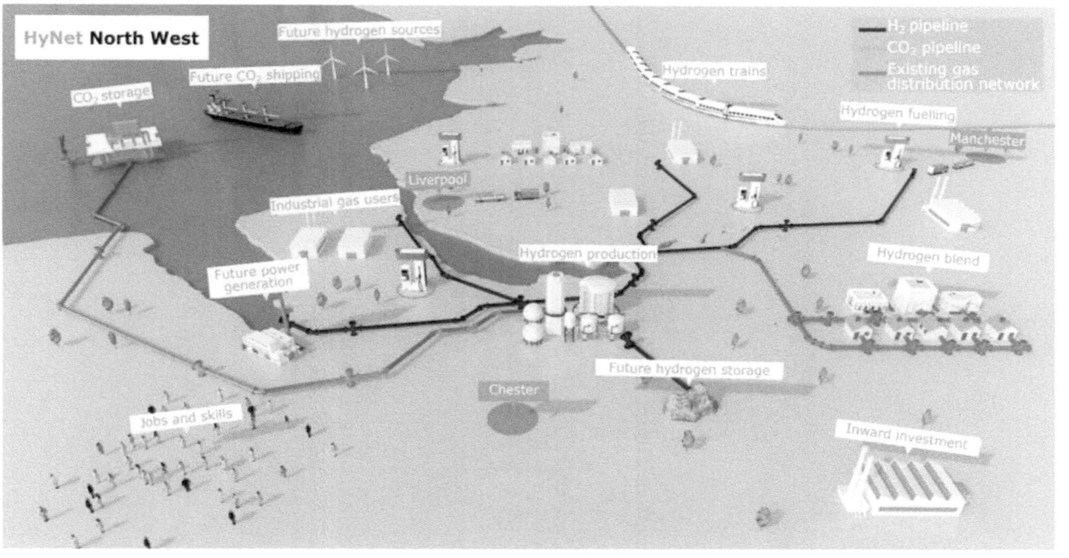

FIGURE 10.45 A schematic of the HyNet project.

Source: French (2020).

gas sector in raising awareness about the energy transition and net-zero energy and how the industry has adapted to the dynamic energy scenario.

10.6.2 Contribution to Net-Zero and Latest Advances

Due to their predominant role in the investment of projects and in social initiatives that raise awareness about the use of cleaner sources of energy, oil companies now appear to be in the business of being perceived as energy companies rather than just oil companies (Pickl, 2019). With this sort of rebranding, there are also opportunities to position themselves in this new environmentally friendly space as leaders of the pack. In this sense, oil companies can invest in multiple phases of the hydrogen production cycle, from those that clean fossil fuel–based processes to completely new endeavors based on hydrogen to generate electricity for different uses (see Figure 10.46).

Net-zero targets are based on treaties like the Paris Agreement, where most countries pledged to set a target for a 1.5°C global warming cap for the year 2050. Because of the level that these pledges have been reproduced in media outlets and the public's response to these messages, major companies have taken this as a signal of important future environmental changes, and most of them have shifted their behavior toward cleaner technologies. Additionally, the regulatory process has advanced to generate strong incentives to invest in renewable energies while also creating penalties for more polluting technologies. Figure 10.47 shows a schematic for a structure to strengthen the incentives for environmental policies that pursue a net-zero goal and another to penalize technologies that are more established but potentially damaging to our climate.

Some examples of policies that penalize existing energy technologies are related to emissions pricing strategies, such as a carbon tax, applied to localities that exceed a certain emission threshold. Also, a carbon cap-and-trade scheme is another proposal where the entire system is allowed a given level of emissions and allowances can be traded between emitters to work out their emissions levels together, creating a small market of sorts. Meanwhile, direct legislation can draft specific standards for technologies, outright banning the use of some of them. An example is the prohibition in the United Kingdom of the sale of coal and wet wood for domestic heating use (Quarton and Samsatli, 2021).

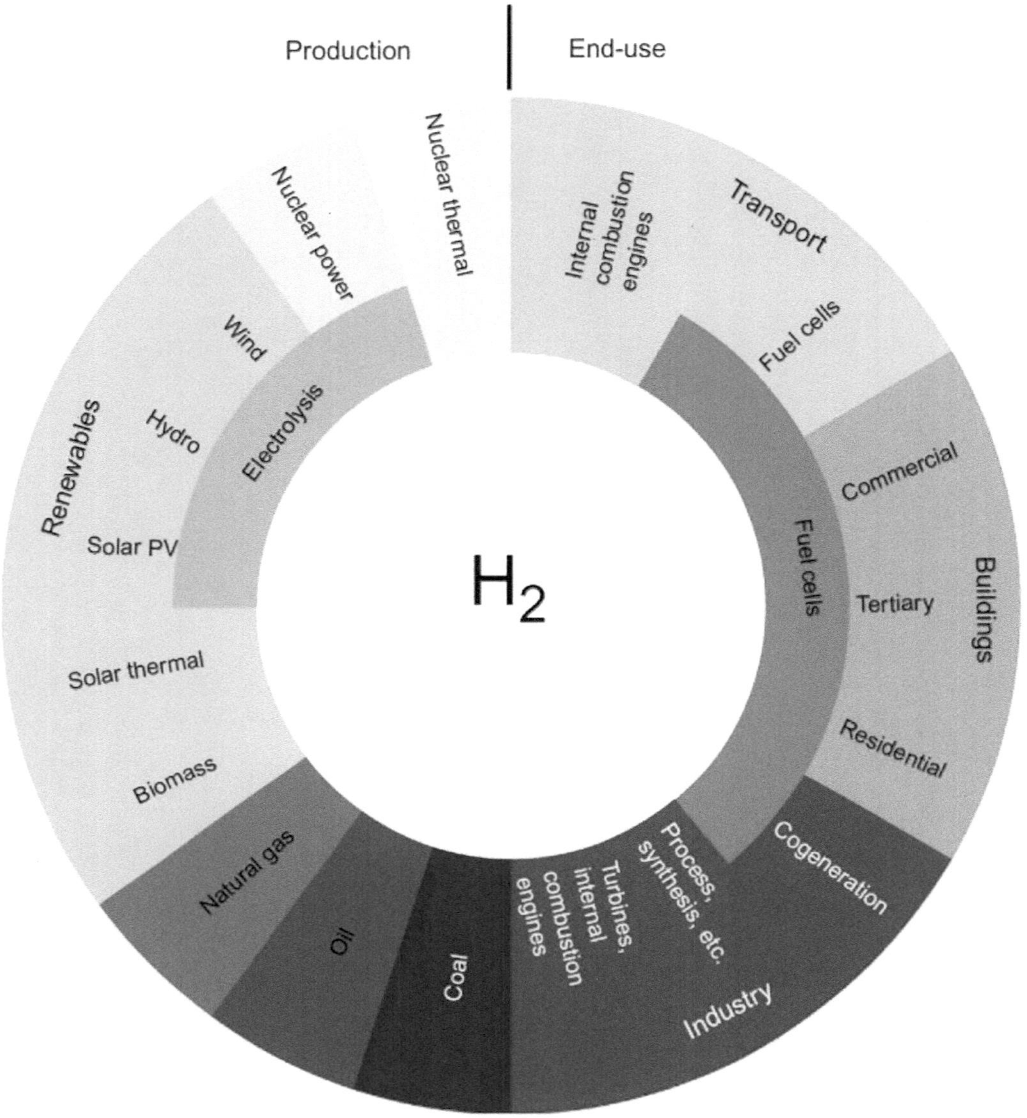

FIGURE 10.46 Hydrogen production and uses.

Source: Mansilla et al. (2018).

As we can see, many governments are laying the conditions for the development of a hydrogen economy, and effort is also being made from the international organization's standpoint. With that said, a complete move requires many industry players, including oil companies and actors like steel, chemicals, and electricity producers, to support, invest, and design clever medium- and long-term strategies to implement practical solutions. For example, legislation may exist to switch mass transportation buses to hydrogen, but without the quick coordination of industry players, a large-scale rollout could take years to see the light of day (van der Spek et al., 2022).

All these moves seem to have planted the seeds of a significant change. In support of the Paris Agreement, the Oil and Gas Climate Initiative (OGCI) was created in 2014, with the participation of CEOs of the most important companies in the sector. Currently, the OGCI has 12 member companies: Aramco, BP, Chevron, CNPC, Eni, Equinor, ExxonMobil, Occidental (better known as

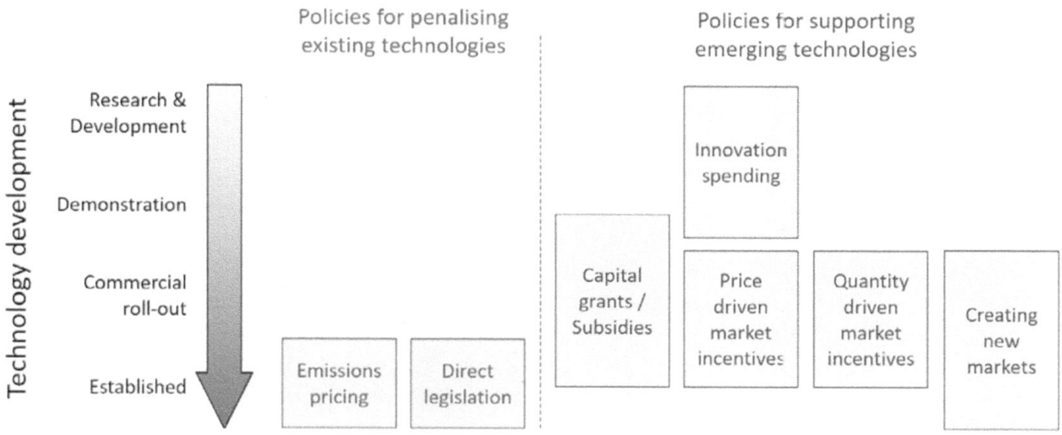

FIGURE 10.47 Policy options for penalizing existing energy technologies or supporting emerging ones. The vertical positions of the policy types indicate the stage of technology development at which they are typically applied.

Source: Quarton and Samsatli (2021).

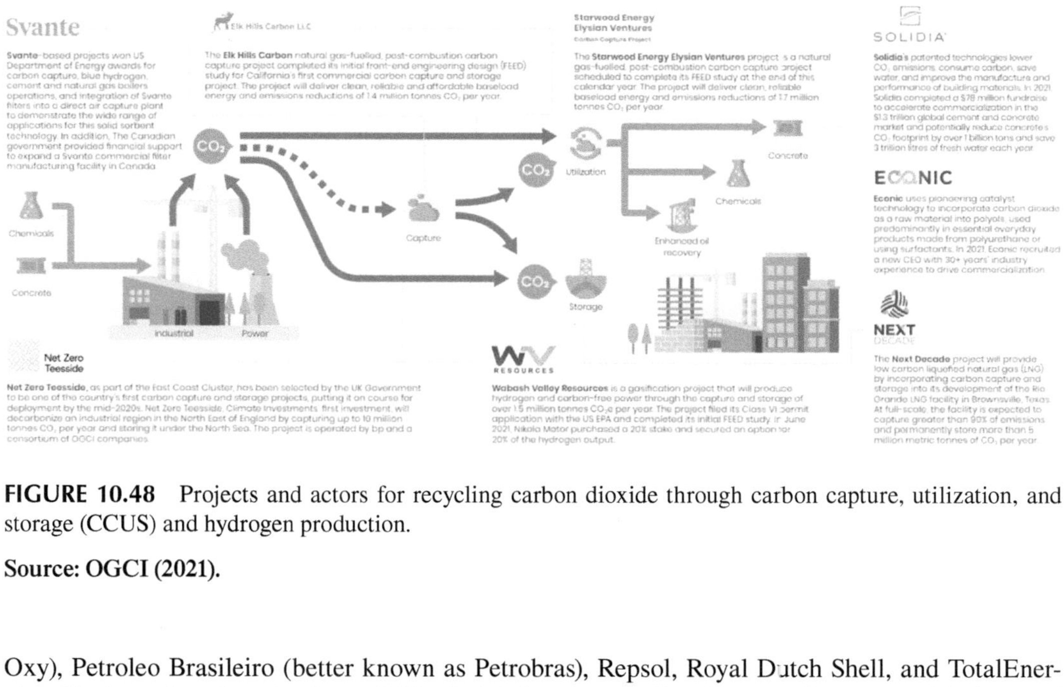

FIGURE 10.48 Projects and actors for recycling carbon dioxide through carbon capture, utilization, and storage (CCUS) and hydrogen production.

Source: OGCI (2021).

Oxy), Petroleo Brasileiro (better known as Petrobras), Repsol, Royal Dutch Shell, and TotalEnergies. Their market share is almost 30% of global-operated oil and gas production, and their stated goals are investing in technologies and projects that speed up decarbonization in the oil and gas sector and in industry and commercial transport (OGCI, 2021).

Regarding hydrogen development and net-zero commitments, the OGCI is currently focused on decarbonizing industrial processes by encouraging low-carbon hydrogen production and carbon and storage projects. Figure 10.48 shows a series of collaborations between innovators and companies which have generated a few CCS/CCUS projects to capture CO_2 and store it to avoid release into the atmosphere. Similarly, some of these endeavors include hydrogen production from the captured carbon, like the Wabash Valley Resources gasification project featured.

Closing in the first decade after big oil and gas players decided to join the quest for the energy transition to renewables shows some progress, but also some areas of improvement. On the one hand, these companies have embraced several avenues for investment and communication of their plans for a more environmentally friendly future. But on the other hand, hydrogen projects still appear to be somewhat tame, with a clear margin for improvement. For example, a welcome addition could be a guideline for a set of investment plans in mass transportation systems based on hydrogen, refueling stations that provide hydrogen for different types of vehicles, and the expansion of hydrogen to help power maritime and aviation operations, to name a few.

With that said, there are promising signs for moving toward a transition to clean energies in the next few decades. There has been hardly another point where so many collaborations and deals have been made between influential energy players, governments, and policymakers to tackle the potential negative consequences of an increase in global temperatures. The next decade should be one of many transformations, incentives, and environmental progress, provided these actors continue offering their expertise, resources, and influence to drive research and development and a road to decarbonization and clean energies.

10.7 CONCLUSIONS

Hydrogen has reappeared as a promising energy carrier that generates low emissions and can revamp fossil fuel–based processes. If the main driver for innovation and environmental policymaking is net-zero GHG emissions by 2050, hydrogen technologies can take the driver's seat and carry most of the burden, especially when powered by renewable sources of energy like solar and wind. In the main, hydrogen is a clean technology that generates mostly water as emission, and if storage and transportation issues improve, it can become the center of the new energy paradigm of the 21st century.

This chapter serves as a primer on hydrogen technologies, their current state of development, the challenges that keep some of these innovations from reaching a more widespread status, and the various opportunities for players to invest in them. As a set of more novel clean methods, they suffer from a lack of maturity when compared to fossil fuel–based approaches, so most of them still need a few years to be fully functional at a larger scale. However, they have a significant upside in their potential for decarbonizing the industrial sector and their prospects for taking over sectors like transportation and residential heating and cooling.

A main consideration with these methods is that their costs may still be somewhat prohibitive to some countries due to the materials required for some of its processes and the lack of economies of scale that lower general costs after a given size and scope is reached. So we are currently in a situation that leaves more prosperous countries with a better opportunity to invest in these methods first, becoming the initial actors to decarbonize. With that said, this constitutes a window for other players to innovate. Big oil and gas players have taken significant steps to invest and raise awareness about these new technologies, striking deals with governments to expand the reach of some of these projects.

Similar to other novel clean energy methods, investments in hydrogen technologies require concerted attempts from multiple actors and a coordinated effort from top to bottom, going from governments to international organizations, to oil majors, and finally, to researchers and academia. Some of the initial cost curves may be too steep for individual actors to overcome on their own, so policymakers should develop flexible regulations that understand these endeavors as long-term strategies to improve the environment, which are also competing in a market where fossil fuel methods are more accessible yet have a larger carbon footprint.

All hands are on deck in the search for lowering greenhouse gas emissions that has allowed for the opportunity to integrate different energy systems into the same general grid, which means that solar can help wind, nuclear can assist these two sources when they are lagging, and so on. Consequently, hydrogen technologies can be of assistance in delivering increasing amounts of energy for different uses, or they can help store it for future use. However, storage is still in a developmental phase, with potential large-scale projects continuing to evaluate the benefits of compressed hydrogen versus the advantages and costs of liquid hydrogen.

The transportation sector is a particularly interesting field of application for storing hydrogen, which, in ideal situations, can produce the necessary energy for many types of vehicles to compete with conventional fossil fuel–based internal combustion engines. Current trade-offs for these newer methods relate to the size and weight of storage tanks, which cannot surpass a certain threshold, lest it becomes an efficiency risk. Additionally, the cost of insulating tanks for both compressed and liquid hydrogen and the potential loss of fuel within the cells are two other aspects currently being studied and improved.

Meanwhile, oil and gas players have developed strategies to participate in the energy transition. By creating the Oil and Gas Climate Initiative (OGCI), these actors publish their progress reports and state-of-the-art clean energy projects, where hydrogen is a relatively new addition to these plans. The most substantial area of innovation is currently centered on carbon and storage/utilization (CCS/CCUS) strategies, which seek to prevent CO_2 from industrial processes from being released into the atmosphere while generating hydrogen from it through processes like coal gasification.

Moreover, the current picture for the development of hydrogen technologies cannot be complete without mentioning the role that governments, policymakers, and international organizations have in pursuing and implementing the energy transition. Thanks to the work of organizations like the United Nations, we now have a set of international environmental regulations that make most countries follow a strict set of rules that aim to mitigate the increase in global temperatures. Similarly, at a country level, policies have been developed to ban the use of specific polluting sources like wood and coal and to establish limits to emissions from industries before a tax is placed on companies.

All in all, hydrogen offers a very enticing potential to overtake many current processes dominated by fossil fuels. A joint effort between actors at all levels of society is required to advance these technologies to a level where they can compete in production costs and eventually replace the more polluting processes. International regulations and policymakers need to become more flexible with these initially more volatile procedures, but since they offer the chance of decarbonization, these methods appear to be a step in the right direction. Then, the future for these technologies seems to be bright, so an economy based on hydrogen, whether blue or green, is a genuine possibility within the next couple of decades.

REFERENCES

Aadhithiyan, S., R. Raja, J. Alzabut, Q. Zhu, and M. Niezabitowski. 2022. Robust non-fragile Mittag-Leffler synchronization of fractional order non-linear complex dynamical networks with constant and infinite distributed delays. *Mathematical Methods in the Applied Sciences* 45(4): 2166–2189.

Aakko-Saksa, P. T., C. Cook, J. Kiviaho, and T. Repo. 2018. Liquid organic hydrogen carriers for transportation and storing of renewable energy – review and discussion. *Journal of Power Sources* 396: 803–823.

Abdalla, A. M., S. Hossain, O. B. Nisfindy, A. T. Azad, M. Dawood, and A. K. Azad. 2018. Hydrogen production, storage, transportation and key challenges with applications: A review. *Energy Conversion and Management* 165: 602–627.

Abdelkareem, M. A., K. Elsaid, T. Wilberforce, M. Kamil, E. T. Sayed, and A. Olabi. 2021. Environmental aspects of fuel cells: A review. *Science of the Total Environment* 752: 141803.

Abdin, Z., A. Zafaranloo, A. Rafiee, W. Mérida, W. Lipiński, and K. R. Khalilpour. 2020. Hydrogen as an energy vector. *Renewable and Sustainable Energy Reviews* 120: 109620.

Abe, J. O., A. P. I. Popoola, E. Ajenifuja, and O. M. Popoola. 2019. Hydrogen energy, economy, and storage: Review and recommendation. *International Journal of Hydrogen Energy* 44: 15072–15086.

Acar, C., A. Beskese, and G. T. Temur. 2018. Sustainability analysis of different hydrogen production options using hesitant fuzzy AHP. *International Journal of Hydrogen Energy* 43: 18059–18076.

Acar, C., and I. Dincer. 2020. The potential role of hydrogen as a sustainable transportation fuel to combat global warming. *International Journal of Hydrogen Energy* 45: 3396–3406.

Akhlaghi, N., and G. Najafpour-Darzi. 2020. A comprehensive review on biological hydrogen production. *International Journal of Hydrogen Energy* 45(43): 22492–22512.

Al Ghafri, S., S. Munro, U. Cardella, T. Funke, W. Notardonato, J. P. M. Trusler, J. Leachman, . . ., and E. F. May. 2022. Hydrogen liquefaction: A review of the fundamental physics, engineering practice and future opportunities. *Energy & Environmental Science*, Accepted Manuscript.

Ali Khan, M. H., R. Daiyan, P. Neal, N. Haque, I. MacGill, and R. Amal. 2021. A framework for assessing economics of blue hydrogen production from steam methane reforming using carbon capture storage & utilisation. *International Journal of Hydrogen Energy* 46(44): 22685–22706.

Al-Khori, K., Y. Bicer, and M. Koç. 2019. Integration of solid oxide fuel cells into oil and gas operations: Needs, opportunities, and challenges. *Journal of Cleaner Production*: 118924.

Ban, S., W. Lin, F. Wu, and J. Luo. 2018. Algal-bacterial cooperation improves algal photolysis-mediated hydrogen production. *Bioresource Technology* 251: 350–357.

Boateng, E., and A. Chen. 2020. Recent advances in nanomaterial-based solid-state hydrogen storage. *Materials Today Advances* 6: 100022.

Brendelberger, S., J. Vieten, M. Roeb, and C. Sattler. 2019. Thermochemical oxygen pumping for improved hydrogen production in solar redox cycles. *International Journal of Hydrogen Energy* 44: 9802–9810.

Brestovič, T., N. Jasminská, and M. Lázár. 2022. Measurements of operating parameters of a metal hydride compressor with a heat pump. *Applied Sciences* 12: 3302.

Broom, D. P., C. J. Webb, G. S. Fanourgakis, G. E. Froudakis, P. N. Trikalitis, and M. Hirscher. 2019. Concepts for improving hydrogen storage in nanoporous materials. *International Journal of Hydrogen Energy* 44: 7768–7779.

Chen, C., Q. Xia, S. Feng, and Q. Liu. 2021. A novel solar hydrogen production system integrating high temperature electrolysis with ammonia based thermochemical energy storage. *Energy Conversion and Management* 237: 114143.

Chen, Z., X. Duan, W. Wei, S. Wang, and B.-J. Nie. 2019. Recent advances in transition metal-based electrocatalysts for alkaline hydrogen evolution. *Journal of Materials Chemistry A* 7: 14971–15005.

Chi, J., and H. Yu. 2018. Water electrolysis based on renewable energy for hydrogen production. *Chinese Journal of Catalysis* 39: 390–394.

Chuayboon, S., and S. Abanades. 2020. An overview of solar decarbonization processes, reacting oxide materials, and thermochemical reactors for hydrogen and syngas production. *International Journal of Hydrogen Energy* 45(48), September: 25783–25810.

Colbertaldo, P., S. B. Agustin, S. Campanari, and J. Brouwer. 2019. Impact of hydrogen energy storage on California electric power system: Towards 100% renewable electricity. *International Journal of Hydrogen Energy* 44: 9558–9576.

Czakiert, T., J. Krzywanski, A. Zylka, and W. Nowak. 2022. Chemical looping combustion: A brief overview. *Energies* 15: 1563.

David, M., C. Ocampo-Martínez, and R. Sánchez-Peña. 2019. Advances in alkaline water electrolyzers: A review. *Journal of Energy Storage* 23, June: 392–403.

Dawood, F., M. Anda, and G. M. Shafiullah. 2020. Hydrogen production for energy: An overview. *International Journal of Hydrogen Energy* 45: 3847–3869.

Dubouis, N., and A. Grimaud. 2019. The hydrogen evolution reaction: From material to interfacial descriptors. *Chemical Science* 10: 9165–9181.

Egeland-Eriksen, T., A. Hajizadeh, and S. Sartori. 2021. Hydrogen-based systems for integration of renewable energy in power systems: Achievements and perspectives. *International Journal of Hydrogen Energy* 46: 31963–31983.

EIA. 2022. *Hydrogen explained*. United States Energy Information Administration. Available in www.eia.gov/energyexplained/hydrogen [accessed March 27th, 2022].

Elberry, A. M., J. Thakur, A. Santasalo-Aarnio, and M. Larmi. 2021. Large-scale compressed hydrogen storage as part of renewable electricity storage systems. *International Journal of Hydrogen Energy* 46: 15671–15690.

El-Shafie, M., S. Kambara, and Y. Hayakawa. 2019. Hydrogen production technologies overview. *Journal of Power and Energy Engineering* 7: 107–154.

Endo, N., E. Shimoda, K. Goshome, T. Yamane, T. Nozu, and T. Maeda. 2019. Simulation of design and operation of hydrogen energy utilization system for a zero emission building. *International Journal of Hydrogen Energy* 44: 7118–7124.

Fasolini, A., R. Cucciniello, E. Paone, F. Mauriello, and T. Tabanelli. 2019. A short overview on the hydrogen production via aqueous phase reforming (APR) of cellulose, C6-C5 sugars and polyols. *Catalysts* 9: 917.

Fonseca, J. D., M. Camargo, J.-M. Commenge, L. Falk, and I. D. Gil. 2019. Trends in design of distributed energy systems using hydrogen as energy vector: A systematic literature review. *International Journal of Hydrogen Energy* 44(19): 9486–9504.

French, S. 2020. The role of zero and low carbon hydrogen in enabling the energy transition and the path to net zero greenhouse gas emissions: With global policies and demonstration projects hydrogen can play a role in a net zero future. *Johnson Matthey Technology Review* 64(3): 357–370.

Fu, Q., D. Wang, X. Li, Q. Yang, Q. Xu, B.-J. Ni, Q. Wang, and X. Liu. 2021. Towards hydrogen production from waste activated sludge: Principles, challenges and perspectives. *Renewable and Sustainable Energy Reviews* 135: 110283.

Galvis E., A. R., F. Leardini, J. R. Ares, F. Cuevas, and J. F. Fernandez. 2018. Simulation and design of a three-stage metal hydride hydrogen compressor based on experimental thermodynamic data. *International Journal of Hydrogen Energy* 43(13): 6666–6676.

Gao, Y., J. Jiang, Y. Meng, F. Yan, and A. Aihemaiti. 2018. A review of recent developments in hydrogen production via biogas dry reforming. *Energy Conversion and Management* 171: 133–155.

Genovese, M., and P. Fragiacomo. 2022. Hydrogen station evolution towards a poly-generation energy system. *International Journal of Hydrogen Energy* 47(24): 12264–12280.

Gray, E. M. 2021. Alloy selection for multistage metal-hydride hydrogen compressors: A thermodynamic model. *International Journal of Hydrogen Energy* 46(29): 15702–15715.

Groppi, D., D. Astiaso Garcia, G. Lo Basso, F. Cumo, and L. De Santoli. 2018. Analysing economic and environmental sustainability related to the use of battery and hydrogen energy storages for increasing the energy independence of small islands. *Energy Conversion and Management* 177: 64–76.

Guo, Y., G. Li, J. Zhou, and Y. Liu. 2019. Comparison between hydrogen production by alkaline water electrolysis and hydrogen production by PEM electrolysis. *IOP Conference Series: Earth and Environmental Science* 371: 042022.

Habibollahzade, A., E. Gholamian, P. Ahmadi, and A. Behzadi. 2018. Multi-criteria optimization of an integrated energy system with thermoelectric generator, parabolic trough solar collector and electrolysis for hydrogen production. *International Journal of Hydrogen Energy* 43(31): 14140–14157.

Hanley, E. S., J. P. Deane, and B. P. Ó. Gallachóir. 2018. The role of hydrogen in low carbon energy futures: A review of existing perspectives. *Renewable and Sustainable Energy Reviews* 82: 3027–3045.

Hegnsholt, E., F. Klose, J. Burchardt, and S. Schönberger. *The real promise of hydrogen*. Boston Consulting Group. Available in http://boston-consulting-group-brightspot.s3.amazonaws.com/img-src/BCG-The-Real-Promise-of-Hydrogen-July-2019_tcm9–225426.pdf [accessed May 17th, 2022].

Hosseini, S. E., and M. A. Wahid. 2020. Hydrogen from solar energy, a clean energy carrier from a sustainable source of energy. *International Journal of Energy Research* 44(6): 4110–4131.

Hu, Q., Y. Shen, J. W. Chew, T. Ge, and C.-H. Wang. 2020. Chemical looping gasification of biomass with Fe2O3/CaO as the oxygen carrier for hydrogen-enriched syngas production. *Chemical Engineering Journal*: 122346.

IEA. 2019. *The future of hydrogen: Seizing today's opportunities*. International Energy Agency. Report prepared by the IEA for the G20, Japan. Available in https://iea.blob.core.windows.net/assets/9e3a3493-b9a6-4b7d-b499-7ca48e357561/The_Future_of_Hydrogen.pdf [accessed May 12th, 2022].

Iida, S., and K. Sakata. 2019. Hydrogen technologies and developments in Japan. *Clean Energy* 3(2): 105–113.

Jalil, Z., A. Rahwanto, Ismail, H. Sofyan, and E. Handoko. 2018. The use of nano-silicon carbide and nickel as catalyst in magnesium hydrides (MgH2) for hydrogen storage material application. *Materials Research Express* 5(6): 064002.

Jenne, S. P., S. Jana, and M. Palanisamy. 2020. Experimental investigation on absorption and desorption characteristics of La0.9Ce0.1Ni5 for hydrogen storage application. *International Journal of Energy Research*: 1–12.

Kaplan, R., and M. Kopacz. 2020. Economic conditions for developing hydrogen production based on coal gasification with carbon capture and storage in Poland. *Energies* 13: 5074.

Khan, M. A., C. Young, and D. B. Layzell. 2021. The techno-economics of hydrogen pipelines. *Transition Accelerator Technical Briefs* 1(2): 1–40.

Kim, M., C. Kim, and Y. Sohn. 2018. Application of metal foam as a flow field for PEM fuel cell stack. *Fuel Cells* 18(2): 123–128.

Kolawole Muritala, I., D. Guban, M. Roeb, and C. Sattler. 2020. High temperature production of hydrogen: Assessment of non-renewable resources technologies and emerging trends. *International Journal of Hydrogen Energy* 45(49), October: 26022–26035.

Kovač, A., M. Paranos, and D. Marciuš. 2021. Hydrogen in energy transition: A review. *International Journal of Hydrogen Energy* 46(16): 10016–10035.

Krane, P., D. Ziviani, J. E. Braun, N. Jain, and A. Marconnet. 2022. Techno-economic analysis of metal-hydride energy storage to enable year-round load-shifting for residential heat pumps. *Energy and Buildings* 256, February: 111700.

Lewandowska-Śmierzchalska, J., R. Tarkowski, and B. Uliasz-Misiak. 2018. Screening and ranking framework for underground hydrogen storage site selection in Poland. *International Journal of Hydrogen Energy* 43(9): 4401–4414.

Likkasit, C., A. Maroufmashat, A. Elkamel, H.-M. Ku, and M. Fowler. 2018. Solar-aided hydrogen production methods for the integration of renewable energies into oil & gas industries. *Energy Conversion and Management* 168: 395–406.

Łukajtis, R., I. Hołowacz, K. Kucharska, M. Glinka, P. Rybarczyk, A. Przyjazny, and M. Kamiński. 2018. Hydrogen production from biomass using dark fermentation. *Renewable & Sustainable Energy Reviews* 91: 665–694.

Luo, M., Y. Yi, S. Wang, Z. Wang, M. Du, J. Pan, and Q. Wang. 2018. Review of hydrogen production using chemical-looping technology. *Renewable and Sustainable Energy Reviews* 81(2), January: 3186–3214.

Lyu, J., V. Kudiiarov, and A. Lider. 2020. An overview of the recent progress in modifications of carbon nanotubes for hydrogen adsorption. *Nanomaterials* 10: 255.

Mahmood, N., Y. Yao, J.-W. Zhang, L. Pan, X. Zhang, and J.-J. Zou. 2018. Electrocatalysts for hydrogen evolution in alkaline electrolytes: Mechanisms, challenges, and prospective solutions. *Advanced Science* 5: 1700464.

Makepeace, J. W., T. He, C. Weidenthaler, T. R. Jensen, F. Chang, T. Vegge, . . ., and W. I. F. David. 2019. Reversible ammonia-based and liquid organic hydrogen carriers for high-density hydrogen storage: Recent progress. *International Journal of Hydrogen Energy* 44: 7746–7767.

Manoharan, Y., S. E. Hosseini, B. Butler, H. Alzhahrani, B. T. F. Senior, T. Ashuri, and J. Krohn. 2019. Hydrogen fuel cell vehicles; current status and future prospect. *Applied Sciences* 9: 2296.

Mansilla, C., C. Bourasseau, C. Cany, B. Guinot, A. Le Duigou, and P. Lucchese. 2018. Hydrogen applications: Overview of the key economic issues and perspectives. *Hydrogen Supply Chains*: 271–292.

Martin, A., M.-F. Agnoletti, and E. Brangier. 2020. Users in the design of hydrogen energy systems: A systematic review. *International Journal of Hydrogen Energy* 45(21): 11889–11900.

Masoudi Soltani, S., A. Lahiri, H. Bahzad, P. Clough, M. Gorbounov, and Y. Yan. 2021. Sorption-enhanced steam methane reforming for combined CO2 capture and hydrogen production: A state-of-the-art review. *Carbon Capture Science & Technology* 1: 100003.

Mayyas, A., M. Wei, and G. Levis. 2020. Hydrogen as a long-term, large-scale energy storage solution when coupled with renewable energy sources or grids with dynamic electricity pricing schemes. *International Journal of Hydrogen Energy* 45(33), June: 16311–16325.

McKinsey & Company. 2021. *Hydrogen Insights: A perspective on hydrogen investment, market development and cost competitiveness*. Hydrogen Council. Available in https://hydrogencouncil.com/wp-content/uploads/2021/02/Hydrogen-Insights-2021-Report.pdf [accessed May 10th, 2022].

Mehrjerdi, H. 2019. Optimal correlation of non-renewable and renewable generating systems for producing hydrogen and methane by power to gas process. *International Journal of Hydrogen Energy* 44(18): 9210–9219.

Mohan, M., V. K. Sharma, E. A. Kumar, and V. Gayathri. 2019. Hydrogen Storage in Carbon Materials—A Review. *Energy Storage 1*.

Moradi, R., and K. M. Groth. 2019. Hydrogen storage and delivery: Review of the state of the art technologies and risk and reliability analysis. *International Journal of Hydrogen Energy* 44: 12254–12269.

Moreno-Blanco, J., G. Petitpas, F. Espinosa-Loza, F. Elizalde-Blancas, J. Martinez-Frias, and S. M. Aceves. 2019. The storage performance of automotive cryo-compressed hydrogen vessels. *International Journal of Hydrogen Energy* 44(31), June: 16841–16851.

Niermann, M., A. Beckendorff, M. Kaltschmitt, and K. Bonhoff. 2019. Liquid organic hydrogen carrier (LOHC) – assessment based on chemical and economic properties. *International Journal of Hydrogen Energy* 44: 6631–6654.

Niermann, M., S. Timmerberg, S. Drünert, and M. Kaltschmitt. 2021. Liquid organic hydrogen carriers and alternatives for international transport of renewable hydrogen. *Renewable and Sustainable Energy Reviews* 135: 110171.

Nurdiawati, A., I. N. Zaini, A. R. Irhamna, D. Sasongko, and M. Aziz. 2019. Novel configuration of supercritical water gasification and chemical looping for highly-efficient hydrogen production from microalgae. *Renewable and Sustainable Energy Reviews* 112: 369–381.

OGCI. 2021. *Accelerating Ambition & Action: A progress report from the oil and gas climate initiative, December 2021*. Oil and Gas Climate Initiative. Available in www.ogci.com/ogci-annual-reports/ [accessed May 23rd, 2022].

Ogden, J., A. Myers Jaffe, D. Scheitrum, Z. McDonald, and M. Miller. 2018. Natural gas as a bridge to hydrogen transportation fuel: Insights from the literature. *Energy Policy* 115: 317–329.

Oni, A. O., K. Anaya, T. Giwa, G. Di Lullo, and A. Kumar. 2022. Comparative assessment of blue hydrogen from steam methane reforming, autothermal reforming, and natural gas decomposition technologies for natural gas-producing regions. *Energy Conversion and Management* 254: 115245.

Osman, A. I., N. Mehta, A. M. Elgarahy, M. Hefny, A. Al-Hinai, A. H. Al-Muhtaseb, and D. W. Rooney. 2021. Hydrogen production, storage, utilisation and environmental impacts: A review. *Environmental Chemistry Letters*: 1–36.

Oudejans, D., M. Offidani, A. Constantinou, S. Albonetti, N. Dimitratos, and A. Bansode. 2022. A comprehensive review on two-step thermochemical water splitting for hydrogen production in a redox cycle. *Energies* 15: 3044.

Pachiega, R., M. Franco Rodrigues, C. Varella Rodrigues, I. Kimiko Sakamoto, M. B. A. Varesche, J. E. De Oliveira, and S. Imaculada Maintinguer. 2019. Hydrogen bioproduction with anaerobic bacteria consortium from brewery wastewater. *International Journal of Hydrogen Energy* 44: 155–163.

Pandey, B., Y. K. Prajapati, and P. N. Sheth. 2019. Recent progress in thermochemical techniques to produce hydrogen gas from biomass: A state of the art review. *International Journal of Hydrogen Energy* 44: 25384–25415.

Pandiyan, A., A. Uthayakumar, R. Subrayan, S.-W. Cha, and S. B. Krishna Moorthy. 2019. Review of solid oxide electrolysis cells: A clean energy strategy for hydrogen generation. *Nanomaterials and Energy* 8(1): 2–22.

Park, M.-J., J.-H. Kim, Y.-H. Lee, H.-M. Kim, and D.-W. Jeong. 2020a. System optimization for effective hydrogen production via anaerobic digestion and biogas steam reforming. *International Journal of Hydrogen Energy* 45(55): 30188–30200.

Park, S., S. Nam, M. Oh, I.-J. Choi, and J. Shin. 2020b. Preference structure on the design of hydrogen refueling stations to activate energy transition. *Energies* 13: 3959.

Parra, D., L. Valverde, F. J. Pino, and M. K. Patel. 2019. A review on the role, cost and value of hydrogen energy systems for deep decarbonisation. *Renewable and Sustainable Energy Reviews* 101: 279–294.

Pickl, M. J. 2019. The renewable energy strategies of oil majors – from oil to energy? *Energy Strategy Reviews* 26: 100370.

Qiu, Y., H. Yang, L. Tong, L. Wang. 2021. Research progress of cryogenic materials for storage and transportation of liquid hydrogen. *Metals* 11: 1101.

Quarton, C. J., and S. Samsatli. 2021. How to incentivise hydrogen energy technologies for net zero: Whole-system value chain optimisation of policy scenarios. *Sustainable Production and Consumption* 27: 1215–1238.

Ramirez-Vidal, P., R. L. S. Canevesi, G. Sdanghi, S. Schaefer, G. Maranzana, A. Celzard, and V. Fierro. 2021. A step forward in understanding the hydrogen adsorption and compression on activated carbons. *ACS Applied Materials & Interfaces* 13(10): 12562–12574.

Rose, P. K., and F. Neumann. 2020. Hydrogen refueling station networks for heavy-duty vehicles in future power systems. *Transportation Research Part D* 83: 102358.

Sadeghi, S., S. Ghandehariun, and M. A. Rosen. 2020. Comparative economic and life cycle assessment of solar-based hydrogen production for oil and gas industries. *Energy* 208: 118347.

Safari, F., and I. Dincer. 2020. A review and comparative evaluation of thermochemical water splitting cycles for hydrogen production. *Energy Conversion and Management* 205: 112182.

Sazali, N. 2020. Emerging technologies by hydrogen: A review. *International Journal of Hydrogen Energy* 45(38), July: 18753–18771.

Shafiee, S. 2020. Operational principles and effect of operating parameters on performance of metal hydride heat pumps. *International Journal of Refrigeration* 120: 22–30.

Shimizu, T., K. Hasegawa, M. Ihara, and Y. Kikuchi. 2020. A region-specific environmental analysis of technology implementation of hydrogen energy in Japan based on life cycle assessment. *Journal of Industrial Ecology* 24: 217–233.

Shiva Kumar, S., and V. Himabindu. 2019. Hydrogen production by PEM water electrolysis – a review. *Materials Science for Energy Technologies* 2: 442–454.

Shojaeddini, E., S. Naimoli, S. Ladislaw, and M. Bazilian. 2019. Oil and gas company strategies regarding the energy transition. *Progress in Energy* 1: 012001.

Singh, R., A. Altaee, and S. Gautam. 2020. Nanomaterials in the advancement of hydrogen energy. *Heliyon* 6: e04487.

Southall, E., and L. Lukashuk. 2022. Hydrogen storage and transportation technologies to enable the hydrogen economy: Liquid organic hydrogen carriers. *Johnson Matthey Technology Review*. Forthcoming.

Stamatakis, E., E. Zoulias, G. Tzamalis, Z. Massina, V. Analytis, C. Christodoulou, and A. Stubos. 2018. Metal hydride hydrogen compressors: Current developments & early markets. *Renewable Energy* 127: 850–862.

Tanç, B., H. Turan Arat, E. Baltacıoğlu, and K. Aydın. 2019. Overview of the next quarter century vision of hydrogen fuel cell electric vehicles. *International Journal of Hydrogen Energy* 44(20), April: 10120–10128.

Tarasov, B. P., M. S. Bocharnikov, Y. B. Yanenko, P. V. Fursikov, K. B. Minko, and M. V. Lototskyy. 2020. Metal hydride hydrogen compressors for energy storage systems: Layout features and results of long-term tests. *Journal of Physics: Energy* 2: 024005.

Tarkowski, R. 2019. Underground hydrogen storage: Characteristics and prospects. *Renewable and Sustainable Energy Reviews* 105: 86–94.

Tian, M.-W., H.-C. Yuen, S.-R. Yan, and W.-L. Huang. 2019. The multiple selections of fostering applications of hydrogen energy by integrating economic and industrial evaluation of different regions. *International Journal of Hydrogen Energy* 44: 29390–29398.

Tian, Y., Y. Tian, Z. Qu, T. Guo, Y. Shi, and L. Zhang. 2018. A new experimental simulation method for the development of non-renewable hydrogen energy—oil and gas resources. *International Journal of Hydrogen Energy* 44(11): 5220–5229.

U.S. Department of Energy. 2022a. *Hydrogen production: Electrolysis*. Office of Energy Efficiency & Renewable Energy. Available in www.energy.gov/eere/fuelcells/hydrogen-production-electrolysis [accessed March 30th, 2022].

U.S. Department of Energy. 2022b. *Hydrogen production: Photobiological*. Office of Energy Efficiency & Renewable Energy. Available in www.energy.gov/eere/fuelcells/hydrogen-production-photobiological [accessed March 30th, 2022].

U.S. Department of Energy. 2022c. *Hydrogen production: Thermochemical water splitting*. Office of Energy Efficiency & Renewable Energy. Available in www.energy.gov/eere/fuelcells/hydrogen-production-thermochemical-water-splitting [accessed March 30th, 2022].

U.S. Department of Energy. 2022d. *Hydrogen storage*. Office of Energy Efficiency & Renewable Energy. Available in www.energy.gov/eere/fuelcells/hydrogen-storage [accessed April 20th, 2022].

U.S. Department of Energy. 2022e. *Physical hydrogen storage*. Office of Energy Efficiency & Renewable Energy. Available in www.energy.gov/eere/fuelcells/physical-hydrogen-storage [accessed April 27th, 2022].

Van de Graaf, T., I. Overland, D. Scholten, and K. Westphal. 2020. The new oil? The geopolitics and international governance of hydrogen. *Energy Research & Social Science* 70: 101667.

van der Spek, M., C. Banet, C. Bauer, P. Gabrielli, W. Goldthrope, M. Mazzotti, . . ., and M. Gazzani. 2022. Perspective on the hydrogen economy as a pathway to reach net-zero CO_2 emissions in Europe. *Energy and Environmental Science* 15: 1034.

Wan, Z., Y. Tao, J. Shao, Y. Zhang, and H. You. 2021. Ammonia as an effective hydrogen carrier and a clean fuel for solid oxide fuel cells. *Energy Conversion and Management* 228: 113729.

Wang, D., M. Muratori, J. Eichman, M. Wei, S. Saxena, and C. Zhang. 2018a. Quantifying the flexibility of hydrogen production systems to support large-scale renewable energy integration. *Journal of Power Sources* 399: 383–391.

Wang, H., J. Xu, L. Sheng, X. Liu, Y. Lu, and W. Li. 2018b. A review on bio-hydrogen production technology. *International Journal of Energy Research* 42(11): 3442–3453.

Wang, M., G. Wang, Z. Sun, Y. Zhang, and D. Xu. 2019. Review of renewable energy-based hydrogen production processes for sustainable energy innovation. *Global Energy Interconnection* 2(5), October: 436–443.

Wang, Q., J. Li, Y. Bu, L. Xu, Y. Ding, Z. Hu, . . ., and Z. Qin. 2020a. *Technical assessment and feasibility validation of liquid hydrogen storage and supply system for heavy-duty fuel cell truck*. 2020 4th CAA International Conference on Vehicular Control and Intelligence (CVCI).

Wang, Y., H. Yang, X. Zhang, F. Han, W. Tu, and W. Yang. 2020b. Microalgal hydrogen production. *Small Methods*: 1900514.

Yang, E., H. Omar Mohamed, S.-G. Park, M. Obaid, S. Y. Al-Qaradawi, P. Castaño, K. Chon, and K.-J. Chae. 2021a. A review on self-sustainable microbial electrolysis cells for electro-biohydrogen production via coupling with carbon-neutral renewable energy technologies. *Bioresource Technology* 320: 124363.

Yang, Q., M. Yan, L. Zhang, X. Xia, Y. Zhu, C. Zhang, B. Zhao, X. Ma, and X. Wang. 2021b. Thermodynamic analysis of chemical looping coupling process for coproducing syngas and hydrogen with *in situ* CO2 utilization. *Energy Conversion and Management* 231: 113845.

Yanxing, Z., G. Maoqiong, Z. Yuan, D. Xueqiang, and S. Jun. 2019. Thermodynamics analysis of hydrogen storage based on compressed gaseous hydrogen, liquid hydrogen and cryo-compressed hydrogen. *International Journal of Hydrogen Energy* 44: 16833–16840.

Yu, Z.-Y., Y. Duan, X.-Y. Feng, X. Yu, M.-R. Gao, and S.-H. Yu. 2021. Clean and affordable hydrogen fuel from alkaline water splitting: Past, recent progress, and future prospects. *Advanced Materials*: 2007100.

Yue, M., H. Lambert, E. Pahon, R. Roche, S. Jemei, and D. Hissel. 2021. Hydrogen energy systems: A critical review of technologies, applications, trends and challenges. *Renewable and Sustainable Energy Reviews* 146: 111180.

Zhang, T., D. Jiang, H. Zhang, Y. Jing, N. Tahir, Y. Zhang, and Q. Zhang. 2020. Comparative study on bio-hydrogen production from corn stover: Photo-fermentation, dark-fermentation and dark-photo co-fermentation. *International Journal of Hydrogen Energy* 45(6), February: 3807–3814.

11 Biomass and Biofuels

Jorge Saldaña, Mesut Yurukcu, Narendra Boppana, Sepehr Arbabi, Jim Henry, and Sebahattin Ziyanak

11.1 INTRODUCTION

In the category of alternatives to fossil fuels, we have seen the marked advancement of renewable sources like solar and wind, the assessment of ocean and tides, the potential of hydropower when considering water, and the cautious exploration of nuclear power. Another of these alternatives is presented in biomass, which can be understood as materials from plants, trees, algae, and various other organic and renewable sources. When these materials are used as a source of energy, we can refer to them as *biofuels*. In the first generation of biofuel technology, its two most common types are ethanol and biodiesel (US Department of Energy, 2023).

Compared to traditional energy sources, biomass offers the comparative advantage of being the by-product of the refinement of organic leftovers or waste products, such as the remnants and discarded vegetable materials from agricultural production. Thus, unused crops, municipal solid waste, or wood from thinning woodlots can turn into valuable energy (EIA, 2022). In this sense, biopower helps in two ways: first, by decreasing humanity's dependence on fossil fuels and, second, by reutilizing material that had no other productive use until recently. Because of significant technological advances, the study of renewable biomass energy has gained increasing attention in the past decade.

To generate biofuels, biomass is refined following different methods. Despite producing CO_2 when burned, the carbon dioxide captured by living biomass through photosynthesis processes that allow it to grow offsets these emissions. Because of this, biomass can be considered a potentially carbon-neutral source of energy (EESI, 2023; EIA, 2022). Other methods to produce biofuels are illustrated in Figure 11.1, consisting mainly of thermochemical and biochemical treatments. While the former method uses biomass in the presence of a heat source and controllable oxygen atmosphere, the latter relies on enzymes, bacteria, and other organisms to transform biomass into liquid fuels (Osman et al., 2021).

In the energy transition process context, assessing the potential of biomass and biofuels as viable alternatives to fossil fuels is imperative. More research over decades has led to academics finding strengths of biomass, such as being a renewable source with reduced carbon emissions compared to fossil fuels. Additionally, the development of biopower can help local communities grow independently from larger urban centers, decentralizing energy production and creating jobs in areas such as farming, processing, and distribution. Moreover, waste management could be improved by developing waste-to-energy plants, which would decrease the amount of waste in landfills while also producing valuable energy (EIA, 2022).

At the present moment, it is estimated that biomass produces around 10% of the energy consumed around the world, at approximately 5×10^9 kJ (kilojoules) of energy a year. Thanks to novel production, collection, and conversion technologies, this value can be increased close to 30 times by 2050 (Alper et al., 2020; Guo et al., 2015a). Additionally, the pressures placed by a changing world and the climate considerations raised by institutions like the Intergovernmental Panel on Climate Change (IPCC) invite the exploration of increasing alternatives to fossil fuels, and biopower is a source that develops from repurposing waste and unused organic materials to produce clean energy.

In this chapter, we review the potential for using biomass on a large scale to produce biofuels that can keep the world running. To this effect, we highlight the strengths and the limitations of current technologies, as well as explore the paths forward enabled by the role of the oil and gas industry in the search for biopower's fullest potential. Additionally, the close synergies and collaborations

 DOI: 10.1201/9781003282273-11

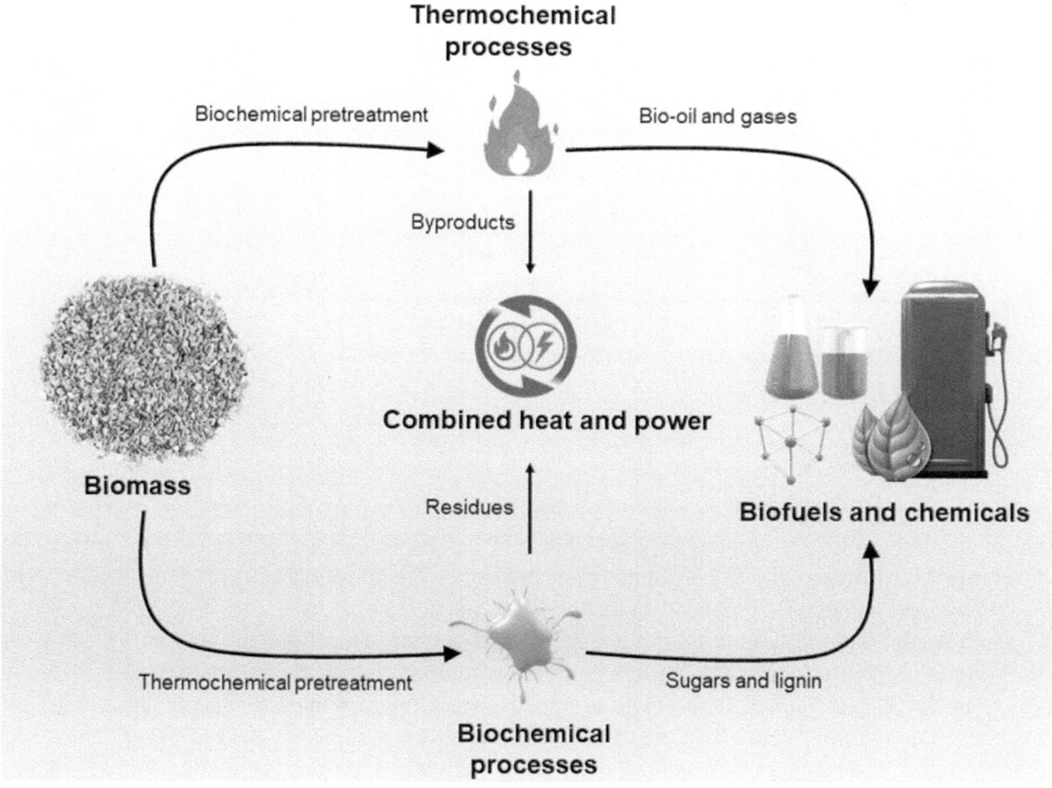

FIGURE 11.1 Integration of hydrothermal and biochemical routes in biomass utilization from a circular economy approach. First, biomass is pre-treated using a biochemical process for the following thermochemical route, or vice versa, by thermochemical pre-treatment for the following biochemical route, eventually producing biofuel or chemicals.

Source: Osman et al. (2021).

between regulatory actors and international organizations to legislate a friendly environment are analyzed closely so that we can provide a framework for expanding these technologies across many more locations in the world that are used by engineers and academics alike.

11.2 SCREENING, SELECTION, DESIGN, AND ECONOMICS OF BIOMASS/BIOFUELS

The choosing of an optimal site to develop biomass energy follows a series of factors that can include the geographical location, the quality of the input material and the soil where it is grown, the potential to utilize this organic material to feed people, social considerations related to the tourism potential of the area, and the level of subsidies provided by local and national authorities. In this section, we review the parameters best suited for selecting a site to develop biomass energy and design criteria for installations. Likewise, we touch on the economic aspects of these decisions and the expected outcomes for locations that seek to capitalize on biopower.

In developing a thriving biopower sector, many concerns need to be tackled, including the building of appropriate infrastructure in the optimal location, the logistics of energy production, and the supply chain that delivers the produced electricity to the final user. Given that producing biomass is not particularly expensive, the success or failure of a large sector that uses excess biomass

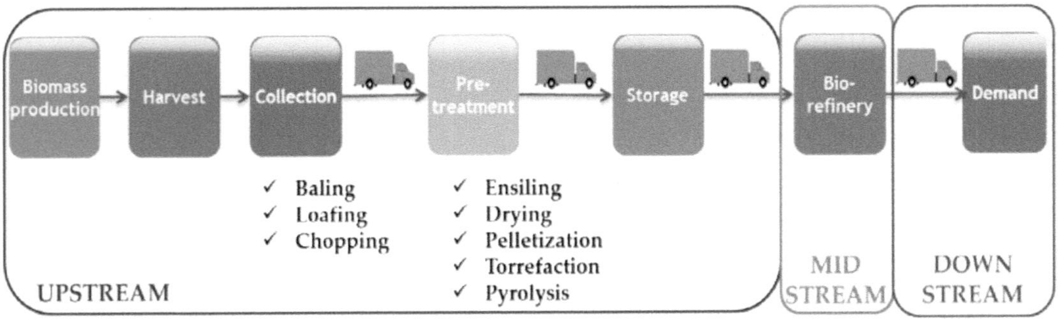

FIGURE 11.2 Major segments and activities in a biomass supply chain.

Source: Zandi Atashbar et al. (2017).

production to generate biofuels centers on lowering operation and management costs and establishing an efficient logistics infrastructure that does not increase these expenditures too significantly. As Figure 11.2 illustrates, efficient midstream and downstream operations are likely the key to a prosperous and sustainable biopower sector.

An example of optimization methods used to find optimal cost strategies is seen in the work of Zhang et al. (2016), where the authors combine simulation modeling with geographic information systems (GIS) technologies to create extended frameworks for biomass supply chains and site selection. The authors utilize GIS methods to assess siting parameters, including variables like connectivity to the railroad and state roads, county boundaries, census population, forest biomass production, water bodies, community dispersion, and co-fired power plant locations. A list of the parameters is illustrated in Figure 11.3.

The rationale for evaluating some of these variables is that similar power plants near the potential location can be seen as a negative aspect due to the competition for biomass that would ensue between installations with similar objectives. Meanwhile, the population count is included to assess the potential for job availability in a given location. Then, water access is evaluated as a factor that could lower variable operating expenses, while lower costs are also expected from locations with a transportation infrastructure already built (Zhang et al., 2016).

Once those variables are evaluated, a second stage in the optimization process consists of a mathematical model that accurately characterizes the process of supplying biofuel facilities so that the costs of providing the required biomass are minimized as much as possible. In the work of Zhang et al. (2016), this supply chain optimization model relies on a mixed-integer linear program (MILP) that identifies the number, size, and location of potential refineries that could process the biomass available in a particular region, while also developing simulations to improve the gathering, handling, and transportation of the materials, to maintain costs on the lower end of the spectrum.

Different quantitative tools and methods have advanced to arrive at optimal supply chain and logistics scenarios informed by site selection parameters. As Ba et al. (2016) point out, GIS-based methods can be coupled with simulation models, while simpler models based on cost calculations are a popular and reliable framework. Moreover, the development of more advanced mathematical programs has allowed for the more widespread use of deterministic, stochastic, and multi-objective models that feature decision variables, different levels of constraints, and objective functions, among other elements.

In deterministic models, linear programs feature variables with linear combinations of continuous variables, which can be solved by simplex algorithms powered by commercial tools like MATLAB. The most common linear program type is the mentioned MILP framework, while non-linear programs are also used, albeit in fewer instances. Meanwhile, stochastic optimization frameworks

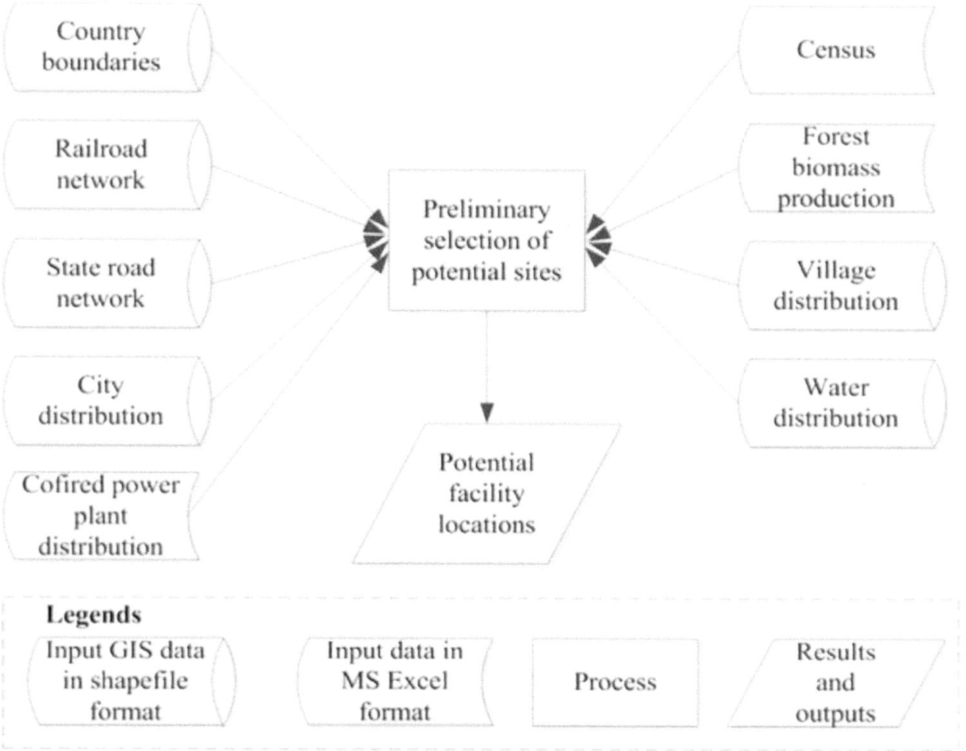

FIGURE 11.3 Overview of the GIS-based methodology proposed in the study.

Source: Zhang et al. (2016).

use random variables to arrive at an objective function, using Monte Carlo simulations to estimate their mean and standard deviation values. Lastly, multi-objective models offer solutions that compromise in different areas, leaving the decision-maker to choose one of the Pareto-optimal alternatives (Ba et al., 2016).

These latter models include diverse variables and arrive at scenarios driven by parameters that move in various ways given their context. As seen in other chapters, multi-criteria decision-making (MCDM) methods are a growing set of tools that help in site selection and cost reduction for different energy alternatives. Additionally, similar models have been applied to improve the location selection of biofuel facilities. As an example, Kheybari et al. (2019) list multi-objective decision-making (MODM) and analytical hierarchical processes (AHPs), along with MCDM, as mathematical frameworks that improve the siting of bioethanol installations.

Moreover, as an improvement to AHP setups, the authors propose a novel MCDM tool called the best-worst method (BWM), which relies on multiple pairwise criteria comparisons performed to find the best scenario, given previous preferences set by decision-makers and experts in searching for the best location for a bioethanol installation in Iran. A series of parameters was established and classified into three areas: economic, social, and environmental. Then, sub-criteria were given different weights following a survey from academic experts who ranked different the criteria on a 10-point Likert scale. Ultimately, the region with the most potential was Khuzestan (Kheybari et al., 2019).

Despite developing criteria that can be classified into one dimension without many problems, some factors can affect multiple dimensions, demanding a more rigorous analysis of the correlates for siting a bioethanol plant. Table 11.1 features a series of considerations that touch on more than

Table 11.1
Criteria for Bioethanol Site Selection That Affects Multiple Dimensions

Criteria	Dimensions		
	Economic	Social	Environmental
Protection of human health (distance to the residential area or density of the population)		*	*
Effect on agriculture		*	*
Effect on tourism (distance to historically important areas)		*	*
Provincial finance subsidies	*	*	
Intensity of natural disasters	*		*
Climate condition	*		*
Waste disposal	*		*
Coordination among supply chain members	*	*	
Safety and security cost (risk)	*	*	

Source: Kheybari et al. (2019).

one of the dimensions noted previously, like protection of human health, effects on agriculture, and effects on tourism, which can affect both social and environmental dimensions. Meanwhile, the intensity of natural disasters, climate conditions, and waste disposal schemes can simultaneously influence the economic and environmental areas. Lastly, local subsidies, safety and security costs, and coordination among supply chain players can affect financial and social dimensions.

In a similar work applied to the Italian context, Delivand et al. (2015) studied a series of constraints that touch on geographical, social, environmental, and economic dimensions using a hybrid model that combines GIS and multi-criteria analysis (MCA). These considerations are related to protected areas, rivers and waterways, forests and woodlands, cultural heritage areas, archaeological sites, and regional ecological networks. Once the availability of land is established, potential sites are assessed in their suitability, removing locations used for industrial or agricultural purposes or those with significant slopes and water bodies.

After these steps, the GIS-AHP model performs a location-allocation analysis, where several candidate plants are placed in more suitable locations, minimizing the distances between biomass source points and a given facility as much as possible. Lastly, an incorporated logistics cost model evaluates investment and operating costs, including collection, loading, transportation, stacking, and storing of biomass materials. This model includes a greenhouse gas emission (GHG) analysis and a life cycle assessment (LCA). These tools allow for a more refined selection of an environmentally friendly site that can stand the test of time (Delivand et al., 2015).

An essential characteristic of these simulation models centers on their contrast of various parameters, which implies that some of these factors could be at odds with each other at times. Therefore, arriving at the most optimal scenario for site selection may mean sacrificing one aspect in one geographical location and a different aspect in another part of the world. To capture this uncertainty, Hezam et al. (2023) developed a fuzzy set framework to site biofuel production plants (BPPs) that takes into consideration the three features that form the triple bottom line (TBL) theory of sustainability: social, environmental, and economic aspects.

Developing a weighted aggregated sum product assessment (WASPAS) method, an improved MCDM framework that ranks options following a compromise solution, while adding picture fuzzy similarity measures (PiFSM) elements, the authors seek to obtain MCDM options under uncertain criteria and weighting from decision-making experts. Thus, following the list of parameters (Q_n) shown in Table 11.2, the authors weigh the assessments from experts on the field and determine

TABLE 11.2
Parameters for Siting a Biofuel Production Plant

Dimension	Criteria	Type
Social	Job creation (Q_1)	Positive
	Training employees (Q_2)	Positive
	Social effects (Q_3)	Positive
	Land use (Q_4)	Negative
	Energy consumption (Q_5)	Negative
Environmental	Relative humidity (Q_6)	Negative
	Waste generation per capita (Q_7)	Negative
	Flood susceptibility (Q_8)	Negative
	Distance from historic tourist places (Q_9)	Positive
	Land price (Q_{10})	Negative
Economic	Pollution (Q_{11})	Negative
	Transportation cost (Q_{12})	Negative
	Operations and maintenance cost (Q_{13})	Negative

Source: Hezam et al. (2023).

that location 2 (P_2) is the most suitable for developing a biofuel installation, given the information provided.

A similar framework is used by Krishankumar et al. (2023), who proposed a q-Rung fuzzy method that weights parameters based on the CRITIC (criteria importance through intercriteria correlations) structure, along with a ranking procedure that is based on the combined compromise solution (CoCoSo) approach. This model weights experts' criteria and ranks biomass locations using an algorithm that evaluates the information from authorities, closely resembling human-driven decision-making. Using Tamil Nadu in India as their case study, the authors found promising sites, such as Hosur, Kongu, Dharmapuri, Kancheepuram, and Mettupalayam.

While developed nations have more leeway when exploring renewable energy installations, developing countries have a more immediate need to provide their populations with alternatives that are not only efficient but also sustainable. Because of this, we see many siting schemes applied in the latter set of countries. As an example, Erdin and Ozkaya (2019) explored the use of ELECTRE. This MCDM tool allows for energy options selection within a group of potential choices analyzed by the framework, similar to the mechanism for PROMETHEE models that work with siting variables. Their study in Turkey finds that biomass is one of two strong choices to pursue in Central and Eastern Anatolia while being one of four promising alternatives for the Marmara region.

In addition to parameters for site selection, biomass and biofuels have the added feature of necessitating very efficient supply chains so that the economic viability of the conversion plants is maximized to its fullest extent. A *biomass supply chain* can be understood as the flow of material between the lands where biomass is created and the biorefinery where it ends up turned into biofuels (Zandi Atashbar et al., 2017). Due to its many moving parts, the optimization of a supply chain can touch on procedures like preprocessing operations, distribution installations, transportation needs, and storage capabilities, among others.

To arrive at the point of proposing novel approaches for optimizing supply chains for biomass resources, some literature starts with site selection methods, arriving later at supply chain improvement proposals through techno-economic assessments. For example, Martinkus et al. (2018) explored using a multi-criteria decision tool (DST) that finds optimal locations for a wood-based supply chain from depots to store biomass to biorefineries used to generate fuels. These locations are found in already-built industrial facilities with a high degree of repurposing, cutting building costs while transforming buildings for clean energy use.

Similarly, Roni et al. (2019) assessed the potential for improvement from preprocessing centers located at biorefineries to be located in distributed biomass installations, also known as "depots," near the biomass sources instead. The argument for a distributed network, as opposed to a centralized one, is to place depots near the biomass sources to produce high-density pellets, which can later be more efficiently transported to the biorefinery instead of the untreated biomass. To analyze this proposed move, the authors relied on a mixed-integer linear programming (MILP) method that helps identify appropriate locations and cost reductions in the blends produced from biomass.

A different scheme is seen in the work of Allman et al. (2021), who tackled supply chain problems with mobile production modules that convert biomass waste into energy in different places, depending on the varying needs of the suppliers. Studying the American states of Minnesota and North Carolina, the authors found supply chain cost savings of 1–4% with this strategy, translating into millions of dollars in the medium term. Additionally, this added flexibility helps tackle some of the uncertainty that varying production degrees from different suppliers in different regions can generate.

Optimization under uncertainty is a crucial aspect of improving biofuel supply chains. Addressing changing factors in different regions of the world, which also feature diverse quantities of resources and levels of biomass production depending on the season, is vital for the economic viability of a bioenergy endeavor. Moved by the variability in the relationships within supply chains, Aboytes-Ojeda et al. (2022) developed a two-stage stochastic programming model, with first-stage variables being the location of depots and refineries and the number of trains to transport biomass, and second-stage factors being third-party bioethanol supply and the flow of biomass between network nodes.

Using the state of Texas as a case study, the authors advanced a hub-and-spoke network model where the supply chain is formed by three major players: farms, preprocessing depots, and biorefineries. In this design, the players are connected by nodes, meaning, transportation from farms to depots by trucks and from depots to biorefineries by train cars. Through an L-shaped algorithm with connectivity constraints, they arrived at an optimized scenario that improves from previous benchmark L-shaped frameworks (Aboytes-Ojeda et al., 2022).

Meanwhile, an optimization framework applied to forest biomass is provided by Acuna et al. (2019), who advanced a strategy of efficient transportation of roots, stems, branches, leaves, and all other materials not used by other industries and can be turned into valuable energy. The dynamics from different sectors can compound the difficulties in finding an appropriate optimization strategy, which invites authors to try methods that combine various strategies that account for uncertainty and diverse potential sources and suppliers.

A review from Lo et al. (2021) examines the trends in biomass supply chain studies, finding that most of these analyses follow a mathematical modeling approach with an optimization tool, while a few works performed stochastic evaluations incorporating biomass supply chain uncertainties. Moreover, these fluctuating considerations can be classified into six main areas: biomass availability, biomass quality, transportation costs, market demand, price fluctuations, and workers' wages. Also, the increasing role of big data and predicting models can improve techno-economic analyses for supply chains, expanding their application across the globe.

One last effort we mention in this section is the work of Gital Durmaz and Bilgen (2020), who sought to find the optimal design for a biomass supply chain from poultry farms to biogas facilities. As Figure 11.4 shows, coupling GIS-based methods with an analytic hierarchy process (AHP) helped the authors narrow down potential sites for a biogas installation. At the same time, a multi-objective mixed-integer linear programming (MOMILP) tool was used to make strategic and tactical decisions on optimal facility locations and transportation network flows, respectively. In other words, this framework helps maximize profits while minimizing the distance traveled between farms and biogas installations.

The First Stage: GIS

Objective:	Achieving environmental objective by considering ecological criteria.
Inputs/data:	Ecological units and distances in table.
Outcome:	Ecological appropriate locations to establish biogas facility.

The Second Stage: AHP

Objective:	Eliminating and assessing appropriate locations with main criteria.
Inputs/data:	Ecological appropriate locations obtained by GIS.
Outcome:	Obtaining more improved results to use as candidate biogas facility locations.

The Third Stage: MOMILP

Objective:	Optimizing biomass supply chain network design by minimizing total cost and total distance of entire supply chian.
Inputs/data:	Candidate biogas facility locations. obtained by AHP.
Outcome:	Locations and capacities of biogas facilities, the amount of biomass transported between farms and facilities, the amount of electricity and fertilizer produced in facilities.

FIGURE 11.4 Multi-stage solution framework for biomass supply chain network design optimization.

Source: Gital Durmaz and Bilgen (2020).

To summarize, the criteria for site selection of biomass facilities also come with the challenge of assessing supply chain optimization features. Therefore, geographical, environmental, and social variables need to be weighed along with economic considerations that minimize costs in several areas in the energy production cycle. These calculations are necessary because biomass production costs are low compared to other energy sources, so most investments are related to building bioenergy installations and the transportation networks established to supply those facilities.

Because of this, the economics of biomass and biofuels are much more closely related to the screening and selection of sites than other energy sources. As a whole, supply chain and biomass logistics operations can constitute one single significant endeavor for researchers. Figure 11.5 shows a diagram of these two areas, with collection, storage, conversion, and delivery to the final users as the main areas of the supply chain. Despite the systematic character of these frameworks, variables like seasonal crop availability, scattered geographical distribution, and variations in quality add significant complexity to optimizing biomass logistics (Malladi and Sowlati, 2018).

Another critical aspect in developing a sustainable biomass and biofuels sector is the effective allocation of resources through the achievement of a balance for using materials for food, feed, or fuel. Thus, a healthy industry must tackle trade-offs and embrace synergies in biomass use to improve people's lives worldwide. As Figure 11.6 shows, decisions regarding the use of organic materials have to cover three areas, with food being the primary purpose of the optimization of

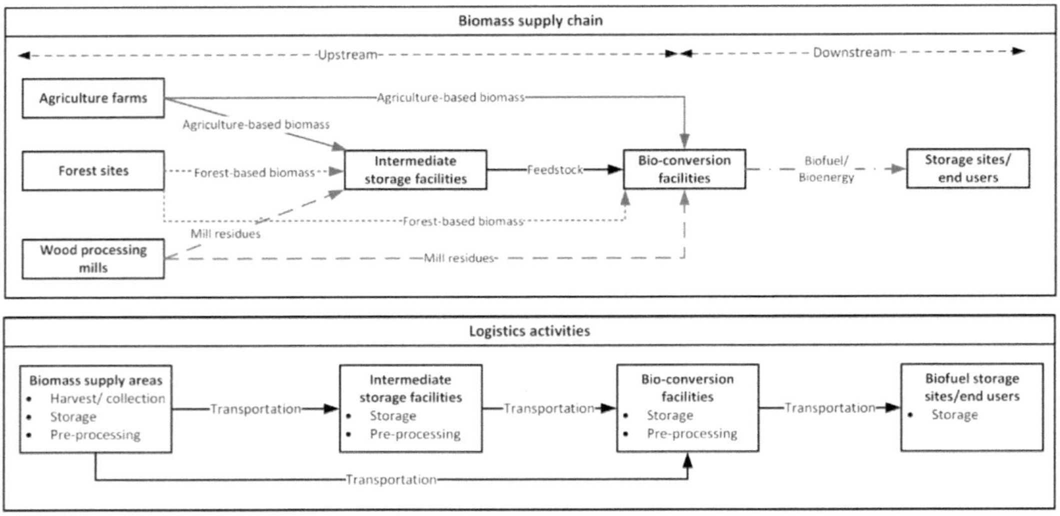

FIGURE 11.5 Biomass supply chain and logistics activities.

Source: Malladi and Sowlati (2018).

crops. Once food is allocated, waste can be minimized by feeding back into the system and relying on livestock to boost virtuous cycles. Then, the remaining available organic materials can be used for fuel and bioenergy purposes (Muscat et al., 2020).

Moreover, in their analysis of the potential of biomass to become a viable alternative to fossil fuels in the European Union, Popp et al. (2021) also highlighted the main priority for food security in the use of biomass. With that said, organic materials' chemical and energy production sectors have increasingly growing markets, especially under a scenario of low-carbon economies and regulations. For the EU's targets regarding greenhouse emissions and net-zero processes, feedstocks of sugar, starch, vegetable oil, and ethanol are required to compete with a fossil fuel–based economy.

This economic transformation requires, among other things, a valorization of the biomass utilized for energy purposes, which does not necessarily take away from organic materials grown and developed for food needs. In this sense, advancements in genetic engineering that enhance specific yeasts' capacities for ethanol production have shown significant promise in recent years (Favaro et al., 2019). Additionally, fourth-generation modified algal biomass has shown exciting potential for use as feedstock, lowering costs and increasing quality and efficacy (Shokravi et al., 2021).

With that said, these endeavors can take very real tolls not only when land is limited but also when labor, capital, funding, and resources like water are not abundant. Therefore, decision-making regarding the use of crops requires models that can confront this ambiguity. As seen in this section, stochastic mathematical models assessing uncertainty have become popular in recent years. Yet the discipline could benefit from biomass logistics experience, a dimension that would add depth to the analysis. For example, Zandi Atashbar et al. (2017) noted the lack of non-dominated sorting generic algorithms (NSGA-II) methods, which have been used in industrial logistic optimization studies with multiple objectives.

Closing this section, we observe the need for encompassing methods that not only observe several basic parameters but also account for seasonal variations, resource quality, transportation needs, economic contexts, and even geopolitical considerations for the use of land and crops. Because of this, we perceive some significant limitations to introducing bioenergy facilities to developing countries. However, technological advancements have allowed for the optimization of fields and the year-round use of land for multiple uses, which can include a more significant share of biomass to be converted to biofuels.

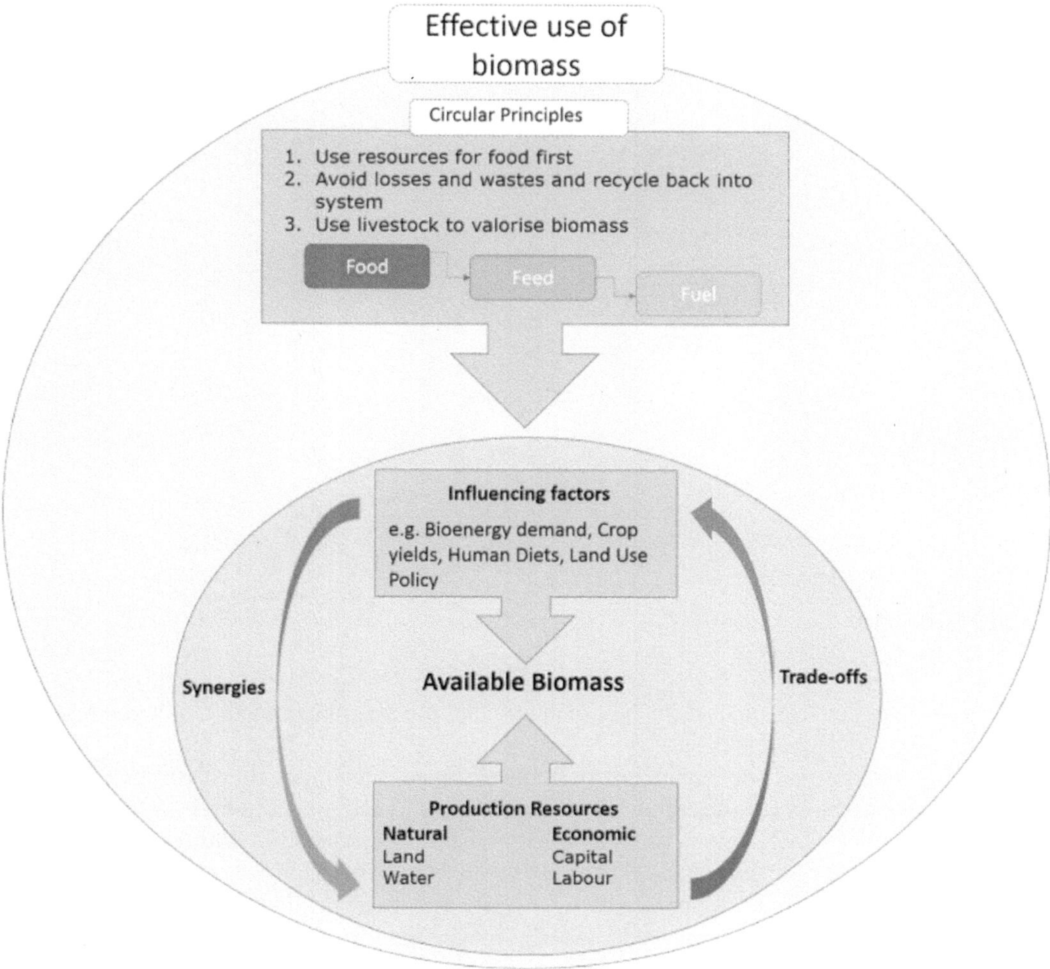

FIGURE 11.6 Framework for an effective use of biomass, with relevant factors, trade-offs, and synergies.

Source: Muscat et al. (2020).

The following sections deal with the different types of biomass resources that are optimal for developing a sustainable biofuel industry and the various principles that have driven some locations in the world to be leaders in the sector. Moreover, a series of best practices is outlined so actors looking to invest in these endeavors can do so with a safer footing. Then, we explore the role of the oil and gas industry in developing bioenergy, their commitments to improving technologies and methods, and the potential of collaboration in the context of the energy transition to renewable energy.

11.3 BIOMASS/BIOFUEL TYPES, PRINCIPLES, AND BEST PRACTICES

As noted, *biomass* is understood as any organic material produced by photosynthesis. In the energy transition context, biomass can be turned into fuels, thus allowing humans to extract energy from it. In the following section, we review the different types of biomass materials that can be successfully converted into biofuels, the potential widespread use of these technologies all over the world, the implications of these operations in the global energy context, the players and regulatory frameworks

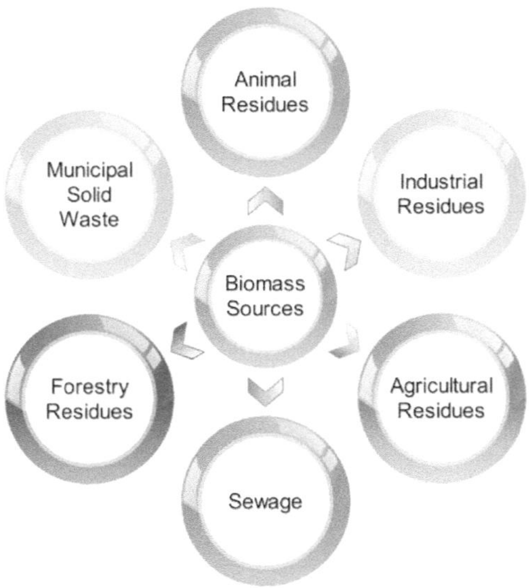

FIGURE 11.7 Key biomass sources.

Source: Tursi (2019).

required to make this happen, and the best practices that can be prescribed from our vantage point of having analyzed the sector closely.

An initial classification of biomass materials touches upon their sources of origin. As Figure 11.7 shows, biomass can come mainly from animal, industrial, forestry, agricultural residues, sewage, and municipal solid waste. Remnants from wood industrial processes like sawdust and shavings, those from algae, sewage, and aquatic crops, and waste from animal farms are the most important sources of biomass. Meanwhile, municipal solid waste and waste originating from anthropogenic activities are both significant sources of biomass, as long as these are not reusable for other purposes after subsequent processing (Tursi, 2019).

Also, biomass sources can be separated according to their chemical composition, resulting in varying properties and processes when converting them into energy for human consumption. Some examples can be seen in different quantities of plant residues, which are made of cellulose, hemicellulose, and lignin. Then, cattle manure mainly comprises starch, featuring some proteins and cereals. The exploration of these types of biomass inputs has generated an opportunity to exploit abundant resources that are obtained at a low cost and that are also not competing for food usage, as is the case with some agricultural and animal products (Brethauer and Studer, 2015).

Meanwhile, biofuels have been classified into different types, with a typology regarding different generations being mainstream in the literature. Following this classification, first-generation biofuels consist of energy from edible resources, such as starch, sugars, animal fats, and vegetable oils. Second-generation biofuels are based on non-edible crops, like crop residues, forest by-products, and industrial waste, while third-generation materials revolve around algae and similar resources. Lastly, fourth-generation biofuels center on advancements on synthetic biology, such as genetically modified crops for energy demands (Vignesh et al., 2020; Zandi Atashbar et al., 2017).

In this section, we delve deeper into the different categories of biomass, the implications of these boundaries, the various processes to generate biofuels, their uses, potential, and expectations for the upcoming years. Additionally, we offer a framework for future research with best practices and principles for academics and engineers who are starting their paths into these energy sources. Knowing all the components of these intricate processes will help facilitate the energy transition process that will phase out fossil fuels in the following decades.

11.3.1 Biomass as Fuel

The understanding of organic materials being used to produce some fuel has accompanied humanity since very early in our existence. In prehistoric times, tribes and early humans burned materials like wood and leaves to produce heat and light. Around 260°C, a combustion process starts for wood, generating solid char and gaseous fumes. Meanwhile, more modern civilizations used crop residues for cooking and heating, among other industrial processes. Archaeologists have found evidence of controlled fires from groups of people who lived more than a million years ago (Guo et al., 2015a).

With the advent of the Industrial Revolution of the 18th and 19th centuries, biomass gained more popularity, serving as feedstock for steam engines and early industrial processes. However, in the 20th century, interest moved to more dense forms of energy, as coal, oil, and natural gas became cost-effective and dominated the energy market in developed countries. That said, developing countries continued using biomass sources, especially in cooking and heating. Due to modern applications requiring significant levels of capital, skills, technology, and a given level of development, modern applications have not been as easy to replicate in developing nations (Rosillo-Calle, 2016).

Despite these differences, interest in bioenergy has been rekindled around the world with similar intensity, mainly due to its potential as a renewable and sustainable approach that would assist in the phasing out of fossil fuels for many industrial, commercial, and everyday uses (Rodionova et al., 2017). In the context of the energy transition, it becomes vital to explore the most optimal ways to convert renewable biomass materials into efficient fuels and how these procedures can be replicated in more places worldwide, both at a national and local scale.

Thanks to constantly improving technologies, biomass materials can now be converted, among other types, into bio-oils, biodiesel, bioethanol, and biogas, pivotal biofuels that can be used for road transportation, in aviation, to boost other niche markets, and for other energy needs. Figure 11.8 shows the different methods of production of bioenergy, as various classes of biofuels are produced from bio-feedstock depending on the degree of technological advancement, the available feedstock, and the energy requirements of the particular location that stands to benefit (Adams et al., 2018).

In what follows, we look at the evolution in the production of biofuels, the primary methods for their generation, and the potential for future technologies to enhance these processes. Additionally, we look at the newer-generation compounds from algae, microalgae, and genetically modified inputs and their potential to become viable alternatives to fossil fuels in the medium term. These dynamics will be later assessed in the context of the energy transition, a process involving actors like governments, international organizations, and the oil and gas industry, with investment, regulations, and other schemes meant to expedite the move to renewable energy sources.

11.3.2 Bio-Oil

One of the main by-products of biomass is in bio-oils, fuels known as pyrolysis oils. Here, lignocellulosic materials are subject to pyrolysis, where thermal decomposition is performed in the absence of oxygen, producing bio-oils, some gases like syngas and char, the solid by-product of the procedure (Bergvall et al., 2021; Guedes et al., 2018; Pinheiro Pires et al., 2019). Usually conducted over temperatures of 500°C, the heat produced deconstructs the polymers bound in the biomass materials, yielding permanent gases, a solid, and a refinable liquid that serves as bio-oil (Lee et al., 2019). Their chemical composition depends on the feedstock and the conditions used to obtain the crude.

In the long run, pyrolysis processes could become self-sustained, as the same bio-oils and char can be used as fuel to drive subsequent reactions. Additionally, due to the absence of oxygen, combustion does not occur in the pyrolysis process, which aids in lowering potential greenhouse gas emissions that result from thermal processes dealing with other materials (US Agricultural Research Service, 2021). Because of this, more attention has been paid to pyrolysis, which has yielded new methods, like the one illustrated in Figure 11.9. A two-step process of hydrotreating and hydrocracking leads to products like gasoline, jet fuel, diesel, and heavy oil.

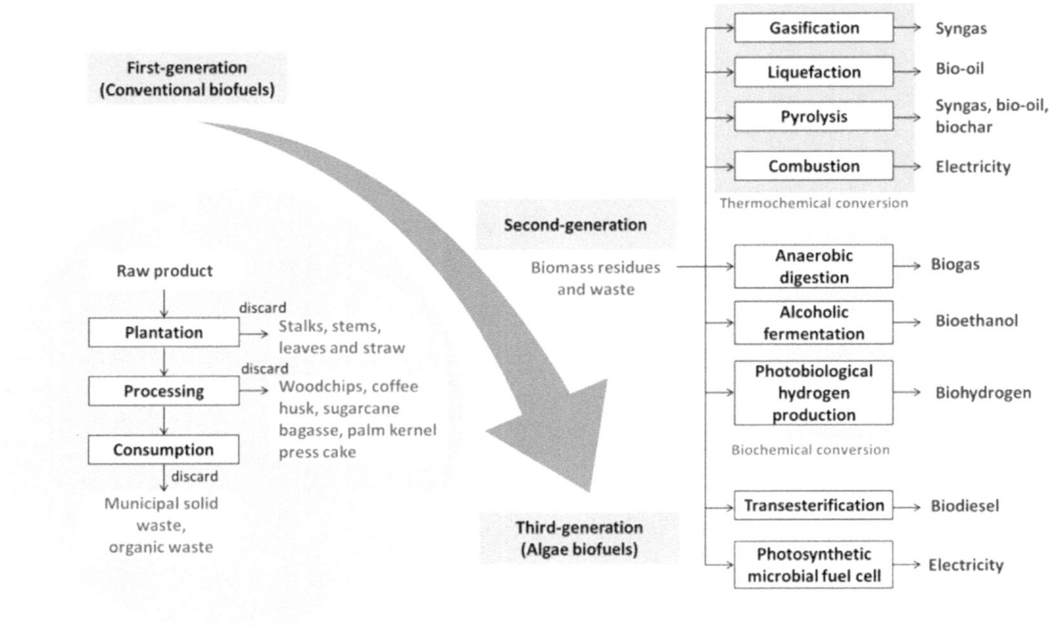

FIGURE 11.8 Diagram for producing different generations of biofuels from biomass residues, emphasizing second-generation compounds.

Source: Lee et al. (2019).

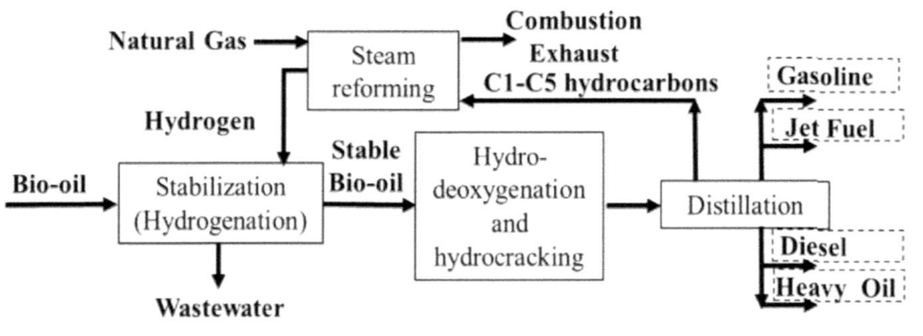

FIGURE 11.9 Hydrotreating and hydrocracking of bio-oil process flow. Products are marked in dashed boxes.

Source: Pinheiro Pires et al. (2019).

This two-stage method has been patented by the Pacific Northwest National Lab (PNNL). It works as follows: A first reactor, known as the stabilization reactor, is operated in temperatures between 150 and 300°C, where some reactions like repolymerization and hydrotreating take place along with converting carbonyl and carboxyl functional groups into alcohols. Then, a second reactor is operated in temperatures between 300 and 500°C, and 10 to 14 MPa (megapascals) of pressure is used, with the goal of hydrocracking and hydrodeoxygenating, requiring hydrogen and a catalyst to obtain lighter molecules from the initial inputs (Bergvall et al., 2021; Pinheiro Pires et al., 2019).

There is ample promise in the exploration of bio-oils through pyrolysis. However, markets continue to be scarce worldwide due to unresolved problems, such as lower profits and more complex operations compared to other energy sources. According to Inayat et al. (2022), the current price for a gallon of bio-oil obtained from algae using fast pyrolysis methods was around US$2.23. This price, while acceptable and only thought to decrease with the eventual enhancement of economies of scale, is still prohibitive for some markets where exploration of biomass is limited by infrastructure and technology constraints.

Some authors have advanced one alternative to pyrolysis methods, which consists of hydrothermal liquefaction (HTL), where chemical and physical conversions are performed in supercritical water, with temperatures over 250°C and pressures over 5 MPa, and the use of a catalyst being optional. This process, also known as hydrous pyrolysis, works well with wet biomass, such as algae, allowing both for the polymerization of small molecules into more significant compounds that form bio-oils and for the degradation of the macromolecules found in biomass (Baloch et al., 2018; Guo et al., 2015b).

Compared to fast pyrolysis, HTL methods can help obtain larger bio-oil yields, lower coke or carbon buildup residues, and less energy consumption. Also, the products obtained from HTL have a higher energy density and superior quality in terms of thermal and storage stability. Because the yields obtained from fast pyrolysis have a more significant concentration of oxygen, subsequent upgrading of the bio-oils is a more challenging endeavor than for HTL-obtained crude. Lastly, pyrolysis produces volatile matter as a by-product and cannot work with wet feedstock, limiting its degree of utility only to dry materials (Baloch et al., 2018; Guo et al., 2015b).

Bio-oils have a promising path forward, yet it is crucial to acknowledge a few limitations to their immediate widespread use. First, the processes we just reviewed continue to be more costly than those procedures performed on established sources of energy. Then, the high oxygen concentration in bio-oils affects their thermal stability during storage, handling, and upgrading procedures. Additionally, not many refining plants exist today to treat bio-oils and convert them into biofuels, which means that production remains at a lower level than desired for a technology that is expected to take over for fossil fuels in the medium term (Pinheiro Pires et al., 2019).

11.3.3 Biodiesel

Biodiesel is produced from vegetable oils, animal fats, used cooking oils, or yellow grease. The primary method to produce biodiesel is through transesterification, which converts these materials into biodiesel and glycerin after mechanical stirring between 50 and 60°C and after the mixture settles at room temperature for several hours. Figure 11.10 illustrates this process, with an esterification phase added if the feedstock requires a pre-treatment to remove water and particulates, such as with yellow grease or used vegetable oil (Guo et al., 2015a).

The term "biodiesel" originated from 19th-century German inventor Rudolf Diesel, who developed a revolutionary combustion engine that could work with some vegetable oils (Tabatabaei et al., 2019). Because of his invention, all fuels that can run that type of engine are known as diesel fuels, where *biodiesel* is the group of fluids produced directly from biomass. Besides transesterification, three other popular methods for biodiesel generation are summarized in Table 11.3, along with their strengths and weaknesses. These are blending, microemulsion, and pyrolysis processes that vary in production, cost-effectiveness, and by-product generation.

As we can see, transesterification is preferred due to the recycling of unreacted feedstock, and glycerol can be used for other valuable products in different sectors. However, the complexity of the equipment required for these processes could be prohibitive for some players. Thus, blending or dilution might be the best option in locations with fewer resources, even though the quality of the output may impede its use as an engine fuel due to its higher viscosity, acid value, gum formation, and free fatty acid (FFA) content (Verma and Sharma, 2016).

With these trade-offs, we are confronted with the many decisions that must be made to develop and flourish a biomass market that sometimes goes against one another. Therefore, funding for equipment could improve the quality of the final biodiesel product. Still, if a crop is bad in a given

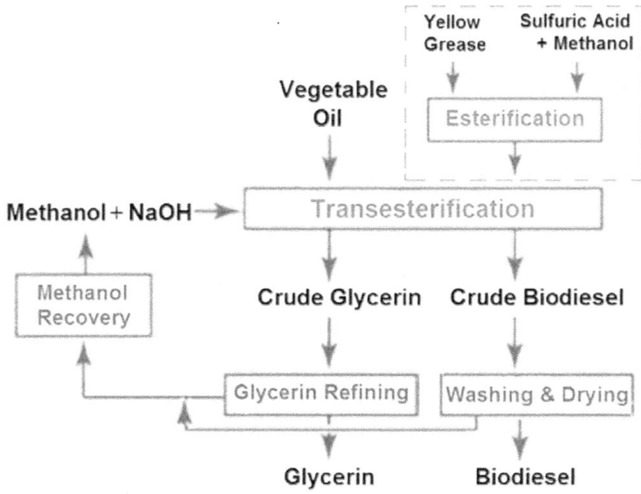

FIGURE 11.10 Technical flowchart for biodiesel production from used vegetable oil.

Source: Guo et al. (2015a).

year, some of the expenditures in infrastructure will be wasted. This raises the question of whether these risky investments are worth it for developing countries, where resources allocated for one area could detract from funding another. The potential role of international energy companies and organizations in helping alleviate the stakes from those decisions will be assessed in later sections.

One strategy that could be taken before considering jumping to biofuel production is to exploit the comparative advantage of a location to produce a given crop and explore the potential to use residues as an energy source once they are abundant. In the case of corn-producing countries, a market has developed where the edible content is used to feed the population, and the co-products, such as corn germ and corn distiller oil, can be used to produce biodiesel. In their work, Veljković et al. (2018) explored the different methods to generate crude from corn germ and the technological advancements to separate corn germ from the crop before the ethanol fermentation process.

Figure 11.11 shows four different methods to recover corn germ to produce crude. These methods are described as follows: First, pressing is the more straightforward method used for feedstocks with low oil content and is recommended for a high volume of the material. The main downside of pressing is its low oil yield and the frequent cleaning of the screw press. Solvent extraction is an efficient procedure that yields high quantities of oil. Still, it remains an unsafe method with potential health risks, and the low quality of the yield may not justify its widespread use. The third mechanism is an economical and green process. Still, aqueous extraction has high costs of low-quality oil, an overall inefficiency, and a comparatively high price for the assisting enzymes (Veljković et al., 2018).

The last method involves using supercritical conditions to extract corn oil, which implies pressures and temperatures higher than the critical point of the input and using carbon dioxide and an optional solvent to obtain it. Using this procedure, the oil yields produced are more significant and of better quality while ensuring a safe and simple operation to generate this biofuel. However, the drawbacks of this process relate to its expensive equipment for high-pressure conditions and its high capital and operational costs (Veljković et al., 2018).

As one of the most common biofuels, biodiesel has received significant governmental backing from countries like the United States, where current yearly production reaches over 2 thousand million gallons as of August of 2023 (EIA, 2023). Meanwhile, the European Union has placed progressively more ambitious goals in their quest to increase their share of biofuels in their energy mix, with

TABLE 11.3
Advantages and Disadvantages of the Four Main Biodiesel Production Processes

Method	Advantage	Disadvantage
Direct use and blending	• Low capital and production costs • Simple production	• Solidification of blend at cold temperatures • Impractical and inappropriate for direct use in diesel engines • High viscosity • Gum formation • Lubricating oil thickening • Incomplete fuel combustion • Oil deterioration • High free fatty acid (FFA) • Low volatility • Unsaturated hydrocarbon chains reactivity • Injector nozzles plugging • Poor atomization • Engine durability reduction, • Higher air pollution emission • Higher engine maintenance costs • Higher engine wear
Microemulsion	• Biodiesel formation with lower viscosity and higher liquidity • Lower nitrogen oxide emissions • No by-product or waste formation • Clear, single-phase, and thermodynamically stable colloidal equilibrium • Dispersion of biodiesel fluid	• Heavy deposition of carbon residue • Inadequate combustion • Lubricating oil thickening • Random injector needle sticking
Pyrolysis	• Suitable for areas with well-established hydro-processing industry • Generation of value-added by-products like syngas • Biofuel with satisfactory physical and chemical properties	• High production cost • Complex equipment requirement • Biofuel has no oxygenated value • Producing short-chain molecules with more similarities to gasoline than diesel fuel
Transesterification	• The most common method for production of biodiesel • Unreacted feedstock can be recycled • The by-product (i.e., glycerol) can be converted into value-added products	• Dry alcohol and oil must be used to increase biodiesel yield by avoiding saponification • Glycerol must be efficiently separated to avoid generation of hazardous gases (i.e., acetaldehyde, formaldehyde) • Expertise requirement • Complex equipment requirement

Source: Tabatabaei et al. (2019).

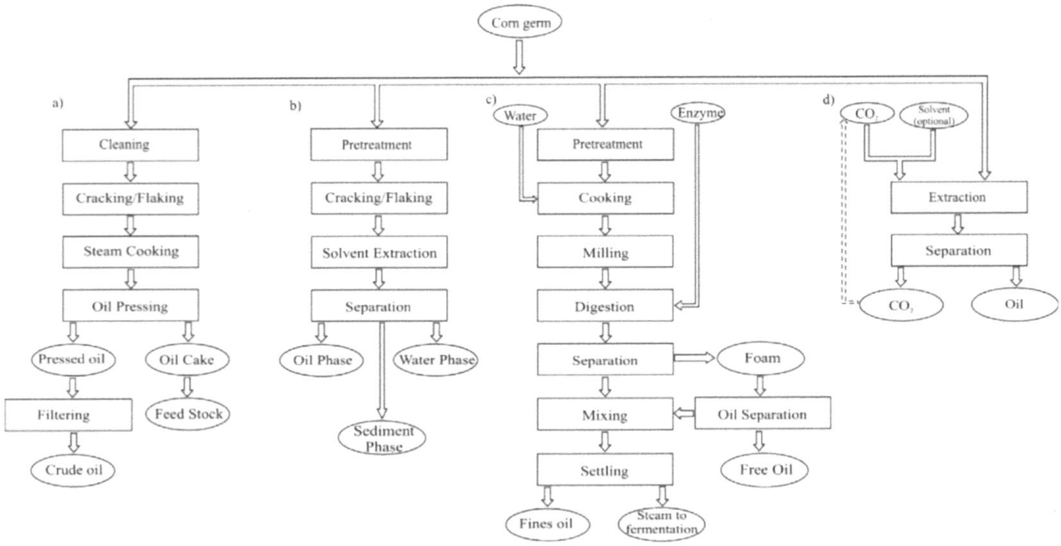

FIGURE 11.11 Schematic presentation of corn germ recovery using (a) pressing, (b) solvent extraction, (c) enzyme-assisted aqueous extraction, and (d) superficial fluid extraction.

Source: Veljković et al. (2018).

new rules implemented in June 2023 that direct member states to increase the share of renewable sources of energy to 14% by 2030, and a 3.5% of advanced biofuels (European Commission, 2023).

11.3.4 Bioethanol

Another liquid fuel that holds tremendous potential to replace transportation fossil fuels is bioethanol, considered a favorable alternative to conventional gas fuel due to its high-octane content of 108 and its lower impact on the environment when going through combustion (Aditiya et al., 2016). One of the main methods to commercially produce bioethanol is fermenting plant biomass–derived simple sugars, such as glucose, fructose, and other saccharides. Moreover, given that cellulose, hemicellulose, and lignin are integral parts of the cell walls of most plant materials, virtually all of them can be used to generate bioethanol, provided they go through effective pre-treatment methods (Guo et al., 2015a).

Figure 11.12 summarizes the leading processes for producing cellulosic ethanol, where pre-treatment consists primarily of steam explosion techniques (SE) that help break down the recalcitrant lignocellulosic matrix. This process uses saturated steam at high pressures in two different stages, with the steam boiling phase provoking a breakdown of the lignocellulosic matrix at temperatures of 170–210°C. Then, the explosion phase converts thermal energy into mechanical energy following a sudden pressure drop that leads to a vapor expansion that disrupts the biomass's fibrous structure (Ziegler-Devin et al., 2021).

Once pre-treatment is done, a conditioning stage may be necessary to deal with inhibitory compounds that can negatively impact the enzymatic hydrolysis and fermentation stages. When the biomass slurry is ready, it is exposed to a mix of cellulolytic enzymes in moderately hot temperatures between 45 and 50°C and an acidity of 4.8 pH, which helps the materials degrade. Then, a fermentation process follows, triggered by the metabolic activity of the agent used to cause it, which could consist of glucose-fermenting bacteria or different types of yeast. Lastly, various distillation methods separate the resulting ethanol from water contents that remain in the mix (Aditiya et al., 2016; Brethauer and Studer, 2015).

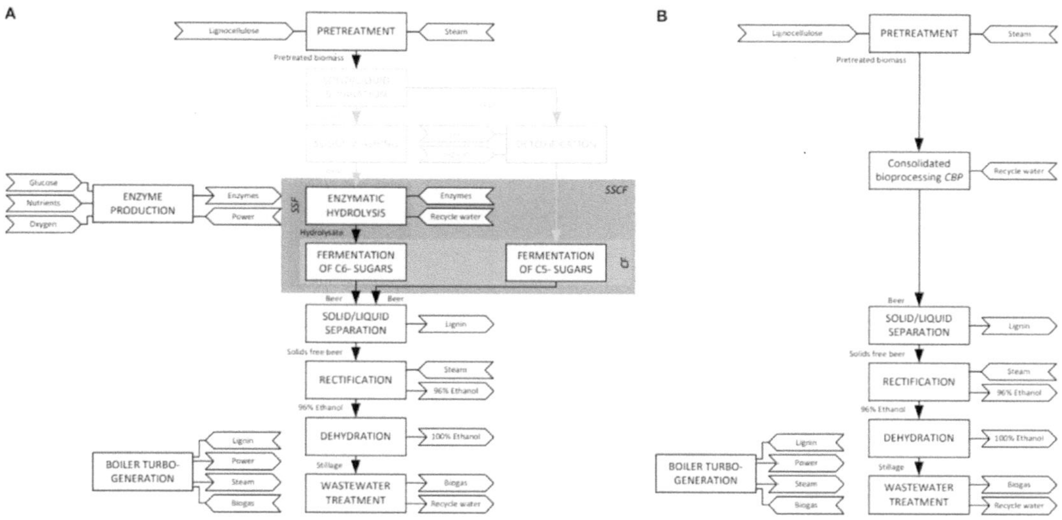

FIGURE 11.12 Process flow sheet for the production of cellulosic ethanol. (a) Cellulosic ethanol production's three primary unit operations: pre-treatment, enzymatic hydrolysis of biomass carbohydrates, and fermentation of released sugars to ethanol. An optional procedure consists of the integration of some unit operations, as depicted in boxes CF (co-fermentation), SSF (simultaneous saccharification and fermentation), and SScF (simultaneous saccharification and co-fermentation). The three main steps are optionally supplemented by several separation, washing, and polishing steps (depicted in gray) to deal with fermentation and enzymatic hydrolysis inhibitors released in the pre-treatment step. Further unit operations include product purification, steam, power generation, enzyme production, and wastewater treatment. (b) Integration of enzyme production, enzymatic hydrolysis, and co-fermentation in one step under the most advanced processing option, known as consolidated bioprocessing (CBP).

Source: Brethauer and Studer (2015).

Features like the quality and yield of bioethanol are directly related to things like the enzyme mix used to degrade the biomass material, the overall production environment, factors like temperature and pH, and the type of feedstock used to make the fuel. Table 11.4 shows a list of sugary biomass materials that can serve as a promising feedstock for first-generation ethanol production, with sugarcane offering a high yield, while crops like dates and watermelons providing varying types of sugars that may achieve increased quality, with the added benefit of waste reduction if used to produce biofuels.

Exploring the nuances in the choosing of a given fermentation agent to produce bioethanol from biomass, Favaro et al. (2019) explored the potential of the *Saccharomyces cerevisiae* microorganism, the preferred yeast strain to achieve high ethanol yields, in aiding a bio-based economy that replaces fossil fuels. The authors find that the great promise shown by these strains with first-generation biomass can be replicated for the new generations of organic materials, provided the possibilities for their industrial improvement are explored. Thus, genetic engineering and smart evolutionary engineering can provide yeasts with higher levels of inhibitor tolerance and a wider range of sugar conversion.

In contrast, starch-based materials like corn, cassava, wheat, and rice are put through liquefaction and saccharification to hydrolyze the carbohydrates into the corresponding sugar monomers. Moreover, corn can go through either dry or wet milling, with the former being an exclusive bioethanol-producing procedure. At the same time, the latter is performed to obtain other by-products, such as high-fructose corn syrup, dextrose, and glucose syrup. These processes can yield up to 400 L of ethanol per ton of corn, around four times higher than the yield from sugar feedstocks (Susmozas et al., 2020).

TABLE 11.4
Potential of Some Sugary Biomasses as Ethanol Feedstock

Biomass	Major Influential Features
Sugarcane (*Saccharum officinarum*)	• High biomass yield • High sucrose content • High efficiency to accumulate solar energy • Global production is 360 million ton/year • Crop yield is 60–79.5 t/ha • Residues are good source for generating electricity and second-generation bioethanol
Sweet sorghum (*Sorghum bicolor*)	• More drought-tolerant crop than sugarcane • Can grow in arid land • Stalk juice a promising feedstock for bioethanol, while grains used for food and starch-based ethanol production • Stalks contain significant amounts of sucrose, glucose, and fructose • Average syrup yield from stalk is 1,900 L/ha
Sugar beet (*Beta vulgaris*)	• Major source of sugars in Europe and North America • Crop yield is 79.1 t/ha • Average syrup yield is 60–120 t/ha • Average ethanol yield is 95 L/ton of stalk
Watermelon (*Citrullus lanatus*)	• Around 20% of waste watermelons during or after harvesting of crops can be used as bioethanol feedstock • Juice can be used as a diluent for molasses that increases overall sugar content in the feedstock and reduces water consumption
Dates (*Phoenix dactylifera*)	• A good amount of dates wasted each year in date-producing countries, which can be used as ethanol feedstock • Contains high amounts of individual sugars (sucrose, glucose, and fructose) as estimated for roughly 70% of total sugars • Longtime storage
Molasses	• A renowned ethanol feedstock enriched with sucrose, glucose, and fructose • An industrial waste and thus no debate of food versus fuel • Valorization of the waste material • Molasses yield approximately 3–7 t/100 t of sugarcane

Source: Zabed et al. (2017).

Even though today most of the bioethanol produced comes from food crops, advanced ethanol can be obtained from second-generation sources, such as industrial waste, lignocellulosic feedstocks, and non-food crops, as well as third-generation feedstock like algae (Susmozas et al., 2020; Zabed et al., 2017). Additionally, bioethanol produced from 2G materials such as forestry and agricultural residues has been shown to reduce greenhouse gas emissions (GHG) to a significant degree when compared to fossil fuels, particularly when the fuel blends are tilted favorably toward using bioethanol, such as the case with E85 mixtures, where only 15% of the blend consists of gasoline (Sharma et al., 2020).

Third- and fourth-generation bioethanol production processes are at a very early stage. Still, they are one of the paths forward to combat the potential scarcity of agricultural resources and the recalcitrance that some materials have to extract the sugars from the biomass. Advanced methods like genetic modification and adaptive evolution are being explored to tackle seasonal and access constraints. Meanwhile, an emphasis on reducing both the energy consumption in the ethanol production process and the costs of enzymes, yeasts, and those incurred in the production cycle are other avenues toward expanding bioethanol use around the world (Brethauer and Studer, 2015; Devi et al., 2023).

11.3.5 Biomass-to-Liquid (BtL) Fuels

These biofuels are obtained through a thermochemical process, such as pyrolysis, gasification, supercritical fluid extraction, and direct liquefaction, using non-food and waste as feedstock. Compared to some of the biochemical processes that we have reviewed, the thermochemical route is significantly faster. These methods also allow for a more seamless integration of the fuels into the conventional fossil fuel distribution system, along with their use in standard engines that rely on gasoline or diesel (Dimitriou et al., 2018; Ibarra-Gonzalez and Rong, 2018).

To produce BtL biofuel, the biomass collected is first converted into synthesis gas, or syngas, and later processed into synthetic biofuels. The method used to produce the fuel will depend on the type of biomass, the energy demand, and the application in industry or laboratory settings. Compared to fossil fuels, BtL fuels have better quality, lower CO_2 emissions, and low emissions of nitrite oxide (NO) and nitrite dioxide (NO2), air-polluting agents often referred to as NOx. Additionally, BtL fuels have a flexible product quality, blending with other fuels without engines having to be modified to work efficiently (Ibarra-Gonzalez and Rong, 2018).

The gasification process is a crucial element in the thermochemical transformation of biomass into biofuels. Using a gasifying agent (GA), biomass is converted into a mixture that produces energy, chemicals, and biofuels. Usually, gasification follows four steps: *drying*, *pyrolysis*, *oxidation*, and *reduction* – phases that generate the gas along with tar and char. To produce light hydrocarbons from large tar molecules, a process of tar reforming can be added to the procedure. As seen with other biomass conversion methods, the composition of syngas ultimately depends on factors such as feedstock quality, technology used, and operating parameters (Molino et al., 2018).

An analogous technique to convert syngas into higher hydrocarbons is the Fischer–Tropsch (FT) method. This catalytic process transforms CO and H_2 into organic compounds using a cobalt-based or iron-based catalyst. The main products of this synthesis method are linear paraffins (alkanes) and olefins (alkenes) of different length chains. Relatedly, the amount of greenhouse gas emissions from fuel-producing FT processes using waste wood as feedstock is negligible, trailing bioethanol, biodiesel, and biogas methods and having at least one-tenth of the impact of fossil fuel techniques that produce gasoline, diesel, or natural gas (Mesfun, 2021; Snehesh and Dasappa, 2016). Figure 11.13 illustrates the FT synthesis method in more detail.

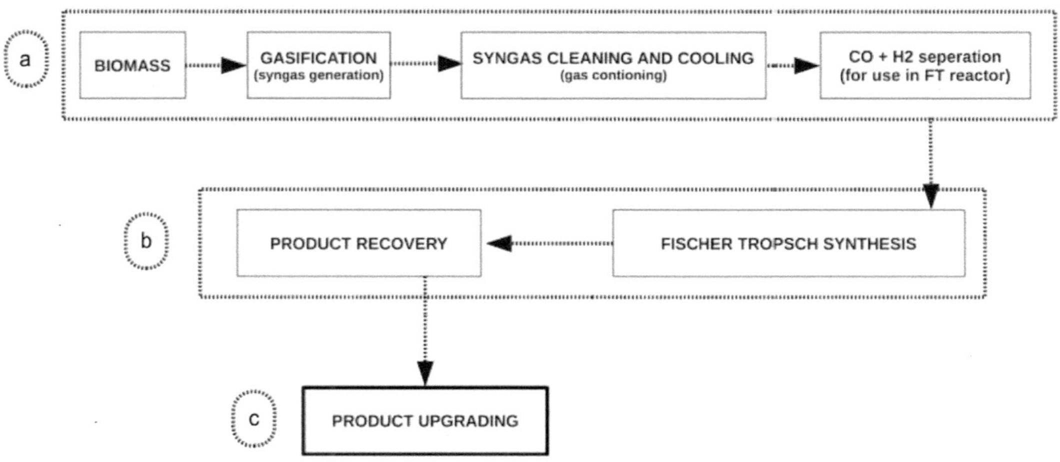

FIGURE 11.13 Fischer–Tropsch schematic: (a) syngas generation via biomass gasification, (b) syngas conversion to higher hydrocarbons via FT reaction, and (c) separation and refining of FT yield to useful products.

Source: Snehesh and Dasappa (2016).

Despite its promise, BtL installations remain in a demonstration or experimental state, with interest failing to take off mostly due to studies showing very ambitious levels of throughput required to achieve economic sustainability, particularly from FT installations. However, a techno-economic analysis performed by Snehesh et al. (2017) shows that sustainability can be reached at levels of 1,000 kg/h, provided that a biomass source is economical and consistent and some of the fuel gas that remains after combustion is used to generate electricity for subsequent fuel generation processes.

As is the case with other fuels reviewed in this chapter, economic efficiency has not been reached for BtL fuels, and many details remain unresolved to expand these technologies across more areas worldwide successfully. In the coming sections, we will review the role that the oil and gas industry, government subsidies, oil prices, legislation, and international organizations can have in increasing the share of biofuel in the energy mix of many countries and the paths that these actors can delineate for the replication of these strategies in multiple places.

11.3.6 Biogas

As part of the group of gaseous biofuels, biogas stands out as a promising alternative to natural gas, with the advantage of being a cleaner and more renewable energy source. In its raw state, biogas is mainly formed of two-thirds of methane (CH_4), approximately one-third of carbon dioxide (CO_2), and small percentages of water vapor, hydrogen gas (H_2), and hydrogen sulfide (H_2S). After going through purification techniques to remove CO_2 and H_2S, upgraded biogas that is of pipeline quality is referred to as *biomethane* (Antar et al., 2021; Guo et al., 2015a; Lee et al., 2019; Rosillo-Calle, 2016).

In nature, biogas is generated after biomass is decomposed gradually by microorganisms. If there is sufficient air supply, decomposition by-products will mainly be CO_2 and H_2O, but in an anaerobic environment like a landfill or a manure lagoon, the by-products will primarily be CH_4 and CO_2. This latter process is known as anaerobic digestion (AD) and consists of four steps: First, *hydrolysis* occurs when bacteria break down molecules into smaller particles, like sugars, fatty acids, and amino acids. Then, *acidogenesis* consists of converting simple organic molecules into CO_2, hydrogen, ammonia, and organic acids. *Acetogenesis* follows, converting organic acids into acetic acids. Lastly, *methanogenesis* is observed after acetic acids are converted into methane and carbon dioxide (Guo et al., 2015a).

Historically, AD has been used in farms, wastewater treatment plants, and households to manage waste or generate energy. To adapt it for widespread industrial use, enhancements to the methanogenesis stage have been developed so that the main components of biogas are retained more efficiently. Examples are using up-flow anaerobic sludge blankets (UASB), membrane bioreactors (MBRs), or coupling membrane techniques with high-rate systems. Membranes help retain selected elements without altering the biological process significantly, while high-rate approaches like UASB can achieve high pathogen removal with an efficient methane production degree (Sárvári Horváth et al., 2016).

Moreover, a co-digestion process has been advanced by some authors, where the first two phases of AD are placed in one part of a reactor and another part of the system performs the last two phases. As seen in Figure 11.14, a separate methanogenesis stage can, among other things, increase stability in pH environment control, achieve a higher loading rate, potentially increase methane yield, and improve pathogen control (Hagos et al., 2017). Grasping parameters like temperature, pH, and volatile fatty acids (VFAs), the ratio of carbon to nitrogen, the retention time to degrade the organic matter entirely, the organic loading rate, along with enhanced pre-treatment options, can significantly affect methane yield and the overall efficiency of the biogas generation process (Pramanik et al., 2019).

One advantage of modern biogas production strategies is that the produced biofuel can use existing natural gas infrastructure, such as pipelines and transportation chains (Rosillo-Calle, 2016).

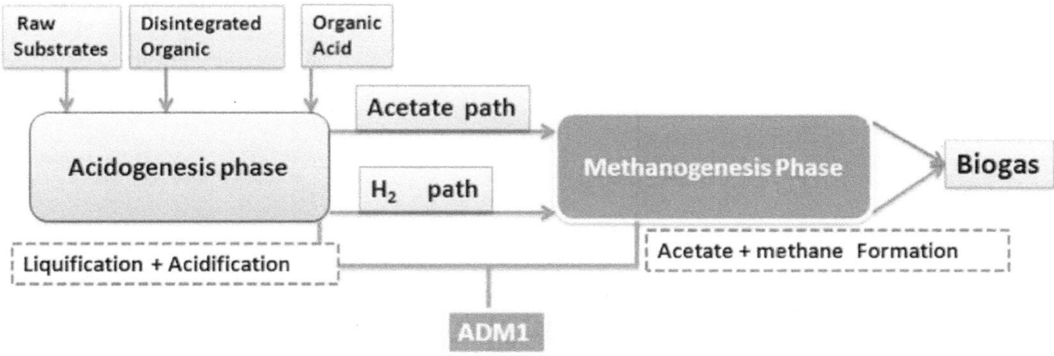

FIGURE 11.14 The phase separation of anaerobic co-digestion process and modeling.

Source: Hagos et al. (2017).

A country that has explored the possibilities for a fossil-free energy sector is Denmark, where water scrubbing and physical absorption are two methods used to upgrade biogas, enhancing its methane contents. In their assessment, Aryal and Kvist (2018) found significant cost-cutting opportunities when supplying biogas directly to food processing industries, with extensive customer mapping as another promising strategy to lower upgrading and compression costs.

As has been the case with other biofuels, the opportunities for the increase in the use of biomethane depend on improving economies of scale, reducing costs across conversion steps, and efficient use of current natural gas infrastructure. These goals can be achieved with subsidies and government incentives, which can be obtained by improving the current legislation that handles the production and upgrading of biogas. The upsides of these moves will help lower GHG emissions and lower waste generation, both attractive targets sought by the energy transition to renewables.

11.3.7 Butanol

Another by-product of biomass, *biobutanol* (or butyl alcohol) has gained increased attention due to its remarkable fuel features, such as high energy content and low pollutant emissions. The main method of biobutanol production from lignocellulosic biomass is an acetone-butanol-ethanol (ABE) fermentation process, where bacteria from the *Clostridia* strain first convert biomass into organic acids, CO_2, and hydrogen. This phase, known as *acidogenesis*, is followed by *solventogenesis*, where the acids reassimilate into acetone, butanol, and ethanol, generating carbon dioxide and hydrogen as other by-products (Abo et al., 2019; da Silva Trindade and dos Santos, 2017).

Despite ample potential for ABE techniques, some constraints for its wider implementation are related to the high cost of the substrates, insufficient levels of yield and productivity, and prohibitive costs of separation and purification methods. Figure 11.15 lays out the main hurdles in the ABE fermentation process, providing strategies to improve butanol production. To tackle the first barrier, pre-treatment of lignocellulosic biomass can be optimized, or non-cellulosic biomass like molasses can be used as a cheap alternative substrate. Additionally, algae offer cost-effective conversion strategies while offering a first look at third-generation biofuel production (Abo et al., 2019; Guo et al., 2022).

Meanwhile, low yields and productivity can be improved by exploring genetic engineering and modifying microorganisms or with a framework of continuous fermentation that foments cell immobilization. For the first point, antisense RNA technologies can "silence" a gene without changing the regulation of gene expression. In the case of butanol production, deleting the catabolic responsive

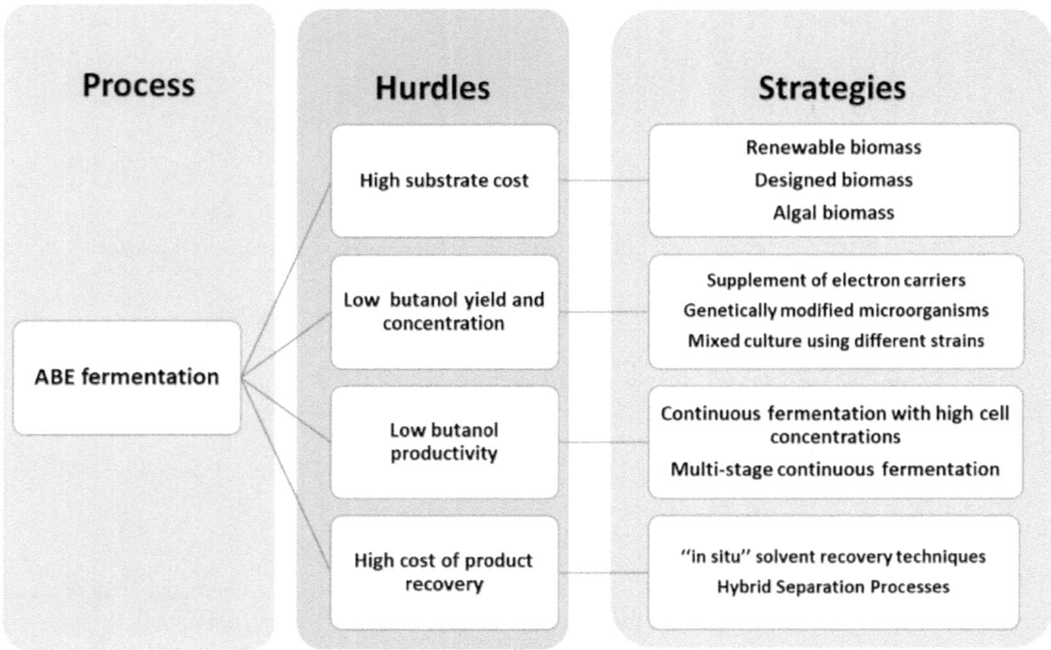

FIGURE 11.15 Hurdles and different strategies to improve butanol production.

Source: Abo et al. (2019).

element (CRE) in *Clostridium acetobutylicum* has led to a seven-and-a-half increase in the butanol yield. Similarly, CRISPR analysis has proven beneficial for butanol production prompted by the *Clostridium beijerinckii* strain (Rao et al., 2016).

Two- or multi-stage continuous fermentation has been explored for the second point. Here, fermenter microbes such as *C. acetobutylicum* or *C. beijerinckii* are fed constantly to the biomass mix across multiple reactors, pumping the resulting broth to the next-stage reactor, to be treated once again, until the mix is eventually pumped to a storage tank as butanol. Compared to batch fermentation techniques, continuous fermentation processes have improved overall ABE production (Zheng et al., 2015). Moreover, novel feedstock like cassava starch has shown promising ABE generation results that warrant further exploration (Lin et al., 2019).

Jang and Choi (2018) simulate a setup with metabolically engineered microbes capable of C6 and C5 sugar co-fermentation to test the commercial viability of a butanol production system with continuous fermentation, placed in a series of fermenters with adsorption columns. Inside this chamber, where temperature and pH are constantly monitored, microbes slowly increase along with the ABE content, which is later adsorbed during circulation. After this, the remaining ABE is desorbed by low-pressure steam, and the cycle restarts. Under these conditions, commercial viability can only be achieved if fixed capital investments and variable operating expenses are optimized as much as possible.

As we can note, technological advancements have opened the door to exploring novel methods to generate butanol from different biomass materials. In time, butanol and ethanol are expected to secure a significant share of the fuels generated to operate combustion engines (da Silva Trindade and dos Santos, 2017). With that said, production costs remain prohibitive for many actors, so until the appropriate players do not tackle these economic considerations, the challenge to expand these methods around the world continues to slow these innovations down.

11.3.8 Photosynthetic Microorganisms and Microalgae

Last, algal and aquatic biomass use holds excellent potential due to the photosynthetic properties that capture CO_2 to produce biomass and the high energy density of its outputs (Barry et al., 2015). *Microalgae* are organisms that convert solar energy into chemical energy, consuming CO_2 for their cultivation and thus helping reduce the overall emission of GHG in fuel production. Also, because of land demands and competition with the food supply, terrestrial vegetation and animal lipids remain constrained as reliable sources for biodiesel, an aspect that increases the potential for microalgae as a raw material for bio-oil production (Guo et al., 2015b; Lee et al., 2019).

As opposed to terrestrial plants, formed of over 95% cellulose, lignin, and hemicellulose, algae are composed of protein, lipids, carbohydrates, and small quantities of cellulose. These differences translate into differences in methods used to generate biofuels and the yield, composition, and ultimate quality of the output (Guo et al., 2015b; Phwan et al., 2018). Thus, one of the most common methods used to break down the solid biomass found in algae is hydrothermal liquefaction, a process that uses water in temperatures between 200 and 350°C and elevated pressures of 10–20 MPa to generate a bio-crude oil (Barry et al., 2015; US Department of Energy, 2023).

Additionally, other methods have been explored to process microalgae biomass with varying degrees of potential. One of these techniques is transesterification, which uses reversible chemical reactions where lipids and triglycerides are reacted with alcohol as a catalyst to achieve a stepwise conversion of the microalgae biomass to generate diglycerides, monoglycerides, and glycerol, along with moieties of fatty acid methyl ester (FAME). This set of reactions, generally triggered by methanol or ethanol as the preferred catalysts, yields a steady release of biodiesel at each process step (Karpagam et al., 2021; Kumar et al., 2020).

Then, fermentation methods rely on microbes to convert microalgae to bioethanol. For these techniques, a pre-treatment process is performed first to break down carbohydrates or starch polymers into monomers like glucose so that bioethanol can be produced. Subsequently, fermentative microorganisms like yeast or bacteria are added to the pre-treated biomass. For the former, *Saccharomyces cerevisiae* and *Schizosaccharomyces pombe* are suitable for producing ethanol in the presence of oxygen. At the same time, for the latter, *Escherichia coli*, *Klebsiella oxytoca*, and *Zymomonas mobilis* have shown the most promise due to higher yield production and amenability to genetic modifications (Phwan et al., 2018).

Depending on the fermenting microorganism selected, a few different paths will follow. One is a separate hydrolysis and fermentation (SHF) process, where the feedstock is degraded into monosaccharides, converting fermentable sugars into bioethanol. Correspondingly, simultaneous saccharification and fermentation (SSF) approaches use a single reactor for hydrolysis and fermentation. A simultaneous saccharification and co-fermentation (SSCF) method is recommended for high-C5 sugar mixes after hydrolysis (Phwan et al., 2018).

Moreover, other methods to process microalgae are proposed, depending on the goals of the synthesis approach. Hence, dark fermentation is tailored for bio-hydrogen production, while photo-fermentation and consolidated bioprocessing are biochemical techniques ideal for bioethanol synthesis (Phwan et al., 2018). Furthermore, as Figure 11.16 illustrates, more straightforward combustion methods are used for power generation, anaerobic digestion approaches are recommended for methane and hydrogen production, and gasification is utilized to produce syngas (Kumar et al., 2020; Voloshin et al., 2016).

Exploring aquatic biomass materials is also at the cutting edge of the move toward more advanced feedstock and conversion technologies. As a third-generation input, microalgae are an ideal testing biomass material for fourth-generation biofuel generation techniques. One important drawback of photosynthetic microorganisms is their comparatively low degree of cell growth, which has encouraged researchers to investigate the properties of light-capturing genes through synthetic biology approaches to improve productivity (Sarkar and Shimizu, 2015).

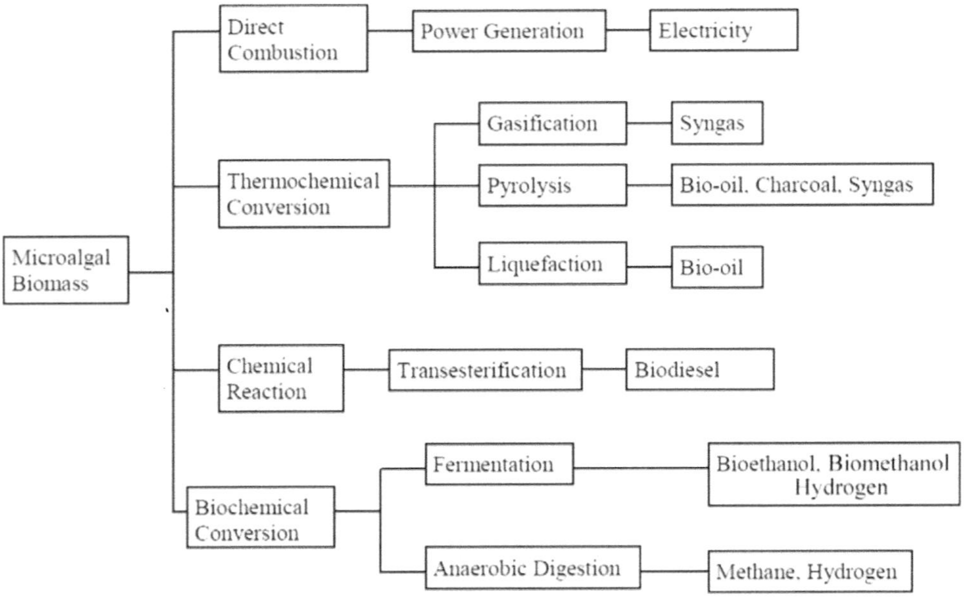

FIGURE 11.16 Methods of processing microalgae biomass.

Source: Voloshin et al. (2016).

To tackle these shortcomings, an emerging trend in the academy has been to explore the possibilities of genetic engineering and modification of algae biomass to produce fuels and improve the manufacturing of other commercial products. As we see in Figure 11.17, microalgae can be grown to produce food, valuable materials, and even medicines. This path is very intricate, but in simple terms, *synthetic biology* involves designing expression vectors, editing genes, transcriptional engineering, and eventually, a genetic transformation that selects for the desired gene expression (Sproles et al., 2021).

Despite the promise of these methods being evident, significant challenges have appeared when putting the potential of genetic engineering into practice. For example, Fajardo et al. (2020) reported a negative effect on algae growth when constitutive promoters are used to express a select gene. To tackle this, gene expression approaches should only be initiated when a certain level of growth has been achieved, and general modifications ought to be taken only after a more general understanding of gene annotation, transformation methods, and metabolic pathways is obtained.

Genome editing (GE) is a technology that has gained popularity for enhancing aquatic biomass to produce fuel and in medicine and food. A multiplicity of GE tools like RNAi, ZNFs, TALENs, and CRISPR/Cas9 is being explored to improve products to their optimal theoretical states, revolutionizing the biology sector (Shokravi et al., 2021). According to Fayyaz et al. (2020), relying on genetic engineering for microalgae development could lower overall costs by 15–20%, helping boost markets and influence actors to invest more in these technologies, creating a virtuous circle based on a renewable energy source.

With that said, genetic engineering has been met with a few roadblocks related to unknown long-term consequences, disruption of natural environments that have taken hundreds of years to evolve, and potential excesses and breaches of ethics when considering the natural expansion of this research to humans or animals. Because of this, governments and international organizations have advanced biosafety regulations to try to find a balance between innovation and evidence-based gene

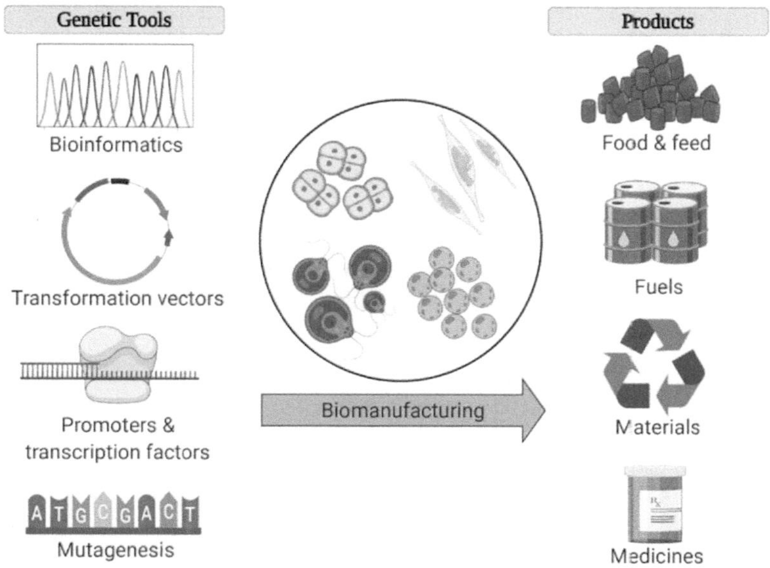

FIGURE 11.17 Genetic tools to develop microalgae as a platform for biomanufacturing various commercial products. Bioinformatic algorithms are used to analyze algal genome sequences, resulting in codon optimization and motif discovery techniques that allow the design of strong expression vectors for the genetic transformation of algae strains. Regulatory elements, such as promoters and transcription factors, allow recombinant gene expression and manipulation of metabolic pathways to obtain products of interest. Random mutagenesis and genome shuffling can further drive the algal production strains toward desired phenotypes.

Source: Sproles et al. (2021).

manipulation. In the next section, we review more of these dynamics in detail and the role the oil and gas industry has played in advancing biomass as an energy alternative.

11.4 BIOFUELS AND THE OIL AND GAS INDUSTRY

In the context of the energy transition to renewables, the oil and gas industry has had a significant role as an investor, as a purveyor of know-how and technology, and ultimately, as a research and development partner to new and established players looking to make the switch. In the case of biofuel development, this is not the exception, as the industry has adapted to the current dynamics by diversifying its energy portfolio, exploring parallel operations for fossil fuels and renewables using its infrastructure, and advocating for policy change that stimulates markets that favor clean energy initiatives.

The new frontiers explored by private and public oil and gas players to facilitate the transition to renewables have one major thing in common: a notable degree of collaboration between several actors that include new energy partners, policymakers, government officials, and international organizations like the United Nations and the European Union. These alliances can be seen, for example, in the search for better regulation for decarbonization efforts, the use of biomass materials within oil and gas operations, the improvement of regulation to stimulate market development and economies of scale, and the inclusion of biomass in carbon capture and storage (CCS) and enhanced oil recovery (EOR) strategies.

Providing an example for one of these areas, we observe the increasing attention that cellulose nanomaterials (CNMs) have received from the oil and gas sector due to their potential to make some processes within the industry more environmentally friendly and sustainable. According to Li et al.

(2023), varying CNMs like cellulose nanofibers (CNFs) and cellulose nanocrystals (CNCs) have properties that suit oil and gas operations well, like attractive hydrophilic, thermal, and rheological features; gas permeability; crystallinity; and elastic moduli. With that said, the authors caution about first performing tests of the materials under harsh conditions so that the switch becomes permanent.

On a related subject, the coupling of carbon capture and storage (CCS) strategies and enhanced oil recovery (EOR) with bioenergy production processes is highlighted by Yasemi et al. (2023). The first joint process offers the prospect of negative emissions due to the production of renewable, clean energy, along with the reduction of the CO_2 that reaches the atmosphere achieved by the CCS procedure. Meanwhile, EOR endeavors have the added challenge of injecting the captured carbon dioxide into reservoirs for production stimulation. Using CO_2 from renewable sources, like in the case with biomass, allows for the creation of closed-loop systems that reduce GHG emissions while improving oil yields.

A similar search for circular economies driven by renewable energies has motivated some authors. According to Basile et al. (2022), these systems where inputs and outputs are contained in closed-loop structures and residues are recycled or refurbished are much more attainable when the energy resources are from sustainable origins. A key role for oil and gas players is in applying their expertise in the decommissioning of depleted offshore installations, exploring repurposing and refurbishing strategies to accommodate a combination of biomass/algae operations, wind turbines, and solar photovoltaic (PV) panels in locations that do not significantly compete with food production.

These alliances are struck for multiple reasons, but two main drivers for these efforts are reduced costs and the potential for knowledge sharing between two or more areas. In the context of the energy transition to renewables, this level of connection is needed to simplify the required transformations. Exploring the Swedish oil refinery segment, Nurdiawati and Urban (2022) noted the concurrent action of international organizations, governments, research institutions, energy companies, and state and private oil and gas players toward decarbonization efforts.

Meanwhile, one state-run oil company that has shown a proactive approach toward renewable energies is Thailand's PTT, which, in 2007, started a series of developments involving wind, solar, hydropower, and biogas projects, along with two associates, Thai Oil Group and Bangchak Petroleum. In their work, Chaipaya et al. (2018) explored the rationale for state and private companies to go green, and as we can see in Figure 11.18, the commitments expressed by Bangchak for increasing the share of biomass in Thailand are related to many social, economic, and environmental areas, such as reducing oil imports, achieving energy security, improving the welfare of farmers, and following government policy.

Meanwhile, international oil and gas players have started their paths toward biofuel exploration. According to a 2023 report from BCC Research, companies like BP and ExxonMobil have invested nearly 3 billion US dollars. At the same time, emerging firms like Manta Biofuel and Innoltek are leading the way in using algae and animal fats as sustainable alternatives to crude oil (PR Newswire, 2023). These moves have been accompanied by favorable legislation, like in California, where a low-carbon fuel standard (LCFS) policy seeks to promote increased biodiesel use despite lingering challenges in cost reduction and technology implementation (Witcover and Williams, 2020).

Likewise, a 2021 report from IEA highlights the sustained recovery of the biofuels sector in that year despite high costs and the health emergency caused by the COVID-19 pandemic. This recovery is led by ethanol and renewable diesel growth in American and European markets, where policies oriented toward GHG reduction have paved the way to more competitiveness with non-renewable sources. On the flip side, demand is driven primarily by Asian and Latin American markets, where mandates like India's 20% ethanol blending by 2025, along with Indonesia's 40% blending requirements, have become incentives that increase the demand (IEA, 2021).

These recoveries and improvements are intrinsically related to the ability of actors to improve the logistics and operations in the biomass sector so that they respond quicker to uncertainties. In the last

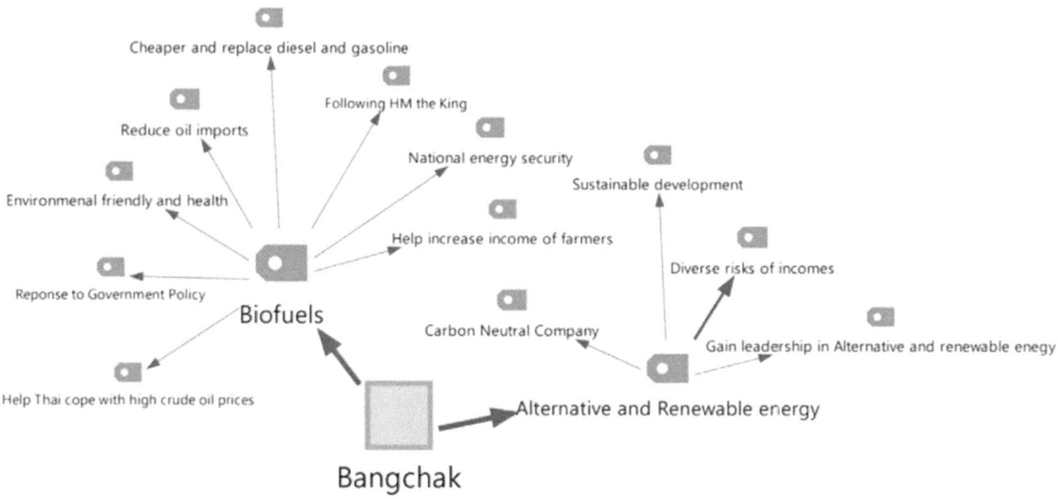

FIGURE 11.18 Bangchak's discourses on the renewable energy investment.

Source: Chaiyapa et al. (2018).

few years, mathematical models have been proposed to enhance bioenergy supply chains, including methods like stochastic programming and robust optimization (RO). Exploring the extent of these alternatives, Mohseni and Pishvaee (2016) propose a sustainable supply chain model that highlights the prospects and limitations of the algae industry and the potential strategies for cost mitigation.

Combating the lack of historical data due to large biomass production not occurring, the authors developed a mixed-integer linear programming (MILP) framework. The model can propose supply chain decisions and guidelines using specific parameters' upper and lower bounds. A few conclusions from model simulations suggest that there are essential trade-offs when considering scaling supply chains and lowering costs, as this relationship is not directly linear. Thus, an optimal scaling level does not always lead to lower costs. In this sense, detecting algae species with a larger lipid content appears more important if the goal is to reduce costs.

Just as many technologies that have been explored and improved for use in varying oil and gas processes are aiding in developing a robust bioenergy sector, the knowledge collected to improve other areas can benefit the biomass economy. This is the case with genetic engineering techniques to increase biofuel production yields and enhance biomass quality. To efficiently achieve these goals, Antar et al. (2021) proposed the novel concept of microbe-to-plant signal compounds, which can potentially increase the growth of diverse species of plants, thus improving agricultural stability and fomenting system sustainability.

As we can see, the role of incumbents in established industries is key to building momentum toward an energy transition process that features a prominent role of biofuels. In particular, the oil and gas sector can provide valuable assistance to help as many countries as possible reach sustainable fourth-generation biofuel generation processes. Working in tandem with governments and policymakers from international organizations, the industry can become an energy player that provides knowledge and expertise and is ready to have a significant position in a post–fossil fuel world.

Despite these dynamics trending toward a positive outcome, some important challenges to a seamless and quick transition remain, particularly when considering the role of biomass and biofuels. First, costs remain prohibitive for many actors worldwide from adopting these methods, with pre-treatment being a significant hurdle. Then, many countries continue developing first-generation biomass, potentially interfering with food production. Therefore, until a substantial number of players do not find it economical to move to non-food or microalgae biofuel development, they will continue relying on conventional energy sources, such as coal and natural gas.

Then, legislation in this area appears to lag behind regulations for other renewable energy sources. An assessment performed by Ebadian et al. (2020) shows that blending mandates only have a modest effect on the increase of biofuels use and that countries that show more success in their efforts to advance biofuels feature a combination of market-pull and technology-push policies. With that said, if economies do not scale properly, the more established oil markets will continue to affect the effectiveness of any policy proposal, regardless of its potential benefits.

11.5 CONCLUSIONS: BIOFUELS IN THE CONTEXT OF THE ENERGY TRANSITION

In this chapter, we assessed the potential of biofuels as the main players in the drive toward a renewable energy economy. As with other clean resources, biofuels are in the sweet spot of a technological revolution and an almost-imperative shift toward their widespread use, triggered by rising temperatures and greenhouse emissions from fossil fuel use. Because of this, biofuels have a very high ceiling and multiple actors have advanced varying ways to enhance their production. With that said, challenges and constraints continue to delay a seamless switch in more locations worldwide.

A primary limitation for biomass is their competition with food supply chains, at least in a subset of countries that have not developed their biofuel production to rely on second-generation biomass inputs onward. Therefore, without significant funding and a policy framework that explores these advancements and encourages parallel industries, widespread use of biofuels will continue to be less likely than conventional energy sources, particularly in countries that depend on food exports or have not developed surplus food supplies for their populations.

Another hurdle is observed in the high costs for production, transportation, and manipulation of biomass materials and the intricate and elaborate operations required to get the most out of these organic residues. For countries without mature markets, optimizing biomass techniques and making them profitable become challenging, and even harder to expect a commitment to sustained use in the short term. Thus, the opportunity costs of developing an R&D sector that excels in biomass production can be too high for many countries that often deal with food security or energy scarcity.

Moreover, a third potential pitfall is related to the limits of genetic engineering methods. Current promising gene editing methods have the flip side of allowing too much manipulation with too little oversight. Laboratory settings are different than nature, and the disruption to natural habitats after gene expression modifications must be better understood before this road is explored much further. In this sense, we do not know enough about any potential trade-offs that a higher yield of biofuels can bring to any organisms in contact with the altered biomass.

These considerations would seem insurmountable, if not for the paths forward provided by the many collaboration efforts between interested parties. Hence, a combination of targeted policies with support from actors with high technological savvy has paved the way for models and policy prescriptions that are increasingly better at predicting uncertainty and providing solutions in highly unknown scenarios. Moreover, in some parts of the world, pilot programs testing low-carbon and renewable fuel standards with minimum percentages of ethanol blending and minimum shares of clean energies are being implemented.

Today, the burden of leading the way toward widespread implementation of biofuels is primarily falling on developed countries, with fewer problems of food scarcity and energy dependence while having a more significant research and development sector. However, we have also observed a few hot spots where more efficient land use and geographical advantages have allowed these locations to lead the development of biofuels. Brazil is one of those countries where sugarcane production and a developed bioethanol sector have produced a powerhouse that has much to offer to the rest of the world.

The lessons from successful cases are steadily replicated in other locations worldwide, yet these efforts may take longer than expected to amount to the energy transition. It is likely that novel technologies will help spread promising methods in advantageous contexts, but the role of policies and legislation cannot be dismissed. Only some limited territories in the United States and the European Union have placed regulations in the books to increase the share of biofuels in the medium term, with most efforts for advancing renewable energies consistently favoring solar, wind, and hydrogen facilities.

We hope that through collaborative efforts, these limitations and seemingly zero-sum estimations turn into diverse choices for energy provision, with more flexible trade-offs and a wider range of action. To achieve this, some players in the oil and gas industry have started to invest in biofuel technologies, keeping in mind their own net-zero pledges and the possibility of becoming energy powerhouses that manage conventional and renewables with equal skill. Despite ample room for improvement, we observe a few robust efforts from the oil and gas industry to place biofuels in the center of the push for a switch to clean energy and decarbonization.

This is just the beginning for biofuels, which have the potential to become widespread around the world due to their ability to foster closed-loop systems where feedstock is used to generate energy to add to the grid and fuels for transportation uses and residues from most operations can be reutilized to convert more feedstock into fuels. Given that these features that boost sustainability and a low-GHG-emissions environment are invaluable, they can be added seamlessly to either carbon capture and storage (CCS) or enhanced oil recovery (EOR) projects, adding synergistic value to the relationship between emerging and established energy sectors.

REFERENCES

Abo, B. O., M. Gao, Y. Wang, C. Wu, Q. Wang, and H. Ma. 2019. Production of butanol from biomass: Recent advances and future prospects. *Environmental Science and Pollution Research* 26: 20164–20182. doi:10.1007/s11356-019-05437-y.

Aboytes-Ojeda, M., K. K. Castillo-Villar, and S. D. Eksioglu. 2022. Modeling and optimization of biomass quality variability for decision support systems in biomass supply chains. *Annals of Operations Research* 314(2): 319–346. doi:10.1007/s10479-019-03477-8.

Acuna, M., J. Sessions, R. Zamora, K. Boston, M. Brown, and M. R. Ghaffariyan. 2019. Methods to manage and optimize forest biomass supply chains: A review. *Current Forestry Reports*. doi:10.1007/s40725-019-00093-4.

Adams, P., T. Bridgwater, A. Lea-Langton, A. Ross, and I. Watson. 2018. Biomass conversion technologies. In *Greenhouse gas balances of bioenergy systems* (pp. 107–139). Academic Press.

Aditiya, H. B., T. M. I. Mahlia, W. T. Chong, H. Nur, and A. H. Sebayang. 2016. Second generation bioethanol production: A critical review. *Renewable and Sustainable Energy Reviews* 66: 631–653. doi:10.1016/j.rser.2016.07.015.

Allman, A., C. Lee, M. Martín, and Q. Zhang. 2021. Biomass waste-to-energy supply chain optimization with mobile production modules. *Computers & Chemical Engineering* 150: 107326. doi:10.1016/j.compchemeng.2021.107326.

Alper, K., K. Tekin, S. Karagöz, and A. J. Ragauskas. 2020. Sustainable energy and fuels from biomass: A review focusing on hydrothermal biomass processing. *Sustainable Energy & Fuels* 4(9): 4390–4414. doi:10.1039/d0se00784f.

Antar, M., P. Gopal, L. A. Msimbira, J. Naamala, M. Nazari, W. Overbeek, . . . and D. L. Smith. 2021. Inter-organismal signaling in the rhizosphere. *Rhizosphere Biology: Interactions Between Microbes and Plants* 255–293. Rhizosphere Biology. Springer, Singapore. https://doi.org/10.1007/978-981-15-6125-2_13

Aryal, N., and T. Kvist. 2018. Alternative of biogas injection into the Danish gas grid system—A study from demand perspective. *ChemEngineering* 2(3): 43. doi:10.3390/chemengineering2030043.

Ba, B. H., C. Prins, and C. Prodhon. 2016. Models for optimization and performance evaluation of biomass supply chains: An Operations Research perspective. *Renewable Energy* 87: 977–989. doi:10.1016/j.renene.2015.07.045.

Baloch, H. A., S. Nizamuddin, M. T. H. Siddiqui, S. Riaz, A. S. Jatoi, . . ., G. J. Griffin. 2018. Recent advances in production and upgrading of bio-oil from biomass: A critical overview. *Journal of Environmental Chemical Engineering* 6(4): 5101–5118. doi:10.1016/j.jece.2018.07.050.

Barry, A. N., S. R. Starkenburg, and R. T. Sayre. 2015. Strategies for optimizing algal biology for enhanced biomass production. *Frontiers in Energy Research* 3: 1. doi:10.3389/fenrg.2015.00001.

Basile, F., L. Pilotti, M. Ugolini, G. Lozza, and G. Manzolini. 2022. Supply chain optimization and GHG emissions in biofuel production from forestry residues in Sweden. *Renewable Energy* 196: 405–421.

Bergvall, N., R. Molinder, A.-C. Johansson, and L. Sandström. 2021. Continuous slurry hydrocracking of biobased fast pyrolysis oil. *Energy & Fuels* 35(3): 2303–2312. doi:10.1021/acs.energyfuels.0c03866.

Brethauer, S., and M. H. Studer. 2015. Biochemical conversion processes of lignocellulosic biomass to fuels and chemicals – A review. *Chimia* 69(10): 572–581. doi:10.2533/chimia.2015.572.

Chaiyapa, W., M. Esteban, and Y. Kameyama. 2018. Why go green? Discourse analysis of motivations for Thailand's oil and gas companies to invest in renewable energy. *Energy Policy* 120: 448–459. doi:10.1016/j.enpol.2018.05.064.

Delivand, M. K., A. R. B. Cammerino, P. Garofalo, and M. Monteleone. 2015. Optimal locations of bioenergy facilities, biomass spatial availability, logistics costs and GHG (greenhouse gas) emissions: A case study on electricity productions in South Italy. *Journal of Cleaner Production* 99: 129–139. doi:10.1016/j.jclepro.2015.03.018.

Devi, A., S. Bajar, P. Sihag, Z. U. D. Sheikh, A. Singh, J. Kaur, N. R. Bishnoi, and D. Pant. 2023. A panoramic view of technological landscape for bioethanol production from various generations of feedstocks. *Bioengineered* 14(1), 81–112. doi:10.1080/21655979.2022.2095702.

Dimitriou, I., H. Goldingay, and A. V. Bridgwater. 2018. Techno-economic and uncertainty analysis of Biomass to Liquid (BTL) systems for transport fuel production. *Renewable and Sustainable Energy Reviews* 88: 160–175. doi:10.1016/j.rser.2018.02.023.

Ebadian, M., S. van Dyk, J. D. McMillan, and J. Saddler. 2020. Biofuels policies that have encouraged their production and use: An international perspective. *Energy Policy* 147: 111906. doi:10.1016/j.enpol.2020.111906.

EESI. 2023. *Bioenergy (biofuels and biomass)*. Available in www.eesi.org/topics/bioenergy-biofuels-biomass/description [accessed September 8th, 2023].

EIA. 2022. *Biomass explained: Biomass and the environment*. Available in www.eia.gov/energyexplained/biomass/biomass-and-the-environment.php [accessed September 8th, 2023].

EIA. 2023. *U.S. biodiesel production capacity*. Available in www.eia.gov/dnav/pet/hist/LeafHandler.ashx?n=pet&s=m_epoordb_8bdpc_nus_mmgl&f=m [accessed October 28th, 2023].

Erdin, C., and G. Ozkaya. 2019. Turkey's 2023 energy strategies and investment opportunities for renewable energy sources: Site selection based on ELECTRE. *Sustainability* 11: 2136. doi:10.3390/su11072136.

European Commission. 2023. *Biofuels*. Available in https://energy.ec.europa.eu/topics/renewable-energy/bioenergy/biofuels_en [accessed October 28th, 2023].

Fajardo, C., M. De Donato, R. Carrasco, G. Martínez-Rodríguez, J. M. Mancera, and F. J. Fernández-Acero. 2020. Advances and challenges in genetic engineering of microalgae. *Reviews in Aquaculture* 12(1): 365–381. doi:10.111/raq.12322.

Favaro, L., T. Jansen, and W. H. van Zyl. 2019. Exploring industrial and natural *Saccharomyces cerevisiae* strains for the bio-based economy from biomass: The case of bioethanol. *Critical Reviews in Biotechnology* 39(6): 800–816. doi:10.1080/07388551.2019.1619157.

Fayyaz, M., K. W. Chew, P. L. Show, T. C. Ling, I.-S. Ng, and J.-S. Chang. 2020. Genetic engineering of microalgae for enhanced biorefinery capabilities. *Biotechnology Advances* 107554. doi:10.1016/j.biotechadv.2020.107554.

Gital Durmaz, Y., and B. Bilgen. 2020. Multi-objective optimization of sustainable biomass supply chain network design. *Applied Energy* 272: 115259. doi:10.1016/j.apenergy.2020.115259.

Guedes, R. E., A. S. Luna, and A. R. Torres. 2018. Operating parameters for bio-oil production in biomass pyrolysis: A review. *Journal of Analytical and Applied Pyrolysis* 129: 134–149. doi:10.1016/j.jaap.2017.11.019.

Guo, M., W. Song, and J. Buhain. 2015a. Bioenergy and biofuels: History, status, and perspective. *Renewable and Sustainable Energy Reviews* 42: 712–725. doi:10.1016/j.rser.2014.10.013.

Guo, Y., T. Yeh, W. Song, D. Xu, and S. Wang. 2015b. A review of bio-oil production from hydrothermal liquefaction of algae. *Renewable and Sustainable Energy Reviews* 48: 776–790. doi:10.1016/j.rser.2015.04.049.

Guo, Y., Y. Liu, M. Guan, H. Tang, Z. Wang, L. Lin, and H. Pang. 2022. Production of butanol from lignocellulosic biomass: Recent advances, challenges, and prospects. *RSC Advances* 12: 18848–18863. doi:10.1039/d1ra09396g.

Hagos, K., J. Zong, D. Li, C. Liu, and X. Lu. 2017. Anaerobic co-digestion process for biogas production: Progress, challenges and perspectives. *Renewable and Sustainable Energy Reviews* 76: 1485–1496. doi:10.1016/j.rser.2016.11.184.

Hezam, I. M., F. Cavallaro, J. Lakshmi, P. Rani, and S. Goyal. 2023. Biofuel production plant location selection using integrated picture fuzzy weighted aggregated sum product assessment framework. *Sustainability* 15: 4215. doi:10.3390/su15054215.

Ibarra-Gonzalez, P., and B.-G. Rong. 2018. A review of the current state of biofuels production from lignocellulosic biomass using thermochemical conversion routes. *Chinese Journal of Chemical Engineering* 27(7): 1523–1535. doi:10.1016/j.cjche.2018.09.018.

IEA. 2021. *Renewables 2021: Analysis and forecast to 2026.* Available in https://iea.blob.core.windows.net/assets/5ae32253-7409-4f9a-a91d-1493ffb9777a/Renewables2021-Analysisandforecastto2026.pdf [accessed October 26th, 2023].

Inayat, A., A. Ahmed, R. Tariq, A. Waris, F. Jamil, . . ., Y.-K. Park. 2022. Techno-economical evaluation of bio-oil production via biomass fast pyrolysis process: A review. *Frontiers in Energy Research* 9: 770355. doi:10.3389/fenrg.2021.770355.

Jang, M.-O., and G. Choi. 2018. Techno-economic analysis of butanol production from lignocellulosic biomass by concentrated acid pretreatment and hydrolysis plus continuous fermentation. *Biochemical Engineering Journal* 134: 30–43. doi:10.1016/j.bej.2018.03.002.

Karpagam, R., K. Jawaharraj, and R. Gnanam. 2021. Review on integrated biofuel production from microalgal biomass through the outset of transesterification route: A cascade approach for sustainable bioenergy. *Science of the Total Environment* 766: 144236. doi:10.1016/j.scitotenv.2020.144236.

Kheybari, S., M. Kazemi, and J. Rezaei. 2019. Bioethanol facility location selection using best-worst method. *Applied Energy* 242: 612–623. doi:10.1016/j.apenergy.2019.03.054.

Krishankumar, R., A. R. Mishra, P. Rani, F. Cavallaro, and K. S. Ravichandran. 2023. A novel integrated q-Rung fuzzy framework for biomass location selection with no apriori weight choices. *Sustainability* 15: 3377. doi:10.3390/su15043377.

Kumar, M., Y. Sun, R. Rathour, A. Pandey, I. S. Thakur, and D. C. W. Tsang. 2020. Algae as potential feedstock for the production of biofuels and value-added products: Opportunities and challenges. *Science of the Total Environment* 137116. doi:10.1016/j.scitotenv.2020.137116.

Lee, S. Y., R. Sankaran, K. W. Chew, C. H. Tan, R. Krishnamoorthy, D.-T. Chu, and P.-L. Show. 2019. Waste to bioenergy: A review on the recent conversion technologies. *BMC Energy* 1(4): 1–22. doi:10.1186/s42500-019-0004-7.

Li, M.-C., X. Liu, K. Lv, J. Sun, C. Dai, B. Liao, . . ., M. Hubbe. 2023. Cellulose nanomaterials in oil and gas industry: Current status and future perspectives. *Progress in Materials Science* 101187. doi:10.1016/j.pmatsci.2023.101187.

Lin, Z., H. Liu, J. Wu, P. Patakova, B. Branska, and J. Zhang. 2019. Effective continuous acetone–butanol–ethanol production with full utilization of cassava by immobilized symbiotic TSH06. *Biotechnology for Biofuels* 12: 219. doi:10.1186/s13068-019-1561-1.

Lo, S. L. Y., B. S. How, W. D. Leong, S. Y. Teng, M. A. Rhamdhani, and J. Sunarso. 2021. Techno-economic analysis for biomass supply chain: A state-of-the-art review. *Renewable and Sustainable Energy Reviews* 135: 110164. doi:10.1016/j.rser.2020.110164.

Malladi, K. T., and T. Sowlati. 2018. Biomass logistics: A review of important features, optimization modeling and the new trends. *Renewable and Sustainable Energy Reviews* 94: 587–599. doi:10.1016/j.rser.2018.06.052.

Martinkus, N., G. Latta, K. Brandt, and M. Wolcott. 2018. A multi-criteria decision analysis approach to facility siting in a wood-based depot-and-biorefinery supply chain model. *Frontiers in Energy Research* 6: 124. doi:10.3389/fenrg.2018.00124.

Mesfun, S. A. 2021. *Biomass to liquids (BtL) via Fischer-Tropsch – a brief review.* European Technology and Innovation Platform. Available in www.etipbioenergy.eu/images/ETIP_B_Factsheet_BtL_2021.pdf [accessed October 30th, 2023].

Mohseni, S., and M. S. Pishvaee. 2016. A robust programming approach towards design and optimization of microalgae-based biofuel supply chain. *Computers & Industrial Engineering* 100: 58–71. doi:10.1016/j.cie.2016.08.003.

Molino, A., V. Larocca, S. Chianese, and D. Musmarra. 2018. Biofuels production by biomass gasification: A review. *Energies* 11: 811. doi:10.3390/en11040811.

Muscat, A., E. M. De Olde, I. J. M. de Boer, and R. Ripoll-Bosch. 2020. The battle for biomass: A systematic review of food-feed-fuel competition. *Global Food Security* 25: 100330.

Nurdiawati, A., and F. Urban. 2022. Decarbonising the refinery sector: A socio-technical analysis of advanced biofuels, green hydrogen and carbon capture and storage developments in Sweden. *Energy Research & Social Science* 84: 102358. doi:10.1016/j.erss.2021.102358.

Osman, A. I., N. Mehta, A. M. Elgarahy, A. Al-Hinai, A. H. Al-Muhtaseb, and D. W. Rooney. 2021. Conversion of biomass to biofuels and life cycle assessment: A review. *Environmental Chemistry Letters* 19: 4075–4118. doi:10.1007/s10311-021-01273-0.

Phwan, C. K., H. C. Ong, W.-H. Chen, T. C. Ling, E. P. Ng, and P. L. Show. 2018. Overview: Comparison of pretreatment technologies and fermentation processes of bioethanol from microalgae. *Energy Conversion and Management* 173: 81–94. doi:10.1016/j.enconman.2018.07.054.

Pinheiro Pires, A. P., J. Arauzo, I. Fonts, M. E. Domine, A. Fernández Arroyo, A., M. E. García-Pérez, . . ., M. García-Perez. 2019. Challenges and opportunities for bio-oil refining: A review. *Energy & Fuels* 33: 4683–4720. doi:10.1021/acs.energyfuels.9b00039.p

Popp, J., S. Kovács, J. Oláh, Z. Divéki, and E. Balázs. 2021. Bioeconomy: Biomass and biomass-based energy supply and demand. *New Biotechnology* 60: 76–84. doi:10.1016/j.nbt.2020.10.004.

PR Newswire. 2023. *Major oil companies are investing in biofuels, according to BCC research report*. Available in www.prnewswire.com/news-releases/major-oil-companies-are-investing-in-biofuels-according-to-bcc-research-report-301968724.html [accessed October 25th, 2023].

Pramanik, S. K., F. B. Suja, S. M. Zain, and B. K. Pramanik. 2019. The anaerobic digestion process of biogas production from food waste: Prospects and constraints. *Bioresource Technology Reports* 100310. doi:10.1016/j.biteb.2019.100310.

Rao, A., A. Sathiavelu, and S. Mythili. 2016. Genetic engineering in biobutanol production and tolerance. *Brazilian Archives of Biology and Technology* 59. doi:10.1590/1678-4324-2016150612.

Rodionova, M. V., R. S. Poudyal, I. Tiwari, R. A. Voloshin, S. K. Zharmukhamedov, H. G. Nam, . . ., S. I. Allakhverdiev. 2017. Biofuel production: Challenges and opportunities. *International Journal of Hydrogen Energy* 42(12): 8450–8461. doi:10.1016/j.ijhydene.2016.11.125.

Roni, M. S., D. N. Thompson, and D. S. Hartley. 2019. Distributed biomass supply chain cost optimization to evaluate multiple feedstocks for a biorefinery. *Applied Energy* 254: 113660. doi:10.1016/j.apenergy.2019.113660.

Rosillo-Calle, F. 2016. A review of biomass energy—shortcomings and concerns. *Journal of Chemical Technology & Biotechnology* 91(7): 1933–1945. doi:10.1002/jctb.4918.

Sarkar, D., and K. Shimizu. 2015. An overview on biofuel and biochemical production by photosynthetic microorganisms with understanding of the metabolism and by metabolic engineering together with efficient cultivation and downstream processing. *Bioresources and Bioprocessing* 2: 17. doi:10.1186/s40643-015-0045-9.

Sárvári Horváth, I., M. Tabatabaei, K. Karimi, and R. Kumar. 2016. Recent updates on biogas production-a review. *Biofuel Research Journal* 3(2): 394–402. doi:10.18331/brj2016.3.2.4.

Sharma, B., C. Larroche, and C.-G. Dussap. 2020. Comprehensive assessment of 2G bioethanol production. *Bioresource Technology* 313: 123630. doi:10.1016/j.biortech.2020.123630.

Shokravi, H., Z. Shokravi, M. Heidarrezaei, H. C. Ong, S. S. Rahimian Koloor, M. Petrů, W. J. Lau, and A. F. Ismail. 2021. Fourth generation biofuel from genetically modified algal biomass: Challenges and future directions. *Chemosphere* 285: 131535. doi:10.1016/j.chemosphere.2021.131535.

da Silva Trindade, W. R. and R. G. dos Santos. 2017. Review on the characteristics of butanol, its production and use as fuel in internal combustion engines. *Renewable and Sustainable Energy Reviews* 69: 642–651. doi:10.1016/j.rser.2016.11.213.

Snehesh, A. S., and S. Dasappa. 2016. Biomass to liquid transportation fuel via Fischer Tropsch synthesis – Technology review and current scenario. *Renewable and Sustainable Energy Reviews* 58: 267–286. doi:10.1016/j.rser.2015.12.143.

Snehesh, A. S., H. S. Mukunda, S. Mahapatra, and S. Dasappa. 2017. Fischer-Tropsch route for the conversion of biomass to liquid fuels—Technical and economic analysis. *Energy* 130: 182–191. doi:10.1016/j.energy.2017.04.101.

Sproles, A. E., F. J. Fields, T. N. Smalley, C. H. Le, A. Badary, and S. P. Mayfield. 2021. Recent advancements in the genetic engineering of microalgae. *Algal Research* 53: 102158. doi:10.1016/j.algal.2020.102158.

Susmozas, A., R. Martín-Sampedro, D. Ibarra, M. E. Eugenio, R. Iglesias, P. Manzanares, and A. D. Moreno. 2020. Process strategies for the transition of 1G to advanced bioethanol production. *Processes* 8(10): 1310. doi:10.3390/pr8101310.

Tabatabaei, M., M. Aghbashlo, M. Dehhaghi, H. K. S. Panahi, A. Mollahosseini, M. Hosseini, and M. M. Soufiyan. 2019. Reactor technologies for biodiesel production and processing: A review. *Progress in Energy and Combustion Science* 74: 239–303. doi:10.1016/j.pecs.2019.06.001.

Tursi, A. 2019. A review on biomass: Importance, chemistry, classification, and conversion. *Biofuel Research Journal* 6(2): 962–979. doi:10.18331/brj2019.6.2.3.

U.S. Agricultural Research Service. 2021. *What is Pyrolysis?* Available in www.ars.usda.gov/northeast-area/wyndmoor-pa/eastern-regional-research-center/docs/biomass-pyrolysis-research-1/what-is-pyrolysis/ [accessed September 18th, 2023].

U.S. Department of Energy. 2023. *Biofuel basics*. Available in www.energy.gov/eere/bioenergy/biofuel-basics [accessed September 8th, 2023].

Veljković, V. B., M. O. Biberdžić, I. B. Banković-Ilić, I. G. Djalović, M. B. Tasić, Z. B. Nježić, and O. S. Stamenković. 2018. Biodiesel production from corn oil: A review. *Renewable and Sustainable Energy Reviews* 91: 531–548. doi:10.1016/j.rser.2018.04.024.

Verma, P., and M. P. Sharma. 2016. Review of process parameters for biodiesel production from different feedstocks. *Renewable and Sustainable Energy Reviews* 62: 1063–1071. doi:10.1016/j.rser.2016.04.054.

Vignesh, P., A. R. P. Kumar, N. S. Ganesh, V. Jayaseelan, and K. Sudhakar. 2021. Biodiesel and green diesel generation: An overview. *Oil & Gas Science and Technology – Revue d'IFP Energies Nouvelles* 76: 6. doi:10.2516/ogst/2020088.

Voloshin, R. A., M. V. Rodionova, S. K. Zharmukhamedov, T. N. Veziroglu, and S. I. Allakhverdiev. 2016. Biofuel production from plant and algal biomass. *International Journal of Hydrogen Energy* 41(39): 17257–17273. doi:10.1016/j.ijhydene.2016.07.084.

Witcover, J., and R. B. Williams. 2020. Comparison of "Advanced" biofuel cost estimates: Trends during roll-out of low carbon fuel policies. *Transportation Research Part D: Transport and Environment* 79: 102211. doi:10.1016/j.trd.2019.102211.

Yasemi, S., Y. Khalili, A. Sanati, and M. Bagheri. 2023. Carbon capture and storage: Application in the oil and gas industry. *Sustainability* 15: 14486. doi:10.3390/su151914486.

Zabed, H., J. N. Sahu, A. Suely, A. N. Boyce, and G. Faruq. 2017. Bioethanol production from renewable sources: Current perspectives and technological progress. *Renewable and Sustainable Energy Reviews* 71: 475–501. doi:10.1016/j.rser.2016.12.076.

Zandi Atashbar, N., N. Labadie, and C. Prins. 2017. Modelling and optimisation of biomass supply chains: A review. *International Journal of Production Research* 56(10): 3482–3506. doi:10.1080/00207543.2017.1343506.

Zhang, F., D. Johnson, M. Johnson, D. Watkins, R. Froese, and J. Wang. 2016. Decision support system integrating GIS with simulation and optimisation for a biofuel supply chain. *Renewable Energy* 85: 740–748. doi:10.1016/j.renene.2015.07.041.

Zheng, J., Y. Tashiro, Q. Wang, and K. Sonomoto. 2015. Recent advances to improve fermentative butanol production: Genetic engineering and fermentation technology. *Journal of Bioscience and Bioengineering* 119(1): 1–9. doi:10.1016/j.jbiosc.2014.05.023.

Ziegler-Devin I, L. Chrusciel, and N. Brosse. 2021. Steam explosion pretreatment of lignocellulosic biomass: A mini-review of theorical and experimental approaches. *Frontiers in Chemistry* 9: 705358. doi:10.3389/fchem.2021.705358.

12 Ocean Energy

Jorge Saldaña, Mesut Yurukcu, Narendra Boppana, Sepehr Arbabi, Jim Henry, and Sebahattin Ziyanak

12.1 INTRODUCTION: THE ENERGY OF THE OCEAN

The search for alternatives to fossil fuels that are clean and can be applied in a widespread fashion has invited researchers to explore all corners of the world. Along with the energy of the sun and the wind, oceans provide an exciting and promising option through the harnessing of tides, waves, and currents to provide electricity that can be added to the grid. In this chapter, we review the main advances to capture the energy of the ocean, the economic aspects of ocean energy endeavors, and the many efforts made to transform this energy to help materialize the transition process to renewables.

Ocean energy is observed mainly in currents and tides that are caused by the gravitational pull of both the moon and the sun, as well as by the rotation of the Earth (EIA, 2022a). To capture the kinetic energy of tides and currents, researchers have devised a few methods where two stand out as the most prevalent. On the one hand, *tidal barrages* are structures like hydroelectric dams built on estuaries or tidal basins and have submerged gates to control water levels during the day (MIT, 2022). Meanwhile, *underwater turbines* are built in zones with strong currents, individually or on arrays, in a fixed structure or a mobile assembly, and capture the energy produced by the constant ebbs and flows of the tides, producing electricity (European Commission, 2017).

The most well-known tidal barrage structure is located in the French river Rance and was built in 1966. This plant features an 8.5 m mean tide, has an installed capacity of 240 MW, and produces an annual output of 520 GWh, suitable to cover 4% of the homes in the Brittany area (Cochet and Lambert, 2017; Melikoglu, 2018; Wang and Wang, 2019). Because water is denser than air, tidal energy is more potent than wind energy, producing a larger output with identical turbine and rotor characteristics. Thus, despite requiring more resources, tidal energy is also a more reliable and consistent energy source than both wind and solar (EIA, 2022a; PNNL, 2021).

Furthermore, similar methods have been advanced to harness ocean energy, such as wave energy and ocean thermal energy conversion (OTEC) technologies. The former seeks to capture the energy of the pressure differential created by the action of the wind on the crest and the trough of a wave. At the same time, the latter takes advantage of the difference in temperature at different depth layers to generate energy (von Jouanne and Brekken, 2017). A variation of the latter method is implemented through ocean thermoelectric generators (OTEGs), which need a more significant temperature gradient to produce electric power (Khan et al., 2017).

Relying on the energy of the oceans is not a new development, as early 19th-century societies used the power of bodies of water to turn waterwheels that helped mill grain. Steadily, the energy needs of increasingly more people were fulfilled as wave energy converters (WECs) were advanced in the 1800s, and the kinetic energy of water was harnessed by large hydroelectric dams that provided electricity for millions of people a few decades later. Today, the power of the ocean and the waves constitutes a predictable and consistent renewable source that can become a staple in the energy grids of many countries (Adcock et al., 2015; Lehmann et al., 2017; Pecher and Kofoed, 2017).

We are currently at a time when technologies have evolved to a point where researchers will scour the Earth in search of reliable alternatives to fossil fuels, just at a time when the environmental situation is getting harder to manage due to those same conventional sources of energy that have

 DOI: 10.1201/9781003282273-12

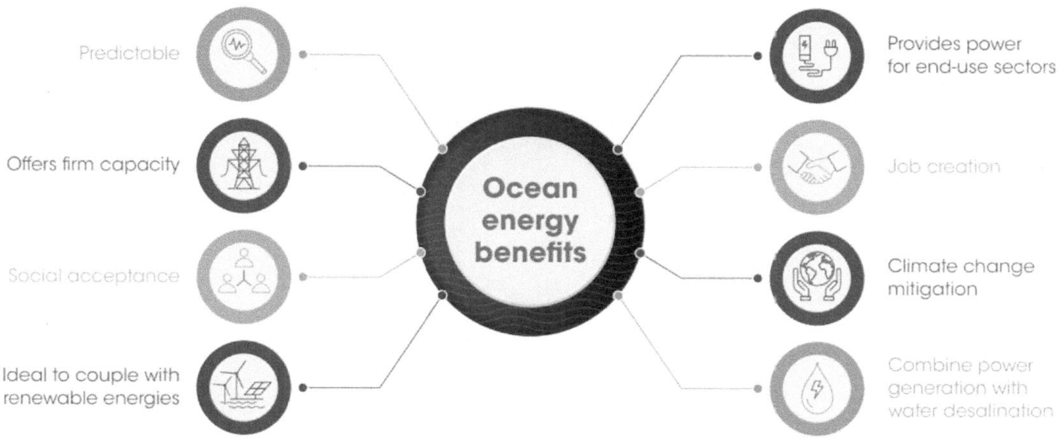

FIGURE 12.1 Benefits of ocean energy technologies.

Source: IRENA-OEE (2023).

contributed to the warming of the climate. Moving forward in harnessing the power of the ocean decisively can provide humanity with a source of energy that has many upsides, as Figure 12.1 illustrates. Unlike fossil fuels, ocean energy helps with climate change mitigation and enjoys a greater degree of social acceptance. Additionally, unlike solar and wind, ocean energy is based on cycles that have a predictable intermittence so that other sources can cover low-energy periods.

At the time, however, ocean energy projects remain largely as prototypes, in the experimental stage, or waiting for proper funding opportunities from governments and interested stakeholders. Part of this apparent skepticism appears to be that most of the criteria for site selection, as well as the economics that would yield viable projects around the world, continue to be parameters that require refinement and further exploration. Yet tidal, wave, and ocean thermal endeavors have the potential to revolutionize the energy sector while providing jobs and improving the prosperity of millions of people worldwide. In this chapter, we provide a guide for engineers to grasp the challenges and opportunities of ocean energy, the paths forward, limitations, and potential as leaders of the energy transition.

12.2 PHYSICS AND THE PRINCIPLES OF SCREENING AND SITE SELECTION FOR OCEAN ENERGY PROJECTS

In a similar fashion to solar and wind projects, the screening and selection of ocean energy sites must assess, first and foremost, the geographical areas that have privileged wave and current power. Having the most advanced theoretical understanding of ocean power utilization will not benefit all stakeholders if that power source is not accessible in the selected location. Therefore, the very first condition to choose a site for future development of an ocean energy facility is the advantage of holding solid currents and wave potential.

Figure 12.2 shows a map of the global distribution of wave power density, where we observe that the most substantial areas for wave generation are located close to the poles, especially in the Southern Hemisphere, between South Africa and Australia. Also, parts of Canada and Alaska, as well as the North of Europe, are located in a naturally advantaged wave density zone. This has encouraged researchers, particularly in the Northern Hemisphere, to develop centers for investigating hydrokinetic energy and to establish test sites for technologies like wave energy converters (WEC). For instance, some initiatives advanced are the Alaska Hydrokinetic Research Center and the European Marine Energy Center (EMEC) (Lehmann et al., 2017; Pecher and Kofoed, 2017).

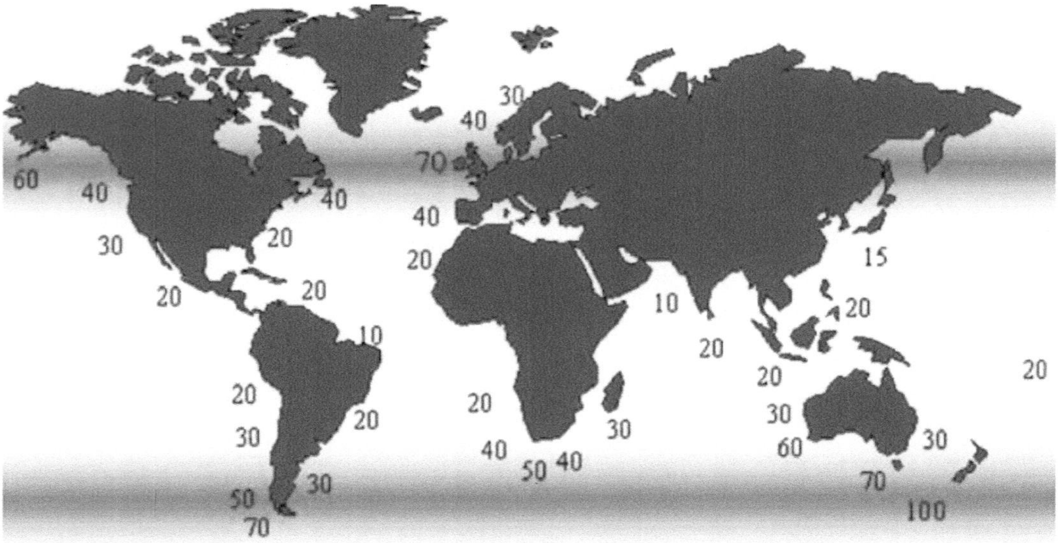

FIGURE 12.2 Wave power density world map, measured in yearly average values of kilowatts per meter (kW/m).

Source: Bracco et al. (2014).

Moreover, some authors have independently explored the potential of sites to provide ocean energy. For example, Coles et al. (2017) assessed the tidal strength around the Channel Islands, an area that covers northern France waters and the English Channel, through a 2D hydrodynamic model that simulates tidal flows. Their findings show that Alderney Race has an average power potential of 5.1 GW, much greater than that of the Pentland Firth, considered the best site for tidal energy in the United Kingdom. Similarly, Moore and Boyle (2014) explored New Zealand's avenues for energy diversification, estimating that the currents at the Manukau Harbor may provide outputs close to those of a small wind farm.

Then, studies have been made to assess the potential of power plant installation projects in less-favored locations. For instance, Buhali et al. (2012) studied the case of the Philippines and their likelihood of hosting tidal power plants using a multi-attribute decision analysis (MDA), finding that attributes like levelized cost of energy, social contribution, risks, environmental impacts, and proximity to the power grid are all critical parameters to decide on green-lighting an ocean energy project. While risks may involve other geological phenomena like earthquakes or tropical storms, social benefits are related to delivering more socially acceptable energy alternatives, providing jobs in a growing clean energy sector, and improving the prosperity and wealth of significant sectors of the population.

These trade-offs can be particularly overwhelming when selecting the first site to harness the energy of the ocean in a given location. The lack of comparable empirical results in similar locations, coupled with very high expectations for the fulfillment of environmental pledges and the economic viability of the site, can be a very challenging task. In their work looking at implementing a novel wave farm, Iglesias and Carballo (2011) developed a framework that expands the data analysis from a few wave patterns in limited time periods to include three other critical parameters: wave height, energy period, and mean wave direction. This trivariate analysis provides a more complete picture that improves from conventional evaluations.

As we can witness, the multiple criteria and the myriad challenges to operationalize different variables to include in the assessment models for ocean energy site selection endeavors can become a significant burden for researchers. In addition to the previously discussed criteria, actors should also take into account the economic feasibility of a project. Other relevant considerations might encompass the presence and potential disruption of existing navigation routes, the appropriate decommissioning or closure procedures for a facility, the possible repurposing of infrastructure for alternative projects, or the conversion of a site into, for instance, a carbon capture and storage (CCS) reservoir.

To simplify the tackling of these challenges, more novel analyses rely on tools such as geographic information system (GIS) software or neural networks supported by artificial intelligence. The assessment of the potential of renewable energy projects has increasingly relied on the systematization of data so that outcomes have the least bias while considering the largest number of cases and variables. Geographic information systems (GIS) improve on previous models by adding the extra dimension of detailed mapping and real-life constraints to arrive at the most optimal locations that are also within the bounds of what is permitted by local legislation and site protection.

For instance, the work of Vasileiou et al. (2017) explores the selection of sites for hybrid projects combining offshore wind energy with wave energy systems, assisted by GIS in Greece. These efforts are greatly simplified thanks to the use of geographical information that helps discard unsuitable sites that are restricted by economic, technical, or social constraints. Then, a second stage allows for the removal of eligible marine areas that meet the exclusion criteria set by the authors, such as areas for military exercise, hydrocarbon exploration areas, protected marine zones, or places with other renewable energy projects already scheduled. These stages are shown in more detail in Figure 12.3.

Then, Cradden et al. (2016) provided more depth in the analysis of wind–wave projects by adding bespoke tools that help with assessing multiple sites using multiple criteria. Analyzing five different cases, the authors found that the most suitable sites for hybrid wind–wave endeavors are European spots with Atlantic-facing coasts. Table 12.1 summarizes the main parameters considered for these five sites, where we see that variables like mean wave power density, wave height, depth, and wind speeds are all good omens for the future use of a site as a wave energy facility.

Meanwhile, a review analysis performed by Shao et al. (2020) regarding site selection criteria for a wide array of renewable energy projects shows that GIS techniques are a well-established tool to arrive at relevant screening and selection criteria, particularly when coupled with multi-criteria frameworks. In this sense, GIS can also be coupled with fuzzy theory, which provides more leeway when dealing with decision-making under real-world conditions. Then, a further set of parameters is mentioned by the authors as relevant for selecting sites dealing with wave energy projects, and these parameters are the distance from ports and shipping density.

In turn, factors for the optimal selection of tidal energy sites are explored by Lee et al. (2017), who focused on finding the most optimal dam location to explore. They started with four main criteria, which are related to business drive, potential opportunity, overall costs, and uncertainty and conflict. Then, each category has multiple parameters, where *tidal data* and *investment return* are part of the first group, while *financial support* and *technological improvements* belong to the second. Lastly, *levelized cost of energy* and *operation and maintenance (O&M) costs* highlight the third category, whereas *cultural conflicts* and *technological uncertainty* are part of the fourth group of variables.

Similarly, Wu et al. (2017) sought to optimize the site selection of a tidal barrage project, starting from the baseline that any tidal energy endeavor will have considerable environmental impacts in its construction phase. Because of this, all considerations that follow come with that cost included, and the final selection of a site must be the most environmentally conscious, something that can be achieved following the three criteria outlined in Figure 12.4. In the image, we see the paths that eventually lead to implementing a tidal barrage project, conditioned by hydrodynamic, biological, and morphological considerations.

Concerning the screening and selection of sites for ocean thermal projects, Zhang et al. (2019) explored the case of China, applying an extended multi-criteria decision model that assessed 16 different parameters first, followed by a political, economic, social, and technological (PEST) analysis. In their framework, the authors focused on various sustainability considerations, such as energy,

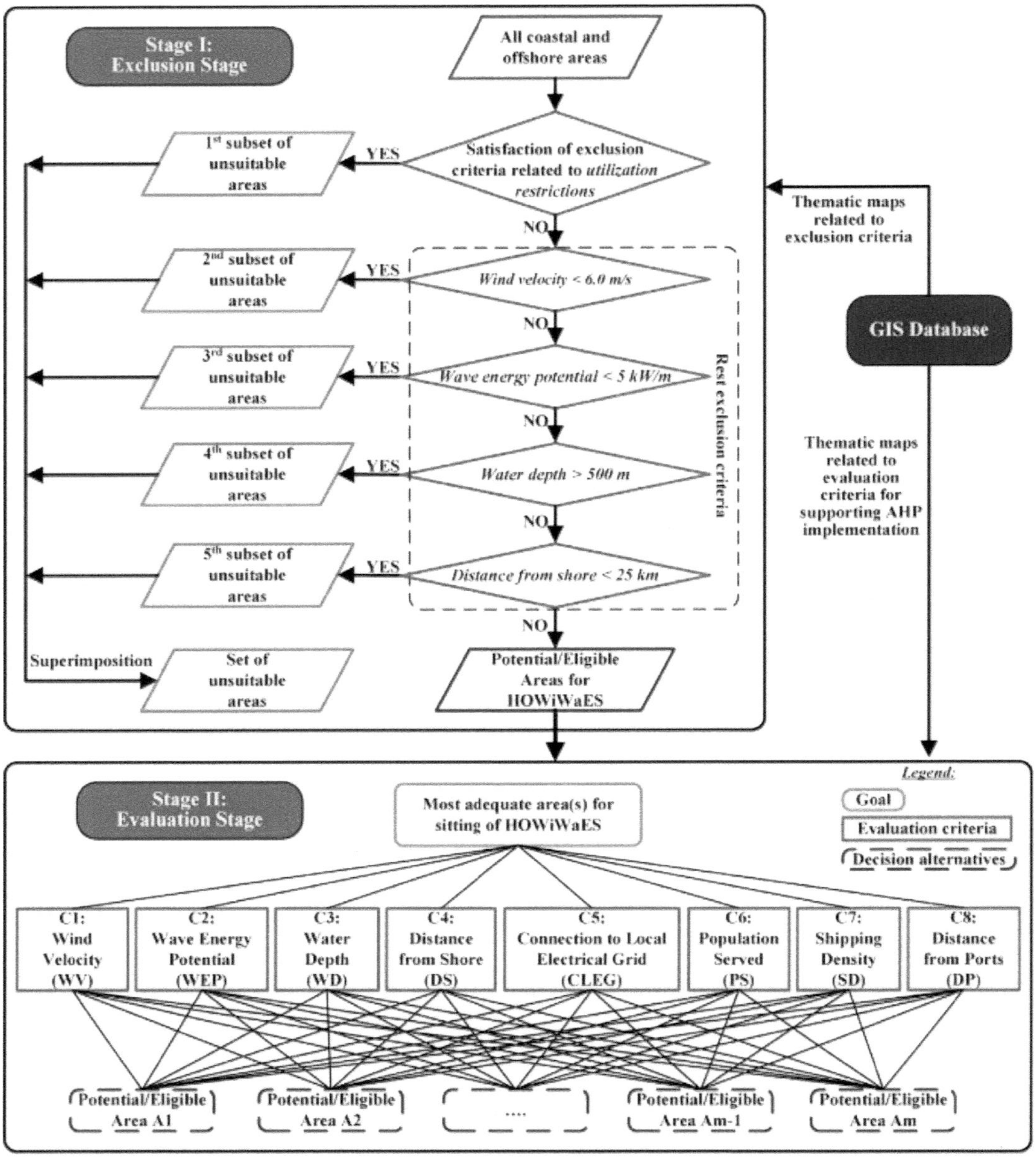

FIGURE 12.3 Framework for site selection of hybrid offshore wind and wave energy systems (HOWiWaES) in the Greek marine environment.

Source: Vasileiou et al. (2017).

construction, and maintenance, economic and environmental factors that help narrow down the selection of the most optimal OTEC installation site. A sensitivity analysis, where changes in outcome variables are analyzed after changes made to input variables, was performed to detect changes in the suggested location. Yet the most potent option remained in Weizhou Island, Guangxi Province.

In turn, studies performed for the reality in Mexico and the Pacific area surrounding the country show that relevant technical variables are related to distance to cold water intake at significant depths, general temperature differences, extreme events, and distance to the electricity grid

TABLE 12.1
Physical, Wind–Wave (Metocean), and Production Characteristics for Sites Selected in the Study

	Shetland Offshore	Crozon Offshore	Norway 1	Norway 3	Sybill Head
Latitude (°)	60.2	48.7	58.25	61.85	52.25
Longitude (°)	−2.85	−5.75	4.45	4.25	−10.7
Depth (ETOPO1) (m)	150	114	178	202	103
Distance to shore (km)	65	75	79	30	17
Mean wind power density (W/m^2)	1,126	795	1,079	1,084	946
Mean wave power density (kW/m)	67	50	28	47	71
95% wind speed at 80 m a.g.l. (m/s)	18.83	17.12	18.9	19.06	18.15
95% significant wave height (m)	6.36	5.66	4.85	5.46	6.52
Wind–wave correlation at time = 0	0.70	0.66	0.78	0.67	0.67
Max wind–wave correlation	0.73	0.69	0.81	0.70	0.70
Time lag to max (h)	4	4	3	3	4
Platform 1 rank (%)	0.77	0.36	n/a	0.27	0.73
Platform 1 capacity factor (%)	40	32	n/a	33	38
Platform 2 rank (%)	0.87	n/a	0.34	0.32	n/a
Platform 2 capacity factor (%)	46	n/a	42	42	n/a
Platform 3 rank (%)	0.81	0.19	0.41	0.32	0.39
Platform 3 capacity factor (%)	58	50	55	54	53
% of hours inaccessible at Hs > 2 m, wind speed >10 m/s	74	60	48	65	72

Source: Cradden et al. (2016).

(Garduño-Ruiz et al., 2021). With that said, sophisticated techno-economic assessments can sometimes leave the realities of marginalized populations, such as local fishermen, outside of the analysis. This omission could lead to some projects being interrupted or to a negative social perception of ocean energy that may not allow future projects of similar scale to move forward (García-Huante et al., 2020).

These studies employ geographical information systems or a blend of morphological and biological evaluations to determine the ocean energy site that causes the least environmental harm. With the arrival of more sophisticated techniques to various sectors of the oil and gas industry, however, many authors have now turned to tools like deep learning neural networks, assisted by artificial intelligence. These techniques allow researchers to estimate the potential of sites with a greater degree of reliability, generating models that ponder multiple criteria, thus helping researchers rank the different parameters in order of importance so that the most optimal site is selected for each individual assessment.

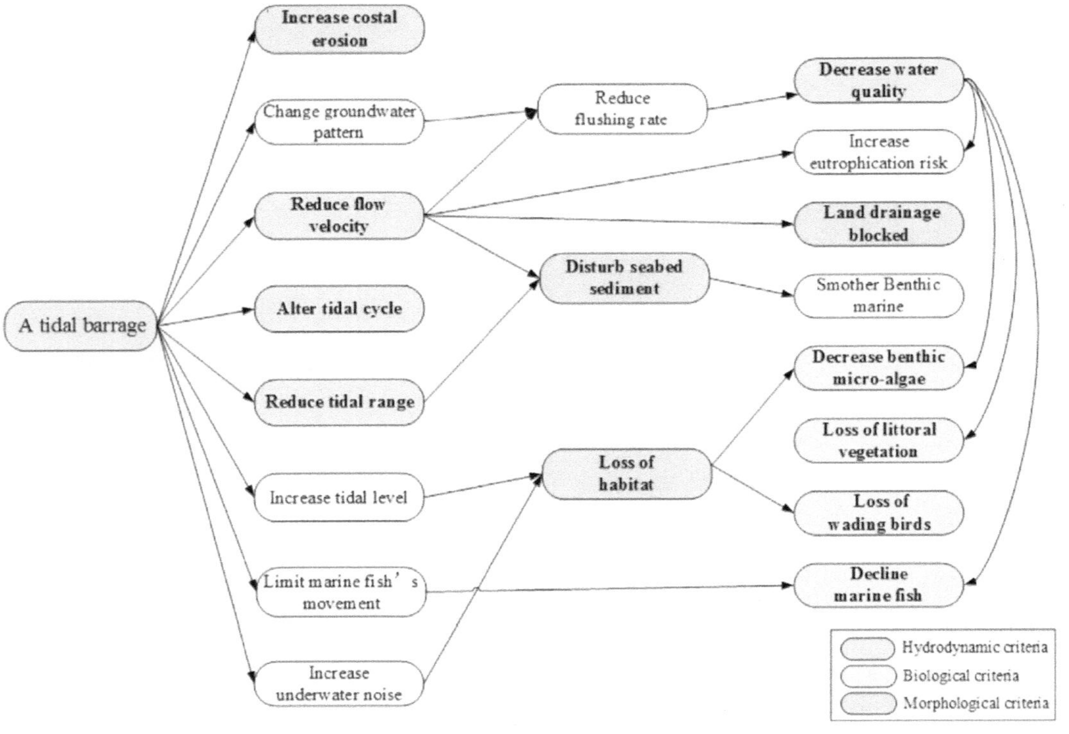

FIGURE 12.4 Correlation relationships among criteria involved in tidal barrage scheme environmental impact assessments (EIAs).

Source: Wu et al. (2017).

One example of these more novel techniques is in the work of Abaei et al. (2017), who expanded the conventional decision-making frameworks to include Bayesian networks, measuring parameters in their probabilistic influence while also helping estimate the expected utility of a wave energy converter (WEC) in a selected location. Relatedly, James et al. (2018) used a machine learning framework trained in the physics-based wave conditions from Monterey Bay, California, to better estimate future sites for wave energy. Through a multi-layer perceptron (MLP) conceptual model, the authors analyzed variables like significant wave heights and characteristic periods, improving the forecast prospects of a WEC site.

Also, Ghosh et al. (2016) combined a multi-criteria decision-making (MCDM) framework with an artificial neural network (ANN) model to arrive at a suitability index for the most optimal wave energy production site. Focusing on variables like wave height and speed, as well as more novel factors like distance between two waves, number of waves, turbulence, coastal erosion, and even tourism potential, the authors can construct an optimal feasibility index for a wave energy site. Figure 12.5 shows the stand-alone ANN framework that is used to estimate the potential for a wave energy location, while findings from their case study analysis show that the most potent factor is wind speed, and coastal erosion is the least important in determining an optimal suitability index.

As this literature review suggests, critical parameters for an optimal site selection for ocean energy projects revolve around strong conditions for generating recurrent and robust waves or tides, or a reasonable degree of temperature difference between surface and seabed in the case of an ocean thermal undertaking. Moreover, social and socio-political constraints, such as protected or military exercise areas, can also become relevant in countries with varied land uses or projects under development. With that said, another suitable set of parameters is those related to the economics of a project, from its funding sources to the potential of achieving a viable and profitable endeavor.

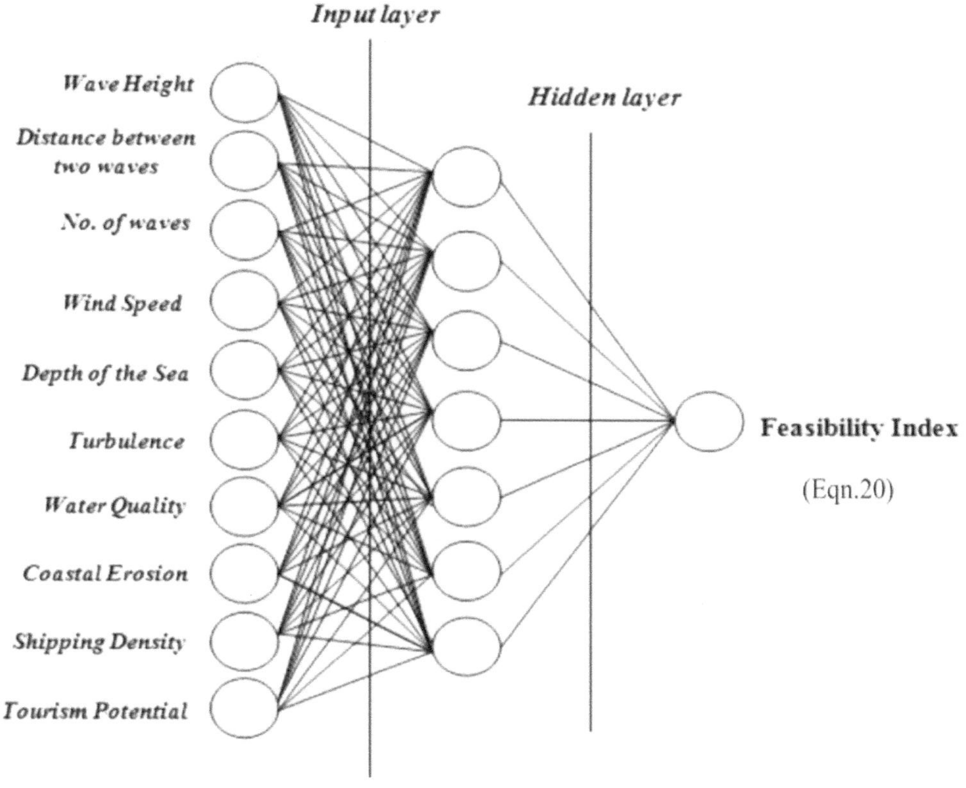

FIGURE 12.5 Artificial neural network (ANN) model architecture developed for feasibility index attainment.
Source: Ghosh et al. (2016).

These factors and the dynamics associated with the economics of ocean energy projects are the subjects of the following section.

12.3 ECONOMICS OF OCEAN ENERGY PROJECTS

Geographical, social, and political considerations constrain wave, tidal, and ocean thermal energy projects. All those limits are established by actors with varying interests or by nature itself. Similarly, different economic considerations constrain ocean energy projects, altering the dynamics and adding one more degree of complexity to the entire effort. For example, stakeholders in the industry and the community may have opposite objectives, which could hinder the future of a project due to cost or previously overlooked externalities. Hence, even though both parties share the common aim of optimizing advantages for the majority while minimizing expenses, there may be instances where they hold divergent perspectives on the same issue.

Despite relying on physics and principles tried over many decades across many distinct areas of the world, harnessing the energy of the ocean to produce mass-level electricity to add to regional or national grids is a very recent enterprise. Because of this, many actors are still in uncharted territory, and prudence may become their best strategy while technologies mature and become more mainstream. However, this caution can delay significant advancements, as well as postpone all potential benefits of a renewable and predictable source of energy that can assist wind and solar and perhaps replace many operations that currently use fossil fuel energy.

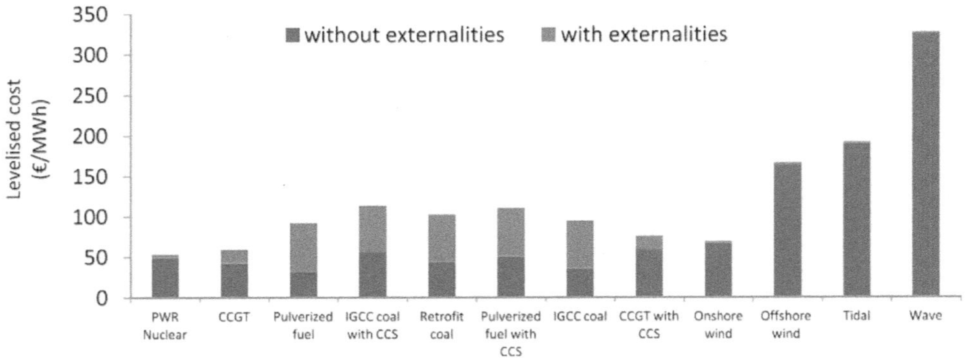

FIGURE 12.6 Levelized cost of energy (€/MWh) for different technologies, including external costs. (*PWR*, pressurized water reactor; *CCGT*, combined cycle gas turbine; *IGCC*, integrated gasification combined cycle; *CCS*, carbon capture and storage.)

Source: Astariz et al. (2015).

In addition, other practical limitations may hinder a more streamlined development of marine technologies. First, funding from governments and international organizations toward renewable energies has historically centered more on solar and onshore wind, given their lower operation costs. This, in turn, creates a path-dependency conundrum where current prices are now lower partly because many wind and solar projects were funded in the past, sometimes to the detriment of marine energy projects. Figure 12.6 shows a comparative measure for the levelized cost of energy for different sources, where we see that tidal and wave energy are still not competitive enough to compete with conventional sources or onshore wind.

Another economic challenge faced by ocean energy projects might arise from the absence of legislation aimed at promoting a market that expands the role of marine resources within the energy portfolios of multiple countries. According to IRENA-OEE, a step forward for achieving greater visibility for ocean energy projects consists of national deployment targets. These large processes seek to include marine projects in broader planning processes, increasing access to finance and reducing perceived market risks. For example, Ireland, Spain, and Portugal have committed to similar targets of deployment of marine energy in their grid, ranging between 30 and 70 MW by 2030, while establishing a path to develop economies of scale for the source in the medium term (IRENA-OEE, 2023).

Lastly, a third roadblock may be exercised by other industry actors with their lack of funding for rollout projects. In theory, providing a "top-up" through revenue support should fund the initial stage of development for an ocean energy endeavor, absorbing initial costs. At the same time, investors are sought and guaranteed a share of the returns when the project becomes financially profitable. Canada and the United Kingdom provide two examples where it has been the government that has stepped up and offered industrial rollout initiatives that helped launch tidal stream pilot projects, establishing ring-fenced budgets with tight deadlines that can be extended if these goals are met (IRENA-OEE, 2023).

A problem that we find regularly when studying the economic dynamics around novel sources of energy is the disparity that developed countries have when compared to developing nations in funding schemes and technological advancements. Given their levels of wealth and their more stable energy grids, the former set of countries can allow themselves to explore potentially less-dense sources of energy, developing economies of scale where the endeavor becomes profitable after a market achieves enough maturity. For example, Allan et al. (2014) studied the potential of government funding for marine energy and the short-term higher costs that this move would mean in Scotland, which would later be compensated by a greater gross regional product and more jobs.

However, most countries worldwide do not have the resources to initiate those projects without the aid of developed countries or international organizations like the United Nations. A review from Magagna and Uihlein (2015) shows that the European Union has demonstrated a very high interest in the development of marine energy projects, hosting half of the ocean energy projects and a little under half of marine energy developers and creating the Ocean Energy Forum, a platform to bring stakeholders and ocean energy actors together. More recently, the European Commission has stated its goal of reaching a 10% share of energy contribution for the Union by 2050 (European Commission, 2023).

One way that actors have found to cut some costs in project development consists of developing hybrid platforms, such as those devised for offshore wind and tidal energy. Astariz and Iglesias (2015) explored this possibility, arguing that despite higher costs for offshore plants generally, hybrid projects can harness the power of two different sources with similar equipment on the same infrastructure, which translates into lower costs passed to the final user of the electricity generated. Similarly, Jeffrey et al. (2013) explored the transfer of technology required across industries to cut costs, achieve EU pledges, and eventually expand the role of ocean energy around the world.

Marine energy is still in the early adopter phase, with some players continuing to be reluctant to invest in it while others have exited the market entirely. Currently, developing tidal arrays and hooking them up to the grid are prohibitive for most actors, given the extra challenges of installing undersea cables and optimizing marine life with large industry operations (Chen et al., 2018; PNNL, 2021). This has maintained very high costs when compared to other renewable sources of energy, as estimations show that commercial-scale tidal energy costs between US$130 and US$250 per megawatt-hour, up to 20 times more than energy coming from wind farms (MIT, 2022).

All this shows the sector is in a very early stage, in need of significant help from established players, such as oil and gas companies looking to invest in renewables, governments aiming to fund more clean energy projects, and other stakeholders looking to respect environmental pledges and expand the reach of the energy source. However, there is promise in the exploration and funding of ocean energy in the medium term. One instance of stakeholders looking for greater involvement, the United Nations developed the Ocean Stewardship Coalition, a group to bring partners, non-governmental organizations (NGOs), companies, and governments together to coordinate projects and remain committed to environmental pledges like the Paris Agreement.

Moreover, the UN Global Compact estimates that given its 15-fold increase in potential for 2040, ocean energy could become 1-trillion-dollar industry, with many opportunities for players that join early (UN Global Compact, 2023). Additionally, in their Ocean Stewardship 2030 Report, the organization proposes a series of recommendations that seek to align policy through strictly targeted strategies and to develop a clear and coordinated regulatory system that advances pathways to healthy ocean energy markets (UN Global Compact, 2020). This framework, where technological aspects and costs at different time frames are laid out, is vital for the blossoming of the sector.

To advance current marine energy systems, the United States Department of Energy announced by the end of 2022 that they would be allocating $35 million, which was later increased by a further $10 million. This money will be used to improve current technologies and to expand the reach of present tidal and river energy. Additionally, it is expected that this funding will attract other players in the industry to develop their projects, generating competition in an emerging market and emphasizing local communities and the environment (EERE, 2022).

As we can see, the economics of ocean energy are currently contingent on a few issues, where actors' coordination toward the development of economies of scale is a pivotal consideration. Because of this challenge, marine energy projects cannot currently take off without considerable help from industrial players, governments, and international organizations, who, through funding schemes and the drafting of friendly legislation, can advance the necessary paths to creating markets that are robust and competitive (Anderlini et al., 2020; Chowdhury et al., 2021). If those steps are carefully followed, optimist scenarios show a potential sector growth from a couple of terawatts-hour before 2020 to 15 TWh of electricity by 2030, as Figure 12.7 suggests.

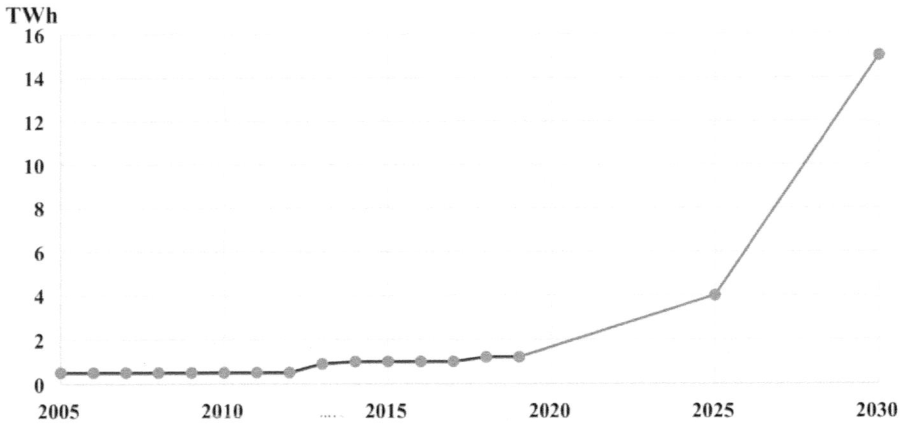

FIGURE 12.7 Ocean power generation scenario, 2000–2030.

Source: Chowdhury et al. (2021).

In summary, the present economic scenario for ocean energy endeavors is one of prudence and waiting. The belief appears to be that there are enough positive aspects to these technologies to be applied in enough favorable contexts eventually and that the momentum created by multiple actors pursuing an energy transition to renewable sources will also drag the more costly ocean energy methods into a position of self-sustainability. In the meantime, however, technologies must achieve a level of maturity that makes it worthwhile for investors to participate. So strategies for improved research and development (R&D) are required, along with a long-term vision that allows for the deployment of many marine projects worldwide. These dynamics will be the focus of the following section.

12.4 OCEAN ENERGY APPLICATIONS, PRINCIPLES, AND BEST PRACTICES

Having reviewed the dynamics for screening and selection of sites to develop ocean energy, we move on now to the basics of ocean energy technologies, current applications, and best practices in the industry. There are many elements that we still do not grasp adequately about the extent of these technologies. Still, early results from experimental research and pilot projects show that harnessing the energy of the ocean to serve humanity can become a permanent staple for many countries. In the context of the energy transition to renewables, marine sources can provide a stable and predictable supply that helps many countries achieve energy independence while phasing out fossil fuels.

12.4.1 Tidal Energy

One promising method to harness the energy of the ocean is to capture the kinetic energy that is generated after the interaction between the wind, the moon, and the sun, coupled with the centrifugal force produced by the rotation of the Earth (von Jouanne and Brekken, 2017). According to Khan et al. (2017), spring (high) tides are expected to occur on new and full moon cycles, while neap (low) tides are observed in waning half-moons that happen due to the misalignment of the Earth with the moon and the sun. Figure 12.8 illustrates these cycles, which have a desirable degree of predictability for researchers, as this offers a more reliable source of inexhaustible energy when compared to solar and wind (IEF, 2021).

As a response to the increasing concerns about the state of the environment, which are partly due to fossil fuel exploitation, engineers and researchers have explored the potential to harness the energy of the oceans as a renewable, clean alternative. This motivation has led to the advancement of two main methods for energy generation from the ocean: tidal barrages and tidal turbines.

Additionally, similar techniques that remain in the very early stages of development are tidal fences and tidal lagoons. In what follows, we review these technologies, their challenges and potential, and the best practices regarding their utilization.

Table 12.2 offers a list of the locations possessing the highest tidal elevations around the world. In places with a considerable ocean energy potential, this power can be directed to operate turbine systems that drive a generator, creating electrical energy that can be added to the grid. Human ingenuity has developed different techniques to capture an increasing fraction of the 3,000 GW of tidal power that are estimated worldwide, offering another renewable source of energy that helps humanity fulfill the energy transition away from fossil fuels.

The French area of La Rance has hosted a tidal barrage facility since 1966, having an installed capacity of 240 MW of electricity while generating close to 500 GWh (gigawatts hour) annually (Cochet and Lambert, 2017; Mendi et al., 2016). This type of facility features a dam-like barrier that connects both sides of an inlet, establishing a basin or reservoir where the water is forced into a smaller area, creating tides over 5 m, then generating pressure to drive turbines that respond to two directional water flows (von Jouanne and Brekken, 2017). More recently, the Sihwa Lake Tidal Station was built in South Korea in 2011, an illustration of which we see in Figure 12.9. This plant has an installed capacity of 254 MW, making it the largest tidal power station in operation today (IEF, 2021).

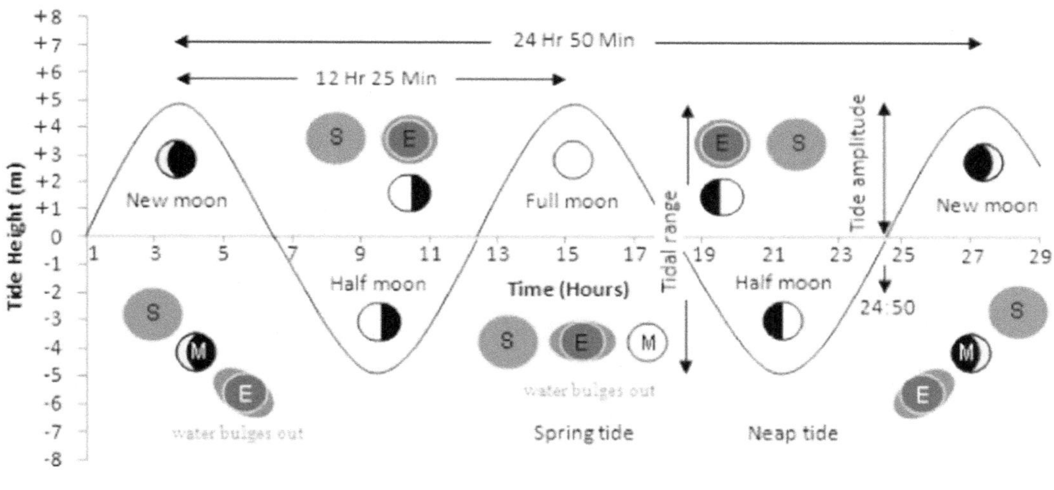

FIGURE 12.8 Natural evolution of spring and neap tides at sea and ocean surfaces.

Source: Khan et al. (2017).

TABLE 12.2
Highest Tides of the Global Ocean

Site	Country	Tidal Elevation (m)
Bay of Fundy	Canada	16.2
Severn Estuary	England	14.5
Port of Granville	France	14.7
La Ranee	France	13.5
Puerto Rio Gallegos	Argentina	13.3
Bay of Mezen (White Sea)	Russia	10
Penzhinskaya Guba (Sea of Okhotsk)	Russia	13.4

Source: Mendi et al. (2016).

The literature on tidal barrage methods highlights a few mechanisms in which electricity is produced in these structures, involving the same phases of filling the basin, maintaining a constant water level in the barrage basin, and operating the turbines to produce power. The first method, ebb generation, consists of a single mode of energy production obtained when the tide flows back to the sea. The sluice gates close when the optimum basin level is achieved and are held until the tidal level outside the basin has decreased enough for the turbines to operate. Here, the water flows out to the sea through the turbines, which generate power (Waters and Aggidis, 2016).

Meanwhile, flood generation is another method of tidal barrage turbine operation, where power is created only twice per day as the tide flows toward the shore. When the tide approaches the facility, the sluice gates are kept closed until the head difference between the basin and the sea exceeds the minimum hydrostatic head difference for the turbines. At this moment, the gates are open, and power is generated after the tide activates the turbines. Lastly, two-way or bidirectional generation methods occur when electricity is generated through all four tidal movements daily, activating the optimized turbines when the tide comes in and goes out (Warak and Goswami, 2020; Waters and Aggidis, 2016).

Moreover, a few methods have been tested to enhance electricity generation when water goes through the turbines. For example, as Figure 12.10 shows, water pumping techniques increase the tidal range between the basin and the sea. These methods use hydro pumps to alter the tide that

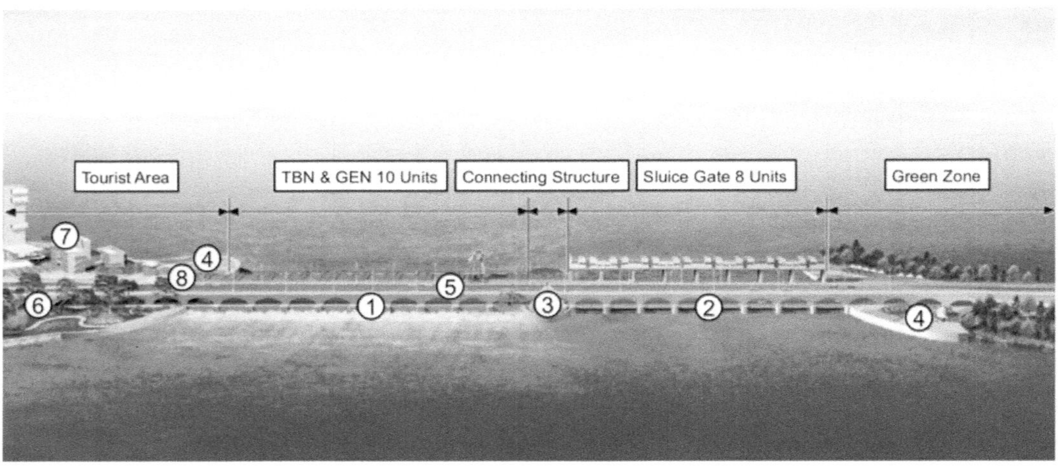

FIGURE 12.9 Construction plan for Sihwa Lake Tidal Station in South Korea, opened in 2011.

Source: von Jouanne and Brekken (2017); Wang and Wang (2019).

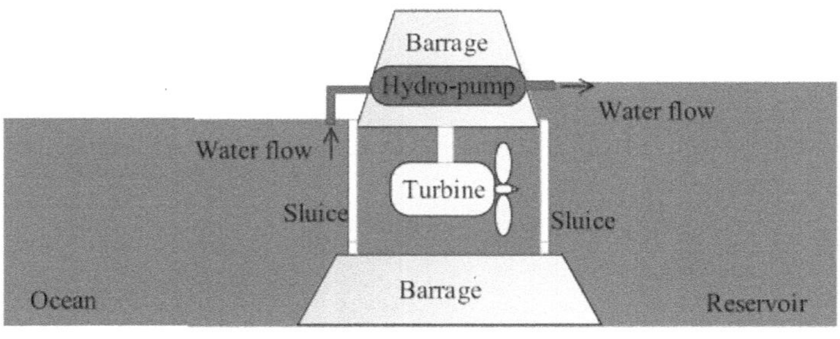

FIGURE 12.10 Structure of the basin-based tidal unit equipped with hydro pump for enhanced power generation.

Source: Negahdari et al. (2023).

reaches the turbines, thus producing more energy (Negahdari et al., 2023; Waters and Aggidis, 2016). Additionally, Ghaedi and Gorginpour (2021) mentioned that an optimized plant design, an optimized operation of turbines and gates, and optimized modes of operation obtained through machine learning methods are three other factors that can enhance production in ocean energy facilities.

Moving on to the second main way to capture the energy of tides, tidal turbine methods rely on the conversion of kinetic energy, acting in a similar fashion as wind turbines yet under more demanding circumstances. Therefore, tidal turbines have blades that turn a rotor, which instead powers a generator, just like wind structures. With that said, given that water is over 800 times denser than air, ocean turbines must be sturdier and heavier than wind structures. This significantly increases the costs of building ocean energy facilities, but an array of underwater turbines can capture more energy than its wind counterpart (EIA, 2022a; Warak and Goswami, 2020).

Depending on the location and specific goals of the project, the layout of tidal turbines can follow two main patterns. On the one hand, horizontal-axis tidal turbines (HATs/HATTs) work when the rotor blades in the structure, which are parallel to the tidal currents, convert the kinetic energy of the ocean into mechanical energy to operate a generator, which in turn generates electricity for the grid (Melikoglu, 2018; Nachtane et al., 2020; NREL, 2011). Figure 12.11 shows a series of different HAT designs developed for different depth ranges. Currently, HAT technologies are the most mature set of methods to capture tidal energy, having already achieved megawatt-level power capacity in several projects (Zhou et al., 2017).

The MCT SeaGen device, located in Strangford Narrows, Northern Ireland, and developed in 2008, was the first commercial-scale tidal turbine deployed to generate electricity for the grid outside a test center (Roberts et al., 2016). In its heyday, the SeaGen facility generated 1.2 MW of electricity between 18 and 20 hours a day, and its system was capable of adding 20 MWh per day, reaching 6,000 MWh for a year (Warman et al., 2020). However, despite having a design life of 20 years, the plant was decommissioned between 2016 and 2019. The main reason for the shutdown was the technology being sold from Siemens to Atlantis Resources and the focus of the latter on a new generation of turbines based on the life cycle of this installation.

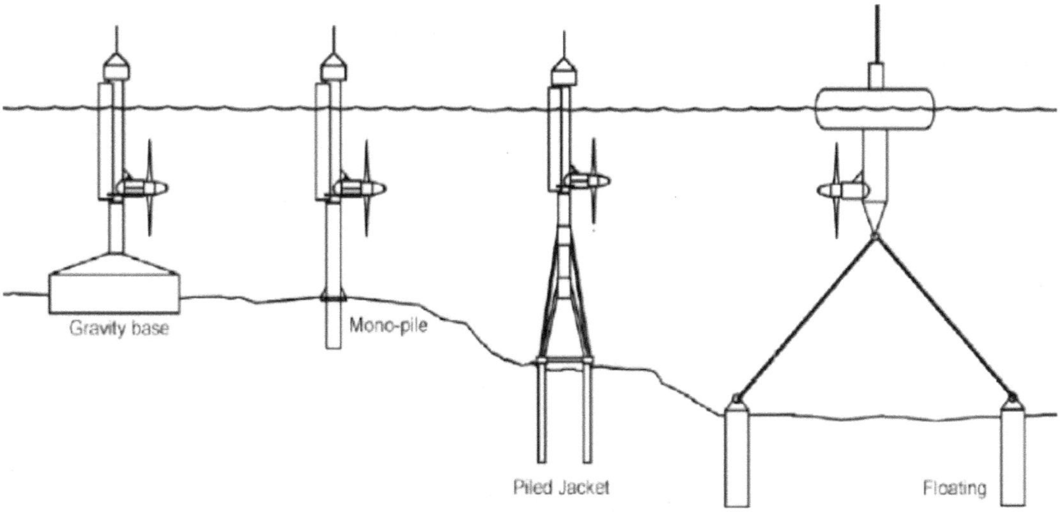

FIGURE 12.11 Examples of horizontal-axis tidal current turbine configurations. Each setup is best suited for different depth ranges.

Source: NREL (2011).

In contrast, vertical-axis tidal turbines (VATs/VATTs), also known as cross-flow turbines, have a vertical rotor featuring blades that are perpendicular to the flow. Due to their lower sensitivity to changes in tidal direction, along with their compact design, vertical-axis turbines can work efficiently in shallow waters. Figure 12.12 shows the different configurations of VATs that vary depending on the shape of the blade and their lift/drag properties. Some advantages of VATs are their simple design, inexpensive manufacture, simple installation, and ability to adapt omnidirectional flows. Conversely, their main challenge is their lower efficiency, explained by the irregular current velocities that limit the RPMs to which these structures can rotate (Roberts et al., 2016; Satrio et al., 2016).

Perhaps the most significant current endeavor in vertical axis tidal turbine technology is the VATTS project in Northern Ireland, a joint venture between GKinetic Energy Ltd., GES Group, the Electric Storage Co., and Queen's University Belfast, funded by the Center for Advanced Sustainable Energy (CASE) and Invest NI. This project is located in the same ideal testing area of Strangford Lough, where the MCT SeaGen device operated for close to a decade. The first turbines of this installation were placed in early 2022, and it is expected that its partners help in their specific areas of hydrokinetic turbines, battery solutions, and maximizing the value of final power delivery and sale of electricity to end users (GES-Group, 2022; The Electric Storage Company, 2023).

Then, oscillating hydrofoils are a variation of these two types of platforms, consisting of a hydrofoil wing attached to a lever arm, as shown in the left side of Figure 12.13. When the tide comes, a lift is generated, which helps raise the lever. Once the lever reaches its highest point, lift is produced on the underside, reversing the motion. This series of movements produces energy to drive a generator, which produces electricity (Roberts et al., 2016). To improve the efficiency of hydrofoils, the work of Ma et al. (2017) concludes that the angle of attack is a key parameter that conditions the success of other variables, like heaving motion and pitching amplitude, in the generation of energy.

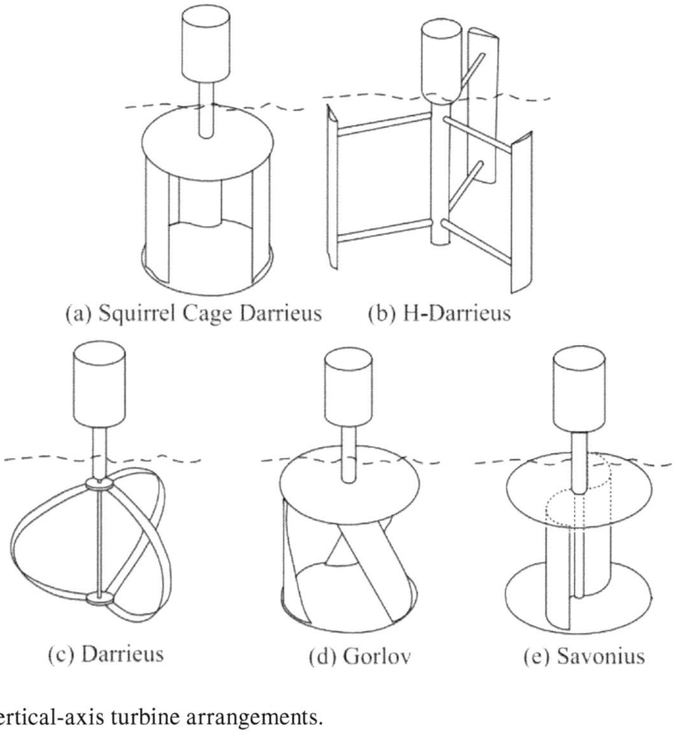

FIGURE 12.12 Vertical-axis turbine arrangements.

Source: Satrio et al. (2016).

Meanwhile, tidal kites are a group of smaller turbines attached to a hydrofoil wing, tethered to the seabed by a rope. Apart from the turbine and the wing, other parts of a kite are a nacelle, a rudder, struts, and the top joint that is attached to the tether (Kaddoura et al., 2020). Kites operate through tidal currents, which lift the turbines and push them forward through the water in a "figure-of-eight" trajectory, as the right side of Figure 12.13 shows. The motion from kites generates kinetic energy that is transformed into electrical energy, which is then transferred through the subsea cable to the electrical grid. In early 2023, the Swedish company Minesto launched operations for a kite power plant in the Faroe Islands, delivering up to 200 kWh per tide with their Dragon Class kites (Offshore Energy, 2023).

Lastly, other variations of these methods have been advanced in the literature, mainly to enhance energy production. First, ducted turbines (or enclosed tips) are horizontal-axis turbines placed in ducts which are contoured to increase the speed of the fluid passing through. Additionally, turbulence and fluid movement are reduced around the structure, which helps guide the water toward the turbines more efficiently. Then, Archimedes screws are a variation of vertical-axis turbines with an added helix, whose goal is to direct the tidal stream so its absorbed energy is optimized when turning the turbines (Nachtane et al., 2020; Wilberforce et al., 2019).

A third, less-preferred method to capture tidal energy is observed in tidal lagoons. These are similar to barrages but consist of entirely artificial basins with turbines embedded within their walls (Roberts et al., 2016). In perhaps the most well-known tidal lagoon project proposal, the Swansea Bay Tidal Lagoon was developed as a pilot initiative by the United Kingdom government in 2015, being approved by the Welsh administration in 2018 with a 200-million-pound investment. However, because of the greater costs of building these installations for comparable energy production, along with potential interruptions in the transport of sand and a small yet significant reduction of upstream water levels, the British government eventually rejected this initial proposal.

The work of Petley and Aggidis (2016) goes over the potential energy generation from a lagoon facility in Swansea Bay, finding that ebb generation techniques that are enhanced with pumping mechanisms can provide the largest amount of annual energy, even more than turbines that are optimized for dual generation of energy (ebb/flow), and that energy can reach over 700 GW/h. Moreover, according to Melikoglu (2018), the energy generated in this lagoon installation can power over 150,000 homes for the next 120 years, perhaps due to these studies and promising figures that, in October 2021, a new project at the same location was announced, this time funded by the private sector (Zhang et al., 2022).

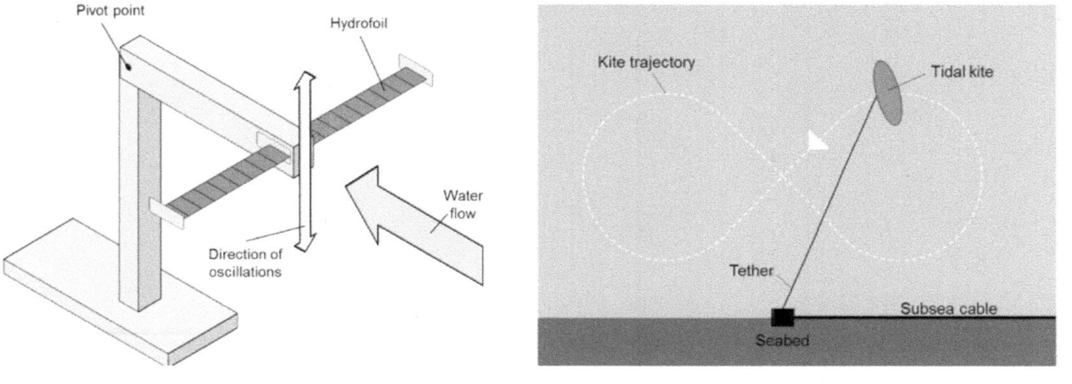

FIGURE 12.13 Schematic of an oscillating hydrofoil (left) and a tidal kite (right).

Source: Roberts et al. (2016).

12.4.2 Wave Energy

Then, a second main method to capture the energy of the ocean consists of harnessing the potential of the wind to generate waves. Waves are another renewable source of energy, with the advantages of being much more dense than solar or wind while also being both reasonably predictable and less volatile than those sources (Lehmann et al., 2017). Waves are mainly formed due to a pressure differential between the crest and the trough caused by airflow over the water's surface. Moreover, due to the propagation of the waves, there is a transfer of energy from air to water, which generates both kinetic energy and potential energy (von Jouanne and Brekken, 2017).

The main way to capture the energy from the ocean is by utilizing a wave energy converter (WEC), which can take a few different forms. Some well-known examples of WECs are oscillating water columns (OWCs), oscillating bodies, attenuators, overtopping systems, and point absorbers, as illustrated in Figure 12.14 (Aderinto and Li, 2018; Melikoglu, 2018; von Jouanne and Brekken, 2017). These methods can be classified into various subcategories, such as their operational concepts, their location with respect to the shore, their water depth positioning of buoyancy, or their orientation concerning the incident wave (Doyle and Aggidis, 2019). We will now review these technologies, their strengths and weaknesses, and the potential for expansion in the future.

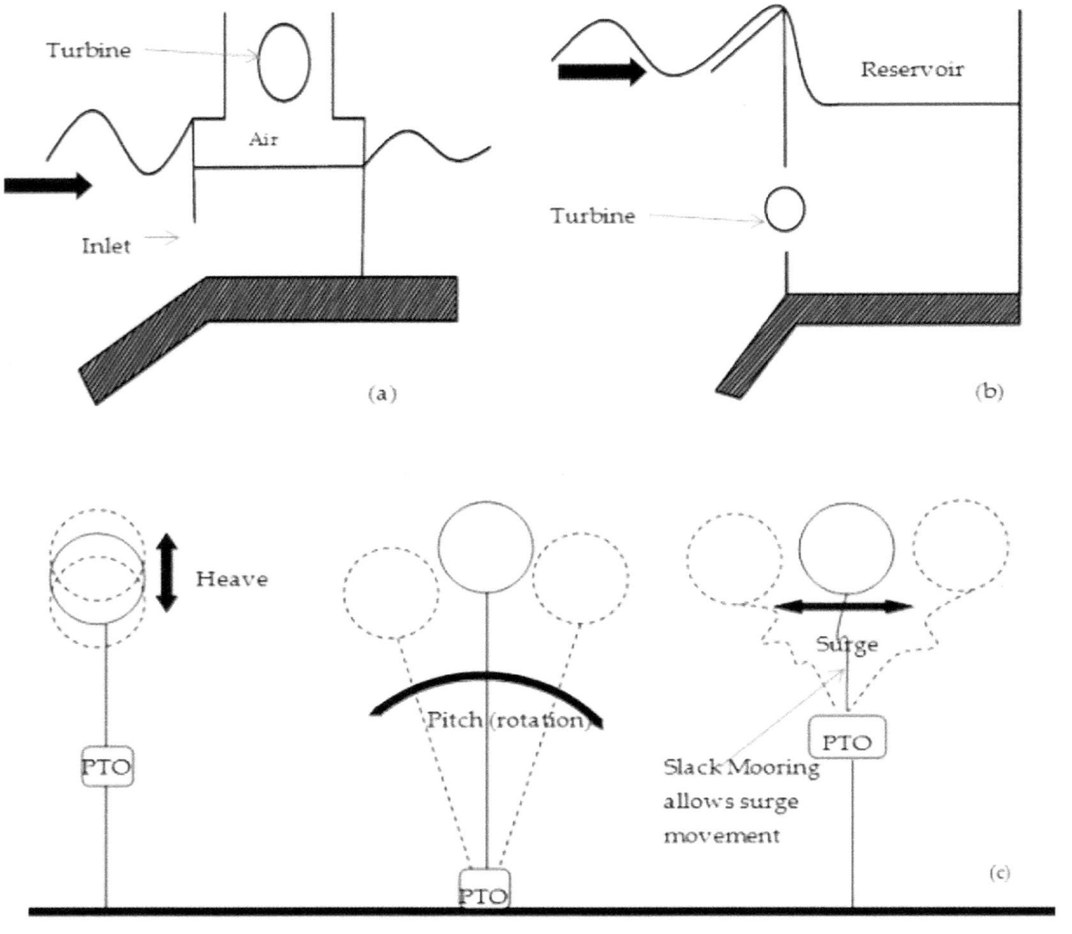

FIGURE 12.14 Classification of wave energy converter (WEC) extraction technologies: (a) oscillating water column, (b) overtopping devices, and (c) oscillating bodies. (*PTO*, power take-off system.)

Source: Aderinto and Li (2018).

To improve the current capacity and planned projects highlighted in Table 12.3, countries with natural advantages and robust funding schemes have shifted heavily toward developing oscillating water column (OWC) systems. These structures have a partially submerged section that traps waves inside a chamber, compressing the air inside and forcing it to go through the turbine to run the rotor, thus generating energy (Falcão et al., 2017; Wilberforce et al., 2019). These devices can produce between 15 and 25 kW/m a year, and some examples are the Limpet, a grid-connected British structure decommissioned in 2018; the Mutriku wave power plant in Spain; the Oceanlinx; and the Ocean Energy OWC (Melikoglu, 2018; von Jouanne and Brekken, 2017).

Then, oscillating bodies move with the motion of the water, under the expectation that they will move cyclically with respect to a fixed point. This motion can be done in six different ways: heave (translational), surge, pitch (rotational), roll, yaw, and sway, with the first three being the most common. Energy is converted by placing an electromagnetic or hydraulic mechanism between the oscillating structure and the fixed point, or if there are multiple structures, it is instead placed between them. Some examples of oscillating bodies are the SeaRay from Columbia Power Technologies, the Powerbouy from Ocean Power Technologies, the Weptos WEC, and the Northwest Energy Innovations (NWEI) Azura (Aderinto and Li, 2018; von Jouanne and Brekken, 2017).

For the most part, the previous set of structures is set up to be perpendicular to the incident wave. When this occurs, the systems are referred to as *terminators*, while *attenuators* are extended in a parallel fashion with respect to the incident wave as if they were riding them (Aderinto and Li, 2018). Meanwhile, *point absorbers* are smaller setups generally placed in large arrays or multiple units, one next to the other, focusing on a smaller area of the incident wavelength (Noad and Porter, 2017). Some examples of attenuators are the McCabe Wave Pump, the Pelamis, Anaconda, Wave Star, and Salter's Duck model (Melikoglu, 2018; Wimalaratna et al., 2022).

Despite being classified as a terminator device by some authors, Salter's Nodding Duck is a model that moves along with the waves, which would place it closer to the attenuator type. The left side of Figure 12.15 illustrates the model that allows for generating energy from the pitch rocking motion that simulates a duck nodding while it sails across the water. Meanwhile, the right side shows an array of these structures in the middle of the sea (Curto et al., 2021). According to Terrero Gonzalez et al. (2021), *attenuators* are a technology with a medium level of development, and many of the groundbreaking projects have already been decommissioned, with the DexaWave still remaining off the coast of Malta.

TABLE 12.3
Notable Wave Energy Converters (WECs) and Their Relevant Parameters

Name	Category	Location	Data	HDE (%)	Scale (m)	Capacity (kW)
AquaBuoy	Point absorber	Canada	2000	20	6	250
Wavebob	Point absorber	Ireland	2007	40	15	1,000
Pelamis	Attenuator	UK	2007	15	150	750
DEXA	Attenuator	Denmark	2011	8	22	160
Biowave	Terminator	Australia	2008	45	16	250
Oyster	Terminator	UK	2005	40	18	315
Mutriku	OWC	Norway	2011	7	180	300
Pico	OWC	Portugal	2000	20	48	400
Wave Dragon	OWEC	Denmark	2009	26	300	1,200
SSG	OWEC	Norway	2008	23	54	350
3-DOF WEC	MDWEC	NA	NA	80	10	152
6-DOF WEC	MDWEC	NA	NA	54	10	831

Source: Wimalaratna et al (2022).

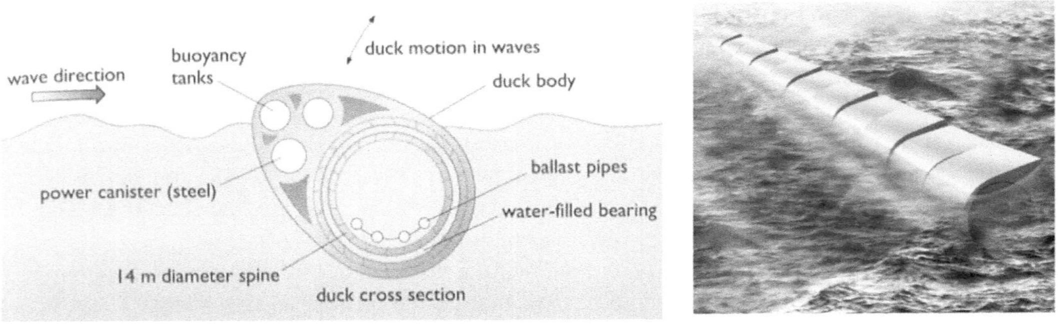

FIGURE 12.15 Salter's Nodding Duck cross-sectional schematic (left) and rendering view of structures placed in an array (right).

Source: Curto et al. (2021).

In turn, *point absorbers* are buoy-like WECs with a smaller size that can be submerged or float. They can take the form of an oscillating body if floating and have served as a starting point to test novel ideas on wave energy harnessing because of their adaptability to different types of seabeds and wave climates (Al Shami et al., 2018; Guo et al., 2022). Lastly, overtopping structures allow for waves to spill over a retaining wall that helps fill a basin at a level higher than the average seawater level. These structures generate energy by running the reservoir water through a low-head hydro turbine. Norway, Denmark, and Spain are the main areas of deployment for these structures (Aderinto and Li, 2018; Terrero Gonzalez et al., 2021; von Jouanne and Brekken, 2017).

A set of technologies with promising potential, wave energy converters are versatile and can adapt to different funding schemes. In this sense, small-scale projects can serve given communities while providing a few jobs in clean energy. At the same time, larger-scale endeavors can transform the energy sector of a particular location. Additionally, the lower carbon footprint that these methods generate constitutes a gateway for the industry to move toward the energy transition process to renewable energy sources. In what remains of this section, we move on now to the third main method to harness marine energy, focused on managing thermal energy differentials.

12.4.3 Ocean Thermal Energy Conversion (OTEC)

A third main way to convert the energy of the oceans into electricity consists of utilizing the differential in temperature observed between the water in the surface and the water in the depths of the sea. The mechanism to generate power relies on an organic Rankine cycle (ORC), where warm water heats up a working liquid and converts it to steam, which later expands in the turbine section of the cycle structure, where the generator converts this mechanical energy into electrical energy (Wang et al., 2018). Simulations for the potential of OTEC methods show that they could provide up to 30 TW, with locations like the Florida Keys generating close to 100 MW (Langer et al., 2020; VanZwieten et al., 2017).

To provide more detail into the primary mechanism for electricity generation in OTEC frameworks, Figure 12.16 illustrates the Rankine cycle process. The main sections in this cycle process are an evaporator, a condenser, a turbine, and a working fluid, along with a connection pipe. In the first step, the working fluid sent through the pump is heated up in the evaporator and converted into vapor. Next, the vapor is sent through the turbine to operate the generator that creates electricity, and once the vapor is out, it is sent to the condenser, where the cold seawater helps it return to a liquid state. Finally, this liquid is sent again through the evaporator, where the process restarts (Ikegami et al., 2018).

Despite ample promise, OTEC methods suffer from low thermal efficiency and a high investment cost. Thus, temperature differentials are not as high to yield very efficient processes, which increases research and development costs. Because of this, different working fluids have been tried as catalysts after the phasing out of R22 due to its harmful impact on the ozone layer. The more well-known fluids studied in the literature are refrigerant-grade ammonia (R717), difluoroethane R152a, and hydrofluorocarbon R134a (Wang et al., 2018). Furthermore, Table 12.4 lists the potential working fluids for OTEC processes, along with their principal properties for evaporation temperature and critical pressure.

Additionally, Ikegami et al. (2018) proposed two distinct OTEC enhancement methods, first through a double-stage Rankine cycle, and then through a Kalina cycle flow. The former technique involves two independent cycles, one high-temperature and one low-temperature sequence, where warm seawater passes from the former to the latter through the evaporators, while cold seawater goes through the condensers in the opposite direction. Meanwhile, the Kalina cycle adds a regenerator and a separator section after the evaporator phase so that the working fluid is separated into saturated vapor and saturated liquid. Then, a novel absorber receives the vapor from the turbine along with the liquid that was not sent through the turbine, and the mix goes back to the condenser to restart the cycle.

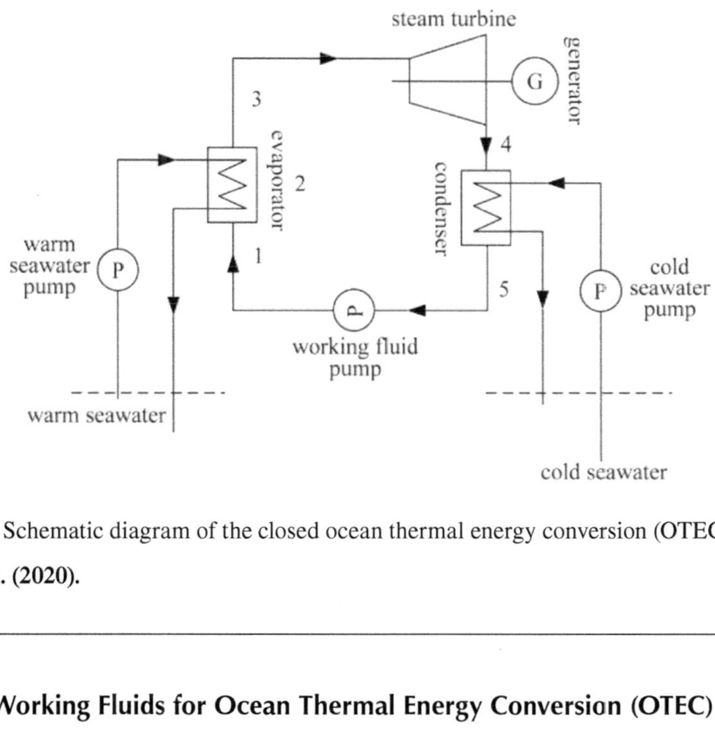

FIGURE 12.16 Schematic diagram of the closed ocean thermal energy conversion (OTEC) system.

Source: Wu et al. (2020).

TABLE 12.4
Properties of Working Fluids for Ocean Thermal Energy Conversion (OTEC) Processes

	Chemical Formula	Critical Temperature (K)	Critical Pressure (kPa)	GWP	Remark
R717	NH_3	405.91	11,380	0	Wet
R152a	CHF_2CH_3	386.41	4,517	140	Wet
R134a	CF_3CH_2F	374.21	4,059	1,200	Wet
R227ea	CF_3CHFCF_3	374.80	2,926	2,050	Dry
R600a	$CH_3CH_3CHCH_3$	407.85	3,629	20	Dry
R601	$CH_3CH_2CH_2CH_2CH_3$	469.36	3,374	11	Dry

Source: Wang et al. (2018).

Compared to the simple ORC, these two processes have shown better efficiency by diminishing the irreversible losses in the heat exchange process. Additionally, ammonia offers the most significant promise in heat transfer performance across all procedures, higher than when using either a water/ammonia mixture or the tetrafluoroethane HFC-134a (Ikegami et al., 2018). Similarly, to assess an improvement in thermal efficiency and power output, Wu et al. (2020) analyzed the improvement of the condenser section, applying constructal theory to optimize the heat transfer plate (HTP). Their findings show that length has a more significant effect than width and effective number, and this dimension ought to be chosen as the main parameter to improve the efficiency of the condenser.

The techniques we have reviewed thus far follow a closed-cycle method for thermal energy conversion. However, two other types of procedures have been advanced in the literature, as Table 12.5 summarizes. First, open-cycle techniques were introduced around 1930, close to 50 years after closed-cycle methods, and their main characteristic is that seawater is used as the fundamental working medium to generate electricity. Thus, warm seawater is pumped into the evaporator, which turns it into the steam that later drives the turbine. Compared to closed-cycle processes, open-cycle methods are more straightforward, are less costly, and can produce by-products like desalinated freshwater (Zhang et al., 2018; Wilberforce et al., 2019).

Meanwhile, hybrid devices combine features from both cycles. So in the first stage, a vacuum chamber evaporates warm seawater, which allows for a working fluid to reach its boiling point after being exposed to the water vapor. Then, electricity is generated by expanding the refrigerant in the turbine, while the vaporized fluid is condensed inside a heat exchanger, which yields desalinated water (Herrera et al., 2021). This system has been studied as an optimized framework to increase both power output and efficiency while also providing an opportunity to produce fresh water for populations near the seaside (EIA, 2022b).

A potential upgrade on OTEC methods that would serve a larger group of end users consists of ocean thermoelectric generators (OTEGs). These plants, in general terms, can transform the waste heat from flue gases, exhaust from vehicles, and thermal power plants into usable electricity. These

TABLE 12.5
Ocean Thermal Energy Conversion Devices

OTEC Energy Devices	Characteristics of the Device
Open-cycle	In a low-pressure environment, warm surface water is flash-evaporated, and the vapor generated helps run a generator. Vapor then condenses when in contact with seawater with lower temperature. Desalinated water is being generated through this process.
Closed-cycle	Ammonia, propane, or chlorofluorocarbons (CFCs) are used as the working fluid at a lower boiling point that is flash-evaporated using warm water by allowing it to flow through a heat exchanger. The resulting vapor is used to run the turbine, where it expands and partially condenses. The two-phase mixture is then passed to a heat exchanger, where it is condensed, pressurized, before being sent back to the evaporator for the entire cycle to commence all over again. These power generation cycles tend to function better than in open-cycles, but they are fairly small due to the fact that the secondary working fluid works at higher pressure.
Hybrid	This combines the characteristics of both the open- and closed-cycle. Warm water, which is used as a primary fluid, enters a low-pressure evaporator, where it is evaporated then passed to a heat exchanger, where its latent heat is used to evaporate the secondary working fluid that is then used to drive the turbine. The secondary working fluid is then condensed and compressed before being sent back to the heat exchanger to repeat the cycle. The condensed water from the heat exchanger is collected as desalinated water.

Source: Wilberforce et al. (2019).

installations commonly require a temperature variation greater than 100°C to produce electric power that would justify the optimization and installation costs. However, novel advances in nanotechnology have created new possibilities in the thermoelectric front, potentially allowing smaller temperature differentials to still produce efficient outcomes (Khan et al., 2017; Khan et al., 2022).

Currently, OTEC methods are the least developed of the three techniques featured in this chapter. However, the United States has been a proactive player in thermal energy conversion research, having operated a 250 kW demonstration plant in Hawaii between 1993 and 1998, supported by the Natural Energy Laboratory of Hawaii Authority. Additionally, the US Navy–supported Makai Ocean's Engineering OTEC plant is currently one of the leading installations for this technology in the world, having been connected to the electrical grid in 2015, providing electricity for 120 homes (EIA, 2022b; Hasan and Dincer, 2020; Khan et al., 2022).

Furthermore, Table 12.6 shows a list of different OTEC projects around the world in various stages of development and operation, as well as featuring different output levels. Updating some of these projects, Herrera et al. (2021) highlighted that the Japan OTEC installation has been enhanced with a water desalination functionality, while the closed-cycle Kona plant in Hawaii also provides around 7,000 gal of fresh water daily. Additionally, the New Energy for Martinique and Overseas (NEMO) project continues to be in the design phase but hopes to take advantage of the favorable 20°C temperature differential to produce 16 MW of electricity to be carried onshore via a submarine cable.

TABLE 12.6
Notable Worldwide Installed or Planned OTEC Power Plants

Location	Nominal or Nameplate Power Output	Description
Hawaii	50 kW	One of the oldest plant, 1979; closed cycle; Lockheed Missile and Space Co.
Japan/Nauru	120 kW	Installed in 1982 by Tokyo Electric Power Services Company.
Hawaii	1 MW	A land-based plant, open-cycle, operated between 1993 and 1998 by LLC and NELHA.
Japan/Okinawa	50 kW	A land-based plant used for power generation and research on other applications of OTEC; installed in 2013 by Xenesys, IHI, and Yokogawa.
Hawaii	10 MW	Working with the US Navy and the Department of Energy, Lockheed Martin has invested $15 million over the past three years toward the technology need for and the design of a 10 MW prototype plant to validate the technologies necessary for small- to large-scale (100 MW or greater) commercial-sized OTEC power plants.
India/Tuticorin, South India	1 MW	A floating closed-cycle plant was attempted by the National Institute of Ocean Technology, India. Difficulties in connecting the 1 km cold water pipe due to lack of marine infrastructure led to closure of the project.
Southern China	10 MW	The 10 MW prototype offshore plant will be the largest planned OTEC project until 2017. Like the Hawaii project, which was also to be a 10 MW facility, the China OTEC plant is designed to pave the way for higher-capacity plants ranging from 10 to 100 MW.
Martinique/Bellefontaine	10 MW	Floating platform; planned to operate from 2016; DCNS France.

Source: Hamedi and Sadeghzadeh (2017).

Another two technologies that use similar methods to produce energy from the oceans are salinity gradient techniques and ocean current methods. The former, also known as osmotic energy, focuses on the differential in salt concentration in the ocean, using membranes to produce usable energy. Due to the largest differential in salt concentration, freshwater–saltwater systems are the most efficient. Currently, two methods are the most used to generate osmotic energy, pressure retarded osmosis (PRO) and reversed electrodialysis (RED), where the former mixes the two types of water in a chamber to increase pressure and operate a turbine, while the latter relies on membranes to separate ions and increase the chemical potential difference, generating voltage and, thus, energy (IRENA, 2020).

Then, ocean current methods explore the interplay between wind, temperature, and salinity across sea bodies around the world. These methods have been the least explored on the list, yet how they operate can be comparable to tidal energy frameworks. The area with the largest potential for current analysis and eventual energy generation is the Gulf Stream, initiating in the Gulf of Mexico and moving east to cover Florida and Virginia, with an estimated potential of 20 GW. Other areas with potential are the East Australian Current, the Agulhas Currents in South Africa, and the Kuroshio Current, off East Asia. Despite currents being slower than tides, their magnitude is more prominent, and their unidirectional flow is considered an advantage for baseload power generation (IRENA, 2020).

This segment has shown the potential and degree of development of the three principal methods to harness the energy of the ocean. The energy transition process requires technologies like those explored here to be in their fullest capacity to begin a successful process of fossil fuel phaseout in the next couple of decades. As we can see, however, most projects continue in a research-and-demonstration stage, with a few installations providing energy and fresh water to seaside communities. In the following section, we review a set of best practices that can help researchers capitalize on the significant technological advancements of the past years, maximizing the contribution of ocean energy endeavors.

12.4.4 Best Practices

The ocean energy sector holds tremendous potential as a clean and renewable source with a predictable supply, offering a promising alternative to traditional fossil fuels. In this sense, harnessing the power of tides, waves, and underwater thermal differentials can make a significant difference in the efforts to reduce greenhouse gas emissions that are linked to global warming. Hence, ensuring a fruitful deployment of these technologies requires the creation and implementation of a set of widespread best practices. This section reviews important considerations regarding the best use of resources and maximizing benefits for marine sector projects.

A first essential consideration to improve ocean energy projects and establish standard practices consists of evaluating current projects and estimating ideal values for relevant parameters when considering future endeavors. In an example for tidal power plants, Ghaedi and Gorginpour (2021) advance a machine learning–assisted optimization strategy that seeks optimal numbers of turbines and gates, as well as efficient measurements for sluice width and the diameters for both the turbine tip and the hub. Also, hydro pumps and the operation modes for barrage installations are evaluated, with findings suggesting that ebb generation aided by pumps is the most optimal method.

Moreover, as the literature dealing with site selection and screening advanced, the success of an ocean energy project is partly explained by a thorough evaluation of various aspects surrounding the projects, such as their life cycle, the impact they could have on the environment, and the installation, operation, and maintenance costs over different periods of time. First, life cycle assessments (LCAs) are scarce and incomplete for ocean energy projects, with an evaluation of the Pelamis wave energy project being one of the most comprehensive due to their inclusion of carbon payback periods and climate change impact measured in CO_2 emissions (Thomson et al., 2019; Uihlein and Magagna, 2016).

Environmental impact assessments (EIAs) are similar evaluations performed to study the potential ecological, economic, and social impacts of a marine energy installation. The work of Mendoza et al. (2019) highlights the lack of assessments performed on the effects of ocean technologies on the physical and biotic marine systems, proposing a methodology that includes five dimensions to bridge this gap. The first evaluation area measures the changes in water turbulence and circulation (hydrodynamic), while the second topic (geomorphologic) considers the potential changes in local sediment properties. Meanwhile, reviewing changes in behavior, noise, or vibrations is the third area (biotic), changes in gas distributions are the fourth (chemical), and impact on fishing is the last one (socio-economic).

The potential impact on the environment around an ocean facility can also be observed before they begin formal operations. Along with its operational impacts, Adcock et al. (2015) explored the likely environmental impacts surrounding the construction phase of the La Rance tidal barrage installation. Because of their findings, some recommendations can also be made regarding the type of installation selected, given the location and the environmental context around them. Thus, regarding ecological impacts and keeping economic costs constant, a tidal lagoon may be preferable to a tidal barrage. In contrast, an array of offshore tidal turbines may be preferable to both options.

As we can see, an emphasis on site specificity allows for a tailored approach to assessments of the potential effects of an ocean energy plant. In this sense, a consideration that the literature has not assessed systematically is the potential effect of cost reductions through learning effects. Exploring more on the path dependence that learning follows, Langer et al. (2020) showed that technological understanding can have a significant impact in reducing operational costs for older plants, as well as installation costs for newer projects. This development could help expand these methods across more locations with advantaged currents and waves suitable for an ocean energy project.

Additionally, developing hybrid technologies that can harness two sources of energy at the same time is a recommended practice, particularly with higher-cost offshore installations. Yilmaz (2019) explored this enhancement opportunity, coupling an OTEC isobutane working fluid system with a wind energy turbine. Findings show that these setups have greater efficiency and lower costs than single-use plants while providing a promising alternative to fossil fuel structures. A similar effort is proposed by Assareh et al. (2021), who explored a triple combination of OTEC, wind turbine, and a solar panel. This configuration could generate almost 450 kW of power at its optimal point.

Likewise, Li et al. (2018) explored the increasing opportunities to couple marine energy installations with seawater desalination systems. Given that current desalinating plants are primarily relying on conventional sources of energy to function and their energy consumption is significant, this pairing can both decrease operational costs while also lowering carbon emissions from overall desalination efforts. Figure 12.17 shows a diagram of the structure of this hybrid system that can be implemented seamlessly across most OTEC projects. Then, Kim et al. (2021) expanded this framework to explore the opportunities for OTEC facilities to provide other types of clean technologies, such as green hydrogen, ocean minerals, and different kinds of water for various aquaculture uses, to name a few.

Similarly, other best practices can be implemented industry-wide regarding turbine design optimization, component improvement, and data collection and management. For example, Zhu et al. (2020) built on the blade element momentum (BEM) theory to propose an enhanced design for a horizontal-axis turbine blade aided by a neural network model. Findings show that optimizing the length of the blade chord, as well as the pitch angle, can increase the power coefficient of the turbine by 2%. Then, Clark and DuPont (2018) highlighted the role of the SuperGen database, which collects reliability and safety information used by international organizations to develop standards in the ocean energy sector.

The use of systematized information can lead to the optimization of operation parameters across the field, which could help lower costs and achieve economies of scale quicker, as well as seamlessly integrate the electricity generated in an ocean energy plant into the grid. Tackling some of these objectives, Hasan and Dincer (2020) looked at the main operation parameters for an OTEC

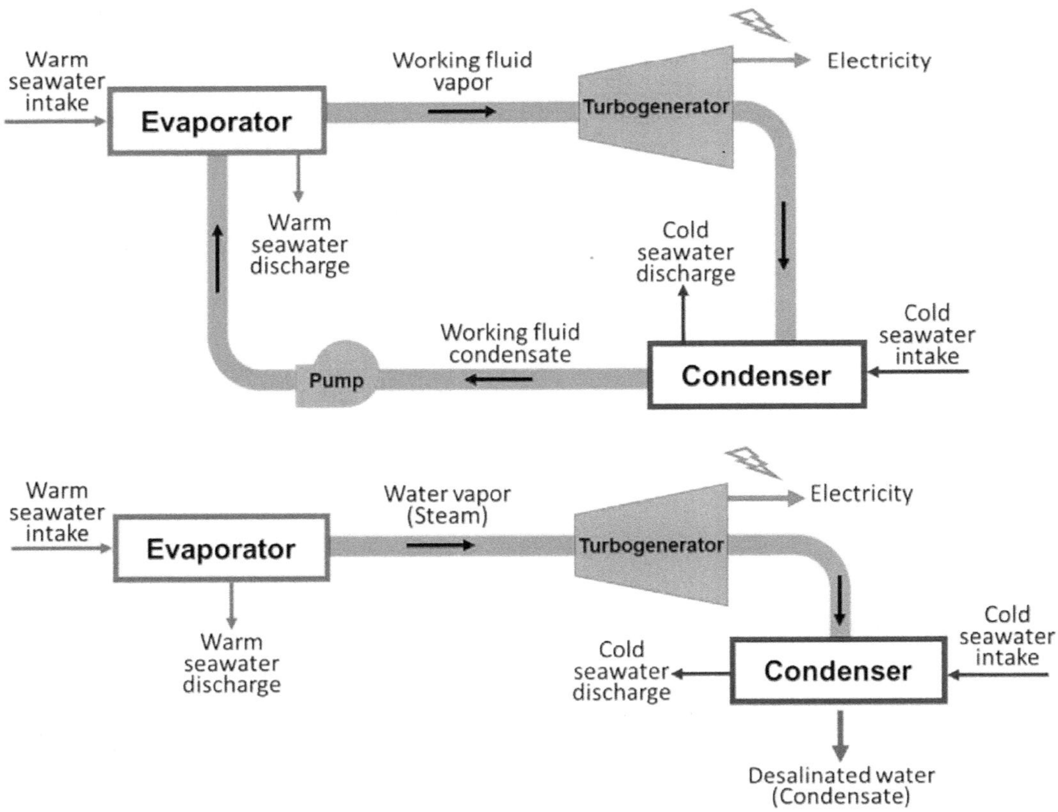

FIGURE 17 SHOWS A diagram of the structure of this hybrid system that can be implemented seamlessly across most OTEC projects.

Figure 12.17 Schematic diagram of OTEC and integrated seawater desalination processes. Upper: close-cycle. Lower: open-cycle using seawater.

Source: Li et al. (2018).

installation so that future simulations contain a standard set of factors to analyze and assess. Taking information from the Makai OTEC power plant in Hawaii, the authors enumerate variables, such as the temperature of both the warm and cold waters, their flow rates, the working fluid, and the maximum and minimum cycle pressure values, to arrive at a standard list of comparable factors across units of study.

Regarding challenges to efficient grid integration, Uihlein and Magagna (2016) explored infrastructure and operation solutions. First, subsea cable connections and an electrical system need to be devised for plants to be serviceable to the community, and some intermittence is to be expected from marine sources. They then implement a standard grid code that establishes minimum requirements for energy facilities to meet after a period of time so that the grid operates safely – also guaranteeing operation and maintenance standards despite the harsh environments for which offshore ocean plants can provide support for more projects in the future. And lastly, improving resource forecasting tools and building battery storage or pumped hydro installations are hailed as other potential solutions.

Similarly, Tran and Smith (2018) studied a series of improvements to levelized cost of energy (LCOE) calculations, which are done under uncertain input contexts and mostly assume a uniform distribution for relevant parameters. In the case of wave and tidal energy, much variability can be found in factors like fixed and variable O&M costs, lifetime of the facility, greenhouse emissions over the life cycle of the plant, capacity factor, and fuel costs. Meanwhile, placing carbon pricing schemes, which limit emissions from fossil fuel plants by charging a fee to companies or incentivizing them monetarily to lower emissions, increases LCOE calculations for coal plants, which increases the competitiveness of renewable projects like ocean energy endeavors (Ma et al., 2022).

Relatedly, deciding between building an onshore or offshore plant can also pave the way for standards that follow a series of optimized practices. As noted in Figure 12.18, an OTEC installation can provide small farms with fresh water to irrigate their crops, another benefit of desalination efforts. Moves like this require higher costs for building infrastructure to transport the water inland yet diminish building costs for an onshore facility, which are found to be much lower than in offshore installations. Additionally, assessing the potential for other electrolysis processes that generate products like oxygen, hydrogen, and ammonia and their use in various fields can be another area of interest.

Mentioned in passing earlier but not comprehensively explored yet is the potential role of machine learning (ML) technologies in the development of suitable standards for the ocean energy sector. For example, the work of Lettenmaier et al. (2017) explores the use of algorithms to observe and track the performance of wave energy converters. Advances in this area can help researchers test WECs without their own generator so that prototypes can be enhanced before they are fully deployed. Relatedly, Wu et al. (2020) explored the development of a genetic algorithm to optimize the size of a plate condenser, achieving Pareto-optimal outcomes for length, width, and effective number.

Then, Moreira et al. (2023) sought to validate an AI-driven model for the La Rance station so that a representation of a tidal barrage structure improved against the on-site data from a constructed installation exists for engineers to utilize and enhance. Using a deep reinforcement learning (DRL)

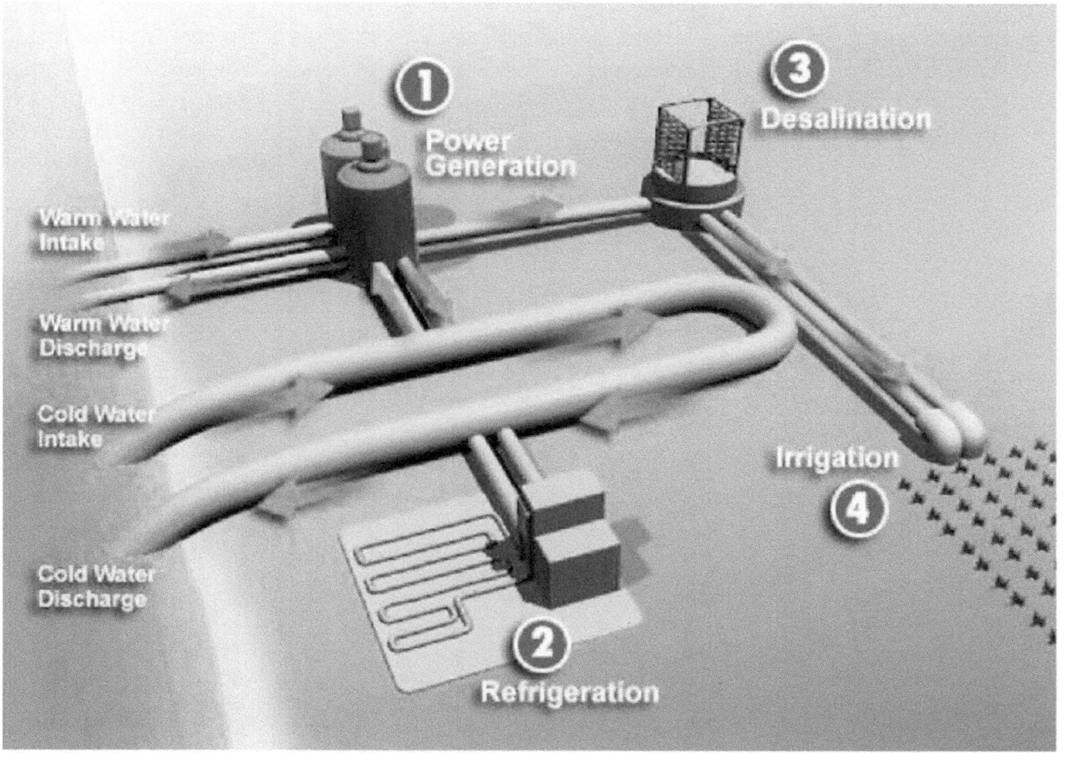

FIGURE 12.18 Diagram of a land-based OTEC plant.

Source: von Jouanne and Brekken (2017).

technique, the authors began with a 0D model, a minimally complex structure only determined by time, and found a series of variables to increase the complexity of the model and increase its practicality. Thus, the parameterization of an ideal structure with variables like sluice gates, turbines, ocean water level, pump power, and lagoon water level is performed for general use in novel projects.

Meanwhile, Zhou (2022) summarizes the uses of AI in ocean energy project analysis, mainly dividing the work encountered in three different areas, as shown in Figure 12.19. First, ML methods have been used for performance estimation, such as deep neural networks (DNN) and extreme learning machines (ELM), for short-term wave energy prediction, improving results from statistical or physics-based approaches. Then, AI models were used for sensitivity and uncertainty assessments, such as controlling a wave converter, evaluating thrust force, and determining changes in energy absorption. Lastly, optimization strategies include using analytical, semi-analytical, and computational intelligence methods to maximize the power supply of an energy converter or to arrive at more efficient structural designs.

In addition, the literature shows that support vector regression (SVR) methods have been used to forecast tide velocity and direction, while techniques like least squares support vector machines (LS-SVM) have been employed to improve the accuracy in predicting both tidal current speed and tidal current direction. Then, hybrid models combining empirical mode decomposition and extreme learning machines (EMD-ELM) were used to predict wave heights. Similarly, enhanced models with Bayesian frameworks have been relied upon to forecast wave height and wave energy flux, given the priors of a given target ocean. Lastly, long short-term memory (LSTM) encoder–decoder methods have been used to contrast past geothermal energy production with forecasts for future production (Lai et al., 2020).

Using novel computational techniques simplifies the work in the field, as researchers can optimize exploration efforts to sites that exclusively show viable potential to be developed. However, those techniques do not replace the work required to convince local communities to build facilities on their land. Therefore, collaboration with coastal communities is critical to advance ocean energy projects and, ultimately, the entire energy transition process. Touching on this, the work of Lange and Cummins (2021) finds that the cultural fabric of a locality is pivotal in the approval of a project, and if trust in institutions is low, a mediator can diffuse tensions and reconcile the interests of all parties involved.

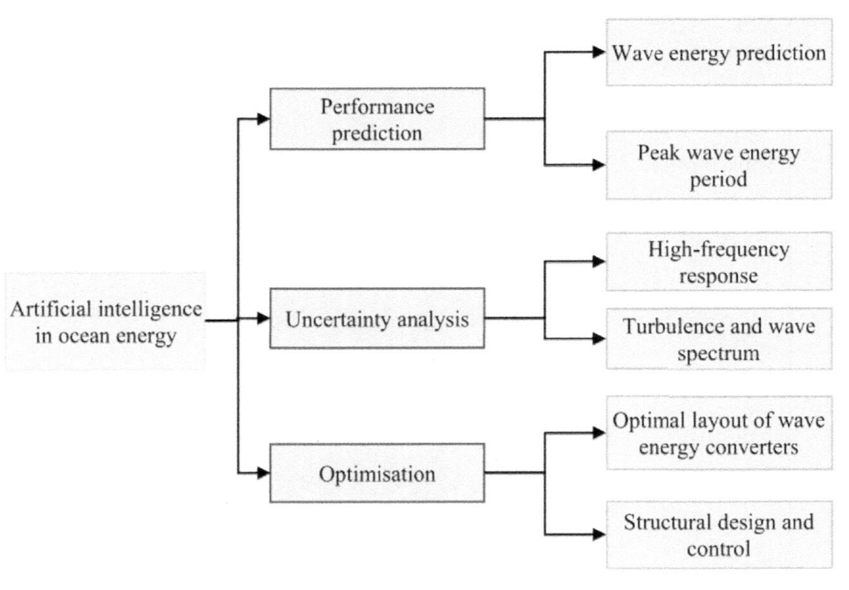

FIGURE 12.19 Summary of the use of artificial intelligence (AI) in developing the ocean energy sector.

Source: Zhou (2022).

All in all, the success or failure of these endeavors is also heavily dependent on the degree of collaboration between stakeholders, like companies, communities, governments, and international organizations. In this sense, enhancing communication channels between actors constitutes another practice to build standards and trust, ideally maximizing benefits for all players involved. An example of a medium that is looking to centralize communication and make it available more widely is the Ocean Thermal Energy Corporation, a team of scientists, entrepreneurs, and business-savvy actors who share the common objective of building and operating sustainable clean water systems (OTEC Corporation, 2023).

A consequence of the strengthening of relationships between actors is also reflected in an improved legal framework that is friendly toward ocean energy projects. Thus far, it appears that non-technological challenges continue to stifle many efforts to harness marine energy. For example, Apolonia et al. (2021) highlighted a series of barriers and enabling features for the development of ocean energy, where the lack of ocean energy legislation, unrealistic ocean energy targets, insufficient funding schemes, governance fragmentation, lack of political ambition, and clarity about procedures are part of the hindering aspects of the current rules framework.

Meanwhile, Ramos et al. (2021) explored the advances in the legal framework for marine renewable energy for the Atlantic region of Europe, emphasizing the delineation of clear marine resources to exploit and the establishment of clear preference criteria with other marine activities. With room for improvement, the authors proposed simplifying licensing procedures, particularly in Spain and France, developing a one-stop-shop process for more quickly advancing marine energy projects. Additionally, implementing sectoral plans that focus on areas like port, grid, technology, and supply chain development can streamline processes and facilitate the approval of ocean energy projects.

As we can see, best practices for ocean energy endeavors comprise a sequence of events that involve the scientific process, but also a series of essential actions performed at the local, regional, national, and even supranational levels. Investment is obtained through the balanced combination of the two sets of practices, ideally in a context of streamlined legislation and competitive funding. As Figure 12.20 shows, the path to an industrial rollout for an ocean energy project is related to successful modeling and experimentation, funding mechanisms, and an increasing readiness of technologies to be applied in different contexts. In addition, social acceptance and positive environmental assessments ensure that these projects are tailored to the communities they seek to serve.

In short, best practices for ocean energy projects involve a synergy between many actors, which adds many more degrees of complexity yet is necessary for the flourishing of a marine technology endeavor over the long run. In this same vein, another actor who can provide significant help in terms of funding, technical *know-how*, and operations and maintenance (O&M) expertise is the oil

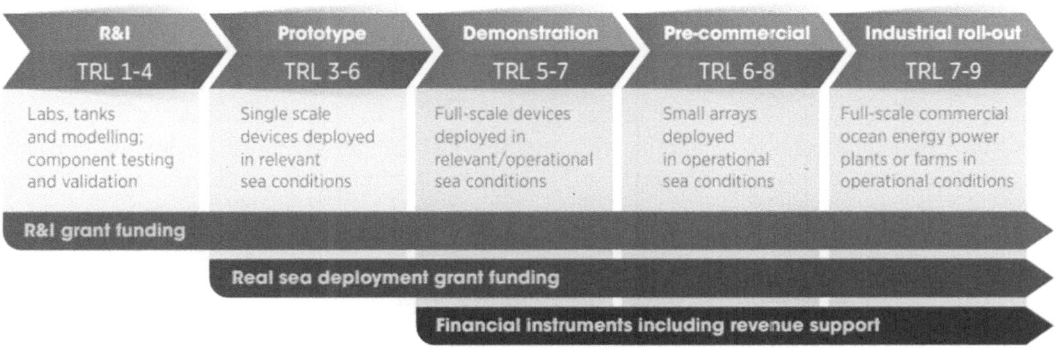

FIGURE 12.20 Ocean energy investment cycle. (*TRL*: technology-readiness level.)

Source: IRENA-OEE (2023).

and gas industry. In the following section, we review the assets this sector can provide to emerging marine energy projects, projects and technologies oil players are supporting, and the paths that open under a collaboration scenario between industries.

12.5 OCEAN ENERGY AND THE OIL AND GAS INDUSTRY

The significant advances in ocean technologies have invited actors to collaborate in funding strategies, regulation improvement, and cost-reduction schemes. In the middle of these developments, the question to be asked now concerns the role of the oil and gas companies in advancing the ocean energy sector. From what can be gathered, oil and gas companies have embraced the general banner of investment in renewables but have stayed mainly on solar, wind, and green hydrogen strategies. With that said, despite ocean energy being a secondary priority for big players in the sector, there are essential assets that these actors offer, which will be reviewed in this section.

First, their unique expertise in exploration and technology development is a valuable asset, and the projects they have initiated in renewable energies feature a set of procedures and standards that can be replicated in the ocean energy sector. For example, the planning and development of offshore wind facilities can be exported without too many modifications to various marine technology endeavors, even being coupled with hybrid platforms that harness two sources of energy in wind and waves. Thus, the human capital advantages from a consolidated industry can serve this emerging sector to achieve similar levels of economic development.

Then, the opportunity to collaborate with rising players in the renewable sector can benefit both industries to a great degree. On the one hand, diversification in cleaner projects allows for environmental responsibility and emissions curbing for the oil and gas sector. Also, the writing has been on the wall for a couple of decades about the pressure on the industry to devise strategies that help them adapt to the changing scenario and the push for an energy transition process (Yun et al., 2019). On the other hand, ambitious players in the renewable sector can receive invaluable funding and development opportunities to increase the reach of their technologies, making marine energy projects more accessible to implement and deploy worldwide.

Another potential area of joint work between the two sectors is the repurposing of existing infrastructure to projects that include cleaner processes. Satterlee et al. (2018) explored this opportunity in the Gulf of Mexico and the work that the Marine Monitoring, Energy and Environmental Research, Science, Education, and Training (MMEERSET) project is doing in the area. The objective of this plan, advanced by the United States, is to convert offshore plants used for fossil fuel operations into research, monitoring, and technology testing installations. In the initial stage, these facilities will assist with data collecting and modeling, observing ocean trends while exploring the possibilities of offshore energy production. With time, various ocean energy projects would be explored and implemented.

The exploration for the repurposing of oil and gas facilities for clean energy use in the future has a lower cost than a complete plant dismantling. Because of this, several avenues for an optimized decommissioning of an offshore platform at the end of its productive period have been explored. Figure 12.21 shows a decision tree regarding the shutdown of an offshore structure, with potential green development options as an alternative to toppling and complete removal. In the case of ocean energy projects, an offshore facility can be left in place but repurposed for initial research and eventual clean energy operations.

Moreover, Figure 12.22 shows the significant degree of commonality in offshore structures between oil and gas facilities and those required for optimal operations of renewable technologies like wind and ocean. The work of Robinson and Engelmann (2018) also provides a series of areas where the work required to advance energy projects is nearly identical, like in the case with leasing proposals, techno-economic assessments and cost estimations, construction and development surveys, calculation of net present value and internal rate of return, installation, and following up with infrastructure integrity once facilities have been built.

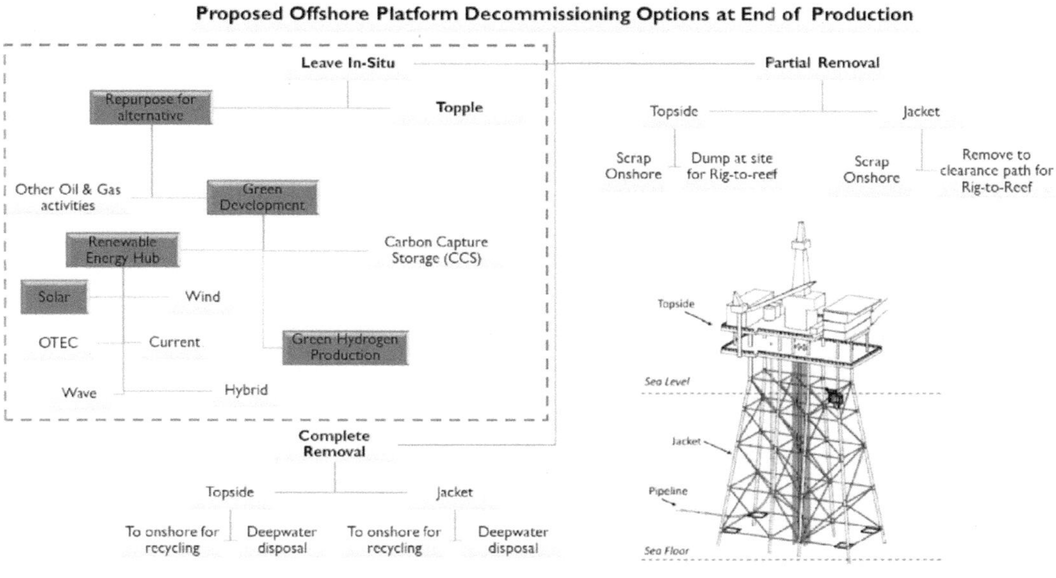

FIGURE 12.21 Offshore platform decommissioning options and the end of production.

Source: Alias and Go (2023).

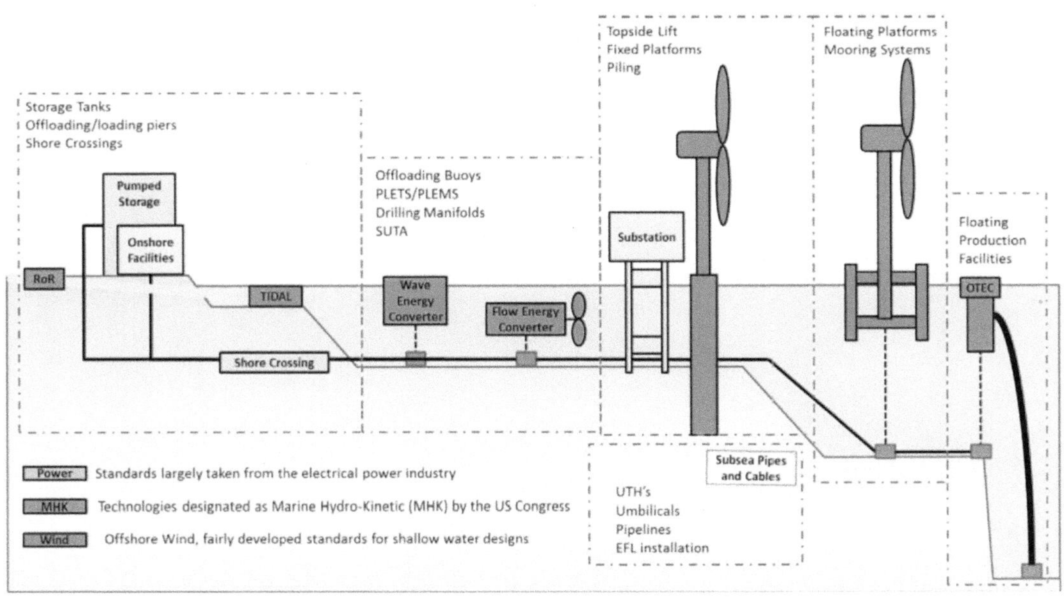

FIGURE 12.22 Offshore renewable energy (ORE) commonalities with oil and gas installations and activities.

Source: Robinson and Engelmann (2018).

Generally, the attitude of oil and gas companies regarding the energy transition has been proactive, with big players setting climate targets that center on net-zero pledges or significantly lowering their greenhouse gas (GHG) emissions by divesting their most polluting operations (Feller, 2019). Additionally, the sector has used technological advancements to their favor, promoting carbon capture and storage (CCS) strategies to avoid polluting the atmosphere with CO_2 from their processes.

In other sections of this book, we review CCS projects in detail, where the reader can evaluate their strengths and opportunities in the energy transition.

Despite not sending powerful signals about the development of ocean energy in particular, oil companies have increased their general messaging about climate change, the energy transition, and low-carbon energy initiatives. Li et al. (2022) analyzed 12 years of reports from different oil companies and found that they have taken a discourse tone more in tune with a move toward the energy transition. In the last couple of years of analysis, Shell has been the most proactive in messaging, followed by BP and ExxonMobil, though, as Figure 12.23 shows, the frequency of use of those terms remains very low, not surpassing 1% of all mentions.

At the same time, electric companies have lacked the incentives and a proactive nature to explore ocean energy endeavors. Narula (2018) points out this deficit, noting the complex dynamic for an early adopter who will experience higher technical and economic costs. However, opening the door to these technologies can be very advantageous once those hurdles are cleared. Those dynamics appear to affect oil majors as well, and if we couple those sentiments with a public perception that barely registers ocean energy projects as part of a potential suite of renewable sources to replace fossil fuels, the delay in moving toward advancing these projects can be easily elucidated.

In the meantime, it appears that ocean energy projects are to be explored, highlighted, and funded, first by governments, and then by independent and relatively new entrepreneurs. At the same time, oil companies will continue their clean energy projects involving primarily solar, wind, and hydrogen technologies. This may delay the impact that marine energy endeavors could bring to the fold amid the energy transition process, but still, the present remains a fertile ground to sow the

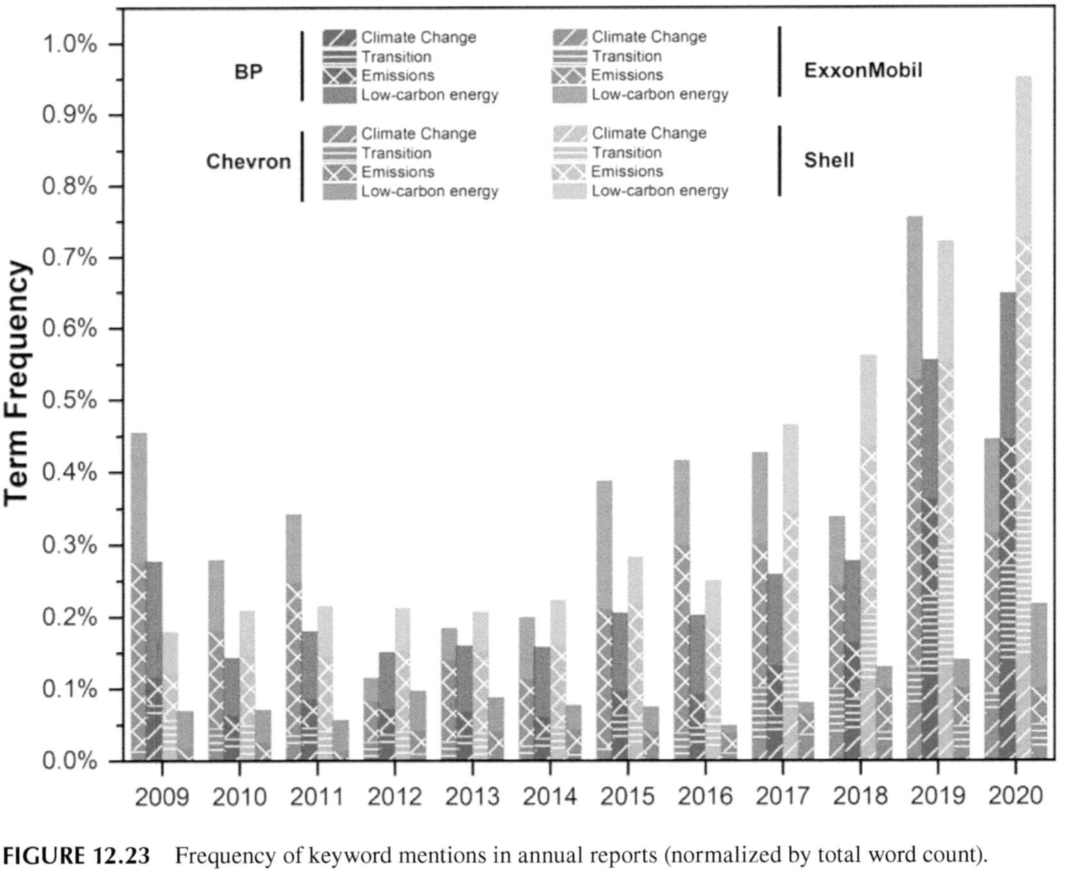

FIGURE 12.23 Frequency of keyword mentions in annual reports (normalized by total word count).

Source: Li et al. (2022).

seeds of future development. In our conclusions, we discuss the present and future opportunities of ocean energy, all in the middle of humanity's quest to gradually phase out fossil fuels and transition to renewable sources.

12.6 CONCLUSIONS: THE PRESENT AND FUTURE OF OCEAN ENERGY

This chapter explored ocean energy methods and processes, their stage of development, and the strengths and weaknesses of current technologies in various parts of the world. The present is full of promise and potential, yet various hurdles remain that are currently limiting the emerging sector from reaching widespread adoption. First, ocean energy methods are still in their infancy, with much more research to optimize site selection, anticipate energy production, build infrastructure to add the produced electricity to the national or regional grid, and guarantee reliable and efficient operation and maintenance standards for multiple years.

In part driven by the previous point, ocean energy costs remain prohibitively high for many players to invest. Perhaps more than all other technologies we have reviewed in this book, marine technologies remain a challenging investment opportunity due to the offshore nature of most of its operations, the limited locations in the world with advantageous tidal and current conditions, and the limited or non-existent auxiliary infrastructure currently in place. Relatedly, the concentration of most actors in renewable sources like solar and wind may also be hindering the growth of ocean energy exploration, a more reliable yet also more costly source of renewable energy.

Then, despite ocean energy projects' status as a clean and renewable source of energy, their environmental impacts are not fully understood as of yet. The current state of ecological impact assessments for marine methods continues to be basic, as the potential disruption to marine ecosystems through potentially concerning interventions that could alter coastal settings and generate noise pollution has not been adequately assessed. Therefore, without a proper set of monitoring and mitigation strategies, it remains too early to prescribe a widespread deployment of ocean energy projects.

With all that said, it is only a matter of time before we see an increase in ocean energy facilities, particularly in locations with advantageous environments and rich funding schemes approved and implemented. Slowly but surely, development costs will be lowered enough to be competitive with those of other clean energies, and the infrastructure will exist to deliver electricity coming from marine methods. Also, initial subsidies from stakeholders may also help lower levelized costs of energy so that building the next offshore plant is not as costly as the current ones. This way, a sustainable business can be achieved, creating a virtuous cycle where dependence on subsidies eventually ceases.

We saw that a few European countries with ample coastlines are actively exploring ocean energy, as is the case with France, Spain, the United Kingdom, and Ireland. Along with favorable winds and currents, these countries have a head start in terms of a friendly legal framework and a set of funding schemes to explore novel methods for ocean energy production. Over time, it is expected that other countries like China and India will replicate these policymaking strategies, while the United States will most likely expand its research and development framework to set ambitious targets for ocean energy generation and integration to the grid, as well as improved energy storage approaches.

As Table 12.7 shows, several tidal and wave energy endeavors are being developed in countries with natural advantages, and these actors have advanced joint efforts to centralize practices and homogenize processes. Thus, an intergovernmental technology collaboration program was formed as part of the Ocean Energy Systems (OES) framework, facilitated by the International Energy Agency (IEA). Currently, this program has 25 members, and some of its objectives are to provide an international base of information, share experience and knowledge, and group together a diversity of interests like universities, research organizations, government departments, energy agencies, and industry associations so that more projects like the ones featured are replicated around the world (OES, 2018).

TABLE 12.7
Notable Ocean Energy Projects in OES Member Countries

1. Korea	Sihwa Tidal Power Plant	Tidal
2. China	LHD Tidal Current Energy Demonstration project	Tidal current
3. Canada	Cape Sharp Tidal project	
4. France	Sabella D10 Tidal Turbine	
5. Italy	GEM "The Ocean's Kite"	
6. UK	Scotrenewables floating tidal system	
7. UK	Shetland Tidal Array	
8. Canada	Water Wall Turbine	
9. Norway	Deep River Power Plant	River current
10. Belgium	Laminaria Wave Energy Converter	Waves
11. Denmark	Resen Waves Smart Ocean Buoy	
12. Ireland	Ocean Energy 035 Buoy	
13. Portugal	WaveRoller	
14. Spain	Oceantec MARMOKA-5 project	
15. Sweden	Sea-based Sotenäs project	
16. USA	Columbia Power Technologies Wave Energy Generator	
17. India	Wave Power Navigational Buoy	
18. Netherlands	Blue Energy Reverse ElectroDialysis project	Salinity gradient
19. Japan	Okinawa OTEC plant	OTEC
20. Germany	StEnSea project	Other uses

Source: OES (2018).

Additionally, the notion of ocean energy coupled with energy from other renewable sources as a set of viable alternatives to fossil fuel power will be commonplace in the short term. The development of hybrid platforms of production, where wind and ocean, solar and ocean, or all three types of energy can be harnessed simultaneously, should be a natural path forward given the technological advancements achieved thus far. Moreover, a significant share of these installations will come from repurposed fossil fuel plants, which now serve as research, testing, and renewable energy production stations.

Another path that has the potential to open in the future is that of global regulations and policy support across the world. As such, increased collaboration is expected from actors like governments, industry stakeholders, non-governmental organizations (NGOs), and supranational organizations like the United Nations and the European Union, with the objective of writing legislation that incentivizes countries to pursue ocean energy projects. Thanks to these plans, other naturally advantaged countries in Oceania, South America, or South Africa can fund their own marine technology schemes and provide clean energy to their citizens.

To conclude, there is ample promise for ocean energy technologies. These methods can deliver reliable, clean energy to millions of people worldwide, neatly complementing sources like wind, solar, hydrogen, hydropower, and even nuclear. Currently, however, we lack the technological and infrastructure development necessary to seamlessly integrate these methods into the renewable energy generation projects already operating around the world. Only through more research, collaboration between actors, and both infrastructure and policymaking efforts can we pave the way for marine technologies to one day be a vibrant energy sector with much to offer to humanity.

REFERENCES

Abaei, M. M., E. Arzaghi, R. Abbassi, V. Garaniya, and I. Penesis. 2017. Developing a novel risk-based methodology for multi-criteria decision making in marine renewable energy applications. *Renewable Energy* 102: 341–348. doi:10.1016/j.renene.2016.10.054.

Adcock, T. A. A., S. Draper, and T. Nishino. 2015. Tidal power generation – A review of hydrodynamic modelling. *Proceedings of the Institution of Mechanical Engineers, Part A: Journal of Power and Energy* 229(7): 755–771. doi:10.1177/0957650915570349.

Aderinto, T., and H. Li. 2018. Ocean wave energy converters: Status and challenges. *Energies* 11: 1250. doi:10.3390/en11051250.

Al Shami, E., R. Zhang, and X. Wang. 2018. Point absorber wave energy harvesters: A review of recent developments. *Energies* 12(1): 47. doi:10.3390/en12010047.

Alias, N. D. B., and Y. I. Go. 2023. Decommissioning platforms to offshore solar system: Road to green hydrogen production from seawater. *Renewable Energy Focus* 46: 136–155. doi:10.1016/j.ref.2023.05.003.

Allan, G. J., P. Lecca, P. G. McGregor, and J. K. Swales. 2014. The economic impacts of marine energy developments: A case study from Scotland. *Marine Policy* 43: 122–131. doi:10.1016/j.marpol.2013.05.003.

Anderlini, E., S. Husain, G. G. Parker, M. Abusara, and G. Thomas. 2020. Towards real-time reinforcement learning control of a wave energy converter. *Journal of Marine Science and Engineering* 8(11), 845. doi:10.3390/jmse8110845.

Apolonia, M., R. Fofack-Garcia, D. R. Noble, J. Hodges, and F. X. Correia da Fonseca. 2021. Legal and political barriers and enablers to the deployment of marine renewable energy. *Energies* 14: 4896. doi:10.3390/en14164896.

Assareh, E., M. Assareh, S. M. Alirahmi, S. Jalilinasrabady, A. Dejdar, and M. Izadi. 2021. An extensive thermo-economic evaluation and optimization of an integrated system empowered by solar-wind-ocean energy converter for electricity generation – Case study: Bandar Abas, Iran. *Thermal Science and Engineering Progress* 25: 100965. doi:10.1016/j.tsep.2021.100965.

Astariz, S., and G. Iglesias. 2015. The economics of wave energy: A review. *Renewable and Sustainable Energy Reviews* 45: 397–408. doi:10.1016/j.rser.2015.01.061.

Astariz, S., A. Vazquez, and G. Iglesias. 2015. Evaluation and comparison of the levelized cost of tidal, wave, and offshore wind energy. *Journal of Renewable and Sustainable Energy* 7(5): 053112. doi:10.1063/1.4932154.

Bracco, G., M. Casassa, E. Giorcelli, G. Giorgi, M. Martini, G. Mattiazzo, . . . and G. Vissio. 2014. Application of sub-optimal control techniques to a gyroscopic wave energy converter. *Renew. Energ. Offshore* 1: 265–269. https://books.google.com/books?hl=en&lr=&id=nNCYCgAAQBAJ&oi=fnd&pg=PA265&dq=Bracco,+G.,+2014.+Application+of+sub-optimal+control+techniques+to+a+gyroscopic+wave+energy+converter.+Renew.+Energ.+Offshore++&ots=on_BSLTRQf&sig=pfVvuSj2benbfmsk8yW6B4vJbm8

Buhali, M. L., M. L. S. Abundo, and M. R. C. O. Ang. 2012. Site selection procedures for tidal in-stream energy in the Philippines: A Preliminary Study. *10th International Power & Energy Conference (IPEC)*. 110–114. doi:10.1109/asscc.2012.6523248.

Chen, H., T. Tang, N. Aït-Ahmed, M. E. H. Benbouzid, M. Machmoum, and M. E.-H. Zaim. 2018. Attraction, challenge and current status of marine current energy. *IEEE Access* 6: 12665–12685. doi:10.1109/access.2018.2795708.

Chowdhury, M. S., K. S. Rahman, V. Selvanathan, N. Nuthammachot, M. Suklueng, A. Mostafaeipour, . . ., K. Techato. 2021. Current trends and prospects of tidal energy technology. *Environment, Development and Sustainability* 23: 8179–8194. doi:10.1007/s10668-020-01013-4.

Clark, C. E., and B. DuPont. 2018. Reliability-based design optimization in offshore renewable energy systems. *Renewable and Sustainable Energy Reviews* 97: 390–400. doi:10.1016/j.rser.2018.08.030.

Cochet, C., and M. Lambert. 2017. The Rance tidal power plant model. *Proceedings of the 24th Telemac-Mascaret User Conference* 191–196.

Coles, D. S., L. S. Blunden, and A. S. Bahaj. 2017. Assessment of the energy extraction potential at tidal sites around the Channel Islands. *Energy* 124: 171–186. doi:10.1016/j.energy.2017.02.023.

Cradden, L., C. Kalogeri, I. Martinez Barrios, G. Galanis, D. Ingram, and G. Kallos. 2016. Multi-criteria site selection for offshore renewable energy platforms. *Renewable Energy* 87: 791–806. doi:10.1016/j.renene.2015.10.035.

Curto, D., V. Franzitta, and A. Guercio. 2021. Sea wave energy. A review of the current technologies and perspectives. *Energies* 14: 6604. doi:10.3390/en14206604.

Doyle, S., and G. A. Aggidis. 2019. Development of multi-oscillating water columns as wave energy converters. *Renewable and Sustainable Energy Reviews* 107: 75–86. doi:10.1016/j.rser.2019.02.021.

EERE. 2022. *U.S. department of energy announces $35 million in funding to advance tidal and river current energy systems.* Available in www.energy.gov/eere/articles/us-department-energy-announces-35-million-funding-advance-tidal-and-river-current [accessed June 8th, 2023].

EIA. 2022a. *Hydropower explained: Tidal power.* Available in www.eia.gov/energyexplained/hydropower/tidal-power.php [accessed May 28th, 2023].

EIA. 2022b. *Hydropower explained: Ocean thermal energy conversion.* Available in www.eia.gov/energyexplained/hydropower/ocean-thermal-energy-conversion.php [accessed June 8th, 2023].

European Commission. 2017. *The tide is turning for underwater turbines.* Available in https://ec.europa.eu/research-and-innovation/en/horizon-magazine/tide-turning-underwater-turbines [accessed May 28th, 2023].

European Commission. 2023. *Ocean energy.* Available in https://research-and-innovation.ec.europa.eu/research-area/energy/ocean-energy_en [accessed June 4th, 2023].

Falcão, A. F. O., J. C. C. Henriques, and L. M. C. Gato. 2017. Rotational speed control and electrical rated power of an oscillating-water-column wave energy converter. *Energy* 120: 253–261. doi:10.1016/j.energy.2016.11.078.

Feller, F. 2019. Oil companies and the transition to a low carbon future overview and opportunities. *Offshore Mediterranean Conference and Exhibition.*

García-Huante, A., Y. Rodríguez-Cueto, E. Garduño-Ruiz, and R. Hernández-Contreras. 2020. General criteria for optimal site selection for the installation of ocean thermal energy conversion (OTEC) plants in the Mexican Pacific". In Kim, A. S., and H.-J. Kim (eds.). *Ocean thermal energy conversion (OTEC): Past, present, and progress.* London, United Kingdom, Ocean Thermal Energy Conversion (OTEC), 129.

Garduño-Ruiz, E. P., R. Silva, Y. Rodríguez-Cueto, A. García-Huante, J. Olmedo-González, M. L. Martínez, . . ., S. Cerdeira-Estrada. 2021. Criteria for optimal site selection for ocean thermal energy conversion (OTEC) plants in Mexico. *Energies* 14(8): 2121. doi:10.3390/en14082121.

GES-Group. 2022. *The new tidal turbine technology from the VATTS project.* Available in https://ges-group.com/2022/03/the-new-tidal-turbine-technology-from-the-vatts-project/ [accessed June 7th, 2023].

Ghaedi, A., and H. Gorginpour. 2021. Generated power enhancement of the barrage type tidal power plants. *Ocean Engineering* 226: 108787.

Ghosh, S., T. Chakraborty, S. Saha, M. Majumder, and M. Pal. 2016. Development of the location suitability index for wave energy production by ANN and MCDM techniques. *Renewable and Sustainable Energy Reviews* 59: 1017–1028. doi:10.1016/j.rser.2015.12.275.

Gonzalez, A. T., P. Dunning, I. Howard, K. McKee, and M. Wiercigroch. 2021. Is wave energy untapped potential? *International Journal of Mechanical Sciences* 205: 106544.

Guo, B., T. Wang, S. Jin, S. Duan, K. Yang, and Y. Zhao. 2022. Review of point absorber wave energy converters. *Journal of Marine Science and Engineering* 10: 1534. doi:10.3390/jmse10101534.

Hamedi, A.-S., and S. Sadeghzadeh. 2017. Conceptual design of a 5 MW OTEC power plant in the Oman Sea. *Journal of Marine Engineering & Technology* 16(2): 94–102. doi:10.1080/20464177.2017.1320839.

Hasan, A., and I. Dincer. 2020. An ocean thermal energy conversion based system for district cooling, ammonia and power production. *International Journal of Hydrogen Energy* 45(32): 15878–15887. doi:10.1016/j.ijhydene.2020.03.173.

Herrera, J., S. Sierra, and A. Ibeas. 2021. Ocean thermal energy conversion and other uses of deep sea water: A review. *Journal of Marine Science and Engineering* 9: 356. doi:10.3390/jmse9040356.

Iglesias, G., and R. Carballo. 2011. Choosing the site for the first wave farm in a region: A case study in the Galician Southwest (Spain). *Energy* 36(9): 5525–5531. doi:10.1016/j.energy.2011.07.022.

Ikegami, Y., T. Yasunaga, and T. Morisaki. 2018. Ocean thermal energy conversion using double-stage Rankine cycle. *Journal of Marine Science and Engineering* 6: 21. doi:10.3390/jmse6010021.

International Energy Agency (IEF). 2021. *The role of critical minerals in clean energy transitions: Mineral requirements for clean energy transitions.* Available in https://www.iea.org/reports/the-role-of-critical-minerals-in-clean-energy-transitions/mineral-requirements-for-clean-energy-transitions

IRENA. 2020. *Innovation outlook: Ocean energy technologies.* International Renewable Energy Agency. Available in www.irena.org/publications [accessed June 12th, 2023].

IRENA-OEE 2023. Scaling up investments in ocean energy technologies. *International Renewable Energy Agency.* Available in www.irena.org/publications [accessed May 28th, 2023].

James, S. C., Y. Zhang, and F. O'Donncha. 2018. A machine learning framework to forecast wave conditions. *Coastal Engineering* 137: 1–10. doi:10.1016/j.coastaleng.2018.03.004.

Jeffrey, H., B. Jay, and M. Winskel. 2013. Accelerating the development of marine energy: Exploring the prospects, benefits and challenges. *Technological Forecasting & Social Change* 80: 1306–1316. doi:10.1016/j.techfore.2012.03.004.

Kaddoura, M., J. Tivander, and S. Molander. 2020. Life cycle assessment of electricity generation from an array of subsea tidal kite prototypes. *Energies* 13(2): 456. doi:10.3390/en13020456.

Khan, M. Z. A., H. A. Khan, and M. Aziz. 2022. Harvesting energy from ocean: Technologies and perspectives. *Energies* 15(9): 3456. doi:10.3390/en15093456.

Khan, N., A. Kalair, N. Abas, and A. Haider. 2017. Review of ocean tidal, wave and thermal energy technologies. *Renewable and Sustainable Energy Reviews* 72: 590–604. doi:10.1016/j.rser.2017.01.079.

Kim, H.-J., H.-S. Lee, S.-T. Lim, and M. Petterson. 2021. The suitability of the pacific islands for harnessing ocean thermal energy and the feasibility of OTEC plants for onshore or offshore processing. *Geosciences* 11: 407. doi:10.3390/geosciences11100407.

Lai, J.-P., Y.-M. Chang, C.-H. Chen, and P.-F. Pai. 2020. A survey of machine learning models in renewable energy predictions. *Applied Sciences* 10(17): 5975. doi:10.3390/app10175975.

Lange, M., and V. Cummins. 2021. Managing stakeholder perception and engagement for marine energy transitions in a decarbonising world. *Renewable and Sustainable Energy Reviews* 152: 111740. doi:10.1016/j.rser.2021.111740.

Langer, J., J. Quist, and K. Blok. 2020. Recent progress in the economics of ocean thermal energy conversion: Critical review and research agenda. *Renewable and Sustainable Energy Reviews* 130: 109960. doi:10.1016/j.rser.2020.109960.

Lehmann, M., F. Karimpour, C. A. Goudey, P. T. Jacobson, and M. R. Alam. 2017. Ocean wave energy in the United States: Current status and future perspectives. *Renewable and Sustainable Energy Reviews* 74: 1300–1313.

Lee, A. H. I., H. H. Chen, and H.-Y. Kang. 2017. A conceptual model for prioritizing dam sites for tidal energy sources. *Ocean Engineering* 137: 38–47. doi:10.1016/j.oceaneng.2017.03.039.

Lettenmaier, T., A. von Jouanne, and T. Brekken. 2017. A new maximum power point tracking algorithm for ocean wave energy converters. *International Journal of Marine Energy* 17: 40–55. doi:10.1016/j.ijome.2017.01.006.

Li, M., G. Trencher, and J. Asuka. 2022. The clean energy claims of BP, Chevron, ExxonMobil and Shell: A mismatch between discourse, actions and investments. *PLoS ONE* 17(2): e0263596. doi:10.1371/journal.pone.0263596.

Li, Z., A. Siddiqi, L. Diaz Anadon, and V. Narayanamurti. 2018. Towards sustainability in water-energy nexus: Ocean energy for seawater desalination. *Renewable and Sustainable Energy Reviews* 82: 3833–3847. doi:10.1016/j.rser.2017.10.087.

Ma, C., X. Wang, and B. Jiang. 2022. Ocean energy development under the background of carbon emissions peak and carbon neutrality in China. *IOP Conference Series: Earth and Environmental Science* 966(1): 012003. IOP Publishing. doi:10.1088/1755-1315/966/1/012003.

Ma, P., Z. Yang, Y. Wang, H. Liu, and Y. Xie. 2017. Energy extraction and hydrodynamic behavior analysis by an oscillating hydrofoil device. *Renewable Energy* 113: 648–659. doi:10.1016/j.renene.2017.06.036.

Magagna, D., and A. Uihlein. 2015. Ocean energy development in Europe: Current status and future perspectives. *International Journal of Marine Energy* 11: 84–104. doi:10.1016/j.ijome.2015.05.001.

Melikoglu, M. 2018. Current status and future of ocean energy sources: A global review. *Ocean Engineering* 148: 563–573. doi:10.1016/j.oceaneng.2017.11.045.

Mendi, V., S. Rao, and J. K. Seelam. 2016. Tidal energy: A review. *Proceedings of International Conference on Hydraulics, Water Resources and Coastal Engineering (Hydro2016)* 2320–2329.

Mendoza, E., D. Lithgow, P. Flores, A. Felix, T. Simas, and R. Silva. 2019. A framework to evaluate the environmental impact of OCEAN energy devices. *Renewable and Sustainable Energy Reviews* 112: 440–449. doi:10.1016/j.rser.2019.05.060.

MIT. 2022. *Why don't we use tidal power more?* Available in https://climate.mit.edu/ask-mit/why-dont-we-use-tidal-power-more [accessed May 28th, 2023].

Moore, T., and C. Boyle. 2014. The tidal energy potential of the Manukau Harbour, New Zealand. *Sustainable Energy Technologies and Assessments* 8: 66–73. doi:10.1016/j.seta.2014.07.001.

Moreira, T. M., J. G. de Faria Jr., P. O. S. Vaz-de-Melo, and G. Medeiros-Ribeiro. 2023. Development and validation of an AI-driven model for the la Rance tidal barrage: A generalisable case study. *Applied Energy* 332: 120506. doi:10.1016/j.apenergy.2022.120506.

Nachtane, M., M. Tarfaoui, I. Goda, and M. Rouway. 2020. A review on the technologies, design considerations and numerical models of tidal current turbines. *Renewable Energy* 157: 1274–1288. doi:10.1016/j.renene.2020.04.155.

Narula, K. 2018. "Renewable Energy from Oceans". In *The maritime dimension of sustainable energy security*. Springer Singapore. doi:10.1007/978-981-13-1589-3.

Negahdari, M. R., A. Ghaedi, M. Nafar, and M. Simab. 2023. Optimal operation of basin-based tidal units equipped with hydro-pumps. *The Journal of Engineering* 2: e12225. doi:10.1049/tje2.12225.

Noad, I. F., and R. Porter. 2017. Modelling an articulated raft wave energy converter. *Renewable Energy* 114: 1146–1159. doi:10.1016/j.renene.2017.07.077.

NREL. 2011. Structural design of a horizontal-axis tidal current turbine composite blade. *ASME 30th International Conference on Ocean, Offshore, and Arctic Engineering*.

OES. 2018. *Spotlight on ocean energy projects: 20 projects + 5 policy initiatives*. Ocean Energy Systems-International Energy Agency. Available in www.ocean-energy-systems.org/publications/search/ [accessed June 14th, 2023].

Offshore Energy. 2023. *Minesto's tidal energy kite reaches new heights in Faroe Islands*. Available in www.offshore-energy.biz/minestos-tidal-energy-kite-reaches-new-heights-in-faroe-islands/ [accessed June 8th, 2023].

OTEC Corporation. 2023. *Ocean thermal energy corporation*. Available in https://otecorporation.com/ [accessed June 13th, 2023].

Pecher, A., and J. P. Kofoed. 2017. *Handbook of ocean wave energy*. Springer Nature.

Petley, S., and G. Aggidis. 2016. Swansea Bay tidal lagoon annual energy estimation. *Ocean Engineering* 111: 348–357. doi:10.1016/j.oceaneng.2015.11.022.

PNNL. 2021. *What is tidal energy*" Available in www.pnnl.gov/explainer-articles/tidal-energy [accessed May 28th, 2023].

Ramos, V., G. Giannini, T. Calheiros-Cabral, P. Rosa-Santos, and F. Taveira-Pinto. 2021. Legal framework of marine renewable energy: A review for the Atlantic region of Europe. *Renewable and Sustainable Energy Reviews* 137: 110608. doi:10.1016/j.rser.2020.110608.

Roberts, A., B. Thomas, P. Sewell, Z. Khan, S. Balmain, and J. Gillman. 2016. Current tidal power technologies and their suitability for applications in coastal and marine areas. *Journal of Ocean Engineering and Marine Energy* 2(2): 227–245. doi:10.1007/s40722-016-0044-8.

Robinson, R., and G. Engelmann. 2018. Synergy in the standards and regulations of offshore renewable energy and offshore oil & gas. *Offshore Technology Conference*. OTC-28630-MS. doi:10.4043/28630-ms.

Satterlee, K., S. Watson, and E. Danenberger. 2018. New opportunities for offshore oil and gas platforms – efficient, effective, and adaptable facilities for offshore research, monitoring, and technology testing. *Oceans MTS/IEEE Charleston* 1–5. doi:10.1109/oceans.2018.8604935.

Satrio, D., I. K. A. P. Utama, and M. Mukhtasor. 2016. Vertical axis tidal current turbine: Advantages and challenges review. *Proceedings of Ocean, Mechanical and Aerospace—Science and Engineering* 3.

Shao, Z., M. E. Huq, B. Cai, O. Altan, and Y. Li. 2020. Integrated remote sensing and GIS approach using Fuzzy-AHP to delineate and identify groundwater potential zones in semi-arid Shanxi Province, China. *Environmental Modelling & Software* 134: 104868.

The Electric Storage Co. 2023. *VATTS project*. Available in https://theelectricstoragecompany.com/projects/vatts-project/ [accessed June 7th, 2023].

Thomson, R. C., J. P. Chick, and G. P. Harrison. 2019. An LCA of the Pelamis wave energy converter. *The International Journal of Life Cycle Assessment* 24: 51–63. doi:10.1007/s11367-018-1504-2.

Tran, T. T. D., and A. D. Smith. 2018. Incorporating performance-based global sensitivity and uncertainty analysis into LCOE calculations for emerging renewable energy technologies. *Applied Energy* 216: 157–171. doi:10.1016/j.apenergy.2018.02.024.

Uihlein, A., and D. Magagna. 2016. Wave and tidal current energy – A review of the current state of research beyond technology. *Renewable and Sustainable Energy Reviews* 58: 1070–1081. doi:10.1016/j.rser.2015.12.284.

UN Global Compact. 2020. *Ocean Stewardship 2030: Ten ambitions and recommendations for growing sustainable ocean business*. New York, UN Global Compact.

UN Global Compact. 2023. *Ocean renewable energy*. Available in https://unglobalcompact.org/take-action/ocean/communication/ocean-renewable-energy [accessed June 4th, 2023].

VanZwieten, J. H., L. T. Rauchenstein, and L. Lee. 2017. An assessment of Florida's ocean thermal energy conversion (OTEC) resource. *Renewable and Sustainable Energy Reviews* 75: 683–691. doi:10.1016/j.rser.2016.11.043.

Vasileiou, M., E. Loukogeorgaki, and D. G. Vagiona. 2017. GIS-based multi-criteria decision analysis for site selection of hybrid offshore wind and wave energy systems in Greece. *Renewable and Sustainable Energy Reviews* 73: 745–757. doi:10.1016/j.rser.2017.01.161.

von Jouanne, A., and T. K. A. Brekken. 2017. Ocean and geothermal energy systems. *Proceedings of the IEEE* 105(11): 2147–2165. doi:10.1109/JPROC.2017.2699558.

Wang, M., R. Jing, H. Zhang, C. Meng, N. Li, and Y. Zhao. 2018. An Innovative Organic Rankine Cycle (ORC) based Ocean Thermal Energy Conversion (OTEC) system with performance simulation and multi-objective optimization. *Applied Thermal Engineering* 145: 743–754. doi:10.1016/j.applthermaleng.2018.09.075.

Wang, Z. J., and Z. W. Wang. 2019. A review on tidal power utilization and operation optimization. *IOP Conference Series: Earth and Environmental Science* 240(5): 052015. doi:10.1088/1755-1315/240/5/052015.

Warak, P., and P. Goswami. 2020. Overview of generation of electricity using tidal energy. *IEEE First International Conference on Smart Technologies for Power, Energy and Control (STPEC)* 1–6. doi:10.1109/stpec49749.2020.9297690.

Warman, E., W. Armi, and F. Fahmi. 2020. A comprehensive study of sea wave tidal power plant (PLTPS). *Proceedings of the International Conference of Science, Technology, Engineering, Environmental and Ramification Researches (ICOSTEERR 2018)Research in Industry 4.0* 280–286. doi:10.5220/0010085302800286.

Waters, S., and G. Aggidis. 2016. Tidal range technologies and state of the art in review. *Renewable and Sustainable Energy Reviews* 59: 514–529. doi:10.1016/j.rser.2015.12.347.

Wilberforce, T., Z. El Hassan, A. Durrant, J. Thompson, B. Soudan, and A. G. Olabi. 2019. Overview of ocean power technology. *Energy* 175: 165–181. doi:10.1016/j.energy.2019.03.068.

Wimalaratna, Y. P., A. Hassan, H. N. Afrouzi, K. Mehranzamir, J. Ahmed, B. M. Siddique, and S. C. Liew. 2022. Comprehensive review on the feasibility of developing wave energy as a renewable energy resource in Australia. *Cleaner Energy Systems* 3: 100021. doi:10.1016/j.cles.2022.100021.

Wu, Y., C. Xu, Y. Ke, K. Chen, and H. Xu. 2017. Multi-criteria decision-making on assessment of proposed tidal barrage schemes in terms of environmental impacts. *Marine Pollution Bulletin* 125(1–2): 271–281. doi:10.1016/j.marpolbul.2017.08.030.

Wu, Z., H. Feng, L. Chen, and Y. Ge. 2020. Performance optimization of a condenser in ocean thermal energy conversion (OTEC) System based on constructal theory and a multi-objective genetic algorithm. *Entropy* 22(6): 641. doi:10.3390/e22060641.

Yilmaz, F. 2019. Energy, exergy and economic analyses of a novel hybrid ocean thermal energy conversion system for clean power production. *Energy Conversion and Management* 196: 557–566.

Yun, P., L. Jia, and Y. JieXin. 2019. International oil companies' low-carbon strategies: Confronting the challenges and opportunities of global energy transition. *IOP Conference Series: Earth and Environmental Science* 237: 042038. doi:10.1088/1755-1315/237/4/042038. DOI 10.1088/1755–1315/237/4/042038

Zhang, J., C. Xu, Z. Song, Y. Huang, and Y. Wu. 2019. Decision framework for ocean thermal energy plant site selection from a sustainability perspective: The case of China. *Journal of Cleaner Production* 225: 771–784. doi:10.1016/j.jclepro.2019.04.032.

Zhang, T., C. Williams, R. Ahmadian, and M. Qadrdan. 2022. Optimal operation of a tidal lagoon as a flexible source of electricity. *IEEE Power & Energy Society General Meeting (PESGM)* 1–5. doi:10.1109/PESGM48719.2022.9916952.

Zhang, W., Y. Li, X. Wu, and S. Guo. 2018. Review of the applied mechanical problems in ocean thermal energy conversion. *Renewable and Sustainable Energy Reviews* 93: 231–244. doi:10.1016/j.rser.2018.05.048.

Zhou, Y. 2022. Ocean energy applications for coastal communities with artificial intelligence–a state-of-the-art review. *Energy and AI* 10: 100189. doi:10.1016/j.egyai.2022.100189.

Zhou, Z., M. Benbouzid, J.-F. Charpentier, F. Scuiller, and T. Tang. 2017. Developments in large marine current turbine technologies – A review. *Renewable and Sustainable Energy Reviews* 71: 852–858. doi:10.1016/j.rser.2016.12.113.

Zhu, F.-W., L. Ding, B. Huang, M. Bao, and J.-T. Liu. 2020. Blade design and optimization of a horizontal axis tidal turbine. *Ocean Engineering* 195: 106652. doi:10.1016/j.oceaneng.2019.106652.

13 Nuclear Energy

Jorge Saldaña, Mesut Yurukcu, Narendra Boppana, Sepehr Arbabi, Jim Henry, and Sebahattin Ziyanak

13.1 INTRODUCTION

In the expanding list of energy resources developed to tackle the environmental shortcomings of fossil fuels, nuclear energy remains one of the top potential options to generate large amounts of low-carbon energy, without the perils of intermittence that solar and wind energy present regularly. In short, nuclear power consists of the use of sustained nuclear fission to generate heat and electricity, with a core set of technologies that have been used for more than 60 years (US Department of Energy, 2022e). Additionally, nuclear energy can be produced through fusion reactions after the nuclei of atoms fuse together (IAEA, 2021c), though this technology remains underdeveloped.

The beginning of nuclear energy methods can be placed in the early 1950s, when the first nuclear reactor (the Experimental Breeder Reactor I) became operational at the Argonne National Laboratory in Idaho, the United States. In the years that followed, countries like the United Kingdom, Russia, France, and Germany began using nuclear energy for commercial purposes, and over 20 other countries have followed suit since (Prăvălie and Bandoc, 2018). Figure 13.1 shows the evolution in the development of nuclear reactors in the last 60 years, and the 1970s was the pivotal juncture that allowed for further growth in the following decade. With that said, the number of reactors has remained close to 450, with a cumulative capacity of 450,000 MW, for most of the last 20 years.

According to Sadekin et al. (2019), two countries have led the way in the expansion of nuclear technologies in the last few years. First, Russia has become the leader in fast neutron reactor technology, which increased the energy obtained from natural uranium when compared to thermal reactors (IAEA, 2022). Meanwhile, France is another country that has embraced nuclear energy, with almost three-quarters of their energy production coming from nuclear sources. According to data from the International Energy Agency (IEA), seen in Figure 13.2, France is significantly ahead of all other countries in terms of its share of nuclear power in their total energy output, more than doubling that of advanced democracies like the United Kingdom, the United States, Canada, and Germany.

Some skepticism has been leveled at developing nuclear facilities due to the much-publicized reactor accidents in Three Mile Island in 1979, Chernobyl in 1986, and Fukushima in 2011, as well as the potential problems handling nuclear waste. The shortcomings revealed by these events have led to strong retrofitting efforts for existing plants and a complete evolution in technology that has led to decommissioning several projects, considered obsolete after safety concerns were raised from those events (IEA, 2019; Knapp and Pevec, 2018). These dynamics have created roadblocks for some investors but have also pushed for the improvement in nuclear technologies, which are now developing what are known as Generation III/III+ projects around the world (Wealer et al., 2021).

Data from the Power Reactor Information System (PRIS), collected by the International Atomic Energy Agency (IAEA, 2021a), shows that there are about 450 total nuclear reactors in operation as of today. Table 13.1 summarizes the top ten countries, including total electrical capacity in MW, the electricity supplied by nuclear sources, and the share of nuclear energy from their total capacity. The country with the most reactors producing the largest amount of electricity is the United States, almost doubling the capacity of the second country in the list, France. However, this country's share of nuclear power is by far the largest in the list, followed only by Ukraine, which features 55% of nuclear power as its share of total power.

 DOI: 10.1201/9781003282273-13

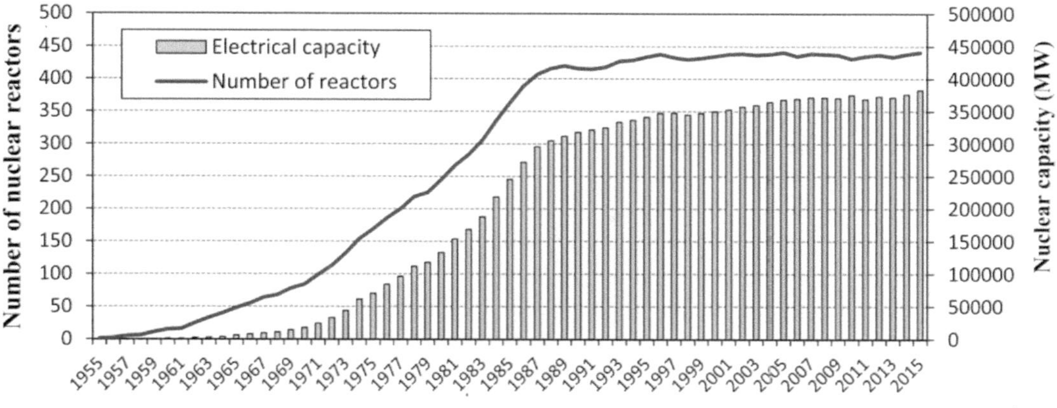

FIGURE 13.1 Worldwide variation of operational nuclear reactors and their net capacity (MW), 1955–2015.

Source: Prăvălie and Bandoc (2018).

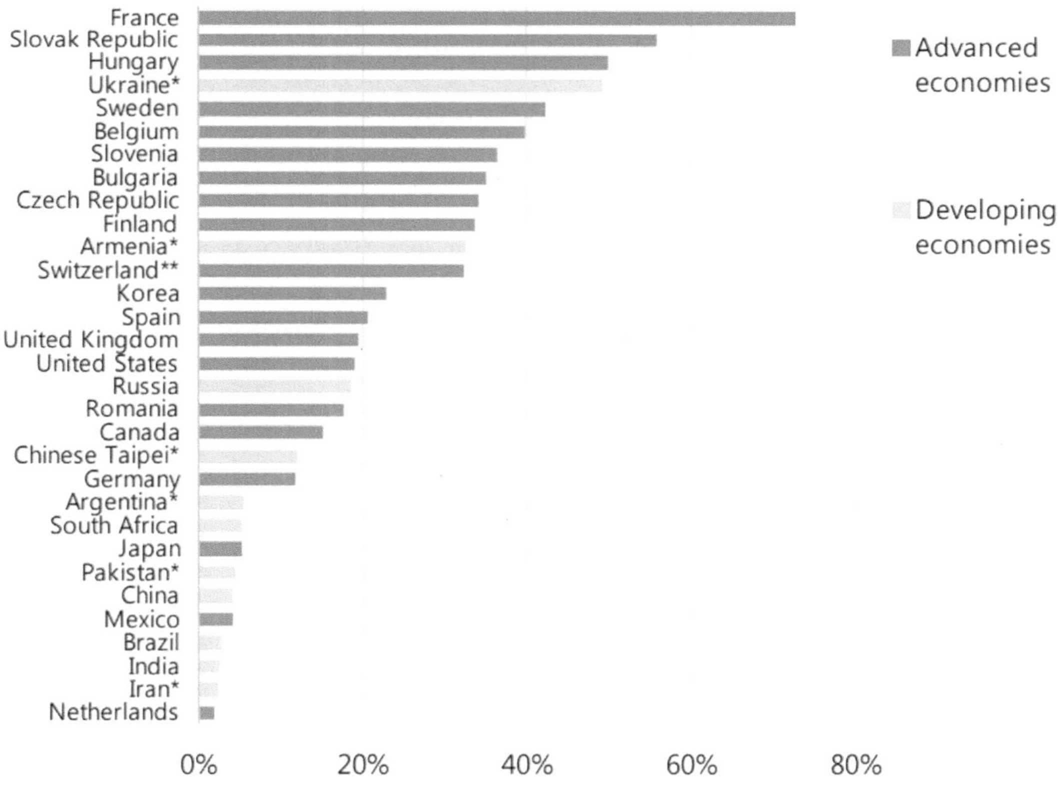

FIGURE 13.2 Share of nuclear power in total electricity generation by country, 2018 (*2016 data; **2017 data).

Source: IEA (2019).

TABLE 13.1
List of Countries with More Than 10 Nuclear Reactors in Operation

Country	Total Net Electrical Capacity (MW)	Number of Operated Reactors	Nuclear Electricity Supplied (GWh)	Nuclear Share (%)
United States	96,553	94	771,637.57	19.6
France	61,370	56	363,394.15	69
China	50,034	53	383,205.32	5
Russia	28,652	38	208,442.8	20
Japan	31,679	33	61,304.12	7.2
Korea	23,091	24	150,456.38	28
India	6,795	22	39,757.85	3.2
Canada	13,624	19	86,780.11	14.3
Ukraine	13,107	15	81,125.58	55
United Kingdom	8,923	15	41,788.5	14.8

Source: IAEA (2021a).

Meanwhile, China is the third ranking in total reactors and in total plant capacity, but the second in supplied electricity, which reflects the efforts that China is aiming for more nuclear energy. Furthermore, China is currently building close to 25 additional nuclear reactors, to provide the largest global energy user with an expanding list of non–fossil fuel energy sources (Wang et al., 2018). Lastly, and despite the worry that some actors have expressed after the nuclear reactor accidents mentioned earlier, Japan, Russia, Ukraine, and the United States continue to rely significantly on nuclear plants to supply energy to its citizens, and the advent of Gen III/III+ technologies only appears to boost the interest to establish nuclear as a regular part of these efforts. Emerging powers like China and India also appear to be paying close attention to these developments.

The last few decades have proved vital for the advancement of nuclear projects, especially in terms of reactor safety and of their potential to deliver large amounts of clean fission energy to millions of people worldwide. The moment may be just right to deploy more nuclear projects around the world, but important limitations and constraints remain, particularly for fusion technologies, which need to be accurately assessed before moving further forward.

Moreover, the role of the oil and gas industry in the development of nuclear energy for the energy transition is explored in more detail. Currently, important players like ENI and Chevron have taken steps to invest in nuclear fusion methods, with the objective to harness the energy of the sun and the stars in generating energy. Additionally, we explore the potential for oil and gas players to occupy the space created after the convergence of climate pledges enforced by international organizations, the advancement of newer generations of safe nuclear energy that could guarantee energy independence for several countries, and the collaboration of countries to share knowledge about nuclear energy and about future reactor projects.

13.2 SCREENING, SELECTION, DESIGN, AND ECONOMICS OF NUCLEAR ENERGY

The multiplicity of trade-offs that appear after an energy need is identified is an aspect that needs important consideration from engineers and researchers. Not every region of the world is equally prepared to host a power plant, and the success of the project depends on many variables, like funding schemes, degree of government or private sector support, positive natural or geographical conditions, public support, and medium- to long-term studies, to name a few. Therefore, the process of screening and selecting a site for the development of an energy facility is another time-consuming endeavor that precedes any conversation about the benefits that any potential plant could have on the lives of people or on the environment (see Figure 13.3).

For the case of nuclear power plants (NPPs), the process of screening and evaluation of potential sites involves extra precautions, like radiological impact assessments and seismic activity. Baskurt and Aydin (2018) advanced a workflow to detect the most suitable regions to host an NPP using geographical information systems (GIS), reaching two main sets of criteria to inform their decision to choose a site. First, exclusionary factors center on geographical and demographic limitations, like closeness to an earthquake fault, availability of cooling water, elevation and flood level, and population density. Then, discretionary criteria relate to infrastructure and geopolitical constraints, like closeness to a national border, proximity to hazardous facilities or transport infrastructure, and closeness to the electrical grid (see Figure 13.4 for an illustration).

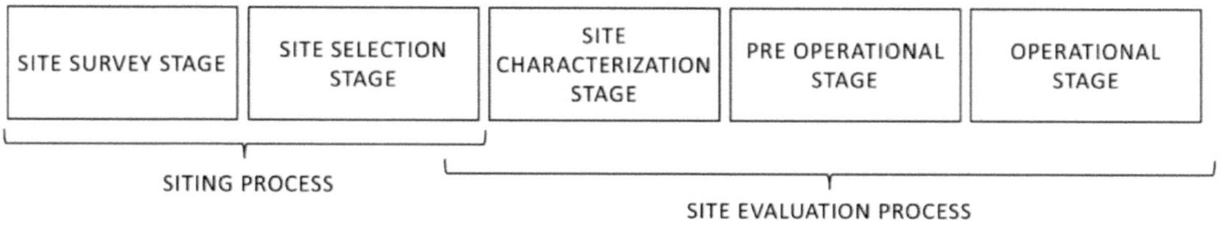

FIGURE 13.3 The stages of siting and site evaluation processes.

Source: Baskurt and Aydin (2018).

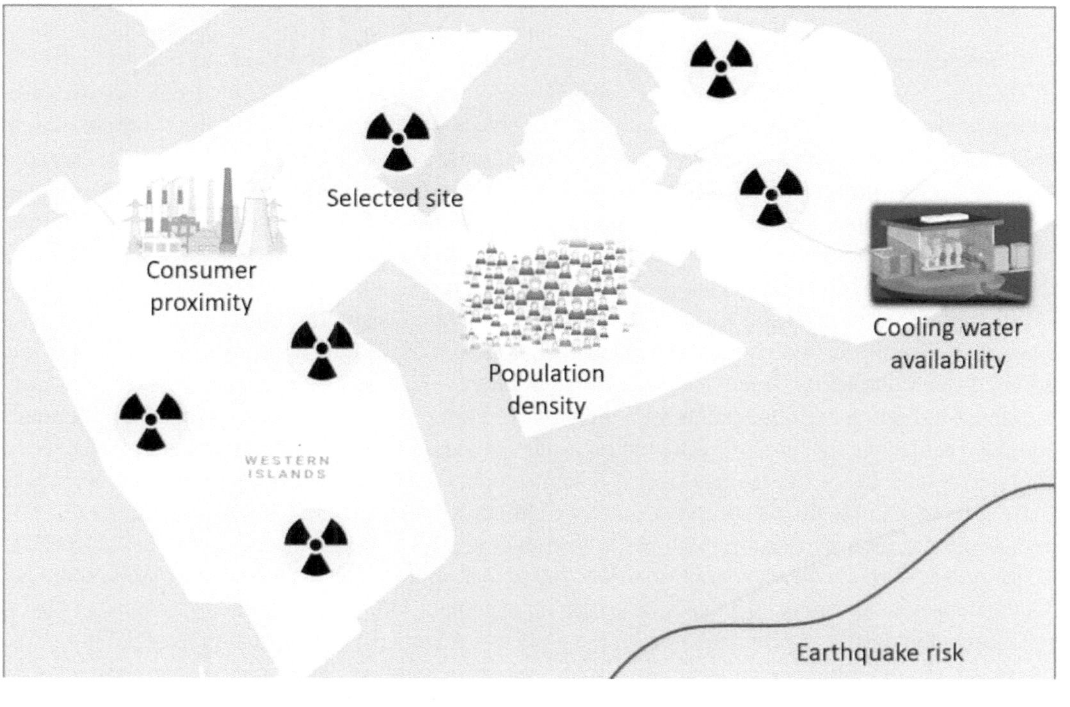

FIGURE 13.4 Some criteria for optimal site selection for small modular reactors (SMR).

Source: Devanand et al. (2019).

TABLE 13.2
Standards for Some Constraints Set by IAEA, DOE, and US NRC for Nuclear Power Plants (NPPs) under Boolean Logic

SNO	Constraint Variables	Criteria Limits
1	Distance from faults, m	8,000
2	Distance from earthquake epicenters, m	8,000
3	Percentage slope	<12
4	Flooding	Areas not included in a floodplain
5	Population density	194 persons per square kilometers
6	Distance from mines, m	8,000
7	Distance from gas pipelines	8,000
8	Distance from protected areas, m	8,000
9	Distance from major tourism spots, m	16,000
10	Distance from the national boundary, m	16,000
11	Coastal areas flooding	Areas not included in a floodplain
12	Distance from power plants, m	8,000
13	Distance from dams, m	8,000
14	Distance from conflict points, m	16,000
15	Distance from major airports, m	16,000
16	Distance from major hospitals, m	16,000
17	Distance from major ports, m	8,000

Source: Agyekum et al. (2021).

Other relevant factors when selecting a site to build a nuclear power plant are related to the use of the land, whether it is a military or culturally protected zone, or if the land belongs to a protected group, like an Indigenous population. Moreover, Table 13.2 shows a comprehensive list of constraints considered by organizations and departments as reasonable standards for the development of nuclear power plants (NPPs). As we can see, distance from different buildings with large gatherings of people, such as major airports and hospitals, and as well as distance from mines, dams, and gas pipelines are two more sets of parameters that need close attention when pondering the construction of an NPP.

To capture more potential variables and improve the analysis, more novel techniques can integrate both quantitative and qualitative information, applying different weights to each variable and adding more complexity to the analysis. These improved techniques consist of, for example, committees of experts providing feedback on the variables selected, along with their relative importance for given projects, or an analytical hierarchy (AHP) process that helps rank the variables used in the analysis. Examples of multi-stage decision frameworks have been used by Chinese researchers to find the most suitable inland locations, and by Ghanian scholars to help with their country's quest to explore nuclear energy starting in 2029 (Agyekum et al., 2021; Peng et al., 2020; Wu et al., 2020).

Moreover, selection involves not only sites but also designs for plants which are based on the latest technologies for safety, efficiency, and cost management. For example, Buongiorno et al. (2019) recommended design improvements with the objective to cut costs, such as a more serial manufacturing and less site-dependent structures, with standardized plant designs which are controlled by a single manager who shows a proven expertise in handling multiple subcontractors. Figure 13.5 shows the use of factory-made sections with standard features that could be applied to multiple projects. This would lower the cost of building nuclear plants, which opens the door to erecting more of these installations over time.

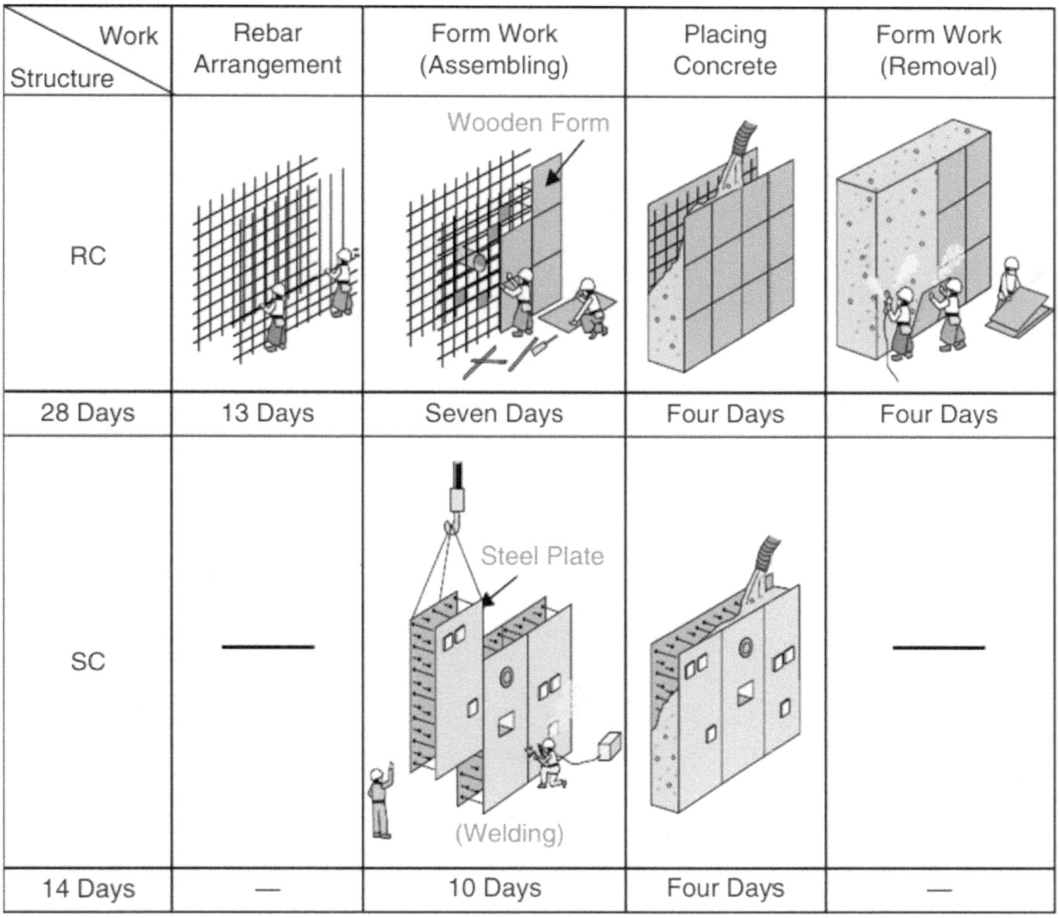

FIGURE 13.5 Construction and layout approach with potential to reduce costs and schedules of new nuclear builds. Factory-fabricated steel plate composites eliminate the need for rebar and formwork in concrete structures, thus reducing expensive labor at the site.

Source: Buongiorno et al. (2019).

With that said, efforts for standardization must include improvements achieved in new generations of research on nuclear energy. Therefore, the effort to optimize is met with the need to adapt all construction to the newest safety considerations, which may also differ across countries with different thresholds of operation for certain facilities. Because of these compromises, then, we could observe important differences in the design of the reactor core, its coolant system, its pressure vessel (RPV), the fuel rods, the fuel composition, or the power conversion system (PCS), among other sections of a nuclear energy facility (Wilding et al., 2019).

One way to deal with these complexities has been to explore the role of virtual reality (VR) technologies. Using VR methods, the processes of design and operation of a nuclear fusion reactor can be integrated into a computer simulation system, where the users can see the development and detect any potential problems in the implementation of a facility before starting their construction (Qin et al., 2020). The proliferation of *in silico* procedures has become increasingly common in areas with large degrees of complexity, because costs related to design and implementation can be lowered through a previous stage of modeling and simulation.

An example of this is seen in the work of Dong and Pan (2018), who proposed a dynamic model for a nuclear heating reactor cogeneration plant. Using a mathematical simulation software like

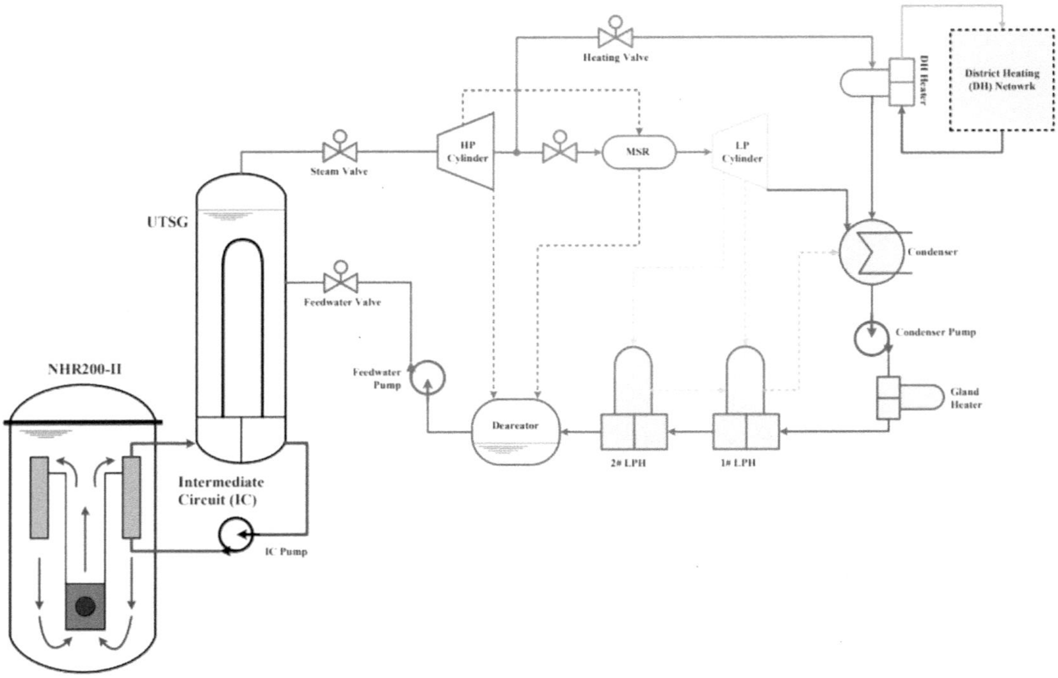

FIGURE 13.6 Simplified diagram of NHR200-II cogeneration plant.

Source: Dong and Pan (2018).

MATLAB®/Simulink®, the authors analyzed the dynamics of a *small modular reactor* (SMR), defined as a fission reactor with electrical power lower than 300 MWe. This NHR200-II facility is designed with advanced features, like a primary circuit integral arrangement, full power range natural circulation, self-pressurization, and a built-in hydraulic control rod system, where a pivotal design feature is that the reactor core is always covered by coolant. Figure 13.6 shows a schematic of the cogeneration plant, which can serve for electricity production as well as district heating.

Lastly for this section, the economics of nuclear energy projects are closely related to those observed for renewable energy projects. Thus, countries are presented with a series of incentives, trade-offs, and opportunities for the implementation of these sorts of projects, which often are constrained by the degree of support from governments and supranational organizations, the degree of innovation and investment from private actors, and natural or geographical limitations. Consequently, countries are placed in many different starting points for their energy endeavors, and their expectations also may vary due to the ambition in each country's environmental and carbon reduction targets.

For example, advanced European countries have already reached a level of economic development that allows them to undertake more ambitious pledges, because these commitments will not mean shortages of power in the short term, as they could mean for some developing countries. In this sense, the surge of China as an energy power in the 21st century may have to do with reaching an acceptable point in the theorized environmental Kuznets curve (EKC), which advances that countries have a curve for the use of more polluting energy strategies, with these being required more under inferior economic conditions (Dong et al., 2018). But as the country improves its economic situation, more environmentally friendly projects can be embraced, while the use of fossil fuels can be questioned with more vigor across more levels of society.

For countries that have reached those levels of economic prosperity, the expectation seems to be a progressive shift toward clean energies, something that is reinforced by international organizations and current market trends. In the middle of these dynamics, nuclear energy can be explored as an alternative to fossil fuels, or as a relief source for renewable energies, when these go through intermittence issues. To understand the role of nuclear technologies in the current scenario, some techno-economic analyses have been performed to assess the potential uses of nuclear technologies to aid in the energy transition process, and to explore the current possibilities under new generation facilities.

One of these analyses is seen in the work of El-Emam and Khamis (2019), who reviewed the results of a collaborative research project on the potential for nuclear hydrogen production in 11 members of the International Atomic Energy Agency (IAEA). The project generated four case studies with different systems that allow for hydrogen production using nuclear energy, each having varying construction periods, decommissioning strategies as well as costs, and expected years of operation. These studies open the door for the expansion in the use of nuclear power to explore an alternative to steam reforming, coal gasification, or water electrolysis, the most common methods to produce hydrogen.

With that said, these costs are highly variable across countries and are also dependent on a prior degree of technological advancements, a strong human capital development, and a relative independence that does not condemn a country to uranium scarcity after some time. Figure 13.7 shows a breakdown of some of the typical generation costs for a nuclear energy project once those prior costs have been incurred. As we can see, almost 80% of the cost is related to construction, while 10% comes from the process to generate fuel. Meanwhile, operation and maintenance costs are expected to take 8%, while decommissioning efforts take 3%.

To tackle some of these burdens, particularly for developing nations, economic assessments have explored the possibility of combining investment on nuclear and renewable sources of energy in the middle to long term. The work of Saidi and Omri (2020) analyzes the impact of nuclear and renewable energies in CO_2 emissions for 15 OECD countries over 20 years, finding that investment in either source of energy helps reduce carbon emissions, which makes them complementary instead of competing sources. This invites countries to explore the many different possibilities of combining investment in these two sources, following constraints like their own government budgets, the amount of private money attracted, the degree of public support, or the feasibility from current infrastructure, to name a few.

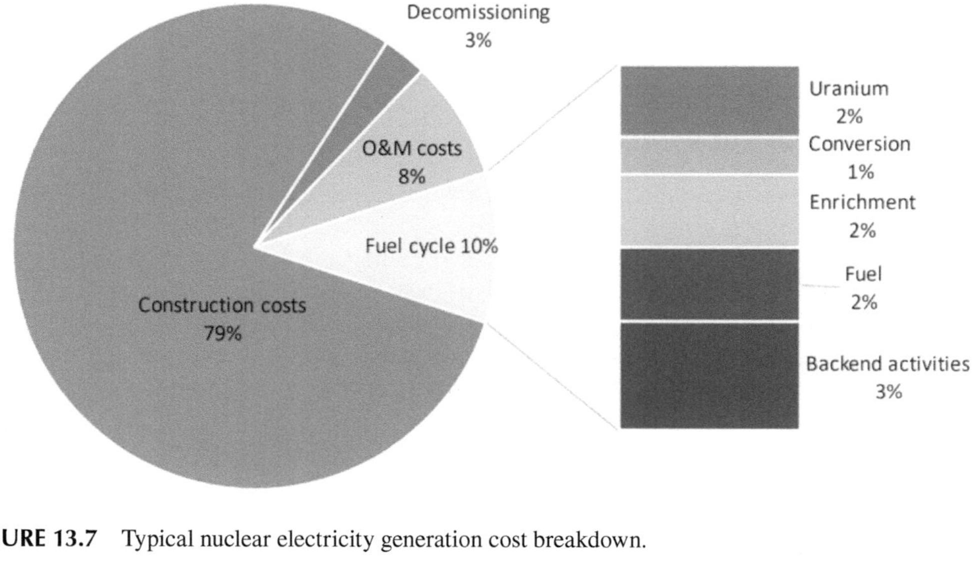

FIGURE 13.7 Typical nuclear electricity generation cost breakdown.

Source: Haas et al. (2019).

As advanced previously, the development of nuclear technologies is also constrained to a significant degree by the human capital inside the country that is seeking to deploy these technologies. Given that nuclear energy presents significant challenges and an important safety profile that needs to be well understood before projects are even started, a limited human capital can thwart any progress that a country is looking to make in the development of cleaner energies. Alam et al. (2019) explored this constraint which affects emerging countries more than advanced economies and stated the problem that developing nations face when trying to abide by international pledges and expectations while yet not having the complete capabilities for developing projects that advanced democracies can tackle much more easily.

Current-generation nuclear installations are set to deliver energy for more than five decades, and the steps of developing, building, and implementing a nuclear plant also take a considerable amount of time and resources, with the IAEA setting an estimated time frame for inexperienced countries of at least 15 years to complete a project of those characteristics. Likewise, these processes cannot occur without a robust human resources development scheme, one that some countries significantly lack. Table 13.3 shows a list of jobs and the time required to train a person to become specialized in nuclear-related occupations, where we see that the simplest job requires three years of training, while the hardest could take up to ten years.

These constraints undoubtedly place important burdens on several countries that may be considering a switch to nuclear energy projects. Therefore, some nations that still desire to switch to cleaner sources of energy will choose less-costly projects – like a limited number of renewable energy facilities – while others will drop their plans entirely when confronted with the limitations, thus sticking to the more polluting sources of energy that have helped them thus far. This suggests that the economic aspects of nuclear endeavors may have a *high-risk/high-reward* feature to them, where some players do not meet the criteria to start a serious nuclear project unless several high-cost measures have been completed before. But countries that are able to start these projects can achieve energy independence through a cleaner source of energy than fossil fuels.

All in all, the challenge to deploy more nuclear energy projects is multifaceted and calls for a greater degree of participation from researchers and engineers, but also from private actors. In this sense, the oil and gas industry can become a very important partner to governments that need both capital and human resources to make their endeavors a reality. Before exploring this relationship in more detail, however, the coming section takes a closer look at nuclear energy, the different types of technologies that exist today, as well as the shortcomings and opportunities that these advancements have when thinking about improving the environment and lowering CO_2 emissions.

13.3 NUCLEAR ENERGY TYPES, PRINCIPLES, BEST PRACTICES, AND ECONOMICS

In a world in need of alternatives to fossil fuels for environmental reasons and sustainability, and in the middle of an increasing demand for electrical energy for demographic growth reasons, nuclear power has become a reasonable alternative for a growing number of countries. Nuclear energy does not generate carbon emissions to the level that fossil fuels do, it does not produce greenhouse gas emissions, and its energy density allows for a greater output for similar levels of production (see Figure 13.8). Thus, nuclear power has positioned itself as a very promising alternative to coal and natural gas, or as a complement in places where it is still too costly to discontinue these sources of energy (Alwaeli and Mannheim, 2022).

On the other hand, the well-established fears about nuclear power going wrong are one important aspect that keeps many countries away from developing such projects. High construction and human capital costs to build and operate a nuclear facility are another reason that limits the interest. Due to these contrasting notions, it becomes important to understand the current strengths of nuclear technologies to be the potential energy supply option for both developing and advanced

TABLE 13.3
Types of Jobs and Training Lead Time Required for Work Functions at a Nuclear Power Plant

Nuclear Work Task	Job Responsibility	Time Required to Train
Project management	• Responsible for bid process and award of contracts • Direct, control, and monitor contractor and in-house design packages • Establish and monitor project milestones	3 years
Maintenance construction	• Perform maintenance and construction work within the power block • Responsible for job package development and work documentation	3-year apprentice training program Planners: 5 years
Operations	• On-shift staff, supervisors, and shift managers responsible for operating primary, secondary, and liquid radwaste systems • Non-shift personnel support operations staff as procedure writers, scheduling coordinators, technical specialists, and training coordinators	Shift: 2–5 years Non-shift: 6–10 years
Reactor engineering	• Analyze fuel performance, monitor core performance • Provide technical direction to operations during refueling, start-up, and shutdown	2 years
Nuclear fuels	• Perform core analysis, reload safety evaluation, reload design analyses, and thermal hydraulic/transient analyses • Responsible for nuclear fuel management, licensing, and acquisition	5 years
Radiological engineering	• Plan and control the radiation as low as reasonably achievable (ALARA) program • Calculate radiation dose and shielding, and review radiation work permits	5 years
Outage management	• Plan and coordinate all outage activities • Central contact point for refueling, maintenance, and forced outage management	5 years
Nuclear safety review	• Responsible for off-site and on-site safety review activities, such as operating abnormalities, license and technical specification changes, and human performance program	5 years
Quality assurance	• Implement approved QA program through periodic audits and surveillances • Establish QA policy, develop QA procedures, and review organizational self-assessments	6 years
Training	• Provide formal training for nuclear staff and instructors • Coordinate training schedules and operate plant simulators	5 years on-the-job experience

Source: Alam et al. (2019).

nations. This section seeks to provide a road map of the basic aspects of nuclear energy, as well as the economics and best practices around the implementation of nuclear power endeavors.

Before we move further, it is important to review the basic aspects of a nuclear power facility, from the operation of its reactor, to the waste generation process. In a few words, nuclear power is generated through physical processes where atoms interact to generate energy. These two processes are called fission and fusion. The *fission* process is forcing larger atoms to split into two smaller atoms, a process that releases large amounts of energy. Meanwhile, *fusion* occurs after two atoms combine to form a larger atom, like when two hydrogen atoms form a helium atom after a fusion

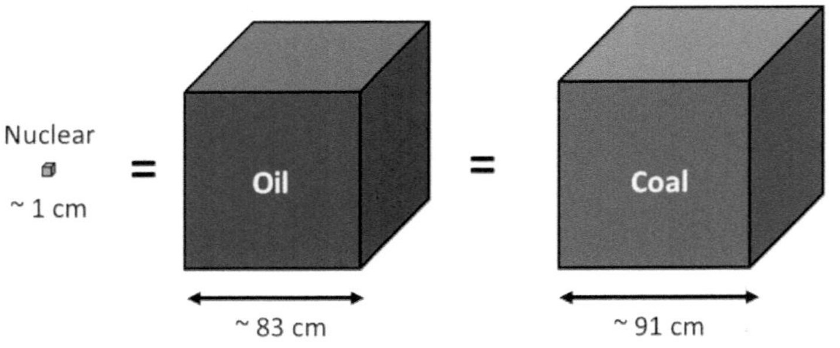

FIGURE 13.8 Nuclear fuel pellet energy content compared to an equivalent amount of energy from oil and coal (figure not to scale).

Source: INL (2022).

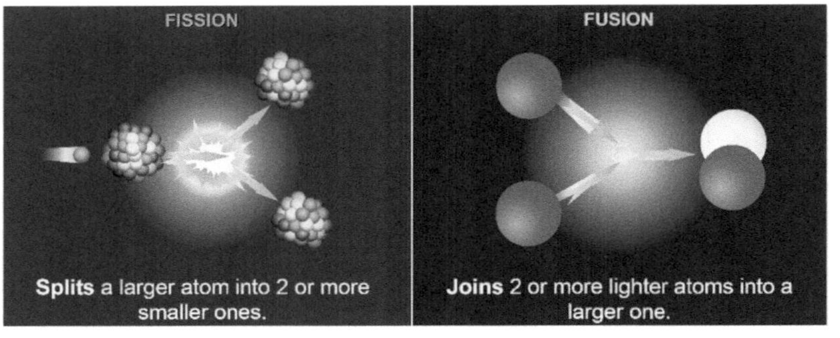

FIGURE 13.9 Fission and fusion energy generation processes.

Source: US Department of Energy (2021).

process that creates an even larger amount of energy than fission (US Department of Energy, 2021). See Figure 13.9 for an illustration of both processes.

These facilities control and contain the nuclear reactions obtained through fission, which produce the heat that allows a turbine to spin, thus creating electricity (US Department of Energy, 2021). To achieve these interactions between atoms, small pellets of enriched uranium are put in fuel rods that are placed in the reactor core. Because it is one of the components adapted to help generate the heat to power the turbines, this enriched uranium is considered *nuclear fuel* (US Energy Information Administration, 2022b; World Nuclear Association, 2022a).

Fuel rods can be used without being replaced for many years, but when they reach the end of their useful life and they can no longer be efficient in producing electricity, this "spent" fuel becomes highly radioactive and needs to be handled with extreme care. Waste of this type, considered high-level waste, consists mostly of isotopes with very long half-lives – thus, they take very long to decay to half their original radioactive levels – that can emit fatal radiation doses in a short time period of exposure. These "transuranic" (TRU) elements are stored in big "spent fuel pools" (see Figure 13.10) which are made of concrete with steel liners, where water up to 40 ft high serves to shield the radiation as well as to cool the fuel rods (US NRC, 2019).

Meanwhile, low-level waste consists of elements that have been contaminated with radioactive material, such as clothing, filters, various equipment and tools, swabs, or syringes. Despite comprising almost 90% of the total amount of radioactive waste, low-level waste accounts for only 1% of the total radioactivity present in the residue (Alwaeli and Mannheim, 2022). Typically, this waste is stored on-site until decay has been significant, or until there is a quantity large enough to ship to a low-level

FIGURE 13.10 Spent nuclear fuel at reactor sites in pool storage (left) and dry storage (right).

Source: Bonano et al. (2018).

waste disposal site. Currently, the United States has eight facilities licensed to dispose of these types of remnants, while burial of TRU is limited across all states in the country (US NRC, 2019).

Furthermore, there is a greater degree of specialization needs from researchers and engineers, and the offshoots of developing these projects contain elements that have been underexplored in the literature. This means that, despite the ample potential of nuclear, there is still not a complete understanding and integration of processes like handling and disposing of nuclear waste, and important limitations plague fusion methods.

Because of this, nuclear technologies may be the set of energy advancements that require the greatest amount of caution to move forward, while perhaps requiring the largest amount of time to develop and implement to a satisfactory degree. Therefore, the potential for ample, reliable energy for an increasing number of people across the world is certainly attainable, but it may require a greater degree of coordination between academics and researchers, governments, and private actors. This means that a successful transition to nuclear will take significant complementary processes of clean energy generation and a steady – albeit lengthy – phaseout of more polluting sources that does not require falling back on older technologies.

The energy transition may be a more challenging process to put in practice than estimated in earlier stages, but the potential of clean technologies like nuclear to provide energy to people all over the world is already being fulfilled in some locations, thanks to new generations of research as well as the success of current facilities to provide energy safely. Thus, the current challenges for the sector are related to expanding access to technologies to more nations, lowering initial costs of nuclear projects, allocating funds for human capital improvement, and disposing of waste safely and definitively once nuclear plants are up and running. The synergy that private actors like the oil and gas industry achieves with governments and international organizations will be key for tackling these challenging times.

13.3.1 Fusion

Like the name implies, *fusion* occurs when two nuclei – or the cores of two atoms – merge to form a new nucleus, releasing the amount of energy equivalent to the loss of mass that results from the difference between the mass of the two separate nuclei and the new nucleus, which will always be lower. Then, the leftover mass becomes energy that can be used for human consumption. Figure 13.11 illustrates a fusion reaction between deuterium (D) and tritium (T) – hydrogen's two naturally occurring isotopes – that produces much more energy than most other fusion reactions while forming a neutron and a helium (He) nucleus (US Department of Energy, 2022a).

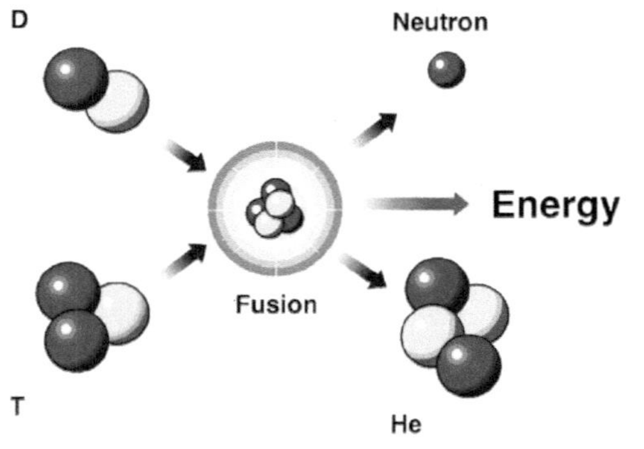

FIGURE 13.11 Depiction of the deuterium (D) and tritium (T) fusion reaction, which produces a helium nucleus (or alpha particle) and a high-energy neutron.

Source: US Department of Energy (2021).

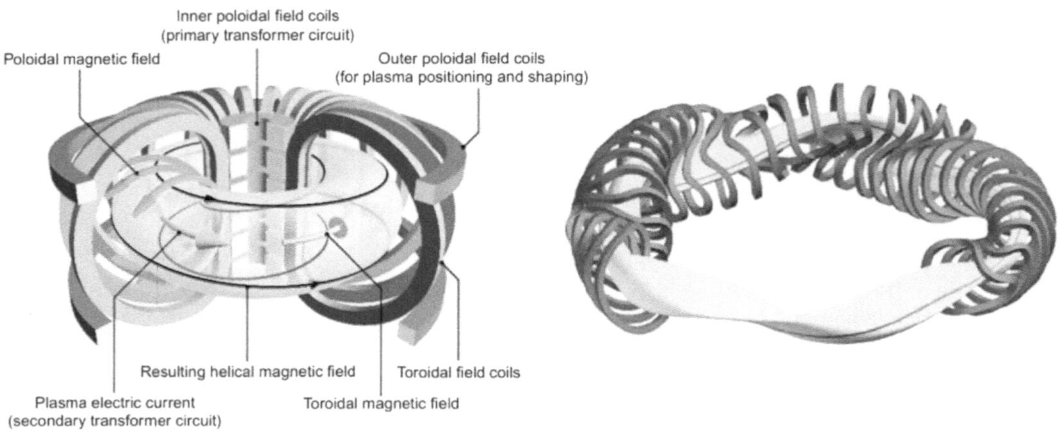

FIGURE 13.12 The main magnets of a tokamak system (left) and a stellarator (right).

Source: Kembleton (2019).

Fusion holds tremendous promise as a method to produce copious amounts of energy. However, some important physical limitations remain which keep fusion more as a theoretical endeavor than a practical effort, at least to a widespread degree. With that said, scientists have explored different technologies that could lead to consistent fusion power. Two of the most well-known mechanisms to achieve this are *tokamaks* and *stellarators*. Both machines work through containing plasma at the right extreme conditions so that thermonuclear fusion reactions can occur. One of the best ways to contain it is through creating a magnetic field with fusion reactors that have toroidal fields, or a donut-shaped structure with a plasma center that has temperatures hotter than the center of the sun (Kembleton, 2019; US Department of Energy, 2022b).

Inside these magnetic field coils, the conditions are ideal for the fusion reactions to occur. Figure 13.12 shows the basic structure of tokamaks and stellarator systems, where we see both generating a field that goes around the torus, or a loop that is shaped like a donut. Stellarators have a few advantages over their counterpart, like the ability to require less injected power to sustain the plasma, their greater design flexibility, and the simplification of some plasma control facets. On the other hand, these structures have a greater degree of complexity than tokamaks, requiring millimetric precision in the manufacturing of wire coils, as well as significant advancements in high-performance computing and plasma theory (US Department of Energy, 2022c).

Another form to achieve fusion reactions consists of a method known as *inertial confinement*, where high-powered lasers are targeted onto a frozen D-T fuel pellet – in what is called "direct drive" – or onto a casing around the pellet, which yields an array of X-rays to heat the pellet, in what is referred to as "indirect drive." Through intense heating, the outer case of the fuel pellet is expanded, while the interior compresses quickly, which results in an implosion that generates a burst of fusion that allows for nuclear reactions (Kembleton, 2019). Ideally, the implosion will cause an ignition which generates more energy than that was applied for the process to occur. However, ignition of this sort has still not been achieved (Osthus et al., 2019).

Because nuclear fusion reactions need to occur in the plasma state of matter, generating widespread conditions for a constant fusion energy output, while also being able to deliver this electrical energy to millions of people, continues to be very challenging. Important limitations are the stability of the plasma to facilitate fusion nuclear reactions, complications to confine the plasma long enough for reactions to occur, the difficulty to maintain equipment and materials exposed to extremely harsh conditions, and the uncertainties in prediction between models and experiment results, which have led to incomplete pictures and unrealistic expectations based on faulty models (IAEA, 2021b; Kembleton, 2019; Osthus et al., 2019).

To confront some of these practical constraints, a set of countries envisioned a joint nuclear fusion endeavor more than three decades ago, which has taken the form of the largest fusion experiment project in history. The ITER initiative – which stands for "the way" in Latin – is a joint effort of 35 nations to develop nuclear fusion energy, consisting of a large tokamak facility situated in Southern France that has the goal of proving the feasibility of fusion as a carbon-free source of energy with the potential to be deployed to a large scale. The current participants in the project are the 27 European Union member nations, plus Switzerland, the United Kingdom, China, India, Japan, Korea, the Russian Federation, and the United States (ITER, 2022c).

Table 13.4 shows a summary of the ITER proposed timeline, without incorporating delays related to the COVID-19 pandemic. With that said, their current level of development toward the completion to first plasma shows to be 77%, as of May 31, 2022 (ITER, 2022a). Moreover, the project has shared virtual tours of its facilities, with Figure 13.13 capturing a still from the project's tokamak pit, which shows important progress. According to the project, their five main objectives are to produce 500 MW of fusion power, show the integrated operation of technologies in a fusion plant, achieve a deuterium–tritium plasma where the reaction is sustained through internal heating, test tritium breeding, all while demonstrating the safety of the whole process (ITER, 2022c).

With all the promise shown by this initiative, fusion is still a very complex technology that is seeking to imitate processes performed by the sun and the stars, under more constrained conditions that are suitable to our planet. Because of this, some voices have raised an alarm regarding potential problems with ITER's design, which may not tolerate too many disturbances, like a plasma disruption or an edge-localized mode (ELM). Using the HEIGHTS simulation package, Hassanein and Sizyuk (2021) computed the behavior of ITER components during exceptional events, warning that divertor plates – made to remove waste material from the plasma while the reactor is still operating – may not withstand too many of these disrupting occurrences, unless important design changes are made.

As Figure 13.14 shows, the current structure of the divertor may not be sufficiently prepared for a plasma leak, especially at extremely high temperatures. Generally, limitations of this type are considered under the moniker RAMI, which stands for *reliability*, *availability*, *maintainability*, and *inspectability*. The work of Maisonnier (2018) identifies the first element as the ability of systems to perform under stated conditions for a specific period of time, and the second as the proportion of time that a system remains in a functioning condition. Meanwhile, *maintainability* and *inspectability* refer to repair rates and preventive maintenance, with corrections made to the functioning of a system if they become necessary.

TABLE 13.4
ITER Timeline

2005	**Decision to site the project in France**
2006	Signature of the ITER agreement
2007	Formal creation of the ITER organization
2007–2009	Land clearing and leveling
2010–2014	Ground support structure and seismic **foundations**① for the tokamak
2012	Nuclear licensing milestone: ITER becomes a basic nuclear installation under French law
2014–2021	Construction of the tokamak building (access for assembly activities in 2019)
2010–2021	Construction of the ITER plant and auxiliary buildings for first plasma
2008–2021	Manufacturing of principal first plasma components
2015–2023	Largest components are transported along the ITER itinerary
2020–2025	Main assembly phase I (*dates according to the 2016 baseline*)
2022	Torus completion*
2024	Cryostat closure*
2024–2025	Integrated commissioning phase (commissioning by system started several years earlier) *
Dec. 2025	**First plasma***
2025–2035	Progressive ramp-up of the machine
2035	**Deuterium–tritium operation begins**

Source: From the ITER 2016 baseline. ITER (2022c).

FIGURE 13.13 Image of ITER's tokamak pit, April 2022 VR tour.

Source: ITER (2022b).

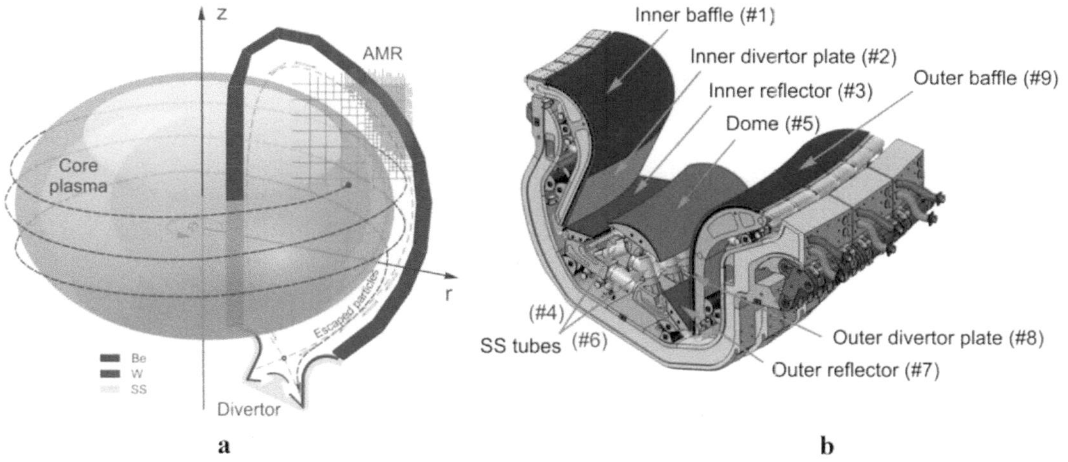

FIGURE 13.14 (a) 3D schematic illustration of ITER components and coordinate system and (b) divertor and nearby component surfaces implemented in HEIGHTS package.

Source: Hassanein and Sizyuk (2021).

When ITER becomes functional, it is expected to feature a low degree of availability. The experimental character of the project may allow for this, but in contrast, its degree of reliability should be very high, given that the success of this endeavor in demonstrating the feasibility of fusion methods will decide the fate of other projects with similar characteristics. The eyes of the world will be looking for projects like ITER to succeed, not only due to all the resources spent on generating the necessary conditions for an endeavor of this magnitude to perform adequately, but also because the current perilous environmental picture could use a breakthrough that finally allows humanity to transition to clean sources of energy permanently, without the fear of having to fall back to more polluting sources in case the former do not perform as expected.

To conclude, *fusion* offers an enticing avenue of exploration for the generation of abundant energy for billions of people. The work done and the expectations created around this technology are tremendous, and once the lock is opened, the world could be looking at the future of energy for centuries to come. However, the steps taken into the development of fusion are still the very initial-phase, and we should not fool ourselves in not recognizing the very steep challenges that remain for this method to be fully operational. Because of this, fission currently offers a more realistic path to energy production, through a process that we will review in the upcoming section.

13.3.2 Fission

All power plants that currently rely on nuclear reactions use fission as the way to obtain energy. As opposed to fusion, *fission* consists of the split of the nucleus of an atom into lighter elements and other neutrons that generate a chain reaction, while also releasing a large amount of energy in the form of heat and radiation. As Figure 13.15 shows, the preferred atom to cause the nuclear fission reaction is a uranium atom, due to their tendency to break apart easily when exposed to the right conditions (US Energy Information Administration, 2022a). To enable and control these nuclear reactions so that they produce the desired amount of heat, facilities known as nuclear reactors have been constructed and developed in many countries around the world.

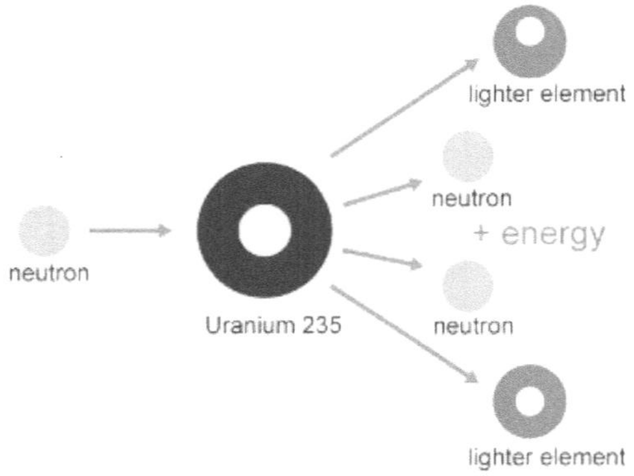

FIGURE 13.15 How fission splits the uranium atom.

Source: US Energy Information Administration (2022a).

Despite their different types, a nuclear reactor will generally have these components: the *fuel*, usually found in the form of pellets of enriched uranium; a *moderator*, material in the core that controls the neutrons released from fission so they can produce more of it (this is usually water, but it can also be heavy water or graphite), and *control rods/blades*, which are neutron-absorbing structures usually made of cadmium, hafnium, or boron inserted or removed from the core to control the rate of reaction. Meanwhile, *coolant fluid* circulates through the core to transfer the heat from it, while a *steam generator* is a feature in the cooling system of pressurized water reactors (PWRs). Lastly, *containment* structures like thick concrete and steel walls are built to protect the contents inside and the people outside (World Nuclear Association, 2022b).

One of the first designs for a nuclear reactor was that of a pressurized water reactor (PWR), developed more than 60 years ago by the US Navy and later put to use on land for electricity generation. Figure 13.16 shows the main elements of a PWR, characterized by two separate coolant loops which are connected by a heat exchanger (HX), also called a steam generator. Due to the heat released by nuclear fission, PWRs heat water to over 300°C at 150 atm, which is transferred to the HX to create steam that powers the turbine that transforms the energy into electricity. A few years later, the boiling water reactor (BWR) was developed, simplifying the structure to a single coolant loop while also operating at a lower pressure (Ho et al., 2019).

These two types of reactors are classified as light water reactors (LWRs) and use nuclear fuel in the previously described pellet form, where uranium has been enriched to feature a greater percentage of uranium-235 (^{235}U), a fissile isotope that allows for nuclear chain reactions to occur consistently (Pioro and Duffey, 2019). Moreover, Canada developed a reactor to work without enriched uranium, known as the Canada Deuterium Uranium (CANDU) reactor (see Figure 13.17), while the United Kingdom created their own versions, the now-defunct Magnox Reactors and the Advanced Gas Reactors (AGRs), some of which remain operational today (Ho et al., 2019).

In turn, Russia created their own low-enriched uranium reactor, known as the RBMK – short for *Reaktor Bolshoy Moshchnosty Kanalny*, meaning, “high-power channel-type reactor.” This structure consists of a light water reactor that is moderated by graphite (Ho et al., 2019). Further, fast neutron reactors (FNR) do not have a moderator and use fast neutrons, generating power mainly from plutonium. Table 13.5 features a summary of these different classes of reactors. In all these iterations developed over time, some clear distinctions can be made in the components and functioning of a nuclear reactor. Because of this, researchers have referred to these classifications as different generations of nuclear reactors (World Nuclear Association, 2022b).

Generally, first-generation reactors consist of those installations developed in the 1950s and the early part of the 1960s as the early prototypes of this technology. These installations mostly used

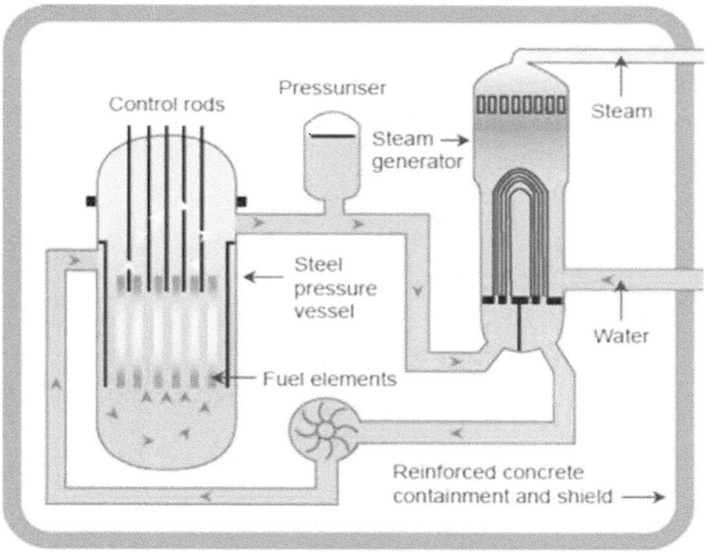

FIGURE 13.16 Diagrammatic layout of a pressurized water reactor (PWR).

Source: Ho et al. (2019).

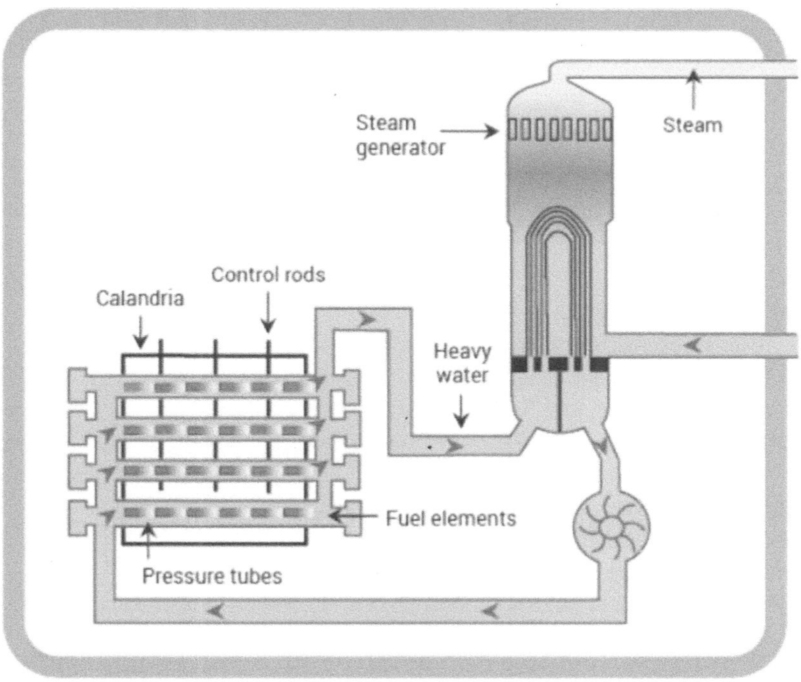

FIGURE 13.17 Illustration of a CANDU nuclear reactor.

Source: World Nuclear Association (2022b).

TABLE 13.5
Operable Nuclear Power Plants

Reactor Type	Main Countries	Number	GWe	Fuel	Coolant	Moderator
Pressurized water reactor (PWR)	USA, France, Japan, Russia, China, South Korea	307	293.3	Enriched UO_2	Water	Water
Boiling water reactor (BWR)	USA, Japan, Sweden	61	61.8	Enriched UO2	Water	Water
Pressurized heavy water reactor (PHWR)	Canada, India	47	24.3	Natural UO2	Heavy water	Heavy water
Advanced gas-cooled reactor (AGR)	UK	9	5.2	Natural U (metal), enriched UO_2	CO_2	Graphite
Light water graphite reactor (LWGR)	Russia	11	7.4	Enriched UO_2	Water	Graphite
Fast neutron reactor (FBR)	Russia	2	1.4	PuO_2 and UO_2	Liquid sodium	None
High-temperature gas-cooled reactor (HTGR)	China	1	0.2	Enriched UO_2	Helium	Graphite
TOTAL		438	393.7			

Source: World Nuclear Association (2022b).

natural uranium as fuel and graphite as moderator, and the last one of them – Wylfa 1 in Wales, United Kingdom – shut down its main reactor in late 2015. Meanwhile, Generation II reactors were developed from the middle of the 1960s onward, with an emphasis in commercial production that featured improvements from the previous set of reactors. Thus, Gen II facilities use enriched uranium as fuel and mostly water as a coolant and moderator (Fernández-Arias et al., 2020; World Nuclear Association, 2022b). Despite these improvements, accidents in Fukushima and Three Mile Island stopped or significantly halted Gen II production in Japan and the United States (Nian, 2018).

These incidents had the silver lining of accelerating the development of Generation III/III+ reactors, which featured greater structural safeguards that further reduce the possibility of a core meltdown accident, more passive safety features, as well as a longer operating life. Japan, China, Russia, and the United Arab Emirates have led the way in building these facilities, which are natural evolutions from PWR, BWR, and CANDU designs (Fernández-Arias et al., 2020; World Nuclear Association, 2021a, 2022b). Table 13.6 lists the current operational Gen III/III+ nuclear reactors as of early 2020, where we see that China has built half of the facilities with the latest technology, far outpacing countries like Japan and Russia.

Finally, Generation IV consists of the group of proposed nuclear reactors of the future that have an emphasis on improving the handling of nuclear waste through a decrease in the amount of spent fuel generated after repeated fission reactions. Figure 13.18 summarizes the development of the different generations over the years, as well as their main prototypes and designs. Logically, the evolution of new generations of nuclear fission technologies follows a shift from a more experimental phase, where fission was obtained through methods that would have potential unintended consequences which were unknown at that time, to an empirically driven stage, where safety and efficiency become the main priorities.

Most of Gen IV designs use a closed fuel cycle so that the resource base is maximized while the production of high-level nuclear waste is minimized. Meanwhile, three of these projects feature a fast reactor, which has no moderator and only relies on fast neutrons for fission. Because of this,

TABLE 13.6
Overview of Operational Gen III/III+ Reactors by Reactor Design, as of March 13, 2020

Reactor Designs	China	Japan	Russia	India	Korea	Iran	Total	Supplier
Gen III	**6**	**4**	**0**	**2**	**2**	**1**	15	
ABWR		4					4	GE/Hitachi/ Toshiba
ACP-1000	2						2	China
AES-91	4						4	Rosatom
AES-92				2		1	3	Rosatom
APR-1400					2		2	KEPCO
Gen III+	**6**	**0**	**3**	**0**	**0**	**0**	9	
AES-2006			3				3	Rosatom
AP1000	4						4	Westinghouse
EPR-1750	2						2	Framatome
Total	12	4	3	2	2	1	24	

Source: Wealer et al. (2021).

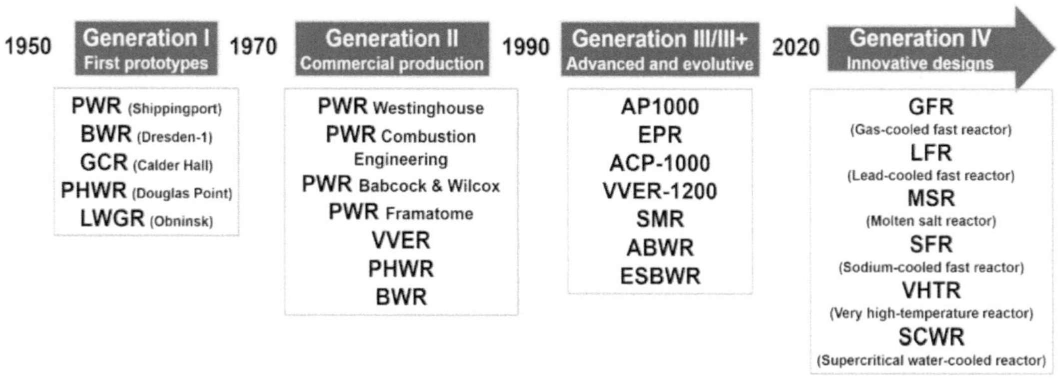

FIGURE 13.18 Different generations of nuclear reactors throughout the past 70 years.

Source: Fernández-Arias et al. (2020).

FNRs use plutonium as fuel, something that improves fission with fast neutrons compared to previous generations of reactors (World Nuclear Association, 2020, 2021b).

Table 13.7 presents a summary of the six Gen IV reactor projects being developed by GIF, with a seventh option having appeared from a variation in the design of molten salt reactors. This offshoot consists of a thermal plant that has the single goal of producing clean hydrogen through nuclear means. These advancements follow the stated general goals from GIF (see Table 13.8) that center on sustainability of the fuel cycle and an improved availability of reactors, better economics, safety and reliability enhancements, as well as an upgraded physical protection for plants, both from outside influences like attacks from bad faith actors as well as from leakage of toxic material from inside to the environment (GIF, 2020).

With the advent of Gen IV technologies, the European Commission launched the joint project called the European Sustainable Nuclear Industrial Initiative (ESNII) in 2010. As is the case with the ITER project for fusion technologies, ESNII has the stated goal of demonstrating fission

TABLE 13.7
Generation IV Reactor Designs under Development by GIF

	Neutron Spectrum (Fast/ Thermal)	Coolant	Tempe-rature (°C)	Pressure*	Fuel	Fuel Cycle	Size (MWe)	Use
Gas-cooled fast reactors	Fast	Helium	850	High	U-238+	Closed, on-site	1,200	Electricity and hydrogen
Lead-cooled fast reactors	Fast	Lead or Pb-Bi	480–570	Low	U-238+	Closed, regional	20–180** 300–1,200 600–1,000	Electricity and hydrogen
Molten salt fast reactors	Fast	Fluoride salts	700–800	Low	UF in salt	Closed	1,000	Electricity and hydrogen
Molten salt reactor **– advanced high-temperature reactors**	Thermal	Fluoride salts	750–1,000	–	UO_2 particles in prism	Open	1,000–1,500	Hydrogen
Sodium-cooled fast reactors	Fast	Sodium	500–550	Low	U-238 and MOX	Closed	50–150 600–1,500	Electricity
Supercritical water–cooled reactors	Thermal or fast	Water	510–625	Very high	UO_2	Open (thermal) Closed (fast)	300–700 1,000–1,500	Electricity
Very high-temperature gas reactors	Thermal	Helium	900–1,000	High	UO_2 prism or pebbles	Open	250–300	Hydrogen and electricity

Source: World Nuclear Association (2020).

technologies and their value to humanity through improving research infrastructures, fuel facilities, and R&D work. Concretely, the alliance wants to advance in four different projects, all with varying degrees of progress and prioritization: a sodium fast reactor (SFR) prototype called ASTRID, a flexible fast spectrum irradiation facility (MYRRHA), a demonstrator for a lead-cooled fast reactor (ALFRED), and a gas-cooled fast reactor dubbed ALLEGRO (ESNII, 2021).

In the middle of these high-tech dynamics, current advancements also open the door for some countries to start their own nuclear programs using small modular reactors (SMRs). These types of reactors rely on more standardized sections and parts, and their energy output is lower than that of an average plant; however, they also boast lower up-front costs while requiring fewer human capital development to operate successfully (Mignacca and Locatelli, 2020). According to Budnitz et al. (2018), SMR prototypes have yet to reach a widespread interest, but there are currently a few projects being developed in Argentina, Russia, and China.

Ultimately, the expectation is that SMRs would provide relief to a series of developing countries that are searching for cleaner sources of energy without putting those nations further in debt due to skyrocketing costs. Figure 13.19 illustrates the contrast between construction of a nuclear power plant versus delivery of a factory-built module that is assembled on-site. The logic behind the widespread use of SMRs is a sort of *plug-and-play* rationale for nuclear energy, where the burden of

TABLE 13.8
Goals for Generation IV Nuclear Energy Systems

Sustainability 1	Generation IV nuclear energy systems will provide sustainable energy generation that meets clean air objectives and provides long-term availability of systems and effective fuel utilization for worldwide energy production.
Sustainability 2	Generation IV nuclear energy systems will minimize and manage their nuclear wastes and notably reduce the long-term stewardship burden, thereby improving protection for the public health and the environment.
Economics 1	Generation IV nuclear energy systems will have a clear life cycle cost advantage over other energy sources.
Economics 2	Generation IV nuclear energy systems will have a level of financial risk comparable to those of other energy projects.
Safety and reliability 1	Generation IV nuclear energy systems operations will excel in safety and reliability.
Safety and reliability 2	Generation IV nuclear energy systems will have a very low likelihood and degree of reactor core damage.
Safety and reliability 3	Generation IV nuclear energy systems will eliminate the need for off-site emergency response.
Proliferation resistance and physical protection	Generation IV nuclear energy systems will increase the assurance that they are very unattractive and the least desirable route for diversion or theft of weapons-usable materials and provide increased physical protection against acts of terrorism.

Source: Pioro and Duffey (2019).

FIGURE 13.19 Factory-built SMR modules delivered to site versus on-site construction.

Source: INL (2022).

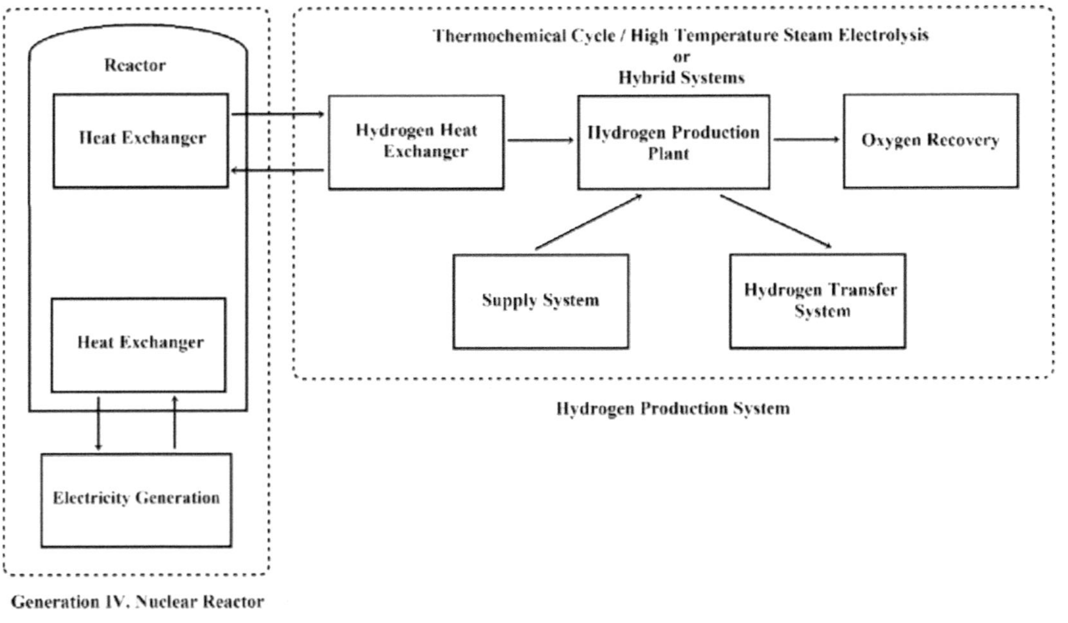

FIGURE 13.20 Concept diagram of the hydrogen generation system integrated in a Generation IV nuclear reactor.

Source: Şahin and Şahin (2021).

designing and building parts goes to one – or very few – actors, economies of scale are reached after making this technology widespread enough, and small countries do not need to spend years arriving at the same point that these developers already have reached.

Along with the goal to provide energy to the final user, the emphasis on nuclear fission technologies is also related to the mitigation of the effects of climate change. As noted previously, there is at least one project in the GIF list that is tailored to the production of clean hydrogen through Gen IV reactors. Şahin and Şahin (2021) explored the potential of hydrogen being produced from nuclear processes, and Figure 13.20 illustrates the integration of the system within a nuclear reactor. As we can see, along with electricity and heat, Gen IV reactors can use the water inside to produce hydrogen via electrolysis, a carbon-free process that also contributes in the quest toward an energy transition to cleaner sources of energy.

With that said, there are important constraints that appear when relying too heavily on new generations of technologies to solve problems that seem long-standing. Knapp and Pevec (2018) analyzed the potential and promising future of nuclear technologies, simultaneously weighing the very real limitations that these new technologies will face in the next couple of decades as to whether fusion technology is feasible in the medium to long term. Significant trade-offs appear when considering the very large sums of money that taxpayers are putting into very ambitious, *high-risk, high reward* ventures, when limited resources and a ticking environmental clock are pushing for more immediate plans.

These challenges may invite some researchers to speed up development of some projects, potentially cutting corners and skipping important safety considerations. Even though newer generations of reactors have helped constrain rogue players, accidents in reactors previously considered safe still haunt nuclear energy deliberations to this day. Three Mile Island occurred in 1979, affecting a PWR; the aftermath of the earthquake and tsunami in Fukushima targeted a BWR; and the Chernobyl plant incident affected a RBMK reactor (Knapp and Pevec, 2018). All those installations had initially passed rigorous safety checks but were blindsided by factors previously discarded, like natural disasters or human error.

TABLE 13.9
Total Uranium Production by Countries in 2018

Country	Uranium Tons	Weight Percentage (%)
Kazakhstan	21,705	40.57
Canada	7,001	13.09
Australia	6,517	12.18
Namibia	5,525	10.33
Niger	2,911	5.44
Russia	2,904	5.43
Uzbekistan	2,404	4.49
China	1,885	3.52
Ukraine	1,180	2.21
USA	582	1.09
India	423	0.79
South Africa	346	0.65
Iran	71	0.13
Pakistan	45	0.08
World total	53,499	100.00

Source: Alwaeli and Mannheim (2022).

Furthermore, expanding the field too much or too quickly may place an unexpected degree of pressure on uranium-producing countries to continue delivering. Table 13.9 shows a list with the largest producers of uranium in the world in 2018 who could end up incurring in costly activities that could have their own unintended consequences, like worsening the climate or exploiting workers. Currently, research has shown that a uranium production can withstand an expansion in nuclear plants, especially if newer generations of reactors can switch from ^{235}U to ^{238}U or ^{232}Th – thorium-232, the element's main isotope – that, despite not being fissile, can be paired with plutonium as fissile fuel (Knapp and Pevec, 2018).

All in all, fission offers a promising avenue to explore the development of clean energy, and the experience of several countries already benefitting from nuclear power plants providing electricity to their power grids should be a strong argument in favor of expanding the reach of nuclear plants across nations. With that said, high economic and human capital costs remain prohibitive for many countries, something that leaves a gap open for interested parties to explore. In this sense, the oil and gas industry has taken proactive steps to invest in renewable energies and appears poised to replicate some of these steps regarding nuclear energy.

The following section explores the role of the oil and gas sector in the development of nuclear initiatives across countries, as well as the pledges made by industry actors to invest and provide a greater visibility to nuclear projects.

13.4 NUCLEAR ENERGY AND THE OIL AND GAS INDUSTRY

After a period of skepticism from governments, private players, and citizens aimed at nuclear power, at times as a response to tragic accidents that cannot be repeated, the advent of safer reactor technologies has brought up a new group of actors interested in investing and exploring the possibilities of nuclear methods for the future of energy. Coupled with the environmental regulations placed by international organizations and the pledges for net-zero carbon emissions made by other industries, the oil and gas sector has not remained idle.

The industry may have received a boost after the recent determination from European Union legislators to declare both nuclear as well as natural gas as "green" technologies (Bloomberg, 2022).

This move could open the door to more investments in projects from these sources, which could, in turn, translate into a larger degree of energy independence for countries in the region. This is especially important under the current scenario of conflict after the war in Ukraine, where Russia may decide to cut their oil supply to countries that heavily depend on it, causing disruptions and potentially risking the lives of millions of people.

Because of these new dynamics, the oil and gas industry is in the middle of a period of adaptation to the new environment, with the energy transition process causing the industry to rethink their business models as well as their investment strategies. In this sense, most oil majors have taken a very proactive stance regarding renewable energies, having struck partnership deals with emergent wind and solar companies for development and research endeavors. Similarly, nuclear provides another opportunity to pivot toward a more diverse suit of sources to provide energy, contributing to the global effort to lower carbon emissions and thus avoiding the more negative scenarios projected by groups like the Intergovernmental Panel on Climate Change (IPCC).

A few examples of collaboration between start-ups and oil majors are the alliances between Chevron and Zap Energy, Equinor and Temasek, as well as Italian company Eni and Commonwealth Fusion Systems (CFS), all forged in 2020, to explore nuclear fusion technologies (Bloomberg, 2020; Eni, 2021; Reuters, 2020). More recently, Norwegian corporation Equinor has gone as far as acquiring a minority shareholding in CFS, a group that spun out of the Plasma Science and Fusion Center at the Massachusetts Institute of Technology (MIT), while Chevron has joined tech giant Google in investing in the nuclear fusion start-up TAE Technologies, with the objective of reaching the goal of producing the first prototype for a hydrogen–boron fusion power plant (CNBC, 2022; Equinor, 2022).

Similar to what has been occurring with renewable sources of energy, trailblazers in the nuclear realm need sponsors that are willing to provide not only money but also research and development (R&D) experience, and even name recognition, all things that oil majors appear ready to provide. This dynamic of collaboration also responds to the need for oil companies to adapt to the new changes in regulation and the subsequent public perception about the continued use of more polluting sources of energy. This challenge is a similar one for oil-exporting countries, who need to contemplate alternative funding strategies under an energy transition framework (OIES, 2018).

According to Biancardi and Di Castelnuovo (2021), oil majors can choose one of four main strategies relative to their degree of innovation. The two extremes in this spectrum (shown in Figure 13.21) are the "last man standing" strategy and the "energy innovator" scheme, which reflect the status quo and a complete revamping of the energy objectives of a company, respectively.

Meanwhile, the two main strategies in the middle reflect some changes, but mostly operating under the same business model. The "gas company" option is taken by players that decide to expand their portfolio to include participation in the growing market of liquefied natural gas (LNG), while the "energy major" option is followed by actors that commit to a widespread portfolio, investing in newer sources of energy, like renewables and nuclear fusion, but keeping their investments in oil and gas. Of course, these categories are referential, as most oil majors can be located in different points of this spectrum and can change their strategies over time to adapt to new market dynamics.

With that said, all oil majors appear to be moving toward the right, with different degrees of speed and intent. Companies that function more in Europe appear to have moved quicker, signaling a stronger allegiance to climate pledges, some of which come from the European Union.

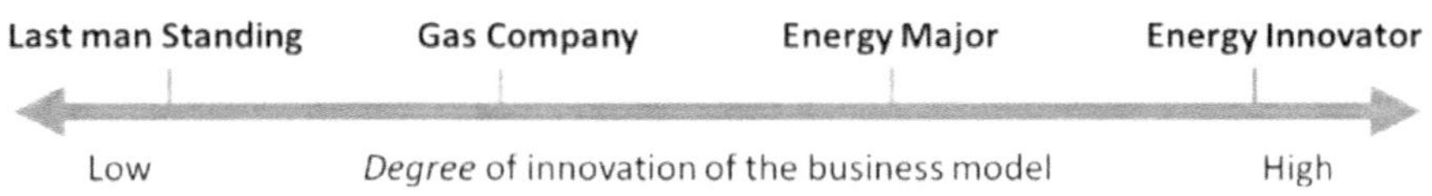

FIGURE 13.21 Four strategic options for "big oil" and their impact on the traditional business model.

Source: Biancardi and Di Castelnuovo (2021).

Meanwhile, companies that focus more exclusively on the United States have been slower to adopt more groundbreaking strategies, both because of the success of fracking methods and the less-rigid regulation. Thus, longtime players like ExxonMobil, Chevron, and Conoco Philips have shown some interest in cleaner technologies but have maintained their business models relatively intact. On the other hand, BP, Royal Dutch Shell, Total, and Eni have been more aggressive in their strategies to become pioneers in the clean energy market (Biancardi and Di Castelnuovo, 2021).

Similarly, national oil companies are also presented with strategies that could significantly shape the future of their businesses. For example, energy companies run by the Chinese government now include a larger set of hydrogen and nuclear technologies that have expanded their reach essentially worldwide. These companies have relative independence but still share the objective of expanding relations and establishing geopolitical influence in disparate regions like Africa or South America. At times, however, profit-seeking companies have gone against official foreign policy directives, as shown by a few moves from national nuclear companies into Europe.

Meanwhile, Sweden has had its own path to the diversification of its public energy sector. Kaijser and Högselius (2019) assessed the role of energy companies in the struggle for independence in the country, where nuclear energy has played a significant role since the days of the Cold War. Characterized by an acute energy dependence in the early years of the 20th century, Sweden took aggressive steps to import coal, particularly from Germany. World War II proved a challenge for the Swedes, who saw their imports diminish due to the conflict, while also raising some eyebrows due to the source of the material being imported. This led to heavy innovation in nuclear technologies, a move that ended up with the first commercial water reactor being built in Stockholm in 1964, used for heat and electricity production.

As we can see, private actors and national companies have developed varying strategies regarding the inclusion of nuclear technologies into their portfolios. With varying degrees of advancements in mind, the general trajectory of virtually all actors has been toward increasingly relying on clean energies, adapting to the new environmental regulations, the public sentiment, and the new market opportunities provided by the changing dynamics. While most players, both private and public, have invested heavily on renewable energies through solar and wind energy facilities, some have also used their resources to fund nuclear energy endeavors, with a particular interest in long-term fusion methods.

These strategies may follow a strictly profit-seeking rationale, where fusion becomes the main new source of energy for decades to come, benefitting those actors that invested first on the technology, just as economic gains of oil and gas have benefitted the "early movers" who explored sites and developed the technology to exploit them. In turn, and using a more balanced approach, we can place these players as rational actors that recognize the changes in their environment, at times driven by governments or by international regulation, and develop a strategy that combines their current way of doing business with an expanded set of sources of energy to explore and exploit.

The following section takes a closer look at how the environment around oil majors and national companies has contributed to their most recent business decisions. Current public perceptions, environmental pledges, and even some nudging actions from governments and international organizations can have a role in the business decisions of oil majors, who also have medium- and long-term profit strategies. Different combinations of actions and behaviors from multiple actors with varying cost and benefit analysis cause a series of different strategies at the country level, with some governments pushing for more investment in clean energies and others relying more on private efforts to reach similar goals.

13.5 NUCLEAR ENERGY IN ENERGY TRANSITION IN THE OIL AND GAS INDUSTRY

The oil and gas industry has chosen a proactive role in the energy transition to cleaner sources of energy. Whether that is due to profit-seeking reasons, an instinct of survival, a rationale to follow

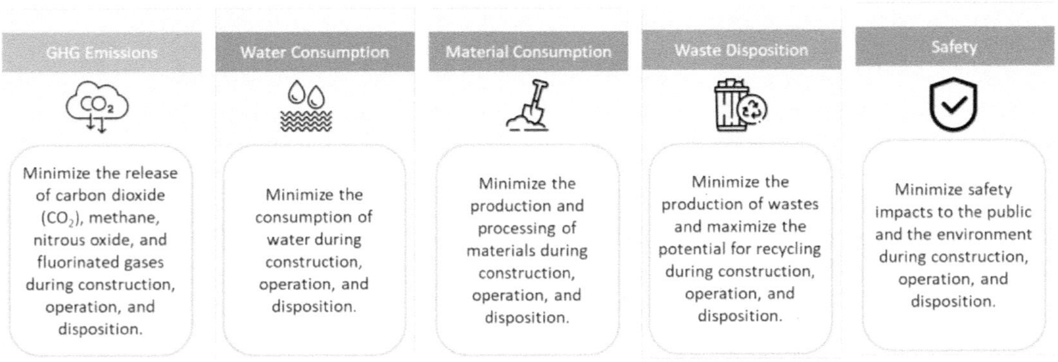

FIGURE 13.22 The net-zero pillars used to evaluate potential energy production technologies.

Source: INL (2022).

novel international regulations and government incentives, or a genuine sentiment of environmental protection, the role of this sector is key in a successful transition. For example, the *know-how* obtained by the industry throughout its work of several decades breaking new ground is a very valuable asset for the current pioneers seeking to break new ground in a different but related area. Thus, in this changing scenario, current innovators still need the toolbox with which old trailblazers opened a path.

The current environmental pledges made by governments and private actors, which are checked regularly by international organizations like the United Nations, have created a framework that is now being followed by multiple other sectors. The most salient commitment being made by businesses like aviation, the automobile industry, and even central banking is the net-zero objective. This goal, at times also referred to as "carbon neutrality," involves the reduction of greenhouse emissions (GHG) to zero through balancing the amount of carbon released into the atmosphere with the amount removed and stored, thanks to technologies like carbon capture and storage (CCS), or the switching to cleaner sources of energy (Dikau et al., 2021).

For many decades, most industrial processes did not account for the unintended consequences that these would generate in the environment. The net-zero pledge is one way to tackle the shortsightedness of many actors seeking to clean up procedures while still trying to maintain current levels of production and service to the public. Figure 13.22 shows the main pillars of the net-zero commitments, which center on reducing GHG emissions, lowering the consumption of water and materials when building and operating, improving waste disposal, and improving the safety of all these processes while reducing their impact on the public and the environment.

As we have reviewed throughout this chapter, nuclear energy is one very promising way to reduce carbon emissions and maintain current levels of production for many industries. However, due to the bad memories from the reactor accidents that have captured the public sentiment for generations, some actors may still be restrained to invest in nuclear technologies to a degree that would help toward a rapid switch to cleaner sources of energy. Additionally, costs may still be prohibitive for a large group of actors, despite their best intentions of making the jump from coal or gas to hydrogen or nuclear.

Therefore, we are confronted with a set of notions that appear to be in contradiction. On the one hand, there is almost complete agreement that there needs to be a quick switch to renewable and clean sources of energy. But on the other, there exists significant reticence to give the opportunity to a source of energy that can become the key to unlock a clean future and provide carbon-free energy for millions of people (see Figure 13.23). The bridging of this gap seems perfectly suited for the oil and gas sector, which can help in raising awareness of the potential benefits of both fission and

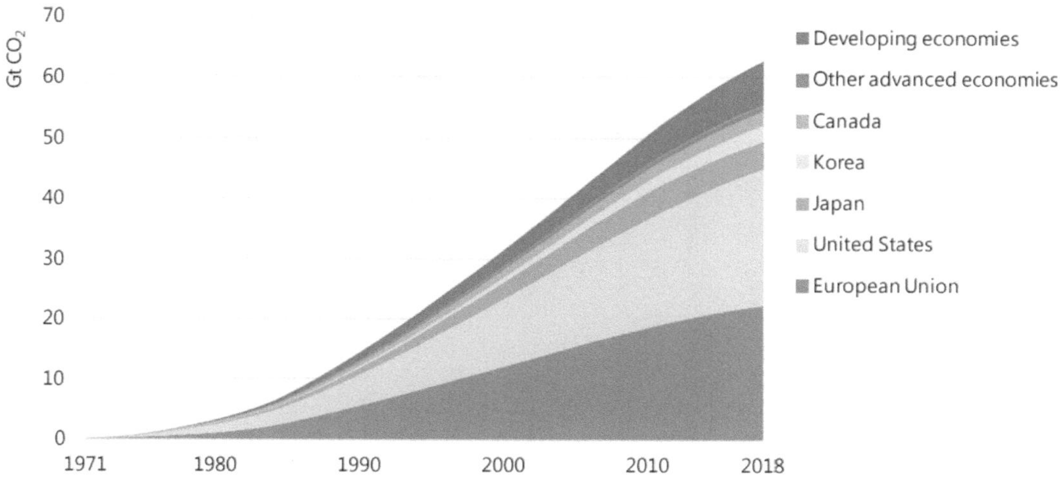

FIGURE 13.23 Cumulative CO_2 emissions avoided by global nuclear power to date.

Source: IEA (2019).

fusion technologies, while also working on achieving economies of scale for some methods like, for example, small modular reactors.

Besides investment in projects like fusion, an adequate degree of coordination should also be established between oil and gas companies and governments around the world to promote nuclear technologies and help with the dissemination of cost-effective opportunities. As an example of the positive feedback loop that can be created thanks to these dynamics, Dinçer et al. (2020) showed that investment in nuclear technologies is associated with the financial development of countries, mainly through the security brought by energy independence. Hence, when countries are seen as energy-robust, industrial production increases, which in turn can help attract investment, strengthening the financial market.

Perhaps due to findings like these, some countries have begun to develop their own networks of association regarding nuclear energy. Jewell et al. (2019) developed a framework to identify an international nuclear collaboration, which they define as an activity where at least two states share, exchange, or combine material resources, knowledge, or information about the development of nuclear energy technologies. According to their degree of specificity, these approaches are further categorized as *agreements*, *partnerships*, *non-binding*, *uranium-related*, or *technological*, with two subclasses for this last label: *technological concrete* and *technological supportive*, depending on whether this is related to matters of construction or knowledge exchange, respectively.

Table 13.10 shows the different types of nuclear cooperation initiatives that countries have been developing between 2000 and 2015. A large number of these approaches are mostly related to knowledge exchange and regarding nuclear safety and security, while construction and operation issues occur fairly regularly as well. Moreover, the countries that participate in these exchanges tend to be the same group of active players. Figure 13.24 shows the network that is generated after mapping the nuclear collaborations, which show that the United States, Russia, France, Canada, Japan, and South Korea are the nations with the greatest degree of activity (Jewell et al., 2019).

Given the significant infrastructure constraints as well as the high degree of preparation and expertise required for countries to start nuclear projects, collaborations are a key tool for the advancement of the technology. These approaches allow for a shared burden that could increase the efficiency in arriving at a successful design, while also forming new generations of capable researchers that could take countries forward into the nuclear path. A consequence of this exchange

TABLE 13.10
The Extent of Different Types of International Nuclear Cooperation

Type of Cooperation	Number of Cooperation of Each Type			
	All	Directional	Partnerships	Tripartite
Total	**737 (100%)**	**346**	**377**	**14**
Technological concrete	**252 (34%)**	**170**	**71**	**11**
Nuclear power plant construction and operation	111 (15%)	86	18	7
Reactor supply	71 (10%)	61	8	2
Nuclear fuel cycle	98 (13%)	63	32	3
Decommissioning and waste management	64 (9%)	35	28	1
Technological supportive	**558 (75%)**	**209**	**344**	**5**
Knowledge exchange and training	384 (52%)	142	240	2
Nuclear safety and security	213 (29%)	37	174	2
Planning, regulation, and supportive infrastructure	102 (14%)	61	39	2
No information	63 (9%)	39	24	–
Uranium	**67 (9%)**	**62**	**3**	**1**

Note: The counts sum to more than the total and the percentages sum to more than 100% because the categories were coded in a non-exclusive manner.

Source: Jewell et al. (2019).

of ideas and knowledge is the development of country-wide nuclear programs in nations that are seeking energy independence or are seeking to galvanize their reliance on clean energy.

The World Association of Nuclear Operators (WANO) has published a road map for countries and private actors to develop their own nuclear reactors, with an emphasis on operational readiness. For the case of a country seeking to start their own nuclear program, the organization highlights IAEA's 19 criteria, shown in Figure 13.25.

The convergence of climate commitments that target carbon neutrality, a safer generation of nuclear reactors that promise abundant clean energy, and collaboration between countries to exchange knowledge about nuclear technology creates a valuable opportunity for private actors to become the financial backers of the many projects that could see the light of day thanks to these dynamics. Seeking to diversify their energy portfolio and adapt to the energies that will power the 21st century, oil and gas players have medium- and long-term windows to invest in fission and fusion projects, with varying degrees of expectations for each method.

Despite fusion having immense potential, so far as to overtake the entire energy system, making every other source used so far seem irrelevant, it still appears way ahead of its time. Results from projects like ITER or the work of TAE Technologies will serve as a compass to project the solid ground where fusion technologies actually stand, and the investment from oil majors like Chevron can be a big part of this development. However, the current climate pledges and expectations call

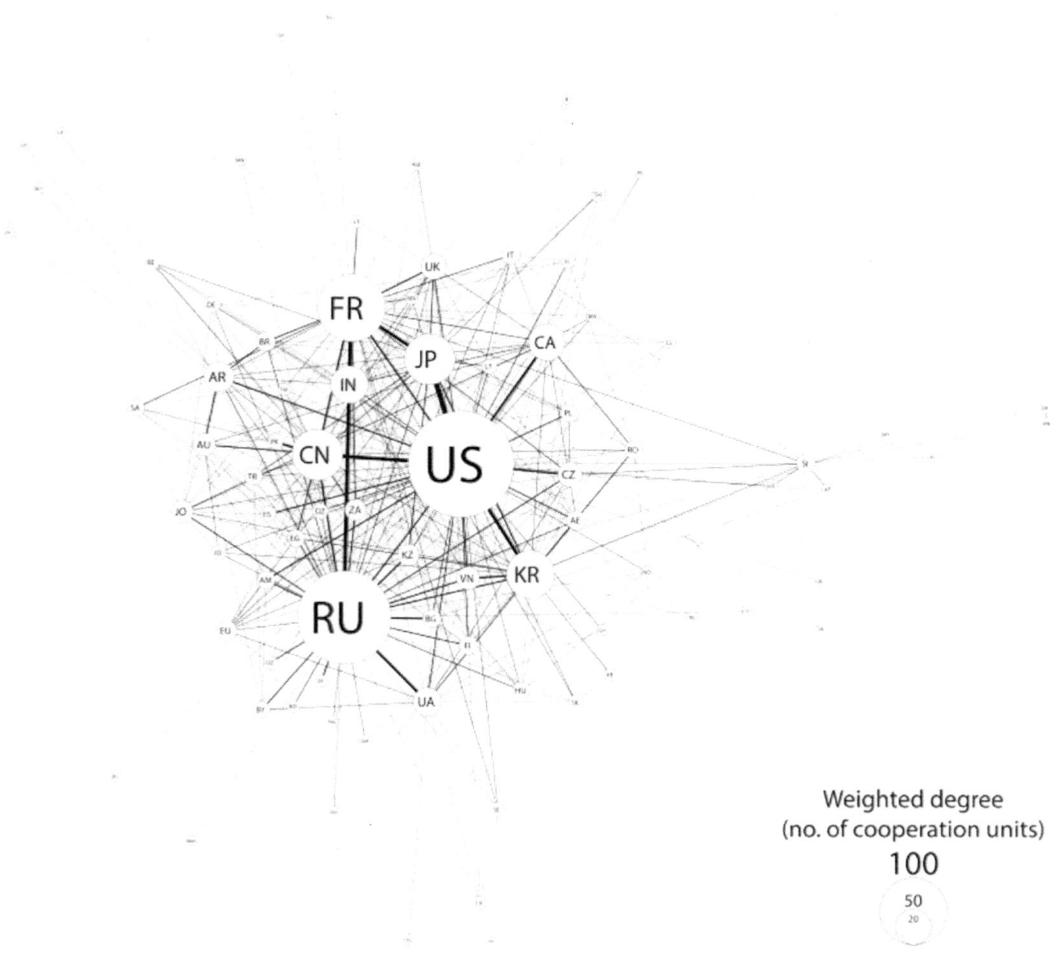

FIGURE 13.24 Aggregate network of all international technological nuclear cooperation.

Source: Jewell et al. (2019).

FIGURE 13.25 Nineteen national criteria and infrastructure elements for countries seeking to begin a nuclear program.

Source: WANO (2020).

for a more urgent set of strategies, where oil and gas actors have leverage to maintain their role as energy providers, this time of clean and renewable sources.

Due to their maturity and readiness to serve the public, fission technologies will become increasingly more important in the next few years. Despite public sentiment staying fairly stable regarding skepticism toward the use of nuclear energy, the previous two years have provided us with many unprecedented situations that could call for a re-evaluation of those feelings. First, coming out of the COVID-19 pandemic has given people a new perspective on global events, and the climate could be the next point of focus. And second, the current conflict in Ukraine may make some people reconsider the role of energy independence, where nuclear can be the missing piece of the puzzle.

Looking at the development of the latest generation of nuclear reactors, choosing to invest in their construction and implementation appears perhaps more enticing than ever before. Following the push for the development of hydrogen as an energy carrier for uses in, for example, the transportation sector across many countries, nuclear reactors can adapt to produce significant quantities of hydrogen at times of lower electricity demand. Similarly, as Figure 13.26 illustrates, the cogeneration capacities of nuclear reactors can also help meet district and building heating needs or produce isotopes for industrial, medical, and research purposes (The Royal Society, 2020).

We are now at a point in time where the many decades of exploration and experimentation with nuclear technologies, marked at times by a difficult road with tragic accidents, can now bring

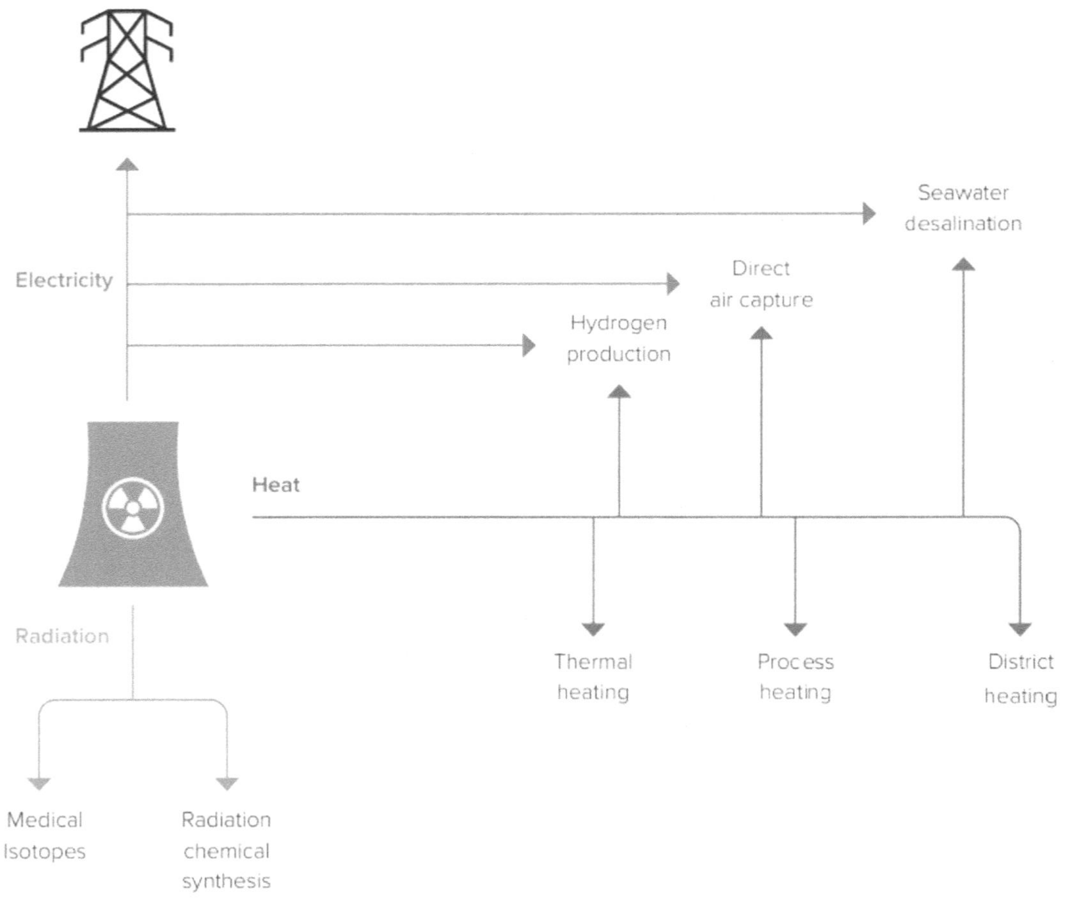

FIGURE 13.26 Potential nuclear cogeneration options for nuclear reactors.

Source: The Royal Society (2020).

exponential benefits just at the right moment. In the middle of explicit efforts to mitigate carbon emissions from governments and international organizations through the pursuit of net-zero strategies, nuclear methods can serve as an excellent complement to renewable energies to eliminate intermittence issues, to effectuate the switch to clean sources that the energy transition requires.

To accelerate this much-needed move, the oil and gas sector can play a key role through investment and R&D initiatives. Using their experience and *know-how*, the industry can help materialize many projects in countries that need more financial backing to implement them, or in developed nations that are seeking to solidify their energy independence. To achieve the ambitious net-zero objectives made by industries and international organizations, the energy transition needs the contribution of all actors across society in the next couple of decades. The window of opportunity has presented itself for the oil and gas industry to revamp itself in the clean energy world, and we can all win if it takes advantage of it.

13.6 CONCLUSIONS

This chapter focused on nuclear energy technologies, its opportunities, challenges, benefits, and the state of the art of current methods. There are two distinct nuclear technologies currently being advanced, with varying degrees of implementation. First, fusion offers the largest degree of potential, but researchers are still considerably away from widespread real-world implementation and successful replication. Then, the technology that is in the center of nuclear reactors, fission, has been implemented for several decades, albeit with methods that have now been updated to bring more safety and to avoid accidents like those in Three Mile Island, Fukushima, and Chernobyl.

The energy transition process requires that researchers advance in these two fronts, but the emergency facing the climate also demands that current applied technologies are able to be expanded. Thus, the challenge of allocating limited resources affects nuclear technologies very significantly. On the one hand, fusion can solve the puzzle of bringing copious amounts of clean energy to millions of people without the need to ever look back to sources like coal or crude oil. But this requires heavy investment, a long time (ten years plus), and human capital development to materialize. On the other hand, fission offers a more incremental way to provide clean energy, which can currently serve more as a complement to coal and oil in some countries, instead of a leading driver.

Moreover, due to many governments lacking the necessary funding schemes to initiate their own nuclear programs or their policymakers not placing enough weight on the potential of nuclear energy altogether, many opportunities to invest and fund projects of this magnitude appear. Here is where the oil and gas industry can become a leader and a strong protagonist in the energy transition, providing capital for the construction of nuclear reactor facilities, helping with community outreach through raising awareness of the potential of these technologies, as well as highlighting the current advancements of Generation III/III+ and Generation IV reactors.

In the first couple of decades of the 21st century, much progress has been made regarding nuclear energy technologies. Reactors have been made safer than ever before, significant improvements have been made in the area of managing nuclear waste, and more cost-effective ways have been developed concurrently so that developing countries can also initiate their path toward nuclear without breaking the bank. Coupled with active collaboration between participating and interested nations and a changing public perception about the potential of nuclear fission technologies, there is ripe soil for the development of many nuclear projects in the near future that could help provide a strong foundation for a successful energy transition process.

Current environmental regulation, enforced by international organizations, is looking to continue the economic development obtained in the last century, but now with climate neutrality in mind. Under this scenario of net-zero pledges and large industries striking deals to lower their carbon emissions, nuclear energy offers a path that covers both current dynamics as well as future ways of generating energy. Despite fusion being the current preferred option for oil majors who have invested large sums of money in start-ups that are exploring it, fission can be the stepping stone that many countries need to achieve an increasingly important state of energy independence.

As we can see, many notions intersect in the search for improved nuclear technologies. Geopolitical elements cross paths with economic considerations, while technical limitations are met with encouraging adaptation strategies. Therefore, the scenario is set for significant medium- and long-term change, and actors are still deciding which cards to play at which moment. But all point toward a strong development and acceptance of nuclear energy, with governments and private actors investing in fission and fusion projects, with the objective of relying permanently on clean sources of energy. The path forward is visible, and now it is time to continue the journey.

REFERENCES

Agyekum, E. B., F. Amjad, F. Aslam, and A. Ali. 2021. Application of Weighted Linear Combination approach in a Geographical Information System environment for nuclear power plant site selection: The case of Ghana. *Annals of Nuclear Energy* 162: 108491.

Alam, F., R. Sarkar, and H. Chowdhury. 2019. Nuclear power plants in emerging economies and human resource development: A review. *Energy Procedia* 160: 3–10.

Alwaeli, M., and V. Mannheim. 2022. Investigation into the current state of nuclear energy and nuclear waste management—A state-of-the-art review. *Energies* 15: 4275.

Baskurt, Z. M., and C. C. Aydin. 2018. Nuclear power plant site selection by Weighted Linear Combination in GIS environment, Edirne, Turkey. *Progress in Nuclear Energy* 104: 85–101.

Biancardi, A, and M. Di Castelnuovo. 2021. No country for "oil" men. What will the future hold for big oil? *European Energy & Climate Journal* 33: 35–45.

Bloomberg. 2020. *Nuclear fusion startup gets $84 million to enter next phase*. Available at www.bloomberg.com/news/articles/2020-05-26/nuclear-fusion-project-backed-by-investors-to-enter-next-phase [accessed July 14th, 2022].

Bloomberg. 2022. EU lawmakers remove last hurdle to label gas, nuclear as green. Available at www.bloomberg.com/news/articles/2022-07-06/eu-lawmakers-remove-last-hurdle-for-gas-nuclear-as-green [accessed July 14th, 2022].

Bonano, E. J., E. A. Kalinina, and P. N. Swift. 2018. The need for integrating the back end of the nuclear fuel cycle in the United States of America. *MRS Advances* 3(19): 991–1003.

Budnitz, R. J., H.-H. Rogner, and A. Shihab-Eldin. 2018. Expansion of nuclear power technology to new countries – SMRs, safety culture issues, and the need for an improved international safety regime. *Energy Policy* 119: 535–544.

Buongiorno, J., M. Corradini, J. Parsons, and D. Petti. 2019. Nuclear energy in a carbon-constrained world: Big challenges and big opportunities. *IEEE Power and Energy Magazine* 17(2): 69–77.

CNBC. 2022. *Google and Chevron invest in nuclear fusion startup that's raised $1.2 billion*. Available at www.cnbc.com/2022/07/19/google-chevron-invest-in-fusion-startup-tae-technologies.html [accessed July 21st, 2022].

Devanand, A., I. A. Karimi, and M. Kraft. 2019. Optimal Site selection for modular nuclear power plants. *Cambridge Centre for Computational Chemical Engineering*. Preprint N° 221.

Dinçer, H., S. Yüksel, Ç. Çağlayan, and G. S. Uluer. 2020. The contribution of nuclear energy investment on sustainable financial and economic development. *Journal of Financial Economics and Banking* 1(1): 39–51.

Dikau, S., N. Robins, and U. Volz. 2021. *Net-zero central banking: A new phase in greening the financial system*. London: Grantham Research Institute on Climate Change and the Environment and Centre for Climate Change Economics and Policy, London School of Economics and Political Science, and Centre for Sustainable Finance, SOAS, University of London.

Dong, K., R. Sun, H. Jiang, and X. Zeng. 2018. CO2 emissions, economic growth, and the environmental Kuznets curve in China: What roles can nuclear energy and renewable energy play? *Journal of Cleaner Production* 196: 51–63.

Dong, Z., and Y. Pan. 2018. A lumped-parameter dynamical model of a nuclear heating reactor cogeneration plant. *Energy* 145: 638–656.

El-Emam, R. S., and I. Khamis. 2019. Advances in nuclear hydrogen production: Results from an IAEA international collaborative research project. *International Journal of Hydrogen Energy* 44(35, July): 19080–19088.

Eni. 2021. *Eni and commonwealth fusion systems*. Available at www.eni.com/en-IT/operations/collaboration-commonwealth-fusion-systems.html [accessed July 18th, 2022].

Equinor. 2022. *Equinor invests in fusion energy*. Available at www.equinor.com/news/ev/invests-fusion-energy [accessed July 21st, 2022].

ESNII. 2021. *ESNII vision paper*. European Sustainable Nuclear Industrial Initiative. Available at https://snetp.eu/wp-content/uploads/2021/07/ESNII-Vision.pdf [accessed July 8th, 2022].

Fernández-Arias, P., D. Vergara, and J. A. Orosa. 2020. A global review of PWR nuclear power plants. *Applied Sciences* 10: 4434.

GIF. 2020. *Annual report 2020*. Gen IV International Forum. Available at www.gen-4.org/gif/jcms/c_178286/gif-2020-annual-report [accessed July 4th, 2022].

Haas, R., S. Thomas, and A. Ajanovic. 2019. The historical development of the costs of nuclear power. In Haas, R., L. Metz, and A. Ajanovic (eds.). *The technological and economic future of nuclear power*. Vienna: Springer.

Hassanein, A., and V. Sizyuk. 2021. Potential design problems for ITER fusion device. *Scientific Reports* 11(1): 1–11.

Ho, M., E. Obbard, P. A. Burr, and G. Yeoh. 2019. A review on the development of nuclear power reactors. *Energy Procedia* 160: 459–466.

IAEA. 2021a. *Nuclear share of electricity generation in 2021*. International Atomic Energy Agency. Available at https://pris.iaea.org/PRIS/WorldStatistics/NuclearShareofElectricityGeneration.aspx [accessed June 4th, 2022].

IAEA. 2021b. *What is fusion, and why is it so difficult to achieve?* Available at www.iaea.org/bulletin/what-is-fusion-and-why-is-it-so-difficult-to-achieve [accessed June 21st, 2022].

IAEA. 2021c. *What is nuclear energy? The science of nuclear power*. International Atomic Energy Agency. Available at www.iaea.org/newscenter/news/what-is-nuclear-energy-the-science-of-nuclear-power [accessed June 1st, 2022].

IAEA. 2022. *Fast reactors*. International Atomic Energy Agency. Available at www.iaea.org/topics/fast-reactors [accessed June 4th, 2022].

IEA. 2019. *Nuclear power in a clean energy system*. International Energy Agency, May. Available at https://iea.blob.core.windows.net/assets/ad5a93ce-3a7f-461d-a441-8a05b7601887/Nuclear_Power_in_a_Clean_Energy_System.pdf [accessed June 1st, 2022].

INL. 2022. *Characteristics of U.S. energy production using nuclear fission*. Idaho National Laboratory. Available at https://inldigitallibrary.inl.gov/sites/sti/sti/Sort_54709.pdf [accessed July 20th, 2022].

ITER. 2022a. *Building ITER* Available at www.iter.org/construction/construction [accessed June 21st, 2022].

ITER. 2022b. *ITER worksite—April 2022 VR Tour*. Available at https://static.iter.org/com/360/2022-04/index.html [accessed June 21st, 2022].

ITER. 2022c. *What is ITER?* Available at www.iter.org/proj/inafewlines [accessed June 21st, 2022].

Jewell, J., M. Vetier, and D. Garcia-Cabrera. 2019. The international technological nuclear cooperation landscape: A new dataset and network analysis. *Energy Policy* 128: 838–852.

Kaijser, A., and P. Högselius. 2019. Under the Damocles Sword: Managing Swedish energy dependence in the twentieth century. *Energy Policy* 126: 157–164.

Kembleton, R. 2019. Nuclear fusion: What of the future? *Managing Global Warming* 199–220.

Knapp, V., and D. Pevec. 2018. Promises and limitations of nuclear fission energy in combating climate change. *Energy Policy* 120: 94–99.

Maisonnier, D. 2018. RAMI: The main challenge of fusion nuclear technologies. *Fusion Engineering and Design* 136(Part B, November): 1202–1208.

Mignacca, B., and G. Locatelli. 2020. Economics and finance of Small Modular Reactors: A systematic review and research agenda. *Renewable and Sustainable Energy Reviews* 118: 109519.

Nian, V. 2018. Technology perspectives from 1950 to 2100 and policy implications for the global nuclear power industry. *Progress in Nuclear Energy* 105: 83–98.

OIES. 2018. *The rise of renewables and energy transition: What adaptation strategy for oil companies and oil-exporting countries?* The Oxford Institute for Energy Studies. Available at https://ora.ox.ac.uk/catalog/uuid:c4bee7fd-305f-4ec9-a681-6f46fb6ec07e/download_file?file_format=application%2Fpdf&safe_filename=The-rise-of-renewables-and-energy-transition-what-adaptation-strategy-for-oil-companies-and-oil-exporting-countries-MEP-19.pdf [accessed July 21st, 2022].

Osthus, D., S. A. Vander Wiel, N. M. Hoffman, and F. J. Wysocki. 2019. Prediction uncertainties beyond the range of experience: A case study in inertial confinement fusion implosion experiments. *SIAM/ASA Journal on Uncertainty Quantification* 7(2): 604–633.

Peng, H.-M., Wang, X.-K., Wang, T.-L., Liu, Y.-H., and Wang, J.-Q. 2020. A multi-criteria decision support framework for inland nuclear power plant site selection under Z-information: A case study in Hunan Province of China. *Mathematics* 8: 252.

Pioro, I., and R. Duffey. 2019. Current and future nuclear power reactors and plants. *Managing Global Warming* 117–197.

Prăvălie, R., and G. Bandoc. 2018. Nuclear energy: Between global electricity demand, worldwide decarbonisation imperativeness, and planetary environmental implications. *Journal of Environmental Management* 209: 81–92.

Qin, S., Q. Wang, and X. Chen. 2020. Application of virtual reality technology in nuclear device design and research. *Fusion Engineering and Design* 161: 111906.

Reuters. 2020. *Oil major Chevron invests in nuclear fusion startup Zap Energy*. Available at www.reuters.com/article/us-chevron-investment-nuclear-idUSKCN25831E [accessed July 18th, 2022].

Saidi, K., and A. Omri. 2020. The impact of renewable energy on carbon emissions and economic growth in 15 major renewable energy-consuming countries. *Environmental Research* 186: 109567.

Sadekin, S., S. Zaman, M. Mahfuz, and R. Sarkar. 2019. Nuclear power as foundation of a clean energy future: A review. *Energy Procedia* 160: 513–518.

Şahin, S., and H. M. Şahin. 2021. Generation-IV reactors and nuclear hydrogen production. *International Journal of Hydrogen Energy* 46(57, August): 28936–28948.

The Royal Society. 2020. *Nuclear cogeneration: Civil nuclear energy in a low-carbon future – policy briefing*. Available at https://royalsociety.org/-/media/policy/projects/nuclear-cogeneration/2020-10-7-nuclear-cogeneration-policy-briefing.pdf [accessed July 21st, 2022].

United States Department of Energy. 2021. *NUCLEAR 101: How does a nuclear reactor work?* Available at www.energy.gov/ne/articles/nuclear-101-how-does-nuclear-reactor-work [accessed June 18th, 2022].

United States Department of Energy. 2022a. *DOE explains . . . nuclear fusion reactions*. Available at www.energy.gov/science/doe-explainsnuclear-fusion-reactions [accessed June 21st, 2022].

United States Department of Energy. 2022b. *DOE explains . . . plasma confinement*. Available at www.energy.gov/science/doe-explainsplasma-confinement [accessed June 21st, 2022].

United States Department of Energy. 2022c. *DOE explains . . . stellarators*. Available at www.energy.gov/science/doe-explainsstellarators [accessed June 21st, 2022].

United States Department of Energy. 2022e. *Nuclear*. Available at www.energy.gov/science-innovation/energy-sources/nuclear [accessed June 1st, 2022].

United States Energy Information Administration. 2022a. *Nuclear explained*. Available at www.eia.gov/energyexplained/nuclear/ [accessed June 21st, 2022].

United States Energy Information Administration. 2022b. *Nuclear explained: The nuclear fuel cycle*. Available at www.eia.gov/energyexplained/nuclear/the-nuclear-fuel-cycle.php [accessed June 18th, 2022].

United States Nuclear Regulatory Commission. 2019. *Backgrounder on radioactive waste*. Available at www.nrc.gov/reading-rm/doc-collections/fact-sheets/radwaste.html [accessed June 19th, 2022].

Wang, Q., R. Li, and G. He. 2018. Research status of nuclear power: A review. *Renewable and Sustainable Energy Reviews* 90: 90–96.

WANO. 2022. *Roadmap to Operational Readiness: A new entrant's guide to preparing for the challenge of safe and efficient nuclear power plant operation*. World Association of Nuclear Operators. Available at www.wano.info/getmedia/346d2b79-2c11-4b09-962f-840ef279e6d0/R2OR-Revision-0-Final-for-Publication-September-2020.pdf.aspx [accessed June 28th, 2022].

Wealer, B., S. Bauer, C. v. Hirschhausen, C. Kemfert, and L. Göke. 2021. Investing into third generation nuclear power plants—Review of recent trends and analysis of future investments using Monte Carlo Simulation. *Renewable and Sustainable Energy Reviews* 143: 110836.

Wilding, P. R., N. R. Murray, and M. J. Memmott. 2019. The use of multi-objective optimization to improve the design process of nuclear power plant systems. *Annals of Nuclear Energy* 107079.

World Nuclear Association. 2020. *Generation IV nuclear reactors*. Available at https://world-nuclear.org/information-library/nuclear-fuel-cycle/nuclear-power-reactors/generation-iv-nuclear-reactors.aspx [accessed July 6th, 2022].

World Nuclear Association. 2021a. *Advanced nuclear power reactors*. Available at https://world-nuclear.org/information-library/nuclear-fuel-cycle/nuclear-power-reactors/advanced-nuclear-power-reactors.aspx [accessed July 6th, 2022].

World Nuclear Association. 2021b. *Fast neutron reactors*. Available at https://world-nuclear.org/information-library/current-and-future-generation/fast-neutron-reactors.aspx [accessed July 6th, 2022].

World Nuclear Association. 2022a. *How is uranium made into nuclear fuel?* Available at www.world-nuclear.org/nuclear-essentials/how-is-uranium-made-into-nuclear-fuel.aspx [accessed June 18th, 2022].

World Nuclear Association. 2022b. *Nuclear power reactors*. Available at https://world-nuclear.org/information-library/nuclear-fuel-cycle/nuclear-power-reactors/nuclear-power-reactors.aspx [accessed June 28th, 2022].

Wu, Y., F. Liu, Y. Huang, C. Xu, B. Zhang, Y. Ke, and W. Jia. 2020. A two-stage decision framework for inland nuclear power plant site selection based on GIS and type-2 fuzzy PROMETHEE II: Case study in China. *Energy Science & Engineering* 8: 1941–1961.

14 Carbon Capture and Storage (CCS)

Jorge Saldaña, Mesut Yurukcu, Narendra Boppana, Sepehr Arbabi, Jim Henry, and Sebahattin Ziyanak

14.1 INTRODUCTION

Confronted with the environmental challenges brought by human-caused climate change, researchers and engineers have had to think not only of ways to decarbonize activities and processes around oil and gas extraction operations but also for existing energy projects to be stripped of some of the carbon that they generate. In this sense, and while we rely on cleaner renewable energy sources like solar and wind to power the world more consistently, a promising mitigation strategy for cutting carbon emissions is carbon capture and storage (CCS).

Also known as carbon sequestration or as carbon capture, utilization, and storage (CCUS), this method consists of the retention of carbon dioxide generated at power plants before reaching the atmosphere, which is then injected underground into reservoirs like deep saline aquifers for long-term storage (Boot-Handford et al., 2014; IPCC, 2005; Szulczewski et al., 2012). CCS methods have the potential to decarbonize many industries, thus helping countries meet climate change targets (Bui et al., 2018). Figure 14.1 illustrates the concept of CCS in its most basic form and all the dynamics involved in recovering and storing CO_2 from electricity plants.

In the processes where different energy sources are brought into power plants for electricity generation or in operations from chemical plants or refineries, a certain amount of carbon dioxide is produced, which in normal circumstances would go into the atmosphere, contributing to the warming of the climate. However, CCS techniques use different approaches to limit the quantity of CO_2 released by these processes, repurposing it for other uses like mineral carbonation or industrial needs or delivering it for storage in onshore or offshore underground sites (Karimi and Khalilpour, 2015).

The procedure has three distinct stages. The first is the separation of carbon dioxide from a gas stream before it can be released entirely into the atmosphere, where it can contribute to warmer temperatures. Then, the separated CO_2 is compressed to ready it for transportation, the second stage. This process can occur using pipelines, ships, trucks, or trains, which move it to injection facilities, where the third phase happens. The gas is injected into geological formations for long-term storage (Koornneef et al., 2008).

This chapter reviews the different carbon capture and storage techniques engineers and companies use to mitigate the carbon emissions generated by their energy processes. Moreover, we will review the various rationales for screening, selection, and the economics of CCS methods, as well as the evolution of these techniques over the years. Additionally, challenges and opportunities are analyzed, and the role of the oil and gas industry is explored, as these methods can help the industry with an enhanced oil recovery (EOR) of depleted reservoirs.

As we have advanced throughout this work, a successful energy transition process that translates into greater reliance on renewable energy sources can only be achieved when different actors across industries come together, recognize the challenges ahead, perform the necessary academic research, and adhere to the pledges advocated. Carbon capture and storage is a promising technique to mitigate the extent of the excessive release of carbon dioxide to the atmosphere from current

 DOI: 10.1201/9781003282273-14

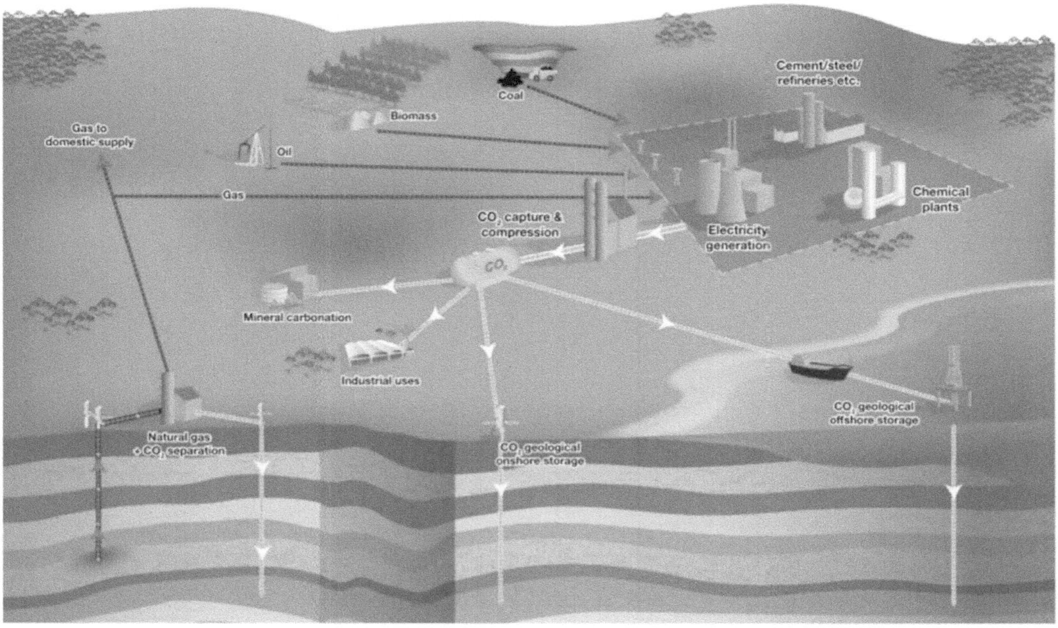

FIGURE 14.1 Overall schematic of the carbon capture and storage (CCS) concept.

Source: Karimi and Khalilpour, 2015].

energy processes and is another piece of the puzzle in the search for a cleaner environment to leave for future generations.

14.2 SCREENING, SELECTION, DESIGN, AND ECONOMICS OF CCS TECHNOLOGIES

Techniques for carbon capture and storage are varied and can be applied to different settings. Thus, for example, a cement production plant will go under a different CCS process than a power plant. Likewise, CO_2 storage projects are substantially costly, requiring private or government subsidies that allocate resources for projects of this type. With these constraints, some countries will feature more CCS projects than others, with prohibitive costs. This section will review the main criteria for screening, selection, and design of CCS technologies and the economics behind site selection, adoption, and execution of these procedures.

With high costs like those from capital and installation (Bui et al., 2018), not all processes can use CCS methods, and all industries will not have the same potential for success after using a CCS technology. Therefore, it becomes crucial to screen for the most favorable criteria to determine which sites, and at what time, will be the most likely to be selected for a carbon capture and storage method. Additionally, with the optimization and professionalization of these procedures, screening workflows can now be performed even for materials based on their absorption capabilities (Farmahini et al., 2021).

Figure 14.2 shows the world map with the current CCS projects in the industry, with one (the Lost Cabin Gas Plant in Wyoming) being suspended. Most of the ventures are in North America, where the United States is leading in applying CCS methods. Meanwhile, Figure 14.3 shows the growth of CCS projects over time, with 2021 featuring a sharp increase in early and announced developments and projects in more advanced stages. Still, according to the International Energy

FIGURE 14.2 Operating CO_2 capture projects in the industry sector.

Source: IEA (2021b).

Agency, the current pace of growth is not on track to achieve the Net Zero Emissions by 2050 Scenario (IEA, 2021b).

In the United States, the Department of Energy (DOE) is the leading agency in charge of deploying CCS methods and site screening for carbon dioxide storage sites. Their phase of analysis of potential sites starts with analyzing copious amounts of datasets in a stage called project analysis. Plans developed in this stage consist of six elements: (1) scope, (2) CO_2 strategy, (3) evaluation criteria, (4) resources, (5) schedule, and (6) risk assessment (see Table 14.1 for a summary). In these stages, the projects are envisioned and delineated with the help of data and analyses being constantly updated, similarly to how oil and gas reservoirs are explored and developed for future use (Rodosta et al., 2011).

Then, the following phase, that of exploration, consists of three sets of independent evaluations: site screening, site selection, and initial characterization. Figure 14.4 shows the site screening workflow, including technical, geographical, and social analyses. First, studies about injection formation, depth, confining, and storage capabilities are performed. Second, regional limitations are assessed, so protected and sensitive sites, population centers, and pipeline rights-of-way (ROWs) are left out of the selection process. Finally, sites are screened for their demographic trends, the public perception of CCS technologies, and the land's previous use (NETL, 2017; Rodosta et al., 2011).

Selecting a site to store CO_2 taken from polluting procedures has become a sub-discipline that requires technical knowledge that goes from geological considerations to land use and public opinion issues. These aspects make CCS technologies prohibitive for some countries that do not have resources allocated for these purposes, so CCS methods may not propagate as fast as expected by international organizations unless there is some participation by private actors. In this sense, oil and gas companies can contribute actively to mitigating their carbon footprints through investment or innovation.

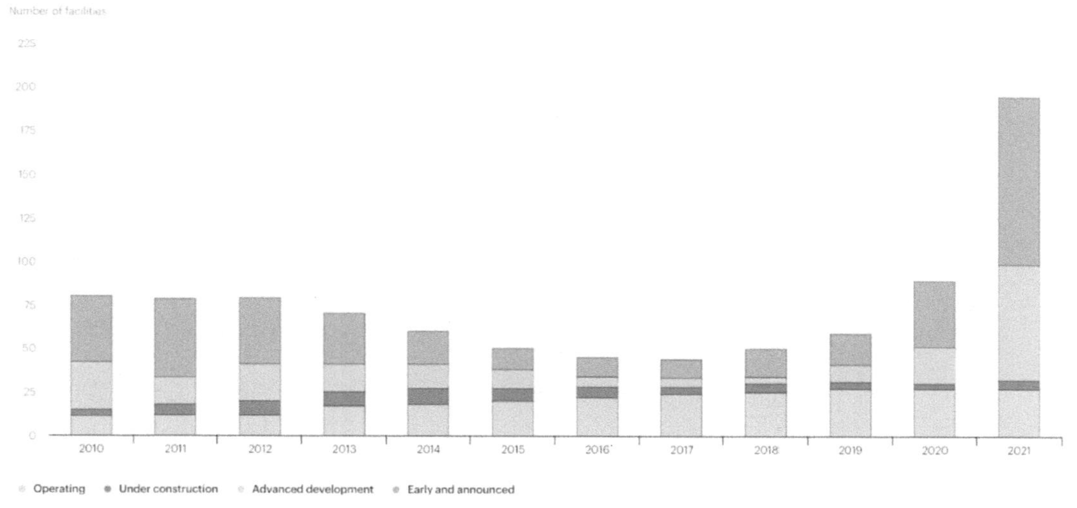

FIGURE 14.3 Global pipeline of commercial CCUS facilities operation and development, 2020–2021.
Source: IEA (2021a).

TABLE 14.1
Guidelines for Project Definition

Project Definition/ Management	Project Analysis	Scope	Define overall project and describe processes for site screening, site selection, and site characterization. Identify project objectives and criteria to evaluate project success or failure.
		Co_2 strategy	Develop a CO_2 management strategy that identifies CO_2 sources, volumes, rates of delivery, and target injection rates. Assess feasibility of several implementation options, with associated risks and mitigation options.
		Evaluation criteria	Establish criteria to be used for qualifying and ranking sites during site screening, site selection, and site characterization. Criteria should include technical, economic, and social parameters.
		Resources	Identify personnel, equipment, and funding resources necessary to complete site screening, site selection, and site characterization processes. Identify necessary areas of expertise, financial thresholds, potential contingencies, and other resource risks.
		Schedule	Develop realistic project schedule for site screening, site selection, and site characterization. Include milestones and contingency plans to mitigate schedule delays.
		Risk assessment	Conduct risk assessment to identify scenarios that could prevent the project from achieving commerciality. Define mitigation options, and develop implementation plan with go/no-go decision gates.

Source: NETL (2017).

Regarding the latter point, another type of screening is that performed on materials for a better carbon capture process. Depending on the type of CCS method, different materials can be utilized to retain the CO_2 emanating from additional energy-generating operations. For example, work has been done to assess the potential for tropical microalgae to serve as efficient CO_2 removers through photosynthesis (Rinanti et al., 2013), as well as the prospects of zeolite and zeolitic imidazolate structures

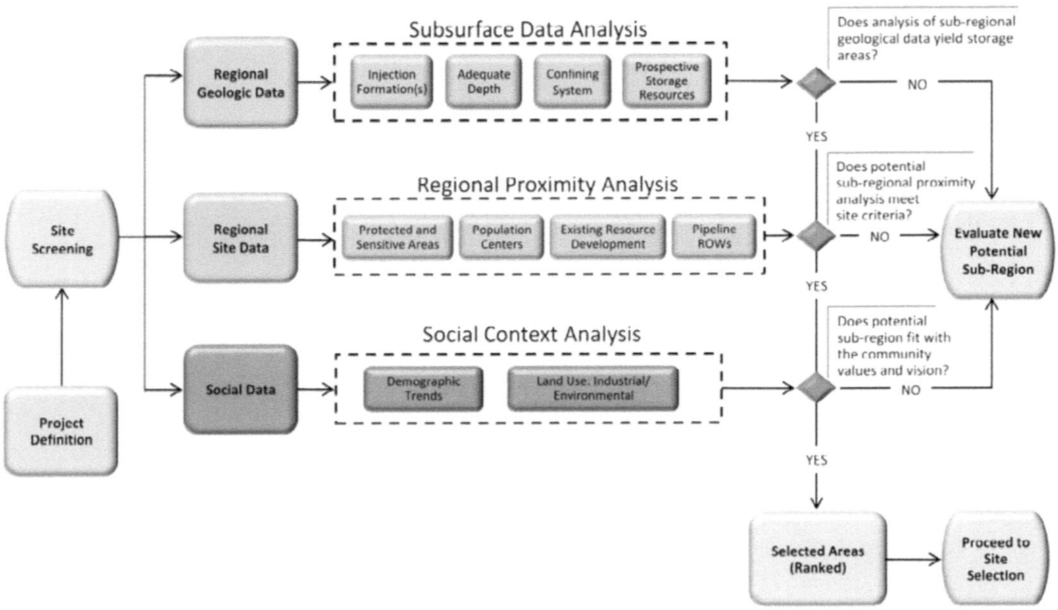

FIGURE 14.4 Process flowchart for site screening evaluation on potential sub-regions.

Source: Rodosta et al. (2011).

TABLE 14.2
Typical Costs of CO_2 Capture for Industrial Plants

Facility	€/MgCO_2		€/MgCO_2
Cement plants	28	Refineries	29–42
Iron and steel plants	29	Hydrogen (flue gas)	36
Ammonia plants (flue gas)	36	Hydrogen (pure CO_2)	3
Ammonia plants (pure CO_2)	3	Petrochemical plants	32–36

Source: Balat and Öz (2007).

to better separate CO_2 from flue gasses emanating from varying combustion procedures (Lin et al., 2012). However, this type of screening will be discussed in more detail in the following section.

Investment in CCS technologies has economic considerations, where market forces are another variable in this dynamic. For example, the cost of removing 1 t of carbon dioxide fluctuates according to technological developments, widespread adoption of CCS procedures, and competition between actors that render associated services. Anderson and Newell (2004) established that the cost of eliminating 1 t of CO_2 was between US$200 and US$250. Conversely, McKinsey & Company (2008) advances that if trends accelerate and industries adapt to steep learning curves, these costs could come down to €30–45 by 2030.

We need to break down the costs for each component to determine the cost of any given CCS method. Thus, estimations need to be made separately for capturing CO_2, the transport of carbon, and its storage in a long-term facility. Table 14.2 shows the different costs of capturing 1 t of CO_2 depending on the type of plant. So refineries and both ammonia and hydrogen plants where flue

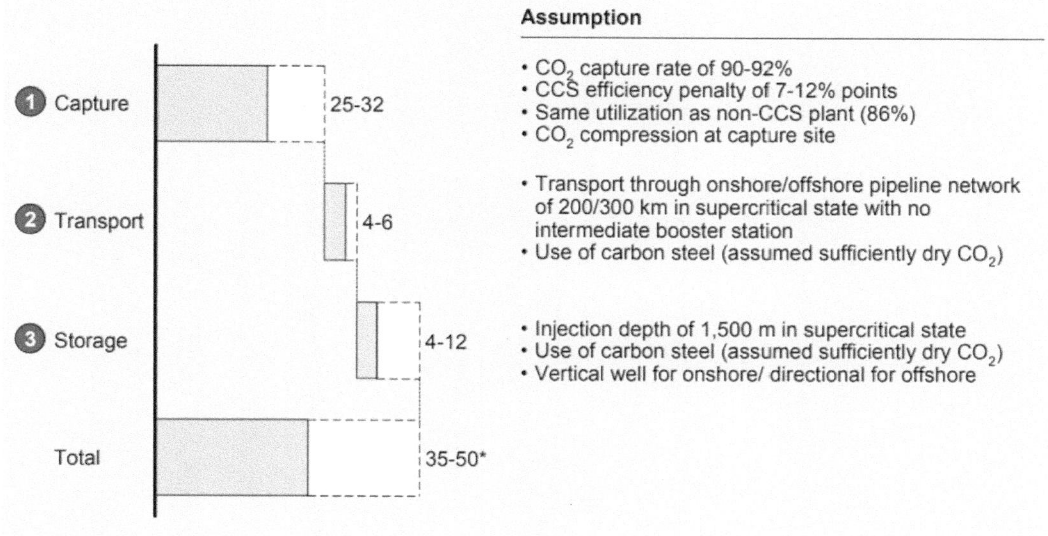

FIGURE 14.5 Total cost of early CCS commercial projects (in €/ton CO_2 abated; ranges include on- and offshore facilities).

Source: McKinsey & Company (2008).

gasses are an offshoot are the places where capturing 1 t of carbon is most expensive, due to the techniques needed to achieve this.

Then, the costs of carbon transportation are subject to economies-of-scale, or "learning rate," considerations (Rai et al., 2010). Thus, there comes a moment when widespread use of these technologies brings overall costs down due to repeated uses of the same already-built pipeline or the same ship routes and the professionalization of the crew in charge of diverse operations and maintenance (O&M) practices (Lau et al., 2021). Similarly, pipeline construction costs are not linear, so constructing much longer pipelines is not much more expensive after a point than building a shorter standard route to transport CO_2 from a refinery to a long-term storage site (Balat and Öz, 2007).

Finally, storage costs for the injection of CO_2 are mainly from drilling and operational costs. It is important to note here that onshore storage is generally less expensive than offshore (Hendricks et al., 2004), while storage capacity is significantly increased when residual and solubility trapping capabilities are better (Szulczewski et al., 2012). Figure 14.5 shows the estimated costs for each of the three stages in CCS projects, with capture being the most expensive step, around five times as expensive as transport and storage. Moreover, carbon transport can be cheaper for the stages, depending on factors like the distance of transportation and the materials for the pipelines. Meanwhile, if the storage phase is done onshore, the injection site is not very deep, and materials are optimized, it can be the least-expensive stage for a particular project.

Meanwhile, the dimension of the cost does not only include monetary expenditures, especially for technologies that require the intervention of sites and the participation of governments through different types of subsidies, which, in the end, means using taxpayers' money to fund local and national policies. Because of this, several social and political considerations must also be assessed. Table 14.3 lists factors relevant to low-carbon initiatives, where CCS belongs. Besides the monetary and learning rate costs, the core potential of the technology and its pillars of support are two other important sets of factors.

TABLE 14.3
The Political Economy of Government Technological Support: A Checklist of Factors to Be Considered When Allocating Government Support for Low-Carbon-Emission Energy Technologies

Important Factors	Key Questions	Considerations
Core Potential		
Energy provision potential	How much energy can this provide?	An option that meets a large share of energy needs is generally desirable. Technologies that come in large units have large capital costs but fit centralized generation grids well.
Emission mitigation potential	How much emission reduction could this technology deliver?	Dependent on characteristics and scale of deployment. Higher life cycle reductions and larger scale of application are preferable.
Timing	On what time scale would it become commercially deployable?	The sooner, the better. But the potential for large gains even on a long time horizon can be attractive.
Costs		
Present cost	What is the current cost of the technology?	Must be seen in relation to current capacity investment (experience curve stage).
Learning rate	How fast can costs come down?	Higher learning rates are desirable, but rates are uncertain.
Learning potential co-benefits and concerns	How far might costs eventually come down?	Higher learning potential is desirable but uncertain.
Development potential	To what extent can it generate economic benefits: jobs, markets, profits, taxes?	Investments are expected to generate economic benefits as well as energy/mitigation.
Energy security	Does it enhance energy security?	Relates to import dependence, diversity of supply, and robustness of infrastructure.
Regional implications	Is it significant for regional energy, mitigation, or development interests?	Regional interests must be accommodated.
Environmental perspective	What are the associated environmental benefits/costs and opportunities/risks?	Fossil fuel CCS, nuclear, hydro, and new renewables have different cost/benefit profiles.
Foreign markets and international competitiveness	What is the potential to develop export markets? How appealing would the technology be to other countries?	Technologies that appeal to other countries with growing energy and mitigation needs may represent important future markets
Compatibility	To what extent is this option compatible with other favored technologies?	Some energy/GHG mitigation options do not "fit" well with others.
Pillars of Support		
Technical/industrial/resource foundations	Does it draw on existing strengths or potentials: resources, industries, innovation clusters?	The importance of exploiting existing knowledge/resource bases and relying on comparative advantages.

Co-funding of RD&D	To what extent will private interests and firms and/or international partners contribute to funding RD&D?	Collaborative engagement can reduce development costs, spread risk, and cover more technological options.
Public receptivity	To what extent is the public supportive/resistant to the technology?	Awareness of public and stakeholder groups; communications; familiarity, accidents.
System Perspectives		
Transition to carbon-emission-free energy system within half-century	How does the technology fit with transition pathways to a carbon-emission-free energy system?	The roles the technology might play; how its potential is distributed over time; potential interactions with other technologies; long-term visions; "backcasting."
Sustainable energy policy	How does it appear from the vantage of sustainable energy policy?	The provision of energy for sustainable development; economic/social and environmental dimensions; holistic assessment.

Source: Torvanger and Meadowcroft (2011).

In the case of CCS methods, this translates into the potential for carbon reduction to reach an environmentally beneficial threshold, given the level of funding and support from governments and private actors, provided that other technologies will eventually replace these polluting processes and industries. Additionally, the population should be made aware of how CCS technologies affect their daily lives and how these methods can benefit their quality of life. Thus, if CCS projects are to become a vital part of the toolkit to mitigate the carbon release of many industrial and energy processes, it is essential that all these moving parts are effectively in motion without missing too many beats.

When one of these branches falls behind, we run the risk of augmented skepticism toward new technologies, and in some cases, its long-term effects have not been documented thoroughly. Hence, the process must target sustainability (Hardisty et al., 2011). Table 14.4 shows a list of potential barriers to the implementation of CCS methods, following keywords mentioned in the literature. As we can see, the biggest issue is the lack of social awareness about these techniques, related directly to the second issue, a lack of a solid communication strategy about these methods' reach and potential benefits.

Similarly, other important social considerations are encapsulated in the *not in my backyard* (or NIMBY) notion, which states that neighbors will prefer initiatives like CCS that present potential benefits to the population, but they would rather not have them implemented close to where they live. A similar notion appears when considering the lack of long-term effects of the technology and the potential for a negative public perception that could be fed by these failures in communication. Thus, improving public perception is one of the main priorities in reducing future costs, implying monetary expenses if public relations do not adequately work with these groundbreaking techniques.

As we note, the decisions for screening and selecting sites require the joint work of many moving parts in unison, where the social and policymaking spheres are closely connected to the scientific and technological domains. In this sense, when researchers make discoveries about new technologies to capture carbon from polluting industries, policymakers need to assess any potential changes in regulations while also informing the public about future projects. Here, the benefits provided by these endeavors and the potential costs some neighbors will have to pay must be reported transparently.

Different projects have different targets for the retention of CO_2, and these distinctions also translate into additional costs for other actors. Therefore, the economy of CCS technologies is highly variable and dependent on variables like geographical context, social acceptance and awareness, policies implemented, the extent of the project, and even materials used. The following section delves more profoundly into the different types of carbon capture and storage technologies

TABLE 14.4
Key Barriers to CCS Technology Implementation, According to Saliency in the Academic Literature

Barriers for CCS implementation	Number of Studies
Lack of public knowledge about CCS, misconceptions	95
Lack of or poor communication strategy	82
Competition between alternative technologies	70
Lack of long-term policy of CCS implementation	57
Controversial economic efficiency, capital intensity, weak market-based mechanism	55
Not enough studied the long-term effects of the technology	52
Lack of trust in some stakeholders	54
NIMBY reaction	38
Site selection and project design without taking into account the specific of locals	34
Appearance of protest potential due to negative public perception	25
Increase in price of energy	22

Source: Tcvetkov et al. (2019).

while introducing the oil and gas industry's role in developing these projects on a larger scale. Additionally, best practices and ways forward are analyzed in detail.

14.3 TYPES OF CCS TECHNOLOGIES, PRINCIPLES, AND BEST PRACTICES

Carbon capture and storage techniques are varied, depending on the process they are trying to aid. Therefore, different procedures, such as cement manufacture, iron, steelmaking, and natural gas treatment, can be fitted with other CO_2 capture technologies (Boot-Handford et al., 2014). In this section, we will review the types of carbon capture and storage methods that feature predominantly in the literature and in which situations are best applied. A guide with principles and best practices is later provided for a better understanding of the use of these methods and their optimal integration in society.

In general terms, carbon capture and storage is done to avoid releasing excessive amounts of carbon dioxide into the atmosphere, which can contribute to climate warming. Recent technological advancements have led to innovations in these techniques and preemptive screening of materials to be used in these processes so that capture is optimized (see Figure 14.6). With these simulations done before any resource is invested in building plants or transporting carbon to storage locations, we can know which materials will provide a better capture, translating into an *ex ante* optimized procedure that helps capture more CO_2.

Likewise, there are currently various methods of separating and capturing carbon dioxide, the most common being absorption, adsorption, cryogenic separation, membrane-based, and microbial/algal systems (Pires et al., 2011; Wilberforce et al., 2018). As we will review shortly, these methods differ from the types of applications of CCS most used in the industrial sector today. Therefore, these same technologies can be applied differently, depending on the moment of the "CO_2 chain" researchers face and the process engineers are trying to intervene. Hence, carbon can be captured before or after the combustion process to generate energy.

Before we move on to those CCS applications, however, Table 14.5 and Figure 14.7 show the main features of some of these separation technologies. We will review them in more depth, but for now it is good to have a general idea of how these technologies are closely related to the CCS applications, which we will analyze shortly. For example, membrane-based capture processes can be used in pre-combustion applications, but due to hydrocarbon loss being very high for procedures

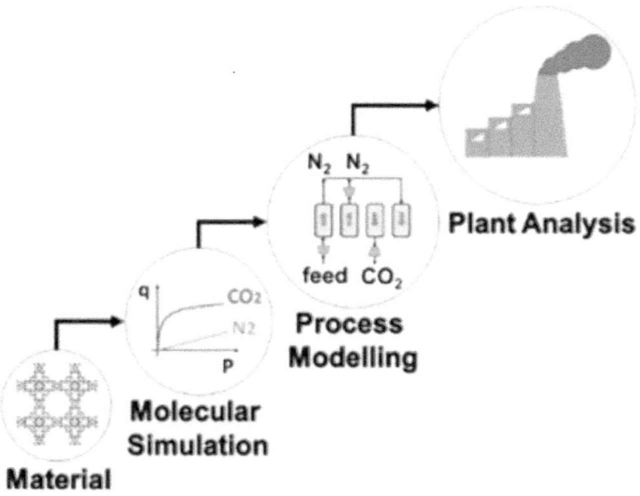

FIGURE 14.6 Performance-based screening of porous materials for carbon capture.

Source: Farmahini et al. (2021).

using membrane techniques, cryogenic methods appear to be a better-suited technique at that stage of combustion. Thus, optimization is critical to taking advantage of these mitigation methods.

Figure 14.8 summarizes this series of processes and technologies through a hierarchical lens. As we can see, the CO_2 chain starts at the level of energy sources, where we know the industrial processes that generate a large amount of carbon dioxide. As a response, carbon capture approaches or applications are developed, where we see the post-, pre-, and oxy-combustion methods. Meanwhile, these applications are performed through different processes, including absorption, adsorption,

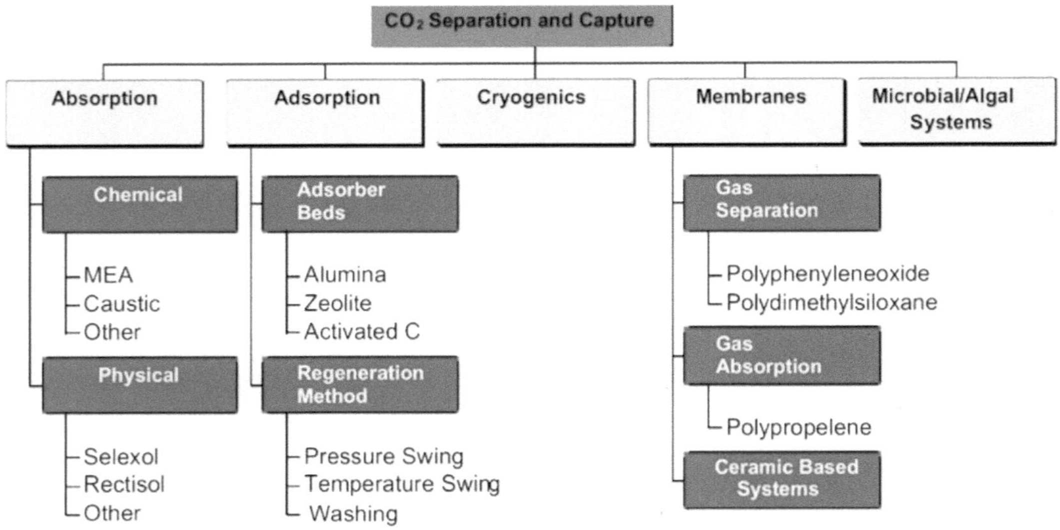

FIGURE 14.7 Various technologies used in carbon capture and storage.

Source: Wilberforce et al. (2018).

TABLE 14.5
Features of Technologies for Carbon Dioxide Capture

Technological Advancement	Merit	Obstacle and Literature Gap
Absorption via chemical means	It is considered a matured kind of technology for post-combustion. It is also suitable for power plants fired by carbon. The efficiency for capturing the carbon dioxide is very high, and losses with respect to hydrocarbons are low.	The capture ratio and heat ratio are very high. For power plants operated using coal, there is high capture energy penalty of approximately 20–30%. Challenges relating to corrosion are also a major obstacle. Solvent challenge relating to stability, reduction in capture ratio, heat ratio, and stripping temperatures to facilitate the usage of waste heat.
Physical absorption	Has high capture efficiency. Very suitable for power plants fired by coal. The capture efficiency is very high, but the heat ratio is low for regeneration. This is also considered a matured technology for processing of natural gas and post-combustion.	The selectivity is low, with high hydrocarbon losses.

(Continued)

TABLE 14.5 (*Continued*)
Features of Technologies for Carbon Dioxide Capture

Technological Advancement	Merit	Obstacle and Literature Gap
Membrane penetration	Suitable for natural gas processing on a large scale. Does not require any chemicals.	Confining of natural gas is needed. There is high hydrocarbon losses.
Pre-combustion	Appropriate for power plants fired by coal. Cost-effective, suitable for hydrogen production in commercial quantities. Has highly efficient, approximately 10–15%, low capture energy penalty.	It is very complex; requires new materials for high carbon dioxide capture at high temperature, huge capital expenses; still undergoing developmental processes. The experience for large-scale hydrogen-fired power plant is still inadequate.
Cryogenic distillation	This is also a matured technology for natural gas with high carbon dioxide composition, high selectivity, and little hydrocarbon losses. There is no need for compression, as the carbon dioxide is obtained in liquid state; hence, transportation is easy and simple. Suitable for high carbon dioxide composition.	Avoiding the carbon dioxide freeze-out is very necessary, and also refrigeration energy penalties.

Source: Wilberforce et al. (2018).

membrane-based, and cryogenic separation. And lastly, the captured carbon is managed mainly in two ways: stored in geological sites or aquifers or utilized for other processes, like enhanced oil recovery (EOR) or carbonation (Olabi et al., 2022).

Currently, the three most used approaches for carbon separation are *post-combustion*, where carbon dioxide (CO_2) is separated from diatomic nitrogen (N_2); *pre-combustion*, where CO_2 is separated from hydrogen gas (H_2); and *oxy-fuel combustion*, where CO_2 is separated from water molecules (H_2O) (see Figure 14.9). Meanwhile, in natural gas (NG) processing, CO_2 is separated from methane (CH_4) and other more complex hydrocarbons (Fernandes Araújo and de Medeiros, 2017). Along with more established methods of scrubbing carbon via absorption or adsorption, more recent developments include membrane-based CO_2 separation (Ji and Zhao, 2017), which we will review in detail in the following pages.

As we can infer, *carbon capture*, *transport*, and *storage* are umbrella terms for different technologies for CO_2 capture applied at other times within an energy process. As a result of the many distinctions between methods, there is no one-size-fits-all alternative to mitigate the release of carbon dioxide into the atmosphere due to energy processes. Additionally, costs vary significantly between these procedures, so not all jurisdictions that have on their minds to apply these will be able to implement them, either due to prohibitive costs, due to public opposition, or because exploratory stages have deemed a project as impractical.

These clarifications are essential due to the complex nature of proposed mitigations. Despite global interest in implementing many of these solutions, we go back to a notion advanced in previous chapters that to execute some of these well-intentioned environmental strategies, a basic level of prosperity is required within the country that is about to invest in these technologies. Hence, impoverished countries seeking a way out of economic despair by burning more fossil fuels will not have the resources to deploy mitigation strategies, at least at the beginning. It is here where private actors and international organizations can carve a role through subsidies and independent projects, an aspect that we will review in more detail in subsequent sections of this chapter.

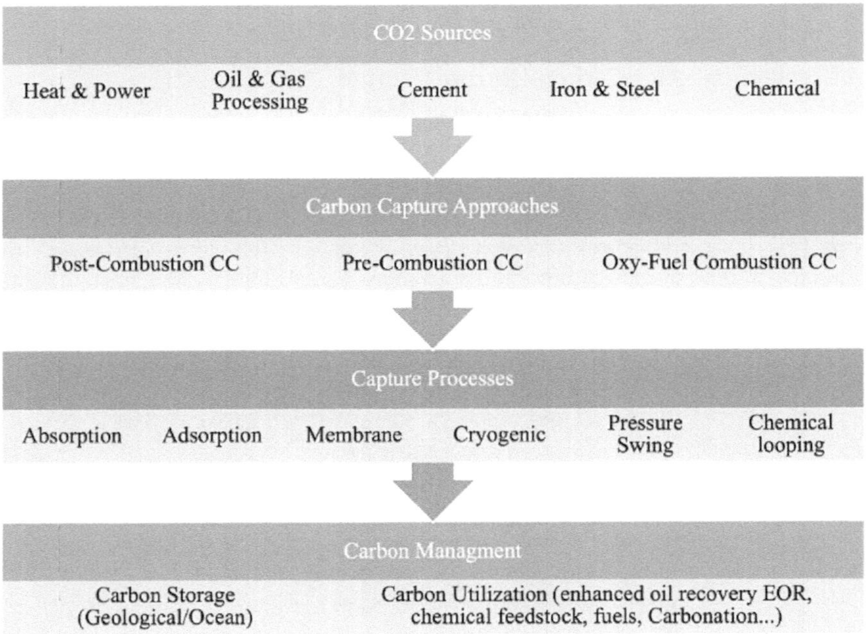

FIGURE 14.8 Carbon capture and utilization approaches.

Source: Olabi et al. (2022).

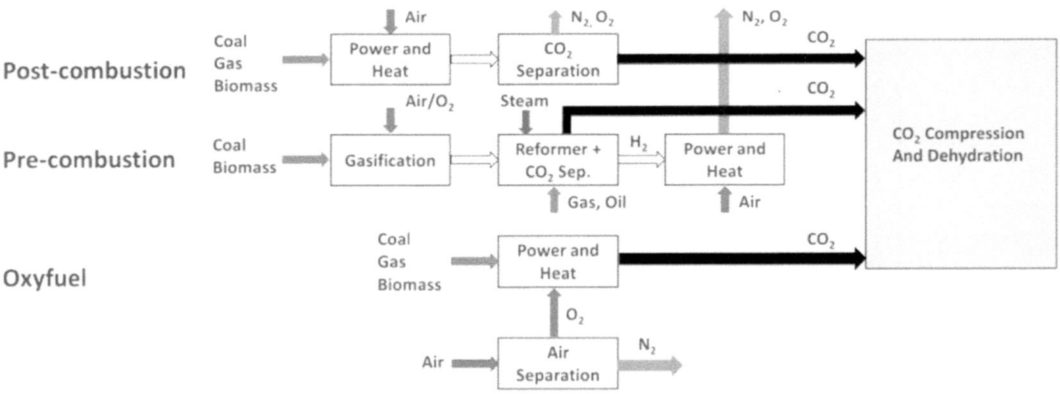

FIGURE 14.9 Different routes to carbon capture from power plants.

Source: Farmahini et al. (2021).

14.3.1 Post-Combustion

Post-combustion CCS techniques are the most basic type of CCS technology because they are performed after energy generation, mostly on flue gases of existing power plants (IPCC, 2005; RFF, 2020). This process allows for the continuation of the work of existing power plants which are needed to power the needs of millions of people, while new capacity is added to meet the demand for cleaner sources of energy (Office of Fossil Energy and Carbon Management, 2022). These methods use a liquid solvent, typically monoethanolamine (MEA), to capture the carbon in flue gas, around 3–15% of the composition of the emanating gas, which separates it from the predominant nitrogen in the air (Balat and Öz, 2007; IPCC, 2005).

Since this process occurs after energy production, post-combustion is the only method that can be retrofitted onto existing plants to mitigate their carbon emissions (Farmahini et al., 2021; Lau et al., 2021; Tan et al., 2016; Vega et al., 2020). Therefore, this is an excellent alternative for existing facilities, which can continue to function to most of their capacity, providing electricity to industries, businesses, and residences. In contrast, renewable energy continues to become the predominant energy source in the 21st century without having to experience significant power shortages. The process of a power plant with a post-combustion CCS method applied to it is illustrated in Figure 14.10.

Given that the Intergovernmental Panel on Climate Change has estimated that current post-combustion systems could capture up to 95% of the CO_2 produced by these processes, the current pace of carbon emissions under status quo conditions can be curbed significantly (IPCC, 2005). This, added to the push from governments and international organizations to achieve the energy transition from fossil fuels to renewables in the next couple of decades, places CCS methods as a helpful bridge technology to close the pollution gap between the present, with a high reliance on fossil fuels, and the future, where renewables are the to-go sources of energy.

Post-combustion methods are the most utilized decarbonization efforts, contributing to the technology's maturity through extensive research and various real-life applications at full-scale commercial plants. With that said, important limitations of this technique are its significant energy penalties for its amine scrubbing process, which can go up to 30% of power output loss, a large water consumption that could impact some locations significantly, and high capital and operational costs due to large equipment required to intervene in the flue gas (Theo et al., 2016).

To tackle problems like those encountered with significant energy penalties, Farmahini et al. (2021) found that a few combinations of elements and materials used in post-combustion techniques have superior productivity and energy consumption values than the industry benchmark, zeolite 13X. Likewise, their findings suggest that the role of machine learning (ML) techniques that are tailored to simulate and find the next best materials for capturing a more significant amount of CO_2 from energy processes is about to become more predominant. In conclusion, *in silico* (or using computers/computational simulations) analyses may be one promising way of cutting costs and making these technologies attainable to more countries worldwide.

14.3.2 Pre-Combustion

In turn, pre-combustion methods separate carbon dioxide from other fuel gases before the gas combustion is completed (Ji and Zhao, 2017). For example, coal ready for a gasification process is

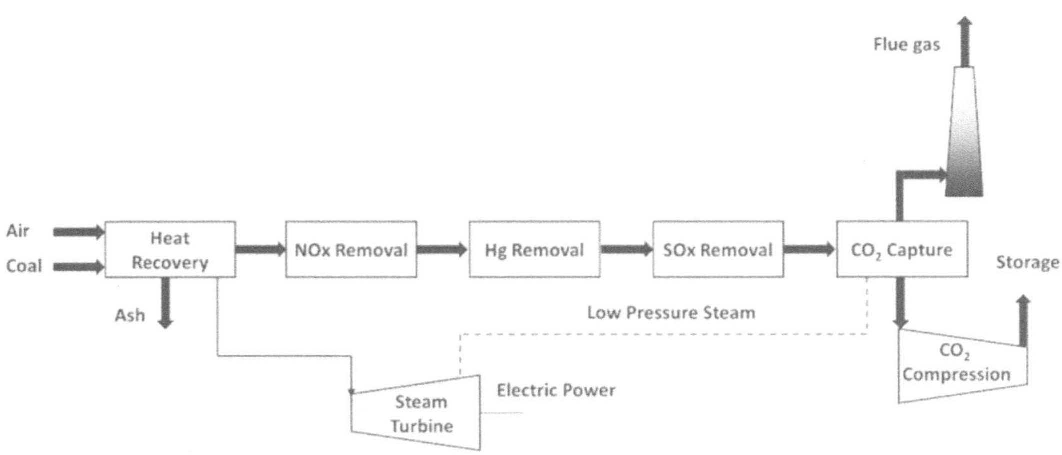

FIGURE 14.10 Schematic illustration of post-combustion CCS plant.

Source: Farmahini et al. (2021).

oxidized in steam and oxygen/air under high temperature and pressure to form synthesis gas, or syngas, a mixture of H_2 and CO_2 formed after a water–gas shift (WGS) reaction (see Figure 14.11). This carbon dioxide can be detached, separated, captured, and sent for storage, while the fuel rich in hydrogen gas can be safely combusted (Farmahini et al., 2021; Fernandes Araújo and de Medeiros, 2017; Office of Fossil Energy and Carbon Management, 2022b; Olabi et al., 2022; Raza et al., 2019).

For these processes, the carbon separation aspect is significantly less energy-intensive; thus, the penalties are not as steep as post-combustion strategies. Similarly, water usage is much lower than in the previous method, and the generation of excess quantities of hydrogen/syngas that can serve as an alternative fuel is a welcome advantage. Pre-combustion techniques still feature high-energy loss due to sorbent regeneration, while CO_2 and H_2 separation is costly and complex (Olabi et al., 2022; Theo et al., 2016; Wilberforce et al., 2018).

A method that is gaining more notoriety, pre-combustion carbon capture is still a distant second from post-combustion processes. As Table 14.6 shows, however, there is a promising avenue for the destination of most of the captured CO_2 using this approach, which is directly related to the oil and gas industry: enhanced oil recovery (EOR). To be explored later in this chapter, EOR methods are a tertiary method of oil recovery used to obtain more oil from reservoirs deemed depleted after conventional extraction techniques (Carter, 2011). Injecting captured CO_2 into these sites can help remove remnant oil from the surface, optimizing the reservoir.

Not only EOR methods can benefit from a more efficient CO_2 capture. If less carbon is released into the atmosphere, environmental decarbonization goals can be achieved and climate warming can be mitigated. The current literature has explored different ways to capture carbon, where gas–liquid absorption has had a predominant role. The work of Padurean et al. (2012) explores the capabilities of four different solvents to capture CO_2 in pre-combustion processes: a mixture of dimethyl ethers of polyethylene glycol (Selexol), methanol (Rectisol), N-methyl-2-pyrolydone (Purisol), and the chemical solvent methyl diethanolamine (MDEA).

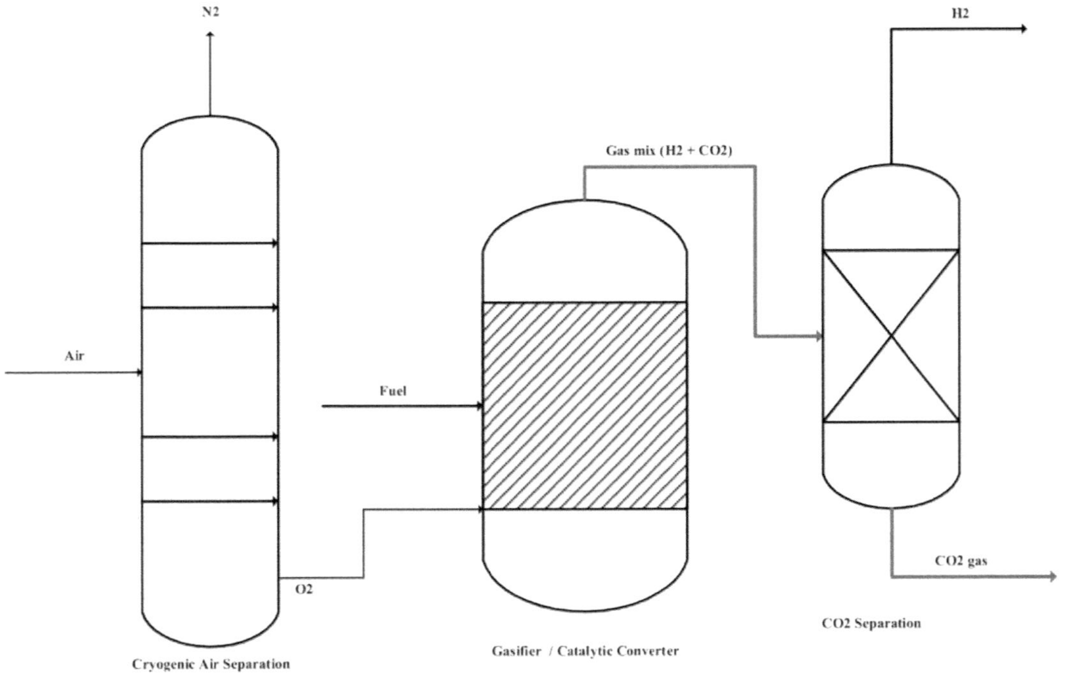

FIGURE 14.11 Pre-combustion carbon capture process.

Source: Olabi et al. (2022).

TABLE 14.6
Summary of Commercial Pre-Combustion Facilities Worldwide by 2016

Commercial Facility	Status	Capacity (TPD)	Capture Technology	Final CO_2 Fate
Pre-combustion capture plants				
Hydrogen Energy California project, a hydrogen/petroleum coke–fueled integrated gasification combined cycle (IGCC) power plant (California, USA)	Constructed in 2015 and operational by 2020	8,219	Rectisol AGR system	Onshore enhanced oil recovery (EOR) at Occidental's Elk Hills Oil Field
Texas Clean Energy Project, a coal-fired power plant (Texas, USA)	Operational by 2019	5,479–8,219	Rectisol AGR system	EOR at Bermian Basin
Kemper County IGCC, a pre-combustion IGCC plant (Mississippi, USA)	Operational by 2016	9,589	TRIG™ Technology	Onshore EOR
Killingholme Project, an IGCC power plant (North Lincolnshire, United Kingdom)	Operational by 2019	6,849	Selexol process	Storage at deep saline formation
Don Valley Power Project, a coal-fueled IGCC power plant (South Yorkshire, United Kingdom)	Constructed in 2013 and operational by 2019	13,425	Selexol process	Offshore storage at deep saline formation
Dongguan Project, a coal-fired 800 MW IGCC power plant (Dongguan, China)	Constructed in 2013 and operational by 2015	2,740	TRIG™ Technology	EOR in Shandong Province
Huaneng GreenGen project, a coal-fired 400 MW IGCC power plant (Bohai Rim, China)	Operational by 2020	5,479	N/A	EOR
Lianyungang project, a coal-fired 1,200 MW IGCC power plant (Jiangsu, China)	Operational by 2015	2,740	N/A	Oceanic storage at Binhai and EOR at North Jiangsu Oil Field

Source: Theo et al. (2016).

TABLE 14.7
Comparisons of Performance and Costs of Integrated Gasification Combined Cycle (IGCC) Power Plants with and without CCS

Parameter	IGCC w/o CCS	IGCC w/ Selexol-CCS	IGCC w/IL-CCS
Gross power output (MW)	681	656	656
Net power output (MW) CO_2 capture system	595	533	534
Power use (MW)		54	53
TCR ($2,011/kWnet)		365	315
Net plant efficiency (HHV, %)	44.3	34.9	35.0
Plant CO_2 emission rate (kg/kWh)	0.710	0.131	0,130
Plant TCR ($2,011/kWnet)	3211	4381	4323
Plant LCOE ($2,011/MWh)[a]	*86.6*	124.4	123.2
Added LCOE for *CCS*[b]			
($2,011/MWh)		37.8	36.6
(Relative %)		44	42
Cost of CO_2 avoided ($2,011/ton CO_2)[b]		65	63

Source: Zhai and Rubin (2018).
[a] The assumed coal price is $1.62/GJ.
[b] This also includes CO2 transport and storage costs.

Using computational simulations, the authors found that the four agents have comparable rates of CO_2 capture, but the physical solvent Selexol is the most energy-efficient of the lot (also see Theo et al., 2016). Meanwhile, Zhai and Rubin (2018) explored the role of ionic liquids (ILs) in pre-combustion approaches. They found that the IL [hmim][Tf2N] delivers comparable results to physical and chemical solvents not only in terms of carbon recovery but also in the penalties for levelized cost of energy (LCOE), power outputs, and plant efficiency that are experienced after implementing these capture strategies in power plants with gasification cycles (see Table 14.7).

Lastly, the exploration of CCS pre-combustion methods has expanded from coal to natural gas plants. However, results thus far have not shown a better performance at these facilities than post-combustion approaches. On the other hand, the steel industry would greatly benefit from pre-combustion methods due to the little interference of the intervention with the iron-making process. All these suggest that research and development (R&D) efforts should be increased to assess the potential for pre-combustion approaches. China, the United States, and countries within the Gulf Cooperation Council (GCC) have become the first movers to develop this carbon capture technology (Jansen et al., 2015).

14.3.3 Oxy-Fuel Combustion

As its name suggests, oxy-fuel combustion systems utilize oxygen instead of air for the main fuel's combustion process, which produces a flue gas that mainly consists of water vapor and carbon dioxide (Li et al., 2021; IPCC, 2005). A plant that uses this CCS approach includes an air separation unit (ASU) that generates an oxygen stream that is combined with the recycled flue gas (RFG) to produce an oxygen-rich gas for the oxidant, which is also reduced in nitrogen (Boot-Handford et al., 2014; Wilberforce et al., 2018). And as Figure 14.12 shows, the resulting CO_2 of high purity can be compressed, transported, and put to storage with no significant interruptions (Lau et al., 2021).

Currently, oxy-combustion processes are not yet commercial, posing more significant risks than the more conventional approaches of post- and pre-combustion due to a lack of maturity (Fernandes

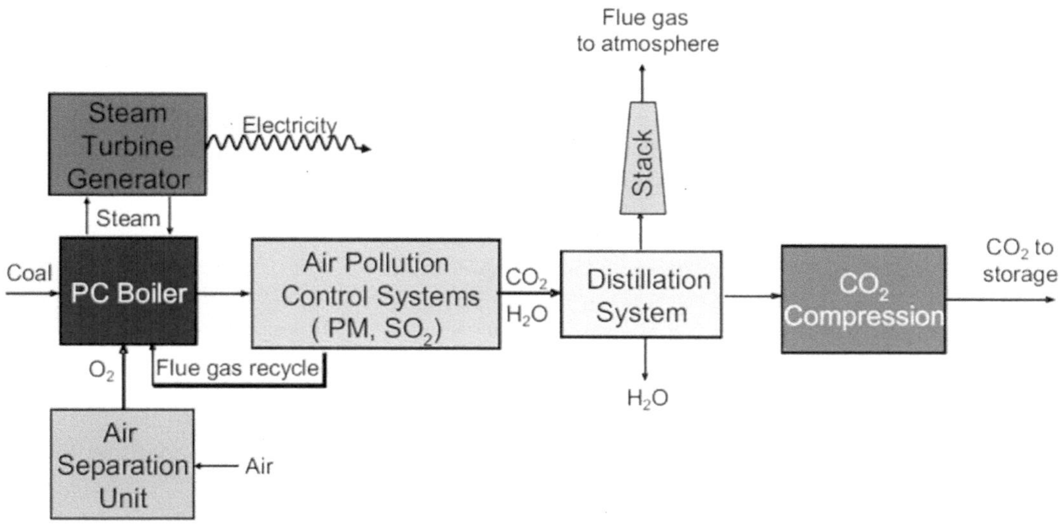

FIGURE 14.12 Oxy-combustion technology utilized in a coal-fired power plant.

Source: Wilberforce et al. (2018).

Araújo and de Medeiros, 2017). Other shortcomings are a net power output reduction because of an energy-intensive air separation unit, a carbon compression process, unresolved issues with full-scale deployment, and the requirement of airtight installations to avoid air or CO_2 leakages. However, oxy-fuels methods are compatible with many coal fuels, require less equipment, and have a very high carbon capture efficiency (Theo et al., 2016).

The choice of replacing air with oxygen to combust fuel follows the goal of reducing the amount of nitrogen in the process. When this is done, flue gases consist almost entirely of CO_2 and water vapor, which can be easily separated through condensation (Anderson and Newell, 2004). Additional impurities like different types of sulfurs (SO_X) can be removed via a process of flue gas desulfurization (FGD), which leaves the CO_2 ready to be stored without the need for more expensive capturing schemes (Wall et al., 2011).

The oxygen used to burn the fuel is produced via cryogenic processes, which separate CO_2 from flue gases using their different condensation and desublimation properties. These methods recover more carbon while yielding a much higher CO_2 purity than other separation techniques (Nemitallah et al., 2017; Song et al., 2019). With that said, the costs of these technologies can be prohibitive for many potential actors interested in using them. But newer research has explored membrane-based separation's role in lowering costs and making these technologies more widely available.

Each of the three CCS applications has its own strengths and weaknesses and better-suited scenarios for their application. Table 14.8 summarizes the benefits and shortcomings of each of these technologies. We can see that high costs and high energy penalties are two of the most salient considerations before choosing one of these approaches. Research has investigated potential new avenues to lower these two dimensions via more efficient capture processes, and we will now turn our attention to the different ways that CO_2 is captured.

14.3.4 Conventional Separation Methods: Absorption, Adsorption, and Cryogenic

This section reviews the more conventional carbon separation techniques used to date. The more mature separation technologies are absorption, adsorption, and cryogenic distillation, all different methods with the same goal of separating CO_2 from flue gases or other elements. *Absorption* is a carbon scrubbing technique using chemical or physical solvents to remove carbon dioxide from

TABLE 14.8
Comparative Benefits of Post-Combustion, Pre-Combustion, and Oxygen-Combustion

Technology	Advantages	Drawbacks
Post-combustion	• Mature technology for other applications (e.g., separation of CO_2 from natural gas) • Standard retrofit of existing power generation capability • Technology improvements and cost reductions possible with additional development	• High energy penalty (~30%) • High cost
Pre-combustion	• Lower costs than post-combustion capture • Lower energy penalties than post-combustion capture • High pressure of CO_2 reduces compression costs • Combine with H_2 production for transportation sector • Technology improvements and cost reductions possible with additional development	• Complex chemical process required for gasification • Repowering of existing capacity needed • Large capital investment needed for repowering
Oxygen-combustion	• Avoid the need for complex post-combustion capture • Potentially higher generation efficiencies • Technology improvements and cost reductions possible with additional development	• New high-temperature materials needed for optimal performance • On-site oxygen separation unit needed • Repowering of existing capacity needed

Source: Balat and Öz (2007).

flue gases or pre-combustion streams. One of the most mature absorption methods uses the solvent monoethanolamine (MEA), which separates CO_2 from flue gases inside an absorption vessel. The solvent, now rich in carbon, is then passed through a second vessel (a stripper column) where the absorption reactions are reversed, and the pure CO_2 is then compressed for transport and future storage (Davison and Thambimuthu, 2005).

Figure 14.13 shows a schematic of the MEA capture process of the considered reference case, which has a 30 wt% (or weight-percent) concentration of MEA as solvent. This process consists of five stages: (1) a washing section that pre-conditions the flue gas; (2) the absorption section, where CO_2 is exposed to the solvent and is then separated from the flue gas; (3) a second washing section that clears flue gases from unwanted pollutants; (4) the regeneration section, where CO_2 is again separated, this time from the solvent; and (5) the compression section, where carbon dioxide is arranged for transport and later storage (Raynal et al., 2011).

Other absorption processes involve using different solvent mixtures with varying degrees of success in removing CO_2 and other impurities from flue gases (see Figure 14.14). Likewise, these methods differ in their operational temperatures, corrosion potential, and other chemical considerations. Table 14.9 lists the strengths and weaknesses of the techniques that feature the most in the literature. In general, more novel solvents are less-volatile, are non-flammable, and utilize recyclable materials, while also trying to tackle high costs, corrosion dangers, and temperature operation issues (Theo et al., 2016).

Meanwhile, adsorption occurs when gas is fed to a bed of solids that can adsorb (or attach to its surface) CO_2, allowing the other gases in the mixture to pass through. When the bed's surface is loaded with carbon, it is replaced by a clean bed, while the original is regenerated to remove the carbon attached to its walls. Desorption is achieved following one of these methods: pressure swing adsorption (PSA), temperature swing adsorption (TSA), electric swing adsorption (ESA), vacuum swing adsorption (VSA), combining pressure and vacuum techniques (PVSA), among others (Davison and Thambimuthu, 2005; Hussin and Aroua, 2020).

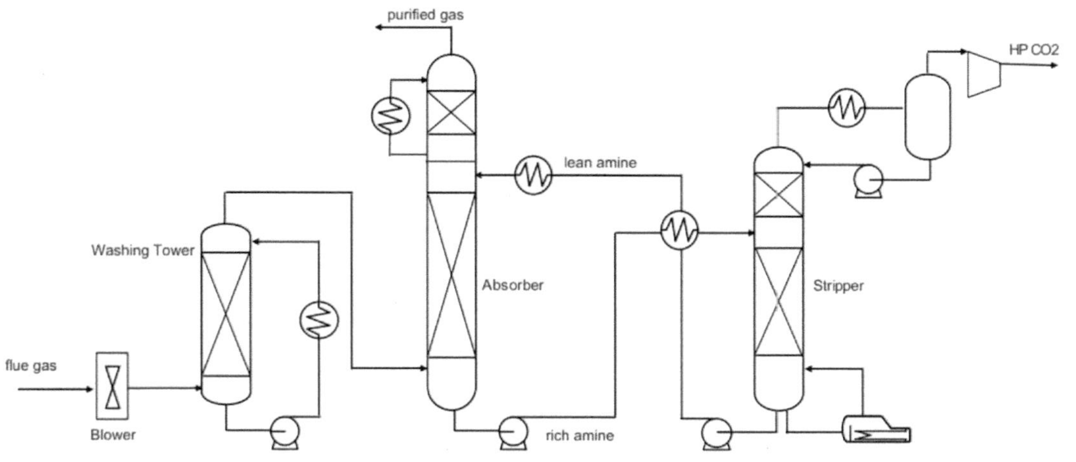

FIGURE 14.13 Simplified process flow diagram of the MEA post-combustion carbon capture process.

Source: Raynal et al. (2011).

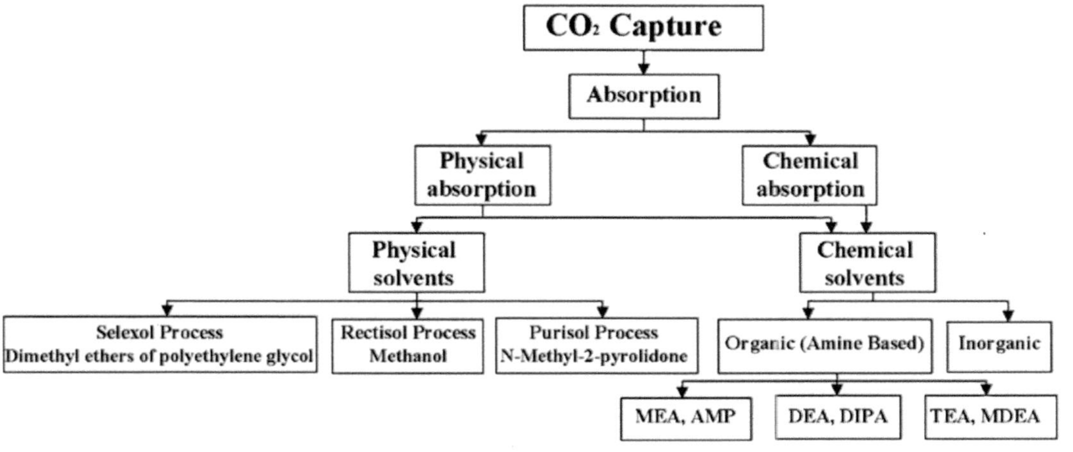

FIGURE 14.14 Carbon dioxide capture process by gas–liquid absorption.

Source: Padurean et al. (2012).

TABLE 14.9
Advantages and Disadvantages of Different Commercial Physical Absorption Processes

Physical Absorption Process	Advantages	Disadvantages
Selexol process	High selectivity for H_2S Does not require water wash for solvent recovery Chemically and thermally stable Low volatility (i.e., minimal solvent loss) Flexible (i.e., compatible with combined removal of H_2S and CO_2, and specific H_2S removal) [370] Capable of moisture removal (due to hydrophilicity of solvent) Minimal capital and operational expenditures (i.e., carbon steel construction) [378]	High viscosity at low side of operational temperature (i.e., 0–175°C), with reduced mass transfer rate and tray efficiency [275] Only suitable for CO_2 removal if CO_2 is more concentrated than H_2S [378]

(Continued)

TABLE 14.9 (*Continued*)
Advantages and Disadvantages of Different Commercial Physical Absorption Processes

Physical Absorption Process	Advantages	Disadvantages
Rectisol process	Good CO_2 and H_2S removal efficiency (i.e., <0.1 ppm) to avoid catalyst poisoning (suitable for synthesis of methanol, hydrogen, ammonia, synthetic natural gas [SNG], and Fischer–Tropsch products) Enable simultaneous capture of H_2S and CO_2 Reasonable viscosity of solvent Minimal solvent loss [370] Lower energy demand than amine-based absorption processes Minimal foaming and good miscibility with water High thermal and chemical stability Low corrosivity Minimal solvent degradation [378]	Low temperature operation (i.e. −30 to −80°C) [379], requiring expensive and complicated cryogenic process stages Possibility of amalgam formation (due to absorption of mercury) at low temperature [378]
Purisol process	High selectivity for H_2S Flexible (i.e., compatible with combined removal of H_2S and CO_2, and specific H_2S removal) [370]	Volatile solvent and requirement of water scrubbing to avoid excessive solvent loss [370]
Morphysorb process	High solvent loading capacity Selective removal of H_2S and CO_2 Reduced corrosion Reduced environmental hazard Low capital and operating costs [380] Lower energy requirement [381] Lower recirculation requirement Lower co-absorption of hydrocarbon [382]	Relatively new process with low maturity level (i.e., at laboratory trial and pilot testing stages) [381]
Fluor process	Non-corrosive and low-viscosity solvent Enable selective removal of H_2S High solubility of carbon dioxide No requirement of make-up water [378]	Uneconomical to achieve high product purity Requirement of cryogenic operational condition Requirement of more efficient gas–liquid contactor [373] High solvent recirculation requirement Expensive solvent [378]

Source: Theo et al. (2016).

The first technique reduces the pressure of the environment to regenerate the adsorbent, while the second restores it by raising the temperature to break the physical or chemical bond, therefore releasing the adsorbed species. PSA and TSA methods are usually combined in large-scale applications. Then, ESA consists of passing a low-voltage current through the adsorbent to regenerate it, and VSA techniques utilize deep vacuum pressure pumps. ESA is perhaps the least-studied method of this group, in part due to high energy penalties for the electric consumption required for the heating and fast cooling of the adsorbent (Davison and Thambimuthu, 2005; Hussin and Aroua, 2020; Ling et al., 2016; Theo et al., 2016).

Figure 14.15 illustrates a typical adsorption-based carbon capture process, with a pre-treatment section that prepares the flue gas to be separated. Then, each of the three sorber sections featured contains a solid adsorbent, like activated carbon, zeolites, metal–organic frameworks (MOFs),

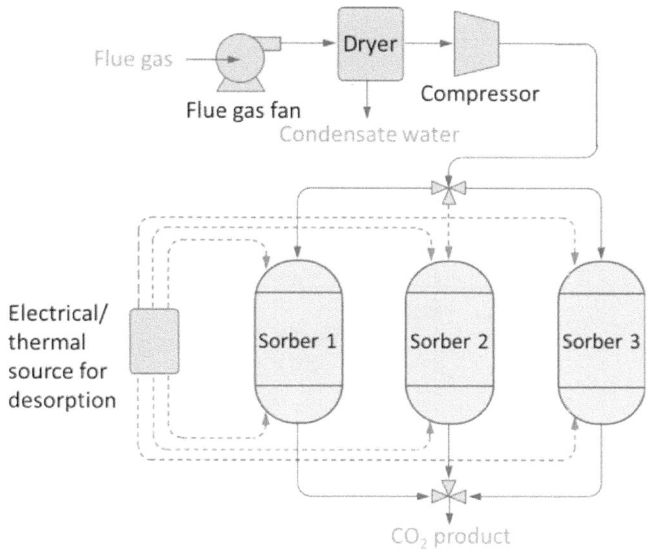

FIGURE 14.15 Process flow diagram for the CO_2 capture from flue gas by adsorption process.

Source: Song et al. (2019).

amine sorbents, or others. A continually functioning process should have at least two adsorbent chambers, so while one is adsorbing, the other is desorbing. After these procedures, the separated CO_2 can be compressed and prepared for transportation and storage or use in EOR or other processes (Song et al., 2019).

A key consideration for adsorption methods is the sustainability of the solid used to separate carbon, along with its capacity to capture CO_2. Therefore, the adsorbent materials must withstand significant temperature changes and multiple processes of sorption and desorption. At the same time, researchers should work on improving efficiency, lowering costs, and abiding by environmental pledges on using clean technologies to advance new capture methods (Ling et al., 2016). Thus, despite its promising future as a CCS method, adsorption technologies must resolve issues like adsorbent regeneration, cost, and lifetime. These issues can be tackled by the advancement in simulations, mathematical modeling (Hussin and Aroua, 2020), or artificial intelligence (AI) tools, like algorithms or machine learning (ML) techniques.

Lastly, cryogenic methods use low temperatures to capture CO_2, separating it from flue gases by taking advantage of its condensation and desublimation chemical properties. The most studied way is cryogenic distillation, a vapor–liquid (V-L) process where separation is achieved thanks to the different boiling points for CO_2 and the rest of the gases in the mixture. Because carbon has to be in liquid form to be removed, cryogenic distillation is an appropriate method in natural gas purification processes. It is a very energy-intensive scheme due to the high pressures required at all stages (Font-Palma et al., 2021).

As potential solutions to the significant energy costs, unconventional vapor–solid (V-S) separation methods have been explored in recent years. Among these processes, we can mention cryogenic liquid, heat exchangers, packed bed, external cooling loop cryogenic carbon capture (CCCECL), anti-sublimation, and controlled freezing zone, to name a few (Font-Palma et al., 2021; Song et al., 2019). Figure 14.16 illustrates a packed bed technique that occurs when H_2O and CO_2 are simultaneously separated from the flue gas in packed beds where the cold energy is provided by liquefied natural gas (LNG) or nitrogen gas. When the flue gas is cold enough, the CO_2 will frost and saturate the bed, and a regeneration process for the bed to remove the captured carbon is required (Font-Palma et al., 2021; Song et al., 2019).

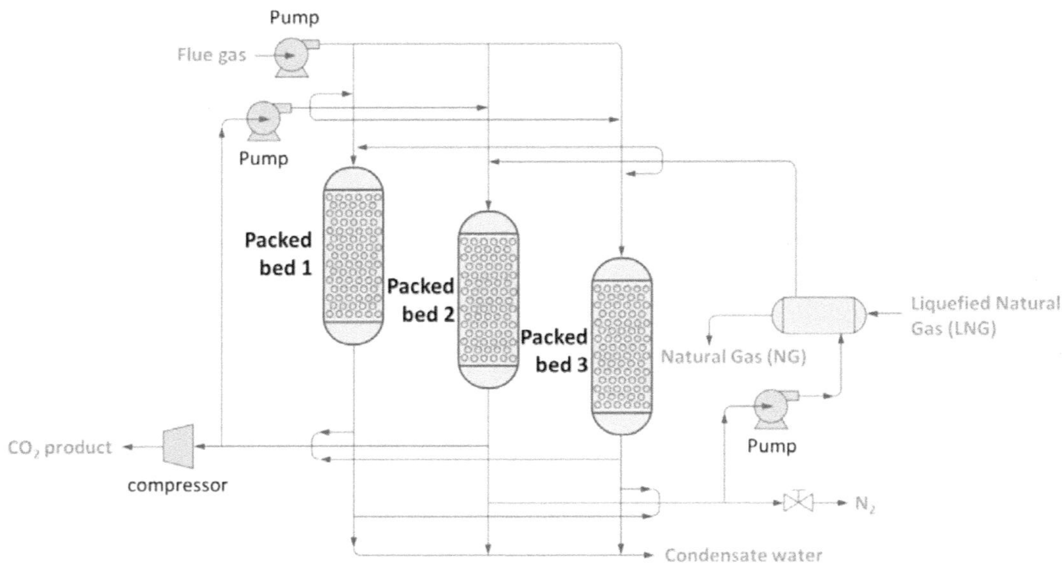

FIGURE 14.16 Schematic diagram for a low-temperature CO_2 capture process by packed bed.

Source: Song et al. (2019).

These low-temperature processes provide a few advantages over amine scrubbing techniques. For example, corrosion and foaming issues, excessive water consumption, and pollution associated with solvent handling are not significant considerations. Likewise, producing liquid CO_2 enables a more economical material transport (Theo et al., 2016). However, their limitations appear, like the potential of stream blockages due to water frosting and the potential to form dry ice inside packed beds (Babar et al., 2021). As a response to these shortcomings, newer carbon capture technologies like membranes, calcium looping, and microalgae-based capture have appeared. We proceed to review them in the next section.

14.3.5 Membrane-Based Gas Separation and More Novel Technologies

Less-mature technologies have appeared as responses to the shortcomings of more conventional CCS methods. One technology that has gained more notoriety in the past few years is membrane-based separation techniques. Applicable in post-combustion, pre-combustion, and oxy-fuel combustion, membrane methods have great potential and compare favorably versus more conventional amine-based absorption methods. Figure 14.17 shows a basic schematic of the separation process through a physical barrier that keeps unwanted gas particles on one side while allowing the targeted gas to go through the membrane (Ji and Zhao, 2017).

The mechanism that allows separation is the difference in the physical or chemical interactions between gases and the membrane materials, which causes distinct components to pass through the membrane at different times (Davison and Thambimuthu, 2005). Scholes et al. (2010) have reviewed different types of membranes used in integrated gasification combined cycle (IGCC) processes, separating them into barriers that permeate H_2, the most common type, and those that permeate CO_2 while retaining H_2. For the former type, metallic, porous inorganic, and polymeric are the most used, while for the CO_2-selective membranes, polymeric and facilitated transport are the ones relied upon the most.

While membranes are a promising avenue for research, they are still not on par with solvent absorption processes. High capital costs, which hinder the development of pilot plant trials, are one

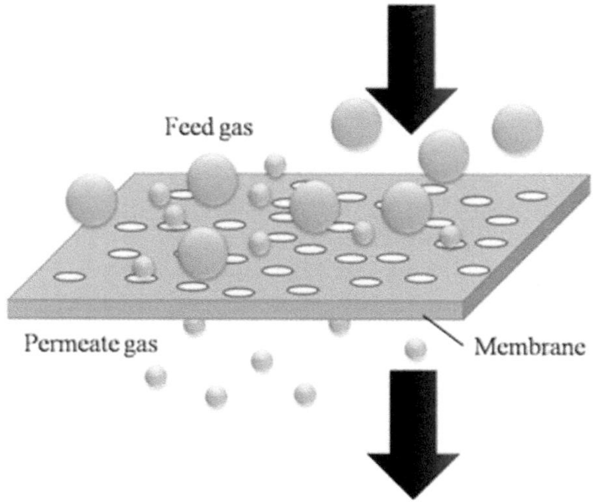

FIGURE 14.17 The schematic of membrane separation for binary gas mixtures.

Source: Ji and Zhao (2017).

TABLE 14.10
Summary of Membranes Used in CCS Processes

Membrane Type	Application	Advantages	Disadvantages
Metal membrane	Pre-combustion	Infinite H_2/CO_2 selectivity	High cost; poisoning; low driving force
Carbon membrane	Pre-combustion	Size sieving effect; high H_2/CO_2 selectivity	High cost; susceptible to oxygen, brittleness
Alumina membrane	Pre-combustion	Low cost; chemical and physical stability	Low H_2/CO_2 selectivity
Zeolite membrane	Pre-combustion and Post-combustion	Low cost; chemical and physical stability	Low H_2/CO_2 selectivity
MOF membrane	Pre-combustion and post-combustion	Large pore volume and surface area	High cost
Silica membrane	Pre-combustion	Proper pore size; low cost; high thermal stability	Poor hydrothermal stability
Polymeric membrane	Post-combustion	Low cost; high CO_2/N_2 selectivity	Low chemical and physical stability; too thick
Fluorite membrane	Oxy-fuel combustion	High O_2/N_2 selectivity	Energy-intensive; hard to seal
Perovskite membrane	Oxy-fuel combustion	High O_2/N_2 selectivity	Energy-intensive; hard to seal; poisoning

Source: Ji and Zhao (2017).

of the most significant shortcomings impeding the collection of information about this new technology (Scholes et al., 2010). On the other hand, Ji and Zhao (2017) highlighted the success of the few pilot programs implemented in the past decade and developed a detailed summary of the different types of membranes used in CCS processes (see Table 14.10). In the last ten years, membrane-based capture methods have advanced tremendously to compete with the industry standard of amine-based solvent scrubbing.

Meanwhile, calcium looping is a technology that removes carbon dioxide from the exhaust of a power plant, creating a stream of CO_2 that is ready for sequestration. Figure 14.18 shows two different models of calcium looping processes that act on the same principles, namely, the generation of calcium carbonate through the exploration of the reversible reaction between calcium oxide and carbon dioxide. Thus, a calcium oxide (CaO) sorbent, often derived from limestone, is cycled between two vessels. The first performs the carbonation of CaO, where the flue gas is stripped of CO_2, and the formed $CaCO_3$ is passed onto the calcination cycle. Here, the CaO formed is sent back to the carbonator, which leaves a stream of CO_2 ready for capture and storage (Blamey et al., 2010).

Despite being in an early stage of research, solid CaO-based sorbents' success in separating carbon dioxide was demonstrated many decades ago. Coal combustion could significantly improve using this technology since these sorbents are not affected by other pollutants like ash or acid gases (Blamey et al., 2010). Additionally, mainstream techniques are plagued with considerable energy penalties, leading to a drop in energy efficiency and affecting electricity prices negatively. Calcium looping processes excel in their lower loss of power plant efficiency and their potential even to increase the system's power output (Erans et al., 2016; Hanak et al., 2015).

Moving on to one of the fastest-growing novel techniques, CO_2 bio-fixation using microalgae has gained significant momentum in the last few years. The efficient photosynthesizing capabilities of microalgae compared to terrestrial plants have taken research deep into the oceans (Song et al., 2019). Also, CO_2 can be used to cultivate microalgae that can be used as feedstock to produce biofuels (Adu et al., 2018). Hence, algae can feature in a solo effort to produce energy while also mitigating pollution from carbon processes. And as an alternative to storing the captured CO_2, algae consume it and grow so that later oil is extracted from them to power the world, keeping a cycle where clean energy is its center.

Figure 14.19 shows this cycle, where algae mitigate anthropogenic sources of carbon emissions by utilizing the CO_2 generated in combustion processes and growing through photosynthesis. Then, high-density oil is extracted from these same algae, serving as a reliable alternative to fossil fuels. Lastly, the processed algae can be fermented to produce methane, which helps run the oil production process all over again. So in addition to mitigating CO_2 emissions from polluting sources and providing an alternative source of energy through the extraction of its oil, the entire process has the potential to be sustainable (Sayre, 2010).

In a discipline aptly named *algae culture*, researchers study the potential of different farming species of marine plants for industrial and commercial uses, where microalgae are a promising

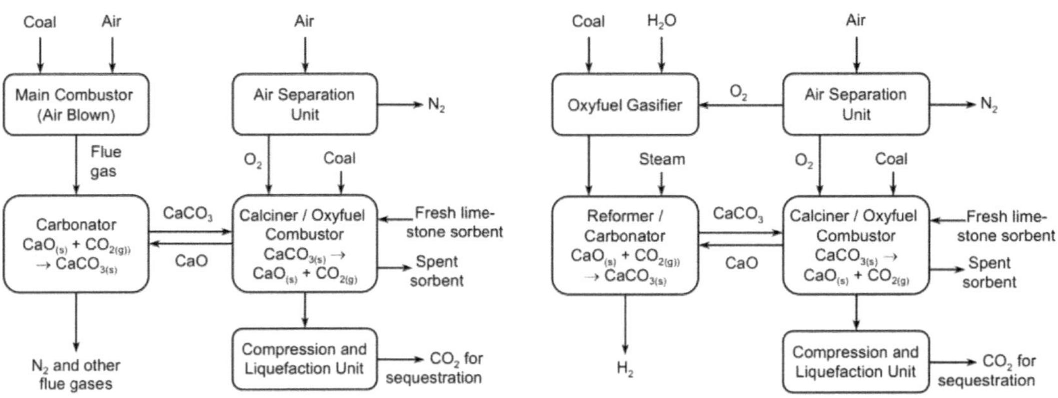

FIGURE 14.18 Two potential combustion processes using the calcium looping cycle.

Source: Blamey et al. (2010).

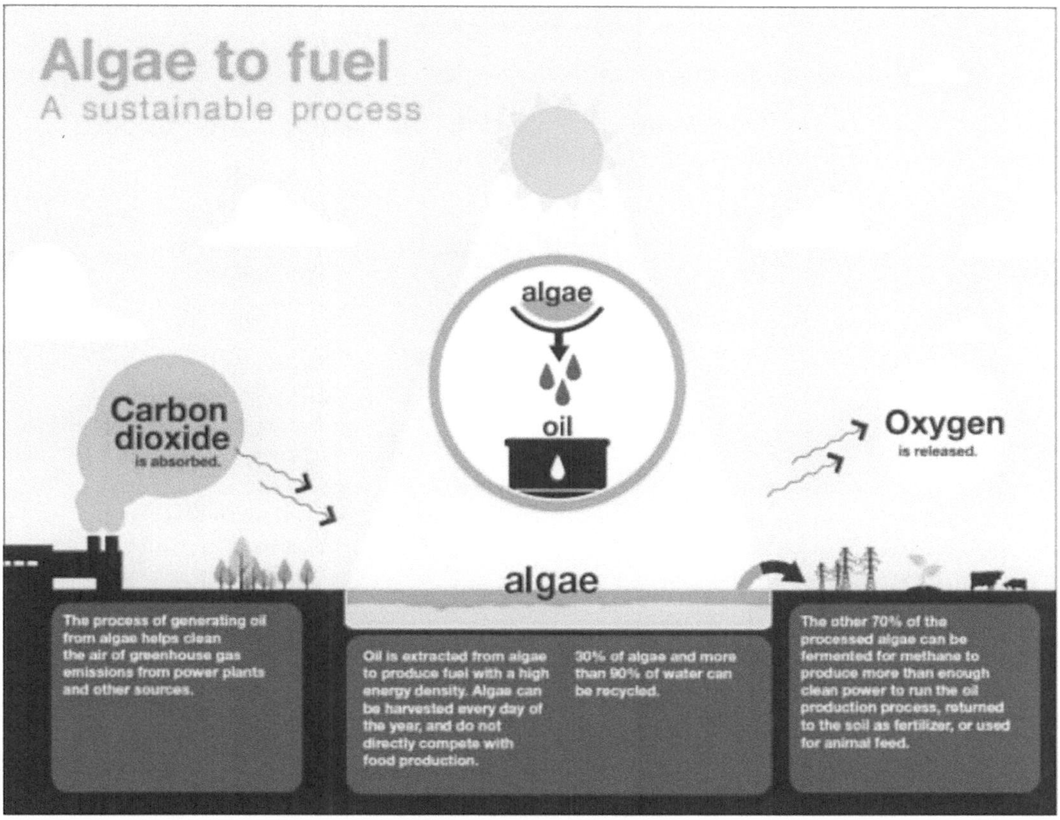

FIGURE 14.19 Microalgal carbon capture and biomass production.

Source: Sayre (2010).

energy alternative. Research has found that seaweed may not be suitable for these endeavors, but phototropic microalgae are idoneous for them due to their ability to absorb sunlight and assimilate CO_2 and other nutrients to develop. Further, some microalgae species have shown potential in some industries, like *Chlorella vulgaris* in steelmaking plants, *Aphanothece microscopica* in refineries, and algae and bacteria combinations for hazardous wastewater and industrial black oil processes (Singh and Ahluwalia, 2012).

As we can see, we are in the middle of promising developments in areas that have not been thoroughly explored regarding the capture of CO_2 from anthropogenic sources. To summarize what we have reviewed in the last two sections, Table 14.11 shows the primary materials utilized in carbon capture approaches, and their strengths and disadvantages. With those trade-offs, therefore, researching more novel technologies to capture carbon dioxide from energy processes continues to be an avenue for improving methods tackling climate change, and several actors have supported these efforts.

From international organizations to governments, research institutions and universities, and private actors like oil and gas majors, a positive feedback loop now exists where academia develops frameworks for the advancement of cleaner technologies and governments subsidize more studies on those topics. Meanwhile, international organizations analyze the literature and propose environmental pledges with marked deadlines and models with climate trajectories that governments and private actors commit to follow. The following sections will explore these relationships in more detail, particularly the role of the oil and gas industry in the decarbonization effort and their plans to explore CCS technologies.

TABLE 14.11
Main Materials Used in Carbon Capture Technologies

Type	Approach	Materials	Features
Oxygen carrier	Oxy-combustion (chemical looping combustion)	CaO, Fe_2O_3, NiO, $FeTiO_2$	• Low cost • Energy efficiency • Inactivation by impurities
Metal–organic framework	Pre-/post-/ oxy-combustion	Cu-BTC, ZIFs, MIL-102	• Chemical diversity • High moldability • Large surface area • High cost (expensive)
Membranes	–	Pd alloys, modified graphene, MOFs	• Flexibility • Energy efficiency • Tradeoff between permeability and selectivity
Zeolites	Pre-/post-combustion	Na-A and alkaline zeolite	• Molecular sieving property • High capacity at moderate conditions • High regeneration energy
Ionic liquid	–	Amino acid ionic liquids	• High thermal stability • Low adsorption rate
Porous carbon	Pre-combustion	Nanotube, functionalized carbon, graphene	• Large adsorption capacity • Low cost • Low selectivity
Mesoporous silica nanoparticle	Pre-combustion	SBA-15, MCM-41	• Tunable surface • Low cost
Amine-based solvent	Post-combustion	Ethanolamines (EAs), piperazine	• Commercial maturation of process • High regeneration energy

Source: Olabi et al. (2022).

14.4 EFFICIENCY, CHALLENGES, AND LATEST ADVANCES IN CCS TECHNOLOGIES

The various processes reviewed in this chapter are tailored toward mitigating carbon dioxide emissions from diverse energy processes. Some of the findings have come from the experience at full-scale power plants. In contrast, other conclusions have come from mathematical simulations and projections where known values generate future scenarios of unknown values. While both avenues have helped broaden the scope of research, the improvement of simulation technologies has opened the door to exploring new materials and techniques *in silico*. This technique has shown very promising results.

In this sense, the potential role that artificial intelligence and machine learning methods could have in the development of models that predict, for example, energy loss after implementation of a pre-combustion CCS method or the changes in power output after switching from one CO_2 adsorption material to another constitutes a game changer. Given the real-life constraints of attempting to test all models in full-scale conditions in completely functional power plants, simulations can help solve some of these challenges, arriving at predictions based on large quantities of data collected from similar plants under similar conditions. Moreover, newer models that include latent, geographical, and other variables have improved the simulation capabilities even further.

These kinds of improvements in computational simulations can help tackle one of the biggest challenges for the widespread implementation of CCS technologies. As mentioned before, despite

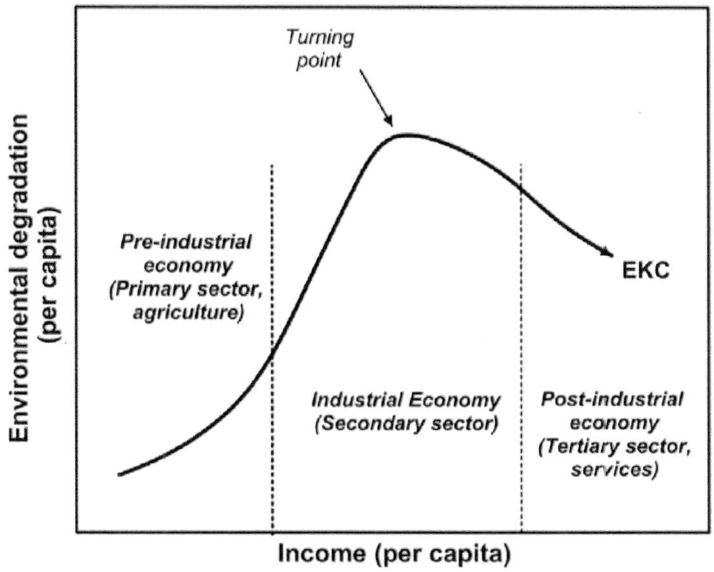

FIGURE 14.20 Environmental Kuznets curve (EKC).

Source: Rashid Gill et al. (2018).

significant support, carbon capture and storage methods are costly for governments and private actors to implement broadly (Middleton et al., 2011). Although research has helped improve the efficiency of current methods, while continuing to advance cheaper and cleaner technologies, the conundrum remains. Countries that do not have robust systems of government subsidies for clean technologies are less prone to implement policies that encourage the development of methods like CCS. And most of the time, a lack of government funding for these projects is caused by a lack of prosperity or economic growth within a country.

In other words, countries that invest in cleaner technologies have reached a level of economic development that other countries are still trying to attain, mostly relying on less-clean energy to achieve it. Therefore, countries that are behind economically are still placing most of their resources for energy spending on the task of developing first and then on a potential switch to cleaner, less-dense energy sources. This notion, based on the econometric Kuznets curve (see Figure 14.20), is explored for European Union countries by Neagu (2019), who advanced that instead of prosperity measured in GDP per capita, countries with a higher "economic complexity" will reach a point where rising carbon emissions will stop, decreasing at every point after that moment of inflection.

This hypothesis relies on the assumption that all countries eventually move toward cleaner technologies only after reaching a level of development that benefits most of its citizens (Rashid Gill et al., 2018). Therefore, countries that have not arrived at that point will not have technologies like carbon capture and storage in their most urgent policy considerations. With that said, it is at this point where private actors, including oil and gas companies, can contribute with research experience, funding, and other needed resources. These actors can help in intermediate stages, where development has not yet been achieved, but some decarbonization efforts can still be implemented.

For example, some efforts have been made to advance natural processes for climate mitigation, like increasing the vegetation of some parts of the globe through afforestation endeavors that could increase the carbon sequestration rate of some regions (Niu and Duiker, 2006; Yosef et al., 2018). Along with this, the improvement of degraded agricultural soils through the addition of carbon dioxide, soil organic carbon (SOC) sequestration processes, biochar as a soil conditioner, and the general treatment of croplands to consume more CO_2 are all cheaper, more immediate ways

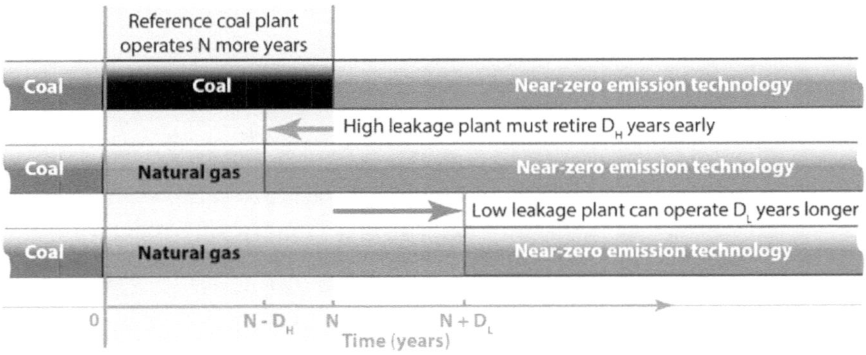

FIGURE 14.21 Concept diagram for transitions from coal energy systems to near-zero emission energy systems.

Source: Zhang et al. (2016).

to jump-start decarbonization efforts in some countries that have been explored in the literature (Smith, 2004; Tan, 2019; VijayaVenkataRaman et al., 2012).

Two other negative emission technologies (NETs) with varying costs that some countries may incur are bioenergy with carbon capture and storage (BECCS) and direct air carbon capture and storage (DACCS) methods. While less-efficient, the former is less-costly. It works similarly to microalgae, with biomass capturing CO_2 and producing several final energy carriers, like electricity, transport fuels, and hydrogen. Meanwhile, DACCS are methods that extract carbon directly from the air with chemicals, with the two main types being a high-temperature (HT) system that uses a liquid sorbent and a low-temperature (LT) one that relies on a solid sorbent to capture the CO_2 (Creutzig et al., 2019; IEA, 2021c; Lehtveer and Emanuelsson, 2021).

Moreover, another alternative that still relies on fossil fuels but constitutes a step away from more polluting energy sources like coal is using natural gas as a "bridge fuel." If wood or coal – sources of energy still preferred in countries without the resources or infrastructure that more developed countries have constructed over the years – is gradually replaced by natural gas to produce electricity, the countries that make the switch can diminish their greenhouse emissions without having to commit fully to a net-zero agenda, at least in the short term. Thus, some middle-of-the-road positions may benefit some countries' search for economic development while still considering the environment.

With that said, there is the risk of natural gas installations not being as efficient as expected, which would make the change from coal to a natural gas one in name only, with little or no of the supposed benefits brought by a switch. As Figure 14.21 shows, a change to natural gas installations that suffer from high degrees of leakage can lead to these facilities being retired earlier than expected, which could translate into falling back on more polluting energy sources to cover a potential energy loss. Therefore, assuming future replacement of these energy systems by a near-zero emissions infrastructure, variations in the condition of natural gas facilities over time can signify either a delay in the switch or a momentary but regrettable regression to a more polluting system to cover for efficiency losses (Zhang et al., 2016).

Currently, the literature has not yet developed this notion of natural gas as a bridge fuel to a great degree (Levi, 2013). Starting with considerations regarding the difference in methane leakage, which could nullify all the supposed benefits of a change to natural gas, researchers have only recently also inquired about the destination of this bridge, the length of this journey, and other less-tangible variables that make the bridge more attractive than the point of departure (Delborne et al., 2020). Accordingly, the discussion is still developing, with some authors positing that this strategy is a "bridge to nowhere" (Howarth, 2014), while others see promising avenues of progress, albeit with regulation and sensible policies put in place (Pierce, 2012).

With that said, the main industry in the middle of all these dynamics is the oil and gas sector, which has been a key to the technological advancements brought by economic development and prosperity in developed countries. At the same time, however, the industry is a significant contributor to the greenhouse emissions we see today affecting our planet. Consequently, we are currently amid significant efforts from oil majors to invest in mitigation strategies like CCS technologies while also working to advance renewable energy research and methods. The following section reviews the industry's efforts in more detail.

14.5 CCS APPLICATIONS IN THE OIL AND GAS INDUSTRY

The O&G sector has been one of the most proactive players in advancing and developing CCS technologies to mitigate the effects of many polluting processes, some of which are performed within the industry. As a sector with a proven experience in energy production, achieved through successful experiences in building lasting infrastructure, the industry is positioned in a unique place to contribute to the new technologies of the 21st century. Using multiple decades of expertise in developing structures to capture and store energy, the industry can respond to the various challenges regarding the capture and sequestration of carbon dioxide.

Figure 14.22 illustrates how captured CO_2 can be utilized for other industrial processes, like producing plastics, chemicals, refrigerants, and fuels through biological conversion. The oil and gas industry has found one promising avenue for storing carbon dioxide, which also helps recover oil from almost-depleted sites. Enhanced oil recovery (EOR) is injecting CO_2 into the pore spaces of subsurface oil reservoirs to move crude out. Due to the miscibility of oil with carbon dioxide – their

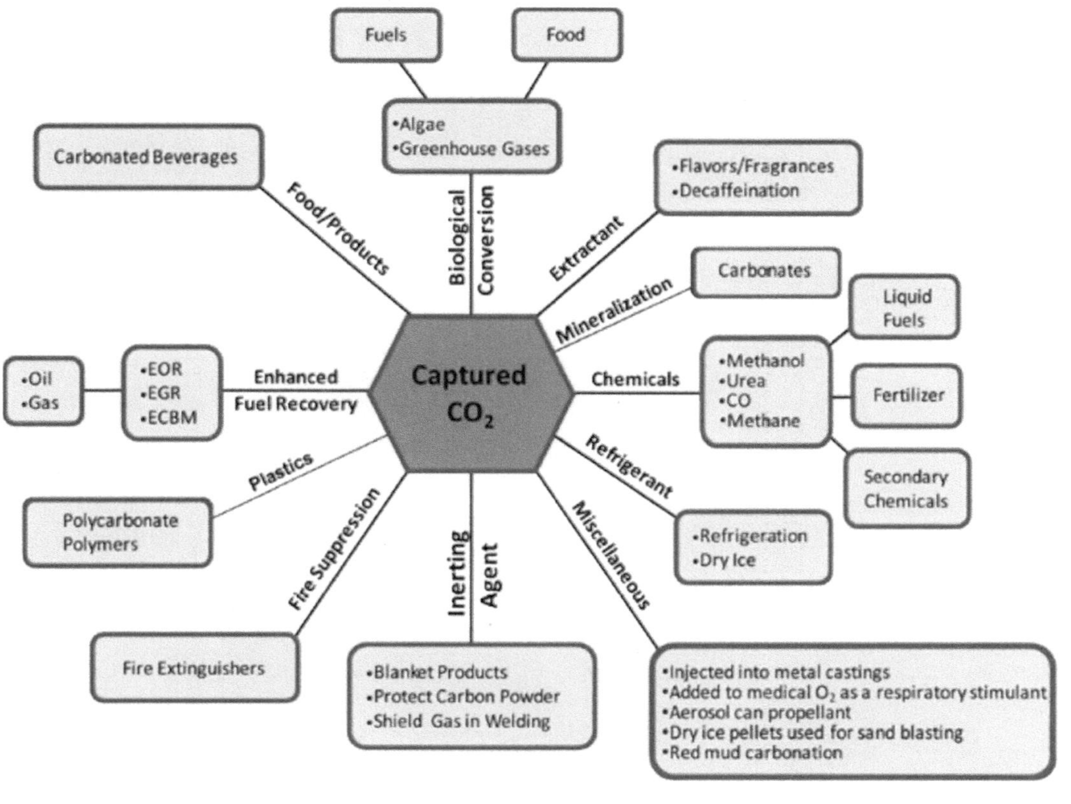

FIGURE 14.22 Different captured CO_2 industrial utilization options.

Source: Adu et al. (2018).

ability to mix in all proportions – the injection of CO_2 in depleted reservoirs helps obtain residual oil that would not be recovered with conventional methods (NETL, 2010).

14.5.1 Enhanced Oil Recovery (EOR)

As Figure 14.23 shows, EOR methods are deployed when an injection well is added to oil reservoirs and previously captured CO_2 is shot through the subsoil so that it captures trapped oil deep underground, increasing domestic production (Ferguson et al., 2009). Due to the mixing of oil and carbon, the former expands and moves toward the well that recovers and extracts the oil, improving the productivity levels of the site (NETL, 2010). Due to large quantities of carbon dioxide being stored deep below the Earth in the reservoir, which prevents them from being released into the atmosphere, EOR technologies also contribute to the reduction of greenhouse gas (GHG) emissions (Eide et al., 2019; Kolster et al., 2017; Sweatman et al., 2009).

As advanced previously, EOR methods are deployed as a tertiary recovery method. EOR techniques are mainly applied after primary and secondary oil recovery methods. The primary recovery method uses the intrinsic energy of the fluid while the pressure in the reservoir is still high. Then, the secondary recovery method requires an injection of water and/or natural gas into the well to alter the pressure underground so that remnant oil is displaced and can be retrieved (Adu et al.,

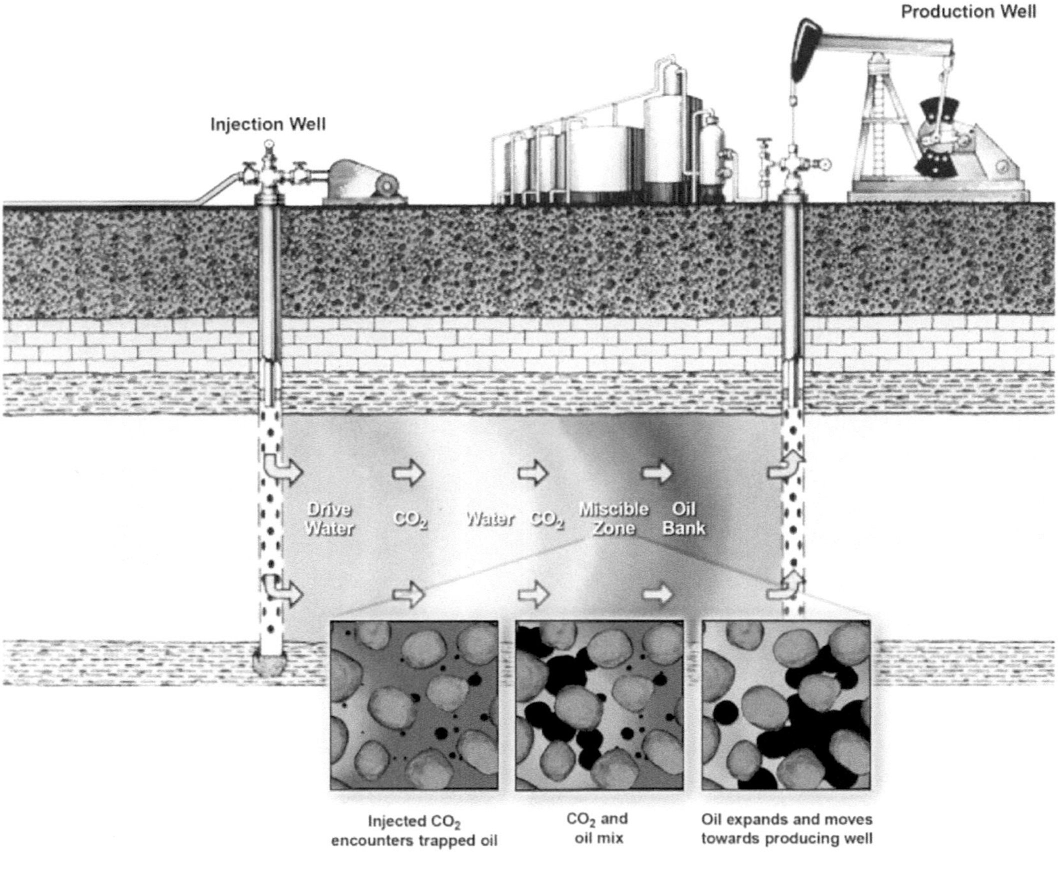

FIGURE 14.23 Cross section illustrating how carbon dioxide and water can flush residual oil from a subsurface rock formation between wells.

Source: NETL (2010).

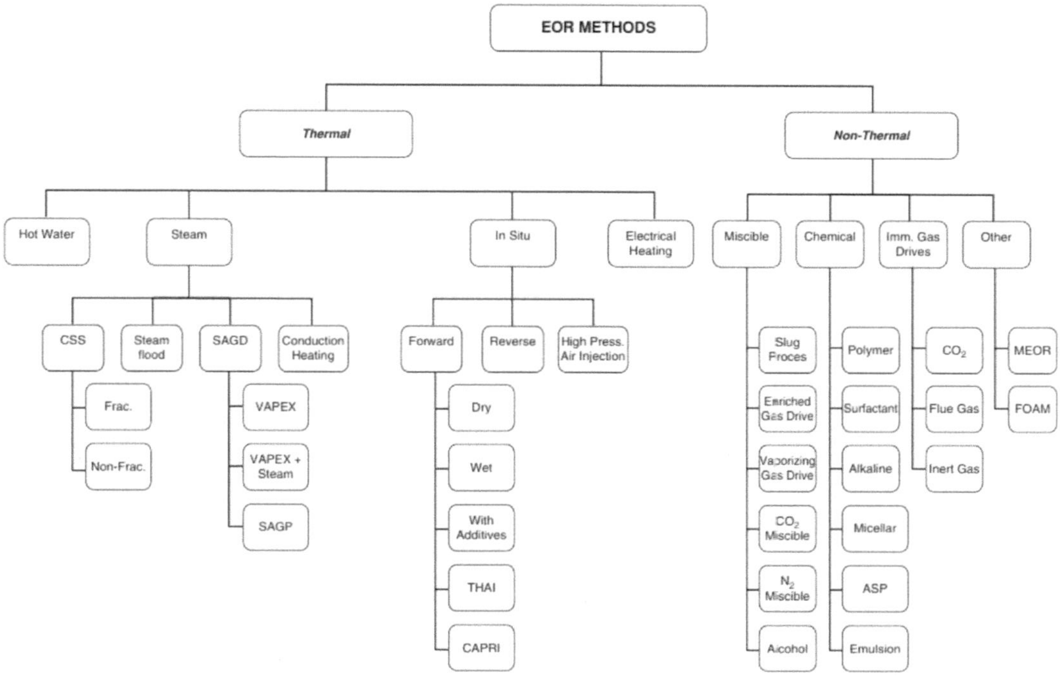

FIGURE 14.24 Classification of EOR methods.

Source: Thomas (2008).

2018). If companies decide to access more of the oil in the reservoir, various EOR methods have been explored in the literature, with CO_2 injection having a proven record after more than four decades of use, dating as far back as the early 1970s (Melzer, 2012).

Moreover, with time, significant improvements have been made to recovery technologies, with important distinctions between onshore and offshore operations and different types of oils seeking to be recovered. For the former, offshore EOR reveals more challenges than onshore operations, like space and weight restrictions and distribution and transportation issues that increase capital (CAPEX) and operational (OPEX) expenditures (Eide et al., 2019). Meanwhile, depending on the oil retrieved, EOR methods are generally divided into thermal and non-thermal approaches (see Figure 14.24). Cyclic steam stimulation (CSS), steam flooding, steam-assisted gravity drainage (SAGD), and fireflooding belong to the former category. In contrast, miscible flooding (which includes CO_2 injection methods), chemical flooding, and other techniques like microbial EOR and foam flooding are part of the latter (Alvarado and Manrique, 2010; Thomas, 2008).

As one of the most promising techniques for enhanced recovery, CO_2 injection has also been studied as a potential decarbonization technique. But as noted previously, some issues in the chain of capture, transportation, and utilization can still be optimized further to offer this solution across a more significant number of projects around the world. For example, the work of Zhang et al. (2018) developed a mixed-integer linear programming (MILP) model to optimize the chain of operations of CCS methods applied in Northeast China, from the cost of the dehydration of the flue gas, to the carbon capture process, to transportation, pipeline, and storage issues, as well as for potential revenue stemming from CO_2 utilization in EOR methods (see Figure 14.25).

Meanwhile, Guo et al. (2019) developed a framework that optimizes the integration between CCS and EOR techniques, focusing on the operational costs for CCS and the expected profit from deploying EOR methods. These two parameters can be obtained by paying close attention to capital

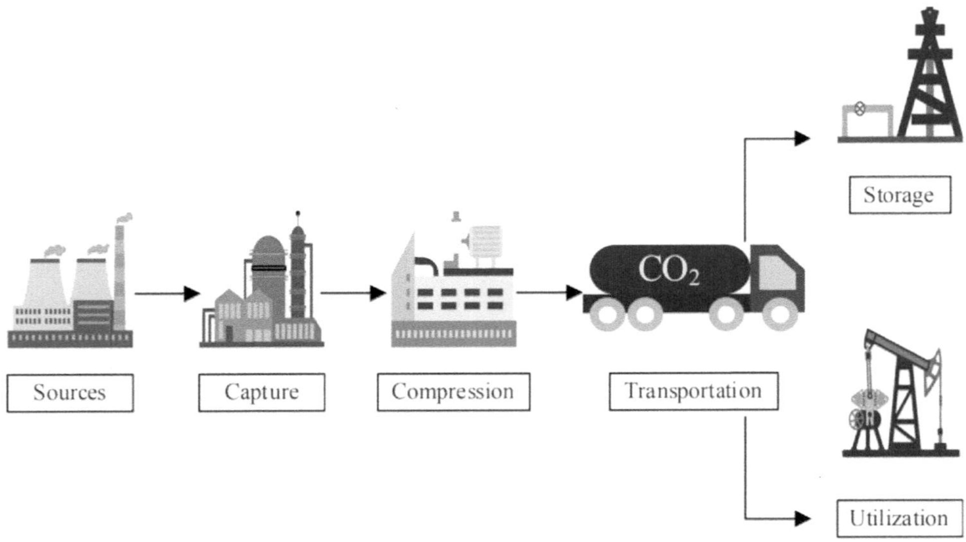

FIGURE 14.25 CCUS supply chain scheme.

Source: Zhang et al. (2018).

expenditures for CCS projects, capture efficiency, depreciation and compound rates, the price of a barrel of oil, and EOR operation and maintenance costs, among others. With that said, these costs can be further diminished after governments promote the development of CO_2-EOR technologies through subsidies and policies that reflect the development of national strategies toward an energy transition away from fossil fuels (Shannak and Malov, 2021).

China is one of the leading countries in innovation, optimization, and government involvement in developing CCS/CCUS strategies. Yang et al. (2019) analyzed the impacts of Chinese government subsidies on the decision to utilize a full-chain carbon capture project, concluding that governments can benefit from targeted policies focusing on CO_2 utilization. Furthermore, CCUS-related research and development schemes can have a high return on investment for governments due to increased academic competition and innovation, which leads to technological advancements that can transform the energy market and solidify carbon capture and storage mitigation techniques.

As Figure 14.26 shows, a business model that integrates government subsidies must work in different areas of the chain. Thus, funding for retrofitting carbon capture and storage installations improves only one area of the chain. At the same time, subsidies for electricity tariffs tackle another, with EOR financing covering a third aspect. Yet, when looking at the entire picture, one can see that retrofitting leads to carbon capture, which later helps with deploying EOR techniques that not only diminish greenhouse emissions but also improve oil production. This excess generates revenue for the parties involved, which leads to more innovation and a potentially improved scenario for the end user of the electricity produced.

These virtuous cycles require many actors to coordinate and work in tandem. As noted in previous chapters, international organizations are another set of actors that can contribute to their development. When these bodies develop regulatory frameworks and establish environmental targets, governments have a clearer idea of how to proceed with their national energy strategies. According to their means, the environmental goals may vary from country to country, so some places will pursue a net-zero energy economy with a complete switch from fossil fuels to renewables. In contrast, others will seek a low-carbon sector with the ability to move forward in the future but a reliance on conventional sources.

These same distinctions across countries will lead to different combinations of policies and regulations being enacted and followed, but with the direction of international organizations, these paths

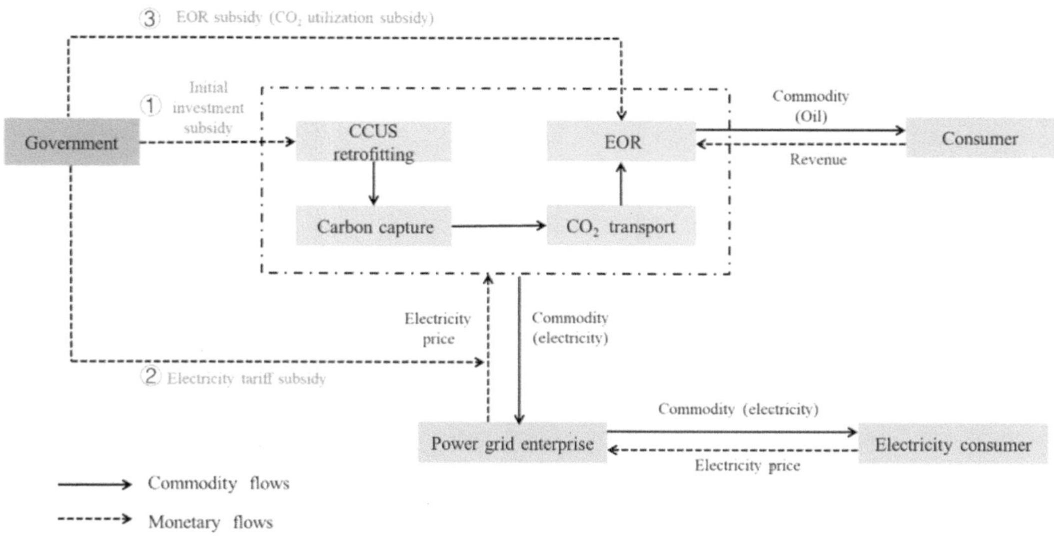

FIGURE 14.26 The vertically integrated business model and corresponding government subsidies for CCUS projects.

Source: Yang et al. (2019).

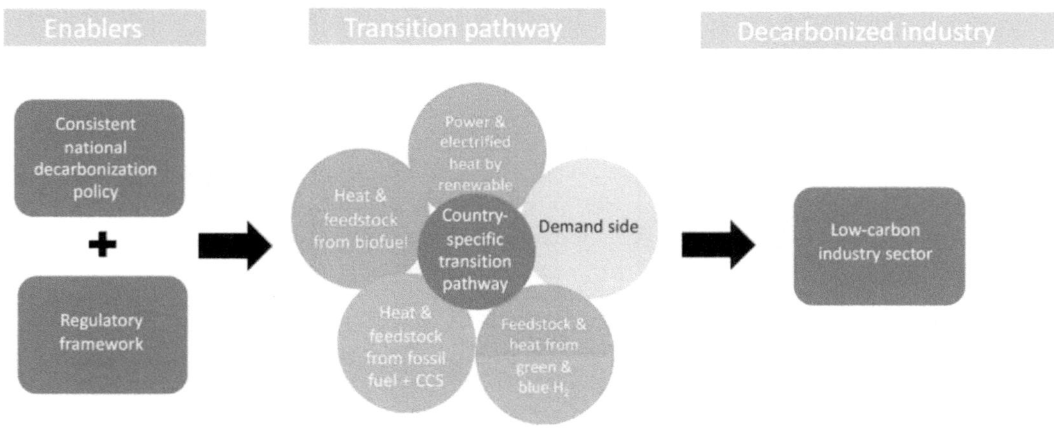

FIGURE 14.27 Country-specific pathways to decarbonize the industry sector.

Source: Lau et al. (2021).

should not be too dissimilar. In the end, the decarbonization efforts for different industry sectors across different countries can look like the illustration in Figure 14.27, where we observe a proposed framework with a clear goal and a specific set of enablers, which trigger a series of standard technologies being implemented with the expectation of achieving the stated goal. However, as noted earlier, a successful transition endeavor depends on multiple actors actively participating, researching, and sharing knowledge about the potential benefits of a lower-carbon industry sector for humanity.

From a revision of the current CCS techniques used in the industrial sector to the role that the oil and gas industry can play through enhanced oil recovery, we have seen how the world can adapt to a scenario where the full transition to zero-carbon is not yet feasible. Given our significant reliance on petroleum, however, a realistic path involves a considerable degree of adaptation to current technologies and the retrofitting of existing infrastructure to lower their carbon emissions or to dispose of the generated CO_2 in ways that cause less harm. In the past few decades, oil producers have joined this endeavor and invested in cleaner technologies while adapting their current methods to more environmentally friendly practices. The following section will review this in more detail.

14.6 POTENTIAL FOR CCS TECHNOLOGIES AND LATEST ADVANCES IN THE WORLD

Throughout this chapter, we reviewed the dynamics and evolution of several carbon capture and storage technologies which can reutilize the CO_2 generated in energy processes or store it deep underground before it reaches the atmosphere. More sustainable development can be achieved by coordinating international organizations, government agencies, oil and gas major players, academia, and informed citizens. Figure 14.28 shows the potential contributions of CCS technologies to different areas, like the environment, the economic sector, and even the social realm. By creating jobs and improving human health, CCS methods can be a strong force for good in the world.

To maximize the potential of these technologies, big players in the oil and gas industry have invested large sums of money to develop partnerships in CCS developments. In 2021, ExxonMobil launched a new project, ExxonMobil Low Carbon Solutions (ELCS), which has announced 20 new projects worldwide. Similarly, Siemens and Aker Carbon Capture have struck a deal to develop CCS technology and business ventures together, as have done G2, NETPower, Siemens, and EJM, who have committed to work on a liquefied natural gas (LNG) plant in Louisiana. Moreover, LaFarge Holcim and Schlumberger have partnered to retrofit cement plants in Europe and the United States. Global player Eni has also joined efforts with local partners in several countries to develop CCS across multiple continents (Global CCS Institute, 2021).

Other essential partnerships highlighted by the Global CCS Institute in their Global Status of CCS 2021 report are Shell's expansion of activities in CCS projects after becoming a founding partner of the Porthos network blue hydrogen project, BP's work in the United Kingdom's Net Zero Teesside network, along with Eni, Equinor, Shell, and Total. Meanwhile, Valero has partnered with Black Rock and Navigator to develop a CO_2 pipeline to transport carbon from bioethanol plants. Lastly, Mitsubishi and the South Pole developed a carbon dioxide removal (CDR) purchasing facility to help companies achieve their net-zero commitments.

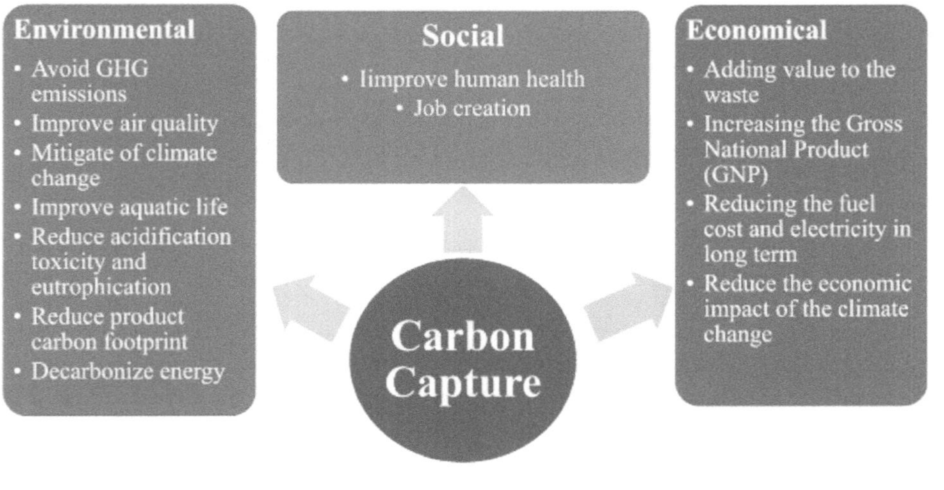

FIGURE 14.28 Contribution of carbon capture and storage in sustainable development.

Source: Olabi et al. (2022).

Despite the increasing value of partnerships and environmental pledges on the international stage, the deployment of CCS projects continues to be slow. It is offtrack of what has been estimated as an ideal pace to offset the potentially negative consequences of climate change (IEA, 2021b). To increase the number of opportunities to catch up to the estimations, some applications which are cost-effective center on gas processing, where natural gas reservoirs can be retrofitted to separate CO_2, which comes out as very concentrated and ready to transport and store. Other potential CCS applications with varying degrees of cost-effectiveness are ethanol production, refining, gas-fired power generation, direct air capture (DAC/DACCS), and low-carbon hydrogen, touted as a promising energy carrier (Turan, 2020).

Regarding this latter technology, CCS technologies can contribute significantly to the production of *blue hydrogen*, which comes from fossil fuels and carbon emissions reduced. Along with *aqua hydrogen*, blue hydrogen relies on non-renewable sources of energy, while a third type, *green hydrogen*, comes from renewable energy and therefore does not emit carbon. Figure 14.29 illustrates the sources of these low-carbon hydrogen solutions. Blue hydrogen is less-expensive than green hydrogen while offering the possibility to continue fossil fuel extraction with a significant CO_2 reduction. Blue hydrogen can help countries achieve climate goals at an acceptable cost while allowing them to explore other low-carbon options (Yu et al., 2021).

Since 2000, nine facilities have started capturing CO_2 from industrial hydrogen production that is based on fossil fuels in North America, France, Japan, and other developed countries. The Quest Carbon Capture and Storage project in Western Canada has stored 5 million tons of CO_2 from 2015 to 2020 (Yu et al., 2021). Additionally, oil companies have also dipped their toes in blue hydrogen production waters, with the following projects developed in the United Kingdom: Equinor's Saltend project, BP's hydrogen plant within the Net Zero Teesside network, and Phillips 66 launching its blue hydrogen venture at the Humber refinery (Global CCS Institute, 2021).

Table 14.12 shows a list of blue hydrogen facilities currently in development as of June 2021. As we can witness, the list is dominated by European countries, with only a couple of projects in the

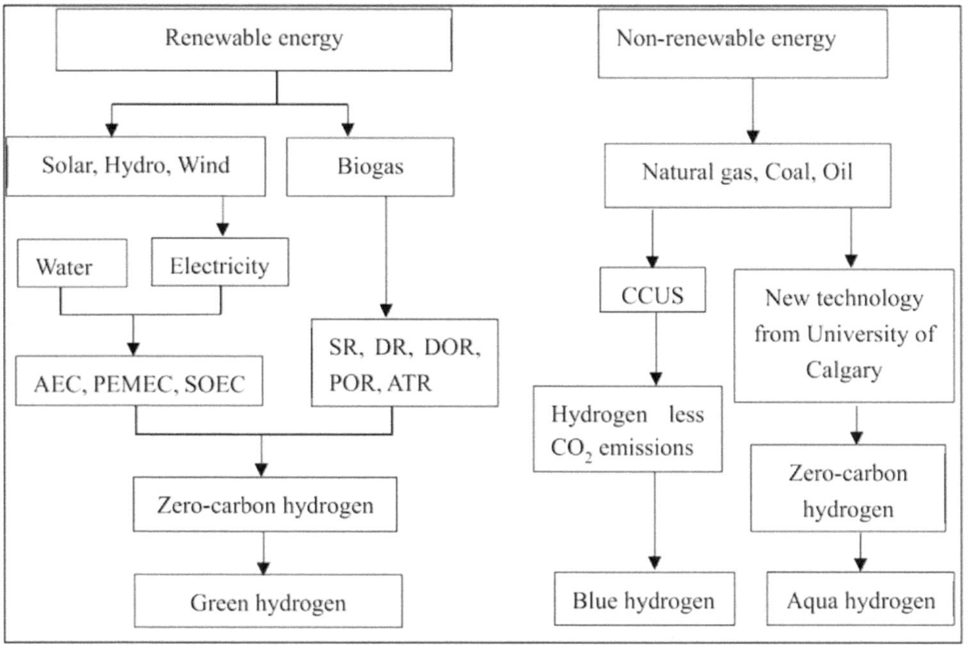

FIGURE 14.29 Low-carbon hydrogen production methods.

Source: Yu et al. (2021).

TABLE 14.12
Blue Hydrogen Production Facilities in Development As of June 2021

Facility	Country	Announced Operational Commencement
Wabash Valley Resources Hydrogen Plant	United States	2022
Air Liquide Refinery Rotterdam	Netherlands	2024
Project Pouakai Hydrogen Production	New Zealand	2024
Shell Refinery Rotterdam	Netherlands	2024
ExxonMobil Benelux Refinery	Netherlands	2024
Air Products Refinery Rotterdam	Netherlands	2024
Acorn Hydrogen	United Kingdom	2025
Clean Energy Systems Carbon Negative Energy Plant – Central Valley	United States	2025
Preem Refinery	Sweden	2025
Barents Blue Clean Ammonia with CCS	Norway	2025
Northern Gas Network H21	United Kingdom	2026
Ravenna Hub – ENI Hydrogen	Italy	2026
Hydrogen to Humber Saltend	United Kingdom	2026–2027
HyNet North West	United Kingdom	Mid-2020s
Polaris CCS Project	Canada	Mid-2020s
Net Zero Teesside – BP H2Teesside	United Kingdom	2027
Humber Zero – Phillips 66 Humber Refinery	United Kingdom	2028
PAU Central Sulawesi Clean Ammonia with CCS	Indonesia	Late 2020s

Note: Excludes facilities that are operating or in construction. Includes facilities where blue hydrogen is an interim product or a final product.

Source: Global CCS Institute (2021).

North American region, one in New Zealand, and another in Indonesia, completing the list. Despite the promising nature of blue hydrogen technology as a carbon emissions mitigation strategy, which we will go deeper into in a separate chapter, there are still essential cost barriers for several countries. Because of this, we expect public–private partnerships between governments and oil and gas producers and joint efforts between novel entrepreneurs and these oil majors to become more widespread over the coming years.

Thanks to all these dynamics, the oil and gas sector has strengthened its position as a global leader in energy technologies by providing resources and *know-how* to CCS projects worldwide. Knowing that there can be a considerable time between our current reliance on fossil fuels and the energy transition that will take us to an energy world centered on clean sources of energy like solar, wind, biomass, and nuclear, private actors have stepped up in providing resources and experience to tackle current environmental challenges. Adaptation is an excellent short-term option to achieve decarbonization pledges without neglecting our existing infrastructure.

14.7 CONCLUSIONS

Strategies of sequestration and capture of carbon are part of a series of measures that are compatible with the current state of things in the industry and energy sectors. At the same time, polluting actors work to mitigate their effects in the medium term, and a process of energy transition to cleaner sources is concurrently underway. Given that around 90% of the emissions associated with the oil and gas industry come from the combustion of hydrocarbons to produce energy, the sector has invested heavily in CCS technologies to mitigate these emissions while maintaining the level of energy production necessary to power the world. Therefore, CCS technologies are not an end but a means to improved processes and a door to the energy transition of the 21st century.

This chapter reviewed the main carbon capture and storage methods used across industrial sectors and the prospects for screening and selecting sites to deploy these technologies. Given the economic constraints and the high installation costs and capital for investment, CCS technologies require a significant amount of work before installation. When considering building a CO_2 capturing station, various elements need to be included in the equation, like the captured method used and the combination of elements needed to interact with the flue gas. This pipeline system will be required to transport the carbon, among other factors. Also, the installation where carbon will be put for long-term storage must be assessed in detail.

Recognizing the important limitation of a potentially prohibitive cost for some countries and actors, CCS technologies are still very promising methods that help decarbonize current energy generation processes, allowing humanity to maintain current levels of electricity production with a significantly lower carbon footprint. CCS methods differ, but there are three different types of technologies to isolate carbon: post-combustion, pre-combustion, and oxy-fuel combustion. These approaches can use different processes, like absorption, adsorption, cryogenic distillation, or more novel ways of separating carbon, like membrane-based separation, calcium looping, or the work of specific microalgae.

The boost in the amount of research performed regarding CCS technologies has been promoted in part by the oil and gas industry, which has been in the middle of all these dynamics and can lend its know-how and experience toward achieving environmental goals that lead the world to an energy transition. In this sense, the sector has struck sensible partnerships with emerging actors and several governments worldwide to establish and maintain a framework that keeps actors in check with their environmental commitments. To help with these, international organizations and think tanks like the United Nations and the Global CCS Institute are key players in the process.

One area where the oil and gas industry can contribute is in the reutilization of large quantities of captured CO_2 to stimulate depleted petroleum reservoirs so that these recover oil thought to be hard, or nearly impossible, to retrieve. In an enhanced oil recovery (EOR) process, carbon is injected deep into reservoirs, which serves two functions: recover more oil from existing reservoirs while also storing CO_2 deep underground so that it cannot reach the atmosphere. With this virtuous cycle of maximizing the efficiency of old petroleum sites while achieving the decarbonization of conventional energy processes, the oil and gas sector is on good footing to be a primary actor in the energy transition process of the next couple of decades.

As is the constant feature with many of these forces, a concerted effort from many actors is required for the energy transition to see the light of day. CCS technologies are one important palliative measure to clean current industrial processes from excessive carbon dioxide generation. Yet more work and awareness are needed to lower costs and extend the technology so more sectors and countries can fulfill their environmental pledges. Consequently, government subsidies and financing of some operations are other important pieces of this puzzle. Incentives of this kind can help industries retrofit energy plants while maintaining or even increasing their electricity generation without increasing their greenhouse emissions.

Carbon capture and storage techniques serve as a mitigation strategy, but success in achieving various decarbonization goals can help galvanize their presence as a set of bridge technologies that connect conventional energy sources with cleaner sources. In a world that still needs to rely on more dense forms of energy for at least a couple more decades, particularly in developing countries, CCS methods have the potential to be deployed worldwide to help diminish the carbon footprint of the many pivotal processes that power the world. Utilizing technological advancements to improve our current situation while we work on longer-term solutions is another key aspect of the human adaptation needed for the future.

REFERENCES

Adu, E., Y. Zhang, and D. Liu. 2018. Current situation of carbon dioxide capture, storage, and enhanced oil recovery in the oil and gas industry. *The Canadian Journal of Chemical Engineering* 9999: 1–29.

Alvarado, V., and E. Manrique. 2010. Enhanced oil recovery: An update review. *Energies* 3: 1529–1575.

Anderson, S., and R. Newell. 2004. Prospects for carbon capture and storage technologies. *Annual Review of Environment and Resources* 29: 109–142.

Babar, M., A. Mukhtar, M. Mubashir, S. Saqib, S. Ullah, A. H. Ali Quddusi, M. A. Bustam, and P. L. Show. 2021. Development of a novel switched packed bed process for cryogenic CO2 capture from natural gas. *Process Safety and Environmental Protection* 147: 878–887.

Balat, H., and C. Öz. 2007. Technical and economic aspects of carbon capture and storage – a review. *Energy Exploration & Exploitation* 25(5): 357–392.

Blamey, J., E. J. Anthony, J. Wang, and P. S. Fennell. 2010. The calcium looping cycle for large-scale CO2 capture. *Progress in Energy and Combustion Science* 36: 260–279.

Boot-Handford, M. E., J. C. Abanades, E. J. Anthony, M. J. Blunt, S. Brandani, N. Mac Dowell, . . ., P. S. Fennell. 2014. Carbon capture and storage update. *Energy & Environmental Science* 7(1): 130–189.

Bui, M., C. S. Adjiman, A. Bardow, E. J. Anthony, A. Boston, S. Brown, . . ., N. Mac Dowell. 2018. Carbon capture and storage (CCS): The way forward. *Energy & Environmental Science* 11: 1062–1176.

Carter, L. D. 2011. *Enhanced oil recovery & CCS*. United States Carbon Sequestration Council.

Creutzig, F., C. Breyer, J. Hilaire, J. Minx, G. P. Peters, and R. Socolow. 2019. The mutual dependence of negative emission technologies and energy systems. *Energy & Environmental Science* 12: 1805–1817.

Davison, J., and K. Thambimuthu. 2005. Technologies for capture of carbon dioxide. *Greenhouse Gas Control Technologies* 7: 3–13.

Delborne, J. A., D. Hasala, A. Wigner, and A. Kinchy. 2020. Dueling metaphors, fueling futures: "Bridge fuel" visions of coal and natural gas in the United States. *Energy Research & Social Science* 61: 101350.

Eide, L. I., M. Batum, T. Dixon, Z. Elamin, A. Graue, S. Hagen, . . ., R. A. Mello Vieira. 2019. Enabling large-scale carbon capture, utilisation, and storage (CCUS) using offshore carbon dioxide (CO2) infrastructure developments—A review. *Energies* 12(10): 1945.

Erans, M., V. Manovic, and E. J. Anthony. 2016. Calcium looping sorbents for CO2 capture. *Applied Energy* 180: 722–742.

Farmahini, A. H., S. Krishnamurthy, D. Friedrich, S. Brandani, and L. Sarkisov. 2021. Performance-based screening of porous materials for carbon capture. *Chemical Reviews* 121: 10666–10741.

Ferguson, R. C., C. Nichols, T. Van Leeuwen, and V. A. Kuuskraa. 2009. Storing CO2 with Enhanced Oil Recovery. *Energy Procedia* 1: 1989–1996.

Fernandes Araújo, O. de Q., and J. L. de Medeiros. 2017. Carbon capture and storage technologies: Present scenario and drivers of innovation. *Current Opinion in Chemical Engineering* 17: 22–34.

Font-Palma, C., D. Cann, and C. Udemu. 2021. Review of cryogenic carbon capture innovations and their potential applications. *C* 7(3): 58.

Global CCS Institute. 2021. *The global status of CCS: 2021*. Australia.

Guo, J.-X., C. Huang, J.-L. Wang, and X.-Y. Meng. 2019. Integrated operation for the planning of CO2 capture path in CCS–EOR project. *Journal of Petroleum Science and Engineering* 106720.

Hanak, D. P., E. J. Anthony, and V. Manovic. 2015. A review of developments in pilot-plant testing and modelling of calcium looping process for CO2 capture from power generation systems. *Energy & Environmental Science* 8: 2199–2249.

Hardisty, P. E., M. Sivapalan, and P. Brooks. 2011. The environmental and economic sustainability of carbon capture and storage. *International Journal of Environmental Research and Public Health* 8(5): 1460–1477.

Hendricks, C., W. Graus, and F. van Bergen. 2004. *Global carbon dioxide storage potential and costs*. Ecofys/TNO-NITG.

Howarth, R. W. 2014. A bridge to nowhere: Methane emissions and the greenhouse gas footprint of natural gas. *Energy Science and Engineering* 2(2): 47–60.

Hussin, F., and M. K. Aroua. 2020. Recent trends in the development of adsorption technologies for carbon dioxide capture: A brief literature and patent reviews (2014–2018). *Journal of Cleaner Production* 253: 119707.

IEA. 2021a. *Carbon capture in 2021: Off and running or another false start?* International Energy Agency: Paris. Available in www.iea.org/commentaries/carbon-capture-in-2021-off-and-running-or-another-false-start [Accessed February 3rd, 2022].

IEA. 2021b. *CCUS in industry and transformation*. Available in www.iea.org/reports/ccus-in-industry-and-transformation [Accessed February 2nd, 2022].

IEA. 2021c. *Net Zero by 2050: A roadmap for the global energy sector*. International Energy Agency.

IPCC. 2005. *Special report on carbon capture and storage*. Intergovernmental Panel on Climate Change: Cambridge, UK, New York.

Jansen, D., M. Gazzani, G. Manzolini, E. van Dijk, and M. Carbo. 2015. Pre-combustion CO2 capture. *International Journal of Greenhouse Gas Control* 40: 167–187.

Ji, G., and M. Zhao. 2017. Membrane separation technology in carbon capture. In Yun, Y. (Ed.). *Recent advances in carbon capture and storage*. IntechOpen: London.

Karimi, F., and R. Khalilpour. 2015. Evolution of carbon capture and storage research: Trends of international collaborations and knowledge maps. *International Journal of Greenhouse Gas Control* 37: 362–376.

Kolster, C., M. S. Masnadi, S. Krevor, N. Mac Dowell, and A. R. Brandt. 2017. CO2 enhanced oil recovery: A catalyst for gigatonne-scale carbon capture and storage deployment? *Energy & Environmental Science* 10: 2594–2608.

Koornneef, J., A. Faaij, and W. Turkenburg. 2008. The screening and scoping of Environmental Impact Assessment and Strategic Environmental Assessment of Carbon Capture and Storage in the Netherlands. *Environmental Impact Assessment Review* 28(6): 392–414.

Lau, H. C., S. Ramakrishna, K. Zhang, and A. V. Radhamani. 2021. The role of carbon capture and storage in the energy transition. *Energy & Fuels* 35(9): 7364–7386.

Lehtveer, M., and A. Emanuelsson. 2021. BECCS and DACCS as Negative emission providers in an intermittent electricity system: Why levelized cost of carbon may be a misleading measure for policy decisions. *Frontiers in Climate* 3: 647276.

Levi, M. 2013. Climate consequences of natural gas as a bridge fuel. *Climatic Change* 118(3–4): 609–623.

Li, X., Z. Peng, Y. Pei, T. Ajmal, K.-J. Rana, A. Aitouche, and R. Mobasheri. 2021. Oxy-fuel combustion for carbon capture and storage in internal combustion engines – A review. *International Journal of Energy Research*: 1–18.

Lin, L.-C., A. H. Berger, R. L. Martin, J. Kim, J. A. Swisher, K. Jariwala, . . ., B. Smit. 2012. In silico screening of carbon-capture materials. *Nature Materials* 11(7): 633–641.

Ling, J., P. Xiao, A. Ntiamoah, D. Xu, P. Webley, and Y. Zhai. 2016. Strategies for CO2 capture from different CO2 emission sources by vacuum swing adsorption technology. *Chinese Journal of Chemical Engineering* 24(4): 460–467.

McKinsey & Company. 2008. *Carbon capture & storage: Assessing the economics*. McKinsey Climate Change Initiative.

Melzer, L. S. 2012. *Carbon dioxide enhanced oil recovery (CO2 EOR): Factors involved in adding carbon capture, utilization and storage (CCUS) to enhanced oil recovery*. Melzer Consulting.

Middleton, R. S., J. M. Bielicki, G. N. Keating, and R. J. Pawar. 2011. Jumpstarting CCS using refinery CO_2 for enhanced oil recovery. *Energy Procedia* 4: 2185–2191.

Neagu, O. 2019. The link between economic complexity and carbon emissions in the European Union countries: A model based on the Environmental Kuznets Curve (EKC) approach. *Sustainability* 11: 4753.

Nemitallah, M. A., M. A. Habib, H. M. Badr, S. A. Said, A. Jamal, R. Ben-Mansour, E. M. A. Mokheimer, and K. Mezghani. 2017. Oxy-fuel combustion technology: Current status, applications, and trends. *International Journal of Energy Research* 41(12): 1670–1708.

NETL. 2010. *Carbon dioxide enhanced oil recovery: Untapped domestic energy supply and long term carbon storage solution*. U.S. Department of Energy, National Energy Technology Laboratory.

NETL. 2017. *Best practices: Site screening, site selection, and site characterization for geologic storage projects: 2017 Revised edition*. U.S. Department of Energy, National Energy Technology Laboratory.

Niu, X., and S. W. Duiker. 2006. Carbon sequestration potential by afforestation of marginal agricultural land in the Midwestern U.S. *Forest Ecology and Management* 223(1–3): 415–427.

Office of Fossil Energy and Carbon Management. 2022a. *Post-combustion carbon capture research*. United States Department of Energy. Available in www.energy.gov/fecm/science-innovation/carbon-capture-and-storage-research/carbon-capture-rd/post-combustion-carbon [Accessed February 22nd, 2022].

Office of Fossil Energy and Carbon Management. 2022b. *Pre-combustion carbon capture research*. United States Department of Energy. Available in www.energy.gov/fecm/science-innovation/carbon-capture-and-storage-research/carbon-capture-rd/pre-combustion-carbon [Accessed February 22nd, 2022].

Olabi, A. G., K. Obaideen, K. Elsaid, T. Wilberforce, E. T. Sayed, H. M. Maghrabie, and M. A. Abdelkareem. 2022. Assessment of the pre-combustion carbon capture contribution into sustainable development goals SDGs using novel indicators. *Renewable and Sustainable Energy Reviews* 153 (January): 111710.

Padurean, A., C.-C. Cormos, and P.-S. Agachi. 2012. Pre-combustion carbon dioxide capture by gas–liquid absorption for Integrated Gasification Combined Cycle power plants. *International Journal of Greenhouse Gas Control* 7: 1–11.

Pierce Jr., R. J. 2012. Natural gas: A long bridge to a promising destination. *Utah Environmental Law Review* 32: 245.

Pires, J. C. M., F. G. Martins, M. C. M. Alvim-Ferraz, and M. Simões. 2011. Recent developments on carbon capture and storage: An overview. *Chemical Engineering Research and Design* 89: 1446–1460.

Rai, V., D. G. Victor, and M. C. Thurber. 2010. Carbon capture and storage at scale: Lessons from the growth of analogous energy technologies. *Energy Policy* 38: 4089–4098.

Rashid Gill, A., K. K. Viswanathan, and S. Hassan. 2018. The Environmental Kuznets Curve (EKC) and the environmental problem of the day. *Renewable and Sustainable Energy Reviews* 81: 1636–1642.

Raynal, L., P.-A. Bouillon, and A. Gomez, and P. Broutin. 2011. From MEA to demixing solvents and future steps, a roadmap for lowering the cost of post-combustion carbon capture. *Chemical Engineering Journal* 171(3): 742–752.

Raza, A., R. Gholami, R. Rezaee, V. Rasouli, and M. Rabiei. 2019. Significant aspects of carbon capture and storage – A review. *Petroleum* 5(4): 335–340.

RFF. 2020. *Carbon capture and storage 101*. Resources for the Future. Available in https://media.rff.org/documents/CCS_101.pdf [Accessed February 23rd, 2022].

Rinanti, A., E. Kardena, D. Indriani Astuti, and K. Dewi. 2013. Screening of potential photosynthetic microalgae from wastewater treatment plant for carbon dioxide capture and storage (CCS). *Asian Transactions on Science & Technology* 3(1): 1–8.

Rodosta, T. D., J. T. Litynsky, S. I. Plasynski, S. Hickman, S. Frailey, and L. Myer. 2011. U.S. department of energy's site screening, site selection, and initial characterization for storage of CO2 in deep geological formations. *Energy Procedia* 4: 4664–4671.

Sayre, R. 2010. Microalgae: The potential for carbon capture. *BioScience* 60(9, October): 722–727.

Scholes, C. A., K. H. Smith, S. E. Kentish, and G. W. Stevens. 2010. CO2 capture from pre-combustion processes – Strategies for membrane gas separation. *International Journal of Greenhouse Gas Control* 4(5): 739–755.

Shannak, S., and A. Malov. 2021. Could China meet its emission reduction goal by CO2-EOR. *Journal of Science and Technology Policy Management* 12(1): 24–40.

Singh, U. B., and A. S. Ahluwalia. 2012. Microalgae: A promising tool for carbon sequestration. *Mitigation and Adaptation Strategies for Global Change* 18(1): 73–95.

Smith, P. 2004. Carbon sequestration in croplands: The potential in Europe and the global context. *European Journal of Agronomy* 20(3): 229–236.

Song, C., Q. Liu, S. Deng, H. Li, and Y. Kitamura. 2019. Cryogenic-based CO2 capture technologies: State-of-the-art developments and current challenges. *Renewable and Sustainable Energy Reviews* 101: 265–278.

Sweatman, R. E., M. E. Parker, and S. L. Crookshank. 2009. Industry experience with CO2-enhanced oil recovery technology. *SPE International Conference on CO2 Capture, Storage, and Utilization*.

Szulczewski, M. L., C. W. MacMinn, H. J. Herzog, and R. Juanes. 2012. Lifetime of carbon capture and storage as a climate-change mitigation technology. *Proceedings of the National Academy of Sciences* 109(14): 5185–5189.

Tan, R. R. 2019. Data challenges in optimizing biochar-based carbon sequestration. *Renewable and Sustainable Energy Reviews* 104: 174–177.

Tan, Y., W. Nookuea, H. Li, E. Thorin, and J. Yan. 2016. Property impacts on Carbon Capture and Storage (CCS) processes: A review. *Energy Conversion and Management* 118: 204–222.

Tcvetkov, P., A. Cherepovitsyn, and S. Fedoseev. 2019. Public perception of carbon capture and storage: A state-of-the-art overview. *Heliyon* 5(12), e02845.

Theo, W. L., J. S. Lim, H. Hashim, A. A. Mustaffa, and W. S. Ho. 2016. Review of pre-combustion capture and ionic liquid in carbon capture and storage. *Applied Energy* 183: 1633–1663.

Thomas, S. 2008. Enhanced oil recovery – An overview. *Oil & Gas Science and Technology – Rev. IFP* 63(1): 9–19.

Torvanger, A., and J. Meadowcroft. 2011. The political economy of technology support: Making decisions about carbon capture and storage and low carbon energy technologies. *Global Environmental Change* 21(2): 303–312.

Turan, G. 2020. *CCS: Applications and opportunities for the oil and gas industry*. Global CCS Institute Brief.

Vega, F., F. M. Baena-Moreno, L. M. Gallego Fernández, E. Portillo, B. Navarrete, and Z. Zhang 2020. Current status of CO2 chemical absorption research applied to CCS: Towards full deployment at industrial scale. *Applied Energy* 260: 114313.

VijayaVenkataRaman, S., S. Iniyan, and R. Goic. 2012. A review of climate change, mitigation and adaptation. *Renewable and Sustainable Energy Reviews* 16(1): 878–897.

Wall, T., R. Stanger, and S. Santos. 2011. Demonstrations of coal-fired oxy-fuel technology for carbon capture and storage and issues with commercial deployment. *International Journal of Greenhouse Gas Control* 5S: S5–S15.

Wilberforce, T., A. Baroutaji, B. Soudan, A. H. Al-Alami, and A. G. Olabi. 2018. Outlook of carbon capture technology and challenges. *Science of the Total Environment* 657: 56–72.

Yang, L., M. Xu, Y. Yang, J. Fan, and X. Zhang. 2019. Comparison of subsidy schemes for carbon capture utilization and storage (CCUS) investment based on real option approach: Evidence from China. *Applied Energy* 255: 113828.

Yosef, G., R. Walko, R. Avisar, F. Tatarinov, E. Rotenberg, and D. Yakir. 2018. Large-scale semi-arid afforestation can enhance precipitation and carbon sequestration potential. *Scientific Reports* 8: 996.

Yu, M., K. Wang, and H. Vredenburg. 2021. Insights into low-carbon hydrogen production methods: Green, blue and aqua hydrogen. *International Journal of Hydrogen Energy* 46(41): 21261–21273.

Zhai, H., and E. S. Rubin. 2018. Systems analysis of physical absorption of CO2 in ionic liquids for pre-combustion carbon capture. *Environmental Science & Technology* 52: 4996–5004.

Zhang, S., L. Liu, L. Zhang, Y. Zhuang, and J. Du. 2018. An optimization model for carbon capture utilization and storage supply chain: A case study in Northeastern China. *Applied Energy* 231: 194–206.

Zhang, X., N. P. Myhrvold, Z. Hausfather, and K. Caldera. 2016. Climate benefits of natural gas as a bridge fuel and potential delay of near-zero energy systems. *Applied Energy* 167: 317–322.

15 Electrical Transportation and Vehicles

Mesut Yurukcu, Jorge Saldaña, Narendra Boppana, Sepehr Arbabi, Jim Henry, and Sebahattin Ziyanak

15.1 INTRODUCTION

The rising demand for traditional vehicles that use non-renewable sources like fossil fuels has intensified the current environmental and energy crises. In recent years, the extensive usage of gasoline and diesel has caused a massive surge in the number of greenhouse gases (carbon dioxide and nitrogen oxide), becoming a significant factor in global warming and climate change (Ju et al., 2021; Zhao et al., 2022). Transportation is one of the major economic sectors that has contributed to air pollution (Higueras-Castillo et al., 2019).

Over the period between 1990 and 2019, the transportation sector's carbon dioxide (CO_2) emissions increased by about 80% (Statistics & Facts, 2022). By 2020, passenger cars were the most significant source of CO_2 emissions, accounting for 41% of the emissions from the transportation sector globally (Tiseo, 2020). Experts forecast that the number of private cars worldwide will double by 2050 (Bauer, 2018), meaning, the number of private cars on the roads could reach 2–2.5 billion by 2050 (Ghadikolaei et al., 2021). In turn, the rise in the number of cars with internal combustion engines will increase the dependence on oil, as the transport sector makes up roughly 60% of the total oil demand (International Energy Agency, 2021a). These facts pose a severe challenge to global energy security and supply.

Wang, Z. (2014), proposed a model and control algorithms for electric vehicle charging within a multi-agent system to assess its impact on the power grid. The popularity of electric vehicles is due to reduced operational costs and potential CO_2 emission reductions. Simulation results show that uncontrolled EV charging can threaten power system stability.

The diffusion of electric vehicles (EVs) provides a sustainable answer to environmental and energy crises. It assists in accomplishing the goals for achieving carbon neutrality under the Paris Agreement and the Green Deal. EVs are powered by electricity, a secondary energy source that replaces fossil fuels and does not release pollutants. The environmental detriment can be minimized, especially when electricity is generated from renewable sources such as wind, sun, biomass, and others (European Environment Agency, 2018; Lucian, 2021).

Thus, the move to electric mobility contributes to the decline of environmental pollution, which is also beneficial for public health and helps reduce fossil energy consumption (Almeida Neves et al., 2019; He et al., 2018; Wu et al., 2017; Yuan et al., 2015); this marks a new paradigm of sustainable energy (Higueras-Castillo et al., 2019). Since 2010, the growth rate of EVs has been higher than that of conventional vehicles (Song et al., 2022). In recent years, international sales of electric cars have risen drastically. In 2018, the global stock of electric passenger cars reached 5 million, a 63% increase from the previous year. At this point, China dominated with about 45%, Europe followed with 24%, and the United States achieved 22% of the global fleet (International Energy Agency, 2019). By the end of 2021, there will be approximately 16.5 million electric cars on the world's roads, three times the amount in 2018 (International Energy Agency, 2022).

Therefore, as demonstrated in Figure 15.1, electric vehicle sales bucked the trend, increasing to thousands and representing 4.1% of total car sales in 2020 (Paoli et al., 2022). For the first time,

DOI: 10.1201/9781003282273-15

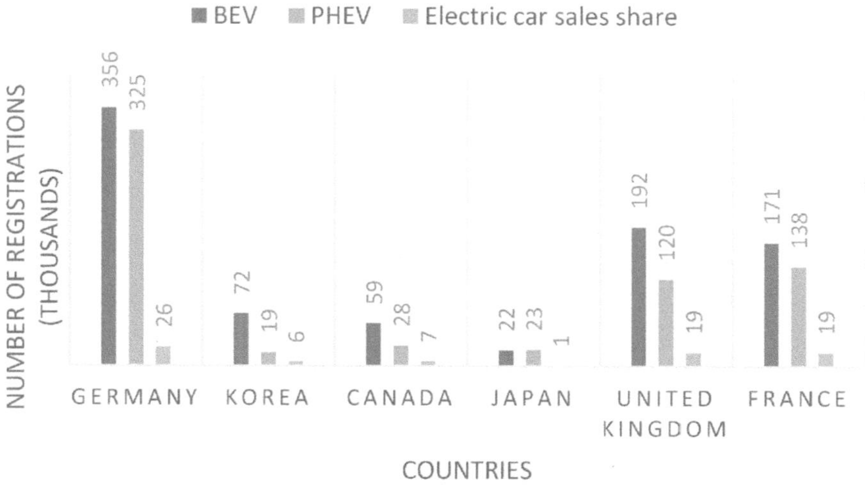

FIGURE 15.1 EV registrations for 2021.

Source: Apata et al. (2023).

Europe surpassed China to become the largest electric vehicle market in the world (International Energy Agency, 2021b). In 2021, electric vehicle sales more than doubled compared to the sales in 2020, totaling 6.6 million new registrations. Nonetheless, their market share was only about 9% of the global car market in 2021. Regarding significant markets, China and Europe drive electric vehicle sales worldwide (Paoli, 2022).

Electric transportation and vehicles have rapidly developed in recent years with the advances in battery technology, motor design, and electric vehicle infrastructure (Albatayneh et al., 2023; Alarcón et al., 2023). Electric transportation also offers an efficient and reliable way to transport goods and services and can be used to reduce traffic congestion (Apata et al., 2023; Garg et al., 2023).

Studying electrical transportation and vehicles is crucial because it explains the industry's current state and the technology being used. It also explains the industry's challenges, such as environmental impacts and cost-effectiveness. Additionally, understanding electrical transportation and vehicles will provide the necessary background information to create an engaging chapter that covers the history and development of the industry, as well as its current challenges, potential solutions, and prospects.

This chapter will explore the current state of electric transportation and vehicles, their challenges, and the opportunities they present. It will discuss the range of vehicle types available, their different types of power sources, and the range of electric vehicle infrastructure being developed. It will also look at the potential for electric transportation to reduce emissions, increase efficiency, and provide more convenient and affordable transportation options. Finally, this chapter will examine the current regulatory framework for electric vehicles and discuss how it could be improved. By delving into this subject, this chapter will give an overview of the current state of electric transportation and vehicles and their challenges and opportunities.

15.2 SCREENING, SELECTION, DESIGN, AND ECONOMICS

Electrical transportation and vehicles (EVs) are important in our society for many reasons. EVs provide a more efficient way of getting around and reducing emissions significantly compared to gasoline-powered vehicles by being cheaper and more reliable (Ali et al., 2023). EVs are becoming increasingly popular due to their environmental and economic benefits, which include reducing air pollution and greenhouse gas emissions, providing a more cost-effective and efficient means of

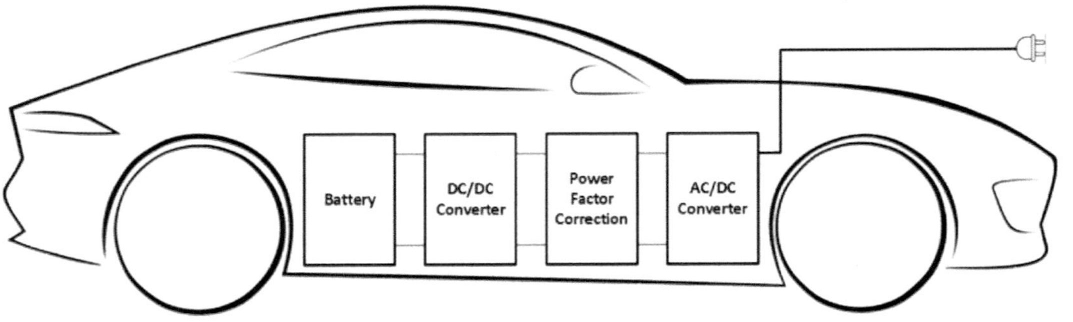

FIGURE 15.2 Onboard EV charger components.

Source: Nour et al. (2020).

TABLE 15.1
Benefits and Challenges of EVs

Benefits	Challenges
Emissions localized at power stations and not in cities, which improves citizens' health	Limited electric range
High efficiency because of high electric motor efficiency	Limited number of charging stations and long charging (refueling) time
Low (for PHEV) or no emissions (for BEV)	Negative impacts on the power grid if uncontrolled charging is used
Fuel or energy independence from countries which produce fuel	Consumer acceptance
Simple transmission, fewer moving parts, and low maintenance	High prices
Less noise pollution	Battery low energy density
Provides electrical services to the power system and helps RESs integration	Expensive battery and needs replacement after few years

Source: Nour et al. (2020).

travel, and being better equipped with the latest technology (Figure 15.2), making them more reliable and safer than ever before (Alanazi, 2023).

Electric vehicles (EVs) are powered by electricity rather than gasoline, which makes them more efficient and cost-effective than traditional vehicles. EVs can reach higher speeds and have greater acceleration than traditional vehicles, allowing for smoother and quicker trips. Additionally, because EVs require less maintenance than traditional vehicles, they can help save money in the long run (Nour et al., 2020).

The importance of electrical transportation and vehicles is becoming increasingly apparent in today's ever-changing world. With the rising fuel costs and the global focus on sustainability, an alternative means of transportation is more necessary than ever. Electric vehicles (EVs) offer numerous benefits (Table 15.1) over gas-powered vehicles, including reduced emissions, lower operating costs, and improved performance (Nour et al., 2020).

In addition to their environmental and economic benefits, EVs can offer improved performance (Table 15.2). EVs have instant acceleration, accelerating faster than conventional gas-powered vehicles. This improved performance can be especially helpful in city driving scenarios, allowing drivers to navigate busy streets quickly and safely (Chan, 2002).

TABLE 15.2
Performance Goal of EV Batteries As Set by USABC

	Parameters	Mid-Term	Long-Term
Primary goals	Energy density (C/3 discharge rate) (Wh/L)	135	300
	Specific energy (C/3 discharge rate) (Wh/kg)	80 (desired: 100)	200
	Power density (W/I)	250	600
	Specific power (80% DOD/30 s) (W/kg)	150 (desired: 200)	400
	Lifetime (year)	5	10
	Cycle life (80% DOD) (cycles)	600	1,000
	Price (USD/kWh)	<150	<100
	Operating temperature (°C)	−30 to 65	−40 to 84
	Recharging time (hour)	<6	3 to 6
	Fast recharging time (40– 80% SOC) (hour)	0.25	
Secondary goals	Self-discharge (%)	<15 (48 h)	<15 (month)
	Efficiency (C/3 discharge, 6 h charge) (%)	75	80
	Maintenance	No maintenance	No maintenance
	Resistance to abuse	Tolerance	Tolerance
	Thermal loss	3.2 W/kWh	3.2 W/kWh

Source: Chan (2002).

Overall, the advantages of electric transportation and vehicles are reduced emissions, lower operating costs, and improved performance. As the world continues to focus on sustainability and reducing emissions, the importance of EVs will only increase.

15.3 TYPES OF ELECTRICAL TRANSPORTATION/VEHICLES

Various types of electric vehicles (EVs) can be classified based on their energy converter (internal combustion engine or electric motor), power source (battery, fuel cell, or gasoline), and charging method (charging station or home charger) (Shaukat et al., 2018). A concise overview of these EV types is provided in the subsequent subsections. The visual representation of the diverse EV structures is depicted in Figure 15.3 (Nour et al., 2020).

In their study, Monteiro et al. 2011 demonstrated that due to the significant reliance on imported fossil fuels and the increasing oil prices, there is a critical need to explore alternatives for traditional people transportation systems. Electric mobility, particularly electric vehicles (EV) and plug-in hybrid electric vehicles (PHEV), emerged as a natural and promising solution. They focused on analyzing the potential impacts of battery charging systems on grid power quality within the context of a smart grid. The study considered the current consumption based on a typical electric consumption profile and investigated voltage degradation for many households. Two EV battery chargers are considered: traditional and smart chargers with sinusoidal current consumption and a unitary power factor. The paper presented simulation results regarding the integration of EVs and PHEVs in terms of power quality, along with experimental findings of a smart charger specially designed for EV charging (Monteiro et al., 2011).

Gray et al. (2014) studied the impact of electric vehicles on power quality in electric distribution systems. Computational studies were conducted using a Monte Carlo approach (Figure 15.4) to evaluate voltage deviations, including under- and/or overvoltage and voltage imbalance. Additionally, the assessment covered distribution transformer overload and unbalance for various vehicle types (plug-in hybrid and battery electric) and vehicle penetration rates, ranging up to 50%, while

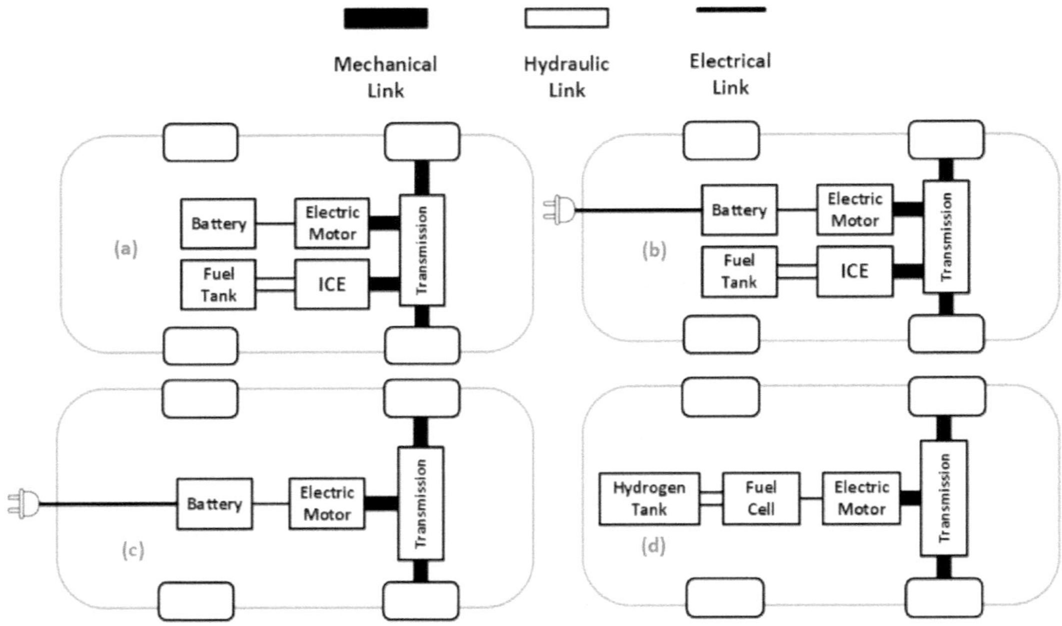

FIGURE 15.3 Basic structure of different electric vehicle (EV) types: (a) hybrid electric vehicle (HEV); (b) plug-in hybrid electric vehicle (PHEV); (c) battery electric vehicle (BEV); and (d) fuel cell electric vehicle (FCEV).

Source: Nour et al. (2020).

considering level 1 and level 2 charging. The findings from the Monte Carlo analysis indicated that battery electric vehicles could cause more overload to distribution transformers than plug-in hybrid electric vehicles. Moreover, it was observed that level 2 and level 1 charging methods could pose specific issues, with level 2 charging leading to secondary bus undervoltage and level 1 charging causing the transformer to be unbalanced (Gray et al., 2014).

As shown in Figure 15.5, there are several types of electric vehicles, including battery electric vehicles (BEVs), plug-in hybrid electric vehicles (PHEVs), hybrid electric vehicles (HEVs), and fuel cell electric vehicles (FCEVs).

Battery electric vehicles (BEVs) are powered entirely by an electric motor and rechargeable batteries (Figure 15.6). They have no internal combustion engine and must be plugged in to recharge.

With continuous advancements in battery technology and price reductions, battery electric vehicles (BEVs) are projected to become the dominant force in the EV market. BEVs rely exclusively on electric motors (EM) powered by a battery and do not incorporate internal combustion engines (ICE). The driving range of BEVs is contingent upon the capacity of their batteries. A significant advantage of BEVs is their ability to produce zero emissions locally, making them particularly vital in large urban centers (Nour et al., 2020).

Plug-in hybrid electric vehicles (PHEVs) have an electric motor and an internal combustion engine. They can be charged from an external power source, and they can also be fueled with gasoline or diesel. Plug-in hybrid electric vehicles are designed with the capability to charge their battery through multiple sources, including regenerative braking, the internal combustion engine (ICE), and an EV charger connected to the distribution network. PHEVs are distinguished by their larger electric motor power, smaller ICE, and bigger battery capacity compared to conventional hybrid electric vehicles (HEVs), which aim to extend their electric range. This enables PHEVs to operate in an all-electric mode, utilizing the electric motor exclusively and resulting in zero greenhouse gas

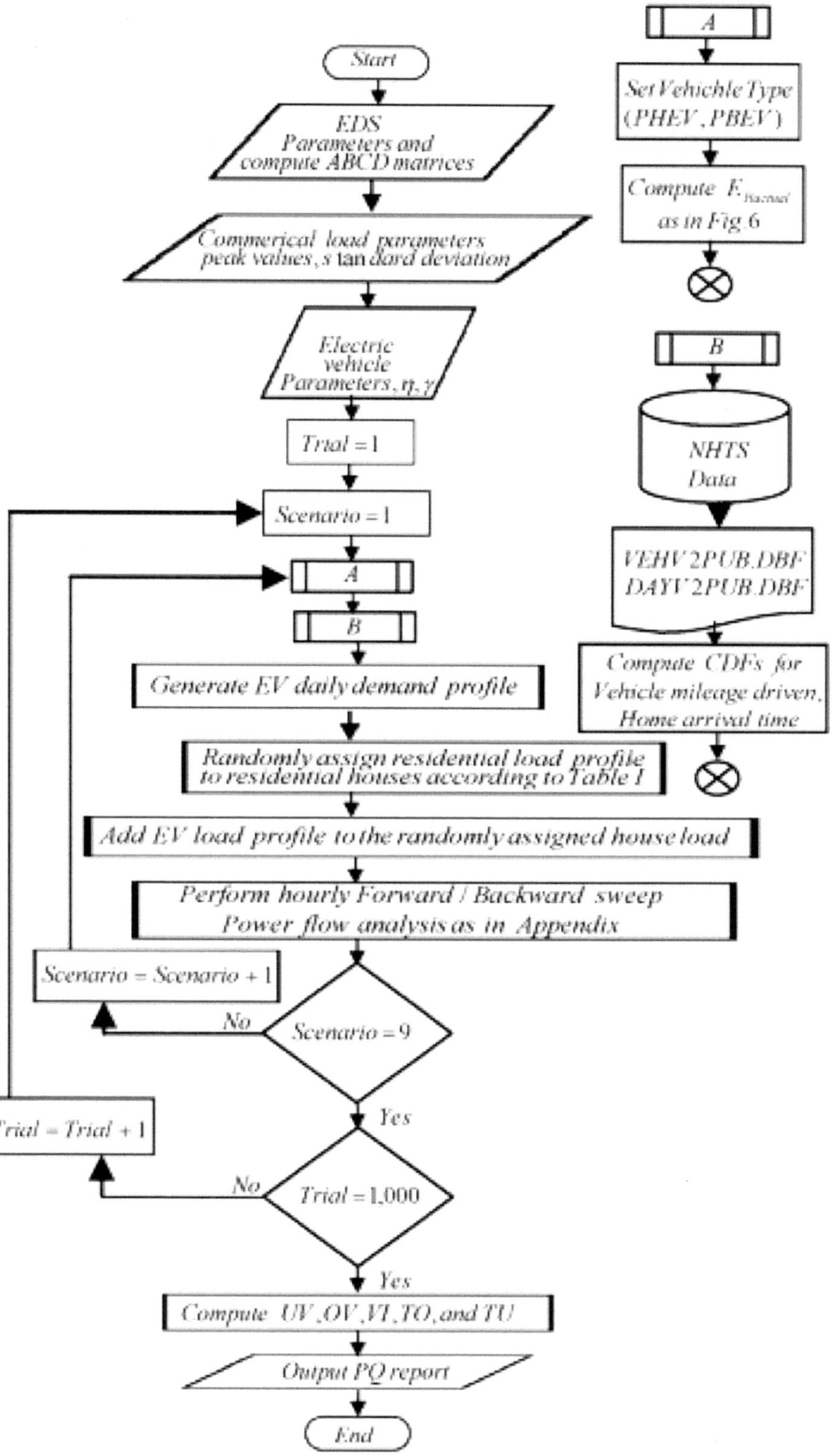

FIGURE 15.4 Flowchart of the Monte Carlo simulation.

Source: Gray et al. (2014).

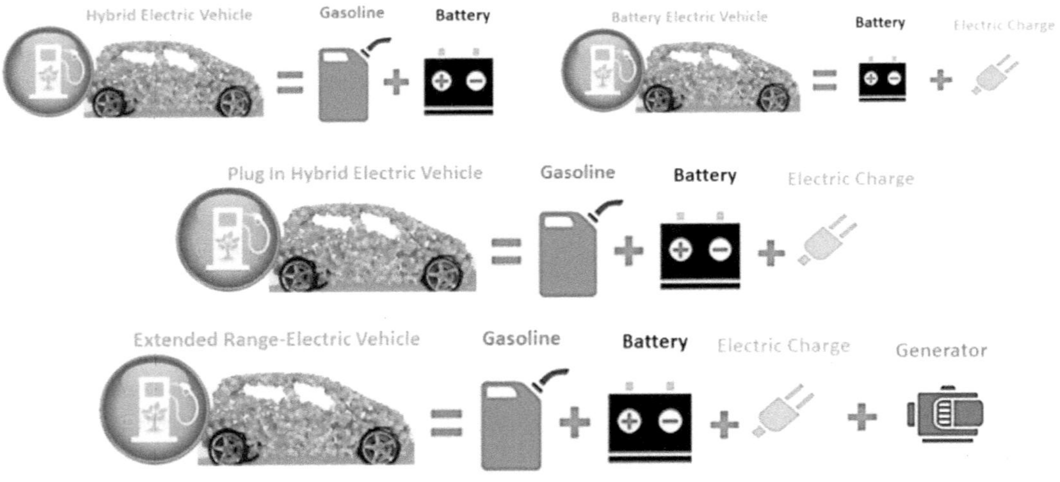

FIGURE 15.5 Types of EV.

Source: Shaukat et al. (2018).

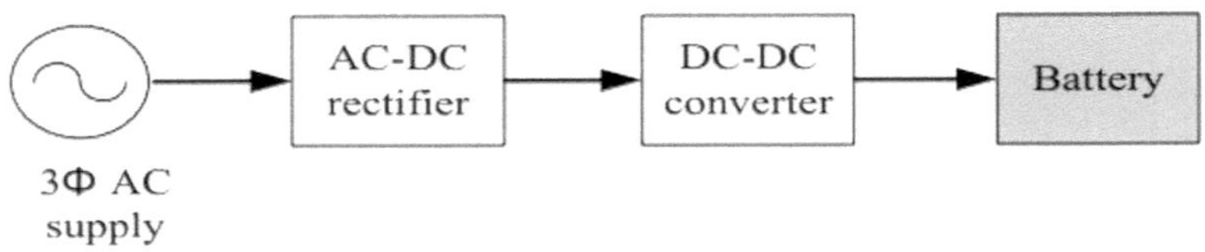

FIGURE 15.6 Basic components of a battery charger.

Source: Shareef et al. (2016).

(GHGs) emissions. The battery's capacity determines the range of electric operation (Table 15.3) (Tie et al., 2013; Un-Noor et al., 2017).

Because of having a smaller battery capacity, PHEVs typically have a limited negative impact on electric power systems and possess restricted capabilities to provide electrical services. PHEVs can be configured in various hybrid setups, offering flexibility in their operation (Nour et al., 2020).

Hybrid electric vehicles (HEVs) have an electric motor and an internal combustion engine but cannot be charged from an external power source. Instead, they rely on regenerative braking and the internal combustion engine to recharge their batteries. Hybrid electric vehicles (HEVs) are similar to internal combustion engine vehicles (ICEVs) but incorporate larger electric motors (EM) and batteries. The battery in HEVs can be charged through regenerative braking and the ICE during light loads. Typically, the battery and electric motor drive the vehicle at lower speeds, while the ICE takes over at higher speeds. The electric motor can also assist the ICE during high loads, improving overall vehicle performance and efficiency. HEVs demonstrate lower greenhouse gas emissions and fuel consumption compared to ICEVs. Notably, they do not rely on charging from the distribution network through EV chargers (Tie et al., 2013; Un-Noor et al., 2017). Consequently, they do not adversely impact the power system due to battery charging, nor can they provide any electrical services (Nour et al., 2020).

Fuel cell electric vehicles (FCEVs) use hydrogen fuel cells to produce electricity, which powers an electric motor (Table 15.4). They emit only water vapor and have longer driving ranges than BEVs. Like battery electric vehicles (BEVs), fuel cell electric vehicles (FCEVs) (Figure 15.7) are also powered by electric motors (EM). However, instead of using batteries, FCEVs employ fuel

TABLE 15.3
Market Penetration and Specification of Plug-In Hybrid Electric (PHEV) and Plug-In Battery Electric Vehicles (PBEV)

PBEV/PHEV	Market Penetration	Electric Range (Miles)	Battery Capacity (kWh)	Specific Energy (kWh/mile)
Chevy Volt (PHEV)	52.45%	40	16	0.36
Nissan Leaf (PBEV)	26.79%	109	24	0.24
Navistar Electric Truck (PBEV)	20.76%	100	80	0.8

Source: Gray et al. (2014).

FIGURE 15.7 Pininfarina H2 Speed is a supercar employing hydrogen fuel cells.

Source: Un-Noor et al. (2017).

cells. These vehicles are refueled with hydrogen, and the fuel cell converts the chemical energy in hydrogen gas into electric energy, which then powers the electric motor. Hydrogen can be produced through different methods, such as natural gas or water electrolysis. FCEVs offer a short refueling time, like internal combustion engine vehicles (ICEVs) (Tie et al., 2013; Un-Noor et al., 2017).

In general, various electric vehicles with their own advantages and disadvantages are currently accessible. As technology progresses, we anticipate the emergence of even more diverse types of electric vehicles.

15.3.1 Land

Battery electric vehicles (BEVs) and plug-in hybrid electric vehicles (PHEVs) are the two main types of EV cars under this category. Table 15.5 summarizes the advantages and disadvantages of different motor types and shows some vehicles using different motor technologies (Un-Noor et al., 2017).

TABLE 15.4
Comparison of Different Vehicle Types

EV Type	Driving Component	Energy Source	Features	Problems
BEV	• Electric motor	• Battery • Ultracapacitor	• No emission • Not dependent on oil • Range depends largely on the type of battery used • Available commercially	• Battery price and capacity • Range • Charging time • Availability of charging stations • High price
HEV	• Electric motor • ICE	• Battery • Ultracapacitor • ICE	• Very little emission • Long range • Can get power from both electric supply and fuel • Complex structure having both electrical and mechanical drivetrains • Available commercially	• Management of the energy sources • Battery and engine size optimization
FCEV	• Electric motor	• Fuel cell	• Very little or no emission • High efficiency • Not dependent on supply of electricity • High price • Available commercially	• Cost of fuel cell • Feasible way to produce fuel • Availability of fueling facilities

Source: Un-Noor et al. (2017).

Electric buses have become a desirable alternative for urban transit systems as societies emphasize environmentally friendly and sustainable modes of transportation. Compared to their regular diesel equivalents, these buses have substantially lower carbon emissions and noise pollution thanks to their electric propulsion systems, which are usually battery-powered. Electric buses, which provide a quieter and cleaner form of public transportation, are becoming more feasible and practical due to battery technology and charging infrastructure developments.

The phenomenon known as "bus bunching" occurs when several buses traveling on the same route arrive at a stop simultaneously, frequently due to delays, which throws off the regular spacing between stops. Bus bunching is especially common in urban centers (Tsoi et al., 2023), where buses can generate significant traffic noise due to the continuous stop-and-go patterns along the bus queues. Electric buses can help mitigate the effects of bus bunching and improve air quality by producing zero emissions and eliminating exhaust fumes, a health hazard in big cities.

Electric buses (Figure 15.8) are also highly energy-efficient and can reduce greenhouse gas emissions. They can also improve air quality by eliminating particulate pollution and pollutants contributing to the ground-level ozone layer. Electric buses can also reduce traffic noise. When all bus fleets are electrified, around 60% of the population can benefit from reducing 1 dBA in the street environment, 15.3% for 1–2 dBA, and 4.3% for more than 2 dBA (Table 15.6). The estimated reduction of preventable deaths and preventable cases of diseases per 100,000 population are 4.15 and 112.99, respectively (Tsoi et al., 2023).

In 2014, Seneca became the first city in the world to launch an all-electric bus fleet. The buses have outperformed their diesel equivalents in fuel and maintenance costs and exceeded expectations regarding charging time, range, and battery life (Horrox, & Casale, 2019). Later, the Chicago Transit Authority's (CTA) rollout of electric buses was one of the first major tests in a cold winter

TABLE 15.5
Different Motor Types: Advantages, Disadvantages, and Usage

Motor Type	Advantage	Disadvantage	Vehicles Used In
Brushed DC motor	• Maximum torque at low speed	• Bulky structure • Low efficiency • Heat generation at brushes	Fiat Panda Elettra (series DC motor), Conceptor G-Van (separately excited DC motor)
Permanent magnet brushless DC motor (BLDC)	• No rotor copper loss • More efficiency than induction motors • Lighter • Smaller • Better heat dissipation • More reliability • More torque density • More specific power	• Short constant power range • Decreased torque with increase in speed • High cost because of PM	Toyota Prius (2005)
Permanent magnet synchronous motor (PMSM)	• Operable in efferent speed ranges without using gear systems • Efficient • Compact • Suitable for in-wheel app cation • High torque even at very low speeds	• Huge iron loss at high speeds during in-wheel operation	Toyota Prius, Nissan Leaf, Soul EV
Induction motor (IM)	• The most mature commutatorless motor drive system • Can be operated like a separately excited DC motor by employing field orientation control		Tesla Model S, Tesla Model X, Toyota RAV4, GM EV1
Switched Reluctance motor (SRM)	• Simple and robust construction • Low cost • High speed • Less chance of hazard • Long constant power range • High power density	• Very noisy • Low efficiency • Larger and heavier than PM machines • Complex design and control	Chloride Lucas
Synchronous reluctance motor (SynRM)	• Robust • Fault-tolerant • Efficient • Small	• Problems in controllability and manufacturing • Low power factor	
PM-assisted synchronous reluctance motor	• Greater power factor than SynRMs • Free from demagnetizing problems observed in IPM		BMW i3
Axial flux ironless permanent magnet motor	• No iron used in outer rotor • No stator core • Lightweight • Better power density • Minimized copper loss • Better efficiency • Variable-speed machine • Rotor capable of being fitted to the lateral side of the wheel		Renovo Coupe

Source: Un-Noor et al. (2017).

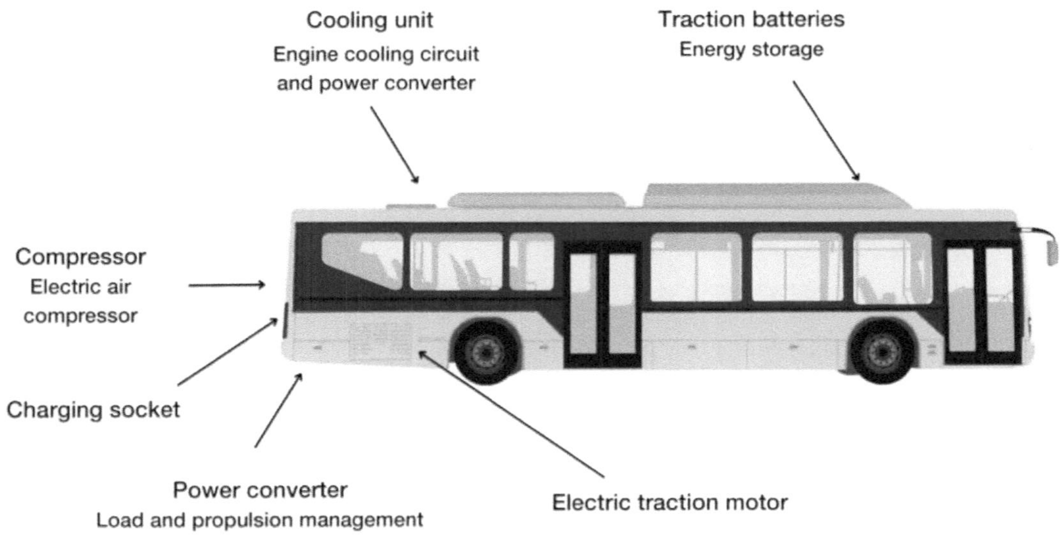

FIGURE 15.8 The different components of an electric bus.

Source: Comeco Group (2023).

TABLE 15.6

(a) Maximum L_{16} Reduction and (b) Maximum L_{night} Reduction in Scenario 1 and Scenario 2

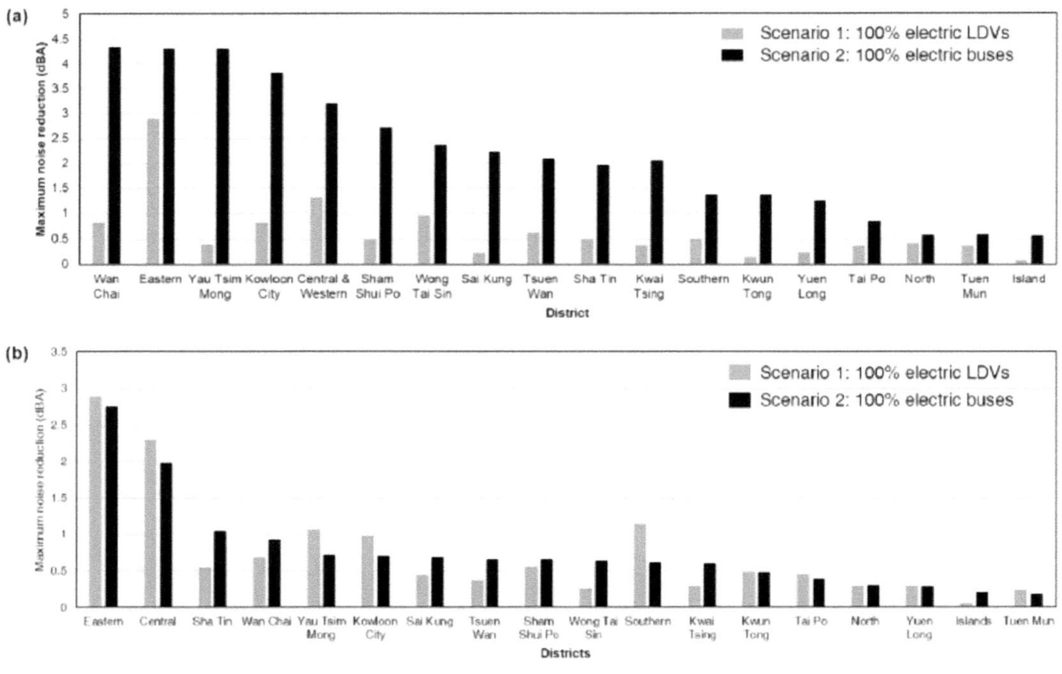

Source: Tsoi et al. (2023).

FIGURE 15.9 King County Metro Proterra electric bus at Eastgate Park and Ride, Bellevue, WA.

Source: SounderBruce, Flickr, via Wikimedia (CC BY-SA 2.0); Horrox and Casale (2019).

climate. The buses have performed well, had no difficulty with extreme temperatures, and saved the CTA more money than fuel and maintenance (Horrox & Casale, 2019). Last but not the least, from the success stories in America, King Country Metro's tests of electric buses have led the agency to commit to transitioning to a full zero-emission fleet by 2040 (Horrox & Casale, 2019).

The electric buses (Figure 15.9) are still an emerging technology. Urban areas, transit agencies, and school districts that are considering implementing electric buses ought to know what to expect and, more importantly, how to optimize the benefits of their investment.

Electric bicycles (e-bikes) have an electric motor and battery that help the rider pedal. The motor is usually powered up to 750 W. Many e-bikes have pedals that propel the bike with or without the help of the electric motor (Qian & Shi, 2023).

E-bikes can help riders:

- Ride farther and faster than a traditional bicycle
- Manage hills
- Travel longer distances without getting tired
- Keep up with traffic
- Pull a small trailer

E-bikes use brushless DC motors, or BLDC motors, which have been the standard for over a decade. These motors do not use brushes to alternate the direction of current flowing to the motor. Pedal-assist sensors activate the motor when the rider pedals the e-bike. There are two types of pedal sensors: cadence sensors and torque sensors. E-bikes typically come with a 48 V or a 52 V battery. A higher-voltage battery provides greater efficiency, with the battery using less electricity to provide the same or better power for the bike (Johnson & Rose, 2015).

Electric scooters are similar to e-bikes but have a smaller frame and are designed for shorter trips. They are also an affordable and efficient option for urban commuting. When shopping for an electric scooter, it is crucial to consider the following factors (Felipe-Falgas et al., 2022):

- *Speed and range.* The *range* of an electric scooter refers to how far it can travel on a single charge. Most commuter scooters have a minimum range of 8 mi, while higher-performance models can go up to 15 mi (Felipe-Falgas et al., 2022).
- *Battery.* The battery of an electric scooter is responsible for powering the motor, lights, controller, and display. It also determines the scooter's speed and range (Felipe-Falgas et al., 2022).
- *Suspension and tires.* A suspension on an electric scooter helps absorb shocks from uneven roads or surfaces. It also provides extra grip to the wheels, making the scooter more manageable (Felipe-Falgas et al., 2022).
- *Power and performance.* The size of the motor on an electric scooter is the most significant factor in determining its top speed, acceleration, and ability to climb hills. Motors range from 200 W to over 6,000 W, with a motor between 350 and 500 W capable of handling most people's daily commute (Felipe-Falgas et al., 2022).

15.3.2 Sea

Electric boats are vessels that are powered by electric motors. Battery packs, solar panels, or generators can power them. Electric boats can be used for many purposes, including as tugboats, ferries, cargo ships, barges, tour boats, fishing trawlers, and cruising yachts (Begovic et al., 2021).

Electric boats can be cheaper to run than marine gasoline because electricity is less expensive than marine gasoline. They also have lower maintenance costs. Some of the most advanced electric boats can sail above 200 nm at cruising speed and 30–50 nm at planing speed. Some disadvantages of electric ships include:

1. Electronic components are prone to corrosion-caused failure.
2. Parts are not readily available because they are not typical components.
3. Skilled and qualified laborers are hard to find.
4. Battery technology has a long way to go to provide the amount of range of a fuel tank.

Some electric boat builders include Duffy, Frauscher, Hinckley Yachts, Nautique, Green Dream Boats, Twin Vee, and X-Shore (Begovic et al., 2021).

15.3.3 Air

Electric airplanes and drones are a new technology that could significantly impact the aviation industry. Electric airplanes use lithium-ion batteries, which are rechargeable and can be charged using solar power or grid electricity, to power their motors. The batteries can be more than half the plane's weight. Electric airplanes are designed to reduce the environmental effects of aviation by providing zero emissions and quieter flights. They have one or two propellers but may have up to four. The electric motor gives them a quiet operation and high efficiency during takeoff and landing (Kocsis Szürke et al., 2023).

Smaller models, such as drones, fly entirely with electricity. Larger aircraft for more than 15 people still rely on a second energy source. Drones are still a new technology that is being used for aerial surveying, monitoring work progress, and measuring stockpiles. Electric airplanes could start to appear in the commercial market around 2026. Air Canada recently ordered 30 battery-powered passenger aircraft (Duan, 2023).

Overall, the use of electric vehicles is increasing rapidly, with new developments and advances being made in various areas.

15.4 EFFICIENCY, CHALLENGES, LATEST ADVANCES

As electric vehicles (EVs) become more integrated into smart city systems, they produce vast amounts of data, including personally identifiable information (PII), vehicle telemetry, and location data, among other things. Recognizing the importance of securing and preserving privacy in the smart city context, it is crucial to understand the sensitivity of this data. Different EV data types require varying protection levels based on their sensitivity, making it necessary to assess their sensitivity levels (Muhammad et al., 2023).

Data classification is the process of categorizing EV data according to its sensitivity, which allows organizations to implement appropriate security measures and control access effectively. This technique identifies and differentiates various information types based on their potential impact if exposed to unauthorized individuals. The Muhammad et al. (2023) conducted a study that employed data classification techniques to determine the sensitivity of EV data. The data were grouped into four categories: confidential, sensitive, unclassified, and secret, as demonstrated in Figure 15.10 (Muhammad et al., 2023).

Integrating electric vehicles (EVs) into smart cities marks a transformative change in urban transportation and sustainability. EVs offer numerous benefits for smart cities, including the potential to reduce greenhouse gas emissions, decrease reliance on fossil fuels, and enhance energy efficiency. However, this integration also brings about complex challenges that must be addressed to ensure the successful adoption and integration of EVs within the urban landscape (Apata et al., 2023).

Electric vehicles are highly efficient, providing up to 90% efficiency compared to traditional gasoline-powered vehicles. They also require less maintenance than gasoline-powered vehicles, making them more cost-effective. Additionally, electric vehicles produce zero emissions, helping to reduce pollution and improve air quality. The efficiency of electric vehicles can be evaluated in several ways.

Here are some of the key metrics that are commonly used to measure the efficiency of electric vehicles.

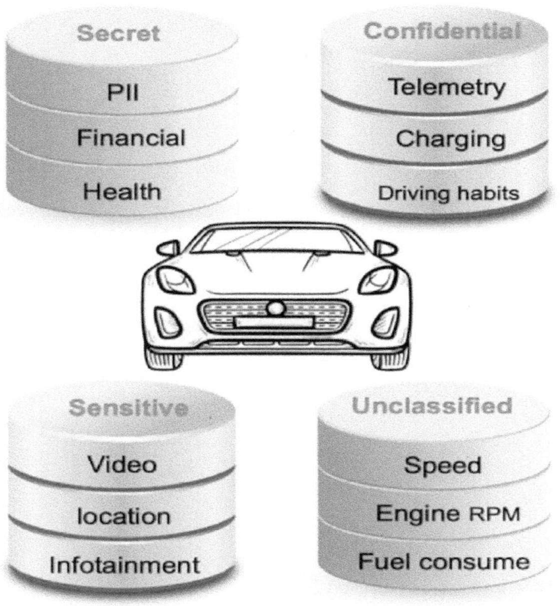

FIGURE 15.10 Data classification in EV.

Source: Muhammad et al. (2023).

Energy efficiency measures how much energy an electric vehicle uses to travel a certain distance. It is usually expressed in miles per kilowatt-hour (miles/kWh) or kilometers per kilowatt-hour (km/kWh).

In a study conducted by Melo and colleagues in 2011, they found that plug-in electric vehicles (PEVs) hold the potential to enhance energy efficiency, decrease CO_2 emissions, and facilitate the integration of renewable energy sources. The authors also discussed the legal framework for advancing electric mobility in Portugal. Furthermore, they provided a brief overview of the methods and modes of charging PEVs and introduced the concept of harmonic theory.

This research involved measuring the performance of normal power and high-power DC charging methods. They specifically focused on the ISR's Nissan Leaf, which was charged through a regular socket outlet in an office building, and the EDP's Mitsubishi i-MiEV, which was charged at a high-power DC charging station located at a Galp gas station.

In a separate study by Shaukat and colleagues in 2018, they aimed to achieve a target of reducing CO_2 emissions to 50.7240 million metric tons per year by harnessing renewable energy resources (RER) to generate 325 GW of power and attaining 90% penetration of electric vehicles (EVs). Figure 15.11 provides a visual representation of the target for CO_2 emissions according to EO S-21-09 for the year 2050.

The *range* of an electric vehicle (Table 15.7) refers to the distance it can travel on a single charge of its battery. Range is a critical factor in determining the practicality and usability of electric vehicles.

Charging efficiency measures how much energy is lost during the charging process. The higher the charging efficiency, the less energy is wasted during the charging process.

Nguyen et al. (2013) studied the connection of many electric vehicles (EVs) to the grid, particularly for fast charging, and its potential to raise several technical problems or significantly

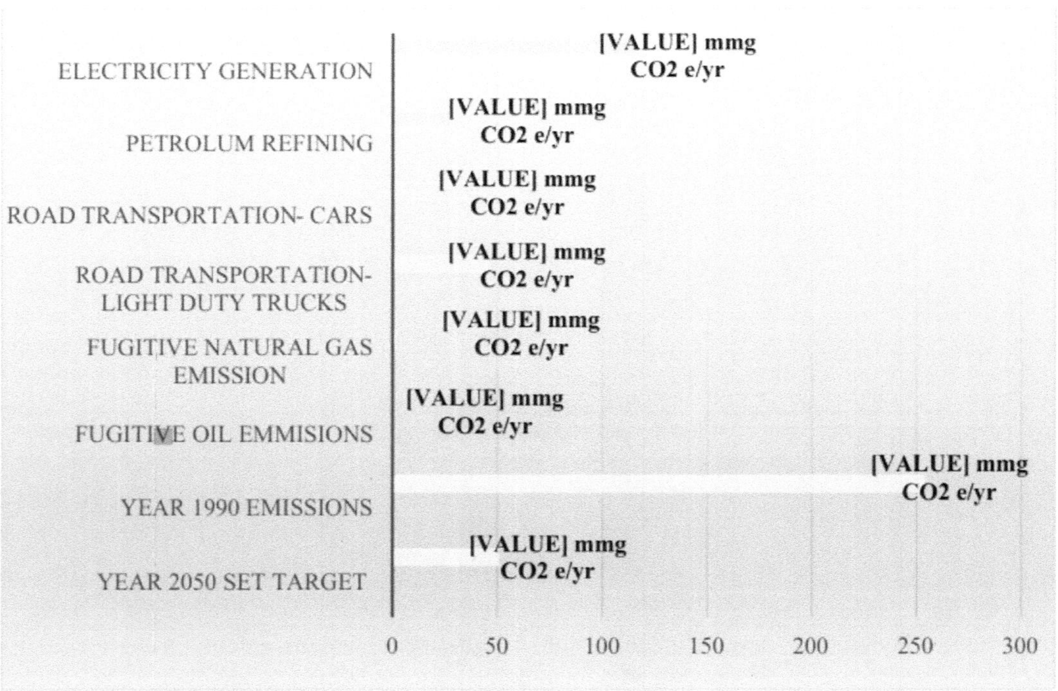

FIGURE 15.11 Goal set by EO S-21-09 for 2050 GHG emission.

Source: Shaukat et al. (2018).

TABLE 15.7
Classification of PEV Charging Methods

	PEV Charging Methods			
Classification	**Mains Connection**	**Power (kW)**	**Current (A)**	**Range[a] (km)**
Normal power	1-phase AC	≤3.7	10–16	<20
Medium power	1-phase or 3-phase AC	3.7–22	16–32	20–110
High-power AC	3-phase AC	>22	>32	>110
High-power DC	DC	>22	>32	>110

Source: Melo et al. (2011).
[a] Assuming an average consumption of 20 kWh/100 km.

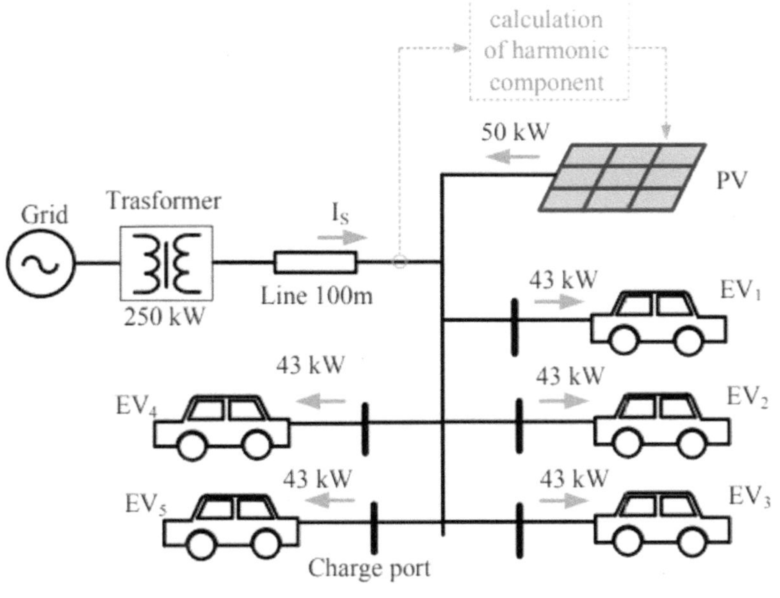

FIGURE 15.12 Fleet of five electric vehicles in the fast charging station and photovoltaic (PV) system.

Source: Nguyen et al. (2013).

impact power systems, including the injection of harmonic currents. They pointed out that the voltage distortion may exceed the admissible limit when multiple electric vehicles undergo fast charging simultaneously. To tackle this issue, they explored using photovoltaic (PV) systems (Figure 15.12) as a viable solution for charging EVs. Their study proposed an advanced control approach for PV inverters utilized as active filters. The effectiveness of the proposed control method was verified through simulations, and the results demonstrated that it successfully reduced the total harmonic distortion (THD) of current and voltage, bringing them within the standard limits (Nguyen et al., 2013).

Regenerative braking efficiency is a process where the vehicle's kinetic energy is converted into electrical energy and stored in the battery. Regenerative braking efficiency measures how much kinetic energy is recovered and stored as electrical energy (Garg et al., 2023).

TABLE 15.8
Charging Characteristics of Different EVs

Model	Vehicle Type	Battery Size (kW h)	Energy Available (kW h)	Range (km)	Energy consumption (kW h/kW)	AC Level 1 Demand (kW)	AC Level 1 Charge time (h)	AC Level 2 Demand (kW)	AC Level 2 Charge time (h)	DC Level 3 Demand (kW)	DC Level 3 Charge time (h)
Nissan Leaf	EV	24	19.2	160	0.12	1.8	11	6.6	2.9	50	0.4
Mitsubishi i-MiEV	EV	16	12.8	150	0.086	1.5	9	3.6	4	50	0.26
Toyota RAV4	EV	41.8	32.18	160	0.20	1.9	17	9.6	3.35	50	0.64
Cooper (BMW)	EV	28	21.5	160	0.13	1.9	11	7.6	2.8	50	0.43
Subaru Stella	EV	9.2	7.1	80	0.09	1.8	4	3.8	2	47	0.15
Tesla Roadster	EV	53	37.1	340	0.11	1.8	21	16.8	22	100	0.37
Chevrolet Volt	PHEV	17.1	13.7	64	0.21	1.4	10	3.6	4	N/A	N/A
Toyota Prius	PHEV	5.2	4.1	25	0.16	1.8	3	3.2	1.5	N/A	N/A
Buick	PHEV	8	6.4	16	0.40	1.4	4.6	3.2	2	N/A	N/A
Fisker Karma	PHEV	22	17.6	80	0.22	1.8	10	3.6	5	N/A	N/A

Source: Shareef et al. (2016).

Overall efficiency is a measure of the efficiency of an electric vehicle over its entire lifetime, including manufacturing, transportation, use, and disposal. It considers all the factors contributing to a vehicle's environmental impact, including energy use, greenhouse gas emissions, and waste (Ahmadi, 2019).

Electric vehicles are generally more energy-efficient than conventional gasoline– or diesel-powered vehicles, as they convert a higher percentage (Table 15.8) of the energy stored in their batteries into motion. However, the efficiency of electric vehicles can vary depending on the specific make and model, as well as driving conditions, battery age, and other factors (Cunanan et al., 2022).

While electric vehicles offer many benefits over conventional gasoline–powered vehicles, several challenges must be addressed to increase their adoption and viability. One of the biggest challenges electric vehicles face is their limited driving range. Although the range of electric vehicles has been improving steadily, they still typically have a shorter range than gasoline-powered vehicles. This can make it difficult for drivers to take long trips without stopping and recharging their vehicles. Electric vehicles need more charging infrastructure. Although there has been significant investment in charging stations in recent years, there are still relatively few in many parts of the world, making it difficult for drivers to find a charging station when they need one.

Hilshey et al. (2012) proposed an algorithm for PEV smart charging and assessed its impact on transformer aging. The study employed a Monte Carlo simulation of a 25 kVA distribution transformer, utilizing ambient temperature data from Burlington, VT, and Phoenix, AZ, to calculate transformer aging under uncoordinated and smart charging conditions. The results revealed that AC level 2 charging caused more significant aging than AC level 1 charging, while smart charging proved effective in mitigating these effects. The research also introduced a more decentralized approach to smart charging, wherein two distributed automaton-based charge management strategies were compared to prevent transformer overloading. These methods empower vehicle owners to prioritize charging options as the vehicles autonomously manage their charging in the given environment.

The cost and durability of electric vehicle batteries are also a challenge. Although batteries have been decreasing steadily, they are still relatively expensive, and the batteries in electric vehicles may need to be replaced after several years of use, which can be costly. According to a study by Zhou and colleagues in 2019, they have developed a method to model how electric vehicles (EVs) affect the operational temperature of medium-voltage (MV) cables and transformers, considering various charging scenarios and harmonic currents. They conducted simulations within a typical urban residential distribution system. The study's findings revealed that when the penetration of EVs exceeded 27.25% with 6.6 kW EV chargers and they were charging in a domestic setting without any control over charging patterns, it caused the 10 kV 400 mm² three-core cable to become overloaded when considering harmonic currents. However, this penetration rate increased to 30.74% when they did not factor in harmonic currents (Zhou et al., 2019).

Additionally, the research determined that the highest acceptable level of EV penetration for a 35/10 kV 16 MVA transformer was 50.22%, while for a 10 kV 1,250 kVA transformer, it was 26.19%. Charging EVs during off-peak hours proved to be a solution to prevent overloading of the cable and transformer, resulting in lower charging costs. Electric vehicles are often more expensive than conventional gasoline–powered vehicles, making them less accessible to consumers. While electric vehicles' costs have decreased, they typically have a higher up-front cost than conventional vehicles (Apata et al., 2023).

As the number of electric vehicles on the road increases, there is a concern about the capacity of the power grid (Figure 15.13) to handle the increased demand for electricity. This could lead to power outages or other disruptions if the grid is not upgraded to handle the increased demand (Das et al., 2020).

The efficiency of electric vehicle batteries declines in cold weather, which can reduce their driving range and performance. This decline can make it more difficult for drivers to use electric vehicles in colder climates, such as in Europe (Figure 15.14) and Canada (Apata et al., 2023).

Overall, while electric vehicles offer many benefits over conventional gasoline–powered vehicles, several challenges still need to be addressed to increase their adoption and viability.

Battery technology has been improving rapidly, with new types of batteries being developed that offer higher energy density, faster charging, and longer lifetimes (Figure 15.15). Solid-state batteries, for example, have the potential to offer higher energy density and faster charging than traditional lithium-ion batteries (Shaukat et al., 2018).

Battery technology (Table 15.9) is constantly improving, allowing electric vehicles to have longer ranges and faster charging times. The availability of charging stations is crucial for adopting electric vehicles in industrial settings. Companies are partnering with charging station providers to develop networks supporting multiple electric vehicle models, making it easier for businesses to switch to electric vehicles.

15.4.1 USA Examples

The availability of charging stations is a key factor in adopting electric vehicles.

Across the United States (Figures 15.16, 15.17), states are taking significant steps to bolster the electric vehicle (EV) infrastructure. California leads the pack with the highest number of electric cars on the road and a massive network of over 25,000 charging stations, featuring more than 2,000 fast chargers capable of refilling a battery up to 80% in just half an hour (Hopkins et al., 2023). New York is also ambitious, targeting installing 10,000 charging stations by the end of 2021 and already boasting over 2,500, including more than 120 fast chargers (Koolman et al., 2021). Texas, another significant market for EVs, has invested in over 1,700 charging stations, of which over 150 are fast DC stations (Brown et al., 2019). Florida is not far behind with its 1,500 -plus charging stations that include over 100 fast DC options (Liu et al., 2023). Massachusetts has set an ambitious goal of its own, aiming for 300,000 electric vehicles on its roads by 2025, supported by a growing infrastructure of over 1,300 charging stations, including more than 50 fast DC stations (Wood et al., 2017).

FIGURE 15.13 Schematic of (a) charging ports and (b) charging connectors.

Source: Das et al. (2020).

Overall, states are making significant investments in charging station infrastructure to support the growing adoption of electric vehicles. With more charging stations available, electric vehicle owners can travel further and more confidently, making electric vehicles a more viable option for businesses and consumers.

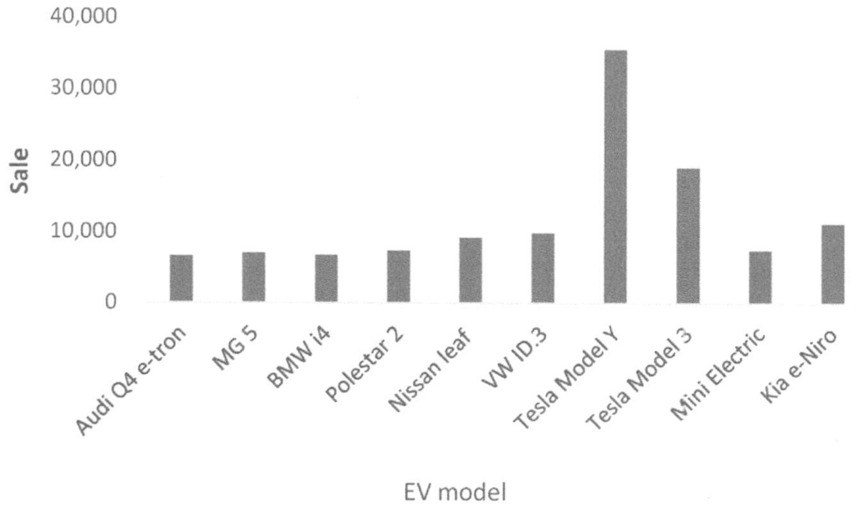

FIGURE 15.14 Models of EV sold in 2022.

Source: Apata et al. (2023).

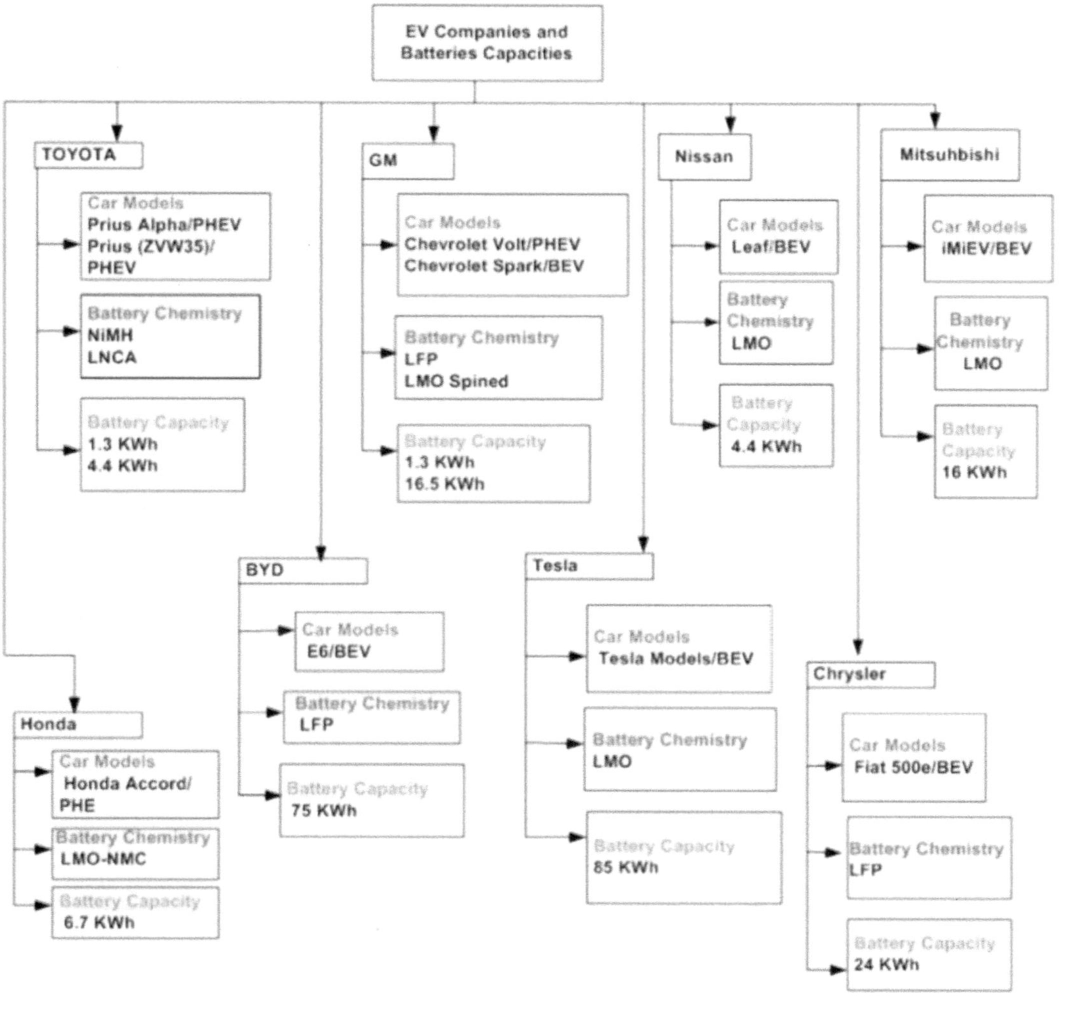

FIGURE 15.15 Companies for EVs with produced battery capacities.

Source: Shaukat et al. (2018).

TABLE 15.9
Common Battery Types, Their Essential Construction Components, Advantages, and Disadvantages

Battery Type	Components	Advantage	Disadvantage
Lead-acid	• Negative active material: spongy lead • Positive active material: lead oxide • Electrolyte: diluted sulfuric acid	• Available in production volume • Comparatively low in cost • Mature technology as used for over 50 years	• Cannot discharge more than 20% of its capacity • Has a limited life cycle if operated on a deep rate of SOC (state of charge) • Low energy and power density • Heavier • May need maintenance
NiMH (nickel–metal hydride)	• Electrolyte: alkaline solution • Positive electrode: nickel hydroxide • Negative electrode: alloy of nickel, titanium, vanadium, and other metals	• Double energy density compared to lead-acid • Harmless to the environment • Recyclable • Safe operation at high voltage • Can store volumetric power and energy • Cycle life longer • Operating temperature range long • Resistant to over-charge and discharge	• Reduced lifetime of around 200–300 cycles if discharged rapidly on high load currents • Reduced usable power because of memory effect
Li-ion (lithium-ion)	• Positive electrode: oxidized cobalt material • Negative electrode: carbon material • Electrolyte: lithium salt solution in an organic solvent	• High energy density, twice of NiMH • Good performance at high temperature • Recyclable • Low memory effect • High specific power • High specific energy • Long battery life, around 1,000 cycles	• High cost • Recharging still takes quite a long time, though better than most batteries
Ni-Zn (nickel-zinc)	• Positive electrode: nickel oxyhydroxide • Negative electrode: zinc	• High energy density • High power density • Uses low-cost material • Capable of deep cycle • Friendly to environment • Usable in a wide temperature range, from −10°C to 50°C	• Fast growth of dendrite, preventing use in vehicles
Ni-Cd (nickel-cadmium)	• Positive electrode: nickel hydroxide • Negative electrode: cadmium	• Long lifetime • Can discharge fully without being damaged • Recyclable	• Cadmium can cause pollution in case of not being property disposed of • Costly for vehicular application

Source: Un-Noor et al. (2017).

15.4.2 Worldwide Examples

Nour et al. 2020 presented a comprehensive review of the increasing adoption of electric vehicles (EVs) in many countries. The rise in EV popularity can be attributed to reduced EV prices, government incentives, the pursuit of energy independence, and environmental concerns. EVs are anticipated to dominate the private cars market in the coming years. However, charging EV batteries

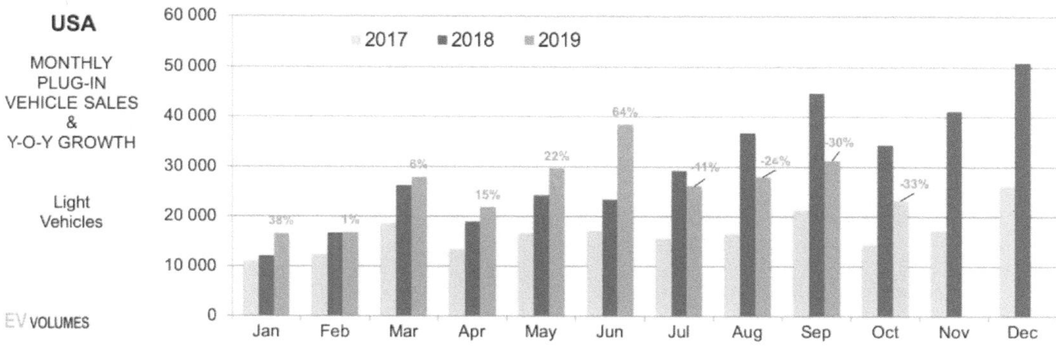

FIGURE 15.16 USA plug-in sales yearly.

Source: ev-volumes.com (2023).

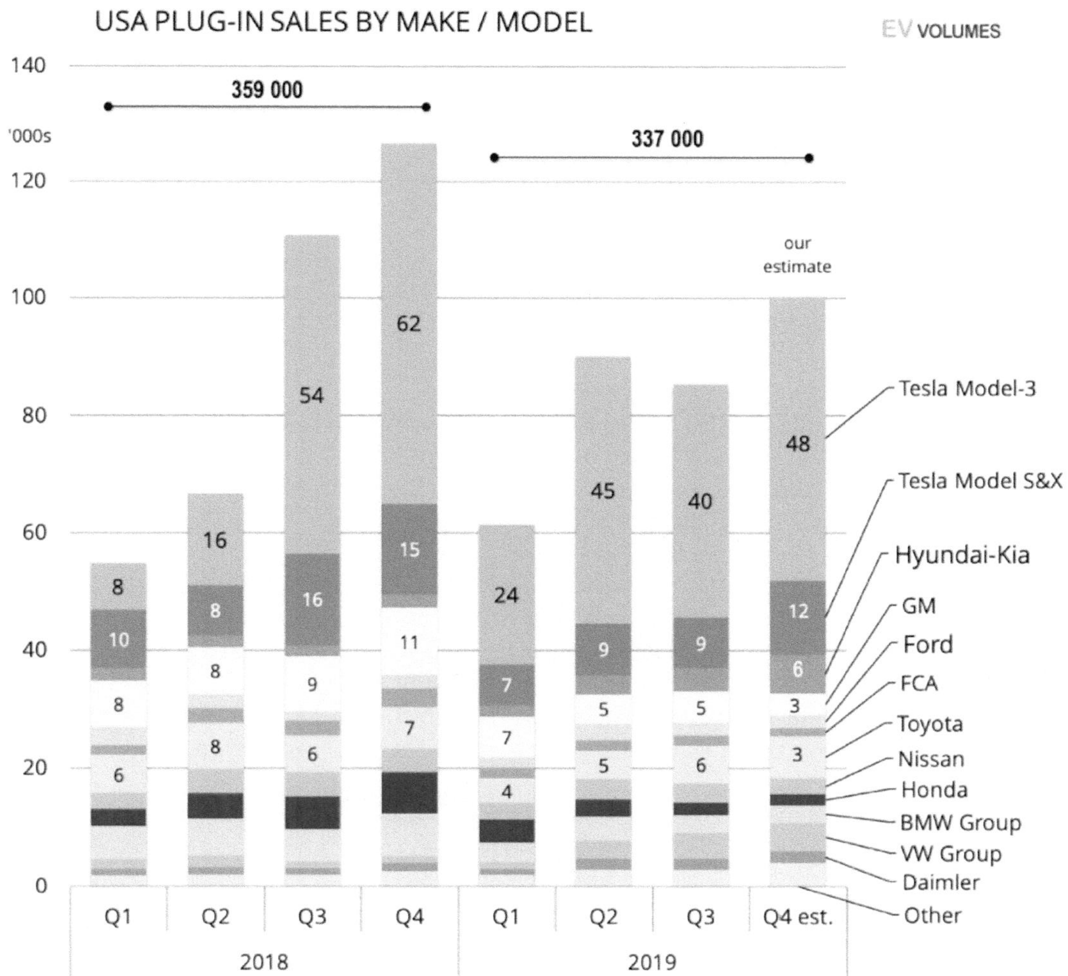

FIGURE 15.17 USA plug-in sales by make and model.

Source: ev-volumes.com (2023).

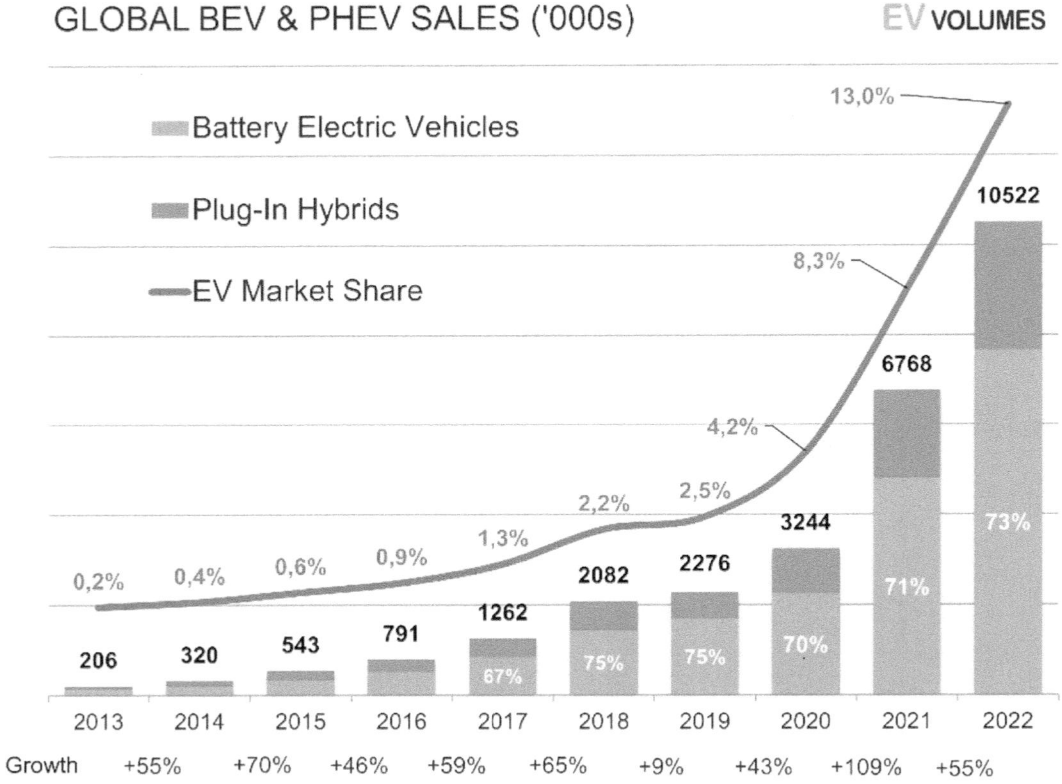

FIGURE 15.18 Top ten best-selling EVs globally in 2022.

Source: Ev-volumes.com (2023).

from the power grid can have severe consequences if not properly managed. Conversely, with proper management, EVs can substantially benefit the power grid and generate revenues for EV owners.

Global EV sales continued to be strong. A total of 10.5 million new BEVs and PHEVs were delivered during 2022, an increase of 55% compared to 2021 figures (Figure 15.18). The regional growth pattern is shifting (Ev-volumes.com, 2023).

Chan (2002) discussed the accelerated development of electric vehicles (EVs) and hybrid electric vehicles (HEVs) in a world (Figure 15.18) where environmental protection and energy conservation are a growing concern. The once-distant dream of commercially viable EVs and HEVs is becoming a reality, with these vehicles gradually becoming available in the market. Many countries worldwide have invested significantly in charging station infrastructure to support the growing adoption of electric vehicles. Here are some examples of countries that have made efforts to improve their charging station infrastructure.

Norway has the highest rate of electric vehicle adoption globally, and the country has made significant investments in charging station infrastructure to support them. The country has over 14,000 charging stations, including over 1,000 fast charging stations. Germany is the largest market for electric vehicles in Europe, and the country has made significant investments in charging station infrastructure to support them. The country has over 35,000 charging stations, including over 1,200 DC fast charging stations (Ibatov et al., 2023). Japan is a significant market for electric vehicles, and the country has made significant investments in charging station infrastructure to support them. The country has over 20,000 charging stations, including over 1,700 fast charging stations (Guzman-Anaya, 2023).

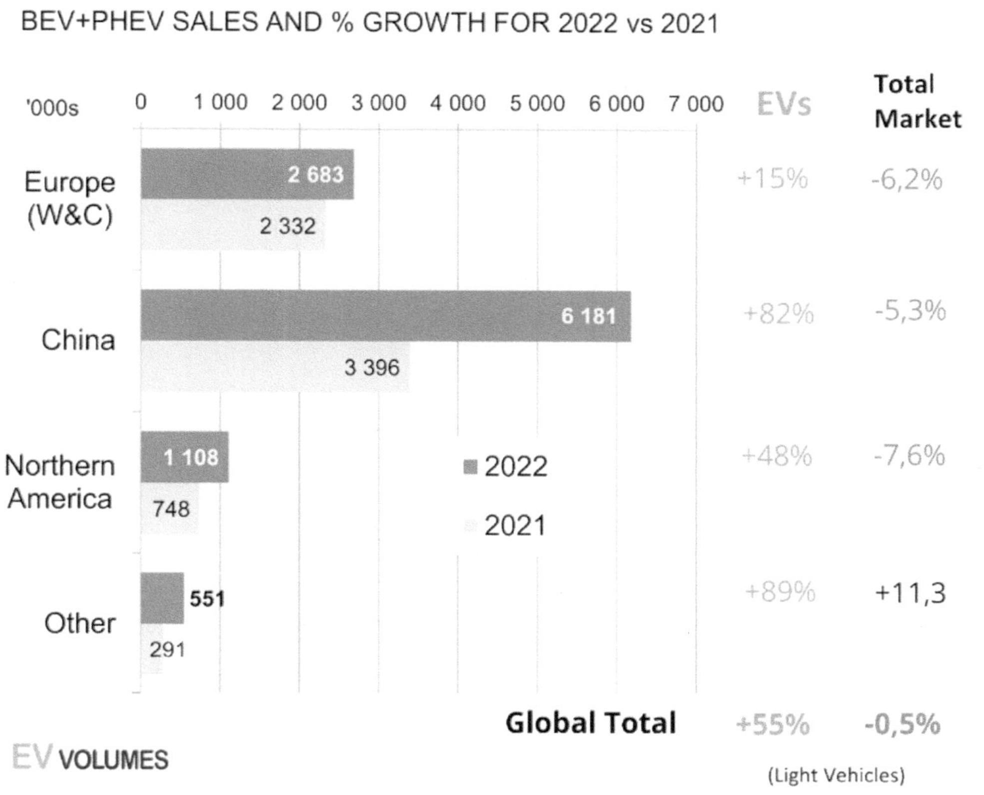

FIGURE 15.19 BEV and PHEV global sales.

Source: Ev-volumes.com (2023).

As electric vehicles gain traction worldwide (Figures 15.19, 15.20), countries are addressing unique climate-related challenges. Norway, the global leader in electric vehicle adoption, has implemented a robust infrastructure of weather-resistant charging stations, complete with heated cables and shelters to protect against harsh conditions (Jain et al., 2023). Similarly, Canada has been focusing on cold-weather solutions for its growing electric vehicle market, with government-backed R&D and innovative approaches like heated pads under charging stations to prevent freezing (Khan et al., 2023). Although Finland's electric vehicle market is smaller than others', it is also getting ahead by investing in cold-weather charging technologies, including heated cables and covered parking spaces (Kelly et al., 2023). In Sweden, the focus has been on urban areas, where the government and private sector are working to provide cold-weather solutions, such as heated cables and shelters, making it easier for residents to switch to EVs (Srithapon et al., 2023). Overall, countries with cold climates have recognized the need for improved cold-weather charging options and have invested in research, development, and infrastructure to support the growing adoption of electric vehicles in these regions. By improving cold weather charging options, electric vehicles can become a more viable option for businesses and consumers in colder regions of the world.

15.5 DEMAND AND SUPPLY

In recent years, there has been a consistent increase in the popularity of electric vehicles (EVs), partially driven by incentives and tax credits available to purchasers. According to data from the International Energy Agency, global sales of EVs rose from 1 million in 2015 to 2 million in 2018.

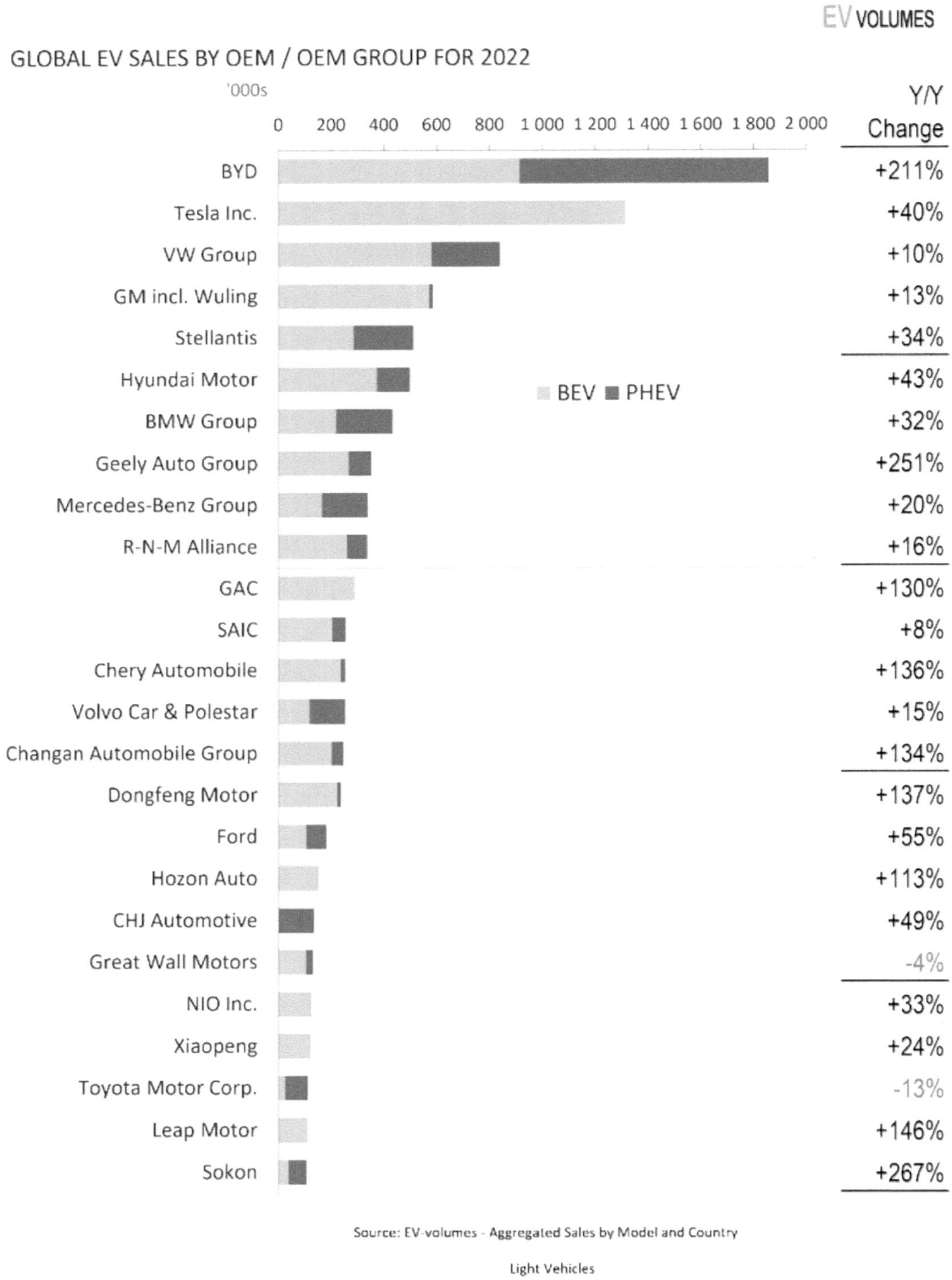

FIGURE 15.20 Global EV sales by OEM.

Source: Ev-volumes.com (2023).

Concurrently, the supply side of the EV market has seen substantial growth in recent years. As of 2019, there were approximately 7.2 million EVs worldwide, significantly rising from the mere 500,000 in 2012 (González & de Haan, 2020). Several factors, including technological advancements, increased investments in industry, and rising consumer demand, have fueled this expansion. As technology advances, the supply of EVs is expected to continue its upward trajectory.

However, forecasting the future demand and supply of EVs remains challenging due to the potential for rapid changes in technology and market conditions. Nevertheless, the demand for EVs is expected to continue its upward trend in the coming years as more individuals become environmentally conscious and recognize the benefits of electric transportation. Furthermore, continued government support through incentives, subsidies, and ongoing technological innovations will likely stimulate the supply of electric vehicles further.

15.6 POTENTIAL TECHNOLOGY CHALLENGES AND LATEST ADVANCES IN THE WORLD

The world has undergone significant changes in technology, politics, and society. Technological advances continue to present both challenges and opportunities. The transportation sector, a major source of greenhouse gas emissions, is transitioning to cleaner options, with electric vehicles (EVs) at the forefront. EVs are being integrated into smart cities globally to reduce emissions and improve urban life. Several cities, including Amsterdam, San Diego, Shenzhen, Yokohama, and Oslo, have successfully embraced EVs. These cities offer charging infrastructure, smart solutions, renewable energy integration, and incentives to drive EV adoption.

15.7 FUTURE TRENDS AND OUTLOOK

An emerging trend in integrating electric vehicles (EVs) into smart cities is deploying high-power charging infrastructure. With the increasing popularity of EVs and advancements in battery technology, there is a growing need for high-power charging solutions to address concerns about charging times. These chargers offer faster and more convenient charges for EV owners. As EV battery technology improves, high-power chargers will become essential in meeting the demand for rapid and efficient charging, significantly reducing charging times from hours to minutes. This advancement will alleviate concerns about charging convenience and range, ultimately encouraging more people to adopt EVs. High-power charging infrastructure can also be customized for specific EV models, optimizing charging efficiency and ensuring compatibility. Placing high-power charging hubs in strategic locations, such as shopping centers and airports, will further enhance the accessibility and convenience of fast charging. Additionally, this infrastructure can support battery-swapping systems, which is particularly beneficial for commercial and fleet applications.

Wireless charging tech (Tie et al., 2013) holds immense promise for revolutionizing EV charging in smart cities by eliminating cables and connectors and enabling easy, automated charging through pads or road-embedded lanes. This integration reduces clutter and is expected to expand in public spaces, residential areas, and fleet operations, enhancing the user experience and boosting EV adoption. Future developments aim to increase power transfer rates for faster, more efficient charging and implement dynamic wireless charging on roadways, allowing EVs to charge while moving during long journeys. Standardization and interoperability between systems will be essential for widespread adoption.

Integrating electric vehicles (EVs) effectively into smart cities hinges on vehicle-to-infrastructure (V2I) communication (Shaukat et al., 2018), which entails information exchange between EVs and urban elements like traffic lights, charging stations, and the power grid. V2I communication empowers EVs with real-time traffic data, enabling them to optimize routes, avoid congestion, and enhance overall traffic flow. Standardizing V2I communication protocols and connectivity standards is crucial for seamless implementation, ensuring compatibility between EV models, infrastructure providers, and cities.

15.8 CONCLUSION

In summary, technology is a double-edged sword. On the one hand, it holds the power to make our lives easier and solve some of our most urgent problems; on the other, the fast-paced evolution of technology presents its own set of issues that need to be thoughtfully managed. There is much hope for what the future could hold, especially when looking at advances like electric vehicles (EVs).

Speaking of electric cars, they are game changers in how we think about transportation. While they have been around for a while, the technology is still fresh. They have some hurdles to overcome – such as limited driving range, steep initial costs, and the need for more charging stations – but recent breakthroughs in battery tech are making them more appealing. As batteries get better and cheaper, it seems inevitable that EVs will become a regular part of our travel routines.

However, it is not just about batteries; the infrastructure supporting these vehicles also needs a reboot. Future advancements in charging systems and renewable energy sources will make it easier for people to own and operate electric cars. Plus, as self-driving technology continues to evolve, electric vehicles stand to benefit. Self-driving cars promise to be safer and more efficient, making the transition to electricity much smoother.

Technology also plays a role in making our roads smarter and safer. As electric cars become common, we will likely see the integration of smart tech like intelligent traffic management systems and vehicle-to-vehicle communications. These innovations could alleviate traffic congestion and increase road safety, facilitating a faster adoption of electric vehicles.

As the icing on the cake, electric vehicles are bound to become more budget-friendly. As technology matures and scales up, the cost of owning an electric car will likely decrease, making it an even more attractive option for consumers.

So electric vehicles are not just a fad – they are set to become a cornerstone of how we get around. Between the ongoing advancements in battery life, autonomous driving, and smart technology, the road ahead for electric vehicles looks promising.

REFERENCES

Ahmadi, P. (2019). Environmental impacts and behavioral drivers of deep decarbonization for transportation through electric vehicles. *Journal of Cleaner Production*, *225*, 1209–1219.

Alanazi, F. (2023). Electric vehicles: Benefits, challenges, and potential solutions for widespread adaptation. *Applied Sciences*, *13*(10), 6016.

Alarcón, F. E., Mac Cawley, A., & Sauma, E. (2023). Electric mobility toward sustainable cities and road-freight logistics: A systematic review and future research directions. *Journal of Cleaner Production*, 138959.

Albatayneh, A., Juaidi, A., Jaradat, M., & Manzano-Agugliaro, F. (2023). Future of electric and hydrogen cars and trucks: An overview. *Energies*, *16*(7), 3230.

Ali, Z. M., Calasan, M., Gandoman, F. H., Jurado, F., & Aleem, S. H. A. (2023). Review of batteries reliability in electric vehicle and E-mobility applications. *Ain Shams Engineering Journal*, 102442.

Almeida Neves, S., Cardoso Marques, A., & Alberto Fuinhas, J. (2019). Technological progress and other factors behind the adoption of electric vehicles: Empirical evidence for EU countries. *Research in Transportation Economics*, *74*, 28–39. [Google Scholar] [CrossRef]

Apata, O., Bokoro, P. N., & Sharma, G. (2023). The risks and challenges of electric vehicle integration into smart cities. *Energies*, *16*(14), 5274.

Bauer, G. (2018). The impact of battery electric vehicles on vehicle purchase and driving behavior in Norway. *Transportation Research, Part D: Transport and Environment*, *58*, 239–258. [Google Scholar] [CrossRef]

Begovic, E., Bertorello, C., De Luca, F., & Rinauro, B. (2021). KISS (Keep It Sustainable and Smart): A research and development program for a zero-emission small crafts. *Journal of Marine Science and Engineering*, *10*(1), 16.

Brown, M. A., & Zhou, S. (2019). Smart-grid policies: An international review. *Advances in Energy Systems: The Large-scale Renewable Energy Integration Challenge*, 127–147.

Chan, C. C. (2002). The state of the art of electric and hybrid vehicles. *Proceedings of the IEEE*, *90*(2), 247–275.

Comeco Group (2023). *How do electric buses work?* Available online: https://bus-news.com/how-do-electric-buses-work (accessed on 2 October 2023).

Cunanan, C. J., Elorza Casas, C. A., Yorke, M., Fowler, M., & Wu, X. Y. (2022). Design and analysis of an offshore wind power to ammonia production system in Nova Scotia. *Energies*, *15*(24), 9558.

Das, H. S., Rahman, M. M., Li, S., & Tan, C. W. (2020). Electric vehicles standards, charging infrastructure, and impact on grid integration: A technological review. *Renewable and Sustainable Energy Reviews*, *120*, 109618.

Duan, F. L. (2023). Why electric aviation—versatile, smarter, and green. In *When AIAA Meets IEEE: Intelligent Aero-engine and Electric Aircraft* (pp. 167–207). Singapore: Springer Nature Singapore.

European Environment Agency (2018). *Electric vehicles from life cycle and circular economy perspectives*. Available online: www.eea.europa.eu//publications/electric-vehicles-from-life-cycle (accessed on 16 April 2022).

EV-Volumes—The electric vehicle world sales database. Available online: www.ev-volumes.com/country/total-world-plug-in-vehicle-volumes/ (accessed on 25 July 2023).

Felipe-Falgas, P., Madrid-Lopez, C., & Marquet, O. (2022). Assessing environmental performance of micromobility using LCA and self-reported modal change: The case of shared e-bikes, e-scooters, and e-mopeds in Barcelona. *Sustainability*, *14*(7), 4139.

Garg, V., Niranjan, S., Prybutok, V., Pohlen, T., & Gligor, D. (2023). Drones in last-mile delivery: A systematic review on Efficiency, Accessibility, and Sustainability. *Transportation Research Part D: Transport and Environment*, *123*, 103831.

Ghadikolaei, M. A., Wong, P. K., Cheung, C. S., Zhao, J., Ning, Z., Yung, K.-F., Wong, H. C., & Gali, N. K. (2021). Why is the world still need to be ready to use alternative fuel vehicles? *Heliyon*, *7*, e07527. [Google Scholar] [CrossRef]

González, A., & de Haan, E. (2020). *The Battery Paradox. How the Electric Vehicle Boom Is Draining Communities and the Planet*. Amsterdam: SOMO.

Gray, M. K., & Morsi, W. G. (2014). Power quality assessment in distribution systems embedded with plug-in hybrid and battery electric vehicles. *IEEE Transactions on Power Systems*, *30*(2), 663–671.

Guzman-Anaya, L. (2023). The role of Japanese cooperation in the transition to electric vehicle production in Mexico. In *Japanese Cooperation and Supporting Industry in Mexico's Automotive Sector: USMCA, Covid-19 Disruptions, and Electric Vehicle Production* (pp. 31–52). Singapore: Springer Nature Singapore.

He, X., Zhan, W., & Hu, Y. (2018). Consumer purchase intention of electric vehicles in China: The roles of perception and personality. *Journal of Cleaner Production*, *204*, 1060–1069. [Google Scholar] [CrossRef]

Higueras-Castillo, E., Molinillo, S., Coca-Stefaniak, J. A., & Liébana-Cabanillas, F. (2019). Perceived value and customer adoption of electric and hybrid vehicles. *Sustainability*, *11*, 4956. [Google Scholar] [CrossRef]

Hilshey, A. D., Rezaei, P., Hines, P. D., & Frolik, J. (2012, July). Electric vehicle charging: Transformer impacts and smart, decentralized solutions. In *2012 IEEE Power and Energy Society General Meeting* (pp. 1–8). IEEE.

Hopkins, E., Potoglou, D., Orford, S., & Cipcigan, L. (2023). Can the equitable roll out of electric vehicle charging infrastructure be achieved? *Renewable and Sustainable Energy Reviews*, *182*, 113398.

Horrox, J., & Casale, M. (2019). *Electric Buses in America: Lessons from Cities Pioneering Clean Transportation*. US PIRG Education Fund. Accessed on 6 June 2021.

Ibatov, N., Chinchalinov, K., & Karim, B. (2023). *Factors of Development of the Electric Vehicle in the EU Countries*. International School of Economics.

International Energy Agency (2019). *Global EV Outlook 2019: Scaling Up the Transition to Electric Mobility*. Available online: www.iea.org/publications/reports/globalevoutlook2019/ (accessed on 20 April 2022).

International Energy Agency (2021a). *Global EV Outlook 2021*. Available online: https://iea.blob.core.windows.net/assets/ed5f4484-f556-4110-8c5c-4ede8bcba637/GlobalEVOutlook2021.pdf (accessed on 26 April 2022).

International Energy Agency (2021b). *Global Energy Review 2021*. Available online: https://iea.blob.core.windows.net/assets/d0031107-401d-4a2f-a48b-9eed19457335/GlobalEnergyReview2021.pdf (accessed on 15 April 2022).

International Energy Agency (2022). *Global EV Outlook 2022. Securing Supplies for an Electric Future*. Available online: https://iea.blob.core.windows.net/assets/ad8fb04c-4f75-42fc-973a-6e54c8a4449a/GlobalElectricVehicleOutlook2022.pdf (accessed on 16 December 2022).

Jain, A., Agrawal, N., & Agrawal, V. (2023, May). Electric vehicle adoption in Rajasthan: Trends, challenges and strategies. In *2023 IEEE Renewable Energy and Sustainable E-Mobility Conference (RESEM)* (pp. 1–5). IEEE.

Johnson, M., & Rose, G. (2015). Extending life on the bike: Electric bike use by older Australians. *Journal of Transport & Health*, *2*(2), 276–283.

Ju, N., Lee, K.-H., & Kim, S. H. (2021). Factors affecting consumer awareness and the purchase of eco-friendly vehicles: Textual analysis of Korean market. *Sustainability*, *13*, 5566. [Google Scholar] [CrossRef]

Kelly, N. J., Flett, G. H., & Hand, J. W. (2023). Developing a statistical electric vehicle charging model and its application in the performance assessment of a sustainable urban charging hub. *Journal of Building Performance Simulation*, 1–20.

Khan, S., Sudhakar, K., & bin Yusof, M. H. (2023). Comparison of mono and bifacial modules for building integration and electric vehicle charging: A case study in Sweden. *Energy Conversion and Management: X*, *20*, 100420.

Kocsis Szürke, S., Perness, N., Földesi, P., Kurhan, D., Sysyn, M., & Fischer, S. (2023). A risk assessment technique for energy-efficient drones to support pilots and ensure safe flying. *Infrastructures*, *8*(4), 67.

Koolman, G., Stecca, M., & Bauer, P. (2021, June). Optimal battery energy storage system sizing for demand charge management in EV fast charging stations. In *2021 IEEE Transportation Electrification Conference & Expo (ITEC)* (pp. 588–594). IEEE.

Liu, Y., Francis, A., Hollauer, C., Lawson, M. C., Shaikh, O., Cotsman, A., . . . Asensio, O. I. (2023). Reliability of electric vehicle charging infrastructure: A cross-lingual deep learning approach. *Communications in Transportation Research*, *3*, 100095.

Lucian, J., Can, R., Stolten, D., & Peters, R. (2021). How to reduce the greenhouse gas emissions and air pollution caused by light and heavy duty vehicles with battery-electric, fuel cell-electric, and catenary trucks. *Environment International*, *152*, 106474. [Google Scholar] [CrossRef]

Melo, N., Mira, F., de Almeida, A., & Delgado, J. (2011, October). Integration of PEV in Portuguese distribution grid: Analysis of harmonic current emissions in charging points. In *11th International Conference on Electrical Power Quality and Utilisation* (pp. 1–6). IEEE.

Monteiro, V., Gonçalves, H., & Afonso, J. L. (2011, October). Impact of Electric Vehicles on power quality in a Smart Grid context. In *11th international conference on electrical power quality and utilisation* (pp. 1–6). IEEE.

Muhammad, Z., Anwar, Z., Saleem, B., & Shahid, J. (2023). Emerging cybersecurity and privacy threats to electric vehicles and their impact on human and environmental sustainability. *Energies*, *16*(3), 1113.

Nguyen, V. L., Tran-Quoc, T., & Bacha, S. (2013, June). Harmonic distortion mitigation for electric vehicle fast charging systems. In *2013 IEEE Grenoble Conference* (pp. 1–6). IEEE.

Nour, M., Chaves-Ávila, J. P., Magdy, G., & Sánchez-Miralles, Á. (2020). Review of positive and negative impacts of electric vehicles charging on electric power systems. *Energies*, *13*(18), 4675.

Paoli, L., & Gül, T. (2022). *Electric Cars Fend Off Supply Challenges to More Than Double Global Sales*. International Energy Agency. Available online: www.iea.org/commentaries/electric-cars-fend-off-supply-challenges-to-more-than-double-global-sales (accessed on 25 April 2022).

Qian, Q., & Shi, J. (2023). Comparison of injury severity between E-bikes-related and other two-wheelers-related accidents: Based on an accident dataset. *Accident Analysis & Prevention*, *190*, 107189.

Shareef, H., Islam, M. M., & Mohamed, A. (2016). A review of the stage-of-the-art charging technologies, placement methodologies, and impacts of electric vehicles. *Renewable and Sustainable Energy Reviews*, *64*, 403–420.

Shaukat, N., Khan, B., Ali, S. M., Mehmood, C. A., Khan, J., Farid, U., . . . Ullah, Z. (2018). A survey on electric vehicle transportation within smart grid system. *Renewable and Sustainable Energy Reviews*, *81*, 1329–1349.

Song, M. R., Chu, W., & Im, M. (2022). The effect of cultural and psychological characteristics on the purchase behavior and satisfaction of electric vehicles: A comparative study of US and China. *International Journal of Consumer Studies*, *46*, 345–364. [Google Scholar] [CrossRef]

Srithapon, C., & Månsson, D. (2023). Predictive control and coordination for energy community flexibility with electric vehicles, heat pumps and thermal energy storage. *Applied Energy*, *347*, 121500.

Statista Research Department Transportation Emissions Worldwide—Statistics & Facts. 2022. Available online: www.statista.com/topics/7476/transportation-emissions-worldwide/#topicHeader__wrapper (accessed on 5 April 2022).

Tie, S. F., & Tan, C. W. (2013). A review of energy sources and energy management system in electric vehicles. *Renewable and Sustainable Energy Reviews*, *20*, 82–102.

Tiseo, I. *Breakdown of CO2 Emissions in the Transportation Sector Worldwide 2020, by Subsector*. Available online: www.statista.com/statistics/1185535/transport-carbon-dioxide-emissions-breakdown/ (accessed on 2 April 2022).

Tsoi, K. H., Loo, B. P., Li, X., & Zhang, K. (2023). The co-benefits of electric mobility in reducing traffic noise and chemical air pollution: Insights from a transit-oriented city. *Environment International*, *178*, 108116.

Un-Noor, F., Padmanaban, S., Mihet-Popa, L., Mollah, M. N., & Hossain, E. (2017). A comprehensive study of key electric vehicle (EV) components, technologies, challenges, impacts, and future direction of development. *Energies*, *10*(8), 1217.

Wang, Z., & Paranjape, R. (2014, November). An evaluation of electric vehicle penetration under demand response in a multi-agent based simulation. In *2014 IEEE Electrical Power and Energy Conference* (pp. 220–225). IEEE.

Wood, E., Raghavan, S., Rames, C., Eichman, J., & Melaina, M. (2017). *Regional Charging Infrastructure for Plug-in Electric Vehicles: A Case Study of Massachusetts* (No. NREL/TP-5400–67436). National Renewable Energy Lab.(NREL), Golden, CO (United States).

Wu, Y., & Zhang, L. (2017). Can the development of electric vehicles reduce the emission of air pollutants and greenhouse gases in developing countries? *Transportation Research, Part D: Transport and Environment*, *51*, 129–145. [Google Scholar] [CrossRef]

Yuan, X., Li, L., Gou, H., & Dong, T(2015). Energy and environmental impact of battery electric vehicle range in China. *Applied Energy*, *157*, 75–84. [Google Scholar] [CrossRef]

Zhao, X., Ma, Y., Shao, S., & Ma, T. (2022). What determines consumers' acceptance of electric vehicles: A survey in Shanghai, China. *Energy Economics*, *108*, 105805. [Google Scholar] [CrossRef]

Zhou, C., Wang, H., Zhou, W., Qian, K., & Meng, S. (2019, October). Determination of maximum level of EV penetration with consideration of EV charging load and harmonic currents. In *IOP Conference Series: Earth and Environmental Science* (Vol. 342, No. 1, p. 012010). IOP Publishing.

16 Energy Storage and Transportation

Mesut Yurukcu, Jorge Saldaña, Narendra Boppana, Sepehr Arbabi, Jim Henry, and Sebahattin Ziyanak

16.1 INTRODUCTION

Energy storage and transportation systems play a crucial role in moving toward the future of new energy demands. These systems not only help use energy more efficiently but also improve the stability and reliability of energy networks. In the last era, the Paris Agreement drew a new road map for our energy sector by reducing carbon emissions while using more renewable energies (Nadeem et al., 2018). With the changing energy landscape, integrating storage technologies becomes essential to manage the intermittent nature of renewable energy sources and meet the growing demand for energy in different sectors (Nadeem et al., 2018; Mitali et al., 2022; Bahloul et al., 2024).

Figure 16.1 illustrates that between 2017 and 2022, energy growth of 43%, amounting to 920 GW, is expected in the global renewable electricity capacity. In its essence, *energy storage* involves capturing produced energy for use, enabling practices that enhance grid efficiency and access to energy (Nadeem et al., 2018). This section explores energy storage and transportation aspects, from selecting technologies to their design and economic evaluation. By looking into storage systems like thermal, electrochemical, and electromagnetic ones, this section offers a detailed overview of

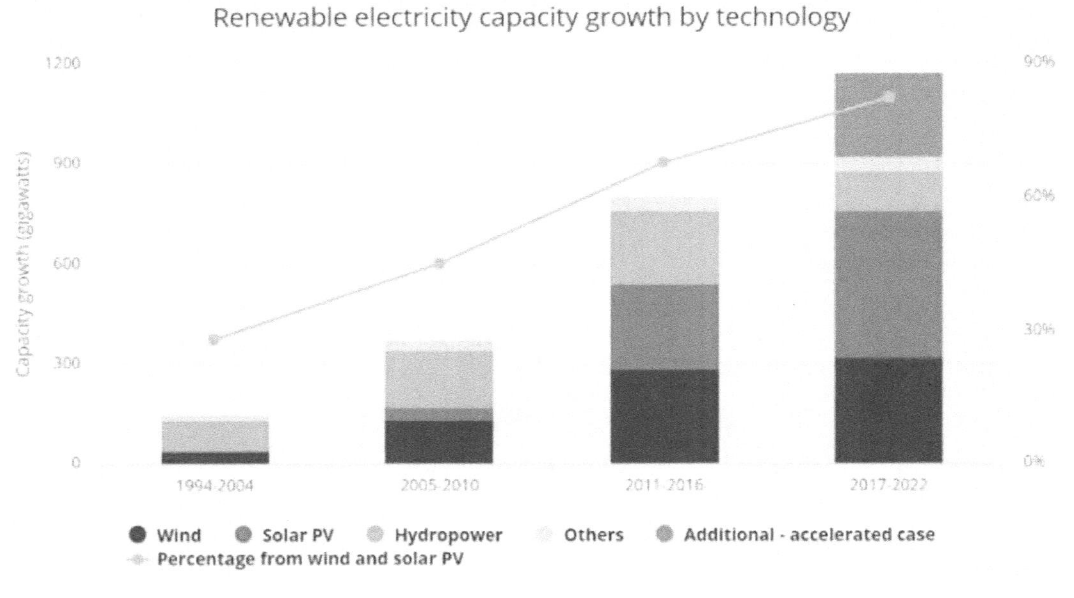

FIGURE 16.1 The capacity of renewables entering the new era.

Source: Nadeem et al. (2018).

DOI: 10.1201/9781003282273-16

operational principles, application areas, life cycle considerations, efficiencies, challenges, and best practices (Aghmadi et al., 2024).

Moreover, this chapter discusses progress in storage systems, such as compressed air and flywheel storage systems, along with innovative methods, like gravity-based stacking using concrete blocks. Thermal storage solutions are explored, ranging from pumped hydro storage to phase-change materials that improve heat retention capacity (Bui et al., 2021; Zhang et al., 2024).

It also discusses storage solutions such as supercapacitors, superconducting magnets, and the growing field of hydrogen and natural gas storage systems, including methods for liquefying hydrogen and utilizing ammonia and salt caverns for storage. Furthermore, the chapter examines the effects, resource constraints, and energy requirements associated with these storage technologies.

In wrapping up this exploration, the chapter delves into the uses of energy storage systems in the oil and gas sector, emphasizing their potential benefits, technological progressions, and global challenges. By offering an analysis of technologies and future trends, this chapter aims to provide readers with a comprehensive understanding of the complexities that influence energy storage and transportation practices. This knowledge can support informed decision-making and innovative problem-solving within the energy industry.

16.2 SCREENING, SELECTION, DESIGN, AND ECONOMICS OF ENERGY STORAGE

While the world is shifting through sustainable and resilient energy, the significance of energy storage systems (ESS) in energy management is emerging. These systems' evaluation, choice, design, and economic evaluation play a role in determining their practicality and efficiency in real-world scenarios. This thorough analysis should incorporate environmental and social viewpoints to ensure that the selected systems offer optimal performance and align with broader energy policy objectives (Gholami et al., 2024).

The screening and selection phases mark the stages where potential technologies are assessed based on criteria tailored to project goals and environmental factors. This process involves evaluating storage options' feasibility, cost-effectiveness, environmental impact, and regulatory adherence. Energy storage technologies differ significantly from systems like pumped hydro and flywheels to solutions such as lithium-ion batteries and flow batteries. Each technology possesses energy density, efficiency, cycle life, and scalability characteristics that need evaluation to meet the application's specific requirements (Carnevali et al., 2024).

Factors including response time, capacity, scalability, and life cycle costs are evaluated along with resilience, conditions, and maintenance needs. Sophisticated analytical models and decision-making frameworks, such as multi-criteria decision analysis (MCDA), are commonly utilized to assist in this intricate selection process, offering a structured approach that considers multiple, sometimes conflicting, factors (Punyavathi et al., 2024).

In their analysis, Cheng and colleagues (2015) discussed the crucial screening and selection processes for implementing energy storage systems. The initial step involves an evaluation of storage technologies based on key performance indicators, like energy density, power capacity, longevity, and cost-efficiency. A critical aspect of the selection process is aligning the technology features with the requirements of the intended application, which can vary from managing large-scale grid power to integrating renewable energy systems at an individual level (Figure 16.2). This phase also entails an assessment of the impact of adherence to regulations and potential risks associated with the technology to ensure sustainable and secure energy storage solutions. The review emphasizes the need for continued research and development to enhance existing technologies and explore innovations that can meet increasing demands for energy storage while minimizing impact (Cheng et al., 2015).

During the design phase, the focus lies on integrating the chosen energy storage technology into the energy infrastructure or creating a system that optimizes the storage capacity. This stage demands comprehending how the grid operates and how different energy sources and storage

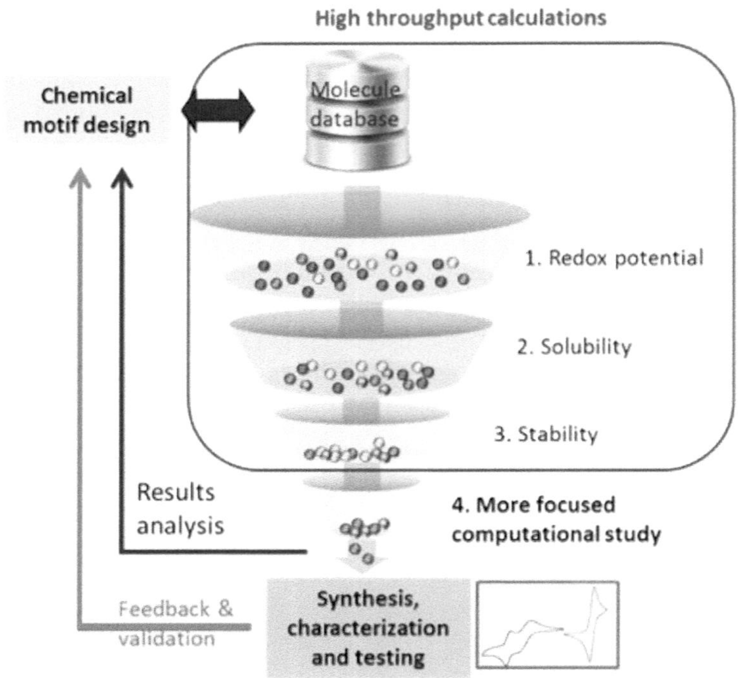

FIGURE 16.2 Schematic of down-selection of candidate molecules for electrical energy storage applications based on high-throughput computations using quantum chemical calculations of specific properties.

Source: Cheng et al. (2015).

technologies interact. Design considerations must not only cover the details of the storage system, like power ratings, energy capacity, and discharge duration, but also anticipate energy demand–supply patterns (Close et al., 2024).

Simulation tools and software are essential in modeling the performance of energy storage systems under diverse scenarios and fine-tuning system configurations for optimal efficiency and reliability. The design process also involves devising safety measures and control systems to mitigate risks associated with energy storage concerning high-energy-density systems such as lithium-ion batteries (Al Wahedi & Bicer, 2024).

Economic assessment is also vital in evaluating energy storage systems, considering capital expenses, operating costs, and potential revenue streams. The financial feasibility of energy storage depends on factors like the cost of energy sources, how long the storage system lasts, upkeep expenses, and how depreciation is calculated. Additionally, economic incentives such as subsidies, tax advantages, and regulatory frameworks all play roles in shaping the economics of energy storage solutions (Vazquez et al., 2010; Rugolo et al., 2012; Hirsch et al., 2018; Gao et al., 2024).

The levelized cost of storage (LCOS) compares the cost-efficiency of storage technologies based on the amount of energy stored. This calculation helps stakeholders grasp when investments in energy storage will break even and how profitable they can be. Moreover, financial models need to account for market dynamics like energy prices, demand fees, and grid services such as peak shaving, load leveling, and frequency regulation that can generate income and shorten the investments' payback period (Figure 16.3) (Lyden et al., 2018; Gao et al., 2024).

In essence, evaluating, selecting, designing, and analyzing the economics of energy storage systems are procedures that require a meticulous approach. With an increasing demand for financially sustainable energy storage solutions comes a greater need for these processes to guide the development and implementation of efficient systems that bolster the sustainability and resilience

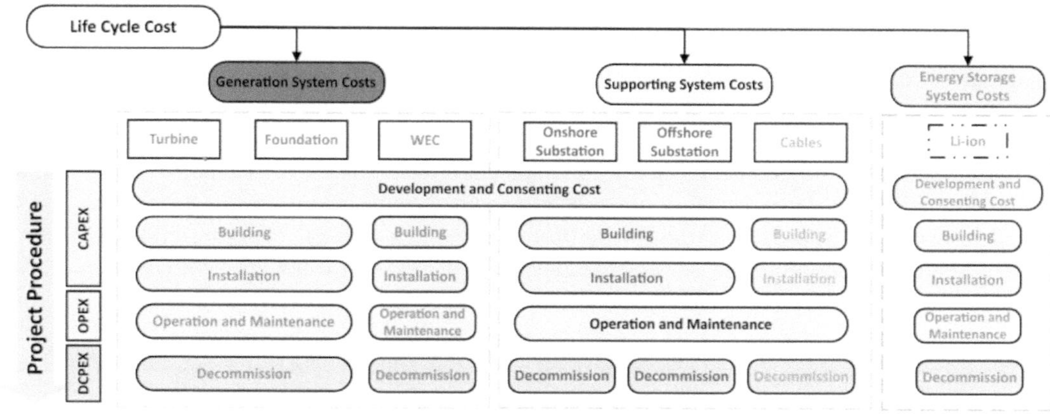

FIGURE 16.3 A diagram of the life cycle cost model.

Source: Gao et al. (2024).

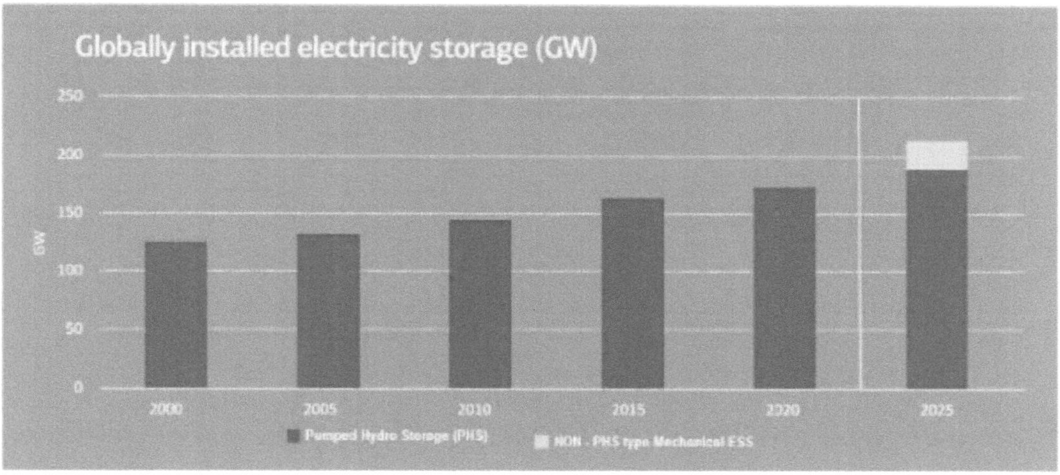

FIGURE 16.4 Annual increase of pumped hydro storage systems.

Source: Nadeem et al. (2018).

of global energy networks. Advancements in technology and better regulations are paving the way for energy storage to transform toward a reliable, efficient, and sustainable grid system.

16.3 ENERGY STORAGE SYSTEMS

Energy storage systems are one of the other important areas for improving energy management, enhancing grid reliability, and supporting the integration of energy sources. These systems come in types based on how they store energy, such as options like pumped hydroelectric storage (PHS) flywheels (Figure 16.4) (Nadeem et al., 2018) and compressed air energy storage (CAES) (Jankowski et al., 2024); thermal solutions using materials with high heat capacities (Tetteh et al., 2024) or phase-change properties like molten salt, hot water tanks, and ice storage (Prieto et al., 2024); electrochemical choices, including batteries like lithium-ion and flow batteries; electrical options with capacitors and supercapacitors; and chemical methods like hydrogen storage through chemical reactions (Aghmadi et al., 2024).

These systems have applications across sectors, including residential and commercial settings for managing peak demand and backup power needs, industrial use for stabilizing loads and improving energy efficiency, utility-scale operations for grid stability and renewable energy integration, and as well as in transportation, primarily in electric and hybrid vehicles. The lifespan of these systems varies depending on the technology used. While some systems, like pumped hydro, can last for decades, certain batteries may have varying lifespans due to wear from charging cycles. Efficiency levels also differ across systems, with supercapacitors reaching up to 95% efficiency, while batteries typically operate between 80 and 90% (Aghmadi et al., 2024).

Energy storage systems (ESS) encounter obstacles and challenges, like enhancing durability and energy capacity; financial barriers, such as steep initial investments and uncertain returns; environmental issues involving resource depletion and disposal problems; and the absence of standardized regulatory frameworks. Each type of energy storage functions based on principles essential for maximizing performance; for example, batteries operate through reversible electrochemical reactions, while mechanical systems like flywheels store kinetic energy (Saka, 2024).

Experts in the energy storage sector emphasize the importance of integrating systems with infrastructure, conducting maintenance and safety checks, and adhering to safety guidelines to minimize potential risks. The economics of energy storage not only evaluates the capital outlay and operating expenses but also considers potential revenues from offering services like frequency regulation and load balancing to the grid. Progress in technology and enhancements in systems that recognize and promote the benefits of energy storage are expected to impact the viability and adoption of ESS. These improvements will likely enhance sustainability and efficiency in energy systems (García-Miguel et al., 2024).

These systems are classified into types, each with its own function and use. Storage technologies like pumped storage (PHS) flywheels and compressed air energy storage (CAES) transform electrical energy into mechanical forms – either potential or kinetic – that can be stored and converted back into electricity when needed (Boruah et al., 2024).

Electrochemical storage mainly consists of lithium-ion, lead-acid, and flow batteries that store energy through chemical reactions that release electrons. Thermal energy storage methods capture heat or cold using materials with heat capacities or phase-change materials (PCMs) for use in heating/cooling applications or power generation. Electrical storage involves capacitors and supercapacitors storing energy in fields for quick discharge/recharge cycles. Lastly, chemical storage, illustrated by hydrogen storage, utilizes chemical processes, like electrolysis and methanation, to store energy in chemical bonds (Nadeem et al., 2018).

Each system possesses attributes regarding its lifespan, discharge duration, energy loss, during-discharge energy density, and power rating. These characteristics significantly influence their suitability for applications, as depicted in Figure 16.5. Each of these systems has a role to play in tackling the issues presented by the rising need for dependable energy systems, aiding in the development of a more eco-friendly and effective energy framework (Nadeem et al., 2018; Aghmadi et al., 2024).

16.3.1 Application Areas

Energy storage systems (ESS) are integral to advancing energy infrastructures' efficiency, reliability, and sustainability across various sectors. Their application areas are diverse, reflecting their ability to mitigate the challenges associated with the intermittency of renewable energy sources, enhance grid stability, and provide flexible energy solutions (Aghmadi et al., 2024; Chatzigeorgiou et al., 2024).

In a study conducted in 2024, Chatzigeorgiou, N. G., studied the applications of energy storage systems. The research explored the suitability of erythritol, a type of sugar alcohol, as a

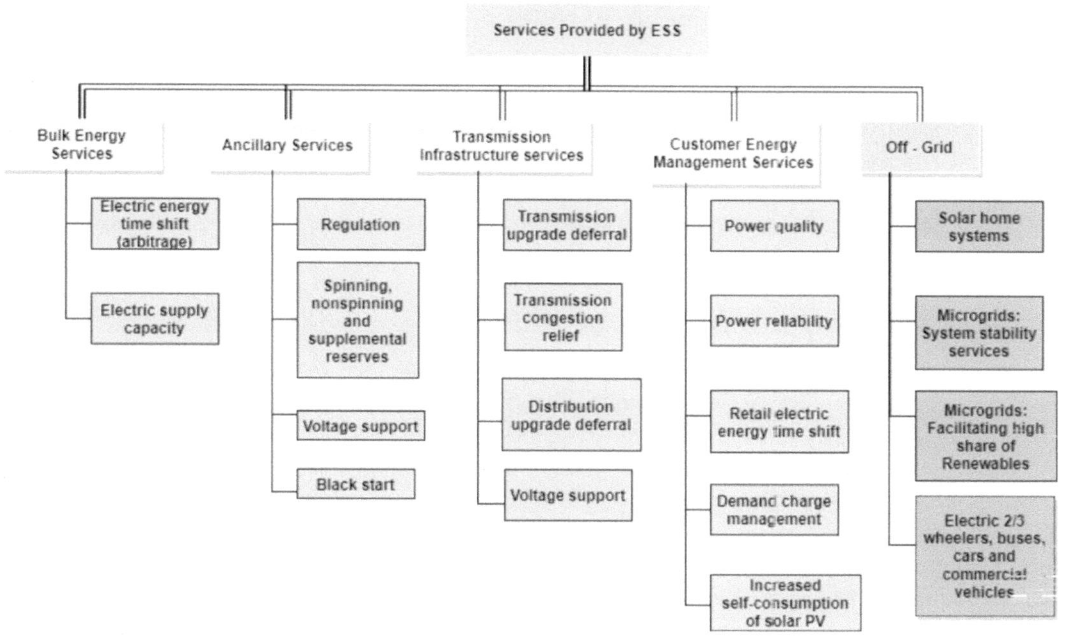

FIGURE 16.5 Different services based on energy storage system.

Source: Nadeem et al. (2018).

phase-change material (PCM) due to its high latent heat capacity and stable thermal properties. Erythritol is deemed ideal for medium-temperature applications in sectors that make use of waste heat recovery and solar thermal energy, such as manufacturing processes and power generation (Chatzigeorgiou et al., 2024).

Energy storage systems (ESS) play a role in improving grid stability, managing demand and supply variations, and integrating different energy sources. By adapting to grid changes, ESS helps regulate frequency, control voltage, and enhance grid reliability. They address the variability of wind power by storing energy during peak production times and releasing it when demand is high or production is low, facilitating greater integration of renewable energy sources (Rana et al., 2023). Additionally, ESS supports load management by releasing stored energy during peak demand periods to lessen the burden on power plants and reduce energy expenses. In the industrial sectors, businesses use ESS to optimize energy usage, shift loads to cost-effective off-peak hours, and enhance energy security – creating precious benefits in regions with expensive electricity or unreliable grids. For power needs and emergency readiness, ESS offers assistance to vital services such as hospitals and data centers during outages. As elements of microgrids, ESS provides energy self-sufficiency for areas that aid in rapidly charging electric vehicles and help meet the high power demands of charging stations to promote electric vehicle adoption (Hamidieh et al., 2022).

Moreover, energy storage systems (ESS) play a role in homes by enabling residents to save solar power for later use when energy demand is low, thereby optimizing solar energy usage and reducing reliance on the grid. ESS also provides utility companies with services such as restarting power after a blackout and managing power quality. They also engage in energy trading by purchasing electricity at prices and selling it at prices to take advantage of market fluctuations. Together, these various applications highlight the importance of ESS in moving toward sustainable, efficient, and reliable energy systems that cater to diverse energy requirements and improve market dynamics (Faisal et al., 2018).

16.4 MECHANICAL STORAGE SYSTEMS

In energy storage systems, energy is transformed between electrical forms. During peak hours, electricity is drawn from the grid and stored mechanically using principles like potential energy, kinetic energy, pressurized gas, and compressed springs until needed. Mechanical energy storage systems are widely used worldwide because they can efficiently convert and utilize stored energy for various tasks. The common types of ESS include pumped hydro storage (PHS) (de Moura Altea et al., 2024), flywheel energy storage (FES) (Wang et al., 2024), compressed air energy storage (CAES) (Tayyeban et al., 2024), and gravity-based energy storage systems (GES) (Kropotin et al., 2024).

16.4.1 Compressed Air Energy Storage (CAES)

The CAES system stores energy by compressing a gas, typically air, into a reservoir. Later, the compressed gas is expanded to turn the turbine and connect it to a generator to generate electricity when needed. CAES systems are considered an alternative to pumped hydroelectric storage (PHS) systems because they have enough capacity, are less dependent on specific locations, have a longer lifespan, and have lower costs per kilowatt. CAES systems are categorized into three types. Isothermal, adiabatic, and diabatic storage systems are based on the processes involved in compressing and expanding air and heat exchange. Isothermal and adiabatic systems are suitable for high-power-density requirements, while diabatic storage systems are commonly used due to their flexibility. In CAES systems, an external power source is utilized to regulate the air temperature, to maintain system efficiency, as shown in Figure 16.6 (Tayyeban et al., 2024; Nabat et al., 2024).

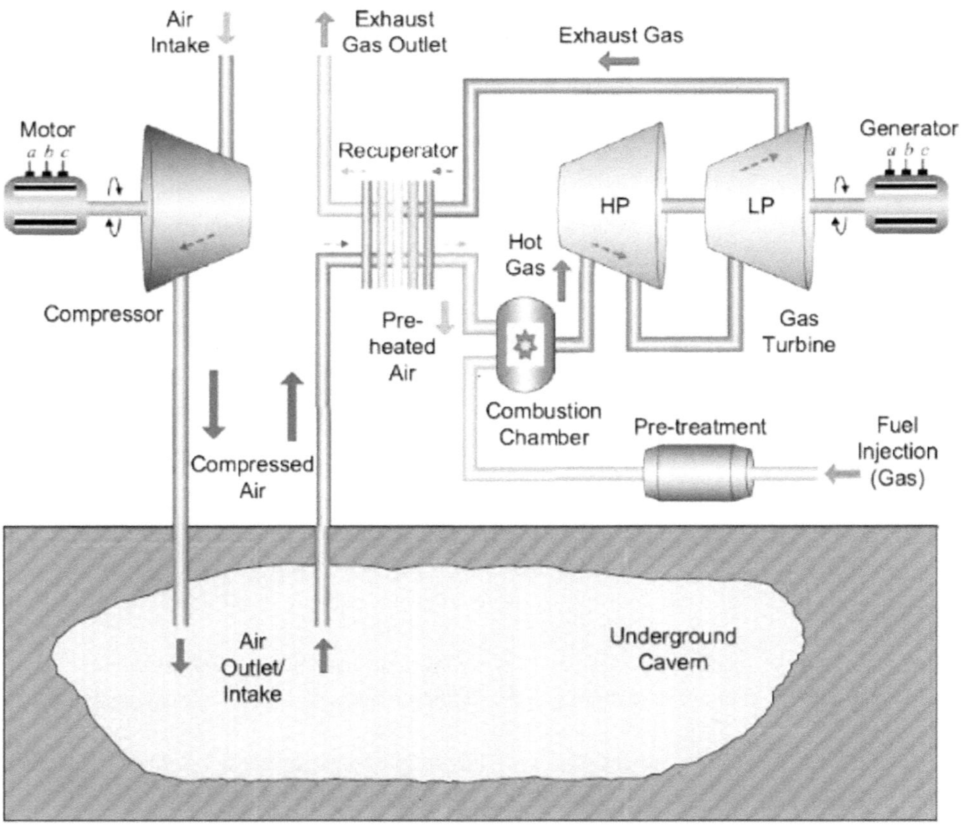

FIGURE 16.6 Large-scale diabatic in-ground cavern CAES system.

Source: Nadeem et al. (2018).

16.4.2 Flywheel Storage

Flywheel energy storage systems store energy in a mass, often a cylinder that rotates rapidly within a sealed container to reduce friction (Figure 16.7) (Nadeem et al., 2018).

The process involves using electricity to speed up the rotation of the cylinder, which then converts this energy into electricity when needed through a generator (Figure 16.8) (Nadeem et al., 2018).

These systems are known for quickly storing and releasing energy. They are suitable for tasks requiring power over specific periods, like maintaining frequency stability and providing backup power during brief interruptions (Table 16.1) (Nadeem et al., 2018).

They are valued for their efficiency, minimal environmental impact, and durability over usage cycles without degradation.

16.4.3 Gravity-Based Stacking Concrete (Energy Vault)

The *energy vault* showcases a gravity-based energy storage system that employs a method involving arranging concrete blocks for energy storage. This approach involves raising blocks with cranes when energy demand is low and excess generation is present, effectively storing potential gravitational energy. When energy is required, these blocks are lowered back. The kinetic energy generated from their descent is converted back into energy using regenerative braking systems in the cranes. This technology promotes sustainability by utilizing easily accessible materials like concrete. It can be customized for various sizes. Its appeal lies in its impact utilization of waste materials for block construction and comparatively lower setup and maintenance costs in comparison to other large-scale energy storage technologies (Cava et al., 2016; Rimpel et al., 2020; Li et al., 2023)

Mechanical storage systems play a role in today's energy setup, providing dependable and practical storage options essential for incorporating renewable energy sources, ensuring grid stability, and improving energy management in different industries. Different types of storage systems come with their advantages and are tailored for particular uses. They can manage energy on a large scale or quickly respond to demands. This flexibility highlights the benefits and significance of energy storage as we move toward more sustainable energy setups (Sandoval Bustamante et al., 2023)

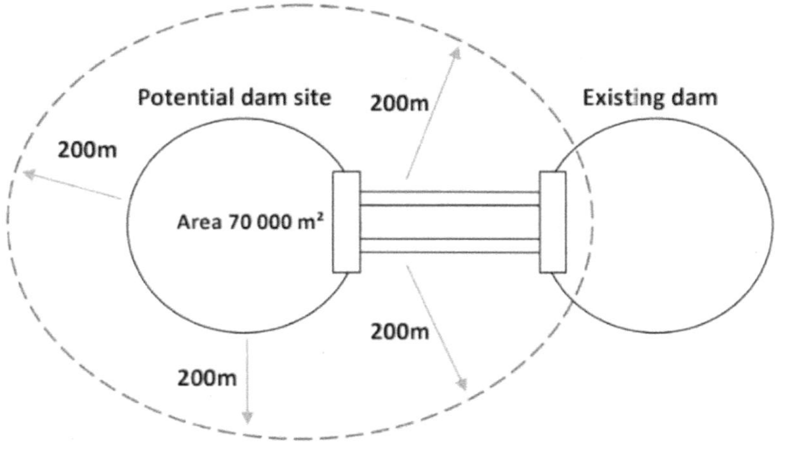

FIGURE 16.7 Pumped hydro storage minimum distance to inhabited areas.

Source: Nadeem et al. (2018).

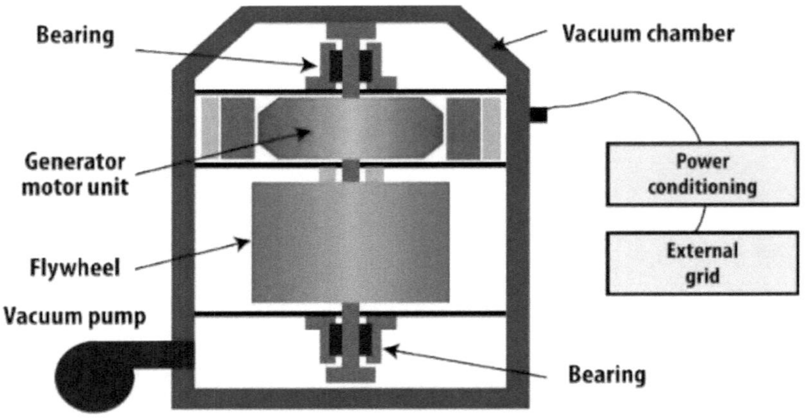

FIGURE 16.8 Structure and components of a pumped hydro storage flywheel energy system.

Source: Nadeem et al. (2018).

TABLE 16.1
Comparison of Low-Speed and High-Speed Flywheels

Specifications	Low-Speed FES	High-Speed FES
Material	Steel	Composite
Electrical machine	Induction, permanent magnet synchronous and reluctance machine	Permanent magnet synchronous and reluctance machine
Quarantine atmosphere	Partial vacuum and partial gas	Absolute vacuum
Required weight of enclosure	Double of flywheel weight	Half of flywheel weight
Applications	Power quality improvement	Aerospace and traction
Economy	Low cost and commercial	High cost and specific usage

Source: Nadeem et al. (2018).

16.5 THERMAL ENERGY STORAGE SYSTEMS

Thermal energy storage (TES) systems store heat or cold, mainly when used alongside energy sources and for heating and cooling purposes (Hirsch et al., 2018; Patel et al., 2024).

Thermal energy can be stored in latent and thermochemical forms or as a mix of those. Latent heat thermal energy storage (LHTES) stands out because it stores and releases energy within a temperature range (Fu et al., 2024). Research indicates that storing the heat using a phase-change material (PCM) can significantly reduce storage volume compared to traditional heat storage materials. The choice of PCM for a system depends on factors like thermal, physical, kinetic, and chemical properties, as well as economic considerations (Brückner et al., 2018). While the selection process typically starts with the phase-change temperature for an application, other factors, such as heat of fusion, thermal conductivity, sub-cooling chemical stability, and thermal stability, are crucial in determining technical feasibility (Steinmann, et al., 2008; Mostafavi, et al., 2016; Niyas, et al., 2017; Ali et al., 2024).

Chandel and Agarwal provided insights into the state of research on PCM commercialization, toxicity issues, energy storage applications, and health risks. Their review highlights PCMs as materials for thermal energy storage solutions but emphasizes the need for increased awareness and engineering innovations to tackle existing challenges (Aneke et al., 2016; Chandel et al., 2017)

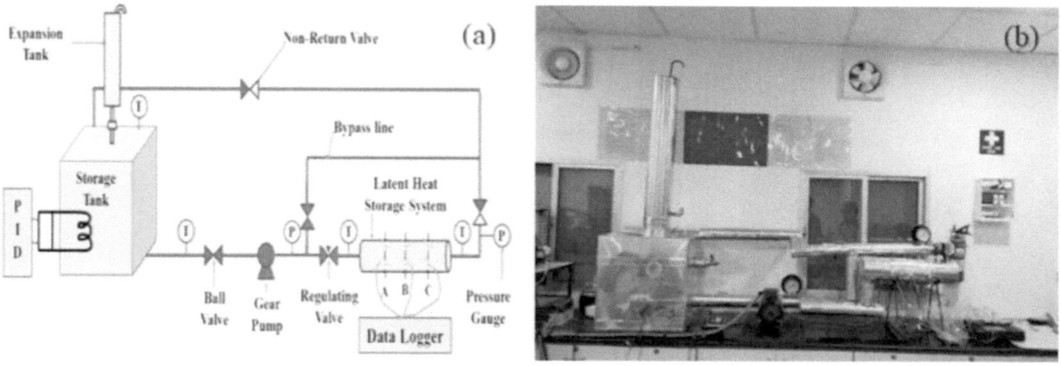

FIGURE 16.9 (a) Schematic diagram and (b) real image of an experimental setup.

Source: Patel et al. (2024).

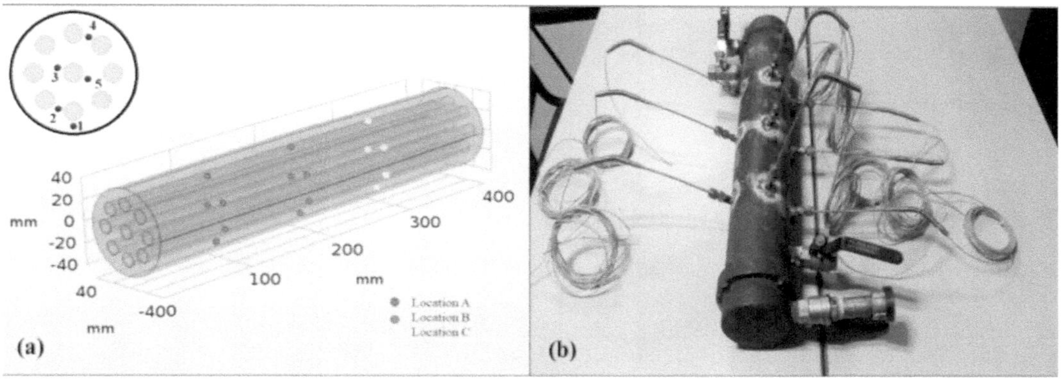

FIGURE 16.10 (a) Dimensions of storage unit, location of thermocouples inside storage unit, and cross-sectional position of five thermocouples (inset); (b) real image of heat storage unit.

Source: Patel et al. (2024).

To develop the system effectively, it is crucial to focus on two main aspects: thoroughly understanding the thermophysical behavior of heat storage materials and comprehending the heat exchange process between these materials and the heat transfer fluid (HTF). While thermophysical properties are innate characteristics of a material, accurately measuring these parameters for materials and studying their performance during various cycles or at elevated temperatures are essential (Figure 16.9). Additionally, selecting the type of heat exchanger is vital to ensure a specific heat transfer rate (Jegadheeswaran et al., 2009; Hérault et al., 2010; Mohamed et al., 2017; Ali et al., 2024).

Among the types of heat exchangers, the shell-and-tube configuration stands out as a good choice for designing an LHTES system. This configuration involves placing the PCM in the shell side while allowing the HTF to flow inside the tubes. It is favored for its ease of manufacturing and implementation, high storage density, and effective heat exchange capabilities (Figure 16.10). Depending on application requirements, optimizing the heat exchange can be achieved through methods such as using tubes, incorporating metal matrices or foam, employing high-conductive composite materials, utilizing carbon fiber cloth, and encapsulating PCM (Zhang et al., 1996; Patel et al., 2024).

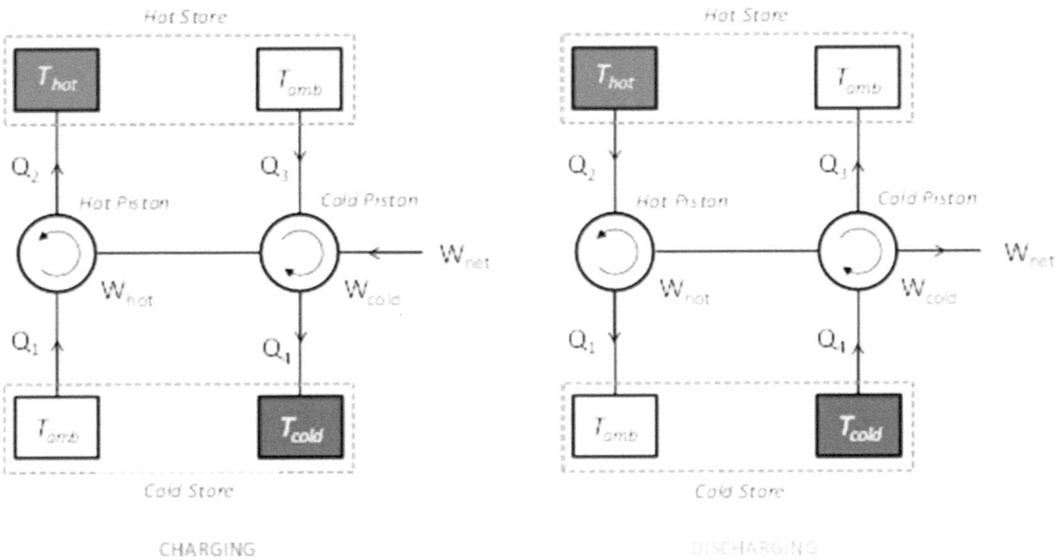

FIGURE 16.11 Schematic of a pumped heat energy storage system.

Source: Smallbone et al. (2017).

16.5.1 Pumped Hydro Storage (PHS)

Pumped hydro energy storage (PHES) has been an accepted technology for storing utility-scale electricity since the 19th century. Not only is hydropower a sustainable energy source, but its flexibility and storage capabilities also contribute to enhancing grid stability and supporting the integration of other intermittent renewables like wind and solar energy. This benefit has led to a growing interest in PHES, with calls for the restoration of small-scale hydro plants (seen in Figure 16.11). Improvements in turbine design are crucial for boosting plant efficiency and adaptability along with developing approaches to optimize storage capacity and maximize profitability in today's deregulated energy market (Smallbone et al., 2017; Javed et al., 2020; Mahfoud et al., 2023).

In the 2000s, PHES re-emerged as a solution for managing peak loads and storing wind and solar energy to ensure power quality. Given the nature of the sources, there is an increasing need for large-scale electricity storage solutions to maintain a steady power supply. While mainly used for storing energy, PHS (Figure 16.12) also deals with other considerations, particularly in controlling how heat is distributed in reservoirs. The process includes pumping water to higher elevations when demand is low and then releasing it through turbines to produce electricity when demand is high (Agyenim et al., 2010; Rehman et al.,2015; Das et al., 2021).

16.5.2 Sensible Heat Storage

This technique entails retaining heat or cold in a medium such as water, sand, or molten salts. The excess energy is stored for later use and released as required. It is used in home climate control systems, industrial heating processes, and industrial solar energy facilities (Agyenim et al., 2010; Raul et al., 2018; Sathishkumar et al., 2024).

Thermal energy storage (TES) systems come in types, such as heat storage (SHS), latent heat storage (LHS) thermochemical reactions, or a combination of these. Among these categories, most thermal energy storage systems rely on heat storage. The reason is that sensible heat materials

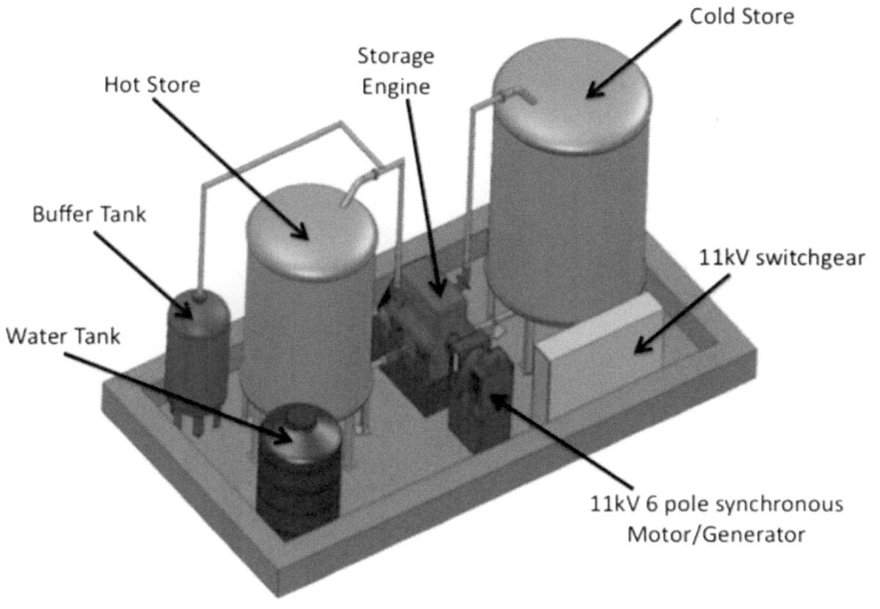

FIGURE 16.12 Image showing the key elements of a pumped heat energy storage system.

Source: Smallbone et al. (2017).

(SHMs) generally have higher conductivities than phase-change materials (PCMs). The higher thermal conductivity of SHMs facilitates heat transfer during charging and cooking processes. On the other hand, PCMs have generally lower thermal conductivity, which results in a longer time needed to absorb and release energy for activities like cooking (Figure 16.13) (Medrano et al., 2010; Lugolole et al., 2018; Rao et al., 2018).

Research by Foong et al. (2011) indicates that SHMs are more cost-effective than PCMs and thermochemical reactions. Therefore, scale-sensitive heat thermal energy storage systems are more suitable for developing countries as they are simpler to build and maintain. These systems can make use of materials like sand, rocks, and concrete charged with air, water, or thermal oils as heat transfer fluids (Cabeza, 2014; Wang et al., 2015).

16.5.3 Latent Heat Storage

Latent heat storage (LHS) is the process of storing thermal energy using substances that change phases at a given temperature, usually from solid to liquid. These materials absorb or release substantial heat without changing temperature during the phase change, making LHS a practical thermal energy storage technique. These properties make them perfect for controlling temperatures or aiding industrial operations (Abhat, 1983; Sharma et al., 2009; Ruwa et al., 2016; Khuda et al., 2024).

The dead zone, which is the area where solid phase-change material (PCM) melts or liquid PCM solidifies, occurs naturally in energy storage systems that use latent heat. This phenomenon happens because the melting or solidifying of PCM in the system affects the charging and discharging rates. The dead zone location varies based on the shape of the heat storage enclosure and how much heat is added. Using fins can help distribute heat evenly, improving the dead zone's phase-change process and temperature distribution (Zalba et al., 2003; Wei et al., 2018; Low et al., 2024).

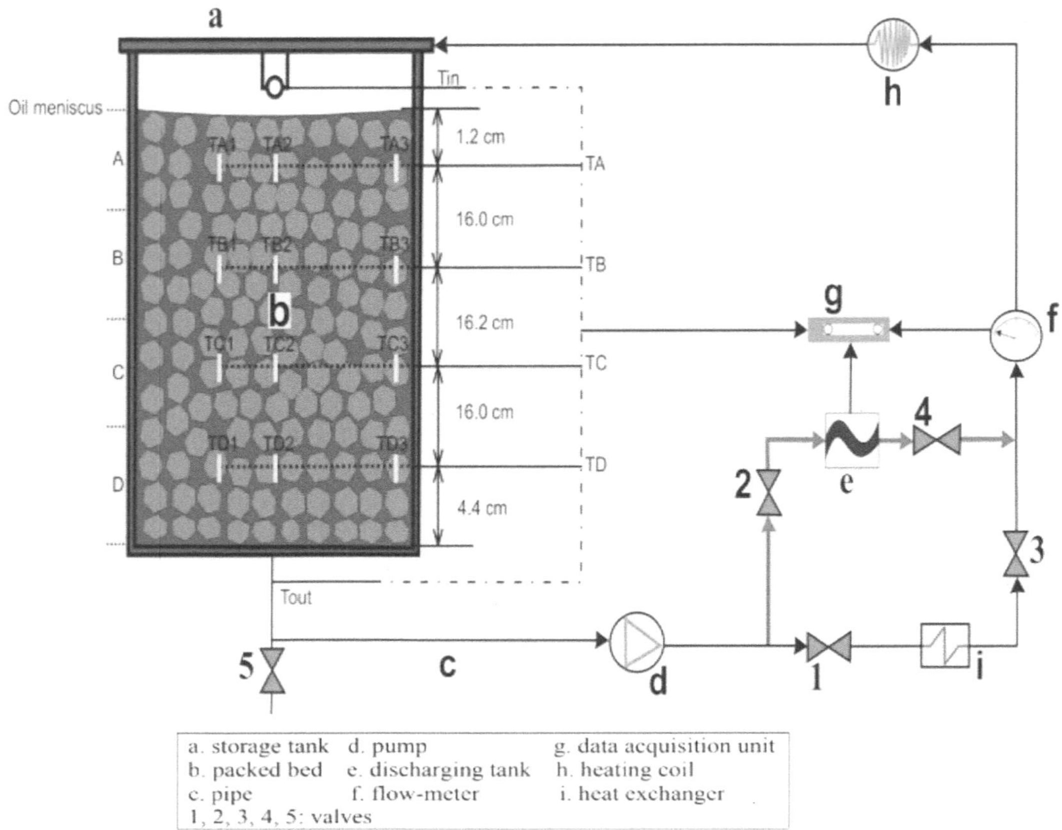

FIGURE 16.13 A schematic diagram of the small-scale TES system.

Source: Lugolole et al. (2018).

16.6 ELECTROCHEMICAL ENERGY STORAGE (ECES)

Battery technologies are the mainstay of electrochemical energy storage systems. They store electrical energy by converting it into chemical energy through electrochemical reactions (Anastasovski et al., 2024).

16.6.1 Flow Batteries

These batteries use liquid electrolytes stored in external tanks and pumped through an electrochemical cell. They are well-suited for applications requiring long discharge times and have a long life cycle (Ouyang et al., 2024).

Flow batteries are energy storage systems that use electrolytes and operate near room temperature. Unlike batteries, where the active electrode material is involved in energy conversion, flow batteries store energy in separate external tanks containing active electrolytes. This design prevents self-discharge (Nguyen et al., 2010). Electrolytes are pumped by a cell into a reactor of battery cells stacked together, where chemical energy is converted into energy and vice versa. A microscopic membrane separates the two electrolytes, allowing only specific ions to pass through into the cell reactor for reaction (Schubert et al., 2023). Flow batteries can be fully recharged by applying a DC voltage to the electrodes during charging. One of the advantages of flow batteries is their ability to decouple power rating from energy storage capacity. The power output of a flow battery depends on

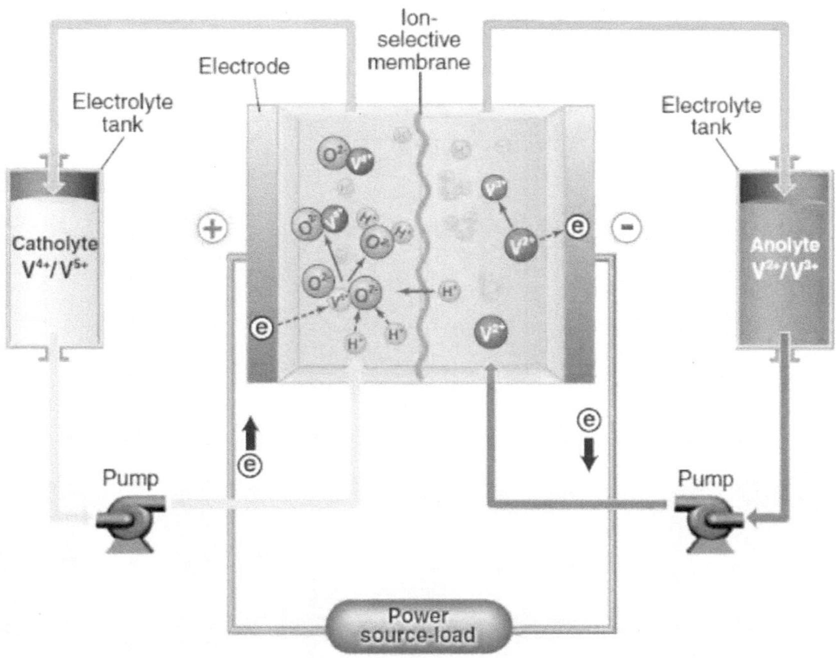

FIGURE 16.14 Redox flow (vanadium) batteries.

Source: Nadeem et al. (2018).

size and stack cell count, while the reservoir size and electrolyte concentration directly influence energy storage capacity (Ouyang et al., 2024).

In the illustration in Figure 16.14, there is a depiction of redox flow batteries using vanadium. The scale of a battery is directly related to its energy storage capability. As a result, they are suitable for energy and power storage applications. Flow batteries share similarities with fuel cells (such as having power and energy ratings). However, fuel cells are not reversible, have lower round-trip efficiency, are economically inefficient, and have life cycles compared to flow batteries (Nadeem et al., 2018).

Therefore, flow batteries are becoming a choice for various stationary uses. Flow batteries can achieve shorter discharge times due to the system's sizing for power and energy. Flow batteries' other benefits include handling overload levels with millisecond response time, no self-discharge in sealed storage units, operation at room temperature, long-lasting charge/discharge cycles, and high round-trip efficiency.

16.6.2 Lead-Based Batteries

Including flooded lead-acid (FLA), valve-regulated lead-acid (VRLA), and lead calcium batteries. Due to their reliability and cost-effectiveness, these batteries are widely used for various applications, from automotive to grid storage (Nadeem et al., 2018).

Lead-acid batteries, first developed in 1859 by physicist G. Plante, are considered one of the best-known battery technologies. They are known for their reliable energy storage capabilities and reasonable cost. These batteries come in sizes ranging from 1 Ah to around 1,000 Ah and have a low daily self-discharge rate of less than 0.3%. They are preferred for many applications due to their durability, safety features, ability to withstand temperatures, high cycle efficiencies

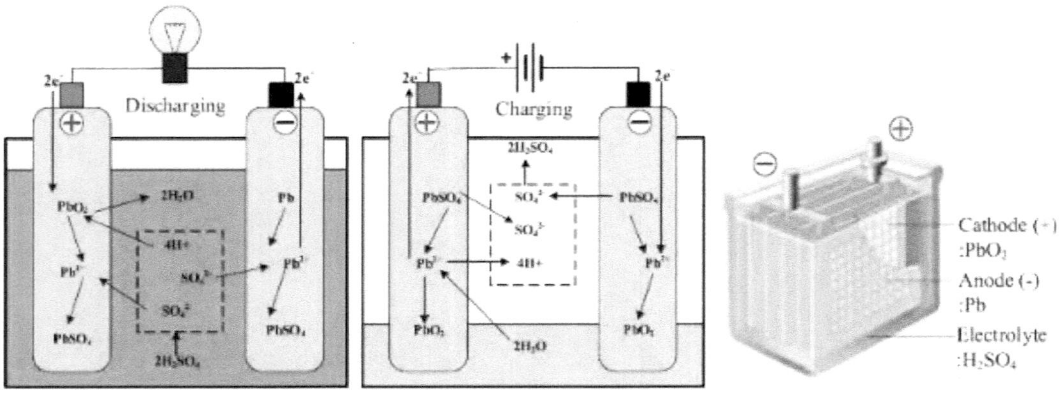

FIGURE 16.15 Lead-acid battery prototype during charging and discharging.

Source: Nadeem et al. (2018).

ranging from approximately 63% to 90%, minimal maintenance requirements, and affordable initial costs. In industries such as data and telecommunications systems, emergency power backup systems, grid utility applications, renewable energy integration projects, and hybrid electric vehicles, lead-acid batteries are commonly used for power supply ranging from kilowatts to tens of megawatts. The lifespan of these batteries is influenced by factors such as temperature charging/discharging rates and the number of charge–discharge cycles they undergo (Figure 16.15) (Nadeem et al., 2018).

16.6.3 Nickel-Based Batteries

Nickel–cadmium (NiCad) and nickel–metal hydride (NiMH) are well-known for being durable and performing effectively across a wide range of temperatures (Aharon et al., 2024; Nyamathulla et al., 2024). In their study conducted in 2024, Wang and colleagues delved into the advancements in nickel-based batteries, explicitly honing in on how surface-amorphized nickel sulfide (Ni_3S_2) could enhance energy storage. They introduced a material that featured unique crystalline, amorphous Ni_3S_2 heterostructure solutions (Wang et al., 2024; Gilbert Zequera et al., 2023).

This innovation significantly boosted battery performance by improving ion and electron transport and reducing resistance at the interface for charge transfer. The material displayed higher capacitance and durability over multiple cycles when tested in supercapacitors and Ni-Zn batteries. The study underscored the potential of incorporating new structures into crystalline hosts to address challenges faced by nickel sulfide batteries, such as low conductivity and sluggish electrochemical processes, thereby paving the way for more effective and long-lasting energy storage solutions (Wang et al., 2024; Gilbert Zequera et al., 2023).

16.6.4 Zinc-Based Batteries

Rechargeable batteries using zinc as a component have gained prominence in the energy storage industry over the decade due to their safety, abundance, and cost-effectiveness. They serve as a complementing factor to lithium-ion batteries. They stand out for their energy density, stable voltage output, efficient charging, safety, and eco-friendliness. These batteries find applications in civilian sectors, such as in aerospace missions, water exploration, and devices like watches and hearing aids. Despite their benefits, zinc silver batteries still face challenges, such as shorter lifespan and problematic underutilization of components (Wang et al., 2024).

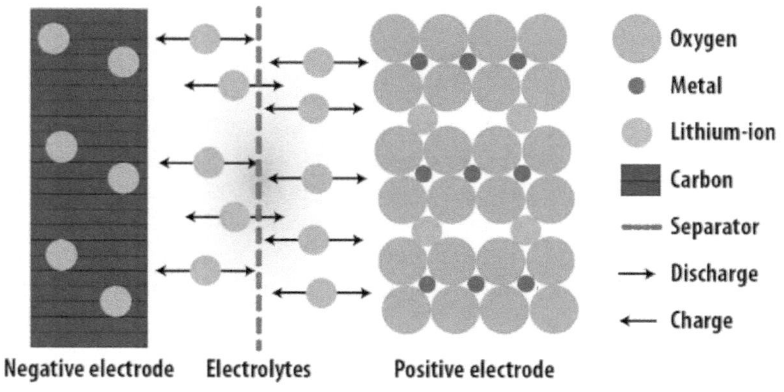

FIGURE 16.16 Schematic of Li-ion cell components and principle.

Source: Nadeem et al. (2018).

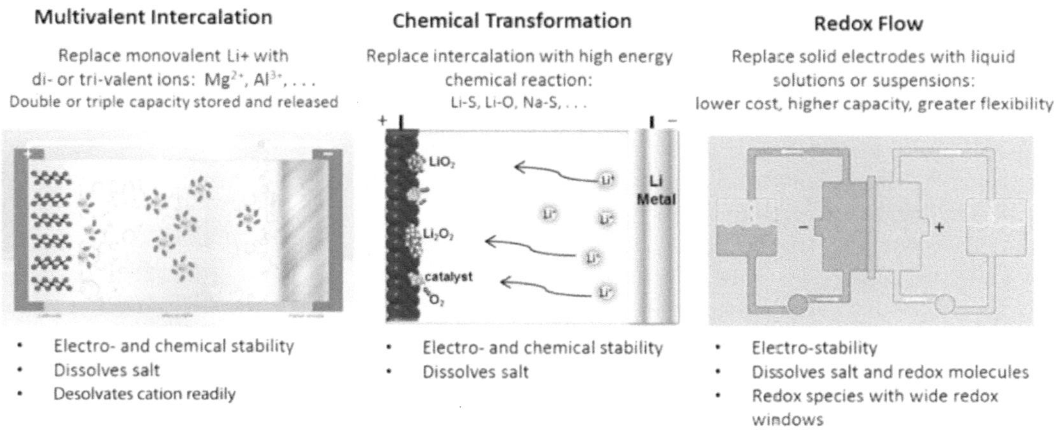

FIGURE 16.17 Illustration of beyond Li-ion battery concepts and basic requirements.

Source: Cheng et al. (2015).

16.6.5 Lithium-Based Batteries

Lithium-ion batteries operate based on the concept of ion extraction within electrodes containing a foil. Figure 16.16 shows the schematic of Li-ion cell components and their operating principle. These electrodes and foil are placed in a solution of Li+ ions before extraction to ensure isolation. Thanks to their energy density (six to eight times that of VLA batteries) per unit volume (75,200 Wh/kg) cycles, low self-discharge rate (<8% per month), extended lifespan, lightweight design, quick charge/discharge capabilities, rapid response time (in milliseconds), and high efficiency, Li-ion batteries have become the primary choice for portable electronics across various market applications (Nadeem et al., 2018).

Over the years, there has been a decrease in costs as the energy density and specific power of Li-ion batteries have increased, resulting in less material usage for manufacturing energy cells. Additionally, modern compact Li-ion batteries are tight. They require no maintenance since there is no gas evolution or electrolysis due to the absence of water in their chemistry. Li-ion batteries consist of lithium-ion and lithium polymer (Figure 16.17) (Nadeem et al., 2018).

16.7 ELECTROMAGNETIC ENERGY STORAGE SYSTEMS

Electromagnetic energy storage systems involve technologies that store energy in electric and magnetic fields, produced by currents, enabling quick energy delivery and high efficiency for applications needing rapid power bursts. The primary types are supercapacitors and superconducting energy storage (SMES) systems. *Supercapacitors*, also called ultracapacitors, store energy using mechanisms allowing for speedy charging and discharging cycles, making them suitable for uses like regenerative braking in vehicles or managing power quality in grid systems (Yadav & Saravanan, 2023). On the other hand, SMES systems store electricity in a field created by the flow of direct current in a superconducting coil, allowing faster energy release, which is beneficial for maintaining grid stability during power outages, enhancing power reliability, and aiding scientific studies that demand intense energy bursts. These technologies contribute to energy management by integrating various energy sources and improving the reliability and effectiveness of power grids and transportation systems (Aghmadi et al., 2024).

16.7.1 SUPERCAPACITORS

They are also known as capacitor energy storage units or electric double-layer capacitors (EDLC). These devices store energy by creating a field through a DC voltage applied across two electrodes separated by a thin insulating layer or dielectric material (Figure 16.18). Unlike capacitors that rely on reactions, these supercapacitors can be rapidly charged with high currents and offer benefits such as high specific energy, low internal resistance, and suitability for use in a wide range of temperatures. This versatility makes them a viable alternative to capacitors used in power electronics and storage batteries. By implementing control strategies, supercapacitor energy storage systems can meet the demanding power requirements of spacecraft's electromagnetic launch systems and high-power radar-equipped ships without causing power disruptions in the microgrid (Nadeem et al., 2018).

While sharing the underlying principles with classical capacitors, supercapacitors differ structurally from them due to their remarkably large specific surface area caused by the need to use liquid electrolytes and porous carbon electrodes. The narrow gap (around 1 nm) between the electrode and electrolyte enables supercapacitor systems to achieve capacitance values in the range of thousand farads (Hepel, 2023).

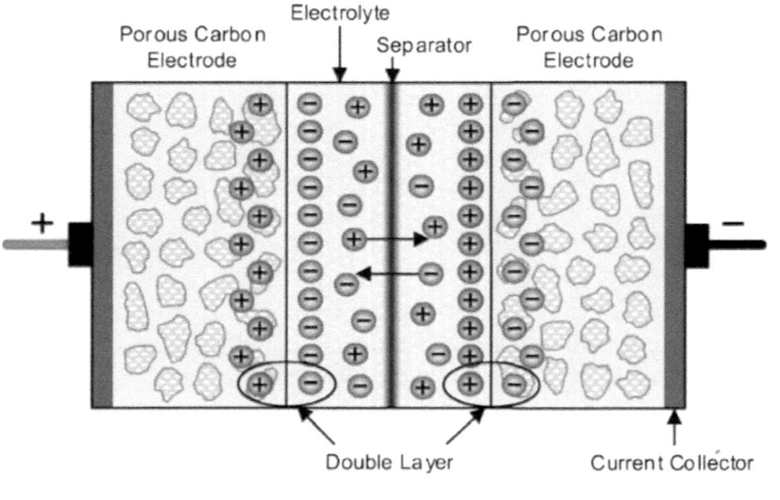

FIGURE 16.18 Graphical representation of symmetrical SCES.

Source: Nadeem et al. (2018).

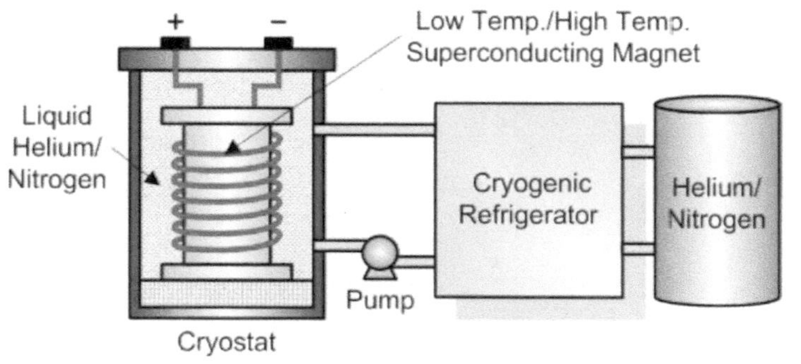

FIGURE 16.19 Super magnetic energy storage (SMES) system design.

Source: Nadeem et al. (2018).

Based on the types of electrodes used, there are two categories of SCES systems: symmetrical and asymmetrical. In *symmetrical systems*, both electrodes are made from the same material, such as activated carbon. On the other hand, *asymmetrical systems* have different materials used in their electrode. Usually, one electrode is made of metal, like nickel hydroxide, and the other is made of superconductor-like materials, such as activated carbon. Asymmetrical systems boost energy density and reduce current leakages. Symmetrical setups are more suitable for smaller applications, while asymmetrical configurations are preferable for larger-scale uses (Khan et al., 2024).

These store energy electrostatically and can charge and discharge much faster than batteries, making them ideal for applications requiring quick bursts of energy.

16.7.2 Superconducting Magnetic Energy Storage

Such systems store energy in the magnetic field generated when direct current flows through a superconducting coil cooled to a temperature higher than its critical superconducting temperature. SMES systems generally provide better response rates and a high power storage capacity.

The stored electrical energy in SMES exists as a field within a superconducting coil generated by the flow of direct current. Once charged, SMES experiences lower losses due to the extremely low resistance of the coil while it is superconducting (Figure 16.19). Cryogenic cooling lowers the coil's temperature so it can perform at its point for superconductivity. During discharge, SMES can rapidly release electrical energy (megawatts) into the grid within milliseconds. Advanced superconductor materials like niobium-titanium (NbTi) and liquid helium or superfluid helium coolant at 4.2 K are commonly used in SMES design for their temperature and high ductile properties. Following the concept of SMES, storing energy in a superconductor while maintaining temperatures below critical values ensures near-permanent energy storage without any decay in current flow (Nadeem et al., 2018; Sesay et al., 2024).

16.8 HYDROGEN/NATURAL GAS STORAGE SYSTEMS

Today's energy storage landscape includes underground salt cavern storage, ammonia, and hydrogen storage, all of which have different uses. Liquid hydrogen is used in fuel cells and power generation for clean energy applications; it is generally stored at high pressures or low temperatures. Ammonia is a hydrogen carrier that can be used as fuel on its own or to liberate hydrogen for various energy systems. Large-scale storage requirements can be met by storing hydrogen, natural gas, or other gases underground in salt caverns (Figure 16.20). These caverns are ideal for various applications, ranging from compact portable devices to large industrial and grid operations, because they offer sufficient capacity and effective withdrawal and injection rates (Małachowska et al., 2022).

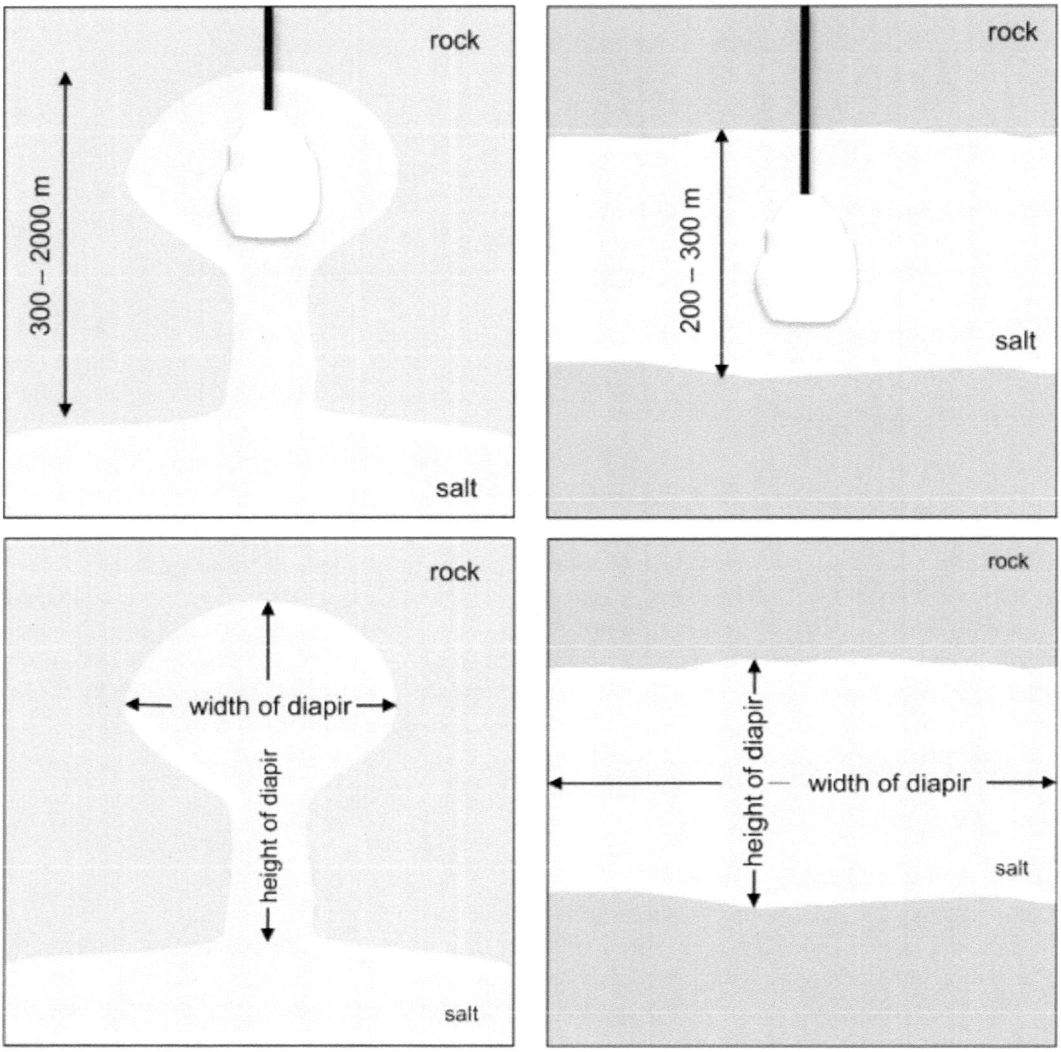

FIGURE 16.20 Salt caverns in (left) diapiric deposits and (right) salt layers.

Source: Małachowska et al. (2022).

16.9 ENERGY INTENSITY, ENVIRONMENTAL IMPACT, AND RESOURCE LIMITATIONS

The discussion about energy storage systems (ESS) extends beyond their aspects to encompass wider environmental and resource-related considerations. Understanding these systems' energy efficiency, environmental impacts, and resource limitations is crucial to assessing their sustainability and long-term viability (Abu et al., 2023).

Energy intensity is the term used to describe the energy required to create, maintain, and use energy storage systems concerning their storage and delivery capacities. It affects these systems' effectiveness and environmental impact. For instance, the high energy intensity seen during production in some battery types may outweigh the advantages of these technologies. Thus, research and development efforts are focused on reducing the energy intensity of energy storage systems by employing advancements in materials and manufacturing techniques (Nadeem et al., 2018; Huang et al., 2022).

The environmental impact of energy storage systems encompasses factors such as emissions, pollution, and disruptions to ecosystems during construction and operation. For instance, batteries may contain substances that pose challenges when it comes to disposal and recycling. Similarly, large-scale systems like pumped hydro storage can significantly impact ecosystems by altering land use patterns and water management practices. Stringent environmental regulations and designs that account for impacts are increasingly being put into practice to tackle these environmental issues (Nadeem et al., 2018; Ngidi et al., 2024).

The sustainability of energy storage technologies heavily depends on the availability of materials required for their manufacturing. Contemporary batteries, such as lithium-ion, rely on relatively rare minerals that can affect the environment and cause conflicts during extraction. Some storage technologies encounter challenges in scalability due to the availability of these resources. Consequently, increasing interest is in developing technologies that utilize environmentally friendly materials. Additionally, there is a growing emphasis on the idea of an economy within the energy storage sector, underscoring the significance of recycling systems to reclaim materials and diminish dependence on raw material extraction (Table 16.2) (Nadeem et al., 2018).

To fully incorporate and broaden the use of energy storage systems within our energy framework, it is crucial to grasp their energy efficiency, environmental consequences, and resource constraints. Tackling these obstacles by leveraging technologies, regulatory actions, and eco-friendly approaches is key to unlocking energy storage systems' capacity and fostering a robust and environmentally friendly energy landscape (Table 16.3) (Nadeem et al., 2018).

16.10 ENERGY STORAGE APPLICATIONS IN THE OIL AND GAS INDUSTRY

Energy storage systems (ESS) are gaining recognition as a technology in the oil and gas sector, providing advantages in terms of operational efficiency, environmental friendliness, and cost savings. Incorporating energy storage technologies into the oil and gas industry supports applications from energy management to improved operational safety (Li et al., 2024).

The research covers a range of topics, from exploring materials in material science to enhancing the capabilities and cost-effectiveness of storage technologies to examining locations and sizes for large-scale energy storage systems to improve stability. An array of research areas also focuses on how energy storage systems operate. Policy and market trends studies delve into issues such as ownership rights, investments, and usage rights that arise from the increasing use of energy storage systems in both utility and residential settings. The development of control and communication solutions for managing these systems are crucial for unlocking the benefits of a smart grid. Establishing communication protocols is particularly important as it enables various other advantages offered by energy storage systems (Khalid, 2024).

Novel optimization algorithms, transactive market mechanisms, and control strategies all rely on having communication channels in place. These channels are essential for monitoring the status of the power network and transmitting operational instructions based on optimization schemes or market mechanisms. To standardize data collection in power systems, it is necessary to have an information model that defines the syntax, semantics, and structure of data exchanges. Various standards have been proposed in the literature for facilitating data exchange within power systems (Table 16.4) (Nadeem et al., 2018).

In the realm of oil and gas operations, it is crucial to ensure a consistent and reliable power supply in offshore locations. Energy storage systems such as batteries or flywheels offer backup power for such cases (Aghmadi et al., 2024). They help balance loads, smoothing out fluctuations in power caused by varying energy sources or grid instability. This is especially important for facilities with safety systems that rely on a power source. The oil and gas industry increasingly turns to renewable energy sources to power their operations, responding to the growing demand for such practices. Energy storage enables the incorporation of sources like wind and solar, ensuring a stable

TABLE 16.2
Technical and Economical Specifications of Various ESS Technologies

ESS Technology	Power Range (MW)	Energy Rating (kWh)	Energy Density (Wh/kg)	Power Density (W/kg)	Discharge Time (ms-h)	Response Time (ms-h)	Round-Trip Efficiency (%)
Mechanical Energy Storage Systems							
PHS	10–5,000	2×10^5 to 5×10^5	0.5–1.5	–	1–24+ hrs	Min	70–85
FES	0.01–0.25	25–5k	5–80	700–12,000	Sec–15 min	Sec	90–05
CAES (underground)	5–300	2×10^5 to 10^6	30–60	–	1–24+ hrs	Min	41–75
CAES (overground)	3–15	2–8.3	140–300 bar	–	2–4 hrs	Sec–min	70–90
Electrical Energy Storage Systems							
SCES	0.01–0.3	10^{-3}–5	0.05–15	10–10^6	Ms–min	Ms	85–95
SMES	0.01–10	15–100, 5×10^6	0.5–5	500–2k	Ms–sec	Ms	90–97
Thermal Energy Storage Systems							
Low-temp TES	0.01–300	–	100–200	10–30	1–8 hrs	Slower	30–50
High-temp TES	1–60	–	80–250	–	1–24+ hrs	Not rapid	80
Chemical Energy Storage Systems							
Hydrogen (FC)	0.3–50	<200k	600–1.2k	5–800	Sec–24 hrs	Sec	30–50
Electrochemical Energy Storage Systems							
Conventional							
Pb-A	<20	18k–100k	30–50	200–400	Sec–5 hrs	Ms	70–90
Ni-Cd	<40	6.75k	15–55	150–350	1–8 hrs	Ms	75
Ni-MH	<0.03	0.01–500	60–80	150–460	Hrs	Ms	70–80
Advanced							
Li-ion	0.05–100	250–25k	120–230	150–2k	Min–1 hr	Ms	85–95
Na-S	10–34	245k	150–240	150–230	6–7.2 hrs	Ms	75–90
Na-$MeCl_2$ (ZEBRA)	0.005–1	120–5k	86–140	180–245	Sec–hr	Ms	90
Metal air							
Zn-air	0.01–1	60–150	1k–1.3k	90–105	10–15 hrs	Ms	50–65
Flow battery							
VRFB	0.01–10	4k-40k	25–35	–	5–10 hrs	Ms	60–75
HFB (ZBB)	2–10	50–500	65–75	–	8–10 hrs	Ms	60–80

Source: Nadeem et al. (2018).

TABLE 16.3
Technical and Economical Specifications of Various ESS Technologies 2

ESS Technology	Lifetime (yr)	Life Cycle (Cycles)	Total Capital Cost: Power Cost SAW	Total Capital Cost: Energy Cost S/kWh	Daily Self-Discharge (%)	Operating Temperature (⁰c)	Nominal Voltage (V)	Technical Maturity
Mechanical Energy Storage Systems								
PHS	30–60	–	500–2k	5–100	Null	Ambient	–	Mature
FES	15–20	20k–100k	100–300	1,000–5,000	1.3–100	20–50+	–	Commercial
CAES (underground)	20–40	8k–13k	500–1.8k	50–400	~0	Ambient	–	Developed
CAES (overground)	~20	0.5k–1.8k	1k–1.55k	200–250	~0	Ambient	–	Developed
Electrical Energy Storage Systems								
SCES	25–30	100k–500k	100–300	300–2k	10–40	–40 to 85	2.3–400	Developing
SMES	20–30	20k–100k	200–350	1k–10k	10–15	–162 to –253	–	Demo/early
Thermal Energy Storage Systems								
Low-temp TES	10–40	–	200–300	20–50	0.5–1	–12 to –150	–	Developing
High-temp TES	5–15	–	200–300	30–60	0.05–1	400–1k	–	Demo/early
Chemical Energy Storage Systems								
Hydrogen (FC)	5–15	1k–10k	400–2k	1–15	~0	50–100, 600–1k	0.2–1.1	Developing/Demo
Electrochemical Energy Storage Systems								
Conventional								
Pb-A	5–15	500–2k	175–600	150–400	0.1–0.4	–30 to 50	2.0–2.35	Mature
Ni-Cd	10–20	800–2.5k	500–1.5k	600–2.4k	0.2–0.3	–45 to 60	1.2	Mature
Ni-MH	5–10	800–3k	120% of Ni-Cd	120% of Ni- Cd	0.4–1.2	–20 to 45	1.2	Mature
Advanced								
Li-ion	20–25	1k–10k	1.2k–4k	400–2.5k	0.15–0.3	10 to 60, –20 to 60	3.6–4.2	Commercial
Na-S	10–15	2.5k–4k	3.2k–4k	300–500	0.05–20	300–350	2.1	Commercial
Na-MeCl$_2$ (ZEBRA)	5–15	1k–1.2k	150–300	230–345	15	270–350	2.58	Commercial
Metal air								
Zn-air	>1	1k–2k	100–250	60–160	Very small	0–50	1.0–1.3	Demo
Flow battery								
VRFB	10–20	>12k	1.4k–3.7k	500–800	Small	10–40	1.4–1.5	Developing

Source: Nadeem et al. (2018).

TABLE 16.4
Storage Services, Applications, and Remarks

Storage Service	Service Applications	Suitable Options	
		Technology	**Remarks**
Long duration (frequent)	• Load leveling • Hot reserve • Arbitrage • Bulk energy storage • Peak shaving • Frequency regulation • Intermittent seasonal balancing • Renewable penetration • Area control	PHS	Site-specific and habitat species limitation, large construction time, and high capital cost
		CAES (underground)	Long construction time, site-specific plant, gas required as fuel, low round-trip efficiency, and high capital cost
		TES	Low round-trip efficiency, slow response, and high operational cost
		VRFB/HFB	Relatively low energy density, expensive than other ECSS, toxic remains in last, electrolyte leakage protection required, too heavy, not yet proven cycle life, and developing technology
		FC	High cost, low efficiency, combustion of fossil fuel, and developing technology
		Pb-A	Too heavy weight, risk of hazardous gas emissions, relatively small lifetime, potential lead pollution, and thermal management system required
		Ni-Cd/Ni-MH	Toxic materials, intolerant to temperature extremes, memory effect, costly recycling, safety process, and low cell voltage
		Na-S	Very high operating temperature, safety concerns, light-weighted, and high initial cost
Medium duration (fast response)	• Black start • Valley filling • Demand-side management • Distributed storage • Loss reduction • Investment deferral • Contingency service • Area control • Spinning reserve • Capacity credit • T&D support	CAES (aboveground)	Special site requirements, low response time, high capital cost, relatively low discharge time, and combustion of gas fuel
		Pb-A	*Same as mentioned for "long duration"*
		Na-MeCl$_2$ (ZEBRA)	Safety concerns, very high operating temperature, light-weighted, and high initial cost
		Li-ion	High capital cost, sophisticated charge control electronics required, light-weighted, fire protection scheme required, safety and recycling of toxic lithium oxide and salt required
		VRFB/HFB	*Same as mentioned for "long duration"*
		FC	*Same as mentioned for "long duration"*
		Ni-Cd/Ni-MH	*Same as mentioned for "long duration"*
		TES	*Same as mentioned for "long duration"*
		FES	Low energy density, constructional limitations, long recharging time due to frictional losses, large standby losses, short-term power, and high capital cost
		Metal-air (Zn-air)	Relatively new and untested, limited shelf and cycle life, voltage drop at the beginning of discharge, electric charging difficult, and only few options available
		SMES	Low energy density, adverse health impact due to extensive magnetic field, manufacturing complexity, and high capital cost

Short duration (highly frequent)	• Power quality • Intermittency mitigation • Transient stability • End user application • Frequency regulation • Renewable integration • Fluctuations smoothing • Emergence power bridging	FES	*Same as mentioned for "long duration"*
		SCES	High operational cost, infancy technology, short-term power, relatively low energy density, toxic compound, and high capital cost
		Pb-A	*Same as mentioned for "long duration"*
		Na-MeCl$_2$ (ZEBRA)	*Same as mentioned for "medium duration"*
		Li-ion	*Same as mentioned (or "medium duration")*
		Ni-Cd/Ni-MH	*Same as mentioned for "long duration"*
		SMES	*Some as mentioned for "medium duration"*

Source: Nadeem et al. (2018).

and dependable energy supply while reducing reliance on fossil fuel–based power generation and decreasing greenhouse gas emissions (Al Wahedi & Bicer, 2024).

Additionally, energy storage plays a role in cutting down on energy costs, by employing peak-shaving strategies during periods of electricity demand, which helps avoid tariff rates and alleviates strain on local grids. Embracing drilling and extraction processes supported by energy storage technology allows for efficient and precise operations compared to traditional diesel-powered equipment, leading to cost savings and reduced environmental impact by minimizing emissions and noise pollution. Moreover, energy storage systems enhance safety by providing power for lighting, fire suppression units, and safety mechanisms during emergencies (Close et al., 2024).

They can also help decrease gas flaring by storing gas that is usually wasted by burning to show a way to use resources efficiently and minimize impact. The intentional inclusion of energy storage systems in the oil and gas sector is crucial for improving efficiency, cutting costs, and promoting the industry's transition to sustainability (Olabi et al., 2021).

16.11 POTENTIAL, TECHNOLOGY, CHALLENGES, AND LATEST ADVANCES IN THE WORLD

Energy storage systems (ESS) play a significant role in the energy transition by marking a significant step forward in meeting future energy demands. ESS offers benefits, such as improving grid reliability, seamlessly incorporating energy sources, and providing backup power to reduce dependence on fossil fuels and lower carbon emissions. The advancement of related technologies has expanded the range of energy storage solutions, with lithium-ion batteries dominating the market due to their energy density and efficiency across applications, from small devices to large-scale grid storage (Aghmadi et al., 2024).

Flow batteries offer scalability and extended discharge periods; hydrogen storage addresses long-term needs. Thermal storage systems allow solar power plants to continue electricity generation during low sunlight conditions or at night. Despite these advancements, challenges like high costs, technological constraints, regulatory issues, and environmental impacts from ESS component production continue to pose obstacles. Recent developments focus on creating materials that reduce resource reliance, enhancing battery recycling techniques, and exploring alternatives such as sodium ion and zinc-based batteries. Moreover, there is a growing trend toward integrating energy storage with energy setups to optimize energy utilization and cost-efficiency (Yang et al., 2024).

It is crucial to understand how these systems work, their abilities, and their impact on power system operations. It helps select strategies to ensure voltage and frequency stability, handle changes in load, and match supply with demand as renewable energy sources grow (Hazra et al., 2024).

16.12 CONCLUSION AND FUTURE WORKS

This study examines cutting-edge energy storage system (ESS) technologies and upcoming trends. It explores application areas based on their capacities and operational characteristics within power system operations. It can be said with confidence that energy storage systems (ESS) are becoming increasingly prevalent with the emergence of technologies. Regarding devices, current research focuses on two aspects: discovering new chemicals or configurations that can store higher energy levels per unit volume and reducing the construction costs to enable more cost-effective energy storage. At the system level, ESS research is geared toward developing algorithms or applications that enhance power systems' resilience, efficiency, and overall performance. Various innovative approaches are being explored, such as using new methods to transport energy to their required destination or implementing inverters with storage for voltage and frequency support in distribution networks. Further groundbreaking solutions will emerge, facilitating the transition to fully digitalized power grids.

In conclusion, as energy storage technology advances, its integration into energy systems is becoming increasingly feasible and economically viable. Overcoming challenges through innovation and supportive policies will play a role in unlocking the full potential of energy storage to meet the world's expanding energy needs sustainably. The ongoing progress in this field presents an outlook for the future of our energy landscape, positioning energy storage as an element in achieving both energy security and environmental sustainability.

REFERENCES

Abhat, A. (1983). Low temperature latent heat thermal energy storage: Heat storage materials. *Solar Energy*, *30*, 313–332. https://doi.org/10.1016/0038-092X(83)90186-X.

Abu, S. M., Hannan, M. A., Ker, P. J., Mansor, M., Tiong, S. K., & Mahlia, T. I. (2023). Recent progress in electrolyser control technologies for hydrogen energy production: A patent landscape analysis and technology updates. *Journal of Energy Storage*, *72*, 108773.

Aghmadi, A., & Mohammed, O. A. (2024). Energy storage systems: Technologies and high-power applications. *Batteries*, *10*(4), 141.

Agyenim, F., Hewitt, N., Eames, P., & Smyth, M. (2010). A review of materials, heat transfer and phase change problem formulation for latent heat thermal energy storage systems (LHTESS). *Renewable and Sustainable Energy Reviews*, *14*, 615–628. https://doi.org/10.1016/j.rser.2009.10.015.

Aharon, I., Shmaryahu, A., Sitbon, M., Dagan, K. J., Baimel, D., & Amar, N. (2024). Premeditated generic energy storage model for sources rating selection in grid applications. *International Journal of Electrical Power & Energy Systems*, *157*, 109837.

Al Wahedi, A., & Bicer, Y. (2024). Comprehensive risk assessment of a renewables-based stand-alone electric vehicle charging station with multiple energy storage technologies. *Energy Storage*, *6*(2), e587.

Ali, H. M., Rehman, T. U., Arıcı, M., Said, Z., Duraković, B., Mohammed, H. I., . . . Teggar, M. (2024). Advances in thermal energy storage: Fundamentals and applications. *Progress in Energy and Combustion Science*, *100*, 101109.

Anastasovski, A., Andreucci, M. B., Kádár, J., & Delli Paoli, M. (2024). Energy storage in urban areas: The role of energy storage facilities, a review. *Energies*, *17*(5), 1117.

Aneke, M., & Wang, M. (2016). Energy storage technologies and real life applications – a state of the art review. *Applied Energy*, *179*, 350–377. https://doi.org/10.1016/j.apenergy.2016.06.097.

Bahloul, M., Trivedi, R., Cardo-Miota, J., Papadimitriou, C., Efthymiou, V., Nouri, A., & Khadem, S. (2024). Analysis of barriers and key enablers toward citizen ESS successful integration. *Journal of Energy Storage*, *86*, 111166.

Boruah, D., & Chandel, S. S. (2024). Techno-economic feasibility analysis of a commercial grid-connected photovoltaic plant with battery energy storage-achieving a net zero energy system. *Journal of Energy Storage*, *77*, 109984.

Brückner, S., Liu, S., Miro, L., Radspieler, M., Cabeza, L. F., Lävemann, E., Jouhara, H., Khordehgah, N., Almahmoud, S., Delpech, B., Chauhan, A., & Tassou, S. A. (2018). Waste heat recovery technologies and applications. *Thermal Science and Engineering Progress*, *6*, 268–289. https://doi.org/10.1016/j.apenergy.2015.01.147.

Bui, V. G., Bui, T. M. T., Hoang, A. T., Nižetić, S., Sakthivel, R., Engel, D., & Hadiyanto, H. (2021). Energy storage onboard zero-emission two-wheelers: Challenges and technical solutions. *Sustainable Energy Technologies and Assessments*, *47*, 101435.

Cabeza, L. F. (2014). *Advances in Thermal Energy Storage Systems: Methods and Applications*. Elsevier Science. https://books.google.co.in/books?id=_590AwAAQBAJ.

Carnevali, A., Palacin, M. R., Grey, C. P., & Franco, A. A. (2024). VOLTA: A tool for battery screening bridging the gap between virtual electrode materials and practical applications. *Energy Storage Materials*, 103254.

Cava, F., Kelly, J., Peitzke, W., Brown, M., & Sullivan, S. (2016). Advanced rail energy storage: Green energy storage for green energy. In *Storing Energy* (pp. 69–86). Elsevier.

Chandel, S. S., & Agarwal, T. (2017). Review of current state of research on energy storage, toxicity, health hazards and commercialization of phase changing materials. *Renewable and Sustainable Energy Reviews*, *67*, 581–596. https://doi.org/10.1016/j.rser.2016.09.070.

Chatzigeorgiou, N. G., Theocharides, S., Makrides, G., & Georghiou, G. E. (2024). A review on battery energy storage systems: Applications, developments, and research trends of hybrid installations in the end-user sector. *Journal of Energy Storage*, *86*, 111192.

Cheng, L., Assary, R. S., Qu, X., Jain, A., Ong, S. P., Rajput, N. N., . . . Curtiss, L. A. (2015). Accelerating electrolyte discovery for energy storage with high-throughput screening. *The Journal of Physical Chemistry Letters*, *6*(2), 283–291.

Close, J., Barnard, J. E., Chew, Y. J., & Perera, S. (2024). A holistic approach to improving safety for battery energy storage systems. *Journal of Energy Chemistry*, *92*, 422–439.

Das, P., Das, B. K., Mustafi, N. N., & Sakir, M. T. (2021). A review on pump-hydro storage for renewable and hybrid energy systems applications. *Energy Storage*, *3*(4), e223.

de Moura Altea, C., & Yanagihara, J. I. (2024). Energy, exergy and environmental impacts analyses of Pumped Hydro Storage (PHS) and Hydrogen (H2) energy storage processes. *Journal of Energy Storage*, *76*, 109713.

Faisal, M., Hannan, M. A., Ker, P. J., Hussain, A., Mansor, M. B., & Blaabjerg, F. (2018). Review of energy storage system technologies in microgrid applications: Issues and challenges. *IEEE Access*, *6*, 35143–35164.

Foong, C. W., Nydal, O. J., & Løvseth, J. (2011). Investigation of a small scale double-reflector solar concentrating system with high temperature heat storage. *Applied Thermal Engineering*, *31*(10), 1807–1815.

Fu, W., Sun, W., & Huo, X. (2024). Design and thermodynamic investigation of a waste heat-assisted compressed air energy storage system integrating thermal energy storage and organic Rankine cycle. *Energy Technology*, *12*(2), 2300838.

Gao, Q., Bechlenberg, A., Jayawardhana, B., Ertugrul, N., Vakis, A. I., & Ding, B. (2024). Techno-economic assessment of offshore wind and hybrid wind–wave farms with energy storage systems. *Renewable and Sustainable Energy Reviews*, *192*, 114263.

García-Miguel, P. L. C., Alonso-Martinez, J., Gómez, S. A., & Rodríguez-Amenedo, J. L. (2024). Impact of risk measures and degradation cost on the optimal arbitrage schedule for battery energy storage systems. *International Journal of Electrical Power & Energy Systems*, *157*, 109883.

Gholami, M., Muyeen, S. M., & Lin, S. (2024). Optimizing microgrid efficiency: Coordinating commercial and residential demand patterns with shared battery energy storage. *Journal of Energy Storage*, *88*, 111485.

Gilbert Zequera, R., Rassõlkin, A., Vaimann, T., & Kallaste, A. (2023). Overview of battery energy storage systems readiness for digital twin of electric vehicles. *IET Smart Grid*, *6*(1), 5–16.

Hérault, D., Rodembusch, F., Campo, L., Gingras, M., Cerveau, G., & Corriu, R. J. P. (2010). Valorization of by-products of the sugar industry: New nanostructured hybrid materials containing sugar derived structures. *Comptes Rendus Chimie*, *13*, 566–574. https://doi.org/10.1016/j.crci.2010.04.001.

Hamidieh, M., & Ghassemi, M. (2022). Microgrids and resilience: A review. *IEEE Access*, *10*, 106059–106080.

Hazra, S. K., Kim, H., Meskher, H., Singh, P., Kansara, S., Thakur, A. K., . . . Hwang, J. Y. Lithium-free redox flow batteries: Challenges and future prospective for safe and efficient energy storage. *Batteries & Supercaps*, e202400100.

Hepel, M. (2023). Advances in micro-supercapacitors (MSCs) with high energy density and fast charge-discharge capabilities for flexible bioelectronic devices—A review. *Electrochemical Science Advances*, *3*(3), e2100222.

Hirsch, A., Parag, Y., & Guerrero, J. (2018, January). Microgrids: A review of technologies, key drivers, and outstanding issues. *Renewable and Sustainable Energy Reviews*, *90*, 402–411.

Huang, Y., & Li, J. (2022). Key Challenges for grid-scale lithium-ion battery energy storage. *Advanced Energy Materials*, *12*(48), 2202197.

Jankowski, M., Pałac, A., Sornek, K., Goryl, W., Żołądek, M., Homa, M., & Filipowicz, M. (2024). Status and development perspectives of the compressed air energy storage (CAES) technologies—A literature review. *Energies*, *17*(9), 2064.

Javed, M. S., Ma, T., Jurasz, J., & Amin, M. Y. (2020). Solar and wind power generation systems with pumped hydro storage: Review and future perspectives. *Renewable Energy*, *148*, 176–192.

Jegadheeswaran, S., & Pohekar, S. D. (2009). Performance enhancement in latent heat thermal storage system: A review. *Renewable and Sustainable Energy Reviews*, *13*, 2225–2244. https://doi.org/10.1016/j.rser.2009.06.024.

Khalid, M. (2024). Smart grids and renewable energy systems: Perspectives and grid integration challenges. *Energy Strategy Reviews*, *51*, 101299.

Khan, S. A., Rahman, M. A., Khraisheh, M., & Hassan, I. G. (2024). Advances in periodic lattice structures for energy Research: Energy Storage, transport and conversion applications. *Materials & Design*, 112773.

Khuda, M. A., Yan, L., Sarunac, N., & Romero, C. (2024). Design, analysis, and testing of a prototype-scale latent heat thermal energy storage (LTES) system. *Journal of Energy Storage*, *89*, 111796.

Kropotin, P., & Marchuk, I. (2024). Analytical and quantitative assessment of capital expenditures for construction of an aboveground suspended weight energy storage. *Renewable Energy*, *220*, 119587.

Li, H., Chen, D., Dong, L., Li, Z., Zhang, W., & Deng, C. (2024, February). Summary of research on operation control of electrochemical energy storage power plants for offshore wind power. In *Ninth International Conference on Energy Materials and Electrical Engineering (ICEMEE 2023)* (Vol. 12979, pp. 680–689). SPIE.

Li, J., Wang, W., Yuan, Z., Chen, J., & Zhang, Y. (2023). Optimal multi-market operation of gravity energy storage and wind power producer using a hybrid stochastic/robust optimization. *Journal of Energy Storage*, *68*, 107760.

Low, Z. H., Qin, Z., & Duan, F. (2024). A review of fin application for latent heat thermal energy storage enhancement. *Journal of Energy Storage*, *85*, 111157.

Lugolole, R., Mawire, A., Lentswe, K. A., Okello, D., & Nyeinga, K. (2018). Thermal performance comparison of three sensible heat thermal energy storage systems during charging cycles. *Sustainable Energy Technologies and Assessments*, *30*, 37–51.

Lyden, A., Pepper, R., & Tuohy, P. G. (2018). A modelling tool selection process for planning of community scale energy systems including storage and demand side management. *Sustainable Cities and Society*, *39*, 674–688.

Mahfoud, R. J., Alkayem, N. F., Zhang, Y., Zheng, Y., Sun, Y., & Alhelou, H. H. (2023). Optimal operation of pumped hydro storage-based energy systems: A compendium of current challenges and future perspectives. *Renewable and Sustainable Energy Reviews*, *178*, 113267.

Małachowska, A., Łukasik, N., Mioduska, J., & Gębicki, J. (2022). Hydrogen storage in geological formations—The potential of salt caverns. *Energies*, *15*(14), 5038.

Medrano, M., Gil, A., Martorell, I., Potau, X., & Cabeza, L. F. (2010). State of the art on high-temperature thermal energy storage for power generation. Part 2-Case studies. *Renewable and Sustainable Energy Reviews*, *14*, 56–72. https://doi.org/10.1016/j.rser.2009.07.036.

Mitali, J., Dhinakaran, S., & Mohamad, A. A. (2022). Energy storage systems: A review. *Energy Storage and Saving*, *1*(3), 166–216.

Mohamed, S. A., Al-Sulaiman, F. A., Ibrahim, N. I., Zahir, M. H., Al-Ahmed, A., Saidur, R., Yılbas, B. S., & Sahin, A. Z. (2017). A review on current status and challenges of inorganic phase change materials for thermal energy storage systems. *Renewable and Sustainable Energy Reviews*, *70*, 1072–1089. https://doi.org/10.1016/j.rser.2016.12.012.

Mostafavi, S. S., Taylor, R. A., Saberi, P., Diarce, G., Tehrani, S. S. M., Taylor, R. A., Saberi, P., & Diarce, G. (2016). Design and feasibility of high temperature shell and tube latent heat thermal energy storage system for solar thermal power plants. *Renewable Energy*, *96*, 120–136. https://doi.org/10.1016/j.renene.2016.04.036.

Nabat, M. H., Yu, H., & Arabkoohsar, A. (2024). A comprehensive and comparative study of an innovative constant-pressure compressed air energy storage (CP-CAES) system. *Sustainable Cities and Society*, *102*, 105163.

Nadeem, F., Hussain, S. S., Tiwari, P. K., Goswami, A. K., & Ustun, T. S. (2018). Comparative review of energy storage systems, their roles, and impacts on future power systems. *IEEE Access*, *7*, 4555–4585.

Ngidi, N. P., Koekemoer, A. F., & Ndlela, S. S. (2024). Recent advancement in the electrochemical performance of electrochemical capacitors based on biomass-derived porous carbon: A review. *Journal of Energy Storage*, *89*, 111638.

Nguyen, T., & Savinell, R. F. (2010). Flow batteries. *The Electrochemical Society Interface*, *19*(3), 54.

Niyas, H., Rao, C. R. C., & Muthukumar, P. (2017). Performance investigation of a lab-scale latent heat storage prototype – experimental results. *Solar Energy*, *155*, 971–984. https://doi.org/10.1016/j.solener.2017.07.044.

Nyamathulla, S., & Dhanamjayulu, C. (2024). A review of battery energy storage systems and advanced battery management system for different applications: Challenges and recommendations. *Journal of Energy Storage*, *86*, 111179.

Olabi, A. G., Onumaegbu, C., Wilberforce, T., Ramadan, M., Abdelkareem, M. A., & Al–Alami, A. H. (2021). Critical review of energy storage systems. *Energy*, *214*, 118987.

Ouyang, T., Zhang, M., Qin, P., & Tan, X. (2024). Flow battery energy storage system for microgrid peak shaving based on predictive control algorithm. *Applied Energy*, *356*, 122448.

Patel, C. P., Pavankumar, T., Narla, A., Bhaskar, A., Mondal, S., Anwer, N., & Malmquist, A. (2024). Experimental study of a latent heat thermal energy storage system using erythritol for medium temperature applications. *Case Studies in Thermal Engineering*, *53*, 103907.

Prieto, C., Tagle-Salazar, P. D., Patiño, D., Schallenberg-Rodriguez, J., Lyons, P., & Cabeza, L. F. (2024). Use of molten salts tanks for seasonal thermal energy storage for high penetration of renewable energies in the grid. *Journal of Energy Storage*, *86*, 111203.

Punyavathi, R., Pandian, A., Singh, A. R., Bajaj, M., Tuka, M. B., & Blazek, V. (2024). Sustainable power management in light electric vehicles with hybrid energy storage and machine learning control. *Scientific Reports*, *14*(1), 5661.

Rana, M. M., Uddin, M., Sarkar, M. R., Meraj, S. T., Shafiullah, G. M., Muyeen, S. M., . . . Jamal, T. (2023). Applications of energy storage systems in power grids with and without renewable energy integration—A comprehensive review. *Journal of Energy Storage*, *68*, 107811.

Rao, C. R. C., Niyas, H., & Muthukumar, P. (2018). Performance tests on lab–scale sensible heat storage prototypes. *Applied Thermal Engineering, 129*, 953–967. https://doi.org/10.1016/j.applthermaleng.2017.10.085.

Raul, A. K., Bhavsar, P., & Saha, S. K. (2018). Experimental study on discharging performance of vertical multitube shell and tube latent heat thermal energy storage. *Journal of Energy Storage*, *20*, 279–288. https://doi.org/10.1016/j.est.2018.09.022.

Rehman, S., Al-Hadhrami, L. M., & Alam, M. M. (2015). Pumped hydro energy storage system: A technological review. *Renewable and Sustainable Energy Reviews*, *44*, 586–598.

Rimpel, A., & Broerman, E. (2020). 4.3 Gravity and buoyancy-based energy storage systems. *Thermal, Mechanical, and Hybrid Chemical Energy Storage Systems*, 235.

Rugolo, J., & Aziz, M. J. (2012). Electricity storage for intermittent renewable sources. *Energy & Environmental Science*, *5*(5), 7151–7160.

Ruwa, T. L., Adun, H. H., & Abbasoglu, S. (2016). Thermal energy storage for solar power plant applications. *13th HONET-ICT Int. Symp. Smart MicroGrids Sustainable Energy sources Enabled by Photonics IoT Sensors, HONET-ICT*, *1*, 170–174. https://doi.org/10.1109/HONET.2016.7753444, 2016.

Saka, S. (2024). An overview of large-scale energy storage systems. *Advanced Redox Flow Technology*, 199–243.

Sandoval Bustamante, J., Singh, A., Soto Crisanto, M. I., Joao, I., & Mohsin, B. (2023, August). Comparative analysis of mechanical energy storage systems. In *ASME Power Conference* (Vol. 87172, p. V001T04A001). American Society of Mechanical Engineers.

Sathishkumar, A., Sundaram, P., Cheralathan, M., & Kumar, P. G. (2024). Effect of nano-enhanced phase change materials on the performance of cool thermal energy storage system: A review. *Journal of Energy Storage*, *78*, 110079.

Schubert, C., Hassen, W. F., Poisl, B., Seitz, S., Schubert, J., Oyarbide Usabiaga, E., . . . Pettinger, K. H. (2023). Hybrid energy storage systems based on redox-flow batteries: Recent developments, challenges, and future perspectives. *Batteries*, *9*(4), 211.

Sesay, S. F., Wekesa, C. W., & Ngoo, L. M. (2024). Optimal placement of superconducting magnetic energy storages in a distribution network with embedded wind power generation. *Engineering, Technology & Applied Science Research*, *14*(2), 13416–13424.

Sharma, A., Tyagi, V. V., Chen, C. R., & Buddhi, D. (2009). Review on thermal energy storage with phase change materials and applications. *Renewable and Sustainable Energy Reviews*, *13*, 318–345. https://doi.org/10.1016/j.rser.2007.10.005.

Smallbone, A., Julch, V., Wardle, R., Roskilly, A. P. (2017). Levelised Cost of Storage for Pumped Heat Energy Storage in comparison with other energy storage technologies. *Energy Conversion and Management*, *152*, 221–228.

Steinmann, W.-D., & Tamme, R. (2008). Latent heat storage for solar steam systems. *Journal of Solar Energy Engineering*, *130*, 011004. https://doi.org/10.1115/1.2804624.

Tayyeban, E., Deymi-Dashtebayaz, M., & Farzaneh-Gord, M. (2024). Multi-objective optimization for reciprocating expansion engine used in compressed air energy storage (CAES) systems. *Energy*, *288*, 129869.

Tetteh, S., Juul, G., Järvinen, M., & Santasalo-Aarnio, A. (2024). Improved effective thermal conductivity of sand bed in thermal energy storage systems. *Journal of Energy Storage*, *86*, 111350.

Vazquez, S., Lukic, S. M., Galvan, E., Franquelo, L. G., & Carrasco, J. M. (2010). Energy storage systems for transport and grid applications. *IEEE Transactions on Industrial Electronics*, *57*(12), 3881–3895.

Wang, H., Liang, M., Li, M., Qu, Y., & Miao, Z. (2024). Surface-amorphized nickel sulfide with boosted electrochemical performance for aqueous energy storage. *Battery Energy*, *3*(1), 20230035.

Wang, K., Di, A., Zhang, S., Ni, L., Wang, H., Liu, H., . . . Ji, X. (2024). Zinc anode based alkaline energy storage system: Recent progress and future perspectives of Zinc–Silver battery. *Energy Storage Materials*, 103385.

Wang, W. W., Wang, L. B., & He, Y. L. (2016). Parameter effect of a phase change thermal energy storage unit with one shell and one finned tube on its energy efficiency ratio and heat storage rate. *Applied Thermal Engineering*, *93*, 50–60. https://doi.org/10.1016/j.applthermaleng.2015.08.108.

Wei, G., Wang, G., Xu, C., Ju, X., Xing, L., Du, X., & Yang, Y. (2018). Selection principles and thermophysical properties of high temperature phase change materials for thermal energy storage: A review. *Renewable and Sustainable Energy Reviews, 81*, 1771–1786. https://doi.org/10.1016/j.rser.2017.05.271.

Yadav, V. V., & Saravanan, B. (2023). Multimachine stability improvement with hybrid renewable energy systems using a superconducting magnetic energy storage in power systems. *Journal of Energy Storage*, *57*, 106255.

Yang, Y., Wu, Z., Yao, J., Guo, T., Yang, F., Zhang, Z., . . . Li, B. (2024). An overview of application-oriented multifunctional large-scale stationary battery and hydrogen hybrid energy storage system. *Energy Reviews*, 100068.

Zalba, B., Marín, J. M., Cabeza, L. F., & Mehling, H. (2003). Review on thermal energy storage with phase change: Materials, heat transfer analysis and applications. *Applied Thermal Engineering*, *23*(3), 251–283. https://doi.org/10.1016/S1359-4311(02)00192-8.

Zhang, Q., Liu, T., Cheng, X., Liu, Y., Zhang, W., & Zhang, Q. (2024). Experimental research of photovoltaic-valley power hybrid heating system with phase change material thermal storage. *Journal of Building Engineering*, *87*, 108788.

Zhang, Y., & Faghri, A. (1996). Heat transfer enhancement in latent heat thermal energy storage system by using the internally finned tube. *International Journal of Heat and Mass Transfer*, *39*, 3165–3173. https://doi.org/10.1016/0017-9310(95)00402-5.

17 Petrochemical Catalysts

Mohammed A. Mohammed, Aliyu Mohammed Kaura, and Adamu Salihu Rabo

17.1 INTRODUCTION

Catalysts are essential in the field of chemistry, driving processes that are vital for modern life. By reducing the activation energy of chemical reactions, catalysts speed up reaction rates without undergoing chemical changes. This unique ability is crucial in producing a range of goods, including fuels, plastics, pharmaceuticals, and fine chemicals (Lower, 2020).

The petrochemical industry relies on catalysts to transform materials like crude oil and natural gas into valuable products. These products, such as ethylene, propylene, and butadiene, serve as the foundation for polymers, synthetic rubbers, and other important materials. Improving processes not only enhances the efficiency and selectivity of these reactions but also promotes sustainability, by reducing energy consumption and waste production (Ranade & Joshi, 2016a).

Catalysts also play a role in transitioning to cleaner energy sources from fossil fuels. They facilitate the conversion of resources like biomass, solar power, and wind energy into usable forms, such as hydrogen and synthetic fuels. Catalysts are key in electrolysis processes for water splitting for hydrogen production and fuel cells that efficiently convert hydrogen into electricity. Moreover, catalysts play a role in the advancement of technologies for capturing and utilizing carbon, contributing to the mitigation of greenhouse gas emissions (Taylor, 2023).

A *catalyst* is a substance that speeds up a chemical reaction without changing itself. This means that the catalyst can be reused after the reaction to help with reactions (Lower, 2020). Catalysts offer an easier activation energy path for transformations, allowing control over selectivity (see Figure 17.1) (Ranade & Joshi, 2016a). They play a role in chemical technology, as 95% of industrial chemical processes involve catalyst types. These substances have uses such as in petroleum products, fertilizers, pharmaceuticals, and plastics. Moreover, *enzymes* – a type of protein – act as catalysts that facilitate most biochemical processes in living beings. Enzymes enable processes that would otherwise be too slow for life functions. The speed of reactions – the rates at which they happen – depends on factors like the nature of the reacting species and the external conditions they encounter (Taylor, 2023). Understanding catalysis is vital because many industrial procedures rely on catalysts for their success. It is crucial to note that a catalyst only impacts the kinetics of a reaction; it does not change the tendency of the reaction to occur (Lower, 2020).

Hence, the change in enthalpy (ΔH) remains constant for both uncatalyzed pathways. When a reaction is catalyzed, the catalyst interacts with the reactants through a chemical process. Eventually, it reverts to its state, ensuring a consistent amount of catalyst throughout. Since the catalyst is not consumed, each molecule can facilitate the conversion of molecules. Choosing the catalyst plays a role in determining process feasibility and cost-effectiveness (Ranade & Joshi, 2016; Taylor, 2023). In addition to selecting a catalyst, it is vital to identify the reactor type and employ effective process intensification strategies to transition laboratory processes into practical applications successfully. Within the petrochemical sector, catalysts are components of production processes as they accelerate the chemical reactions that transform materials into desired end products. Catalysts enhance reaction efficiency by reducing the activation energy needed for initiation. They also boost product yields by minimizing side reactions and by-products (Varshouee & Zarand, 2020). However, choosing petrochemical catalysts depends on the reaction type, reactants involved, and desired outputs. This intricate selection

 DOI: 10.1201/9781003282273-17

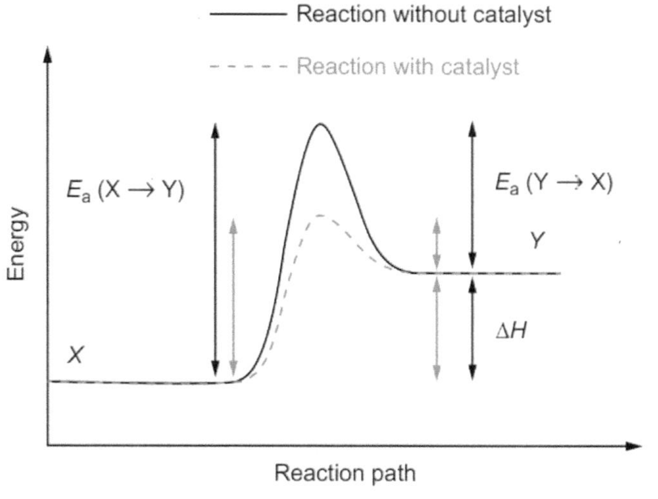

FIGURE 17.1 Activation energy Ea, case of an exothermic reaction.

Source: Ranade and Joshi (2016a).

process involves reactions where utilizing catalysts becomes imperative for achieving success in catalytic cracking, reforming, hydrogenation, and polymerization. Catalysts play a role in improving the quality and purity of products, as highlighted by Varshouee and Zarand (2020).

The introduction of catalysts has transformed the chemical and petrochemical sector by offering a cost-efficient method for producing materials needed for various applications, as discussed by Chaudhari (2016) and Rostrup-Nielsen (1997). The innovation of catalysts has opened doors to discovering chemical reactions and processes that were previously unattainable. To ensure efficiency and sustainability in production, the petrochemical industry must invest in research to uncover catalysts. Petrochemical catalysts exhibit functionalities depending on the reaction at hand, including adsorption of molecules, activation of chemical bonds, stabilization of reaction intermediates, and alteration of reaction pathways (Chaudhari, 2016). For instance, catalytic cracking, a used process in petroleum refining, relies on catalysts to facilitate the conversion of larger hydrocarbon molecules into more valuable products, like gasoline and olefins. Common catalysts used in the petrochemical industry include catalysts for ethylene oxide production, platinum on zeolite for making aromatics, alumina for hydrocracking, and different specialty materials tailored for reactions (Matar et al., 1989; Rostrup-Nielsen, 1997; Ma et al., 2015).

In this chapter, we will explore the catalyst types used in the petrochemical sector, including heterogeneous, heterogenized, homogeneous, and biocatalysts. We will delve into processes like hydrocarbon hydrogenation, alcohol manufacturing, and hydrocarbon dehydrogenation. Each segment will showcase the importance of these processes, the catalysts utilized, and how they are applied to create items.

Furthermore, we will touch upon the catalysis technologies set to transform the industry by providing sustainable and cost-efficient solutions. The chapter will end with an examination of how catalysts contribute to the energy transition, underscoring their role in developing more effective energy production methods. Through this overview, we aim to offer insight into the vital role of catalysts in advancing chemical technology and their influence on the petrochemical industry and the global shift toward sustainable energy.

Traditionally, catalysts were distinguished into *homogeneous* and *heterogeneous*. This distinction is linked to the fact that the catalyst operates in the same phase where the reaction occurs (homogeneous catalysts) or in a different phase (heterogeneous or heterogenized catalysts).

17.1.1 Homogeneous Catalysts

Homogeneous catalysis involves a system where the substances and catalysts needed for a reaction are combined in one phase. Regarding kinetics and mechanisms, homogeneous catalysis can be divided into two categories: *species* and *complex* catalysis. In single-species catalysis, a molecule or ion interacts with a molecule to kick-start the reaction and then returns to its original state. An example of this is hydrogen ion catalysis. On the other hand, complex catalysis involves the catalyst that creates a chemical system that interacts with the reactants. Acid-based catalysis and metal–organic complexes are examples of this type. Practical homogeneous catalysts include acids and bases, metal salts, organometallic complexes with different ligands, nonaqueous ionic liquids (NAILs), metal clusters, and enzymes (Chaudhari, 2016). The central metal atom is connected to inorganic ligands (Ranade & Joshi, 2016b).

Adjusting ligand changes can be made easily to the catalyst's environment to modify its properties. Transition metals play a role in forming these complexes because they have d-orbitals that enable ligands to bond in a way that allows for further chemical reactions. Some common examples of catalysts are the Wilkinson catalyst, known as RhCl(PPh3)3 , used for hydrogenation; $HRh(CO)(PPh_3)_3$, for olefin hydroformylation; [Rh(CO)2I2]^/HI , for methanol carbonylation; $Co(OAc)_2/Co(OAc)_2/HBr$, for xylene oxidation; Pd(pyca) (PPh3) (PPh3(OTs)) / TsOH / LiCl , for alcohol and olefin carbonylation; and Rh(1)R, RDIPAMP, for the asymmetric hydrogenation of olefins (L-dopa), as shown in Figure 17.2 (Chaudhari, 2016).

Another type of homogeneous catalyst is the Ziegler–Natta catalyst (Figure 17.3), discovered by Ziegler and Natta in the 1960s, which is used in polymerization reactions to produce high-density polyethylene (HDPE) and polypropylene (PP) (Behr, 2012). The catalyst is typically composed of a transition metal compound, such as titanium or vanadium, and an organoaluminum compound, such as triethylaluminum (Khanna & Anand Rishi, 2016). The catalyst has high activity and selectivity, making it ideal for industrial-scale polymerization.

Other homogeneously catalyzed reactions include ester hydrolysis, Diels–Alder, Cannizzaro, and enzymatic processes. They are also capable of initiating reactions at lower temperatures and pressures. However, one of the major challenges of using homogeneous catalysts is separating them from the products, which can be difficult and often requires additional purification steps (Stock, 2023).

In the next section, we will explain the realm of catalysis, which functions based on unique principles and provides specific benefits and uses in different chemical processes.

17.1.2 Heterogeneous Catalysts

A *heterogeneous catalyst* is a type of catalyst that operates in a state from the substances involved in a reaction. This usually means a catalyst interacting with gases (like the oxidation of SO_2 with V_2O_5) or two liquid phases (such as the emulsion copolymerization of styrene and butadiene). These types of catalysts are commonly used in the petrochemical industry to assist in reactions like hydrogenation, oxidation, and reforming processes. Heterogeneous catalysis primarily occurs on the surface of materials. The catalyst's surface area should be extensive, allowing reactants to easily reach it (Missen et al., 1999; Khanna & Anand Rishi, 2016). Reactions involving catalysis progress through both physical stages, as seen in Figure 17.4. The fundamental steps involved in reactions include (Unnikrishnan & Srinivas, 2016);

1. *External diffusion.* Reactants move from the phase to the catalyst surface and then internally diffuse toward active sites.
2. *Adsorption.* Reactants bind to sites, on the surface.
3. *Surface reaction* takes place.
4. *Desorption.* Products are released from sites before diffusing out of the catalyst.

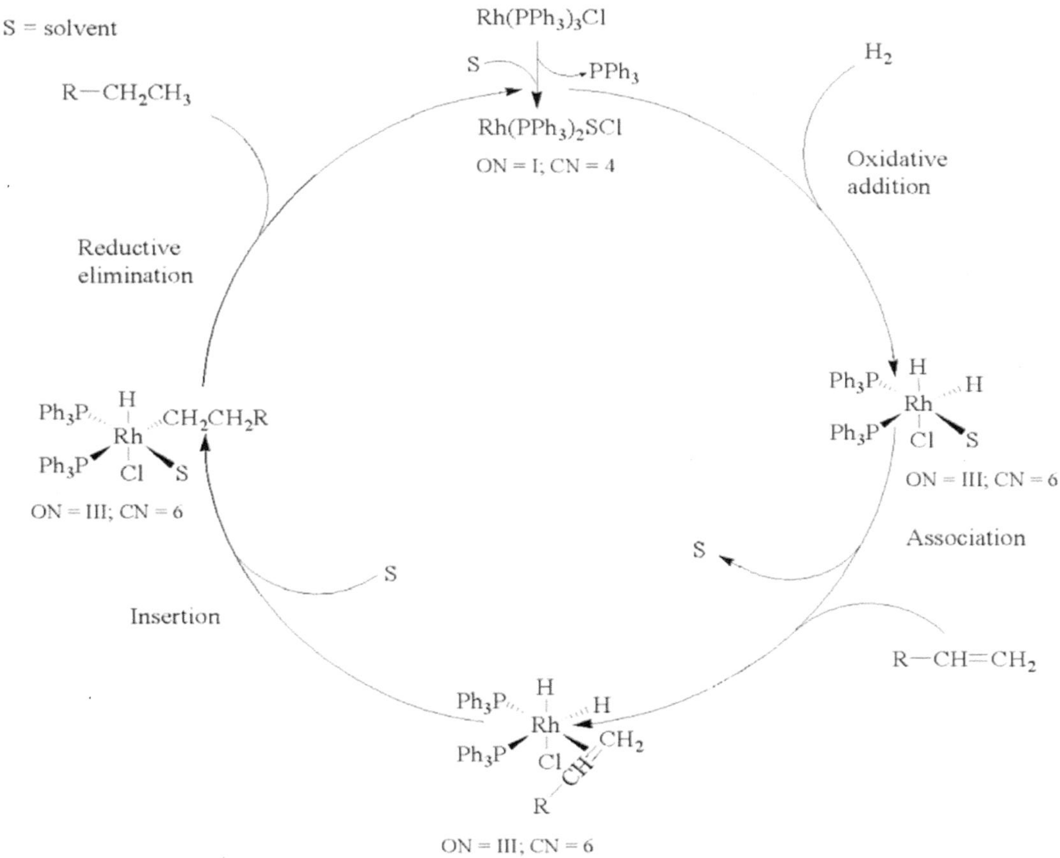

FIGURE 17.2 Hydrogenation mechanism using the Wilkinson catalyst.

Source: Augustine (2016); Behr (2012); Chaudhari (2016).

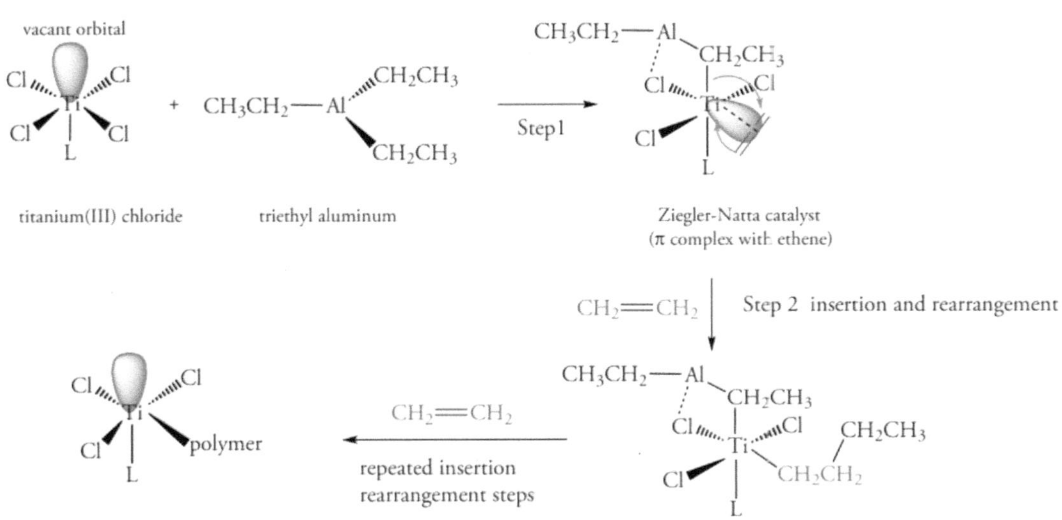

FIGURE 17.3 Schematic diagram of steps in Ziegler–Natta catalysis.

Source: Janice (2020).

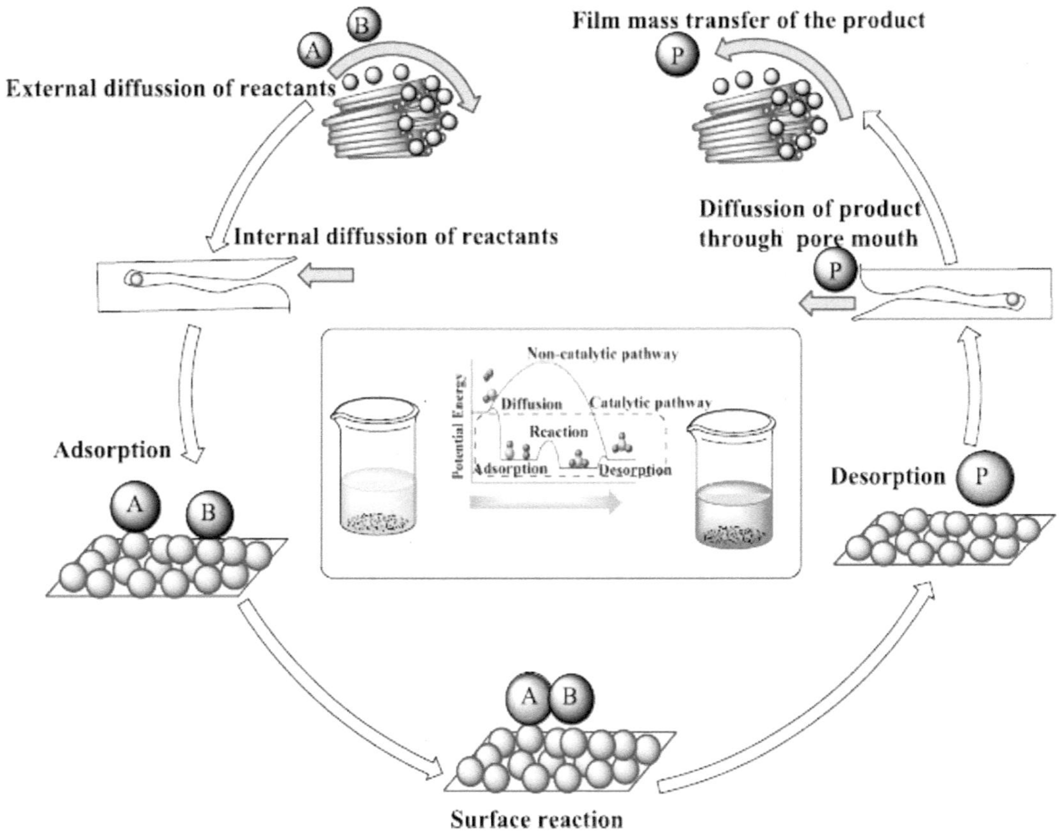

FIGURE 17.4 Reaction steps involved in heterogeneous catalysis.

Source: Unnikrishnan and Srinivas (2016).

Each of these stages plays a role in influencing the pace of the process. Any alteration in the speed of these stages can result in a shift in the reaction rate. The next part delves deeper into these steps involved in reactions (Unnikrishnan & Srinivas, 2016).

Heterogeneous catalysts are essential in the production of petrochemical products like plastics, fuels, and solvents (Taylor, 2023). They function by creating a surface for reactions, reducing the energy required for activation and speeding up chemical transformations. These catalysts can be customized for reactions to show high selectivity, proving valuable in industrial processes by achieving desired product outcomes while minimizing unwanted by-products (Khanna & Anand Rishi, 2016).

Utilizing catalysis is a sustainable approach in industrial settings. The advancement of these catalysts centers on improving feedstock conversion rates and product selectivity (Glotov & Karakhanov, 2021). This entails selecting materials, managing surface characteristics, and comprehending catalytic mechanisms. Researchers continually refine these catalysts to enhance efficiency, durability, and environmental sustainability. Moving on from discussing catalysis, we will explore heterogenized catalysis, blending elements from both homogeneous and heterogeneous setups to provide distinct advantages across different uses.

17.1.3 Heterogenized Homogeneous Catalysts

Homogeneous catalysts possess appealing qualities, like selectivity and accessibility to all catalytic sites. Yet they come with notable drawbacks, such as corrosion, toxicity, handling complexity, high expenses, and solid waste generation. These challenges have led to a decrease in the acceptance of catalysts in today's aware and economically pressured era (Zhao et al., 2006). Consequently, the primary focus in catalysis science and engineering has shifted toward replacing catalysts with alternatives. One approach involves anchoring the catalyst onto a support to create what is known as heterogenization of the homogeneous catalyst (Hagen, 2006; Zhao et al., 2006).

The process of heterogenization entails fixing a catalyst onto a solid support to facilitate easier separation and reuse in different reactions. This method merges the advantages of catalysis, offering benefits like simplified catalyst separation and recycling. The transformation of a catalyst into a heterogenized form typically includes attaching or immobilizing the metal complex or catalytic species onto a solid support or carrier material. This immobilization can occur through methods such as modifying the support with ligands that interact with the catalyst or enclosing the species within a porous or structured material.

Various techniques are utilized to transform catalysts into ones each presenting its advantages and obstacles. These methods encompass immobilization through covalent bonding, entanglement, encapsulation, ionic heterogenization or ionic liquid support, and surface functionalization (Hagen, 2006; Zhao et al., 2006). After examining the idea and approaches of making homogeneous catalysis heterogeneous, we will now explore biocatalysts, which bring advantages for usage due to their specificity and environmentally friendly nature.

17.1.4 Biocatalysts

A *biocatalyst* refers to a substance or agent utilized to kick-start, assist, or hasten reactions. These biocatalysts can take the form of enzymes, microorganisms, or other biological catalysts that facilitate chemical processes in industries. Enzymes are particularly effective and specific in speeding up reactions under mild conditions, offering notable advantages compared to traditional chemical catalysts (Hagen, 2006).

Biocatalysis is a field that combines biochemistry and organic chemistry with biology, enzymology, protein chemistry, catalysis, transport phenomena, and reaction engineering. It has garnered attention in sectors like pharmaceuticals, food and beverage production, agriculture, and biofuels due to its potential for cost-sustainable processes. The use of enzymes as biocatalysts aligns with the increasing demand for sustainable technologies. Biocatalysts are sourced from a variety of origins, such as cells from microbial, plant, or animal sources, as well as cell-free extracts and enzymes. Despite the abundance of known biocatalysts available, only a small percentage is utilized on a scale. For example, out of around 4,000 identified enzymes, approximately 400 are commercially accessible; however, only about 40 are employed in operations (Hagen, 2006).

Enzymes are typically more effective than catalysts, often displaying turnover numbers exceeding 100,000 s^{-1} in contrast to the rate of 0.01 s^{-1} observed for heterogeneous and homogeneous catalysts (Hagen, 2006). Enzymes work under conditions that showcase high substrate specificity as well as chemo-, regio-, and enantioselectivity (Hagen, 2006).

Biocatalysts are also utilized in the production of petrochemicals by transforming materials like oil or natural gas into valuable items, such as plastics, solvents, and other chemical products. They are preferred for their precision, efficiency, and potential friendly attributes when compared to chemical catalysts. Their application can aid in promoting eco-conscious approaches within the petrochemical sector (Hagen, 2006). Enzymes are frequently the biocatalysts employed in petrochemical procedures. They offer management over production processes and enable the alteration of chemical compositions to produce specific final goods. The enzymes utilized in petrochemical operations encompass lipases, proteases, and amylases, among others (Hagen, 2006).

Microorganisms are also utilized as biocatalysts in petrochemical manufacturing. Specific strains of bacteria, yeast, and fungi are leveraged for their capability to generate enzymes or facilitate chemical reactions. For instance, certain microorganisms can decompose hydrocarbons into molecules – a process in petroleum refining and creating different chemical compounds (Hagen, 2006). In the next section, we will explore specialty and custom catalysts tailored to meet needs and provide distinct benefits in specific applications.

17.2 HYDROCARBON HYDROGENATION

Hydrocarbon hydrogenation involves the addition of hydrogen to hydrocarbon molecules with the help of catalysts like nickel, palladium, or platinum to create hydrocarbons. This process aims to saturate compounds by adding hydrogen to triple bonds in hydrocarbons. When carbon–carbon or carbon–heteroatom bonds are broken during the reaction, it is known as *hydrogenolysis* (Stock, 2023).

The process of hydrogenation is commonly used in the production of bulk and fine organic chemicals and plays a key role in purification procedures. For instance, it can selectively reduce amounts of acetylene in ethylene streams or butadiene in streams, eliminate unwanted oxygen or carbon oxides, and decrease levels of heteroatoms like sulfur (hydrodesulfurization, HDS), nitrogen (hydrodenitrification, HDN), and metals (hydrodemetallation, HDM) from refinery streams. These reactions are generally categorized as hydrotreating (HDT) processes (Stock, 2023). Hydrogenation is limited by a thermodynamic equilibrium with the reverse dehydrogenation reaction:

$$A + nH_2 \rightleftharpoons \mathbf{B}\Delta H < 0 \qquad \text{(Eq. 17.1)}$$

A notable limitation of hydrogenation is its equilibrium with the dehydrogenation reaction. This process releases an amount of heat and follows Le Chatelier's principle: At lower temperatures, the equilibrium shifts toward favoring hydrogenation due to its exothermic nature (Busca, 2014; Stock, 2023). When the temperature rises, the balance shifts to the left, which makes dehydrogenation a valuable synthetic method (Stock, 2023). In hydrogenation, managing the temperature rise resulting from the reaction is crucial. While higher temperatures can boost reaction speeds, they might also have an impact on conversion, selectivity, and catalyst lifespan (Stock, 2023). Achieving a reaction rate at temperatures involves selecting an appropriate catalyst based on the hydrogenated compound's activation energy. Higher operating pressures promote the reaction since there are more product moles than reactant moles. Therefore, increasing hydrogen partial pressure tilts the equilibrium toward the product side (Stock, 2023).

Here are some examples of hydrogenation (Eq. 2–5)(Busca, 2014; Stock, 2023):

1. Ethene (C_2H_4) reacts with hydrogen at $150^{\circ}C$ to produce ethane (C_2H_4).

$$C_2H_4 + H_2 \rightarrow C_2H_6 \qquad \text{(Eq. 17.2)}$$

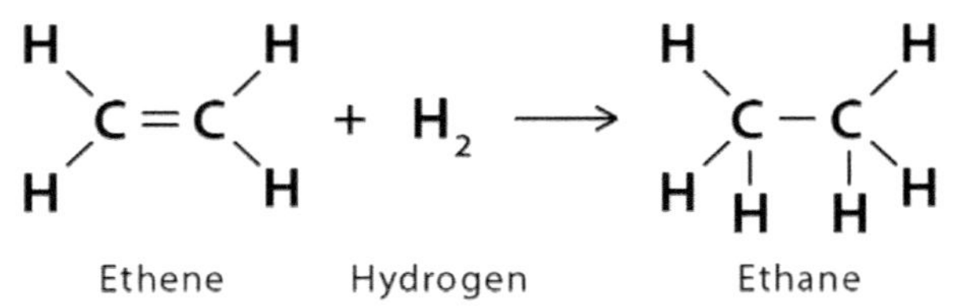

(Eq. 17.3)

2. Cyclohexene (C_6H_{10}) converts into cyclohexane (C_6H_{12}) by reacting with hydrogen in the presence of a palladium (Pd) catalyst.

$$C_6H_{10} + H_2 \rightarrow C_6H_{12} \qquad \text{(Eq. 17.4)}$$

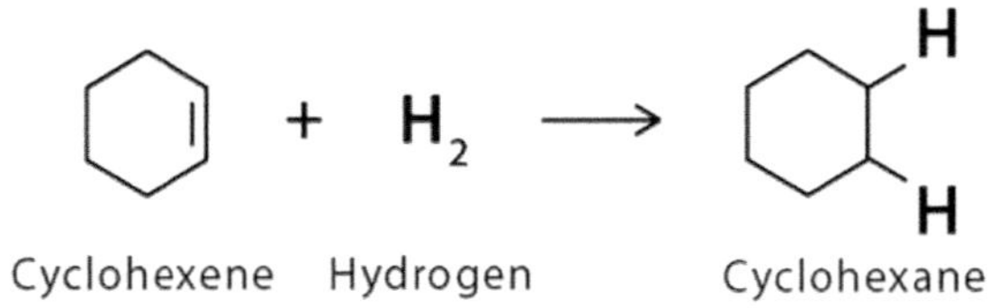

(Eq. 17.5)

3. Hydrogenation of olefins using the Wilkinson catalyst.

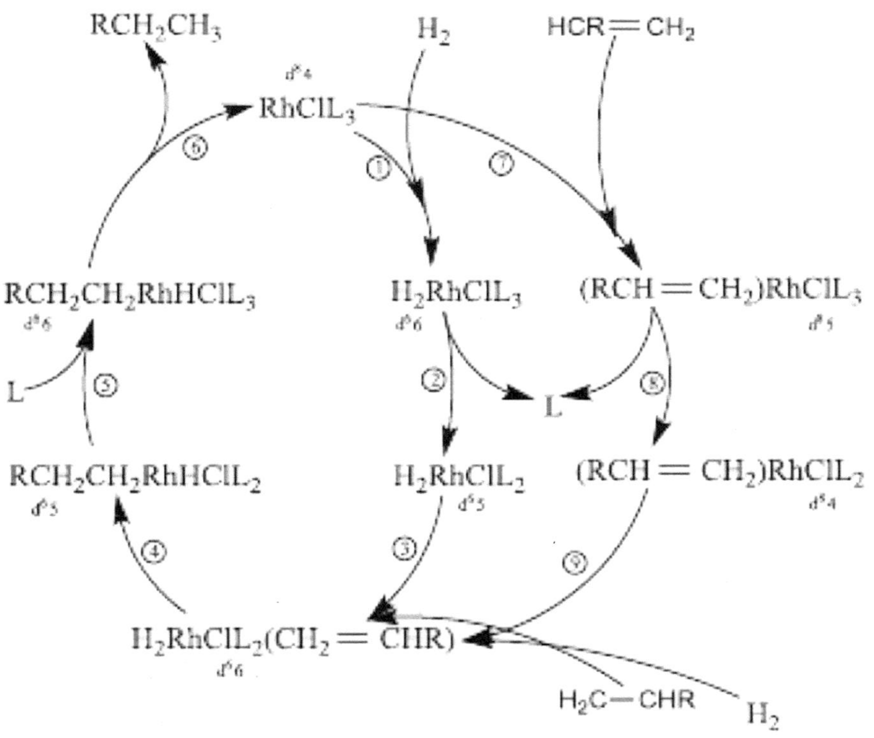

(Eq. 17.6)

17.2.1 Selective Hydrogenations (C_2 to Pygas)

Selective hydrogenation is a process that decreases the amount of compounds in a mixture by focusing on unsaturated bonds, like carbon–carbon double (C=C) and triple bonds (C=C). It is carried out with a catalyst that minimizes unwanted reactions, as mentioned by Huang and Liu (2018).

This method plays a role in the production of petrochemical components, such as ethylene, propylene, butadiene, and aromatics, as highlighted in the hydrocarbons report from 2021. It is commonly utilized in refinery operations to improve the coordination between refining and petrochemical processes. Pyrolysis gasoline (pygas), which is produced during ethylene manufacturing, contains aromatics along with impurities, like sulfur, nitrogen, and olefins, that need to be removed through hydrogenation. The hydrogenation process for pygas typically involves two steps: hydrogenation of diolefins in the phase, followed by gas phase hydrogenation to eliminate the remaining impurities. The purified pygas can then serve as a material for the production of chemicals and high-octane gasoline (Huang & Liu, 2018; Hydrocarbon, 2021).

The selective hydrogenation of acetylene (C_2H_2) to ethylene (C_2H_2) during the conversion of C_2 to pygas helps enhance the quality of pyrolysis gasoline. This method utilizes palladium-based catalysts known for their selectivity in converting acetylene while leaving ethylene untouched, according to a Nitriles study from 1975 (Nitriles, 1975).

17.2.2 Dearomatization

The realm of chemistry has long been intrigued by dearomatization reactions, a topic that traces its origins back to the 19th century (Zhang & You, 2016). Early studies on dearomatization encompassed methods such as reactions, transition metal–mediated processes, hydrogenation techniques, and enzyme-catalyzed pathways. Dearomatization essentially involves the removal of aromaticity from a compound, leading to the creation of compounds or chemical bonds with unique properties (Christopher, 2015; Nemoto, 2023). This reaction holds significance across domains including natural product synthesis and medicinal chemistry. Within the petrochemical sector, dearomatization plays a role in yielding products like high-octane gasoline, diesel fuel, and petrochemical intermediates. It is commonly accomplished through conversions like hydrocracking or hydrotreating processes that transform compounds into intricate three-dimensional structures (Aleksiev & García, 2023).

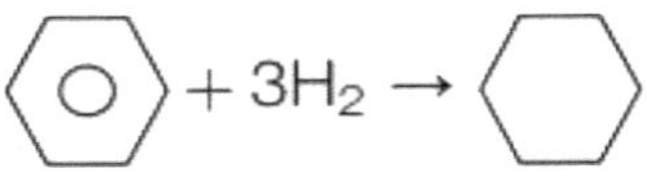

(Eq. 17.7)

Catalysts are essential in the process of removing aromas by aiding the reactions. Successful dearomatization catalysts are crafted to deliver selectivity, activity, and durability in petrochemical settings. Crafting and refining these catalysts necessitate a comprehension of the chemical reactions at play and the characteristics of both the input materials and desired outcomes (Zheng & You, 2021). Furthermore, selecting the catalyst and reaction parameters must be done with precision to align with needs. There are various types of dearomatization reactions, each with its mechanism and application. These types include hydrogenative (Birch reduction), alkylative, photochemical, thermal, oxidative, transition metal–assisted, and enzymatic (Zhu et al., 2023).

17.2.2.1 Oxidative Dearomatization

Oxidative dearomatization reactions are a method that transforms compounds into aromatic systems through oxidative processes. One strategy involves utilizing transition metal catalysts, such as iridium, ruthenium, or palladium, to facilitate oxidation reactions. These catalysts aid in transferring oxygen atoms to the ring, resulting in the formation of dearomatized products. Organocatalytic approaches employing organic molecules as catalysts have also proven to be effective. Various oxidizing agents like DDQ, PIDA, IBX, and other hypervalent iodine reagents are commonly employed for this purpose (Christopher, 2015; Roche & Porco, 2011).

17.2.2.2 Birch Reduction

Birch reduction (Eq. 9), or hydrogenative reaction, is a widely used chemical reaction that involves the reduction of aromatic rings. Typically, this reaction is carried out by using alkali metals, such as sodium or lithium, in the presence of ammonia or alcohol solvents. This classic dearomatization reaction has been studied extensively and is known for its ability to convert aromatic compounds into dienes or cyclohexadiene (Zhu et al., 2023).

The reaction mechanism involves the transfer of an electron from the metal to the aromatic ring, leading to the formation of a radical anion intermediate that can undergo further reactions. Birch reduction is a very powerful tool in organic chemistry, and its potential for further functionalization makes it an indispensable method for the synthesis of complex molecules (Eq. 8–9) (Zhu et al., 2023).

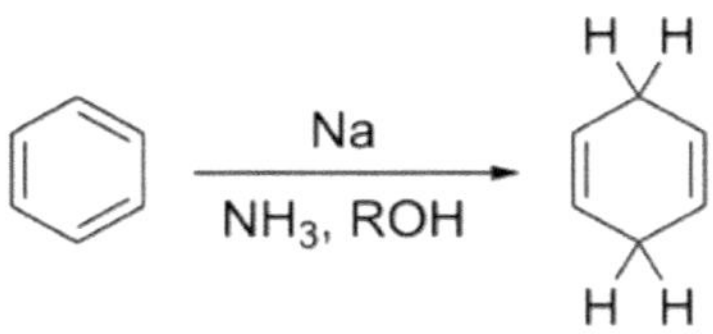

(Eq. 17.8)

An example is the reduction of naphthalene in ammonia and ethanol.

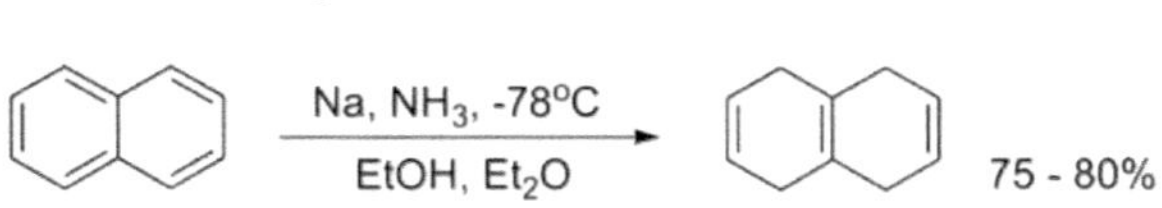

(Eq. 17.9)

17.2.2.3 Aromatic Substitution Reactions

Aromatic substitution reactions are a class of chemical reactions that involve the substitution of an atom or group of atoms in an aromatic compound. These reactions are widely used in organic synthesis to introduce functional groups or to form fused ring systems. One of the most extensively used aromatic substitution reactions is the Friedel–Crafts reaction, which can proceed via acylation or alkylation. This reaction has been employed for the dearomatization of aromatic substrates, leading to the formation of cyclohexadiene (Ashenhurst, 2020; Ye et al., 2023; W. Y. Zhang et al., 2023).

17.2.2.4 Enantioselective Dearomatization

Enantioselective dearomatization is a type of asymmetric dearomatization reaction that selectively converts an aromatic compound into a non-aromatic compound with the introduction of chirality (Ashenhurst, 2020). This process is achieved by using chiral reagents or catalysts (Ye et al., 2023). Enantioselective dearomatization reactions have emerged as an important tool for the synthesis of chiral compounds from readily available starting materials. Such reactions have been widely studied and have found applications in the synthesis of natural products, pharmaceuticals, and materials (Ashenhurst, 2020; Ye et al., 2023; W. Y. Zhang et al., 2023).

17.2.2.5 Cycloaddition Dearomatization Reactions

Cycloaddition dearomatization reactions refer to a class of chemical reactions that utilize Diels–Alder and related cycloaddition reactions to transform aromatic compounds. Specifically, these reactions involve the formation of bridged or fused ring systems that result in the dearomatization of the compound. This process enables the construction of polycyclic structures, which are often used in the synthesis of natural products and complex molecules. In other words, these reactions offer a powerful tool for chemists to build complex molecular architectures through the creation of multiple rings, which can facilitate the formation of diverse and novel chemical entities (Ashenhurst, 2020; Ye et al., 2023; W. Y. Zhang et al., 2023).

17.2.2.6 Electrophilic Aromatic Substitution

Electrophilic aromatic substitution is a reaction mechanism that involves the introduction of an electrophile to the aromatic ring, which is then attacked by the aromatic pi electrons. The result of this reaction is the formation of a new carbon–carbon bond, which can lead to the incorporation of a functional group or the formation of a fused ring system (Ashenhurst, 2020; Ye et al., 2023; W. Y. Zhang et al., 2023).

17.2.2.7 Transition Metal–Catalyzed Dearomatization

Dearomatization reactions catalyzed by transition metals involve transforming aromatic compounds, such as benzene derivatives and indoles, into non-aromatic intermediates (Nemoto, 2023). These intermediates can then be further modified to produce a variety of valuable products. Transition metal catalysts facilitate dearomatization reactions in multiple ways, including oxidative addition, migratory insertion, and reductive elimination. By selectively activating specific bonds in the aromatic compounds, these catalysts enable the formation of new carbon–carbon bonds and the creation of intricate three-dimensional structures. This approach has proven highly effective in synthesizing complex organic molecules, including pharmaceuticals and natural products. A classic example of transition metal–assisted dearomatization is the Buchner ring expansion (Christopher, 2015).

17.2.3 Resins Hydrogenation

Petroleum resins are compounds with low molecular weight obtained from hydrocarbon pyrolysis tars or coal tars. These resins can exist in semi-liquid forms, depending on their makeup. Aliphatic resins are typically derived from the C5 fraction of pyrolysis tar, while aromatic resins come from polymerizing hydrocarbons in the C8–C10 fractions. Copolymer hydrocarbon resins are created by copolymerizing aliphatic monomers. In resin hydrogenation, hydrogen is added to bonds in the resin with the help of a catalyst. This process selectively reduces bonds while leaving functional groups unaffected. The choice of catalyst – nickel (Ni) or palladium (Pd) – depends on the resin type and desired product characteristics. Hydrogenation enhances resin properties and stability, making them suitable for use in adhesives, coatings, and plastics. Hydrogenation occurs at temperatures and pressures using hydrogen gas. The specific operating conditions vary based on the desired product outcome and the level of hydrogenation required. Full, partial, and selective hydrogenation methods produce resins with compatibility and good stability by eliminating double bonds (Antonov et al., 2017; Petrukhina et al., 2018).

17.2.3.1 Alcohols

Organic compounds known as alcohols contain a hydroxyl (-OH) group attached to a saturated carbon atom, according to the IUPAC guidelines from 2019. These alcohols vary from forms, like methanol and ethanol, to intricate ones, such as sucrose and cholesterol. The inclusion of an OH group alters the characteristics of hydrocarbons significantly, rendering them hydrophilic and offering a platform for reactions (IUPAC, 2019).

17.2.3.2 Oxo Alcohols

Oxo alcohols, also referred to as aldehyde alcohols, are created by reacting alkenes (olefins) with synthesis gas (a mix of carbon monoxide and hydrogen) using a rhodium or cobalt catalyst (Wagner et al., 2000). The resulting aldehyde is then transformed into oxo alcohols. These substances, as chemicals, play a role in making plasticizers, coatings, detergents, and various industrial and consumer goods. Moreover, oxo alcohols are solvents used in producing fragrances and flavoring esters (Wagner et al., 2000).

In Figure 17.5, oxo alcohols are made by combining carbon monoxide (CO) and hydrogen (as synthesis gas) with an olefin in the hydroformylation process to produce an aldehyde. This aldehyde is then transformed into alcohol through hydrogenation (Weber & Falbe, 1970). In an intermediate stage, the aldol condensation reaction involves merging two aldehydes to form an aldehyde before the hydrogenation process. When creating chain oxo alcohols, alpha olefins from the Shell higher olefin process are commonly used, resulting in alcohols, like isodecyl alcohol (Wagner et al., 2000).

17.2.4 Butanediol and Equivalents (THF and GBL)

Butanediol is a term used for chemical compounds called diols or glycols which have two hydroxyl groups attached to a butane structure. The main types include 1,4 butanediol and 2,3 butanediol; 1,4 butanediol (BDO) is classified as an alcohol and an organic compound, represented by the formula

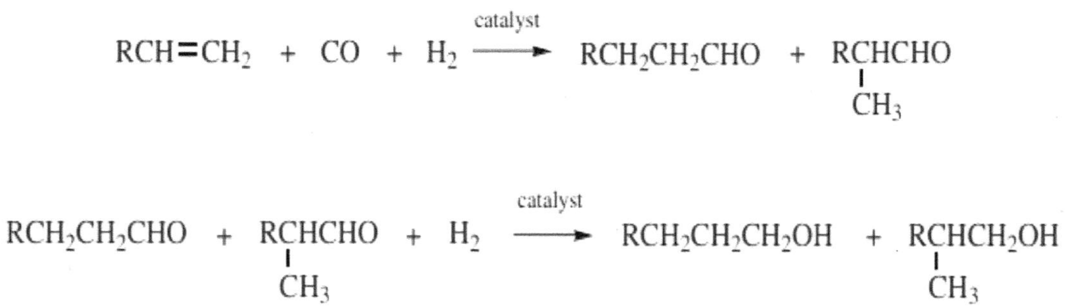

FIGURE 17.5 The hydroformylation reaction and then hydrogenation of the aldehyde to obtain the alcohol.

Source: Wagner et al. (2000).

$HOCH_2CH_2CH_2CH_2OH$. This clear, thick liquid was initially created in 1890 through the breakdown of N,N'-dinitro-1,4-butanediamine by Dutch scientist Pieter Johannes Dekkers, who named it "glycol" (Dekkers, 1890; Johnson, 2020). They can be derived from various sources, including maleic anhydride (MAH), succinic acid (SAC), and butane. MAH can be obtained from the oxidation of either butane or benzene, while SAC can be sourced from biomass feedstock. The production process involves two stages: esterification and hydrogenolysis (see Figure 17.6).

17.2.4.1 SMPO (Acetophenone to Phenyl Ethanol to Styrene)

The SMPO (monomer propylene oxide process) involves a procedure where acetophenone is transformed into phenylethanol and then converted into styrene. This method is commonly utilized in the production of styrene, a component in plastics and synthetic materials. The conversion of acetophenone to phenyl ethanol usually occurs through a reduction reaction, while the transformation of phenyl ethanol to styrene is achieved via a dehydration reaction (Buijink et al., 2008; Cochran & Dubner, 1992). The SMPO process comprises four main reaction steps that are schematically drawn in Figure 17.7; the simplified process flow sheet is found in Figure 17.8.

The SMPO process consists of four reaction stages. Initially, ethylbenzene (EB) undergoes oxidation to form ethylbenzene hydroperoxide (EBHP). This hydroperoxide then oxidizes propylene to generate propylene oxide (PO) and methyl phenyl carbinol (MPC), also known as phenylethanol. The phenylethanol is subsequently dehydrated to produce a monomer (SM). Furthermore, methyl phenyl ketone (MPK), a by-product of the reaction, is hydrogenated to MPC for conversion to SM (Cochran & Dubner, 1992).

17.3 INORGANICS

17.3.1 Hydrogen Peroxide (H_2O_2)

Hydrogen peroxide, also known as H_2O_2, was first identified by Thénard in 1818 as a result of the interaction between barium peroxide and nitric acid (Campos Martin et al., 2006). It serves as an oxidizing agent that remains effective across a wide pH spectrum, boasting a high oxidation potential and producing water as its by-product. Equations 10–14 are various possible reactions of hydrogen peroxide, depending on the substrate; these include decomposition, oxidation, molecular addition, reduction, and substitution.

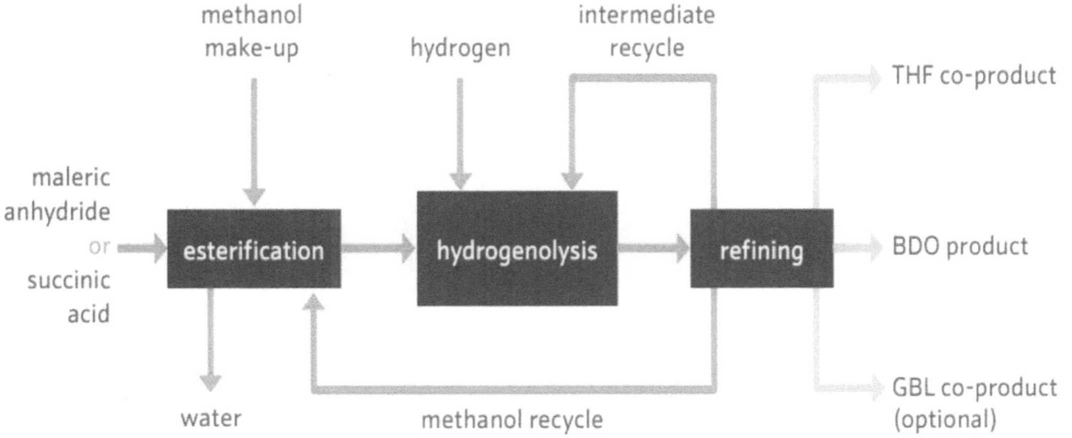

FIGURE 17.6 Workflow for the production of BDO and its derivatives, THF and GBL.

Source: Johnson (2020).

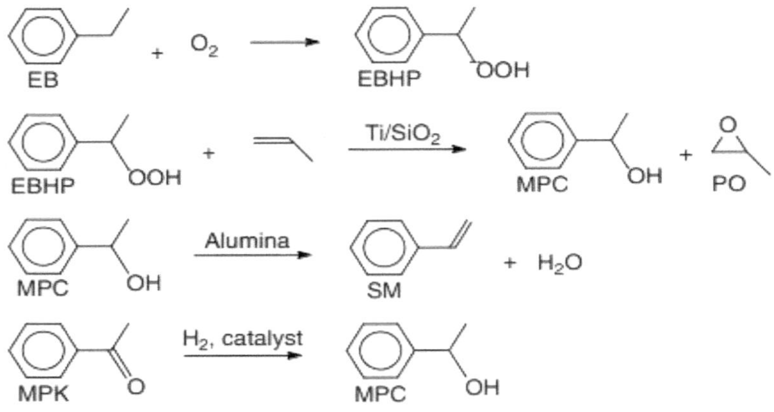

FIGURE 17.7 Main reaction steps of the SMPO process.

Source: Buijink et al. (2008).

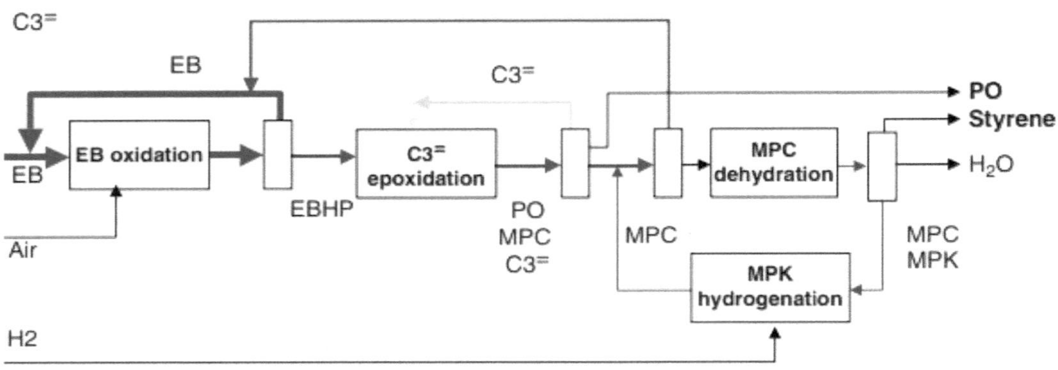

FIGURE 17.8 Simplified process flow scheme for the SMPO process.

Source: Buijink et al. (2008).

$$Decompostion: 2H_2O_2 \rightarrow 2H_2O + O_2 \quad \text{(Eq. 17.10)}$$

$$Oxidation: H_2O_2 + M \rightarrow MO + H_2O \quad \text{(Eq. 17.11)}$$

$$Addition: H_2O_2 + A \rightarrow AH_2O_2 \quad \text{(Eq. 17.12)}$$

$$Reduction: H_2O_2 + R \rightarrow RH_2 + O_2 \quad \text{(Eq. 17.13)}$$

$$Substitution: H_2O_2 + RX \rightarrow ROOH + HX \quad \text{(Eq. 17.14)}$$

Hydrogen peroxide, a liquid known for its powerful oxidizing characteristics, finds extensive application in various industries, such as in chemical, pharmaceutical, and paper production. Its production involves methods, including the anthraquinone process, alcohol oxidation, and electrochemical synthesis, as outlined by Campos Martin et al. (2006) and Hans Joachim and Pfiefderer (2002).

The anthraquinone process is a widely used industrial process for producing hydrogen peroxide (Figures 17.9 and 17.10). It involves the reaction of anthraquinone, a yellow crystalline solid, with hydrogen gas in the presence of a catalyst, such as palladium or platinum. This reaction results in the formation of hydrogen peroxide, a clear and colorless liquid, and anthrahydroquinone, a brownish solid.

Another way of producing hydrogen peroxide is via alcohol oxidation. This process involves heating and pressurizing isopropyl alcohol with air over a silver-based catalyst (Campos-Martin et al., 2006). As shown in Equations 15–17, electrochemical synthesis uses electrolysis to create hydrogen peroxide from a dilute solution of either sulfuric acid or sodium hydroxide (Campos-Martin et al., 2006). During this process, an electrical current is passed through the solution, which results in the formation of hydrogen peroxide at the anode. This method of producing hydrogen peroxide is a cost-effective and efficient alternative to traditional methods and has found applications in various industries, including pharmaceuticals, electronics, and water treatment.

$$Anode: 2OH^- \rightarrow H_2O + H_2O + \frac{1}{2}O_2 + 2e^- \quad \text{(Eq. 17.15)}$$

$$Cathode: H_2O + O_2 + 2e^- \rightarrow HO_2^- + OH^- \quad \text{(Eq. 17.16)}$$

$$Over\,all\,reaction: HaOH + \frac{1}{2}O_2 \rightarrow NO_2Na \quad \text{(Eq. 17.17)}$$

Hydrogen peroxide (H_2O_2) is utilized in the petrochemical industry as a reactant or oxidizing agent in various catalytic processes (Eq. 18) (Organic, 2003; Xu et al., 2004). It can be used in combination with a catalyst to selectively oxidize specific functional groups within organic molecules.

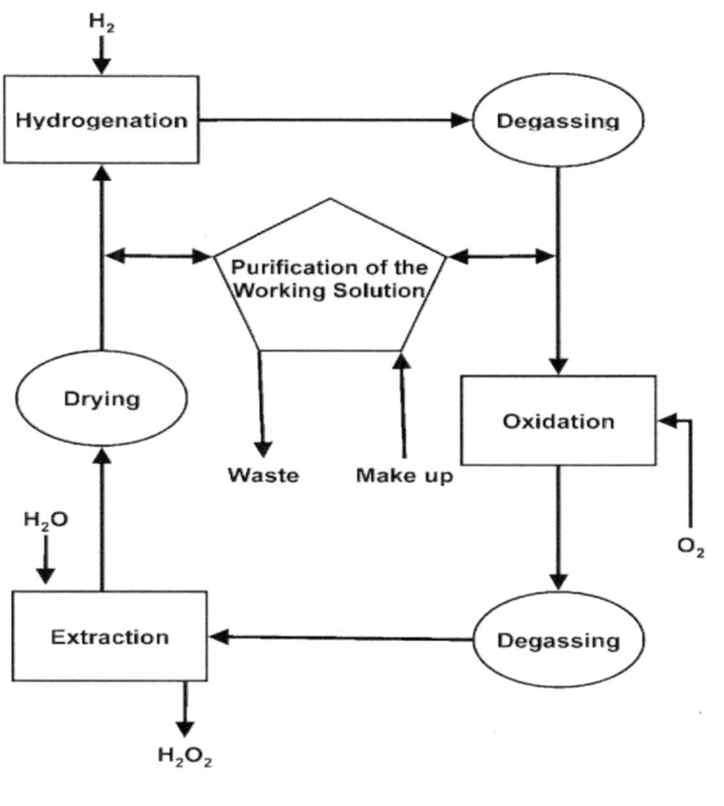

FIGURE 17.9 Flowchart involved in the anthraquinone method for the synthesis of hydrogen peroxide.

Source: Campos-Martin et al. (2006).

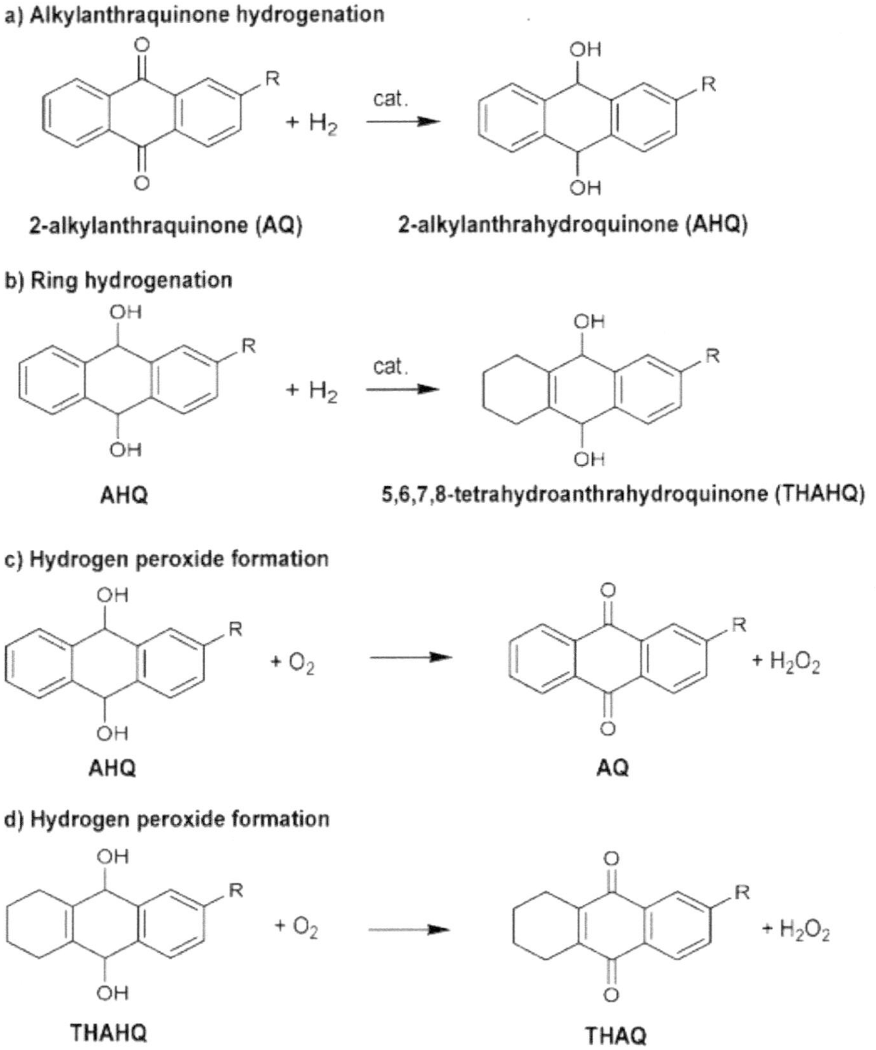

FIGURE 17.10 Anthraquinone autoxidation reactions.

Source: Campos-Martin et al. (2006).

The catalytic decomposition of hydrogen peroxide can be employed for in situ generation of reactive oxygen species for petrochemical transformations (Mayer & Ofial, 2018).

$$P - S - CH_3 + H_2O_2 \rightarrow Ph - S(O) - CH_3 + H_2O \quad \text{(Eq. 17.18)}$$

17.3.2 CFC Catalysts

Chlorofluorocarbons (CFCs) are compounds with fluorine and hydrogen known for their carbon–fluorine bonds. They were commonly used in refrigerants and propellants, although they have been phased out due to their damaging effects on the ozone layer, a concern addressed by the Montreal Protocol. The asymmetry of CFCs is influenced by variations in size and charge among the atoms of fluorine, chlorine, and hydrogen. While they do not serve as catalysts themselves, the term "CFC catalyst" refers to catalysts specifically created to break down CFC molecules, thereby reducing their harm. Various metals, such as ZrO_2, Al_2O_3, TiO_2, ZrO_2–TiO_2, and Cr_2O_3, supported on surfaces are examples of materials used in oxidation processes to transform harmful CFCs into less-detrimental substances (Siegemund et al., 2000).

17.3.3 Silicones

Silicones are man-made materials made up of alternating silicon and oxygen atoms mixed with hydrogen and carbon to create groups (Moretto et al., 2000). The characteristics of silicone polymers differ based on the groups attached and the length and flexibility of the polymer chain. Renowned for their ability to withstand temperatures, water, and chemicals, silicones find application in a range of uses, from medical equipment to automotive components. In contrast to industrial polymers, silicones have an inorganic structure that lacks carbon atoms. Polysiloxanes, also known as siloxanes, exhibit a backbone composed of silicon–oxygen bonds (O-Si-O) with two organic groups linked to each silicon atom, like vinyl (CH_2), methyl (CH_3), or phenyl (C_6H_5) (Moretto et al., 2000). The general formula for silicone is $(R_2SiO)x$, where *R* signifies a group. Figure 17.11 displays polydimethylsiloxane, one of the utilized silicone polymers featuring methyl groups bonded covalently to silicon atoms (Moretto et al., 2000; Lucchesini et al., 2016; Xometry, 2022).

17.3.4 Brine Polishing

Refining salt water through brine polishing is essential for enhancing its suitability for purposes. This process involves techniques such as filtration, ion exchange, and reverse osmosis. Brine polishing plays a role in industries like desalination, chemical manufacturing, and mining by boosting effectiveness, minimizing fouling, and prolonging the lifespan of equipment. It guarantees that the brine adheres to quality guidelines, thus averting challenges like corrosion (Lucchesini et al., 2016; Moretto et al., 2000; Xometry, 2022).

17.3.5 Chlorinated Organics

Chlorinated organics, also known as *organochlorines*, are substances consisting of carbon, hydrogen, and chlorine atoms. They can be found in sources and are present in many biomolecules, as highlighted by Kjeld (1986) and Wagner et al. (2009). Examples of these compounds include dioxins, which are produced in high-temperature settings, like forest fires, and simple chlorinated hydrocarbons, like dichloromethane, as mentioned by Gribble (1994).

Within the petrochemical industry, chlorinated organics play a role in the production of plastics, solvents, and intermediates. For instance, vinyl chloride is created by chlorinating ethylene and serves as a precursor for PVC. These substances are also utilized in agrochemicals and pharmaceuticals. Effective management practices are essential due to the health impacts associated with chlorinated organics. Some of these compounds can persist and accumulate in organisms over time, posing risks to both ecosystems and human health (Gribble, 1999; Wagner et al., 2009).

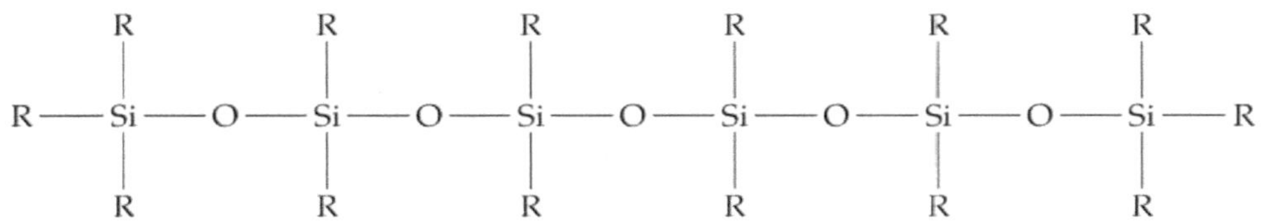

FIGURE 17.11 Silicon–oxygen backbone of silane polymers. *R* stands for a CH3 group.

Source: Lucchesini et al. (2016).

17.4 DEHYDROGENATION, DEHYDRATION, OXIDATION

The process of dehydrogenating hydrocarbons involves breaking carbon–hydrogen bonds to create hydrogen and carbon–carbon double bonds. This results in the production of materials used in petrochemical products, such as polymers and fuels (Stock, 2023). By converting alkanes into olefins, this process serves as a foundation for the creation of aldehydes, alcohols, polymers, and aromatic compounds (Wittcoff et al., 2004).

A significant obstacle faced in dehydrogenation is the fouling and deactivation of catalysts caused by coking, which involves the polymerization of substances through dehydrogenation (Guisnet & Magnoux, 2001). Dehydrogenation plays a role in the petrochemical sector for producing compounds that necessitate high temperatures exceeding (RCO_3H) (Wittcoff et al., 2004). One prominent example of a large-scale dehydrogenation reaction is the production of styrene from ethylbenzene. Conventional catalysts employed in this process typically consist of iron(III) oxide enhanced with potassium oxide or potassium carbonate (Guisnet & Magnoux, 2001; Airaksinen et al., 2002; Wittcoff et al., 2004; James & Castor, 2000; Vora, 2012; Stock, 2023).

17.4.1 BASF–Linde Propane Dehydrogenation Process

The BASF–Linde propane dehydrogenation process (Figure 17.12) represents a leap in the field of petrochemicals enabling the transformation of propane into propylene, a component in the production of plastics, chemicals, and fuels. This innovative technology, introduced in 1995, employs an approach wherein the dehydrogenation reactor alternates between operations such as purging and regeneration (Monai et al., 2021; S. Chen et al., 2021).

Within this process, propane is introduced into the reactor and undergoes dehydrogenation to yield propylene along with hydrogen by-products. The utilization of Pt-based and chromium oxide–based catalysts supported on alumina and hydrotalcite has proven to offer selectivity and activity. Operating at temperatures ranging between 550 and 650°C under pressures ensures a propylene selectivity exceeding 90% with periodic catalyst regeneration being carried out. Key benefits associated with this method include conversion efficiencies, a straightforward fixed-bed reactor design, and reduced reactor dimensions attributed to the absence of dilution. These advantages translate into equipment investment and heightened safety measures (Monai et al., 2021; S. Chen et al., 2021).

17.4.2 SNAP Catalyst for Butane and Propane Dehydrogenation

The SNAP catalyst consists of a noble metal catalyst supported on alumina pellets. Noble metals, known for their effectiveness and precision, are commonly utilized in catalysis. The SNAP catalyst is

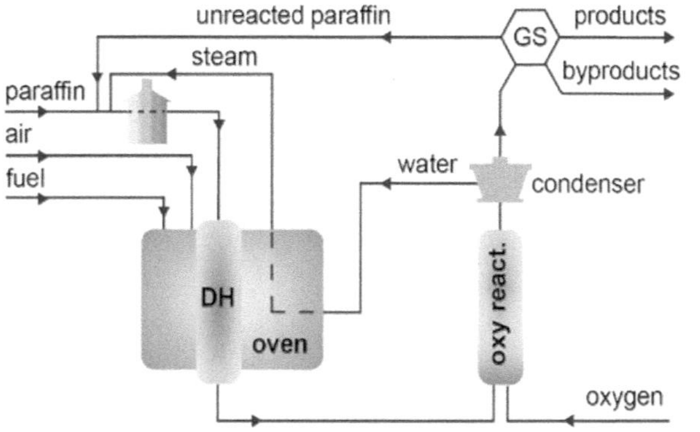

FIGURE 17.12 Linde-BASF-Statoil propane dehydrogenation (PDH) reactors.

Source: Monai et al. (2021).

employed in the dehydrogenation process of butane and propane, resulting in the production of substances such as butadiene, propylene, and isoamylene. This procedure plays a role in the petrochemical sector by generating building blocks for polymers and chemical intermediates (Huang et al., 2021).

Within the dehydrogenation reactor, the SNAP catalyst aids in extracting hydrogen atoms from butane and propane to form olefin products. The composition and structure of the catalyst are engineered to enhance dehydrogenation while minimizing side reactions like cracking or coke formation. Typically, platinum or palladium is dispersed on alumina pellets to create a surface area for interaction with hydrocarbons (Huang et al., 2021).

The SNAP catalyst is specifically designed to function at elevated temperatures and under conditions commonly found in butane and propane dehydrogenation units. Its catalytic efficiency, selectivity, and durability are crucial for achieving high product yields and ensuring term effectiveness (Huang et al., 2021).

17.4.3 Ethanol to Ethylene

Ethylene, a key component in plastics, is primarily produced by steam-cracking hydrocarbons but can also be made by dehydrating ethanol from renewable sources (Fan et al., 2013). The catalytic dehydration of ethanol uses solid acidic catalysts like alumina or silica-alumina at 200–300°C. This process removes water from ethanol, forming ethylene, which is then purified for applications such as polyethylene and ethylene oxide production. The process involves introducing ethanol into a reactor with a dehydrating catalyst, maintaining the catalyst in a fluidized state at temperatures above 700°F, and withdrawing ethylene and regenerating the catalyst by heating it in a fluidized state (Fan et al., 2013).

The conversion of ethanol to ethylene holds significance in the context of sustainable and renewable feedstock utilization, as shown in Figure 17.13, as ethanol can be derived from biomass sources such as sugarcane, corn, and cellulosic materials. This process represents a potential pathway for the production of ethylene from renewable feedstocks, contributing to the diversification of ethylene sources and reducing reliance on fossil fuels.

17.4.4 Vinyl Acetate Monomer (VAM)

Vinyl acetate monomer (VAM) is a substance known for its slightly sweet and strong scent, recognized for its high flammability and ability to create explosive combinations with air. In terms of composition, VAM is denoted as $CH_3COOCH{=}CH_2$ and contains a vinyl group along with a

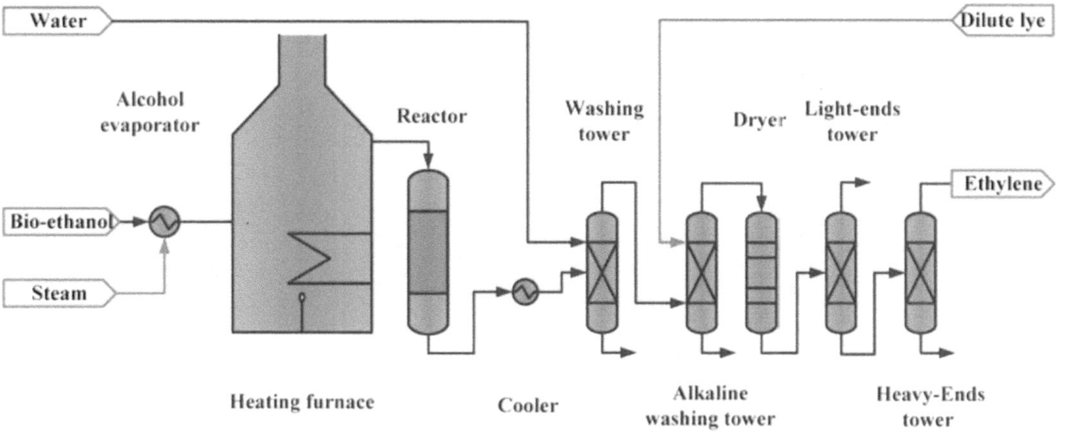

FIGURE 17.13 A simple process diagram of a bioethanol-based ethylene plant.

Source: Y. Chen (2007); M. Zhang and Yu (2013).

reactive acetate group, giving it some solubility in water but full solubility in most organic solvents. VAM is mainly utilized in the manufacturing of polymers like polyvinyl acetate (PVA), polyvinyl alcohol (PVOH), and ethylene vinyl acetate (EVA), serving purposes in adhesives, paints, fabrics, coatings, and food packaging materials. These items benefit from VAMs' characteristics of resilience, potency, clarity, and improved adhesion. All factors for their use in packaging materials, adhesives, and foams (Dow, 2021).

17.4.5 Amines

Amines are compounds derived from ammonia with aromatic variations (Figure 17.14). Similar to ammonia, amines exhibit properties due to the presence of an unshared electron pair, on the nitrogen atom (Charlesworth, 2022).

Quaternary (4°) amine, on the other hand, refers to an ion in which a nitrogen atom in a compound is bonded to four alkyl or aryl groups and bears a positive charge (Jiao & Stahl, 2019). With the substituent group attached identified, numbering should start from the carbon atom nearest to the amine group. Substituents, which are attached to the nitrogen atom instead of the carbon of the chain, are designated by a capital *N* (Lappert et al., 2008; Arshad et al., 2020).

17.4.6 Aliphatic Amines

Aliphatic amine compounds are created by replacing hydrogen atoms in ammonia (NH_3) with alkyl or alkanol radicals. While the higher homologs of these compounds are not soluble in water, the lower homologs, such as ammonia, exist as gases, easily dissolved in water. In solution, all aliphatic amines exhibit certain properties and can form salts that are nonvolatile, odorless solids that readily dissolve in water (Altarawneh et al., 2016).

17.4.7 Aromatic Amines

Amines which contain aromatic rings are called *aromatic amines*. The amine group ($-NH_2$) is directly attached to the aromatic ring (also called arylamines) (Figure 17.15) (Turesky, 2011).

17.4.7.1 Aniline

Aniline (Figure 17.16), also referred to as phenylamine or aminobenzene, consists of the chemical formula $C_6H_5NH_2$ or C_6H_7N. Due to the presence of the amino group within its structural formula, it is considered as an aromatic amine. This compound is known to possess all the characteristics of an aromatic compound; they are toxic and are used in a wide variety of industries (Anjalin et al., 2020).

17.4.7.2 Toluene Diamine (TDA)

Toluene diamine (TDA), also referred to as 2,4-diaminotoluene, is an organic compound with the formula $C_6H_3(NH_2)_2CH_3$. It is produced by using a nickel catalyst to hydrogenate 2,4-dianitrotoluene. Commercial samples frequently have up to 20% of the 2,6-isomer (Abou Hammad et al., 2020).

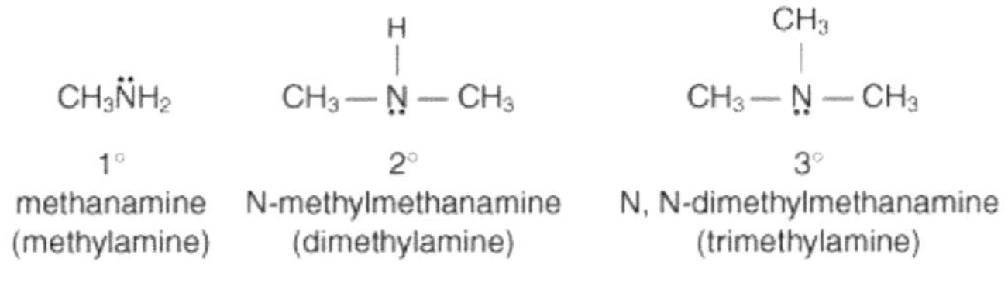

FIGURE 17.14 Classification of amine molecules.

Source: Charlesworth (2022).

FIGURE 17.15 Examples of arylamines.

Source: Turesky (2011).

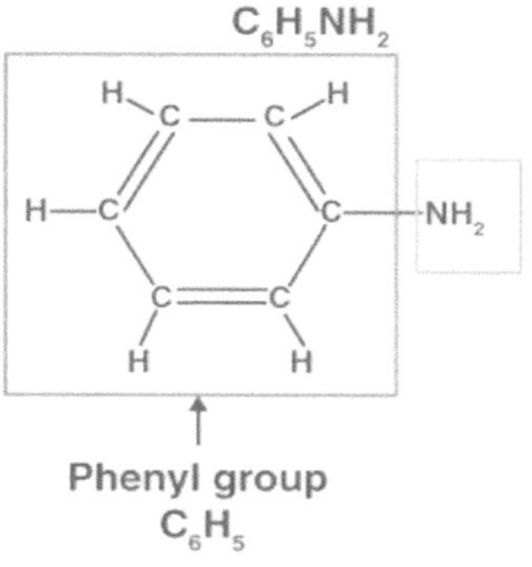

FIGURE 17.16 Aniline composition.

Source: Anjalin et al. (2020).

17.4.8 Acrylamide

Acrylamide, often referred to as acrylic amide, is an organic compound with the chemical formula $CH_2 = CHC(O)NH_2$. It is an odorless white material that dissolves in several organic solvents and water (Friedman, 2015). Acrylamide is a chemical compound formed naturally from starchy food materials during baking, roasting, high-temperature cooking (including frying), and industrial processing at $+120°C$ and low moisture. It is the same reaction that "browns" food and affects its taste, and the main chemical process that causes this is known as the Maillard reaction. Sugars and amino acids are found naturally in a variety of foods and produce acrylamide (Lineback et al., 2012; Sennakesavan et al., 2020).

17.5 THERMOPLASTICS

The term "thermoplastics" refers to any type of plastic that can be heated until it melts and becomes soft then solidifies once it cools down. There are two categories of polymers: crystalline and amorphous. Thermoplastics have some level of crystallinity that makes them resistant to chemical attacks from things like fuel, oil, and paint remover, which is why they are used as matrices for high-performance composites (Biron, 2015). Polymers are divided into thermosetting or thermoplastic categories based on how they react to temperatures. *Thermoplastic polymers* can be either branched or linear in structure; they harden upon cooling and soften when heated. In contrast, *thermosetting polymers* do not soften when heated because of their cross-linked structures once they have hardened (Peters, 2002).

17.5.1 Acrylonitrile Butadiene Styrene

Acrylonitrile butadiene styrene (ABS) (Figure 17.17) is made of three monomers: acrylonitrile, butadiene, and styrene (Ramesan, 2014; Ramlan & Shuaib, 2018). It is an amorphous polymer with an impact-resistant engineering thermoplastic. Due to its physical properties, such as strain, resistance to impact, high rigidity, and abrasion, it is a preferred choice for structural applications. ABS is used in Lego toys, auto parts, electronic housings, pipe fittings, and consumer products (Kesarla et al., 2018).

17.5.2 Polycarbonate

A transparent thermoplastic with carbonate functional groups is called polycarbonate (PC), an excellent alternative to glass because of its light weight and, similarly, resistance to impact and fracture due to its high strength. To give the PC an appropriate shape, it is heated and then forced firmly into a mold. Owing to its recyclability and eco-friendly processing, it is widely used. PC is a transparent thermoplastic polymer, a high-performance, tough, and amorphous polymer (Kyriacos, 2017; Olabisi & Adewale, 2016; Kausar, 2018).

17.5.3 Polyethylene

Ethane, an important component of natural gas, undergoes cracking to produce ethylene (C_2H_4), a gaseous hydrocarbon also obtained from petroleum (Ronca, 2017). Ethylene molecules have the formula $CH_2=CH_2$, consisting of two methylene units (CH_2) joined by a double bond (Gahleitner et al., 2019). This double bond can be broken with polymerization catalysts, forming long chain-like molecules with hydrogen atoms connected to a carbon backbone (Maddah, 2016; Gahleitner & Paulik, 2017; Ronca, 2017; Gahleitner et al., 2019).

17.5.4 Polyethylene Terephthalate

Polyethylene terephthalate (PET) is created through the condensation polymerization of ethylene glycol and terephthalic acid. This method, referred to as step growth polymerization, results in the production of water as a side product (Bomgardner et al., 2015; Crowther & Quoquab, 2023; Sin & Tueen, 2023).

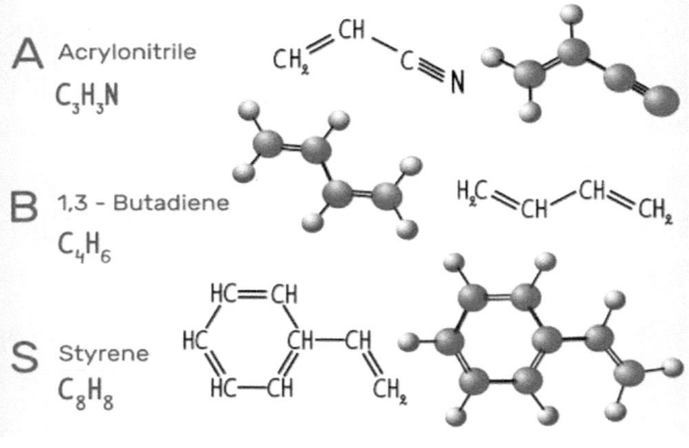

FIGURE 17.17 Acrylonitrile butadiene styrene molecules composition.

Source: Kesarla et al. (2018).

17.5.5 Polytetrafluoroethylenes

The process of combining results in polytetrafluoroethylene, a man-made material that is fire-resistant, durable, strong, and of a waxy texture. When chlorodifluoromethane is heated between 600 and 700°C, it transforms into a scentless gas known as tetrafluoroethylene (C_2F_4). Additionally, the reaction of hydrogen fluoride (HF) with chloroform ($CHCl_3$) leads to the creation of chlorodifluoromethane. By applying pressure and utilizing radical initiators, tetrafluoroethylene monomers are mixed with water before being polymerized (Wang et al., 2002; Wang et al., 2017; Dhanumalayan & Joshi, 2018; Primc, 2020).

17.5.6 Polyvinyl Chloride (PVC)

A synthetic resin obtained from the polymerization of vinyl chloride is called polyvinyl chloride (PVC). Chloroethylene, also known as (CH_2=CHCl), is produced by the reaction of hydrogen chloride, ethylene, and oxygen in the presence of a copper catalyst (Amobonye et al., 2022; Novotný et al., 2022).

17.5.7 Polymethyl Methacrylate

Polymethyl methacrylate is a man-made resin created through the process of polymerizing methyl methacrylate, making it part of the group of acrylic resins. It is derived from propylene, a substance extracted from the components of oil (Mateos-Timoneda, 2009; Gohil et al., 2017). Cumene, or isopropyl benzene, is produced by reacting benzene and propylene. After the cumene is oxidized to cumene hydroperoxide, it is treated with acid to produce acetone (Puska et al., 2011).

17.5.8 Polypropylene

Polypropylene is a synthetic resin produced by propylene polymerization. The thermal cracking of ethane, propane, butane, and the naphtha fraction of petroleum produces propylene. Propylene, like ethylene, belongs to hydrocarbon molecules with single carbon atoms connected by a double bond, called olefins (Rani et al., 2022).

17.5.9 Polystyrene

Polystyrene is a rigid resin that is highly transparent. Its strength and durability contribute to the quality of rubber and plastic items through copolymerization with polymers. The process involves the reaction of benzene with ethylene to form ethylbenzene, which is then hydrogenated to create a hydrocarbon called phenylethylene, or styrene (Tripathi et al., 2016; Sharma et al., 2022; Gu et al., 2024).

17.5.10 Expanded Polystyrene

Expanded polystyrene (EPS) is a sealed cell foam material made from solid polystyrene beads. It includes an expansion gas incorporated into the polystyrene bead through the polymerization process of monomers (Adhikary & Ashish, 2022; Giuliani et al., 2020).

17.6 THERMOSETS

Thermosets are a polymer that forms a rigid structure when heated or treated with a curing agent. They are widely used in applications such as epoxy adhesives, coatings, electrical equipment, transportation, construction, and building industries (Koltzenburg et al., 2017). Additionally, thermosets play a role in producing materials like glass and reinforced carbon fibers, making them essential in advanced sectors, such as aerospace and military industries (AlMaadeed et al., 2020).

17.6.1 Epoxide

A cyclic ether with a three-membered ring is called an epoxide. An epoxide basic structure consists of two adjacent carbon atoms of a hydrocarbon attached to an oxygen atom (Tanaka, 2018). An epoxide is much more reactive than a typical acyclic ether because of the strain of the three-membered ring. The most important epoxide produced by the oxidation of ethylene over a silver catalyst is ethylene oxide, generally used to produce ethylene glycol, to make antifreeze, and as a fumigant and in other useful compounds. Epoxidation of alkenes with the transfer of oxygen atoms by peroxy acid (RCO_3H) is used to produce more-complex epoxides (Södergren et al., 2000; Magnus et al., 2002; Jana et al., 2023).

17.6.2 Phenol-Formaldehyde

Phenol-formaldehyde resin is a type of material made by combining formaldehyde with phenol. In settings, polymers can be transformed into resins through two main methods. The first approach involves creating a low-molecular-weight prepolymer called a resole by mixing formaldehyde and phenol in a water solution with the help of a base catalyst. This resole can then be solidified into a thermosetting network polymer for plywood production by layering it between wood veneers and subjecting it to pressure during assembly (Hocking, 2006).

17.6.3 Polyurethane

Polyurethane, or PUR, or PU, is a polymer made up of organic units connected by urethane molecules. Most polyurethanes belong to the category of thermosetting polymers because they do not melt when heated. However, it is worth noting that certain types of polyurethanes exhibit distinct properties and can be melted and reshaped with heat. Traditional methods for producing polyurethanes involve chemical reactions between polyols and di- or tri-isocyanates. Due to the presence of two types of monomers that undergo repeated polymerization, polyurethanes can be seen as alternating copolymers. Additionally, each molecule of the polyols and isocyanates used as monomers in the production of polyurethanes typically contains two or more groups (Zafar & Sharmin, 2012; Giuliani et al., 2020; De Souza et al., 2021).

17.6.4 Unsaturated Polyester Resins

Unsaturated polyester refers to a type of thermosetting resin formed by blending a weight unsaturated polyester with a vinyl monomer and then copolymerizing them. This process results in the creation of a plastic material known for its toughness. These unsaturated polyesters are commonly used in the production of components like boat hulls, pipes, and countertops reinforced with fiberglass or minerals. When an unsaturated dicarboxylic acid or its anhydride is combined with one to produce polyesters, they are called copolyesters. The distinctive ester groups formed during the reaction between these acids and dialcohols like ethylene glycol or propylene glycol play a role in creating long-chain polyester molecules (Dholakiya, 2012).

17.7 ELASTOMERS

Elastomers are materials made up of chain-like molecules known as polymers. They can regain their shape after being stretched, which is why they are called elastomers, derived from polymer. In conditions, the long molecules in materials are coiled irregularly. When a force is applied, these molecules align themselves in the direction of the force. Upon release, they spontaneously return to their random arrangement (Massey, 2007). The used elastomer is polyisoprene, which is the polymer found in natural rubber obtained from the milky latex of various trees, predominantly *Hevea* rubber trees. Despite facing competition from alternatives, like butadiene rubber and styrene butadiene rubber produced from gas and petroleum products, natural rubber remains a significant industrial polymer (Rosato et al., 2004).

17.7.1 Unsaturated Rubbers That Can Be Cured by Sulfur Vulcanization

Sulfur vulcanization is a chemical process that transforms rubber or similar polymers into materials of varying hardness, flexibility, and strength by heating them with sulfur or compounds containing sulfur. The sulfur creates connections between parts of the polymer chains, impacting their electrical characteristics. Many everyday items, like tires, shoe soles, hoses, and conveyor belts, are produced using vulcanized rubber (Council, 1995; Osswald & Menges, 1996; Cheremisinoff, 2012).

17.7.2 Saturated Rubbers That Cannot Be Cured by Sulfur Vulcanization

A *saturated compound* refers to a compound where bonds connect a series of carbon atoms. Such rubber types cannot undergo sulfur vulcanization alone. Rubbers containing a saturated chain (with heteroatoms) exhibit strong resistance to oxygen, heat, and ozone exposure. Listed in what follows are some of the common types of saturated rubbers (Council, 1995; Osswald & Menges, 1996; Cheremisinoff, 2012):

- Ethylene propylene rubber (EPM/EPDM)
- Epichlorohydrin rubber (ECO)
- Polyacrylic rubber (ACM, ABR)
- Silicone rubber (SI, Q, VMQ)
- Fluorosilicone rubber (FVMQ)
- Fluoroelastomers (FKM and FEPM): Viton, Tecnoflon, Fluorel, Aflas, and Dai-El
- Perfluoroelastomers (FFKM): Tecnoflon PFR, Kalrez, Chemraz, Perlast
- Polyether block amides (PEBA)
- Chlorosulfonated polyethylene (CSM): Hypalon
- Ethylene-vinyl acetate (EVA)

17.7.2.1 Ethylene-Propylene Rubber (EPM/EPDM)

Ethylene-propylene rubber, also referred to as ethylene-propylene diene monomer (EPDM), is a man-made rubber material primarily consisting of ethylene and propylene, along with an amount of a monomer that aids in vulcanization by providing cross-linking points. This vulcanization process enhances EPDM's ability to withstand temperatures, making it suitable for temperature conditions. EPDM rubber, commonly known as ethylene propylene diene monomer rubber, is widely used in industries. Dienes such as ethylidene norbornene (ENB), dicyclopentadiene (DCPD), and vinyl norbornene (VNB) are utilized in the production of EPDM rubbers at concentrations ranging from 4 to 8% (Whelan, 1994; AZO, 2003; Costa et al., 2008; Osborne, 2004; Winters et al., 2002; W. Zhang & You, 2016).

17.8 EMERGING TECHNOLOGIES

New technologies are continually shaping the realm of petrochemical catalysts, leading to advancements in efficiency, selectivity, and environmental friendliness. Here are some noteworthy emerging technologies in petrochemical catalysts:

1. *Singular-atom catalysts (SACs).* Singular-atom catalysts represent a cutting-edge innovation where individual metal atoms are dispersed on a supporting material. SACs offer activity and selectivity by maximizing the utilization of metal atoms. In petrochemical processes, SACs hold promise for improving performance in conversions while reducing metal usage and enhancing sustainability.
2. *Plasmonic catalysts.* By utilizing interactions between light and matter, plasmonic catalysts can elevate reactions, leading to enhanced efficiency and selectivity. In applications within the petrochemical industry, plasmonic catalysts provide opportunities for driven catalytic processes like photocatalysis for organic transformations.
3. *Catalysts based on MOFs.* Metal frameworks (MOFs) are a category of materials with adjustable structures and surface areas. Due to their surface-to-volume ratio and tailored functionalities, MOFs are increasingly being considered as supports for catalysts. Catalysts based on MOFs show potential in petrochemical processes for enhancing activity through the management of active sites and pore structures.
4. Inspired by nature, biomimetic catalysts imitate enzyme structures and mechanisms to drive reactions. By mimicking the processes found in nature, these catalysts offer precise and selective outcomes in petrochemical transformations, promoting more sustainable production methods within the industry.
5. The trend of combining homogeneous catalysts is gaining momentum in catalysis research. Hybrid catalyst systems merge the benefits of both types, harnessing the effectiveness of catalysis alongside the reusability of catalysis. This strategy shows promise in improving efficiency and sustainability in petrochemical processes.
6. The realm of catalyst design and optimization is undergoing a revolution through machine learning and computational modeling techniques. By utilizing data-driven methodologies and predictive modeling, researchers can accelerate the discovery of catalyst materials and enhance their performance. Machine learning tools are being employed within the petrochemical sector to speed up catalyst innovation efforts.

17.9 SIGNIFICANCE OF CATALYSTS IN ENERGY TRANSITION

Catalysts play a role in transitioning to energy by making energy production more efficient, eco-friendly, and cleaner. Here are some important ways in which catalysts contribute to this shift.

Catalysts are crucial for converting energy sources like wind power into usable forms, such as hydrogen or synthetic fuels. They assist in processes like water splitting through electrolysis and the creation of fuels from sources, helping integrate renewable energy into current energy systems. Additionally, catalysts are players in developing the hydrogen economy by supporting the production, storage, and application of hydrogen as an energy source. They are used in processes including steam methane reforming, water–gas shift reactions, electrolysis for hydrogen production, and fuel cells that efficiently convert hydrogen into electricity.

Catalysts play a role in devices such as batteries and fuel cells in electrochemical energy storage. They improve the efficiency and performance of these storage tools by enabling charge–discharge cycles, boosting energy conversion efficiencies, and facilitating the use of energy sources for transportation and grid functions.

In the realm of carbon capture and utilization (CCU), catalysts play a role in reducing carbon dioxide emissions by transforming them into products. They speed up the process of converting CO_2 into chemicals, fuels, or materials, aiding in the efforts to lower carbon footprints and achieve carbon neutrality across industries.

Moreover, catalysts are vital in the production of biofuels, where plant-based materials are transformed into fuels through reactions. They facilitate the conversion of biomass components like sugars, lignocellulose, and fats into biofuels such as ethanol, biodiesel, and biojet fuel. This promotes the adoption of alternatives to fossil fuels.

Furthermore, catalysts find applications in optimizing industrial processes, including petrochemicals, chemicals, and refining. They enhance reaction efficiencies, decrease energy consumption, and reduce impacts. By promoting more sustainable production methods, catalysts enhance energy efficiency and environmental performance within industrial settings.

REFERENCES

Abou Hammad, A. B., Elzwawy, A., Mansour, A. M., Alam, M. M., Asiri, A. M., Karim, M. R., Rahman, M. M., & El Nahrawy, A. M. (2020). Detection of 3, 4-diaminotoluene based on Sr 0.3 Pb 0.7 TiO 3/CoFe 2 O 4 core/shell nanocomposite via an electrochemical approach. *New Journal of Chemistry*, *44*(19), 7941–7953.

Adhikary, S. K., & Ashish, D. K. (2022). Turning waste expanded polystyrene into lightweight aggregate: Towards sustainable construction industry. *Science of the Total Environment*, *837*. https://doi.org/10.1016/j.scitotenv.2022.155852

Airaksinen, S. M. K., Harlin, M. E., & Krause, A. O. I. (2002). Kinetic modeling of dehydrogenation of isobutane on chromia/alumina catalyst. *Industrial and Engineering Chemistry Research*, *41*(23), 5619–5626. https://doi.org/10.1021/ie020371j

Aleksiev, M., & García, M. O. (2023). Enantioselective dearomatization reactions of heteroarenes by anion-binding organocatalysis. *Chemical Communications*, *59*(23), 3360–3372. https://doi.org/10.1039/d2cc07101k

AlMaadeed, M. A. A., Ponnamma, D., & El-Samak, A. A. (2020). Polymers to improve the world and lifestyle: Physical, mechanical, and chemical needs. *Polymer Science and Innovative Applications: Materials, Techniques, and Future Developments*, 1–19. https://doi.org/10.1016/B978-0-12-816808-0.00001-9

Altarawneh, M., Almatarneh, M. H., Marashdeh, A., & Dlugogorski, B. Z. (2016). Decomposition of ethylamine through bimolecular reactions. *Combustion and Flame*, *163*, 532–539. https://doi.org/10.1016/j.combustflame.2015.10.032

Amobonye, A. E., Bhagwat, P., Singh, S., & Pillai, S. (2022). Biodegradability of Polyvinyl chloride. *Biodegradability of Conventional Plastics: Opportunities, Challenges, and Misconceptions*, 201–220. https://doi.org/10.1016/B978-0-323-89858-4.00017-8

Anjalin, M., Kanagathara, N., & Baby Suganthi, A. R. (2020). A brief review on aniline and its derivatives. *Materials Today: Proceedings*, *33*, 4751–4755. https://doi.org/10.1016/j.matpr.2020.08.358

Antonov, S. V., Petrukhina, N. N., Pakhmanova, O. A., & Maksimov, A. L. (2017). Hydrogenation process for producing light petroleum resins as adhesive and hot-melt components (review). *Petroleum Chemistry*, *57*(12), 983–1001. https://doi.org/10.1134/S0965544117120027

Arshad, M., Zubair, M., Rahman, S. S., & Ullah, A. (2020). Polymers for advanced applications. In *Polymer Science and Nanotechnology*. Elsevier. https://doi.org/10.1515/ci.2006.28.6.34a

Ashenhurst, J. (2020). *Reactions of Aromatic Molecules*. www.masterorganicchemistry.com/2017/07/11/electrophilic-aromatic-substitution-introduction/

Augustine, R. L. (2016). Whither Goest Thou, catalysis. *Catalysis Letters*, *146*(12), 2393–2416. https://doi.org/10.1007/s10562-016-1865-8

AZO. (2003). Ethylene propylene rubbers – properties and applications of ethylene propylene diene (EPDM) and ethylene propylene copolymers (EPM). *Ethylene Propylene Rubbers – Properties and Applications of Ethylene Propylene Diene (EPDM) and Ethylene Propylene Copolymers (EPM)*. www.azom.com/article.aspx?ArticleID=1822

Behr, A. (2012). Catalysis, homogeneous. *Ullmann's Encyclopedia of Industrial Chemistry*, *7*, 255–269. https://doi.org/10.1002/14356007.o05_o01

Biron, M. (2015). Material selection for thermoplastic parts: Practical and advanced information for plastics engineers. *Material Selection for Thermoplastic Parts: Practical and Advanced Information for Plastics Engineers*, 1–682. https://doi.org/10.1016/C2013-0-18801-0

Bomgardner, M., Andrady, A., Southerton, D., & Fossi, C. (2015). Plastics in a sustainable society. *Mistra.Org*, 1–40. www.mistra.org/wp-content/uploads/2018/01/Background-paper-Plastics-in-a-sustainable-society.pdf

Buijink, J. K. F., Lange, J.-P., Bos, A. N. R., Horton, A. D., & Niele, F. G. M. (2008). Propylene epoxidation via shell's SMPO process. *Mechanisms in Homogeneous and Heterogeneous Epoxidation Catalysis*, 355–371. https://doi.org/10.1016/b978-0-444-53188-9.00013-4

Busca, G. (2014). Metal catalysts for hydrogenations and dehydrogenations. *Heterogeneous Catalytic Materials*, 297–343. https://doi.org/10.1016/b978-0-444-59524-9.00009-2

Campos-Martin, J. M., Blanco-Brieva, G., & Fierro, J. L. G. (2006). Hydrogen peroxide synthesis: An outlook beyond the anthraquinone process. *Angewandte Chemie—International Edition*, *45*(42), 6962–6984. https://doi.org/10.1002/anie.200503779

Charlesworth, N. (2022). *New Methods for the Construction of Fluorinated Cyclic Amines and Amides*. PhD thesis, University of Nottingham.

Chaudhari, R. V. (2016). Fundamentals of homogeneous catalysis. In *Industrial Catalytic Processes for Fine and Specialty Chemicals* (pp. 17–39). Elsevier. https://doi.org/10.1016/B978-0-12-801457-8.00002-1

Chen, S., Chang, X., Sun, G., Zhang, T., Xu, Y., Wang, Y., Pei, C., & Gong, J. (2021). Propane dehydrogenation: Catalyst development, new chemistry, and emerging technologies. *Chemical Society Reviews*, *50*(5), 3315–3354. https://doi.org/10.1039/d0cs00814a

Chen, Y. (2007). Simulation and synthesis of heat exchange networks for producing ethylene from ethanol. *Tianjin University*, 2007.

Cheremisinoff, N. P. (2012). Condensed encyclopedia of polymer engineering terms. *Condensed Encyclopedia of Polymer Engineering Terms*, 1–362. https://doi.org/10.1016/C2009-0-25687-X

Christopher, P. F. (2015). Dearomatization reactions: An overview. *Arene Chemistry: Reaction Mechanisms and Methods for Aromatic Compounds*, 399–423. https://doi.org/10.1002/9781118754887.ch15

Cochran, R. N., & Dubner, W. S. (1992). Propylene oxide-styrene monomer process. *US 5,210,354*. 1–6.

Costa, F. R., Dutta, N. K., Choudhury, N. R., & Bhowmick, A. K. (2008). Thermoplastic elastomers. In *Current Topics in Elastomers Research* (pp. 101–164). Wiley. https://doi.org/10.1002/polc.5070260104

Council, N. R. (1995). Polymer science and engineering: The shifting research frontiers. *Choice Reviews Online*, *32*(07), 32-3901-32-3901. https://doi.org/10.5860/choice.32-3901

Crowther, D., & Quoquab, F. (2023). The dilemma of plastic. *Developments in Corporate Governance and Responsibility*, *19*, 1–8. https://doi.org/10.1108/S2043-052320230000019001

De Souza, F. M., Kahol, P. K., & Gupta, R. K. (2021). Introduction to polyurethane chemistry. *ACS Symposium Series*, *1380*, 1–24. https://doi.org/10.1021/bk-2021-1380.ch001

Dekkers, M. P. J. (1890). Le glycol tétraméthylénique. *Recueil Des Travaux Chimiques Des Pays-Bas*, *9*(4), 92–102. https://doi.org/10.1002/recl.18900090403

Dhanumalayan, E., & Joshi, G. M. (2018). Performance properties and applications of polytetrafluoroethylene (PTFE)–a review. *Advanced Composites and Hybrid Materials*, *1*(2), 247–268. https://doi.org/10.1007/s42114-018-0023-8

Dholakiya, B. (2012). Unsaturated polyester resin for specialty applications. In *Polyester*. https://doi.org/10.5772/48479

Dow. (2021). *Vinyl Acetate Monomer (VAM)*. www.dow.com

Fan, D., Dai, D. J., & Wu, H. S. (2013). Ethylene formation by catalytic dehydration of ethanol with industrial considerations. *Materials*, *6*(1), 101–115. https://doi.org/10.3390/ma6010101

Friedman, M. (2015). Acrylamide: Inhibition of formation in processed food and mitigation of toxicity in cells, animals, and humans. *Food & Function*, *6*(6), 1752–1772.

Gahleitner, M., & Paulik, C. (2017). Polypropylene and other polyolefins. *Brydson's Plastics Materials: Eighth Edition*, 279–309. https://doi.org/10.1016/B978-0-323-35824-8.00011-6

Gahleitner, M., Tranninger, C., & Doshev, P. (2019). Polypropylene copolymers. *Polypropylene Handbook: Morphology, Blends and Composites*, 295–355. https://doi.org/10.1007/978-3-030-12903-3_6

Giuliani, F., Autelitano, F., Garilli, E., & Montepara, A. (2020). Expanded polystyrene (EPS) in road construction: Twenty years of Italian experiences. *Transportation Research Procedia*, *45*, 410–417. https://doi.org/10.1016/j.trpro.2020.03.033

Glotov, A., & Karakhanov, E. (2021). Heterogeneous catalysts for petrochemical synthesis and oil refining. In *Heterogeneous Catalysts for Petrochemical Synthesis and Oil Refining*. https://doi.org/10.3390/books978-3-0365-1429-1

Gohil, S. V., Suhail, S., Rose, J., Vella, T., & Nair, L. S. (2017). Polymers and composites for orthopedic applications. *Materials and Devices for Bone Disorders*, 349–403. https://doi.org/10.1016/B978-0-12-802792-9.00008-2

Gribble, G. W. (1994). The natural production of chlorinated compounds. *Environmental Science & Technology*, *28*(7), 310A–319A. https://doi.org/10.1021/es00056a712

Gribble, G. W. (1999). The diversity of naturally occurring organobromine compounds. *Chemical Society Reviews*, *28*(5), 335–346. https://doi.org/10.1039/a900201d

Gu, J. D., Ford, T., Mitton, B., & Mitchell, R. (2024). Research on biodeteriotration of plastics. *International Biodeterioration and Biodegradation*, *186*. https://doi.org/10.1016/j.ibiod.2023.105699

Guisnet, M., & Magnoux, P. (2001). Organic chemistry of coke formation. *Applied Catalysis A: General*, *212*(1–2), 83–96. https://doi.org/10.1016/S0926-860X(00)00845-0

Hagen, J. (2006). Industrial catalysis: A practical approach. In *Focus on Catalysts* (Vol. 2006, Issue 5). Wiile. https://doi.org/10.1016/s1351-4180(06)71636-1

Hans-Joachim, R., & Pfiefderer, G. (2002). Production of hydrogen peroxide. *Membrane Technology*, *2002*(6), 12. https://doi.org/10.1016/s0958-2118(02)80183-5

Hocking, M. B. (2006). *Handbook of Chemical Technology and Pollution Control*. Elsevier.

Huang, L., & Liu, G. (2018). Optimization of two-stage pygas hydrogenation reactor based on hydrogen network integration. *Industrial and Engineering Chemistry Research*, *57*(31), 10539–10550. https://doi.org/10.1021/acs.iecr.8b01703

Huang, Y., Wang, B., Yuan, H., Sun, Y., Yang, D., Cui, X., & Shi, F. (2021). The catalytic dehydrogenation of ethanol by heterogeneous catalysts. *Catalysis Science & Technology*, *11*, 1652. https://doi.org/10.1039/d0cy02479a

Hydrocarbon, P. (2021). *Axens SeLene Selective Hydrogenation Catalysts New Family*. www.hydrocarbonprocessing.com/news/

IUPAC. (2019). Alcohol. In *IUPAC Compendium of Chemical Terminology* (3rd ed.). International Union of Pure and Applied Chemistry. https://doi.org/https://doi.org/10.1351/goldbook.A00204

James, D. H., & Castor, W. M. (2000). Styrene. In *Ullmann's Encyclopedia of Industrial Chemistry*. Wiley-VCH Verlag GmbH & Co. KGaA. https://doi.org/10.1002/14356007.a25_329

Jana, A., Chakraborty, S., Sarkar, K., & Maji, B. (2023). Ruthenium-catalyzed reductive coupling of epoxides with primary alcohols via hydrogen transfer catalysis. *Journal of Organic Chemistry*, *88*(1), 310–318. https://doi.org/10.1021/acs.joc.2c02354

Janice, G. S. (2020). 28 – Synthetic polymers. In *Organic Chemistry* (6th ed., pp. 1260–1285). https://lccn.loc.gov/2018032811

Jiao, N., & Stahl, S. S. (2019). Green oxidation in organic synthesis. *Green Oxidation in Organic Synthesis*, i–iv. https://doi.org/10.1002/9781119304197

Johnson Matthey group. (2020). *DAVY Butanediol and Co-products Process*. www.jmprotech.com/licensed-processes-butanediol

Kausar, A. (2018). A review of filled and pristine polycarbonate blends and their applications. *Journal of Plastic Film and Sheeting*, *34*(1), 60–97. https://doi.org/10.1177/8756087917691088

Kesarla, H., Rohit, K., Mohod, A., Tanji, S., Mane, O., & Venkatachalam, G. (2018). Study on tensile behavior of fly ash reinforced hybrid polymer matrix composite. *Materials Today: Proceedings*, *5*(5), 11922–11932. https://doi.org/10.1016/j.matpr.2018.02.166

Khanna, A., & Anand Rishi. (2016). The essential chemical industry – Online. *2016 International Conference on Internet of Things and Applications (IOTA)*, 266–270. www.essentialchemicalindustry.org/chemicals.html

Kjeld, C. Engvild. (1986). Chlorine-containing natural compounds in higher plants. *Phytochemistry*, *25*(4), 781–791.

Koltzenburg, S., Maskos, M., & Nuyken, O. (2017). *Polymer Chemistry* (pp. 477–491). Springer.

Kyriacos, D. (2017). Polycarbonates. *Brydson's Plastics Materials: Eighth Edition*, 457–485. https://doi.org/10.1016/B978-0-323-35824-8.00017-7

Lappert, M., Protchenko, A., Power, P., & Seeber, A. (2008). Metal amide chemistry. *Metal Amide Chemistry*, 1–355. https://doi.org/10.1002/9780470740385

Lineback, D. R., Coughlin, J. R., & Stadler, R. H. (2012). Acrylamide in foods: A review of the science and future considerations. *Annual Review of Food Science and Technology*, *3*, 15–35. http://ovidsp.ovid.com/ovidweb.cgi?T=JS&PAGE=reference&D=emed10&NEWS=N&AN=22136129

Lower, S. (2020). Chem1 virtual textbook. *LibreTexts*, 885. https://chem.libretexts.org/Bookshelves/General_Chemistry/Book%3A_Chem1_(Lower)/16%3A_Electrochemistry/16.04%3A_The_Nernst_Equation

Lucchesini, A., Gozzini, S., Marinelli, C., & Marmugi, L. (2016). Low energy atomic photodesorption from organic coatings. *Coatings*, *6*(4). https://doi.org/10.3390/coatings6040047

Ma, Z., Wei, L., Zhou, W., Jia, L., Hou, B., Li, D., & Zhao, Y. (2015). Overview of catalyst application in petroleum refinery for biomass catalytic pyrolysis and bio-oil upgrading. *RSC Advances*, *5*(107), 88287–88297. https://doi.org/10.1039/c5ra17241a

Maddah, H. A. (2016). Polypropylene as a promising plastic: A review. *American Journal of Polymer Science*, *6*(1), 1–11.

Magnus, A., Bertilsson, S. K., & Andersson, P. G. (2002). Asymmetric base-mediated epoxide isomerisation. *Chemical Society Reviews*, *31*(4), 223–229. https://doi.org/10.1039/b104372m

Massey, L. K. (2007). The effect of UV light and weather: On plastics and elastomers. *The Effect of UV Light and Weather: On Plastics and Elastomers*, 1–456.

Matar, S., Mirbach, M. J., & Tayim, H. A. (1989). Types of catalysts used in petrochemical processes. *Catalysis in Petrochemical Processes*, 19–34. https://doi.org/10.1007/978-94-009-1177-2_2

Mateos-Timoneda, M. A. (2009). Polymers for bone repair. *Bone Repair Biomaterials*, 231–251. https://doi.org/10.1533/9781845696610.2.231

Mayer, R. J., & Ofial, A. R. (2018). Nucleophilic reactivities of bleach reagents. *Organic Letters*, *20*(10), 2816–2820. https://doi.org/10.1021/acs.orglett.8b00645

Missen, R. W., Mims, C. A., & Saville, B. A. (1999). *Introduction to Chemical Reaction Engineering and Kinetics*. J. Wiley.

Monai, M., Gambino, M., Wannakao, S., & Weckhuysen, B. M. (2021). Propane to olefins tandem catalysis: A selective route towards light olefins production. *Chemical Society Reviews*, *50*(20), 11503–11529. https://doi.org/10.1039/d1cs00357g

Moretto, H., Schulze, M., & Wagner, G. (2000). Silicones. In *Ullmann's Encyclopedia of Industrial Chemistry*. Wiley. https://doi.org/10.1002/14356007.a24_057

Nemoto, T. (2023). Development of transition-metal-catalyzed dearomatization reactions. *Chemical and Pharmaceutical Bulletin*, *71*(8), 624–632. https://doi.org/10.1248/CPB.C23-00227

Nitriles, S. (1975). Selective hydrogenation catalyst for furfurol. *Journal Chemical Abstracts*, *40*(2), 267–268.

Novotný, Č., Fojtík, J., Mucha, M., & Malachová, K. (2022). Biodeterioration of compost-pretreated polyvinyl chloride films by microorganisms isolated from weathered plastics. *Frontiers in Bioengineering and Biotechnology*, *10*. https://doi.org/10.3389/fbioe.2022.832413

Olabisi, O., & Adewale, K. (2016). Handbook of thermoplastics. In O. Olabisi & K. Adewale (Eds.), *Handbook of Thermoplastics*. CRC Press. https://doi.org/10.1201/b19190

Organic, S. (2003). Mild and selective oxidation of sulfur compounds in trifluoroethanol: Diphenyl disulfide and methyl phenyl sulfoxide. *Organic Syntheses*, *80*, 184. https://doi.org/10.15227/orgsyn.080.0184

Osborne, D. (2004). Organic chemistry principles and industrial practice. *Synthesis*, *2004*(15), 2594–2594. https://doi.org/10.1055/s-2004-835255

Osswald, T. A., & Menges, G. (1996). Materials science of polymers for engineers. *Choice Reviews Online*, *33*(11), 33-6326-33-6326. https://doi.org/10.5860/choice.33-6326

Peters, E. N. (2002). Plastics: Thermoplastics, thermosets, and elastomers. *Handbook of Materials Selection*, 335–355. https://doi.org/10.1002/9780470172551.ch11

Petrukhina, N. N., Zakharyan, E. M., Korchagina, S. A., Nagieva, M. V., & Maksimov, A. L. (2018). Hydrogenation of petroleum resins in the presence of supported sulfide catalysts. *Petroleum Chemistry*, *58*(1), 48–55. https://doi.org/10.1134/S0965544118010127

Primc, G. (2020). Recent advances in surface activation of polytetrafluoroethylene (PTFE) by gaseous plasma treatments. *Polymers*, *12*(10), 1–7. https://doi.org/10.3390/polym12102295

Puska, M., J., A., & Vallittu, P. (2011). Polymer composites for bone reconstruction. In *Advances in Composite Materials—Analysis of Natural and Man-Made Materials*. https://doi.org/10.5772/20657

Ramesan, M. T. (2014). Flammability, oil resistance, and interaction of petroleum fuels with dichlorocarbene modified styrene butadiene rubber/fly ash composites. *Petroleum Science and Technology*, *32*(15), 1775–1783. https://doi.org/10.1080/10916466.2012.662254

Ramlan, M. I. S., & Shuaib, N. A. (2018). Environmental assessment and mechanical properties of acrylonitrile butadiene styrene (ABS) reinforced with fly ash filler. *AIP Conference Proceedings*, *2030*. https://doi.org/10.1063/1.5066659

Ranade, V. V., & Joshi, S. S. (2016a). Catalysis and catalytic processes. *Industrial Catalytic Processes for Fine and Specialty Chemicals*, 1–14. https://doi.org/10.1016/B978-0-12-801457-8.00001-X

Ranade, V. V., & Joshi, S. S. (2016b). Catalysis and catalytic processes. In *Industrial Catalytic Processes for Fine and Specialty Chemicals* (pp. 1–14). Elsevier. https://doi.org/10.1016/B978-0-12-801457-8.00001-X

Rani, M., Meenu, & Shanker, U. (2022). The role of nanomaterials in plastics biodegradability. *Biodegradability of Conventional Plastics: Opportunities, Challenges, and Misconceptions*, 283–308. https://doi.org/10.1016/B978-0-323-89858-4.00012-9

Roche, S. P., & Porco, J. A. (2011). Dearomatization strategies in the synthesis of complex natural products. *Angewandte Chemie—International Edition*, *50*(18), 4068–4093. https://doi.org/10.1002/anie.201006017

Ronca, S. (2017). Polyethylene. *Brydson's Plastics Materials: Eighth Edition*, 247–278. https://doi.org/10.1016/B978-0-323-35824-8.00010-4

Rosato, D. V., Rosato, D. V., & Rosato, M. V. (2004). Plastic product material and process selection handbook. *Plastic Product Material and Process Selection Handbook*, 1–618. https://doi.org/10.1016/B978-1-85617-431-2.X5000-2

Rostrup-Nielsen, J. (1997). Catalytic processes in the petrochemical industry-present and future challenges. *World Petroleum Congress Proceedings*, *1997-Octob*, 767–772.

Sennakesavan, G., Mostakhdemin, M., Dkhar, L. K., Seyfoddin, A., & Fatihhi, S. J. (2020). Acrylic acid/acrylamide based hydrogels and its properties—A review. *Polymer Degradation and Stability*, *180*. https://doi.org/10.1016/j.polymdegradstab.2020.109308

Sharma, R., Sharma, A. K., Sharma, B., & Sarkar, A. (2022). Biodegradability of synthetic plastics: Effective degradation mechanisms. *Biodegradability of Conventional Plastics: Opportunities, Challenges, and Misconceptions*, 101–120. https://doi.org/10.1016/B978-0-323-89858-4.00005-1

Siegemund, G., Schwertfeger, W., Feiring, A., Smart, B., Behr, F., Vogel, H., & McKusick, B. (2000). Fluorine compounds, organic. In *Ullmann's Encyclopedia of Industrial Chemistry*. https://doi.org/10.1002/14356007.a11_349

Sin, L. T., & Tueen, B. S. (2023). Plastics and environmental sustainability issues. *Plastics and Sustainability*, 1–43. https://doi.org/10.1016/b978-0-12-824489-0.00006-4

Södergren, M. J., Bertilsson, S. K., & Andersson, P. G. (2000). Allylic alcohols via catalytic asymmetric epoxide rearrangement. *Journal of the American Chemical Society*, *122*(28), 6610–6618. https://doi.org/10.1021/ja000545t

Stock, C. (2023). Hydrogenation and dehydrogenation. *Ullmann's Encyclopedia of Industrial Chemistry*, 1–20. https://doi.org/10.1002/14356007.a13_487.pub3

Tanaka, Y. (2018). Synthesis and characteristics of epoxides. *Epoxy Resins*, 9–283. https://doi.org/10.1201/9780203756713-2

Taylor, H. S. (2023). *Catalysis | Chemistry, Classification, & Chemical Reactions | Britannica*. www.britannica.com/science/catalysis

Tripathi, S., Yadav, A., & Devendra, M. T. (2016). Plastic waste: Environmental pollution, health hazards and biodegradation strategies. *Plastic Waste: Environmental Pollution Health Hazards & Biodegradation Strategies*, 152–177.

Turesky, R. J. (2011). Heterocyclic aromatic amines: Potential human carcinogens. *Current Cancer Research*, *6*, 95–112. https://doi.org/10.1007/978-1-61737-995-6_5

Unnikrishnan, P., & Srinivas, D. (2016). Heterogeneous catalysis. In *Industrial Catalytic Processes for Fine and Specialty Chemicals* (pp. 41–111). Elsevier. https://doi.org/10.1016/B978-0-12-801457-8.00003-3

Varshouee, G. H., & Zarand, S. M. G. (2020). Investigating the effect of co-catalyst on the final product properties and the yield of ziegler–natta catalyst in propylene polymerization in optimum temperature reaction in absence of hydrogen via mathematical model. *SN Applied Sciences*, *2*(4). https://doi.org/10.1007/s42452-020-2436-6

Vora, B. V. (2012). Development of dehydrogenation catalysts and processes. *Topics in Catalysis*, *55*(19–20), 1297–1308. https://doi.org/10.1007/s11244-012-9917-9

Wagner, C., Omari, M. El, & König, G. M. (2009). Biohalogenation: Nature's way to synthesize halogenated metabolites. *Journal of Natural Products*, *72*(3), 540–553. https://doi.org/10.1021/np800651m

Wagner, J. D., Lappin, G. R., & Zietz, J. R. (2000). Alcohols, higher aliphatic, synthetic processes. In *Kirk-Othmer Encyclopedia of Chemical Technology*. https://doi.org/10.1002/0471238961.1925142023010714.a01

Wang, R., Xu, G., & He, Y. (2017). Structure and properties of polytetrafluoroethylene (PTFE) fibers. *E-Polymers*, *17*(3), 215–220. https://doi.org/10.1515/epoly-2016-0059

Wang, X. Q., Chen, D. R., Han, J. C., & Du, S. Y. (2002). Crystallization behavior of polytetrafluoroethylene (PTFE). *Journal of Applied Polymer Science*, *83*(5), 990–996. https://doi.org/10.1002/app.2279

Weber H, & Falbe, J. (1970). Oxo synthesis technology. *Industrial & Engineering Chemistry*, *62*(4), 33–37. https://doi.org/10.1021/ie50724a006

Whelan, T. (1994). Polymer technology dictionary. In *Polymer Technology Dictionary*. Springer. https://doi.org/10.1007/978-94-011-1292-5

Winters, R., Heinen, W., Verbruggen, M. A. L., Lugtenburg, J., Van Duin, M., & De Groot, H. J. M. (2002). Solid-state 13C NMR study of accelerated-sulfur-vulcanized 13C-labeled ENB-EPDM. *Macromolecules*, *35*(5), 1958–1966. https://doi.org/10.1021/ma001716h

Wittcoff, H. A., Reuben, B. G., & Plotkin, J. S. (2004). Industrial organic chemicals. In *Choice Reviews Online* (Vol. 34, Issue 01). Wiley. https://doi.org/10.1002/0471651540

Xometry. (2022). 4 types of silicone and their differences. *Xometry*. www.xometry.com/resources/materials/types-of-silicone/

Xu, W. L., Li, Y. Z., Zhang, Q. S., & Zhu, H. S. (2004). A selective, convenient, and efficient conversion of sulfides to sulfoxides. *Synthesis*, *2004*(02), 227–232. https://doi.org/10.1055/S-2004-44387

Ye, Z., Liu, W., Gu, H., & Yang, X. (2023). Enantioselective dearomatization of substituted phenols via organocatalyzed electrophilic amination. *Organic Letters*, *25*(31), 5838–5843. https://doi.org/10.1021/acs.orglett.3c02100

Zafar, F., & Sharmin, E. (2012). Polyurethane: An introduction. In *Polyurethane*. www.intechopen.com/books/polyurethane

Zhang, M., & Yu, Y. (2013). Dehydration of ethanol to ethylene. *Industrial and Engineering Chemistry Research*, *52*(28), 9505–9514. https://doi.org/10.1021/ie401157c

Zhang, W., & You, S.-L. (2016). Miscellaneous Asymmetric dearomatization reactions. In *Asymmetric Dearomatization Reactions* (pp. 379–389.). Wiley. https://doi.org/10.1002/9783527698479

Zhang, W. Y., Wang, H. C., Wang, Y., Zheng, C., & You, S. L. (2023). Enantioselective dearomatization of indoles via SmI2-mediated intermolecular reductive coupling with ketones. *Journal of the American Chemical Society*, *145*(18), 10314–10321. https://doi.org/10.1021/jacs.3c01994

Zhao, X. S., Bao, X. Y., Guo, W., & Lee, F. Y. (2006). Immobilizing catalysts on porous materials. *Materials Today*, *9*(3), 32–39. https://doi.org/10.1016/S1369-7021(06)71388-8

Zheng, C., & You, S. L. (2021). Advances in catalytic asymmetric dearomatization. *ACS Central Science*, *7*(3), 432–444. https://doi.org/10.1021/acscentsci.0c01651

Zhu, H., Fujimori, S., Kostenko, A., & Inoue, S. (2023). Dearomatization of C6 aromatic hydrocarbons by main group complexes. *Chemistry – A European Journal*, *29*(59). https://doi.org/10.1002/chem.202301973

18 Latest Advances in Technology

Aliyu Mohammed Kaura, Mohammed Aliyu, Labaran Jamilu Ari, and Usman Yusuf

18.1 INDUSTRY REVOLUTION 4.0

Industry 4.0, or the Fourth Industrial Revolution, is a transformational shift in the manufacturing sector that integrates digital technologies. This revolution, driven by various technologies, including the Internet of Things (IoT), artificial intelligence (AI), robotics, and big data analytics, holds immense potential for the energy sector. The primary objective of Industry 4.0 is to increase the general interconnection and cooperation of resources, including people and IT systems, to create a more efficient and intelligent manufacturing process. The industry has undergone various revolutions, progressing from Industrial Revolution (IR) 1.0 to 4.0, as illustrated in Figure 18.1 (Wibowo et al., 2020; Tran, 2023).

The First Industrial Revolution began in the 18th century with steam power and mechanization of production, leading to significant changes in human productivity. The Second Industrial Revolution, in the 19th century, introduced electricity in assembly-line production to transform automobile production. The Third Industrial Revolution (20th century) introduced partial automation using memory-programmable controls and computers, allowing for the automation of entire production processes without human intervention (Lu et al., 2019; Hussain & Di Sia, 2020). The beginning of the 21st century saw the emergence of Industry 4.0, also known as the Fourth Industrial Revolution, which brought about a transformation through the digitalization and automation of industrial operations. *Industry 4.0* is defined as using information and communication technologies to create "cyber-physical production systems" and smart factories (Buła & Niedzielski, 2021). The era brought remarkable technological advancements, such as the Internet of Things (IoT), autonomous robotics, artificial intelligence (AI), big data analytics, cloud computing, and sensors (Buła & Niedzielski, 2021; Munirathinam, 2020; Sackschewsky et al., 2018; Tsaramirsis et al., 2022). The progress of these technologies has impacted the production, delivery, and utilization of goods and services. As a result, business approaches have been reassessed and reorganized, giving rise to what is now known as Entrepreneurship 4.0 (Đorđević et al., 2023).

Industry 4.0 involves ten steps: mechanizing, sensorizing, transmitting, integrating, analyzing, visualizing, augmenting, robotizing, crafting, and virtualizing. These steps collect and analyze data to make faster decisions and manage assets effectively. However, key technologies of IR 4.0 (Figure 18.2) include the Internet of Things (IoT), artificial intelligence and machine learning, big data analytics, cyber-physical systems (CPS), additive manufacturing (3D printing), augmented reality and virtual reality, cloud computing, robotics and automation, digital twins, and edge computing. These technologies work together to revolutionize manufacturing processes, improving efficiency, productivity, quality, and flexibility. Collaboration enhances industry efficiency, productivity, quality, and flexibility. Some key technologies associated with Industry 4.0 include the following (Lu et al., 2019; Elijah et al., 2021):

- *Internet of Things (IoT).* Devices and sensors are interconnected to collect and exchange data in real time, enabling the monitoring and control of manufacturing processes.
- *Artificial intelligence (AI) and machine learning (ML).* AI and ML algorithms analyze data to optimize processes, predict maintenance needs, and improve efficiency.

DOI: 10.1201/9781003282273-18

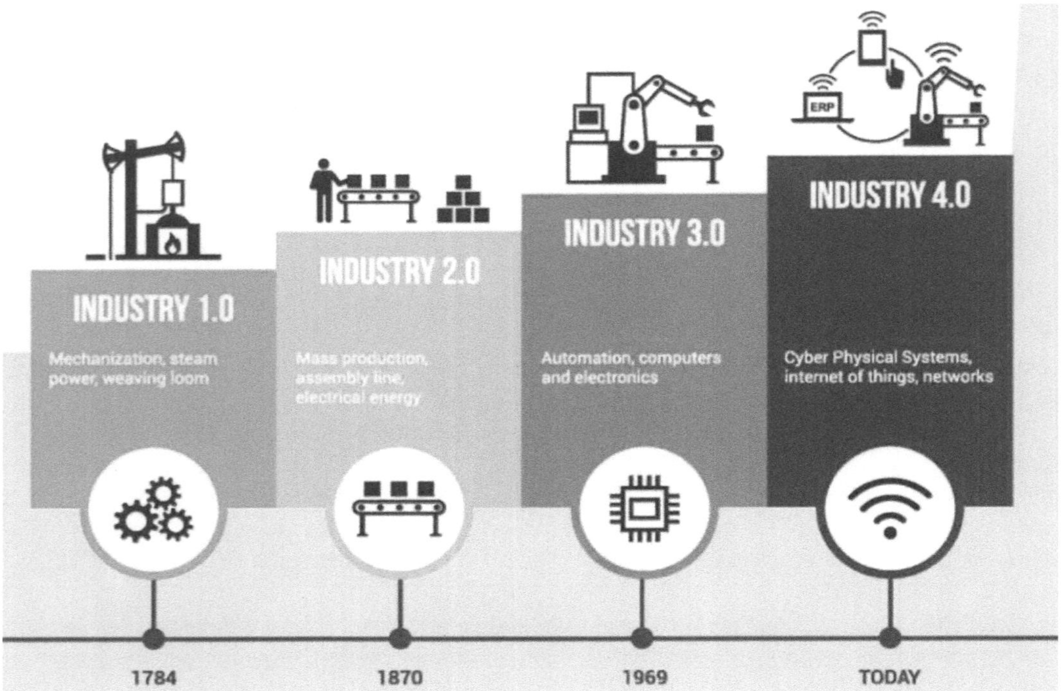

FIGURE 18.1 Industrial Revolution from 1.0 to 4.0.

Source: Wibowo et al. (2020).

- *Big data analytics.* Large volumes of data generated by IoT devices are processed and analyzed to uncover insights, patterns, and trends, aiding decision-making and process optimization.
- *Cyber-physical systems (CPS).* Physical systems are integrated with computational and networking capabilities, enabling real-time monitoring, control, and optimization.
- *Additive manufacturing (3D printing).* Allows the production of complex and customized parts with reduced waste, shorter lead times, and greater flexibility than traditional manufacturing methods.
- *Augmented reality (AR) and virtual reality (VR).* AR and VR technologies are used for training, maintenance, and remote assistance, enhancing productivity and reducing downtime.
- *Cloud computing.* Provides scalable and flexible computing resources for data storage, processing, and analysis, facilitating collaboration and access to resources from anywhere.
- *Robotics and automation.* Advanced robots and automated systems perform tasks with precision, speed, and flexibility, increasing productivity and safety in manufacturing environments.
- *Digital twin.* Virtual replicas of physical assets, processes, or systems are created to simulate and analyze their behavior, enabling optimization and predictive maintenance.
- *Edge computing.* Data processing and analysis occur closer to the data source (e.g., IoT devices), reducing latency and enabling faster decision-making in real-time applications.

Industry 4.0 has significantly influenced the oil and gas supply chain. Digital technology has transformed the entire value chain and altered best practices across all industries. Advancements in seismic imaging and data processing are improving efficiency in manufacturing and production within the oil and gas industry. Industry 4.0 is driving transformation across various corporate sectors,

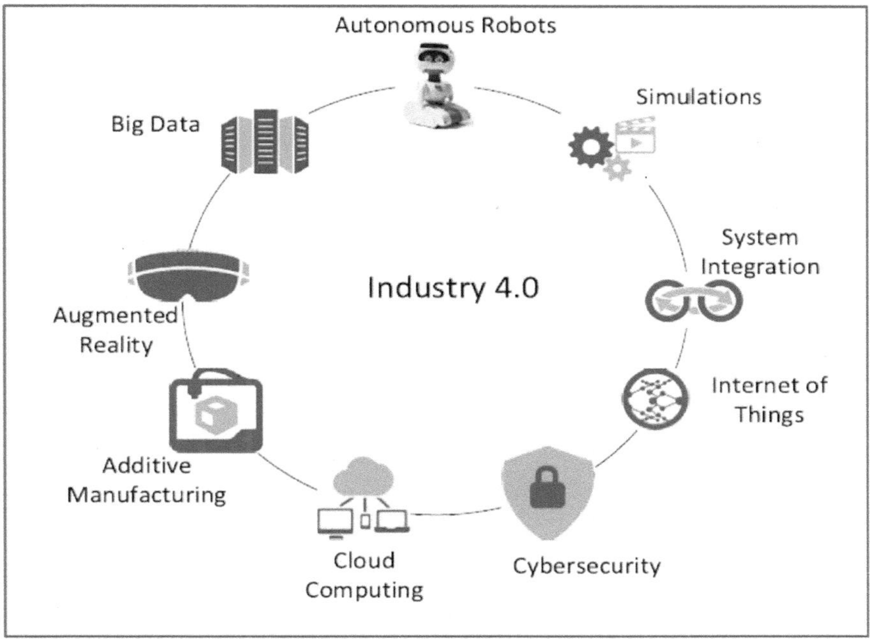

FIGURE 18.2 Industry 4.0 technologies.

Source: Elijah et al. (2021).

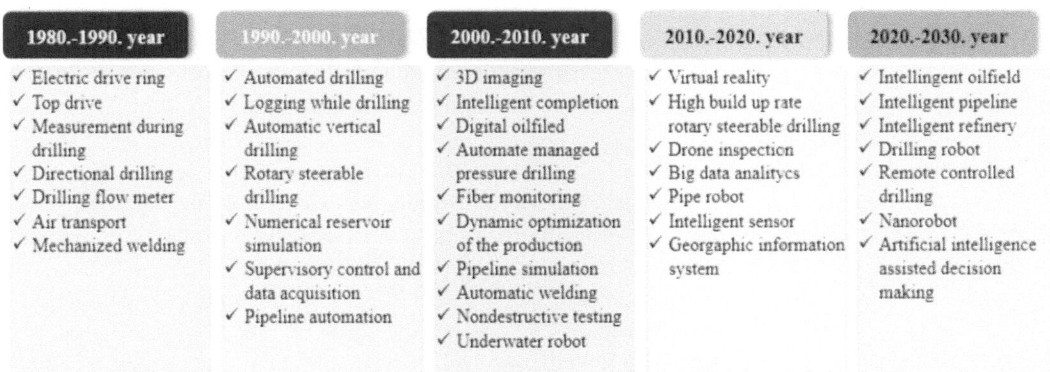

FIGURE 18.3 Key elements of development and application of technological innovations in the oil and gas industry in the last half-century.

Source: Majstorović (2022).

including research and development and finance, through the continuous advancement of technology. Technology is at the forefront of advancements in the oil and gas industry (Majstorović, 2022).

As big data analytics improve drilling and exploration, autonomous vehicles will shorten production times and reduce waste. The transformation will increase profits, improve safety and decision-making procedures, and secure the industry's long-term viability. In addition to enhancing environmental reporting and lowering carbon footprints, the deployment of Industry 4.0 must be prioritized. Figure 18.3 illustrates key elements of developing and applying technological innovations in the oil and gas industry (Majstorović, 2022).

Every Industrial Revolution has resulted in unparalleled advantages, including heightened production, greater living standards, and improved global interconnectedness. Nevertheless, they have also posed obstacles. The First Industrial Revolution witnessed severe working conditions in factories. The Second Industrial Revolution caused environmental pollution. The Third Industrial Revolution raised issues around data privacy and cybersecurity. The Fourth Industrial Revolution explores issues related to employment displacement caused by automation and the ethical application of AI (Majstorović, 2022).

18.1.1 Benefits of Adapting to IR 4.0

The Fourth Industrial Revolution, Industry 4.0, has brought business advantages. It boosts efficiency, cuts cost, enhances quality, offers flexibility and customization options, and enables maintenance. Additionally, it streamlines supply chains using technologies such as blockchain and RFID. Automation reduces tasks by fostering creativity and innovation within companies. Industry 4.0 sets businesses apart by attracting customers who appreciate innovation and sustainability practices. It supports sustainability by encouraging eco-production methods, efficient energy usage, and waste reduction initiatives. Moreover, it ensures competitiveness on a scale by keeping up with advancements and market trends. Therefore, embracing Industry 4.0 is crucial for businesses to excel in a connected world (Lu et al., 2019).

Implementing Industry 4.0 in the oil and gas sector involves integrating sensors and IoT devices that utilize data analytics and AI technologies, establishing cybersecurity protocols, leveraging cloud computing alongside edge computing solutions for data management, and employing collaborative robots for automation tasks. These technologies provide real-time insights into equipment performance metrics, maintenance requirements, and operational efficiencies within the industry domain. Data analytics can identify trends, predict equipment malfunctions, optimize processes, and enhance decision-making capabilities. Cybersecurity measures are in place to safeguard networks through encryption methods and access controls. Cloud computing combined with edge computing facilitates data storage, data processing, and analytical functions. Collaborative robots and automation strategies streamline operations and enhance safety standards in high-risk environments (Georgiou et al., 2021).

Embracing the principles of Industry 4.0 is essential for companies looking to thrive in today's landscape to enhance their competitiveness, drive innovation, and meet evolving customer needs and market trends. Leveraging Industry 4.0 technologies enables organizations to tackle challenges, capitalize on opportunities, and create a business path forward (Zakeri et al., 2022).

Industry 4.0 processes and tools are characterized by their integration capabilities, sustainability, agility, significance, efficiency, and transparency. The primary goal of Industry 4.0 is to address optimization challenges in aspects of life by reducing human-related impacts and obstacles in areas like energy management, time efficiency, and cost-effectiveness. In the past, consumers were largely unaware of the advancements shaping their products because they operated behind the scenes in production processes. However, with Industry 4.0, advancements have become more accessible and visible to consumers (Lim et al., 2021).

Digital transformation technology plays a role in Industry 4.0 by establishing data gathering and analysis infrastructure. The Internet of Things (IoT) connects devices while interacting with intelligence (AI) systems; blockchain technology stores and manages this data flow, contributing to self-learning AI systems' development. The proliferation of devices has introduced sensors into our daily lives, resulting in vast amounts of data known as big data; effectively analyzing big data is vital for feeding accurate information into AI systems (Borysiak et al., 2022).

18.1.2 Current Status and Key Applications of IR 4.0

The oil and gas industry is experiencing a profound transformation, thanks to the integration of Industry 4.0 technologies. This digital revolution, incorporating IoT, AI, and data analysis, is not just a buzzword but a game changer for enhancing efficiency, safety, and sustainability. The adoption of these technologies is revolutionizing energy production and distribution, leading to increased productivity, improved safety measures, and enhanced sustainability efforts. This shift is also reducing downtime and the risks of accidents. Companies in the oil and gas sector are progressively embracing these technologies to boost efficiency, safety standards, and sustainability practices. This involves the use of automation, IoT sensors, big data analytics, AI applications, and robotics across various operations. These advancements enable functions like maintenance activities, remote monitoring capabilities for operations' efficiency improvement, and real-time decision-making support. The adoption rates vary among companies and regions due to factors such as costs, challenges encountered during implementation, and existing infrastructure limitations. However, the industry is increasingly moving toward this transformation to stay competitive and meet evolving demands (Wang et al., 2023).

In the realm of Industry 4.0, notable applications include robots for inspecting and maintaining pipelines and other infrastructure. Furthermore, implementing analytics and machine learning algorithms aids in optimizing production processes and minimizing downtime. Incorporating sensors and Internet of Things (IoT) devices facilitate real-time monitoring of equipment and operations, enabling issue detection and proactive maintenance measures. These applications are further expounded upon later (Pérez et al., 2019).

18.1.3 Digital Twins

A *digital twin* is a cutting-edge technology that has revolutionized the way we understand and manage physical objects and systems. It is a virtual replica of a physical object or system created using data from sensors, machines, and other sources (Almessabi et al., 2023; Whaley, 2020). This virtual representation simulates the behavior of the real-world object or system in real time, enabling businesses and organizations to gain valuable insights into their operations, predict maintenance needs, ensure safety, and prepare for emergencies (Wanasinghe et al., 2021).

Since their inception, digital twins have come a long way and have rapidly gained popularity in different industries. Initially used by NASA to simulate space missions and test spacecraft designs before launching them into space (China, 2023), digital twins are now used in a wide range of applications, from manufacturing and construction to health care and transportation. They offer a level of accuracy and precision that was previously unattainable and can potentially transform the way we work and live (Sheveleva et al., 2021). Digital twin technology is a crucial aspect of Industry 4.0 that provides a powerful and disruptive strategy with great potential for huge organizations (Almessabi et al., 2023). The oil and gas industry is among the industries incorporating modern technologies to enhance safety, operational efficiency, production, and decision-making processes (China, 2023; Wanasinghe et al., 2020).

18.1.4 Industrial Internet of Things (IIoT)

The Industrial Internet of Things (IIoT) is another element of IR 4.0 that has revolutionized the manufacturing industry by providing real-time data collection from equipment and components. Smart sensors embedded in drilling equipment gather data on drilling methods and well output optimization, allowing quick adjustments. Using AI-driven systems, automated drillers enhance production efficiency by monitoring operations in real time. This technology has reduced costs and increased productivity, driven innovation, and improved efficiency in the manufacturing industry (Munirathinam, 2020).

18.1.5 Predictive Analytics

Predictive analytics is an umbrella term for analytics that utilize data, statistical modeling, data mining techniques, and machine learning to forecast future outcomes (Schwab, 2020). It aids companies

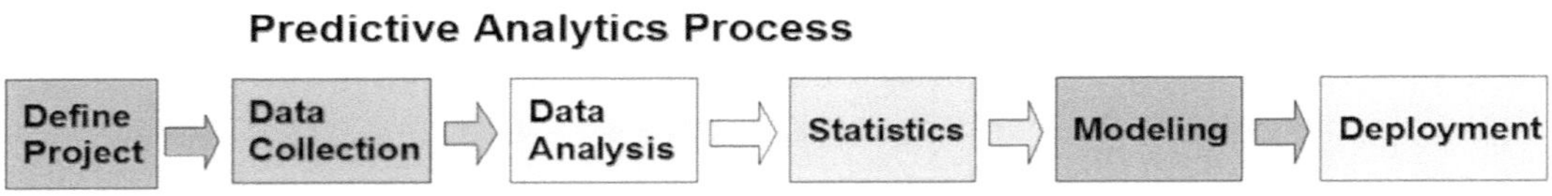

FIGURE 18.4 Predictive analytics process flow.

Source: Patresearch (2013).

in recognizing risks and opportunities by pinpointing patterns within data. Data scientists employ learning and machine learning algorithms like regression, neural networks, and decision trees to analyze historical data and predict forthcoming trends. Predictive analytics is used across sectors to anticipate and address business challenges. In the banking sector, it assists in forecasting credit risk and detecting fraud, enabling companies to adjust to market shifts. In health care, it aids in identifying and managing patients with conditions while monitoring infections such as sepsis. HR teams leverage analytics to align job candidates, reduce turnover rates, and boost engagement levels, thereby cutting recruitment expenses while enhancing employee satisfaction. In marketing and sales, it facilitates client interactions by predicting customer churn rates and implementing selling strategies. Within supply chain management operations, predictive analytics plays a role in overseeing product inventory levels and establishing pricing strategies to help companies evaluate the costs versus returns of their products over time (Fan et al., 2020).

In the oil and gas sector, predictive analytics ensures efficient operations. Augmented reality (AR) and virtual reality (VR) models can conduct analysis by leveraging data analysis and existing knowledge of process procedures. This method involves identifying patterns and trends in data to forecast issues or failures in operations. Such insights aid in prioritizing parameters and shutdown sequences, ultimately improving efficiency. Simply put, predictive analytics help anticipate issues in advance, allowing for actions to mitigate costly downtime or safety risks (see Figure 18.4) (Kumar et al., 2019).

18.1.6 Cybersecurity

In the age of Industry 4.0, where interconnected systems are prevalent, cybersecurity has emerged as a concern. Growing reliance on technologies necessitates cybersecurity practices to defend against various cyber threats. Safeguarding critical infrastructure such as power grids, transportation networks, and health-care facilities is crucial for the operation of society. It is vital to prevent cyber threats, like hacking, data breaches, and identity theft, to uphold the privacy and security of information. Data integrity is also important since any unauthorized manipulation or modification of data can have repercussions. Therefore, implementing cybersecurity measures is essential to ensure the safety and security of interconnected systems. In the energy sector encompassing oil and gas and power utilities, power plants implementing stringent cybersecurity measures is imperative to protect against cyberterrorism (Progoulakis et al., 2021).

18.1.7 Autonomous Operations

IR 4.0 enables the deployment of autonomous technologies in oil and gas operations, such as autonomous drones and unmanned aerial vehicles (UAVs). These drones can perform pipeline inspections, flare stack monitoring, and facility surveillance more efficiently and safely than traditional methods. They also collect high-resolution data that enhances decision-making and reduces operational risks. Autonomous operations lead to better asset management, reduced costs, and increased safety, particularly in hazardous environments (Devold & Fjellheim, 2019).

18.1.8 Remote Monitoring and Control

Industry 4.0 empowers the supervision and management of oil and gas assets, enabling operators to oversee operations from control centers. By utilizing devices and cloud-based platforms, operators can remotely track equipment performance, adjust parameters, and address any irregularities in real time, enhancing operational efficiency and flexibility. Remote monitoring and control also facilitate identifying and responding to safety risks or security vulnerabilities (Sayed et al., 2020).

A significant advantage of Industry 4.0 lies in its capacity to support real-time monitoring and optimization, leading to energy management. This is achieved through the implementation of grids, sophisticated sensors, and data-informed decision-making processes. Intelligent grids facilitate the distribution of energy while minimizing wastage. Advanced sensors aid in measuring and monitoring energy usage, enabling the pinpointing of areas where energy conservation measures can be implemented. With data-driven decision-making practices, energy consumption patterns can be analyzed for optimization. Ultimately, Industry 4.0 technologies serve as a resource for achieving energy efficiency enhancements and overall energy conservation (Masudin et al., 2024).

18.1.9 Blockchain Technology

Blockchain technology is being investigated within the oil and gas sector to improve transparency, security, and efficiency in managing the supply chain, trading activities, and asset transactions. Organizations can simplify transactions, minimize disagreements, and uphold data integrity throughout the value chain using blockchain-powered systems. Additionally, blockchain technology allows for effective monitoring of oil and gas resources, from production to usage, lowering the chances of activities while enhancing transparency (Al-Ajwad et al., 2024).

18.1.10 Advanced Data Analytics

Oil and gas companies leverage big data analytics and machine learning algorithms to extract actionable insights from vast operational data. These insights enable operators to optimize production processes, identify cost-saving opportunities, and improve safety and environmental performance. Advanced data analytics also helps operators optimize supply chains, improve procurement processes, and enhance customer service. Advanced big data analytics require computing and energy-intensive data centers with the highest adoption of cybersecurity (Digitemie & Ekemezie, 2024).

18.2 INDUSTRIAL INTERNET OF THINGS (IIOT) IN THE ENERGY SECTOR

The Internet of Things (IoT) is a network of internet-connected objects capable of collecting and sharing data. It comprises two main components. These are the *internet*, which serves as the connectivity infrastructure, and *things*, which refer to objects or devices (Bhavana et al., 2022). IoT involves objects that gather data and transmit it to the internet, allowing other connected devices to access it. IoT devices can range from sensors and cameras to household appliances, electric vehicles, humanoid or industrial robots, and industrial machinery, all communicating with each other to enhance automation, efficiency, and fast and smart decisions (Abdullahi et al., 2024).

Several organizations have created their systems of classification and categorization for Internet of Things (IoT) applications and scenarios (McEwen & Cassimally, 2013). "Industrial IoT" is commonly used by enterprises and associations to refer to IoT applications in manufacturing and utilities. Others analyze IoT based on device categories, like wearables and appliances. Some individuals concentrate on IoT within integrated location-based applications, like "smart homes" or "smart cities." IoT use cases can potentially encompass almost every area of our lives, regardless of the application (Angurala & Rustamovich, 2024).

The Industrial Internet of Things (IIoT) is a subclass of IoT that covers use cases for IoT technologies in industrial applications and processes, including but not limited to manufacturing, logistics, operational and production efficiency, and energy management. The Industrial Internet of Things (IIoT) is a network of manufacturing machines and advanced analytics used to enhance industrial processes. IIoT integrates machine learning, big data, and machine-to-machine communication to execute manufacturing efforts efficiently. Microsensors on pre-existing machinery provide unlimited access to a data pool describing system health in real time, making operators intuitive to the network's ebb and flow and making it possible to pinpoint skewed numbers. Figure 18.5 shows the elements of IIoT (Gomes et al., 2020).

Thanks to the Internet of Things (IoT) technology, the oil and gas industry is transforming. IoT is helping automate numerous business activities, making the industry more efficient and cost-effective. With the help of IoT, oil and gas companies can reduce costly downtime in urban areas, automate several processes, and improve efficiency across all manufacturing phases. This

FIGURE 18.5 Elements of the Industrial Internet of Things.

Source: Gomes et al. (2020).

technology also allows business owners to control various aspects of their operations from their desks, giving them greater control and flexibility. By leveraging IoT, oil and gas enterprises can streamline their operations, reduce costs, and ensure the smooth functioning of their businesses (Elsharief et al., 2024).

18.2.1 Cases and Current Applications

Digitalization has enhanced many activities in every industry, leading to the emergence of Industry 4.0. This Internet of Things (Figure 18.6) is a crucial component that enhances corporate performance by boosting the efficiency of different activities. It enhances a company's worth by increasing earnings and delivering high-quality service to its clients. The oil and gas business utilizes IoT to manage its demand and supply cost-effectively and proactively. Advanced data analytics, cloud computing, and IoT have greatly enhanced business performance in the oil and gas sector by emphasizing critical factors like decision-making, revenue generation, and operational procedures. Data is collected extensively to improve these factors using SCADA systems, Internet of Things systems, drones, and satellites (Rishabhsoft, 2023).

IoT sensors are not just revolutionizing the oil and gas industry but also enhancing safety protocols. These sensors, including temperature, proximity, pressure, humidity, electro-optical, level, and gas, are vigilant guardians of safety. Temperature sensors keep track of equipment temperatures and alert operators when they exceed a limit, ensuring equipment safety. They also monitor pressure and humidity levels, maintaining optimal conditions. Pressure sensors measure pressure levels and generate signals based on oil or gas pressure, preventing potential hazards. Humidity sensors detect air temperature and atmospheric elements, while electro-optical sensors convert light into signals, enhancing safety in various conditions. Level sensors assess temperature, vibration, and electromagnetic interference; gas sensors detect the presence of gases, leaks, and potential explosions, further bolstering safety protocols (Sharma & Villányi, 2024).

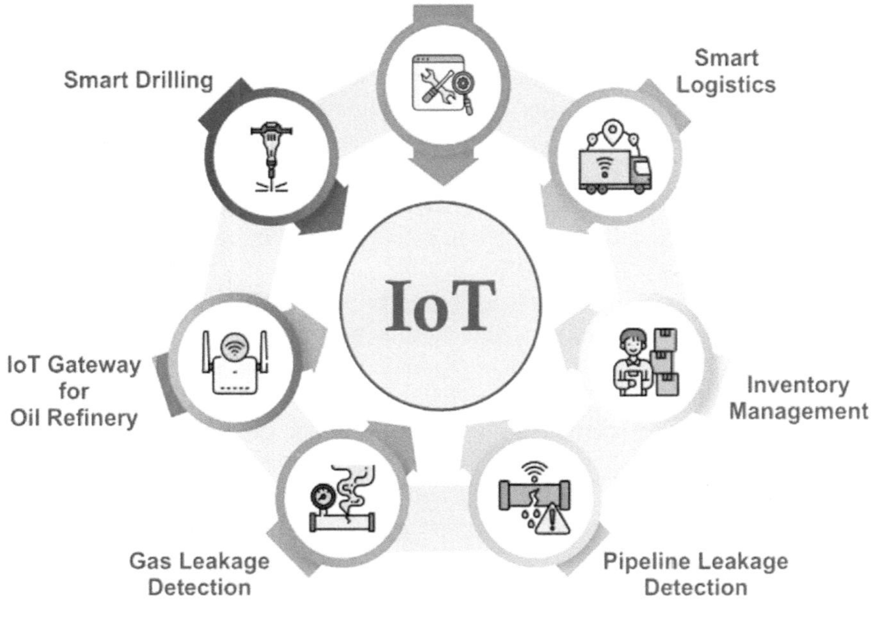

FIGURE 18.6 IoT application in the oil and gas industry.

Source: Rishabhsoft (2023).

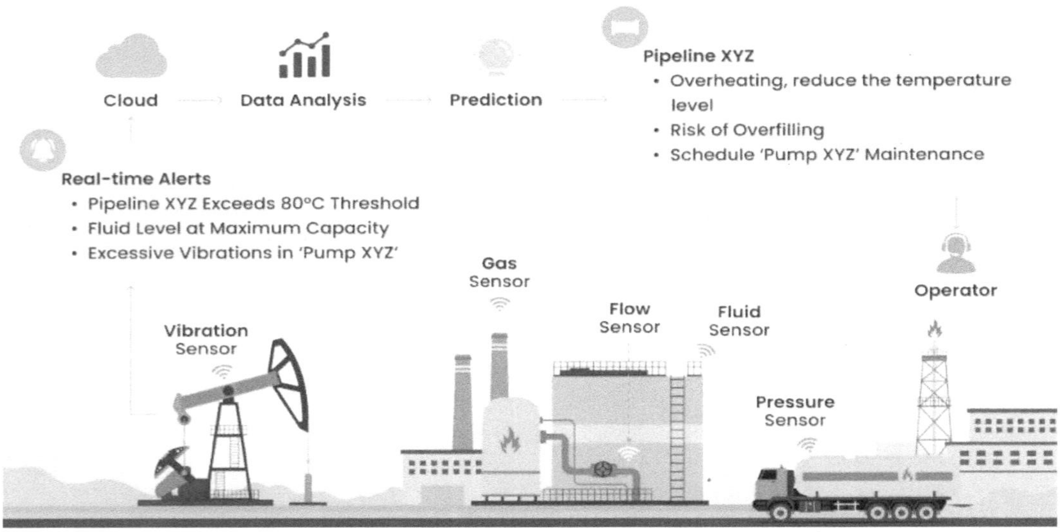

FIGURE 18.7 IoT in predictive maintenance for the oil and gas industry.

Source: Nilay (2021).

Industrial Internet of Things (IIoT) (Figure 18.7) technologies are not just improving but revolutionizing seismic interpretation by enhancing data acquisition, processing, analysis, and decision-making processes in the oil and gas industry. These technologies, including real-time data acquisition, remote monitoring and control of seismic equipment, data integration and fusion, edge computing for real-time analysis, predictive maintenance, and enhanced collaboration and visualization, pave the way for a more efficient and productive industry. IIoT sensors, by collecting data like ground vibrations, temperature, humidity, and atmospheric pressure, enable continuous monitoring of environmental conditions, further enhancing the efficiency of operations. Remote monitoring allows operators to track and adjust parameters, optimizing data collection processes. IIoT devices also enable real-time seismic data analysis, detecting anomalies and identifying events without relying on centralized processing resources (Naghneh et al., 2024).

IIoT is not just a technology but a game changer that can revolutionize oil well drilling processes. By collecting real-time seismic data, IoT devices empower the decision-making process in drilling and extraction operations. The ability to monitor drilling parameters in real time is a significant step toward improving efficiency and preventing equipment issues, giving operators a greater sense of control and confidence (Temer & Pehl, 2017).

It offers real-time monitoring and control, predictive maintenance, automated drilling systems, remote operations, data analytics, safety and environmental monitoring, and integration with digital twins. These technologies allow operators to monitor drilling parameters, detect early signs of equipment wear, schedule repairs, and optimize drilling trajectories. They also enable data analytics and decision support, enabling operators to make informed decisions about wellbore design, drilling strategies, and reservoir characterization. Additionally, IoT sensors can monitor environmental parameters, enabling early detection of hazards and compliance with regulatory requirements. By leveraging IIoT technologies, operators can enhance efficiency, reduce costs, improve safety, and maximize productivity (Kamdjou et al., 2024).

IIoT is an essential tool in reservoir management, enabling real-time monitoring, data analysis, and predictive maintenance. It allows remote monitoring and control of water levels, temperature, pressure, flow rates, and quality. Sensors in oil and gas wells offer real-time data on reservoir dynamics, fluid characteristics, and production rates. This helps operators monitor reservoir behavior, improve production tactics, and maximize hydrocarbon recovery. Operators can enhance production and reservoir management by combining sensor data with reservoir simulation models and predictive analytics, enabling data-driven decisions to increase oil and gas reserves (Allahloh & Mohammad, 2018).

The Industrial Internet of Things (IIoT) is essential for optimizing pipelines through its ability to offer real-time monitoring and control features. IIoT allows for remote monitoring of parameters, including temperature, pressure, flow rate, leakage, and corrosion levels, using sensors throughout the pipeline network. Drones are also used for leakage inspection on pipeline corridors, speeding up maintenance and improving reliability significantly. Operators can use this data to identify abnormalities, forecast probable malfunctions, and enhance operations to achieve better efficiency and safety. IIoT enables predictive maintenance, asset tracking, and integration with other systems to improve decision-making in pipeline management. IoT solutions may examine numerous data points simultaneously, guaranteeing pipeline integrity and reducing environmental effects (Dickerson & Worthen, 2024).

Predictive maintenance plays a role across various industries, including oil and gas, utilities, manufacturing, and transportation. In *predictive maintenance*, IIoT sensors gather real-time data on equipment health, performance, and environmental variables. This data enables operators to anticipate equipment failures, plan maintenance in advance, and reduce unexpected downtime. Organizations can use predictive analytics algorithms to analyze sensor data, enabling them to identify maintenance needs, diagnose problems from a distance, and enhance asset performance to boost dependability and lower operational expenses (Nilay, 2021, Abdullahi et al., 2024). IoT technology transforms oil and gas logistics and distribution through real-time fleet and asset tracking, remote monitoring in storage facilities, and smart billing. It enhances route efficiency and asset usage and minimizes fuel consumption while prioritizing safety and security. IoT sensors offer real-time data to streamline billing procedures (Naghneh et al., 2024). IIoT allows remote monitoring of assets, equipment, and infrastructure in oil and gas fields, enabling operators to analyze performance, identify abnormalities, and optimize maintenance plans. Installed sensors on machinery like pumps, compressors, and pipes send live data to central platforms, allowing for predictive maintenance, minimizing downtime, and prolonging the life of assets (Khan et al., 2024).

IIoT sensors are utilized for environmental monitoring and compliance in oil and gas operations to track air quality, water quality, and emissions levels, ensuring adherence to regulations and reducing environmental impact. Real-time environmental monitoring systems using IIoT sensors gather and analyze pollutant levels, leak detection, and environmental performance data. This allows operators to address issues, reduce risks, and show environmental responsibility (Sharma et al., 2024).

18.3 CLOUD COMPUTING IN THE ENERGY SECTOR

The industry is a sophisticated, data-driven platform with business-, computing-, and data-intensive features. The requirement for more storage capacity grows along with the volume of data. Big data and cloud computing have many potential uses in industrial settings. With the help of cloud computing, users can construct infrastructure, link services, upload data, and carry out analyses on demand (Zhifeng, 2019). Cloud computing is an integrated solution that combines grid computing, distributed computing, network storage, virtualization, load balancing, and other classic computer technologies (Hussain et al., 2022). The primary aspect of cloud computing is the distribution of services to many dispersed computers. Cloud computing utilizes an enterprise data center to facilitate resource migration to meet application needs and grant access to all storage devices. Cloud computing offers essentially customizable storage with elastically extensible and reducible size, which helps address the volume challenge of big data by providing unlimited storage support (Guo et al., 2020). Google was the pioneer in introducing cloud computing services. IBM introduced the "BlueCloud platform schedule" in 2007. Microsoft developed

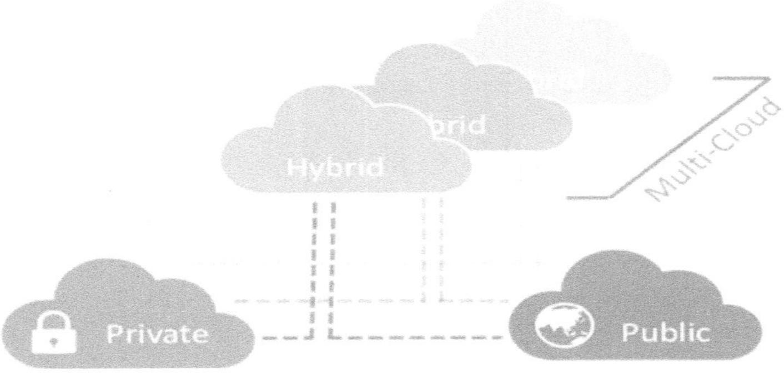

FIGURE 18.8 Public, private, hybrid, and multi-cloud deployment models.

Source: Hussain et al. (2022).

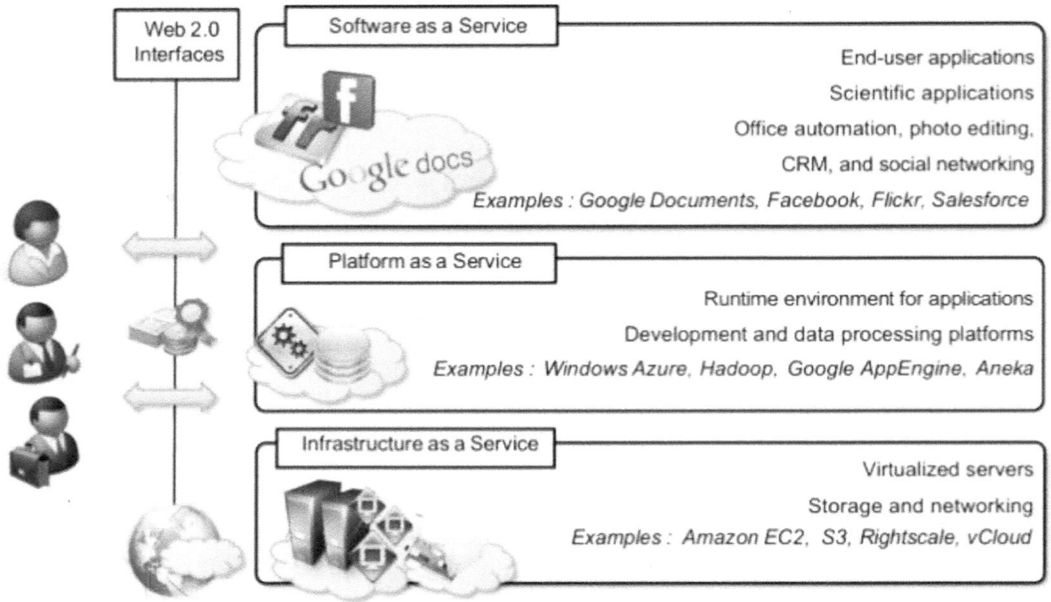

FIGURE 18.9 Cloud computing delivery models.

Source: Kwek et al. (2019).

Windows Live online services (Zhifeng, 2019). There are four main forms of cloud architecture (shown in Figure 18.8): public cloud, private cloud, hybrid cloud, and multi-cloud (Mohammad & Rajeswari, 2019; Zhifeng, 2019; Hussain et al., 2022).

As shown in Figure 18.9, cloud computing provides various services through three layers: software as a service (SAAS), platform as a service (PAAS), and infrastructure as a service (IAAS). These firms offer a variety of services to customers at a reasonable cost. *Software as a service* (SaaS) is a program intermediating the cloud and the user. *Platform as a service* (PAAS) provides tools and resources for building software applications. *Infrastructure as a service* (IAAS) provides customers hardware components, such as CPU, monitor, and network tools, but does not offer services beyond that (Hussain et al., 2022; Mohammad & Rajeswari, 2019; Zhifeng, 2019).

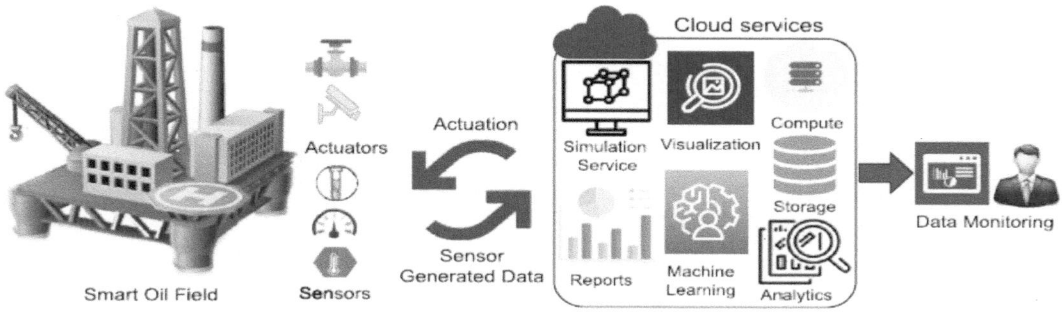

FIGURE 18.10 Diverse cloud services can be utilized to store, process, and analyze sensor-generated data and manage industrial equipment in a smart oil and gas sector.

Source: Sackschewsky et al. (2018); Hussain et al. (2022).

Cloud computing provides various benefits, including scalability, on-demand self-service, resource pooling, elasticity, pay-per-use billing, broad network access, multi-tenancy, high availability and reliability, managed services, security, and disaster recovery capabilities (Hussain et al., 2022). These features allow organizations to focus on core business objectives while leveraging cloud computing's agility, flexibility, and cost-effectiveness. Users can provision computing resources without human intervention, and providers can automatically adjust resource allocation to match workload demands. Pay-per-use billing allows users to pay for resources they consume, while broad network access allows flexibility and mobility.

The smart oil and gas business relies increasingly on cloud-based services housed in remote internet servers, referred to as cloud data centers, as depicted in Figure 18.10. Data centers are facilities designed for the storage and processing of data. Sensor-generated data is transferred to cloud providers to access various cloud services (Hussain et al., 2022). Some services provide intelligent instructions to actuators to complete the automation control loop in the smart oil field. Cloud technology enables O&G organizations to utilize data-related and computational services like machine learning and visualization without overseeing any computing infrastructure. Firms are concerned about data privacy and security when contemplating fully adopting cloud services. Security concerns have caused a minor delay in adopting cloud services, particularly among important business stakeholders. A more dependable approach involves storing the data in an on-premises computer facility, sometimes called a private cloud or fog computing (Sackschewsky et al., 2018; Hussain et al., 2022).

18.3.1 Cases and Current Applications

Big data is a term that refers to vast amounts of data that are characterized by their volume, variety, and velocity. This data can be structured, semi-structured, or unstructured, requiring advanced tools and techniques to manage, process, and analyze (Berisha et al., 2022; Gaurav, 2021). Big data analytics aim to extract valuable insights and knowledge from this data, which can help organizations make informed decisions, improve their performance, and gain a competitive advantage. Big data sources are diverse and can include user interactions with websites or applications, social media activity, IoT sensors, machine logs, transactional data, and more. By leveraging big data analytics, businesses can gain a deeper understanding of their customers, products, and operations and use this knowledge to optimize their strategies and achieve their goals. Figure 18.11 shows the flow in the processing of big data (Berisha et al., 2022).

Big data and cloud computing are two prominent technologies that are currently the focus of attention in the IT industry and are mainstream in many sectors of our modern economy. Big data

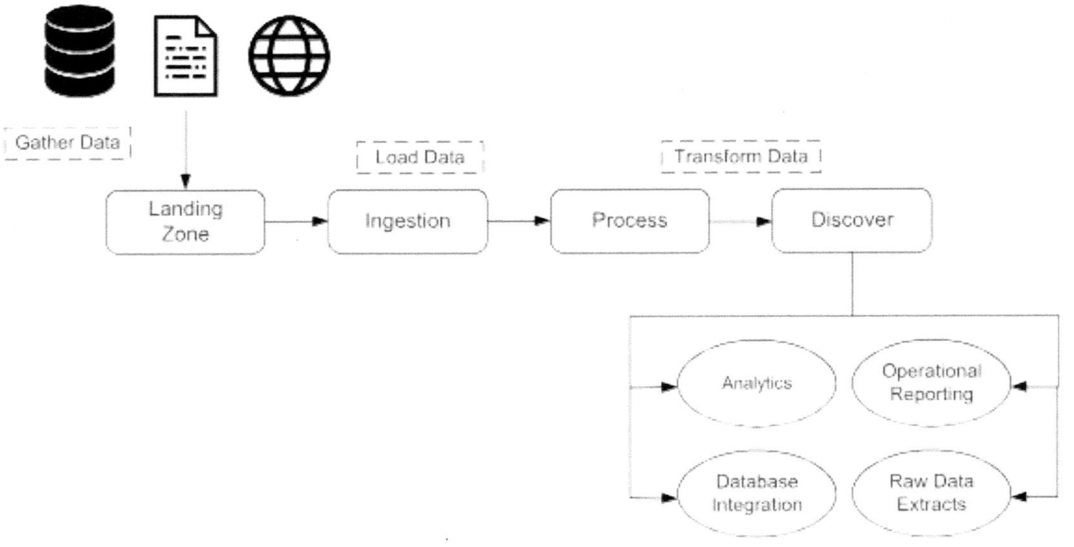

FIGURE 18.11 Flow in the processing of big data.

Source: Berisha et al. (2022).

represents content, whereas cloud computing offers on-demand storage, computation resources, and tools to store and analyze big data (Gaurav, 2021; Yang et al., 2017). The two technologies can work together to produce outstanding outcomes, but they are different. Big data and cloud computing have revolutionized businesses and industries, enhancing efficiency, data-driven decision-making, and customer experiences across multiple sectors, demonstrating their transformative potential, including different industries, such as health care, finance, education, transportation and logistics, energy, telecommunications, and government.

In health-care applications, patient records, medical pictures, and genomic data are being analyzed using big data analytics, changing health care by improving diagnosis, treatment, and personalized medicine. Thanks to cloud computing, huge volumes of medical data can be securely stored and processed, and doctors and other medical staff can access this data from anywhere in the world (Angurala et al., 2024).

Financial institutions use big data analytics to detect fraud, manage risk, and make investment decisions based on data. Cloud computing allows them to scale their infrastructure as needed and access powerful computing resources for complex analytics, quickly analyzing vast amounts of data and making decisions (Khan et al., 2024).

Retailers use big data to analyze customer behavior, optimize pricing and promotions, and manage inventory. Cloud computing enables them to store and process large volumes of sales data, run predictive analytics, and deploy e-commerce platforms with high scalability and reliability, leading to better customer experience (Chhotaray et al., 2024).

In the education sector, big data analytics and cloud computing can track student performance, personalize learning experiences, improve academic outcomes, and enhance administrative processes through data-driven decision-making. This has resulted in improved student outcomes and increased efficiency in educational institutions (Rajamohan et al., 2024).

Big data optimizes supply chain operations, route planning, and fleet management. Cloud computing enables transportation companies to leverage data from GPS devices, RFID tags, and other sources to improve efficiency and reduce costs, leading to optimized logistics and increased profitability (Kayvanfar et al., 2024).

Telecom companies use big data analytics to analyze network performance, predict customer churn, and personalize marketing campaigns. Cloud computing allows them to store and process large amounts of detailed call records and other cost-effective data, leading to improved customer retention and increased revenue (Abdugafur & Rustamovich, 2024).

Energy companies utilize big data and cloud computing to optimize power generation, predict equipment failures, and manage energy consumption. This leads to reduced costs and increased environmental sustainability. The oil and gas industry uses big data for exploration, production, supply chain management, safety monitoring, cost-saving opportunities, and collaboration among geographically dispersed teams (Anwar et al., 2024).

Governments use big data analytics for various purposes, including public safety, health care, urban planning, and disaster response. Cloud computing allows them to share data securely across agencies and leverage advanced analytics tools without heavy up-front investment, leading to improved decision-making and increased efficiency (Rath et al., 2024).

Marketers can leverage big data analysis and cloud computing to understand better customer behavior, segment audiences, optimize advertising campaigns, and measure the effectiveness of marketing strategies through real-time data insights. This leads to improved targeting and increased return on investment (Rath et al., 2024).

The oil and gas industry is at the forefront of leveraging cloud computing technology. It uses cloud storage and integration of hardware and software resources to meet informatization requirements, enabling unified data management across various departments. The industry generates massive data from hydrocarbon reservoirs, complex fluids, and environmental risks (Hussain et al., 2022). Big data analytics streamlines essential processes in upstream, midstream, and downstream sectors, including exploration, drilling, production, and delivery. Research and desktop cloud applications are improving work efficiency, collaboration, and remote visualization. Key applications include remote monitoring, data analytics, predictive maintenance, reservoir simulation, digital oil field solutions, supply chain management, environmental monitoring, compliance, and safety and risk management (Oguanobi & Joel, 2024). These applications are discussed in what follows.

One of the most significant contributions of cloud computing to the oil and gas industry is data management and storage. Cloud computing platforms offer scalable storage solutions because of the vast amounts of data generated from exploration, production, drilling, and field operations. These solutions empower operators to efficiently store, manage, and access seismic data, well logs, production data, and other critical information. By leveraging cloud storage services, operators can ensure secure data storage, reduce infrastructure costs, and streamline data management processes (Buyya et al., 2024).

Another application is remote monitoring and control. Cloud-based systems are not just about enhancing efficiency and productivity in remote oil and gas operations, such as drilling rigs, production platforms, and pipelines; they also play a crucial role in ensuring safety and managing risks. By leveraging cloud technologies, companies can collect, analyze, and visualize operational data from geographically dispersed assets, enhancing safety and reducing risks. Cloud platforms offer secure document sharing, project management tools, videoconferencing capabilities, and real-time collaboration features that enable stakeholders to collaborate on projects, share information, and coordinate activities across the value chain, improving teamwork and decision-making. This emphasis on safety and risk management is a key aspect of cloud computing in the energy sector (Li et al., 2024).

Cloud computing empowers oil and gas companies to leverage advanced data analytics and machine learning algorithms to predict equipment failures, optimize maintenance schedules, and reduce downtime. By analyzing historical and real-time data from sensors and equipment, companies can proactively address maintenance needs, enhance asset reliability, and minimize operational disruptions (Ebirim et al., 2024).

Moreover, cloud-based reservoir simulation and modeling tools play a critical role in the oil and gas industry. They support exploration and production activities by simulating fluid flow, predicting reservoir behavior, and optimizing production strategies. These tools utilize a cloud infrastructure, which provides several benefits to companies, including the ability to run complex simulations, evaluate multiple scenarios, and make informed decisions to maximize hydrocarbon recovery and reservoir performance (Wang et al., 2024).

The cloud-based reservoir simulation and modeling tools offer a range of features, such as advanced visualization, data management, and collaboration capabilities. The advanced visualization feature helps users better understand the reservoir behavior by providing 3D visualization of the data, which helps identify potential drilling locations and optimize production strategies. The data management feature ensures data integrity, security, and accessibility, while the collaboration capability enables team members to work together seamlessly, regardless of location (Mustapha, 2024).

Cloud-based reservoir simulation and modeling tools have revolutionized the oil and gas industry by enabling companies to optimize their production strategies and improve their reservoir performance. These tools allow companies to minimize risks, reduce costs, and increase efficiency, ultimately leading to higher profitability (Sidqi et al., 2024).

In addition to another application, the adoption of cloud computing technology has significantly improved the oil and gas industry. With the implementation of digital oil field solutions, data from various sources, such as sensors, SCADA systems, and historical records, can be integrated. This integration helps optimize oil and gas operations by providing real-time data visualization, ensuring collaborative decision-making, and automating workflows (Ershaghi et al., 2024).

Digital oil field platforms have transformed operations, enabling oil and gas companies to reduce operational costs and enhance production efficiency. With real-time data visualization, decision-makers can make informed decisions in real time, and automation of workflow processes ensures that operations are carried out more efficiently (Haouel & Nemeslaki, 2024).

Another application is digital twin modeling, and cloud computing enables developing and deploying digital twin models for oil and gas assets, such as wells, pipelines, refineries, and platforms. *Digital twins* are virtual replicas of physical assets that integrate real-time sensor data, historical performance data, and physics-based models to simulate asset behavior, predict potential issues, and optimize asset performance. Cloud-based digital twin platforms enable operators to monitor asset health, simulate scenarios, and optimize maintenance strategies to prolong asset lifespan and maximize production efficiency (Hashmi et al., 2024).

Cloud computing providers offer robust cybersecurity measures and data protection protocols to safeguard the oil and gas industry's sensitive data, intellectual property, and operational information. Cloud platforms implement multi-layered security controls, encryption mechanisms, access management policies, and compliance certifications to protect data from cyber threats, unauthorized access, and data breaches, ensuring data confidentiality, integrity, and availability for oil and gas operators (Rekeraho et al., 2024).

In addition, cloud-based supply chain management systems streamline the oil and gas industry's procurement, logistics, inventory management, and vendor relationships. By leveraging cloud technologies, companies can optimize supply chain operations, reduce long lead times, improve visibility into the sourcing and distribution of materials, and enhance supplier collaboration to ensure timely and cost-effective delivery of equipment and services (Fraihat et al., 2024).

Moreover, cloud computing supports environmental monitoring and compliance in the oil and gas sector by enabling companies to track and report on environmental performance metrics, emissions data, and compliance with regulatory standards. Cloud-based environmental management systems help oil and gas companies minimize environmental risks, enhance sustainability efforts, and ensure adherence to environmental regulations (Tirlangi et al., 2024).

Finally, cloud-based safety and risk management solutions assist oil and gas companies in proactively identifying, assessing, and mitigating operational risks and safety hazards. These solutions centralize safety data, conduct risk assessments, and implement safety protocols through the cloud (Jin, 2024).

18.4 DATA ANALYTICS IN THE ENERGY SECTOR

Data analytics is a field that uses a mix of techniques, like math, statistics, and computer science, to uncover insights from datasets (Coursera, 2023). It involves tasks like analyzing data, designing data collection methods, and setting up storage systems. Data analytics helps businesses gain understanding on and improve their processes and services (Verma & Marchette, 2020). It offers insights into customer experiences and challenges. By going beyond data and linking insights to action, companies can create customer experiences, develop relevant digital products, streamline operations, and enhance employee efficiency. Big data analytics involves identifying patterns, trends, and connections in datasets. These sophisticated analyses require tools, technologies, computational resources, and scalable data storage solutions (Mathur et al., 2024).

Inside data analytics, machine learning, a subset of artificial intelligence (AI), aims to create algorithms to analyze data and automate tasks without instructions. It has decision-making, forecasting, and prediction applications in health care, business, and finance. *Machine learning* is broadly classified into learning, where models are trained on labeled data, and *unsupervised learning*, which involves identifying patterns in data. These different types of machine learning play a role in building systems capable of making informed decisions based on complex datasets (Bellodi et al., 2022; Kumar & Prabhu, 2020).

Data analytics encompasses a sequence of steps to deliver precise analysis. Data analysts collaborate with scientists and engineers to develop data pipelines and establish models. Data collection is the initial stage, where data is identified and gathered from various source systems or extracted from specific subsets. The second stage entails data quality adjustment, which includes identifying and rectifying data quality issues, preparing data for the analytical model based on business requirements, and enforcing data governance policies. The next step involves constructing an analytical model, testing it with an experimental dataset, and finally, executing it against the required dataset in production mode. The last stage involves showcasing the outcomes of the models to end users and company executives. The big data analytics process flowchart is shown in Figure 18.12 (Aksahan, 2021).

There are five types of data analytics: descriptive, diagnostic, predictive, prescriptive, real-time, and augmented. *Descriptive* data analytics examines past data to explain what happened, using statistical analysis techniques to compare past results and identify anomalies. *Diagnostic* data analytics aims to answer why an anomaly happened using techniques like drill-down, data discovery, data

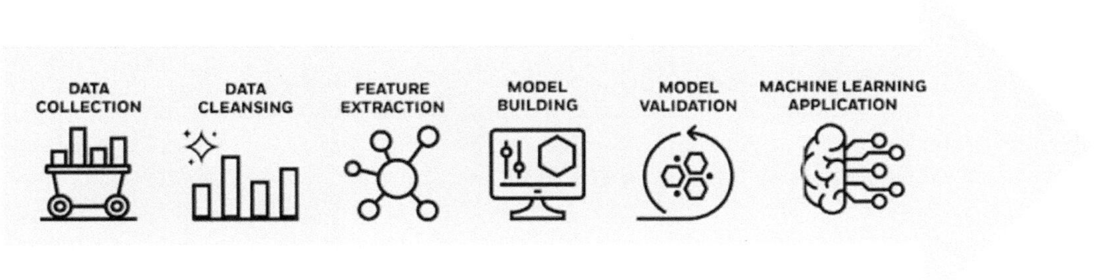

FIGURE 18.12 Big data analytics process flow diagram.

Source: Aksahan (2021).

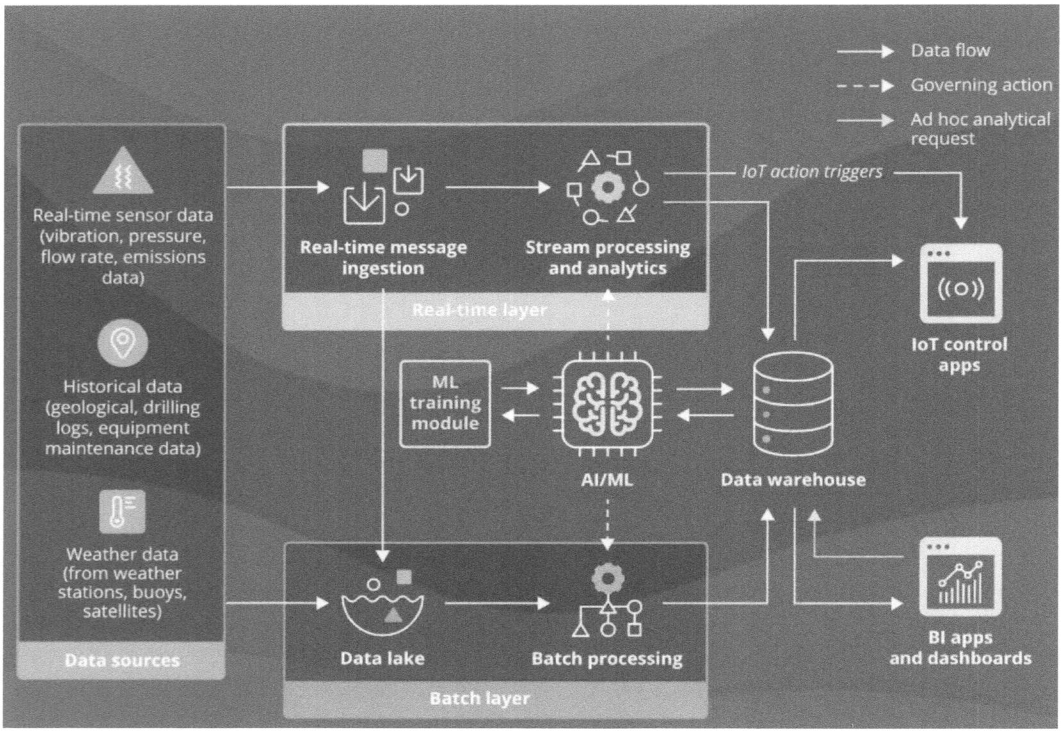

FIGURE 18.13 Big data software architecture: Big data software for the oil and gas company analyzes multi-source data in real time and in batches, following a specific computation schedule. The real-time component responds to events in real time, while the batch layer provides complete historical analytics. Users can get insights and alerts through business intelligence apps and dashboards. The data warehouse holds highly structured data that can be analyzed and queried by the BI layer.

Source: ScienceSoft (2022).

mining, and correlations. *Predictive* data analytics use current or historical data to predict future actions, using algorithms and methodologies like a linear or logistic regression model. *Prescriptive* data analytics selects the best solution for a problem from available options, using tools like recommendation engines and loan approval engines to examine decision options and personalize the process. *Real-time* data analytics use data immediately when entered in the database, analyzing new data from customers or external sources. *Augmented* data analytics uses machine language (ML) and natural language processing (NLP) to analyze data, automate code-based data exploration, and reduce errors. Most analytics software integrates augmented data analytics into leverage machine language and other features. Understanding each type of data analytics is essential for companies to make informed decisions and improve their overall performance (Rao et al., 2024).

The relationship between big data, cloud computing, data analytics, and machine learning is synergistic and crucial for driving business value and innovation. *Big data* provides large datasets for analysis. *Cloud computing* offers the essential infrastructure and resources to efficiently store, process, and analyze big data on a huge scale. *Data analytics* utilizes methods and technologies to extract practical insights from big data, often utilizing machine learning algorithms to automate analysis and decision-making. These technologies foster innovation and help organizations extract value from their data assets, improving decision-making, products, services, and customer experiences (Ochuba et al., 2024).

The oil and gas sector has grown data-intensive due to the introduction of data recording devices in exploration, drilling, and production processes. Big data in the oil and gas industry is utilized to analyze seismic and micro-seismic data, enhance reservoir characterization and simulation, decrease drilling time and safety, optimize production pump performance, improve petrochemical asset management, enhance shipping and transportation, and enhance occupational safety. The oil and gas industry has shown increased interest in employing big data analytics (Figure 18.13). The influence of data analytics is significant, ranging from enhancing operational efficiency to enabling improved decision-making (ScienceSoft, 2022).

18.4.1 Current Applications

In today's moving world, data analytics has become a tool for businesses of all sizes to gain valuable insight into their operations and make well-informed decisions. Whether it is the oil and gas industry aiming to optimize production, health-care providers striving to enhance patient outcomes, financial institutions working on risk management, or retailers focusing on improving customer experiences, data analytics is widely used across sectors. Common applications include predicting equipment failures, recognizing disease trends, uncovering activities, analyzing consumer behavior, and enhancing supply chain efficiency. With analytics tools and methods, companies can harness the power of their data and maintain a competitive edge in today's data-centric environment (Konya & Nematzadeh, 2024).

Data analytics holds immense importance in the oil and gas energy sector because it can extract valuable insights from vast amounts of data generated throughout the exploration, production, refining, and distribution processes. The recent technological advances have caused oil and gas exploration and production industries to create massive daily datasets. Oil and gas data analytics algorithms are crucial in various sectors, including upstream, midstream, and downstream exploration, drilling, production, and delivery. Upstream analytics gather seismic data, while customized predictive models detect potential equipment issues. Downhole sensors such as temperature, sound, and pressure increase reservoir output. Oil and gas data analytics improve reservoir modeling and management through production data analysis (Nguyen et al., 2024).

Seismic data interpretation of data analytics. Interpreting seismic reflection data typically requires high-performance multi-core computers, sophisticated visualization methods, and the creation of various seismic data formats and characteristics (Roden, 2016). Data analytics approaches assist in recognizing patterns, trends, and anomalies in seismic data, which helps detect underlying structures and geological characteristics. Algorithms quantitatively assess seismic features to precisely characterize reservoir qualities, including porosity, permeability, and fluid saturation (Roden, 2016).

Machine learning algorithms may be trained to identify seismic facies, fault patterns, and lithology types, helping geoscientists with interpretation duties. This involves utilizing computer algorithms that iteratively learn from the data and autonomously adjust to generate dependable, consistent outcomes. Principal component analysis (PCA) and self-organizing maps (SOM) are unsupervised machine learning techniques used to analyze large datasets and enhance geological interpretation (Roden, 2016).

Data analytics technologies combine seismic data with additional geographical and geological variables to provide insight into underlying formations and reservoir characteristics. By examining past seismic data and well logs, predictive models can anticipate reservoir activity, optimize drilling locations, and reduce exploration risks. Sophisticated image processing methods improve seismic images to provide clearer visualization and comprehension of underground structures, faults, and fluid flow paths. Data analytics simplifies seismic interpretation by automating routine processes, enabling geoscientists to concentrate on more advanced research and decision-making (see Figure 18.14). (Roden, 2016).

Applications for data analytics in drilling management. Drilling for oil and gas involves a wide range of equipment and sensors (Khare, 2022). It is crucial for monitoring and controlling drilling operations. These sensors are installed by mud logging, logging while drilling (LWD), and

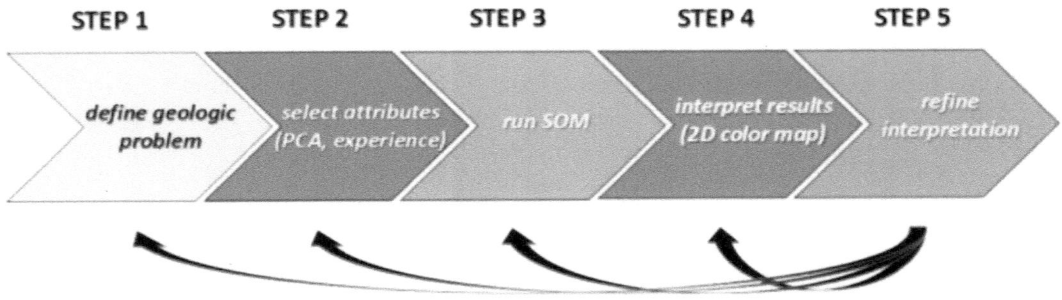

FIGURE 18.14 Interpretation steps employing machine learning (PCA and SOM).

Source: Roden (2016).

measurement while drilling (MWD) by service companies (Li et al., 2011). The information gathered from these sensors is then analyzed by a big data solution, which provides real-time insights into the drilling process (Mohammadpoor & Torabi, 2020; Abdullah et al., 2023).

The big data software also collects operational data, such as the rotation speed of the drilling bit, drilling fluid temperature, and equipment metadata, like model and operational settings. This information is then used to build accurate machine learning (ML) and artificial intelligence(AI) models which help generate alerts on abnormal events and identify failure-causing equipment usage patterns. This allows oil and gas companies to minimize non-productive time, optimize inventory management processes and maintenance schedules, extend equipment lifespan, and achieve significant operational efficiencies (Mohammadpoor & Torabi, 2020; Abdullah et al., 2023).

The conditions on a drilling rig can be quite harsh, with heavy lifting, vibrations, and harsh weather conditions causing significant wear and tear on the electronic sensors. As a result, it can be challenging to keep these sensors in optimum condition and maintain the accuracy and quality of the data they generate. Sensor failures are common on rigs, leading to downtime and non-productive time (NPT) in drilling operations (Mohammadpoor & Torabi, 2020; Abdullah et al., 2023). Furthermore, the excess vibrations and noise associated with drilling operations make sensors installed on the rig floor, mast, pumps, and mud pit particularly prone to fatigue, wear, and tear. This can result in obtaining inaccurate data, which can have serious consequences for drilling operations. Therefore, monitoring and replacing the sensors are essential to ensure that drilling operations proceed smoothly and efficiently.

For data analytics in optimization log interpretation, the art of well log interpretation is a constantly evolving field that requires creativity and the integration of modern technologies and techniques. Data analytics improves well log interpretation within the oil and gas sector. It aids in integrating and refining data from multiple sources, identifying inconsistencies and errors, and enhancing reservoir characterization, drilling operations, and production performance. Advanced analytic techniques, such as machine learning (ML) and artificial intelligence (AI), can detect patterns, trends, and anomalies in data, streamlining the interpretation process. Predictive modeling techniques, including regression analysis and neural networks, can estimate reservoir parameters and forecast fluid behavior (Wong et al., 2003). Data visualization tools and interactive dashboards can help understand complex geological systems. Integrated data analytics systems can integrate well log data with geological models and reservoir simulations, optimizing well placement, production plans, and reservoir management practices (Dorrington & Link, 2004; Hsieh et al., 2005; Ilkhchi et al., 2006; Deng et al., 2023; Odimarha, A. C., et al., 2024).

Last but not the least, in AI technology, computer vision is a field of artificial intelligence and computer science that enables computers to interpret and understand the visual world. It involves developing algorithms and techniques that allow machines to analyze, process, and extract information from images or videos. This technology enables applications such as object detection, image classification, facial recognition, and autonomous navigation (Gonzalez, 2024, April).

18.5 AUTOMATION IN THE ENERGY SECTOR

Automation uses technologies to manufacture and provide goods and services with little human involvement. Implementing automation technologies enhances the efficiency, dependability, and speed of jobs traditionally done by humans (Kanade, 2024; Rouse, 2023). Automation is a vital component of contemporary workplaces, employing control systems and machinery such as computers and robots to carry out activities that were formerly done by hand (Concept, 2024; Rouse, 2023).

Today's industry expects high-quality products that are consistent and competitively priced due to increased competitiveness. Industries are exploring novel product designs and integrated manufacturing procedures alongside the usage of automated technologies to tackle this challenge (Concept, 2024; Fogg, 2020). Industrial automation is a significant and prominent strategy for addressing the aforementioned difficulty. By implementing new, innovative, and integrated technologies and services, industrial automation enhances product quality, reliability, and production rate while decreasing manufacturing and design costs (Concept, 2024; Fogg, 2020).

Automation in the oil and gas (O&G) industry is a growing trend that uses digital technologies to improve energy production and global market competitiveness. It emerged as a powerful catalyst for transformation within the industry. Common automation O&G includes drilling automation, production optimization, asset management, safety systems, remote operations, data analytics, robotics, supply chain management, and environmental monitoring. These systems use sensors, algorithms, and robotics to optimize drilling processes, monitor production in real time, and ensure early detection of potential hazards (Abdalla, 2024). Automation also helps companies adapt to changing market conditions and regulatory requirements, reducing costs and improving safety (Abdalla, 2024).

18.5.1 Cases and Examples of Automation

Automation in the oil and gas industry is revolutionizing various aspects, including pipeline inspection drones, autonomous underwater vehicles (AUVs), robotic drilling systems, robotic maintenance and repair, automated tank gauging systems, remote monitoring and control, and drone-based environmental monitoring. These technologies provide high-resolution imagery, autonomous navigation, leak detection, structural monitoring, and real-time parameter adjustments. They also improve accuracy and efficiency in monitoring crude oil levels in storage tanks, reduce manual measurements, and improve safety and production. Integrating automation, robots, and drones transforms the industry from exploration to maintenance and environmental monitoring (Ershaghi et al., 2024).

One area of using automation is crucial in maintaining pipeline integrity in the oil and gas industry. It allows continuous monitoring, leak detection, re-routing, corrosion monitoring, pipeline inspection, data analysis, and remote monitoring and control. Automation uses sensors to detect changes in pressure, temperature, flow rate, and other parameters, allowing operators to identify potential issues promptly. It also helps predict maintenance needs and prevents costly downtime. Inspection tools like drones and robotic crawlers capture high-resolution images and data, allowing operators to assess pipeline conditions without disrupting operations. Automation also enables operators to monitor and control pipeline operations remotely, ensuring safety and compliance with regulatory requirements (Arinze et al., 2024).

Another area of automation is increasingly being used in the oil and gas industry for transportation logistics and work order generation. These systems optimize route planning, scheduling, and vehicle allocation, while real-time tracking and dispatching systems assign tasks based on proximity, capacity, and urgency. Automation platforms integrate transportation logistics with other supply chain processes, ensuring better coordination and optimization. Work order generation is automated through asset management systems which track equipment and infrastructure performance and use predictive maintenance algorithms to identify potential issues. Integration with ERP systems streamlines workflow processes and communication between departments, while mobile workforce management allows field technicians to receive and manage work orders directly on mobile devices. This approach optimizes resource utilization, reduces operational costs, and enhances efficiency and productivity.

Moreover, procurement automation in the oil and gas industry involves using technology to streamline and optimize the procurement process for goods and services. Automation platforms include electronic procurement systems, supplier relationship management software, automated purchase requisition approval workflows, strategic sourcing tools, inventory management and replenishment automation, integration with enterprise resource planning (ERP) systems, contract life cycle management (CLM) automation, and data analytics and reporting tools. These tools digitize and centralize the procurement process, reduce manual paperwork, and improve compliance. They also enable the creation of predefined approval workflows for purchase requisitions, support strategic sourcing initiatives, monitor inventory levels in real time, integrate with ERP systems, and analyze procurement data for cost savings and supplier optimization (Gallegos et al., 2024).

The latest automation technology in the oil and gas industry is drone inspection, which offers improved safety, cost-effectiveness, and efficiency. High-resolution cameras and sensors can inspect tanks for corrosion, leaks, or structural damage, providing detailed imagery and data. Thermal imaging cameras can detect temperature variations, identifying potential leaks or insulation problems. Drones can access hard-to-reach areas of tanks quickly and safely, reducing personnel risk and downtime. Flare stack inspections can be done using drones, capturing detailed visual and thermal data without shutting down operations or using costly scaffolding. Drones can also enter confined spaces within vessels, providing visual inspection data and ensuring safety and environmental regulations compliance (Spadoro et al., 2024).

18.6 DISRUPTIVE TECHNOLOGIES IN THE ENERGY SECTOR

The oil and gas industry embraces disruptive technologies to optimize operations, reduce costs, and mitigate environmental impacts. These include digital oil fields, advanced analytics, automation and robotics, enhanced oil recovery, renewable energy integration, digital twins, and blockchain. These technologies enhance efficiency and resource extraction and shift toward a more sustainable and digitally driven future. They enable real-time monitoring, predictive maintenance, remote inspection, and improved safety. Adopting these technologies is transforming the industry and paving the way for a more sustainable future (Ekemezie & Digitemie, 2024).

The oil and gas sector utilizes 3D printing, virtual reality (VR), and augmented reality (AR) to enhance operations across several areas: 3D printing allows for creating personalized parts for drilling equipment as needed, while AR applications offer support for remote maintenance. VR simulations provide immersive training for personnel, while AR superimposes digital information onto actual training surroundings. Digital twins generate digital representations of assets, enabling continuous performance monitoring. Remote inspection decreases safety hazards and operating expenses. Hazard simulators replicate dangerous situations, whereas AR applications offer immediate safety notifications. These technologies improve operational efficiency, safety results, and asset management strategies, leading to cost savings and sustainable growth (Hussain et al., 2024).

Disruptive technology (Figure 18.15) is an innovation that radically alters existing industry practices, leading to the creation of new markets that transform consumer and corporate operations (Green & Newman, 2017; IFC, 2019). These innovations put conventional wisdom to the test and upend established business paradigms, causing profound changes in the sector. Disruptive technologies are innovations that replace a process, a product, or a technology already well-established, giving rise to a new way to operate, be it for consumers, organizations, or both. Discussed in what follows are some examples of disruptive technology (Olajiga et al., 2024).

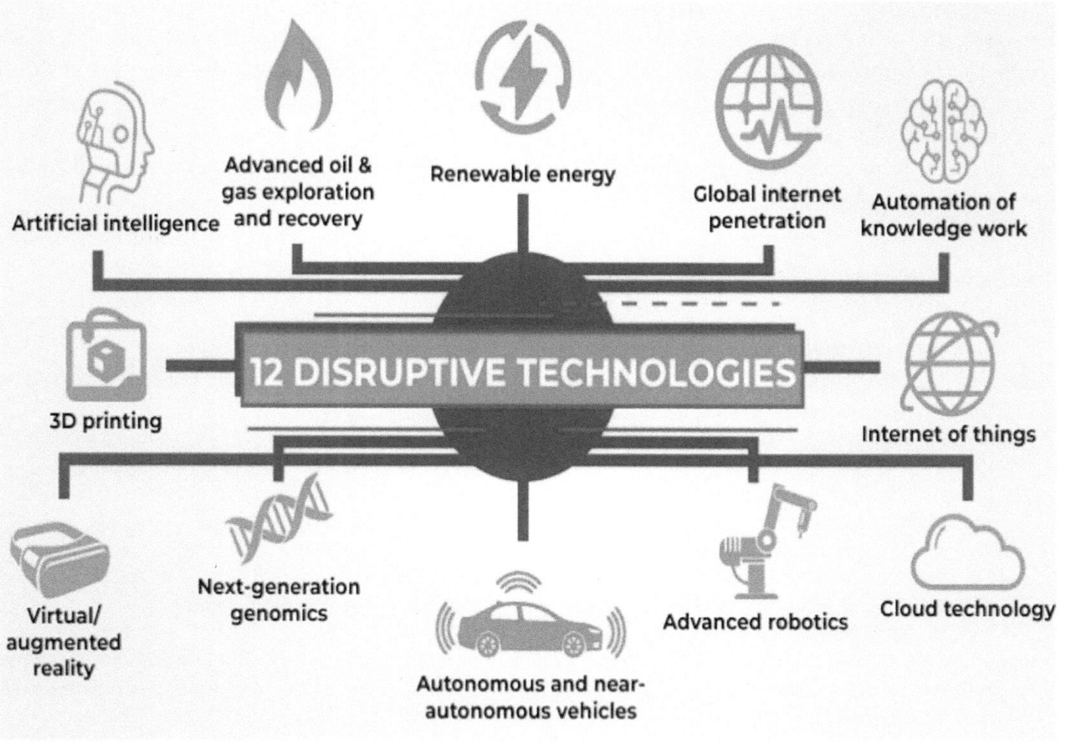

FIGURE 18.15 Disruptive technology source.

Source: The ASEAN Post (2018).

18.6.1 Current Applications

18.6.1.1 Safety Training

Disruptive technologies are revolutionizing safety training in the oil and gas industry by providing immersive and interactive learning experiences. These include virtual reality (VR) simulations which recreate hazardous scenarios, and augmented reality (AR) training modules that overlay digital information onto physical environments. Interactive e-learning platforms use AI and machine learning algorithms to personalize safety training modules, making them more engaging and effective. Remote training and collaboration tools enable remote employees to participate in safety training programs, fostering teamwork and knowledge sharing. Advanced analytics tools analyze training performance metrics to identify trends and improve safety training programs. These technologies help organizations adapt to evolving safety regulations and industry's best practices while minimizing operational disruptions (Oyewole et al., 2024).

18.6.1.2 Immersion Training

Immersion training is a comprehensive learning experience in the oil and gas industry, covering topics such as exploration, production, refining, and distribution processes (Amazon, 2018). It can take various formats, such as on-site visits, classroom lectures, workshops, simulations, and hands-on exercises. These programs can range from a few days to several weeks or months. Participants may include engineers, geologists, technicians, and managers. The benefits of immersion training include accelerated learning, networking opportunities, exposure to real-world challenges, and practical skills. Industry associations, educational institutions, and specialized training providers often organize these programs (Smirnova et al., 2020).

The push for achieving net-zero greenhouse gas emissions by 2050 is spurring technological progress across industries. Within the oil and gas sector, digitalization, IoT, and advanced data analysis are improving the effectiveness and decision-making processes. The energy transition toward energy sources and eco-friendly practices fundamentally transforms how we generate and use energy. These technological advancements and the strategic move toward renewables play a role in meeting the climate goals outlined in global agreements, securing a sustainable and robust energy landscape for the future.

REFERENCES

Abdalla, R. (2024). Transforming the Industry: Digitalization and Automation in Oil and Gas Wells. *Advances in Oil and Gas Well Engineering*. https://doi.org/10.5772/intechopen.112512

Abdugafur, K., & Rustamovich, R. T. (2024). Internet of Things (IoT) and Its Role in Telecommunications. *Genius Repository*, *28*, 59–63.

Abdullah, M. A., Aseel, A., Roy, R., & Sunil, P. (2023). Predictive Big Data Analytics for Drilling Downhole Problems: A Review. *Energy Reports*, *9*, 5863–5876. https://doi.org/10.1016/j.egyr.2023.05.028

Abdullahi, I., Longo, S., & Samie, M. (2024). Towards a Distributed Digital Twin Framework for Predictive Maintenance in Industrial Internet of Things (IIoT). *Sensors*, *24*(8), 2663.

Aksahan, F. (2021). *Industry 4.0 in the Oil and Gas Industry* %7C. www.honeywell.com%7C

Al-Ajwad, M. N. K., Ahmad, R., & Mohammed, F. (2024). Blockchain for Oil and Gas Industry Asset Management: Pioneering Industry 5.0 Solutions. In *The Future of Human-Computer Integration* (pp. 25–38). CRC Press.

Allahloh, A. S., & Mohammad, S. (2018, October). Development of the Intelligent Oil Field with Management and Control Using IIOT (Industrial Internet of Things). In *2018 2nd IEEE International Conference on Power Electronics, Intelligent Control and Energy Systems (ICPEICES)* (pp. 815–820). IEEE.

Almessabi, A., Al-Saedi, K., Ddin, M. U., & Al Naqbi, M. (2023). Transformation of Operations Through Digital Twin Application. *Society of Petroleum Engineers—ADIPEC, ADIP, 2023*. https://doi.org/10.2118/216714-MS

Amazon. (2018). *What Is Immersion Reading?* https://audible.custhelp.com/app/answers/detail/a_id/5176/~/what-is-immersion-reading%3F

Angurala, M., Sharma, Y., & Mishra, S. (2024, February). Empowering Industries: Unveiling the Potential of Industrial Internet of Things. In *2024 11th International Conference on Computing for Sustainable Global Development (INDIACom)* (pp. 166–170). IEEE.

Anwar, T., Khan, G. A., Ashraf, Z., Ansari, Z. A., Ahmed, R., & Azrour, M. (2024). The Combination of Blockchain and the Internet of Things (IoT): Applications, Opportunities, and Challenges for Industry. *Blockchain and Machine Learning for IoT Security*, 56–76.

Arinze, C. A., Ajala, O. A., Okoye, C. C., Ofodile, O. C., & Daraojimba, A. I. (2024). Evaluating the Integration of Advanced IT Solutions for Emission Reduction in the Oil and Gas Sector. *Engineering Science & Technology Journal*, *5*(3), 639–652.

The ASEAN Post. (2018). *How Disruptive Technologies Are Transforming Southeast Asia, 2023*. The ASEAN Post.

Bellodi, E., Zese, R., Riguzzi, F., & Lamma, E. (2022). Introduction to Machine Learning. *Machine Learning and Non-Volatile Memories*, 1–21. https://doi.org/10.1007/978-3-031-03841-9_1

Berisha, B., Mëziu, E., & Shabani, I. (2022). Big Data Analytics in Cloud Computing: An Overview. *Journal of Cloud Computing*, *11*(1). https://doi.org/10.1186/s13677-022-00301-w

Bhavana, B. C., Gowda, C. . T. V., & Rakshitha, B. H. (2022). A Survey: Internet of Things (IOT) Technologies, Applications. *International Journal for Research in Applied Science and Engineering Technology*, *10*(8), 640–644. https://doi.org/10.22214/ijraset.2022.42595

Borysiak, O., Wołowiec, T., Gliszczyński, G., Brych, V., & Dluhopolskyi, O. (2022). Smart Transition to Climate Management of the Green Energy Transmission Chain. *Sustainability*, *14*(18), 11449.

Buła, P., & Niedzielski, B. (2021). Industrial Revolution – from Industry 1.0 to Industry 4.0. *Management, Organisations and Artificial Intelligence*, 1–34. https://doi.org/10.4324/9781003184027-1

Buyya, R., Ilager, S., & Arroba, P. (2024). Energy-Efficiency and Sustainability in New Generation Cloud Computing: A Vision and Directions for Integrated Management of Data Centre Resources and Workloads. *Software: Practice and Experience*, *54*(1), 24–38.

Chhotaray, P., Behera, B. C., Moharana, B. R., Muduli, K., & Sephyrin, F. T. R. (2024). Enhancement of Manufacturing Sector Performance with the Application of Industrial Internet of Things (IIoT). In *Smart Technologies for Improved Performance of Manufacturing Systems and Services* (pp. 1–19). CRC Press.

China, C. (2023). *10 Ways the Oil and Gas Industry Can Leverage Digital Twin Technology*. IBM. www.ibm.com/blog/10-ways-the-oil-and-gas-industry-can-leverage-digital-twin-technology/

Concept, S. (2024). *What Is Industrial Automation & Types of Industrial Automation*. https://conceptsystemsinc.com/what-is-industrial-automation-types-of-industrial-automation/

Coursera. (2023). *Data Analytics: Definition, Uses, Examples, and More*. https://www.Coursera.Org/Articles/Data-Analytics.

Deng, T., Ambía, J., & Torres-Verdín, C. (2023). Interpretation of Well Logs and Core Data via Bayesian Inversion. *Geophysics*, *88*(2), D49–D67. https://doi.org/10.1190/geo2022-0164.1

Devold, H., & Fjellheim, R. (2019, November). Artificial Intelligence in Autonomous Operation of Oil and Gas Facilities. In *Abu Dhabi International Petroleum Exhibition and Conference* (p. D022S176R002). SPE.

Dickerson, P., & Worthen, J. (2024, May). Optimizing Pipeline Systems for Greater Precision, Efficiency & Safety Using Emerging Technologies. In *PSIG Annual Meeting* (pp. PSIG-2426). PSIG.

Digitemie, W. N., & Ekemezie, I. O. (2024). Enhancing Carbon Capture and Storage Efficiency in the Oil and Gas Sector: An Integrated Data Science and Geological Approach. *Engineering Science & Technology Journal*, *5*(3), 924–934.

Đorđević, L., Ćoćkalo, D., Bakator, M., & Novaković, B. (2023). Entrepreneurship in Society 5.0: Integrating technology and humanistic values for a sustainable future. *Industrija*, *51*(1), 29–45. https://doi.org/10.5937/industrija51-45802

Dorrington, K. P., & Link, C. A. (2004). Genetic-Algorithm/Neural-Network Approach to Seismic Attribute Selection for Well-Log Prediction. *Geophysics*, *69*(1), 212–221. https://doi.org/10.1190/1.1649389

Ebirim, W., Usman, F. O., Olu-lawal, K. A., Ninduwesuor-Ehiobu, N., Ani, E. C., & Montero, D. J. P. (2024). Optimizing Energy Efficiency in Data Center Cooling Towers Through Predictive Maintenance and Project Management. *World Journal of Advanced Research and Reviews*, *21*(2), 1782–1790.

Ekemezie, I. O., & Digitemie, W. N. (2024). Climate Change Mitigation Strategies in the Oil & Gas Sector: A Review of Practices and Impact. *Engineering Science & Technology Journal*, *5*(3), 935–948.

Elijah, O., Ling, P. A., Rahim, S. K. A., Geok, T. K., Arsad, A., Kadir, E. A., Abdurrahman, M., Junin, R., Agi, A., & Abdulfatah, M. Y. (2021). A Survey on Industry 4.0 for the Oil and Gas Industry: Upstream Sector. *IEEE Access*, *9*, 144438–144468. https://doi.org/10.1109/ACCESS.2021.3121302

Elsharief, M., Emran, A., Hassan, H., Sabuj, S. R., & Jo, H. S. (2024). Energy-Efficient Synchronization in Industrial Internet of Things: An Intelligent Neighbor-Knowledge Approach. *IEEE Transactions on Industrial Informatics*, *20*(6), 8548–8558.

Ershaghi, I., Paul, D. L., & Popa, A. (2024, April). Digital Decarbonization in the Oil and Gas Industry. In *SPE Western Regional Meeting* (p. D021S008R001). SPE.

Fan, J., Li, R., Zhang, C. H., & Zou, H. (2020). *Statistical Foundations of Data Science*. Chapman and Hall, CRC.

Fogg, E. (2020). *Industrial Automation: How It Works, Types, and Benefits*. MachineMetrics.

Fraihat, B. A. M., Zowid, F., Ayasrah, F. T. M., Rababah, A., Ahmad, A. Y., & Othman, O. H. O. (2024). *The Impact of Cloud Computing on Supply Chain Performance the Mediating Role of Knowledge Sharing in Utilities and Energy Sectors*. Growing Science.

Gallegos, J., Quevedo, H., Etcheverry, K., Romero, J., Vázquez, H., Rueda, P., . . . & Carrera, J. (2024, February). From Drilling to Production: Digital Solutions for Automation and Remote Operations in the Amazon Region. In *International Petroleum Technology Conference* (p. D021S068R003). IPTC.

Gaurav, S. (2021). Big Data & Cloud Computing: The Roles & Relationships. *Big Data & Cloud Computing: The Roles & Relationships*. Opportunities and Challenges. *Periodica Polytechnica Social and Management Sciences*, *32*(1), 1–16.

Georgiou, K., Mittas, N., Mamalikidis, I., Mitropoulos, A., & Angelis, L. (2021). Analyzing the Roles and Competence Demand for Digitalization in the Oil and Gas 4.0 Era. *IEEE Access*, *9*, 151306–151326.

Gomes, P., Magaia, N., & Neves, N. (2020). Industrial and Artificial Internet of Things with Augmented Reality. *Internet of Things*, 323–346. https://doi.org/10.1007/978-3-030-44907-0_13

Gonzalez, S. (2024, April). Generative Humane-Machine Interaction in Oil & Gas. In *Offshore Technology Conference* (p. D041S049R008). OTC.

Green, J., & Newman, P. (2017). Disruptive Innovation, Stranded Assets and Forecasting: The Rise and Rise of Renewable Energy. *Journal of Sustainable Finance and Investment*, *7*(2), 169–187. https://doi.org/10.1080/20430795.2016.1265410

Guo, H., Goodchild, M. F., & Annoni, A. (2020). Manual of Digital Earth. In H.Guo, M. F. Goodchild, & A. Annoni (Eds.), *Manual of Digital Earth*. Springer. https://doi.org/10.1007/978-981-32-9915-3

Hashmi, R., Liu, H., & Yavari, A. (2024). Digital Twins for Enhancing Efficiency and Assuring Safety in Renewable Energy Systems: A Systematic Literature Review. *Energies*, *17*(11), 2456.

Haouel, C., & Nemeslaki, A. (2024). Digital Transformation in Oil and Gas Industry: Opportunities and Challenges. *Periodica Polytechnica Social and Management Sciences*, *32*(1), 1–16.

Hsieh, B. Z., Lewis, C., & Lin, Z. S. (2005). Lithology Identification of Aquifers from Geophysical Well Logs and Fuzzy Logic Analysis: Shui-Lin Area, Taiwan. *Computers and Geosciences*, *31*(3), 263–275. https://doi.org/10.1016/j.cageo.2004.07.004

Hussain, C. M., & Di Sia, P. (2020). Handbook of Smart Materials, Technologies, and Devices. In C. M.Hussain & P. Di Sia (Eds.), *Handbook of Smart Materials, Technologies, and Devices*. Springer International Publishing. https://doi.org/10.1007/978-3-030-58675-1

Hussain, M., Alamri, A., Zhang, T., & Jamil, I. (2024). Application of Artificial Intelligence in the Oil and Gas Industry. In *Engineering Applications of Artificial Intelligence* (pp. 341–373). Springer Nature.

Hussain, R. F., Mokhtari, A., Ghalambor, A., & Salehi, M. A. (2022). IoT for Smart Operations in the Oil and Gas Industry: From Upstream to Downstream. *IoT for Smart Operations in the Oil and Gas Industry: From Upstream to Downstream*, 1–245. https://doi.org/10.1016/C2020-0-04614-1

IFC. (2019). *Reinventing Business Through Disruptive Technologies: Sector Trends and Investment Opportunities for Firms in Emerging Markets*. EM Compass Notes. IFC.

Ilkhchi, A. K., Rezaee, M., & Moallemi, S. A. (2006). A Fuzzy Logic Approach for Estimation of Permeability and Rock Type from Conventional Well Log Data: An Example from the Kangan Reservoir in the Iran Offshore Gas Field. *Journal of Geophysics and Engineering*, *3*(4), 356–369. https://doi.org/10.1088/1742-2132/3/4/007

Jin, W. (2024). Security and Privacy of Digital Economic Risk Assessment System Based on Cloud Computing and Blockchain. *Soft Computing*, *28*(3), 2753–2768.

Kamdjou, H. M., Baudry, D., Havard, V., & Ouchani, S. (2024). Resource-Constrained Extended Reality Operated With Digital Twin in Industrial Internet of Things. *IEEE Open Journal of the Communications Society*, *5*, 928–950.

Kanade, V. (2024). *What Is Automation? Definition, Types, Benefits, and Importance*. www.spiceworks.com/tech/artificial-intelligence/articles/what-is-automation/

Kayvanfar, V., Elomri, A., Kerbache, L., Vandchali, H. R., & El Omri, A. (2024). A Review of Decision Support Systems in the Internet of Things and Supply Chain and Logistics Using Web Content Mining. *Supply Chain Analytics*, 100063.

Khan, N., Solvang, W. D., & Yu, H. (2024). Industrial Internet of Things (IIoT) and Other Industry 4.0 Technologies in Spare Parts Warehousing in the Oil and Gas Industry: A Systematic Literature Review. *Logistics*, *8*(1), 16.

Khare, S. K. (2022). Machine Vision for Drill String Slip Status Detection. *Petroleum Research*, *7*(1), 115–122. https://doi.org/10.1016/j.ptlrs.2021.08.001

Konya, A., & Nematzadeh, P. (2024). Recent Applications of AI to Environmental Disciplines: A Review. *Science of the Total Environment*, *906*, 167705.

Kumar, A., Ramasree, R. J., & Faisal, M. (2019, April). Performing Predictive Analysis Using Machine Learning on the Information Retrieved from Production Data of Oil & Gas Upstream Segment. In *2019 International Conference on Communication and Signal Processing (ICCSP)* (pp. 0385–0391). IEEE.

Kumar, M. S., & Prabhu, J. (2020). Recent Development in Big Data Analytics: Research Perspective. *Research Anthology on Artificial Intelligence Applications in Security*, *3–4*, 1640–1663. https://doi.org/10.4018/978-1-7998-7705-9.ch072

Kwek, D., Maulana, D., Kaburuan, E. R., & Legowo, N. (2019). Enterprise Architecture Planning Information System Based on Cloud Computing Using Togaf (Case Study: Pandi. Id Registry). *International Journal of Scientific and Technology Research*, *8*(9), 1167–1178.

Li, C. S., Chen, Y. Y., & Sun, W. (2011). Lithology Identification Method While Drilling Based on Element Mudlogging Information. *Zhongguo Shiyou Daxue Xuebao (Ziran Kexue Ban)/Journal of China University of Petroleum (Edition of Natural Science)*, *35*(6), 66–70. https://doi.org/10.3969/j.issn.1673-5005.2011.06.010

Li, H., Wang, X., Feng, Y., Qi, Y., & Tian, J. (2024). Driving Intelligent IoT Monitoring and Control Through Cloud Computing and Machine Learning. *arXiv preprint arXiv:2403.18100*.

Lim, C. H., Lim, S., How, B. S., Ng, W. P. Q., Ngan, S. L., Leong, W. D., & Lam, H. L. (2021). A Review of Industry 4.0 Revolution Potential in a Sustainable and Renewable Palm Oil Industry: HAZOP Approach. *Renewable and Sustainable Energy Reviews*, *135*, 110223.

Lu, H., Guo, L., Azimi, M., & Huang, K. (2019). Oil and Gas 4.0 Era: A Systematic Review and Outlook. *Computers in Industry*, *111*, 68–90. https://doi.org/10.1016/j.compind.2019.06.007

Majstorović, V. (2022). Application of Industry 4.0 Model in Oil and Gas Companies. *Journal of Engineering Management and Competitiveness*, *12*(1), 77–84. https://doi.org/10.5937/jemc2201077m

Masudin, I., Tsamarah, N., Restuputri, D. P., Trireksani, T., & Djajadikerta, H. G. (2024). The Impact of Safety Climate on Human-Technology Interaction and Sustainable Development: Evidence from Indonesian Oil and Gas Industry. *Journal of Cleaner Production*, *434*, 140211.

Mathur, S., Hemachandran, K., & Shanmugarajah, D. (2024). AI in Energy Sector. In *Artificial Intelligence for Business* (pp. 294–309). Productivity Press.

McEwen, A., & Cassimally, H. (2013, October). The Internet of Things: An Overview. *Designing the Internet of Things*, *2019*, 8.

Mohammad, S., & Rajeswari, M. N. (2019). Overview of Cloud Computing and Its Types. *Jetir*, *6*(3). www.jetir.org

Mohammadpoor, M., & Torabi, F. (2020). Big Data Analytics in Oil and Gas Industry: An Emerging Trend. *Petroleum*, *6*(4), 321–328. https://doi.org/10.1016/j.petlm.2018.11.001

Munirathinam, S. (2020). Industry 4.0: Industrial Internet of Things (IIOT). In *Advances in Computers* (Vol. 117, No. 1, pp. 129–164). Elsevier.

Mustapha, H. (2024). *U.S. Patent No. 11,886,784*. U.S. Patent and Trademark Office.

Naghneh, H. M., Amani, M., Farhadi, A., & Isaai, M. T. (2024). Application of the Closed Loop Industrial Internet of Things (IIoT)-Based Control System in Enhancing the Oil Recovery Factor and the Oil Production. *IET Cyber-Physical Systems: Theory & Applications*, *9*(1), 72–86.

Nguyen, T. H., Paramasivam, P., Le, H. C., & Nguyen, D. C. (2024). Harnessing a Better Future: Exploring AI and ML Applications in Renewable Energy. *JOIV: International Journal on Informatics Visualization*, *8*(1), 55–78.

Nilay, D. (2021). *Handy Tips to Leverage Industrial IoT for Oil and Gas Fleet Management*. www.intuz.com/blog/iiot-for-oil-and-gas-fleet-management

Ochuba, N. A., Amoo, O. O., Okafor, E. S., Akinrinola, O., & Usman, F. O. (2024). Strategies for Leveraging Big Data and Analytics for Business Development: A Comprehensive Review Across Sectors. *Computer Science & IT Research Journal*, *5*(3), 562–575.

Odimarha, A. C., Ayodeji, S. A., & Abaku, E. A. (2024). Machine Learning's Influence on Supply Chain and Logistics Optimization in the Oil and Gas Sector: A Comprehensive Analysis. *Computer Science & IT Research Journal*, *5*(3), 725–740.

Oguanobi, V. U., & Joel, O. T. (2024). Scalable Business Models for Startups in Renewable Energy: Strategies for Using GIS Technology to Enhance SME Scaling. *Engineering Science & Technology Journal*, *5*(5), 1571–1587.

Olajiga, O. K., Olu-Lawal, K. A., Usman, F. O., & Ninduwezuor-Ehiobu, N. (2024). Conceptual Framework for Effective Communication Strategies in High-Risk Industries: Insights from the Energy Sector. *World Journal of Advanced Engineering Technology and Sciences*, *11*(2), 080–090.

Oyewole, A. T., Okoye, C. C., Ofodile, O. C., Odeyemi, O., Adeoye, O. B., Addy, W. A., & Ololade, Y. J. (2024). Human Resource Management Strategies for Safety and Risk Mitigation in the Oil and Gas Industry: A Review. *International Journal of Management & Entrepreneurship Research*, *6*(3), 623–633.

Patresearch. (2013). *Predictive Analytics in Oil and Gas*. www.predictiveanalyticstoday.com/predictive-analytics-oil-gas/rishabhsoft.

Pérez, M. D. L. E. M., Scholten, D., & Stegen, K. S. (2019). The Multi-Speed Energy Transition in Europe: Opportunities and Challenges for EU Energy Security. *Energy Strategy Reviews*, *26*, 100415.

Progoulakis, I., Nikitakos, N., Rohmeyer, P., Bunin, B., Dalaklis, D., & Karamperidis, S. (2021). Perspectives on Cyber Security for Offshore Oil and Gas Assets. *Journal of Marine Science and Engineering*, *9*(2), 112.

Rajamohan, K., Rangasamy, S., Manoj, C. N., Mary, L., Sabu, R., & Jose, S. S. (2024). Internet of Things (IoT) as a Game Changer to the Education Sector. In *Advanced IoT Technologies and Applications in the Industry 4.0 Digital Economy* (pp. 251–274). CRC Press.

Rao, A. P., Nagireddi, B., Jha, S., Perumal, R., & Pothabattula, L. K. B. (2024). *U.S. Patent Application No. 18/508,919*. U.S. Patent and Trademark Office.

Rath, K. C., Khang, A., & Roy, D. (2024). The Role of Internet of Things (IoT) Technology in Industry 4.0 Economy. In *Advanced IoT Technologies and Applications in the Industry 4.0 Digital Economy* (pp. 1–28). CRC Press.

Rekeraho, A., Cotfas, D. T., Cotfas, P. A., Bălan, T. C., Tuyishime, E., & Acheampong, R. (2024). Cybersecurity Challenges in IoT-Based Smart Renewable Energy. *International Journal of Information Security*, *23*(1), 101–117.

Rishabhsoft. (2023). *IoT in Oil and Gas Sector: Benefits, Use Cases, Challenges*. www.rishabhsoft.com/blog/iot-in-oil-and-gas-industry

Roden, R. (2016). Seismic Interpretation in the Age of Big Data. *SEG Technical Program Expanded Abstracts*, *35*, 4911–4915. https://doi.org/10.1190/segam2016-13612308.1

Rouse, M. (2023). Automation. In *Techopedia*. Technopedia. www.techopedia.com/definition/32099/automation

Sackschewsky, C., Yavar, E., & Cascini, J. (2018). *How Industry 4.0 Is Transforming the Oil & Gas Supply Chain What Is Industry 4.0?* 1–4. www.bdo.com

Sayed, R. A. F., Hashim, S. M., Shahin, E. S., & Abd-Elmaksoud, M. H. (2020, May). Effective Use of Real-Time Remote Monitoring System in Offshore Oil Wells Optimization, Case Study from Gulf of Suez-July Field. In *Offshore Technology Conference* (p. D031S032R001). OTC.

Schwab, J. H. (2020). Predictive Analytics. *Spine Journal*, *20*(7), 1152–1153. https://doi.org/10.1016/j.spinee.2020.03.003

ScienceSoft. (2022). *Big Data in Oil & Gas: Market Trends, Use Cases, and Toolset*. www.scnsoft.com/analytics/big-data/oil-gas

Sharma, R., & Villányi, B. (2024). Safe and Secure Oil and Gas Pipeline Transportation System Based on Industrial Internet of Things. *IEEE Sensors Journal*, *24*(5), 6834–6845.

Sharma, S. K., Rani, A., Bakhariya, H., Kumar, R., Tomar, D., & Ghosh, S. (2024). The Role of IoT in Optimizing Operations in the Oil and Gas Sector: A Review. *Transactions of the Indian National Academy of Engineering*, 1–20.

Sheveleva, A., Tyaglov, S., & Khaiter, P. (2021). Digital Transformation Strategies of Oil and Gas Companies: Preparing for the Fourth Industrial Revolution. *Digital Strategies in a Global Market: Navigating the Fourth Industrial Revolution*, 157–171.

Sidqi, M., Prabata, W., Situmorang, J., Iqbal, B. M., Nugroho, I., & Wiliam, H. (2024). *JIWA DBase: Integrated Geothermal Data Management and Analysis in Cloud System*. Forthcoming Proceedings.

Smirnova, A., Zaychenko, I., Bagaeva, I., & Gorshechnikova, P. (2020). Digital Technologies in the Industry: Application of Immersive Training Technologies in the Oil and Gas Complex. *SHS Web of Conferences*, *84*, 03003. https://doi.org/10.1051/shsconf/20208403003

Spadoro, J., Vombergar, K., & Hoempler, M. (2024). Flying High: The Role of Drones in Enhancing Utility Services. *Climate and Energy*, *40*(9), 1–12.

Temer, E., & Pehl, H. J. (2017, November). Moving Toward Smart Monitoring and Predictive Maintenance of Downhole Tools Using the Industrial Internet of Things IIoT. In *Abu Dhabi International Petroleum Exhibition and Conference* (p. D031S065R004). SPE.

Tirlangi, S., Teotia, S., Padmapriya, G., Kumar, S. S., Dhotre, S., & Boopathi, S. (2024). Cloud Computing and Machine Learning in the Green Power Sector: Data Management and Analysis for Sustainable Energy. In *Developments Towards Next Generation Intelligent Systems for Sustainable Development* (pp. 148–179). IGI Global.

Tran, T. M. (2023). *Revolution in the Pipeline – Part 1: Disruptive Technologies in Oil and Gas Industry*. https://fptsoftware.com/resource-center/blogs/revolution-in-the-pipeline-part-1-disruptive-technologies-in-oil-and-gas-industry

Tsaramirsis, G., Kantaros, A., Al-Darraji, I., Piromalis, D., Apostolopoulos, C., Pavlopoulou, A., Alrammal, M., Ismail, Z., Buhari, S. M., Stojmenovic, M., Tamimi, H., Randhawa, P., Patel, A., & Khan, F. Q. (2022). A Modern Approach towards an Industry 4.0 Model: From Driving Technologies to Management. *Journal of Sensors*, *2022*. https://doi.org/10.1155/2022/5023011

Verma, R. M., & Marchette, D. J. (2020). What Is Data Analytics? *Cybersecurity Analytics*, 9–33. https://doi.org/10.1201/9780429326813-2

Wanasinghe, T. R., Trinh, T., Nguyen, T., Gosine, R. G., James, L. A., & Warrian, P. J. (2021). Human Centric Digital Transformation and Operator 4.0 for the Oil and Gas Industry. *IEEE Access*, *9*, 113270–113291.

Wanasinghe, T. R., Wroblewski, L., Petersen, B. K., Gosine, R. G., James, L. A., De Silva, O., Mann, G. K. I., & Warrian, P. J. (2020). Digital Twin for the Oil and Gas Industry: Overview, Research Trends, Opportunities, and Challenges. *IEEE Access*, *8*, 104175–104197. https://doi.org/10.1109/ACCESS.2020.2998723

Wang, T., Wei, Q., Xiong, W., Wang, Q., Fang, J., Wang, X., . . . & Wang, J. (2024). Current Status and Prospects of Artificial Intelligence Technology Application in Oil and Gas Field Development. *ACS Omega*, *9*(3), 3173–3183.

Wang, Z., Li, S., Jin, Z., Li, Z., Liu, Q., & Zhang, K. (2023). Oil and Gas Pathway to Net-Zero: Review and Outlook. *Energy Strategy Reviews*, *45*, 101048.

Whaley, J. (2020). *What Is Digital Twin Technology – And How Is It Used in Oil and Gas?* GEOExPro. www.geoexpro.com/articles/2020/10/what-is-digital-twin-technology-and-how-is-it-used-in-oil-and-gas

Wibowo, E. B., Legionosuko, T., Mahroza, J., & Chandra Jaya, Y. (2020). Industry 4.0: Challenges and Opportunities in Competency Development for Defense Apparatus' Human Resources. *International Journal of Advanced Science and Technology*, *29*(7), 45–60.

Wong, S. W., van der Bas, F., Groenenboom, J., & Zuiderwijk, P. (2003). Near Wellbore Stimulation by Acoustic Waves. In *SPE European Formation Damage Conference and Exhibition* (pp. SPE-82198). SPE.

Yang, C., Huang, Q., Li, Z., Liu, K., & Hu, F. (2017). Big Data and Cloud Computing: Innovation Opportunities and Challenges. *International Journal of Digital Earth*, *10*(1), 13–53. https://doi.org/10.1080/17538947.2016.1239771

Zakeri, B., Paulavets, K., Barreto-Gomez, L., Echeverri, L. G., Pachauri, S., Boza-Kiss, B., . . . & Pouya, S. (2022). Pandemic, War, and Global Energy Transitions. *Energies*, *15*(17), 6114.

Zhifeng, Y. (2019). Cloud Computing and Big Data for Oil and Gas Industry Application, China. *Journal of Computers*, *14*(4), 268–282. https://doi.org/10.17706/jcp.14.4.268-282

19 Future Directions in Oil and Gas – Renewables and Energy Transition

Deniz Yağmur Kayhan

19.1 INTRODUCTION

The Paris Agreement signed at the United Nations Climate Change Conference has aimed to keep the global average temperature increase at pre-industrial levels – in other words, below 2°C, and to limit this increase specifically to 1.5°C (United Nations, 2015; European Commission, n.d.a.). The common goal of Net-Zero Emissions According to the 2050 Scenario published by the International Energy Agency (IEA), the 1.5°C Global Warming Special Report of the Intergovernmental Panel on Climate Change (IPCC), and the Green Deal of the European Commission is to achieve net-zero emissions of greenhouse gases by 2050 (IPCC, 2018; IEA, 2021a, 2022a; European Commission, n.d.b.).

The energy sector has been significantly impacted by the decisions, reports, and recommendations put forth by agreements and commissions. These initiatives have shaped both future energy transitions, guiding expectations and actions toward a decarbonized, clean, friendly, and sustainable future. This influence spans areas including fossil fuels, renewables, and low-emissions technologies, with a focus on reducing or eliminating CO_2 emissions and keeping temperature increases below thresholds. According to Figure 19.1 in the study by Davis et al. (2018), energy services are being provided without contributing to CO_2 levels in the atmosphere. The illustration showcases electricity generation using sources like solar, wind, hydropower, and nuclear power alongside carbon management technologies, such as carbon capture and storage facilities. It also includes processes for producing hydrogen and ammonia utilizing hydrocarbons as biofuels or synthetic fuels as end uses tailored to meet demands.

In order to achieve a decarbonized future, plans that include the steps should be designed for pathways based on years. For example, Figure 19.2 represents decarbonization pathways starting from 2022, 2030, and 2050. Each year comprises various primary decarbonization methods, such as low-carbon technologies and lifestyles, carbon capture, utilization and storage (CCUS), and negative emission technologies (NETs). Every step has a certain CO_2 mitigation amount. The study by Vitillo et al. (2022) is just one example for them.

This chapter starts by examining the energy scenarios put forth by organizations and companies. Most of these scenarios consider both pessimistic outcomes, providing insights into the potential global challenges based on their strategies. The analysis delves into the expected impacts of energy transitions on sectors such as oil and gas, as well as renewables, drawing from these scenario projections. Subsequently, it elaborates on perspectives and models that outline the pathways in energy transition, along with supporting factors.

19.2 FUTURE SCENARIOS

Foresight designed based on certain aspects and future expectations of energy plays important roles in the direction of technology, industry, human life, and environment. In this sense, several scenarios

 DOI: 10.1201/9781003282273-19

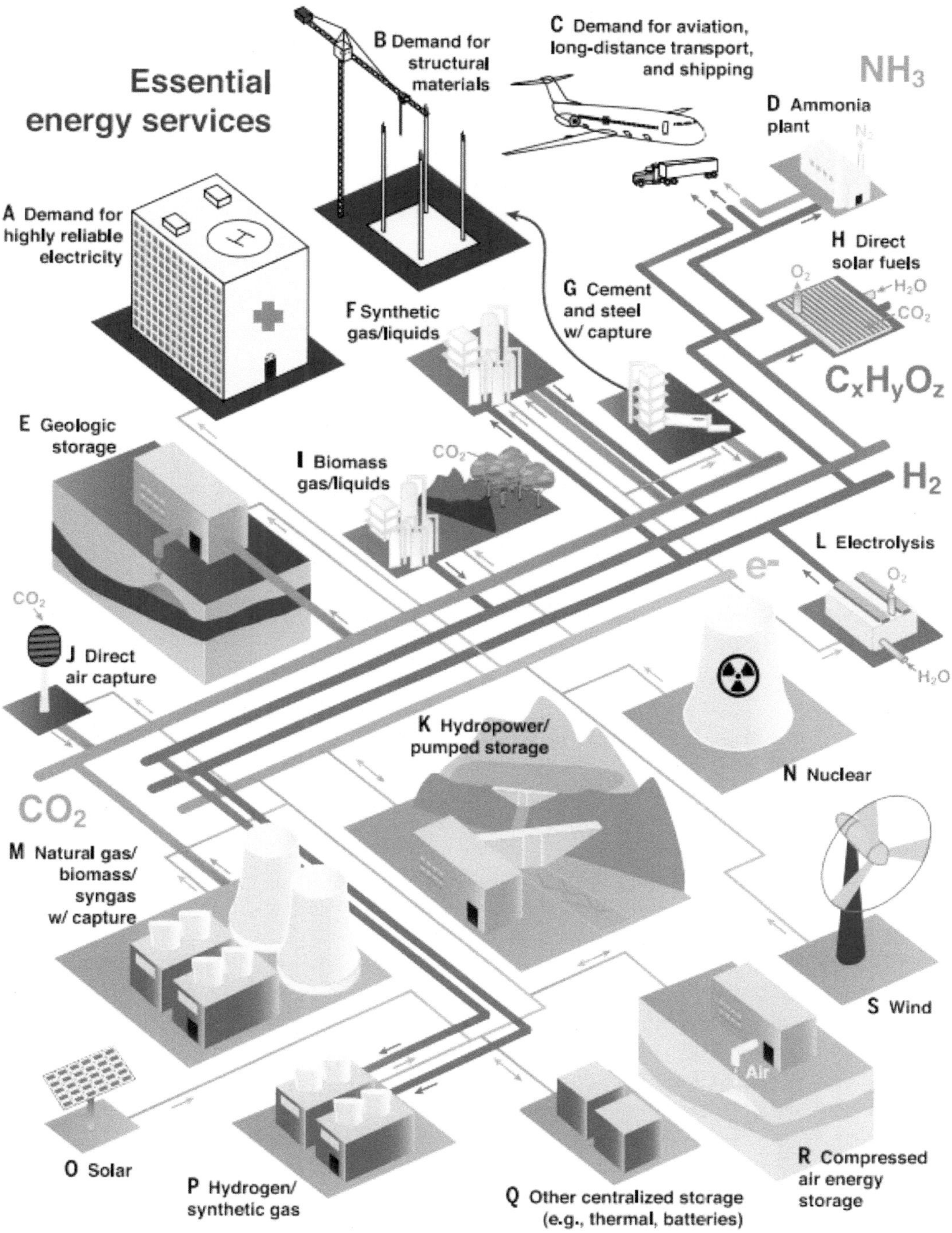

FIGURE 19.1 Systems that provide energy without adding any CO_2 to the atmosphere (red for *carbon management*, purple for *hydrocarbon*, blue for *hydrogen*, green for *electricity generation and transmission*, orange for *ammonia*, and black for *end uses*).

Source: Davis et al. (2018).

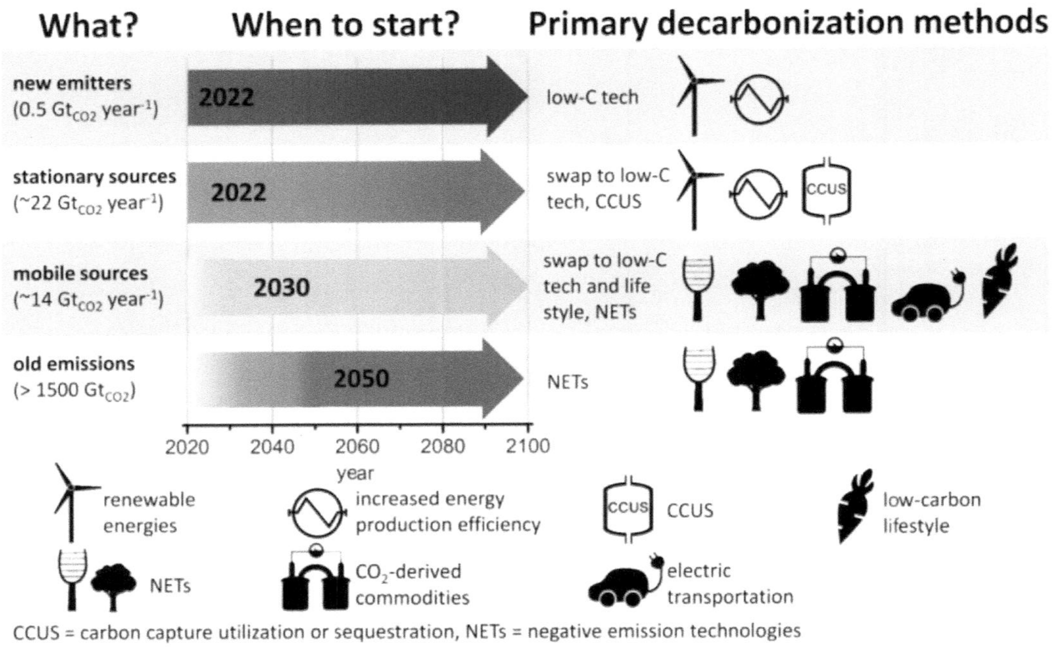

FIGURE 19.2 Decarbonization pathways with the amount of emitted carbon (gigatons CO_2 per year, Gt_{CO2}year^{-1}), start time, and primary decarbonization methods.

Source: Vitillo et al. (2022).

about the future of energy resources and technologies have been released in reports from various sources, such as the International Energy Agency (IEA), International Renewable Energy Agency (IRENA), and companies from a variety of fields, including Barclays, BP, Equinor, ExxonMobil, McKinsey, Shell, TotalEnergies, and World Energy Council. Even though many of these reports have been prepared to meet the target of limiting global temperature rise at certain degrees by 2050, the majority of them contain both relatively strong and weak cases for the foresight on energy future. They comprise expectations on the future of both fossil fuels and renewables with carbon mitigation and low-emissions technologies via evaluating the world's estimated population, climate change, economic growth, and energy demand.

There are also some publications which looked at and compared these future energy scenarios in several ways, like energy demand, total consumption, electricity generation by fuels, and mitigation of CO_2 emissions (Newell et al., 2019, 2021; Davies and Simmons, 2021). Figure 19.3 shows the global primary energy fuel shares by comparisons of 1990, 2000, 2015, and 2019 data, with energy future scenarios of various organizations and companies by 2040 (Newell et al., 2021).

According to Figure 19.3, the aforementioned and categorized scenarios include different actions and share of energy resources from each other. In 2019, global primary energy demand is 572 qBtu (quadrillion British thermal units). The reference and evolving policies scenarios show a rise in the global energy demand, which may increase higher than expected. Specifically, recent advancements in artificial intelligence (AI) will inevitably reshape all industries, which will require significant power and energy for data processing. According to 2040 primary energy fuel projections, fossil fuels continue to have similar or even higher demand compared to 2019 data. On the other hand, most of the ambitious climate scenarios represent a fall in global energy consumption, since energy efficiency measures reduce the energy intensity of the economy, which is highly unlikely due to further adoption of artificial intelligence and electrified transportation. Except liquid fossil fuels, natural gas demand is expected to grow even in some of the ambitious climate scenarios. In contrast

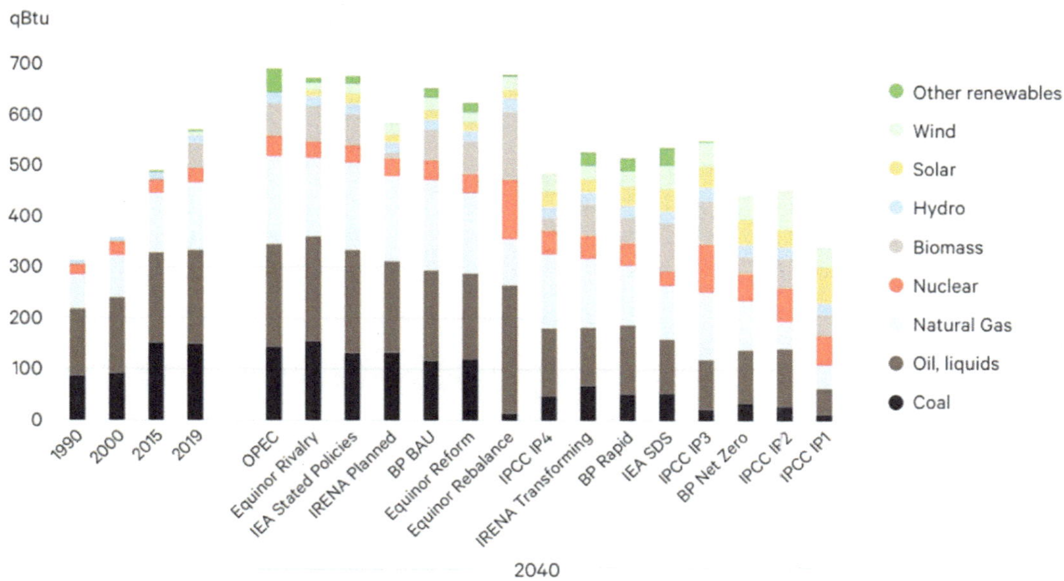

FIGURE 19.3 Global primary energy fuel shares (quadrillion British thermal units, qBtu) in 1990, 2000, 2015, 2019, and 2040, based on energy future scenarios of various organizations and companies.

Source: Newell et al. (2021).

to all the differences between reference, evolving policies, and ambitious climate scenarios, renewable energies have an increasing trend in all these scenarios (Newell et al., 2021). Nevertheless, the highest share of these energies is in ambitious climate scenarios.

IEA announced a series of future energy scenarios which are named as Stated Policies Scenario (STEPS), Announced Pledges Scenario (APS), Sustainable Development Scenario (SDS), and Net-Zero Emission by 2050 Scenario (NZE) (IEA, 2021a, 2021b, 2022b, 2023a). STEPS is characterized by current policy settings according to sectoral assessments that have been announced by governments. APS presumes climate commitments have been organized by governments, containing nationally determined contributions (NDCs) and longer-term net-zero targets that are achieved in full and on time. NZE is a narrow but achievable pathway designed for the global energy sector to reach net-zero in CO_2 emissions and limit global warming well below 1.5°C by 2050. SDS is similar to NZE at some points, and it aims to reach the United Nation's Sustainable Development Goals (SDGs) and global net-zero emissions by 2070 (IEA, 2021b, 2022b).

While focusing on three scenarios of IEA, STEPS, APS, and NZE, comparing global electricity demand, supply, and generation data is a good way to understand the differences in their pathways for the future. Electricity comprised 20% of the total energy consumption of the world, which is equivalent to one-third of energy-related CO_2 emissions in 2021. Table 19.1 represents global electricity demand and supply in 2010 and 2021, with three IEA scenarios for 2030 and 2050. STEPS, APS, and NZE show nearly 80%, 120%, and 150% increases, respectively, in electricity demand through 2050 compared to 2021 data. Rapid electrification of end uses and widespread usage of electrical vehicles accelerate this demand, especially in APS and NZE.

Figure 19.4 contains global electricity generation by sources in 2010 and 2021 data and based on three IEA scenarios by 2030 and 2050. For the electricity supply, renewables seem to be the dominant source in the future according to the forenamed scenarios. Their share in global electricity generation was 28% in 2021. It is expected to be 43% by 2030 and 65% by 2050 in STEPS, 49% by 2030 and 80% by 2050 in APS, and 61% by 2030 and 88% by 2050 in NZE. Solar photovoltaics (PV) and wind provide the main contributors to this development according to the scenarios.

TABLE 19.1
Global Electricity Demand in Sectors and Supply (TWh) by Sources in 2010 and 2021 Data and According to STEPS, APS, and NZE by 2030 and 2050

			STEPS		APS		NZE	
	2010	2021	2030	2050	2030	2050	2030	2050
Buildings	9,637	12,594	15,383	21,940	14,889	19,623	13,293	15,850
Industry	7,450	10,166	12,036	15,073	12,471	18,332	13,776	21,697
Transport	295	441	1,169	3,607	1,570	7,845	2,236	10,243
Hydrogen production	–	2	159	663	879	5,714	2,464	11,433
Global electricity demand	18,548	24,700	30,621	43,672	31,752	53,810	33,733	62,159
Unabated coal	8,670	10,201	9,044	5,892	8,076	1,580	4,666	0
Unabated natural gas	4,855	6,552	6,848	6,658	6,100	3,577	4,977	82
Unabated oil	969	682	432	312	363	175	180	3
Fossil fuels with CCUS	–	1	5	133	75	1,338	282	1,317
Nuclear	2,756	2,776	3,351	4,260	3,547	5,103	3,896	5,810
Hydropower	3,449	4,327	5,078	6,809	5,213	7,543	5,725	8,251
Wind	342	1,870	4,604	10,691	5,816	17,416	7,840	23,486
Solar PV	32	1,003	4,011	12,118	4,838	18,761	7,551	27,006
Other renewables	411	859	1,380	2,833	1,707	5,153	1,948	5,762
Hydrogen and ammonia			9	44	79	567	603	1,467
Global electricity supply	21,539	28,334	34,834	49,845	35,878	61,268	37,723	73,232
Renewables share	20%	28%	43%	65%	49%	80%	61%	88%

Source: IEA (2022b).

Notes: TWh = terawatt-hours; *CCUS* = carbon capture, utilization, and storage; *PV* = photovoltaics; *STEPS* = Stated Policies Scenario; *APS* = Announced Pledges Scenario; *NZE* = Net Zero Emissions by 2050 Scenario. *Electricity demand* is defined as total gross electricity generated less own use generation, plus imports, less exports and transmission and distribution losses. Other sources are included in electricity supply.

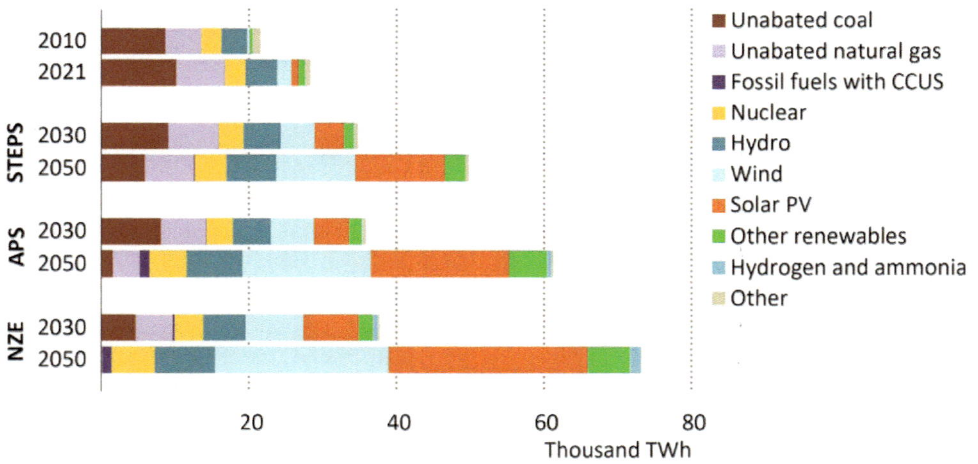

FIGURE 19.4 Global electricity generation (thousand terawatt hours, TWh) by sources in 2010 and 2021 data and based on STEPS, APS, and NZE by 2030 and 2050.

Source: IEA (2022b).

The role of hydrocarbons declines in global electricity generation through 2050. The share of unabated coals is expected to fall from 36% in 2021 to 26% by 2030 and 12% by 2050 in STEPS, 23% by 2030 and 3% by 2050 in APS. Unabated natural gas had 23% share in electricity generation in 2021; it declines to 20% by 2030 and 13% by 2050 in STEPS, 17% by 2030 and 6% by 2050 in APS. The shares of both unabated coal and natural gas decrease faster in NZE, reaching global net-zero emissions in electricity by 2040. Unabated oil had 2% share in electricity generation in 2021. It also declines in the defined three scenarios of IEA (IEA, 2022b).

The other significant concern is CO_2 emissions in future energy scenarios. Based on STEPS, global CO_2 emissions had a quick rise in 2021, then increased to 36 Gt (gigatons) by 2030. On the other hand, APS expects global CO_2 emissions to reach 34 Gt by 2030 and to fall into current values by 2050. According to NZE, CO_2 emissions are expected to decrease to 21 Gt by 2030 and to reach net-zero by 2050. According to the IEA Model for the Assessment of Greenhouse Gas–Induced Climate Change (MAGICC), the global temperature rise is expected to be 2.6°C in STEPS and 2.1°C in APS by 2100, and then they show an increasing trend through the future. In contrast, both SDS and NZE estimate temperature rise at 1.7°C and 1.5°C until 2050, then to transition into a decreasing trend. Briefly, the key differences between the weak and strong scenarios, STEPS and NZE, result from lower emissions (by renewables, CCUS, low-emissions fuels), more contingent in energy area (by electrification of end uses, electricity networks), and effective energy transition (by reducing fossil fuel usage, investments for less-polluting alternative energies, and retirement of unabated fossil fuels) in NZE (IEA, 2021b).

In summary, NZE provides the optimum conditions for essential climate action between the energy future scenarios of the IEA. The key priorities of further CO_2 emissions reduction in NZE are clean electrification, energy efficiency, clean energy innovation, and preventing methane release from fossil fuel operations. Figure 19.5 compares the strong scenarios of IEA and NZE on different shares of energy sources and technologies in global energy system pathways for clean energy transition between today and 2050 (IEA, 2021b).

IRENA launched two scenarios, which are the Planned Energy Scenario (PES) and the 1.5°C Scenario (IRENA, 2023). PES has been prepared according to governments' energy plans and targets and policies planned based on G20 countries. The 1.5°C Scenario is coherent with limiting global warming to 1.5°C by the end of the present century. A total of six key performance indicators are used for the PES and the 1.5°C Scenario. These are renewables for electricity, direct usage of renewables, improvements in energy intensity, direct electrification of end uses, production and supply of clean hydrogen and derivative fuels, and capturing and removing CO_2 (IRENA, 2023). Table 19.2 provides PES and 1.5°C Scenario projections for 2030 and 2050 based on 2020 data. According to Table 19.2, the 1.5°C Scenario provides the optimum contribution through the 1.5°C pathway for both global and G20 countries scale, whereas PES supplies insufficient support at this point. Nevertheless, PES has still relatively better foresight for future needs compared to the 2020 data.

Figure 19.6 shows electricity generation and capacity PES and 1.5°C Scenario projections for 2030 and 2050 starting from 2020 data. According to the figure, renewable energy capacity increases to 6,773 GW by 2030 and 15,835 GW by 2050 in PES. Even though the share of renewables in electricity generation is expected to have a significant increase from 28% in 2020 to 46% in 2030 and 73% in 2050 in PES, this is not enough to meet limiting global temperature increases well below 1.5°C. That means decarbonization in the power sector by 2050 is essential to reach the climate action target of the Paris Agreement. On the contrary, the renewables' share in electricity generation increases to 68% in 2030 and 91% in 2050 for the 1.5°C Scenario. Solar PV and wind energy supply are the majority of electricity generation and capacity in both PES and 1.5°C Scenarios (IRENA, 2023). In addition, 1.5°C Scenario by 2050 does not contain any share of coal and oil in electricity generation and capacity foresight. It only contains a minor share of natural gas. That situation is quite compatible with the pathway that has been aimed for the future.

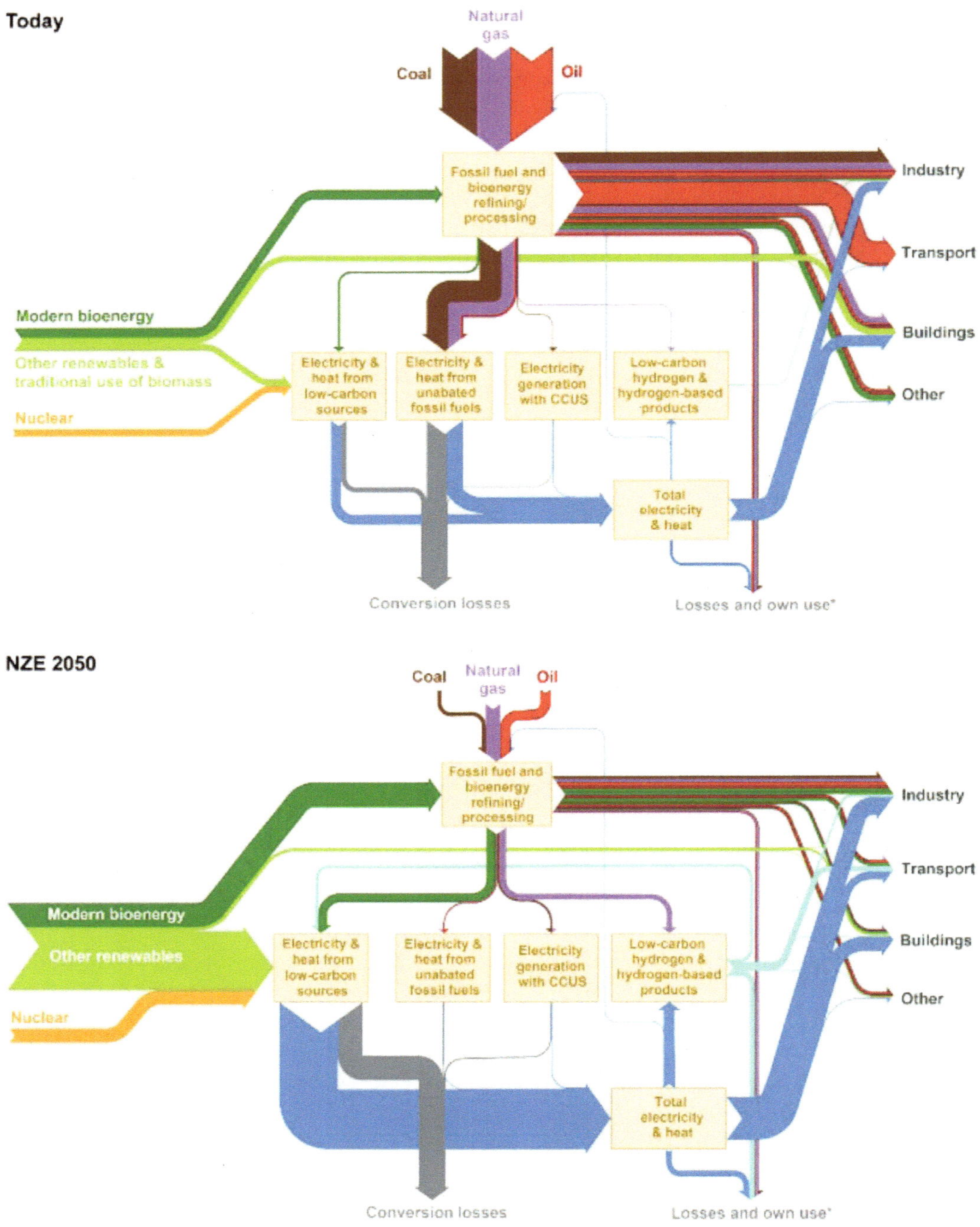

FIGURE 19.5 Global energy system of today and NZE by 2050.

Source: IEA (2021b).

The major difference between two IRENA scenarios, PES and 1.5°C Scenario, can be explained via their potential results, like CO_2 emissions. Global CO_2 emissions under PES and the 1.5°C Scenario between 2023 and 2050 based on sectors are shown in Figure 19.7. For 2050 PES projections, annual CO_2 emissions slightly decrease to 34 Gt of CO_2 ($GtCO_2$), which is primarily based on 12.7 industry 21%, power and heat plants with 9.8 $GtCO_2$, and transport with 7.9 $GtCO_2$ emissions.

TABLE 19.2
Six Key Performance Indicators Prepared for the 1.5°C Scenario with 2020 Numbers, and with Expected Numbers of 2030 and 2050 in PES and 1.5°C Scenario for Both Global and G20 Countries Separately

			Recent Years	2030		2050	
			2020	PES	1.5°C Scenario	PES	1.5°C Scenario
KPI.01 Renewables (Power)	Electricity generation (TWh/yr)	Global	7,468	16,504	27,358	38,118	82,148
		G20	6,237	14,269	22,397	31,071	60,547
	Renewable energy share in electricity generation (%)	Global	28%	46%	68%	73%	91%
		G20	28%	48%	69%	74%	91%
KPI.02 Renewables (Direct Uses)	Renewable energy share in TFEC (%)	Global	18%	23%	35%	33%	82%
		G20	16%	22%	36%	35%	82%
	Modern use of bioenergy (EJ)[1]	Global	21	30	50	41	64
		G20	19	26	36	33	42
KPI.03 Energy Intensity	Energy intensity improvement rate (%)	Global	1.7%	1.8%	3.3%	2.0%	2.8%
		G20	2.1%	2.1%	3.6%	2.3%	3.1%
KPI.04 Electrification in End Use Sectors (Direct)	Electrification rate in TFEC (%)	Global	22%	23%	29%	28%	51%
		G20	24%	26%	31%	32%	55%
KPI.05 Clean Hydrogen and Derivatives	Production of clean hydrogen (Mt)	Global	0.7 Mt/yr[2]	2	125	21	523
		G20	0.5 Mt/yr[2]	2	94	20	373
KPI.06 CCS, BECCS, and Others	CO_2 captured from CCS, BECCS, and other removal measures (Gt)	Global	0.04 $GtCO_2$/yr[3]	0.1	2.2	0.5	7.0
		G20	0.03 $GtCO_2$/yr[3]	0.1	2.1	0.4	4.9

Source: IRENA (2023).

In the 1.5°C Scenario, they reach net-zero by 2050 (even reaching negative emissions of 0.2 $GtCO_2$/year), with a steep decreasing trend through 2030 and following a fall. The 1.5°C Scenario 2050 is performed by the contribution of the advancement of new technologies. The largest shares of this decrease come from renewables, including power and direct uses, and energy conservation and efficiency, each with a 25% share. Then, they are followed by direct electrification in end use sectors (19%), hydrogen and its derivatives (12%), bioenergy with carbon capture and storage (BECCS) and other carbon removal measures (11%), and direct carbon capture and sequestration (CCUS) in industry (8%) (IRENA, 2023).

Barclays proposed three scenarios for the future of energy and called them 3D Scenarios, which are Dynamism, Development, and Deadlock (Barclays, 2019). The energy mix in 2017 and 3D Scenarios in 2050 are represented in Figure 19.8. Between them, Dynamism is the best-case energy scenario for mitigating carbon emissions that contains aggressive policies and rapid electrification in technology development. It is compatible with the aim of limiting the global temperature increase to 2°C in this century with the 1.5°C ultimate goal so that its targets are in line with the Paris Agreement decisions. It aims to adopt energy-efficient technologies; use

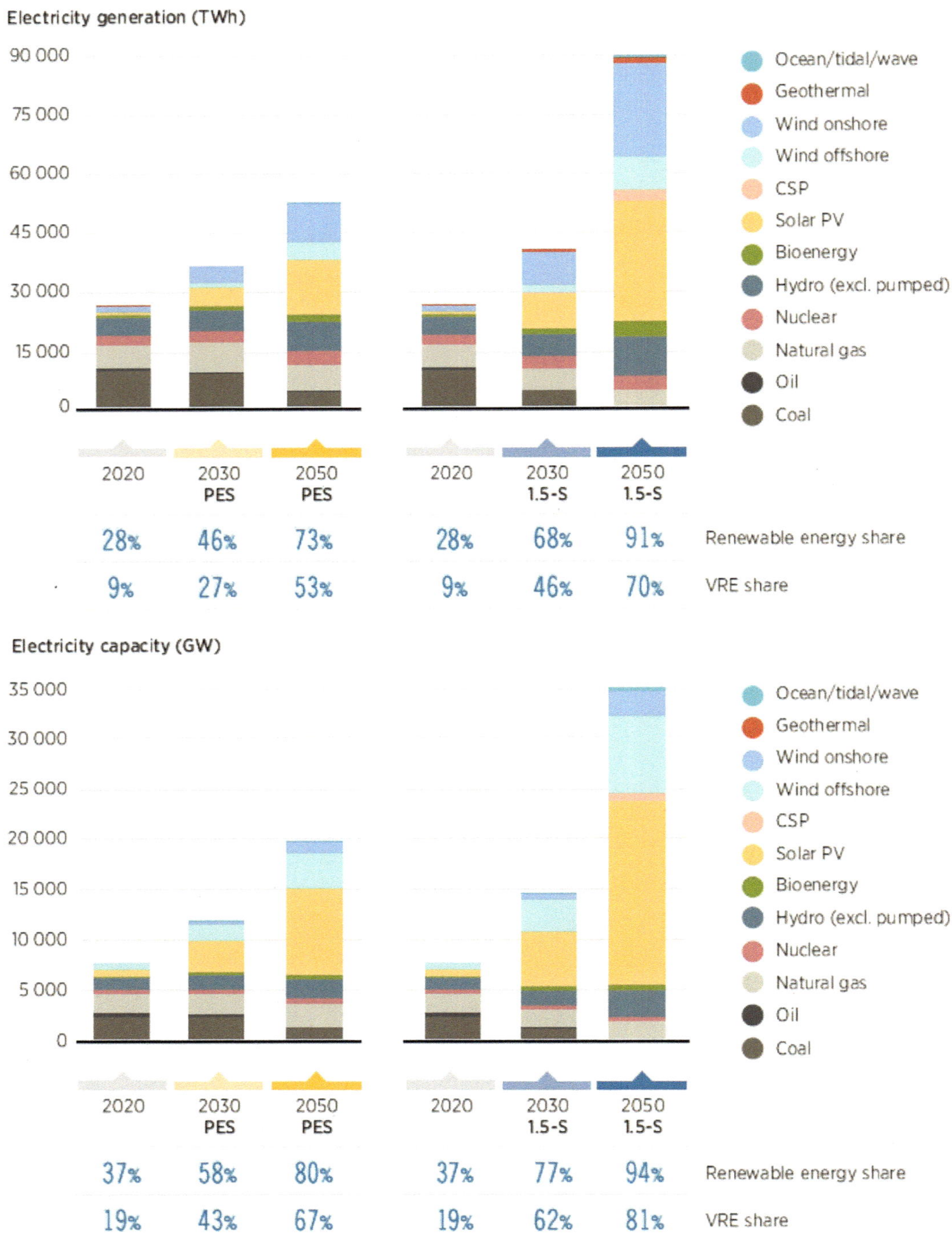

FIGURE 19.6 Electricity generation (terawatt hour, TWh) and capacity (gigawatt, GW) in PES and 1.5°C Scenario in 2020, 2030, and 2050. (*VRE*: variable renewable energy; *CCS*: carbon capture and storage.)

Source: IRENA (2023).

low-carbon alternatives in different areas, like industrial materials, buildings, and infrastructure; reduce petrochemical products with plastic recycling and alternatives; and aggressively transition to electric and hybrid vehicle engines. It also assumes significant actions in carbon capture and storage (CCS) from 2040 onward, and developments in jet fuel efficiency and alternative airplane technologies (Barclays, 2019).

Development is called a "business-as-usual" scenario, which assumes more tightened environmental policies in the industry. However, this scenario is not enough to meet the 2°C target. In this scenario, it is expected to use less oil in industrial materials, infrastructure, and cars. On the contrary, this might lead to an increase in demand for airlines, transport by truck, and petrochemical

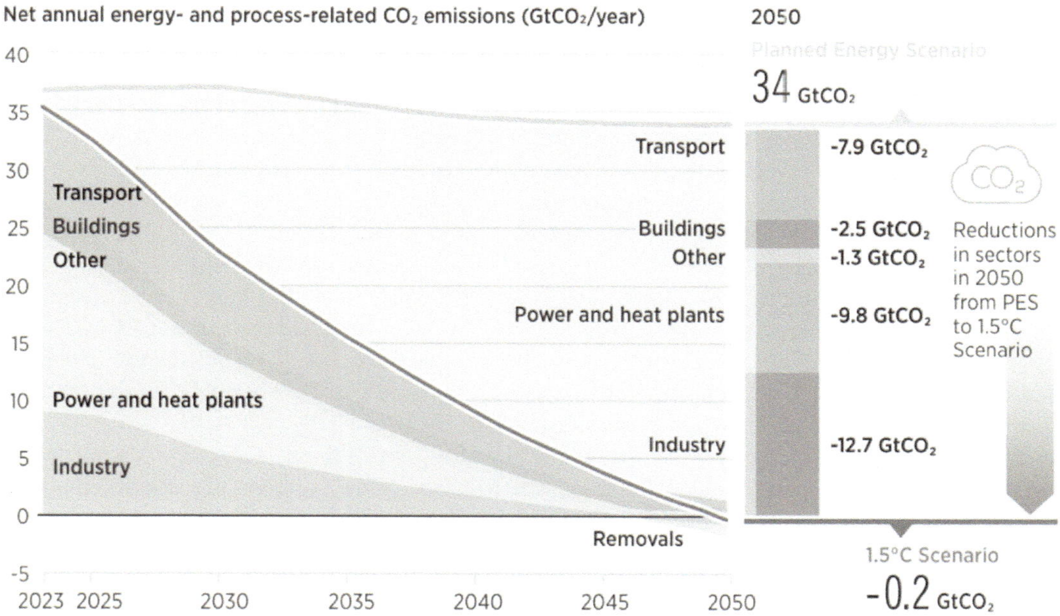

FIGURE 19.7 Global CO_2 emissions (gigatons of CO_2, $GtCO_2$) under PES and the 1.5°C Scenario between 2023 and 2050 based on sectors.

Source: IRENA (2023).

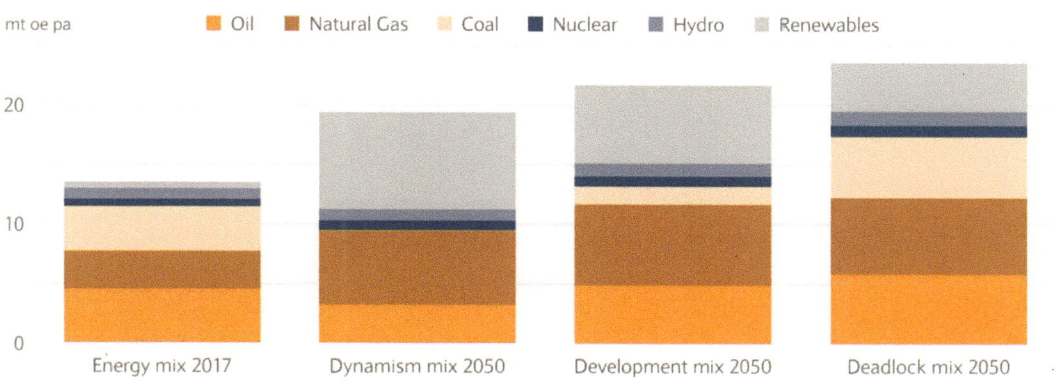

FIGURE 19.8 Energy mix in 2017 compared to Barclays's 3D Scenarios (Dynamism, Development, and Deadlock) in 2050 based on the type of sources (million tons of oil equivalent per annum, mtoe pa).

Source: Barclays (2019).

materials. Deadlock is the worst-case scenario on the subject of reducing carbon emissions. This scenario offers a limited focus on low-carbon policies due to a polarized world, little technological adoption, and trade wars. It still shows modest economic growth (Barclays, 2019).

When we focus on key assumptions of Barclays's 3D Scenarios, there are important disparities between these three scenarios. Gross domestic product (GDP) growth initially starts at a slow pace, then accelerates to 3.1% per annum (pa) in Dynamism, a historical average of 2.6% per annum (pa) in Development, and lower than the historical average of 2.4% per annum (pa) in Deadlock. For energy intensity, the energy–GDP growth ratio will fall from 65% to 30% in Dynamism, 50% in Development, and 60% in Deadlock by 2050. The electrification rate will increase from 17% to 35% in Dynamism, 25% in Development, and 20% in Deadlock in the energy mix by 2050. The share of renewables in the energy mix will rise from 5% to 40% in Dynamism, 30% in Development, and 15% in Deadlock by 2050 (Figure 19.8).

BP announced three main scenarios, which are Accelerated, Net-Zero, and New Momentum (BP, 2023). Accelerated and Net-Zero are aligned with a range of IPCC scenarios, including limiting global temperature rise well below 2°C and 1.5°C above pre-industrial temperature by 2100. With reference to 2019 levels, CO_2-equivalent emissions in the Accelerated scenario, Net-Zero scenario, and New Momentum scenario aim to reduce by 75%, 95%, and 30%, respectively, by 2050. Figure 19.9 illustrates the total final consumption of these scenarios through 2050. According to 2019 figures, total final consumptions in the Accelerated scenario, Net-Zero scenario, and New Momentum scenarios are expected to be 15% lower, 30% lower, and 10% higher, respectively, by 2050. All three scenarios propose a fall in consumption of fossil fuels, which is from 65% in 2019 to 20–50% by 2050. Especially, the rate of coal use has declined dramatically since the use of electrification and low-carbon hydrogen adoption increases in the industry. The share of oil also decreases due to electrification in the transportation industry (BP, 2023).

Equinor proposed two main scenarios: Walls and Bridges (Equinor, 2023). Walls is related to current market trends, policy, and technology, which shows a slow accelerating pace at climate in the future. Contrary to Walls, Bridges is a normative back-cast scenario that is 50% compatible with no more than the 1.5°C global temperature rise target. Energy intensity increased by 1.2% per year between 1990 and 2020. It decreases by 1.9% per year in Walls and 3.2% per year in

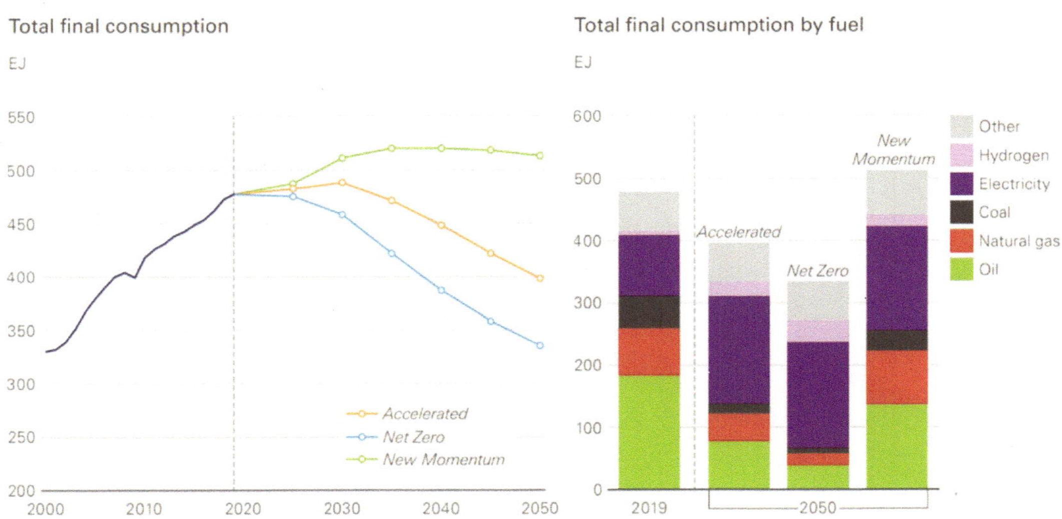

FIGURE 19.9 Total final energy consumption from 2000 to 2050 and the one by fuel (exajoule, EJ) in Accelerated, Net-Zero, and New Momentum scenarios compared with 2019 data.

Source: BP (2023).

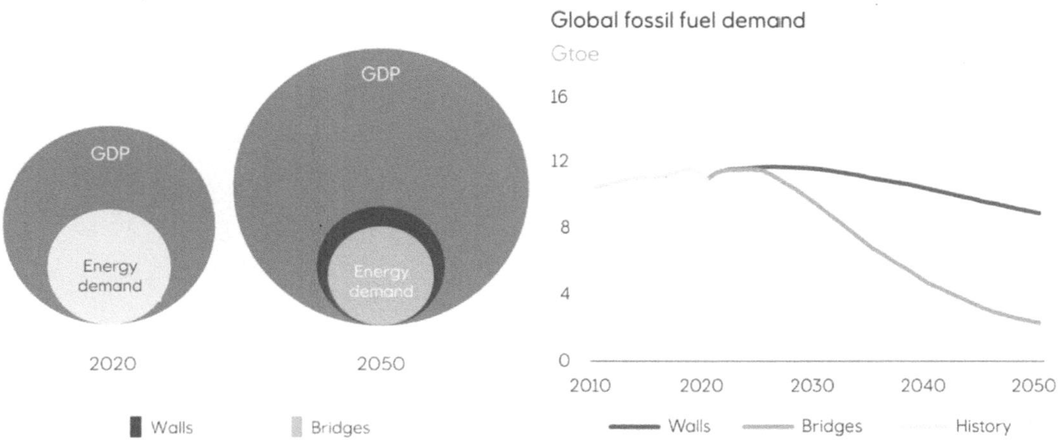

FIGURE 19.10 Global energy demand and global fossil fuel demand (gigatons of oil equivalent, Gtoe) based on Walls and Bridges scenarios.

Source: Equinor (2023).

Bridges toward 2050 with increasing electrification. Figure 19.10 shows global energy and fossil fuel demand. According to the figure, global fossil fuel demand will peak in 2026 in Walls, followed by a gentle decrease. On the other hand, this demand will decrease rapidly after 2025 in Bridges, and the effect of all fossil fuel use will be abated or recovered by carbon removal. Gas demand will peak 14 years earlier in Bridges by 2025 than in Walls. Energy consumption is higher in Bridges compared to Walls, with massive acceleration before 2030. Wind capacity is five times greater in Walls and eight times greater in Bridges by 2050 according to 2020 data. Solar photovoltaic (PV) capacity will be 9 times higher in Walls and 13 times higher in Bridges by 2050. Even though electrification and hydrogen have important roles in the decarbonization of transportation in both scenarios, Bridges shows further decarbonization. CCUS operations start to accelerate after 2030 in Walls. Furthermore, Equinor scenarios contain foresight about mineral demands. In Walls, demand for minerals will double by the early 2030s according to average annual demand between 2016 and 2020 (Equinor, 2023).

ExxonMobil (2019) prepared energy projections for 2040 based on three drivers: technology, policy, and consumer preferences. The global energy demand is expected to be mainly driven by non-OECD (Organisation for Economic Co-operation and Development) countries (nearly 70%). This demand has the highest share in electricity generation by sector, which is followed by transportation and industry in 2040. Figure 19.11 includes the global energy mix in demand that shifts to lower-carbon fuels mostly. According to ExxonMobil (2019), natural gas will grow faster than other energy sources through 2040. Moreover, renewables and nuclear have increasing trends through this time. Oil seems to have a leading role in energy demand, and coal is still used mostly in developing nations, with a nearly 20% decrease in share by 2040. For climate action, four key areas are pointed, which are increasing energy efficiency, transitioning the energy mix toward low-carbon sources, preparing policies for cost-efficiency, and investing in advanced technologies (ExxonMobil, 2019).

McKinsey announced five scenarios and called them Fading Momentum, Current Trajectory, Further Acceleration, Achieved Commitments, and 1.5°C Pathway, sorted from slower to faster action according to the speed of energy transition (McKinsey, 2022). In Fading Momentum, public sentiment will result in longtime dominance of fossil fuels and a slower pace in climate policies and cost reductions, with an expected global temperature rise higher than 2.4°C in 2050. The Current Trajectory continues to fall in renewable costs continuously, and current policies are insufficient, with around 2.4°C (between 1.9 and 2.9°C) temperature rise linked with emissions. In Further

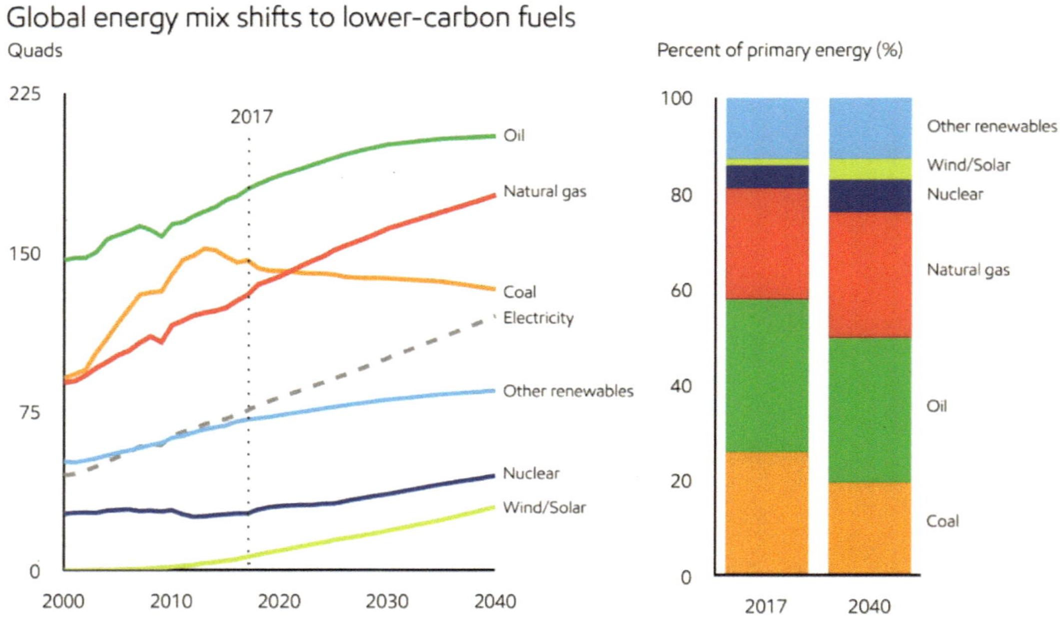

FIGURE 19.11 Global energy mix in demand from 2000 to 2040 and percent of primary energy in 2017 and 2040 by sources.

Source: ExxonMobil (2019).

Acceleration, energy transition will be carried out by country-specific decisions, and technological and financial limitations still continue with an increase in global temperature around 1.9°C (between 1.6 and 2.4°C). Achieved Commitments include net-zero commitments which are achieved by intentional policies of leading countries and a slower pace of followers in energy transition, with around 1.7°C (between 1.4 and 2.1°C) expected global temperature rise. The 1.5°C Pathway is expected to be globally adopted, which requires investments for rapid decarbonization and changes in behaviors with limiting temperature increase below 1.5°C in 2050. Moreover, the CO_2 pricing will be less than €50, €55–130, €75–140, €100–180, and more than €200 in the mentioned five scenarios, going from slower to faster action one (McKinsey, 2022).

The global power generation based on the Current Trajectory, Further Acceleration, and Achieved Commitments scenarios through 2050 is illustrated in Figure 19.12. These three scenarios are expected to reach between 79 and 89% in renewable energy share in 2050. In Further Acceleration, renewable energy share is anticipated to increase from 29% to 60% between 2020 and 2035, which means double within 15 years. Mainly, solar and onshore wind will be responsible for most of this growth, with 43% and 26% share due to further cost and technology improvements. On the contrary, nuclear energy still requires public acceptance and economic support from policies in these three scenarios (McKinsey, 2022).

Shell (2018) announced a new scenario named Sky and compared it with the company's original New Lens Scenarios, which are Mountains and Oceans (Figure 19.13). Whereas Mountains is a government-led scenario with a top-down approach, Oceans is a bottom-up scenario with a market-driven outcome. Both Mountains and Oceans are not enough to limit global temperature rise well below 2°C. Different from these two scenarios, Sky contains better multi-lateral collaboration for greenhouse gas emissions and climate effect issues, and it helps to reach well below the 2°C target. Between these three scenarios, Sky targets the highest CO_2 emissions reduction by achieving net-zero by 2070. Then, the second one is Mountain, and the third one is Oceans, in this aspect. Sky presumes that consumers preferentially select low-carbon, high-efficiency options for their energy

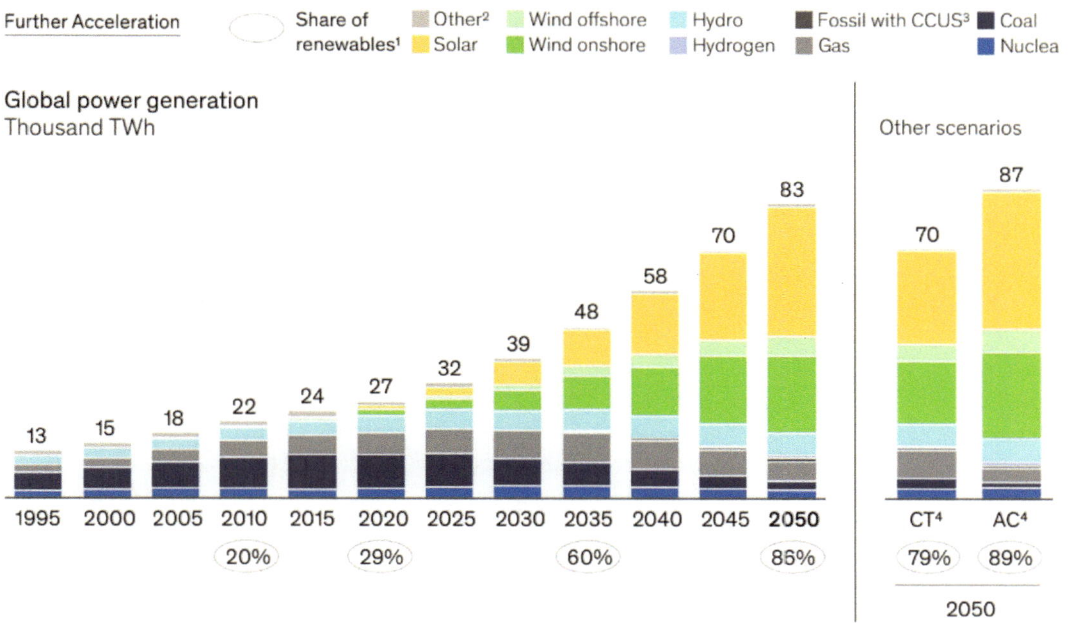

FIGURE 19.12 Global power generation (thousand terawatt hour, TWh) in Further Acceleration from 1995 to 2050, and in Current Trajectory (CT) and Achieved Commitments (AC) in 2050.

Source: McKinsey (2022).

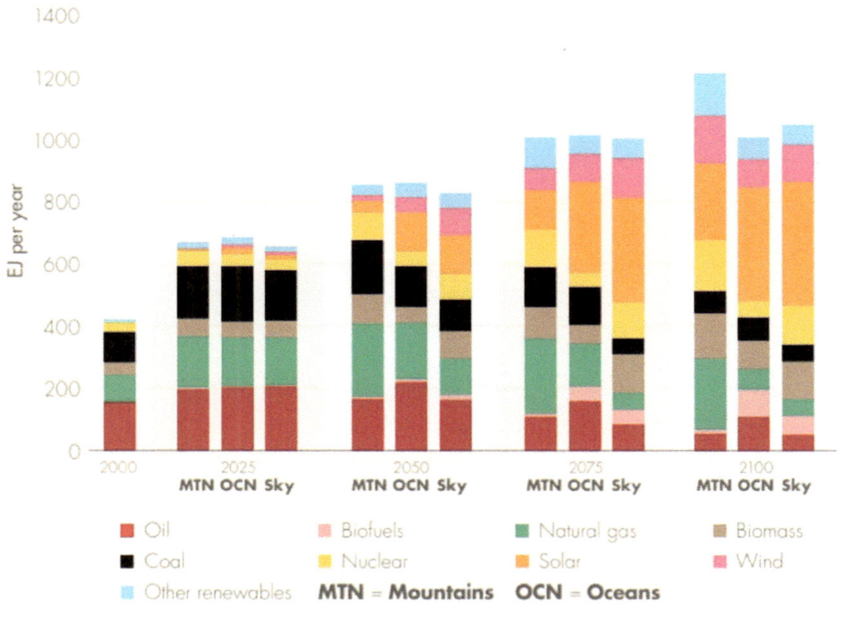

FIGURE 19.13 Primary energy by source (exajoule, EJ per year) in 2000, and in Mountains, Oceans, and Sky scenarios in 2025, 2050, 2075, and 2100.

Source: Shell (2018).

service needs. Carbon pricing is expected to be adopted globally by governments in the 2020s with appropriate costs of CO_2 for services and goods. Electrification of end users has increased by more than three times, and the generation of electricity is more than five times the 2018 numbers. Primary energy from renewables overcomes fossil fuels in the 2050s, while new energy sources increase by 50-fold (Figure 19.13). Solar, bioenergy, and wind are the most-used renewables; oil will dominate the fossil fuel sector by 2070. In addition, nearly 10,000 large-scale CCS facilities are constructed (Shell, 2018).

TotalEnergies (2022a, 2022b) proposed two main energy scenarios for 2050, which are Momentum and Rupture. Momentum is a forward-looking scenario related to existing decarbonization strategies of 40 Net-Zero by 2050 countries. The global temperature increase will be between 2.1 and 2.3°C by 2100. GDP and energy growth are expected to be +3.0% and +0.4% per year, respectively. This scenario starts with a high demand for fossil fuels that leads to high greenhouse gas emissions. Revolution in transportation contains electrification of light and heavy-duty vehicles and sustainable fuels for aviation and marine transportation, but they are below the targets. Usage of renewable energy sources and electrification of end uses have growing trends. Natural gas has an important role in energy transition with biofuels and biogas in specific areas. Hydrogen (H_2) energy grows after 2030. Moreover, polymer recycling is aimed at this scenario (TotalEnergies, 2022a).

Rupture is a back-casting scenario that aims to catch the well below 2°C objective of the Paris Agreement by 2050. Different from Momentum, Rupture is expected to be applied in all emerging economies in order to reach its objectives. GDP and energy growth are expected to be +3.0% and +0.2% per year, respectively. Energy efficiency is increased, and GHG emissions are reduced. Further development in renewable electricity is necessary. Road transport is evolved by the widespread usage of zero-emission vehicles. New energy carriers, such as biofuels, biogas, e-fuels, and clean H_2, are highly preferred (TotalEnergies, 2022a).

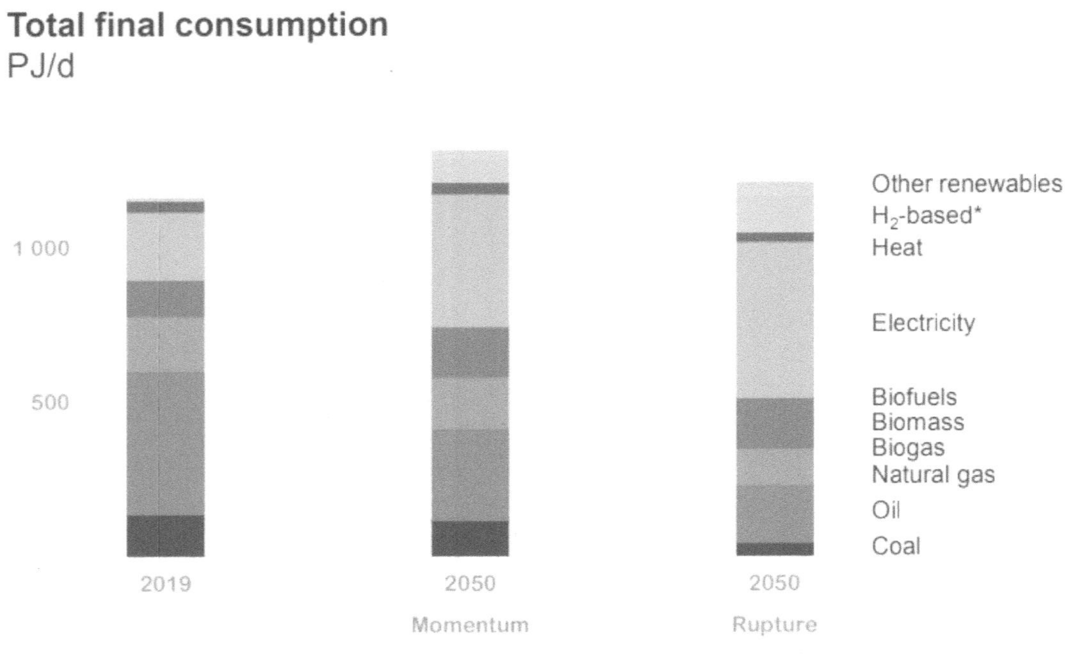

FIGURE 19.14 Total final consumption (petajoule per day, PJ/d) in 2019, and in Momentum and Rupture in 2050.

Source: TotalEnergies (2022a).

Figure 19.14 shows the total final consumption in 2019 and expected consumption in 2050 for both Momentum and Rupture. In Rupture, coal is almost phased out completely and oil has a dramatic fall. Natural gas and green gases seem to see a continuation in usage. Bioenergies (biofuels, biogas, biomass), electricity, and H_2-based fuels are nearly two-thirds of consumption. In 2050, 6.5 Gt of CO_2 emissions will be eliminated mainly in power generation and industry, and this amount might reach 7 Gt with technologies like direct air capture (DAC) (TotalEnergies, 2022a).

In addition, Rupture+ has been announced as the improved version of Rupture. It aims to limit the global temperature rise to 1.5°C and reach net-zero CO_2 emissions in 2050. Similar to Rupture, Rupture+ has a higher energy demand in 2050 than the one in 2019. Oil demand decreases to 23 million barrels per day (Mb/d), which is close to IEA's NZE scenario (24 Mb/d). Power generation from green H_2 is accelerated. Low-carbon H_2 and electricity have rising trends in transportation, and 7.6 Gt CO_2 emissions are removed by CCS operations (TotalEnergies, 2022a).

The World Energy Council launched three scenarios, named Modern Jazz, Unfinished Symphony, and Hard Rock (World Energy Council, 2019). *Modern Jazz* is a digitally disruptive and innovative scenario which has energy plus services, dominant market mechanisms, and is dominated by data. *Unfinished Symphony* is a long-term scenario with strong policies and significant action for climate and broad agendas. *Hard Rock* is characterized by low global cooperation and more national focus that has a low success rate different from the previous ones (World Energy Council, 2019).

The comparisons of Modern Jazz, Unfinished Symphony, and Hard Rock scenarios are explained in Figure 19.15 with primary and final energies. These scenarios show growth in the primary energy demand at the global level from 2020 to 2040. Unfinished Symphony has only a 4% increase in energy demand, with advantages of policies and regulations, while Modern Jazz has 13% growth via innovative and market-driven systems, and Hard Rock has 21%, the highest growth, without strong policy and market-driven innovations. Fossil fuels comprise more than two-thirds of primary energy demand in all the defined scenarios. Coal and oil have a volumetric decrease in Modern Jazz and Unfinished Symphony through 2040, but they increase in Hard Rock. *Gas*, defined as less harmful hydrocarbons compared to oil and coal, has a volumetric rise in all three scenarios, especially in Modern Jazz, with more availability and relatively low cost. The final energy consumption also increases in all scenarios, with differences mainly in industry and transportation

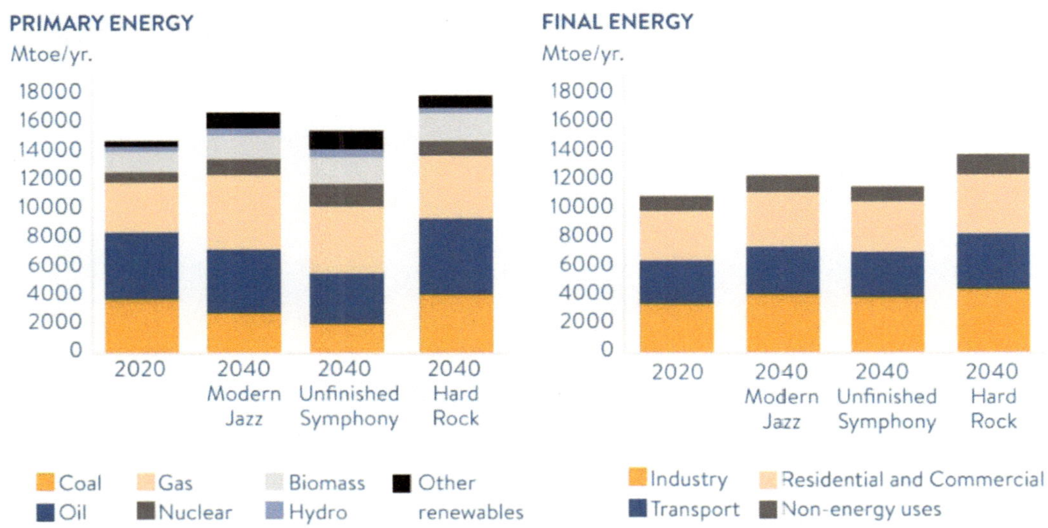

FIGURE 19.15 Primary energy by source and final energy by use (million tons of oil equivalent per year, Mtoe/yr) in 2020, and in Modern Jazz, Unfinished Symphony, and Hard Rock in 2040.

Source: World Energy Council (2019).

(Figure 19.15). Unfinished Symphony has the strongest growth in electrification with 60% increase, followed by Modern Jazz with 28%, and Hard Rock with 4% between 2020 and 2040. CCUS technologies have significant progress in Unfinished Symphony with good support in carbon pricing mechanisms. Modern Jazz also has progress in CCUS with several full-scale commercial projects. On the other hand, deployment of this technology fails in Hard Rock. Hydrogen production for use has similar trends, like CCUS, for each scenario. Furthermore, all scenarios cannot meet the well below 2°C target of the Paris Agreement (World Energy Council, 2019).

19.3 FUTURE DIRECTIONS IN OIL AND GAS

In light of the majority of the reports explained in the previous part, oil and gas will mostly have a lower share in the global energy mix in future scenarios. In some cases, gas will continue to grow even until 2050 since it is a lower-carbon alternative to coal that can decrease carbon intensity and improve air quality (ExxonMobil, 2023).

Figure 19.16 illustrates the projections of hydrocarbon production based on IEA's Net-Zero Emissions (NZE) scenario. All hydrocarbon production trends have a significant downward shift – after reaching peak production. According to an IEA (2021a) report, oil demand will never reach its 2019 peak value again. It will decrease from 88 million barrels per day (mb/d) in 2020 to 72 mb/d in 2030 and 24 mb/d in 2050. On the other hand, gas will show an increasing trend through the mid-2020s. Its peak is expected to be around 4,300 bcm (billion cubic meters). Later, the demand is expected to drop to 3,700 bcm by 2030 and 1,750 bcm by 2050, respectively. A major reduction will be observed in coal. Its usage will reduce from 5,250 Mtce (million tons of coal equivalent) in 2020 to 2,500 Mtce by 2030 and less than 600 Mtce by 2050. As a result, oil, gas, and coal demand reduction is expected to reach 75%, 55%, and 90% by 2050, respectively (IEA, 2021a).

Figure 19.17 shows the oil and gas production trend through 2050 based on world geographies. During the downtrend, the Middle East and North America will have higher oil production output throughout 2050. Gas production is more evenly distributed across Eurasia, Asia Pacific, the Middle East, and North American regions. Hence, oil and gas supply will be concentrated in a few major countries that produce them (IEA, 2021a).

In order to provide significant future foresight for both oil and gas, the mean values of several future scenarios were calculated by Davies and Simmons (2021). It is called the Paris Mean. Based on these calculations, both oil and gas demands with the Paris Mean and required demand on the

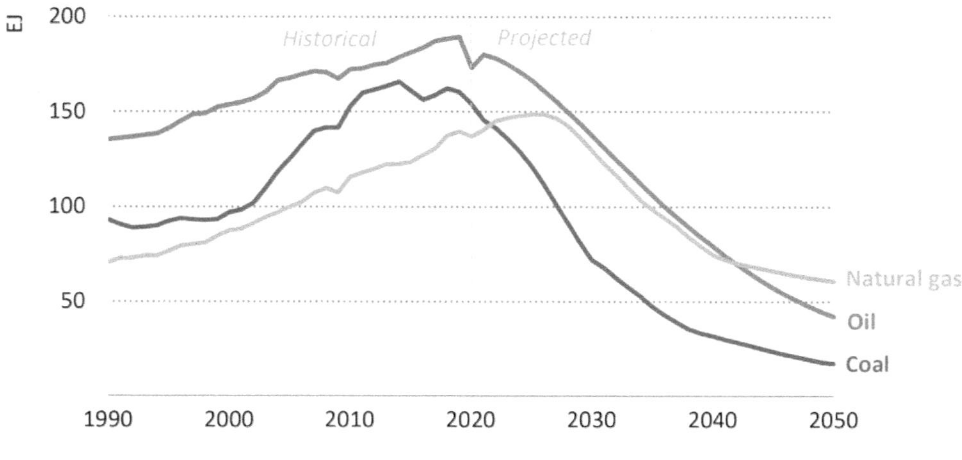

FIGURE 19.16 Production of hydrocarbons (exajoule, EJ) from 1990 to 2050 according to NZE.

Source: IEA (2021a).

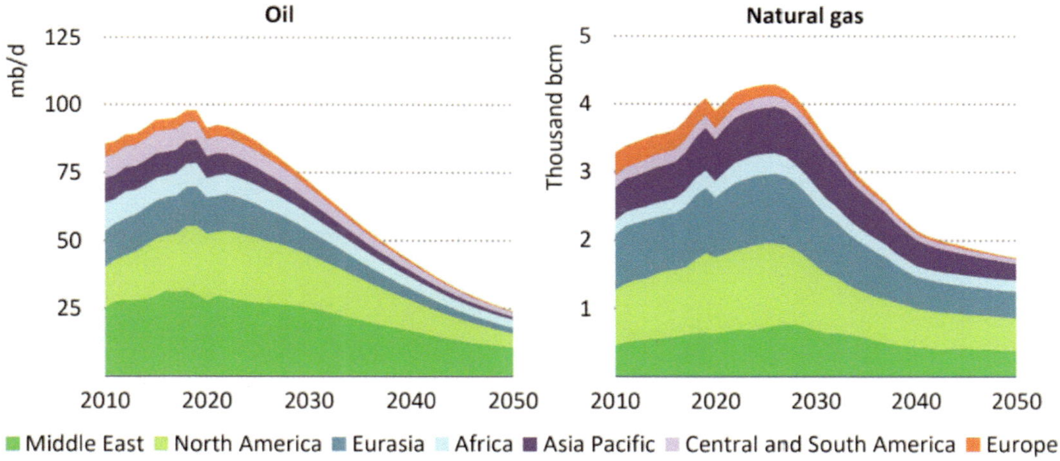

FIGURE 19.17 Production of oil (million barrels per day, mb/d) and gas (thousand billion cubic meters, bcm) between 2010 and 2050 based on regions of the world.

Source: IEA (2021a).

existing fields in the world have been mentioned. While global oil demand is expected to be between 1,115.8 and 805.4 Bbbl (billion barrels) by 2050, the equivalent Paris Mean demand will be 943.3 Bbbl, with 104.5 Bbbl standard deviation. To meet oil demand from 2020 to 2050, the total increase in oil production from the existing fields is between 586.1 and 659.1 Bbbl, which translates into an annual production increase of between 9.2 and 11.5 Bbbl.

Between 2020 and 2050, global gas demand ranges from 5,374.1 to 4,017.03 Tcf (trillion cubic feet). The Paris Mean demand is 4,733.49 Tcf, with 424.9 Tcf standard deviation. To meet this demand, the total increase in gas production from the existing fields is between 2,458.91 and 2,562.77 Tcf, so between 70.0 and 73.4 Tcf of gases need to be explored every year.

Between 1860 and 2019, the world consumed 1.52 trillion barrels and 4,347.8 Tcf of oil and gas, respectively. Consequently, the Paris Mean calculations equal 62% of oil and 108.9% of gas consumed (Davies and Simmons, 2021).

In addition, "advantaged hydrocarbons" is used as a new term in the petroleum industry, which originally indicates a large volume of hydrocarbons extracted at low cost (Tischuk and Latham, 2021). Nowadays, this term has been used to define the hydrocarbons that are needed to be explored and produced with minimized greenhouse gas emissions. The key features of advantaged hydrocarbons are minimum risk of exploration failure, comparable simplicity in drilling, lower need for subsurface operations during the production phase, with limited environmental footprint, with strong and predictable long-term production of oil and gas. This concept has been mainly affected by drilling operations, hydrocarbon type, reservoir characteristics, and additional factors, such as geography, policy, and technical developments. The effects of drilling on the advantaged hydrocarbons are directly related to choices (extent of logging and coring), matter (onshore and offshore), location (onshore or offshore), and nature of subsurface (pressure, lithology, depth, hazards) (Davies and Simmons, 2021). The cost of operations influences energy expenditure, with a direct correlation to the depth of the well. In addition, complexities in operations containing unconsolidated, cemented, or faulted formations, high-pressure zones, and sloughing shale raise operational costs and cause disadvantages (Hossain, 2015; Davies and Simmons, 2021). For the hydrocarbon type, gas can be accepted as an advantage since it leads to less greenhouse gas emissions (Tanaka et al., 2019). These emissions are also related to oil's density, extraction method, and depth. For example, extra-heavy oils have 60% more greenhouse gases compared to light oils, and the sour gases are used for end use power generation, which results in an additional 15% emissions on average (Jiang et al., 2011;

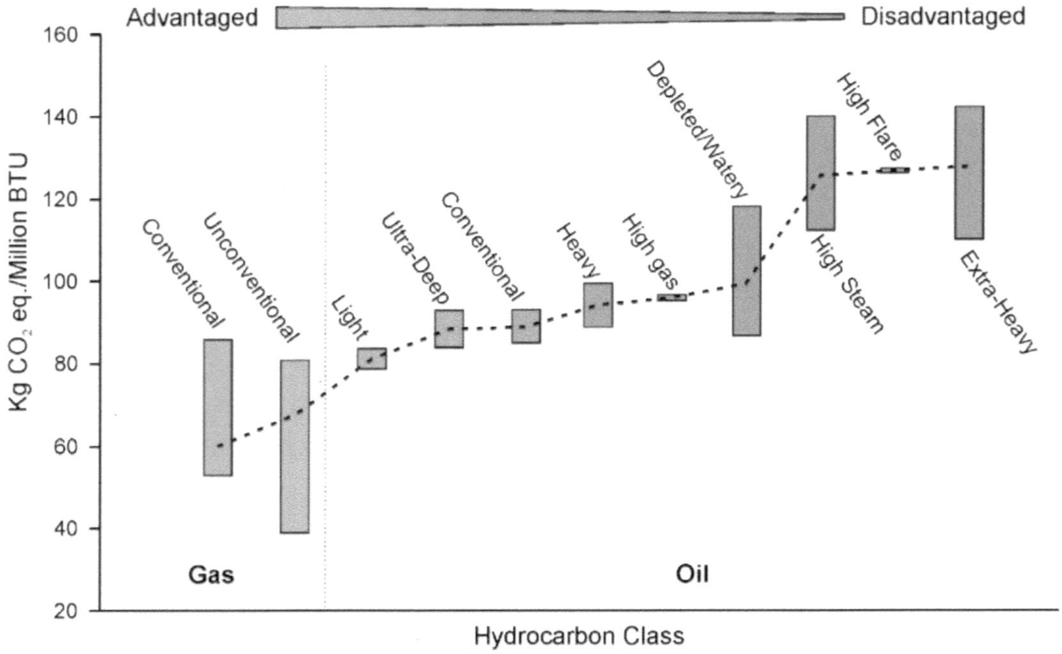

FIGURE 19.18 Hydrocarbon class versus total life cycle GHG emissions (kilograms of CO_2 equivalent per million British thermal units, kg CO_2 eq./million BTU) represented from advantaged to disadvantaged hydrocarbons.

Source: Davies and Simmons (2021).

Burnham et al., 2012; Alkatheri et al., 2018). Reservoir characterization is mainly related to the depth of burial, heterogeneity, average porosity, and diagenesis (Aliyuda et al., 2020), and a better understanding of it is to provide fewer risks during exploration, a lower number of drilling dry wells with high recovery factors that means less wasted energy and CO_2 emissions. Davies and Simmons (2021) have summarized the advantaged and disadvantaged hydrocarbon scale according to hydrocarbon class versus greenhouse gas emissions in Figure 19.18. Based on the figure, gas, especially the conventional one, is the most advantaged, whereas extra-heavy oil is the least-advantaged one.

Abating oil and gas emissions of the hydrocarbon industry is an essential issue in the future. Based on the NZE scenario of IEA, reductions in total oil and gas emissions will depend on some key actions, which are reduction in hydrocarbon demand, decrease in methane emissions, removal of non-emergency flaring, electrification in upstream activities, adaption of CCUS to oil and gas industry, and largely, use of low-emission electrolysis hydrogen in refineries (Figure 19.19). Since the Industrial Revolution, methane has led to nearly 30% of global temperature rise. In the oil and gas industry, estimated methane emissions are 80 Mt (million tons), which is equal to 2.4 Gt of CO_2 equivalent (Gt CO_2-eq). Hence, eliminating methane emissions in a sustainable and rapid way is essential. According to the NZE scenario, the IEA methane reduction target is 60 Mt, nearly 75%, by 2030. Flaring caused 500 Mt CO_2-eq greenhouse gas emissions in 2022 by the combustion of methane and natural gas liquids, while 365 Mt CO_2-eq emissions and 95% of flared volumes will be eliminated by removing non-emergency flaring. Electrification in upstream activities is also important to decrease half of the CO_2 emissions resulting from oil and gas operations. CCUS projects captured and stored 25 Mt CO_2/year from natural gas processing. Low-emission electrolysis hydrogen has the smallest role in reducing carbon emission among the other mentioned key actions. However, it is forecasted that it might have an important role after 2030 (IEA, 2023b).

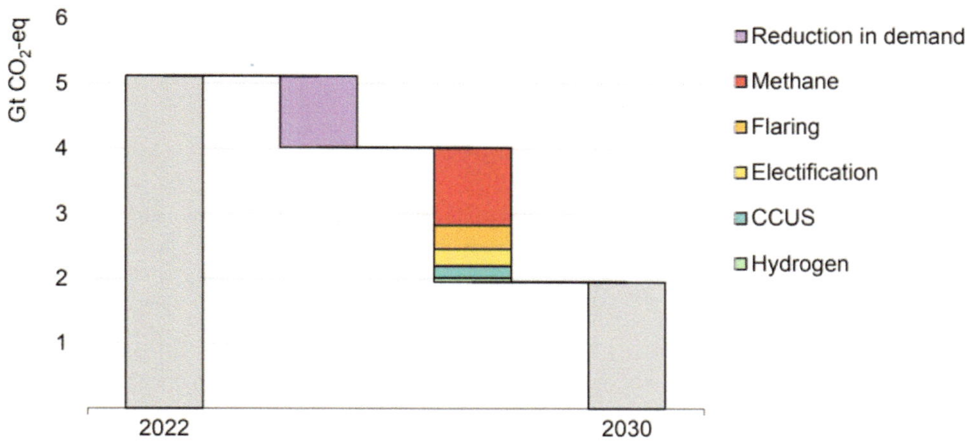

FIGURE 19.19 Oil and gas emissions reductions (gigatons of CO_2 equivalent, Gt CO_2-eq) in NZE to 2030. **Source: IEA (2023b).**

While focusing on the future directions of oil and gas, CCS/CCUS technology has been applied to the fossil fuel industry since the 1970s by injecting CO_2 underground for enhanced oil recovery (EOR). The first CO_2 storage project, which was first developed for the purpose of reducing greenhouse gas emissions other than EOR, was put into operation in the Sleipner gas field in Norway in 1996 (IEA, 2022c). In the Sleipner CCS plant in the North Sea, more than 19 Mt of CO_2 have been stored in the Utsira Formation by the end of 2020 (Equinor, n.d.). Sleipner's CO_2 injection rate is 0.85–1.0 Mt per annum. However, this rate is smaller than many proposed CCS projects worldwide now (IEEFA, 2023).

Approximately 35 commercial facilities are operating globally in 2022 applying CCUS to industrial processes, fuel conversion, and power generation. The total annual CO_2 capture potential of these facilities is approximately 45 Mt. The project developers have announced their targets for more than 200 new carbon capture plants that will be operational by 2030 and are expected to capture more than 220 Mt of CO_2 per year (IEA, 2022d). In addition, CO_2 storage capacity by 2050 can be projected to be around 700 Mt per year, which is just 10% of the required amount. In order to catch the required level for CCUS operations in the future, five aspects are defined by Martin-Roberts et al. (2021) (Table 19.3). These aspects include opening new carbon capture facilities, equipment availability, transportation routes, development of storage sites, and deployment of infrastructures.

Several future energy scenarios have higher targets than today's data for CCUS applications. According to BP's Accelerated and Net-Zero scenarios, CCUS will reach 4–6 Gt CO_2 by 2050 (BP, 2023). McKinsey's Achieved Commitments, Further Acceleration, and Current Trajectories scenarios aim to reach nearly 4.2, 3.6, and 2.1 Gt CO_2, respectively, by 2050 (McKinsey, 2022). In both BP and McKinsey scenarios, the deployment of CCUS with blue hydrogen production takes a large share in CCUS operations (BP, 2023; McKinsey, 2022). TotalEnergies' Momentum and Rupture scenarios are targeting 3 and 6.5 Gt CO_2 emission reductions by 2050 via CCUS operations. In Rupture scenario, this reduction occurs in power generation (50%) and industry (25%) (TotalEnergies, 2022a).

The fossil fuel industry is more prone to apply CCUS of its previous experiences, like EOR techniques (IEA, 2016). Some major oil companies, mainly European companies, such as BP, Shell, Equinor, ENI, Total, and OXY, have shown higher involvement in CO_2 storage to go through a carbon-neutrality future, according to Martin-Roberts et al. (2021).

TABLE 19.3
Aspects for CCUS Deployment

Resource Being Assessed	Amount of Resource Needed	Industry Comparison	Feasibility
Commissioning of new capture facilities	About 75–100 new capture plants/year (worldwide) would be required.	Historic commissioning rates of combined cycle gas turbines (CCGT) (global CCGT build-out peaked in early 2000s, commissioning more than 70 new plants/year between 2000 and 2005, with a maximum of 119 plants/year in 2003).	The required build-out rates thus seem to be possible, but it remains to be seen whether the high rate can be maintained over the decades necessary to bring carbon capture up to speed.
Availability of CO_2 compressors	One compressor per new capture plant would be required.	Current size of CO_2 compressor industry used as comparison.	While production capacity may initially be limited, the industry would be able to adapt and expand to meet CCS requirements.
Creation of CO_2 transport networks	Assumed 10% of CO_2 would be transported by ship, the remainder by pipeline. An estimated 4,500–12,000 km of new CO_2 pipeline/year would be required. CO_2 transport by ship estimated to be required to increase by 15–30 Mtpa.	Development of natural gas transportation pipelines used as a proxy for CO_2 pipeline build-out rates. For shipping estimates, shipping rates of liquefied natural gas (LNG) was used.	Natural gas pipeline construction has maintained a build-out rate of 5,000–9,000 km/year since the 1990s, comparable with estimated requirements for CO_2 transport. Shipping of LNG has an average yearly increase of −~20 Mtpa CO_2 equivalent. Transport of captured CO_2 is not anticipated to block CCS deployment.
Development of CO_2 geological storage sites	Up to 60 storage sites required to be discovered and developed per year (injecting 5–10 Mtpa for 20 years, with final cumulative storage of 100–200 Mt). This equates to discovering 3.0–5.5 Gt storage resource per year, rising to 6.5 Gt by 2050.	Comparison with natural gas discovery rates.	Historically, gas fields discovered at an average rate of ~6 Gt CO_2 storage/year, suggesting that early-stage storage site discovery rates are feasible, but that exploration efficiency will be required to increase in the future to allow continued CCS development.
Deployment of drilling rigs/platforms/well pads	Around 1–4 wells required per 1 Mtpa of CO_2 injection (to account for failed wells, poor injectivity, monitoring, and production wells). A requirement of between 300 and 1,200 new wells per year after 25 years of growth. Around ~40–100 drilling rigs, and installation of 60–120 wellheads or platforms per year.	Comparison with historic global drilling rig counts.	Historic global drilling rig counts increase and decrease on the order of 300 rigs per year. New rigs can be constructed quickly in response to increased demand, suggesting that the required increase in rigs for storage site development is feasible.

Source: Martin-Roberts et al. (2021).

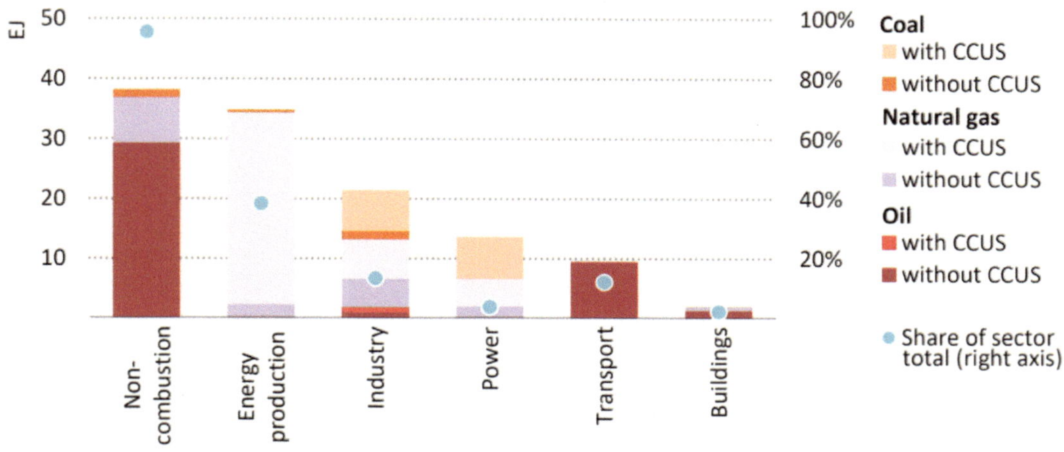

FIGURE 19.20 Usage of fossil fuels (exajoule, EJ) and their share by sector in 2050 in NZE.

Source: IEA (2021a).

Figure 19.20 shows the usage of fossil fuels and their share by sector in 2050 (IEA NZE scenario). Based on IEA (2021a), nearly 120 EJ fossil fuels will be consumed in NZE, which is lower than the value in 2020 (460 EJ). Although the energy areas emit no CO_2 according to the mentioned scenario, usage of fossil fuels does not decrease to zero by 2050. There are three reasons for that. The first one is the usage of fossil fuels in non-energy areas. More than 30% of this usage is non-combustion, such as chemical feedstock, which means it does not lead to direct CO_2 emissions. The second one is that almost half of fossil fuel use in plants will be equipped with CCUS by 2050, which targets for 3.6 Gt CO_2 emissions to be captured by this case. The last one is the usage of fossil fuels in sectors which have no or limited technology options. The unabated usage of fossil fuels will result in 1.7 Gt CO_2 emissions that are aimed to be eliminated by the applications of bioenergy with carbon capture and storage (BECCS) and direct air capture with carbon capture and storage (DACCS) (IEA, 2021a).

Figure 19.21 shows a dramatic drop in investment in hydrocarbon sources through 2030 in IEA APS and NZE scenarios according to IEA (2023c). On the other hand, investments in clean energies, containing clean electrification, low-emission fuels, and energy efficiency, have sharp increases within the same time period.

19.4 FUTURE DIRECTIONS IN RENEWABLES

Despite oil and gas, renewables will have increasing trends in the future based on the reports that have been explained in the previous part. In global energy flows, renewables will meet more than half of the power demand by 2050, according to NZE of IEA. That situation, based on sources and sectors compared to 2021 data, is shown in Figure 19.22. The total electricity production will rise by 3.5% per year through 2050 to catch the expected demand. While focusing on renewable energy sources separately, bioenergy catches around a 15% share in total final consumption by 2050, which has potential usage in power and heat, transport, industry, fuel transformation, buildings, and architecture. Solar PV and wind have the largest increase in share from 2021 to 2050. Hydrogen and its fuels become important, especially in heavy industry and long-distance transport, after 2030. They reach almost 10% share in total final consumption in 2050 (IEA, 2023a).

Moreover, renewables are essential for reducing CO_2 emissions in the energy sector while limiting the global temperature rise well below 1.5°C by 2050. Based on IEA's energy future scenario

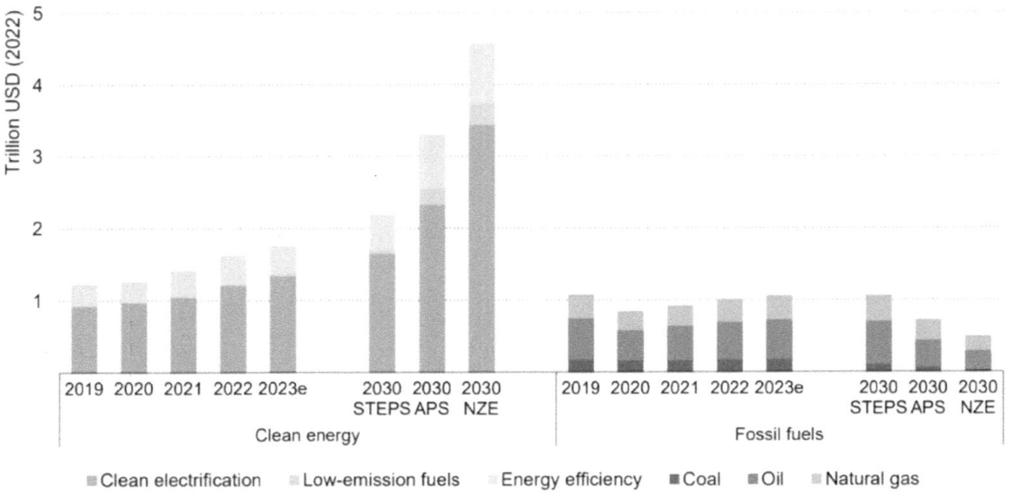

FIGURE 19.21 Investments (trillion United States dollars, USD) in the energy bench market from 2019 to 2023, estimated (2023e), and IEA's scenarios STEPS, APS, NZE in 2030.

Source: IEA (2023c).

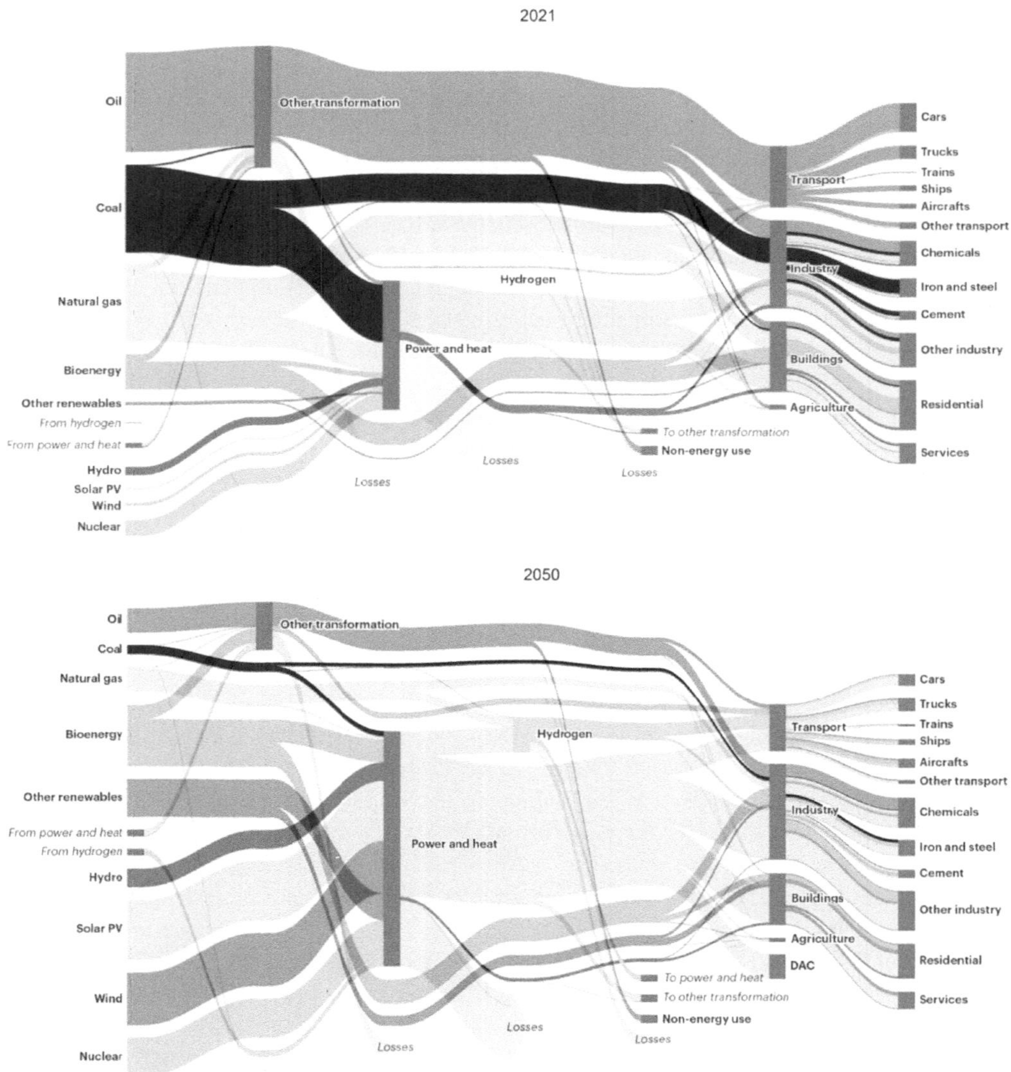

FIGURE 19.22 Global energy flows based on sources and sectors in 2021 and 2050 in IEA's NZE.

Source: IEA (2023a).

NZE, decarbonization of the energy systems by 2050 is held on eight main pillars, which are solar PV and wind, electrification, bioenergy, CCUS, behavioral change and avoided demand, hydrogen, and other fuel shifts (Figure 19.23). Between these pillars, energy efficiency and behavioral change do not need fundamental changes. However, other pillars that account for more than 70% of cumulative CO_2 emission reductions require essential deployment of new technology infrastructures and equipment. In addition, supply chains of some decarbonization pillars are pillar-specific, represented in Figure 19.23. For example, solar PV and wind will be under low-emission electricity supply chain, and electrification will be under heat pumps and electrical vehicles supply chain. Some pillars have cross-cutting supply chains, such as low-emission synthetic hydrocarbon fuels for CCUS, hydrogen, and bioenergy (IEA, 2023a).

Replacing fossil fuels with renewable energies is a significant part of the future pathway. For example, Chen et al. (2022) stated that the new marine energy will replace fossil fuels in the future. The new marine energy comprises wave energy, tidal energy, and temperature difference energy. They can generate nearly 37 trillion kilowatts of electricity per year, which is equal to the heat generated by burning 3,800 barrels of oil. The development of this energy can provide China's carbon-neutrality pathway by 2060. They established a game approach for this potential replacement for the future and, especially, increase of the energy output growth rate of new marine energy while decrease of the one of fossil fuels takes interest (Chen et al., 2022). Moreover, the most energy-efficient and cleanest results can be obtained when low-emissions technologies are used in collaboration with renewables for energy transitions. Hence, the maximum outcome will be reached by combining them with other energy sources and applications, such as wind, solar, geothermal, and electrification of end uses. To illustrate, Wei et al. (2021) mentioned that combining CCUS and renewables increases energy efficiency. Miranda-Barbosa et al. (2017) stated that geothermal energy can be made more useful by combining it with other energy technologies. Combining CCS with geothermal energy is more advantageous as it can reduce CO_2 emissions, reduce construction costs, and use CO_2 instead of water for heat transfer (Miranda-Barbosa et al., 2017). Furthermore, the combination of bioenergy and CCS (BECCS) includes an energy pathway in which CO_2 is captured from a biogenic source and stored permanently. Currently, only about 2 Mt of biogenic CO_2 is sequestered per year, especially in bioethanol applications, and it is targeted to reach 40 Mt CO_2/year by 2030 (IEA, 2022e).

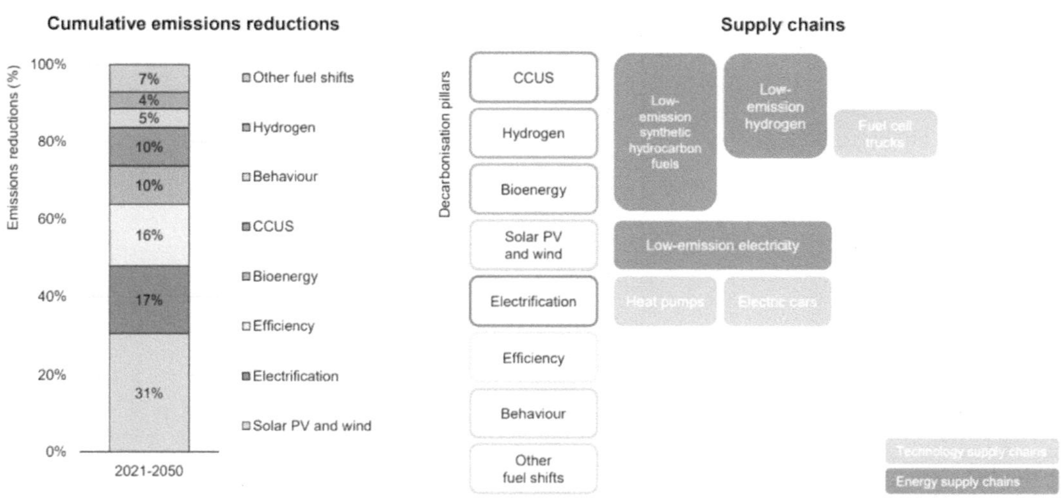

FIGURE 19.23 Cumulative CO_2 emissions reductions and supply chains in the energy sector by 2050 based on IEA's NZE.

Source: IEA (2023a).

Investments in energy transition are likely to have more share of renewable energies. Figure 19.24 shows IRENA in PES and 1.5°C Scenario (2023) for global cumulative energy investments between 2023 and 2050. These expected cumulative amounts are USD 103 trillion in PES, whereas they are USD 150 trillion in the 1.5°C Scenario. The difference between these two scenarios is USD 47 trillion, which means USD 1.7 trillion more investments per year is needed for the 1.5°C Scenario. Both scenarios comprise higher percentages of renewables, electrification, carbon, and hydrogen technologies compared to fossil fuels. The power sector through renewables would require USD 61 trillion in cumulative investments, containing USD 39 trillion for renewable power generation capacity and USD 22 trillion for renewables' infrastructure through 2050. Solar and wind energy would need a higher share of investments. Solar PV, onshore wind, and offshore wind would require USD 333 billion, USD 356 billion, USD 283 billion annual average investments until 2050. Other renewables, including bioenergy, hydropower, ocean technologies, geothermal, and solar thermal, would contain more than USD 400 billion annual investment until 2050 (IRENA, 2023).

With the foresight on a higher share of renewables for power generation and more investment in them in the future, dealing with their challenges becomes essential. One of them is the seasonal and interannual variability of many renewable energy resources. Solar PV, wind turbines, and hydropower depend on the availability of the sun, wind, and hydro, respectively. According to the location and climate conditions, the variability of these energy sources changes significantly. For example, the viability of solar peaks during the summer whereas wind peaks in the winter in a continental (warm summer) system. An arid (cold) system has very flat solar PV potential throughout the

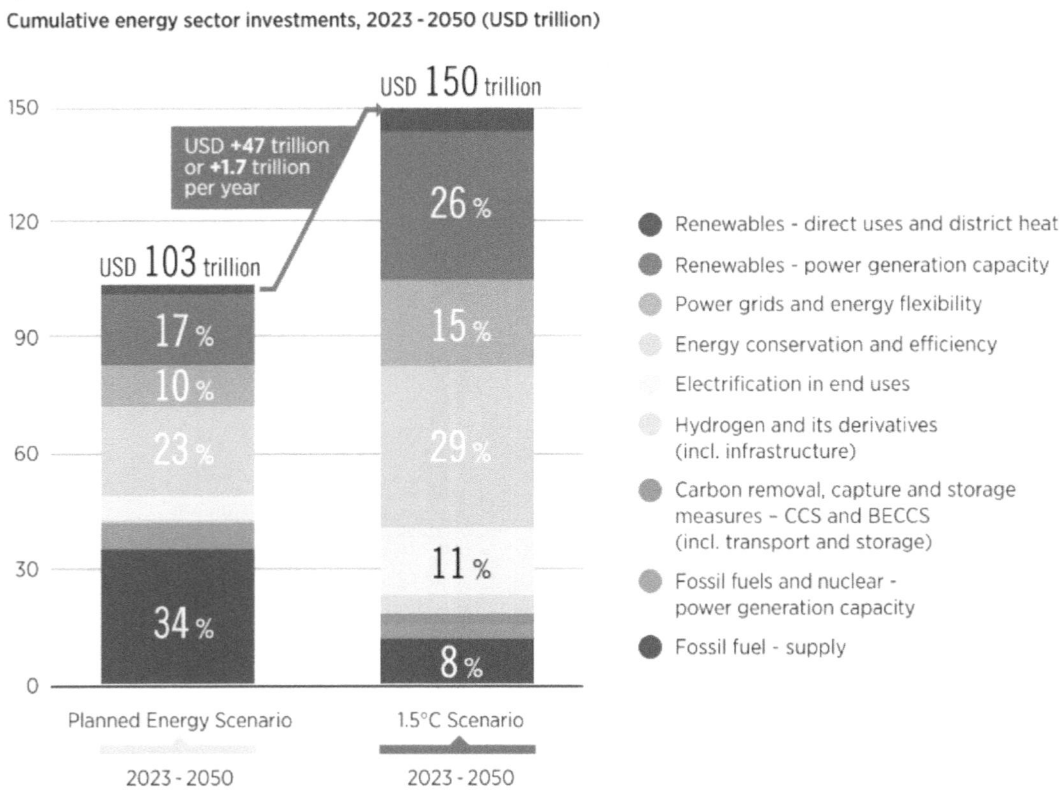

FIGURE 19.24 Global cumulative investments (trillion United States dollars, USD) in the energy sector based on PES and the 1.5°C Scenario of IRENA between 2023 and 2050.

Source: IRENA (2023).

year and a significant fall in wind generation at the beginning of the year. Consequently, systems with variable renewable energies need seasonal flexibility services that include power plants, grids, energy storage, and demand-side response (IEA, 2023d). The other issue is that technologies for clean energies have lower durability compared to other energy technologies. Figure 19.25 represents the lifetime and construction time for energy sources for the purpose of heating and power production. To illustrate, renewables, like solar PV, CSP (concentrating solar power), and onshore and offshore wind, have a significant drop in performance. Most of their technological equipment has a lifetime of around 25 years. On the other hand, the majority of fossil fuel–based power plants can be operated between 30 and 40 years, even more. Hence, this situation increases the requirement to refurbish and recycle the old equipment to decrease their effects on the environment. The last challenge that has been mentioned in this part of the chapter is the necessity of a high number of skilled workers in the renewable energy sector. Some fossil fuel employees have skills which can be transferred to renewable energy sectors (IEA, 2023a). For instance, oil and gas industry workers' knowledge and experiences can be advantageous in the works related to offshore wind sector, such as shipping operations and hydrodynamics (C&S Partners, 2021). To sum up, addressing the challenges in renewables will provide more extensive usage of these clean and environmentally friendly sources and technologies in the future to reach net-zero emissions and limit global temperature rise with 1.5°C targets.

19.5 FUTURE DIRECTIONS IN ENERGY TRANSITION

Energy transition is more complex than just switching from carbon-based to non-carbon-based fuels, according to Craig (2022). Issues such as rapid technological and scientific developments in various sectors, changes in the consumption trends of energy and natural resources over the years, social concerns about climate and environment, energy efficiency, energy demand, and population growth have required energy transformation throughout the history (Craig, 2022). There are three essential transformations in energy: *primary* transformation continues to be from wood to coal, *secondary* energy transformation is from coal to other hydrocarbon sources, and *tertiary* energy transformation is from oil and gas to new energy sources. Their trends and forecasts for global

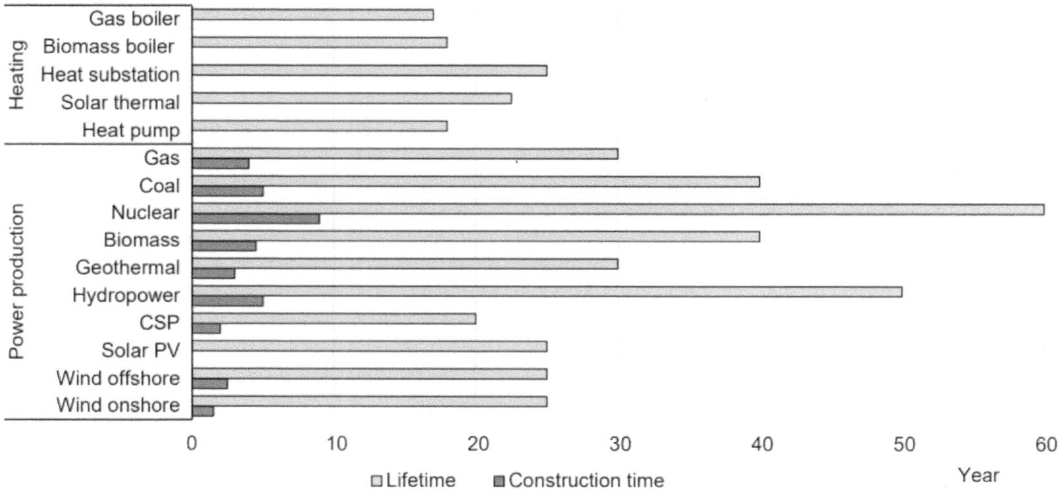

FIGURE 19.25 Operating lifetime and construction time of energy technologies (*CSP*: concentrating solar power).

Source: IEA (2023a).

energy consumption by year are given in Figure 19.26. This chapter focuses on the conditions in the third energy transformation, which started by shifting from coal, oil, and natural gas to a new low-carbon-energy, sustainable, and energy-efficient world. This can also be named the transition from fossil fuels to non–fossil fuels. The driving forces of these development patterns in this current energy transition are mainly the increase in energy demand, the Industrial Revolution, and the need for transport, social civilization, environmental and ecological problems, technological and scientific developments in industry (Zou et al., 2016). Furthermore, it is expected that the current energy transition will take place not only by considering innovations and markets like the previous ones but also by the intensity of the impact of governments seeking urgent solutions to anthropogenic climate change (Gardiner et al., 2023). Natural gas, which has relatively less damage to the environment compared to coal and oil (Zhang et al., 2016; Zou et al., 2016), as well as CCUS from the experiences from EOR technology with reducing CO_2 emissions and using CO_2 in beneficial forms, could provide a bridge for the third energy transition.

Besides all the positive foresight about the future energy transition, there is some consensus about it. Craig (2022) mentioned that governments, businesses, and people in fossil fuel–producing regions will protect their economies and the possibility that the fossil fuel age will not end anytime soon. In some future energy scenarios, it is predicted that despite the rapid growth in the use of renewable energy sources, oil, natural gas, and coal can be used at a high rate in order to meet the social and economic expectations of the world's growing population (IEA, 2021a, 2021b, 2022b; Craig, 2022). World oil demand is expected to increase by 110 million barrels per day, or more, by 2040, and the global temperature rise is expected to be 3.5–5.5°C, unless the necessary emission reductions are realized. In scenarios that limit the temperature increase to below 2°C, it is expected that oil demand will reach its peak in the 2020s and decrease to 70–90 million barrels per day in 2040. In the 1.5°C Scenario, on the other hand, in order to reach the target, this decrease should be up to 65–70 million barrels of daily oil demand in 2040. But even in times of the COVID-19 pandemic, the daily oil consumption in the world was 80 million barrels (Craig, 2022). Considering all the mentioned situations, it is essential to focus on renewables, alternative energy sources and technologies, and emission reduction methods in approaching the determined scenario targets during the energy transition in the future.

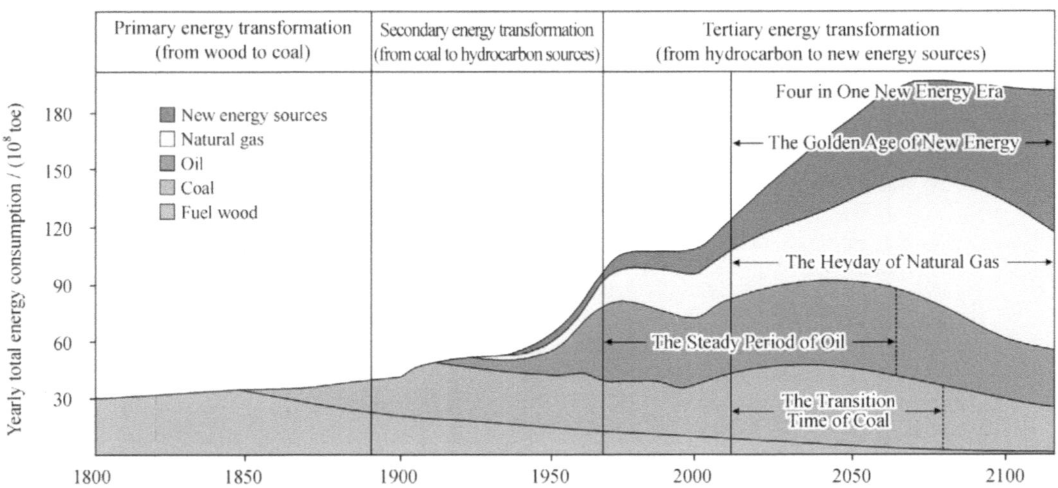

FIGURE 19.26 Global energy consumption trends and forecasts with energy transformations based on source by years.

Source: Zou et al. (2016).

This new energy transition and these environmental concerns point out the necessity of a carbon-neutral future. The term "carbon-neutral" indicates that CO_2 emissions within a year are balanced with CO_2 removed by technologies. This term also means net-zero CO_2 emissions (IPCC, 2018). Major countries have decided to apply some progress and practices for a carbon-neutral future. These are gradual withdrawal from coal-based power generation, acceleration in the usage of new energies, application of CCUS and carbon conversion technologies, and development of carbon pricing mechanisms with an increase in the cost of carbon emissions. In addition, there are four major carbon-neutral approaches, which are carbon replacement, carbon reduction, carbon sequestration, and carbon cycle. *Carbon replacement* is performed by electricity, hydrogen, and heat. *Carbon reduction* aims at efficiency improvement and energy saving. *Carbon sequestration* includes the steps of the capture of CO_2, transportation, utilization of it to usable goods, and storage in large underground geological areas to isolate it from the atmosphere. The *carbon cycle* comprises carbon conversion in an artificial way and forest carbon sink. In carbon reduction and carbon sequestration, black carbon emissions (CO_2 which cannot be used or fixed and remains in the atmosphere) are reduced. On the other hand, gray carbon (CO_2 that is available or fixed in the atmosphere) is recycled in the carbon cycle. Figure 19.27 represents the contribution of four major carbon-neutral approaches based on mitigating CO_2 emissions between 2020 and 2050. According to this figure, carbon replacement will meet almost half of the emission reduction by 2050. It will be followed by carbon reduction, carbon cycle, and carbon sequestration with 21%, 17%, and 15%, respectively. Moreover, CO_2 emissions reduction is relatively slow from 2020 to 2030, but it will accelerate between 2030 and 2050 due to further development in the related technologies, the promotion of renewable energy–related projects, and the cost competition between new energies and fossil fuels (Zou et al., 2021).

Decarbonization is essential in the future energy transition, and fossil fuel power plants are among the biggest priorities for this purpose. For the decarbonization of existing fossil fuel power plants, Zhang et al. (2022) stated three pathways, namely, early retirement, decarbonization retrofit, and emission offsetting. While choosing these pathways, technical, financial, and social dimensions should be thought. These pathways are represented in Figure 19.28. Emissions offsetting supplies an immediate solution, and it is appropriate for easy-to-decarbonize sectors. It does not provide a fundamental solution. Hence, retirement and retrofitting can provide better

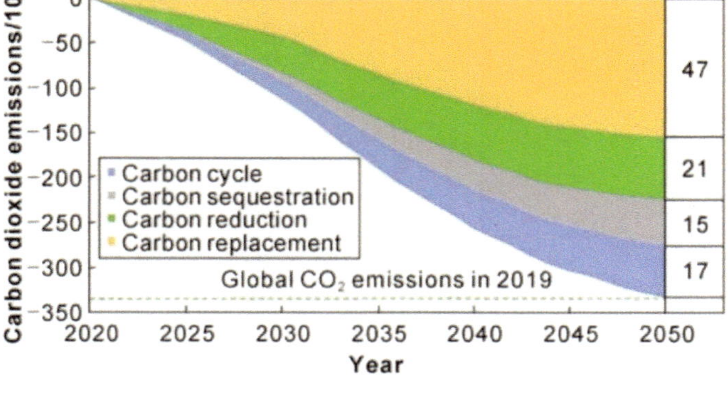

FIGURE 19.27 CO_2 emissions (10^8 tons, *t*) versus year indicating contribution (%) of four carbon-neutral approaches (carbon replacement, carbon reduction, carbon sequestration, and carbon cycle) between 2020 and 2050.

Source: Zou et al. (2021).

ways, containing investments, bioenergy supply, CO_2 storage, and integrating H_2 applications (Zhang et al., 2022).

In addition, several publications have pointed out that the energy transition is closely related to the United Nations Sustainable Development Goals (SDGs). The 17 SDGs with 169 targets were announced under the 2030 Agenda for Sustainable Development by the United Nations (United Nations, 2015). These SDGs and their relevant targets represent very broad areas, including energy, environment, ecology, industry, health, education, equality, and peace. The energy future is directly related to SDGs, especially with SDG13 ("Climate action"), SDG7 ("Affordable and clean energy"), SDG12 ("Responsible consumption and production"), and SDG9 ("Industry, innovation and infrastructure"). The future expectations will support SDG13 by decreasing CO_2 emissions, helping achieve the 1.5°C target, and reducing the effects of industrial actions; SDG7 by supporting energy transitions and collaborating with several renewables to increase energy efficiency; SDG12 by companies' transition to sustainable consumption and production and their conscious usage of natural resources; and SDG9 by integrating conscious energy resources and technologies into the industry (Kayhan, 2022). When social, economic, global, and educational aspects are considered as part of the energy transition, it is also related to some other SDGs, such as SDG1 ("No poverty"), SDG8 ("Decent work and economic growth"), SDG17 ("Partnerships for the goals"), SDG14 ("Life below water"), SDG15 ("Life on land"), SDG11 ("Sustainable cities and communities"), and SDG4 ("Quality education").

In the sustainable development value chain, several transformation indicators should be addressed under the scope of SDGs. These indicators contain more green jobs, innovation, education standards, socio-economic inclusiveness, entrepreneurship, and human development (Leal Filho et al., 2019; Aldieri et al., 2021; Gatto, 2022a). Gatto (2022b) stated that the future energy transition will

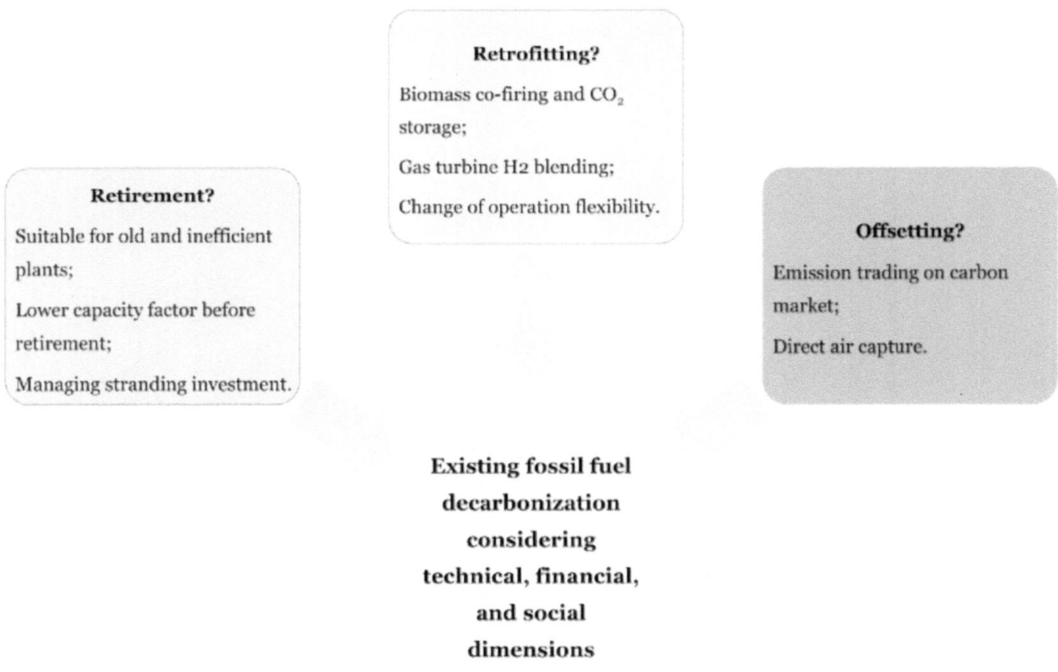

FIGURE 19.28 Decarbonization pathways of existing fossil fuel power plants (retirement, retrofitting, and offsetting).

Source: Zhang et al. (2022).

mainly depend on policymaking, interest groups and organizations, and people's choices. Singh and Singh (2023) recommended four key points for the future of energy transition and climate action. These recommendations are the use of co-benefits in order to take the interests of diverse audiences, less-polarizing discussions, passing from doomism to optimism, and finding the balance between quantitative and qualitative information (Singh and Singh, 2023). These are shown in Figure 19.29 with explanations for each recommendation.

Moreover, the FORECAST model is a bottom-up long-term scenarios model in industry by 2050 that is developed in the scope of energy demand and the amount of greenhouse gases. It considers several mitigation options for decarbonization in 2050, containing CCS, energy efficiency, fuel switching from fossil fuels to renewable and low-carbon options, material efficiency, circular economy, and recycling. This model thinks about key aspects for strategic decision-making for decarbonization in energy transition as a transition pathway, a broad range of mitigation options, costs, policies, and high level of technology. It especially points out the importance of policies for energy efficiency and fuel switching. These policies would cover taxes on CO_2 and energy, technology subsidies, and operational expenditures (OPEX) supports, like feed-in tariffs for renewables (Fleiter et al., 2018).

For the future energy transition, another important issue is the timing of it. Sovacool (2016) suggested four conclusions for the timing of energy transition. Firstly, even though this transition can be fast or slow, definitional patterns are essential, which indicate significance, the scale of society, and resources, carriers, controverts, and services. Secondly, energy transition timing can be subjective. Thirdly, these transitions are complex and related to more than one reason, factor, and plan. Fourthly, many energy transitions depend on pathways, and they are cumulative. According to these results, some transitions will seem like a punctuated equilibrium rather than an exponential line on

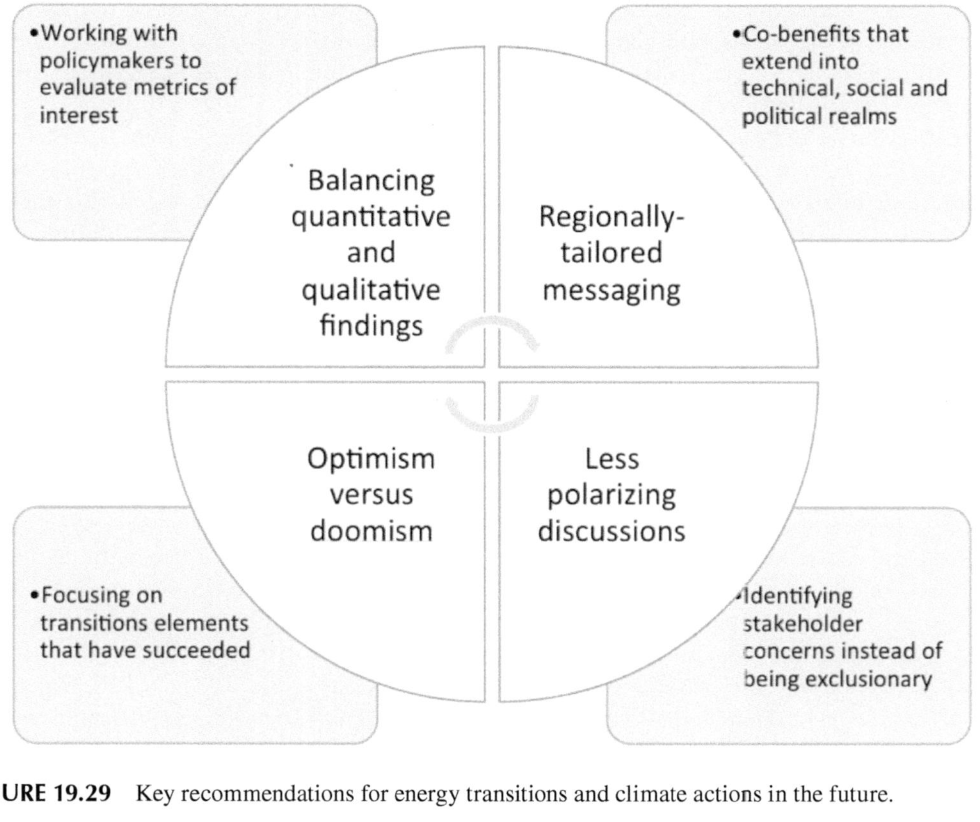

FIGURE 19.29 Key recommendations for energy transitions and climate actions in the future.

Source: Singh and Singh (2023).

a graph. Furthermore, future energy scenarios, forecasts, and research will increase transparency (Sovacool, 2016).

19.6 CONCLUSIONS

To mitigate anthropogenic climate change, reaching net-zero greenhouse gas emissions and limiting global temperature increase well below 1.5°C, or 2°C, have been aimed in light of agreements, reports, and commissions. These decisions have affected the energy sector majorly and influenced the current and future energy transitions. Future directions in energy transition have a close relationship with the oil and gas industry and renewables. Future scenarios and future directions in oil and gas, renewables, and energy transition have been explained in this chapter within the scope of these targets.

This chapter first reviewed the future scenarios of various organizations and companies. They offer strong case scenarios for the best direction in energy for a carbon-neutral future. The strong case scenarios that are mentioned in this chapter are IEA's NZE, IRENA's 1.5°C Scenario, Barclays's Dynamism, BP's Net Zero, Equinor's Bridges, McKinsey's 1.5°C Pathway, Shell's Sky, TotalEnegies' Rupture+, and World Energy Council's Unfinished Symphony. The majority of these scenarios are designed to reach net-zero greenhouse gas emissions and are compatible with limiting global temperature rise well below 1.5°C or 1.5°C by 2050. In addition, weak case scenarios, such as IEA's STEPS, IRENA's PES, Barclays's Deadlock, Equinor's Walls, McKinsey's Fading Momentum, Shell's Mountains and Oceans, TotalEnergies' Momentum, and World Energy Council's Hard Rock, show the potential results of the continuation of the current energy policies and the government-led decisions instead of broad perspectives. For this reason, they are inadequate to catch the significant targets for the future. While comparing the strong and weak case scenarios, differences at certain points draw attention. A higher share of renewables and a lower share of fossil fuels in energy demands and consumptions, electrification of end uses, more carbon mitigation and low-emissions technologies, and higher CO_2 pricing are the main key points between scenarios. These differences have big impacts on the future directions in the energy field that have been implemented by the expected results of the strong and weak scenarios.

Whereas the oil and gas industry has generally a lower share, especially in strong case scenarios, gas has a high share in some future foresight in the global energy mix by 2050. In spite of concerns about leading to greenhouse gas emissions and delaying integration of net-zero emission energies, natural gas seems as one of the bridges for the third energy transition. The general falling trends in share can be explained by both the actions for climate change and the approach of the Hubbert curve. There are essential steps in the future of the oil and gas industry. The first one is choosing advantaged hydrocarbons, which are the hydrocarbons that need to be explored and produced with minimized CO_2 emissions. There are several parameters, such as hydrocarbon type, reservoir characteristics, drilling operations, policies, geography, and techniques, to classify them as advantaged, and taking care of these parameters can provide environmental benefits even in the future directions with a relatively high share of hydrocarbons. The second one is abating oil and gas emissions. It requires key actions, including decreasing demand for hydrocarbon, reducing methane emissions, removing non-emergency flaring, electrification, CCUS applications, and using hydrogen in refineries. Among these actions, CCUS takes attention as a low-emissions technology. With the help of experiences in the oil and gas industry on EOR, CCUS has a higher opportunity to be applied in this field. CCUS applications are also mentioned in several strong case scenarios as a significant pathway for mitigating CO_2 emissions by 2050.

Lastly, renewables have increasing shares in the global energy mix through 2050. Especially, solar PV and wind have the highest increase from today to 2050. Renewables are essential to catch the future targets for the climate. Strong case scenarios of organizations and companies indicated the advantages of widespread usage of renewables to limit the temperature rise and reach net-zero greenhouse gas emissions in the future. In addition, combining renewables and low-emissions technologies and using them instead of fossil fuels in some applications can provide energy-efficient and cleaner results. Besides all these positive points about renewables, dealing with the challenges, like seasonal and inter-annual variability, lower durability compared to other energy sources, and the necessity of a high number of skilled employees, is important.

The current and future energy transitions are expected to be not only by technological developments, population growth, and energy demand but also by concerns about climate and environment. The transition from fossil fuels to non–fossil fuels (transition from oil and gas to low-carbon and energy-efficient sources) is called the third energy transformation. Nevertheless, there is some consensus about this transition because of not meeting the social and economic expectations of the world's growing population only by renewables and low-emission technologies. That situation points out the requirement for more developments and investments in these sources and technologies and the acceleration of their applications. A carbon-neutral future is an important topic in the energy transition which can be performed by carbon replacement, carbon reduction, carbon sequestration, and carbon cycle. Moreover, this topic is in line with several SDGs of the United Nations, especially SDG13, SDG7, SDG12, and SDG9.

In conclusion, the future of the energy field is expected to be cleaner and more sustainable. It will have a higher share of renewables and low-emissions technologies and a lower share of fossil fuels. The strong and weak case scenarios are beneficial to understanding the potential share of future energy sources. The best and most important direction is the one that can provide the maximum help while preventing and mitigating the harmful effects of climate change.

REFERENCES

Aldieri, L., A. Gatto, and C.P. Vinci. 2021. Evaluation of energy resilience and adaptation policies: An energy efficiency analysis. *Energy Policy* 157: 112505.

Aliyuda, K., J. Howell, and E. Humphrey. 2020. Impact of geological variables in controlling oil-reservoir performance: An insight from a machine-learning technique. *SPE Reservoir Evaluation and Engineering* 23(4): 1314–1327.

Alkatheri, M., A. Betancourt-Torcat, and A. Almansoori. 2018. Comparison of life cycle greenhouse gas emissions from unconventional ultra-sour and conventional gas feedstock for power: A case study of the United Arab Emirates. *Journal of Cleaner Production* 197: 908–918.

Barclays. 2019. *Oil in 3D: The demand outlook to 2050.* Impact Series, 6th ed. Barclays: 39 p.

BP. 2023. *Energy outlook*, 2023 ed. BP Plc: 127 p.

Burnham, A., J. Han, C.E. Clark, M. Wang, J.B. Dunn, and I. Palou-Rivera. 2012. Life-cycle greenhouse gas emissions of shale gas, natural gas, coal, and petroleum. *Environmental Science and Technology* 46(2): 619–627.

C&S Partners. 2021. *Materializing leadership: From oil and gas to renewable energies.* Available in https://concretesteelpartners.com/wp-content/uploads/2021/06/Odd-Stromsnes-From-OG-to-Renewable-publish.pdf [Accessed September 12th, 2023].

Chen, Z., Z. He, W. Zhang, and Z. Sun. 2022. Towards carbon neutrality: A dynamic game approach replacing fossil energy with new marine energy. *Energy Reports* 8: 738–755.

Craig, J. 2022. The "energy transition": Myth or reality? *EPJ Web of Conferences* 268: 00004.

Davies, A., and M.D. Simmons. 2021. Demand for "advantaged" hydrocarbons during the 21st century energy transition. *Energy Reports* 7: 4483–4497.

Davis, S.J., N.S. Lewis, M. Shaner, S. Aggarwal, D. Arent, I.L. Azevedo, . . ., and K. Caldeira. 2018. Net-zero emissions energy systems. *Science* 360: eaas9793.

Equinor. 2023. *Energy perspectives 2023: Global macroeconomic and energy market outlook.* Equinor ASA: 15 p.

Equinor. n.d. *Sleipner area.* Available in www.equinor.com/energy/sleipner [Accessed September 12th, 2023].

European Commission. n.d.a. *Paris agreement.* Available in https://climate.ec.europa.eu/eu-action/international-action-climate-change/climate-negotiations/paris-agreement_en [Accessed August 10th, 2023].

European Commission. n.d.b. *A European green deal.* Available in https://ec.europa.eu/info/strategy/priorities-2019-2024/europeangreen-deal_en [Accessed January 3rd, 2023].

ExxonMobil. 2019. *2019 outlook for energy: A perspective to 2040.* Exxon Mobil Corporation: 57 p.

ExxonMobil. 2023. *Energy supply*. Available in https://corporate.exxonmobil.com/what-we-do/energy-supply/global-outlook/energy-supply [Accessed September 28th, 2023].

Fleiter, T., M. Rehfeldt, A. Herbst, R. Elsland, A-L. Klingler, P. Manz, and S. Eidelloth. 2018. A methodology for bottom-up modelling of energy transitions in the industry sector: The FORECAST model. *Energy Strategy Reviews* 22: 237–254.

Gardiner, N.J., J.J. Roberts, G. Johnson, D.J. Smith, C.E. Bond, R. Knipe, S. Haszeldine, S. Gordon, and M. O'Donnell. 2023. Geosciences and the energy transition. *Earth Science, System and Society* 3: 10072.

Gatto, A. 2022a. The energy futures we want: A research and policy agenda for energy transitions. *Energy Research & Social Science* 89: 102639.

Gatto, A. 2022b. Polycentric and resilient perspectives for governing the commons: Strategic and law and economics insights for sustainable development. *Ambio* 51: 1921–1932.

Hossain, M.E. 2015. Drilling costs estimation for hydrocarbon wells. *Journal of Sustainable Energy Engineering* 3: 3–32.

IEA (International Energy Agency). 2016. *20 years of carbon capture and storage: Accelerating future deployment*. IEA Publication: 111 p.

IEA (International Energy Agency). 2021a. *Net zero by 2050: A roadmap for the global energy sector*. IEA Publications: 222 p.

IEA (International Energy Agency). 2021b. *World energy outlook 2021*. IEA Publications: 383 p.

IEA (International Energy Agency). 2022a. *Global energy and climate model: Documentation*. IEA Publications: 124 p.

IEA (International Energy Agency). 2022b. *World energy outlook 2022*. IEA Publications: 522 p.

IEA (International Energy Agency). 2022c. *CO2 storage resources and their development: An IEA CCUS handbook*. IEA Publications: 125 p.

IEA (International Energy Agency). 2022d. *Carbon capture, utilisation and storage*. Available in www.iea.org/reports/carbon-capture-utilisation-and-storage-2 [Accessed September 2nd, 2023].

IEA (International Energy Agency). 2022e. *Bioenergy with carbon capture and storage*. Available in www.iea.org/reports/bioenergy-with-carbon-capture-and-storage [Accessed August 15th, 2023].

IEA (International Energy Agency). 2023a. *Energy technology perspectives 2023*. IEA Publications: 463 p.

IEA (International Energy Agency). 2023b. *Emissions from oil and gas operations in net zero transitions: A world energy outlook special report on the oil and gas industry and COP28*. IEA Publications: 31 p.

IEA (International Energy Agency) 2023c. *World energy investment 2023*. IEA Publications: 179 p.

IEA (International Energy Agency). 2023d. *Managing seasonal and interannual variability of renewables*. IEA Publications: 68 p.

IEEFA (Institute for Energy Economics and Financial Analysis). 2023. *Norway's carbon capture and storage projects augur geological risks in global aspirations to bury carbon dioxide*. Available in https://ieefa.org/articles/norways-carbon-capture-and-storage-projects-augur-geological-risks-global-aspirations-bury [Accessed September 12th, 2023].

IPCC (Intergovernmental Panel on Climate Change). 2018. *Special report: Global warming of 1.5°C*. Cambridge University Press: 562 p.

IRENA (International Renewable Energy Agency). 2023. *World energy transitions outlook: 1.5°C pathway*, Volume 1. IRENA: 175 p.

Jiang, M., W.M. Griffin, C. Hendrickson, P. Jaramillo, J. VanBriesen, and A. Venkatesh. 2011. Life cycle greenhouse gas emissions of Marcellus shale gas. *Environmental Research Letters* 6: 034014.

Kayhan, D.Y. 2022. Net-sıfır emisyona ulaşabilmek adına karbon yakalama, kullanma ve depolama (CCUS) faaliyetlerinin endüstride sürdürülebilir kalkınma açısından önemi. *MTA Doğal Kaynaklar ve Ekonomi Bülteni* 34: 16–26.

Leal Filho, W., S.K. Tripathi, J.B.S.O.D. Andrade Guerra, R. Ginè-Garriga, V. Orlovic Lovren, and J. Willats. 2019. Using the sustainable development goals towards a better understanding of sustainability challenges. *International Journal of Sustainable Development & World Ecology* 26(2): 179–190.

Martin-Roberts, E., V. Scott, S. Flude, G. Johnson, R.S. Haszeldine, and S. Gilfillan. 2021. Carbon capture and storage at the end of a lost decade. *One Earth* 4: 1569–1584.

McKinsey. 2022. *Global energy perspective 2022: Executive summary*. McKinsey & Company: 27 p.

Miranda-Barbosa, E., B. Sigfusson, J. Carlsson, and E. Tzimas. 2017. Advantages from combining CCS with geothermal energy. *Energy Procedia* 114: 6666–6676.

Newell, R.G., D. Raimi, and G. Aldana. 2019. *Global energy outlook 2019: The next generation of energy*. Resources for the Future: 38 p.

Newell, R.G., D. Raimi, S. Villanueva, and B. Prest. 2021. *Global energy outlook 2021: Pathways from Paris*. Resources for the Future: 40 p.

Shell. 2018. *Shell scenarios: Sky*. Meeting the Goals of the Paris Agreement. Shell International B.V.: 69 p.

Singh, U., and S. Singh. 2023. Future research directions to facilitate climate action and energy transitions. *Energy and Climate Change* 4: 100092.

Sovacool, B.K. 2016. How long will it take? Conceptualizing the temporal dynamics of energy transitions. *Energy Research & Social Science* 13: 202–215.

Tanaka, K., O. Cavalett, W.J. Collins, and F. Cherubini. 2019. Asserting the climate benefits of the coal-to-gas shift across temporal and spatial scales. *Natural Climate Change* 9: 389–396.

Tischuk, A., and A. Latham. 2021. The hunt for advantaged oil. *Wood Mackenzie*. Available in www.woodmac.com/news/opinion/the-hunt-for-advantaged-oil/ [Accessed August 19th, 2023].

TotalEnergies. 2022a. *Energy outlook 2022*. TotalEnergies SE: 42 p.

TotalEnergies. 2022b. *TotalEnergies energy outlook 2022*. TotalEnergies Publishes Its Annual Contribution to the Energy Transition Dialogue. TotalEnergies SE, Press Release, September 27.

United Nations. 2015. *Transforming our world: The 2030 agenda for sustainable development. general assembly*. United Nations.

Vitillo, J.G., M.D. Eisaman, E.S.P. Aradottir, F. Passarini, T. Wang, and S.W. Sheehan. 2022. The role of carbon capture, utilization, and storage for economic pathways that limit global warming to below 1.5°C. *iScience* 25: 104237.

Wei, N., X. Li, S. Liu, S. Lu, and Z. Jiao. 2021. A strategic framework for commercialization of carbon capture, geological utilization, and storage technology in China. *International Journal of Greenhouse Gas Control* 110: 103420.

World Energy Council. 2019. *World energy scenarios 2019: Exploring innovation pathways to 2040*. World Energy Council: 148 p.

Zhang, C., H. Zhai, L. Cao, X. Li, F. Cheng, L. Peng, K. Tong, J. Meng, L. Yang, and X. Wang. 2022. Understanding the complexity of existing fossil fuel power plant decarbonization. *iScience* 25: 104758.

Zhang, X., N. P. Myhrvold, Z. Hausfather, and K. Caldeira. 2016. Climate benefits of natural gas as a bridge fuel and potential delay of near-zero energy systems. *Applied Energy* 167: 317–322.

Zou, C., B. Xiong, H. Xue, D. Zheng, Z. Ge, Y. Wang, L. Jiang, S. Pan, and S. Wu. 2021. The role of new energy in carbon neutral. *Petroleum Exploration and Development* 48(2): 480–491.

Zou, C., Q. Zhao, G. Zhang, and B. Xiong. 2016. Energy revolution: From a fossil energy era to a new energy era. *Natural Gas Industry B* 3: 1–11.

Index

Note: numbers in **bold** indicate a table. Numbers in *italics* indicate a figure on the corresponding page

C

G

H

I

J

K

L

M

N